Modern Control Theory

Third Edition

WILLIAM L. BROGAN, Ph.D.
Professor of Electrical Engineering
University of Nevada, Las Vegas

Prentice Hall

PRENTICE HALL, Upper Saddle River, NJ 07458

Library of Congress Cataloging-in-Publication Data

Brogan, William L.
Modern control theory / William L. Brogan. -- 3rd ed.
p. cm.
Includes bibliographical references.
ISBN 0-13-589763-7
1. Control theory. I. Title.
QA402.3.B76 1991
629.8'312--dc20 90-6977
CIP

Editorial/production supervision
and interior design: Patrice Fraccio/Bayani Mendoza de Leon
Cover design: Karen Stephens
Manufacturing buyers: Lori Bulwin/Linda Behrens

Printed in the United States of America
20

ISBN 0-13-589763-7

Prentice-Hall International (UK) Limited, *London*
Prentice-Hall of Australia Pty. Limited, *Sydney*
Prentice-Hall of Canada, Inc., *Toronto*
Prentice-Hall Hispanoamericana, S. A., *Mexico*
Prentice-Hall of India Private Limited, *New Delhi*
Prentice-Hall of Japan, Inc., *Tokyo*
Prentice-Hall Asia Pte. Ltd., *Singapore*
Editora Prentice-Hall do Brasil, Ltda., *Rio de Janeiro*

This book is dedicated to my parents Lloyd and Alice,
and to my grandchildren Jesse and Jacque,
representatives of the next generation.

Contents

Preface

This is a text on state variable modelling, analysis and control of dynamical systems. The analytical approach to many control problems consists of three major steps: (1) develop an idealized mathematical representation of the real physical system, (2) apply mathematical analysis and design techniques to the model and (3) interpret the mathematical results in terms of implications on the real, physical system. If the resulting implications are not acceptable or do not seem to match reality or experimental observations, one or more of the above steps may need to be modified and repeated. This book attempts to illustrate these steps and some of the tools that may be involved. While most of the steps are mathematically based, this is intended as an engineering text. No abstract theorems/proofs are included just to enhance the mathematical elegance.

The first two chapters are intended as a review of prerequisite introductory courses on modelling and control of physical systems using primarily a transfer function approach. These are not intended to be complete, stand-alone treatments. Rather, a distilled summary of the essentials is presented. It has been observed that students often get bogged down in the myriad of details of these prerequisite courses, thus losing sight of the big picture. The intent here was to place the major points in perspective before going on to the state variable approach.

State variables, state vectors, state space and linear system matrices are introduced in Chapter 3. Methods of obtaining state variable models from other system descriptions are provided. In previous editions this chapter came after a series of chapters on matrix theory and linear algebra. The present, inverted approach provides early motivation for the need to master the mathematics of matrices and linear algebra. It also allows earlier introduction of control-related examples, which appear throughout the subsequent chapters. Chapters 4 through 8 provide a thorough development of the needed mathematical tools. The treatment has been considerably revised, based on experience gained with previous editions, as well as helpful comments from reviewers.

The depth to which the mathematical topics need to be pursued depends upon the preparation of the reader and the level of understanding needed. My recent experience has been that most students approach this book after having had a first course in linear algebra. It still appears fruitful to cover Chapters 4 through 8 fairly carefully, and students gain new insights from the large number of engineering-motivated examples. Not all of the more abstract topics need to be covered in an undergraduate course, however. A more advance graduate level approach would merely skim the early sections of these chapters, but put more emphasis on the extensions and proof found in the problems.

In addition to reversing the order of presentation mentioned above, this book deleted some interesting but peripheral topics, to make room for new material which is more central to the controls field. Additional topics now covered include QR decomposition of a matrix, which can be used to iteratively solve for eigenvalues and eigenvectors. More importantly, it is useful in finding the Kalman controllable and/or observable canonical forms. These provide a very satisfactory method of determining minimal realizations in Chapter 11. A portion of the material on matrix fraction description of systems, and application to controller/observer design is now included. This Diophantine equation approach supplements and provides an insightful alternative to the state variable approach. The treatment of optimal control has been revised to emphasize the linear quadratic problem and the associated Riccati equations. The question of robustness is addressed in an introductory manner. Two extensions to the LQ theory are introduced; projective controls is a method of designing low order controllers which preserve the dominant modes of a full order optimal controller, and frequency-weighted cost functions which lead to dynamic controllers capable of coping with modelling approximations.

More emphasis is given to stability of time varying linear systems and a new chapter provides some tools for approaching nonlinear system problems.

As in the previous editions, the problems, especially the illustrative problems for which complete solutions are given, should be considered as an integral part of each chapter. Many useful results are derived and presented *only* in these problems.

This third edition has evolved from the first two, and thus all former users who sent comments or filled out review forms for Prentice Hall have contributed to this work. Those individuals who contributed more directly to the preparation of the previous editions have had a lasting impact here too. During the manuscript preparation, student feedback from several classes was very helpful. Steven Crammer, Saeed Karamooz and L. Lane Sanford deserve special thanks. Professors John Boye and George Schade from Nebraska, Hal Tharp of Arizona and Sahjendra Singh of UNLV provided comments or suggestions. Production editors Patrice Fraccio and Bayani Mendoza de Leon who handled the editorial supervision and interior design of the book, and the five anonymous reviewers who provided detailed comments to Prentice Hall editor Tim Bozik, are especially thanked. Finally, I wish to acknowledge Mailliw Nagorb for typing the manuscript.

William L. Brogan

1

Background and Preview

1.1 INTRODUCTION

Control theory is often regarded as a branch of the general, and somewhat more abstract, subject of systems theory [1].‡ The boundaries between these disciplines are often unclear, so a brief section is included to delineate the point of view of this book.

In order to put control theory into practice, a bridge must be built between the real world and the mathematical theory. This bridge is the process of modeling, and a summary review of modeling is included in this chapter [2, 3].

Control theory can be approached from a number of directions. The first systematic method of dealing with what is now called control theory began to emerge in the 1930s. Transfer functions and frequency domain techniques were predominant in these "classical" approaches to control theory. Starting in the late 1950s and early 1960s a time-domain approach using state variable descriptions came into prominence.

For a number of years the state variable approach was synonymous with "modern control theory." At the present time the state variable approach and the various transfer function–based methods are considered on an equal level, and nicely complement each other. Distinctions exist, and the major one appears to be in the kinds of mathematical tools used. The state variable approach uses linear algebra based on the real or complex number field. The newer multivariable transfer function approaches involve the algebra of polynomial matrices and related concepts. By defining the number field properly, the major mathematical tool is once again linear algebra, but on a somewhat less familiar level (see, for example, Sections 6.3.1 and 6.3.2). Some of these concepts are pointed out and used throughout this book. The older classical control theory point of view, using single-input, single-output transfer functions is

‡ Reference citations are given numerically in the text in brackets. The references are listed at the end of each chapter.

reviewed briefly in Chapter 2. However, for the most part this book is devoted to the state variable point of view.

1.2 SYSTEMS, SYSTEMS THEORY, AND CONTROL THEORY

According to the *Encyclopedia Americana,* a system is "... an aggregation or assemblage of things so combined by nature or man as to form an integral and complex whole...." Mathematical systems theory is the study of the interactions and behavior of such an assemblage of "things" when subjected to certain conditions or inputs. The abstract nature of systems theory is due to the fact that it is concerned with mathematical properties rather than the physical form of the constitutent parts.

Control theory is more often concerned with physical applications. A control system is considered to be any system which exists for the purpose of regulating or controlling the flow of energy, information, money, or other quantities in some desired fashion. In more general terms, a control system is an interconnection of many components or functional units in such a way as to produce a desired result. In this book control theory is assumed to encompass all questions related to design and analysis of control systems.

Figure 1.1 is a general representation of an *open-loop* control system. The input, or control, $\mathbf{u}(t)$ is selected based on the goals for the system and all available a priori knowledge about the system. The input is in no way influenced by the output of the system, represented by $\mathbf{y}(t)$. If unexpected disturbances act upon an open-loop system, or if its behavior is not completely understood, then the output will not behave precisely as expected.

Another general class of control systems is the *closed-loop,* or *feedback,* control system, as illustrated in Figure 1.2. In the closed-loop system, the control $\mathbf{u}(t)$ is modified in some way by information about the behavior of the system output. A feedback system is often better able to cope with unexpected disturbances and uncertainties about the system's dynamic behavior. However, it need not be true that closed-loop control is always superior to open-loop control. When the measured outputs have errors which are sufficiently large and when unexpected disturbances are relatively unimportant, closed-loop control can have a performance which is inferior to open-loop control.

EXAMPLE 1.1 In order to provide financial security for the retirement years, a person arranges to have \$300 per month invested into an annuity account. The system "input" each month is $u(t) = \$300$. The system output $y(t)$ is the accrued value in the account. Since $u(t)$ is not affected by the current economic climate or by $y(t)$, this is an open-loop system. ■

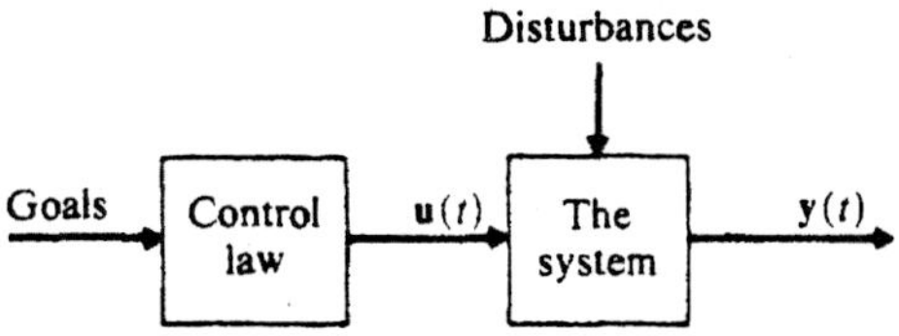

Figure 1.1 An open-loop control system.

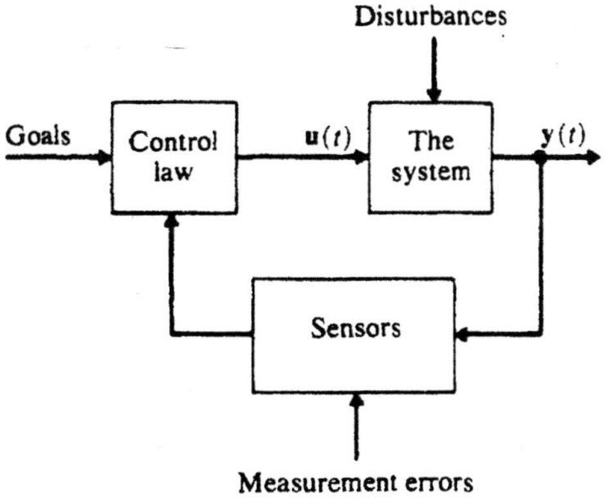

Figure 1.2 A closed-loop control system.

EXAMPLE 1.2 Another person with the same goal of financial security plans to invest in the stock market, by attempting to implement the strategy of buying-low and selling-high. The input $u(t)$ at any given time is influenced by the perceived market conditions, the past success of the stock account, and so forth. This is a feedback or closed-loop system. ∎

EXAMPLE 1.3 A typical industrial control system involves components from several engineering disciplines. The automatic control of a machine shown in Figure 1.3 illustrates this. In this example, the desired time history of the carriage motion is patterned into the shape of the cam. As the cam-follower rises and falls, the potentiometer pick-off voltage is proportional to the desired carriage position. This signal is compared with the actual position, as sensed by another potentiometer. This difference, perhaps modified by a tachometer-generated rate signal, gives rise to an error signal at the output of the differential amplifier. The power level of this signal is usually low and must be amplified by a second amplifier before it can be used for corrective action by an electric motor or a servo valve and a hydraulic motor or some other prime mover. The prime mover output would usually be modified by a precise gear train, a lead screw, a chain and sprocket, or some other mechanism. Clearly, mechanical, electrical, electronic, and hydraulic components play important roles in such a system. ∎

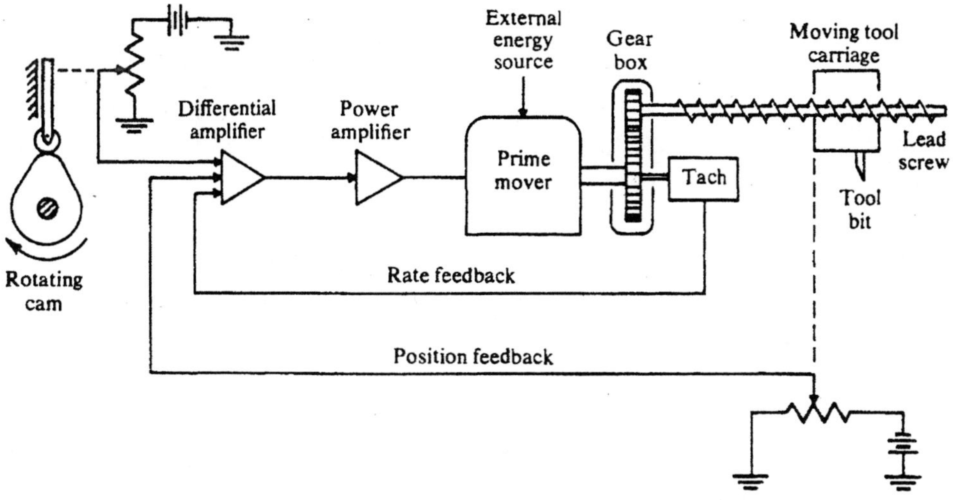

Figure 1.3

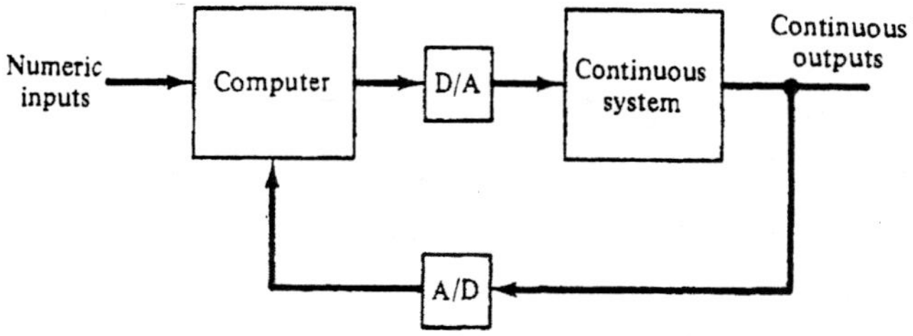

Figure 1.4

EXAMPLE 1.4 The same ultimate purpose of controlling a machine tool could be approached somewhat differently using a small computer in the loop. The continuous-time, or analog, signals for position and velocity must still be controlled. Measurements of these quantities would probably be made directly in the digital domain using some sort of optical pulse counting circuitry. If analog measurements are made, then an analog-to-digital (A/D) conversion is necessary. The desired position and velocity data would be available to the computer in numeric form. The digital measurements would be compared and the differences would constitute inputs into a corrective control algorithm. At the output of the computer a digital-to-analog (D/A) conversion could be performed to obtain the control inputs to the same prime mover, as in Example 1.3. Alternately, a stepper motor may be selected because it can be directly driven by a series of pulses from the computer. Figure 1.4 shows a typical control system with a computer in the loop. ■

1.3 MODELING

Engineers and scientists are frequently confronted with the task of analyzing problems in the real world, synthesizing solutions to these problems, or developing theories to explain them. One of the first steps in any such task is the development of a mathematical model of the phenomenon being studied. This model must not be oversimplified, or conclusions drawn from it will not be valid in the real world. The model should not be so complex as to complicate unnecessarily the analysis.

System models can be developed by two distinct methods. *Analytical modeling* consists of a systematic application of basic physical laws to system components and the interconnection of these components. *Experimental modeling,* or modeling by synthesis, is the selection of mathematical relationships which seem to fit observed input-output data. Analytical modeling is emphasized first. Some aspects of the other approach are presented in Chapter 6 (least-squares data fitting).

1.3.1. Analytical Modeling

An outline of the analytical approach to modeling is presented in Figure 1.5. The steps in this outline are discussed in the following paragraphs.

1. *The intended purposes of the model must be clearly specified.* There is no single model of a complicated system which is appropriate for all purposes. If the purpose is a detailed study of an individual machine tool, the model would be very different from one used to study the dynamics of work flow through an entire factory.

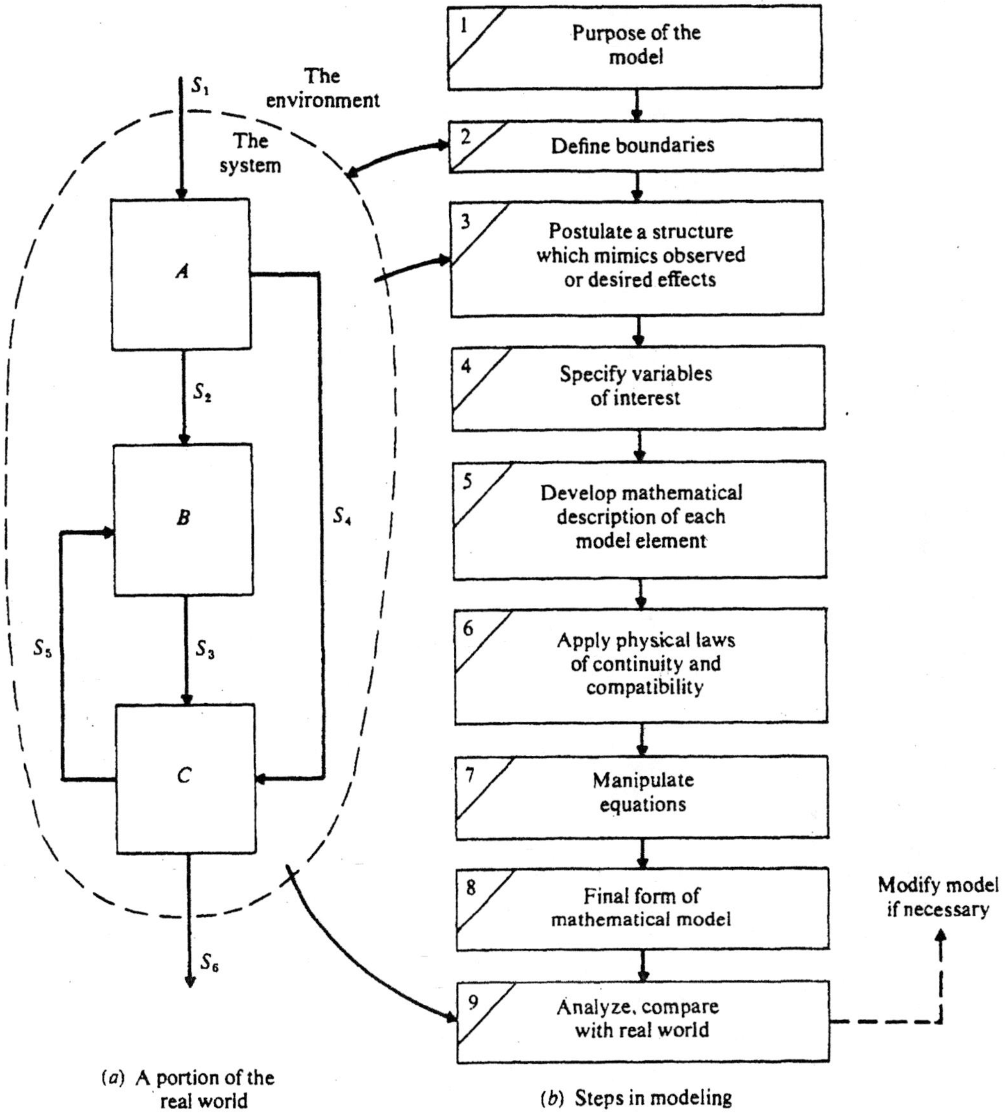

Figure 1.5 Modeling considerations.

2. *The system boundary is a real or imagined separation of the part of the real world under study, called the system, and the rest of the real world, referred to as the environment.* The system boundary must enclose all components or subsystems of primary interest, such as subsystems A, B, and C in Figure 1.5a.

   A second requirement on the selection of the boundary is that all causative actions or effects (called signals) crossing the boundary be more or less one-way interactions. The environment can affect the system, and this is represented by the input signal S_1. The system output, represented by the signal S_6, should not affect the environment, at least not to the extent that it would modify S_1. If there is no interest in subsystem A, then a boundary enclosing B and C, and with inputs S_2 and S_4, could be used. Subsystem C should not be selected as an isolated system because one of its outputs S_5 modifies its input S_3 through subsystem B. The requirement is that all inputs are known, or can be assumed known for the purpose of the study, or can be controlled independently of the internal status of the system.

EXAMPLE 1.5 The purpose of the models of Figure 1.6 is to study the flow of work and information within a production system due to an input rate of orders. These orders could be an input from the environment, as in Model I. If the purpose is to study the effects of an advertising campaign, then orders are determined, at least in part, by a major system variable. In this case the rate of orders should be an internal variable in the feedback system of Model II. ■

3. *All physical systems, whether they are of an electrical, mechanical, fluid, or thermal nature, have mechanisms for storing, dissipating, or transferring energy, or transforming energy from one form to another.* The third step in modeling is one of reducing the actual system to an interconnection of simple, idealized elements which preserve the character of these operations on the various kinds of energy. An electric circuit diagram illustrates such an idealization, with ideal sources representing inputs. In mechanical systems, idealized connections of point masses, springs, and dashpots are often used. In thermal or fluid systems, and to a certain extent in economic, political, and social systems, similar idealizations are possible. This process is referred to as *physical modeling*. The level of detail required depends on the type of information expected from the model.
4. *If the physical model is properly selected, it will exhibit the same major characteristics as the real system.* In order to proceed with development of a mathematical model, variables must be assigned to all attributes of interest. If a quantity of interest does not yet exist and thus cannot be labeled, a modification will be required in Step 3 in order to include it. The classification of system types is discussed in the next section. This book deals mainly with *deterministic lumped-parameter systems*. In all lumped-parameter systems there are basically just two types of variables. They are *through variables* (sometimes called path variables or rate variables) and *across variables* (sometimes called point variables or level

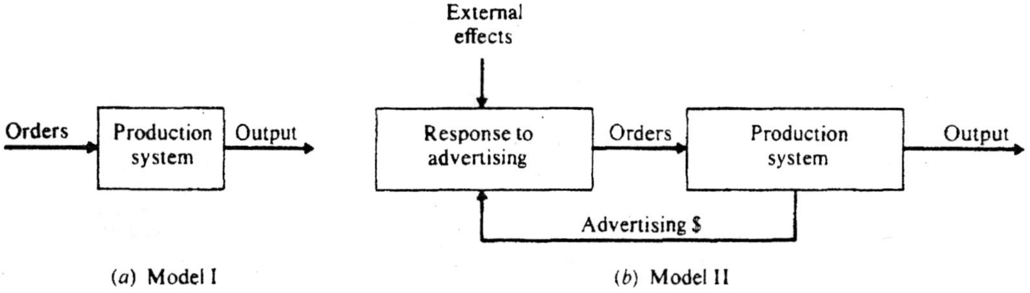

Figure 1.6

variables). Through variables flow through two-terminal elements and have the same value at both terminals. Examples are electric current, force or torque, heat flow rate, fluid flow rate, and rate of work flow through a production element. Across variables have different values at the two terminals of a device. Examples are voltage, velocity, temperature, pressure, and inventory level.

5. *Each two-terminal element in the idealized physical model will have one through and one across variable associated with it.* Multiterminal devices such as transformers or controlled sources will have more. In every device, mathematical relationships will exist between the two types of variables. These relationships, called *elemental equations,* must be specified for each element in the model. This step could uncover additional variables that need to be introduced. This would mean a modification of Step 4. Common examples of elemental equations are the current-voltage relationships for resistors, capacitors, and inductors. The form of these relations may be algebraic, differential, or integral expressions, linear or nonlinear, constant or time-varying.
6. *After a system has been reduced to an interconnection of idealized elements, with known elemental equations, equations must be developed to describe the interconnection effects.* Regardless of the physical type of the system, there are just two types of physical laws that are needed for this purpose. The first is a statement of *conservation* or *continuity* of the through variables at each node where two or more elements connect. Examples of this basic law are Kirchhoff's node equations, D'Alembert's version of Newton's second law, conservation of mass in fluid flow problems, and heat balance equations. The second major law is a *compatibility* condition relating across variables. Kirchhoff's voltage law around any closed loop is but one example. Similar laws regarding relative velocities, pressure drops, and temperature drops must also hold. Both of these laws yield linear equations in through or across variables, regardless of whether the elemental equations are linear or nonlinear. This fact is responsible for the name given to linear graphs, an extremely useful tool in applying these two laws.

EXAMPLE 1.6 Consider the system with six elements, including a source v_0, shown in Figure 1.7. Each element is represented as a branch of the linear graph, and the interconnection points are nodes. Each node is identified by an across variable v_i, and each branch has a through variable, called f_i, with the arrow establishing the sign convention for positive flow. Let

b (number of branches) = 6
s (number of sources) = 1
n (number of nodes) = 4

Two unknowns exist for each branch, except source branches have a single unknown. Thus there are $2b - s = 11$ unknowns, and 11 equations are needed. They are $b - s = 5$ elemental equations, $n - 1 = 3$ continuity equations (node 1 is used as a reference and is redundant)

$$f_0 - f_1 - f_2 = 0, \qquad f_2 - f_3 - f_4 = 0, \qquad f_4 - f_5 = 0$$

and $b - (n - 1) = 3$ compatibility (loop) equations. Letting $v_{ab} = v_a - v_b$, these are

$$v_{21} - v_0 = 0, \qquad v_{23} + v_{31} + v_{12} = 0, \qquad v_{34} + v_{41} + v_{13} = 0$$

These $2b - s$ equations can be used to determine all unknowns. ■

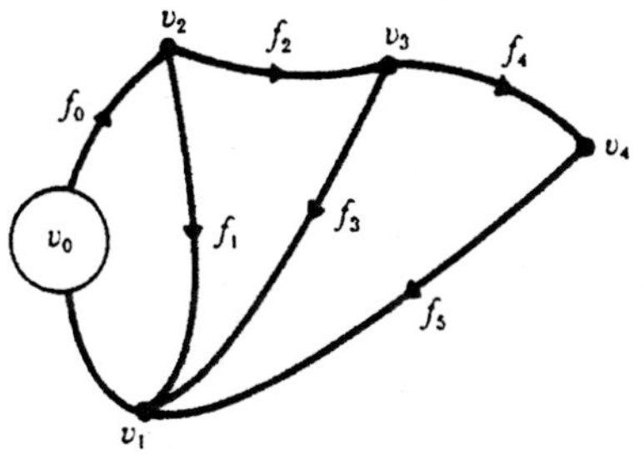

Figure 1.7

7. *This step consists of manipulating the elemental, continuity, and compatibility equations into a desired final form.* A further discussion of this step is presented in Section 1.5.
8. *Item 8 of Figure 1.5 is the end result of the modeling process.* It may be arrived at by a process of iteration, as mentioned in Step 9.
9. *The model developed in the preceding steps should never be confused with the real world system being studied.* Whenever possible, the results produced by the real system should be compared with model results under similar conditions. If unacceptable discrepancies exist, the model is inadequate and should be modified.

1.3.2 Experimental Modeling

Time series models: autoregressive, moving average, and ARMA models. In some experimental modeling situations the system structure may be based solidly on the laws of physics, and perhaps only a few key parameter values are uncertain. Even these unknown parameters may be known to some degree. Upper and lower bounds or the mean and variance or other probabilistic descriptors may be available at the outset.

In other situations, notably in areas of socioeconomic or biological systems, the only thing available is an assumed model form, which is convenient to work with and which does a reasonable job of fitting observations. All the coefficients are usually unknown and must be determined in these cases. When this situation applies, the autoregressive moving average (ARMA) model is frequently used. A brief overview follows.

A large variety of technical applications can be framed in a similar mathematical form. Let $y(k)$ be some variable of interest at a general time t_k. This might be the price of a stock or similar commodity. This is the case of an economic time series problem. In another situation $y(k)$ could be the magnitude of the acoustic waveform of speech, and the interest might be in either speech coding or synthetic speech generation. The major interest here is in the identification of unknown systems, and $y(k)$ is the system output. In all these cases it is assumed that the next value in the time series, $y(k+1)$, is influenced by the present value $y(k)$ and past (lagged) values $y(k-1)$, $y(k-2), \ldots, y(k-n)$. In addition, the signal $y(k+1)$ is normally influenced by one or more current and past input signals $u_i(k+1)$, $u_i(k-j)$ for $j = 0, 1, 2, \ldots, p-1$. It is assumed here that the time series is generated by a linear difference equation with a single input,

$$y(k+1) = a_0 y(k) + a_1 y(k-1) + a_2 y(k-2) + \cdots + a_n y(k-n) + b_0 u(k+1) + b_1 u(k) + \cdots + b_p u(k+1-p) + v(k) \tag{1.1}$$

where $v(k)$ is a random noise term. The identification problem is thus reduced to the estimation of the system coefficients

$$\boldsymbol{\theta} = [a_0 \quad a_1 \cdots a_n \quad b_0 \quad b_1 \cdots b_p]^T \tag{1.2}$$

from a series of measurements of the inputs $u(i)$ and the outputs $y(i)$. Equation (*1.1*) can be recast as

$$\begin{aligned} y(k+1) &= [y(k) \quad y(k-1) \cdots u_1(k+1) \quad u_1(k) \cdots]\boldsymbol{\theta} + v(k) \\ &= \mathbf{C}(k)\boldsymbol{\theta} + v(k) \end{aligned} \tag{1.3}$$

If the measurements of past values of $y(i)$ and $u(i)$ are sufficiently accurate, $\mathbf{C}(k)$ can be assumed known. Equation (*1.3*) is then in a form suitable for recursive least-squares estimation of the unknown parameters $\boldsymbol{\theta}$. A series of equations can be stacked into one larger equation:

$$\begin{bmatrix} y(k+1) \\ y(k+2) \\ \cdot \\ \cdot \\ \cdot \\ y(k+N) \end{bmatrix} = \begin{bmatrix} \mathbf{C}(k) \\ \mathbf{C}(k+1) \\ \cdot \\ \cdot \\ \cdot \\ \mathbf{C}(k+N-1) \end{bmatrix} \boldsymbol{\theta} + \begin{bmatrix} v(k) \\ v(k+1) \\ \cdot \\ \cdot \\ \cdot \\ v(k+N-1) \end{bmatrix} \tag{1.4}$$

Equation (*1.4*) is suitable for use in a batch least-squares estimation procedure for determining approximate values for the unknown constant parameters $\boldsymbol{\theta}$. Both batch and recursive least-squares solutions are presented in Chapter 6.

The performance of such a parameter estimation scheme is dependent upon the input signal. Sometimes, specially selected input signals can be used during the identification process. In other situations only the normal operating signals can be used. It may be tolerable to add a small sinusoidal component, called a *dither signal,* to the input to aid the identification process. It is intuitively clear that the input must excite those modes of the system that are intended for identification. That is, if a constant input is used and if the system has been operating sufficiently long for steady state to be reached, little about the system can be identified, other than its steady-state gain. The input must be "sufficiently exciting" or "sufficiently rich" if the identification is to be successful.

It is informative to take the Z-transform of Eq. (*1.1*). Z-transforms are defined briefly and used in Problems 1.15 through 1.20. For present purposes it suffices to view the variable z^{-1} as a time-delay operator. Then Eq. (*1.1*) can be written in delay operator—i.e., transformed—form as

$$y(z)[z - a_0 - a_1 z^{-1} - a_2 z^{-2} - \cdots - a_n z^{-n}] = [b_0 z + b_1 + b_2 z^{-1} + \cdots + b_p z^{-(p-1)}]u(z)$$

Then the input-output transfer function can be written

$$\frac{y(z)}{u(z)} = \frac{b_0 + b_1 z^{-1} + \cdots + b_p z^{-p}}{1 - (a_0 z^{-1} + \cdots + a_n z^{-(n+1)})} = H(z) \tag{1.5}$$

Some commonly used nomenclature [4] is now defined. If $y(k+1)$ depends only on the u terms and not on past y terms, all the a_i coefficients would be zero and the transfer function then would have zeros, but all p poles would be at the origin—i.e., a pure time delay of p units. This is sometimes referred to as an all-zero model, or a *moving average* (MA) model. If the only input is the random term $\nu(k)$ (or perhaps a single $u(j)$ term), there is at most one non-zero b_i term. This transfer function has poles, but all zeros, if any, are at the origin. It is called an all-pole model, or alternatively, an *autoregressive* (AR) model. The general case involves both poles and zeros and is often referred to as an *autogregressive moving average* (ARMA) model.

In a multiple-input, multiple-output system, $H(z)$ is, of course, a transfer function *matrix*. Scalar transfer functions are used in Chapters 2 and 3. The complete treatment of state variable/transfer function relationships, including the matrix case, begins in Chapter 3 and continues in Chapter 12.

Alternate Model Forms

Matrix fraction description [5]. Let **P**, **N**, and **R** be finite matrix polynomials in the variable z^{-1}, which for present purposes can be treated as a delay operator. Then the ARMA-type models can be simply expressed as

$$\mathbf{P}(z^{-1})\mathbf{y}(k) = \mathbf{N}(z^{-1})\mathbf{u}(k) + \mathbf{R}(z^{-1})\mathbf{v}(k) \tag{1.6}$$

Of course, with just one input and one output, **P**, **N**, and **R** are scalar polynomials, and division by the denominator P puts Eq. (*1.6*) into the transfer function form. In the multivariable case these terms are matrices, as is the transfer function $\mathbf{H}(z)$. **P** will always be square. Taking its inverse gives the so-called left MFD (matrix fraction description) form for the transfer function,

$$\mathbf{H}(z) = \mathbf{P}^{-1}(z)\mathbf{N}(z)$$

This and the alternative right MFD are discussed in future chapters.

State variables. It will be shown in Chapter 3 that the preceding discrete-time models can be written in state variable form as

$$\begin{aligned} \mathbf{x}(k+1) &= \mathbf{A}(\boldsymbol{\theta})\mathbf{x}(k) + \mathbf{B}(\boldsymbol{\theta})u(k) \\ y(k+1) &= \mathbf{C}(\boldsymbol{\theta})\mathbf{x}(k+1) + \mathbf{D}(\boldsymbol{\theta})u(k+1) + v(k+1) \end{aligned} \tag{1.7}$$

where

k denotes a discrete time point
$\mathbf{x}$ is the state
u is the input, deterministic or random
y is the output
v is a random noise
$\boldsymbol{\theta}$ is a vector of unknown parameters
A, **B**, **C**, and **D** are system matrices

Most of the rest of this book will deal with state variable models, under the assumption that the values of $\boldsymbol{\theta}$ have been identified already, and hence $\{\mathbf{A}, \mathbf{B}, \mathbf{C}, \mathbf{D}\}$ will be assumed known. In some adaptive and self-tuning control systems, the least-squares estimation of model parameters, $\boldsymbol{\theta}$, is carried out in real time. This estimation process constitutes an outer loop. The inner control loops then use the estimated models to carry out control functions. Adaptive control is not treated in this book [6, 7].

Impulse response [4, 8]. Let the inverse Z-transform of $H(z)$ be $h(k)$; then the inverse transform of $y(z)$ can be written as a summation convolution:

$$\begin{aligned} y(k) &= \Sigma h(k-j)u(j) + v(k) \\ &= h(0)u(k) + h(1)u(k-1) + h(2)u(k-2) + h(3)u(k-3) \\ &\quad + \cdots + v(k) \end{aligned} \tag{1.8}$$

Normally, as the time parameter k continues to increase, the preceding sum continues to grow in length. Then the system description is called an infinite impulse response (IIR). The coefficients $h(k)$ are values of the impulse response, and they could also be calculated, at least in the scalar case, by long division of the transfer function. That is, $h(i)$ would be the coefficient of z^{-i} when the transfer function is written as a power series in z^{-1}. For a stable system, $h(i) \to 0$ as $i \to \infty$. This fact means that the power series might be truncated after some finite number of terms. A system model with only a finite number of past input terms is called a finite input response (FIR).

EXAMPLE 1.7 A specific second-order example of Eq. (*1.1*) is

$$y(k+1) = 1.3y(k) - 0.4\, y(k-1) + u(k+1) \tag{1.9}$$

The input-output transfer function relation is

$$y(z) = u(z)/[(1-0.5z^{-1})(1-0.8z^{-1})] = H(z)u(z) \tag{1.10}$$

The impulse response $h(kT)$ is given by the inverse Z-transform

$$h(kT) = Z^{-1}\{H(z)\}$$

By using partial fraction expansion,

$$H(z) = \frac{\frac{8}{3}z}{z-0.8} - \frac{\frac{5}{3}z}{z-0.5}$$

Then $h(kT) = \frac{8}{3}(0.8)^k - \frac{5}{3}(0.5)^k$ for any sample time k. A partial tabulation of this impulse response function follows:

k	$h(kT)$
0	1
1	1.3
2	1.29
3	1.157
4	0.9881
5	0.8217
10	0.2847
20	0.0512
30	0.0055

These $h(kT)$ values could also be evaluated as coefficients of z^{-1} in the infinite series obtained by long division of

$$H(z) = \frac{1}{1 - 1.3z^{-1} + 0.4z^{-2}}$$

Suppose that $h(kT)$ is approximated as zero for $k > 30$. Then an approximate expression for generating the outputs is

$$y(k) = u(k) + 1.3u(k-1) + 1.29u(k-2) + 1.157u(k-3) + \cdots + 0.0055u(k-30) \qquad (1.11)$$

The original form of Eq. (*1.10*) (a second-order autoregressive model) has two poles. Equation (*1.11*), a moving-average model, appears to have no nonzero poles but 30 zeros. These types of approximate equivalencies illustrate the difficulty in experimental modeling. The same sequence of measured inputs and outputs could lead to either of these results, or others, depending on the model structure which is assumed. The nonuniqueness of the answer may or may not cause problems, depending on the purpose of the derived model. ■

1.4 CLASSIFICATION OF SYSTEMS

As a result of the modeling discussion of Section 1.3, it can be seen that the types of equations required to describe a system depend on the types of elemental equations and the types of inputs from the environment. System models are classified according to the types of equations used to describe them. The family tree shown in Figure 1.8 illustrates the major system classifications. Combinations of these classes can also occur. The most significant combination is the continuous-time system, digital controller of the type mentioned in Example 1.4. The digital signals are discrete-time in nature, that is, they only change at discrete time points. The most common approach to these problems is to represent the continuous-time part of the system by a discrete-time approximate model and then proceed with a totally discrete problem. The experimentally derived ARMA models of Section 1.3.2 are approximations of this type. Other discrete approximations are given later (Problem 2.18 and Section 9.8).

In Figure 1.8 dashed lines indicate the existence of subdivisions similar to the others shown on the same level.

Distributed parameter systems require partial differential equations [9] for their description, for example, as in the description of currents and voltages at every spatial point along a transmission line. These will not be considered further, but can often be approximated by lumped-parameter models. Lumped-parameter systems are those for which all energy storage or dissipation can be lumped into a finite number of discrete spatial locations. They are described by ordinary difference equations, or in some cases by purely algebraic equations. Discrete component electric circuits fall into this category.

Systems containing parameters or signals (including inputs) which can only be described in a probabilistic fashion (due to ignorance or actual random behavior) are called stochastic, or random, systems. Because random process theory [10] is not an assumed prerequisite for this text, emphasis will be on deterministic (nonrandom)

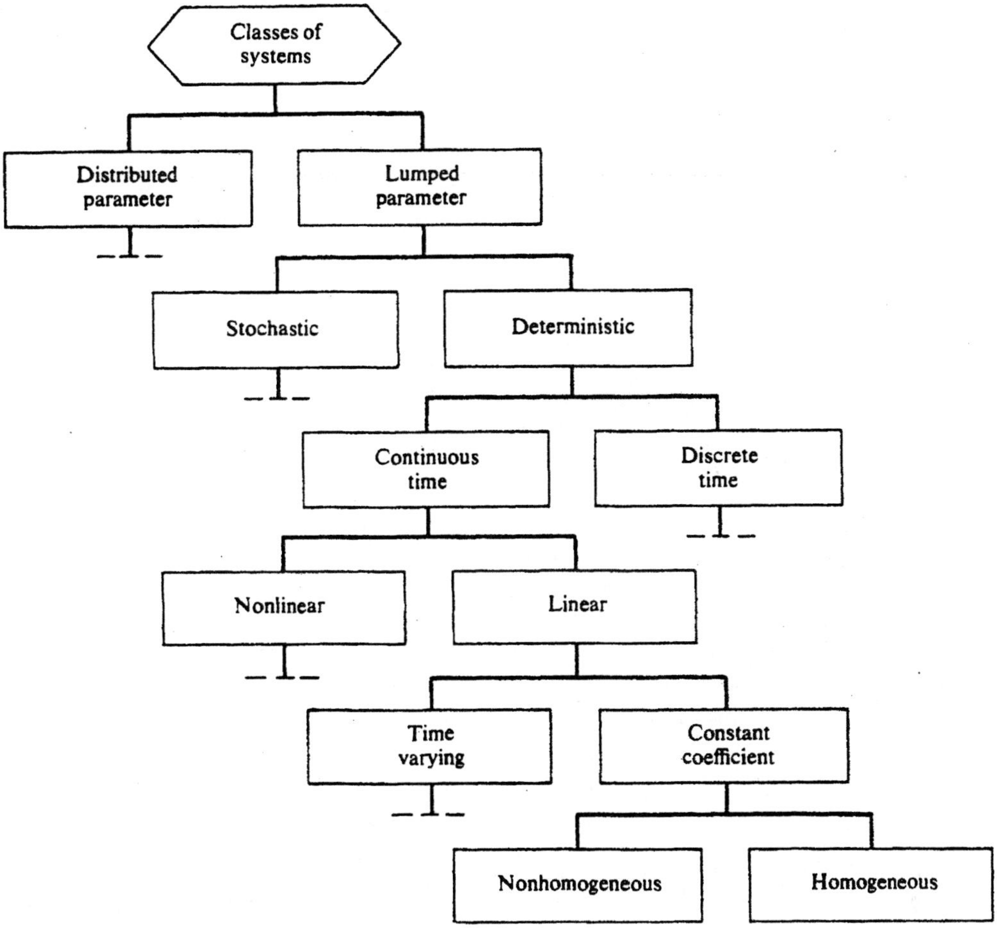

Figure 1.8 Major classes of system equations.

lumped-parameter systems. There will be a few occasions, such as in the discussion of noisy measurements, where the random nature of certain error signals cannot be totally ignored.

If all elemental equations are defined for all time, then the system is a continuous-time system. If, as in sampling or digital systems, some elemental equations are defined or used only at discrete points in time, a discrete-time system is the result. Continuous-time systems are described by differential equations, discrete-time systems by difference equations.

If all elemental equations are linear, so is the system. If one or more elemental equations are nonlinear, as is the case for a diode, then the overall system is nonlinear. When all elemental equations can be described by a set of constant parameter values, as in the familiar *RLC* circuit, the system is said to be stationary or time-invariant or constant coefficient. If one or more parameters, or the very form of an elemental equation, vary in a known fashion with time, the system is said to be time-varying.

Finally, if there are no external inputs and the system behavior is determined entirely by its initial conditions, the system is said to be homogeneous or unforced. With forcing functions acting, a nonhomogeneous system must be considered.

One additional distinction not shown in Figure 1.8 could be made between large-scale (many variables) systems and small-scale systems. The degree of difficulty in analysis varies greatly among these system classifications. These differences have motivated different methods of approach. Modern control theory provides one of the most general approaches.

1.5 MATHEMATICAL REPRESENTATIONS OF SYSTEMS

During the analytical modeling process, equations are developed to describe the behavior of each individual system element and also to describe the interconnections of these elements. These equations, or for that matter the corresponding linear graph, could be taken as the mathematical representation of the system. Normally, however, additional manipulations will be performed before the mathematical representation of the model is in final form.

Many forms are possible, but generally they divide into one of two categories: (*a*) input-output equations, and (*b*) equations which reveal the internal behavior of the system as well as input-output terminal characteristics.

Input-output equations are derived by a process of elimination of all system variables except those constituting inputs and those considered as outputs. For the system shown in Figure 1.5*a*, this would mean expressing signals S_2, S_3, S_4, and S_5 in terms of the input signals S_1 and output signals S_6. This is done using the known dynamic relations of the subsystems A, B, and C. For example, S_2 and S_4 are related to S_1 through the dynamics of subsystem A. The input-output equations could constitute one or more differential or difference equations in any of the classes shown in Figure 1.8. The independent variable is usually time, the dependent variables are the system outputs, and the inputs act as forcing functions. Models developed experimentally from measurements of inputs and outputs are almost invariably of the input-output type. Subsequent conversions to other forms, such as state variable models, are always possible.

When the constituent equations are linear with constant coefficients, Laplace or Z-transforms can be used to define input-output *transfer functions*. When more than one input and more than one output must be treated, matrix notation and the concepts of *transfer matrices* are convenient. It is assumed that the Laplace transform is a tool familiar to most readers. The Z-transform [8] may be less familiar, and so a bare-bones minimum introduction to it is contained in the problems. Although time-domain methods will be stressed in this book, transforms and transfer functions will at times be useful. A reader with no prior exposure should also consult the references.

Integral forms of the input-output equations, using the system *weighting function*, are also widely used. The weighting function, or system impulse response, is obtained from the inverse Laplace or Z-transform of the input-output transfer function.

Equations which reveal the internal behavior of the system, and not just the terminal characteristics, can take several forms. Loop equations, involving all unknown loop currents (through variables), or node equations, involving all unknown nodal voltages (across variables), can be written. The choice between alternatives may depend on the number of loops versus the number of nodes. If everything is linear, transform techniques lead to the concepts of impedance and admittance matrices. Hybrid combinations of voltage and current equations are also used.

The *state space* approach [1] will be developed extensively in the remainder of this book. At this time it is sufficient to say that state variables consist of some minimum set of variables which are essential for completely describing the internal status, i.e., state of the system.

1.6 MODERN CONTROL THEORY: THE PERSPECTIVE OF THIS BOOK

The most effective control theory makes use of good models of the real-world systems being controlled. A goal of this chapter has been to stress the importance of the modeling process. No attempt was made to present a complete, self-contained theory of modeling. Rather, the intent was to build upon knowledge acquired in prior courses on circuit theory, dynamics, kinematics, and so on. Some background knowledge in introductory feedback control is also assumed. A summary review of this topic is included in Chapter 2.

In some industrial control applications, good results are achieved even though only a rudimentary knowledge of a process model is available. The widely used proportional-integral-derivative (PID) controller can be tuned to give satisfactory performance based solely on knowledge of dominant system time constants. This fact does not violate the dictum that good models are required; it simply reinforces the point that models should be suited to the intended purpose.

In other areas of control practice, systems are successfully designed and built without use of mathematical models or analysis. This artisan, or craftsperson, approach can often work reasonably well when the designer is sufficiently experienced and the system is only incrementally different from previous successful designs. However, when problems do arise, the same ad hoc approach to an attempted solution can sometimes compound the difficulties. One such example occurred with an industrial robotic system. The system exhibited unacceptable signal fluctuations on occasion, which were erroneously diagnosed as a noise problem. Compensating capacitors were added between key signal lines and ground to allow the high-frequency noise terms to bleed off. An after-the-fact model revealed the true problem was caused by very low stability margins. The "compensating" capacitors added additional phase lag and only made the problem worse.

Elegant mathematical results derived from an inappropriate model can likewise yield negative results. Exact models of real systems are extremely rare. Therefore, robustness is an important quality in control design. *Robustness* can be defined in various ways, but generally the word implies the maintenance of adequate stability margins or other performance levels in spite of model errors or deliberate oversimplifications. The ability to operate in the presence of disturbance inputs is also important. The widely used PID controllers owe much of the success to what today would be called their robustness. The linear-quadratic (LQ) optimal controllers of Chapter 14 have certain guaranteed robustness properties. The newer H^∞ design techniques, which are basically worst-case analyses, represent another approach to robustness in the face of model error. These are not pursued in this book [11].

Many real systems are nonlinear and/or time-varying. Yet, approximately 80% of this book is devoted to linear, constant systems. Perhaps 10% is devoted explicitly to linear time-varying systems and another 10% (primarily Chapter 15) is devoted to nonlinear systems. The purpose of this book is to build a foundation for specialized study that may follow, and linear systems theory is the major part of that foundation. The treatment of nonlinear systems in Chapter 15 is restricted mainly to extensions of the linear theory that follow easily from earlier developments in this book. Several useful approaches to the control of some classes of nonlinear systems are presented.

Chapters 4 through 8 present a large amount of linear algebra for controls rather than control theory per se. An attempt has been made to motivate the linear algebraic developments by bringing in related control topics, even though the same topics may be developed more fully later. Some algorithmic considerations are also included. Liberal use of computer algorithms has been made throughout the book. One advantage of the state variable approach to control problems is that the structure remains the same whether there are 2, 20, or 100 states. The fact is, however, that only the smallest problems can be solved without a computer. Many solutions provided in this book were carried out using code acquired from the literature [12], perhaps with modifications, or with programs developed during the years of teaching this material. There are now several commercially available packages which have the needed capabilities [13, 14].

A large number of diverse problems and examples are included in each chapter. The intention is to show not only how a given problem is worked but why it is worked in a certain way and what the ramifications are. For example, knowing how to compute the feedback gains to achieve certain closed-loop poles is a mathematical result. Additional engineering insight is needed in order to decide whether a pole placement approach should be used and, if so, what constitutes good pole locations. Alternatives are classical feedback design or an optimal control design. In these, intelligent trade-offs must be made between response time, control effort, or disturbance rejection. The problems are intended to give some insight into these issues, in addition to illustrating the mechanics of a given method.

Closely related technical areas include self-tuning and adaptive control [7, 8], learning systems and artificial intelligence [7, 15], neural networks [16], and robotics [17]. A short chapter or two at the end of the present book could not do justice to any of these important topics. Therefore, this book concentrates on developing a basic foundation which would be useful to the broadest class of readers. Those wishing to pursue one of these special topics later will be able to do so more effectively after mastering the material given here.

REFERENCES

1. Zadeh, L. A. and C. A. Desoer: *Linear System Theory, The State Space Approach,* McGraw-Hill, New York, 1963.
2. Cannon, R. H., Jr.: *Dynamics of Physical Systems,* McGraw-Hill, New York, 1967.

3. Shearer, J. A., A. T. Murphy, and H. H. Richardson: *Introduction to System Dynamics,* Addison-Wesley, Reading, Mass., 1967.
4. Makhoul, J.: "Linear Prediction: A Tutorial Review," *Proceedings of the IEEE,* Vol. 63, No. 4, April 1975.
5. Kailath, T.: *Linear Systems,* Prentice-Hall, Englewood Cliffs, N.J., 1980.
6. Harris, C. J. and S. A. Billings: *Self-Tuning and Adaptive Control,* Peter Peregrinus Ltd. (for IEE), London, 1981.
7. Astrom, K. J. and B. Wittenmark: *Adaptive Control,* Addison-Wesley, Reading, Mass., 1989.
8. Franklin, G. F. and J. D. Powell: *Digital Control of Dynamic Systems,* Addison-Wesley, Reading, Mass., 1980.
9. Brogan, W. L.: "Optimal Control Applied to Systems Described by Partial Differential Equations," *Advances in Control Systems,* Vol. 6, C. T. Leondes, Ed., Academic Press, New York, 1968.
10. Papoulis, A.: *Probability, Random Variables and Stochastic Processes,* McGraw-Hill, New York, 1965.
11. Zames, G.: "Feedback and Optimal Sensitivity: Model Reference Transformations, Multiplicative Seminorms and Approximate Inverses," *IEEE Transactions on Automatic Control,* Vol. AC-26, No. 2, April 1981, pp. 301–320.
12. Melsa, J. L. and S. K. Jones: *Computer Programs for Computational Assistance in the Study of Linear Control Theory,* 2d ed., McGraw-Hill, New York, 1973.
13. Herget, C. J. and A. J. Laub, Eds.: "Computer Aided Design of Control Systems Special Issue," *IEEE Control Systems Magazine,* Vol. 2, No. 4, Dec. 1982.
14. Moler, C. "MATLAB Users' Guide," *Tech. Report CS81-1* (Revised) Dept. Of Computer Science, University of New Mexico, Albuquerque, N.M., Aug. 1982.
15. Grossberg, S.: *The Adaptive Brain I: Cognition, Learning, Reinforcement, and Rhythm,* and *The Adaptive Brain II: Vision Speech, Language and Motor Control,* Elsevier/North Holland, Amsterdam, 1986.
16. Bavarian, B., Ed.: "Special Section on Neural Networks for Systems and Control" (five articles), *IEEE Control Systems Magazine,* Vol. 8, No. 2, Apr. 1988, pp. 3–31. (See also Vol. 9, No. 3, Apr. 1989, pp. 25–59 for five more articles on neural networks in controls.)
17. Klafter, R. D., T. A. Chmielewski, and M. Negin: *Robotic Engineering, an Integrated Approach,* Prentice Hall, Englewood Cliffs, N.J., 1989.
18. Chestnut, H.: "A Systems Approach to the Economic Use of Computers for Controlling Systems in Industry," *General Electric Report no. 70-C-089,* Schenectady, N.Y., Feb. 1970.
19. Kuo, B. C.: *Analysis and Synthesis of Sampled-Data Control Systems,* Prentice-Hall, Englewood Cliffs, N.J., 1963.

ILLUSTRATIVE PROBLEMS

1.1 Explain why it would be inappropriate to consider a single branch of an electric network a system, even if the only variable of interest is the current through that branch.

The current in one branch affects the currents and voltages in other parts of the network. This, in turn, affects the current in the first branch. Because of the two-way coupling, the network must be considered as a whole, and must be solved using simultaneous equations.

1.2 Develop an electromechanical model of the fixed field, armature-controlled dc motor. Consider the voltage supplied to the armature as the input and account for the observed dissipation of electrical energy and mechanical energy.

The dissipation of electrical energy can be accounted for by lumping all armature resist-

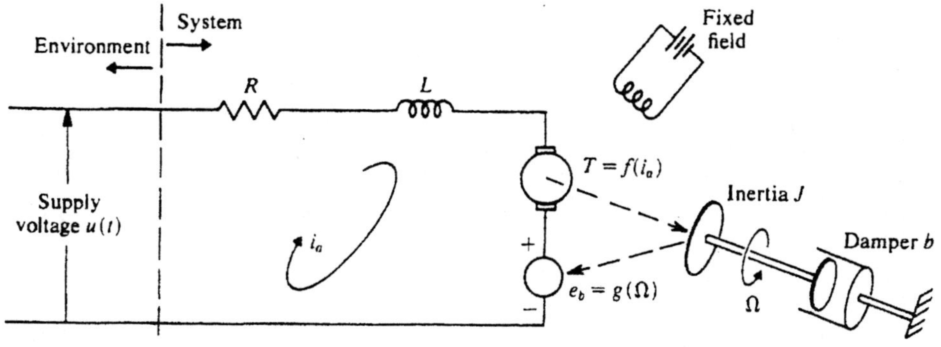

Figure 1.9

ance into a resistor R. A noticeable phase shift between the supply voltage and the current through the armature windings can be accounted for by a lumped inductor L. The load and all rotating parts can be represented as a lumped inertia element J. Mechanical energy losses are accounted for by adding an ideal damper b between the rotating load and some fixed reference. The connection between the electrical and the mechanical aspects is obtained from Maxwell's equations. A moving charge in a magnetic field has a force exerted upon it, so that the armature torque T is a function of the armature current i_a. Likewise, a conductor moving in a magnetic field has a voltage induced in it, the back emf e_b. The model is shown schematically in Figure 1.9.

The torque and the back emf are often approximated by the linear relationships $T = Ki_a$ and $e_b = K\Omega$, where K is a constant for a particular motor. For the linear case the transfer function between the input voltage $u(t)$ and the output angle $y(t) = \int \Omega dt$ is derived as follows. The loop equation for the electrical circuit is $u(t) = L(di_a/dt) + Ri_a + e_b$. The mechanical torque balance is $T(t) = J\ddot{y} + b\dot{y}$. The Laplace transforms of these equations are $u(s) = (Ls + R)i_a(s) + e_b(s)$ and $T(s) = (Js^2 + bs)y(s)$. Solving gives $i_a(s) = [u(s) - e_b(s)]/(R + Ls)$. The electromechanical conversion equation then gives $T(s) = K[u(s) - e_b(s)]/(R + Ls)$. Equating the two forms for $T(s)$ and using $e_b(s) = Ksy(s)$ gives $(Js^2 + bs)y(s) = K[u(s) - Ksy(s)]/(R + Ls)$. The input-output transfer function is found from this:

$$y(s)/u(s) = K/[(Js^2 + bs)(R + Ls) + K^2 s]$$

1.3 Some electronic test gear (Figure 1.10) is mounted near a large tank of liquid gas at −350°F. Develop a simple model which would be useful in estimating the coldest temperature at which the electronic equipment will need to operate.

Because of the insulation material, heat is allowed to flow only in one direction, from the 70° air through the electronic package to the −350° liquid gas. The environment, consisting of the two constant temperatures of 70 and −350, is represented by two ideal sources. There is a single unknown temperature T (across variable), that of the electronic package interior. The

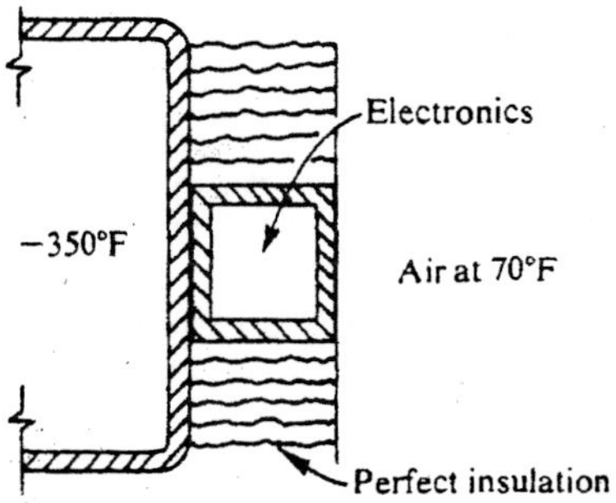

Figure 1.10

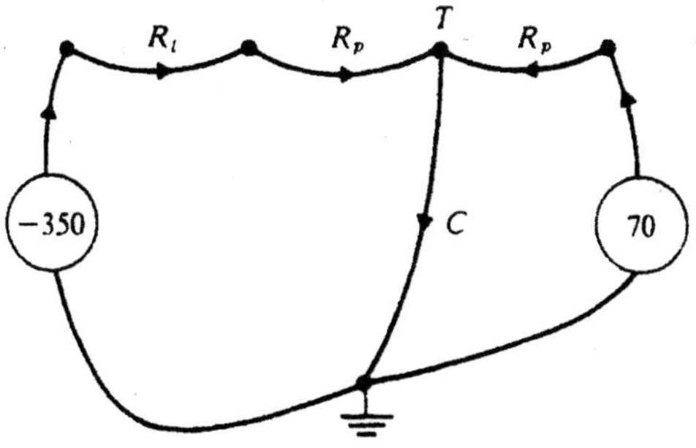

Figure 1.11

package has some thermal capacitance C and its end walls and the tank wall present thermal resistance, R_p and R_t, to heat flow Q. The linear graph is shown in Figure 1.11.

In steady state no heat flows in the branch representing the capacitance. Thus $70 - T = QR_p$ and $T - (-350) = Q(R_p + R_t)$. Eliminating the heat flow Q gives

$$\frac{70 - T}{R_p} = \frac{T + 350}{R_p + R_t} \quad \text{or} \quad T = \frac{70(R_p + R_t) - 350R_p}{2R_p + R_t}$$

If the thermal resistivity R_p of each end of the electronic package is 1°F s/Btu and if the thermal resistivity R_t of the adjacent area of the tank is 2°F s/Btu, then $T = [70(3) - 350(1)]/4 = -35$°F.

1.4 In many ways the flow of work through a factory is similar to fluid flow in a piping network. Figure 1.12 shows such a network. Two pumps deliver fluid at constant pressure P_1 and P_2, respectively. Six lumped approximations for fluid resistance R_i are indicated. They account for the pressure drop in each segment of pipe proportional to the flow Q through the segment. Three fluid capacitances C_i are indicated. The pressure at their base is proportional to the height of the standing fluid, that is, proportional to the integral of the flow into them. Two ideal elements, called fluid inertances I_1 and I_2, are included to account for inertia effects. They cause a pressure drop proportional to the rate of change of flow. a. Draw the linear graph, label all variables; b. write the elemental equations; c. write the continuity equations; d. write the compatibility equations. Neglect all changes in height except in the capacitances, and use atmospheric pressure as the reference node.

(a) Each distinct pressure (across variable) will form a system node. Between each pair of nodes a branch will represent the ideal element that accounts for the pressure change. This allows the construction of the linear graph of Figure 1.13.

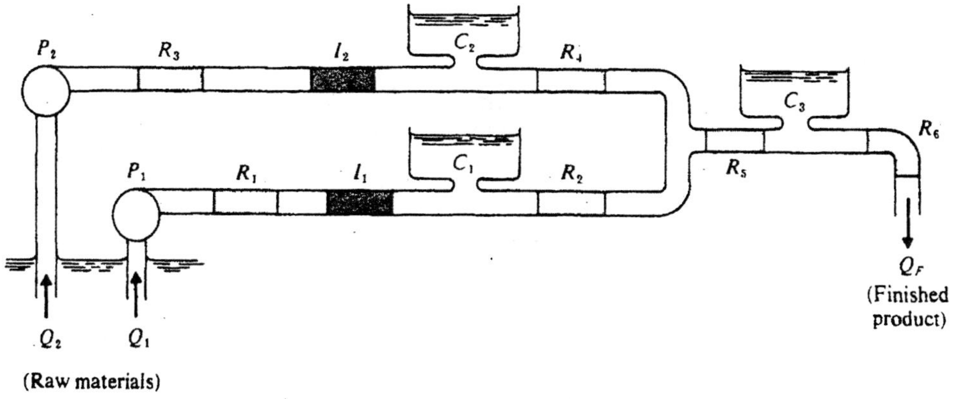

Figure 1.12

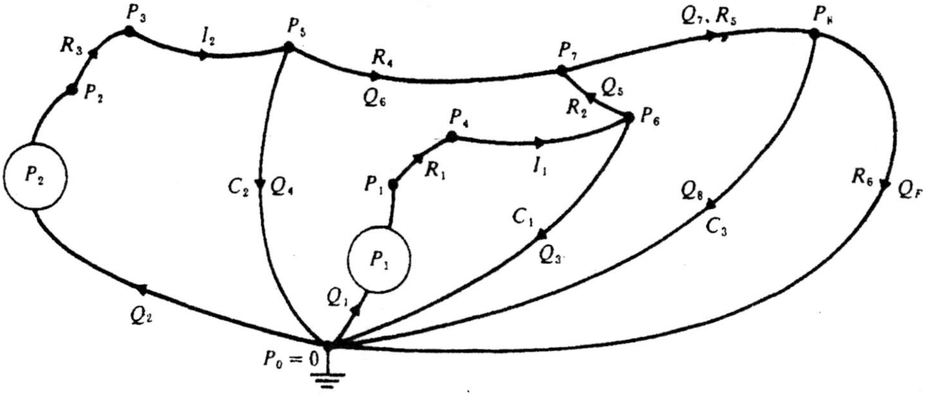

Figure 1.13

(b) There are $n = 9$ nodes, $b = 13$ branches, and $s = 2$ sources. The $b - s = 11$ elemental equations are

$$P_{23} = R_3 Q_2 \qquad P_{14} = R_1 Q_1 \qquad P_{46} = I_1 \frac{dQ_1}{dt} \qquad P_{45} = I_2 \frac{dQ_2}{dt}$$

$$C_2 \frac{dP_{50}}{dt} = Q_4 \qquad C_1 \frac{dP_{60}}{dt} = Q_3 \qquad P_{57} = R_4 Q_6 \qquad P_{67} = R_2 Q_5$$

$$P_{78} = R_5 Q_7 \qquad C_3 \frac{dP_{80}}{dt} = Q_8 \qquad P_{80} = R_6 Q_F$$

(c) The $n - 1 = 8$ continuity equations are

$Q_1 = Q_1$ (twice) since Q_1 flows in three separate branches

$Q_2 = Q_2$ (twice) since Q_2 flows in three separate branches

$$Q_1 - Q_3 - Q_5 = 0 \qquad Q_2 - Q_4 - Q_6 = 0 \qquad Q_5 + Q_6 - Q_7 = 0$$

$$Q_7 - Q_8 - Q_F = 0$$

(d) The $b - (n - 1) = 5$ compatibility equations are

$$P_1 = P_{14} + P_{46} + P_{60} \qquad P_2 = P_{23} + P_{35} + P_{50}$$

$$P_{50} = P_{57} + P_{76} + P_{60} \qquad P_{60} = P_{67} + P_{78} + P_{80}$$

$$P_{08} + P_{80} = 0$$

By eliminating variables in various ways, a set of differential equations involving only flow rates Q_i, or only nodal pressures P_i, or a combination of both could be obtained.

1.5 **(a)** Write equations describing the lumped-parameter approximate model for the transmission line shown in Figure 1.14.

(b) Find the input-output transfer function $y(s)/u(s)$. The input $u(t)$ is the source voltage v_s, and the output $y(t)$ is the load voltage v_L.

(a) For simplicity, the line is segmented into three equal lengths as shown. The leakage conductance G is the reciprocal of the leakage resistance. More segments could be used in the same manner.

The values of R and L are obviously 1/3 that for the entire line, while G and C each have values equal to 1/2 that for the entire line. Since the source voltage is known, there are six unknowns: i_0, i_1, i_2, v_1, v_2, and v_L.

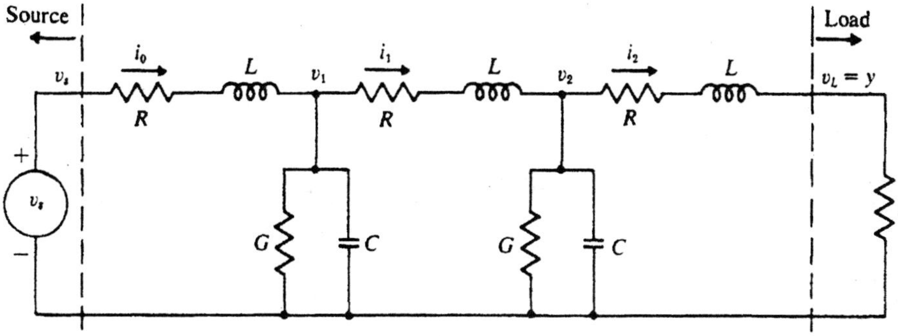

Figure 1.14

(b) Writing loop equations, using the Laplace transforms of the elemental equations, gives

$$v_s = [R + Ls]i_0 + v_1 \qquad v_1 = [R + Ls]i_1 + v_2$$
$$v_2 = [R + Ls]i_2 + v_L \qquad v_L = R_L i_2$$

The nodal equations are $i_0 = (G + Cs)v_1 + i_1$ and $i_1 = (G + Cs)v_2 + i_2$. Letting $A = R + Ls$ and $B = G + Cs$ gives

$$v_s = Ai_0 + v_1 = (AB + 1)v_1 + Ai_1 = (AB + 1)v_2 + (A^2B + 2A)i_1$$

By continuing this process of substitution, a final expression containing only $v_s = u$ and $v_L = y$ is obtained:

$$\frac{y(s)}{u(s)} = \frac{R_L}{(A^3B^2 + 4A^2B + 3A) + R_L(A^2B + 3AB + 1)}$$

1.6 Derive a difference equation for the purely resistive ladder network shown in Figure 1.15 (perhaps a dc version of the lumped approximation for a transmission line).

The difference equation for a typical $(k + 1)$st loop is obtained by writing a loop equation

$$(2r + R)i_{k+1} - ri_k - ri_{k+2} = 0$$

This holds for $1 \le k + 1 \le N - 1$. It is a second-order difference equation, and two boundary conditions are needed in order to uniquely specify the solution. The first and the last loops, which do not satisfy the general equation, provide the two necessary conditions:

$$v_s = (R + r)i_0 - ri_1$$
$$0 = (r + R + R_L)i_N - ri_{N-1}$$

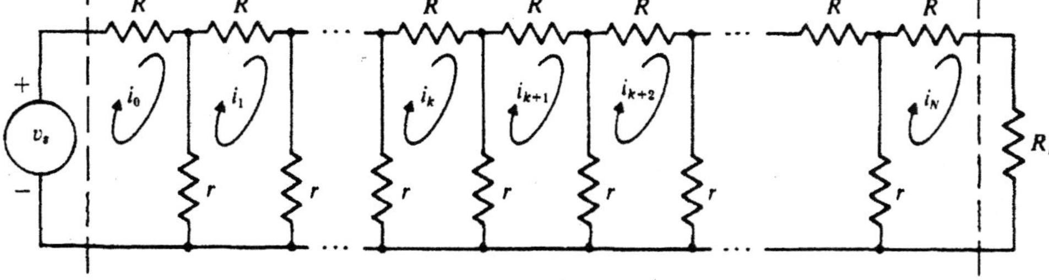

Figure 1.15

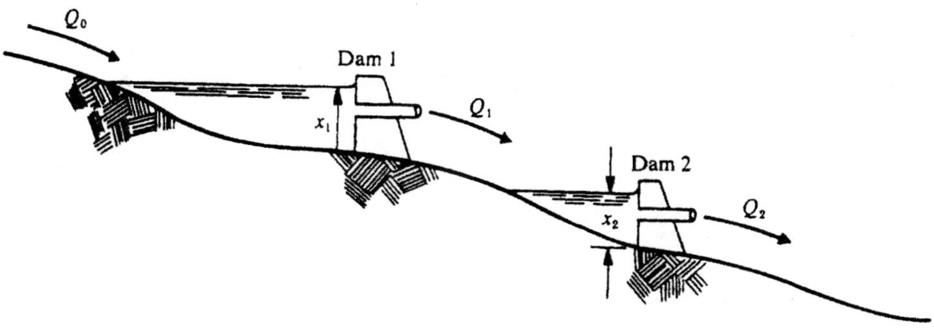

Figure 1.16

1.7 A pair of dams in a flood-control project is shown in Figure 1.16. The water level at dam 1 at a given time t_k is $x_1(k)$, and $x_2(k)$ is the height at dam 2 at the same time. The amount of run-off water collected in reservoir 1 between times t_k and t_{k+1} is $Q_0(k)$. The water released from dams 1 and 2 during this period is denoted by $Q_1(k)$ and $Q_2(k)$. Develop a discrete-time model for this system.

Conservation of flow requires

$$x_1(k+1) = x_1(k) + \alpha[Q_0(k) - Q_1(k)]$$

$$x_2(k+1) = x_2(k) + \beta[Q_1(k) - Q_2(k)]$$

These represent lumped-parameter discrete-time equations. If the amount of controlled spillages Q_1 and Q_2 are selected as functions of the water heights x_1 and x_2, a discrete feedback control system obviously results. Z-transform theory could be used to analyze such a system.

1.8 Draw the linear graph for the ideal transformer circuit of Figure 1.17 noting that the transformer is a four-terminal element.

The linear graph is shown in Figure 1.18*a* with the transformer represented as shown in Figure 1.18*b*.

The equations for this circuit are

Node equations: $i_s = i_1$, $i_3 = -i_2$, $i_3 = i_4$

Elemental equations: $v_2 = Nv_1$, $i_1 = -Ni_2$ (transformer)

$v_2 - v_3 = i_3 R$ (resistance)

$C\dot{v}_3 = i_4$ (capacitance)

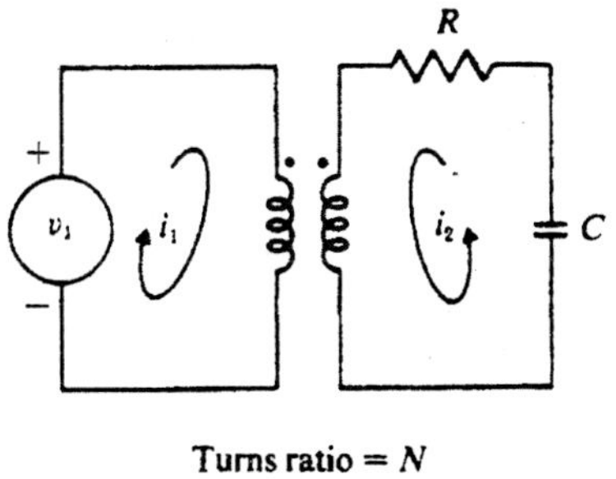

Figure 1.17

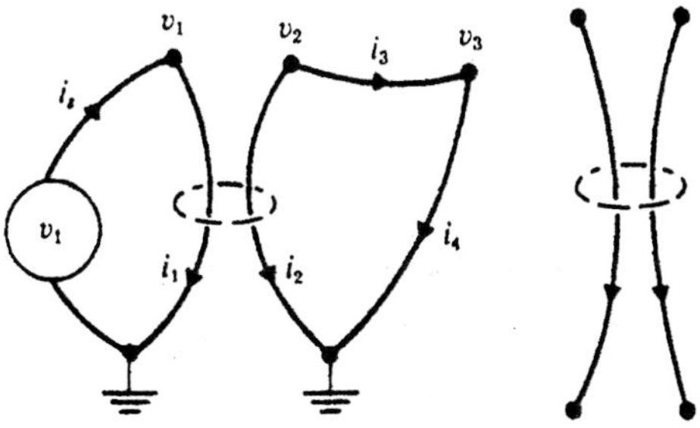

Figure 1.18

Note that the transformer requires two equations for its specification. Similar multiterminal elements are required whenever energy is transformed from one form to another. Transformers, transducers, and gyrators all have similar representation. Note also that an ideal transformer has zero instantaneous power flow into it, i.e.,

$$v_1 i_1 + v_2 i_2 = v_1(-Ni_2) + (Nv_1)i_2 = 0$$

1.9 Develop a model of an automobile which would be appropriate for studying the effectiveness of the suspension system, tire characteristics, and seat design on passenger comfort.

For simplicity, lateral rolling motions are ignored. An idealized model might be represented as shown in Figure 1.19. The displacements x_1 and x_2 are inputs from the environment (road surface). Masses m_1 and m_2 represent the wheels, whereas M and J represent the mass and pitching inertia of the main car body. The seat and passenger mass are represented by m_p. The elasticity and energy dissipation properties of the tires are represented by k_1, k_2, b_1, and b_2. The suspension system is represented by k_3, k_4, b_3, and b_4. The seat characteristics are represented by k_s. Newton's second law is applied to the wheels, giving

$$m_1\ddot{x}_3 = k_1(x_1 - x_3) + b_1(\dot{x}_1 - \dot{x}_3) + k_3(x_5 - x_3) + b_3(\dot{x}_5 - \dot{x}_3)$$

$$m_2\ddot{x}_4 = k_2(x_2 - x_4) + b_2(\dot{x}_2 - \dot{x}_4) + k_4(x_6 - x_4) + b_4(\dot{x}_6 - \dot{x}_4)$$

Letting l_1 be the distance from the left end to the center of gravity cg and letting l_2 be the distance to the seat mount, the following geometric relations can be obtained. Assuming small angles,

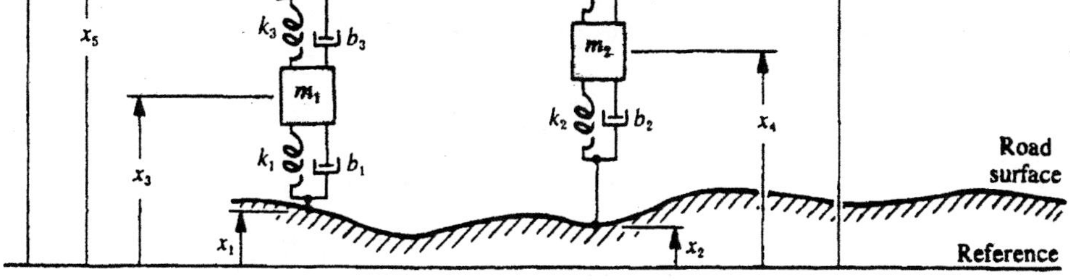

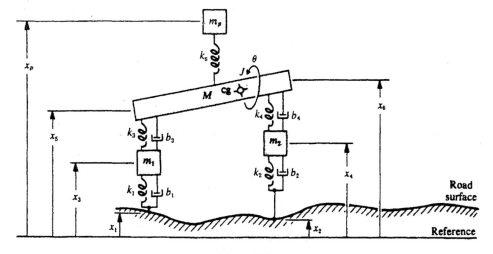

Figure 1.19

$$x_{cg} = x_5 + \frac{l_1}{l}(x_6 - x_5)$$

$$x_s = x_5 + \frac{l_2}{l}(x_6 - x_5) \quad \text{and} \quad \theta = \frac{x_6 - x_5}{l}$$

where l is the total length (wheel base). Summing forces on M gives

$$M\ddot{x}_{cg} = k_3(x_3 - x_5) + k_4(x_4 - x_6) + k_s(x_p - x_s) + b_3(\dot{x}_3 - \dot{x}_5) + b_4(\dot{x}_4 - \dot{x}_6)$$

Summing torques gives

$$J\ddot{\theta} = -l_1 k_3(x_3 - x_5) + (l - l_1)k_4(x_4 - x_6) - (l_1 - l_2)k_s(x_p - x_s)$$
$$-l_1 b_3(\dot{x}_3 - \dot{x}_5) + (l - l_1)b_4(\dot{x}_4 - \dot{x}_6)$$

Finally, summing forces on m_p gives $m_p\ddot{x}_p = k_s(x_s - x_p)$. This set of five coupled second-order differential equations, along with the geometric constraints, constitutes an approximate model for this system.

1.10 A typical common base amplifier circuit, using a *pnp* transistor, is shown in Figure 1.20*a*. The h-parameter equivalent circuit for small signals within the amplifier mid-band frequency range is given in Figure 1.20*b*. Draw the linear graph for the amplifier.

The input signal voltage is replaced by an ideal source v_s in series with the source resistance R_s. The three-terminal transistor device is described by the four hybrid parameters h_{ib}, h_{rb}, h_{fb}, and h_{ob}, which are straight line approximations to the various nonlinear device characteristics in the vicinity of the operating point.

In this example there are two dependent, or controlled, sources, described by $v_d = h_{rb}v_3$ and $i_d = h_{fb}i_2$. Using the equations implied by the linear graph, Figure 1.21,

$$v_1 = i_2 h_{ib} + h_{rb}v_3 \qquad (E\text{-}B \text{ loop equation}) \qquad (1)$$

$$h_{fb}i_2 + i_3 + i_5 + i_7 = 0 \qquad (C \text{ node equation}) \qquad (2)$$

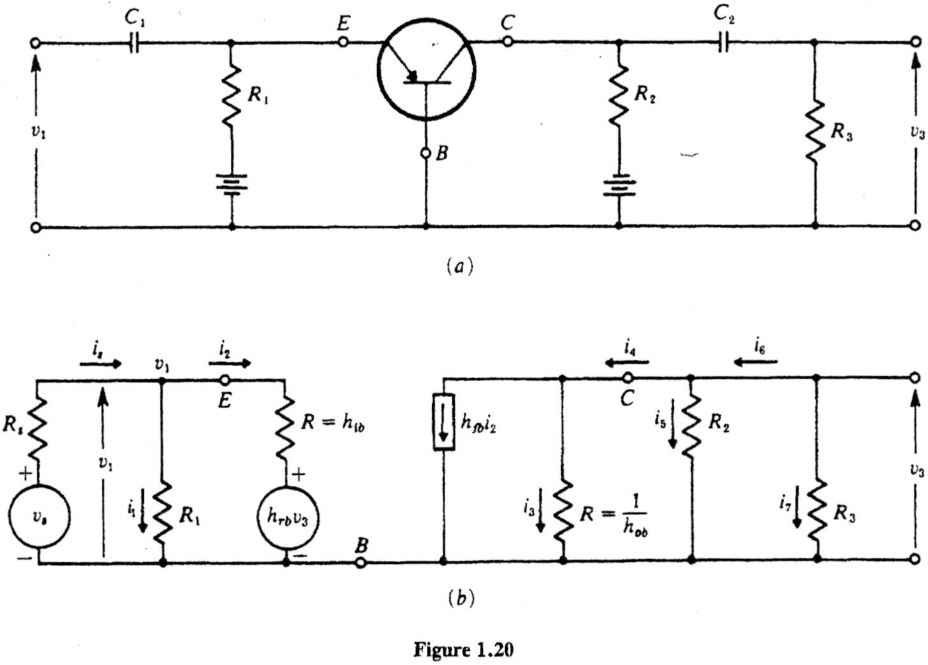

Figure 1.20

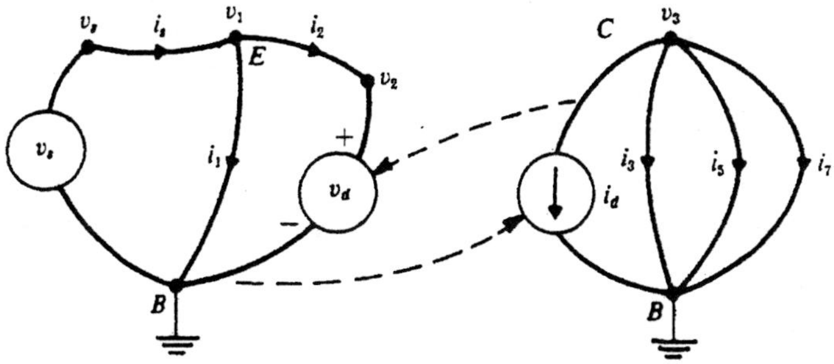

Figure 1.21

or

$$h_{fb}i_2 + v_3 h_{ob} + v_3/R_2 + v_3/R_3 = 0 \qquad \text{(using elemental equations)} \tag{3}$$

From equation (3),

$$i_2 = \frac{-v_3}{h_{fb}}\left(h_{ob} + \frac{1}{R_2} + \frac{1}{R_3}\right) \tag{4}$$

Using equation (4) in equation (1) gives the voltage input-output equation:

$$v_3 = \frac{-h_{fb}v_1}{h_{ib}(h_{ob} + 1/R_2 + 1/R_3) - h_{rb}h_{fb}}$$

1.11 Draw a block diagram for the motor system of Problem 1.2, preserving the individual identity of the electrical, mechanical, and conversion aspects, and illustrating the feedback nature of this system.

The block diagram of Figure 1.22 is drawn directly from the constituent equations of Problem 1.2.

1.12 Classify the systems described by the following equations:

(a) $\ddot{y} + t^2\dot{y} - 6y = u(t)$

(b) $\dddot{y} + \ddot{y}y + 4y = 0$

(c) $\dfrac{\partial^2 y}{\partial t^2} = a\dfrac{\partial^2 y}{\partial x^2}$

(d) $\ddot{y} + a\dot{y} + by^2 = u(t)$

(e) $\dot{y} + ay = u(t)$ if $t < t_1$
$\dot{y} + by = u(t)$ if $t \geq t_1$

(f) $\dot{y} + a \max(0, y) = 0$

(a) Lumped parameter, linear, continuous time, time variable coefficient (t^2), nonhomogeneous.
(b) Lumped parameter, nonlinear (due to $\ddot{y}y$ term), continuous time.
(c) Distributed parameter.
(d) Lumped parameter, nonlinear (due to y^2 term).
(e) Lumped parameter, linear, time variable.
(f) Lumped parameter, nonlinear.

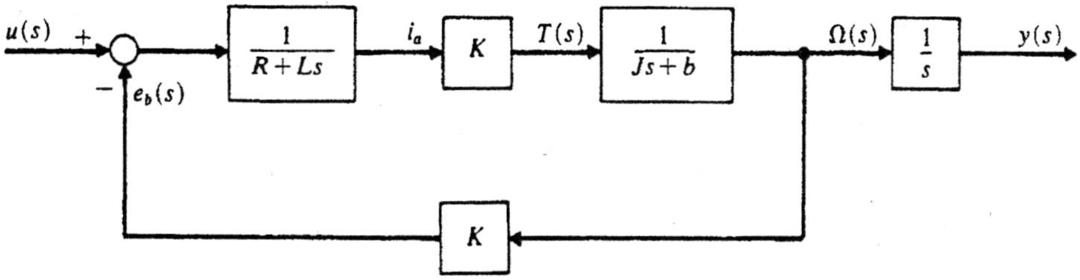

Figure 1.22

Changing of system characteristics, through switching at predetermined times, does not make a system nonlinear. However, if the switching depends on the magnitude of the dependent variable y, the system is nonlinear.

1.13 Use the concepts of Figure 1.5, page 5, to discuss the development of a mathematical model for a rocket vehicle.

1. With such a vague problem statement, many purposes for this model could be considered, such as the structural adequacy of the design, or the temperature history of a component within the vehicle. Suppose that the purpose is to study the trajectory of the vehicle.
2. The boundary of the system is the physical envelope of the vehicle. The inputs from the external environment consist of atmospheric and gravitational effects, as well as a thrust force caused by the gases being expelled across the system boundary. Additional inputs are the mission data, which specify key characteristics that the trajectory should possess. Outputs are the components of position and velocity along the resulting trajectory.
3. The structure of this system consists of several subsystems. One of these is the vehicle dynamics subsystem, which relates forces and torques to the vehicle acceleration. Another is the kinematic subsystem, which relates accelerations to vehicle positions and velocities. A third subsystem is the navigation subsystem, which takes measurements of position, velocity, or acceleration and provides useful signals containing present position and velocity information. A fourth subsystem, the guidance system, accepts the position-velocity data, compares it with mission goals, and computes guidance commands. The final subsystem is the control subsystem. It accepts guidance commands as inputs, and its outputs are commanded body attitude angles or angular rates which will cause the vehicle to steer to the desired trajectory. The control system also turns the thrust on and off. The overall system is shown in Figure 1.23.
4. A wide diversity of models could be developed. If only position and velocity are of interest, the vehicle may be represented as a point mass and the attitude control system might be assumed perfect. If angular attitude information is desired, detailed equations for rotational motion may be required. The description could include elastic vehicle bending, spurious torques due to fuel sloshing, the dynamics of hydraulic control actuators, etc. Each of the other subsystems could be broken down into very fine detail, if required.

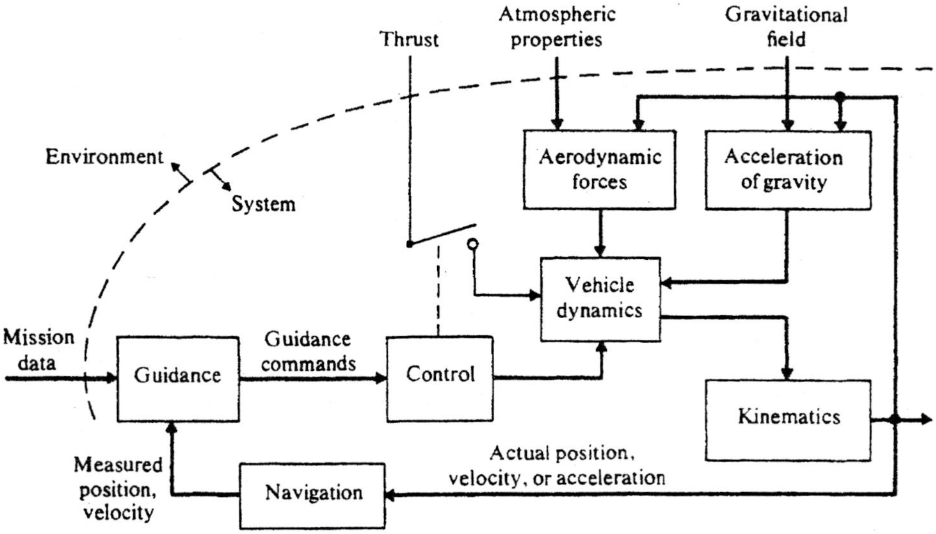

Figure 1.23

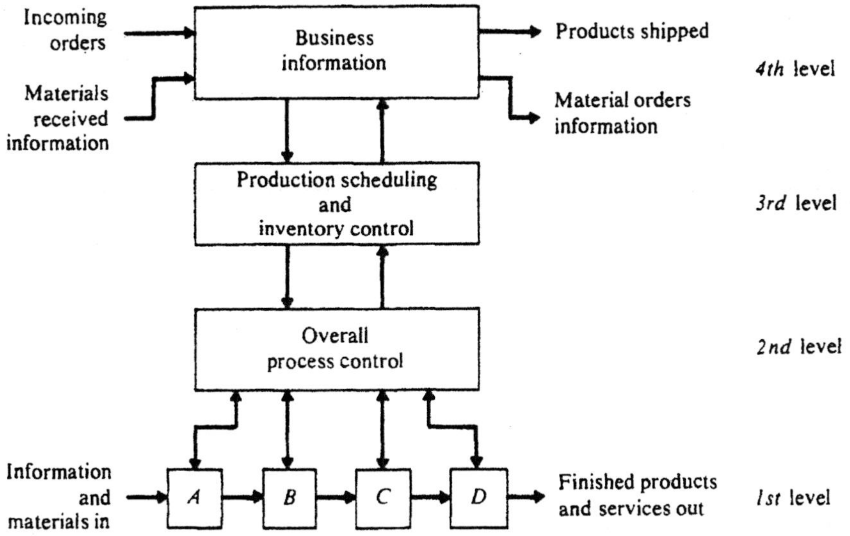

Figure 1.24

5. The remaining steps in Figure 1.5 are relatively straightforward if the preceding steps have been carried out correctly. The resulting equations will be nonlinear and involve many variables, coordinate transformations, etc. Except in certain simple cases, computer simulation will be required.

1.14 Discuss the various levels at which control techniques can be applied in industrial systems.

Harold Chestnut [18] gives the four-level representation to control activities in the business environment shown in Figure 1.24.

Individual operations A, B, C, and D might represent automatic machine tools, as in Example 1.3. At this level, a very complete model is required to study detailed behavior. The overall sequence A, B, C, D could represent a process such as an automated steel mill or chemical plant. The general characteristics of the sequence might be modeled in a way similar to the system of Problem 1.4. The characteristics of the interacting sequence would be described by the level 2 model, but details of individual elements need no longer be apparent. At level 3 a still broader view is taken. The model might be used to determine how work schedules should be set in order to efficiently use the system capability while maintaining optimum inventory, avoiding premium overtime pay, and meeting delivery schedules. At the fourth level, the broadest view is taken. A complete production line might be viewed as a simple time delay. Broader questions regarding market forecasts, new product development, plant expansion, and customer relations become dominant.

According to Mr. Chestnut, "Traditionally the automatic control engineers have focused their attention on the first and second levels of control, which are those associated with the fast control functions in the energy and materials ends of the industrial spectrum. With the current emphasis being developed in the systems aspect of the overall industrial process, more attention is being given to the third and fourth levels of control where significant economies in time and money and resources can and are being realized and can be more readily brought to the attention of the customers." Some of these "big picture" economic systems models are now being facilitated by the various spread sheet and data base management programs which are widely available on management's personal microcomputers.

1.15 Define the Z-transform of a continuous-time signal $y(t)$.

The Z-transform of $y(t)$, written $Z\{y(t)\} = Y(z)$, is defined as the result of a three-step operation:

(i) Modulate $y(t)$ with a periodic train of Dirac delta functions, i.e.,

$$y_s(t) = y(t) \sum_{n=-\infty}^{\infty} \delta(t - nT)$$

where T is the sample period. Since $\delta(\ )$ is zero, except when its argument is zero,

$$y_s(t) = \sum_{n=0}^{\infty} y(nT)\delta(t - nT)$$

assuming $y(t) = 0$ for $t < 0$.

(ii) Laplace transform the impulse-modulated signal

$$Y^*(s) \triangleq \mathcal{L}\{y_s(t)\} = \sum_{n=0}^{\infty} y(nT)e^{-nTs}$$

Note that $y(nT)$ is no longer a function, but just a set of sample values. These act as constants as far as $\mathcal{L}$ is concerned.

(iii) Make a change of variables $z = e^{Ts}$. Thus

$$Y(z) = Y^*(s)|_{z=e^{Ts}} = \sum_{n=0}^{\infty} y(nT)z^{-n}$$

The reason for the change of variables is to allow working with polynomials in z rather than transcendental functions in s.

1.16 What is the significance of the Z-transform as expressed in the previous problem?

The Z-transform of a function can be written in various other forms such as ratios of polynomials in z or z^{-1}. However, if a function's Z-transform can be manipulated into an infinite series in z^{-1}, then we can pick off the function's value at time $t = nT$ as the coefficient multiplying z^{-n}. This series form can be found by long division or by using knowledge of some standard infinite series results.

If $y(t)$ is a unit step, then all $y(nT) = 1$, so

$$Y(z) = \sum_{n=0}^{\infty} z^{-n} = \frac{1}{1 - z^{-1}}$$

Conversely, if $Y(z) = \frac{z}{z - 0.5}$, then long division gives

$$Y(z) = 1 + 0.5z^{-1} + 0.25z^{-2} + \cdots$$

From this it is immediately known that

$$y(0) = 1, y(T) = 0.5, y(2T) = 0.25, \ldots, \text{etc.}$$

1.17 What are some other methods of determining the inverse Z-transform?

(i) There are extensive tables of transform pairs available [19]. It should be pointed out that the a function has unique Z-transform, but the inverse transform is not unique. Many functions have the same sample values, but are different between samples.

(ii) A complicated transform expression can often be written as the sum of several simple terms, using a variation of partial fraction expansion. Then each simple term can be inverted.

(iii) The formal definition of the inverse transform is

$$y(nT) = \frac{1}{2\pi j} \oint Y(z)z^{n-1}\,dz$$

The contour integral is around a closed path that encloses all singularities of $Y(z)$. This integral can be evaluated using Cauchy's residue theory of complex variables.

1.18 Why are Z-transforms useful when dealing with constant coefficient difference equations and sampled-data signals?

The Z-transform possesses all the advantages for these systems as does the Laplace transform with differential equations. It allows much of the solution effort to be carried out with only algebraic manipulations in z. After the transform of the desired output variable $Y(z)$ is isolated algebraically, then its inverse can be calculated to give $y(nT)$.

1.19 Analyze the difference equation

$$y(t_k) + a_2 y(t_{k-1}) + a_1 y(t_{k-2}) + a_0 y(t_{k-3}) = b_0 x(t_k) + b_1 x(t_{k-1})$$

$x(t_k)$ is a known input sequence.

Let $Y(z)$ and $X(z)$ be the Z-transforms of y and x, respectively. Since the Z-transform is a linear operator, it can be applied to each individual term in the sum, giving

$$Y(z) + a_2 z^{-1} Y(z) + a_1 z^{-2} Y(z) + a_0 z^{-3} Y(z) = b_0 X(z) + b_1 z^{-1} X(z)$$

The "delay operator" nature of z^{-1} has been used here. As should be apparent from Problems 1.15 and 1.16, a shift of n sample periods in the time domain is achieved by multiplying by z^{-n} in the Z-domain. Thus

$$Y(z) = \left[\frac{b_0 + b_1 z^{-1}}{1 + a_2 z^{-1} + a_1 z^{-2} + a_0 z^{-3}}\right] X(z)$$

The output transform $Y(z)$ is the input transform $X(z)$ multiplied by a rational function of z^{-1} (or z). This rational function is the Z-domain transfer function $H(z)$.

1.20 What is the significance of the poles and zeros of $H(z)$?

Just as in the Laplace s-domain, the behavior of the system depends very heavily on the roots of the denominator of $H(z)$, i.e., the poles. A stable system must not have any s-plane roots with positive real parts. Since $z = e^{Ts}$ this means that in the z-plane all poles of a stable system must be inside the unit circle.

The zeros of the transfer function affect the magnitude of the various terms in the time-domain output. That is, the poles determine the system modes and the zeros help determine how strongly the modes will contribute to the total response. This is evident if $H(z)$ is expanded in partial fraction form.

PROBLEMS

1.21 Derive the elemental equation for the fluid storage tank of Figure 1.25 and show that it is analogous to an electric capacitance. Let Q be the volume flow rate, P the pressure at the base of the tank of cross sectional area A, and h the height of the fluid.

1.22 Show that an inventory storage unit can be modeled by an elemental equation analogous to an electric capacitance. Let the net flow of goods into inventory be Q items per unit time, and let the number of items in inventory at time t be $v(t)$.

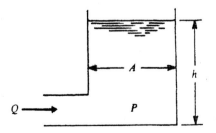

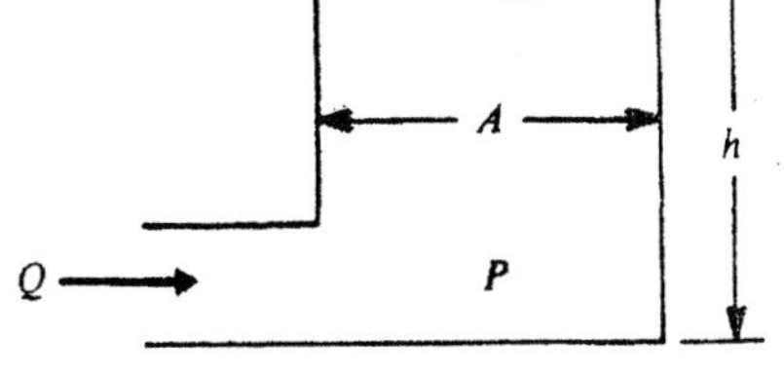

Figure 1.25

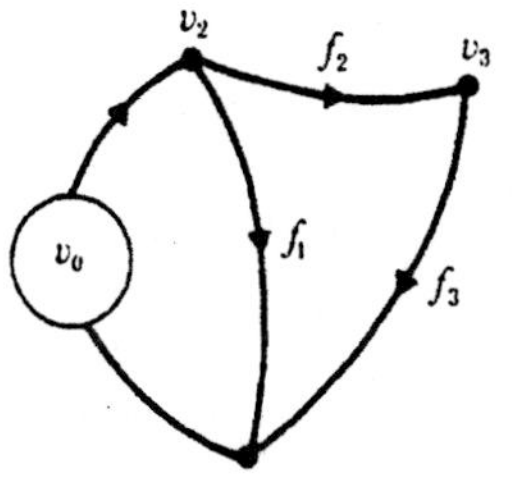

Figure 1.26

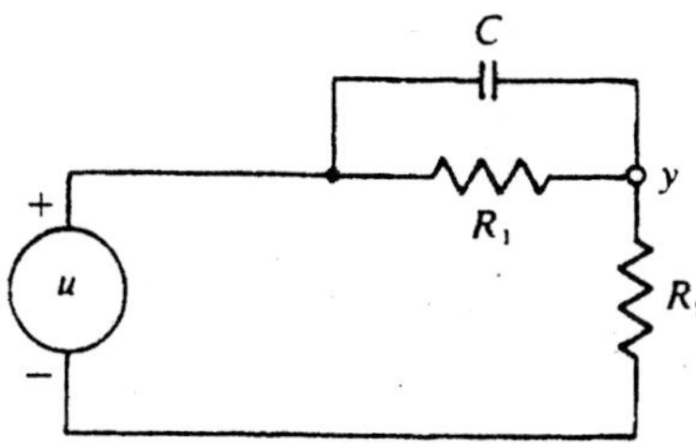

Figure 1.27

1.23 If branch 1 of Figure 1.26 contains an ideal capacitance C, branch 2 an ideal inductance L, and branch 3 an ideal resistance R, find the input-output equation relating v_0 and f_3.

1.24 Derive the input-output differential equation for the network of Figure 1.27. Treat $u(t)$ as the input voltage and $y(t)$ as the output voltage. Also give the input-output transfer function $y(s)/u(s)$.

1.25 A government agency would like a model for studying the effectiveness of its air pollution monitoring and control program. Discuss the factors involved in such a model.

1.26 Derive the Z-transform of $y(t) = e^{-0.1t}$. For $t < 0$, $y(t) = 0$. Use a sample time of $T = 2.0$ s.

1.27 The following values apply to the system in Problem 1.19.

$$a_0 = -0.064, \quad a_1 = 0.56, \quad a_2 = -1.4, \quad b_0 = 10.0, \quad \text{and} \quad b_1 = 5.0$$

Find the poles and zeros of $H(z)$. Does this represent a stable system?

1.28 For the system of Problem 1.19, find $y(nT)$ if the input $x(nT)$ is the sampled version of a unit step function starting at $t = 0$.

2

Highlights of Classical Control Theory

2.1 INTRODUCTION

Classical control theory, at the introductory level, deals primarily with linear, constant coefficient systems. Few real systems are exactly linear over their whole operating range, and few systems have parameter values that are precisely constant forever. But many systems approximately satisfy these conditions over a sufficiently narrow operating range. This chapter reviews the classical methods, which are applicable to linear, constant coefficient systems. More extensive discussions are in References 1 through 5.

2.2 SYSTEM REPRESENTATION

The consideration of linear, stationary systems is greatly simplified by the use of transform techniques and frequency domain methods. For continuous-time systems this means Laplace transforms (or sometimes Fourier transforms) [4]. Z-transforms provide equivalent advantages for discrete-time systems [6, 7]. These methods are basic in classical control systems analysis. Thus algebraic equations in the transformed variables are dealt with rather than the system's differential or difference equations. Manipulation of the algebraic cause and effect relations is facilitated by the use of transfer functions and block diagrams or signal flow graphs [1].

2.3 FEEDBACK

Most systems considered in classical control theory are feedback control systems. A typical single-input, single-output continuous-time (totally analog) system is shown in Figure 2.1*a*. Figure 2.1*b* shows a typical feedback arrangement for controlling a continuous-time process using a digital controller. Notation commonly used in classical

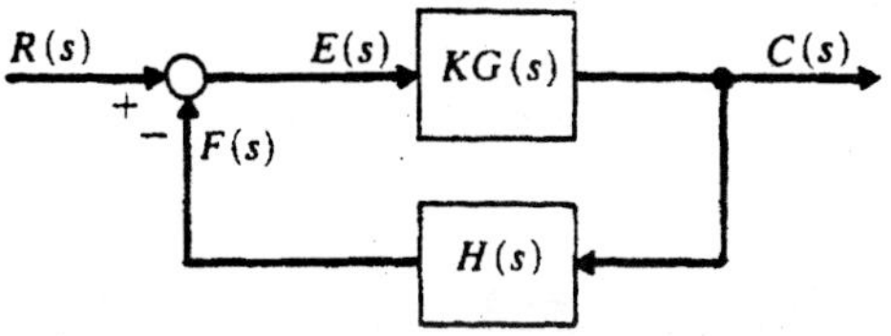

Figure 2.1*a* Elementary feedback control system.

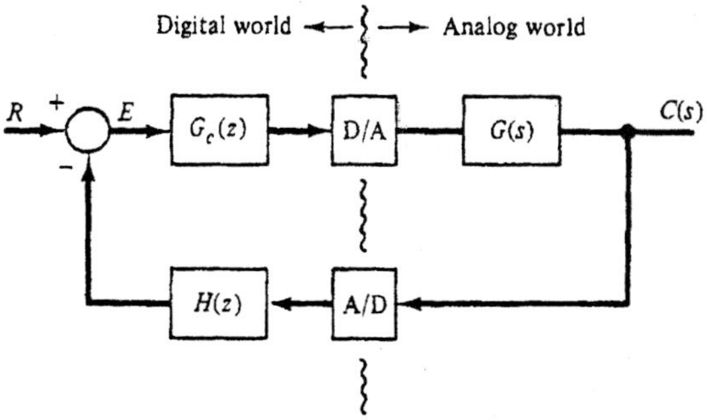

Figure 2.1*b*

control theory will be used in this chapter. In both cases the input or reference signal is R, the output or controlled signal is C, the actuating or error signal is E, and the feedback signal is F. This should cause no confusion with the parameters R and C used in resistance and capacitance networks.

While there are many similarities between these two types of systems, the differences are significant enough to merit a brief separate discussion of each.

Continuous-Time Systems

The forward *transfer function is* $KG(s)$, where K is an adjustable gain. The forward transfer function often consists of two factors $G(s) = G_c(s)G_p(s)$, where $G_p(s)$ is fixed by the nature of the plant or process to be controlled. $G_c(s)$ is a compensation or controller transfer function, which the designer can specify (within certain limits) to achieve desired system behavior. The *feedback transfer function* is $H(s)$. This often represents the dynamics of the instrumentation used to form the feedback signals, but it can also include signal conditioning or compensation networks. The designer may be able to at least partially specify $H(s)$ in some cases, and in other cases it may be totally fixed, or even just unity. In any case, the *open-loop transfer function* is $KG(s)H(s)$. It represents the transfer function around the loop, say from E to F, when the feedback signal is disconnected from the summing junction.

In the feedback system of Figure 2.1*a* and *b* the actuating signal is determined by comparing the feedback signal with the input signal. When $H(s) = 1$, the unity feedback case, the comparison is directly between the output and the input. Then the difference E is truly an error signal.

A major part of classical control theory for continuous-time systems is devoted to the analysis of feedback systems like the one shown in Figure 2.1*a*. Multiple input-output systems and multi-loop systems can also be considered using transfer function techniques (see Problems 4.2 through 4.7), although most of this book is devoted to a state variable approach instead. It is beneficial to have a thorough understanding of single-input, single-output systems before the multivariable case is considered. This chapter provides a review of the methods used in studying the behavior of C and E as influenced by R.

EXAMPLE 2.1 Relations, in the Laplace transform domain, between the input R and the output C and between R and the error E are derived algebraically as follows. At the summing junction, $R - HC = E$. The relation between E and C is $KGE = C$. Elimination of E gives $KGR - KGHC = C$, so that $C = KGR/(1 + KGH)$. Using $E = C/KG$ gives $E = R/(1 + KGH)$. ■

The system of Figure 2.1*a* is the prototype for all continuous system discussions in this chapter. The following terminology will be used frequently. In general, $G(s) = g_n(s)/g_d(s)$ and $H(s) = h_n(s)/h_d(s)$ will be ratios of polynomials in s. The values of s which are roots of the numerator are called *zeros*. Roots of the denominator are called *poles*. In particular, the *open-loop zeros* are values of s which are roots of the numerator of the open-loop transfer function $KG(s)H(s) = Kg_n(s)h_n(s)/[g_d(s)h_d(s)]$. The *open-loop poles* are roots of the denominator of $KG(s)H(s)$. Since the closed-loop transfer function is $C(s)/R(s) = KG(s)/[1 + KG(s)H(s)] = Kg_n(s)h_d(s)/[g_d(s)h_d(s) + Kg_n(s)h_n(s)]$, the *closed-loop zeros* are all the roots of $g_n(s)h_d(s)$. The *closed-loop poles* are roots of $1 + KG(s)H(s) = 0$ or equivalently, roots of $g_d(s)h_d(s) + Kg_n(s)h_n(s) = 0$.

Discrete-Time Systems

There is a richer variety of possibilities when dealing with the digital control of continuous systems. The points of conversion from continuous-time signals to discrete-time signals (A/D) and back again (D/A) can vary from one application to the next. An analog feedback sensor could be used and then its output sampled, or a direct digital measurement may be used. The reference input R could be a continuous-time signal that needs to be sampled before being sent to the control computer, or it might be a direct digital input. Figure 2.1*b* is just one possible arrangement. Other configurations can be analyzed in a similar way. Before proceeding with the analysis, models of the A/D and D/A conversion processes are required. These conversions are also referred to as sampling and desampling or signal reconstruction, respectively.

Sampling. Assume a periodic sampler with period T. A convenient model of the A/D conversion is an impulse modulator, usually shown symbolically as a switch like the one of Figure 2.2, where a general signal $y(t)$ is being sampled. Impulse modulation is not really what physically occurs, since no infinite amplitude signals such as $y^*(t_k)$ actually exist in the system. This series of impulse functions have infinite amplitude at the sample times, but it is their areas or strengths that represent the real signal amplitudes $y(t_k)$ mathematically. This artificial representation is used because

1. It allows the use of Z-transforms, which simplify much of the analysis.
2. The correct answers are obtained (except for quantization effects) as long as it is understood that within the digital portion it is the strengths of the impulses, not their amplitudes, that describe the signals.

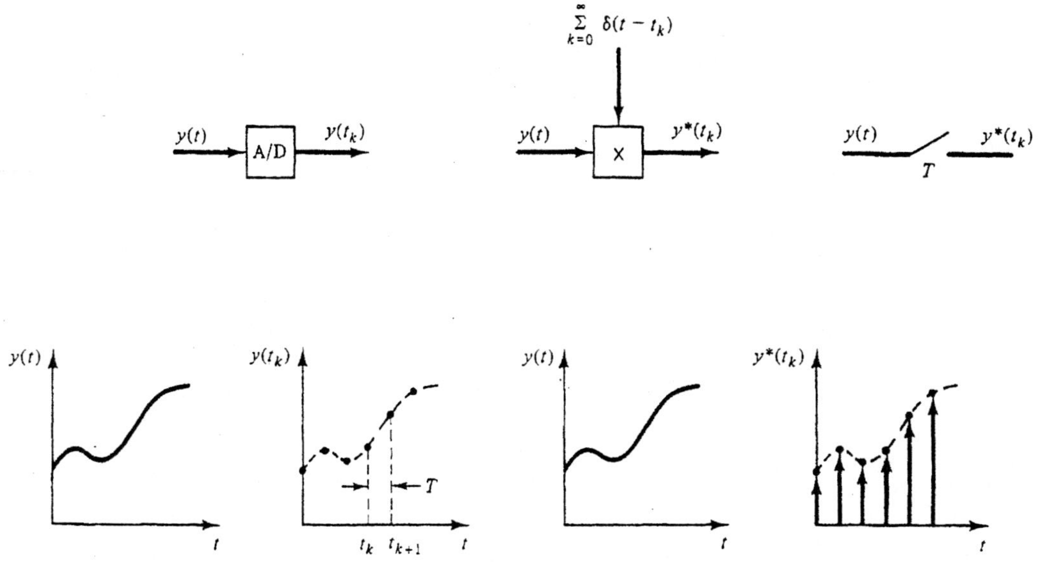

(a) Reality

(b) Impulse modulation model

Figure 2.2

3. The correct effect on the continuous-time part of the system is obtained provided some sort of "hold" circuit is used on the impulse train before the signal reenters the analog world. There are various versions of hold devices. One function common to them all is an integration, which eliminates the impulses and once again gives a finite amplitude physical signal. This is the desampling function of the D/A.

Desampling. The only model of the D/A process to be considered here is another sampler (perfect time synchronization assumed), followed by a zero order hold (ZOH). The zero order hold integrates the difference between two consecutive impulses in the periodic impulse train shown in Figure 2.2 and repeated in Figure 2.3. Therefore, the output is a piecewise constant signal whose value between t_k and t_{k+1} is clamped at $y(t_k)$ (again, ignoring quantization errors). If the computer made no modification to the signal between the A/D and D/A, the end-to-end effect of this sampling-desampling operation would be to create a piecewise constant approximation to the continuous input signal. From Chapter 1, the Laplace transform of an impulse-modulated signal is (after a change of variable) the Z-transform of the signal. Any linear operation that the computer algorithm performs on the signal samples between the A/D and D/A can be represented by a Z-transform domain transfer function, sometimes called a pulse transfer function. $G_c(z)$ and $H(z)$ in Figure 2.1b are examples of this.

Extensive tables of Z-transforms are available [7]. Use of these tables, plus a few simple rules, will allow systems like Figure 2.1b to be analyzed almost as easily as, and

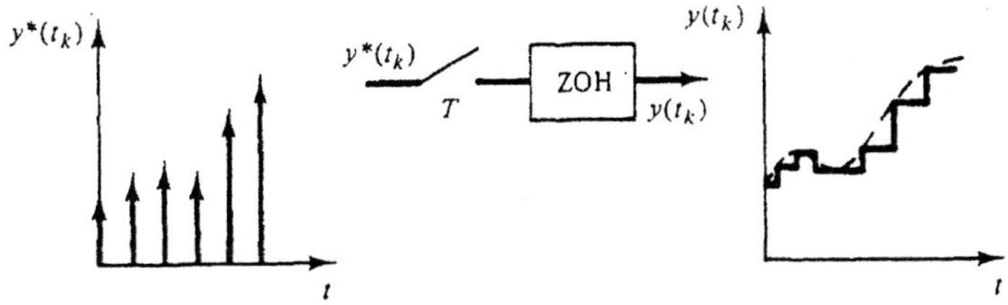

Figure 2.3

with a great deal of similarity to, those of Figure 2.1*a*. Of course, a complete understanding and appreciation will require a more thorough treatment, as can be found in the references. Some key rules of manipulation are:

1. *As a signal passes through the "switch," it is Z-transformed.*

2. *The transform of the signal out of a transfer function block is the product of that transfer function and the transform of the input signal.*

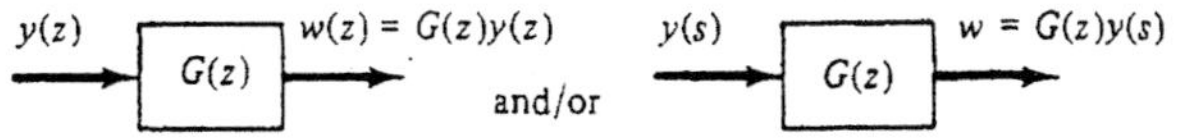

3. *Sampling a signal that is already sampled does not change it.*

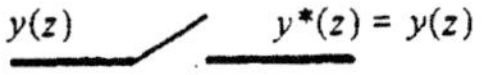

or $Z\{y(z)\} = y(z)$

4. *Pulsed or Z-transformed signals (and transfer functions) and s-domain signals (and transfer functions) will appear together in the same expression at times.* The action of a sampler (or a Z-transform) on these mixed signals is illustrated next.

$$Z\{G(z)y(s)\} = G(z)y(z)$$

5. *The Z-transform operator is not associative for products.* The placement of "switches" in a block diagram is important.

$$w_1(z) = Z\{y(s)G_1(s)G_2(s)\} \neq w_2(z) = Z\{G_1(s)y(s)\}G_2(z)$$

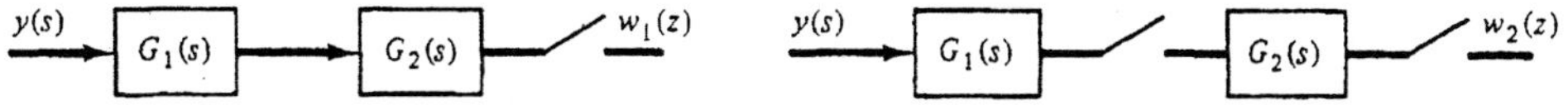

6. *When working with closed-loop systems like Figure 2.1b, it is generally best to follow a two-step process.* First, algebraically solve for the variable at the input of a sampler in terms of external inputs and/or outputs of that sampler. Second, "close the loop" by passing through the sampler, i.e., taking the Z-transform. This will give a result, entirely in the Z-domain, which can be used to solve for the system output sequence at the sample times. If the sampling period is small enough compared to the rate of change of the signal, this approximation may be all that is needed to describe the continuous system output.

EXAMPLE 2.2 The system of Figure 2.1*b* is redrawn in Figure 2.4, using symbolism just introduced for A/D and D/A operations. The transfer function for a ZOH is also used, $G_0 = (1 - z^{-1})/s$.

The input to the forward path sampler is first isolated:

$$\begin{aligned} E_1(z) &= G_c(z)[R(z) - F(z)] \\ &= G_c(z)[R(z) - E_1^*(z)Z\{(1 - z^{-1})G(s)/s\}H(z)] \end{aligned}$$

Define the Z-transform of $G(s)/s$ times $(1 - z^{-1})$ as $G'(z)$. Then

$$E_1(z) = G_c(z)[R(z) - E_1(z)G'(z)H(z)]$$

Solving gives

$$E_1(z) = G_c(z)R(z)/[1 + G_c(z)G'(z)H(z)]$$

In this case the input to the selected sampler was already sampled, so the second step of passing through the sampler has no effect, i.e. $E_1^*(z) = E_1(z)$. The full two-step process is better illustrated by selecting the feedback sampler instead.

Before doing that, note that

$$C(s) = E_1(z)[(1 - z^{-1})G(s)/s)]$$

and that

$$\begin{aligned} C(z) &= E_1(z)G'(z) \\ &= G_c(z)G'(z)R(z)/[1 + G_c(z)G'(z)H(z)], \end{aligned}$$

an expression very similar to the continuous system result.

The same example is reworked by isolating the input to the feedback sampler first:

$$C(s) = [R(z) - H(z)C(z)]G_c(z)(1 - z^{-1})G(s)/s$$

Then the traverse around the loop is completed by passing through the sampler to obtain

$$C(z) = [R(z) - H(z)C(z)]G_c(z)G'(z)$$

Solving this for $C(z)$ gives the same result as above.

In order to dispel the idea that a discrete system result can always be written from the continuous system result by substituting all the individual Z-transfer functions for their Laplace transform counterparts, the reader is urged to work through Problems 2.38, 2.39, and 2.41, which have different sampling arrangements. ■

Once $C(z)$ is found, the values of the output $C(t)$ can be determined *at the sample times* t_k by finding the inverse Z-transform. While this will not give the values of $C(t)$

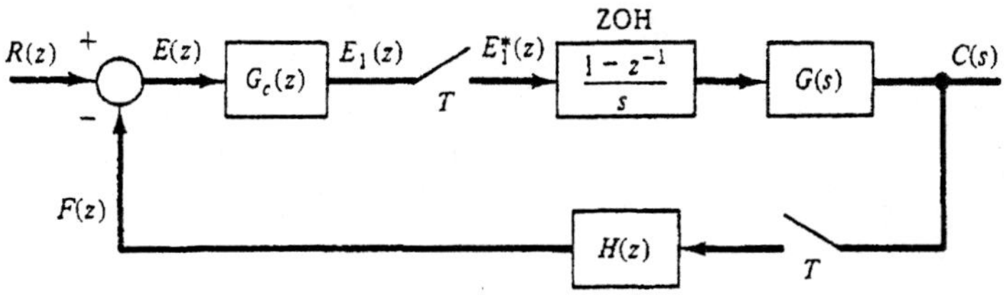

Figure 2.4

between sample points, it often gives a sufficiently accurate representation of the continuous signal.

Control systems like those of Figure 2.1*a* and *b* are used extensively because of the advantages that can be obtained by using feedback. The advantages of feedback control are:

1. The system output can be made to follow or track the specified input function in an automatic fashion. The name automatic control theory is frequently used for this reason.
2. System performance is less sensitive to variations of parameter values (see Problem 2.1).
3. System performance is less sensitive to unwanted disturbances (see Problem 2.4).
4. Use of feedback makes it easier to achieve the desired transient and steady-state response (see Problem 2.5).

The advantages of feedback are gained at the expense of certain disadvantages, the principal ones being:

1. The possibility of instability is introduced and stability becomes a major design concern. Actually, feedback can either stabilize or destabilize a system.
2. There is a loss of system gain, and additional stages of amplification may be required to compensate for this.
3. Additional components of high precision are usually required to provide the feedback signals (see Problems 2.2 and 2.3).

Once the system models are specified in terms of transfer functions and block diagrams, classical control theory is devoted to answering three general questions:

(a) What are appropriate measures of system performance that can be easily applied to feedback control systems?
(b) How can a feedback control system be easily analyzed in terms of these performance measures?
(c) How should the system be modified if its performance is not satisfactory?

2.4 MEASURES OF PERFORMANCE AND METHODS OF ANALYSIS IN CLASSICAL CONTROL THEORY

If the complete solutions for the system output $C(t)$ were available in analytical form for every conceivable input, system performance could be assessed. To obtain an analytical expression for $C(t)$, the inverse Laplace transform of

$$C(s) = \frac{KG(s)R(s)}{1 + KG(s)H(s)} \tag{2.1a}$$

is required. To determine $C(t_k)$ in the discrete-time case or in the mixed discrete-continuous case, the inverse Z-transform of

$$C(z) = \frac{G_c(z)G'(z)R(z)}{1 + G_c(z)G'(z)H(z)} \tag{2.1b}$$

(or a similar expression) is needed. In both cases, if the denominators can be factored, partial fraction expansion can be used to obtain a sum of easily invertible terms. However, the denominator of equation (*2.1*) may be a high-degree polynomial. Also, an infinite number of possible inputs $R(s)$ or $R(z)$ could be considered. Rather than seek complete analytical solutions, classical control theory uses only certain desirable features, which C should possess, in order to evaluate performance. Methods of classical control theory were developed before the widespread availability of digital computers. As a result, all the techniques seek as much information as possible about the behavior of $C(t)$ or $C(t_k)$ without actually solving for them. The methods have been developed for ease of application and stress graphical techniques. They are still useful methods of analysis and design because of the insight they provide.

The problem of infinite variety for possible inputs is dealt with by considering important aperiodic and periodic signals as test inputs. Step functions, ramps, and sinusoids are common examples.

The general characteristics a well-designed control system should possess are (1) stability, (2) steady-state accuracy, (3) satisfactory transient response, (4) satisfactory frequency response, and (5) reduced sensitivity to model parameter variations and disturbance inputs. These requirements are interrelated in various ways and often present conflicting goals. For example, decreasing response times generally requires increasing system bandwidth, which increases suceptibility to high-frequency noise. A key concept in many aspects of feedback system's performance is the so-called return difference function, which happens to be the denominators of Eqs. (*2.1a*) and (*2.1b*). These are both of the form $R_d(\zeta) = 1 + KF(\zeta)$, with the complex variable ζ being either s or z. The name *return difference* arises as follows. If a feedback loop is broken at a given point, the difference between a signal v inserted into the loop at that point and the resulting signal r, which returns to the broken point after traversing the loop, is

$$v - r = [1 + KF(\zeta)]v = R_d(\zeta)v$$

A graphical interpretation of the return difference will be given after polar plots—i.e., Nyquist plots—are introduced.

1. *Stability means that $C(t)$ or $C(t_k)$ must not grow without bound due to a bounded input, initial condition, or unwanted disturbance.* (This intuitive definition is expanded upon in Chapters 10 and 15.) For linear constant coefficient systems, stability depends only on the locations of the roots of the closed-loop characteristic equation. The continuous system's characteristic equation is the denominator of equation (*2.1a*) set to zero. The discrete case uses the denominator of equation (*2.1b*). Both of these are of the form

$$R_d(\zeta) = 1 + KF(\zeta) = 0 \tag{2.2}$$

with the complex variable ζ being either s or z. The principal difference is the stability region. The roots must be in the left-half s-plane for a stable continuous system. Since $z = e^{Ts}$, the entire left-half s-plane maps into the interior of the unit circle in the Z-plane. A stable discrete system must have all roots of its characteristic equation inside the unit circle. Methods of determining stability are discussed below.

a. *Routh's criterion* determines how many roots have positive real parts directly from the coefficients of the characteristic polynomial. The actual root locations are not found. The number of unstable roots of a continuous system are obtained directly. Routh's criterion can also be applied to a discrete system, but first a bilinear transformation $z = (w + 1)/(w - 1)$ is used to map the inside of the unit circle in the Z-plane into the left half of a new complex w-plane. This converts the characteristic equation into a polynomial in w, to which Routh's criterion can be applied.

b. *Root locus* is a graphical means of factoring the characteristic equation (or any algebraic polynomial of similar form). Both continuous and discrete systems can be described simultaneously by using Eq. (2.2) as the characteristic equation. The essence of the method is to consider $KF(\zeta) = -1$. Since $F(\zeta)$ is a complex number with a magnitude and a phase angle, this implies two conditions, which are considered separately. They are, assuming the gain is positive and real,

$$\angle F(\zeta) = (1 + 2m)180^\circ \qquad \text{for any integer } m \tag{2.3a}$$

and

$$K|F(\zeta)| = 1 \tag{2.3b}$$

Thus root locus determines the closed-loop roots (and therefore stability) by working with the open-loop transfer function $KF(\zeta)$, which is normally available in factored form.

c. *Bode plots* are another graphical method which provides stability information for *minimum phase systems* (systems with no open-loop poles or zeros in the unstable region). Magnitude and phase angle are considered separately, as in Eqs. (2.3a) and (2.3b), but the only values of ζ considered are on the stability boundary. This technique is widely used with continuous systems, in which case the stability boundary is defined by $s = j\omega$. This corresponds to the consideration of sinusoidal input functions with frequencies ω. This method is greatly simplified by using decibel units for magnitude and a logarithmic frequency scale for plotting. This allows for rapid construction of straight line asymptotic approximations of the magnitude plot. The critical point for stability, -1, becomes the point of 0 db and -180° phase shift. Bode techniques can also be applied to discrete systems by first using the same bilinear transformation as was mentioned under Routh's criterion. The stability boundary in the w-plane can also be characterized by the purely imaginary values $w = j\omega_w$. This "transformed" frequency ω_w is generally badly distorted from the true sinusoidal frequency, so intuition is of less value in this approach, insofar as stability margins, bandwidths, and similar concepts are concerned. For this reason, Bode methods are probably used less often with discrete systems, and they will not be pursued here. The same is more or less true for the following two frequency domain methods as well, so they are only discussed for continuous systems.

d. *Polar plots and Nyquist's stability criterion.* Polar plots convey much the same information as Bode plots, but the term $KG(j\omega)H(j\omega)$ is plotted as a locus of phasors with ω as the parameter. The critical point is again -1. Note that the return difference $R_d(j\omega)$ (or $R_d(e^{j\omega T})$ in the discrete case) can be represented as a phasor from the critical -1 point to a point on a plot of the loop transfer function $KG(j\omega)H(j\omega)$. (See Problem 2.1 and, in particular, Figure 2.9c.) Nyquist's stability criterion, which applies to nonminimum phase systems as well, states that the number of unstable closed-loop poles is $Z_R = P_R - N$, where N is the number of encirclements of the critical point -1 made by the locus of phasors. Counterclockwise encirclements are considered positive. P_R is the number of open-loop poles in the right-half plane.

e. *Log magnitude versus angle plots* are sometimes used for stability analysis. They contain the same information as Bode plots, but magnitude and angle are combined on a single graph with ω as a parameter.

2. *Steady-state accuracy requires that the signal $E(t)$, which is often an error signal, approach a sufficiently small value for large values of time.* The final value theorem facilitates analyzing the requirement without actually finding inverse transforms. That is, for continuous systems

$$\lim_{t\to\infty}\{E(t)\} = \lim_{s\to 0}\{sE(s)\} \tag{2.4a}$$

For discrete systems, the Z-transform version of the final value theorem is used,

$$\lim_{k\to\infty}\{E(t_k)\} = \lim_{(z\to 1)}\{(z-1)E(z)\} \tag{2.4b}$$

Both versions of the final value theorem are only valid when the indicated limits exist. By considering step, ramp, and parabolic test inputs, the useful parameters called *position, velocity,* and *acceleration* (or step, ramp, and parabolic) *error constants* are developed. These provide direct indications of steady-state accuracy (see Problem 2.8).

3. *Satisfactory transient response means there is no excessive overshoot for abrupt inputs, an acceptable level of oscillation in an acceptable frequency range, and satisfactory speed of response and settling time, among other things.* These are actually questions of relative stability, and depend upon the location of the closed-loop poles in the s-plane or Z-plane and their proximity to the stability boundary. Questions regarding transient response are best studied using root locus, since it is the only classical method which actually determines closed-loop pole locations. Bode, Nyquist, and log-magnitude plot methods also give information regarding transient response, at least indirectly. *Gain margin GM* is a measure of additional gain a system can tolerate with no change in phase, while remaining stable. *Phase margin PM* is the additional phase shift that can be tolerated, with no gain change, while remaining stable. Note that these stability margins are measures of the magnitude of the minimum return difference phasor. Experience has shown that acceptable transient response will usually require stability margins on the order of

$$PM > 30°, \qquad GM > 6 \text{ db}$$

These frequency domain stability margins can often be used to draw conclusions regarding transient performance, because many control systems have their response characteristics dominated by a pair of underdamped complex poles. For this case known correlations exist between frequency domain and time domain characteristics. A few approximate rules of thumb are

$$\text{damping ratio} \cong 0.01\, PM \text{ (in degrees)}$$

$$\% \text{ overshoot} + PM \cong 75$$

$$\text{(rise time)(closed-loop bandwidth in rad/s)} \cong 0.45\, (2\pi)$$

Other response times have similar inverse relationships with bandwidth. The frequency of 0 db magnitude for the open-loop KGH term has an effect similar to bandwidth. Increasing this crossover frequency increases bandwidth and decreases response times.

4. *Satisfactory frequency response implies such things as satisfactory bandwidth, limits on maximum input-to-output magnification, frequency at which this magnification occurs, as well as gain and phase margin specifications.* Bode, Nyquist, and log-magnitude-angle plots all are frequency response methods, and they deal with the open-loop transfer function. If the closed-loop characteristics, such as closed-loop bandwidth, must be determined, then the *Nichol's chart* [1] can be used. The Nichol's chart is a graphical conversion from open-loop magnitude-phase characteristics to closed-loop characteristics. Normally, one of the open-loop graphical methods is first used and the results are then transferred to a Nichol's chart. From this, the closed-loop frequency response characteristics can be read off directly.

5. *Since perfect models are never available, either because of intentional simplifications or because of unavoidable ignorance, time variations, or noise corruption, a good control system must be at least somewhat forgiving of these errors.* Problems 2.1 through 2.4 briefly review how feedback can lead to reduced sensitivity to external disturbances and internal parameter variations. The concept of return difference plays a prominent role [8, 9].

2.5 METHODS OF IMPROVING SYSTEM PERFORMANCE

Whenever the performance of a feedback control system is not satisfactory, the following possible approaches should be considered.

1. A simple adjustment of the gain parameter K. This could be considered by using any of the analysis methods mentioned in the preceding paragraphs. From a consideration of the system's root locus, it is obvious that gain adjustment can only shift the closed-loop poles along well-defined loci. Perhaps no points on these loci give satisfactory results.

2. Minor changes in the system's structure, such as adding additional measurements to be used as feedback signals. Addition of minor feedback loops can alter the loci of possible pole locations as K is varied. The inclusion of a rate feedback loop, using a tachometer for example, is a common means of improving stability.
3. Major changes in the system's structure or components. A hydraulic motor may perform better than an electric motor in some cases. A higher capacity pump or a more streamlined aerodynamic shape may be the answer in other cases.
4. Addition of compensating networks—i.e., $G_c(s)$ or $H(s)$—or digital algorithms—i.e., $G_c(z)$ or $H(z)$—to alter the root locus or to change the magnitude and phase characteristics in a critical frequency range.

Of these four techniques for improvement, only the second and fourth constitute what are usually referred to as *compensation techniques*. The advantages of root locus, Bode, and Nyquist methods of analysis are that compensating changes in the open-loop transfer function can be rapidly taken into account. The modifications may be made in order to reshape the locus, improve gain or phase margins, or increase the error constants. The classical methods thus constitute design techniques as well as analysis techniques. A process of design by analysis is usually used. That is, a compensating network is selected and then analyzed. However, a little experience gives great insight into the kinds of compensation that are needed. If the major problem is to improve relative stability with less concern for error constants, lead compensation networks are usually tried. If the system has acceptable stability margins, but poor steady-state accuracy, lag compensation networks will usually be appropriate. If a combination of both improvements is needed, a lag-lead network may give the desired results. More complicated networks, such as the bridged-T network, Butterworth filters, and so on, can be used to effectively cancel undesirable left-half-plane poles and replace them with more favorable ones. Cancellation compensation should never be used to eliminate unstable poles, because parameter tolerances will preclude exact cancellation. Even an infinitesimal error in cancellation will leave an unstable closed-loop pole. The form of the desired specifications and the personal preference of the designer will influence the choice of the analysis method. Extra insight can usually be gained by looking at a compensation problem from both the root locus and one of the frequency domain techniques.

An alternative method of design, through synthesis rather than analysis, is also possible. In this approach, the design specifications are translated into a desired closed-loop transfer function which satisfies them. Let the closed-loop transfer function be $M(z)$. This can be related to the compensator $G_c(z)$. For example, the system of Example 2.2 has

$$M(z) = G_c(z)G'(z)/[1 + G_c(z)G'(z)H(z)]$$

which can be solved to give

$$G_c(z) = \frac{M(z)}{[1 - M(z)H(z)]G'(z)} \tag{2.5}$$

Because of this result it is clear that certain restrictions must be imposed on $M(z)$ if the resulting compensator is to be physically realizable. This is discussed in Problems 2.23

and 2.24. More details on this method can be found in References 6 and 7. Since discrete system compensators are just computer algorithms, there is no concern about synthesizing the results in terms of passive electrical components R, C, and maybe an occasional L that dominated classical control compensation in the early years. It is perhaps for this reason that the algebraic synthesis methods seem to be more widely used in discrete system design, although the continuous system version was described years earlier by Truxal [10]. Even in the continuous system domain the definition of what is practical now is quite different from the early years because of progress in technology, such as operational amplifiers and large-scale integrated circuit technology.

One final design parameter in discrete systems is the sampling period T. It can have a profound effect on system performance. Nyquist's sampling theorem tells us that a signal must be sampled at least twice per cycle of the highest frequency present in order to avoid losing information about the signal. The highest frequency present is often interpreted as the highest frequency of interest, and the sampler is generally preceded by a low pass filter. This prevents the high-frequency terms from being *aliased* as low-frequency signals as a result of sampling. *Frequencies of interest* are related to system bandwidth, since that is what determines which frequencies the system is capable of passing or responding to. The "twice" is strictly a theoretical limit based on an unachievable ideal low pass filter, which would be needed to reconstruct the original signal from its sampled version. In reality, a cushion is provided by sampling at a considerably higher rate if possible. Frequently a sampling rate of three to five times the Nyquist rate is more appropriate. The reason for the sampling in the first place might be because of time-shared or multiplexed equipment, so possible T values may be restricted in many cases. In closed-loop systems T has another effect beyond the sampling theorem considerations. The value of T interacts with the loop gain K (and, of course, pole-zero locations, too) to determine system stability.

EXAMPLE 2.3 Investigate the system of Figure 2.5 for stability.

The characteristic equation is

$$1+\frac{K}{(s-10)(s+20)(s+100)}=0 \quad \text{or} \quad s^3+110s^2+800s-20{,}000+K=0$$

The Routhian array is a table with one more row than the highest power of s in the characteristic equation. The first two rows are filled in a sawtooth pattern with the coefficients of the characteristic equation. Each succeeding row is computed from terms in the two rows just above it. The pattern for the computed rows is as follows. Suppose two typical rows with a_i and b_i coefficients are available as in Table 2.1a. Then the c_i terms are given by

$$c_1=(b_1a_2-a_1b_2)/b_1, \qquad c_2=(b_1a_3-a_1b_3)/b_1, \qquad c_3=(b_1a_4-a_1b_4)/b_1$$

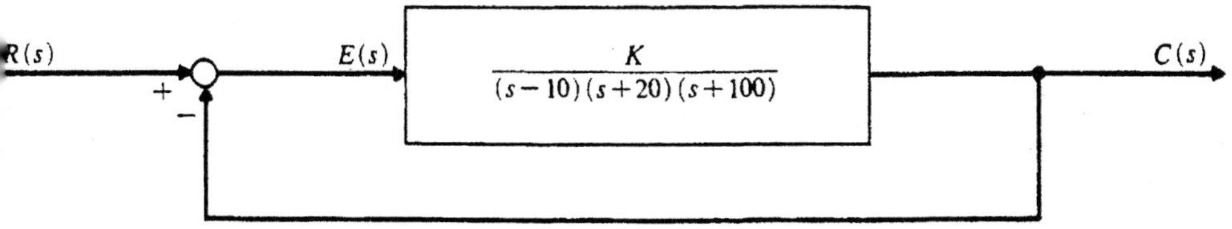

Figure 2.5

TABLE 2.1a

a_1	a_2	a_3	a_4
b_1	b_2	b_3	b_4
c_1	c_2	c_3	c_4

TABLE 2.1b

s^3	1	800
s^2	110	$K - 20{,}000$
s	$\frac{108{,}000 - K}{110}$	
s^0	$K - 20{,}000$	

Each row is filled in from left to right until all remaining terms are zero. In this case c_4 and all higher c_i terms are zero because of blanks in the a_i and b_i rows.

Table 2.1*b* gives the array for the system of Figure 2.5. Routh's criterion states that the number of sign *changes* in the first column is equal to the number of roots in the right-half plane. For stability the first column must have all entries positive. Therefore, the system is stable if $20{,}000 \le K \le 108{,}000$. If $K < 20{,}000$, one sign *change* exists in column one and there will be one unstable root. If $K > 108{,}000$, there are two sign changes and therefore two unstable roots. If $K = 108{,}000$, the s row is zero. Whenever an entire row is zero, the coefficients of the preceding row are used to define the *auxiliary equation*. Roots of the auxiliary equation are also roots of the original characteristic equation. With $K = 108{,}000$, the auxiliary equation is $110s^2 + 88{,}000 = 0$, indicating poles at $s = \pm j\sqrt{800}$. With K at this maximum value, the system oscillates to $\omega = \sqrt{800}$ rad/s. ■

EXAMPLE 2.4 Investigate the steady-state following error for the system of Figure 2.5 if $K = 100{,}000$ and the input is a unit step.

The error is

$$E(s) = \frac{1/s}{1 + \dfrac{K}{(s-10)(s+20)(s+100)}}$$

Using the final value theorem, the steady-state error is

$$E(t)|_{ss} = \frac{1}{1 + \dfrac{K}{(-10)(20)(100)}} = -0.25$$

Since the input is unity, the steady-state output has a 25% error. ■

EXAMPLE 2.5 Add compensation to the previous system in order to achieve a steady-state error of less than 10%. The oscillatory poles should have a damping ratio of $0.7 < \zeta < 0.9$ and a damped natural frequency of about 10 to 20 rad/s.

In order to meet the steady-state error specifications, a gain increase by a factor of 2.2 is required. This would give an unstable system and then the final value theorem cannot be used. Compensation is required, and it will be added in the forward loop. Because of the form of the specifications, root locus will be used. First, the locus of points satisfying Eq. (*2.2*) is found. The following rules greatly simplify this procedure.

1. The number of branches of the root locus equals the number of open-loop poles. One closed-loop pole will exist on each branch.
2. One branch of the locus starts at each open-loop pole. One branch terminates at each open-loop zero and the remaining branches approach infinity.

3. The part of the locus on the real axis lies to the left of an odd number of poles plus zeros.
4. With $K = 0$, open and closed-loop poles coincide. As K increases, the closed-loop poles move along the loci. As $K \to \infty$ each closed-loop pole approaches either an open-loop zero or infinity.
5. Branches that go to infinity do so along asymptotes with angles given by $\phi_i = 180°(1 + 2k)/(n - m)$ for $k = 0, \pm 1, \pm 2, \ldots$, and where n and m are the number of open-loop poles and zeros respectively.
6. The asymptotes emanate from the *center of gravity* given by cg = [(sum of real parts of all open-loop poles) − (sum of real parts of all open-loop zeros)]/$(n - m)$.
7. The loci are symmetric with respect to the real axis.

By using these rules and by testing the angle criterion at a few additional points off the real axis, the uncompensated root locus of Figure 2.6 is obtained. It is obvious that the locus must be reshaped in order to meet the specifications. The angle criterion at points inside the desired region indicates that an additional 30–70° of phase lead is needed if the locus is to pass through this region. By placing a zero at $s = -20$ and a pole at $s = -45$, adequate phase lead is obtained. However, the lead network transfer function $G_{C1}(s) = (s + 20)/(s + 45)$ introduces a decrease in the error constant by a factor of $\frac{20}{45} = 0.445$. This decrease can be made up, and the additional increase gained by using a lag filter, such as

$$G_{C2} = \frac{(s + 0.1)}{(s + 0.01)}$$

This pole-zero pair near the origin will have only a small effect on the locus in the region of interest since their angle contributions almost cancel each other. Using the compensator transfer function $G_C(s) = (s + 0.1)(s + 20)/[(s + 0.01)(s + 45)]$, the compensated root locus of Figure

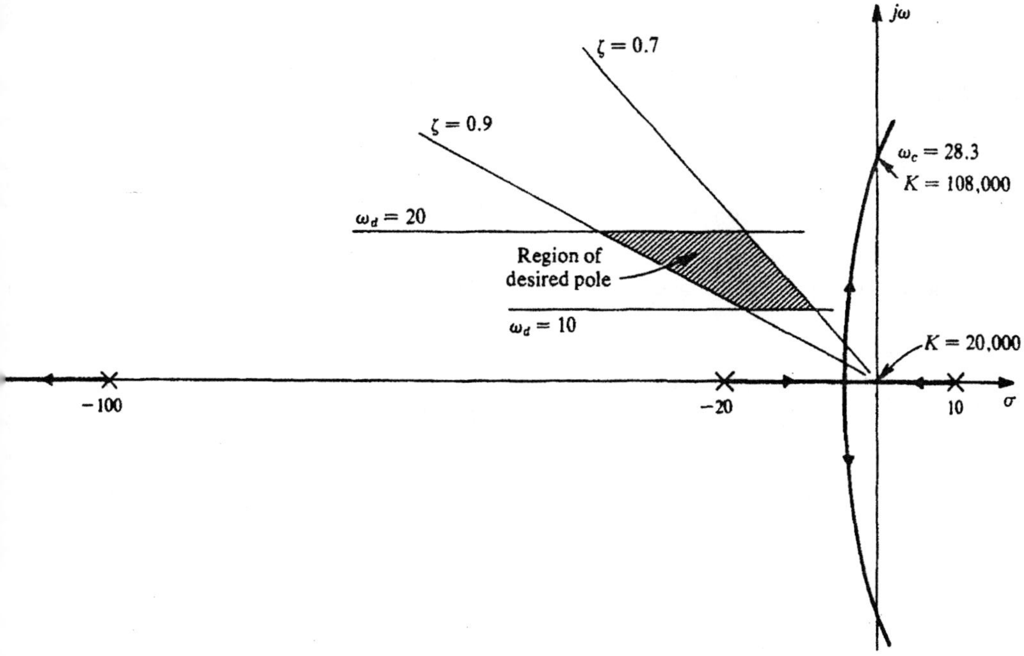

Figure 2.6 Approximate sketch of uncompensated root locus.

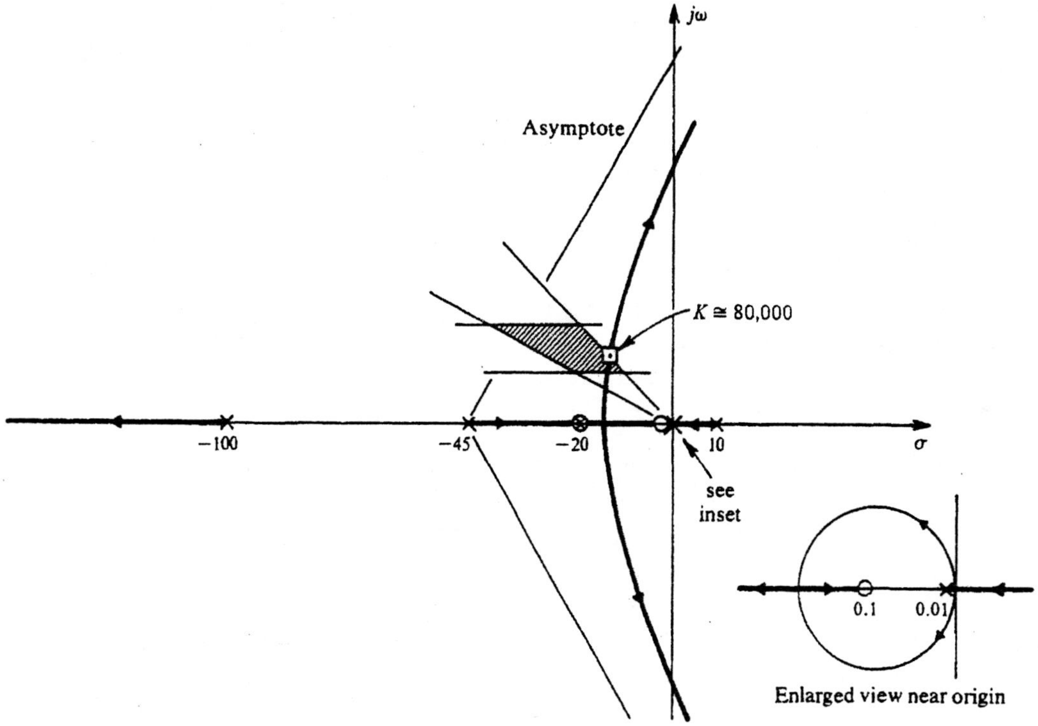

Figure 2.7

2.7 is obtained. Applying Eq. (2.3) at the point ⊡ indicates that the required gain is approximately $K = 80{,}000$. Using this result,

$$E(t)|_{ss} = \frac{1}{1 + \dfrac{K(0.1)}{(-10)(45)(100)(0.01)}} = -0.06 = 6\% \text{ error}$$

■

EXAMPLE 2.6 Consider the error-sampled system of Figure 2.8, which represents a linearized model of a position control system using an armature controlled dc motor with a time constant $\tau = 2$ seconds.

1. Find expressions for $C(s)$, $C(z)$, and $E(z)$.
2. Show that the steady-state error is zero for a step input and approaches a constant for

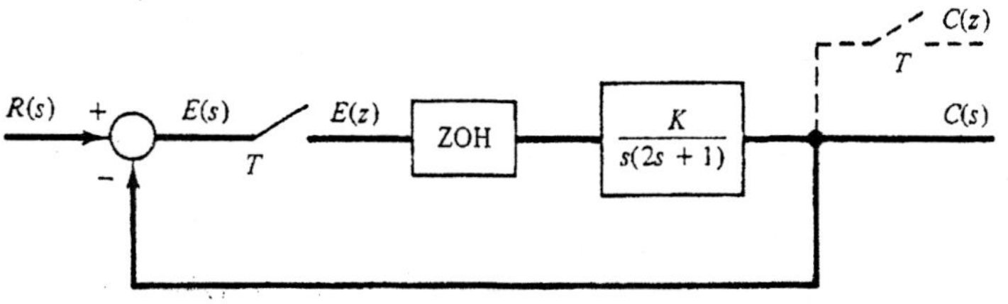

Figure 2.8

ramp inputs, with the constant decreasing inversely with K, so long as the system remains stable.

3. Show that the maximum allowable gain for stability is inversely related to the sampling period T.
4. With $T = 2$ s, find the maximum gain for stability.
5. Set the gain to $K = 0.5$ and find the steady-state error for a ramp input $R(t) = t$. The sampling period is $T = 2$.

1. *Using the two-step procedure of Section 2.3,*

$$E(s) = R(s) - E(z)(1 - z^{-1})K/[s^2(2s + 1)]$$

$$E(z) = R(z) - E(z)(1 - z^{-1})KZ\{1/[s^2(2s + 1)]\}$$

Define $(1 - z^{-1})Z\{.5K/[s^2(s + 0.5)]\} = G'(z)$. Using Z-transform tables,

$$G'(z) = \frac{K\{z[T - 2(1 - e^{-0.5T})] + [2(1 - e^{-0.5T}) - Te^{-0.5T}]\}}{(z - 1)(z - e^{-0.5T})} \tag{2.6}$$

Therefore,

$$E(z) = \frac{R(z)}{1 + G'(z)}$$

$$C(s) = E(z)(1 - z^{-1})K/[s^2(2s + 1)]$$

and

$$C(z) = E(z)G'(z)$$

$$= \frac{G'(z)R(z)}{1 + G'(z)} \tag{2.7}$$

Note that the sampled signal $C(t_k)$ does not exist at any point in this system, so a fictitious sampler is added at the output to facilitate finding it. This sampler does not affect actual system operation.

2. *If $R(t)$ is a unit step, then $R(z) = z/(z - 1)$ and the final value theorem gives*

$$\lim_{k\to\infty} E(t_k) = 1/[1 + \lim_{z\to 1} G'(z)] = 0$$

since $G'(z) \to \infty$ as $z \to 1$.

If $R(t) = t$, then $R(z) = Tz/(z - 1)^2$ and the final value theorem now gives

$$\lim_{k\to\infty} E(t_k) = \lim_{z\to 1} Tz/[(z - 1)G'(z)] = T/[KT] = 1/K$$

This result holds as long as all limits exist, which requires that the system be stable.

3. *Stability requires that the roots of the characteristic equation*

$$1 + G'(z) = 0$$

be inside the unit circle. The characteristic equation can be reduced to

$$F(z) = z^2 + \alpha z + \beta = 0 \tag{2.8}$$

where the α and β coefficients are

$$\alpha = K[T - 2(1 - e^{-0.5T})] - (1 + e^{-0.5T}) \tag{2.9}$$

$$\beta = e^{-0.5T}(1 - KT) + 2K(1 - e^{-0.5T}) \tag{2.10}$$

Rather than factor this quadratic directly, use the bilinear transformation $z = (w + 1)/(w - 1)$ to find a quadratic in w:

$$[1 + \alpha + \beta]w^2 + [2(1 - \beta)]w + [1 + \beta - \alpha] = 0 \tag{2.11}$$

Applying Routh's criterion to the w quadratic shows that for stability,

$$1 + \alpha + \beta > 0, \qquad 2(1 - \beta) > 0, \quad \text{and} \quad 1 + \beta - \alpha > 0$$

are required. In terms of the original z quadratic, these requirements are

$$F(1) > 0, \qquad F(0) < 1, \quad \text{and} \quad F(-1) > 0$$

These requirements are true for any second-order characteristic equation $F(z) = 0$, and are an example of the Schur-Cohn stability test [7]. For this problem $F(1) = KT(1 - e^{-0.5T})$ is positive for all positive T and K.

The second condition states that $\beta < 1$, or

$$K[2 - 2e^{-0.5T} - Te^{-0.5T}] + e^{-0.5T} < 1 \tag{2.12}$$

If $T = 0$, then $\beta = 1$, but T will never be zero. As $T \to \infty$, $\beta \to 2K$, so this condition must be checked in detail when specific values are given in part 4.

Finally, $F(-1) > 0$ leads to

$$K < \frac{2(1 + e^{-0.5T})}{T(1 + e^{-0.5T}) - 4(1 - e^{-0.5T})} \tag{2.13}$$

For very small T this gives $K < 2/T$, and for very large T it gives $K < 2/(T - 4)$. This demonstrates the inverse relationship between K_{max} and T.

4. *With $T = 2$, the requirement of Eq. (2.12) gives $K_{max} = 1.196$, and Eq. (2.13) gives* $K < 13.19$. The most constraining result is the one which is operable.

5. *With $T = 2$*

$$G'(z) = 0.7357588K(z + 0.71828)/[(z - 1)(z - 0.36788)] \tag{2.14}$$

With $K = 0.5$, the steady-state error $\lim_{k \to 1} E(t_k) = 1/K = 2$. This motor control system will follow a commanded ramp in position, but the actual position will be offset by two units from the command. This may not be accurate enough. Even if the gain is increased to near its limit, the error only decreases to around one unit, and the transients will be very slow to die out with the system being that close to the stability limits. This system will need to have some form of compensation if the sampling time cannot be decreased. Note that if $T = 1$, then

$$G'(z) = 0.21306K(z + 0.84675)/[(z - 1)(z - 0.6065)] \tag{2.15}$$

and the maximum allowable stable gain increases to 2.18. The steady-state error due to a ramp input is still $1/K$ independent of T. ■

2.6 EXTENSION OF CLASSICAL TECHNIQUES TO MORE COMPLEX SYSTEMS

When multiple loops or multiple inputs and outputs must be considered, signal flow graph techniques can be used to reduce the problem to one of single-loop analysis. However, the resulting "open-loop" transfer function will usually not be in the convenient factored form. Even the simple techniques can become tedious in this case.

Linear multiple-input, multiple-output systems can be treated systematically using various transfer function matrix representations. Pursuing the subject in that direction quickly leads to the theory of polynomial matrices and the so-called matrix fraction description of systems. In this book, with very few exceptions, the alternate approach of using state space methods is pursued.

Simulation has long served as a supplement and extension to the classical analytical techniques, especially when dealing with complex and nonlinear systems. Analog computers were first historically, then came digital and hybrid methods. At present, the strong trend toward the digital computer continues. Computer-aided design (CAD) tools, which ease or even automate many of the tasks described in this chapter, are now widely available. Further, computers are frequently used as components in the control loop. The controllers or compensators $G_c(z)$ are routinely implemented in microprocessor form.

REFERENCES

1. Dorf, R. C.: *Modern Control Systems, 5th ed., Addison-Wesley, Reading, Mass., 1989.*
2. *Rowland, J. R.: Linear Control Systems,* John Wiley, New York, 1986.
3. Franklin, G. F., J. D. Powell, and A. Emami-Naeini: *Feedback Control of Dynamic Systems,* Addison-Wesley, Reading, Mass., 1986.
4. Kuo, B. C.: *Automatic Control Systems,* 5th ed., Prentice Hall, Englewood Cliffs, N.J., 1987.
5. D'Azzo, J. J. and C. H. Houpis: *Linear Control System Analysis and Design,* Third ed., McGraw-Hill, New York, 1981.
6. Franklin, G. F. and J. D. Powell: *Digital Control of Dynamic Systems,* Addison-Wesley, Reading, Mass., 1980.
7. Kuo, B. C.: *Analysis and Synthesis of Sampled-Data Control Systems,* Prentice Hall, Englewood Cliffs, N.J., 1963.
8. Cruz, J. B., Jr. and W. R. Perkins: "A New Approach to the Sensitivity Problem in Multivariable Feedback Systems," *IEEE Transaction on Automatic Control,* Vol. AC-9, July 1964, pp. 216–223.
9. Cruz, J. B., Jr., Ed.: *Systems Sensitivity Analysis,* Dowden, Hutchinson and Ross, Stroudsburg, Penn., 1973.
10. Truxal, J. G.: *Automatic Control System Synthesis,* McGraw-Hill, New York, 1955.

ILLUSTRATIVE PROBLEMS

Properties of Feedback

2.1 Compare the open-loop and feedback control systems of Figure 2.9 in terms of the sensitivity of the output C to variations in system parameters.

In the open-loop case $C = G_1 R$ and $\partial C/\partial G_1 = R$, so that a change δG produces a change in the output $\delta C = R\,\delta G$. System sensitivity S is defined as percentage change in C/R divided by the percentage change in the process transfer function. For the open-loop system,

$$S = \frac{\delta G/G}{\delta G/G} = 1$$

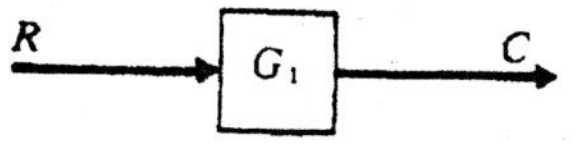

(a) Open loop

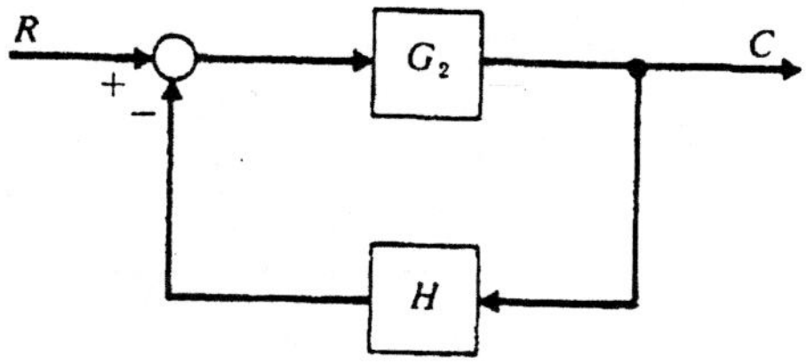

(b) Closed loop

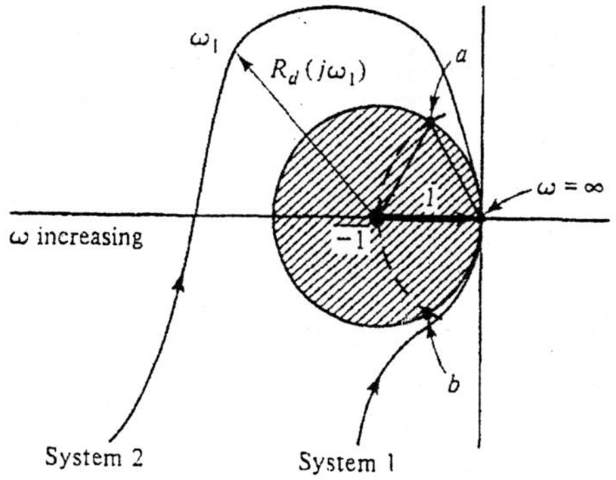

(c)

Figure 2.9

For the closed-loop feedback system $C = G_2R/(1 + G_2H)$,

$$\delta G_2 \frac{\partial C}{\partial G_2} = \frac{R\,\delta G_2}{1 + G_2H} - \frac{G_2RH\,\delta G_2}{(1 + G_2H)^2} = \frac{R\,\delta G_2}{(1 + G_2H)^2}$$

The sensitivity is

$$S = \frac{\delta(C/R)}{C/R}\frac{G_2}{\delta G_2} = \frac{\delta G_2}{(1 + G_2H)^2}\frac{(1 + G_2H)}{G_2}\frac{G_2}{\delta G_2} = \frac{1}{1 + G_2H} = \frac{1}{R_d}$$

If the magnitude of the return difference is greater than unity at all frequencies ω, i.e.,

$$|R_d(j\omega)| > 1 \tag{1}$$

then the closed-loop configuration (*b*) is always less sensitive to parameter variations than the open-loop configuration (*a*). Equation (*1*) requires that the polar plot of $G_2(j\omega)H(j\omega)$ never enters the unit-radius disk centered at -1. Figure 2.9*c* shows polar plots for two possible systems which satisfy this requirement. System 1 has an infinite gain margin (increase or decrease). With system 2 the gain can be increased an infinite amount or decreased by at least 50% while remaining stable. Consideration of the equilateral triangles formed by points $(a, -1, 0)$ or $(b, -1, 0)$ make it clear that systems which do not penetrate the unit disk also have phase margins of at least 60°. Certain linear-quadratic optimal systems are guaranteed to have these desirable properties, as is discussed in Sec. 14.7. For other systems, the design goal is often to shape the polar plot and/or use a high loop gain in order to satisfy Eq. (*1*) at all frequencies within the system bandwidth but perhaps not at all frequencies. Satisfactory reduction of sensitivity to the most problematic parameter variations is thus achieved.

2.2 Show how precise feedback coefficients can give precision feedback control even if gross errors exist in the forward loop system being controlled.

As the loop gain increases, the feedback transfer function becomes

$$\frac{C}{R} = \frac{KG}{1+KGH} \to \frac{1}{H}$$

Thus for very high loop gain the response is largely determined by H rather than G.

2.3 Show that the sensitivity of the output to errors in H approaches unity for high loop gain.

This could easily be shown by starting with the result of the last problem. Alternatively, define sensitivity to H as

$$S_H = \frac{\delta(C/R)/(C/R)}{\delta H/H} = \frac{-G^2\,\delta H}{(1+GH)^2}\,\frac{1+GH}{G}\,\frac{H}{\delta H} = -\frac{GH}{1+GH} \to -1 \quad \text{as} \quad |GH| \to \infty$$

This indicates the need for precision components in the feedback loop.

2.4 Compare performance of the systems shown in Figure 2.10a,b as degraded by the unwanted disturbance D.

In the open-loop case, $C(s) = KG_1 G_2 R(s) + G_2 D(s)$.

In the feedback case, $C(s) = \dfrac{KG_1 G_2 R(s)}{1+KG_1 G_2 H} + \dfrac{G_2 D(s)}{1+KG_1 G_2 H}$.

In the second case the contribution of the disturbance to the output can be made small by increasing the gain K. More generally, this is accomplished by increasing the return difference magnitude over the frequency range of interest, and this should be done by increasing the magnitude of KG_1. Notice that feedback also introduces a loss of useful gain between C and R. For example, if $H = 1$ and G_1 and G_2 are ideal amplifiers with constant gains, the open-loop system gain is $KG_1 G_2$. The feedback system gain $KG_1 G_2/(1 + KG_1 G_2)$ would then be less than unity.

2.5 Use the dc motor of Problem 1.2, page 17, to demonstrate how feedback can favorably improve transient response. Neglect the inductance L.

When $L = 0$, the transfer function can be written as

$$\frac{\Omega(s)}{V(s)} = \frac{K/JR}{s + (bR + K^2)/JR}$$

This is considered as the open-loop system. If a step voltage $V(s) = V/s$ is applied, the open-loop speed response is, letting $K' = K/JR$ and $a = (bR + K^2)/JR$,

$$\Omega(s) = \frac{VK'}{s(s+a)} \quad \text{or} \quad \omega(t) = \frac{VK'}{a}(1 - e^{-at})$$

The speed of response is determined by a, which is determined by the load and motor characteristics. If response is too slow for a given load, a new motor with a larger value of K could be

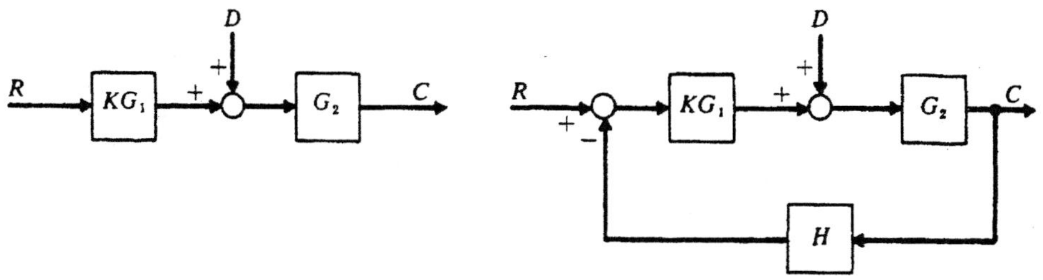

(a) Open loop (b) Feedback

Figure 2.10

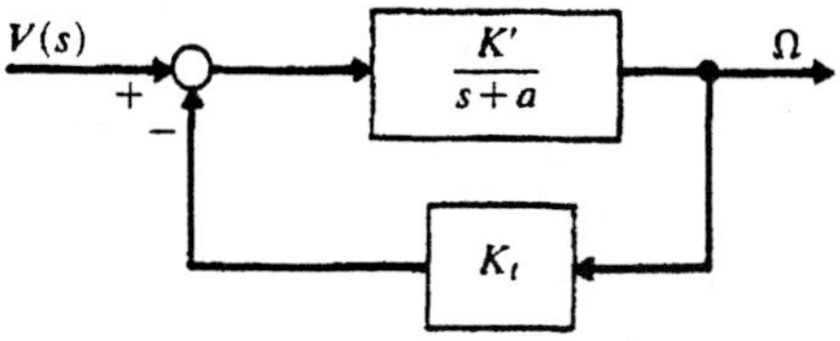

Figure 2.11

installed. Alternatively, consider the tachometer feedback system of Figure 2.11. Here $\Omega(s) = [K'/(s + a + K'K_t)]V(s)$. The system response time is now determined by $a' = a + K'K_t$ and can obviously be improved by proper choice of K_t.

Routh's Criterion

2.6 Use Routh's criterion to determine the number of roots of

$$s^5 + 5s^4 - 2s^3 + 8s^2 + 10s + 3 = 0$$

which have positive real parts.

The Routhian array is given in Table 2.2

Note that any row can be normalized by multiplying or dividing by a positive constant, as in the s^3 row. For the terms in column 1, the sign changes from row s^4 to row s^3 and again from row s^3 to row s^2. Two sign changes indicate there are two right-half-plane roots to this equation.

2.7 Does the following equation have any roots in the right-half plane?

$$s^4 + 2s^3 + 4s^2 + 8s + \alpha = 0$$

The Routhian array is shown in Table 2.3.

Note that the leading term in the s^2 row is zero. Whenever this happens, the zero is replaced by a small number ϵ and the rest of the array is computed as usual. The limiting behavior as $\epsilon \to 0$ is used to determine stability. Here the first column reduces to $\{1, 2, 0, \lim_{\epsilon \to 0} (-2\alpha/\epsilon), \alpha\}$. If $\alpha < 0$, there is just one sign change between the s and s^0 rows, and, therefore, just one right-half-plane root. If $\alpha > 0$, there are two sign changes and two right-half-plane roots.

Steady-State Error and Error Constants

2.8 Derive expressions for the steady-state value of $E(t)$ for the system of Figure 2.1*a* when the input $R(s)$ is a step, ramp, and parabolic function, respectively.

The Laplace transform of the error is $E(s) = R(s)/[1 + KG(s)H(s)]$. For a unit step, $R(s) = 1/s$. Using the final value theorem and assuming a constant steady-state error exists,

TABLE 2.2

s^5	1	-2	10
s^4	5	8	3
s^3	$\frac{5(-2) - (1)(8)}{5} = -18/5 \sim -18$	$\frac{5(10) - 1(3)}{5} = \frac{47}{5} \sim 47$	—
s^2	$\frac{-18(8) - 5(47)}{-18} = \frac{379}{18}$	3	—
s	$\frac{(379/18)(47) - (-18)(3)}{379/18} = \frac{18785}{379}$	—	
s^0	3		

TABLE 2.3

s^4	1	4	α
s^3	2	8	
s^2	$\frac{2(4)-8}{2} = 0^{\epsilon}$	α	
s	$\frac{8\epsilon - 2\alpha}{\epsilon}$		
s^0	α		

$E_{ss} \triangleq \lim_{t\to\infty}\{E(t)\} = 1/[1 + K \lim_{s\to 0}\{GH\}]$. Similarly, for a ramp input, $R(s) = 1/s^2$ and $E_{ss} = 1/K \lim_{s\to 0}\{sGH\}$. If the input is the parabola $t^2/2$, $R(s) = 1/s^3$ and $E_{ss} = 1/K \lim_{s\to 0}\{s^2 GH\}$.

To proceed, we must know the system type. The system type is the number of s terms that factor out of the denominator of $G(s)H(s)$. For a type 0 system there are no such factors, so $K \lim_{s\to 0}\{GH\} = K_b$, the Bode gain. Likewise, for type 0, $K \lim_{s\to 0}\{sGH\} = 0$ and $K \lim_{s\to 0}\{s^2 GH\} = 0$. For type 1 systems $K \lim_{s\to 0}\{GH\} = \infty$, $K \lim_{s\to 0}\{sGH\} = K_b$ and $K \lim_{s\to 0}\{s^2 GH\} = 0$. Similar results hold for type 2 and higher systems. The three limiting values for each system are called the position, velocity, and acceleration *error constants* K_p, K_v, and K_a. These are summarized in Table 2.4, along with the steady-state error values. Note that larger error constants give smaller steady-state error.

Miscellaneous Methods

2.9 Sketch the root locus for a system with

$$KG(s) = \frac{K}{s(s+8)(s^2+8s+32)}, \qquad H(s) = s + 4$$

The open-loop poles, plotted as ×, are located at $s = 0$, -8, $-4+4j$, and $-4-4j$. The open-loop zero ⊙ is at $s = -4$. There are four branches of the loci, and Rule 3 gives the real axis portion, shown in Figure 2.12. Three branches must approach $s = \infty$ as $K \to \infty$. One is on the negative real axis, and the others are at $\pm 60°$, since

$$\phi_i = \frac{(1+2k)180°}{3} = (1+2k)60°$$

$$= +60 \quad (k=0), \qquad -60 \quad (k=-1), \quad \text{and} \quad 180° \quad (k=1)$$

Other values of k give multiples of the same three asymptotic angles. Rule 6 gives

$$\text{cg} = \frac{1}{3}[(0-8-4-4)-(-4)] = -4$$

TABLE 2.4

	Error constants			Steady-state error		
System type	K_p	K_v	K_a	Step input	Ramp input	Parabolic input
0	K_b	0	0	$1/(1+K_b)$	∞	∞
1	∞	K_b	0	0	$1/K_b$	∞
2	∞	∞	K_b	0	0	$1/K_b$

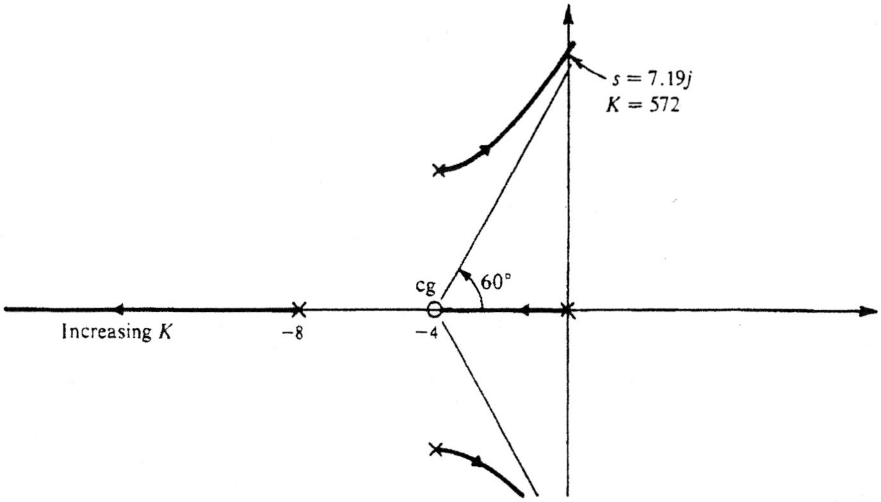

Figure 2.12

In general, the phase angle of the transfer function, equation (2.2), can be written as

$$\angle GH(s) = \phi_{z_1} + \phi_{z_2} + \cdots - \phi_{p_1} - \phi_{p_2} = \cdots = (1 + 2k)180°$$

where ϕ_{z_i} is the angle of the line segment from the zero z_i to a point s and ϕ_{p_i} is the angle from pole p_i to s. In order to determine the angle of departure of the locus from a complex pole, a test point s_1 is used, which is infinitesimally close to the pole. The angles of the vectors from all zeros and poles except one are easily measured. The remaining angle, associated with that complex pole, can be computed. This is the angle of departure and in this case it is 0°. With this information, a few more test points allow an accurate sketch of the complete locus.

2.10 For the system shown in Figure 2.13, select K so that the phase margin is greater than 30° and the gain margin is greater than 10 db.

In Bode form,

$$KGH = \frac{50K}{200}\frac{(0.02s + 1)}{s(0.1s + 1)(0.05s + 1)}$$

The Bode plots are drawn in Figure 2.14 with the Bode gain $K_b = K/4$ set to unity. At $\omega = 10$, the phase is $-150°$ and the gain is -24 db. This means that K_b could be increased from 0 db to +24 db, and the phase margin specification would just be satisfied. If this were done, then at $\omega = 24$ rad/s, where the phase is $-180°$, the gain would increase from -39 db to -15 db. This gain margin of 15 db satisfies the specifications, so $K_b = 24$ db, which converts to a real gain of about 15. This means that $K = 4K_b = 60$ can be used to satisfy both specifications.

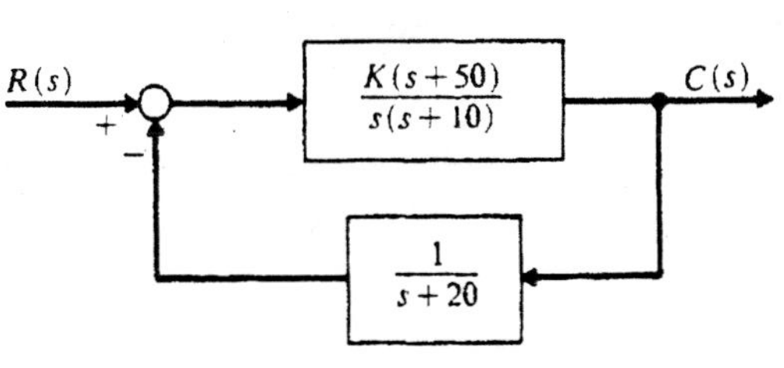

Figure 2.13

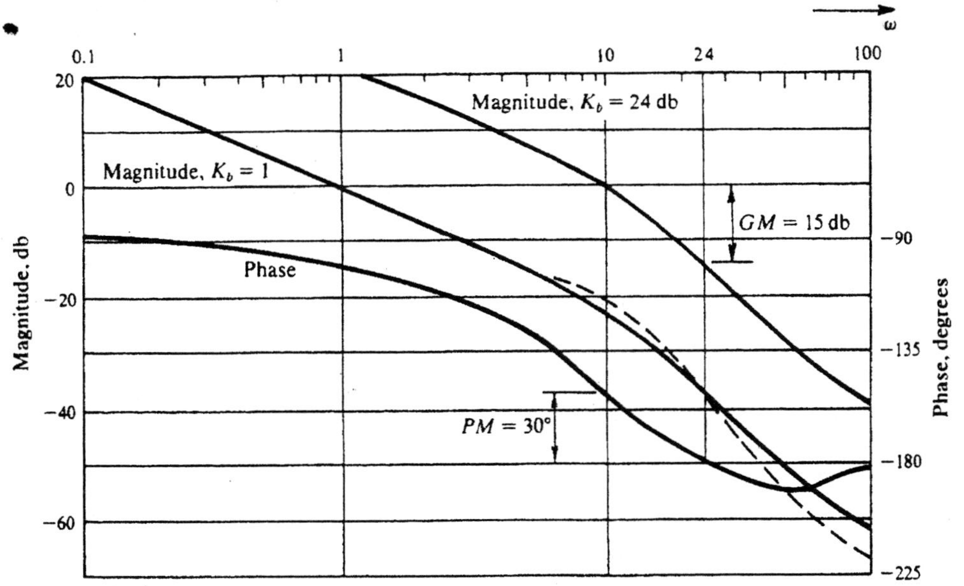

Figure 2.14

2.11 Use root locus to determine the closed-loop poles of Problem 2.10 when $K = 60$ is used.

The upper half of the locus is sketched in Figure 2.15 using the rules of Example 2.4 and a spirule to check Eq. (2.2) at a few additional points. The closed-loop poles are shown as ⊡. Notice that the complex poles lie almost exactly on the $\zeta = 0.3$ damping line. For this example, the rule of thumb $\zeta = 0.01\,PM$ is verified.

Polar Plots and Nyquist's Criterion

2.12 What information is readily available from a polar plot of $KG(j\omega)H(j\omega)$?

The number of closed-loop poles in the right half plane can be determined in terms of the number of encirclements of -1, using Nyquist's criterion. The system type is indicated, assuming a minimum phase system, by the phase angle at $\omega = 0$. Type 0 systems have a finite magnitude and zero phase angle. Type 1 systems approach infinite magnitude at an angle of $-90°$, type 2 systems aproach infinite magnitude at an angle of $-180°$, etc. The excess of open-loop poles compared to zeros is indicated by the behavior as $\omega \to \infty$. If there is an equal number of poles and zeros, the magnitude approaches a finite constant. In all other cases the magnitude approaches zero, but if there is one more pole than zero, the approach is along the $-90°$ axis. For two more poles than zeros, it is along the $-180°$ axis, etc. Relative stability, in terms of gain and phase margins, is also readily apparent. For example, in the plot shown in Figure 2.16 the system is type 1, and it has three more poles than zeros. Assuming no open-loop, right-half-plane poles, the system is stable, since the plot does not encircle the point -1. The phase margin is 60° and the gain margin is 1.25, since the gain could be increased by that factor without causing the plot to encircle the -1 point.

2.13 Draw the polar plot for

$$KGH = \frac{K(s + 0.5)(s + 10)}{(s + 1)(s + 2)(s^2 + 2s + 5)}$$

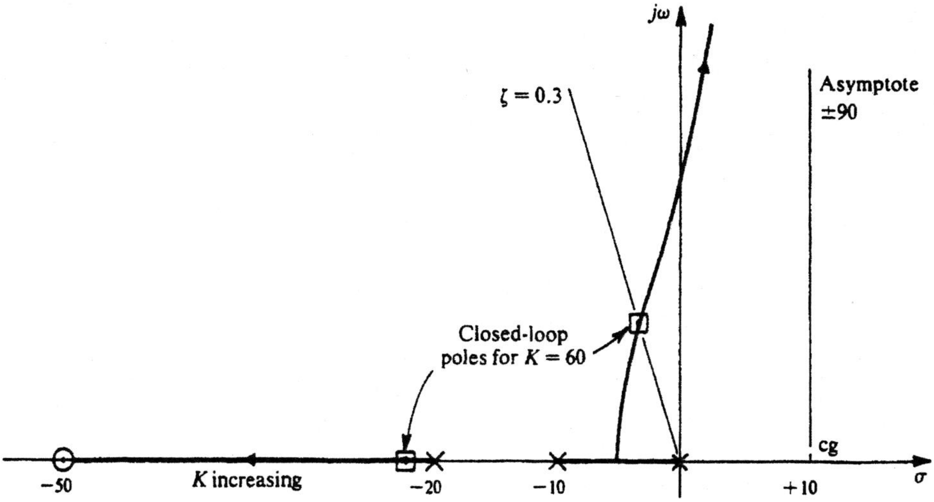

Figure 2.15

Determine the gain and phase margins when $K = 2$.

The polar plot for this type 0 system is given in Figure 2.17. The -1 point is not encircled, so $N = 0$. Since there are no open-loop, right-half-plane poles, $P_R = 0$. Therefore, Nyquist's criterion indicates that there are no unstable closed-loop poles. The phase margin is approximately 50°. The magnitude at 180° phase is 0.32, so the gain margin is $1/0.32 \cong 3.1$.

The smallest magnitude of the return difference, which occurs here at approximately $\omega = 3$, is an excellent measure of stability margins. This is the basis for the constant M-circle concepts in classical control analysis. The larger the minimum return difference is, the more robust the system is to modeling errors and parameter variations, which may cause gain changes or phase shifts.

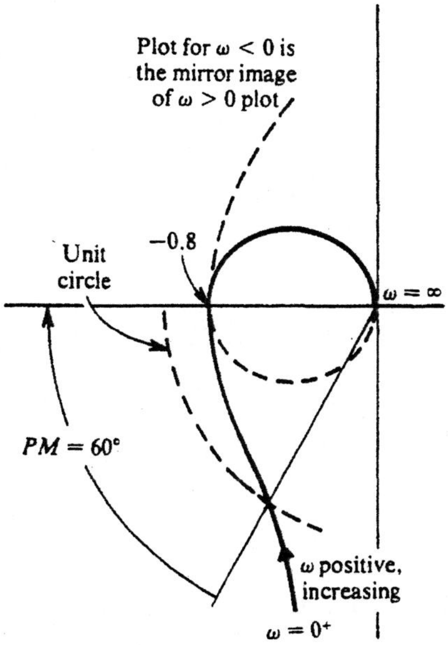

Figure 2.16

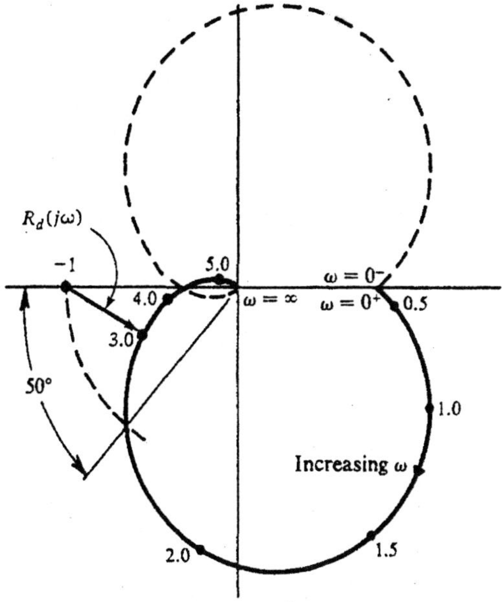

Figure 2.17

2.14 Analyze the polar plot for the nonminimum phase system with

$$KGH = \frac{K(s+10)(s+20)}{s(s-10)(s+40)}$$

Even though this is a type 1 system, the phase angle approaches $-270°$ as $\omega \rightarrow 0$ since the minus sign on the unstable pole contributes $-180°$ phase. The general shape of the polar plot is shown in Figure 2.18*a*.

The -1 point could be encircled by either the *a-d-c-b-a* circuit, counterclockwise, or by

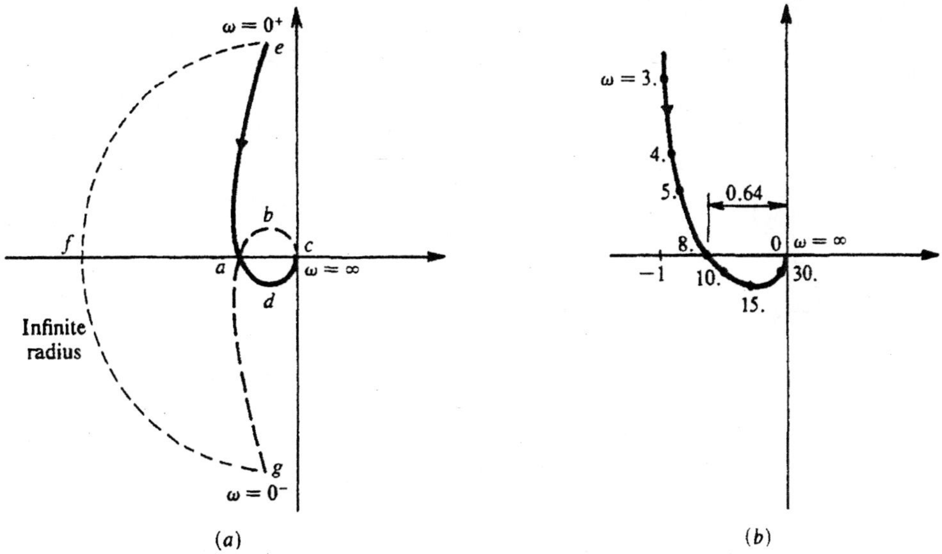

Figure 2.18

the infinite *a-g-f-e-a* circuit, clockwise. Which circumstance prevails depends upon K. If -1 is inside the small circuit, then $N = 1$. Since $P_R = 1$, there would be $Z_R = 1 - 1 = 0$ unstable closed-loop poles. If -1 is to the left of point *a*, then there is one clockwise encirclement so $N = -1$ and, therefore, $Z_R = 1 - (-1) = 2$, indicating two unstable closed-loop poles. An enlarged plot is also given in Figure 2.18*b* for $K = 10$. The magnitude of 0.64 at $-180°$ indicates that the system is unstable for values of $K < 10/0.64 \cong 15.6$, and stable if $K > 15.6$. This problem illustrates that the usual visualization of gain and phase margin is incorrect for nonminimum phase systems. This is why Bode plots should not be used with nonminimum phase transfer functions.

Compensation

2.15   Explain the essence of classical compensation using root locus techniques.

The basis for compensation is a knowledge of the correspondence between closed-loop pole locations and the type of transient time response terms they yield. Some typical closed-loop pole locations are indicated by □ in Figure 2.19, and the corresponding time response terms are shown.

Poles farther to the left of the imaginary axis give terms which die out faster, i.e., faster response times. Complex poles give oscillating terms with a frequency equal to the distance from the real axis and decay time inversely proportional to the real part of the pole, σ. The damping ratio, related to overshoot, is defined in terms of the angle γ as $\zeta = \sin \gamma$. The undamped natural frequency ω_n is the radial distance from the origin, so that $\sigma_i = \zeta \omega_n$ and the damped frequency is $\omega_i = \omega_n \sqrt{1 - \zeta^2}$.

Based on these relations, the desired locations of the most dominant closed-loop poles are selected. Checking the root locus angle criterion at that point indicates whether additional lag or lead is needed and how much. Compensating poles and zeros are then selected to provide this phase shift.

Suppose a damping ratio of $\zeta = 0.707$ and a frequency of 10 rad/s are desired. Then a pair of complex closed-loop poles must be located as shown in Figure 2.20. If the angle for the uncompensated KGH is $-160°$ at that point, then an additional $-20°$ phase must be provided by compensation. It is common practice to place the compensator zero directly below the desired closed-loop pole. Obviously, a single pole zero pair can provide up to 60–65° phase shift. If a greater shift is needed, more complicated compensation is required.

2.16   Discuss the compensation characteristics of the phase-lag and phase-lead circuits of Figure 2.21 using Bode plots.

1. $c_1 e^{-\sigma_1 t}$

2. $c_2 e^{-\sigma_2 t}$

3. $c_3 e^{-\sigma_3 t} \sin (\omega_3 t + \phi_3)$

4. $c_4 e^{-\sigma_4 t} \sin (\omega_4 t + \phi_4)$

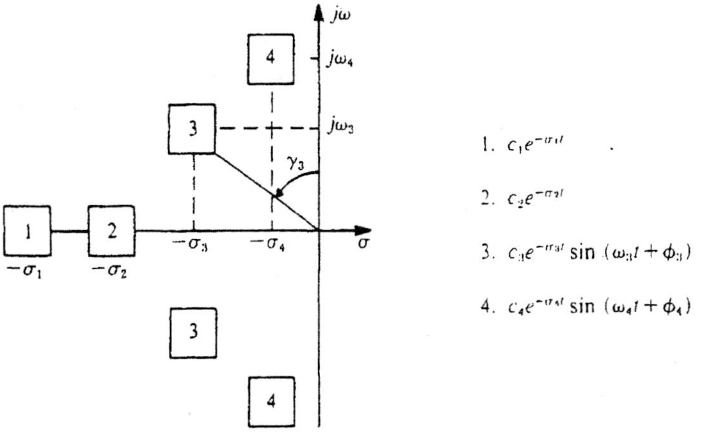

Figure 2.19

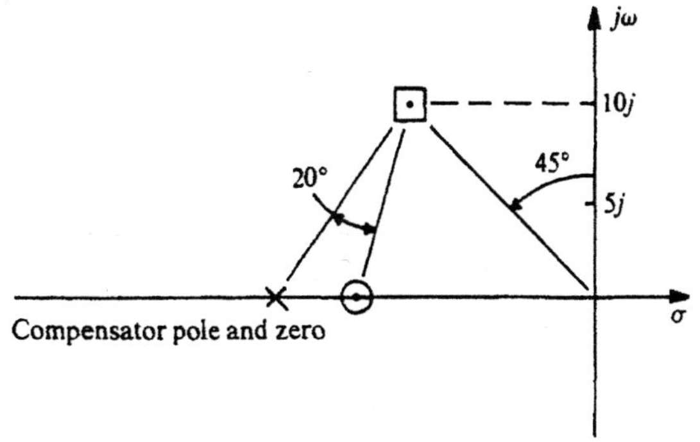

Figure 2.20

The lag circuit lowers the gain at high frequencies and leaves the phase unchanged for $\omega >> 1/\tau_1$ (Figure 2.22*a*). The corner frequencies $1/\tau_1$ and $1/\tau_2$ would be chosen well below the critical crossover frequency. This means the dc gain can be raised in order to improve steady-state accuracy. This increase in gain returns the high frequency gain to its uncompensated level and thus leaves the stability margins relatively unchanged.

The lead circuit Bode plot assumes an additional dc gain has been added to compensate for the τ_2/τ_1 reduction. By proper choice of $1/\tau_1$ and $1/\tau_2$ relative to the cross-over frequencies, the phase lead can be used to increase phase margin. This circuit also delays the 0 db crossover frequency, thus increasing bandwidth and speed of response. It leaves the low frequency characteristics, such as steady-state error, unchanged.

Discrete-Time Problems

2.17 In the process of designing a digital control system, it was determined that the following compensator was desired:

$$G_c(z) = \frac{10(z + 0.4)(z - 0.5)}{(z + 0.5)(z + 1)}$$

Determine an algorithm which can be coded on the computer to implement $G_c(z)$.

There are several possible answers, two of which follow. Writing G_c in expanded polynomial form gives

$$G_c(z) = \frac{10(z^2 - 0.1z - 0.2)}{z^2 + 1.5z + 0.5} = \frac{10(1 - 0.1z^{-1} - 0.2z^{-2})}{1 + 1.5z^{-1} + 0.5z^{-2}}$$

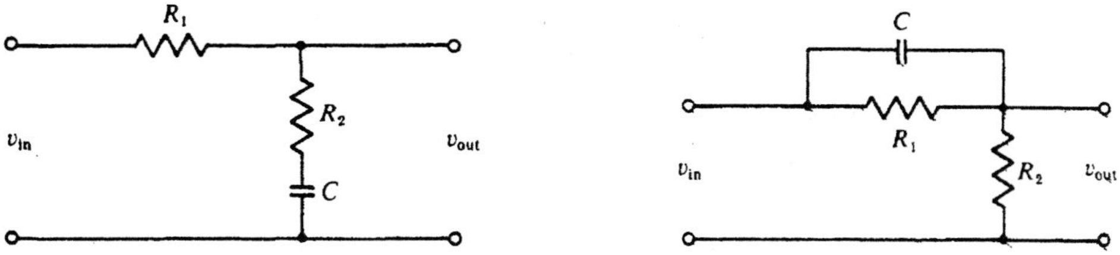

$$\frac{V_{out}}{V_{in}} = \frac{1 + \tau_1 s}{1 + \tau_2 s} \quad \tau_2 > \tau_1$$

(*a*) Phase-lag circuit

$$\frac{V_{out}}{V_{in}} = \left(\frac{\tau_2}{\tau_1}\right)\frac{1 + \tau_1 s}{1 + \tau_2 s} \quad \tau_1 > \tau_2$$

(*b*) Phase-lead circuit

Figure 2.21

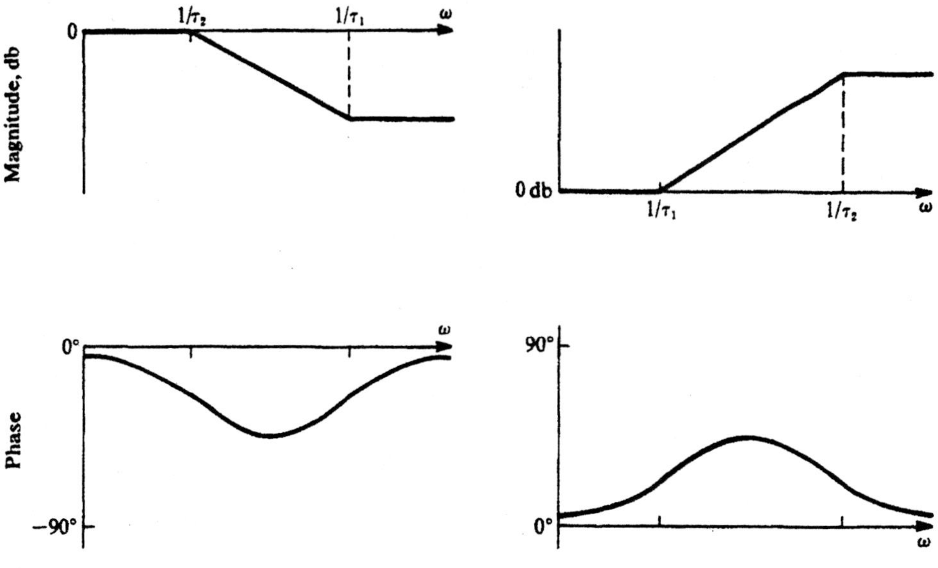

(*a*) Phase-lag circuit (*b*) Phase-lead circuit

Figure 2.22

If the input to G_c is $E(t_k)$ and the output is $y(t_k)$, then using z^{-1} as the delay operator, cross multiplication gives the so-called direct form realization

$$(1 + 1.5z^{-1} + 0.5z^{-2})Y(z) = (10 - z^{-1} - 2z^{-2})E(z)$$

or, in the time domain

$$y(t_k) = 10E(t_k) - E(t_{k-1}) - 2E(t_{k-2}) - 1.5y(t_{k-1}) - 0.5y(t_{k-2})$$

If $G_c(z)/z$ is expanded in partial fractions and the result is then multiplied by z, one obtains

$$G_c(z) = -4 - 4z/(z + 0.5) + 18z/(z + 1)$$

This represents three separate paths through the compensator, as shown in Figure 2.23. This is the so-called parallel realization.

The algorithm thus consists of

$$y_1(t_k) = 4E(t_k)$$

$$y_2(t_k) = E(t_k) - 0.5y_2(t_{k-1})$$

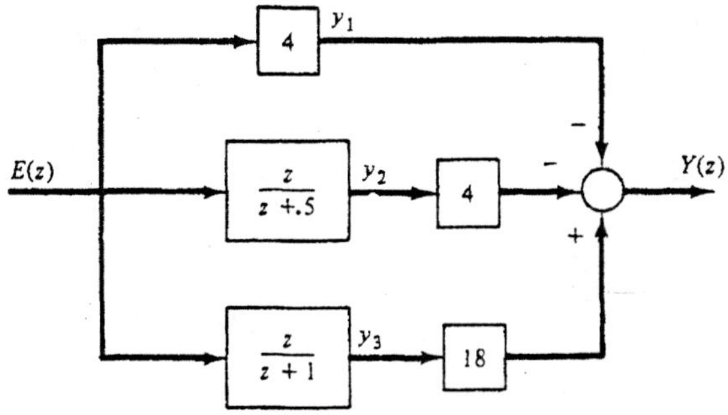

Figure 2.23

$$y_3(t_k) = E(t_k) - y_3(t_{k-1})$$

and

$$y(t_k) = 18y_3(t_k) - y_1(t_k) - 4y_2(t_k)$$

Other forms are possible.

2.18 Derive the transfer function for the device which outputs the piecewise constant approximation to a continuous-time function $E(t)$ as shown in Figure 2.24.

Let the unit step function starting at time $t = 0$ be $u(t)$. Then a shifted unit step function starting at time t_k is $u(t - t_k)$. The piecewise constant function can be written as

$$\begin{aligned} Y^*(t) &= E(0)u(t) + [E(1) - E(0)]u(t - t_1) + [E(2) - E(1)]u(t - t_2) + \cdots \\ &\quad + [E(t_k) - E(t_{k-1})]u(t - t_k) + \cdots \\ &= \sum_{k=0}^{\infty} [E(t_k) - E(t_{k-1})]u(t - t_k) \end{aligned}$$

Since $du(t - t_k)/dt = \delta(t - t_k)$, the derivative of the device output is

$$x(t) = dY^*(t)/dt = \sum_{k=0}^{\infty} [E(t_k) - E(t_{k-1})]\delta(t - t_k)$$

Using the definition of the Z-transform in Problem 1.15, the transform of $x(t)$ is $X(z) = E(z) - z^{-1}E(z) = (1 - z^{-1})E(z)$. The final relationship among these variables is shown in Figure 2.25.

Since $y(t)$ is the integral of $x(t)$, $Y(s) = [(1 - z^{-1})/s]E(z)$. Clearly, then, the transfer function of the zero order hold is $G_0(s) = (1 - z^{-1})/s$.

2.19 Investigate methods of obtaining a Z-transfer function $G'(z)$ which has approximately the same behavior as an s-transfer function $G(s)$. There are several approaches to this question.

(a) One way is to determine the exact Z-transform that corresponds to the product of $G(s)$ and the ZOH transfer function $G_0(s)$, perhaps using transform tables. The ZOH should be included in most cases involving combined continuous and discrete systems, as shown in Figure 2.1*b*. The reason is that Z-transforms are always applied to everything between two samplers (or the same sampler around a complete loop). For most physical systems of interest, a D/A will be included in this segment of the system.

(b) An approximate conversion is obtained by treating s as a derivative operator and using a forward finite difference approximation on the sampled signal:

$$\dot{y}(t) \cong [y(t_{k+1}) - y(t_k)]/T$$

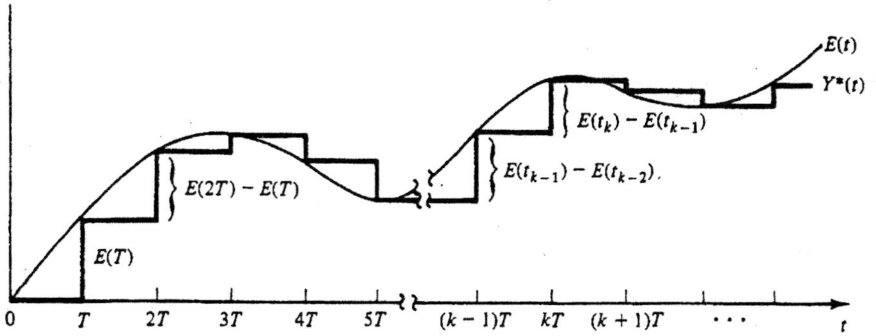

Figure 2.24

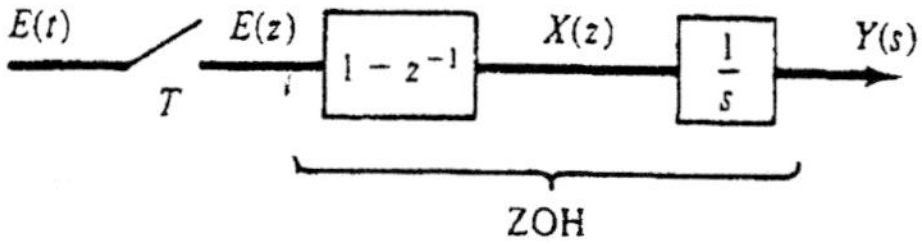

Figure 2.25

In the transform domain this means that $s \cong (z-1)/T$, since z is the advance operator. If this approximation for s is used in $G(s)$, the approximate transfer function $G_b'(z)$ is obtained. Note that this result is also given directly from $z = e^{Ts} \cong 1 + Ts$, which is approximately true for small Ts.

(c) A backward difference approximation is also possible. $\dot{y}(t) \cong [y(t_k) - y(t_{k-1})]/T$ leads to $s \cong (1 - z^{-1})/T$ and to $G_c'(z)$. This result is also obtainable from $z = e^{Ts} = 1/e^{-Ts} \cong 1/[1 - Ts]$.

(d) If z is written as $z = e^{Ts/2}/e^{-Ts/2} \cong [1 + Ts/2]/[1 - Ts/2]$, then s can be solved for as $s \cong (2/T)[z-1]/[z+1]$. Using this gives $G_d'(z)$.

All the above results can also be derived by approximating the integration operator instead of the derivative operator [6].

2.20 Apply the above techniques to determine discrete approximations for $G(s) = 1/(s+a)$.

(a.1) $Z\{(1 - z^{-1})G(s)/s\} = [1 - e^{-aT}]/[a(z - e^{-aT})] = G_a'(z)$

Note that the pole is the exact s- to z-plane mapping of $s = -a$, so stability properties are preserved.

(a.2) For comparison, the transform without the zero order hold is $Z\{G(s)\} = z/[z - e^{-aT}]$. Even though the denominators are the same, there is a one-period delay difference and a gain difference. Unless one is working with true pulsed circuits, form a.1 is the appropriate one to use.

(b) With $s \cong (z-1)/T$, $G_b'(z) = T/[z - 1 + aT]$. The z-plane pole can be in the unstable region (outside the unit circle) even if the s-plane pole is stable.

(c) With $s \cong (z-1)/(Tz)$, $G_c'(z) = [T/(1+aT)]z/\{z - [1/(1+aT)]\}$. Here the z-plane pole is always stable (inside the unit circle) whenever the s-plane pole is stable (and sometimes even when the s-plane pole is unstable).

(d) With $s \cong (2/T)[z-1]/[z+1]$, $G_d'(z) = [T/(aT+1)]z/[z - 1/(1+aT)]$. In this case the z-plane pole is inside the unit circle if $a < 0$, is on the unit circle if $a = 0$, and is outside if $a > 0$. Thus $G_d'(z)$ inherits the exact stability properties of $G(s)$. It can be shown that this is true for all transfer functions formed using approximation (d). That is,

$$z = [1 + Ts/2]/[1 - Ts/2]$$

exactly maps the left-hand s-plane into the interior of the unit circle. Approximations (a) and (d) both preserve stability properties, while (b) and (c) do not. However, the behavior in (c) is preferable to that of (b) in this regard.

2.21 By using the mapping $z = e^{Ts}$, determine how the pole locations discussed in Problem 2.15 map into the Z-plane.

For poles [1] and [2] or any other on the negative real axis, $z = e^{-\sigma T}$ is a positive number between 0 and 1. Therefore, Z-plane poles on the positive real axis correspond to exponential time functions. These are stable (decaying) exponentials for poles inside the unit circle. Poles on the positive real axis in the s-plane also give positive real axis Z-plane poles that are outside the unit circle. These correspond to unstable (growing) exponentials.

Poles like [3] and [4] give $z = e^{-\sigma T}e^{j\omega T} = e^{-\sigma T}[\cos(\omega T) \pm j\sin(\omega T)]$. These will always occur in conjugate pairs, and are on the unit circle if $\sigma = 0$ and give persistent oscillations. If $\sigma > 0$, they are inside the unit circle and give damped oscillations. For $\sigma < 0$, the poles are outside the unit circle and correspond to growing oscillations. The decay or growth rate depends directly on the radial distance of the poles from the unit circle. The frequency of oscillation is directly related to their angular position ωT. Note that if $\omega T = \pi$, the poles are on the negative real axis. No s-plane pole with $\omega > \pi/T$ will occur if the sampling rate satisfies the Nyquist sampling theorem. If such higher frequency poles do occur, they map into the Z-plane at points which also correspond to lower frequency poles. This ambiguity is called aliasing, and can be avoided by sampling at least twice per period of the highest frequency pole in the system.

As any s-plane pole's real part approaches $-\infty$, the Z-plane pole approaches the origin. Poles at $s = 0$ map into $z = 1$.

It is informative to consider several familiar contours in the s-plane and see how they map into the Z-plane. In Figure 2.26*a*, a closed contour is considered, with arrows indicating the traverse direction and numbered points showing the correspondence at eight key points. In Figure 2.26*b*, lines of constant frequency, constant settling time, and lines of constant damping ratio are shown for the s- and Z-planes.

2.22 The dc motor controller of Example 2.6 is to be designed using root locus. The sampling period is $T = 1$ s. Closed-loop Z-plane poles at $z = 0.19877 \pm 0.30956j$ are desired. These are the images of $s = -1 \pm j$. They are selected because of the desirable properties they give to continuous-time systems, namely, a settling time of about 4 seconds and the damping ratio of 0.7, which gives about 5% overshoot. The open-loop transfer function is given in Eq. (*2.15*).

The uncompensated root locus is sketched in Figure 2.27*a*. To achieve the desired root locations, compensation is required to reshape the locus. A forward-path cascade

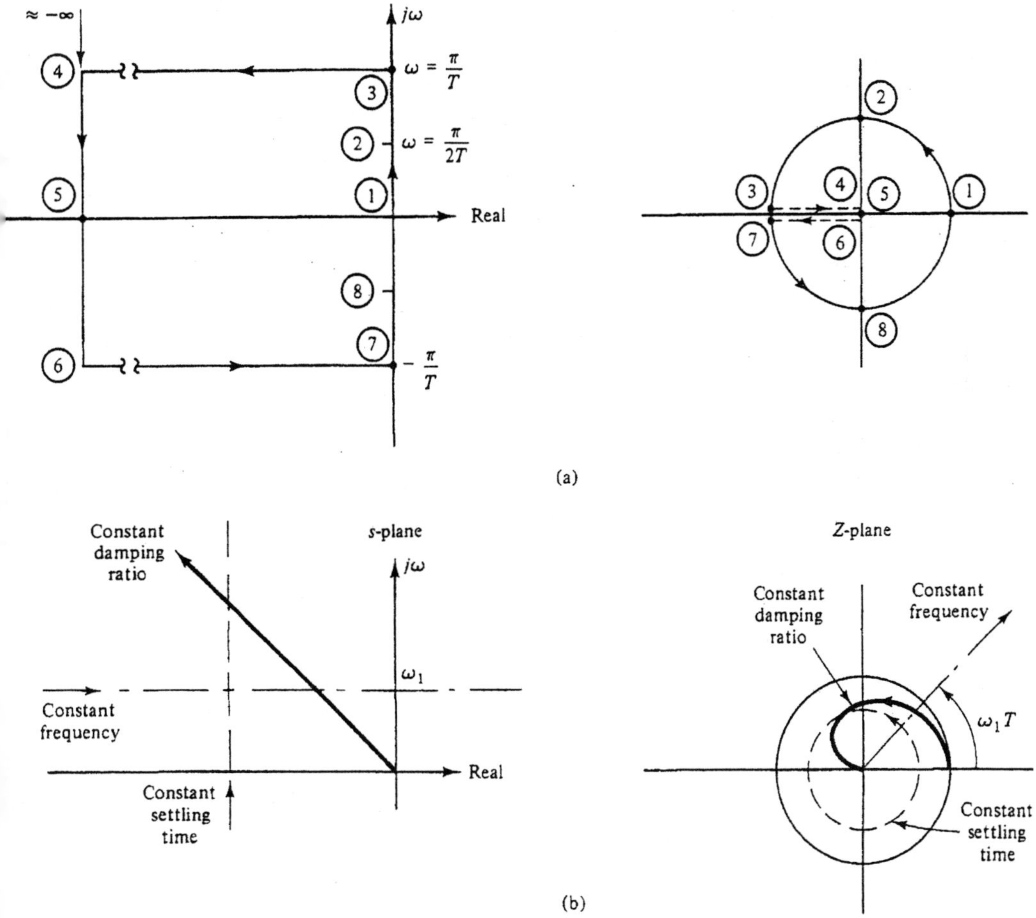

Figure 2.26

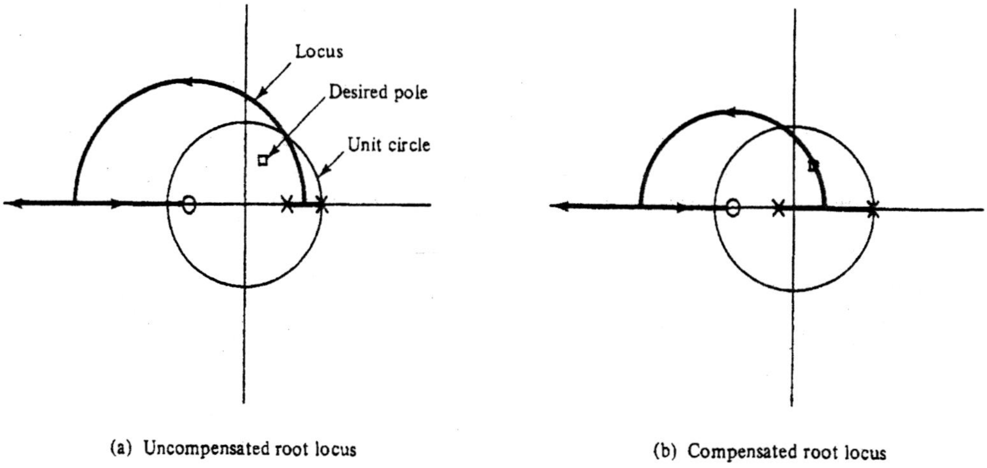

Figure 2.27

compensator of the form $G_c(z) = K_c(z - 0.60653)/(z + a)$ is selected. The value of a will be selected so that the locus will be shifted over, as shown in Fig. 2.27*b*. This is a lead-type compensator. As shown in Figure 2.28, if the desired point is to be on the locus, it must be true that $\phi_3 - \phi_1 - \phi_2 = -180$. But $\phi_1 = 180 - \tan^{-1}[0.30956/(1 - 0.19877)] = 158.88°$. Likewise, $\phi_3 = \tan^{-1}[0.30956/(0.84675 + 0.19877)] = 16.49°$. Therefore, ϕ_2 must be 37.61°. This means that the compensator pole must be at $z = -0.203$. The required root locus gain is computed as the product of the vector lengths from the poles, divided by the product of the vector lengths from the zeros to the desired root location ⊡. This gives $K_{RL} = [(0.7378)(0.2572)/1.1889]^{1/2} = 0.3995$. The total root locus gain for this sytem is $K_{RL} = K_c(0.21306K)$, so if, for example, the gain K in the open-loop transfer function has a value of 0.5, then $K_c = 3.7501$, giving the final compensator

$$G_c(z) = 3.7501(z - 0.6065)/(z + 0.2030).$$

The compensated system's response to a step input is shown in Figure 2.29. It shows about 4% overshoot, just what might have been expected from the s-plane roots. This result is not necessarily typical. Quite often the discrete system will have far more overshoot than expected from the s-plane pole positions. This is due to the fact that the sampled system is essentially running open-loop for the intersample periods T. This allows larger overshoot to build up before corrective feedback action can occur. Also, the location of the zero is a major factor in determining the response [6]. The final value theorem shows that this system will have a steady-state error of 3.26 when the input is a ramp.

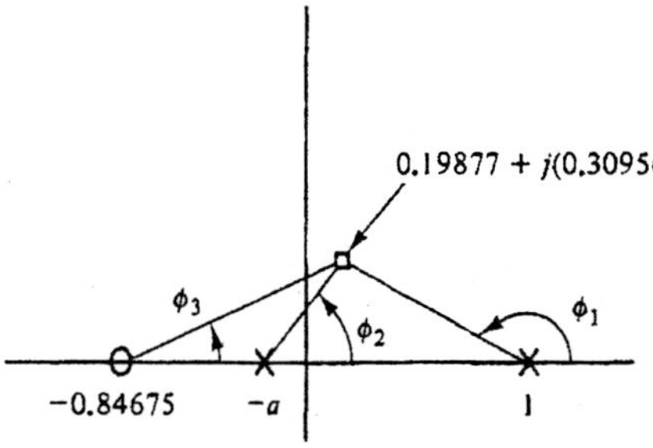

Figure 2.28

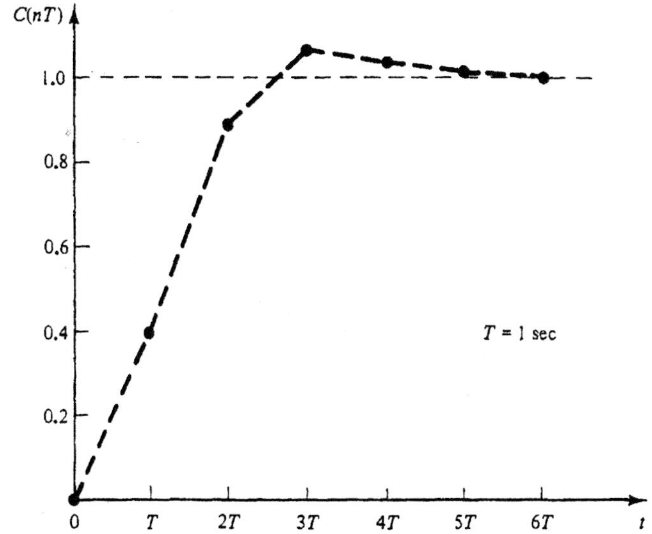

Figure 2.29 Step response.

2.23 Design a cascade compensator for the system of Example 2.6 so that the steady-state value of the error $E(t_k)$ is zero for a ramp input, and such that E goes to zero in the minimum number of sampling periods. This is referred to as the deadbeat response controller.

The transfer function for the error is

$$E(z) = W(z)R(z)$$

where $W(z) = 1/[1 + G_c(z)G'(z)]$.

If $E(t_k)$ is to go to and remain at zero in some finite time, then $W(z)R(z)$ must be a *finite* polynomial in z^{-1}. Since steps, ramps, and other typically used inputs contain a denominator factor of $(1 - z^{-1})^\alpha$, $W(z)$ must contain as a factor $(1 - z^{-1})^\alpha$. Otherwise, an infinite series in z^{-1} would result for $E(z)$. In this particular case of a ramp input, $\alpha = 2$ and $W(z)$ must have the factor $(1 - z^{-1})^2$. It is generally necessary that $W(z)$ contain another factor $F(z^{-1})$ as well. If $W(z)$ is given, then it is easy to show that $G_c(z) = [1 - W(z)]/[W(z)G'(z)]$. The reason why the extra factor F may be required is that the resulting $G_c(z)$ must be *forced* to be physically realizable if it doesn't come out that way initially. The general rules for doing this are:

1. If $G'(z)$ has no poles or zeros on or outside the unit circle (a single pole at $z = 1$ is acceptable), then $F(z^{-1}) = 1$ is all that is required. A leading 1 is always assumed for the polynomial F, and if any unnecessary extra powers of z^{-1} are included, they only delay the time at which E reaches zero. Otherwise, the following three additional dictums must be met:
2. $F(z^{-1})$ must contain as zeros all the unstable poles of $G'(z)$.
3. $1 - W(z)$ must contain as zeros all the zeros of $G'(z)$ on or outside the unit circle.
4. $1 - W(z)$ must contain z^{-1} as a factor.

In the present case $G'(z)$ has no poles or zeros outside the unit circle so it is permissible to use $F = 1$. Then $W(z) = (1 - z^{-1})^2$ and the compensator is

$$G_c(z) = (2z - 1)(z - 0.6065)/[0.21306K(z + 0.84675)(z - 1)]$$

When this compensator is used, the closed-loop transfer function is $M(z) = 2z^{-1} - z^{-2}$. The responses to step and ramp inputs are as shown in Figs. 2.30 and 2.31, respectively. Note that the design objectives are met for the ramp input, but the step response may not be acceptable since it has 100% overshoot. This illustrates one drawback of deadbeat response controllers: they are

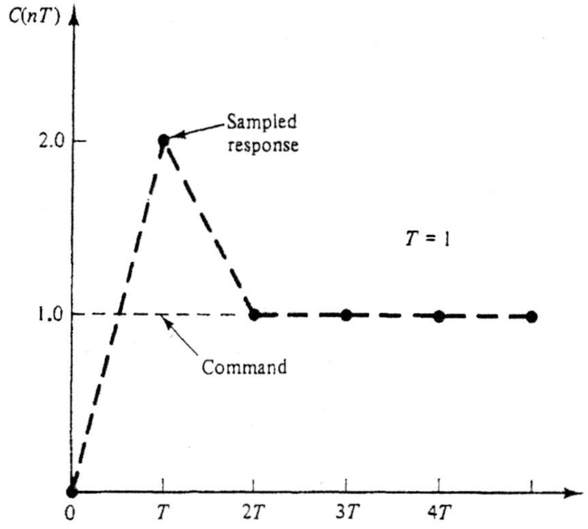

Figure 2.30 Step response.

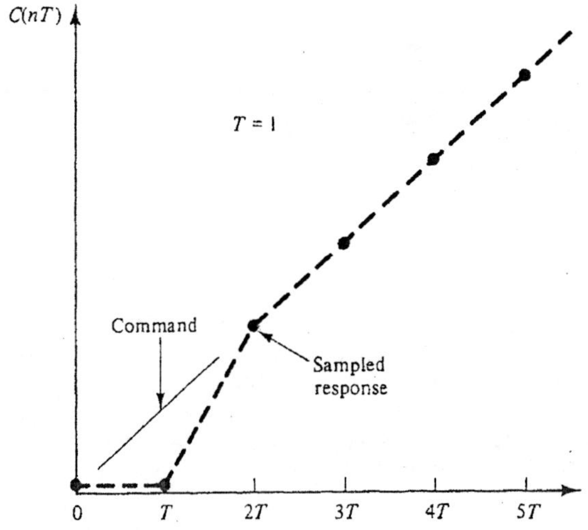

Figure 2.31 Ramp response.

tuned to specific inputs. The other potential drawback is that the intersample error is not necessarily zero just because $E(t_k)$ is zero. There are other methods of suppressing intersample ripple [7].

2.24 Consider the same system as in the previous two problems. This time a cascade compensator is sought which meets the followng specs: (1) closed-loop poles at $z = 0.19877 \pm 0.30956j$, (2) zero steady-state error for a step input, and (3) a velocity error constant K_v of 5 to insure an ability to follow ramp inputs with acceptably small error. This is an example of a direct synthesis method [7], sometimes called the method of Raggazini [6].

A summary of the rules associated with this design method is given before applying them to this specific problem. Let the final closed-loop transfer function be called $M(z)$.

1. In order to ensure that the final compensator is physically realizable, $M(z)$ must have at least as many more poles than zeros as $G'(z)$ does.

2. To ensure a stable closed-loop system, $M(z)$ must contain as zeros all zeros of $G'(z)$ which are on or outside the unit circle. To see this, let the plant transfer function be written $G'(z) = N_{p1}N_{p2}/D_p$, where N_{p2} contains all zeros on or outside the unit circle. Let $G_c = N_c/D_c$. Then

$$M(z) = \frac{N_c N_{p1} N_{p2}}{D_c D_p + N_c N_{p1} N_{p2}}$$

The only way to prevent M from having the zeros in question is to select the compensator denominator to have the factor N_{p2}. But that leaves N_{p2} as a common factor in the denominator of M, thus constituting unstable closed-loop poles. Since perfect cancellation is never possible, this approach is not satisfactory. N_{p2} must be a factor in $M(z)$.

3. Of necessity, if $M(z)$ is to be stable, $1 - M(z)$ must have as zeros all the unstable poles of $G'(z)$. (Those on or outside the unit circle—a single pole at $z = 1$ is not considered unstable.) The reason why this is necessary is clear from a consideration of Eq. (2.5), rearranged as $M = G_c G'(1 - M)$.

4. In order to have zero steady-state error, the final value theorem indicates that $\lim_{z\to 1}\{(z-1)R(z)[1 - M(z)]\} = 0$. The implication of this depends on what $R(z)$ is, but for a step input, since $R(z) = z/(z-1)$, it means that $\lim_{z\to 1} M(z) = 1$.

5. The velocity error coefficient is defined as

$$K_v = (1/T)\lim_{z\to 1}\{(z-1)G_c(z)G'(z)\} = (1/T)\lim_{z\to 1}(z-1)M(z)/[1 - M(z)]$$

For the case where a step input was used in (4), $M(1) = 1$, so the expression for the error constant is indeterminant of the form 0/0. Use of L'Hospital's rule gives $K_v = -1/\{Td[M(z)]/dz\}|_{z=1}$. Solving gives a requirement on $M(z)$ in terms of a specified K_v:

$$\left.\frac{dM(z)}{dz}\right|_{z=1} = \frac{-1}{K_v T}$$

Now for the specific problem here, $M(z)$ must have at least one more pole than zero, since $G'(z)$ does. The open-loop system $G'(z)$ has no poles or zeros outside the unit circle. In view of this and the desired closed-loop poles, a tentative $M(z)$ is selected, which satisfies (1), (2), and (3).

$$M(z) = \frac{A(z+a)}{z^2 - 0.39754z + 0.13534}$$

There are two free design parameters A and a that can be used to satisfy (4) and (5). From (4), $A(1+a)/0.737797 = 1$ and from (5)

$$A/0.737797 - A(1+a)(2 - 0.39754)/(0.737797)^2 = -1/5$$

From these two equations the two unknowns are found to be $A = 1.4549$ and $a = -0.492888$. The compensated closed-loop transfer function is

$$M(z) = 1.4549(z - 0.492888)/(z^2 - 0.39754z + 0.135337)$$

Using this, Eq. (2.5) gives the following compensator, after algebraic simplification:

$$G_c(z) = \frac{6.82859(z - 0.60653)(z - 0.492888)}{K(z - 0.85244)(z + 0.84675)}$$

The transient response of the closed-loop system to step and ramp inputs are shown in Figures 2.32 and 2.33, respectively. Note that the maximum percent overshoot (at the sampling times) due to a step input is 45.5% and the settling time is about $4T$, i.e., 4 s. A different set of design specifications may lead to improved performance.

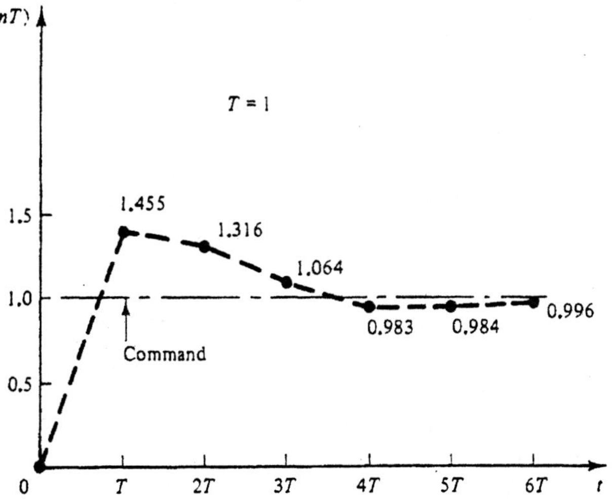

Figure 2.32 Step response.

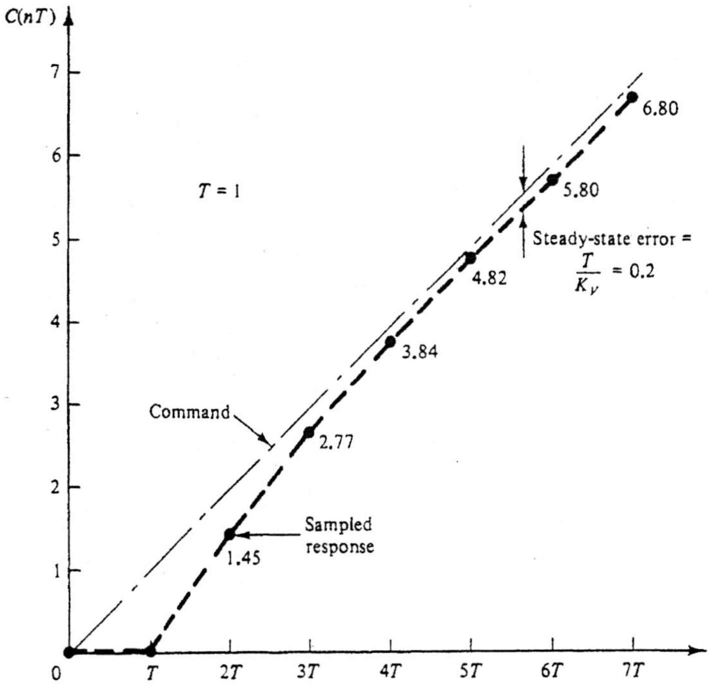

Figure 2.33 Ramp response.

PROBLEMS

2.25 A single-input, single-output system is described by $\dddot{y} + \ddot{y} + 6\dot{y} + (K-3)y = u(t)$. What is the range of values of K for stability?

2.26 Find the gain K and the frequency ω at which the system of Figure 2.34 becomes unstable. Consider only positive gains.

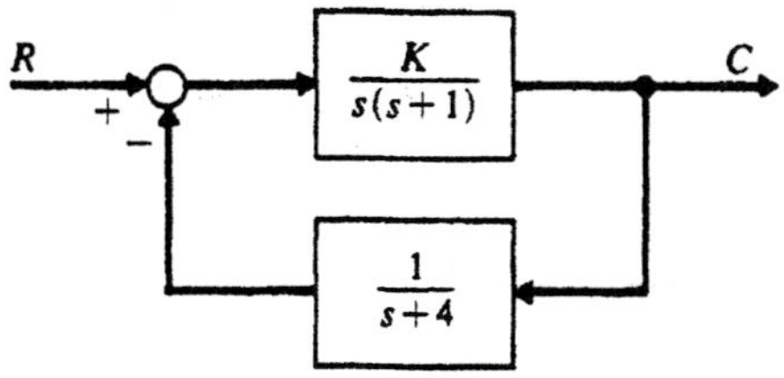

Figure 2.34

2.27 Use Routh's criterion to accurately compute the gain and frequency at the imaginary axis crossover for the system of Problem 2.9.

2.28 The asymptotic gain portions of three Bode plots are shown in Figure 2.35. Identify the system types and their error coefficients.

2.29 Use Bode's method to show that the system with $KGH = K/[s^2(s+1)(s^2+2s+225)]$ is unstable for all positive K.

2.30 Sketch the polar plot and use Nyquist's criterion to investigate the stability of a system with $KGH = K(s+10)/s^2$.

2.31 Sketch the polar plot and use Nyquist's criterion to investigate the stability of a system with $KGH = K(s+10)(s+30)/s^3$.

2.32 Use the polar plot to determine the gain-phase margins for the system described by

$$KGH = \frac{5000(s+2)}{s^2(s+10)(s+30)}$$

2.33 A feedback control system has an open-loop transfer function

$$KGH = \frac{65{,}000K}{s(s+25)(s^2+100s+2600)}$$

Find the value of K such that the exponential envelope of the dominant terms decays to 0.15% of its maximum value in 1 s. Also find the frequency of this damped oscillation.

2.34 Why should Bode plots not be used to infer stability margins for the system in Example 2.3, page 43.

2.35 Give three different algorithms for realizing a digital compensator

$$G_c(z) = (z - 0.5)/[z(z+0.5)]$$

whose input and output are E and Y, respectively.

2.36 Determine an expression for the output sequence $C(nT)$, valid for any nonnegative n, if $C(z) = 10z/[(z-1)(z-0.5)(z+0.5)]$.

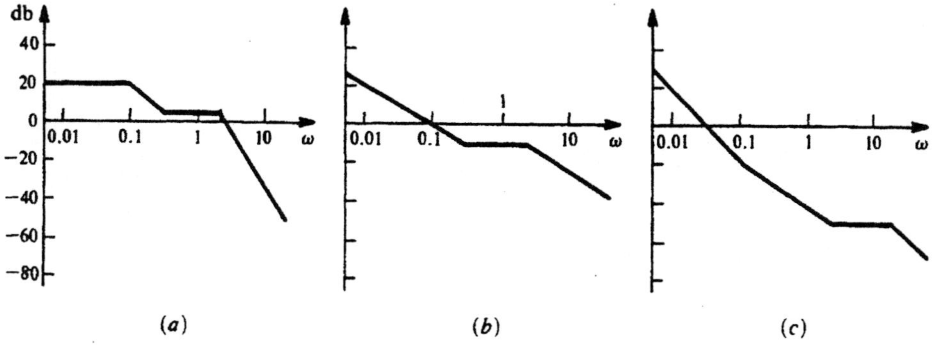

Figure 2.35

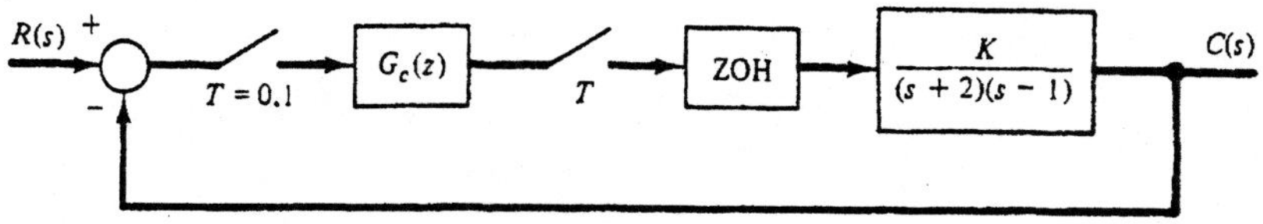

Figure 2.36

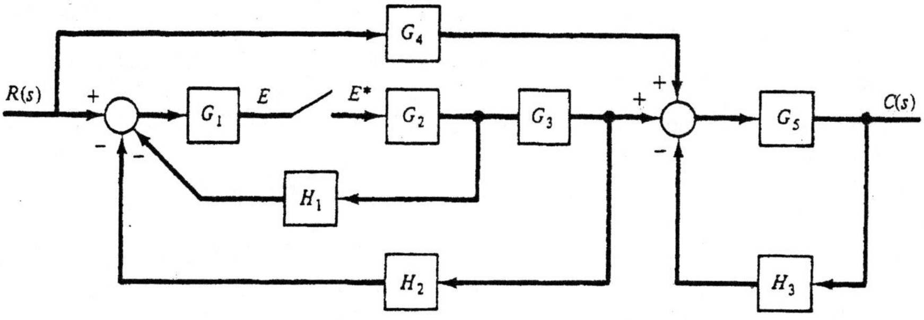

Figure 2.37

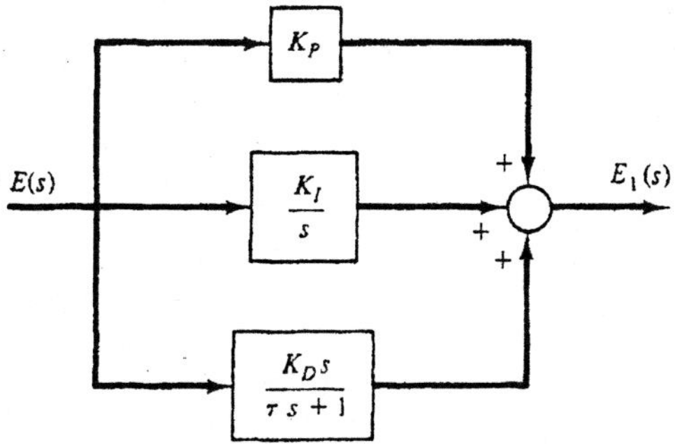

Figure 2.38

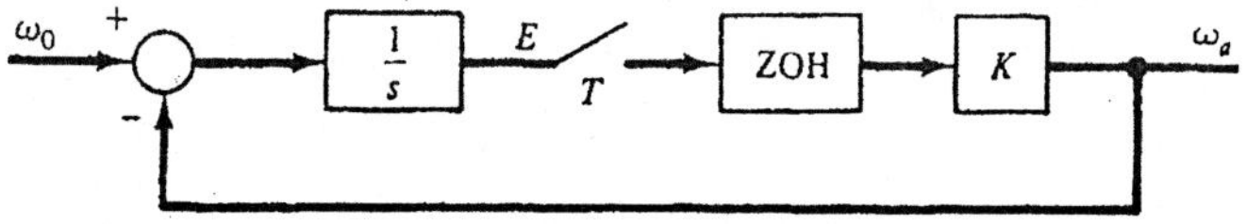

Figure 2.39

2.37 For $C(z) = (z-a)(z-b)/[(z-\alpha)(z-\beta)]$, with a, b, α, and β all distinct:

(a) Find $C(0)$, $C(T)$, and $C(2T)$ using long division.

(b) Find an expression for $C(nT)$ valid for all $n > 0$.

(c) Apply the final value theorem (assume $|\alpha|$ and $|\beta|$ are less than 1), and from its results verify the limiting value of your part (b) answer as $n \to \infty$.

2.38 The system of Figure 2.36 is open-loop unstable.

(a) Determine the closed-loop transfer function $C(z)/R(z)$, assuming that $G_c(z) = 1$.

(b) Sketch the root locus for the uncompensated system. From this sketch, show that the closed-loop system will be stable for some narrow range of K values, but that the resulting system's transient response will not be very desirable. Compensation will probably be required.

2.39 Determine $E(z)$, $C(s)$, and $C(z)$ for the system of Figure 2.37.

2.40 Consider the widely used PID controller (proportional, integral, and derivative control action) shown in Figure 2.38. Use the stable forward difference approximation for s and determine an equivalent digital controller transfer function. Note that a pure derivative does not have a physically realizable transfer function, so a small time-constant term τ is included.

The wide and persistent appeal of PID controllers, especially in the process control industries, can be attributed to their robustness. That is, a properly tuned PID controller can give a good compromise between acceptable time response and disturbance rejection, even with significant model errors present.

2.41 Using the sample and zero-order hold models suggested for A/D and D/A converters, show that the cascade connection of a D/A followed by an A/D has a Z-transfer function of 1, meaning no alteration of a digital signal sequence passing through it.

2.42 A simple one-dimensional model of a digital tracking loop is shown in Figure 2.39. The purpose of the control loop is to keep the angular rate of the antenna ω_a approximately equal to the angular rate ω_0 that the line of sight to the tracked object is making. The integrated angular rate difference is a pointing angle error E. This error angle is sampled, because of a time-shared control computer, and then used to command an antenna rate proportional to the error. The dynamics of the antenna drive system are so rapid that they are neglected, meaning that the actual ω_a is equal to its commanded value. Show that this loop is stable for $0 < KT < 2$.

3

State Variables and the State Space Description of Dynamic Systems

3.1 INTRODUCTION

In the previous two chapters mathematical models of systems have been presented and discussed using differential or difference equations. In the linear, constant-coefficient case, transfer functions were found to be convenient. In either form, these models were directed toward an input-output description of the system. In this chapter the concept of *state* is introduced and methods of writing state variable forms of the system models are presented. The state variable model of a system includes a description of the internal status of that system, in addition to the input-output behavior. Therefore, state variable models represent a more complete description in general. The state variable approach also applies to time-varying and nonlinear systems which cannot easily be described by transfer functions. Before we begin, a few mathematical conventions must be established.

Functions, Transformations, and Mappings

A *function* is a rule by which elements in one set are associated with elements in another set. A function consists of three things: two specified sets of elements $\mathscr{X} = \{x_i\}$ and $\mathscr{Y} = \{y_i\}$ and a rule relating elements $x_i \in \mathscr{X}$ to elements $y_i \in \mathscr{Y}$. The rule must be unambiguous. That is, for every $x \in \mathscr{X}$, there is associated a single element $y \in \mathscr{Y}$. The rule is often written as

$$y = f(x)$$

The words function, transformation, and mapping will be used interchangeably. A common notation which indicates all three aspects of a function is

$$f: \mathscr{X} \rightarrow \mathscr{Y}$$

This states that there is a rule, f, by which every element in $\mathcal{X}$ is mapped into some element in $\mathcal{Y}$. The set $\mathcal{X}$ is called the *domain* of the function, and the set $\mathcal{Y}$ is called the *codomain*. For a particular x, $y = f(x)$ is called the *image* of x, or conversely x is the *pre-image* of y. When the function is applied to every element in $\mathcal{X}$, a set of image points in $\mathcal{Y}$ is generated. This set of images is called the *range* of the function, and is sometimes expressed as $f(\mathcal{X})$.

The words *into* and *onto* are frequently used in conjunction with functions or mappings. The set $f(\mathcal{X})$ is always contained within or possibly equal to $\mathcal{Y}$, written $f(\mathcal{X}) \subseteq \mathcal{Y}$. Thus f is said to map $\mathcal{X}$ *into* $\mathcal{Y}$. If every element in $\mathcal{Y}$ is the image of at least one $x \in \mathcal{X}$, then $f(\mathcal{X}) = \mathcal{Y}$, and the function is said to map $\mathcal{X}$ *onto* $\mathcal{Y}$.

In order that the function be unambiguously defined, there is always just one y associated with each x. However, it is possible that two or more distinct elements $x \in \mathcal{X}$ have the same image point $y \in \mathcal{Y}$. The special case for which distinct elements $x \in \mathcal{X}$ map into distinct elements $y \in \mathcal{Y}$ is called a *one-to-one* mapping. That is, if f is one-to-one, then $x_1 \neq x_2$ implies that $f(x_1) \neq f(x_2)$. Implications such as this are more concisely written as

$$x_1 \neq x_2 \Rightarrow f(x_1) \neq f(x_2)$$

As in all logical arguments, negating both propositions reverses the implication, so that f is one-to-one if

$$f(x_1) = f(x_2) \Rightarrow x_1 = x_2$$

If a function is both one-to-one and onto, then for each $y \in \mathcal{Y}$ there is a *unique* pre-image $x \in \mathcal{X}$. The unique relation between y and x defines the inverse function $g = f^{-1}$ with

$$g: \mathcal{Y} \to \mathcal{X}, \quad \text{where} \quad g[f(x)] = x$$

Vector-Matrix Notation

The bookkeeping conveniences of vector-matrix notation will be used in this chapter. An in-depth and systematic development of these subjects is presented in the following two chapters. A few of the rudimentary notions are now presented. An ordered set of n objects $f_1, f_2, \ldots, f_n$ can be arranged in various array forms, with the position in the array preserving in some way the order of the set. For example, we could write

$$[f_1 \;\; f_2 \;\; f_3 \;\; \cdots \;\; f_n] \quad \text{or} \quad \begin{bmatrix} f_1 \\ f_2 \\ f_3 \\ \vdots \\ f_n \end{bmatrix} \quad \text{or} \quad \begin{bmatrix} f_1 & f_2 & f_3 \\ f_4 & f_5 & f_6 \\ f_7 & \cdots & \end{bmatrix}$$

Which of these forms to use depends on the intended purpose. It is more convenient to refer to the entire array by a single symbol, such as **R**, **C**, or **S**. Ordered arrays of objects (numbers, functions of time, polynomial functions of complex variables, etc.) are called matrices. The first example displayed is a row matrix **R** and the second is a column matrix **C**. **R** and **C** are related by a transpose—i.e., an interchange of rows and columns—written $\mathbf{C}^T = \mathbf{R}$. Either of the first two could be thought of as vectors: a row vector or a column vector. Actually, a vector is more general than just a column of numbers, as is discussed in Chapter 5. The rectangular array **S** in the third representation would normally have its entries designated with two subscripts, a row number i and a column number j. When combining or manipulating arrays in terms of single symbols such as **R**, **C**, or **S**, a logical algebra must be used if the results are to make sense. The reader is probably familiar with the rules of matrix addition, subtraction, multiplication, and transposition. If not, an occasional reference to Chapter 4 may be useful, since those notions will be used in this chapter.

Vector Space

Let the ordered set of three physical position coordinates (with respect to some coordinate system) be considered as a position vector. Then the set of all such possible vectors can be thought of as a vector space, in this case the three-dimensional physical space. This simple example of a vector space as a set of vectors can be generalized. An ordered n-tuple can be thought of as a point in an n-dimensional space. Certain technical requirements, presented in Chapter 5, are necessary in order to qualify as a valid vector space. Certain relationships (linear functions, mappings, or transformations) between two vectors in finite-dimensional vector spaces can be expressed as a product of a matrix and a vector. The form of the transformation matrix depends upon the coordinate system being used. For example, a point in physical space can be represented in spherical or rectangular coordinates. The column of three numbers that represents this position vector would be very different in the two coordinate systems. The form of the transformation matrix that relates two such vectors would also differ greatly.

3.2 THE CONCEPT OF STATE

The concept of state occupies a central position in modern control theory. However, it appears in many other technical and nontechnical contexts as well. In thermodynamics the equations of *state* are prominently used. Binary sequential networks are normally analyzed in terms of their *states*. In everyday life, monthly financial *state*ments are commonplace. The president's *state* of the Union message is another familiar example.

In all these examples the concept of state is essentially the same. It is a complete summary of the status of the system at a particular point in time. Knowledge of the state at some initial time t_0, plus knowledge of the system inputs after t_0, allows the determination of the state at a later time t_1. As far as the state at t_1 is concerned, it makes no difference how the initial state was attained. Thus the state at t_0 constitutes a complete history of the system behavior prior to t_0, insofar as that history affects future behavior. Knowledge of the present state allows a sharp separation between the past and the future.

At any fixed time the state of a system can be described by the values of a set of variables x_i, called *state variables*. One of the state variables of a thermodynamic system is temperature and its value can range over the continuum of real numbers $\mathcal{R}$. In a binary network state variables can take on only two discrete values, 0 or 1. Note that the state of your checking account at the end of the month can be represented by a single number, the balance. The state of the Union can be represented by such things as gross national product, percent unemployment, the balance of trade deficit, etc. For the systems considered in this book the state variables may take on any scalar value, real or complex. That is, $x_i \in \mathcal{F}$. Although some systems require an infinite number of state variables, only systems which can be described by a finite number n of state variables will be considered here. Then the state can be represented by an n component *state vector* $\mathbf{x} = [x_1 \quad x_2 \quad \cdots \quad x_n]^T$.

The state at a given time belongs to an n-dimensional vector space defined over the field $\mathcal{F}$. A general n-dimensional space will be denoted by $\mathcal{X}^n$. However, because of its great importance, the *state space* will be referred to as Σ from now on.

The systems of interest in this book are dynamic systems. Although a more precise definition is given in the next section, usually the word dynamic refers to something active or changing with time. *Continuous-time* systems have their state defined for all times in some interval, for example, a continually varying temperature or voltage. For *discrete-time* systems the state is defined only at discrete times, as with the monthly financial statement or the annual state of the Union message. Continuous-time and discrete-time systems can be discussed simultaneously by defining the times of interest as $\mathcal{T}$. For continuous-time systems $\mathcal{T}$ consists of the set of all real numbers $t \in [t_0, t_f]$. For discrete-time systems $\mathcal{T}$ consists of a discrete set of times $\{t_0, t_1, t_2, \ldots, t_k, \ldots, t_N\}$. In either case the initial time could be $-\infty$ and the final time could be ∞ in some circumstances.

The state vector $\mathbf{x}(t)$ is defined only for those $t \in \mathcal{T}$. At any given t, it is simply an ordered set of n numbers. However, the character of a system could change with time, causing the *number* of required state variables (and not just the values) to change. If the dimension of the state space varies with time, the notation Σ_t could be used. It is assumed here that Σ is the same n-dimensional state space at all $t \in \mathcal{T}$.

3.3 STATE SPACE REPRESENTATION OF DYNAMIC SYSTEMS

A general class of multivariable control systems is considered. There are r *real*-valued inputs or control variables $u_i(t)$, referred to collectively as the $r \times 1$ vector $\mathbf{u}(t)$. For a fixed time $t \in \mathcal{T}$, $\mathbf{u}(t)$ belongs to the real r-dimensional space $\mathcal{U}^r$. There are m real-valued outputs $y_i(t)$, referred to collectively as the $m \times 1$ vector $\mathbf{y}(t)$. For any $t \in \mathcal{T}$, $\mathbf{y}(t)$ belongs to the real m-dimensional space $\mathcal{Y}^m$. Let the subset of times $t \in \mathcal{T}$ which satisfy $t_a \le t \le t_b$ be denoted as $[t_a, t_b]_{\mathcal{T}}$. For continuous-time systems $[t_a, t_b]_{\mathcal{T}} = [t_a, t_b]$, and for discrete-time systems $[t_a, t_b]_{\mathcal{T}}$ is the intersection of $[t_a, t_b]$ and $\mathcal{T}$.

It is necessary to distinguish between an input function (the graph of $\mathbf{u}(t)$ versus $t \in \mathcal{T}$) and the value of $\mathbf{u}(t)$ at a particular time. The notation $\mathbf{u}_{[t_a, t_b]}$ will be used to indicate a *segment* of an input function over the set of times in $[t_a, t_b]_{\mathcal{T}}$. Similarly, a segment of an output function will be indicated by $\mathbf{y}_{[t_a, t_b]}$. The admissible input functions (actually sequences in the discrete-time case) are elements of the input function space $\mathcal{U}$, that is, $\mathbf{u}$: $\mathcal{T} \to \mathcal{U}$. The output functions (or sequences) are elements of the output function space $\mathcal{Y}$, $\mathbf{y}$: $\mathcal{T} \to \mathcal{Y}$. Heuristically, $\mathcal{U}^r$ and $\mathcal{Y}^m$ can be thought of as "cross sections" through $\mathcal{U}$ and $\mathcal{Y}$ at a particular time t. Similarly, the graph of $\mathbf{x}(t)$ versus t, the so-called *state trajectory,* could be considered as an element of a function space $\mathcal{S}$. The state space Σ can be visualized as a cross section through $\mathcal{S}$ at a particular $t \in \mathcal{T}$.

The primary interest in a control system may be in the relationship between inputs, which can be manipulated, and outputs, which determine whether or not system goals are met. This relationship may be thought of as a mapping or transformation $\mathcal{W}: \mathcal{U} \to \mathcal{Y}$, that is, $\mathbf{y}(t) = \mathcal{W}(\mathbf{u}(t))$. Examples of this type of relationship have been mentioned in Sec. 1.5. It is easy to show that a given $\mathbf{u}_{[t_a, t_b]}$ need *not* define a unique output function $\mathbf{y}_{[t_a, t_b]}$.

EXAMPLE 3.1 The input-output differential equation for the circuit of Figure 3.1 is $dy/dt + (1/RC)y = u(t)/RC$. The solution is

$$y(t) = e^{-(t-t_0)/RC}\, y(t_0) + (1/RC)\int_{t_0}^{t} e^{-(t-\tau)/RC}\, u(\tau)\, d\tau$$

A given input function $u(t) \in \mathcal{U}$ defines a *family* of output functions $y(t) \in \mathcal{Y}$. In order to relate one unique output with each input, additional information, such as the value of $y(t_0)$, must be specified. Unless this is done, the nonuniqueness prevents $\mathcal{W}$ from being a transformation in the strict sense. ■

State variables are important because they resolve the nonuniqueness problem illustrated in Example 3.1 and at the same time completely summarize the *internal* status of the system.

Definition 3.1. The state variables of a system consist of a minimum set of parameters which completely summarize the system's status in the following sense. If at any time $t_0 \in \mathcal{T}$, the values of the state variables $x_i(t_0)$ are known, then the output $\mathbf{y}(t_1)$ and the values $x_i(t_1)$ can be *uniquely* determined for any time $t_1 \in \mathcal{T}, t_1 > t_0$, provided $\mathbf{u}_{[t_0, t_1]}$ is known.

Definition 3.2. The state at any time t_0 is a set of the minimum number of parameters $x_i(t_0)$ which allows a *unique* output segment $\mathbf{y}_{[t_0, t]}$ to be associated with each input segment $\mathbf{u}_{[t_0, t]}$ for every $t_0 \in \mathcal{T}$ and for all $t > t_0$, $t \in \mathcal{T}$.

The implications of Definitions 3.1 and 3.2 can be stated in terms of transformations on the input, state, and output spaces. Given (1) a pair of times, t_0 and t_1 in $\mathcal{T}$, (2) $\mathbf{x}(t_0) \in \Sigma$, and (3) a segment $\mathbf{u}_{[t_0, t_1]}$ of an input in $\mathcal{U}$, both the state and the output must be uniquely determinable. This requires that there be a transformation $\mathbf{g}$ which maps the elements $(t_0, t_1, \mathbf{x}(t_0), \mathbf{u}_{[t_0, t_1]})$, which can be treated collectively as a single element of

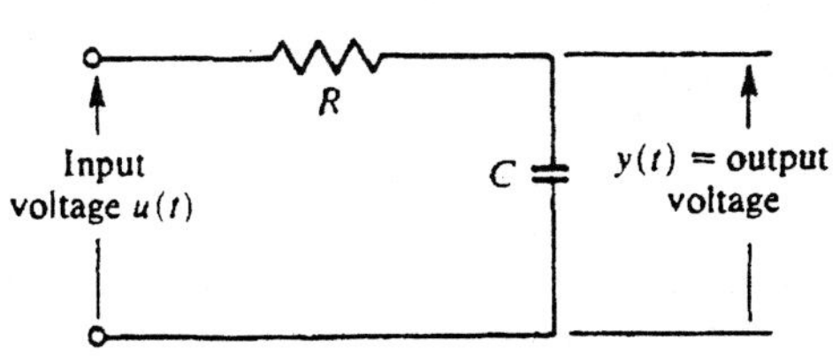

Figure 3.1

the product space (Section 5.10) $\mathcal{T} \times \mathcal{T} \times \Sigma \times \mathcal{U}$, into a unique element in Σ; that is, $\mathbf{g}$: $\mathcal{T} \times \mathcal{T} \times \Sigma \times \mathcal{U} \rightarrow \Sigma$, where

$$\mathbf{x}(t_1) = \mathbf{g}(t_0, t_1, \mathbf{x}(t_0), \mathbf{u}_{[t_0, t_1]}) \tag{3.1}$$

Furthermore, since $\mathbf{y}(t_1)$ is uniquely determined, a second transformation exists, $\mathbf{h}$: $\mathcal{T} \times \Sigma \times \mathcal{U}^r \rightarrow \mathcal{Y}^m$ with

$$\mathbf{y}(t_1) = \mathbf{h}(t_1, \mathbf{x}(t_1), \mathbf{u}(t_1)) \tag{3.2}$$

The transformation $\mathbf{h}$ has no memory, rather $\mathbf{y}(t_1)$ depends only on the instantaneous values of $\mathbf{x}(t_1)$, $\mathbf{u}(t_1)$, and t_1. The transformation $\mathbf{g}$ is *nonanticipative* (also called causal). This means that the state, and hence the output at t_1, do not depend on inputs occurring after t_1.

Definition 3.3. The model of a physical system is called a *dynamical system* [1, 2] if a set of times $\mathcal{T}$, spaces $\mathcal{U}$, Σ, and $\mathcal{Y}$, and transformations $\mathbf{g}$ and $\mathbf{h}$ can be associated with it. The transformations are those of Eqs. (*3.1*) and (*3.2*) and must have the following properties:

$$\mathbf{x}(t_0) = \mathbf{g}(t_0, t_0, \mathbf{x}(t_0), \mathbf{u}_{[t_0, t_1]}) \quad \text{for any } t_0, t_1 \in \mathcal{T} \tag{3.3}$$

If $\mathbf{u} \in \mathcal{U}$ and $\mathbf{v} \in \mathcal{U}$ with $\mathbf{u} = \mathbf{v}$ over some segment $[t_0, t_1]_{\mathcal{T}}$, then

$$\mathbf{g}(t_0, t_1, \mathbf{x}(t_0), \mathbf{u}_{[t_0, t_1]}) = \mathbf{g}(t_0, t_1, \mathbf{x}(t_0), \mathbf{v}_{[t_0, t_1]}) \tag{3.4}$$

If $t_0, t_1, t_2 \in \mathcal{T}$ and $t_0 < t_1 < t_2$, then

$$\begin{aligned} \mathbf{x}(t_2) &= \mathbf{g}(t_0, t_2, \mathbf{x}(t_0), \mathbf{u}_{[t_0, t_2]}) \\ &= \mathbf{g}(t_1, t_2, \mathbf{x}(t_1), \mathbf{u}_{[t_1, t_2]}) \\ &= \mathbf{g}(t_1, t_2, \mathbf{g}(t_0, t_1, \mathbf{x}(t_0), \mathbf{u}_{[t_0, t_1]}), \mathbf{u}_{[t_1, t_2]}) \end{aligned} \tag{3.5}$$

Equation (*3.3*) indicates that $\mathbf{g}$ is the identity transformation whenever its two time arguments are the same. Equation (*3.4*), called the *state transition property*, indicates that $\mathbf{x}(t_1)$ does not depend on inputs prior to t_0 except insofar as the past is summarized by $\mathbf{x}(t_0)$. It also indicates that $\mathbf{x}(t_1)$ does not depend on inputs after t_1. Equation (*3.5*), called the *semigroup property*, states that it is immaterial whether $\mathbf{x}(t_2)$ is computed directly from $\mathbf{x}(t_0)$ and $\mathbf{u}_{[t_0, t_2]}$ or if $\mathbf{x}(t_1)$ is first obtained from $\mathbf{x}(t_0)$ and $\mathbf{u}_{[t_0, t_1]}$, and then this state is used, along with $\mathbf{u}_{[t_1, t_2]}$.

Lumped-parameter continuous-time dynamical systems can be represented in state space notation by a set of first-order differential equations (*3.6*) and a set of single-valued algebraic output equations (*3.7*):

$$\dot{\mathbf{x}} = \mathbf{f}(\mathbf{x}, \mathbf{u}, t) \tag{3.6}$$

$$\mathbf{y}(t) = \mathbf{h}(\mathbf{x}, \mathbf{u}, t) \tag{3.7}$$

In order that $\mathbf{x}$ be a valid state vector, Eq. (*3.6*) must have a unique solution. In essence, this means that the components of $\mathbf{f}$ are restricted so that Eq. (*3.8*) defines a unique $\mathbf{x}(t)$ for $t \geq t_0$:

$$\mathbf{x}(t) = \mathbf{x}(t_0) + \int_{t_0}^{t} \mathbf{f}(\mathbf{x}(\tau), \mathbf{u}(\tau), \tau)\, d\tau \tag{3.8}$$

The theory of differential equations [3] indicates that it is sufficient if $\mathbf{f}$ satisfies a Lipschitz condition with respect to $\mathbf{x}$, is continuous with respect to $\mathbf{u}$, and is piecewise continuous with respect to t. Then there will be a unique $\mathbf{x}(t)$ for any t_0, $\mathbf{x}(t_0)$ provided $\mathbf{u}(t)$ is piecewise continuous. The unique solution of Eq. (*3.6*) defines the transformation g of Eq. (*3.1*).

EXAMPLE 3.2 Consider a point mass falling in a vacuum. The input is the constant gravitational attraction a and the output is the altitude $y(t)$ shown in Figure 3.2. The input-output relation is $\ddot{y} = -a$. Integrating gives

$$\dot{y}(t) = \dot{y}(t_0) - a(t - t_0)$$

Integrating again gives

$$y(t) = y(t_0) + \dot{y}(t_0)(t - t_0) - a(t - t_0)^2/2$$

The variable y does not constitute the state, since its initial value is not sufficient for uniquely determining $y(t)$. The set of variables y, $\dot{y}$, and $\ddot{y}$ does not form the state, since this is not the *minimum* set of variables required to uniquely determine $y(t)$. The parameter $\ddot{y}$ is not needed. A valid state vector is $\mathbf{x}(t) = [y(t) \quad \dot{y}(t)]^T$. It is easily seen that the form of Eqs. (*3.1*) and (*3.2*) for this example is

$$\mathbf{x}(t) = \begin{bmatrix} 1 & t - t_0 \\ 0 & 1 \end{bmatrix} \mathbf{x}(t_0) - \begin{bmatrix} (t - t_0)^2/2 \\ t - t_0 \end{bmatrix} a, \qquad y(t) = [1 \quad 0]\mathbf{x}(t)$$ ■

Lumped-parameter discrete-time dynamical systems can be described in an analogous way by a set of first-order difference equations (*3.9*) and a set of algebraic output equations (*3.10*):

$$\mathbf{x}(t_{k+1}) = \mathbf{f}(\mathbf{x}(t_k), \mathbf{u}(t_k), t_k) \tag{3.9}$$

$$\mathbf{y}(t_k) = \mathbf{h}(\mathbf{x}(t_k), \mathbf{u}(t_k), t_k) \tag{3.10}$$

In general, the functions $\mathbf{f}$ and $\mathbf{h}$ of Eqs. (*3.6*), (*3.7*), (*3.9*), and (*3.10*) can be nonlinear. However, the linear case is of major importance and also lends itself to more detailed analysis. The most general state space representation of a *linear* continuous-time dynamical system is given by Eqs. (*3.11*) and (*3.12*):

$$\dot{\mathbf{x}} = \mathbf{A}(t)\mathbf{x}(t) + \mathbf{B}(t)\mathbf{u}(t) \tag{3.11}$$

$$\mathbf{y}(t) = \mathbf{C}(t)\mathbf{x}(t) + \mathbf{D}(t)\mathbf{u}(t) \tag{3.12}$$

$\mathbf{A}$, $\mathbf{B}$, $\mathbf{C}$, and $\mathbf{D}$ are matrices of dimension $n \times n$, $n \times r$, $m \times n$, and $m \times r$, respectively. Equations (*3.11*) and (*3.12*) are shown in block diagram form in Figure 3.3. The

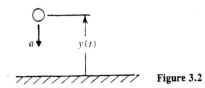

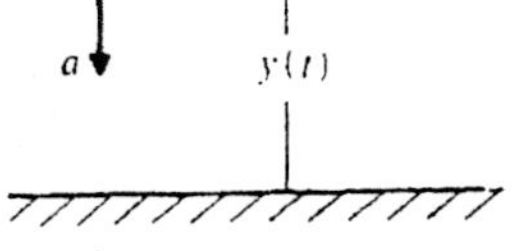

Figure 3.2

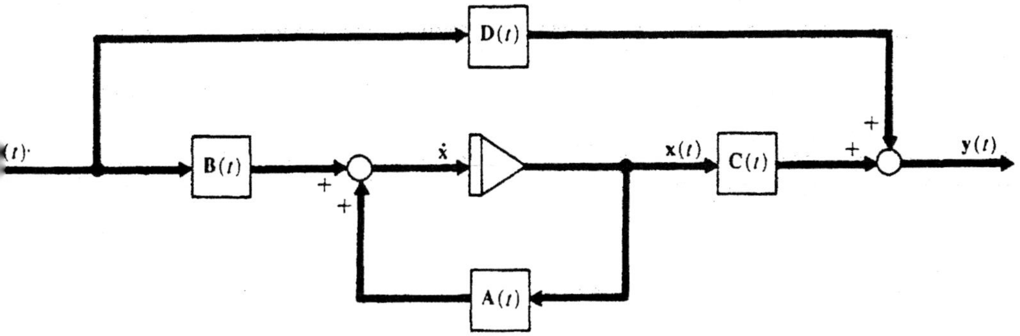

Figure 3.3 State space representation of continuous-time linear system.

heavier lines indicate that the signals are vectors, and the integrator symbol really indicates n scalar integrators.

The state space representation of a discrete-time *linear* system is given by Eqs. (*3.13*) and (*3.14*). The simplified notation k refers to a general time $t_k \in \mathcal{T}$:

$$\mathbf{x}(k+1) = \mathbf{A}(k)\mathbf{x}(k) + \mathbf{B}(k)\mathbf{u}(k) \tag{3.13}$$

$$\mathbf{y}(k) = \mathbf{C}(k)\mathbf{x}(k) + \mathbf{D}(k)\mathbf{u}(k) \tag{3.14}$$

The matrices **A**, **B**, **C**, and **D** have the same dimensions as in the continuous-time case, but their meanings are different. The block diagram representation of Eqs. (*3.13*) and (*3.14*) is given in Figure 3.4. The delay symbol is analogous to the integrator in Figure 3.3 and really symbolizes n scalar delays.

The notational choice of using the same symbols {**A**, **B**, **C**, **D**} in both cases carries some potential for confusion in certain instances. The advantage of using the same symbols is that many continuous and discrete concepts are revealed as being essentially identical, not only in this chapter but throughout the book. One final word regarding notational symbology seems appropriate. A sizable fraction of the literature uses **F**, **G**, and **H** in place of **A**, **B**, and **C** in the continuous case. In the discrete case, $\mathbf{\Phi}$ and $\mathbf{\Gamma}$ are frequently used in the literature in place of **A** and **B**. It is hoped that with this note of caution the concepts here and in other works can be appreciated and integrated without being tied to a rigid notational standard.

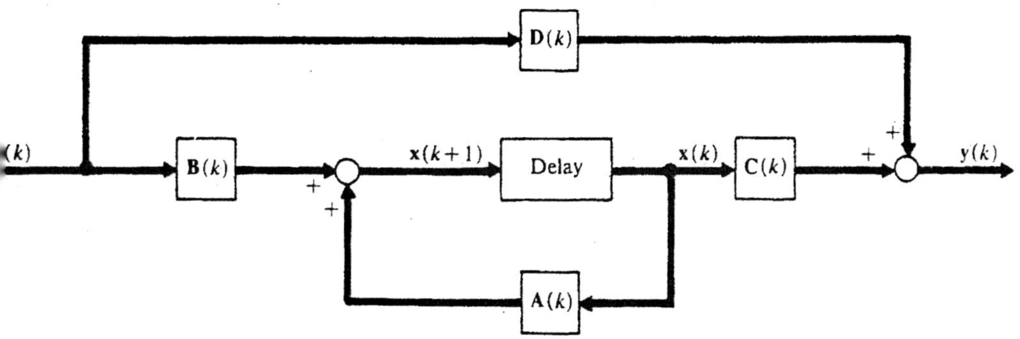

Figure 3.4 State space representation of discrete-time linear system.

Frequently, the discrete-time state variable system models, as in Eqs. (*3.9*) and (*3.10*) or Eqs. (*3.13*) and (*3.14*), are the result of *approximating* a continuous system, or, because of sampling, time-multiplexing of equipment or digital implementation prevents continuous-time operation. Some systems are inherently discrete-time. Other systems are partly discrete and partly continuous, such as a digital controller driving a continuous motor. Some of these sampling and approximation problems will be dealt with in Section 9.8. For present purposes it is assumed that the original system description is in discrete form, perhaps from an empirically obtained ARMA model of Chapter 1 or as a result of taking the Z-transform of a continuous system.

The *analysis* of and *solutions* for linear system state equations are presented in Chapter 9, after the necessary mathematical tools have been developed. The remainder of this chapter develops methods of *obtaining* state space system representations such as those given in Eqs. (*3.6*), (*3.7*) or Eqs. (*3.9*), (*3.10*) for nonlinear systems or Eqs. (*3.11*), (*3.12*) or Eqs. (*3.13*), (*3.14*) for linear systems.

3.4 OBTAINING THE STATE EQUATIONS

3.4.1 From Input-Output Differential or Difference Equations

A class of single-input, single-output systems can be described by an nth-order linear ordinary differential equation:

$$\frac{d^n y}{dt^n} + a_{n-1}\frac{d^{n-1} y}{dt^{n-1}} + \cdots + a_2\frac{d^2 y}{dt^2} + a_1\frac{dy}{dt} + a_0 y = u(t) \tag{3.15}$$

This class of systems can be reduced to the form of n first-order state equations as follows. Define the state variables as

$$x_1 = y, \quad x_2 = \frac{dy}{dt}, \quad x_3 = \frac{d^2 y}{dt^2}, \ldots, \quad x_n = \frac{d^{n-1} y}{dt^{n-1}} \tag{3.16}$$

These particular state variables are often called *phase variables*. As a direct result of this definition, $n-1$ first-order differential equations are $\dot{x}_1 = x_2$, $\dot{x}_2 = x_3, \ldots,$ $\dot{x}_{n-1} = x_n$. The nth equation is $\dot{x}_n = d^n y/dt^n$. Using the original differential equation and the preceding definitions gives

$$\dot{x}_n = -a_0 x_1 - a_1 x_2 - \cdots - a_{n-1} x_n + u(t) \tag{3.17}$$

so that

$$\dot{\mathbf{x}} = \begin{bmatrix} 0 & 1 & 0 & 0 & \cdots & 0 \\ 0 & 0 & 1 & 0 & \cdots & 0 \\ 0 & 0 & 0 & 1 & \cdots & 0 \\ \vdots & & & & & \vdots \\ 0 & 0 & 0 & 0 & \cdots & 1 \\ -a_0 & -a_1 & -a_2 & -a_3 & \cdots & -a_{n-1} \end{bmatrix} \mathbf{x} + \begin{bmatrix} 0 \\ 0 \\ 0 \\ \vdots \\ 0 \\ 1 \end{bmatrix} u(t) = \mathbf{A}\mathbf{x} + \mathbf{B}u(t) \tag{3.18}$$

The output is $y(t) = x_1(t) = [1 \quad 0 \quad 0 \quad \cdots \quad 0]\,\mathbf{x}(t) = \mathbf{C}\mathbf{x}(t)$. In this case the coefficient matrix $\mathbf{A}$ is the companion matrix (see Section 7.9 and Problem 7.32).

A comparable class of discrete-time systems is described by an nth-order difference equation

$$y(k+n)+a_{n-1}y(k+n-1)+\cdots+a_2y(k+2)$$
$$+a_1y(k+1)+a_0y(k)=u(k). \tag{3.19}$$

Phase-variable-type states can be defined as

$$x_1(k)=y(k),\quad x_2(k)=y(k+1),\quad x_3(k)=y(k+2),\ldots,\quad x_n(k)=y(k+n-1)$$

where the discrete time points t_k are simply referred to as k. With these definitions, the first $n-1$ state equations are of the form

$$x_i(k+1)=x_{i+1}(k)$$

The original difference equation becomes

$$y(k+n)=x_n(k+1)=-a_0x_1(k)-a_1x_2(k)-\cdots-a_{n-1}x_n(k)+u(k) \tag{3.20}$$

Comparison of the continuous and discrete systems just considered shows that they have the identical forms for the **A**, **B**, and **C** system matrices and both have $\mathbf{D}=[0]$. The only difference is that $\mathbf{x}(k+1)$ replaces $\dot{\mathbf{x}}$. In both cases the coefficients a_i could be functions of time, yielding time-variable $\mathbf{A}(t)$ or $\mathbf{A}(k)$ $n\times n$ system matrices.

The fact that the equations have the same *form* should *not* be used to conclude that a discrete approximation to a continuous system can be obtained merely by replacing $\dot{\mathbf{x}}(t)$ with $\mathbf{x}(k+1)$. If Eq. (*3.19*) is an approximation of Eq. (*3.15*), the individual a_i coefficients will be quite different in the two cases.

EXAMPLE 3.3 A continuous-time system is described by

$$\ddot{y}+4\dot{y}+y=u(t)$$

so that $a_0=1$ and $a_1=4$. Use the forward difference approximation for derivatives $\dot{y}(t_k)\cong[y(t_{k+1})-y(t_k)]/T$ and $\ddot{y}\cong[\dot{y}(t_{k+1})-\dot{y}(t_k)]/T$, where $T=t_{k+1}-t_k$ is the constant sampling period. Find the approximate difference equation.

It follows that $\ddot{y}(t_k)\cong[y(k+2)-2y(k+1)+y(k)]/T^2$, so that substitution into the differential equation and regrouping terms gives

$$y(k+2)+(4T-2)y(k+1)+(T^2-4T+1)y(k)=T^2u(k)$$

Thus the discrete coefficients are $a_0=T^2-4T+1$ and $a_1=4T-2$.

If a backward difference approximation to the derivatives is used, a very different set of coefficients will be found. Also, the time argument on the u input term will change. ■

3.4.2 Simultaneous Differential Equations

The same method of defining the state variables can be applied to multiple-input, multiple-output systems described by several coupled differential equations if the inputs are not differentiated.

EXAMPLE 3.4 A system has three inputs u_1,u_2,u_3 and three outputs y_1,y_2,y_3. The input-output equations are

$$\ddot{y}_1 + a_1\ddot{y}_1 + a_2(\dot{y}_1 + \dot{y}_2) + a_3(y_1 - y_3) = u_1(t)$$

$$\ddot{y}_2 + a_4(\dot{y}_2 - \dot{y}_1 + 2\dot{y}_3) + a_5(y_2 - y_1) = u_2(t)$$

$$\dot{y}_3 + a_6(y_3 - y_1) = u_3(t)$$

Notice that in the second equation $\dot{y}_3$ can be eliminated by using the third equation. State variables are selected as the outputs and their derivatives up to the $(n-1)$th, where n is the order of the highest derivative of a given output.

Select $x_1 = y_1$, $x_2 = \dot{y}_1$, $x_3 = \ddot{y}_1$, $x_4 = y_2$, $x_5 = \dot{y}_2$, $x_6 = y_3$. Then

$$\dot{x}_1 = x_2, \quad \dot{x}_2 = x_3, \quad \dot{x}_4 = x_5$$

$$\dot{x}_3 = -a_1x_3 - a_2(x_2 + x_5) - a_3(x_1 - x_6) + u_1$$

$$\dot{x}_5 = -a_4(x_5 - x_2 + 2\dot{x}_6) - a_5(x_4 - x_1) + u_2$$

$$\dot{x}_6 = -a_6(x_6 - x_1) + u_3$$

Eliminating $\dot{x}_6$ from the $\dot{x}_5$ equation leads to

$$\begin{bmatrix} \dot{x}_1 \\ \dot{x}_2 \\ \dot{x}_3 \\ \dot{x}_4 \\ \dot{x}_5 \\ \dot{x}_6 \end{bmatrix} = \begin{bmatrix} 0 & 1 & 0 & 0 & 0 & 0 \\ 0 & 0 & 1 & 0 & 0 & 0 \\ -a_3 & -a_2 & -a_1 & 0 & -a_2 & a_3 \\ 0 & 0 & 0 & 0 & 1 & 0 \\ a_5 - 2a_4a_6 & a_4 & 0 & -a_5 & -a_4 & 2a_4a_6 \\ a_6 & 0 & 0 & 0 & 0 & -a_6 \end{bmatrix} \begin{bmatrix} x_1 \\ x_2 \\ x_3 \\ x_4 \\ x_5 \\ x_6 \end{bmatrix} + \begin{bmatrix} 0 & 0 & 0 \\ 0 & 0 & 0 \\ 1 & 0 & 0 \\ 0 & 0 & 0 \\ 0 & 1 & -2a_4 \\ 0 & 0 & 1 \end{bmatrix} \begin{bmatrix} u_1 \\ u_2 \\ u_3 \end{bmatrix}$$

The output equation is

$$\begin{bmatrix} y_1 \\ y_2 \\ y_3 \end{bmatrix} = \begin{bmatrix} 1 & 0 & 0 & 0 & 0 & 0 \\ 0 & 0 & 0 & 1 & 0 & 0 \\ 0 & 0 & 0 & 0 & 0 & 1 \end{bmatrix} \mathbf{x}$$

■

When derivatives of the input appear in the system differential equation, the previous method of state variable selection must be modified. If the method is applied without modification, a set of first-order differential equations is obtained as desired, but the input derivatives will still be present. The state equations must express $\dot{\mathbf{x}}$ as a function of $\mathbf{x}$ and $\mathbf{u}$ (and not $\dot{\mathbf{u}}$). A serious mistake that is sometimes made is to define a new vector $\mathbf{u}$ with components made up of u and its derivatives. This is wrong because the inputs to the state equations must be the actual physical inputs to the system. Arbitrary mathematical redefinitions are not allowed on these (or on the output variables $\mathbf{y}$). This differs from the situation for the internal state variables, which may or may not correspond to real physical signals. The input components must be independently selectable control variables. Clearly if $u(t)$ is specified, there is no freedom left in specifying its derivative $\dot{u}(t)$. The correct method of dealing with input derivatives is to somehow absorb the derivative terms into the definitions of the state variables. In simple cases various ad hoc choices may be apparent. For example, consider $\ddot{y} + a\dot{y} + by = u + c\dot{u}$. Rearranging gives $\ddot{y} - c\dot{u} = -a\dot{y} - by + u$. One could use $x_1 = y$, $x_2 = \dot{y} - cu$ so that, assuming c is constant, $\dot{x}_1 = \dot{y} = x_2 + cu$ and $\dot{x}_2 = -a[x_2 + cu] - bx_1 + u$; or

$$\begin{bmatrix} \dot{x}_1 \\ \dot{x}_2 \end{bmatrix} = \begin{bmatrix} 0 & 1 \\ -b & -a \end{bmatrix} \begin{bmatrix} x_1 \\ x_2 \end{bmatrix} + \begin{bmatrix} c \\ 1 - ac \end{bmatrix} u$$

$$y = [1 \quad 0]\mathbf{x}$$

Since the *selection* of state variables is not a unique process, other choices could be made. For complex higher-order and coupled equations, the ad hoc methods become quite cumbersome. Straightforward systematic methods can be developed, with simulation diagrams being a useful tool.

3.4.3 Using Simulation Diagrams

Equation (*3.8*) indicates that state variables for continuous-time systems are always determined by integrating a function of state variables and inputs. The simulation diagram approach makes use of this fact. Six ideal elements are used as building blocks in the simulation diagrams. They are described in Table 3.1.

Note that a differentiating element $\rightarrow\boxed{d/dt}\rightarrow$ is *not* included. If the equations for a continuous-time system can be simulated using any combination of these elements except the delay, and if no *unnecessary* integrators are used, then the output of each integrator can be selected as a state variable.

An integrator sums up past inputs to form its present output and hence represents a memory element. For discrete-time systems the ideal delay is used instead of the integrator as the ideal memory element. The close analogy between the integrator and the delay is emphasized in Figure 3.5. The transform domain analogy between $1/s$ and $1/z$ is especially clear. The outputs of delays in a discrete-time simulation diagram constitute a valid choice of state variables. In both cases, the outputs of memory or storage devices constitute states. This notion recurs in Section 3.4.5 where states will be associated with energy storage devices.

EXAMPLE 3.5 One possible simulation diagram for $\ddot{y} + a\dot{y} + by = u + c\dot{u}$ is given in Figure 3.6. Selecting the outputs of the integrators as x_1 and x_2 leads to the same state equations given earlier for this system. ■

Ad hoc use of the simulation diagram requires a degree of ingenuity in more complicated systems if unnecessary integrators or delays are to be avoided. A systematic approach which is especially simple for constant coefficient systems is now presented. It can be applied with only a modest amount of extra effort to linear, time-variable systems. (See Problems 3.9 and 3.10.) The insight that the procedure provides is also useful in dealing with certain nonlinear systems (see Problem 3.17).

1. Solve each differential equation for its highest derivative. In the case of a difference equation, solve for the most time-advanced term in each equation. A typical nth-order equation is considered for discussion purposes.
2. Formally integrate each differential equation as many times as the highest derivative, here assumed to be n. For difference equations, n delays replace the n-fold integration. In both cases the goal is to achieve the current $y(t)$ or $y(t_k)$ on the left-hand side.

TABLE 3-1

Element	Symbol	Input-output relation
1. Integrator	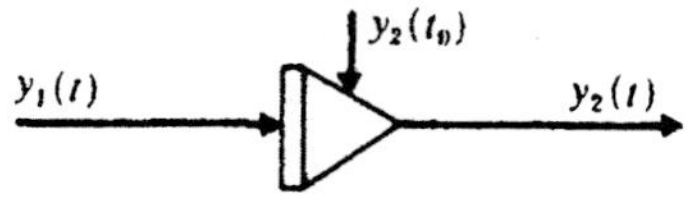	$y_2(t) = y_2(t_0) + \int_{t_0}^{t} y_1(\tau)\, d\tau$
2. Delay		$y_2(t) = y_1(t - T)$
3. Summing junction	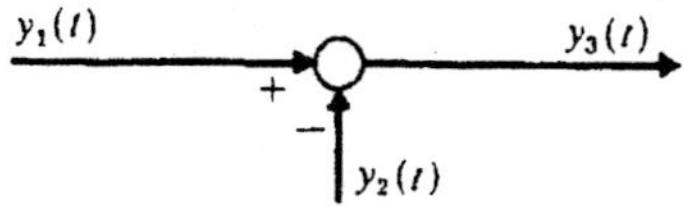	$y_3(t) = y_1(t) - y_2(t)$
4. Gain change	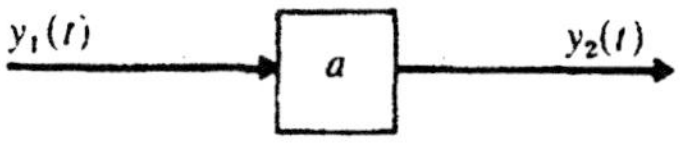	$y_2(t) = ay_1(t)$
5. Multiplier	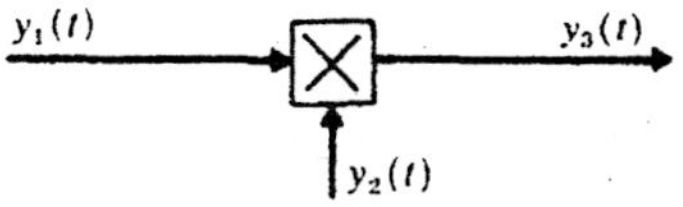	$y_3(t) = y_1(t)y_2(t)$
6. Single-valued function generator		$y_2(t) = f(y_1(t))$

3. Group terms on the right-hand side in the form of a nested sequence of integrations (or delays). Use up integrations to remove all derivatives. Use delay operators to remove all advance terms. In constant coefficient cases the coefficients move freely past either of these operators. Time-variable coefficients require use of integration by parts. (See Problems 3.9 and 10.)
4. Draw the simulation diagram by inspection, using the fact that terms inside of an integral sign or delay operator domain constitute inputs to that integrator or delay. A term already containing an integral or delay is the output of another such operator.
5. When the diagram is completed, there should be n integrators or delays. The

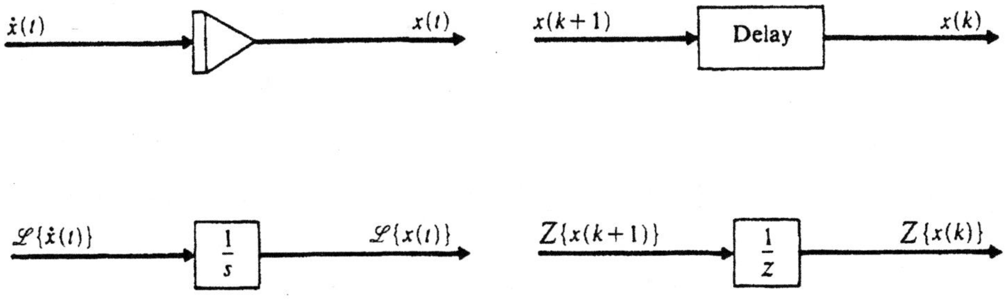

(*a*) Ideal integrator (*b*) Ideal delay

Figure 3.5

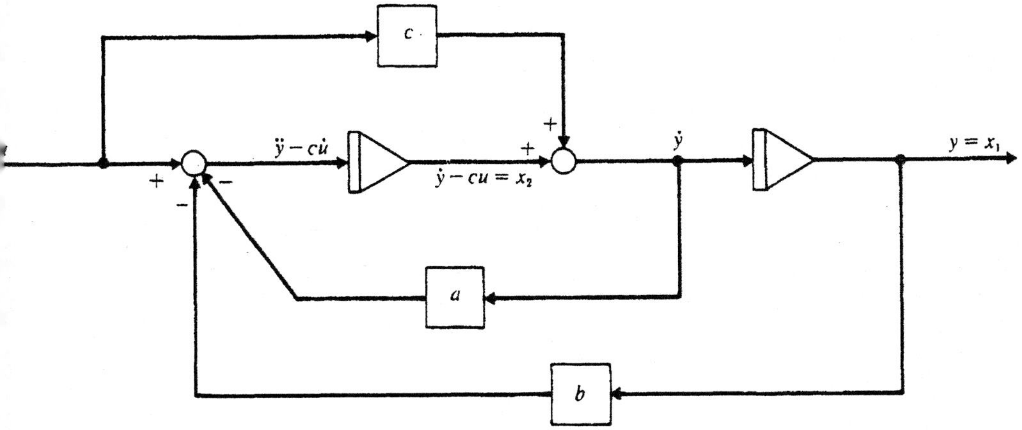

Figure 3.6 ■

output of each integrator is selected as a state $x_i(t)$ in the continuous-time case. The output of each delay in the discrete-time case is selected as a state $x_i(k)$. Then state differential or difference equations can be written directly, using the fact that $\dot{x}_i(t)$ (or $x_i(k+1)$) is the input to that integrator (or delay element).

This procedure leads to state equations in *observable canonical form.* The reason for the name and the importance of the form become apparent in Chapter 11, where properties of state equations are discussed. For now, just note that the **C** output matrix is in an especially simple form, all 0s and 1s, whereas **B** is more complicated. An alternate state variable description is given shortly, in which the input matrix **B** has a similarly simple form, containing all 0s and 1s, but with elements of **C** being more complicated. That form will be called the *controllable canonical form.* Note that the first procedure of Section 3.4, where there were no input derivatives, led to a set of state equations with *both* **B** and **C** having these simple forms.

EXAMPLE 3.6 Consider

$$\frac{d^n y}{dt^n} + a_{n-1}\frac{d^{n-1} y}{dt^{n-1}} + \cdots + a_1\frac{dy}{dt} + a_0 y = \beta_0 u + \beta_1\frac{du}{dt} + \cdots + \beta_m\frac{d^m u}{dt^m}.$$

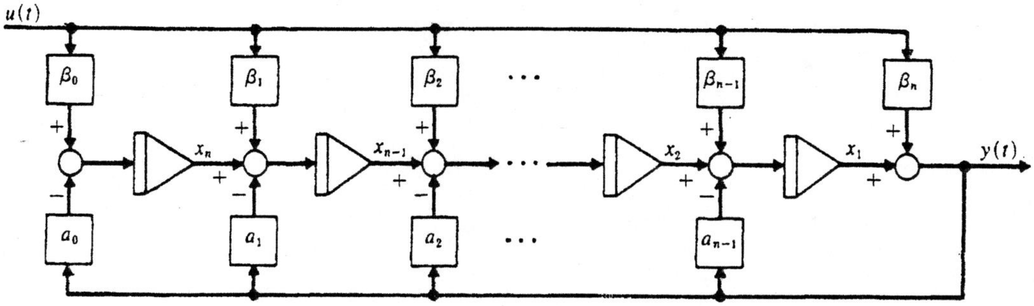

Figure 3.7

The coefficients a_i and β_i are constant. Assume $m = n$. If $m < n$, then some of the β_i terms can be set to zero after the final result is obtained.

(1), (2) $$y(t) = \underbrace{\int\int\cdots\int}_{n \text{ integrals}} \left\{\beta_n \frac{d^n u}{dt^n} + \left(\beta_{n-1}\frac{d^{n-1}u}{dt^{n-1}} - a_{n-1}\frac{d^{n-1}y}{dt^{n-1}}\right) + \cdots + \left(\beta_1 \frac{du}{dt} - a_1 \frac{dy}{dt}\right) + (\beta_0 u - a_0 y)\right\} dt\ldots dt'$$

(3) $$y(t) = \beta_n u + \int \left\{(\beta_{n-1}u - a_{n-1}y) + \int \left[\beta_{n-2}u - a_{n-2}y + \int\left(\cdots + \int \{\beta_1 u - a_1 y + \int (\beta_0 u - a_0 y)\,dt\}\cdots\right) dt'\right] dt''\right\} dt'''$$

(4) The simulation diagram is shown in Figure 3.7.

(5) Numbering outputs of integrators from the right gives

$$\dot{x}_1 = -a_{n-1}[x_1 + \beta_n u] + \beta_{n-1}u + x_2$$
$$= -a_{n-1}x_1 + x_2 + (\beta_{n-1} - a_{n-1}\beta_n)u$$
$$\dot{x}_2 = -a_{n-2}x_1 + x_3 + (\beta_{n-2} - a_{n-2}\beta_n)u$$
$$\vdots$$
$$\dot{x}_{n-1} = -a_1 x_1 + x_n + (\beta_1 - a_1\beta_n)u$$
$$\dot{x}_n = -a_0 x_1 + (\beta_0 - a_0\beta_n)u$$

The output equation is

$$y = x_1 + \beta_n u = [1 \quad 0 \quad 0 \quad \cdots \quad 0]\mathbf{x} + \beta_n u$$

This represents the observable canonical form of the state equations. ■

EXAMPLE 3.7 A system is described by the following equation:

$$y(k+n) + a_{n-1}y(k+n-1) + \cdots + a_1 y(k+1) + a_0 y(k) = \beta_0 u(k) + \beta_1 u(k+1) + \cdots + \beta_m u(k+m)$$

First solve for the most advanced output term (or terms if coupled equations are involved):

$$y(k+n) = -a_{n-1}y(k+n-1) - \cdots - a_1 y(k+1) - a_0 y(k) + \beta_0 u(k) + \beta_1 u(k+1) + \cdots + \beta_m u(k+m)$$

Then delay every term in the equation n times so that $y(k)$ is obtained on the left-hand side. The symbol $\mathfrak{D}$ will be used to represent the delay operation. As in Example 3.6, it is assumed that $m = n$ for convenience. If $m < n$, then some of the coefficients β_i can be set to zero. It is impossible that $m > n$ for physically realizable systems:

$$y(k) = -a_{n-1}\mathfrak{D}(y(k)) - \cdots - a_1\mathfrak{D}^{n-1}(y(k)) - a_0\mathfrak{D}^n(y(k)) + \beta_0\mathfrak{D}^n(u(k)) + \beta_1\mathfrak{D}^{n-1}(u(k)) + \cdots + \beta_n u(k)$$

Rearrange this expression as a nested sequence of delayed terms:

$$y(k) = \beta_n u(k) + \mathfrak{D}\{-a_{n-1}y(k) + \beta_{n-1}u(k) + \mathfrak{D}[-a_{n-2}y(k) + \beta_{n-2}u(k) + \mathfrak{D}(\cdots + \mathfrak{D}\{-a_0 y(k) + \beta_0 u(k)\})]\}$$

The simulation diagram of Figure 3.8 can now be drawn, noting that everything which is operated upon by the delay operator $\mathfrak{D}$ forms the input to that delay.

This diagram is exactly like the one of Figure 3.7 except that the integrators have been replaced by delay elements. The state equations can be written down from the simulation diagram by using the fact that if the output of a delay is $x_i(k)$, then its input signal must be $x_i(k+1)$:

$$\begin{aligned}
x_1(k+1) &= -a_{n-1}[x_1(k) + \beta_n u(k)] + x_2(k) + \beta_{n-1}u(k) \\
&= -a_{n-1}x_1(k) + x_2(k) + [\beta_{n-1} - a_{n-1}\beta_n]u(k) \\
x_2(k+1) &= -a_{n-2}x_1(k) + x_3(k) + [\beta_{n-2} - a_{n-2}\beta_n]u(k) \\
&\ \vdots \\
x_{n-1}(k+1) &= -a_1 x_1(k) + x_n(k) + [\beta_1 - a_1\beta_n]u(k) \\
x_n(k+1) &= -a_0 x_1(k) + [\beta_0 - a_0\beta_n]u(k)
\end{aligned}$$

The output equation is

$$y(k) = x_1(k) + \beta_n u(k) = [1 \quad 0 \quad 0 \quad \cdots \quad 0]\mathbf{x}(k) + \beta_n u(k)$$ ■

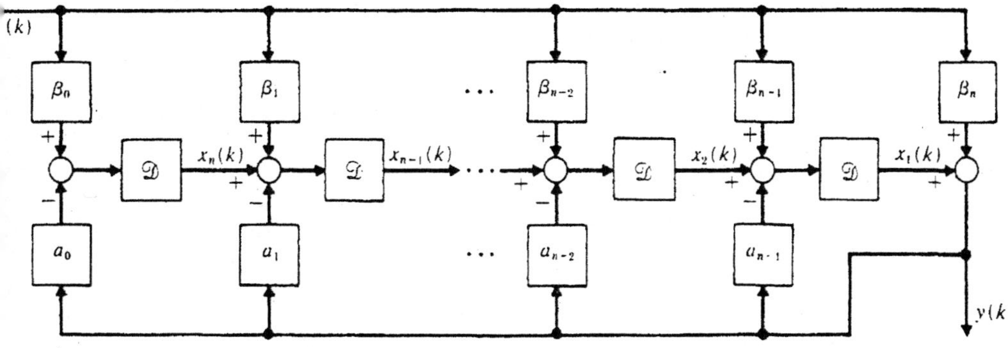

Figure 3.8

3.4.4 State Equations from Transfer Functions

The discussion is restricted to single-input, single-output constant coefficient systems described by a Laplace transform transfer function:

$$\frac{Y(s)}{u(s)} = T(s) = \frac{\beta_m s^m + \beta_{m-1} s^{m-1} + \cdots + \beta_1 s + \beta_0}{s^n + a_{n-1} s^{n-1} + \cdots + a_1 s + a_0}$$

or a Z-transform transfer function

$$\frac{y(z)}{u(z)} = T(z) = \frac{\beta_m z^m + \beta_{m-1} z^{m-1} + \cdots + \beta_1 z + \beta_0}{z^n + a_{n-1} z^{n-1} + \cdots + a_1 z + a_0}$$

For any physical system, causality (physical realizability) requires that $m \le n$. Otherwise, the output $y(k)$ at time t_k would depend upon future inputs $u(j)$ at times t_j with $j > k$. The transfer function can also be written in terms of negative powers of z:

$$T(z) = \frac{z^{m-n}[\beta_m + \beta_{m-1} z^{-1} + \cdots + \beta_1 z^{-(m-1)} + \beta_0 z^{-m}]}{1 + a_{n-1} z^{-1} + a_{n-2} z^{-2} + \cdots + a_1 z^{-(n-1)} + a_0 z^{-n}}$$

Since z^{-1} provides a delay of one sample period, this might be called the delay operator form. It is clear that there is a delay of $n - m$ sample periods from input to output.

The preceding transfer functions correspond exactly to the systems considered in Examples 3.6 and 3.7. Therefore, the previous method of picking states obviously applies. The reason for considering these systems further from the transfer function point of view is that alternative forms of the transfer functions are easily written using algebraic manipulations. The alternative forms can be used to find alternative state variable models. In particular, four major categories of state variable models, called *realizations*, will be presented. Each has its own particular set of advantages, disadvantages, and implications in state variable applications.

1. *Direct realizations* are so named because they derive directly from the expanded polynomial form of the transfer functions. The two major direct realizations are
   (a) Observable canonical form
   (b) Controllable canonical form
2. *Cascade realizations* are so named because they derive from the transfer function written as a product of simple factored terms, which could be represented by a series of cascaded blocks in a block diagram.
3. *Parallel realizations* are so named because they derive from the transfer function written as a sum of partial fraction expansion terms, which would appear as parallel blocks on a block diagram.

Within each of these categories there remains a certain amount of freedom of choice, such as how factors are to be grouped in the cascade form. Actually, the possibilities are infinite because if $\mathbf{x}$ is any valid state vector, then so is $\mathbf{Sx}$ for any nonsingular transformation matrix $\mathbf{S}$. Finally, there are other canonical realizations not discussed here, such as the lattice and ladder network realizations.

The four major realizations discussed in this book are now illustrated for continuous-time systems. The observable canonical form has already been derived using the nested integrator approach. Its simulation diagram is shown in Figure 3.7. The state equation component equations are written in matrix form as

$$\dot{\mathbf{x}}(t) = \begin{bmatrix} -a_{n-1} & 1 & 0 & 0 & \dots & 0 \\ -a_{n-2} & 0 & 1 & 1 & \dots & 0 \\ \vdots & & & & & \\ -a_1 & 0 & 0 & 0 & \dots & 1 \\ -a_0 & 0 & 0 & 0 & \dots & 0 \end{bmatrix} \mathbf{x}(t) + \begin{bmatrix} \beta_{n-1} - a_{n-1}\beta_n \\ \beta_{n-2} - a_{n-2}\beta_n \\ \vdots \\ \beta_1 - a_1\beta_n \\ \beta_0 - a_0\beta_n \end{bmatrix} u(t)$$

and

$$y(t) = [1 \quad 0 \quad 0 \quad \dots \quad 0]\mathbf{x}(t) + \beta_n u(t) \tag{3.21}$$

The controllable canonical form is obtained by artificially splitting the transfer function denominator polynomial—call it $a(s)$—and the numerator polynomial—call it $b(s)$—as shown in Figure 3.9. An intermediate variable g has been introduced. Now the transfer function from u to g is exactly the same as the transfer function from u to y for Eq. (*3.15*) considered earlier using phase variables. So once again the states can be selected as $x_1 = g$, $x_2 = \dot{g}, \dots, x_n = \overset{(n-1)}{g}$. As a result the differential equation part of the controllable canonical state equations is given by Eq. (*3.17*). The block from g to the output y in Figure 3.9 indicates that $y(t)$ is a linear combination of g and its derivatives, $y(t) = \beta_0 g + \beta_1 \dot{g} + \beta_2 \ddot{g} + \cdots + \beta_m \overset{(m)}{g}$. It is assumed that $m = n$ in order to get the most general result. If m is actually less than n, simply set the higher β_i terms to zero in the results to follow. When the state definitions are used, each g term except the last $(d^n g/dt^n)$ is simply replaced by the appropriate state. The nth order derivative of g is now $\dot{x}_n$, and this is a combination of all the states and the a_i coefficients. Regrouping terms gives the state output equation for the controllable canonical form as

$$y(t) = [\beta_0 - a_0\beta_n \quad \beta_1 - a_1\beta_n \quad \beta_2 - a_2\beta_n \quad \cdots \quad \beta_{n-1} - a_{n-1}\beta_n]\mathbf{x}(t) + \beta_n u(t) \tag{3.22}$$

When $m < n$, $\beta_n = 0$ and this takes on a deceptively simple form. In order to illustrate the full generality of the result $m = n$ has been assumed. The simulation diagram for the controllable canonical form is given in Figure 3.10. All the feed-forward β_i terms are from $b(s)$ and all the feedback terms are from $a(s)$.

The nth-order s-domain transfer function can be written in factored form, Eq. (*3.23*). If all poles p_i are distinct and if $m = n$, the partial fraction expansion form of Eq. (*3.24*) can be obtained:

$$T(s) = \frac{\beta_m(s + z_1)(s + z_2)\cdots(s + z_m)}{(s + p_1)(s + p_2)\cdots(s + p_n)} \tag{3.23}$$

$$T(s) = b_0 + \frac{b_1}{s + p_1} + \frac{b_2}{s + p_2} + \cdots + \frac{b_n}{s + p_n} \tag{3.24}$$

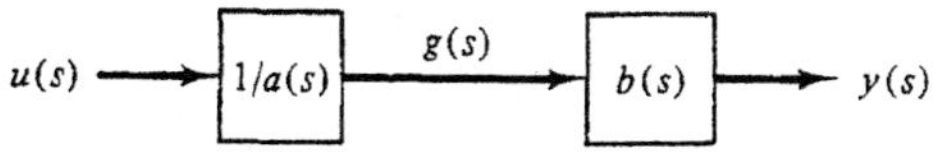

Figure 3.9

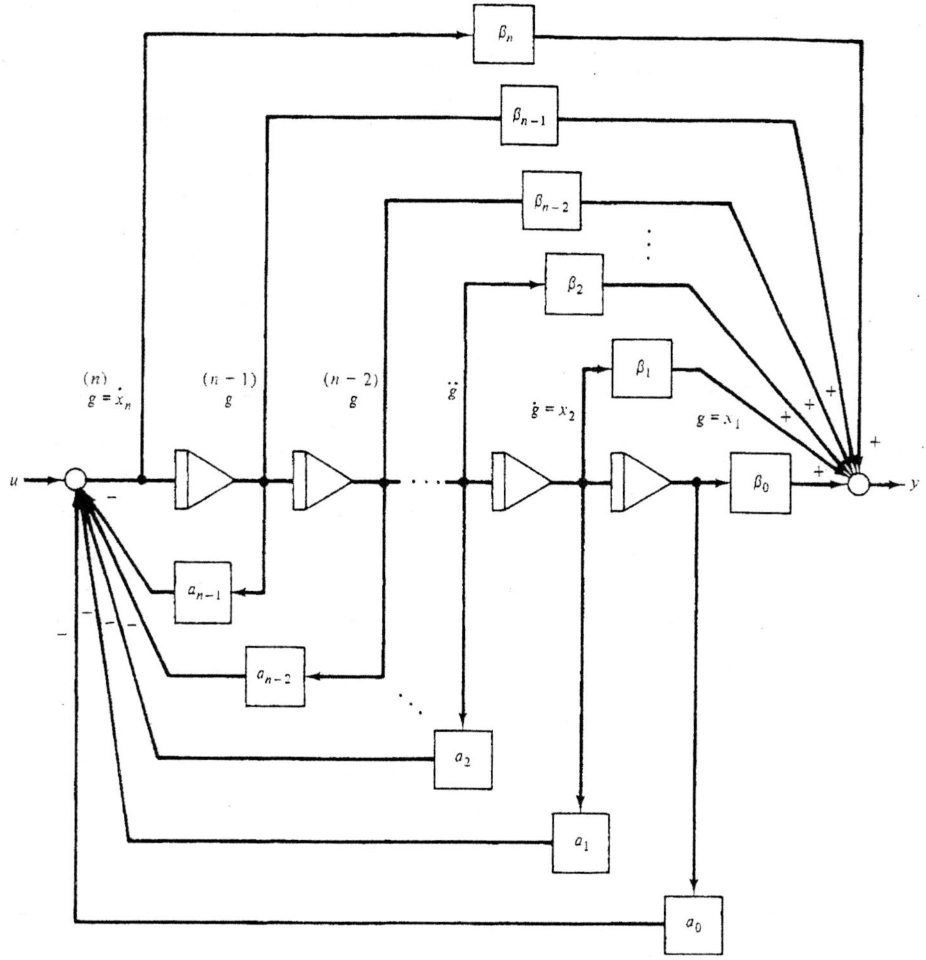

Figure 3.10

If some poles are repeated, the partial fraction expansion will contain additional terms involving powers of the multiple pole $(s + p_i)$ in the denominator (see Problems 3.4 and 3.22). If $m < n$, $b_0 = 0$. For a dynamical system, m can never exceed n.

Equation (3.23) indicates that $T(s)$ can be written as the product of simple factors $1/(s + p_i)$ or $(s + z_j)/(s + p_i)$. Quadratic terms could be considered as well. If a denominator root p_i is complex, it and its conjugate may be kept together in a quadratic factor containing only real coefficients. In any case the system can be represented by a series (cascade) connection of simple first- and/or second-order factors such as those shown in Figure 3.11. The grouping of numerator factors with denominator factors is not discussed here. Each block should be realizable (i.e., the power of the numerator cannot exceed the power of the denominator). Beyond that restriction, how the factors are grouped is left arbitrary, although it does affect the scaling of the signals passed between the blocks. A simulation diagram can be determined for each block by any of

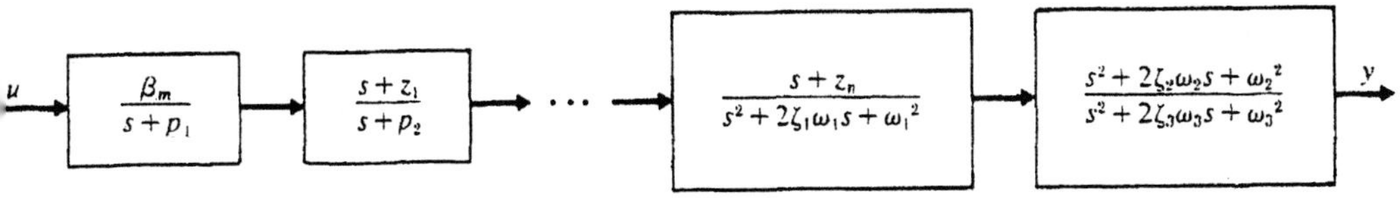

Figure 3.11

the previous methods (see also Problem 3.1). The overall simulation diagram for the cascade realization is the series connection of the diagrams found for each block. The cascade realization of the state equations is then written by selecting integrator outputs as state variables. An example follows shortly.

The partial fraction expanded form in Eq. (*3.24*) indicates that $T(s)$ can be represented as a parallel connection of simple terms (Figure 3.12).

The diagram of Figure 3.12 assumes p_{n-1} is a double pole. The system simulation diagram is also a parallel connection of the individual terms.

EXAMPLE 3.8 Select a suitable set of state variables for the system whose transfer function is

$$T(s) = \frac{s+3}{s^3+9s^2+24s+20}$$

Note that in factored form

$$T(s) = \frac{s+3}{(s+2)^2(s+5)}$$

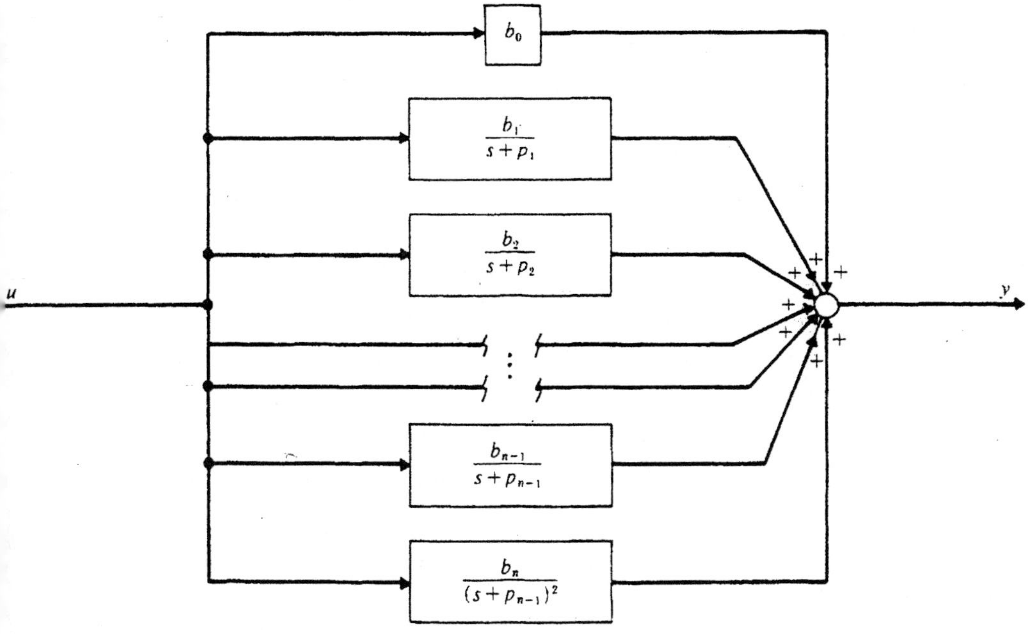

Figure 3.12

and, using a partial fraction expansion,

$$T(s) = \frac{\frac{2}{9}}{s+2} + \frac{\frac{1}{3}}{(s+2)^2} + \frac{-\frac{2}{9}}{s+5}$$

Using the original form of the transfer function will lead to one possible direct realization. First, it is noted that

$$\dddot{y} = -9\ddot{y} - 24\dot{y} - 20y + \dot{u} + 3u$$

or

$$y = \int\left\{-9y + \int\left[-24y + u + \int(-20y + 3u)\,dt\right]dt'\right\}dt''$$

from which the simulation diagram of Figure 3.13 is obtained.

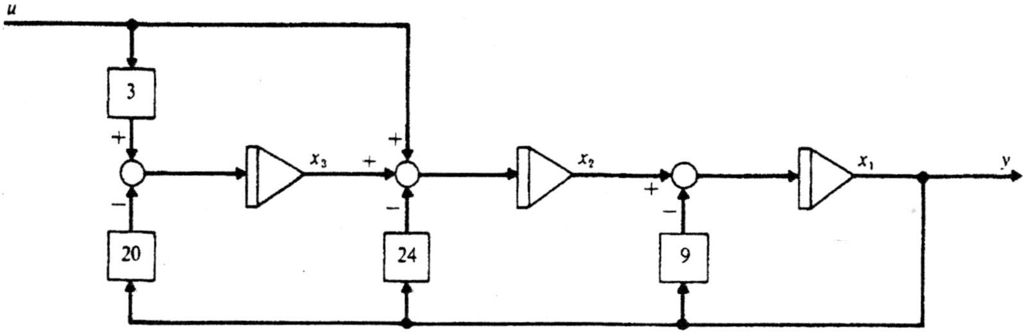

Figure 3.13

Using Figure 3.13, the observable canonical form of the state equations is

$$\begin{bmatrix}\dot{x}_1\\ \dot{x}_2\\ \dot{x}_3\end{bmatrix} = \begin{bmatrix}-9 & 1 & 0\\ -24 & 0 & 1\\ -20 & 0 & 0\end{bmatrix}\begin{bmatrix}x_1\\ x_2\\ x_3\end{bmatrix} + \begin{bmatrix}0\\ 1\\ 3\end{bmatrix}u$$

and

$$y = [1 \quad 0 \quad 0]\mathbf{x}$$

This could have been written directly from the general result of Eq. (*3.21*).

A second direct form of the transfer function will now be developed. The numerator and denominator of the transfer function are separated, as shown in Figure 3.14. The intermediate variable thus created is labeled g for convenience. The relationship between u and g is given by

$$\dddot{g} + 9\ddot{g} + 24\dot{g} + 20g = u$$

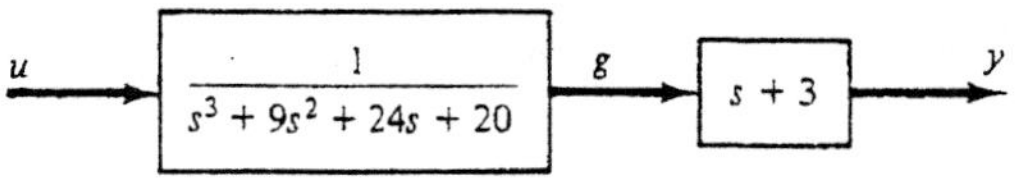

Figure 3.14

This means that the **A** matrix will be in companion form and the **B** matrix will assume its simplest possible form—all 0s or 1s.

The relationship between the fictitious g and the output y depends only on the numerator of $T(s)$. In this simple case

$$y = \dot{g} + 3g$$

Noting that since the s variable indicates differentiation, y is seen to be a linear combination of g and its various derivatives. These derivatives are available as inputs to the various integrators. Using this reasoning leads to the simulation diagram of Figure 3.15. Then, by picking outputs of integrators as states, the controllable canonical form of the state equations is as follows:

$$\dot{\mathbf{x}} = \begin{bmatrix} 0 & 1 & 0 \\ 0 & 0 & 1 \\ -20 & -24 & -9 \end{bmatrix}\mathbf{x} + \begin{bmatrix} 0 \\ 0 \\ 1 \end{bmatrix}u$$

$$y = [3 \quad 1 \quad 0]\mathbf{x} + [0]u$$

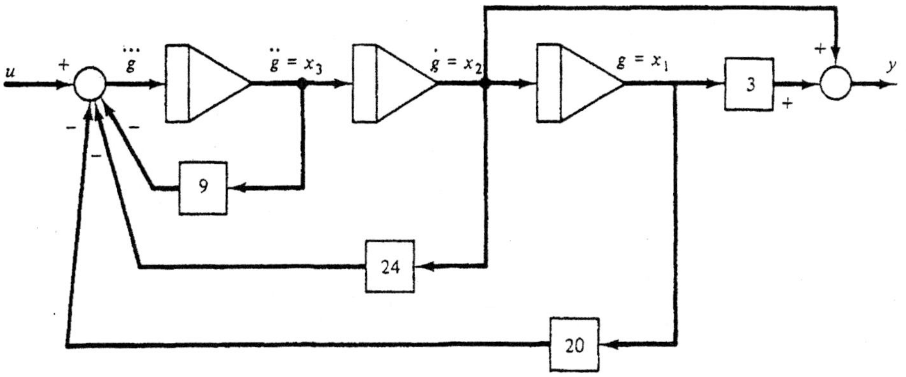

Figure 3.15

These could have been written directly from Eqs. (*3.18*) and (*3.22*).

Using the factored form of $T(s)$, the simulation diagram of Figure 3.16 is obtained. From Figure 3.16, one possible cascade realization is

$$\begin{bmatrix} \dot{x}_1 \\ \dot{x}_2 \\ \dot{x}_3 \end{bmatrix} = \begin{bmatrix} -5 & 1 & 0 \\ 0 & -2 & 1 \\ 0 & 0 & -2 \end{bmatrix}\begin{bmatrix} x_1 \\ x_2 \\ x_3 \end{bmatrix} + \begin{bmatrix} 0 \\ 1 \\ 1 \end{bmatrix}u \quad \text{and} \quad y = [1 \quad 0 \quad 0]\mathbf{x}$$

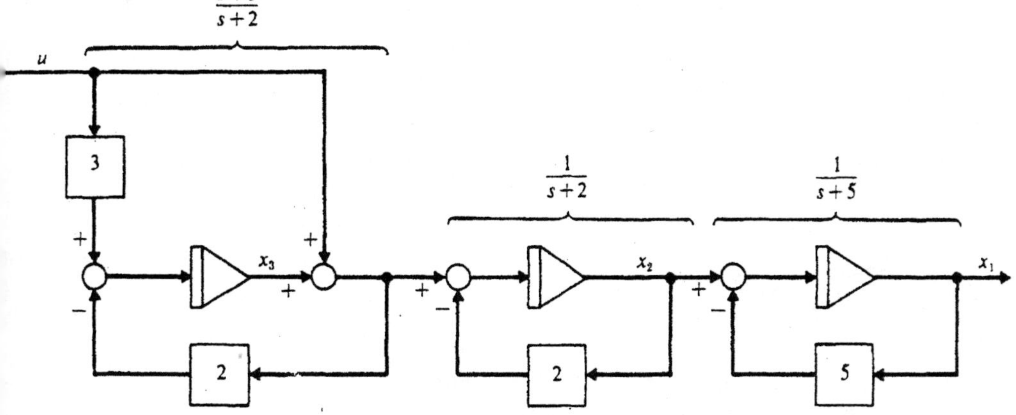

Figure 3.16

Using the partial fraction expansion, the simulation diagram of Figure 3.17 is obtained. From Figure 3.17, the parallel realization is

$$\begin{bmatrix} \dot{x}_1 \\ \dot{x}_2 \\ \dot{x}_3 \end{bmatrix} = \begin{bmatrix} -5 & 0 & 0 \\ 0 & -2 & 1 \\ 0 & 0 & -2 \end{bmatrix} \begin{bmatrix} x_1 \\ x_2 \\ x_3 \end{bmatrix} + \begin{bmatrix} 1 \\ 0 \\ 1 \end{bmatrix} u \quad \text{and} \quad y = \begin{bmatrix} -\frac{2}{9} & \frac{1}{3} & \frac{2}{9} \end{bmatrix} \mathbf{x}$$

■

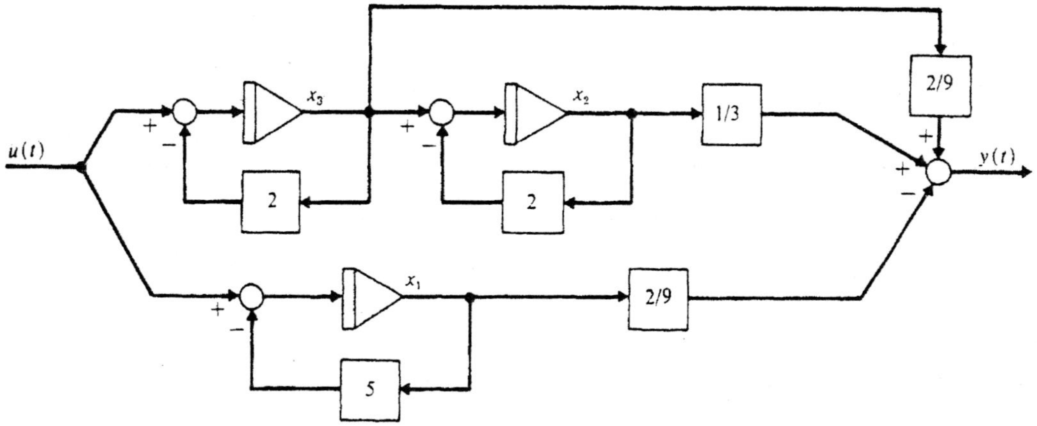

Figure 3.17

Four different sets of valid state equations have been derived for this system. In the third and fourth forms, the diagonal terms of the matrix **A** are the same, the system poles (and, as we see in Chapter 7, the eigenvalues of **A**).The fourth form gives **A** in Jordan canonical form (see Chapter 7). All four representations are different, but all have the same number of state variables, and this is the order of the system. They also all have the same eigenvalues (poles). Note that if separate parallel paths had been used for the terms $1/(s+2)^2$ and $1/(s+2)$, one unnecessary integrator would have been used. This should be avoided, since unnecessary integrators means unnecessary state variables. This relates to the topic of minimal realizations in Chapter 12 and controllability and observability properties of Chapter 11.

The same four realizations can be found for discrete-time systems described by Z-transform transfer functions. Much of the work is exactly the same, with delay operators replacing integrators. Therefore, the major points are presented by way of an example. To give a comparison with the continuous-system results, the third-order system of Example 3.8 is used. When it is preceded by a zero-order hold and then Z-transformed with a sampling period of 0.2, the resulting discrete transfer function is

$$T(z) = \frac{0.013667z^2 + 0.00167z - 0.0050}{z^3 - 1.7085z^2 + 0.9425z - 0.1653} \tag{3.25}$$

EXAMPLE 3.9 Direct Realization, Observable Canonical Form The delay operator form of the transfer function converts immediately to the difference equation considered in Example 3.7. The result of the previous example is a set of observable canonical form state equations

$$\mathbf{x}(k+1)=\begin{bmatrix}-a_{n-1} & 1 & 0 & 0 & \dots & 0 & 0\\ -a_{n-2} & 0 & 1 & 0 & \dots & 0 & 0\\ \vdots & & & & & & \vdots\\ -a_1 & 0 & 0 & 0 & \dots & 0 & 1\\ -a_0 & 0 & 0 & 0 & \dots & 0 & 0\end{bmatrix}\mathbf{x}(k)+\begin{bmatrix}\beta_{n-1}-a_{n-1}\beta_n\\ \beta_{n-2}-a_{n-2}\beta_n\\ \vdots\\ \beta_1-a_1\beta_n\\ \beta_0-a_0\beta_n\end{bmatrix}u(k)$$

$$y(k)=[1\quad 0\quad 0\quad 0\quad \dots\quad 0\quad 0]\mathbf{x}(k)+\beta_n u(k)$$

For the specific third-order example, the observable canonical state equations are therefore

$$\mathbf{x}(k+1)=\begin{bmatrix}1.7085 & 1 & 0\\ -0.9425 & 0 & 1\\ 0.1653 & 0 & 0\end{bmatrix}\mathbf{x}(k)+\begin{bmatrix}0.01361\\ 0.00167\\ -0.0050\end{bmatrix}u(k)$$

$$y(k)=[1\quad 0\quad 0]\mathbf{x}(k)$$

■

EXAMPLE 3.10 Direct Realization, Controllable Canonical Form The previous transfer function is artificially split into two parts with the fictitious variable $g(k)$ in between, as shown in Figure 3.18. The simulation diagram for determining $g(k)$ is first developed using

$$g(k+3)=1.7085g(k+2)-0.9425g(k+1)+0.1653g(k)+u(k)$$

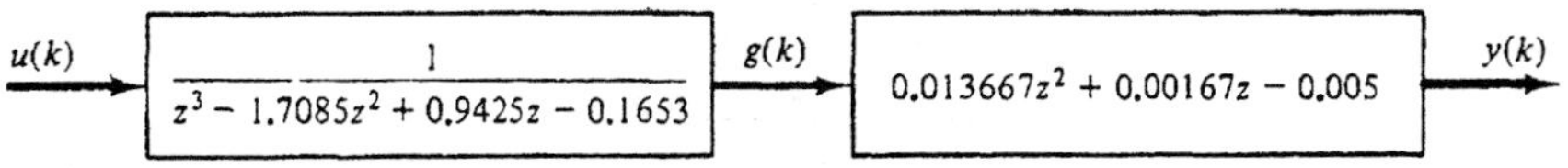

Figure 3.18

Then three successive delay elements give $g(k+2)$, $g(k+1)$, and $g(k)$. This constitutes part of the diagram in Figure 3.19. The second transfer function in Figure 3.18 states that

$$y(k)=-0.0050g(k)+0.00167g(k+1)+0.013667g(k+2)$$

This relationship constitutes the rest of Figure 3.19. Numbering the outputs of the delays as states in the order shown immediately gives a controllable canonical form of the state equations

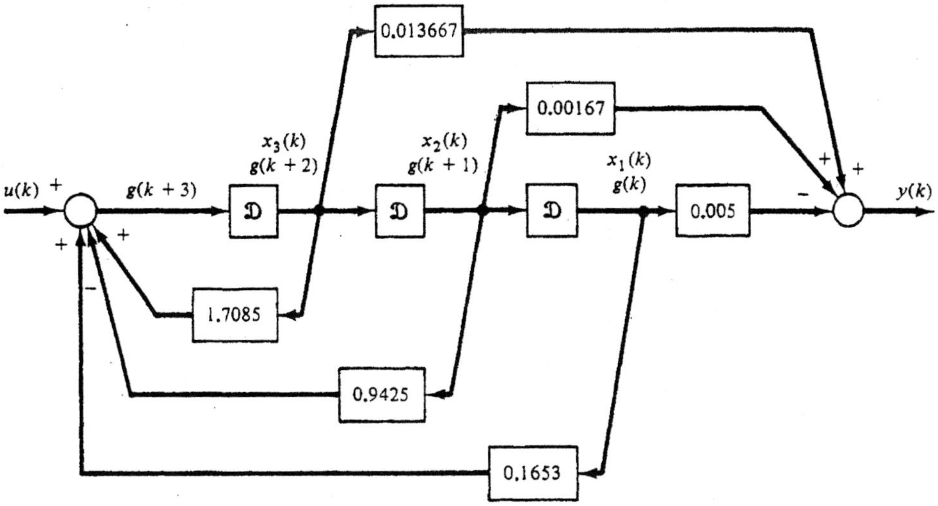

Figure 3.19

$$\mathbf{x}(k+1) = \begin{bmatrix} 0 & 1 & 0 \\ 0 & 0 & 1 \\ 0.1653 & -0.9425 & 1.7085 \end{bmatrix} \mathbf{x}(k) + \begin{bmatrix} 0 \\ 0 \\ 1 \end{bmatrix} u(k)$$

$$y(k) = [-0.0050 \quad 0.00167 \quad 0.013667]\mathbf{x}(k)$$ ■

EXAMPLE 3.11 Cascade Realization The same transfer function is again considered, but this time in factored form (and grouped in the way terms will be cascaded together):

$$T(z) = \left[\frac{1}{z - 0.3679}\right]\left[\frac{z - 0.5488}{z - 0.6703}\right]\left[\frac{z + 0.6714}{z - 0.6703}\right][0.013667]$$

There are at least two valid simulation diagrams for factors such as $\frac{z+a}{z-b}$, as shown in Figure 3.20*a* and *b*. Using the second form, the total simulation diagram for a cascade realization is as shown in Figure 3.21. Using state variables numbered as shown, the state equations are

$$\mathbf{x}(k+1) = \begin{bmatrix} 0.6703 & 0.1215 & 1 \\ 0 & 0.6703 & 1 \\ 0 & 0 & 0.3679 \end{bmatrix} \mathbf{x}(k) + \begin{bmatrix} 0 \\ 0 \\ 1 \end{bmatrix} u(k)$$

$$y(k) = 0.013667[1.3417 \quad 0.1215 \quad 1]\mathbf{x}(k)$$

$$= [0.01834 \quad 0.00166 \quad 0.013667]\mathbf{x}(k)$$ ■

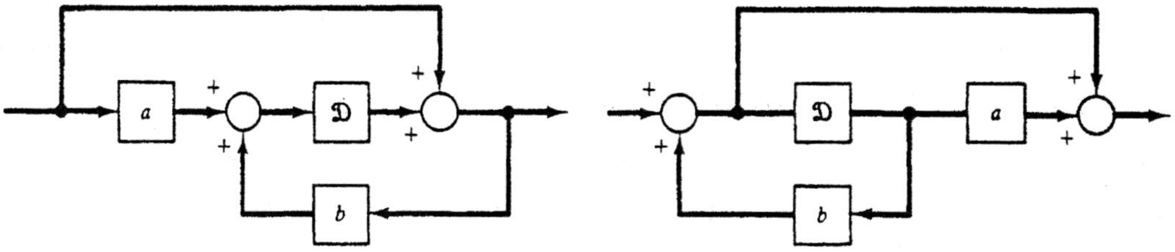

Figure 3.20

EXAMPLE 3.12 Parallel Realization A parallel realization of the same transfer function is now developed using partial fraction expansions. If $T(z)/z$ is expanded and that result is then multiplied by z, the result is

$$T(z) = 0.0305 - 0.07637\frac{z}{(z - 0.3679)} + 0.011\frac{z}{(z - 0.6703)^2} + 0.0459\frac{z}{(z - 0.6703)}$$

Although there is often good reason for going through the extra steps to put an expanded Z-transfer function into this form, there is no reason to do so in the present context. A direct expansion of $T(z)$ in the usual manner gives

$$T(z) = \frac{0.0074}{(z - 0.6703)^2} + \frac{0.0418}{(z - 0.6703)} - \frac{0.0281}{(z - 0.3679)}$$

A simulation diagram of this is shown in Figure 3.22. The state equations are written directly from this as

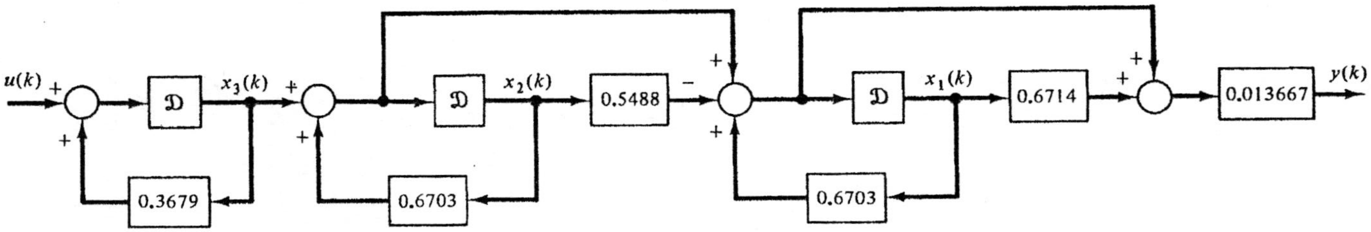

Figure 3.21

$$\mathbf{x}(k+1) = \begin{bmatrix} 0.6703 & 1 & 0 \\ 0 & 0.6703 & 0 \\ 0 & 0 & 0.3679 \end{bmatrix} \mathbf{x}(k) + \begin{bmatrix} 0 \\ 1 \\ 1 \end{bmatrix} u(k)$$

$$y(k) = [0.0074 \quad 0.0418 \quad -0.0281]\mathbf{x}(k)$$

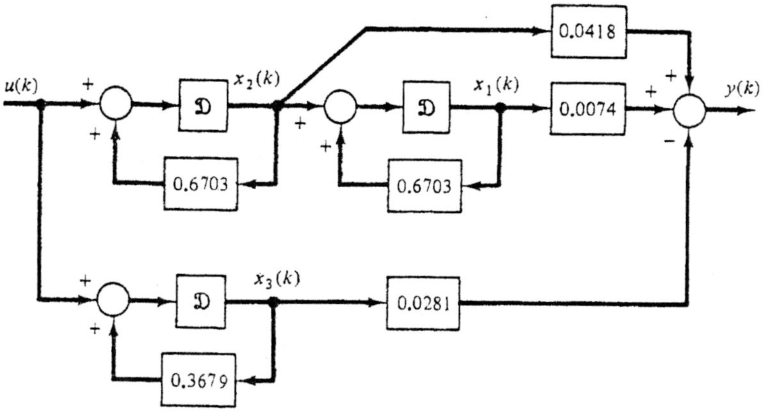

Figure 3.22

As in the continuous-time case, the partial fraction procedure has given an **A** matrix in Jordan canonical form.

The alternative form of the partial fraction expansion equation leads to exactly the same final state equations. It would be a good exercise for the reader to verify this. The *apparent* direct path from u to y through the gain of 0.0305 will cancel out in the manipulations, and a zero D term does in fact result. There is no path from input to output with less than a one sample period delay. ■

3.4.5 State Equations Directly from the System's Linear Graph [4, 5]

Linear graphs were used in Chapter 1 in connection with system modeling. Recall that linear graph techniques are not restricted to linear or constant-coefficient systems. A method of obtaining state equations directly from the system graph is now presented. This is a powerful method because it avoids many of the intermediate manipulations with transfer functions and input-output differential equations, which are restricted to linear, constant systems. Furthermore, the linear graph technique often gives greater engineering insight because the state variables thus obtained are usually related to the energy stored in the system. Before describing the method, a few additional definitions regarding linear graphs are required.

A *tree* is a set of branches of the graph that (1) contains every node of the graph, (2) is connected, and (3) contains no loops. A tree is formed from a graph by removing certain branches. A branch of the graph included in the tree is called a tree branch. Those branches which were deleted while forming a tree are called *links*. Each time a link is added to a tree, one loop is formed. A loop consisting of one link plus a number of tree branches is called the *fundamental loop* associated with that link. For a given tree, if any one tree branch is cut, the tree is separated into two parts. A *fundamental cutset* of a given tree branch consists of that one cut tree branch plus all links that connect between nodes of the two halves of the severed tree. In other words, if a line is drawn through the original graph in such a way as to (1) divide the graph into two parts and (2) cut only one tree branch, then that branch plus all links cut by the dividing line form a fundamental cutset.

The following procedure is a systematic method of obtaining state equations from a linear graph:

1. *Form a tree from the graph which includes:* (*a*) all across variable (voltage) sources; (*b*) as many elements as possible which store energy by virtue of their across variable (capacitors or the analogous elements in other disciplines), that is, elements whose elemental equation has the across variable differentiated; (*c*) elements with algebraic elemental equations (resistors and their analogs); (*d*) as few elements as possible which store energy by virtue of their through variable and have the through variable differentiated in their elemental equation (inductors and their analogs); and (*e*) no through variable (current) sources. (*f*) If ideal transformers are included in the graph, one side of the transformer should be treated like a through variable source and the other side like an across variable source. There will usually be several trees which satisfy these rules.
2. *Choose as state variables the across variables of all capacitor-like elements included in the tree and the through variables of all inductor-like elements not included in the tree.*
3. *The elemental equations for elements involving the selected state variables will be of the form*

   $$\dot{x}_i = \text{function of through or across variables and inputs}$$

   Use the compatibility laws (Kirchhoff's voltage laws) around the fundamental loops and the conservation laws (Kirchhoff's current laws) into the fundamental cutsets to eliminate nonstate variables from these functional equations.

EXAMPLE 3.13 Consider the linear graph shown in Figure 3.23. The symbols L, C, and R_i are used to indicate the type elements in each branch, although they need not be of an electrical nature.

Using the given rules, the tree of Figure 3.24 is selected. The state variables are the voltage x_1 across C and the current x_2 through L. It is assumed that the elemental equations are nonlinear,

$$C\dot{x}_1 = f_1(i_C) \quad \text{and} \quad L\dot{x}_2 = f_2(v_s - v_1)$$

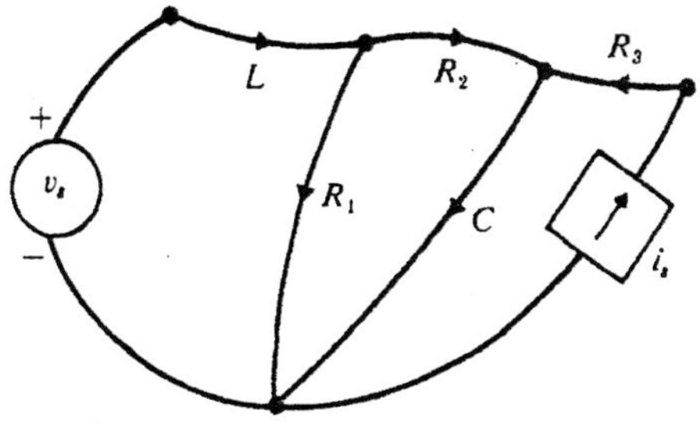

Figure 3.23

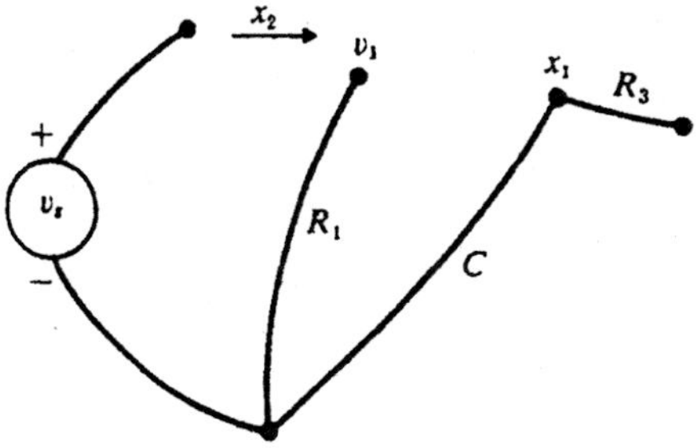

Figure 3.24

To complete the state description, i_C and v_1 must be expressed as functions of x_1, x_2, v_s, and i_s.

The fundamental cutset associated with branch C is given in Figure 3.25 and thus $i_C = i_2 + i_s$. The cutset associated with R_1 is given in Figure 3.26, from which $i_1 = x_2 - i_2$. The fundamental loop formed with link R_2 gives the compatibility equation $v_1 - i_2 R_2 = x_1$, assuming R_2 is a linear resistor. If R_1 is also linear, $v_1 = R_1 i_1$.

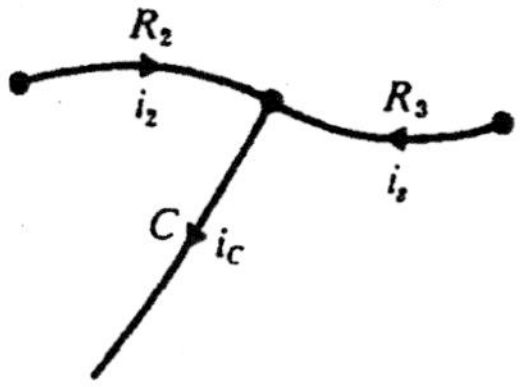

Figure 3.25

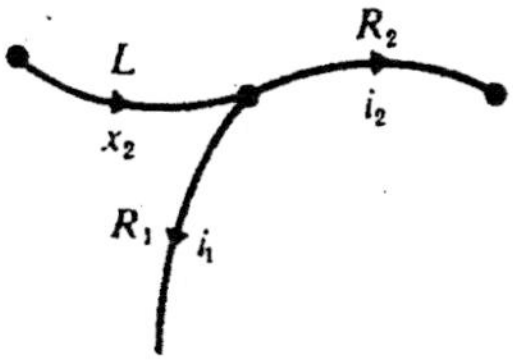

Figure 3.26

These equations are solved simultaneously to give

$$i_C = i_s + \frac{R_1 x_2 - x_1}{R_1 + R_2}$$

$$v_1 = \frac{R_1 R_2 x_2 + R_1 x_1}{R_1 + R_2}$$

so that the state equations are

$$\dot{x}_1 = \frac{1}{C} f_1\left(i_s + \frac{R_1 x_2 - x_1}{R_1 + R_2}\right)$$

$$\dot{x}_2 = \frac{1}{L} f_2\left(v_s - \frac{R_1 R_2 x_2 + R_1 x_1}{R_1 + R_2}\right)$$

If L and C are also linear, then letting $u_1 = v_s$ and $u_2 = i_s$ reduces the equations to

$$\begin{bmatrix} \dot{x}_1 \\ \dot{x}_2 \end{bmatrix} = \begin{bmatrix} -\dfrac{1}{C(R_1 + R_2)} & \dfrac{R_1}{C(R_1 + R_2)} \\ -\dfrac{R_1}{L(R_1 + R_2)} & -\dfrac{R_1 R_2}{L(R_1 + R_2)} \end{bmatrix} \begin{bmatrix} x_1 \\ x_2 \end{bmatrix} + \begin{bmatrix} 0 & 1/C \\ 1/L & 0 \end{bmatrix} \begin{bmatrix} u_1 \\ u_2 \end{bmatrix}$$

If the voltage across R_2 is considered to be the output y, then

$$y = v_1 - x_1 = \begin{bmatrix} -\dfrac{R_2}{R_1+R_2} & \dfrac{R_1R_2}{R_1+R_2} \end{bmatrix} \mathbf{x}$$

■

3.5 INTERCONNECTION OF SUBSYSTEMS

Suppose that two subsystems i and j have been modeled in state variable format:

$$\dot{\mathbf{x}}_i = \mathbf{A}_i\mathbf{x}_i + \mathbf{B}_i\mathbf{u}_i \qquad \dot{\mathbf{x}}_j = \mathbf{A}_j\mathbf{x}_j + \mathbf{B}_j\mathbf{u}_j$$

$$\mathbf{y}_i = \mathbf{C}_i\mathbf{x}_i + \mathbf{D}_i\mathbf{u}_i \qquad \mathbf{y}_j = \mathbf{C}_j\mathbf{x}_j + \mathbf{D}_j\mathbf{u}_j$$

Suppose that the input to subsystem j is the output from subsystem i—that is, $\mathbf{u}_j = \mathbf{y}_i$—and assume that $\mathbf{y}_i$ and $\mathbf{y}_j$ are both considered outputs of the composite system (see Fig. 3.27). The first subsystem's state equations remain unchanged, and by substitution the second set can be written as

$$\dot{\mathbf{x}}_j = \mathbf{A}_j\mathbf{x}_j + \mathbf{B}_j[\mathbf{C}_i\mathbf{x}_i + \mathbf{D}_i\mathbf{u}_i]$$

$$\mathbf{y}_j = \mathbf{C}_j\mathbf{x}_j + \mathbf{D}_j[\mathbf{C}_i\mathbf{x}_i + \mathbf{D}_i\mathbf{u}_i]$$

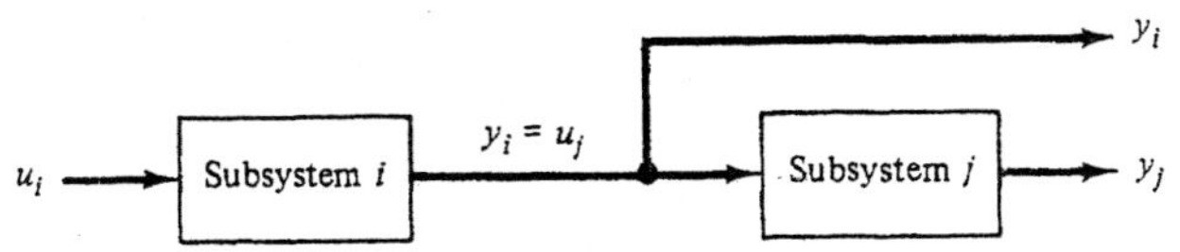

Figure 3.27

The composite system state vector and output vector are

$$\mathbf{x} = \begin{bmatrix} \mathbf{x}_i \\ \mathbf{x}_j \end{bmatrix}, \qquad \mathbf{y} = \begin{bmatrix} \mathbf{y}_i \\ \mathbf{y}_j \end{bmatrix}$$

and they satisfy

$$\dot{\mathbf{x}} = \begin{bmatrix} \mathbf{A}_i & \mathbf{0} \\ \mathbf{B}_j\mathbf{C}_i & \mathbf{A}_j \end{bmatrix}\mathbf{x} + \begin{bmatrix} \mathbf{B}_i \\ \mathbf{B}_j\mathbf{D}_i \end{bmatrix}\mathbf{u}_i$$

$$\mathbf{y} = \begin{bmatrix} \mathbf{C}_i & \mathbf{0} \\ \mathbf{D}_j\mathbf{C}_i & \mathbf{C}_j \end{bmatrix}\mathbf{x} + \begin{bmatrix} \mathbf{D}_i \\ \mathbf{D}_j\mathbf{D}_i \end{bmatrix}\mathbf{u}_i$$

Notice that the upper right partitions of the composite system matrices $\mathbf{A}$ and $\mathbf{C}$ are both zero. This is because there is no path from subsystem j into i. This substitution approach can be extended to various interconnection topologies. Consider the four-subsystem arrangement of Figure 3.28. The overall system has two groups of inputs and four groups of outputs:

$$\mathbf{u} = [\mathbf{u}_1^T \ \ \mathbf{u}_3^T]^T \quad \text{and} \quad \mathbf{y} = [\mathbf{y}_1^T \ \ \mathbf{y}_2^T \ \ \mathbf{y}_3^T \ \ \mathbf{y}_4^T]^T$$

The state vector $\mathbf{x}$ is also the "stacked-up" composite of $\mathbf{x}_1$, $\mathbf{x}_2$, $\mathbf{x}_3$, and $\mathbf{x}_4$. After tedious substitution and rearrangement, it is found that the composite system can be described by the state equations

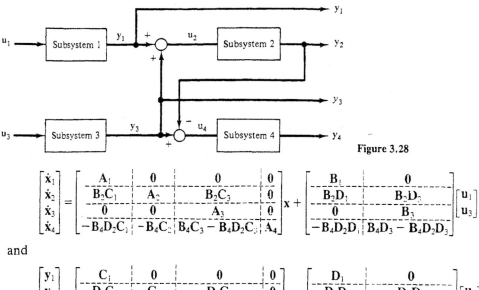

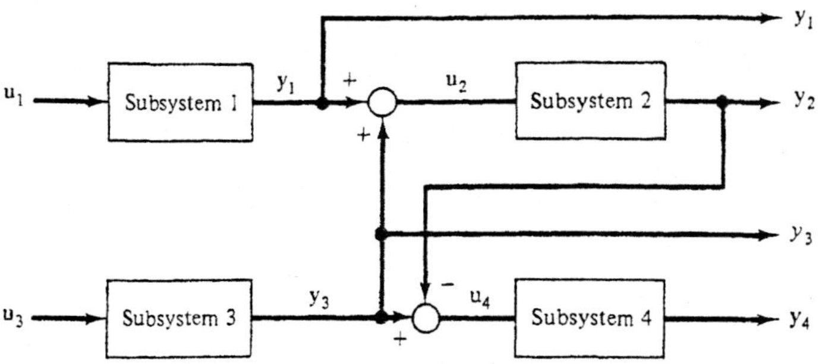

Figure 3.28

$$\begin{bmatrix} \dot{x}_1 \\ \dot{x}_2 \\ \dot{x}_3 \\ \dot{x}_4 \end{bmatrix} = \left[\begin{array}{c|c|c|c} A_1 & 0 & 0 & 0 \\ \hline B_2C_1 & A_2 & B_2C_3 & 0 \\ \hline 0 & 0 & A_3 & 0 \\ \hline -B_4D_2C_1 & -B_4C_2 & B_4C_3 - B_4D_2C_3 & A_4 \end{array}\right] x + \left[\begin{array}{c|c} B_1 & 0 \\ \hline B_2D_1 & B_2D_3 \\ \hline 0 & B_3 \\ \hline -B_4D_2D_1 & B_4D_3 - B_4D_2D_3 \end{array}\right] \begin{bmatrix} u_1 \\ u_3 \end{bmatrix}$$

and

$$\begin{bmatrix} y_1 \\ y_2 \\ y_3 \\ y_4 \end{bmatrix} = \left[\begin{array}{c|c|c|c} C_1 & 0 & 0 & 0 \\ \hline D_2C_1 & C_2 & D_2C_3 & 0 \\ \hline 0 & 0 & C_3 & 0 \\ \hline -D_4D_2C_1 & -D_4C_2 & D_4C_3 - D_4D_2C_3 & C_4 \end{array}\right] x + \left[\begin{array}{c|c} D_1 & 0 \\ \hline D_2D_1 & D_2D_3 \\ \hline 0 & D_3 \\ \hline -D_4D_2D_1 & D_4D_3 - D_4D_2D_3 \end{array}\right] \begin{bmatrix} u_1 \\ u_3 \end{bmatrix}$$

There is obviously a pattern in these matrices that derives from the subsystem interconnection topology, but this is not pursued here. Also note the absence of closed-feedback loops in both these examples. These and other more complicated interconnections are best left until after a more thorough presentation of matrix algebra, including inversion, is presented in the next chapter. Finally, the interconnection procedure just presented can yield a nonminimal state realization in some cases. One such case is when poles and zeros of cascaded blocks cancel. The presumption has been made here that each subsystem represents a real, physical system whose modes and signals are to be preserved in the final model. If doing this results in extra states, so be it. The implications of this shall be made clear in Chapter 11.

3.6 COMMENTS ON THE STATE SPACE REPRESENTATION

The selection of state variables is not a unique process. Various sets of state variables can be used. Some are easier to derive, whereas others are easier to work with once they are obtained. These comments all relate to *mathematics*. Some *physical* considerations also exist. It may be that the starting information about a system is derived from an experimentally obtained transfer function, perhaps by fitting straight-line approximations to a frequency response plot (Bode plot). It could be that an ARMA model has been constructed by fitting to historical data. The procedure you follow in a real situation depends upon what data you have at the start. There is often good reason to select states which have physical significance. These can then at least potentially be measured, perhaps by adding additional instrumentation.

The state equations consist of two generic parts. The differential or difference equation represents the so-called dynamics of the system, and the *algebraic* output equation is often referred to as the output or measurement equation. This can be misleading, since most real sensors have their own inherent dynamical responses. For example, a temperature sensor invariably has some time constant which prevents the instantaneous measurement of the temperature state. Whenever these kinds of instrument dynamics are significant, they must be included in the "dynamical" part of the state equations. That is, they will add states, just as in Section 3.5 when subsystems were cascaded together.

At the risk of oversimplification, it can be said that control theory started and flourished using transfer function methods. Then the state variable approach was developed, and for many years it was synonymous with *modern control.* Some of the advantages of the state variable approach are as follows:

1. It provides a convenient, compact notation and allows the application of the powerful vector-matrix theory, which is developed in the next few chapters of this book.

2. The uniform notation for all systems, regardless of order, makes possible a uniform set of solution techniques and computer algorithms. This is in sharp contrast with, for example, phase-plane methods, which give great insight into the behavior of second-order systems.

3. The state space representation is in an ideal format for computer solution, either analog or digital. In fact, the simulation diagrams used here to write the state equations are ideal starting points for system simulation. This is important because computers are invariably needed in the analysis of all but the most trivial systems.

4. The state space approach originally was able to define and explain more completely many system characteristics and attributes. Currently, most of the advantages and insights gained by use of state variable methods in the early years have been found to have counterparts in the new and expanded input-output transfer function methods of multivariable systems. The two approaches are now both being used. They require somewhat different mathematical tools, but they complement each other in various ways. Some of the newer aspects of transfer function methods appear in later chapters, but this book stresses state variable methods and the linear algebra and matrix theory upon which they depend.

REFERENCES

1. Kalman, R. E.: "Mathematical Description of Linear Dynamical Systems," *Jour. Soc. Ind. Appl. Math-Control Series,* Series A, Vol. 1, No. 2, 1963, pp. 152–192.
2. Desoer, C. A.: *Notes for a Second Course on Linear Systems,* Van Nostrand Reinhold, New York, 1970.
3. Coddington, E. A. and N. Levinson: *Theory of Ordinary Differential Equations,* McGraw-Hill, New York, 1955.
4. Chen, C. T.: *Introduction to Linear System Theory,* Holt, Rinehart and Winston, New York, 1970.

5. Martens, H. R. and D. R. Allen: *Introduction to Systems Theory,* Charles E. Merrill Publishing Co., Columbus, Oh., 1969.
6. Zadah, L. A. and C. A. Desoer: *Linear System Theory, The State Space Approach,* McGraw-Hill, New York, 1963.
7. De Russo, P. M., R. J. Roy, and C. M. Close: *State Variables for Engineers,* John Wiley, New York, 1965.

ILLUSTRATIVE PROBLEMS

Linear, Continuous-Time, Constant Coefficients

3.1 Four input-output transfer functions $y(s)/u(s)$ are given. Describe the systems they represent in state variable form:

(a) $1/(s+\alpha)$ (b) $(s+\beta)/(s+\alpha)$

(c) $(s+\beta)/(s^2+2\zeta\omega s+\omega^2)$ (d) $(s^2+2\zeta_1\omega_1 s+\omega_1^2)/(s^2+2\zeta_2\omega_2 s+\omega_2^2)$

The solutions are obtained by writing the input-output differential equation, drawing the simulation diagram, and selecting the integrator outputs as state variables.

(a) The differential equation is $\dot{y}+\alpha y=u$. This is simulated in Figure 3.29*a*, from which $\dot{x}=-\alpha x+u$ and $y=x$.

(b) The differential equation is $\dot{y}+\alpha y=\beta u+\dot{u}$. This is simulated in Figure 3.29*b*, from which $\dot{x}=-\alpha x+(\beta-\alpha)u$ and $y=x+u$.

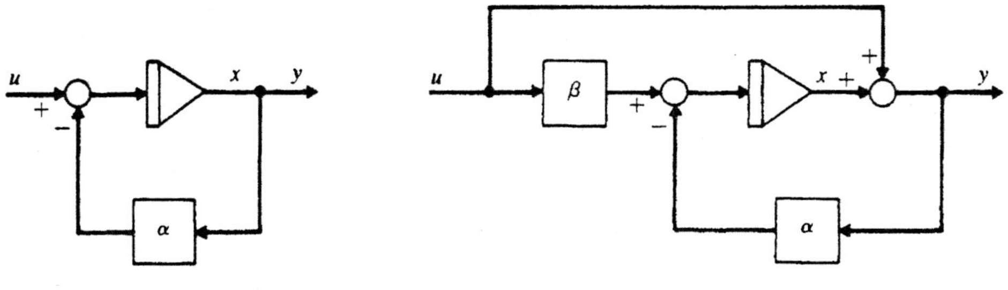

(*a*) (*b*)

Figure 3.29 (a) and (b)

(c) The differential equation is $\ddot{y}+2\zeta\omega\dot{y}+\omega^2 y=\beta u+\dot{u}$, and is simulated in Figure 3.29*c*. The state equations are $\dot{x}_1=-2\zeta\omega x_1+x_2+u$, $\dot{x}_2=-\omega^2 x_1+\beta u$, and $y=[1\quad 0]\mathbf{x}$.

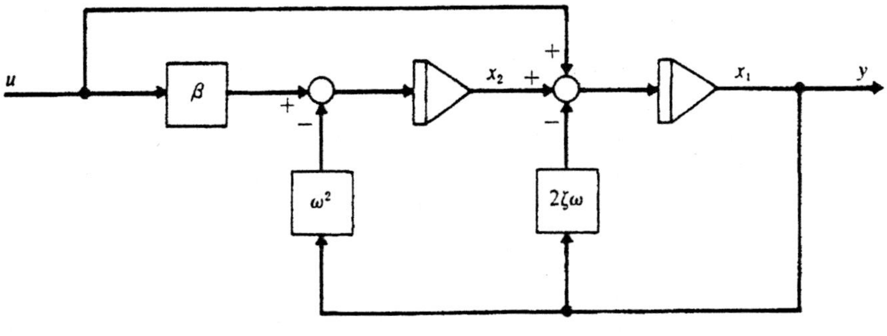

Figure 3.29 (c)

(d) The differential equation is $\ddot{y} + 2\zeta_2\omega_2\dot{y} + \omega_2^2 y = \ddot{u} + 2\zeta_1\omega_1\dot{u} + \omega_1^2 u$. Integrating twice allows this to be written as

$$y = u + \int \left\{ 2\zeta_1\omega_1 u - 2\zeta_2\omega_2 y + \int [\omega_1^2 u - \omega_2^2 y]\, dt \right\} dt'$$

The simulation diagram of Figure 3.29d gives $\dot{x}_1 = -2\zeta_2\omega_2 x_1 + x_2 + (2\zeta_1\omega_1 - 2\zeta_2\omega_2)u$, $\dot{x}_2 = -\omega_2^2 x_1 + (\omega_1^2 - \omega_2^2)u$, and $y = [1 \quad 0]\mathbf{x} + u$.

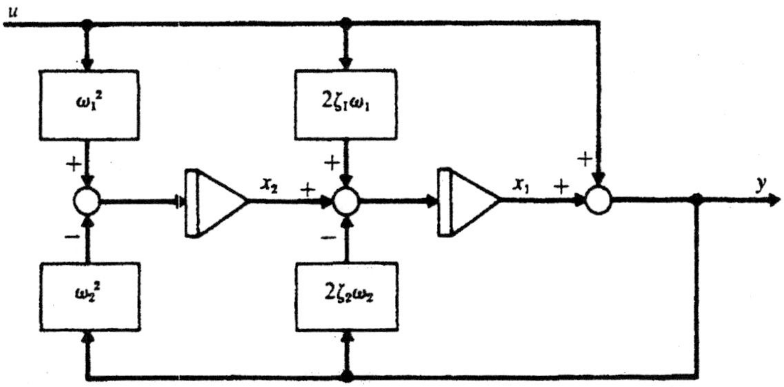

Figure 3.29 (d)

3.2 A system with two inputs and two outputs is described by

$$\ddot{y}_1 + 3\dot{y}_1 + 2y_2 = u_1 + 2u_2 + 2\dot{u}_2 \quad \text{and} \quad \ddot{y}_2 + 4\dot{y}_1 + 3y_2 = \ddot{u}_2 + 3\dot{u}_2 + u_1$$

Select a set of state variables and find the state space equations for this system.

Integrating each equation twice gives

$$y_1 = \iint \{-3\dot{y}_1 - 2y_2 + u_1 + 2\dot{u}_2 + 2u_2\}\, dt\, dt'$$

$$y_2 = \iint \{-4\dot{y}_1 - 3y_2 + \ddot{u}_2 + 3\dot{u}_2 + u_1\}\, dt\, dt'$$

or

$$y_1 = \int \left\{ -3y_1 + 2u_2 + \int [-2y_2 + u_1 + 2u_2]\, dt \right\} dt'$$

$$y_2 = u_2 + \int \left\{ -4y_1 + 3u_2 + \int [-3y_2 + u_1]\, dt \right\} dt'$$

The simulation diagram of Figure 3.30 can now be drawn.

The state equations are

$$\left.\begin{aligned} \dot{x}_1 &= -3x_1 + x_2 + 2u_2 \\ \dot{x}_2 &= -2x_3 + u_1 \\ \dot{x}_3 &= -4x_1 + x_4 + 3u_2 \\ \dot{x}_4 &= -3x_3 + u_1 - 3u_2 \end{aligned}\right\} \quad \text{or} \quad \dot{\mathbf{x}} = \begin{bmatrix} -3 & 1 & 0 & 0 \\ 0 & 0 & -2 & 0 \\ -4 & 0 & 0 & 1 \\ 0 & 0 & -3 & 0 \end{bmatrix}\mathbf{x} + \begin{bmatrix} 0 & 2 \\ 1 & 0 \\ 0 & 3 \\ 1 & -3 \end{bmatrix}\mathbf{u}$$

and

$$\mathbf{y} = \begin{bmatrix} y_1 \\ y_2 \end{bmatrix} = \begin{bmatrix} 1 & 0 & 0 & 0 \\ 0 & 0 & 1 & 0 \end{bmatrix}\mathbf{x} + \begin{bmatrix} 0 & 0 \\ 0 & 1 \end{bmatrix}\mathbf{u}.$$

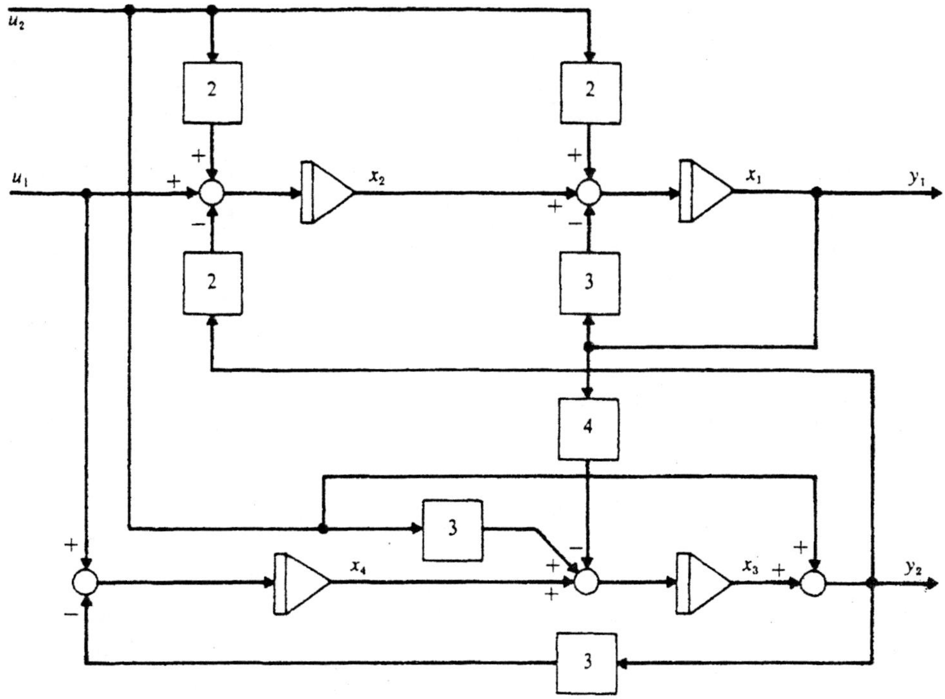

Figure 3.30

3.3 A single-input, single-out system has the transfer function

$$\frac{y(s)}{u(s)} = \frac{1}{s^3 + 10s^2 + 27s + 18} = T(s)$$

Find three different state variable representations.

(a) The transfer function represents the differential equation $\dddot{y} + 10\ddot{y} + 27\dot{y} + 18y = u$. Setting $x_1 = y$, $x_2 = \dot{y}$, $x_3 = \ddot{y}$ gives

$$\begin{bmatrix} \dot{x}_1 \\ \dot{x}_2 \\ \dot{x}_3 \end{bmatrix} = \begin{bmatrix} 0 & 1 & 0 \\ 0 & 0 & 1 \\ -18 & -27 & -10 \end{bmatrix} \begin{bmatrix} x_1 \\ x_2 \\ x_3 \end{bmatrix} + \begin{bmatrix} 0 \\ 0 \\ 1 \end{bmatrix} u \quad \text{and} \quad y = [1 \quad 0 \quad 0]\mathbf{x}$$

(b) In factored form $T(s) = 1/[(s + 6)(s + 1)(s + 3)]$, and a simulation diagram is given in Figure 3.31.

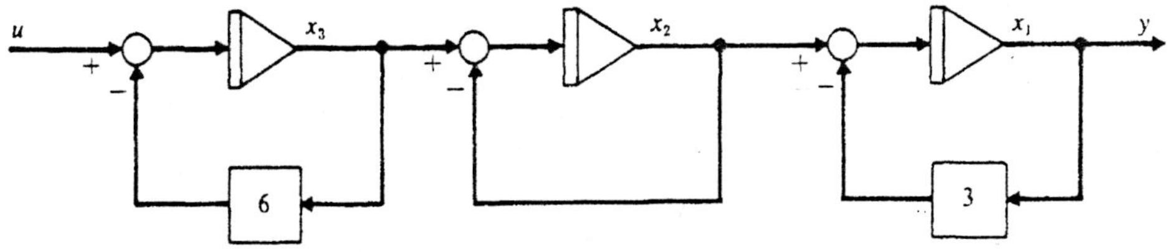

Figure 3.31

Then

$$\begin{bmatrix}\dot{x}_1\\ \dot{x}_2\\ \dot{x}_3\end{bmatrix} = \begin{bmatrix}-3 & 1 & 0\\ 0 & -1 & 1\\ 0 & 0 & -6\end{bmatrix}\begin{bmatrix}x_1\\ x_2\\ x_3\end{bmatrix} + \begin{bmatrix}0\\ 0\\ 1\end{bmatrix}u \quad \text{and} \quad y = [1 \quad 0 \quad 0]\mathbf{x}$$

(c) Using partial fractions, $T(s) = a/(s+6) + b/(s+1) + c/(s+3)$, where

$$a = (s+6)T(s)|_{s=-6} = \tfrac{1}{15} \qquad c = (s+3)T(s)|_{s=-3} = -\tfrac{1}{6}$$

$$b = (s+1)T(s)|_{s=-1} = \tfrac{1}{10}$$

so the simulation diagram of Figure 3.32 is obtained.

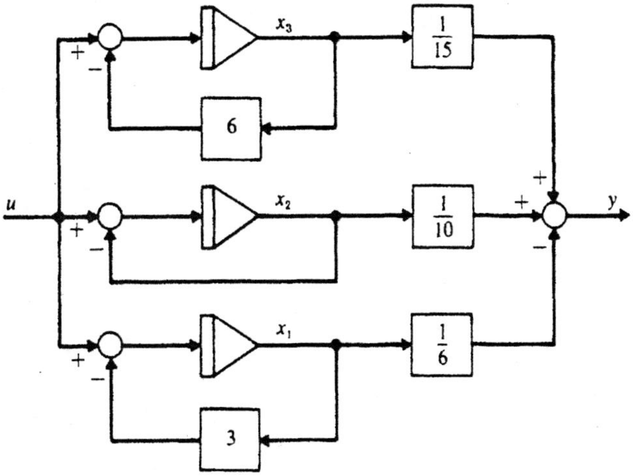

Figure 3.32

The state equations are

$$\dot{\mathbf{x}} = \begin{bmatrix}-3 & 0 & 0\\ 0 & -1 & 0\\ 0 & 0 & -6\end{bmatrix}\mathbf{x} + \begin{bmatrix}1\\ 1\\ 1\end{bmatrix}u$$

$$y = [-\tfrac{1}{6} \quad \tfrac{1}{10} \quad \tfrac{1}{15}]\mathbf{x}$$

Note that since Figure 3.33*a* and *b* are equivalent, an alternative form for the matrices **B** and **C** is $\mathbf{B} = [-\tfrac{1}{6} \quad \tfrac{1}{10} \quad \tfrac{1}{15}]^T$ and $\mathbf{C} = [1 \quad 1 \quad 1]$.

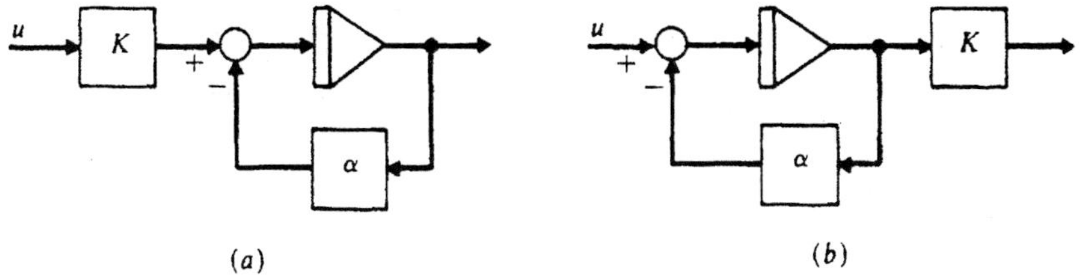

Figure 3.33

3.4 A system input-output transfer function is $T(s) = 1/[s^2(s+3)^3(s+1)]$. Find a state variable representation, using the partial fraction expansion of $T(s)$.

The expansion is

$$T(s) = a_1/s^2 + a_2/s + a_3/(s+3)^3 + a_4/(s+3)^2 + a_5/(s+3) + a_6/(s+1)$$

where

$$a_1 = s^2 T(s)\big|_{s=0} = \frac{1}{27} \qquad a_4 = \frac{d}{ds}\{(s+3)^3 T(s)\}\big|_{s=-3} = -\frac{7}{108}$$

$$a_2 = \frac{d}{ds}\{s^2 T(s)\}\big|_{s=0} = -\frac{2}{27} \qquad a_5 = \frac{1}{2!}\frac{d^2}{ds^2}\{(s+3)^3 T(s)\}\big|_{s=-3} = -\frac{11}{216}$$

$$a_3 = (s+3)^3 T(s)\big|_{s=-3} = -\frac{1}{18} \qquad a_6 = (s+1)T(s)\big|_{s=-1} = \frac{1}{8}$$

Using this expansion gives the simulation diagram of Figure 3.34.

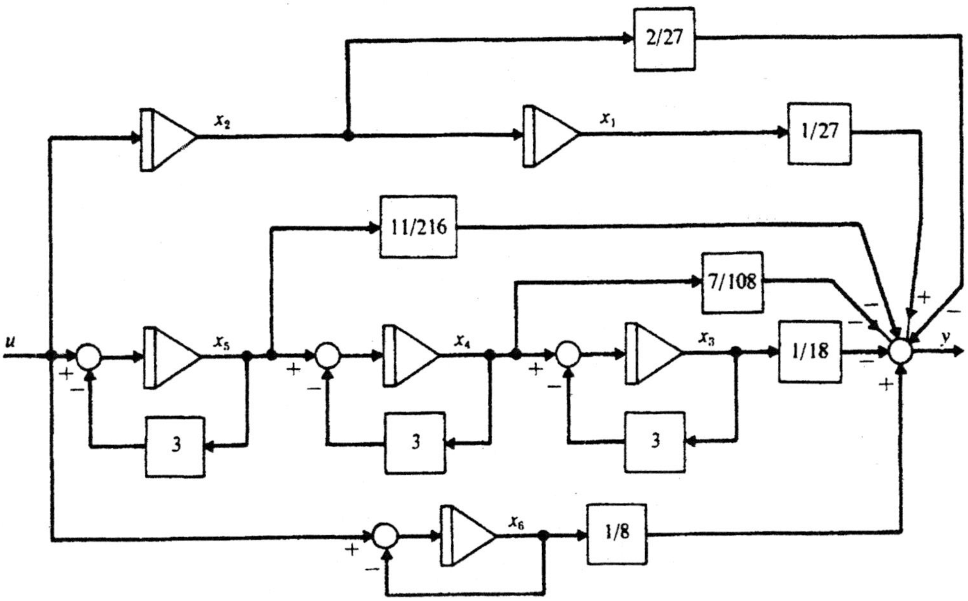

Figure 3.34

From the diagram,

$$\dot{\mathbf{x}} = \begin{bmatrix} 0 & 1 & 0 & 0 & 0 & 0 \\ 0 & 0 & 0 & 0 & 0 & 0 \\ 0 & 0 & -3 & 1 & 0 & 0 \\ 0 & 0 & 0 & -3 & 1 & 0 \\ 0 & 0 & 0 & 0 & -3 & 0 \\ 0 & 0 & 0 & 0 & 0 & -1 \end{bmatrix}\mathbf{x} + \begin{bmatrix} 0 \\ 1 \\ 0 \\ 0 \\ 1 \\ 1 \end{bmatrix} u \qquad \text{(note } \mathbf{A} \text{ is in Jordan form)}$$

$$y = \begin{bmatrix} \frac{1}{27} & -\frac{2}{27} & -\frac{1}{18} & -\frac{7}{108} & -\frac{11}{216} & \frac{1}{8} \end{bmatrix}\mathbf{x}$$

3.5 Find the state space representation for a system described by $T(s) = (s+1)/(s^2+7s+6)$.

Even though $T(s) = (s+1)/[(s+1)(s+6)]$, the common factor should not be canceled, or the system will be mistaken for a first-order system. Rather, use $\ddot{y} + 7\dot{y} + 6y = \dot{u} + u$, from which $\begin{bmatrix} \dot{x}_1 \\ \dot{x}_2 \end{bmatrix} = \begin{bmatrix} -7 & 1 \\ -6 & 0 \end{bmatrix} \begin{bmatrix} x_1 \\ x_2 \end{bmatrix} + \begin{bmatrix} 1 \\ 1 \end{bmatrix} u$ and $y = [1 \quad 0]\mathbf{x}$.

Linear Discrete-Time State Equations

3.6 A system has three inputs, $u_1(k)$, $u_2(k)$, and $u_3(k)$, and three outputs, $y_1(k)$, $y_2(k)$, and $y_3(k)$. The input-output difference equations are

$$y_1(k+3) + 6[y_1(k+2) - y_3(k+2)] + 2y_1(k+1) + y_2(k+1) + y_1(k) - 2y_3(k) = u_1(k) + u_2(k+1) \quad (1)$$

$$y_2(k+2) + 3y_2(k+1) - y_1(k+1) + 5y_2(k) + y_3(k) = u_1(k) + u_2(k) + u_3(k) \quad (2)$$

$$y_3(k+1) + 2y_3(k) - y_2(k) = u_3(k) - u_2(k) + 7u_3(k+1) \quad (3)$$

Draw a simulation diagram, select state variables, and write the matrix state equations.

Delaying each term in equation (*1*) three times gives

$$y_1(k) = \mathcal{D}\{-6[y_1(k) - y_3(k)] + \mathcal{D}[u_2(k) - 2y_1(k) - y_2(k) + \mathcal{D}[u_1(k) - y_1(k) + 2y_3(k)]]\}$$

Delaying each term in equation (*2*) twice gives

$$y_2(k) = \mathcal{D}\{-3y_2(k) + y_1(k) + \mathcal{D}[u_1(k) + u_2(k) + u_3(k) - 5y_2(k) - y_3(k)]\}$$

and from equation (*3*), delayed once,

$$y_3(k) = 7u_3(k) + \mathcal{D}\{u_3(k) - u_2(k) - 2y_3(k) + y_2(k)\}$$

The simulation diagram can be represented as shown in Figure 3.35.

Labeling x_1 through x_6 as shown in Figure 3.35 gives

$$\mathbf{x}(k+1) = \begin{bmatrix} -6 & 1 & 0 & 0 & 0 & 6 \\ -2 & 0 & 1 & -1 & 0 & 0 \\ -1 & 0 & 0 & 0 & 0 & 2 \\ 1 & 0 & 0 & -3 & 1 & 0 \\ 0 & 0 & 0 & -5 & 0 & -1 \\ 0 & 0 & 0 & 1 & 0 & -2 \end{bmatrix} \mathbf{x}(k) + \begin{bmatrix} 0 & 0 & 42 \\ 0 & 1 & 0 \\ 1 & 0 & 14 \\ 0 & 0 & 0 \\ 1 & 1 & -6 \\ 0 & -1 & -13 \end{bmatrix} \begin{bmatrix} u_1(k) \\ u_2(k) \\ u_3(k) \end{bmatrix}$$

and

$$\mathbf{y}(k) = \begin{bmatrix} 1 & 0 & 0 & 0 & 0 & 0 \\ 0 & 0 & 0 & 1 & 0 & 0 \\ 0 & 0 & 0 & 0 & 0 & 1 \end{bmatrix} \mathbf{x}(k) + \begin{bmatrix} 0 & 0 & 0 \\ 0 & 0 & 0 \\ 0 & 0 & 7 \end{bmatrix} \begin{bmatrix} u_1(k) \\ u_2(k) \\ u_3(k) \end{bmatrix}$$

3.7 A system is described by the input-output equation

$$y(k+3) + 2y(k+2) + 4y(k+1) + y(k) = u(k)$$

This case is analogous to the simplest continuous-time problem where the input is not differentiated. Consequently, state variables can be selected as the output $y(k)$ and the output advanced by one and by two time steps. That is,

$$x_1(k) = y(k), \qquad x_2(k) = y(k+1), \qquad x_3(k) = y(k+2)$$

Then

$$x_1(k+1) = x_2(k), \qquad x_2(k+1) = x_3(k)$$

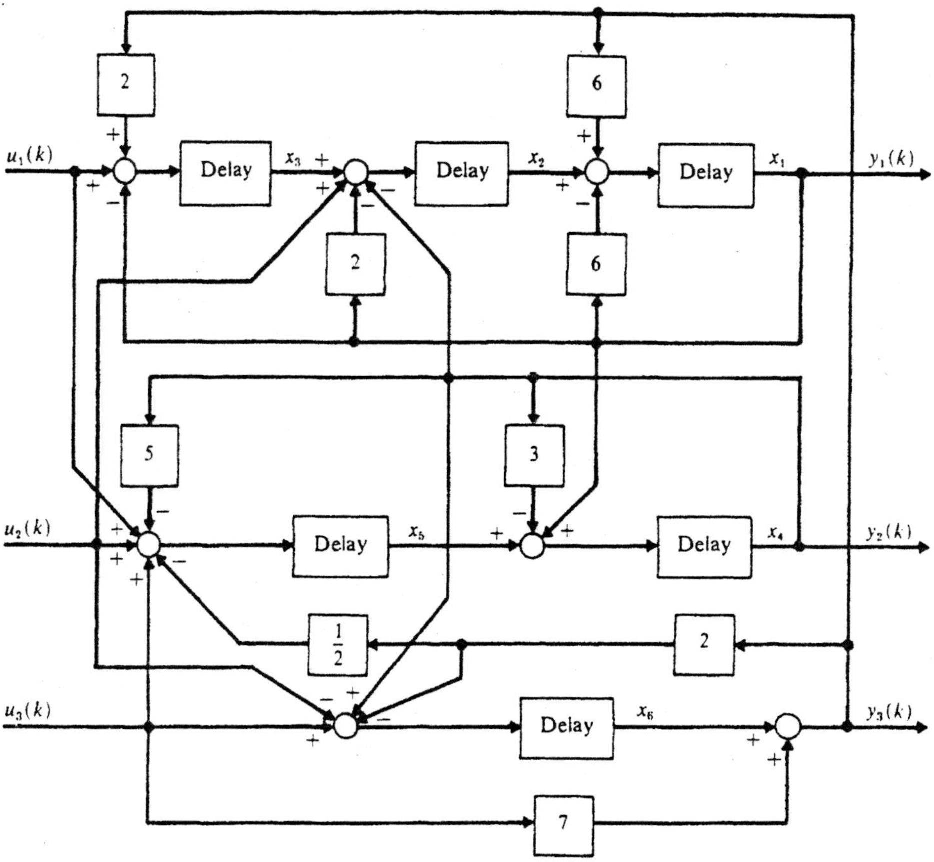

Figure 3.35

The final component of the state vector equation comes from the original difference equation and is

$$x_3(k+1) = -2x_3(k) - 4x_2(k) - x_1(k) + u(k)$$

Although its use is unnecessary in this simple problem, a possible simulation diagram is shown in Figure 3.36.

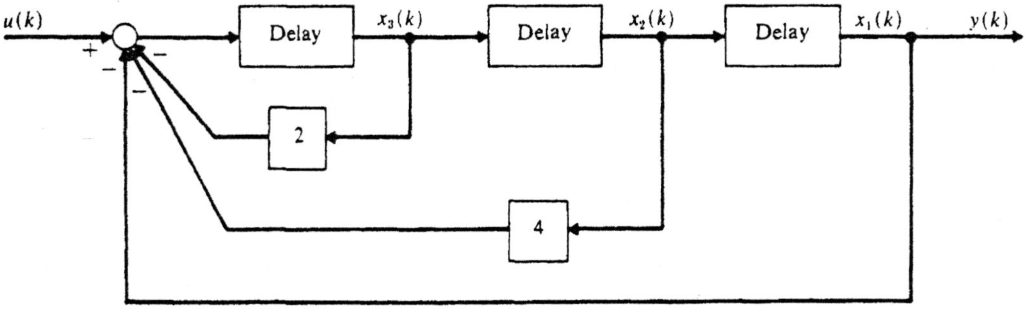

Figure 3.36

The state equations for this example are

$$\begin{bmatrix} x_1(k+1) \\ x_2(k+1) \\ x_3(k+1) \end{bmatrix} = \begin{bmatrix} 0 & 1 & 0 \\ 0 & 0 & 1 \\ -1 & -4 & -2 \end{bmatrix} \begin{bmatrix} x_1(k) \\ x_2(k) \\ x_3(k) \end{bmatrix} + \begin{bmatrix} 0 \\ 0 \\ 1 \end{bmatrix} u(k)$$

$$y(k) = [1 \quad 0 \quad 0]\mathbf{x}(k) + 0u(k)$$

Time-Varying Coefficients

3.8 Obtain a state variable representation for the linear system with time-varying coefficients $\ddot{y} + e^{-t^2}\dot{y} + e^t y = u$.

Let $x_1 = y$, $x_2 = \dot{y}$, then $\dot{x}_2 = \ddot{y}$, so

$$\begin{bmatrix} \dot{x}_1 \\ \dot{x}_2 \end{bmatrix} = \begin{bmatrix} 0 & 1 \\ -e^t & -e^{-t^2} \end{bmatrix} \begin{bmatrix} x_1 \\ x_2 \end{bmatrix} + \begin{bmatrix} 0 \\ 1 \end{bmatrix} u \quad \text{and} \quad y = [1 \quad 0]\mathbf{x}$$

3.9 Apply the integration method of Section 3.4.3 to find a state variable model for

$$\dot{y} + a(t)y = b_0(t)u + b_1\dot{u}$$

Solving for $\dot{y}$ and integrating once gives

$$y(t) = \int b_1 \dot{u}\, dt + \int [b_0 u - ay]\, dt$$

Using formal integration by parts on the first integral gives $\int b_1 \dot{u}\, dt = b_1 u(t) - \int \dot{b}_1 u\, dt$, so that $y(t) = b_1 u + \int [b_0 u - ay - \dot{b}_1 u]\, dt$. The simulation diagram of Figure 3.37 is drawn from this.

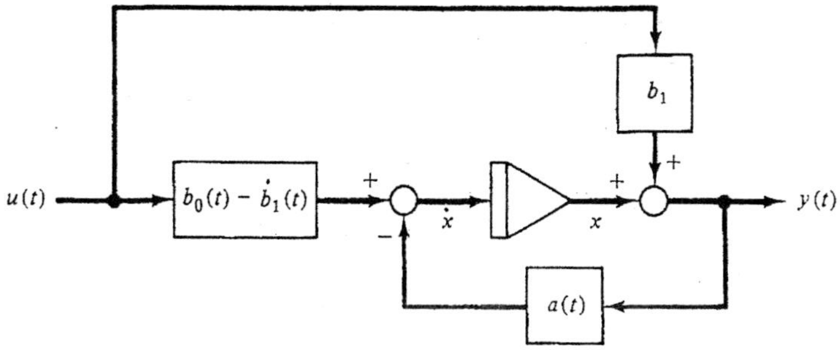

Figure 3.37

Using the integrator output as $x = y - b_1 u$ gives the state equations

$$\dot{x}(t) = -a(t)x(t) + [b_0(t) - \dot{b}_1(t) - b_1(t)a(t)]u(t)$$

$$y(t) = x(t) + b_1(t)u(t)$$

3.10 Find a state variable model for the second-order time-varying system

$$\ddot{y} + a_1(t)\dot{y} + a_0(t)y = b_0(t)u + b_1(t)\dot{u} + b_2(t)\ddot{u}$$

Solving for $\ddot{y}$ and integrating twice gives

$$y(t) = \iint \{b_0 u - a_0 y\}\, dt\, dt' + \int \left\{ \int [b_1 \dot{u} - a_1 \dot{y}]\, dt \right\} dt' + \iint b_2 \ddot{u}\, dt\, dt'$$

The first integrand on the right contains no derivatives and therefore is in final form. The second term on the right contains first derivatives of y and u, so integration by parts is needed to eliminate them. The innermost integral becomes

$$\int [b_1 \dot{u} - a_1 \dot{y}]\, dt = b_1 u - a_1 y - \int [\dot{b}_1 u - \dot{a}_1 y]\, dt$$

Since the third term in the expression for $y(t)$ contains $\ddot{u}$, integration by parts must be used twice.

$$\iint b_2 \ddot{u}\, dt\, dt' = \int \left\{ b_2 \dot{u} - \int \dot{b}_2 \dot{u}\, dt \right\} dt'$$

$$= b_2 u - \int \dot{b}_2 u\, dt - \int \left[\dot{b}_2 u - \int \ddot{b}_2 u\, dt \right] dt'$$

Recombining all terms into one nested integrator equation gives

$$y(t) = b_2 u + \int \left\{ b_1 u - a_1 y - 2\dot{b}_2 u + \int [b_0 u - a_0 y - \dot{b}_1 u + \dot{a}_1 y + \ddot{b}_2 u]\, dt \right\} dt'$$

The simulation diagram in Figure 3.38 is drawn from this. The state equations are

$$\dot{\mathbf{x}} = \begin{bmatrix} -a_1(t) & 1 \\ -[a_0(t) - \dot{a}_1(t)] & 0 \end{bmatrix} \mathbf{x} + \begin{bmatrix} b_1 - 2\dot{b}_2 - a_1 b_2 \\ b_0 - \dot{b}_1 + \ddot{b}_2 - b_2(a_0 - \dot{a}_1) \end{bmatrix} u(t)$$

$$y(t) = [1 \quad 0]\mathbf{x}(t) + b_2(t) u(t)$$

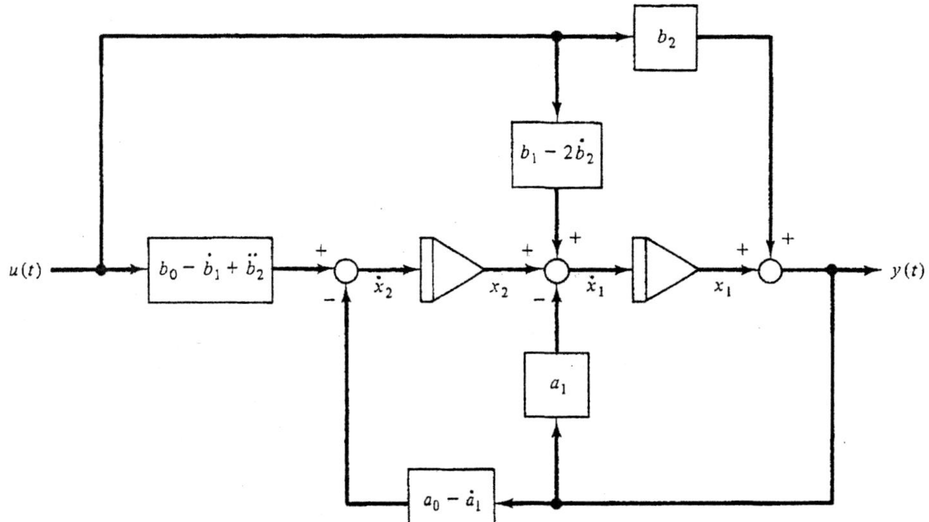

Figure 3.38

Linear Graph Method

3.11 Write the differential equations for the circuit of Figure 3.39 in state variable form. Consider the voltage across R_3 as the output.

The tree selected is shown in Figure 3.40.

The state variables are selected as the capacitor voltage x_1 and the inductor current x_2, so that $\dot{x}_1 = (1/C)i_C$, $\dot{x}_2 = (1/L)(v_1 - v_2)$. To express i_C, v_1, and v_2 in terms of x_1, x_2 and inputs, use

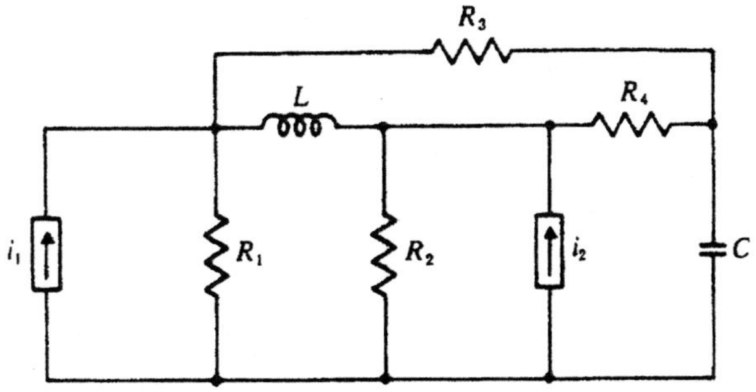

Figure 3.39

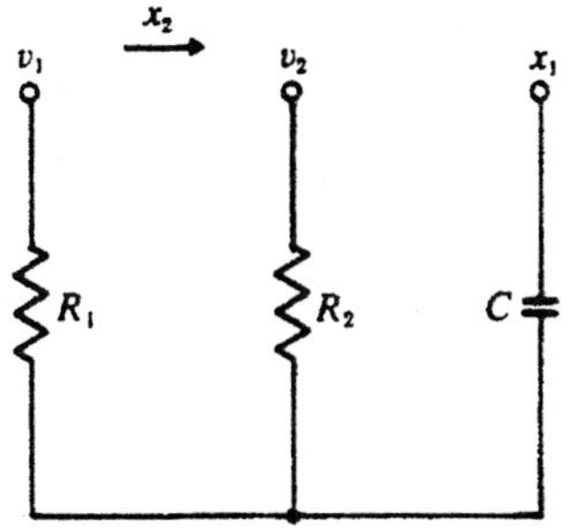

Figure 3.40

$i_C = i_{R_4} + i_{R_3}$ (cutset equation for tree branch C)

$i_{R_1} = i_1 - i_{R_3} - x_2$ (cutset equation for tree branch R_1)

$i_{R_2} = x_2 + i_2 - i_{R_4}$ (cutset equation for tree branch R_2)

$R_3 i_{R_3} = v_1 - x_1$ (fundamental loop equation using link R_3)

$R_4 i_{R_4} = v_2 - x_1$ (fundamental loop equation using link R_4)

$v_1 = R_1 i_{R_1}$

$v_2 = R_2 i_{R_2}$

These are seven equations with seven unknowns. Solving for all the resistor currents gives

$$i_{R_1} = \frac{R_3}{R_1+R_3} i_1 + \frac{x_1}{R_1+R_3} - \frac{R_3 x_2}{R_1+R_3} \qquad i_{R_3} = -\frac{x_1}{R_1+R_3} - \frac{R_1 x_2}{R_1+R_3} + \frac{R_1 i_1}{R_1+R_3}$$

$$i_{R_2} = \frac{x_1}{R_2+R_4} + \frac{R_4 x_2}{R_2+R_4} + \frac{R_4 i_2}{R_2+R_4} \qquad i_{R_4} = -\frac{x_1}{R_2+R_4} + \frac{R_2 x_2}{R_2+R_4} + \frac{R_2 i_2}{R_2+R_4}$$

Using these to determine i_C, v_1, and v_2 gives

$$\begin{bmatrix} \dot{x}_1 \\ \dot{x}_2 \end{bmatrix} = \begin{bmatrix} \dfrac{-(R_1+R_2+R_3+R_4)}{C(R_1+R_3)(R_2+R_4)} & \dfrac{R_2R_3 - R_1R_4}{C(R_2+R_4)(R_1+R_3)} \\ \dfrac{(R_1R_4 - R_2R_3)}{L(R_1+R_3)(R_2+R_4)} & -\dfrac{1}{L}\left[\dfrac{R_1R_3}{R_1+R_3} + \dfrac{R_2R_4}{R_2+R_4}\right] \end{bmatrix} \begin{bmatrix} x_1 \\ x_2 \end{bmatrix}$$

$$+ \begin{bmatrix} \dfrac{R_1}{C(R_1+R_3)} & \dfrac{R_2}{C(R_2+R_4)} \\ \dfrac{R_1R_3}{L(R_1+R_3)} & \dfrac{-R_2R_4}{L(R_2+R_4)} \end{bmatrix} \begin{bmatrix} u_1 \\ u_2 \end{bmatrix}$$

where $u_1 = i_1$, $u_2 = i_2$ are the inputs. The output is

$$y = R_3 i_{R_3} = \begin{bmatrix} \dfrac{-R_3}{R_1+R_3} & \dfrac{-R_1R_3}{R_1+R_2} \end{bmatrix} \mathbf{x} + \begin{bmatrix} \dfrac{R_1R_3}{R_1+R_3} & 0 \end{bmatrix} \mathbf{u}$$

3.12 Describe the hydraulic system of Problem 1.4, page 19, in state variable form.

The linear graph is redrawn as Fig. 3.41 using the analogous RLC symbols for the elements.

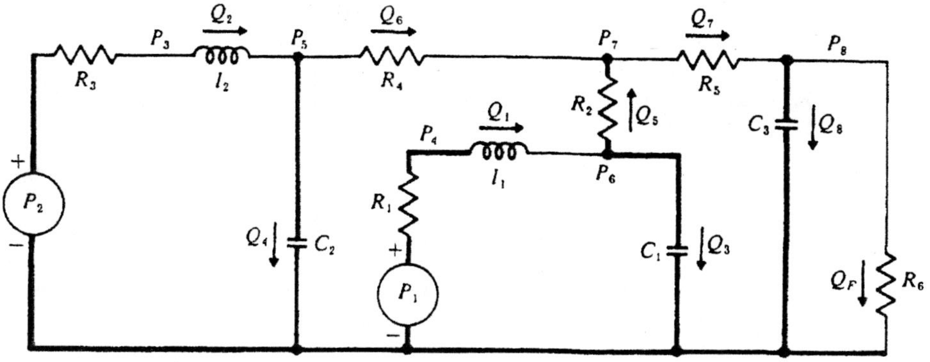

Figure 3.41

The tree is shown in heavy lines. The state variables are chosen as the pressures $x_1 = P_6$, $x_2 = P_5$, $x_3 = P_8$ and the flow rates $x_4 = Q_2$, $x_5 = Q_1$. Then $\dot{x}_1 = (1/C_1)\,Q_3$, $\dot{x}_2 = (1/C_2)\,Q_4$, $\dot{x}_3 = (1/C_3)\,Q_8$, $\dot{x}_4 = (1/I_2)\,P_{35}$, $\dot{x}_5 = (1/I_1)\,P_{46}$.

The last two are the easiest to complete and this is done first: $P_{35} = P_2 - R_3 x_4 - x_2$, $P_{46} = P_1 - R_1 x_5 - x_1$.

In order to express Q_3, Q_4, and Q_8 in terms of the state variables, the following simultaneous equations must be solved:

$$\left[\begin{array}{ccc|ccc} 1 & 0 & 0 & 1 & 0 & 0 \\ 0 & 1 & 0 & 0 & 1 & 0 \\ 0 & 0 & 1 & 0 & 0 & -1 \\ \hline 0 & 0 & 0 & 1 & 1 & -1 \\ 0 & 0 & 0 & 0 & R_4 & R_5 \\ 0 & 0 & 0 & -R_2 & R_4 & 0 \end{array}\right]\left[\begin{array}{c} Q_3 \\ Q_4 \\ Q_8 \\ \hline Q_5 \\ Q_6 \\ Q_7 \end{array}\right] = \left[\begin{array}{c} x_5 \\ x_4 \\ -x_3/R_6 \\ \hline 0 \\ x_2 - x_3 \\ x_2 - x_1 \end{array}\right]$$

The required solutions are

$$Q_3 = x_5 + \frac{1}{\Delta}[-R_4(x_2 - x_3) + (R_4 + R_5)(x_2 - x_1)]$$

$$Q_4 = x_4 + \frac{1}{\Delta}[-R_2(x_2 - x_3) - R_5(x_2 - x_1)]$$

$$Q_8 = -\frac{x_3}{R_6} + \frac{1}{\Delta}[(R_4 + R_2)(x_2 - x_3) - R_4(x_2 - x_1)]$$

where $\Delta = R_2R_4 + R_4R_5 + R_2R_5$. Using these relations allows the system equations to be put in the form $\dot{\mathbf{x}} = \mathbf{Ax} + \mathbf{Bu}$, where $\mathbf{u} = [P_1 \quad P_2]^T$. The output is $Q_F = y$ and is given by $y = [0 \quad 0 \quad 1/R_6 \quad 0 \quad 0]\mathbf{x}$.

Nonlinear State Models

3.13 A schematic of a motor-generator system driving an inertia load J, with viscous damping b, at an angular velocity Ω is shown in Fig. 3.42. Derive a state space model of this system.

The pertinent equations are

1. $L_f \dfrac{di_f}{dt} + R_f i_f = e_f$
2. $e_g = f(i_f)$
3. $e_g - e_m = (R_g + R_m)i_m + (L_g + L_m)\dfrac{di_m}{dt}$
4. $T = K_m i_m$
5. $T = J\dot{\Omega} + b\Omega$
6. $e_m = K_m\Omega$

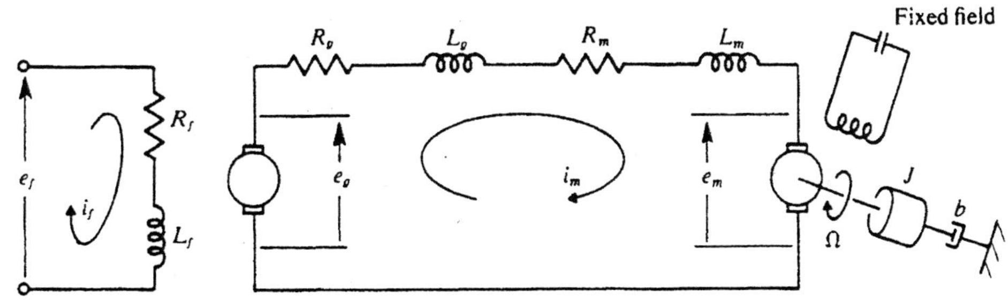

Figure 3.42

The simulation diagram of Fig. 3.43 can be constructed directly from these equations.

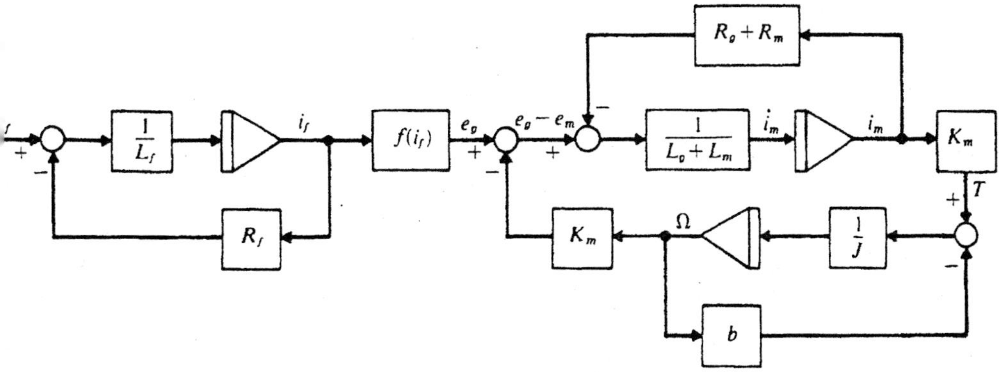

Figure 3.43

Selecting integrator outputs as state variables, $x_1 = \Omega$ (indicates kinetic energy in load), $x_2 = i_m$ (indicates magnetic energy in motor inductance), and $x_3 = i_f$ (indicates magnetic energy in the generator field), and letting the input be $e_f = u(t)$ gives

$$\begin{bmatrix} \dot{x}_1 \\ \dot{x}_2 \\ \dot{x}_3 \end{bmatrix} = \begin{bmatrix} -\frac{b}{J}x_1 + \frac{K_m}{J}x_2 \\ -\frac{K_m}{L_g + L_m}x_1 - \frac{R_g + R_m}{L_g + L_m}x_2 + \frac{f(x_3)}{L_g + L_m} \\ -\frac{R_f}{L_f}x_3 + \frac{u}{L_f} \end{bmatrix}$$

If it can be assumed that the generator characteristics are linear, i.e., if $e_g = K_g i_f$, then the above equations can be written in the standard linear form $\dot{\mathbf{x}} = \mathbf{A}\mathbf{x} + \mathbf{B}u$. If the speed Ω is considered to be the output y, then $y = [1 \quad 0 \quad 0]\mathbf{x}$.

3.14 Develop a state space model of a rocket vehicle (Figure 3.44) moving vertically above the earth. The vehicle thrust is $T = K\dot{m}$, where $\dot{m}$ is the rate of mass expulsion and can be controlled. Assume a drag force is given as a nonlinear function of velocity.

Letting the instantaneous mass of the vehicle be $m(t)$ and letting $D = f(\dot{h})$ be the drag, Newton's second law gives the dynamic force balance $m\ddot{h} = T(t) - f(\dot{h}) - [m(t)k^2 g_0/(k + h)^2]$, where an inverse square gravity law has been assumed. A simulation diagram is given in Figure 3.45.

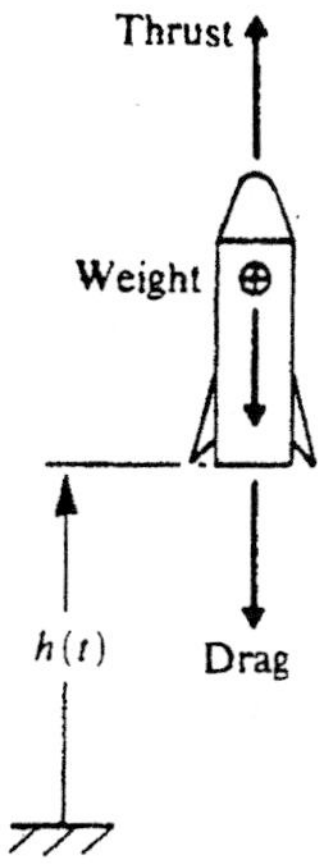

Figure 3.44

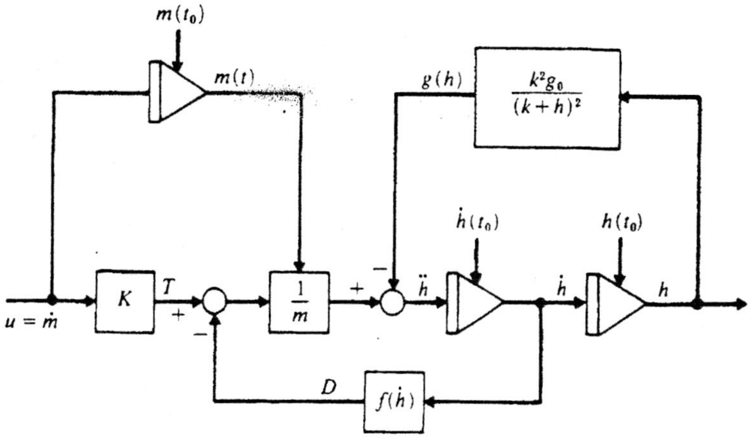

Figure 3.45

This is a nonlinear, time-variable system, but as before, outputs of the integrators are selected as state variables, $x_1 = h$, $x_2 = \dot{h}$, $x_3 = m$. Then

$$\begin{bmatrix} \dot{x}_1 \\ \dot{x}_2 \\ \dot{x}_3 \end{bmatrix} = \begin{bmatrix} x_2 \\ \dfrac{Ku - f(x_2)}{x_3} - g(x_1) \\ u \end{bmatrix}$$

In this case it is not possible to obtain the linear form $\dot{\mathbf{x}} = \mathbf{Ax} + \mathbf{Bu}$, but the preceding result is of the more general form $\dot{\mathbf{x}} = \mathbf{f}(\mathbf{x}, \mathbf{u}, t)$.

3.15 Derive the equations of motion for a satellite rotating in free space under the influence of gas jets mounted along three mutually orthogonal body-fixed axes.

Let $\omega = [\omega_x \quad \omega_y \quad \omega_z]^T$ be the three components of angular velocity expressed with respect to the body-fixed axes. Let $\mathbf{T} = [T_x \quad T_y \quad T_z]^T$ be the three components of input control torques. Newton's second law, as applied to a rotating body, states that $d\mathbf{H}/dt = \mathbf{T}$, where $\mathbf{H}$ is the angular momentum vector, and the time rate of change d/dt is with respect to a fixed inertial reference. The vector $\mathbf{H}$ can be expressed in body coordinates as $\mathbf{H} = [J_x\omega_x \quad J_y\omega_y \quad J_z\omega_z]^T$, where the constants J_i are moments of inertia of the body and x, y, z are assumed to be principal axes of inertia. The inertial rate of change $d\mathbf{H}/dt$ is related to the apparent rate $[\dot{\mathbf{H}}]$ as seen by an observer moving with the body by $d\mathbf{H}/dt = [\dot{\mathbf{H}}] + \boldsymbol{\omega} \times \mathbf{H}$. Therefore,

$$T_x = J_x\dot{\omega}_x + (J_z - J_y)\omega_y\omega_z, \qquad T_y = J_y\dot{\omega}_y + (J_x - J_z)\omega_x\omega_z,$$

$$T_z = J_z\dot{\omega}_z + (J_y - J_x)\omega_x\omega_y$$

These equations are often referred to as Euler's dynamical equations. Rearranging gives

$$\begin{bmatrix} \dot{\omega}_x \\ \dot{\omega}_y \\ \dot{\omega}_z \end{bmatrix} = \begin{bmatrix} \dfrac{J_y - J_z}{J_x}\omega_y\omega_z \\ \dfrac{J_z - J_x}{J_y}\omega_x\omega_z \\ \dfrac{J_x - J_y}{J_z}\omega_x\omega_y \end{bmatrix} + \begin{bmatrix} \dfrac{T_x}{J_x} \\ \dfrac{T_y}{J_y} \\ \dfrac{T_z}{J_z} \end{bmatrix}$$

This obviously is in the state variable form and is linear in the control variables T_i but nonlinear in the state variables ω_i due to the products $\omega_i\,\omega_j$.

3.16 Apply the results of the preceding problem to the satellite of Figure 3.46, which is spin stabilized about the x axis. That is, $\omega_x = S$ is a large value, ω_y and $\omega_z \ll S$ represent small wobbling errors. Assume the satellite is rotationally symmetric about the x axis. Find a linear approximation for the state equation.

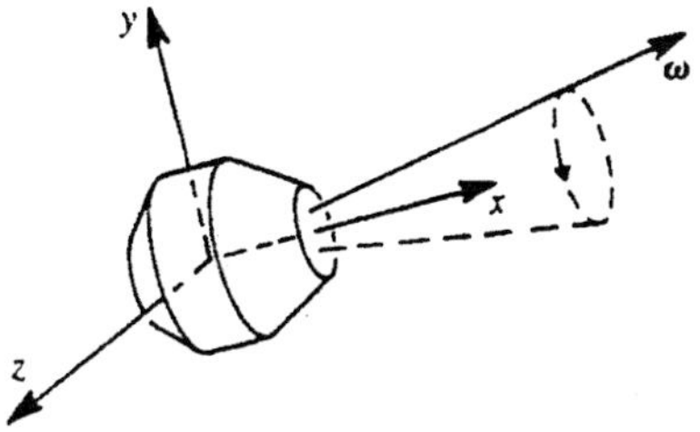

Figure 3.46

Due to symmetry $J_y = J_z$, so $\dot{\omega}_x = T_x/J_x$. If $T_x = 0$, in the absence of any other torques, $\omega_x =$ constant. If the definition $[(J_x - J_y)/J_z]S \triangleq \Omega$ is introduced, then the linear relations between the control torques T_y, T_z and the small wobbling errors ω_y, ω_z are

$$\begin{bmatrix} \dot{\omega}_y \\ \dot{\omega}_z \end{bmatrix} = \begin{bmatrix} 0 & -\Omega \\ \Omega & 0 \end{bmatrix} \begin{bmatrix} \omega_y \\ \omega_z \end{bmatrix} + \frac{1}{J_y} \begin{bmatrix} T_y \\ T_z \end{bmatrix}$$

3.17 A nonlinear time-varying second-order system is described by

$$\ddot{y} + g(y, \dot{y}, u, t) + f(y, t)\dot{u} = 0$$

Assume that the function $f(\ )$ is sufficiently smooth so that the first derivatives with respect to each of its arguments exist and are well behaved. Derive a state variable model for this system.

Solving for $\ddot{y}$ and integrating twice gives

$$y(t) = \iint -g(y, \dot{y}, u, \tau)\, d\tau\, dt' - \iint f(y, \tau)\dot{u}\, d\tau\, dt'$$

$$= \iint -g(y, \dot{y}, u, \tau)\, d\tau\, dt' - \int f(y, t')u\, dt' + \iint [u\, df/dt']\, d\tau\, dt'$$

The chain-rule expanded form $df/dt = [\partial f/\partial y]\dot{y} + [\partial f/\partial t]$ is used in the preceding equation, which is used to draw the simulation diagram shown in Figure 3.47. The blocks in this diagram are general functional evaluation boxes, and the outputs depend upon all the signals shown as inputs in a way described by the equation inside the box. The state equations cannot be written as simple matrix products as in the linear case.

$$\dot{x}_1 = x_2 - f(x_1, t)u(t)$$

$$\dot{x}_2 = u(t)[\partial f/\partial t + x_2\, \partial f/\partial x_1] - g(x_1, x_2, t, u)$$

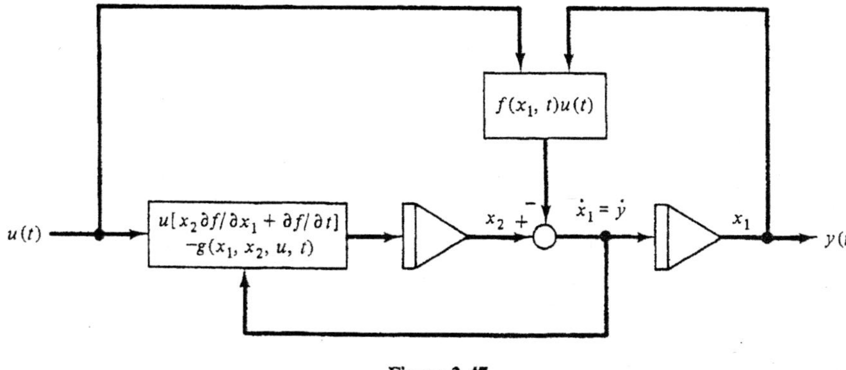

Figure 3.47

and

$$y = [1 \quad 0]\mathbf{x}$$

3.18 Draw a simulation diagram and express the following nonlinear, time-varying system in state space form:

$$y(k+3) + y(k+2)y(k+1) + \alpha \sin(\omega k)y^2(k) + y(k) = u(k) - 3u(k+1)$$

Rewriting the equation as

$$y(k+3) + 3u(k+1) = u(k) - y(k+2)y(k+1) - \alpha \sin(\omega k)y^2(k) - y(k)$$

allows the simulation diagram to be drawn as shown in Figure 3.48.

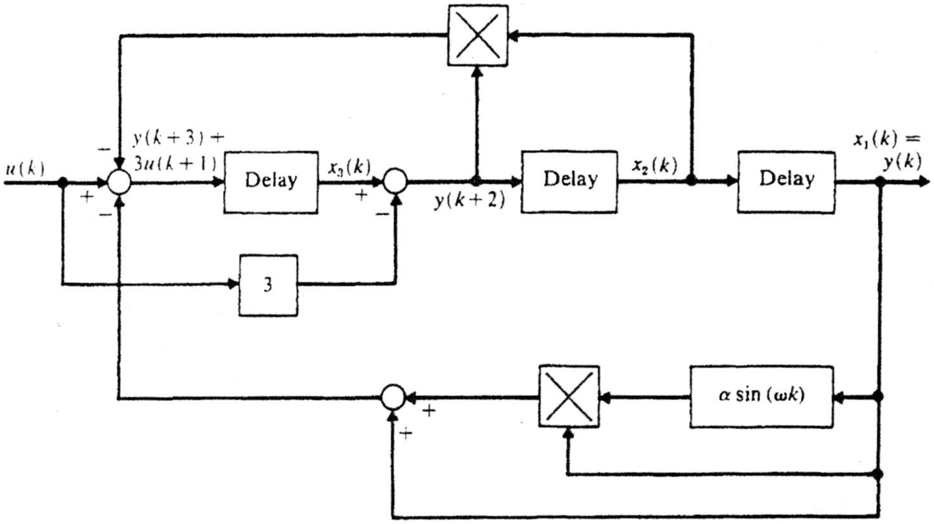

Figure 3.48

Labeling x_1 through x_3 as shown, the state equations are

$$x_1(k+1) = x_2(k)$$

$$x_2(k+1) = x_3(k) - 3u(k)$$

$$x_3(k+1) = -x_2(k)x_3(k) + 3x_2(k)u(k) - \alpha \sin(\omega k)x_1^2(k) - x_1(k) + u(k)$$

and the output equation is $y(k) = [1 \quad 0 \quad 0]\mathbf{x}(k)$.

PROBLEMS

3.19 Convince yourself that both Figure 3.49*a* and *b* represent the system described by

$$\ddot{y} + a\dot{y} + by = u$$

and find the matrices **A, B, C,** and **D** for each case.

3.20 Find a state space representation for the system described by

$$\dot{y}_1 + 3(y_1 + y_2) = u_1$$

$$\ddot{y}_2 + 4\dot{y}_2 + 3y_2 = u_2$$

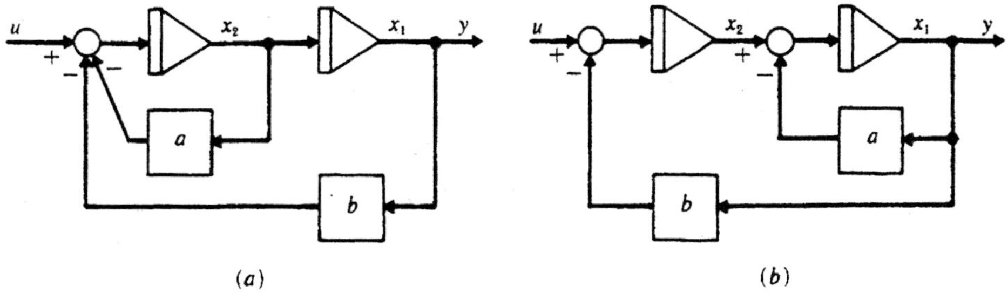

Figure 3.49

3.21 Find a state space representation for the system described by

$$\ddot{y}_1 + 3\dot{y}_1 + 2(y_1 - y_2) = u_1 + \dot{u}_2$$

$$\dot{y}_2 + 3(y_2 - y_1) = u_2 + 2\dot{u}_1$$

3.22 Find the state variable equations for a system described by

$$T(s) = 1/(s^3 + 8s^2 + 13s + 6)$$

using the partial fraction expansion.

3.23 Describe the circuit of Figure 3.50 in state variable form.

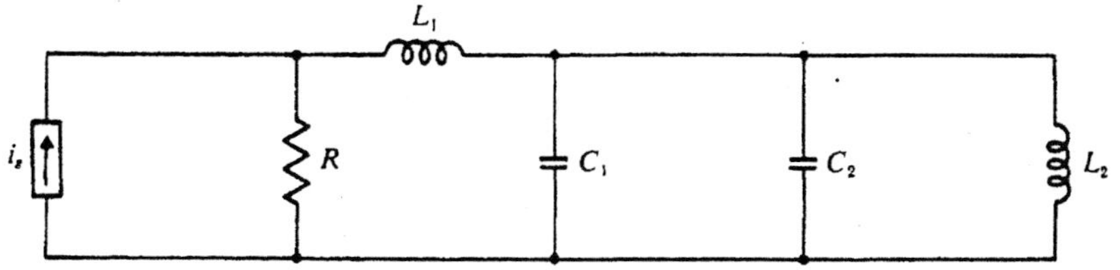

Figure 3.50

3.24 Represent the circuit of Figure 3.51 in state variable form. Assume that the amplifier is an ideal voltage amplifier, that is $v_5 = Kv_4$ and the amplifier draws no current. The transformer ratio is N. The output is the voltage across C_2.

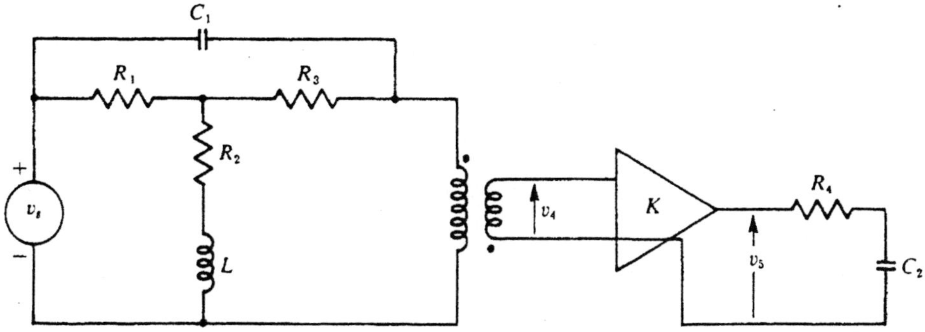

Figure 3.51

3.25 Use as state variables the voltages x_1 and x_2 and the current x_3 as shown in Figure 3.52. Derive the state equations, letting $v_s = u_1$ and $i_s = u_2$.

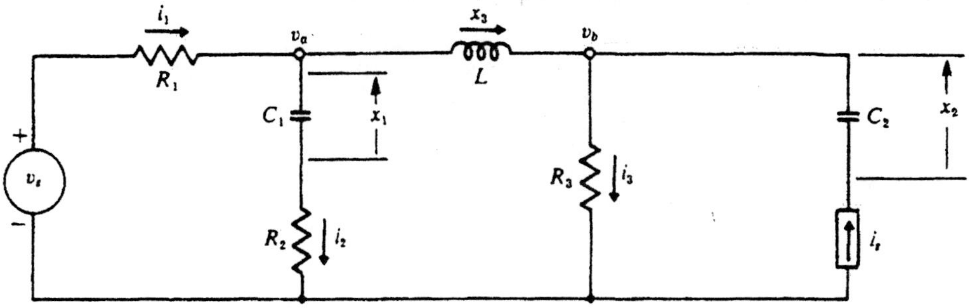

Figure 3.52

3.26 A system has two inputs and two outputs. The input-output equations are

$$y_1(k+2) + 10y_1(k+1) - y_2(k+1) + 3y_1(k) + 2y_2(k) = u_1(k) + 2u_1(k+1)$$

$$y_2(k+1) + 4[y_2(k) - y_1(k)] = 2u_2(k) - u_1(k)$$

Select state variables and write the vector matrix state equations.

3.27 Draw two different simulation diagrams and obtain two different state variable representations for the system described by

$$y(k+2) + 3y(k+1) + 2y(k) = u(k)$$

4

Fundamentals of Matrix Algebra

1 INTRODUCTION

In Chapter 3 matrix notation was used in writing state variable descriptions of dynamic systems. In this chapter a more careful and complete development of matrix algebra and some matrix calculus is presented from first principles. If the matrix usage in the preceding chapter posed no difficulty for the reader, then the introductory parts of this chapter can be treated as a review. The more advanced notions in this chapter, many of which are introduced in the problems, will still be worthwhile. If the earlier introduction of matrices caused the reader some uncertainty, then at least that exposure provided motivation for careful study of the present chapter. Experience shows that the similarities between matrix algebra and scalar algebra have a way of lulling the unwary into a sense of complacency. The usual scalar manipulations often seem to carry over and yield correct results. But ignoring the crucial differences will ultimately cause embarrassing or silly results.

Modeling, design, and analysis of control systems are the major subjects of this book. However, matrix theory and linear algebra are useful in almost every branch of science and engineering. The investment of time and effort that is required to work carefully through this and the next two chapters will pay dividends in deeper understanding, greater insight, and better computational skills later on.

.2 NOTATION

Matrices are rectangular arrays of elements. The elements of a matrix are referred to as *scalars* and will be denoted by lowercase letters, a, b, α, β, etc. In order to define algebraic operations with matrices, it is necessary to restrict these scalar elements to be members of a field. A field $\mathcal{F}$ is any set of two or more elements for which the operations of addition, multiplication, and division are defined, and for which the following axioms hold:

1. If $a \in \mathcal{F}$ and $b \in \mathcal{F}$, then $(a + b) = (b + a) \in \mathcal{F}$.
2. $(ab) = (ba) \in \mathcal{F}$.
3. There exists a unique null element $0 \in \mathcal{F}$ such that $a + 0 = a$ and $0(a) = 0$.
4. If $b \neq 0$, then $(a/b) \in \mathcal{F}$.
5. There exists a unique identity element $1 \in \mathcal{F}$ such that $1(a) = (a)1 = (a/1) = a$.
6. For every $a \in \mathcal{F}$ there is a unique negative element $-a \in \mathcal{F}$ such that $a + (-a) = 0$.
7. The associative, commutative, and distributive laws of algebra are satisfied.

Note that the set of integers does not form a field because axiom 4 is not necessarily true. Some examples of fields are the set of all rational numbers, the set of all real numbers, and the set of all complex numbers. The set of all rational polynomial functions also forms a field. Such functions are ratios of two polynomials $b(s)/a(s)$, where $a(s)$ and $b(s)$ are polynomials in a complex variable s (or z) with real or complex coefficients. Most matrices used in this book are assumed to be defined over the complex number field. For simplicity, many examples will be further restricted to real numbers, with the integers being a special subset. However, many control problems are posed in terms of transfer function matrices, so the field of rational polynomial functions is important. Matrices with polynomial elements also occur. The set of polynomial elements do not form a field because axiom 4 fails. Polynomial elements must be considered as members of the broader class of rational polynomial functions, just as integers are considered as special members of the field of rational numbers.

Boldface uppercase letters will be used to represent matrices, such as

$$\mathbf{A} = \begin{bmatrix} 42 & 16 \\ 5 & 3 \\ 8 & 1 \end{bmatrix}$$

Horizontal sets of entries such as (42 16) and (5 3) are called rows, whereas vertical sets of entries such as (42 5 8) are called columns. It will often be convenient to refer to the element in the ith row and jth column of $\mathbf{A}$ as a_{ij}. Rather than explicitly displaying all elements of $\mathbf{A}$, the shorthand notation $\mathbf{A} = [a_{ij}]$ will sometimes be used. If $\mathbf{A}$ has m rows and n columns, it is said to be an $m \times n$ (or m by n) matrix. In that case, the indices i and j in the shorthand notation indicate collectively the range of values $i = 1, 2, \ldots, m$ and $j = 1, 2, \ldots, n$. In particular, when $m = n = 1$, the matrix has a single element and is just a scalar. The subscripts are then unnecessary. If $n = 1$, the matrix has a single column and is called a *column matrix*. The column index j is then superfluous and is sometimes omitted. Similarly, when $m = 1$, the matrix is called a *row matrix*. Whenever $m = n$, the matrix is called a *square matrix*. In general, m and n can take on any finite integer values.

The four state variable system matrices $\{\mathbf{A}, \mathbf{B}, \mathbf{C}, \mathbf{D}\}$, the input vector $\mathbf{u}$, the output vector $\mathbf{y}$, and the state vector $\mathbf{x}$ were introduced in Chapter 3. These quantities will receive major attention throughout the book. However, the same symbols also will be used in a more generic way in the discussions of matrix algebra.

3 ALGEBRAIC OPERATIONS WITH MATRICES

Matrix Equality

Matrices **A** and **B** are equal, written **A** = **B**, if and only if their corresponding elements are equal. That is, $a_{ij} = b_{ij}$ for $1 \le i \le m$ and $1 \le j \le n$. Of course, this means that equality can exist only between matrices of the same size, $m \times n$ in this case.

Matrix Addition and Subtraction

Matrix addition and subtraction are performed on an element-by-element basis. That is, if $\mathbf{A} = [a_{ij}]$ and $\mathbf{B} = [b_{ij}]$ are both $m \times n$ matrices, then $\mathbf{A} + \mathbf{B} = \mathbf{C}$ and $\mathbf{A} - \mathbf{B} = \mathbf{D}$ indicate that the matrices $\mathbf{C} = [c_{ij}]$ and $\mathbf{D} = [d_{ij}]$ are also $m \times n$ matrices whose elements are given by $c_{ij} = a_{ij} + b_{ij}$ and $d_{ij} = a_{ij} - b_{ij}$ for $i = 1, 2, \ldots, m$ and $j = 1, 2, \ldots, n$.

Matrix Multiplication

Two types of multiplication can be defined. Multiplication of a matrix $\mathbf{A} = [a_{ij}]$ by an arbitrary scalar $\alpha \in \mathcal{F}$ amounts to multiplying every element in **A** by α. That is, $\alpha\mathbf{A} = \mathbf{A}\alpha = [\alpha a_{ij}]$.

Multiplication of an $m \times n$ matrix $\mathbf{A} = [a_{ij}]$ by a $p \times q$ matrix $\mathbf{B} = [b_{ij}]$ is now considered. In forming the product **AB** = **C**, it is said that **A** *premultiplies* **B** or equivalently, **B** *postmultiplies* **A.** This product is only defined when **A** has the same number of columns as **B** has rows. When this is true, that is, when $n = p$, **A** and **B** are said to be *conformable.* The elements of $\mathbf{C} = [c_{ij}]$ are then computed according to

$$c_{ij} = \sum_{k=1}^{n} a_{ik} b_{kj}$$

Clearly, the product **C** is an $m \times q$ matrix.

EXAMPLE 4.1 Let $\mathbf{A} = \begin{bmatrix} 2 & 3 \\ 4 & 5 \end{bmatrix}$, $\mathbf{B} = \begin{bmatrix} 1 & 3 & 5 \\ 2 & 4 & 8 \end{bmatrix}$, and $\mathbf{C} = \begin{bmatrix} 4 \\ -5 \end{bmatrix}$. Then

$$\mathbf{AB} = \begin{bmatrix} 2(1)+3(2) & 2(3)+3(4) & 2(5)+3(8) \\ 4(1)+5(2) & 4(3)+5(4) & 4(5)+5(8) \end{bmatrix} = \begin{bmatrix} 8 & 18 & 34 \\ 14 & 32 & 60 \end{bmatrix}$$

$$\mathbf{AC} = \begin{bmatrix} 2(4)+3(-5) \\ 4(4)+5(-5) \end{bmatrix} = \begin{bmatrix} -7 \\ -9 \end{bmatrix}, \qquad 10\mathbf{A} = \begin{bmatrix} 20 & 30 \\ 40 & 50 \end{bmatrix}$$

The products **BA, CA,** and **CB** are not defined. ■

Once the mechanics of matrix products are mastered, the notational advantages when dealing with simultaneous equations become clear. Matrices with purely numeric entries provide good introductory examples, but the algebra being developed is much more general than this.

EXAMPLE 4.2 Consider the three coupled differential equations of Example 3.4. Ignoring initial conditions, the Laplace transforms are

$$(s^3 + a_1 s^2 + a_2 s + a_3)y_1(s) + a_2 s y_2(s) - a_3 y_3(s) = u_1(s)$$

$$-(a_4 s + a_5)y_1(s) + (s^2 + a_4 s + a_5)y_2(s) + 2a_4 s y_3(s) = u_2(s)$$

$$-a_6 y_1(s) + (s + a_6)y_3(s) = u_3(s)$$

By defining the 3×3 matrix with complex polynomial elements as

$$\mathbf{P}(s) = \begin{bmatrix} (s^3 + a_1 s^2 + a_2 s + a_3) & a_2 s & -a_3 \\ -(a_4 s + a_5) & (s_2 + a_4 s + a_5) & 2a_4 s \\ -a_6 & 0 & (s + a_6) \end{bmatrix}$$

and the 3×1 column vectors as

$$\mathbf{Y}(s) = \begin{bmatrix} y_1(s) \\ y_2(s) \\ y_3(s) \end{bmatrix} \qquad \mathbf{U}(s) = \begin{bmatrix} u_1(s) \\ u_2(s) \\ u_3(s) \end{bmatrix}$$

these equations are compactly written as

$$\mathbf{P}(s)\mathbf{Y}(s) = \mathbf{U}(s)$$ ■

Kronecker Product

Other less frequently used definitions of products of matrices can be defined. One which will be found useful in Chapter 6 is the *Kronecker product,* written $\mathbf{A} \otimes \mathbf{B}$. Each scalar component a_{ij} of the first factor is multiplied by the entire matrix $\mathbf{B}$. There are no conformability-like restrictions on the dimensions of the factors $\mathbf{A}$ and $\mathbf{B}$ that enter into such a product. If $\mathbf{A}$ is $n \times m$ and $\mathbf{B}$ is $p \times q$, then $\mathbf{A} \otimes \mathbf{B}$ will be of dimension $np \times mq$. Note that $\mathbf{A} \otimes \mathbf{B} \neq \mathbf{B} \otimes \mathbf{A}$, although these two products both have the same size. An example illustrates the Kronecker product definition.

$$\begin{bmatrix} a_{11} & a_{12} \\ a_{21} & a_{22} \end{bmatrix} \otimes \begin{bmatrix} b_{11} & b_{12} & b_{13} \\ b_{21} & b_{22} & b_{23} \end{bmatrix} = \begin{bmatrix} a_{11}\mathbf{B} & a_{12}\mathbf{B} \\ a_{21}\mathbf{B} & a_{22}\mathbf{B} \end{bmatrix}$$

$$= \begin{bmatrix} a_{11}b_{11} & a_{11}b_{12} & a_{11}b_{13} & a_{12}b_{11} & a_{12}b_{12} & a_{12}b_{13} \\ a_{11}b_{21} & a_{11}b_{22} & a_{11}b_{23} & a_{12}b_{21} & a_{12}b_{22} & a_{12}b_{23} \\ a_{21}b_{11} & a_{21}b_{12} & a_{21}b_{13} & a_{22}b_{11} & a_{22}b_{12} & a_{22}b_{13} \\ a_{21}b_{21} & a_{21}b_{22} & a_{21}b_{23} & a_{22}b_{21} & a_{22}b_{22} & a_{22}b_{23} \end{bmatrix}$$

Division

Division by a matrix, per se, is not defined. Thus it is not meaningful to "solve for" $\mathbf{Y}(s)$ in Example 4.2 by dividing out the $\mathbf{P}(s)$ matrix. An operation somewhat analogous to division, called *matrix inversion,* is discussed later.

The Null Matrix and the Unit Matrix

As a necessary part of scalar algebra, axioms 3 and 5 for number fields introduce a null element and an identity element. Correspondingly, the null matrix $\mathbf{0}$ is one that has all its elements equal to zero. Then, $\mathbf{A} + \mathbf{0} = \mathbf{A}$ and $\mathbf{0A} = \mathbf{0}$. Note, however, that the null matrix is not unique because the numbers of rows and columns it possesses can be any finite positive integers. Whenever necessary, the dimensions of the null matrix will be indicated by two subscripts, $\mathbf{0}_{mn}$. Another difference of major importance exists between the scalar zero and the null matrix. In scalar algebra, $ab = 0$ implies that either a or b or both are zero. No similar inference can be drawn from the matrix product $\mathbf{AB} = \mathbf{0}$. For a simple verification, let $\mathbf{A} = \begin{bmatrix} 1 & -1 \\ -1 & 1 \end{bmatrix}$ and $\mathbf{B} = \begin{bmatrix} 3 & 4 \\ 3 & 4 \end{bmatrix}$ and form the product $\mathbf{AB}$. Although neither $\mathbf{A}$ nor $\mathbf{B}$ are null matrices, one or the other of these factors possesses properties which are in some sense (to be made clear later) somewhat like a zero. In matrix algebra there are varying degrees of "behaving like zero" of scalar algebra. A null matrix is a very strict, hard zero that has every property to be expected from the scalar zero. It will be seen later that matrix concepts of determinant, rank, trace, eigenvalue, singular value, and matrix norm can all be related to a matrix having some property associated with scalar zero.

The *identity,* or *unit, matrix* $\mathbf{I}$ is a square matrix with all elements zero, except those on the main diagonal ($i = j$ positions) are ones. The unit matrix is not unique because of its dimensions. When necessary, an $n \times n$ unit matrix will be denoted by $\mathbf{I}_n$. The unit matrix has algebraic properties similar to the scalar identity element—namely, if $\mathbf{A}$ is $m \times n$, then $\mathbf{I}_m \mathbf{A} = \mathbf{A}$ and $\mathbf{AI}_n = \mathbf{A}$.

4.4 THE ASSOCIATIVE, COMMUTATIVE, AND DISTRIBUTIVE LAWS OF MATRIX ALGEBRA

Many of the associative, commutative, and distributive laws of scalar algebra carry over to matrix algebra, as summarized next:

$$\mathbf{A} + \mathbf{B} = \mathbf{B} + \mathbf{A}, \qquad \mathbf{A} - \mathbf{B} = \mathbf{A} + (-\mathbf{B}) = -\mathbf{B} + \mathbf{A}$$

$$\mathbf{A} + (\mathbf{B} + \mathbf{C}) = (\mathbf{A} + \mathbf{B}) + \mathbf{C}, \qquad \alpha(\mathbf{A} + \mathbf{B}) = \alpha\mathbf{A} + \alpha\mathbf{B}$$

$$\alpha\mathbf{A} = \mathbf{A}\alpha, \qquad \mathbf{A}(\mathbf{BC}) = (\mathbf{AB})\mathbf{C}$$

$$\mathbf{A}(\mathbf{B} + \mathbf{C}) = \mathbf{AB} + \mathbf{AC}, \qquad (\mathbf{B} + \mathbf{C})\mathbf{A} = \mathbf{BA} + \mathbf{CA}$$

One major difference exists between scalar and matrix algebra. Scalar multiplication is commutative, i.e., $ab = ba$. However, matrix multiplication is not commutative, i.e., $\mathbf{AB} \neq \mathbf{BA}$. In many cases the reversed product is not even defined because the conformability conditions are not satisfied. Even when both $\mathbf{A}$ and $\mathbf{B}$ are square so that $\mathbf{AB}$ and $\mathbf{BA}$ are both defined, they need not be equal. It is for this reason that it is necessary to distinguish between premultiplication and postmultiplication.

EXAMPLE 4.3 Let $\mathbf{A} = \begin{bmatrix} 2 & 3 \\ 1 & 8 \end{bmatrix}$, $\mathbf{B} = \begin{bmatrix} -1 & 1 \\ 0 & 4 \end{bmatrix}$. Then $\mathbf{AB} = \begin{bmatrix} -2 & 14 \\ -1 & 33 \end{bmatrix}$ and $\mathbf{BA} = \begin{bmatrix} -1 & 5 \\ 4 & 32 \end{bmatrix}$. ■

4.5 MATRIX TRANSPOSE, CONJUGATE, AND THE ASSOCIATE MATRIX

The operation of matrix transposition is the interchanging of each row with the column of the same index number. If $\mathbf{A} = [a_{ij}]$, then the *transpose* of $\mathbf{A}$ is $\mathbf{A}^T = [a_{ji}]$. The matrix $\mathbf{A}$ is said to be *symmetric* if $\mathbf{A} = \mathbf{A}^T$. If $\mathbf{A} = -\mathbf{A}^T$, then $\mathbf{A}$ is *skew-symmetric*. An important property of matrix transposition of products is illustrated by

$$(\mathbf{AB})^T = \mathbf{B}^T\mathbf{A}^T, \qquad (\mathbf{ABC})^T = \mathbf{C}^T\mathbf{B}^T\mathbf{A}^T, \ldots$$

The *conjugate of* $\mathbf{A}$, written $\overline{\mathbf{A}}$, is the matrix formed by replacing every element in $\mathbf{A}$ by its complex conjugate. Thus $\overline{\mathbf{A}} = [\overline{a}_{ij}]$. If all elements of $\mathbf{A}$ are real, then $\overline{\mathbf{A}} = \mathbf{A}$. If all elements are purely imaginary, then $\overline{\mathbf{A}} = -\mathbf{A}$.

The *associate matrix* of $\mathbf{A}$ is the conjugate transpose of $\mathbf{A}$. The order of these two operations is immaterial. Matrices satisfying $\mathbf{A} = \overline{\mathbf{A}}^T$ are called *Hermitian matrices*. *Skew-Hermitian* matrices satisfy $\mathbf{A} = -\overline{\mathbf{A}}^T$. For real matrices, symmetric and Hermitian mean the same thing.

4.6 DETERMINANTS, MINORS, AND COFACTORS

Determinants are defined for square matrices only. The determinant of the $n \times n$ matrix $\mathbf{A}$, written $|\mathbf{A}|$, is a scalar-valued function of $\mathbf{A}$. The familiar form of the determinants for $n = 1, 2$, and 3 are

$$n = 1 \qquad |\mathbf{A}| = a_{11}$$

$$n = 2 \qquad |\mathbf{A}| = a_{11}a_{22} - a_{12}a_{21}$$

$$n = 3 \qquad |\mathbf{A}| = a_{11}a_{22}a_{33} + a_{12}a_{23}a_{31} + a_{13}a_{21}a_{32} - a_{13}a_{22}a_{31} - a_{12}a_{21}a_{33} - a_{11}a_{32}a_{23}$$

There is a common pattern which can be generalized for any n. Each determinant has $n!$ terms, with each term consisting of n elements of $\mathbf{A}$, one from each row and from each column. However, the general pattern is inefficient for evaluating large determinants. Usually, a larger-order determinant is first reduced to an expression involving one or more smaller determinants. The methods of *Laplace expansion* and *pivotal condensation* can be used for this purpose. Also, the basic properties of determinants can be used to simplify the evaluation task. Some of these methods are discussed later.

Notice that a square null matrix and the matrix $\mathbf{B} = \begin{bmatrix} 3 & 4 \\ 3 & 4 \end{bmatrix}$, which was used in the discussion of the null matrix, both have zero determinants. Square matrices with zero determinants do possess some of the behaving like zero properties mentioned earlier. For example, the transfer function matrix equation from Example 4.2,

$$\mathbf{P}(s)\mathbf{Y}(s) = \mathbf{U}(s)$$

can have a nonzero output $\mathbf{Y}(s)$ even though $\mathbf{U}(s)$ is zero if the 3×3 matrix $\mathbf{P}$ has a zero determinant (for certain values of the complex variable s). This is in agreement with our concept of transfer function zeros for scalar systems. More often the input-output system transfer function would be expressed as

$$\mathbf{Y}(s) = \mathbf{H}(s)\mathbf{U}(s)$$

and the zeros of the square transfer function matrix $\mathbf{H}(s)$ could be defined as the values of s which make $|\mathbf{H}(s)| = 0$. The implication now is that nonzero inputs $\mathbf{U}(s)$ can cause zero outputs $\mathbf{Y}(s)$. After matrix inversion is introduced, it will be seen that the zeros of $\mathbf{P}(s)$ are the poles of $\mathbf{H}(s)$. However, since determinants are defined only for square matrices, they are not the most general tool for measuring when a matrix behaves in some sense like zero.

Minors

An $n \times n$ matrix $\mathbf{A}$ contains n^2 elements a_{ij}. Each of these has associated with it a unique scalar, called a *minor* M_{ij}. The minor M_{pq} is the determinant of the $n-1 \times n-1$ matrix formed from $\mathbf{A}$ by crossing out the pth row and qth column.

Cofactors

Each element a_{pq} of $\mathbf{A}$ has a *cofactor* C_{pq}, which differs from M_{pq} at most by a sign change. Cofactors are sometimes called signed minors for this reason and are given by $C_{pq} = (-1)^{p+q} M_{pq}$.

Determinants by Laplace Expansion

If $\mathbf{A}$ is an $n \times n$ matrix, any arbitrary row k can be selected and $|\mathbf{A}|$ is then given by $|\mathbf{A}| = \sum_{j=1}^{n} a_{kj} C_{kj}$. Similarly, Laplace expansion can be carried out with respect to any arbitrary column l, to obtain $|\mathbf{A}| = \sum_{i=1}^{n} a_{il} C_{il}$. Laplace expansion reduces the evaluation of an $n \times n$ determinant down to the evaluation of a string of $(n-1) \times (n-1)$ determinants, namely, the cofactors.

EXAMPLE 4.4 Given $\mathbf{A} = \begin{bmatrix} 2 & 4 & 1 \\ 3 & 0 & 2 \\ 2 & 0 & 3 \end{bmatrix}$. Three of its minors are

$$M_{12} = \begin{vmatrix} 3 & 2 \\ 2 & 3 \end{vmatrix} = 5, \qquad M_{22} = \begin{vmatrix} 2 & 1 \\ 2 & 3 \end{vmatrix} = 4, \quad \text{and} \quad M_{32} = \begin{vmatrix} 2 & 1 \\ 3 & 2 \end{vmatrix} = 1$$

The associated cofactors are

$$C_{12} = (-1)^3 5 = -5, \qquad C_{22} = (-1)^4 4 = 4, \qquad C_{32} = (-1)^5 1 = -1$$

Using Laplace expansion with respect to column 2 gives $|\mathbf{A}| = 4C_{12} = -20$. ■

Pivotal Condensation

Pivotal condensation [1], also called the method of Chio [2, 3], reduces an $n \times n$ determinant to a single $(n-1) \times (n-1)$ determinant and thus avoids the long string of determinants encountered with Laplace expansion. Let a_{pq} be any nonzero element of **A**. This is called the *pivot element.* An $(n-1)\times(n-1)$ determinant is formed, with each of its elements obtained from a 2×2 determinant. Each 2×2 determinant contains a_{pq}, one other element from row p, one other element from column q, and the fourth element is from the fourth corner of the rectangle defined by the previous three elements. Let the $(n-1)\times(n-1)$ determinant be called $|\Delta|$. Then $|\mathbf{A}| = [1/(a_{pq})^{n-2}]|\Delta|$. Although the procedure looks complicated, in actual applications the large number of 2×2 determinants easily reduce to their numeric values. The method is best illustrated by an example.

EXAMPLE 4.5 Let **A** be a 4×4 matrix and assume that $a_{23}\neq 0$. Then

$$|\mathbf{A}| = \frac{1}{(a_{23})^2}\begin{vmatrix} \begin{vmatrix} a_{11} & a_{13} \\ a_{21} & a_{23} \end{vmatrix} & \begin{vmatrix} a_{12} & a_{13} \\ a_{22} & a_{23} \end{vmatrix} & \begin{vmatrix} a_{13} & a_{14} \\ a_{23} & a_{24} \end{vmatrix} \\ \begin{vmatrix} a_{21} & a_{23} \\ a_{31} & a_{33} \end{vmatrix} & \begin{vmatrix} a_{22} & a_{23} \\ a_{32} & a_{33} \end{vmatrix} & \begin{vmatrix} a_{23} & a_{24} \\ a_{33} & a_{34} \end{vmatrix} \\ \begin{vmatrix} a_{21} & a_{23} \\ a_{41} & a_{43} \end{vmatrix} & \begin{vmatrix} a_{22} & a_{23} \\ a_{42} & a_{43} \end{vmatrix} & \begin{vmatrix} a_{23} & a_{24} \\ a_{43} & a_{44} \end{vmatrix} \end{vmatrix}$$

Note that the pivot element a_{23} is in the same location relative to the other elements within each 2×2 determinant as it is in the original **A** matrix. ■

Useful Properties of Determinants

1. If **A** and **B** are both $n\times n$, then $|\mathbf{AB}| = |\mathbf{A}||\mathbf{B}|$.
2. $|\mathbf{A}| = |\mathbf{A}^T|$.
3. If all the elements in any row or in any column are zero, then $|\mathbf{A}| = 0$.
4. If any two rows of **A** are proportional, $|\mathbf{A}| = 0$. If a row is a linear combination of any number of other rows, then $|\mathbf{A}| = 0$. Similar statements hold for columns.
5. Interchanging any two rows (or any two columns) of a matrix changes the sign of its determinant.
6. Multiplying all elements of any one row (or column) of a matrix **A** by a scalar α yields a matrix whose determinant is $\alpha|\mathbf{A}|$.
7. Any multiple of a row (column) can be added to any other row (column) without changing the value of the determinant.

4.7 RANK AND TRACE OF A MATRIX

The *rank* of **A**, designated as r_A or rank(**A**), is defined as the size of the largest nonzero determinant that can be formed from **A**. A zero determinant is interpreted in terms of the zero of the number field being used. Therefore, a matrix with rational polynomial entries is considered singular only if its determinant is identically zero and not just if its determinant happens to have a zero value for certain isolated values of s or z. The same notion applies to the determination of rank for these matrices. The maximum possible rank of an $m \times n$ matrix is obviously the smaller of m and n. If $\mathbf{A}$ takes on its maximum possible rank, it is said to be of full rank. If $\mathbf{A}$ is $n \times n$ (square) and has its maximal rank n, then the matrix is said to be *nonsingular*. We see below that nonsingular matrices can be inverted, but singular matrices have no inverse.

Nonsingular matrices do not possess any of the properties of behaving like zero, whereas singular matrices do. The generalization to nonsquare matrices is accomplished by the concept of rank. Note that null matrices always have a zero rank. Heuristically, full-rank matrices will not have zero-like behavior, whereas rank-deficient matrices, those having less than full rank, will. The amount by which they are rank-deficient, i.e.,

$$q = \min(n, m) - r_A$$

is called the *degeneracy*, or *nullity*, of the matrix $\mathbf{A}$. This concept appears repeatedly in later chapters.

The rank of the product of two or more matrices is never more than the smallest rank of the matrices forming the product. For example, if r_A and r_B are the ranks of $\mathbf{A}$ and $\mathbf{B}$, then $\mathbf{C} = \mathbf{AB}$ has rank r_C satisfying $0 \le r_C \le \min\{r_A, r_B\}$.

Let $\mathbf{A}$ be an $n \times n$ matrix. Then the *trace* of $\mathbf{A}$, denoted by $\mathrm{Tr}(\mathbf{A})$, is the sum of the diagonal elements of $\mathbf{A}$, $\mathrm{Tr}(\mathbf{A}) = \sum_{i=1}^{n} a_{ii}$. If $\mathbf{A}$ and $\mathbf{B}$ are conformable square matrices, then $\mathrm{Tr}(\mathbf{A} + \mathbf{B}) = \mathrm{Tr}(\mathbf{A}) + \mathrm{Tr}(\mathbf{B})$ and $\mathrm{Tr}(\mathbf{AB}) = \mathrm{Tr}(\mathbf{BA})$. From the definition of the trace it is obvious that $\mathrm{Tr}(\mathbf{A}^T) = \mathrm{Tr}(\mathbf{A})$. From this it follows that $\mathrm{Tr}(\mathbf{AB}) = \mathrm{Tr}(\mathbf{B}^T\mathbf{A}^T)$.

EXAMPLE 4.6 Let $\mathbf{A} = \begin{bmatrix} 1 & 5 & 8 \\ 3 & -1 & 2 \\ 4 & -4 & 6 \end{bmatrix}$. $\mathbf{B} = \begin{bmatrix} 1 & -1 & 8 \\ 3 & -3 & 2 \\ 4 & -4 & 6 \end{bmatrix}$. Then $|\mathbf{A}| = -112$, so that $r_A = 3$ and $\mathbf{A}$ is nonsingular. Also, $\mathrm{Tr}(\mathbf{A}) = 6$. The matrix $\mathbf{B}$ has $|\mathbf{B}| = 0$, so $r_B < 3$. Crossing out column 2 and row 3 of $\mathbf{B}$ gives a 2×2 determinant with a value -22, so $r_B = 2$. The trace of $\mathbf{B}$ is $\mathrm{Tr}(\mathbf{B}) = 4$. Forming $\mathbf{AB}$ and $\mathbf{BA}$ shows that $r_{AB} = 2$ and $r_{BA} = 2$. Also, $\mathrm{Tr}(\mathbf{A} + \mathbf{B}) = 10 = \mathrm{Tr}(\mathbf{A}) + \mathrm{Tr}(\mathbf{B})$ and $\mathrm{Tr}(\mathbf{AB}) = 100 = \mathrm{Tr}(\mathbf{BA})$. Note that $\mathrm{Tr}(\mathbf{AB}) \ne \mathrm{Tr}(\mathbf{A})\mathrm{Tr}(\mathbf{B})$. ■

4.8 MATRIX INVERSION

The inverse of the scalar element a is $1/a$, or a^{-1}. It satisfies $a(a^{-1}) = (a^{-1})a = 1$. If an arbitrary matrix $\mathbf{A}$ is to have an analogous inverse $\mathbf{B} = \mathbf{A}^{-1}$, then the following must hold:

$$\mathbf{BA} = \mathbf{AB} = \mathbf{I}$$

Because of conformability requirements, this can never be true if $\mathbf{A}$ is not square. In addition, $\mathbf{A}$ must have a nonzero determinant, i.e., $\mathbf{A}$ must be nonsingular. When this is true, $\mathbf{A}$ has a unique inverse given by

$$\mathbf{A}^{-1} = \frac{\mathbf{C}^T}{|\mathbf{A}|}$$

where $\mathbf{C}$ is the matrix formed by the cofactors C_{ij}. The matrix $\mathbf{C}^T$ is called the *adjoint matrix*, Adj(**A**). Thus the inverse of a nonsingular matrix is

$$\mathbf{A}^{-1} = \text{Adj}(\mathbf{A})/|\mathbf{A}|$$

EXAMPLE 4.7 Let $\mathbf{A} = \begin{bmatrix} 1 & 1 \\ 2 & 2 \end{bmatrix}$, $\mathbf{B} = \begin{bmatrix} 1 & 2 \\ 3 & 4 \end{bmatrix}$, $\mathbf{D} = \begin{bmatrix} 4 & 2 & 1 \\ 2 & 6 & 3 \\ 1 & 3 & 5 \end{bmatrix}$. Then $\mathbf{A}^{-1}$ does not exist since $|\mathbf{A}| = 0$. Since $|\mathbf{B}| = -2$, $\mathbf{B}^{-1}$ exists and is given by $\mathbf{B}^{-1} = \begin{bmatrix} -2 & 1 \\ 3/2 & -1/2 \end{bmatrix}$. Similarly, $\mathbf{D}^{-1} = \dfrac{1}{70}\begin{bmatrix} 21 & -7 & 0 \\ -7 & 19 & -10 \\ 0 & -10 & 20 \end{bmatrix}$. ■

The definition of the matrix inverse just given is perfectly general. It applies to matrices whose elements are functions of time or of complex variables, such as s in Example 4.2. Thus

$$\mathbf{H}(s) = \mathbf{P}(s)^{-1} \tag{4.1}$$

As with scalar transfer functions, the poles of $\mathbf{H}(s)$ are those values of s for which elements of $\mathbf{H}$ are unbounded. The definition of the matrix inverse shows that this happens when $|\mathbf{P}(s)| = 0$. The poles of $\mathbf{H}(s)$ are the zeros of $\mathbf{P}(s)$, as mentioned earlier.

Inversion of large matrices by direct application of the above definition is tedious. Numerical techniques such as *Gaussian elimination* are often used. Matrix partitioning can also be employed to obtain a matrix inverse in terms of several smaller inverses. Another method, based on the Cayley-Hamilton theorem, is given in Chapter 8.

In many applications the entries in a matrix to be inverted are complex numbers. Although the general definition of the matrix inverse is valid for complex entries, the actual calculations become much more cumbersome. Some computer algorithms for matrix inversion are restricted to matrices with real numbers for elements. Problem 4.22 gives some partial results on inverting complex matrices using only real numbers. In other cases, the complex inverse is not the desired end result but is only an intermediate quantity that occurs while solving for $\mathbf{X}$ in simultaneous equations of the form

$$\mathbf{AX} = \mathbf{B} \quad \text{or} \quad \mathbf{XA} = \mathbf{B}$$

It is shown in Problem 4.23 that if $\mathbf{A}$ and $\mathbf{B}$ have complex entries which occur in complex conjugate pairs in a certain way, then the solution for $\mathbf{X}$ is purely real and can easily be computed using only real matrix inversion calculations.

The Inverse of a Product

Let $\mathbf{A}, \mathbf{B}, \mathbf{C}, \ldots, \mathbf{W}$ be any number of conformable nonsingular matrices. Then

$$(\mathbf{ABC}\cdots\mathbf{W})^{-1} = \mathbf{W}^{-1}\cdots\mathbf{C}^{-1}\mathbf{B}^{-1}\mathbf{A}^{-1}$$

Some Matrices with Special Relationships to Their Inverses

If $\mathbf{A}^{-1} = \mathbf{A}$, $\mathbf{A}$ is said to be *involutory*.
If $\mathbf{A}^{-1} = \mathbf{A}^T$, $\mathbf{A}$ is said to be *orthogonal*.
If $\mathbf{A}^{-1} = \overline{\mathbf{A}}^T$, $\mathbf{A}$ is said to be *unitary*.

.9 PARTITIONED MATRICES

Any matrix $\mathbf{A}$ can be subdivided or partitioned into a number of smaller submatrices. If conformable matrices are partitioned in a compatible fashion, the submatrices can be treated just as if they were scalar elements when performing the operations of addition and multiplication. Of course, the order of the products is not arbitrary, as it would be with scalars.

EXAMPLE 4.8 $\mathbf{AB} = \mathbf{C}$ can be partitioned in various ways. A few of them are given next:

$$\text{(a)}\quad \begin{bmatrix} \mathbf{A}_1 \\ \hline \mathbf{A}_2 \end{bmatrix} [\mathbf{B}_1 \mid \mathbf{B}_2] = \left[\begin{array}{c|c} \mathbf{A}_1\mathbf{B}_1 & \mathbf{A}_1\mathbf{B}_2 \\ \hline \mathbf{A}_2\mathbf{B}_1 & \mathbf{A}_2\mathbf{B}_2 \end{array}\right] = \left[\begin{array}{c|c} \mathbf{C}_1 & \mathbf{C}_2 \\ \hline \mathbf{C}_3 & \mathbf{C}_4 \end{array}\right]$$

$$\text{(b)}\quad \left[\begin{array}{c|c} \mathbf{A}_1 & \mathbf{A}_2 \\ \hline \mathbf{A}_3 & \mathbf{A}_4 \end{array}\right] \begin{bmatrix} \mathbf{B}_1 \\ \hline \mathbf{B}_2 \end{bmatrix} = \begin{bmatrix} \mathbf{A}_1\mathbf{B}_1 + \mathbf{A}_2\mathbf{B}_2 \\ \hline \mathbf{A}_3\mathbf{B}_1 + \mathbf{A}_4\mathbf{B}_2 \end{bmatrix} = \begin{bmatrix} \mathbf{C}_1 \\ \hline \mathbf{C}_2 \end{bmatrix}$$

$$\text{(c)}\quad \left[\begin{array}{c|c} \mathbf{A}_1 & \mathbf{A}_2 \\ \hline \mathbf{A}_3 & \mathbf{A}_4 \end{array}\right] \left[\begin{array}{c|c} \mathbf{B}_1 & \mathbf{B}_2 \\ \hline \mathbf{B}_3 & \mathbf{B}_4 \end{array}\right] = \left[\begin{array}{c|c} \mathbf{A}_1\mathbf{B}_1 + \mathbf{A}_2\mathbf{B}_3 & \mathbf{A}_1\mathbf{B}_2 + \mathbf{A}_2\mathbf{B}_4 \\ \hline \mathbf{A}_3\mathbf{B}_1 + \mathbf{A}_4\mathbf{B}_3 & \mathbf{A}_3\mathbf{B}_2 + \mathbf{A}_4\mathbf{B}_4 \end{array}\right] = \left[\begin{array}{c|c} \mathbf{C}_1 & \mathbf{C}_2 \\ \hline \mathbf{C}_3 & \mathbf{C}_4 \end{array}\right] \qquad \blacksquare$$

Partitioned matrices were used without comment in Sec. 3.5 where subsystems of state variable systems were combined to obtain an overall composite state variable model. That application illustrates one possible motivation for using partitioned matrices. They allow the clustering together of groups of variables and treating the group by an identifying symbol. It is an intermediate step between displaying *all* the scalar entries and displaying the entire matrix by just a single symbol.

Partitioned matrices can be used to find an expression for the inverse of a nonsingular matrix $\mathbf{A}$. If $\mathbf{A}$ is partitioned into four submatrices, then $\mathbf{A}^{-1} = \mathbf{B}$ will also have four submatrices:

$$\mathbf{AB} = \mathbf{I} \quad \text{or} \quad \left[\begin{array}{c|c} \mathbf{A}_1 & \mathbf{A}_2 \\ \hline \mathbf{A}_3 & \mathbf{A}_4 \end{array}\right] \left[\begin{array}{c|c} \mathbf{B}_1 & \mathbf{B}_2 \\ \hline \mathbf{B}_3 & \mathbf{B}_4 \end{array}\right] = \left[\begin{array}{c|c} \mathbf{I} & \mathbf{0} \\ \hline \mathbf{0} & \mathbf{I} \end{array}\right]$$

The partitioned form implies four separate matrix equations, two of which are $\mathbf{A}_1\mathbf{B}_1 + \mathbf{A}_2\mathbf{B}_3 = \mathbf{I}$ and $\mathbf{A}_3\mathbf{B}_1 + \mathbf{A}_4\mathbf{B}_3 = \mathbf{0}$. These can be solved simultaneously for $\mathbf{B}_1$ and $\mathbf{B}_3$. The remaining two equations give $\mathbf{B}_2$ and $\mathbf{B}_4$ and lead to the result

$$\mathbf{A}^{-1} = \left[\begin{array}{c|c} (\mathbf{A}_1 - \mathbf{A}_2\mathbf{A}_4^{-1}\mathbf{A}_3)^{-1} & -\mathbf{A}_1^{-1}\mathbf{A}_2(\mathbf{A}_4 - \mathbf{A}_3\mathbf{A}_1^{-1}\mathbf{A}_2)^{-1} \\ \hline -\mathbf{A}_4^{-1}\mathbf{A}_3(\mathbf{A}_1 - \mathbf{A}_2\mathbf{A}_4^{-1}\mathbf{A}_3)^{-1} & (\mathbf{A}_4 - \mathbf{A}_3\mathbf{A}_1^{-1}\mathbf{A}_2)^{-1} \end{array}\right]$$

Several matrix identities can be derived by starting with the reversed order, $\mathbf{BA} = \mathbf{I}$, repeating the above process, and then using the uniqueness of $\mathbf{B} = \mathbf{A}^{-1}$ to equate the various terms. One such identity, called the *matrix inversion lemma*, is particularly useful. A general form is

$$(\mathbf{A}_1 - \mathbf{A}_2\mathbf{A}_4^{-1}\mathbf{A}_3)^{-1} = \mathbf{A}_1^{-1} + \mathbf{A}_1^{-1}\mathbf{A}_2(\mathbf{A}_4 - \mathbf{A}_3\mathbf{A}_1^{-1}\mathbf{A}_2)^{-1}\mathbf{A}_3\mathbf{A}_1^{-1}$$

By letting $\mathbf{A}_1 = \mathbf{P}^{-1}$, $\mathbf{A}_2 = \mathbf{H}^T$, $\mathbf{A}_3 = \mathbf{H}$, and $\mathbf{A}_4 = -\mathbf{Q}^{-1}$, an extremely useful special form of the inversion lemma that will be encountered in recursive weighted least squares is obtained:

$$[\mathbf{P}^{-1} + \mathbf{H}^T\mathbf{Q}\mathbf{H}]^{-1} = \mathbf{P} - \mathbf{P}\mathbf{H}^T[\mathbf{H}\mathbf{P}\mathbf{H}^T + \mathbf{Q}^{-1}]^{-1}\mathbf{H}\mathbf{P}$$

Diagonal, Block Diagonal, and Triangular Matrices

If the only nonzero elements of a square matrix $\mathbf{A}$ are on the main diagonal, then $\mathbf{A}$ is called a *diagonal matrix*. This is often written as $\mathbf{A} = \text{diag}[a_{11} \quad a_{22} \quad \ldots \quad a_{nn}]$. For this case, $|\mathbf{A}| = a_{11}a_{22}\cdots a_{nn}$ and $\mathbf{A}^{-1} = \text{diag}[1/a_{11} \quad 1/a_{22} \quad \ldots \quad 1/a_{nn}]$. The unit matrix is a special case with all $a_{ii} = 1$.

A *block diagonal*, or *quasidiagonal*, matrix is a square matrix that can be partitioned so that the only nonzero elements are contained in square submatrices along the main diagonal,

$$\mathbf{A} = \begin{bmatrix} \mathbf{A}_1 & & & \\ & \mathbf{A}_2 & & \\ & & \ddots & \\ & & & \mathbf{A}_k \end{bmatrix} = \text{diag}[\mathbf{A}_1 \quad \mathbf{A}_2 \quad \cdots \quad \mathbf{A}_k]$$

For this case $|\mathbf{A}| = |\mathbf{A}_1||\mathbf{A}_2|\cdots|\mathbf{A}_k|$ and $\mathbf{A}^{-1} = \text{diag}[\mathbf{A}_1^{-1} \quad \mathbf{A}_2^{-1} \quad \cdots \quad \mathbf{A}_k^{-1}]$, provided that $\mathbf{A}^{-1}$ exists.

A square matrix which has all its elements below (above) the main diagonal equal to zero is called an *upper triangular* (*lower triangular*) matrix. The determinant of any triangular matrix is the product of its diagonal elements.

4.10 ELEMENTARY OPERATIONS AND ELEMENTARY MATRICES

Three basic operations on a matrix, called *elementary operations*, are as follows:

1. The interchange of two rows (or of two columns).
2. The multiplication of every element in a given row (or column) by a scalar α.
3. The multiplication of the elements of a given row (or column) by a scalar α, and adding the result to another row (column). The original row (column) is unaltered.

It is stressed that the nature of the scalar α depends upon which number field is in use. For example, if α is a rational polynomial function, the entire discussion of elementary operations still applies without change. When these row operations are applied to the unit matrix, the resultant matrices are called elementary matrices, and are denoted as follows:

$\mathbf{E}_{p,q}$: pth and qth rows of $\mathbf{I}$ interchanged

$\mathbf{E}_p(\alpha)$: pth row of $\mathbf{I}$ multiplied by α

$\mathbf{E}_{p,q}(\alpha)$: pth row of $\mathbf{I}$ multiplied by α and added to qth row

The elementary matrices are all nonsingular. In fact,

$$|\mathbf{E}_{p,q}| = -1, \qquad |\mathbf{E}_p(\alpha)| = \alpha, \qquad |\mathbf{E}_{p,q}(\alpha)| = 1$$

The inverse of each elementary matrix is also an elementary matrix.

Premultiplication (postmultiplication) of a matrix by one of the elementary matrices performs the corresponding elementary row (column) operation on that matrix.

By performing a sequence of elementary row and column operations, any matrix of rank r can be reduced to one of the following normal forms:

$$\mathbf{I}_r, \qquad [\mathbf{I}_r \mid \mathbf{0}], \qquad \begin{bmatrix} \mathbf{I}_r \\ \hline \mathbf{0} \end{bmatrix}, \qquad \left[\begin{array}{c|c} \mathbf{I}_r & \mathbf{0} \\ \hline \mathbf{0} & \mathbf{0} \end{array}\right]$$

These are special cases of, or analogous to, matrices in the *row-reduced echelon* form. They are also sometimes called *Hermite normal forms*. These will be defined more formally in Chapter 6, where their value will be more fully appreciated. Thus elementary operations provide a practical means of computing the rank of a matrix, but they have many other uses as well.

11 DIFFERENTIATION AND INTEGRATION OF MATRICES

When a matrix $\mathbf{A}$ has elements which are functions of a scalar variable (such as time), differentiation and integration of the matrix are defined on an element-by-element basis. If $\mathbf{A}(t) = [a_{ij}(t)]$, then $d\mathbf{A}/dt = \dot{\mathbf{A}} = [\dot{a}_{ij}(t)]$ and $\int \mathbf{A}(\tau)\,d\tau = [\int a_{ij}(\tau)\,d\tau]$. Because of the integration rule, Laplace transforms and inverse Laplace transforms of matrices are also found on element-by-element basis. This is also true for Z-transforms of matrices.

Equation (*4.1*) could be applied to $\mathbf{P}(s)$ given in Example 4.2 to determine $\mathbf{H}(s)$ directly. However, there is an alternative approach which can—and frequently will—be used. In Example 3.4 a state variable model for the system in question was determined. The form of that model is

$$\dot{\mathbf{x}} = \mathbf{A}\mathbf{x} + \mathbf{B}\mathbf{u} \tag{4.2a}$$

$$\mathbf{y} = \mathbf{C}\mathbf{x} + \mathbf{D}\mathbf{u} \tag{4.2b}$$

and the specific form of $\mathbf{A}$, $\mathbf{B}$, $\mathbf{C}$, and $\mathbf{D}$ have been given. Assume that the initial conditions for $\mathbf{x}(t)$ at time $t = 0$ are given by $\mathbf{x}_0$. Taking the Laplace transform of Eq. (*4.2a*) and combining the two $\mathbf{X}(s)$ terms gives

$$(s\mathbf{I}_n - \mathbf{A})\mathbf{X}(s) = \mathbf{x}_0 + \mathbf{B}\mathbf{U}(s)$$

Premultiplying both sides by the inverse of $(s\mathbf{I}_n - \mathbf{A})$ gives

$$\mathbf{X}(s) = (s\mathbf{I}_n - \mathbf{A})^{-1}\mathbf{x}_0 + (s\mathbf{I}_n - A)^{-1}\mathbf{B}\mathbf{U}(s) \tag{4.3}$$

The transform of the state vector consists of the initial condition response plus the forced response. These are often called the zero input response and the zero state response, respectively. When Eq. (*4.3*) is substituted into the Laplace transform of Eq. (*4.2b*), the result is

$$\mathbf{Y}(s) = \{\mathbf{C}(s\mathbf{I}_n - \mathbf{A})^{-1}\mathbf{B} + \mathbf{D}\}\mathbf{U}(s) + \mathbf{C}(s\mathbf{I}_n - \mathbf{A})^{-1}\mathbf{x}_0 \tag{4.4}$$

Ignoring the initial condition term, the input-output transfer function matrix is given by the first term in Eq. (*4.4*),

$$\mathbf{H}(s) = \mathbf{C}(s\mathbf{I}_n - \mathbf{A})^{-1}\mathbf{B} + \mathbf{D} \tag{4.5}$$

This is an alternative to the calculation in Eq. (*4.1*) for the transfer function matrix. In more general problems with input derivatives, the input-output expression in Example 4.2 will take the form

$$\mathbf{P}(s)\mathbf{Y}(s) = \mathbf{N}(s)\mathbf{U}(s)$$

or

$$\mathbf{Y}(s) = \mathbf{P}(s)^{-1}\mathbf{N}(s)\mathbf{U}(s) \tag{4.6}$$

where $\mathbf{N}(s)$ will be an $m \times r$ matrix of polynomials in s. As before, $\mathbf{P}$ will be an $m \times m$ matrix, where the number of inputs components in $\mathbf{U}$ is r and the number of output components in $\mathbf{Y}$ is m. The generalization of Eq. (*4.1*) for the $m \times r$ input-output transfer function matrix is

$$\mathbf{H}(s) = \mathbf{P}(s)^{-1}\mathbf{N}(s) \tag{4.7}$$

This is *one* particular form of the *matrix fraction description* (MFD) of a multiple-input, multiple-output system transfer function. It is an alternative to the state variable form of Eq. (*4.5*) [4].

Differentiation of a Determinant

Two useful rules for differentiating a determinant are

$$\frac{\partial|\mathbf{A}|}{\partial a_{ij}} = C_{ij} \qquad \text{(follows immediately from the Laplace expansion)}$$

If the $n \times n$ matrix $\mathbf{A}$ is a function of t, then $|\mathbf{A}|$ is also a function of t. Then $d|\mathbf{A}|/dt$ is just the sum of n separate determinants. The first determinant has row (or column) one differentiated, the second has row (or column) two differentiated, and so on through all n rows (columns).

4.12 ADDITIONAL MATRIX CALCULUS

4.12.1 The Gradient Operator and Differentiation with Respect to a Vector

Let $f(x_1, x_2, \ldots, x_n)$ be a scalar-valued function of n variables x_i. The variables may be, but need not be, state variables in the present discussion. For notational convenience the dependence on n variables x_i is written as $f(\mathbf{x})$, with $\mathbf{x}$ being a vector with components x_i. The n partial derivatives of $f(\mathbf{x})$, $\partial f/\partial x_i$, will be used frequently. It is convenient to group these partials into an array and give the array a special symbol. A single-rowed array, a row vector, could be—and frequently is—used for this purpose. Here the array is arranged as a single column, that is, a column vector. This convention is an arbitrary choice. Both forms are used in the literature, and both are referred to as the *gradient vector* [1]. Three different symbols are frequently used to identify the gradient of $f(\mathbf{x})$, $\nabla_{\mathbf{x}} f = \text{grad}_{\mathbf{x}} f = df/d\mathbf{x}$. The meaning of these symbols is given by

$$\nabla_{\mathbf{x}} f = \begin{bmatrix} \dfrac{\partial f}{\partial x_1} & \dfrac{\partial f}{\partial x_2} & \cdots & \dfrac{\partial f}{\partial x_n} \end{bmatrix}^T \tag{4.8}$$

The only differences which arise between the row and column vector definitions are the presence or absence of the transpose in various algebraic manipulations. Conformability requirements for matrix multiplication must always be satisfied and can be used to determine whether a row or column definition is implied.

EXAMPLE 4.9 Let $f_1(\mathbf{x}) = \mathbf{x}^T\mathbf{A}\mathbf{y}$, a bilinear function. Expanding this in terms of individual components gives $f_1(\mathbf{x}) = \sum_i \sum_j a_{ij} x_i y_j$. A typical component of the gradient is $\partial f_1/\partial x_k = \sum_i \sum_j a_{ij}(\partial x_i/\partial x_k) y_j$.

Using the independence of the x_i components gives $\partial x_i/\partial x_k = \delta_{ik} = \begin{cases} 1 \text{ if } i = k \\ 0 \text{ if } i \neq k \end{cases}$. This means that only one term in the summation over i is nonzero, so $\partial f_1/\partial x_k = \sum_j a_{kj} y_j$. This is just the kth component of the matrix product $\mathbf{Ay}$, so the gradient vector for this example is $\nabla_{\mathbf{x}}(\mathbf{x}^T\mathbf{A}\mathbf{y}) = \mathbf{A}\mathbf{y}$. ■

EXAMPLE 4.10 If $f_2(\mathbf{x}) = \mathbf{y}^T\mathbf{A}\mathbf{x}$, then $\nabla_{\mathbf{x}} f_2 \neq \mathbf{y}^T\mathbf{A}$. The gradient operator *is not* simply a canceling of the $\mathbf{x}$ vector as might be inferred from Example 4.9. By convention, the gradient is a column vector, so it cannot be equal to the row vector $\mathbf{y}^T\mathbf{A}$. The correct expression for the gradient is $\nabla_{\mathbf{x}} f_2 = \mathbf{A}^T\mathbf{y}$. ■

EXAMPLE 4.11 Let $f_3(\mathbf{x}) = \mathbf{x}^T\mathbf{A}\mathbf{x}$, a quadratic form. In summation notation,

$$f_3(\mathbf{x}) = \sum_i \sum_j a_{ij} x_i x_j \quad \text{and} \quad \frac{\partial f_3}{\partial x_k} = \sum_i \sum_j a_{ij} \left\{ \frac{\partial x_i}{\partial x_k} x_j + x_i \frac{\partial x_j}{\partial x_k} \right\} = \sum_j a_{kj} x_j + \sum_i a_{ik} x_i$$

Returning to matrix notation $\nabla_{\mathbf{x}}(\mathbf{x}^T\mathbf{A}\mathbf{x}) = \mathbf{A}\mathbf{x} + \mathbf{A}^T\mathbf{x}$. If $\mathbf{A} = \mathbf{A}^T$, as is usual when dealing with quadratic forms, then $\nabla_{\mathbf{x}}(\mathbf{x}^T\mathbf{A}\mathbf{x}) = 2\mathbf{A}\mathbf{x}$. ■

The geometrical interpretation of the gradient is often useful. To aid in visualization, the vector $\mathbf{x}$ is restricted to two components. Then for each point $\mathbf{x}$ in the plane, the function $f(\mathbf{x})$ has some prescribed value. Figure 4.1 shows such a function.

The equation $f(\mathbf{x}) = c$, with c constant, specifies a locus of points in the plane. Figure 4.2 shows the locus of points in the plane for several different values of c.

At a given point such as $\mathbf{x}_0$ in Figure 4.2, $\nabla_{\mathbf{x}} f$ is a vector normal to the curve $f(\mathbf{x}) = c$, and it points in the direction of increasing values of $f(\mathbf{x})$. The gradient defines the direction of maximum increase of the function $f(\mathbf{x})$.

The derivative of a scalar function with respect to a vector yields a vector, the gradient vector. If a vector-valued function of a vector, $\mathbf{f}(\mathbf{x}) = [f_1(\mathbf{x}) f_2(\mathbf{x}) \cdots f_m(\mathbf{x})]^T$, is

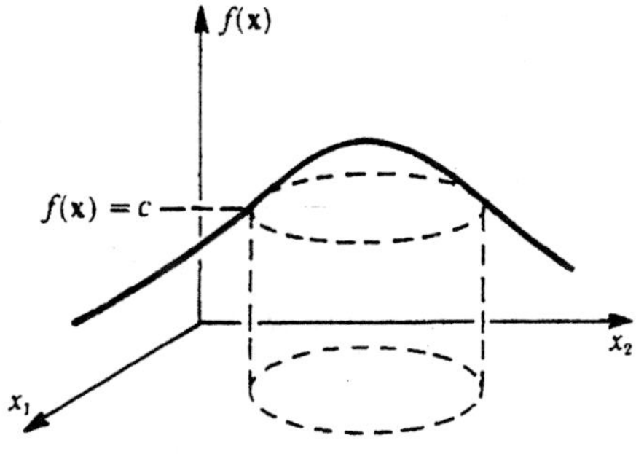

Figure 4.1

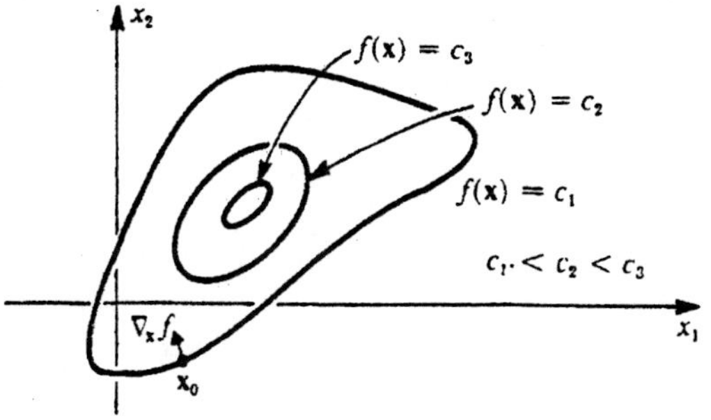

Figure 4.2

considered, the gradient of each component $f_i(\mathbf{x})$ is a column vector with the same dimension as $\mathbf{x}$. If the function $\mathbf{f}(\mathbf{x})$ is transposed to a row vector, then

$$\nabla_{\mathbf{x}}\mathbf{f}^T(\mathbf{x}) = [\nabla_{\mathbf{x}} f_1 \quad \nabla_{\mathbf{x}} f_2(\mathbf{x}) \quad \cdots \quad \nabla_{\mathbf{x}} f_m(\mathbf{x})] \tag{4.9}$$

is an $n \times m$ matrix whose columns are gradients. The transpose of this matrix will be denoted by the symbols $\nabla_{\mathbf{x}}\mathbf{f}(x)$ or simply $d\mathbf{f}/d\mathbf{x}$. That is, $d\mathbf{f}/d\mathbf{x} \triangleq [\partial f_i/\partial x_j]$ and this $m \times n$ matrix is the *Jacobian matrix*. Note that the symbol $d\mathbf{f}/d\mathbf{x}$ is just a suggestive name attached to the prescribed array of first partial derivatives. The symbol $d\mathbf{f}/d\mathbf{x}$ could just as well have been called $\partial\mathbf{f}/\partial\mathbf{x}$, and it *could* have been defined alternatively as $[\partial f_j/\partial x_i]$.

EXAMPLE 4.12 Let $\mathbf{f}(\mathbf{x}) = \mathbf{A}\mathbf{x}$, and let the jth column of $\mathbf{A}^T$ be $\mathbf{a}_j$. Then since $f_j(\mathbf{x})$ can be written as $\mathbf{a}_j^T\mathbf{x} = \mathbf{x}^T\mathbf{a}_j$, it is immediate that $\nabla_x f_j(\mathbf{x}) = \mathbf{a}_j$, so that $d[\mathbf{A}\mathbf{x}]/d\mathbf{x} = \mathbf{A}$. ■

Let $\mathbf{g}(\mathbf{x})$ be a vector-valued function of $\mathbf{x}$ and let $f(\mathbf{g})$ be a scalar-valued function. Then the chain rule gives

$$\begin{aligned} df/dx_i &= \partial f/\partial g_1\, \partial g_1/\partial x_i + \partial f/\partial g_2\, \partial g_2/\partial x_i + \cdots + \partial f/\partial g_n\, \partial g_n/\partial x_i \\ &= [\partial g_j/\partial x_i] df/d\mathbf{g} \end{aligned}$$

By virtue of the convention adopted previously, the total gradient can be written as $\nabla_x f = [d\mathbf{g}/d\mathbf{x}]^T df/d\mathbf{g} = [d\mathbf{g}/d\mathbf{x}]^T\, \nabla_{\mathbf{g}} f$.

EXAMPLE 4.13 Let $f(\mathbf{g}) = \mathbf{g}^T\mathbf{W}\mathbf{g}$ and let $\mathbf{g}(\mathbf{x}) = \mathbf{A}\mathbf{x} - \mathbf{y}$. Then $d\mathbf{g}/d\mathbf{x} = \mathbf{A}$ and $df/d\mathbf{g} = 2\mathbf{W}\mathbf{g}$, so that $d\{[\mathbf{A}\mathbf{x} - \mathbf{y}]^T\mathbf{W}[\mathbf{A}\mathbf{x} - \mathbf{y}]\}/d\mathbf{x} = 2\mathbf{A}^T\mathbf{W}\mathbf{g}$. ■

The preceding extends to scalar functions of several vector functions of $\mathbf{x}$. If $f = f(\mathbf{g}(\mathbf{x}), \mathbf{h}(\mathbf{x}))$, then

$$df/d\mathbf{x} = [d\mathbf{g}/d\mathbf{x}]^T df/d\mathbf{g} + [d\mathbf{h}/d\mathbf{x}]^T df/d\mathbf{h}$$

EXAMPLE 4.14 Let $f = \mathbf{g}^T(\mathbf{x})\mathbf{h}(\mathbf{x})$ and let $\mathbf{g}(\mathbf{x}) = \mathbf{A}\mathbf{x} + \mathbf{b}$ and $\mathbf{h}(\mathbf{x}) = \mathbf{B}\mathbf{x} + \mathbf{c}$. Then

$$df/d\mathbf{x} = \mathbf{A}^T[\mathbf{B}\mathbf{x} + \mathbf{c}] + \mathbf{B}^T[\mathbf{A}\mathbf{x} + \mathbf{b}]$$ ■

The second partial derivatives of a function of a vector also arise on occasion. When $f(\mathbf{x})$ is a scalar-valued function, the matrix of all second partial derivatives, called the *Hessian matrix*, will be denoted by

$$\frac{d^2f}{d\mathbf{x}^2} = \begin{bmatrix} \frac{\partial^2 f}{\partial x_1^2} & \frac{\partial^2 f}{\partial x_1 \partial x_2} & \cdots & \frac{\partial^2 f}{\partial x_1 \partial x_n} \\ \frac{\partial^2 f}{\partial x_2 \partial x_1} & \frac{\partial^2 f}{\partial x_2^2} & \cdots & \frac{\partial^2 f}{\partial x_2 \partial x_n} \\ \vdots & & & \vdots \\ \frac{\partial^2 f}{\partial x_n \partial x_1} & \frac{\partial^2 f}{\partial x_n \partial x_2} & \cdots & \frac{\partial^2 f}{\partial x_n^2} \end{bmatrix} \tag{4.10}$$

It becomes notationally awkward to continue these definitions to the second derivative of a vector function **f** with respect to a vector. This would require a three- dimensional array, with a typical element being $\partial^2 f_i/\partial x_j\, \partial x_k$.

4.12.2 Generalized Taylor Series

The Taylor series expansion is one of the most useful formulas in the analysis of nonlinear equations. The expansion of a scalar-valued function of a scalar is recalled [5]:

$$f(x_0 + \delta x) = f(x_0) + \left.\frac{df}{dx}\right|_0 \delta x + \frac{1}{2!}\left.\frac{d^2f}{dx^2}\right|_0 \delta x^2 + \cdots \tag{4.11}$$

The notation $\left.\frac{df}{dx}\right|_0$ indicates that all derivatives are evaluated at the point x_0.

The Taylor expansion of a function of two variables x and y is

$$\begin{aligned} f(x_0 + \delta x, y_0 + \delta y) = f(x_0, y_0) &+ \left.\frac{\partial f}{\partial x}\right|_0 \delta x + \left.\frac{\partial f}{\partial y}\right|_0 \delta y \\ &+ \frac{1}{2!}\left[\left.\frac{\partial^2 f}{\partial x^2}\right|_0 \delta x^2 + 2\left.\frac{\partial^2 f}{\partial x\, \partial y}\right|_0 \delta x\, \delta y + \left.\frac{\partial^2 f}{\partial y^2}\right|_0 \delta y^2\right] + \cdots \end{aligned} \tag{4.12}$$

If the two variables x and y are used to define the vector $\mathbf{x} = [x \quad y]^T$, the preceding expansion is more compactly written as

$$f(\mathbf{x}_0 + \delta\mathbf{x}) = f(\mathbf{x}_0) + (\nabla_{\mathbf{x}} f|_0)^T \delta\mathbf{x} + \frac{1}{2!}\delta\mathbf{x}^T \left.\frac{d^2f}{d\mathbf{x}^2}\right|_0 \delta\mathbf{x} + \cdots \tag{4.13}$$

This generalized form of the Taylor expansion is valid for any number of components of the vector **x**.

An m component vector-valued function $\mathbf{f}(\mathbf{x})$ can be viewed as m separate scalar functions. The Taylor expansion through the first two terms can be written as

$$\mathbf{f}(\mathbf{x}_0 + \delta\mathbf{x}) = \mathbf{f}(\mathbf{x}_0) + \nabla_{\mathbf{x}} \mathbf{f}|_{\mathbf{x}_0} \delta\mathbf{x} + \cdots \tag{4.14}$$

The slight discrepancy in the gradient terms of Eqs. (*4.13*) and (*4.14*) is due to the definition of the gradient as a column vector and is the reason why the row definition is preferred by some authors. Higher terms in Eq. (*4.14*) cannot be written conveniently in matrix notation. However, the first-order terms in $\delta\mathbf{x}$ are frequently all that are used, and a good approximation results if all components of $\delta\mathbf{x}$ are sufficiently small.

The generalized Taylor series is the standard tool used in linearizing nonlinear system equations. This is utilized in Chapter 15.

4.12.3 Vectorizing a Matrix

When matrices were introduced in the previous chapter they were presented as a convenient way of arranging a number of scalar variables. No certain order was required initially, although later manipulations (e.g., matrix products) did expect certain conventions be followed. In the next chapter it will be seen that no special arrangement of elements in an abstract vector is required. We could arrange the elements around a circle if we wished. The sum of two such "circles" would be a circle, as would the product of a circle with a scalar. That is, the set of such circles is closed under the operations of addition and multiplication by a scalar. There is no compelling reason to use such a strange definition, but it could be done. However, there are times when it is more convenient to arrange elements that are traditionally in a rectangular array, and hence thought of as a matrix, into a linear column array, hence having the characteristics of a vector. The rearrangement from a rectangular array to a column is called *vectorizing* the matrix. It could be done in row order, column order, or perhaps some other scanning order. Here, the column order will be arbitrarily selected but consistently used. The capital letters used to indicate matrices will be retained, but the vectorized column form will be indicated by enclosing the letter in parenthesis. That is, if

$$\mathbf{A} = \begin{bmatrix} a_{11} & a_{12} & a_{13} & \cdots \\ a_{21} & a_{22} & a_{23} & \cdots \\ a_{31} & a_{32} & a_{33} & \cdots \\ \vdots & \vdots & \vdots & \end{bmatrix} \quad \text{then} \quad (\mathbf{A}) = \begin{bmatrix} a_{11} \\ a_{21} \\ a_{31} \\ a_{12} \\ a_{22} \\ a_{32} \\ a_{13} \\ \vdots \end{bmatrix}$$

To save space, the transpose can be written $(\mathbf{A})^T = [a_{11}\ a_{21}\ a_{31}\ \cdots\ a_{12}\ a_{22}\ a_{32}\ \cdots\ a_{13}\ a_{23}\ \cdots]$. The operations of vectorizing and transposition do not commute, that is, $(\mathbf{A}^T) \neq (\mathbf{A})^T$. Why introduce such nonstandard notation? It has not been done just to make the valid point that arrangement order is arbitrary as long as it is used in a logically consistent fashion. The major reason for introducing the concept of vectorizing a matrix is that it is *convenient* in many situations. One such situation is the derivation of matrix gradient expressions, because it allows use of the already familiar vector gradient results. Another convenient application of vectorized matrices appears in Chapter 6 in the solution of a special class of linear matrix equations called Lyapunov equations.

Matrix Gradients

Let $\mathbf{A}$ be an $m \times n$ matrix and let $f(\mathbf{A})$ be a scalar-valued function of $\mathbf{A}$. Then the matrix gradient of f with respect to $\mathbf{A}$ is written as $\partial f(\mathbf{A})/\partial\mathbf{A}$. This is just a symbol for the $m \times n$ rectangular array of scalar derivatives $[\partial f(\mathbf{A})/\partial a_{ij}]$. By vectorizing $\mathbf{A}$ and applying familiar formulas for vector gradients with respect to $(\mathbf{A})$, a column arrangement of the same scalar derivatives is easily derived for many functions f. Then the definition of $(\mathbf{A})$ can be "undone" to write the matrix gradient in the more traditional rectangular matrix form. This process is now used to derive a catalog of useful matrix gradient results. If $\mathbf{A}$ and the unit matrix are both vectorized, then it is easy to see that $\text{Tr}[\mathbf{A}] = (\mathbf{A})^T(\mathbf{I})$. In this example the trace operation is an example of the function f discussed previously, and its matrix argument must now be square in this case. Then $\partial\text{Tr}[\mathbf{A}]/\partial\mathbf{A} = \partial(\mathbf{A})^T(\mathbf{I})/\partial(\mathbf{A}) = (\mathbf{I}) = \mathbf{I}$. A number of gradients of the trace of matrix products are easily derived by noting that $\text{Tr}[\mathbf{AB}] = (\mathbf{A})^T(\mathbf{B}^T) = (\mathbf{B})^T(\mathbf{A}^T) = (\mathbf{B}^T)^T(\mathbf{A}) = (\mathbf{A}^T)^T(\mathbf{B})$. Therefore, $\partial\text{Tr}[\mathbf{AB}]/\partial\mathbf{A} = \mathbf{B}^T$, $\partial\text{Tr}[\mathbf{AB}]/\partial\mathbf{B} = \mathbf{A}^T$, $\partial\text{Tr}[\mathbf{AB}]/\partial\mathbf{A}^T = \mathbf{B}$, and $\partial\text{Tr}[\mathbf{AB}]/\partial\mathbf{B}^T = \mathbf{A}$. Consider the trace of a three-term product $\text{Tr}[\mathbf{ABC}] = \text{Tr}[\mathbf{BCA}] = \text{Tr}[\mathbf{CAB}]$ and define $\mathbf{BC} \equiv \mathbf{D}$. Then, in vectorized form, $\text{Tr}[\mathbf{ABC}] = (\mathbf{A})^T(\mathbf{D}^T)$, so that $\partial\text{Tr}[\mathbf{ABC}]/\partial\mathbf{A} = (\mathbf{D}^T) = \mathbf{D}^T = \mathbf{C}^T\mathbf{B}^T$. Similarly, $\partial\text{Tr}[\mathbf{ABC}]/\partial\mathbf{B} = \mathbf{A}^T\mathbf{C}^T$ and $\partial\text{Tr}[\mathbf{ABC}]/\partial\mathbf{C} = \mathbf{B}^T\mathbf{A}^T$. The general rule that $\partial\text{Tr}[\ \ ]/\partial\mathbf{A}^T = \{\partial\text{Tr}[\ \ ]/\partial\mathbf{A}\}^T$ for any matrix $\mathbf{A}$ allows the transpose of the preceding results to be used to get the gradient with respect to $\mathbf{A}^T$, $\mathbf{B}^T$, or $\mathbf{C}^T$. Now consider two-term or three-term products where a factor is repeated, such as in $\mathbf{A}^T\mathbf{A}$, $\mathbf{ABA}$, or $\mathbf{ABA}^T$. Since $\text{Tr}[\mathbf{AA}^T] = (\mathbf{A})^T(\mathbf{A}^T)$, it is found that $\partial\text{Tr}[\mathbf{AA}^T]/\partial\mathbf{A} = 2(\mathbf{A}) = 2\mathbf{A}$. This suggests a chain-rule-like behavior in which the total gradient is the sum of the factors obtained by treating one factor at a time as variable and treating the other factors as fixed. That is, $\partial\text{Tr}[\mathbf{ABA}^T]/\partial\mathbf{A} = \partial\text{Tr}[\mathbf{AC}]/\partial\mathbf{A} + \partial\text{Tr}[\mathbf{DA}^T]/\partial\mathbf{A}$, where $\mathbf{C} = \mathbf{BA}^T$ and $\mathbf{D} = \mathbf{AB}$ are treated as constants until after the differentiation. Previous formulas can be used on each of the terms in the sum to give

$$\partial\text{Tr}[\mathbf{ABA}^T]/\partial\mathbf{A} = (\mathbf{BA}^T)^T + \mathbf{AB} = \mathbf{AB}^T + \mathbf{AB}$$

Likewise, $\partial\text{Tr}[\mathbf{ABA}]/\partial\mathbf{A} = \mathbf{A}^T\mathbf{B}^T + \mathbf{B}^T\mathbf{A}^T$. Extensions to other variations are almost limitless. For example, $\partial\text{Tr}[\mathbf{ABAC}]/\partial\mathbf{A} = \mathbf{C}^T\mathbf{A}^T\mathbf{B}^T + \mathbf{B}^T\mathbf{A}^T\mathbf{C}^T$. The previous example is a special case of this result with $\mathbf{C} = \mathbf{I}$. Also, $\partial\text{Tr}[\mathbf{ABA}^T\mathbf{C}]/\partial\mathbf{A} = \mathbf{C}^T\mathbf{AB}^T + \mathbf{CAB}$. Some formulas involving $\mathbf{A}^{-1}$ can be derived by using $\mathbf{A}^{-1} = \mathbf{A}^{-1}\mathbf{AA}^{-1}$ and using the previous chain rule to find $\partial\text{Tr}[\mathbf{A}^{-1}]/\partial\mathbf{A} = \partial\text{Tr}[\mathbf{A}^{-1}\mathbf{C}]/\partial\mathbf{A} + \partial\text{Tr}[\mathbf{DA}^{-1}]/\partial\mathbf{A} + \partial\text{Tr}[\mathbf{AE}]/\partial\mathbf{A}$, where $\mathbf{C} = \mathbf{AA}^{-1} = \mathbf{I}$, $\mathbf{D} = \mathbf{A}^{-1}\mathbf{A} = \mathbf{I}$, and $\mathbf{E} = [\mathbf{A}^{-1}]^2 = \mathbf{A}^{-2}$. Therefore, $\partial\text{Tr}[\mathbf{A}^{-1}]/\partial\mathbf{A} = 2\partial\text{Tr}[\mathbf{A}^{-1}]/\partial\mathbf{A} + \mathbf{E}^T$, or $\partial\text{Tr}[\mathbf{A}^{-1}]/\partial\mathbf{A} = -\mathbf{E}^T = -[\mathbf{A}^{-2}]^T$. Similar manipulation can be used to show that

$$\partial\text{Tr}[\mathbf{BA}^{-1}\mathbf{C}]/\partial\mathbf{A} = -[\mathbf{A}^{-1}\mathbf{CBA}^{-1}]^T$$

One final matrix gradient expression where the scalar-valued function f is the determinant of its argument can be derived without use of the intermediate vectorization process. Since $|\mathbf{A}| = \Sigma a_{ij}C_{ij}$, where the Laplace expansion can be along any row or column and where C_{ij} is the ijth cofactor, it is easy to see that

$$\partial|\mathbf{A}|/\partial\mathbf{A} = [C_{ij}] = \text{Adj}[\mathbf{A}]^T$$

Applications of matrix gradients arise in several optimal control and estimation problems. A scalar cost function, such as the trace, is minimized with respect to a selectable matrix by setting the matrix gradient to zero.

REFERENCES

1. DeRusso, P. M., R. J. Roy, and C. M. Close: *State Variables for Engineers,* John Wiley, New York, 1965.
2. Hovanessian, S. A. and L. A. Pipes: *Digital Computer Methods in Engineering,* McGraw-Hill, New York, 1969.
3. Pipes, L. A.: *Applied Mathematics for Engineers and Physicists,* 3rd ed., McGraw-Hill, New York, 1971.
4. Kailath, T.: *Linear Systems,* Prentice Hall, Englewood Cliffs, N.J., 1980.
5. Taylor, A. E. and R. W. Mann: *Advanced Calculus,* Xerox, Boston, 1972.
6. Cruz, J. B., Jr., J. S. Freundenberg, and D. P. Looze: "A Relationship Between Sensitivity and Stability of Multivariable Feedback Systems," *IEEE Transactions on Automatic Control,* Vol. AC-26, No. 1, Feb. 1981, pp. 66–74.
7. Lawson, C. L. and R. J. Hanson: *Solving Least Squares Problems,* Prentice-Hall, Englewood Cliffs, N.J., 1974.
8. Kunz, K. S.: *Numerical Analysis,* McGraw-Hill, New York, 1957.

ILLUSTRATIVE PROBLEMS

Introductory Manipulations

4.1 Let

$$\mathbf{A} = \begin{bmatrix} 1 & 4 \\ 2 & 5 \end{bmatrix}, \quad \mathbf{B} = \begin{bmatrix} 3 & 1 \\ 1 & 3 \end{bmatrix}, \quad \text{and} \quad \mathbf{C} = \begin{bmatrix} 42 & 16 \\ 5 & 3 \\ 8 & 1 \end{bmatrix}$$

Compute $\mathbf{A}+\mathbf{B}$, $\mathbf{A}-\mathbf{B}$, $\mathbf{AB}$, $\mathbf{BA}$, $\mathbf{CA}$, $\mathbf{CB}$, $\mathbf{AC}$, and $\mathbf{AC}^T$.

$$\mathbf{A}+\mathbf{B} = \begin{bmatrix} 1+3 & 4+1 \\ 2+1 & 5+3 \end{bmatrix} = \begin{bmatrix} 4 & 5 \\ 3 & 8 \end{bmatrix}, \qquad \mathbf{A}-\mathbf{B} = \begin{bmatrix} -2 & 3 \\ 1 & 2 \end{bmatrix}$$

$$\mathbf{AB} = \begin{bmatrix} 1(3)+4(1) & 1(1)+4(3) \\ 2(3)+5(1) & 2(1)+5(3) \end{bmatrix} = \begin{bmatrix} 7 & 13 \\ 11 & 17 \end{bmatrix}, \qquad \mathbf{BA} = \begin{bmatrix} 5 & 17 \\ 7 & 19 \end{bmatrix}$$

$$\mathbf{CA} = \begin{bmatrix} 42(1)+16(2) & 42(4)+16(5) \\ 5(1)+\ 3(2) & 5(4)+\ 3(5) \\ 8(1)+\ 1(2) & 8(4)+\ 1(5) \end{bmatrix} = \begin{bmatrix} 74 & 248 \\ 11 & 35 \\ 10 & 37 \end{bmatrix}, \qquad \mathbf{CB} = \begin{bmatrix} 142 & 90 \\ 18 & 14 \\ 25 & 11 \end{bmatrix}$$

$\mathbf{AC}$ is not defined because of the conformability rule: $(2 \times 2)(3 \times 2)$.

$$\mathbf{AC}^T = \begin{bmatrix} 1 & 4 \\ 2 & 5 \end{bmatrix}\begin{bmatrix} 42 & 5 & 8 \\ 16 & 3 & 1 \end{bmatrix} = \begin{bmatrix} 1(42)+4(16) & 1(5)+4(3) & 1(8)+4(1) \\ 2(42)+5(16) & 2(5)+5(3) & 2(8)+5(1) \end{bmatrix}$$

$$= \begin{bmatrix} 106 & 17 & 12 \\ 164 & 25 & 21 \end{bmatrix}$$

Multiple Variable Systems and Transfer Matrices

4.2 Consider the multiple-input, multiple-output feedback system shown in Figure 4.3, where $\mathbf{G}_1$, $\mathbf{G}_2$, $\mathbf{H}_1$, and $\mathbf{H}_2$ are transfer function matrices and $\mathbf{R}$, $\mathbf{E}_1$, $\mathbf{E}_2$, $\mathbf{V}$, $\mathbf{W}$, $\mathbf{C}$, $\mathbf{D}$, $\mathbf{F}_1$, and $\mathbf{F}_2$ are column matrices. If $\mathbf{R}$ has r components, $\mathbf{V}$ has m components, $\mathbf{C}$ has n components, and $\mathbf{D}$ has p components, determine the dimensions of all other matrices in the diagram.

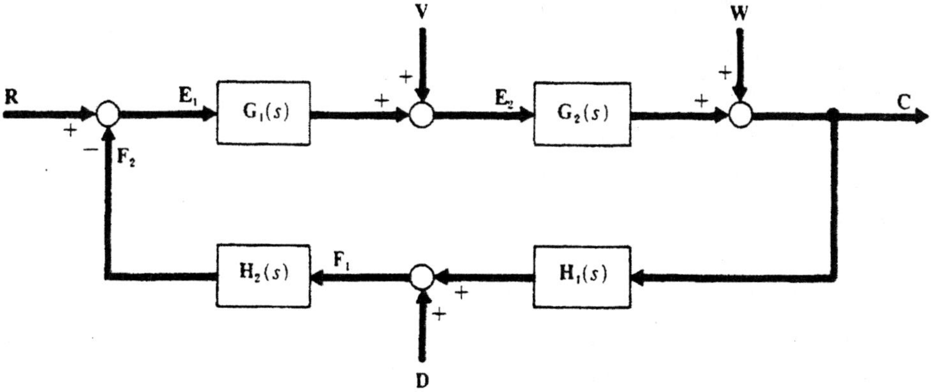

Figure 4.3

In order for $\mathbf{R} - \mathbf{F}_2 = \mathbf{E}_1$ to make sense, $\mathbf{F}_2$ and $\mathbf{E}_1$ must be $r \times 1$ matrices like $\mathbf{R}$. If $\mathbf{G}_1\mathbf{E}_1$ is to be conformable, $\mathbf{G}_1$ must have r columns. Since this product adds to $\mathbf{V}$, it must be an $m \times 1$ matrix. So $\mathbf{G}_1$ must be $m \times r$. Since $\mathbf{G}_1\mathbf{E}_1 + \mathbf{V} = \mathbf{E}_2$, $\mathbf{E}_2$ must be an $m \times 1$ matrix. Conformability requires that $\mathbf{G}_2$ have m columns, and since $\mathbf{G}_2\mathbf{E}_2 + \mathbf{W} = \mathbf{C}$ is an $n \times 1$ matrix, the transfer matrix $\mathbf{G}_2$ must be $n \times m$. Similar reasoning requires that $\mathbf{H}_1$ be a $p \times n$ matrix, $\mathbf{F}_1$ be a $p \times 1$ matrix, and $\mathbf{H}_2$ be an $r \times p$ matrix.

4.3 Referring to the system of Problem 4.2, derive the overall transfer matrix relating the input $\mathbf{R}$ to the output $\mathbf{C}$.

Ignoring all inputs except $\mathbf{R}$, this system is described by five matrix equations:

$$\mathbf{E}_1 = \mathbf{R} - \mathbf{F}_2, \qquad \mathbf{E}_2 = \mathbf{G}_1\mathbf{E}_1, \qquad \mathbf{C} = \mathbf{G}_2\mathbf{E}_2, \qquad \mathbf{F}_1 = \mathbf{H}_1\mathbf{C}, \quad \text{and} \quad \mathbf{F}_2 = \mathbf{H}_2\mathbf{F}_1$$

Depending upon the sequence of algebraic manipulations used to eliminate all terms except $\mathbf{R}$ and $\mathbf{C}$, different forms of the final result are obtained. Four different sequences are presented.

1. Eliminating $\mathbf{E}_2$ gives $\mathbf{C} = \mathbf{G}_2\mathbf{G}_1\mathbf{E}_1$ and eliminating $\mathbf{F}_1$ gives $\mathbf{F}_2 = \mathbf{H}_2\mathbf{H}_1\mathbf{C}$. Combining gives $\mathbf{C} = \mathbf{G}_2\mathbf{G}_1\mathbf{R} - \mathbf{G}_2\mathbf{G}_1\mathbf{H}_2\mathbf{H}_1\mathbf{C}$ or $[\mathbf{I}_n + \mathbf{G}_2\mathbf{G}_1\mathbf{H}_2\mathbf{H}_1]\mathbf{C} = \mathbf{G}_2\mathbf{G}_1\mathbf{R}$ so that

$$\mathbf{C} = [\mathbf{I}_n + \mathbf{G}_2\mathbf{G}_1\mathbf{H}_2\mathbf{H}_1]^{-1}\mathbf{G}_2\mathbf{G}_1\mathbf{R}$$

The similarity with the scalar closed-loop transfer function is apparent.

2. In this sequence $\mathbf{E}_2$ is first isolated and then, in turn, related to $\mathbf{C}$ as follows: $\mathbf{E}_2 = \mathbf{G}_1(\mathbf{R} - \mathbf{F}_2)$, but

$$\mathbf{F}_2 = \mathbf{H}_2\mathbf{H}_1\mathbf{G}_2\mathbf{E}_2 \quad \text{so that } [\mathbf{I}_m + \mathbf{G}_1\mathbf{H}_2\mathbf{H}_1\mathbf{G}_2]\mathbf{E}_2 = \mathbf{G}_1\mathbf{R}$$

or

$$\mathbf{E}_2 = [\mathbf{I}_m + \mathbf{G}_1\mathbf{H}_2\mathbf{H}_1\mathbf{G}_2]^{-1}\mathbf{G}_1\mathbf{R} \quad \text{so that } \mathbf{C} = \mathbf{G}_2\mathbf{E}_2 = \mathbf{G}_2[\mathbf{I}_m + \mathbf{G}_1\mathbf{H}_2\mathbf{H}_1\mathbf{G}_2]^{-1}\mathbf{G}_1\mathbf{R}$$

Note the different arrangement of terms and the size of the matrix to be inverted.

3. If $\mathbf{E}_1$ is first isolated in a similar way, we obtain

$$\mathbf{E}_1 = [\mathbf{I}_r + \mathbf{H}_2\mathbf{H}_1\mathbf{G}_2\mathbf{G}_1]^{-1}\mathbf{R}$$

from which

$$\mathbf{C} = \mathbf{G}_2\mathbf{G}_1[\mathbf{I}_r + \mathbf{H}_2\mathbf{H}_1\mathbf{G}_2\mathbf{G}_1]^{-1}\mathbf{R}$$

4. If $\mathbf{F}_1$ is first isolated, we obtain

$$\mathbf{F}_1 = [\mathbf{I}_p + \mathbf{H}_1\mathbf{G}_2\mathbf{G}_1\mathbf{H}_2]^{-1}\mathbf{H}_1\mathbf{G}_2\mathbf{G}_1\mathbf{R}$$

Premultiplying this by $\mathbf{H}_2$ gives $\mathbf{F}_2$. Subtracting $\mathbf{F}_2$ from $\mathbf{R}$ gives

$$\mathbf{E}_1 = \{\mathbf{I}_r - \mathbf{H}_2[\mathbf{I}_p + \mathbf{H}_1\mathbf{G}_2\mathbf{G}_1\mathbf{H}_2]^{-1}\mathbf{H}_1\mathbf{G}_2\mathbf{G}_1\}\mathbf{R}$$

Premultiplying this by $\mathbf{G}_2\mathbf{G}_1$ gives

$$\mathbf{C} = \{\mathbf{G}_2\mathbf{G}_1 - \mathbf{G}_2\mathbf{G}_1\mathbf{H}_2[\mathbf{I}_p + \mathbf{H}_1\mathbf{G}_2\mathbf{G}_1\mathbf{H}_2]^{-1}\mathbf{H}_1\mathbf{G}_2\mathbf{G}_1\}\mathbf{R}$$

4.4 Use the result of Problem 4.3 to establish some useful matrix identities.

In the previous problem four results of the form $\mathbf{C} = \mathbf{T}_i\mathbf{R}$ were found, with the differences contained in the four $\mathbf{T}_i$ matrices. Equating two forms for the output $\mathbf{C}$ gives

$$\mathbf{T}_i\mathbf{R} = \mathbf{T}_j\mathbf{R}$$

In general, this equation is not sufficient for concluding that $\mathbf{T}_i = \mathbf{T}_j$. However, in this case it must be true for *all* $\mathbf{R}$ that $(\mathbf{T}_i - \mathbf{T}_j)\mathbf{R} = \mathbf{0}$ so that $\mathbf{T}_i - \mathbf{T}_j$ must be the null matrix, or $\mathbf{T}_i = \mathbf{T}_j$. Therefore, the results of Problem 4.3 give the following matrix identities. They are true for any matrices which satisfy the conformability conditions, whenever the indicated inverses exist:

$$[\mathbf{I}_n + \mathbf{G}_2\mathbf{G}_1\mathbf{H}_2\mathbf{H}_1]^{-1}\mathbf{G}_2\mathbf{G}_1 = \mathbf{G}_2[\mathbf{I}_m + \mathbf{G}_1\mathbf{H}_2\mathbf{H}_1\mathbf{G}_2]^{-1}\mathbf{G}_1 = \mathbf{G}_2\mathbf{G}_1[\mathbf{I}_r + \mathbf{H}_2\mathbf{H}_1\mathbf{G}_2\mathbf{G}_1]^{-1}$$
$$= \mathbf{G}_2\mathbf{G}_1 - \mathbf{G}_2\mathbf{G}_1\mathbf{H}_2[\mathbf{I}_p + \mathbf{H}_1\mathbf{G}_2\mathbf{G}_1\mathbf{H}_2]^{-1}\mathbf{H}_1\mathbf{G}_2\mathbf{G}_1$$

4.5 Consider the system of Problem 4.2. a. Find the closed-loop transfer function matrices which relate the following input-output matrix pairs: $\mathbf{R}$ to $\mathbf{E}_1$, $\mathbf{V}$ to $\mathbf{E}_2$, $\mathbf{W}$ to $\mathbf{C}$, and $\mathbf{D}$ to $\mathbf{F}_1$. Also determine the characteristic equations. b. Discuss the matrix generalization of the return difference concept.

(a) Starting with the basic relations in Problem 4.3 and using similar manipulations leads to

$$\mathbf{E}_1 = [\mathbf{I}_r + \mathbf{H}_2\mathbf{H}_1\mathbf{G}_2\mathbf{G}_1]^{-1}\mathbf{R} \qquad \mathbf{C} = [\mathbf{I}_n + \mathbf{G}_2\mathbf{G}_1\mathbf{H}_2\mathbf{H}_1]^{-1}\mathbf{W}$$

$$\mathbf{E}_2 = [\mathbf{I}_m + \mathbf{G}_1\mathbf{H}_2\mathbf{H}_1\mathbf{G}_2]^{-1}\mathbf{V} \qquad \mathbf{F}_1 = [\mathbf{I}_p + \mathbf{H}_1\mathbf{G}_2\mathbf{G}_1\mathbf{H}_2]^{-1}\mathbf{D}$$

Since, for example,

$$[\mathbf{I}_r + \mathbf{H}_2\mathbf{H}_1\mathbf{G}_2\mathbf{G}_1]^{-1} = \frac{\text{Adj}\,[\mathbf{I}_r + \mathbf{H}_2\mathbf{H}_1\mathbf{G}_2\mathbf{G}_1]}{|\mathbf{I}_r + \mathbf{H}_2\mathbf{H}_1\mathbf{G}_2\mathbf{G}_1|}$$

the characteristic equation is obtained by setting the determinant in the denominator to zero. The characteristic equation is a property of the system and not the particular inputs or outputs considered. It is reasonable to expect that the following determinant identities are true (this is proven in Chapter 7, Problem 7.22, by other means):

$$|\mathbf{I}_r + \mathbf{H}_2\mathbf{H}_1\mathbf{G}_2\mathbf{G}_1| = |\mathbf{I}_m + \mathbf{G}_1\mathbf{H}_2\mathbf{H}_1\mathbf{G}_2| = |\mathbf{I}_n + \mathbf{G}_2\mathbf{G}_1\mathbf{H}_2\mathbf{H}_1|$$
$$= |\mathbf{I}_p + \mathbf{H}_1\mathbf{G}_2\mathbf{G}_1\mathbf{H}_2|$$

The roots of any one of these determinants constitute the poles of the multivariable system, provided no cancellation with numerator terms has occurred.

(b) The matrices contained in the preceding determinants and whose inverses appear in the preceding transfer functions are termed *return difference matrices*. These are the multivariable generalization of the return difference function in Chapter 2. In single-loop scalar problems, the return difference $R_d(s)$ is the same regardless of where the loop is broken. In the multivariable case here, the return difference matrix is different at each point in the loop because of the order in which the factors $\mathbf{G}_i$ and $\mathbf{H}_i$ appear. In this matrix case,

$\mathbf{F}_d(s) = \{\mathbf{I} + [$the product of $\mathbf{G}_i$ and $\mathbf{H}_i$ matrices in the order encountered while traversing the loop backward from the point in question$]\}$

The *determinants* of the return difference matrices are all the same, but the matrices themselves differ from point to point. As in the scalar case, a small return difference indicates low stability margins and poor sensitivity to disturbances and model variations. Although the scalar case is unambiguous, the "size" of the return difference matrix can be measured in various ways. What is really needed is a measure of how near these matrices are to being singular. In general, the determinant is a poor measure of near singularity. The singular values of a matrix, developed in Chapter 7, are a much more meaningful measure, and the preceding four matrices will all have different singular values.

Disturbance rejection: The contribution of each of the four inputs to the output of Figure 4.3 are

$$\mathbf{C}_R = \mathbf{G}_2\mathbf{G}_1[\mathbf{I}_r + \mathbf{H}_2\mathbf{H}_1\mathbf{G}_2\mathbf{G}_1]^{-1}\mathbf{R}$$

$$\mathbf{C}_V = \mathbf{G}_2[\mathbf{I}_m + \mathbf{G}_1\mathbf{H}_2\mathbf{H}_1\mathbf{G}_2]^{-1}\mathbf{V}$$

$$\mathbf{C}_W = [\mathbf{I}_n + \mathbf{G}_2\mathbf{G}_1\mathbf{H}_2\mathbf{H}_1]^{-1}\mathbf{W}$$

$$\mathbf{C}_D = \mathbf{G}_2\mathbf{G}_1\mathbf{H}_2[\mathbf{I}_p + \mathbf{H}_1\mathbf{G}_2\mathbf{G}_1\mathbf{H}_2]^{-1}\mathbf{D}$$

Presumably $\mathbf{R}$ is a desired input. If the return differences are made "large," say by increasing $\mathbf{G}_1$, the outputs due to $\mathbf{V}$ and $\mathbf{W}$ disturbance inputs can be made "small." Without getting into exactly what large and small mean here, this is the essential idea behind disturbance rejection in feedback systems. Sensitivity to model errors is also reduced as the return difference matrix is made larger [6]. The return difference matrix, like its scalar counterpart, is frequency-dependent. For any real system the transfer function magnitudes will eventually go to zero as $\omega \to \infty$. Therefore, the return differences will eventually go to $\mathbf{I}$ (or 1). Robust control system design is concerned with maintaining a sufficiently high return difference over the frequency range of interest and then having it approach its asymptotic value in a graceful fashion.

4.6 A single-input, two-output feedback system has the form shown in Figure 4.3, with

$$\mathbf{G}_2\mathbf{G}_1 = \begin{bmatrix} \dfrac{1}{s+1} \\[2ex] \dfrac{1}{s+2} \end{bmatrix} \qquad \mathbf{H}_2\mathbf{H}_1 = [s \quad 1]$$

Find the characteristic equation and the transfer function matrix relating $\mathbf{R}$ to $\mathbf{C}$, using two different formulations.

Using $|\mathbf{I}_2 + \mathbf{G}_2\mathbf{G}_1\mathbf{H}_2\mathbf{H}_1| = 0$ gives

$$\left|\begin{bmatrix} 1 & 0 \\ 0 & 1 \end{bmatrix} + \begin{bmatrix} \dfrac{s}{s+1} & \dfrac{1}{s+1} \\[2ex] \dfrac{s}{s+2} & \dfrac{1}{s+2} \end{bmatrix}\right| = \left(1 + \frac{s}{s+1}\right)\left(1 + \frac{1}{s+2}\right) - \frac{s}{(s+1)(s+2)} = 0$$

Using $|\mathbf{I}_1 + \mathbf{H}_2\mathbf{H}_1\mathbf{G}_2\mathbf{G}_1| = 1 + s/(s+1) + 1/(s+2) = 0$ leads to the same characteristic equation with less effort.

Likewise, $\mathbf{C} = [\mathbf{I}_2 + \mathbf{G}_2\mathbf{G}_1\mathbf{H}_2\mathbf{H}_1]^{-1}\mathbf{G}_2\mathbf{G}_1\mathbf{R}$ gives

$$\mathbf{C} = \frac{\begin{bmatrix} 1+\frac{1}{s+2} & -\frac{1}{s+1} \\ -\frac{s}{s+2} & 1+\frac{s}{s+1} \end{bmatrix}\begin{bmatrix} \frac{1}{s+1} \\ \frac{1}{s+2} \end{bmatrix}\mathbf{R}}{1+\frac{s}{s+1}+\frac{1}{s+2}}$$

Using $\mathbf{C} = \mathbf{G}_2\mathbf{G}_1[1 + \mathbf{H}_2\mathbf{H}_1\mathbf{G}_2\mathbf{G}_1]^{-1}\mathbf{R}$ gives

$$\mathbf{C} = \frac{\begin{bmatrix} \frac{1}{s+1} \\ \frac{1}{s+2} \end{bmatrix}\mathbf{R}}{1+\frac{s}{s+1}+\frac{1}{s+2}}$$

These are identical, but the second form is obtained without the requirement of matrix inversion.

4.7 Consider the four forms of the closed-loop transfer matrix derived in Problem 4.3. What are the dimensions of the matrices that need to be inverted in each form if $\mathbf{G}_1$ is 10×1000, $\mathbf{G}_2$ is 50×10, $\mathbf{H}_1$ is 1×50, and $\mathbf{H}_2$ is 1000×1?

The first form requires inverting $\mathbf{I}_n + \mathbf{G}_2\mathbf{G}_1\mathbf{H}_2\mathbf{H}_1$, which is a 50×50 matrix. Form 2 requires inverting a 10×10 matrix since $m = 10$. The third form requires inversion of a 1000×1000 matrix, since $r = 1000$. The fourth form requires only a scalar division since $p = 1$. The same possibilities exist for the size of the determinant to be used in finding the characteristic equation.

Determinants, Cramer's Rule, Rank

4.8 A 5×5 matrix decomposes into the unit matrix plus a product, as shown. Evaluate its determinant.

$$\mathbf{A} = \begin{bmatrix} 0 & -2 & -3 & -4 & -5 \\ -1 & -1 & -3 & -4 & -5 \\ 4 & 8 & 13 & 16 & 20 \\ 2 & 4 & 6 & 9 & 10 \\ 8 & 16 & 24 & 32 & 41 \end{bmatrix} = \mathbf{I}_5 + \begin{bmatrix} -1 \\ -1 \\ 4 \\ 2 \\ 8 \end{bmatrix}[1 \quad 2 \quad 3 \quad 4 \quad 5]$$

$$|\mathbf{A}| = |\mathbf{I}_5 + \mathbf{GH}| = 1 + \mathbf{HG} = 1 + [1 \quad 2 \quad 3 \quad 4 \quad 5]\begin{bmatrix} -1 \\ -1 \\ 4 \\ 2 \\ 8 \end{bmatrix} = 58$$

4.9 Does $\mathbf{A} = \begin{bmatrix} 16 & 0 & 4 & 7 \\ -3 & 8 & 8 & 2 \\ 1 & 0 & 5 & 2 \\ -7 & 6 & 5 & 4 \end{bmatrix}$ have an inverse?

$\mathbf{A}^{-1}$ exists if and only if $|\mathbf{A}| \neq 0$. To check the determinant, the method of Laplace expansion is used with respect to the second column, $|\mathbf{A}| = 8C_{22} + 6C_{42}$. The two cofactors are

$$C_{22} = (-1)^4 \begin{vmatrix} 16 & 4 & 7 \\ 1 & 5 & 2 \\ -7 & 5 & 4 \end{vmatrix} = 368 \qquad C_{42} = (-1)^6 \begin{vmatrix} 16 & 4 & 7 \\ -3 & 8 & 2 \\ 1 & 5 & 2 \end{vmatrix} = -33$$

Therefore, $|\mathbf{A}| = 8(368) - 6(33) = 2746 \neq 0$ and $\mathbf{A}^{-1}$ does exist.

.10 Check the determinant in the previous problem using a_{31} as the pivot element.

$$|\mathbf{A}| = \frac{1}{1^2} \begin{vmatrix} \begin{vmatrix} 16 & 0 \\ 1 & 0 \end{vmatrix} & \begin{vmatrix} 16 & 4 \\ 1 & 5 \end{vmatrix} & \begin{vmatrix} 16 & 7 \\ 1 & 2 \end{vmatrix} \\ \begin{vmatrix} -3 & 8 \\ 1 & 0 \end{vmatrix} & \begin{vmatrix} -3 & 8 \\ 1 & 5 \end{vmatrix} & \begin{vmatrix} -3 & 2 \\ 1 & 2 \end{vmatrix} \\ \begin{vmatrix} 1 & 0 \\ -7 & 6 \end{vmatrix} & \begin{vmatrix} 1 & 5 \\ -7 & 5 \end{vmatrix} & \begin{vmatrix} 1 & 2 \\ -7 & 4 \end{vmatrix} \end{vmatrix} = \begin{vmatrix} 0 & 76 & 25 \\ -8 & -23 & -8 \\ 6 & 40 & 18 \end{vmatrix}$$

Using the new a_{31} as the pivot gives

$$|\mathbf{A}| = \frac{1}{6} \begin{vmatrix} \begin{vmatrix} 0 & 76 \\ 6 & 40 \end{vmatrix} & \begin{vmatrix} 0 & 25 \\ 6 & 18 \end{vmatrix} \\ \begin{vmatrix} -8 & -23 \\ 6 & 40 \end{vmatrix} & \begin{vmatrix} -8 & -8 \\ 6 & 18 \end{vmatrix} \end{vmatrix} = \frac{1}{6} \begin{vmatrix} -456 & -150 \\ -182 & -96 \end{vmatrix} = \begin{vmatrix} -76 & -25 \\ -182 & -96 \end{vmatrix} = 2746$$

.11 $\mathbf{A}$, $\mathbf{B}$, $\mathbf{I}_p$, and $\mathbf{0}$ are submatrices. Show that

$$\left| \begin{bmatrix} \mathbf{A} & \mathbf{0} \\ \mathbf{B} & \mathbf{I}_p \end{bmatrix} \right| = |\mathbf{A}|$$

Using Laplace expansion p times with respect to the last p columns gives

$$\begin{vmatrix} \mathbf{A} & \mathbf{0} \\ \mathbf{B} & \mathbf{I}_p \end{vmatrix} = 1 \begin{vmatrix} \mathbf{A} & \mathbf{0} \\ \mathbf{B}' & \mathbf{I}_{p-1} \end{vmatrix} = 1 \cdot 1 \begin{vmatrix} \mathbf{A} & \mathbf{0} \\ \mathbf{B}'' & \mathbf{I}_{p-2} \end{vmatrix} = \cdots = |\mathbf{A}|$$

.12 Show that

$$\begin{vmatrix} \mathbf{A} & \mathbf{0} \\ \mathbf{B} & \mathbf{C} \end{vmatrix} = |\mathbf{A}| \cdot |\mathbf{C}|$$

and that

$$\begin{vmatrix} \mathbf{A} & \mathbf{B} \\ \mathbf{C} & \mathbf{D} \end{vmatrix} = |\mathbf{A}| \cdot |\mathbf{D} - \mathbf{C}\mathbf{A}^{-1}\mathbf{B}| \quad \text{if } \mathbf{A}^{-1} \text{ exists}$$

$$= |\mathbf{D}| \cdot |\mathbf{A} - \mathbf{B}\mathbf{D}^{-1}\mathbf{C}| \quad \text{if } \mathbf{D}^{-1} \text{ exists}$$

We have

$$\left| \begin{bmatrix} \mathbf{A} & \mathbf{0} \\ \mathbf{B} & \mathbf{C} \end{bmatrix} \right| = \left| \begin{bmatrix} \mathbf{A} & \mathbf{0} \\ \mathbf{B} & \mathbf{I} \end{bmatrix} \begin{bmatrix} \mathbf{I} & \mathbf{0} \\ \mathbf{0} & \mathbf{C} \end{bmatrix} \right| = \begin{vmatrix} \mathbf{A} & \mathbf{0} \\ \mathbf{B} & \mathbf{I} \end{vmatrix} \cdot \begin{vmatrix} \mathbf{I} & \mathbf{0} \\ \mathbf{0} & \mathbf{C} \end{vmatrix} = |\mathbf{A}| \cdot |\mathbf{C}|$$

using results of the previous problem.

If $\mathbf{A}^{-1}$ exists, the desired determinant can be multiplied by $\begin{vmatrix} \mathbf{I} & \mathbf{0} \\ -\mathbf{C}\mathbf{A}^{-1} & \mathbf{I} \end{vmatrix} = 1$:

$$\begin{vmatrix} \mathbf{A} & \mathbf{B} \\ \mathbf{C} & \mathbf{D} \end{vmatrix} = \begin{vmatrix} \mathbf{I} & \mathbf{0} \\ -\mathbf{C}\mathbf{A}^{-1} & \mathbf{I} \end{vmatrix} \begin{vmatrix} \mathbf{A} & \mathbf{B} \\ \mathbf{C} & \mathbf{D} \end{vmatrix} = \begin{vmatrix} \mathbf{A} & \mathbf{B} \\ \mathbf{0} & \mathbf{D} - \mathbf{C}\mathbf{A}^{-1}\mathbf{B} \end{vmatrix} = |\mathbf{A}| \cdot |\mathbf{D} - \mathbf{C}\mathbf{A}^{-1}\mathbf{B}|$$

If $\mathbf{D}^{-1}$ exists, use $\begin{vmatrix} \mathbf{A} & \mathbf{B} \\ \mathbf{C} & \mathbf{D} \end{vmatrix} = \begin{vmatrix} \mathbf{I} & -\mathbf{B}\mathbf{D}^{-1} \\ \mathbf{0} & \mathbf{I} \end{vmatrix} \begin{vmatrix} \mathbf{A} & \mathbf{B} \\ \mathbf{C} & \mathbf{D} \end{vmatrix}$ and repeat the above procedure.

4.13 Use Cramer's rule to find the solutions for x_1 and x_2, if $3x_1 + 2x_2 = 6$, and $x_1 - 5x_2 = 1$. In matrix form,

$$\begin{bmatrix} 3 & 2 \\ 1 & -5 \end{bmatrix}\begin{bmatrix} x_1 \\ x_2 \end{bmatrix} = \begin{bmatrix} 6 \\ 1 \end{bmatrix} \quad \text{or} \quad \mathbf{AX} = \mathbf{Y}$$

If the coefficient matrix **A** is nonsingular, Cramer's rule gives the solution for the component x_i as

$$x_i = \frac{|\mathbf{B}_i|}{|\mathbf{A}|}$$

where $\mathbf{B}_i$ is formed from **A** by replacing column i by **Y**. In this example

$$x_1 = \frac{\begin{vmatrix} 6 & 2 \\ 1 & -5 \end{vmatrix}}{|\mathbf{A}|} = \frac{32}{17} \quad \text{and} \quad x_2 = \frac{\begin{vmatrix} 3 & 6 \\ 1 & 1 \end{vmatrix}}{|\mathbf{A}|} = \frac{3}{17}$$

4.14 Use Cramer's rule and partitioned matrices to prove the following theorem.

Theorem. Let A be an $n \times n$ matrix and let a_{ij} be any nonzero element. Define $\mathbf{B}_{ij}$ as the $n-1$ by $n-1$ matrix formed by deleting row i and column j of **A**. Let **R** and **C** be $1 \times n-1$ and $n-1 \times 1$ row and column matrices formed by deleting a_{ij} from the ith row and jth column of **A**. Then

$$|\mathbf{A}| = (-1)^{i+j} a_{ij} \left| \mathbf{B}_{ij} - \frac{1}{a_{ij}} \mathbf{CR} \right|$$

Proof. Consider the set of linear equations $\mathbf{AX} = \mathbf{Y}$, where **X** and **Y** are column matrices and **Y** is all zero except $y_i = 1$. Cramer's rule is used to solve for x_j:

$$\mathbf{x}_j = \frac{1}{|\mathbf{A}|}\begin{vmatrix} \mathbf{B}_1 & \begin{matrix} 0 \\ \vdots \end{matrix} & \mathbf{B}_2 \\ a_{i1} \cdots & 1 & \cdots a_{in} \\ \mathbf{B}_3 & \begin{matrix} \vdots \\ 0 \end{matrix} & \mathbf{B}_4 \end{vmatrix} \quad \text{where} \quad \mathbf{B}_{ij} = \begin{bmatrix} \mathbf{B}_1 & \mathbf{B}_2 \\ \mathbf{B}_3 & \mathbf{B}_4 \end{bmatrix}$$

Using the Laplace expansion method with respect to column j and then rearranging gives

$$|\mathbf{A}| = \frac{(-1)^{i+j}}{x_j} |\mathbf{B}_{ij}| \tag{1}$$

In order to determine $1/x_j$, the original equation $\mathbf{AX} = \mathbf{Y}$ can be written as

$$\mathbf{B}_{ij}\mathbf{X}_a + \mathbf{C}x_j = \mathbf{0}$$

$$\mathbf{RX}_a + a_{ij}x_j = 1$$

$\mathbf{X}_a$ is formed from **X** by deleting x_j. Thus

$$\mathbf{X}_a = -\mathbf{B}_{ij}^{-1}\mathbf{C}x_j$$

so that $(-\mathbf{RB}_{ij}^{-1}\mathbf{C} + a_{ij})x_j = 1$. Using this in Eq. (*1*) gives

$$|\mathbf{A}| = (-1)^{i+j}|\mathbf{B}_{ij}| \cdot (a_{ij} - \mathbf{RB}_{ij}^{-1}\mathbf{C}) = (-1)^{i+j} a_{ij} |\mathbf{B}_{ij}| \cdot \left| 1 - \frac{1}{a_{ij}} \mathbf{RB}_{ij}^{-1}\mathbf{C} \right|$$

The determinant identities of Problem 4.5 give the final result:

$$|\mathbf{A}| = (-1)^{i+j} a_{ij} \left| \mathbf{B}_{ij} - \frac{1}{a_{ij}} \mathbf{CR} \right|$$

(A limiting process can be used to show the validity of this proof and the final result even if $|\mathbf{A}| = 0$ or $|\mathbf{B}_{ij}| = 0$.)

15 Use the previous theorem to evaluate $|\mathbf{A}|$, using a_{11} as the divisor.

$$|\mathbf{A}| = \begin{vmatrix} 4 & 8 & 1 & 3 \\ 2 & 5 & -1 & 3 \\ -1 & 6 & 7 & 9 \\ 1 & 1 & -3 & 3 \end{vmatrix} = 4 \left| \begin{bmatrix} 5 & -1 & 3 \\ 6 & 7 & 9 \\ 1 & -3 & 3 \end{bmatrix} - \tfrac{1}{4} \begin{bmatrix} 2 \\ -1 \\ 1 \end{bmatrix} [8 \quad 1 \quad 3] \right|$$

$$= 4 \begin{vmatrix} 1 & -3/2 & 3/2 \\ 8 & 29/4 & 39/4 \\ -1 & -13/4 & 9/4 \end{vmatrix} = 246$$

.16 Find the rank and trace of

$$\mathbf{A} = \begin{bmatrix} 1 & 2 & 3 & 3 \\ 2 & -2 & 1 & 1 \\ 0 & 1 & 5 & 8 \\ 1 & -6 & 4 & -4 \end{bmatrix}$$

The sum of the diagonal terms gives $\text{Tr}(\mathbf{A}) = 1 - 2 + 5 - 4 = 0$. Using any one of several methods gives $|\mathbf{A}| = 338$. Since the determinant is nonzero, the rank of $\mathbf{A}$ is 4.

Matrix Inversion and Related Topics

.17 Given

$$\begin{bmatrix} 2 & 0.5 & 2 \\ 3 & 3 & 0 \\ 1 & 0.5 & 2 \end{bmatrix} \begin{bmatrix} x_1 \\ x_2 \\ x_3 \end{bmatrix} = \begin{bmatrix} 3 \\ 1 \\ 5 \end{bmatrix} \quad \text{or} \quad \mathbf{AX} = \mathbf{Y}$$

find x_1, x_2, and x_3 using the definition of matrix inversion.

The solution is $\mathbf{X} = \mathbf{A}^{-1}\mathbf{Y}$, where $|\mathbf{A}| = 6$ and

$$\text{Adj}\,\mathbf{A} = \begin{bmatrix} 6 & -6 & -1.5 \\ 0 & 2 & -0.5 \\ -6 & 6 & 4.5 \end{bmatrix}^T$$

so that

$$\mathbf{X} = \frac{1}{6} \begin{bmatrix} 6 & 0 & -6 \\ -6 & 2 & 6 \\ -1.5 & -0.5 & 4.5 \end{bmatrix} \begin{bmatrix} 3 \\ 1 \\ 5 \end{bmatrix} = \begin{bmatrix} -2 \\ \frac{7}{3} \\ \frac{35}{12} \end{bmatrix}$$

.18 Explain the method of Gaussian elimination in terms of elementary matrices and partitioned matrices.

Gaussian elimination is used to solve $\mathbf{AX} = \mathbf{Y}$, where $\mathbf{A}$ is known, $n \times n$, and $\mathbf{Y}$ is known, $n \times p$. The $n \times p$ matrix $\mathbf{X}$ is unknown. Often $\mathbf{X}$ and $\mathbf{Y}$ are column vectors, with $p = 1$. Assuming that $\mathbf{A}^{-1}$ exists, $\mathbf{A}^{-1}\mathbf{AX} = \mathbf{A}^{-1}\mathbf{Y}$ or $\mathbf{IX} = \mathbf{A}^{-1}\mathbf{Y}$. The elementary row operations are equivalent to premultiplication by the elementary matrices. A sequence of these operations which reduces the coefficient of $\mathbf{X}$ from $\mathbf{A}$ to $\mathbf{I}$ will simultaneously change $\mathbf{Y}$ to $\mathbf{A}^{-1}\mathbf{Y}$, that is, $\mathbf{X}$. The operations are:

1. Form the $n \times n + p$ matrix $\mathbf{W}_0 = [\mathbf{A} \mid \mathbf{Y}]$.
2. Find the element in column one with maximum absolute value. Interchange that row with row one. This gives $\mathbf{W}_1 = \mathbf{E}_{1,q}\mathbf{W}_0 = [\mathbf{E}_{1,q}\mathbf{A} \mid \mathbf{E}_{1,q}\mathbf{Y}]$.

3. Divide the entire first row of $\mathbf{W}_1$ by w_{11}, assuming $w_{11} \neq 0$. This gives $W_2 = \mathbf{E}_1(1/w_{11})\mathbf{W}_1$. If $w_{11} = 0$, then the entire first column is zero, indicating that $\mathbf{A}$ is singular. If this happens, skip to step 6.
4. Multiply the first row of $\mathbf{W}_2$ by $-w_{21}$ and add to the second row. $\mathbf{W}_3 = \mathbf{E}_{1,2}(-w_{21})\mathbf{W}_2$.
5. Repeat step 4 using $\mathbf{E}_{1,3}(-w_{31}), \ldots, \mathbf{E}_{1,n}(-w_{n1})$ in sequence. This reduces column one to a one followed by $n-1$ zeros.
6. Find the maximum absolute value element $w_{\alpha 2}$ from column 2, rows 2 through n. Interchange that row with row 2, $\mathbf{W}_{k+1} = \mathbf{E}_{2,\alpha}\mathbf{W}_k$.
7. Divide row 2 by the current value of w_{22}. This is analogous to step 3. Repeat steps 4 and 5 until w_{22} is 1 (possibly zero if $\mathbf{A}$ is singular) and all $w_{i2} = 0$ below w_{22}.
8. Repeat steps 6 and 7 until the first n columns form an $n \times n$ upper triangular matrix with $\delta_i = 1$ or 0 for the ith diagonal. From this, $|\mathbf{A}| = (-1)^\nu \mu_1 \mu_2 \mu_3 \cdots \mu_n$, where ν is the number of row interchanges used and μ_i is the divisor used for the ith row, steps 3 and 7. If a zero is encountered on the diagonal, then $|\mathbf{A}| = 0$. When this happens, the rank of $\mathbf{A}$ is still of interest. The triangular form is a convenient starting point for reducing to one of the normal forms to determine rank. If $|\mathbf{A}| \neq 0$, continue to step 9.
9. Multiply the current $\mathbf{W}$ by $\mathbf{E}_{n,1}(-w_{1n})$, then by $\mathbf{E}_{n,2}(-w_{2n}), \ldots$. This reduces the nth column of $\mathbf{W}$ to $n-1$ zeros but leaves the n, n element unity.
10. Repeat step 9 for other columns, until $\mathbf{W}$ has the unit matrix for its first n columns. The last n columns of this final $\mathbf{W}$ matrix contain $\mathbf{A}^{-1}\mathbf{Y}$.

    Note that $\mathbf{A}^{-1}$ can be found by using $\mathbf{Y} = \mathbf{I}$ when setting up $\mathbf{W}_0$.

4.19 Find $\begin{bmatrix} 0 & 3 \\ 4 & 2 \end{bmatrix}^{-1}$.

The sequence of matrices, interchanges. and divisors is

$$\mathbf{W}_0 = \left[\begin{array}{cc|cc} 0 & 3 & 1 & 0 \\ 4 & 2 & 0 & 1 \end{array}\right] \xrightarrow[\nu=1]{\text{interchange}} \left[\begin{array}{cc|cc} 4 & 2 & 0 & 1 \\ 0 & 3 & 1 & 0 \end{array}\right] \xrightarrow{\mu_1 = 4} \left[\begin{array}{cc|cc} 1 & \frac{1}{2} & 0 & \frac{1}{4} \\ 0 & 3 & 1 & 0 \end{array}\right]$$

$$\xrightarrow{\mu_2 = 3} \left[\begin{array}{cc|cc} 1 & \frac{1}{2} & 0 & \frac{1}{4} \\ 0 & 1 & \frac{1}{3} & 0 \end{array}\right] \longrightarrow \left[\begin{array}{cc|cc} 1 & 0 & -\frac{1}{6} & \frac{1}{4} \\ 0 & 1 & \frac{1}{3} & 0 \end{array}\right]$$

This problem demonstrates why row interchanges are required to avoid dividing by zero. The results are

$$|\mathbf{A}| = (-1)^\nu \mu_1 \mu_2 = -12, \qquad r_A = 2, \qquad \mathbf{A}^{-1} = \frac{1}{12}\begin{bmatrix} -2 & 3 \\ 4 & 0 \end{bmatrix}$$

4.20 Let $\mathbf{A} = \begin{bmatrix} 1 & 2 & 3 \\ 2 & 4 & 6 \\ 0 & 1 & 5 \end{bmatrix}$. Determine $|\mathbf{A}|$, r_A, and $\mathbf{A}^{-1}$ if it exists.

$$\mathbf{W}_0 = \left[\begin{array}{ccc|ccc} 1 & 2 & 3 & 1 & 0 & 0 \\ 2 & 4 & 6 & 0 & 1 & 0 \\ 0 & 1 & 5 & 0 & 0 & 1 \end{array}\right] \xrightarrow{\nu = 1} \left[\begin{array}{ccc|ccc} 2 & 4 & 6 & 0 & 1 & 0 \\ 1 & 2 & 3 & 1 & 0 & 0 \\ 0 & 1 & 5 & 0 & 0 & 1 \end{array}\right]$$

$$\xrightarrow{\mu_1 = 2} \left[\begin{array}{ccc|ccc} 1 & 2 & 3 & 0 & \frac{1}{2} & 0 \\ 1 & 2 & 3 & 1 & 0 & 0 \\ 0 & 1 & 5 & 0 & 0 & 1 \end{array}\right] \longrightarrow \left[\begin{array}{ccc|ccc} 1 & 2 & 3 & 0 & \frac{1}{2} & 0 \\ 0 & 0 & 0 & 1 & -\frac{1}{2} & 0 \\ 0 & 1 & 5 & 0 & 0 & 1 \end{array}\right]$$

$$\xrightarrow[\nu=2]{\text{interchange}} \left[\begin{array}{ccc|ccc} 1 & 2 & 3 & 0 & \frac{1}{2} & 0 \\ 0 & 1 & 5 & 0 & 0 & 1 \\ 0 & 0 & 0 & 1 & -\frac{1}{2} & 0 \end{array}\right]$$

At this point we see that $\mathbf{A}$ is singular, that is, $|\mathbf{A}| = 0$ and $\mathbf{A}^{-1}$ does not exist. We can therefore drop the right half of $\mathbf{W}$ and use row or column operations to give

$$\begin{bmatrix}1 & 2 & 3\\ 0 & 1 & 5\\ 0 & 0 & 0\end{bmatrix} \longrightarrow \begin{bmatrix}1 & 2 & -7\\ 0 & 1 & 0\\ 0 & 0 & 0\end{bmatrix} \longrightarrow \begin{bmatrix}1 & 2 & 0\\ 0 & 1 & 0\\ 0 & 0 & 0\end{bmatrix} \longrightarrow \left[\begin{array}{cc|c}1 & 0 & 0\\ 0 & 1 & 0\\ \hline 0 & 0 & 0\end{array}\right] = \left[\begin{array}{c|c}\mathbf{I}_2 & 0\\ \hline 0 & 0\end{array}\right]$$

Therefore, $r_A = 2$.

21 Use Gaussian elimination to solve for x_1 and x_2, if $\begin{bmatrix}2 & -2\\ 3 & 8\end{bmatrix}\begin{bmatrix}x_1\\ x_2\end{bmatrix} = \begin{bmatrix}10\\ -5\end{bmatrix}$.

$$\mathbf{W}_0 = \left[\begin{array}{cc|c}2 & -2 & 10\\ 3 & 8 & -5\end{array}\right] \longrightarrow \left[\begin{array}{cc|c}3 & 8 & -5\\ 2 & -2 & 10\end{array}\right] \longrightarrow \left[\begin{array}{cc|c}1 & \frac{8}{3} & -\frac{5}{3}\\ 2 & -2 & 10\end{array}\right]$$

$$\longrightarrow \left[\begin{array}{cc|c}1 & \frac{8}{3} & -\frac{5}{3}\\ 0 & -\frac{22}{3} & \frac{40}{3}\end{array}\right] \longrightarrow \left[\begin{array}{cc|c}1 & \frac{8}{3} & -\frac{5}{3}\\ 0 & 1 & -\frac{20}{11}\end{array}\right] \longrightarrow \left[\begin{array}{cc|c}1 & 0 & \frac{35}{11}\\ 0 & 1 & -\frac{20}{11}\end{array}\right]$$

Therefore, $x_1 = \frac{35}{11}$ and $x_2 = -\frac{20}{11}$.

.22 $\mathbf{G}$ is an $n \times n$ complex matrix. Find $\mathbf{G}^{-1}$ using only real matrix inversion routines.

Let $\mathbf{G} = \mathbf{A} + j\mathbf{B}$, with $\mathbf{A}$ and $\mathbf{B}$ real. Assuming that $\mathbf{G}^{-1}$ exists, it can also be expressed in terms of two real matrices $\mathbf{C}$ and $\mathbf{D}$, $\mathbf{G}^{-1} = \mathbf{C} + j\mathbf{D}$. The basic requirement of a matrix inverse is that

$$\mathbf{G}\mathbf{G}^{-1} = \mathbf{I} = (\mathbf{A} + j\mathbf{B})(\mathbf{C} + j\mathbf{D}) = (\mathbf{AC} - \mathbf{BD}) + j(\mathbf{AD} + \mathbf{BC})$$

Therefore, equating real parts to real parts, $\mathbf{AC} - \mathbf{BD} = \mathbf{I}$. Equating imaginary parts $\mathbf{AD} + \mathbf{BC} = \mathbf{0}$. If $\mathbf{A}^{-1}$ exists, the solution can be written as $\mathbf{C} = (\mathbf{A} + \mathbf{BA}^{-1}\mathbf{B})^{-1}$ and $\mathbf{D} = -\mathbf{CBA}^{-1}$. The rearrangement identities of Problem 4.4 are useful in proving this. If $\mathbf{B}^{-1}$ exists but $\mathbf{A}^{-1}$ doesn't, then $[j\mathbf{G}]^{-1}$ can be sought instead. This effectively reverses the roles of $\mathbf{A}$ and $\mathbf{B}$ so the above procedure can again be used. If both $\mathbf{A}$ and $\mathbf{B}$ are singular, but $\mathbf{G}$ is not, further modifications will be necessary.

.23 Given a set of simultaneous linear equations

$$\mathbf{XA} = \mathbf{B} \tag{1}$$

where $\mathbf{A}$ and $\mathbf{B}$ are known complex-valued matrices of size $n \times n$ and $m \times n$, respectively, and where $\mathbf{X}$ is the unknown $m \times n$ matrix. Assume that columns i and $i + 1$ of $\mathbf{A}$ are complex conjugates. Assume the same for columns of $\mathbf{B}$. Show that $\mathbf{X}$ is purely real and can be computed using only real numbers from

$$\mathbf{X} = \mathbf{B}_* \mathbf{A}_*^{-1} \tag{2}$$

where $\mathbf{A}_*$ and $\mathbf{B}_*$ are formed from $\mathbf{A}$ and $\mathbf{B}$ by replacing their two complex columns by the real part and imaginary part of their respective column i.

Postmultiplying Eq. (*1*) by any nonsingular $n \times n$ matrix $\mathbf{T}$ and solving gives

$$\mathbf{X} = (\mathbf{BT})(\mathbf{AT})^{-1} = \mathbf{BA}^{-1}$$

A particular $\mathbf{T}$ is selected that differs from the unit matrix only in the four elements defined by the intersections of rows i and j with columns i and j. The four exceptional elements form the 2×2 block

$$\mathbf{T}_i = \begin{bmatrix}\dfrac{1}{2} & \dfrac{-j}{2}\\[2ex] \dfrac{1}{2} & \dfrac{j}{2}\end{bmatrix}$$

Clearly $\mathbf{T}$ satisfies the nonsingular condition. Its determinant is just $j/2$. Furthermore, $\mathbf{BT} = \mathbf{B}_*$ and $\mathbf{AT} = \mathbf{A}_*$, as demonstrated by a 2×2 case:

$$\begin{bmatrix} \alpha + j\beta & \alpha - j\beta \\ \gamma + j\delta & \gamma - j\delta \end{bmatrix} \begin{bmatrix} \frac{1}{2} & \frac{-j}{2} \\ \frac{1}{2} & \frac{j}{2} \end{bmatrix} = \begin{bmatrix} \alpha & \beta \\ \gamma & \delta \end{bmatrix}$$

Result (2) applies to matrices with any number of conjugate-pair columns, and they need not be adjacent. However, the conjugate pairs must appear in the same column numbers in both **A** and **B**. Form $\mathbf{A}_*$ and $\mathbf{B}_*$ by replacing one member of each pair by the real part and the second member by the imaginary part. Two equal real columns in **B** qualify as conjugate pairs, meaning $\mathbf{B}_*$ will have an all-zero column. This causes no problem, but the same cannot be done in **A** because two equal columns (or an all-zero column) means that **A** and $\mathbf{A}_*$ are singular. No matrix-inversion method will solve that problem.

If the original problem is to solve $\mathbf{AX} = \mathbf{B}$, then *everything* said above about *columns* must be changed to *rows*. This is equivalent to solving the transposed problem $\mathbf{X}^T\mathbf{A}^T = \mathbf{B}^T$ for $\mathbf{X}^T$.

Cholesky Decomposition

4.24 If **A** is symmetric and positive definite (see Chapter 7), it can be uniquely (except for signs) factored into $\mathbf{A} = \mathbf{S}^T\mathbf{S}$, where **S** is an upper triangular matrix. **S** is called the square root matrix of **A**. The procedure for factoring **A** is most commonly called Cholesky decomposition [7] (although it is sometimes called the method of Banachiewicz and Dwyer [8]. Deduce the algorithm for finding **S**.

The algorithm for finding the s_{ij} entries in **S** is as follows:

$$s_{11} = [a_{11}]^{1/2}; s_{1j} = a_{1j}/s_{11} \quad \text{for } j = 2, \ldots, n$$

$$s_{22} = [a_{22} - (s_{12})^2]^{1/2}$$

$$s_{2j} = [a_{2j} - s_{12}s_{1j}]/s_{22} \quad \text{for } j = 3, \ldots, n$$

$$\vdots$$

$$s_{ii} = \left[a_{ii} - \sum_{k=1}^{i-1} (s_{ki})^2\right]^{1/2} \quad \text{for } i = 2, \ldots, n$$

$$s_{ij} = \left[a_{ij} - \sum_{k=1}^{i-1} s_{ki}s_{kj}\right]/s_{ii} \quad \text{for } j = i+1, \ldots, n$$

4.25 Show how Cholesky decomposition can be used in solving simultaneous equations of the form $\mathbf{Ax} = \mathbf{y}$. Assume **y** is known and that **A** is known, symmetric, and positive definite.

Assume **S** has been found such that $\mathbf{A} = \mathbf{S}^T\mathbf{S}$. Then $\mathbf{S}^T\mathbf{Sx} = \mathbf{y}$. Define $\mathbf{Sx} = \mathbf{v}$. Then $\mathbf{S}^T\mathbf{v} = \mathbf{y}$. Because $\mathbf{S}^T$ is lower triangular, the elements of **v** can easily be found one-by-one by back substitution,

$$v_1 = y_1/s_{11}, \qquad v_2 = (y_2 - s_{12}v_1)/s_{22}, \qquad v_3 = (y_3 - s_{13}v_1 - s_{23}v_2)/s_{33}, \ldots$$

Once **v** is determined, a similar procedure can be used to find the components of **x**:

$$x_n = v_n/s_{nn}, x_{n-1} = (v_{n-1} - s_{n-1,n}v_n)/s_{n-1,n-1},$$

$$x_{n-2} = (y_{n-2} - s_{n-2,n}v_n - s_{n-2,n-1}v_{n-1})/s_{n-2,n-2}, \ldots$$

Linearizing Nonlinear Equations

4.26 The two-port nonlinear electrical device shown in Figure 4.4 is characterized by the four quantities i_1, i_2, v_1, and v_2. Use Taylor series to develop a linear model for small signal variations about a nominal operating point.

Figure 4.4

There are a variety of linear models that can be developed, depending on the form assumed for the functional relations among the four variables. There is one relationship at the input port and another at the output port. One possibility is the pair $v_1 = f_0(v_2, i_1, i_2)$ and $i_2 = g_0(i_1, v_2, v_1)$. These can be combined to yield $v_1 = f_0(v_2, i_1, g_0(i_1, v_2, v_1))$ and $i_2 = g_0(i_1, v_2, f_0(v_2, i_1, i_2))$. This shows that only two independent variables i_1 and v_2 suffice to determine v_1 and i_2. These relationships are rewritten more simply as $v_1 = f(i_1, v_2)$ and $i_2 = g(i_1, v_2)$. Let $v_2 = v_{2n} + \delta v_2$ and $i_1 = i_{1n} + \delta i_1$, where v_{2n} and i_{1n} define the nominal operating point and δv_2 and δi_1 are small variations from the nominal. Then Taylor series expansion gives

$$v_1 = f(i_{1n}, v_{2n}) + \frac{\partial f}{\partial i_1}\bigg|_n \delta i_1 + \frac{\partial f}{\partial v_2}\bigg|_n \delta v_2$$

$$i_2 = g(i_{1n}, v_{2n}) + \frac{\partial g}{\partial i_1}\bigg|_n \delta i_1 + \frac{\partial g}{\partial v_2}\bigg|_n \delta v_2$$

Obviously, $v_{1n} = f(i_{1n}, v_{2n})$ and $i_{2n} = g(i_{1n}, v_{2n})$ so that

$$\delta v_1 \triangleq v_1 - v_{1n} = h_{11}\,\delta i_1 + h_{12}\,\delta v_2$$

$$\delta i_2 \triangleq i_2 - i_{2n} = h_{21}\,\delta i_1 + h_{22}\,\delta v_2$$

where the h_{ij} terms are the partial derivatives evaluated at the nominal point. These are the hybrid or h parameters commonly used in small signal, linearized analysis of transistors and other nonlinear devices.

.27 A tracking station measures the azimuth angle α, the elevation angle β, and the range r to an earth satellite as shown in Figure 4.5.

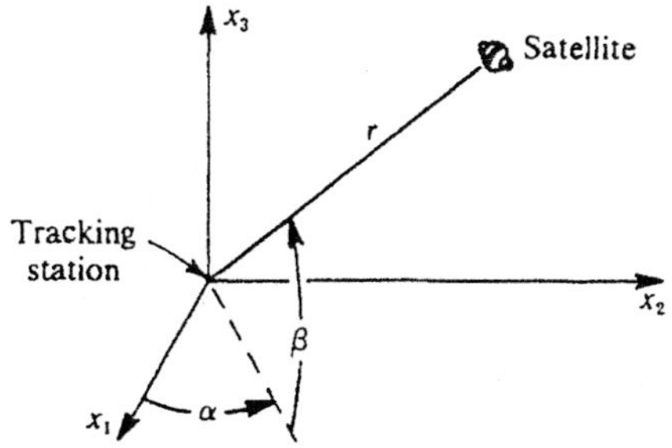

Figure 4.5

(a) Derive the nonlinear equations which relate the satellite's relative position $[x_1 \;\; x_2 \;\; x_3]'$ to the measured quantities.
(b) Obtain linear equations which relate small perturbations in satellite location to small perturbations in the measurements.

(a) The station-to-satellite range magnitude is $r = \sqrt{x_1^2 + x_2^2 + x_3^2}$ and the tracking antenna angles are $\alpha = \tan^{-1}(x_2/x_1)$ and $\beta = \tan^{-1}(x_3/\sqrt{x_1^2 + x_2^2})$.
(b) Letting $\mathbf{x}(t) = \mathbf{x}_n(t) + \delta\mathbf{x}(t)$, $r(t) = r_n(t) + \delta r(t)$, $\alpha(t) = \alpha_n(t) + \delta\alpha(t)$, and $\beta(t) = \beta_n(t) + \delta\beta(t)$, the Taylor series expansion gives

$$\delta r = \sum_{i=1}^{3} \frac{\partial r}{\partial x_i}\,\delta x_i = \frac{1}{r_n}\mathbf{x}_n^T\,\delta\mathbf{x}$$

$$\delta\alpha = \sum_{i=1}^{3} \frac{\partial \alpha}{\partial x_i}\,\delta x_i = 1/(x_{1n}^2 + x_{2n}^2)[-x_{2n} \quad x_{1n} \quad 0]\,\delta\mathbf{x}$$

$$\delta\beta = \sum_{i=1}^{3} \frac{\partial \beta}{\partial x_i}\,\delta x_i = \frac{1}{r_n^2\sqrt{x_{1n}^2 + x_{2n}^2}}[-x_{1n}x_{3n} - x_{2n}x_{3n} \quad x_{1n}^2 + x_{2n}^2]\,\delta\mathbf{x}$$

Letting $\delta\mathbf{y} = [\delta r \quad \delta\alpha \quad \delta\beta]^T$ allows the preceding results to be expressed as $\delta\mathbf{y}(t) = \mathbf{C}(t)\delta\mathbf{x}(t)$, where $\mathbf{C}$ is a 3×3 matrix.

4.28 A certain process is characterized by a set of parameters $\mathbf{x}$. Measurements $\mathbf{y}$ can be made on this process, and they are related to $\mathbf{x}$ by a nonlinear algebraic equation, $\mathbf{y} = \mathbf{f}(\mathbf{x})$. The nominal values of $\mathbf{x}$ are $\mathbf{x}_n$. Describe a method of estimating the actual values of $\mathbf{x}$ based on the measurements $\mathbf{y}$.

Let $\mathbf{x} = \mathbf{x}_n + \delta\mathbf{x}$. Then $\mathbf{y} = \mathbf{f}(\mathbf{x}_n + \delta\mathbf{x}) \cong \mathbf{f}(x_n) + \left.\frac{d\mathbf{f}}{d\mathbf{x}}\right|_n \delta\mathbf{x}$.
Call $\mathbf{f}(\mathbf{x}_n) \triangleq \mathbf{y}_n$ and $\mathbf{y} - \mathbf{y}_n \triangleq \delta\mathbf{y}$. Since $\mathbf{y}$ is measured and since $\mathbf{y}_n$ can be computed from a knowledge of $\mathbf{x}_n$, $\delta\mathbf{y}$ is a known vector. The Jacobian matrix $\left.\frac{d\mathbf{f}}{d\mathbf{x}}\right|_n \triangleq \mathbf{A}$ can also be computed. The relation $\delta\mathbf{y} = \mathbf{A}\delta\mathbf{x}$ is of the form treated in Chapter 6. Because of measurement inaccuracies, redundant measurements and the least-squares technique are most commonly used to solve for $\delta\mathbf{x}$. Then the estimated parameter values are given by $\mathbf{x} = \mathbf{x}_n + \delta\mathbf{x}$.

4.29 Equation (*4.7*) expressed a matrix transfer function $\mathbf{H}(s)$ in one form of the MFD, with the inverse of a polynomial matrix as the *left* factor. This form naturally arises from systems such as the one of Example 4.2 but with input derivative terms. Show how a second MFD form can be obtained with an inverse of a polynomial matrix as the factor on the *right*.

Assume that $\mathbf{H}(s)$ is $m \times r$. If each element in $\mathbf{H}(s)$ is placed over a common denominator, the scalar polynomial $a(s)$, then

$$\mathbf{H}(s) = \mathbf{N}(s)/a(s) = \mathbf{N}(s)[\mathbf{I}_r\,a(s)]^{-1} = [\mathbf{I}_m\,a(s)]^{-1}\,\mathbf{N}(s)$$

This shows that both the left and right forms of the MFD are possible, but the special forms here are misleading. The numerator $\mathbf{N}(s)$ is not generally the same in both forms, and the inverted matrix is not generally diagonal, as demonstrated by the $\mathbf{P}(s)^{-1}$ in Eq. (*4.7*). To indicate the more general forms which are possible, write

$$\mathbf{H}(s) = \mathbf{P}_1(s)^{-1}\,\mathbf{N}_1(s) = \mathbf{N}_2(s)\mathbf{P}_2(s)^{-1}$$

These factors are clearly nonunique. Rewrite the two expressions as

$$\mathbf{P}_1(s)\mathbf{H}(s) = \mathbf{N}_1(s) \quad \text{and} \quad \mathbf{H}(s)\mathbf{P}_2(s) = \mathbf{N}_2(s)$$

Let $\mathbf{T}_1(s)$ and $\mathbf{T}_2(s)$ be any arbitrary nonsingular $m \times m$ and $r \times r$ polynomial matrices. Then

$$\mathbf{T}_1(s)\mathbf{P}_1(s)\mathbf{H}(s) = \mathbf{T}_1(s)\mathbf{N}_1(s) \quad \text{or} \quad \mathbf{H}(s) = [\mathbf{T}_1\mathbf{P}_1]^{-1}[\mathbf{T}_1\,\mathbf{N}_1]$$

and

$$\mathbf{H}(s)\mathbf{P}_2(s)\mathbf{T}_2(s) = \mathbf{N}_2(s)\mathbf{T}_2(s) \quad \text{or} \quad \mathbf{H}(s) = [\mathbf{N}_2\,\mathbf{T}_2][\mathbf{P}_2\,\mathbf{T}_2]^{-1}$$

Thus new factors $\mathbf{P}_1' = \mathbf{T}_1\,\mathbf{P}_1$ and $\mathbf{N}_1' = \mathbf{T}_1\,\mathbf{N}_1$ or $\mathbf{P}_2' = \mathbf{P}_2\,\mathbf{T}_2$ and $\mathbf{N}_2' = \mathbf{N}_2\,\mathbf{T}_2$ can always be created, and they will also be polynomial matrices. The matrices $\mathbf{T}_1$ and $\mathbf{T}_2$ can sometimes be generalized to include certain rational polynomial factors as long as the primed $\mathbf{P}, \mathbf{N}$ factors are still polynomial matrices (i.e., no fractions). Note that form 2 is the matrix analog of the controllable canonical form procedure, illustrated in Figure 3.9, which is generalized in Figure 4.6. The matrix equations $\mathbf{P}_2\,\mathbf{g} = \mathbf{u}$ and $\mathbf{N}_2\,\mathbf{g} = \mathbf{y}$ define an intermediate vector $\mathbf{g}$, which is often called the partial state vector. MFD form 1 is the matrix analog of the Chapter 3 observable canonical form procedure. In the scalar case, these canonical form names were attached because of the

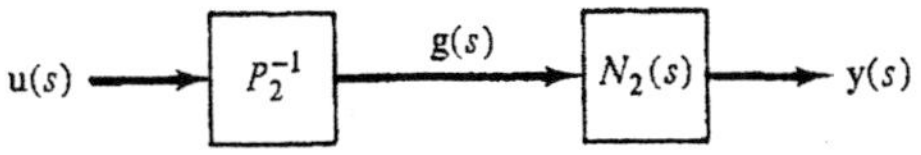

Figure 4.6

controllable or observable properties which were guaranteed to the resulting state models. Those properties are *not* guaranteed to the matrix generalizations mentioned here. Extra work is required to select "good" choices for the generalized denominator and numerator matrix factors **P** and **N**.

.30 The transfer function

$$\mathbf{H}(s) = \begin{bmatrix} \dfrac{1}{s+1} & \dfrac{2}{(s+1)(s+2)} \\ \dfrac{1}{(s+1)(s+3)} & \dfrac{1}{s+3} \end{bmatrix} = \dfrac{\begin{bmatrix} (s+2)(s+3) & 2(s+3) \\ (s+2) & (s+1)(s+2) \end{bmatrix}}{(s+1)(s+2)(s+3)}$$

can be written in either MFD form, with $\mathbf{P} = (s+1)(s+2)(s+3)\mathbf{I}_2$. The degree of the determinant of this denominator matrix is 6 in either case. Start with form 1 and use elementary row operations to find an alternative MFD which has $|\mathbf{P}_1(s)|$ with degree 4.

Elementary row operations are equivalent to premultiplying by one of the elementary matrices, which are all nonsingular. The $\mathbf{T}_1$ matrix of the previous problem is a product of elementary matrices. Premultiplying $\mathbf{P}_1$ and $\mathbf{N}_1$ is accomplished by doing row operations on $[\mathbf{P}_1 \quad \mathbf{N}_1]$. One obvious elementary operation is to divide each row by any common factors that might be present, e.g., $(s+3)$ in row 1 and $(s+2)$ in row 2 of

$$\left[\begin{array}{cc|cc} (s+1)(s+2)(s+3) & 0 & (s+2)(s+3) & 2(s+3) \\ 0 & (s+1)(s+2)(s+3) & (s+2) & (s+1)(s+2) \end{array}\right]$$

The resulting new $P(s)$ has the desired degree of 4 and

$$\mathbf{H}(s) = \begin{bmatrix} (s+1)(s+2) & 0 \\ 0 & (s+1)(s+3) \end{bmatrix}^{-1} \begin{bmatrix} (s+2) & 2 \\ 1 & (s+1) \end{bmatrix}$$

The fact that **P** is still diagonal is a peculiarity of this problem and is not a general result. Can a lower-degree $|\mathbf{P}|$ be found? This relates to the problem of finding minimal state variable realizations and is considered in Chapter 12.

.31 Repeat Problem 4.30, starting with the second MFD form $\mathbf{H} = \mathbf{N}_2\mathbf{P}_2^{-1}$ and use elementary column operations. Postmultiplication of $\mathbf{N}_2$ and $\mathbf{P}_2$ by $\mathbf{T}_2$ is equivalent to carrying out a sequence of elementary column operations on

$$\begin{bmatrix} \mathbf{N}_2 \\ \hline \mathbf{P}_2 \end{bmatrix} = \left[\begin{array}{cc} (s+2)(s+3) & 2(s+3) \\ s+2 & (s+1)(s+2) \\ \hline (s+1)(s+2)(s+3) & 0 \\ 0 & (s+1)(s+2)(s+3) \end{array}\right]$$

The first obvious operation is to cancel a factor of $s+2$ from column 1. Then column 1 times $(s+1)(s+2)$ is subtracted from column 2 (which is the same as postmultiplying by the elementary matrix $\mathbf{E}_{1,2}(\alpha)$ of Sec. 4.10, with $\alpha = -(s+1)(s+2)$). The result of these steps is

$$\left[\begin{array}{cc} (s+3) & -s(s+3)^2 \\ 1 & 0 \\ \hline (s+1)(s+3) & -(s+1)^2(s+2)(s+3) \\ 0 & (s+1)(s+2)(s+3) \end{array}\right]$$

Now a factor of $s+3$ can be canceled from column 2, giving

$$\mathbf{H}(s) = \begin{bmatrix} (s+3) & -s(s+3) \\ 1 & 0 \end{bmatrix} \begin{bmatrix} (s+1)(s+3) & -(s+1)^2(s+2) \\ 0 & (s+1)(s+2) \end{bmatrix}^{-1}$$

It is not generally true that the degree of all possible left- and right-form MFD determinants will be the same. The degree can be *increased* simply by selecting arbitrarily high-order polynomial factors in $\mathbf{T}_i$. There is a limit to how far the degree can be *decreased*. The desired degree 4 is achieved with this $\mathbf{P}_2(s)$. In fact, the minimal-order state variable system which has this transfer function is 4, as is discussed in Chapter 12. The determinants of both the left and right MFD forms of $\mathbf{P}$ hint that the poles or characteristic modes of the fourth-order realization will be at $s = -1, -1, -2$, and -3. The main objective of this and the preceding problem is to demonstrate certain elementary operations on polynomial matrices to achieve alternative MFD forms. The underlying questions of systematic procedures to follow, when to stop, what constitutes good forms, minimal degree, and so on are left to Chapter 12. This same transfer function is considered again in Examples 12.2 and 12.7 using other methods. Section 6.3 gives more on polynomial matrix methods.

4.32 Two subsystems are described in state variable form as

$$\dot{\mathbf{x}}_1 = \mathbf{A}_1\mathbf{x}_1 + \mathbf{B}_1\mathbf{u}_1; \quad \mathbf{y}_1 = \mathbf{C}_1\mathbf{x}_1 + \mathbf{D}_1\mathbf{u}_1; \quad r \text{ inputs, } m \text{ outputs}$$

$$\dot{\mathbf{x}}_2 = \mathbf{A}_2\mathbf{x}_2 + \mathbf{B}_2\mathbf{u}_2; \quad \mathbf{y}_2 = \mathbf{C}_2\mathbf{x}_2 + \mathbf{D}_2\mathbf{u}_2; \quad m \text{ inputs, } r \text{ outputs}$$

They are interconnected in the feedback loop shown in Figure 4.7. Use substitution and matrix algebra to derive the state variable model for the composite system with inputs $\mathbf{u}_a$ and outputs $\mathbf{y}_1$ and $\mathbf{y}_2$.

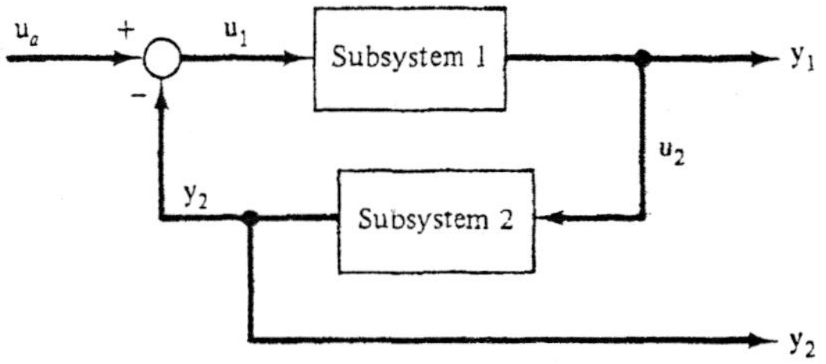

Figure 4.7

Write the output of the summing junction as

$$\mathbf{u}_1 = \mathbf{u}_a - \mathbf{y}_2 = \mathbf{u}_a - \mathbf{C}_2\mathbf{x}_2 - \mathbf{D}_2\mathbf{y}_1$$

$$= \mathbf{u}_a - \mathbf{C}_2\mathbf{x}_2 - \mathbf{D}_2[\mathbf{C}_1\mathbf{x}_1 + \mathbf{D}_1\mathbf{u}_1]$$

Combining the two $\mathbf{u}_1$ terms and premultiplying by the matrix inverse gives $\mathbf{u}_1 = [\mathbf{I}_r + \mathbf{D}_2\mathbf{D}_1]^{-1}\{\mathbf{u}_a - \mathbf{C}_2\mathbf{x}_2 - \mathbf{D}_2\mathbf{C}_1\mathbf{x}_1\}$. For convenience, let $\mathbf{L} = [\mathbf{I}_r + \mathbf{D}_2\mathbf{D}_1]^{-1}$. Then

$$\dot{\mathbf{x}}_1 = \mathbf{A}_1\mathbf{x}_1 + \mathbf{B}_1\mathbf{L}\mathbf{u}_a - \mathbf{B}_1\mathbf{L}\mathbf{C}_2\mathbf{x}_2 - \mathbf{B}_1\mathbf{L}\mathbf{D}_2\mathbf{C}_1\mathbf{x}_1$$

$$\mathbf{y}_1 = \mathbf{C}_1\mathbf{x}_1 + \mathbf{D}_1\mathbf{L}\mathbf{u}_a - \mathbf{D}_1\mathbf{L}\mathbf{C}_2\mathbf{x}_2 - \mathbf{D}_1\mathbf{L}\mathbf{D}_2\mathbf{C}_1\mathbf{x}_1$$

and

$$\dot{\mathbf{x}}_2 = \mathbf{A}_2\mathbf{x}_2 + \mathbf{B}_2\{\mathbf{C}_1\mathbf{x}_1 + \mathbf{D}_1\mathbf{L}\mathbf{u}_a - \mathbf{D}_1\mathbf{L}\mathbf{C}_2\mathbf{x}_2 - \mathbf{D}_1\mathbf{L}\mathbf{D}_2\mathbf{C}_1\mathbf{x}_1\}$$

$$\mathbf{y}_2 = \mathbf{C}_2\mathbf{x}_2 + \mathbf{D}_2\{\mathbf{C}_1\mathbf{x}_1 + \mathbf{D}_1\mathbf{L}\mathbf{u}_a - \mathbf{D}_1\mathbf{L}\mathbf{C}_2\mathbf{x}_2 - \mathbf{D}_1\mathbf{L}\mathbf{D}_2\mathbf{C}_1\mathbf{x}_1\}$$

or

$$\begin{bmatrix} \dot{\mathbf{x}}_1 \\ \dot{\mathbf{x}}_2 \end{bmatrix} = \begin{bmatrix} \mathbf{A}_1 - \mathbf{B}_1\mathbf{L}\mathbf{D}_2\mathbf{C}_1 & -\mathbf{B}_1\mathbf{L}\mathbf{C}_2 \\ \mathbf{B}_2\mathbf{C}_1 - \mathbf{B}_2\mathbf{D}_1\mathbf{L}\mathbf{D}_2\mathbf{C}_1 & \mathbf{A}_2 - \mathbf{B}_2\mathbf{D}_1\mathbf{L}\mathbf{C}_2 \end{bmatrix} \begin{bmatrix} \mathbf{x}_1 \\ \mathbf{x}_2 \end{bmatrix} + \begin{bmatrix} \mathbf{B}_1\mathbf{L} \\ \mathbf{B}_2\mathbf{D}_1\mathbf{L} \end{bmatrix} \mathbf{u}_a$$

and

$$\begin{bmatrix} y_1 \\ y_2 \end{bmatrix} = \begin{bmatrix} C_1 - D_1LD_2C_1 & -D_1LC_2 \\ D_2C_1 - D_1LD_2C_1 & C_2 - D_2D_1LC_2 \end{bmatrix} \begin{bmatrix} x_1 \\ x_2 \end{bmatrix} + \begin{bmatrix} D_1L \\ D_2D_1L \end{bmatrix} u_a$$

PROBLEMS

.33 Show that every real, square matrix **A** can be written as the sum of a symmetric matrix and a skew-symmetric matrix.

.34 Let $\mathbf{E} = [e_1 \quad e_2 \quad \cdots \quad e_n]^T$ be a column of errors in a multivariable control system. Show that the sum of the squares of the errors can be written in several forms, $e_1^2 + e_2^2 + \cdots + e_n^2 = \mathbf{E}^T\mathbf{E} = \mathrm{Tr}(\mathbf{E}\mathbf{E}^T)$.

.35 Consider the h-parameter model of a transistor, which is typical of many two-port devices (Figure 4.8).

$$\begin{bmatrix} v_1 \\ i_2 \end{bmatrix} = \begin{bmatrix} h_{11} & h_{12} \\ h_{21} & h_{22} \end{bmatrix} \begin{bmatrix} i_1 \\ v_2 \end{bmatrix}$$

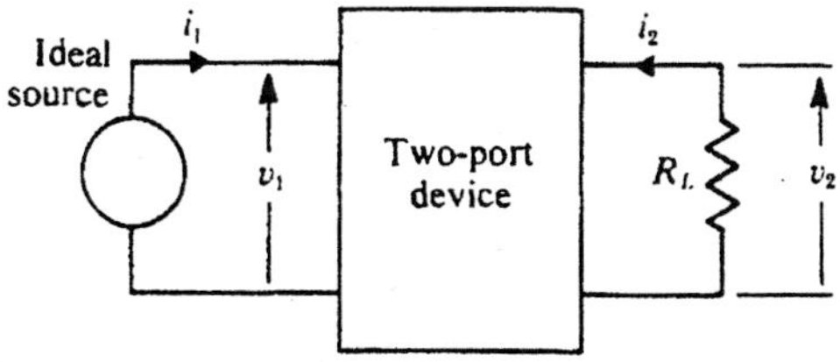

Figure 4.8

Add the third equation $v_2 = -R_L i_2$ and find i_1, i_2, and v_2 if the source is an ideal voltage source v_1.

.36 If the ideal source in the previous problem is a current source i_1, find v_1, i_2, and v_2.

.37 Compute $|\mathbf{A}|$ using Laplace expansion, pivotal condensation, elementary operations, and the method of Problem 4.14. Draw conclusions about the effort required by each method.

$$\mathbf{A} = \begin{bmatrix} 1 & 3 & -1 & 4 \\ 2 & 0 & 1 & 5 \\ -1 & 6 & 10 & -8 \\ 0 & -2 & 7 & 1 \end{bmatrix}$$

.38 Find the inverses of

$$\mathbf{A} = \begin{bmatrix} 4 & 1 & 1 \\ 2 & 0 & 3 \\ 1 & 1 & 5 \end{bmatrix}, \qquad \mathbf{B} = \begin{bmatrix} 0 & -2 & -3 & -4 & -5 \\ -1 & -1 & -3 & -4 & -5 \\ 4 & 8 & 13 & 16 & 20 \\ 2 & 4 & 6 & 9 & 10 \\ 8 & 16 & 24 & 32 & 41 \end{bmatrix},$$

$$\mathbf{C} = \left[\begin{array}{cc|cc|cc} 1 & 2 & 0 & 0 & 0 & 0 \\ -2 & 1 & 0 & 0 & 0 & 0 \\ \hline 0 & 0 & 5 & 7 & 0 & 0 \\ 0 & 0 & 1 & 0 & 0 & 0 \\ \hline 0 & 0 & 0 & 0 & 3 & 8 \\ 0 & 0 & 0 & 0 & 8 & 3 \end{array}\right]$$

(*Hint*: **B** is the matrix of Problem 4.8. Use an identity from Problem 4.4).

4.39 Find the Laplace transforms of

$$\mathbf{A}(t) = \begin{bmatrix} 1 & t \\ e^{-at} & b \sin \beta t \\ t^2 & e^{-t} \cos \beta t \end{bmatrix}, \qquad \mathbf{B}(t) = \begin{bmatrix} \cosh \beta t & \sinh \beta t \\ te^{-at} & \cos \beta t \end{bmatrix}$$

4.40 Find the inverse Laplace transform of

$$\mathbf{A}(s) = \begin{bmatrix} \frac{24}{s^5} & \frac{s^2 - \beta^2}{(s^2 + \beta^2)^2} \end{bmatrix}, \qquad \mathbf{B}(s) = \begin{bmatrix} \frac{1}{(s+1)(s+2)} & \frac{s}{(s+1)(s+2)} \\ 0 & s \end{bmatrix}$$

4.41 Find the upper triangular square root matrix of

$$\text{(a) } \mathbf{A} = \begin{bmatrix} 4 & -1 & 2 \\ -1 & 8 & 4 \\ 2 & 4 & 9 \end{bmatrix} \qquad \text{(b) } \mathbf{A} = \begin{bmatrix} 4 & 6 & 1 \\ 6 & 1 & 2 \\ 1 & 2 & 2 \end{bmatrix}$$

$$\text{(c) } \mathbf{A} = \begin{bmatrix} 16 & 4 & 1 & -1 & 3 \\ 4 & 10 & 4 & 2 & -2 \\ 1 & 4 & 25 & 4 & 1 \\ -1 & 2 & 4 & 11 & 7 \\ 3 & -2 & 1 & 7 & 17 \end{bmatrix}$$

4.42 Find both a left and right MFD form for $H(s) = \begin{bmatrix} 1/(s+1)^2 & 1/(s+1) \\ 0 & 1/(s+1) \end{bmatrix}$ Both should have $\mathbf{P}(s)$ with degree 3.

4.43 Find an MFD form with the inverse on the right and with degree of $|\mathbf{P}(s)| = 3$ for $\mathbf{H}(s) =$

$$\begin{bmatrix} 1/[(s+1)(s+2)] & 1/(s+1) \\ 1/(s+2) & 1/(s+3) \end{bmatrix}.$$

4.44 Find the **A**, **B**, **C**, and **D** state matrices for a composite system of Figure 3.27 of Sec. 3.5. Subsystem 1 has the transfer function $H_1(s) = (s+5)/(s^2 + 3s + 6)$ and subsystem 2 has $H_2(s) = (s+2)/(s^2 + 4s + 3)$.

4.45 Use the same two subsystems as in Problem 4.44 and add two more, $H_3(s) = 10/(s+6)$ and $H_4(s) = (s-2)/(s^2 + s + 1)$, interconnected as in Figure 3.28. Find the state matrices for the composite system, using controllable canonical forms to describe each subsystem.

4.46 Find the composite state variable matrices for a system with the feedback topology of Problem 4.32. Use systems 1 and 2 of Problem 4.44.

4.47 Pressure drop-flow rate relations through many devices are nonlinear. For an orifice the flow rate Q and the pressure drop $P_1 - P_2$ are related by $Q = c\sqrt{P_1 - P_2}$. Derive a linear relation for the flow out of an orifice at the bottom of a tank. The tank is nominally kept filled to a height h_n. The fluid density is ρ lb-sec^2/ft^4. Thus $P_1 = \rho g h$ and $P_2 = 0$.

4.48 A navigation scheme uses a sextant to measure the angle included between the directions to two known landmarks from the position of the sextant. Let the sextant position vector be **x**. The landmark position vectors are $\mathbf{r}_1$ and $\mathbf{r}_2$.
(a) Find the nonlinear expression for the measured angle θ.
(b) Find the linear relation between small perturbations in θ and in **x**.

5 Vectors and Linear Vector Spaces

.1 INTRODUCTION

Every student of introductory physics is familiar with the concept of a vector as a quantity which possesses both a magnitude and a direction. In Chapter 3 the terms state vector and state space were used in parameterizing models of dynamical systems. The state components could very well be a mixed set of physical quantities such as voltages, temperatures, and displacements. The formal procedures for picking states could even yield state components which are linear combinations of these disparate items. Arranging such a mixture of elements in a column matrix and referring to it as a vector seems inconsistent with the physical magnitude and direction concept of a vector. A primary objective of this chapter is to rectify these different notions of vectors and the vector spaces to which they belong. The discussion begins with a review of vector in the more familiar physical sense. The generalizations to the more abstract notions of vectors are then presented.

A second objective is to discuss various kinds of transformations on vectors. These topics have wide applicability in almost every branch of science and engineering. The central focus of this book is modeling and controlling physical systems. Therefore, there is interest in knowing how an initial state vector transforms into a state vector at a later time or how inputs transform to states or to outputs. Certain transformations of coordinates allow greater insight into system behavior and simplify the analysis and calculations. Some intrinsic system properties remain invariant to transformations, just as the length of a physical vector must not change when different coordinate systems are selected. Although all these topics cannot be dealt with completely in this chapter, the conceptual and computational foundations are presented.

5.2 PLANAR AND THREE-DIMENSIONAL REAL VECTOR SPACES

Many physical quantities, such as force and velocity, possess both a magnitude and a direction. Such entities are referred to as *vector* quantities. They are often represented by directed line segments or arrows. The length represents the magnitude, and the orientation indicates the direction.

If one point in the plane is defined as the origin, a unique vector can be associated with every point in the (two-dimensional) plane. The origin is defined as the zero vector, $\mathbf{0}$, and every other point can be associated with the directed line segment from the origin to the point. The same correspondence between points and directed line segments can obviously be made with points along the real line $\mathcal{R}$ (one dimension) and in three dimensions. Thus the terms point and vector can be used interchangeably.

If a coordinate system is defined in the plane, then each point can be identified by a unique pair of ordered numbers. These coordinate numbers can be written as a column matrix. However, since many different coordinate systems could be selected, a given vector could be represented by many different column matrices. A vector is *not* a column matrix but is a more basic entity which may be *represented* by a column matrix once a coordinate system is defined.

Vector Addition, Subtraction, and Multiplication by a Scalar

The coordinate-free description of vector addition is given by the parallelogram law. The sum of two vectors $\mathbf{v}_1$ and $\mathbf{v}_2$ is the diagonal of the parallelogram formed with $\mathbf{v}_1$ and $\mathbf{v}_2$ as sides. Since $-\mathbf{v}_2$ is a vector with the same magnitude and orientation, but the opposite direction of $\mathbf{v}_2$, the vector difference $\mathbf{v}_1 - \mathbf{v}_2$ is just the sum of $\mathbf{v}_1$ and $-\mathbf{v}_2$. Multiplication of a vector by a scalar alters the magnitude but not the orientation. In particular, any nonzero vector $\mathbf{v}$ can be used to form a *unit* vector $\hat{\mathbf{v}}$ with the same direction as $\mathbf{v}$ by multiplying $\mathbf{v}$ by the reciprocal of its magnitude.

Whenever vectors are represented as column matrices with respect to a common coordinate system, the usual rules apply for addition of matrices and multiplication of a matrix by a scalar.

Vector Products

Products such as $\mathbf{vw}$ are not defined because of matrix conformability requirements. Three types of vector products are defined.

The *inner product* (or scalar product or dot product) of $\mathbf{v}$ and $\mathbf{w}$ is defined as $\langle \mathbf{v}, \mathbf{w} \rangle = vw \cos\theta$, where v and w are the vector magnitudes and θ is the angle included between the two vectors. When real vectors are represented in orthogonal cartesian coordinates, the inner product may be computed in terms of the components as $\langle \mathbf{v}, \mathbf{w} \rangle = \mathbf{v}^T\mathbf{w} = \mathbf{w}^T\mathbf{v}$. If $\mathbf{v}$ and $\mathbf{w}$ are perpendicular, then $\theta = \pi/2$ so that $\langle \mathbf{v}, \mathbf{w} \rangle = 0$. Any two vectors which have a zero inner product are said to be *perpendicular*, or *orthogonal*. The zero vector is considered to be orthogonal to every other vector. The magnitude of a vector $\mathbf{v}$ can be expressed as $v = \langle \mathbf{v}, \mathbf{v} \rangle^{1/2}$.

The *outer product* of two real vectors $\mathbf{v} = [v_1 \quad v_2 \quad v_3]^T$ and $\mathbf{w} = [w_1 \quad w_2 \quad w_3]^T$ is defined as

$$\mathbf{v}\rangle\langle\mathbf{w} = \mathbf{v}\mathbf{w}^T = \begin{bmatrix} v_1 w_1 & v_1 w_2 & v_1 w_3 \\ v_2 w_1 & v_2 w_2 & v_2 w_3 \\ v_3 w_1 & v_3 w_2 & v_3 w_3 \end{bmatrix}$$

Since matrix multiplication is not commutative, neither is the outer product; $\mathbf{v}\mathbf{w}^T \neq \mathbf{w}\mathbf{v}^T$.

The *cross product* $\mathbf{v} \times \mathbf{w}$ is defined only in three dimensions. This product yields another vector, perpendicular to the plane of $\mathbf{v}$ and $\mathbf{w}$. It points in the direction a right-hand screw would advance if $\mathbf{v}$ were rotated toward $\mathbf{w}$ through the smaller of the two angles θ between them. The magnitude is equal to the area of the parallelogram formed by $\mathbf{v}$ and $\mathbf{w}$, i.e., $vw \sin\theta$.

5.3 AXIOMATIC DEFINITION OF A LINEAR VECTOR SPACE

Concepts such as directed line segments, lengths, angles, and dimensions of the space are considered to be intuitively obvious in one, two, or three dimensions. In dimensions higher than three, visualization is no longer possible. In cases such as a state vector with components made up of voltages, temperatures, and displacements, it is not yet clear what the vector characteristics are, even if there are only two or three components. For these reasons a more axiomatic definition of vectors and vector spaces is required. It will still be helpful to consider some of the general results for the particular cases of two or three dimensions. Geometrical descriptions of this nature will frequently be of use in gaining understanding of the concepts.

Linear Vector Spaces

A linear vector space $\mathcal{X}$ is a set of elements, called vectors, defined over a scalar number field $\mathcal{F}$, which satisfies the following conditions for addition and multiplication by scalars.

1. For any two vectors $\mathbf{x} \in \mathcal{X}$ and $\mathbf{y} \in \mathcal{X}$, the sum $\mathbf{x} + \mathbf{y} = \mathbf{v}$ is also a vector belonging to $\mathcal{X}$.
2. Addition is commutative: $\mathbf{x} + \mathbf{y} = \mathbf{y} + \mathbf{x}$.
3. Vector addition is also associative: $(\mathbf{x} + \mathbf{y}) + \mathbf{z} = \mathbf{x} + (\mathbf{y} + \mathbf{z})$.
4. There is a zero vector, $\mathbf{0}$, contained in $\mathcal{X}$ which satisfies $\mathbf{x} + \mathbf{0} = \mathbf{0} + \mathbf{x} = \mathbf{x}$.
5. For every $\mathbf{x} \in \mathcal{X}$ there is a unique vector $\mathbf{y} \in \mathcal{X}$ such that $\mathbf{x} + \mathbf{y} = \mathbf{0}$. This vector $\mathbf{y}$ is $-\mathbf{x}$.
6. For every $\mathbf{x} \in \mathcal{X}$ and for any scalar $a \in \mathcal{F}$, the product $a\mathbf{x}$ gives another vector $\mathbf{y} \in \mathcal{X}$. In particular, if a is the unit scalar,

   $$1 \cdot \mathbf{x} = \mathbf{x} \cdot 1 = \mathbf{x}$$

7. For any scalars $a \in \mathcal{F}$ and $b \in \mathcal{F}$, and for any $\mathbf{x} \in \mathcal{X}$, $a(b\mathbf{x}) = (ab)\mathbf{x}$.

8. Multiplication by scalars is distributive,

$$(a + b)\mathbf{x} = a\mathbf{x} + b\mathbf{x}$$

$$a(\mathbf{x} + \mathbf{y}) = a\mathbf{x} + a\mathbf{y}$$

The sets of all real one-, two-, or three-dimensional vectors discussed in the previous section satisfy all of these conditions and, therefore, are linear vector spaces. Elements in these familiar spaces can be represented as ordered sets of real numbers $[\alpha_1]$, $[\alpha_1 \quad \alpha_2]^T$, and $[\alpha_1 \quad \alpha_2 \quad \alpha_3]^T$, respectively. A fairly obvious generalization is to consider spaces whose elements are ordered n-tuples of real numbers $[\alpha_1 \quad \alpha_2 \cdots \alpha_n]^T$, where n is a finite integer. This space is referred to as $\mathcal{R}^n$. If the scalars α_i are allowed to be complex, then the space is referred to as $\mathcal{C}^n$. Both of these possibilities will be simultaneously covered by referring to an ordered set of n-tuples $\alpha_i \in \mathcal{F}$ as belonging to the space $\mathcal{X}^n$. Many other vector spaces can be defined. Some examples which can be verified to satisfy the required axioms are:

1. The set of all $m \times n$ matrices with elements in $\mathcal{F}$. Since valid number fields $\mathcal{F}$ include such possibilities as the real numbers, the complex numbers, or the set of rational polynomial functions with real or complex coefficients, the possibilities here are many. In particular, it is possible to define a vector space of all $m \times n$ transfer functions.
2. The set of all continuous or piecewise continuous time functions $f(t)$ on some interval $a \le t \le b$ or an ordered set $f_1(t), f_2(t), \ldots, f_n(t)$ of such functions.
3. The set of all polynomials of degree less than or equal to n, with coefficients belonging to the real or complex number fields $\mathcal{F}$. The zero element required by axiom 4 would be the $m \times n$ null matrix, the function which is identically zero, or the polynomial which has all its coefficients zero, respectively. A rational polynomial function could also be defined as a vector. The zero element would have all the numerator coefficients identically zero.

This short list provides just some of the possibilities and makes it clear that the notions of magnitude and direction are not required—or at least are not obvious—in the definitions of abstract vectors. The notion of a vector as an ordered n-tuple also seems not to apply in some of these examples, although that will be seen to depend upon how the notion of coordinate systems is generalized. That is, the countable Fourier expansion coefficients can be thought of as ordered components of a periodic function, defined as a vector. When the full generality required by some of these examples is implied, the vector space will be referred to as $\mathcal{X}$. For the most part, the discussions here will deal with vectors in the sense of the previous section and their generalizations to $\mathcal{X}^n$.

One fact emerges from the preceding examples that seems puzzling at first. The set of rational polynomial functions was used as a field and a vector space of $m \times n$ matrices was defined over that field in (1). Then the same rational polynomial function was itself declared a vector in (3). A comparison of the axioms used to define a field in Chapter 4 and those used here to define an abstract vector space show a great deal of similarity. The requirements on a vector space are weaker because no inverse vector element is required. Actually *any* set of elements which qualifies as a field in the sense defined in Chapter 4 can also be used to define a vector space over itself as the field. The real line is an example. It can be considered as a one-dimensional vector space (i.e., ordered one-tuples) defined over the real number field. More complicated vector spaces can be built up as ordered sets of the simpler vectors. The simplest example is that n-tuples of reals are ordered one-tuples. The same is true of the rational polynomial functions. One of them can be treated as a vector defined over itself as the field. Or, an ordered array of them, e.g., a transfer function, can be defined as a vector. It is important to note, however, that not all vector spaces are equivalent to fields because of the requirement of the inverse element. The set of polynomials can define a vector space, but they do not constitute a field because the ratio of two such polynomials is in general not a member of the set of polynomials.

In addition to the fact that vector inverse elements are not required, it is explicitly pointed out that no notion of the *product* of two vector elements is found in the required axioms. Very frequently an additional definition of an *inner product* is imposed upon a vector space. This is extremely useful. It allows the generalization of familiar geometrical concepts, such as length or distance, and angles between vectors. Whenever this extra definition is imposed, a restricted special class of vector spaces, called *inner product spaces,* is being dealt with.

5.4 LINEAR DEPENDENCE AND INDEPENDENCE

Consider three vectors $\mathbf{x}_1$, $\mathbf{x}_2$, and $\mathbf{x}_3$ in three-dimensional space. If there exists a relation among them, such as $\mathbf{x}_3 = \alpha\mathbf{x}_1 + \beta\mathbf{x}_2$, then it is clear that $\mathbf{x}_3$ lies in the plane through $\mathbf{x}_1$ and $\mathbf{x}_2$, no matter what scalar values are attached to α and β. $\mathbf{x}_3$ is said to be dependent on $\mathbf{x}_1$ and $\mathbf{x}_2$. The notions of dependence and independence must be generalized to arbitrary sets of vectors.

Definition 5.1. Let a finite number of vectors belonging to a linear vector space $\mathcal{X}$ be denoted by $\{\mathbf{x}_i\} = \{\mathbf{x}_1, \mathbf{x}_2, \ldots, \mathbf{x}_n\}$. If there exists a set of n scalars, a_i, at least one of which is not zero, which satisfies $a_1\mathbf{x}_1 + a_2\mathbf{x}_2 + \cdots + a_n\mathbf{x}_n = \mathbf{0}$, then the vectors $\{\mathbf{x}_i\}$ are said to be *linearly dependent.*

Definition 5.2. Any set of vectors $\{\mathbf{x}_i\}$ which is not linearly dependent is said to be *linearly independent.* That is, if $a_1\mathbf{x}_1 + a_2\mathbf{x}_2 + \cdots + a_n\mathbf{x}_n = \mathbf{0}$ implies that each $a_i = 0$, then $\{\mathbf{x}_i\}$ is a set of linearly independent vectors.

EXAMPLE 5.1 Consider the set of n vectors $\mathbf{e}_i$, each of which has n components. All components of $\mathbf{e}_i$ are zero, except the ith component, which is unity. Then

$$a_1\mathbf{e}_1 + a_2\mathbf{e}_2 + \cdots + a_n\mathbf{e}_n = a_1\begin{bmatrix}1\\0\\0\\\vdots\\0\end{bmatrix} + a_2\begin{bmatrix}0\\1\\0\\\vdots\\0\end{bmatrix} + \cdots + a_n\begin{bmatrix}0\\0\\\vdots\\0\\1\end{bmatrix} = \begin{bmatrix}a_1\\a_2\\\vdots\\a_n\end{bmatrix}$$

The only way that this sum can give the $\mathbf{0}$ vector is if each and every $a_i = 0$. Thus the set $\{\mathbf{e}_i\}$ is linearly independent. This set of $\mathbf{e}_i$ vectors represents the natural extension of the cartesian coordinate directions often used in two- and three-dimensional spaces. They will be referred to as the natural cartesian coordinates. ■

EXAMPLE 5.2 Let the components of three vectors with respect to the natural cartesian coordinates be

$$\mathbf{x}_1^T = [5 \quad 2 \quad 3], \qquad \mathbf{x}_2^T = [-1 \quad 7 \quad 4], \qquad \mathbf{x}_3^T = [14 \quad 50 \quad 36]$$

These vectors are linearly dependent because $2\mathbf{x}_1 + 3\mathbf{x}_2 - \frac{1}{2}\mathbf{x}_3 = 0$. ■

Lemma 5.1. Let $\mathcal{V} = \{\mathbf{x}_i, i = 1, n\}$ be a set of linearly dependent vectors. Then the set formed by adding any vector $\mathbf{x}_{n+1}$ to $\mathcal{V}$ is also linearly dependent.

Lemma 5.2. If a set of vectors $\{\mathbf{x}_i\}$ is linearly dependent, then one of the vectors can be written as a linear combination of the others.

Tests for Linear Dependence

Consider a set of n vectors $\{\mathbf{x}_i\}$, each having n components with respect to a given coordinate system. Let $\mathbf{A}$ be the $n \times n$ matrix which has the $\mathbf{x}_i$ vectors as columns. The set of vectors is linearly dependent if and only if $|\mathbf{A}| = 0$. The zero in this determinant test for independence must be the zero element of the number field over which the vectors are defined. In particular, if the rational polynomial functions are the field, the determinant must be identically zero for *all* values of the variable s or z or t used in defining the polynomials. It is not sufficient for the polynomial to equal zero for specific isolated values.

EXAMPLE 5.3 Use the preceding test to show that the three vectors of Example 5.2 are linearly dependent.

We find $|\mathbf{A}| = \begin{vmatrix} 5 & -1 & 14 \\ 2 & 7 & 50 \\ 3 & 4 & 36 \end{vmatrix} = 0$. Therefore, the set is linearly dependent. ■

The previous test for linear independence is not applicable when considering a set of n vectors $\{\mathbf{x}_i\}$, each of which has m components, with $m \neq n$. The matrix $\mathbf{A}$ is $m \times n$ and $|\mathbf{A}|$ is not defined. Assume the set is linearly dependent so that $\sum_{i=1}^{n} a_i \mathbf{x}_i = \mathbf{0}$ with at least one nonzero a_i. Premultiplying by $\bar{\mathbf{x}}_1^T$ gives the scalar equation

$$a_1 \bar{\mathbf{x}}_1^T \mathbf{x}_1 + a_2 \bar{\mathbf{x}}_1^T \mathbf{x}_2 + \cdots + a_n \bar{\mathbf{x}}_1^T \mathbf{x}_n = 0$$

Repeated premultiplication by $\bar{\mathbf{x}}_2^T$, then $\bar{\mathbf{x}}_3^T$, and so on gives a set of n simultaneous equations, which can be written in matrix form as

$$[\bar{\mathbf{x}}_i^T \mathbf{x}_j][\mathbf{a}] = \mathbf{0}$$

If the $n \times n$ matrix $\mathbf{G} \triangleq [\bar{\mathbf{x}}_i^T \mathbf{x}_j]$ has a nonzero determinant, then $\mathbf{G}^{-1}$ exists, and solving gives

$$\mathbf{a} = \mathbf{G}^{-1}\mathbf{0} = \mathbf{0}$$

This contradicts the assumption of at least one nonzero a_i. The matrix $\mathbf{G}$ is called the *Grammian matrix*. A necessary and sufficient condition for the set $\{\mathbf{x}_i\}$ to be linearly dependent is that $|\mathbf{G}| = 0$. An alternate means of determining linear independence is to reduce the matrix $\mathbf{A}$ to row-reduced echelon form, as mentioned in Section 4.10. This approach is convenient for computer applications and will be used in Chapter 6.

EXAMPLE 5.4 Consider two vectors defined over the complex number field, $\mathbf{x}_1 = [1+j \quad 6]^T$ and $\mathbf{x}_2 = [5+j \quad 18-12j]^T$. Show that they are linearly dependent.

$$\begin{vmatrix} 1+j & 5+j \\ 6 & 6(3-2j) \end{vmatrix} = 6\begin{vmatrix} 1+j & 5+j \\ 1 & 3-2j \end{vmatrix} = 6[(5+j)-(5+j)] \equiv 0$$

Thus the vectors are dependent. In fact, $\mathbf{x}_2 = (3-2j)\mathbf{x}_1$. Similarly, $\mathbf{x}_1 = \begin{bmatrix} (z+1)/[z(z-0.5)] \\ 1/z \end{bmatrix}$ and $\mathbf{x}_2 = \begin{bmatrix} 1/[(z+1)(z-0.5)] \\ 1/(z^2+2z+1) \end{bmatrix}$ are dependent as can be verified by (1) showing that the determinant formed with these columns is identically zero, (2) by forming the Grammian, (3) by performing elementary row and/or column operations to show that the rank is 1, or (4) by noting that $\mathbf{x}_2 = [z/(z+1)^2]\mathbf{x}_1$. Noticing the linear dependencies by inspection is not nearly so easy when the vectors—and hence the scalar proportionality factors—are defined over the complex numbers or rational polynomial functions. ■

Geometrical Significances of Linear Dependence

Two vectors can normally be used to form sides of a parallelogram. If the vectors are linearly dependent, they have the same direction, so the parallelogram degenerates to a line. It is shown in Problem 5.14 that the 2×2 Grammian determinant is equal to the square of the area of the parallelogram formed by the vectors. Thus $|\mathbf{G}| = 0$ indicates that the parallelogram has degenerated to a single line. Three vectors can normally be used to define the sides of a parallelepiped. If there is one linear dependency relation (i.e., any two of the three vectors are linearly independent but the set of three is linearly dependent), then the parallelepiped has degenerated to a plane figure and hence has zero volume. $|\mathbf{G}| = 0$ indicates this. If there are two dependency relations, the parallelepiped degenerates to a single line. Similar significance can be attached in higher dimensional cases. The number of dependency relationships among a set of vectors (or the columns of a matrix) is called the *degeneracy*, q. For an $n \times n$ matrix $\mathbf{A}$, q, n, and the rank r_A are related by

$$n = r_A + q$$

Often it is easier to determine the rank first and then use that to determine $q = n - r_A$. The degeneracy q is the key to finding eigenvectors and generalized eigenvectors for a matrix with repeated eigenvalues. This is discussed in Chapter 7.

Sylvester's Law of Degeneracy

If $\mathbf{A}$ and $\mathbf{B}$ are square conformable matrices whose product is $\mathbf{AB} = \mathbf{C}$, Sylvester's law of degeneracy can be used to place bounds on the degeneracy of $\mathbf{C}$, q_c in terms of the degeneracy of $\mathbf{A}$, q_A, and of $\mathbf{B}$, q_B:

$$\max\{q_A, q_B\} \le q_C \le q_A + q_B$$

If the relations between n, rank, and degeneracy are used, the following limits on the rank of **C** can be obtained:

$$r_A + r_B - n \leq r_C \leq \min\{r_A, r_B\}$$

A similar result was presented in Chapter 4 for **A** and **B** not necessarily square, but the lower limit there was $0 \leq r_C$.

5.5 VECTORS WHICH SPAN A VECTOR SPACE; BASIS VECTORS AND DIMENSIONALITY

The dimension of a vector space has been referred to several times. In two or three dimensions, the concept is obvious, but in higher dimensions a precise definition must be relied upon rather than intuition. It is first necessary to define what is meant by a set of vectors which *span* a vector space.

Let $\mathscr{X}$ be a linear vector space and let $\{\mathbf{u}_i, i = 1, m\}$ be a subset of vectors in $\mathscr{X}$. The set $\{\mathbf{u}_i\}$ is said to span the space $\mathscr{X}$ if for every vector $\mathbf{x} \in \mathscr{X}$ there is at least one set of scalars $a_i \in \mathscr{F}$ which permits $\mathbf{x}$ to be expressed as a linear combination of the $\mathbf{u}_i$,

$$\mathbf{x} = a_1\mathbf{u}_1 + a_2\mathbf{u}_2 + \cdots + a_m\mathbf{u}_m = \sum_{i=1}^{m} a_i\mathbf{u}_i$$

Note that if the vectors $\mathbf{x}$ and $\mathbf{u}_i$ can be expressed as columns of n scalars with respect to a common coordinate system, then in matrix notation $\mathbf{x} = \mathbf{Ua}$, where $\mathbf{U}$ is the $n \times m$ matrix whose ith column is $\mathbf{u}_i$ and $\mathbf{a} = [a_1 \quad a_2 \ldots a_m]^T$. The scalars a_i are the components of $\mathbf{x}$ in the $\mathbf{u}_i$ coordinate direction. Vectors such as $\mathbf{u}_i$, which merely span the space, do not make a good coordinate system because there may be more vectors than necessary, and as a result the a_i coefficients are not unique.

EXAMPLE 5.5 Consider all vectors in the plane. Then any pair of noncollinear vectors such as $\{\mathbf{x}, \mathbf{y}\}$, $\{\mathbf{x}', \mathbf{y}'\}$, and $\{\mathbf{x}'', \mathbf{y}''\}$ spans the two-dimensional space, since every vector in the plane can be represented as a combination of any one of these pairs. Another set of vectors which spans this space is $\{\mathbf{x}, \mathbf{y}, \mathbf{y}''\}$. Two ways of expressing a vector $\mathbf{w}$ in terms of these three vectors are shown in Figure 5.1. The coefficients in the linear expansion of a vector in terms of a set of spanning vectors need not be unique. There is an infinite number of possibilities in this example. ■

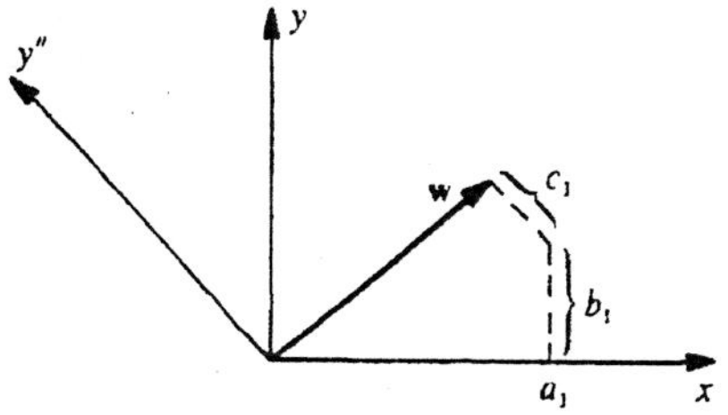

(a) $\mathbf{w} = a_1\mathbf{x} + b_1\mathbf{y} + c_1\mathbf{y}''$

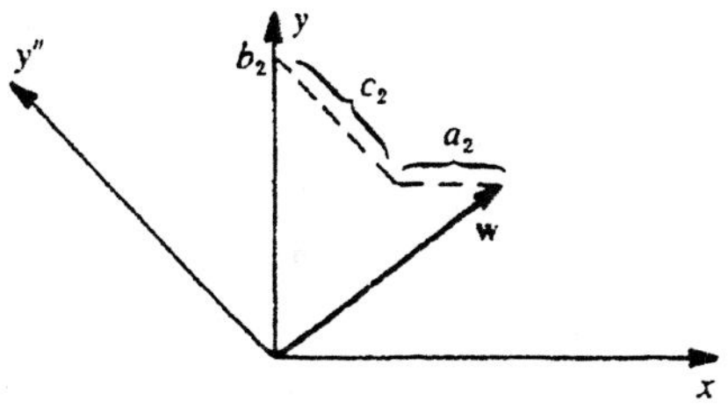

(b) $\mathbf{w} = a_2\mathbf{x} + b_2\mathbf{y} - c_2\mathbf{y}''$

Figure 5.1

Basis Vectors

A set of basis vectors, $\mathcal{B} = \{\mathbf{v}_i\}$, for a space $\mathcal{X}$ is a subset of vectors in $\mathcal{X}$ which (1) spans the space $\mathcal{X}$ and (2) is a linearly independent set. Alternatively, a set of basis vectors is a set consisting of the minimum number of vectors required to span the space $\mathcal{X}$. There are infinitely many choices for basis vectors in a given vector space. For example, in Fig. 5.1 any *two* of the three vectors $\{\mathbf{x}, \mathbf{y}, \mathbf{y}''\}$ or any two other noncollinear vectors in the plane could be selected. Every valid basis set for a given space will contain the same number of vectors, e.g., two for the plane. Once a basis is selected for the space $\mathcal{X}$, every vector $\mathbf{x} \in \mathcal{X}$ has a *unique* representation or expansion with respect to that basis. That is, there is a unique set of scalar coefficients such that $\mathbf{x} = a_1\mathbf{v}_1 + a_2\mathbf{v}_2 + \cdots + a_n\mathbf{v}_n$. As before, if $\mathbf{x}$ and each $\mathbf{v}_i$ can be expressed as a column of scalars, then in matrix notation $\mathbf{x} = V\mathbf{a}$. Because of the uniqueness of the relation between a given $\mathbf{x}$ and a set of coefficients $\{a_i\}$ for a given basis set, basis vectors are the natural generalization of coordinate vectors discussed in two and three dimensions. The column of scalars $\mathbf{a}$ can be viewed as the same vector $\mathbf{x}$ but expressed in a different coordinate system. It should now be clear why a column of scalars is not a vector but is only one representation of the vector in a particular set of coordinates. The vector itself is a more abstract entity, such as the directed line segment used in elementary physics, which exists independent of any coordinate system. The vector appears as a column of scalars only after a coordinate system—i.e., a basis set—is introduced.

Some basis sets are more convenient to work with than others. Most would agree that the mutually orthogonal $\mathbf{x}, \mathbf{y}$ of Fig. 5.1 would be more convenient than the other choices shown there. That set would be even more convenient if both $\mathbf{x}$ and $\mathbf{y}$ had unit length. This generalizes naturally in $\mathcal{X}^n$ to the set $\{\mathbf{e}_i\}$ discussed in Example 5.1. Whenever a specific basis is not mentioned for $\mathcal{X}^n$, this natural cartesian basis set will be implied.

Dimension of a Vector Space

Definition 5.3. The dimension of a vector space $\mathcal{X}$, written $\dim(\mathcal{X})$, is equal to the number of vectors in the basis set $\mathcal{B}$. Thus an n-dimensional linear vector space has n basis vectors.

EXAMPLE 5.6

1. *Let $\mathcal{X}$ be the linear vector space consisting of all n-component vectors*

$$\mathbf{x}^T = [x_1 \quad x_2 \quad x_3 \quad \cdots \quad x_n]$$

which satisfy $x_1 = x_2 = x_3 = \cdots = x_n$. Since the basis set for this space consists of the single vector $\mathbf{v}^T = [1 \quad 1 \quad 1 \quad \cdots \quad 1]$, this space is one-dimensional.

2. *Let $\mathcal{X}$ be the linear space consisting of all polynomials of degree $n-1$ or less, $\{f(t) | f(t) = \alpha_1 + \alpha_2 t + \alpha_3 t^2 + \cdots + \alpha_n t^{n-1}, \alpha_i \in \mathcal{F}\}$.* An obvious basis set is $\{1, t, t^2, \ldots, t^{n-1}\}$. Since the basis contains n elements, $\dim(\mathcal{X}) = n$, but $\mathcal{X}$ is not $\mathcal{X}^n$ as defined earlier. ■

It should be pointed out that a linear vector space can consist of a single element, the zero vector $\mathbf{0}$. Such a space is said to be a zero-dimensional space. A field, by way of contrast, must always have at least two elements, 0 and 1.

5.6 SPECIAL OPERATIONS AND DEFINITIONS IN VECTOR SPACES

In order to generalize many of the useful concepts of familiar two- and three-dimensional spaces to n-dimensional spaces, some additional definitions are required.

Inner Product

Let $\mathscr{X}$ be an n-dimensional linear vector space defined over the scalar number field $\mathscr{F}$. If, to each pair of vectors $\mathbf{x}$ and $\mathbf{y}$ in $\mathscr{X}$, a unique scalar belonging to $\mathscr{F}$, called the inner product, is assigned, then $\mathscr{X}$ is said to be an inner product space. Various definitions for the inner product are possible. Any scalar valued function of $\mathbf{x}$ and $\mathbf{y}$ can be defined as the *inner product,* written $\langle \mathbf{x}, \mathbf{y}\rangle$, provided the following axioms are satisfied:

1. $\langle \mathbf{x}, \mathbf{y}\rangle = \overline{\langle \mathbf{y}, \mathbf{x}\rangle}$ (complex conjugate property)
2. $\langle \mathbf{x}, \alpha\mathbf{y}_1 + \beta\mathbf{y}_2\rangle = \alpha\langle \mathbf{x}, \mathbf{y}_1\rangle + \beta\langle \mathbf{x}, \mathbf{y}_2\rangle$ (linear, homogeneous property)
3. $\langle \mathbf{x}, \mathbf{x}\rangle \geq 0$ for all $\mathbf{x}$ and $\langle \mathbf{x}, \mathbf{x}\rangle = 0$ if and only if $\mathbf{x} = \mathbf{0}$ (nonnegative length)

A commonly used definition of the complex inner product in $\mathscr{X}^n$, which is sufficiently general for our purposes, is

$$\langle \mathbf{x}, \mathbf{y}\rangle = \overline{\mathbf{x}}^T \mathbf{y}$$

If $\mathscr{F}$ is the set of reals, then the real inner product can be defined in the same way, but the complex conjugate on $\mathbf{x}$ is then superfluous. The inner product space defined on the real scalar field is called *Euclidean space.* Unless otherwise stated, the inner product will be assumed to be the complex inner product given above.

Combining axioms 1 and 2, it is easy to show that the inner product also satisfies

$$\langle \alpha\mathbf{x}_1 + \beta\mathbf{x}_2, \mathbf{y}\rangle = \overline{\alpha}\langle \mathbf{x}_1, \mathbf{y}\rangle + \overline{\beta}\langle \mathbf{x}_2, \mathbf{y}\rangle$$

The Grammian matrix introduced in Sec. 5.4 is generally defined in terms of the inner product as $\mathbf{G} = [\langle \mathbf{x}_i, \mathbf{x}_j\rangle]$. Some further definitions of inner products are given in the problems. In particular, see Problem 5.25 for vector spaces of matrices.

Vector Norm

Axioms 1 and 3 for inner products ensure that $\langle \mathbf{x}, \mathbf{x}\rangle$ is a nonnegative real number and is zero if and only if $\mathbf{x} = \mathbf{0}$. Because of these properties, the inner product can be used to define the *length,* or *norm,* of a vector as $\|\mathbf{x}\| = \langle \mathbf{x}, \mathbf{x}\rangle^{1/2}$. This norm will be used throughout this book unless an explicit statement to the contrary is made. It is called the quadratic norm, or in the case of real vector spaces, the Euclidean norm. In two or three dimensions, it is easy to see that this definition for the length of $\mathbf{x}$ satisfies the conditions of Euclidean geometry. It is a generalization to n dimensions of the theorem of Pythagoras.

Many other norms can be defined; the only requirements are that $\|\mathbf{x}\|$ be a nonnegative real scalar satisfying

1. $\|\mathbf{x}\| = 0$ if and only if $\mathbf{x} = \mathbf{0}$
2. $\|\alpha\mathbf{x}\| = |\alpha|\cdot\|\mathbf{x}\|$ for any scalar α
3. $\|\mathbf{x} + \mathbf{y}\| \le \|\mathbf{x}\| + \|\mathbf{y}\|$

The last condition is called the triangle inequality, for reasons which are obvious in two or three dimensions.

An important inequality, called the Cauchy-Schwarz inequality, can be expressed in terms of the norm and the absolute value of the inner product:

$$|\langle \mathbf{x}, \mathbf{y}\rangle| \le \|\mathbf{x}\|\cdot\|\mathbf{y}\|$$

The equality holds if and only if $\mathbf{x}$ and $\mathbf{y}$ are linearly dependent.

Unit Vectors

A unit vector, $\hat{\mathbf{x}}$, is by definition a vector whose norm is unity, $\|\hat{\mathbf{x}}\| = 1$. Any nonzero vector $\mathbf{x}$ can be normalized to form a unit vector.

$$\hat{\mathbf{x}} = \frac{\mathbf{x}}{\|\mathbf{x}\|}$$

Metric or Distance Measure

The concept of distance between two points (vectors are used synonymously with points) in a linear vector space can be introduced by using the norm. The distance between two points $\mathbf{x}$ and $\mathbf{y}$ is defined as the scalar function

$$\rho(\mathbf{x}, \mathbf{y}) = \|\mathbf{x} - \mathbf{y}\|$$

When the quadratic norm is used, this gives

$$\rho(\mathbf{x}, \mathbf{y}) = \langle \mathbf{x} - \mathbf{y}, \mathbf{x} - \mathbf{y}\rangle^{1/2}$$

Generalized Angles in n-Dimensional Spaces

The concept of angles between vectors can be generalized to real n-dimensional spaces by extending the notion of the dot product of two- or three-dimensional spaces,

$$\mathbf{x}\cdot\mathbf{y} = \langle \mathbf{x}, \mathbf{y}\rangle = \|\mathbf{x}\|\cdot\|\mathbf{y}\| \cos\theta$$

Thus, the cosine of the angle between $\mathbf{x} \in \mathscr{X}$ and $\mathbf{y} \in \mathscr{X}$ is

$$\cos\theta = \frac{1}{\|\mathbf{x}\|\cdot\|\mathbf{y}\|}\langle \mathbf{x}, \mathbf{y}\rangle = \langle \hat{\mathbf{x}}, \hat{\mathbf{y}}\rangle$$

It was mentioned earlier that various inner products can be defined. The particular choice of inner product dictates a specific meaning for the geometric concept of angle. Since $\langle \mathbf{x}, \mathbf{y} \rangle$ need not be real in spaces defined over the complex scalars, it is not particularly useful to try to place an interpretation upon angles in complex spaces.

Outer Product

The outer product (sometimes called the dyad product) of two vectors **x** and **y** belonging to $\mathscr{X}^n$ is

$$\mathbf{x}\rangle\langle\mathbf{y} = \mathbf{x}\bar{\mathbf{y}}^T$$

The brackets are motivated by a comparison with the usual definition for the inner product.

Multiplication of a Vector by an Arbitrary, Conformable Matrix

Since a vector in $\mathscr{X}^n$ can be represented as a column matrix with respect to a specific set of basis vectors, all the operations of matrix algebra can then be applied. In particular, premultiplication by a conformable matrix yields another column matrix, which is the representation of a vector. If the matrix multiplier is $n \times n$ and skew-symmetric, then two vectors **x** and **y** in $\mathscr{R}^n$ related by $\mathbf{y} = \mathbf{A}\mathbf{x}$ can be seen to have $\langle \mathbf{x}, \mathbf{y} \rangle = 0$. In two- or three-dimensional spaces, a zero inner product indicates that the two vectors are orthogonal (this is generalized to any vector space in the next section). Thus multiplication by a skew-symmetric matrix is somewhat akin to generalizing the cross product in that the resultant is orthogonal to **x**.

Just as vectors are abstract elements that often can be represented by column matrices, matrices as used in the preceding paragraph are specific coordinate-system dependent representations of a more abstract *transformation* operator, to be discussed later.

5.7 ORTHOGONAL VECTORS AND THEIR CONSTRUCTION

Any two vectors **x** and **y** which belong to a linear vector space $\mathscr{X}$ are said to be *orthogonal* if and only if

$$\langle \mathbf{x}, \mathbf{y} \rangle = 0$$

This is the natural generalization of the geometric concept of perpendicularity. Note that this definition of orthogonality indicates that the zero vector is orthogonal to every other vector. If each pair of vectors in a given set is mutually orthogonal, then the set is said to be an *orthogonal set*. If, in addition, each vector in this orthogonal set is a unit vector, then the set is said to be *orthonormal*. Each pair of orthonormal vectors $\hat{\mathbf{v}}_i$ and $\hat{\mathbf{v}}_j$ satisfies $\langle \hat{\mathbf{v}}_i, \hat{\mathbf{v}}_j \rangle = \delta_{ij}$, where δ_{ij} is the Kronecker delta and equals 1 if $i = j$ and 0 otherwise. Orthonormal vectors are convenient choices for basis vectors. The natural set of cartesian basis vectors $\mathbf{e}_i$ of Example 5.1 is the simplest example of an orthonormal set.

The Gram-Schmidt Process

Orthonormal vectors form a convenient basis set, so it is of interest to know how to construct an orthonormal set. Given any set of n linearly independent vectors $\{\mathbf{y}_i, i = 1, n\}$, an orthonormal set $\{\hat{\mathbf{v}}_i, i = 1, n\}$ can be constructed by using the Gram-Schmidt process. The process consists of two steps. First an orthogonal set $\{\mathbf{v}_i\}$ is constructed, and second, each vector in this set is normalized. Let $\mathbf{v}_1 = \mathbf{y}_1$ and select $\mathbf{v}_2$ as the vector formed from $\mathbf{y}_2$ by subtracting out the component in the direction of $\mathbf{v}_1$. This is equivalent to requiring that $\langle \mathbf{v}_1, \mathbf{v}_2 \rangle = 0$. Let $\mathbf{v}_2 = \mathbf{y}_2 - a\mathbf{v}_1$. Then in order to satisfy orthogonality,

$$a = \frac{\langle \mathbf{v}_1, \mathbf{y}_2 \rangle}{\langle \mathbf{v}_1, \mathbf{v}_1 \rangle}$$

so that

$$\mathbf{v}_2 = \mathbf{y}_2 - \frac{\langle \mathbf{v}_1, \mathbf{y}_2 \rangle}{\langle \mathbf{v}_1, \mathbf{v}_1 \rangle}\mathbf{v}_1$$

the next vector is chosen as

$$\mathbf{v}_3 = \mathbf{y}_3 - a_1 \mathbf{v}_1 - a_2 \mathbf{v}_2$$

and the two scalars a_i are chosen to satisfy

$$\langle \mathbf{v}_1, \mathbf{v}_3 \rangle = 0 \quad \text{and} \quad \langle \mathbf{v}_2, \mathbf{v}_3 \rangle = 0$$

This leads to

$$\mathbf{v}_3 = \mathbf{y}_3 - \frac{\langle \mathbf{v}_1, \mathbf{y}_3 \rangle}{\langle \mathbf{v}_1, \mathbf{v}_1 \rangle}\mathbf{v}_1 - \frac{\langle \mathbf{v}_2, \mathbf{y}_3 \rangle}{\langle \mathbf{v}_2, \mathbf{v}_2 \rangle}\mathbf{v}_2$$

Continuing in this manner leads to the general equation

$$\mathbf{v}_i = \mathbf{y}_i - \sum_{k=1}^{i-1} \frac{\langle \mathbf{v}_k, \mathbf{y}_i \rangle}{\langle \mathbf{v}_k, \mathbf{v}_k \rangle}\mathbf{v}_k$$

After all n of the vectors $\mathbf{v}_i$ are computed, the normalization

$$\hat{\mathbf{v}}_i = \frac{\mathbf{v}_i}{\|\mathbf{v}_i\|}, \qquad i = 1, \ldots, n$$

gives the desired orthonormal set.

EXAMPLE 5.7 Construct a set of orthonormal vectors from

$$\mathbf{y}_1^T = [1 \quad 0 \quad 1], \qquad \mathbf{y}_2^T = [-1 \quad 2 \quad 1], \qquad \mathbf{y}_3^T = [0 \quad 1 \quad 2]$$

Since the Grammian gives $|\mathbf{G}| = 4$, these vectors are linearly independent.

Step 1. Let

$$\mathbf{v}_1 = \mathbf{y}_1 = [1 \quad 0 \quad 1]^T$$

$$\mathbf{v}_2 = \mathbf{y}_2 - \frac{\langle \mathbf{v}_1, \mathbf{y}_2 \rangle}{\langle \mathbf{v}_1, \mathbf{v}_1 \rangle}\mathbf{v}_1 = \mathbf{y}_2$$

In this case $\mathbf{y}_1$ and $\mathbf{y}_2$ are already orthogonal.

$$\mathbf{v}_3 = \mathbf{y}_3 - \frac{\langle \mathbf{v}_1, \mathbf{y}_3 \rangle}{\langle \mathbf{v}_1, \mathbf{v}_1 \rangle}\mathbf{v}_1 - \frac{\langle \mathbf{v}_2, \mathbf{y}_3 \rangle}{\langle \mathbf{v}_2, \mathbf{v}_2 \rangle}\mathbf{v}_2 = \begin{bmatrix} -\frac{1}{3} & -\frac{1}{3} & \frac{1}{3} \end{bmatrix}^T$$

Step 2. Normalize $\mathbf{v}_i$ to get $\hat{\mathbf{v}}_i$:

$$\hat{\mathbf{v}}_i = \frac{\mathbf{v}_1}{\|\mathbf{v}_1\|} = \frac{1}{\sqrt{2}}\begin{bmatrix} 1 \\ 0 \\ 1 \end{bmatrix}, \quad \hat{\mathbf{v}}_2 = \frac{\mathbf{v}_2}{\|\mathbf{v}_2\|} = \frac{1}{\sqrt{6}}\begin{bmatrix} -1 \\ 2 \\ 1 \end{bmatrix}, \quad \hat{\mathbf{v}}_3 = \frac{\mathbf{v}_3}{\|\mathbf{v}_3\|} = \frac{1}{\sqrt{3}}\begin{bmatrix} -1 \\ -1 \\ 1 \end{bmatrix}$$

■

The Modified Gram-Schmidt Process

By modifying the sequence of operations slightly, a modified Gram-Schmidt process is obtained. It has superior numerical properties when the operations are carried out on a computer with finite word size. The benefits are most apparent when some vectors in the set are nearly collinear. As before, select $\mathbf{y}_1 = \mathbf{v}_1$ and $\hat{\mathbf{v}}_1 = \mathbf{v}_1/\|\mathbf{v}_1\|$. Next subtract from every $\mathbf{y}_j$, $j \geq 2$, the components in the direction of $\hat{\mathbf{v}}_1$. That is, $\mathbf{y}_j' = \mathbf{y}_j - \langle \hat{\mathbf{v}}_1, \mathbf{y}_j \rangle \hat{\mathbf{v}}_1$ for $j = 2, 3, \ldots, n$. The unit normalized version of $\mathbf{y}_2$ is selected as $\hat{\mathbf{v}}_2$. Then the components along the direction of $\hat{\mathbf{v}}_2$ are subtracted from all $\mathbf{y}_j'$,

$$\mathbf{y}_j'' = \mathbf{y}_j' - \langle \hat{\mathbf{v}}_2, \mathbf{y}_j \rangle \hat{\mathbf{v}}_2 \quad \text{for } j = 3, 4, \ldots, n$$

This continues until all n orthonormal vectors are found. Theoretically, identical results will be obtained from both versions, but practically, because of finite machine precision, some $\langle \hat{\mathbf{v}}_i, \hat{\mathbf{v}}_j \rangle$ factors will not be precisely zero and the results will differ. Matrix versions of the two construction processes are given in Problems 5.17 and 5.18.

EXAMPLE 5.8 Repeat the previous example using the modified Gram-Schmidt process.

As before $\mathbf{v}_1 = \mathbf{y}_1$ and $\hat{\mathbf{v}}_1$ is unchanged. Then $\mathbf{y}_2' = \mathbf{y}_2 - \langle \hat{\mathbf{v}}_1, \mathbf{y}_2 \rangle \hat{\mathbf{v}}_1$ as before, and $\mathbf{y}_3' = \mathbf{y}_3 - \langle \hat{\mathbf{v}}_1, \mathbf{y}_3 \rangle \hat{\mathbf{v}}_1 = [-1 \quad 1 \quad 1]^T$. Finally, $\hat{\mathbf{v}}_2 = \mathbf{y}_2'/\|\mathbf{y}_2'\|$ and $\mathbf{y}_3'' = \mathbf{y}_3' - \langle \hat{\mathbf{v}}_2, \mathbf{y}_3' \rangle \hat{\mathbf{v}}_2 = [-\frac{1}{3} \quad -\frac{1}{3} \quad \frac{1}{3}]^T$. The final orthonormal set is the same as before, but the intermediate operations are different. ■

Use of the Gram-Schmidt Process to Obtain QR *Matrix Decomposition*

It is frequently useful to express an $n \times m$ matrix $\mathbf{A}$ as a product of an orthogonal matrix $\mathbf{Q}$ (i.e., $\mathbf{Q}^{-1} = \mathbf{Q}^T$) and an upper-triangular matrix $\mathbf{R}$. The Gram-Schmidt process is one way of determining $\mathbf{Q}$ and $\mathbf{R}$ such that $\mathbf{A} = \mathbf{QR}$. Assume first that the m columns $\mathbf{a}_j$ of $\mathbf{A}$ are linearly independent. This requires that $m \leq n$. If the Gram-Schmidt process is applied to the set $\{\mathbf{a}_j\}$ to obtain the orthornormal set $\{\mathbf{t}_j\}$, the construction equations for the $\mathbf{t}_j$ vectors are

$$\mathbf{t}_1 = \alpha_{11}\,\mathbf{a}_1, \qquad \mathbf{t}_2 = \alpha_{12}\,\mathbf{a}_1 + \alpha_{22}\,\mathbf{a}_2, \ldots, \qquad \mathbf{t}_j = \alpha_{1j}\,\mathbf{a}_1 + \alpha_{2j}\,\mathbf{a}_2 + \cdots + \alpha_{jj}\,\mathbf{a}_j$$

Calculation of the α_{ij} scalar coefficients involve inner product and norm operations, as demonstrated earlier. Collectively, these construction equations can be written as the matrix equation

$$[\mathbf{t}_1 \quad \mathbf{t}_2 \quad \dots \quad \mathbf{t}_m] = [\mathbf{a}_1 \quad \mathbf{a}_2 \quad \dots \quad \mathbf{a}_m]\begin{bmatrix} \alpha_{11} & \alpha_{12} & \alpha_{13} & \cdots & \alpha_{1m} \\ 0 & \alpha_{22} & \alpha_{23} & \cdots & \alpha_{2m} \\ 0 & 0 & \alpha_{33} & \cdots & \alpha_{3m} \\ \vdots & & & & \\ 0 & 0 & 0 & \cdots & \alpha_{mm} \end{bmatrix} \tag{5.1}$$

or simply as $\mathbf{T} = \mathbf{AS}$. Although $\mathbf{T}$ need not be square, it has certain orthogonality properties. By construction, $\langle \mathbf{t}_i, \mathbf{t}_j \rangle = \delta_{ij}$. This means that $\mathbf{T}^T\mathbf{T} = \mathbf{I}_m$. However, $\mathbf{T}$ is not orthogonal in the sense defined as the end of Sec. 4.8 because $\mathbf{TT}^T \neq \mathbf{I}_n$ unless $n = m$. The $\mathbf{S}$ matrix is upper-triangular and nonsingular (because by Sylvester's law of degeneracy $\mathbf{S}$ must have rank m). Its inverse is also upper triangular. Therefore, several alternate forms are immediate.

$$\mathbf{I}_m = \mathbf{T}^T\mathbf{AS}, \qquad \mathbf{T}^T\mathbf{A} = \mathbf{S}^{-1}, \quad \text{and} \quad \mathbf{A} = \mathbf{TS}^{-1} \tag{5.2}$$

The last equation is almost in the widely used **QR** decomposition form—but not quite because $\mathbf{T}$ is not generally square and not truly orthogonal. The original columns in $\mathbf{A}$ can always be augmented with additional vectors $\mathbf{v}_k$ in such a way that the matrix $[\mathbf{A} \mid \mathbf{V}]$ has n linearly independent columns. The Gram-Schmidt process can then be applied to construct a full set of n orthonormal vectors $\{\mathbf{t}_j\}$, which can be used to define the columns of the $n \times n$ matrix $\mathbf{Q}$. This is true regardless of the size or rank of $\mathbf{A}$—that is, the earlier assumptions that $m \le n$ and $\text{Rank}(\mathbf{A}) = m$ are no longer required.

Although the expressions in Eq. (5.2) were derived from a Gram-Schmidt construction point of view, the last version has a Gram-Schmidt *expansion* interpretation as well. This point of view is used here. If a given column $\mathbf{a}_j$ is expanded in terms of the orthonormal set $\{\mathbf{t}_j\}$, the kth expansion coefficient is $\langle \mathbf{t}_k, \mathbf{a}_j \rangle$, or $\mathbf{t}_k^T \mathbf{a}_j$. All the expansion coefficients for all $\mathbf{a}_j$ columns are contained in the matrix given by $\mathbf{Q}^T\mathbf{A} = \mathbf{R}$. The matrix $\mathbf{R}$ thus obtained will be upper-triangular (or the nonsquare generalization of upper-triangular), with exactly $r_A = \text{rank}(\mathbf{A})$ nonzero rows. Since $\mathbf{Q}$ is orthogonal, it follows that $\mathbf{A} = \mathbf{QR}$. $\mathbf{R}$ is a generalization of $\mathbf{S}^{-1}$ and $\mathbf{Q}$ is a generalization of $\mathbf{T}$ in Eq. (5.2). The following five matrices illustrate the range of possibilities. Both the original $(\mathbf{A} = \mathbf{TS}^{-1})$ and the augmented $(\mathbf{A} = \mathbf{QR})$ forms of the decomposition are shown for each matrix. Results are rounded approximations.

$$m = n, \text{ not full rank:} \quad \mathbf{A} = \begin{bmatrix} 1 & 1 \\ 1 & 1 \end{bmatrix} = \begin{bmatrix} 0.707 \\ 0.707 \end{bmatrix}[1.414 \quad 1.414]$$

$$= \begin{bmatrix} 0.707 & -0.707 \\ 0.707 & 0.707 \end{bmatrix}\begin{bmatrix} 1.414 & 1.414 \\ 0 & 0 \end{bmatrix}$$

$$m < n, \text{ full rank:} \quad \mathbf{B} = \begin{bmatrix} 1 & 1 \\ 2 & -1 \\ 3 & 2 \end{bmatrix} = \begin{bmatrix} 0.2673 & 0.3132 \\ 0.5345 & -0.8351 \\ 0.8018 & 0.4523 \end{bmatrix}\begin{bmatrix} 3.7417 & 1.3363 \\ 0 & 2.053 \end{bmatrix}$$

$$= \begin{bmatrix} 0.2673 & 0.3132 & 0.9113 \\ 0.5345 & -0.8351 & 0.1302 \\ 0.8018 & 0.4523 & -0.3906 \end{bmatrix}\begin{bmatrix} 3.7417 & 1.3363 \\ 0 & 2.053 \\ 0 & 0 \end{bmatrix}$$

$$m < n, \text{ not full rank:} \quad \mathbf{C} = \begin{bmatrix} 1 & 2 \\ 2 & 4 \\ 3 & 6 \end{bmatrix} = \begin{bmatrix} 0.2673 \\ 0.5345 \\ 0.8018 \end{bmatrix} [3.7417 \quad 7.4833]$$

$$= \begin{bmatrix} 0.2673 & -0.9569 & -0.1139 \\ 0.5345 & 0.2455 & -0.8087 \\ 0.8018 & 0.1553 & 0.5771 \end{bmatrix} \begin{bmatrix} 3.7416 & 7.4833 \\ 0 & 0 \\ 0 & 0 \end{bmatrix}$$

$$m > n, \text{ full rank:} \quad \mathbf{D} = \begin{bmatrix} 2 & 1 & 6 \\ 1 & 4 & 8 \end{bmatrix}$$

$$= \begin{bmatrix} 0.8944 & -0.4472 \\ 0.4472 & 0.8944 \end{bmatrix} \begin{bmatrix} 2.2361 & 2.6833 & 8.9443 \\ 0 & 3.1305 & 4.4721 \end{bmatrix}$$

$$m > n, \text{ not full rank:} \quad \mathbf{E} = \begin{bmatrix} 1 & 4 & 7 & 3 \\ 2 & 0 & 2 & 1 \\ 3 & 4 & 9 & 4 \end{bmatrix}$$

$$= \begin{bmatrix} 0.2673 & 0.7715 \\ 0.5345 & -0.6172 \\ 0.8018 & 0.1543 \end{bmatrix} \begin{bmatrix} 3.7417 & 4.2762 & 1.0156 & 4.543 \\ 0 & 3.7033 & 5.5549 & 2.3156 \end{bmatrix}$$

$$= \begin{bmatrix} 0.2673 & 0.7715 & 0.5774 \\ 0.5345 & -0.6172 & 0.5774 \\ 0.8018 & 0.1543 & -0.5774 \end{bmatrix} \begin{bmatrix} 3.7417 & 4.2762 & 1.0156 & 4.543 \\ 0 & 3.7033 & 5.5549 & 2.3156 \\ 0 & 0 & 0 & 0 \end{bmatrix}$$

The determination of a **QR** decomposition is not generally a hand calculation. It is very worthwhile to have a computer algorithm for this purpose. The **QR** decomposition procedure provides a good way of determining the rank of a matrix. It can be adapted to solving the eigenvalue problem of Chapter 7. Finally, in Chapter 12 it provides an easy way of determining a minimal-dimension state model from an arbitrary state model.

5.8 VECTOR EXPANSIONS AND THE RECIPROCAL BASIS VECTORS

Every vector $\mathbf{x} \in \mathscr{X}$ has a unique expansion

$$\mathbf{x} = \sum_{i=1}^{n} a_i \mathbf{v}_i$$

with respect to the basis set $\mathscr{B} = \{\mathbf{v}_i, i = 1, n\}$. Taking the inner product of $\mathbf{v}_j$ and $\mathbf{x}$ gives

$$\langle \mathbf{v}_j, \mathbf{x} \rangle = \left\langle \mathbf{v}_j, \sum_{i=1}^{n} a_i \mathbf{v}_i \right\rangle = \sum_{i=1}^{n} a_i \langle \mathbf{v}_j, \mathbf{v}_i \rangle$$

If the basis set is orthonormal so that $\langle \mathbf{v}_j, \mathbf{v}_i \rangle = \delta_{ij}$, then the jth expansion coefficient is $a_j = \langle \mathbf{v}_j, \mathbf{x} \rangle$.

EXAMPLE 5.9 Use the set of orthonormal vectors generated in Example 5.7 as basis vectors and find the three coefficients a_i which allow $\mathbf{z} = [4 \quad -8 \quad 1]^T$ to be written as

$$\mathbf{z} = a_1 \hat{\mathbf{v}}_1 + a_2 \hat{\mathbf{v}}_2 + a_3 \hat{\mathbf{v}}_3$$

Because $\{\hat{\mathbf{v}}_i\}$ is an orthonormal set,

$$a_1 = \langle \hat{\mathbf{v}}_1, \mathbf{z}\rangle, \qquad a_2 = \langle \hat{\mathbf{v}}_2, \mathbf{z}\rangle \quad \text{and} \quad a_3 = \langle \hat{\mathbf{v}}_3, \mathbf{z}\rangle$$

so that

$$\mathbf{z} = \frac{5}{\sqrt{2}}\hat{\mathbf{v}}_1 - \frac{19}{\sqrt{6}}\hat{\mathbf{v}}_2 + \frac{5}{\sqrt{3}}\hat{\mathbf{v}}_3$$ ■

Basis vector expansions can be used to gain insight into the state equations of a time-invariant dynamic system

$$\dot{\mathbf{x}} = \mathbf{Ax} + \mathbf{Bu}$$

$$\mathbf{y} = \mathbf{Cx} + \mathbf{Du}$$

At any given time instant, $\mathbf{x} \in \Sigma$, the n-dimensional state space. Let $\{\mathbf{v}_i, i = 1, \ldots, n\}$ be a set of constant basis vectors for Σ. Then at any time instant there exists a set of unique scalars α_i such that $\mathbf{x} = \alpha_1 \mathbf{v}_1 + \cdots + \alpha_n \mathbf{v}_n$, or $\mathbf{x} = \mathbf{V}\boldsymbol{\alpha}$. Since all $\mathbf{v}_i$ are constant, the time variations of $\mathbf{x}$ must be contained in the expansion coefficients α_i, and thus $\dot{\mathbf{x}} = \mathbf{V}\dot{\boldsymbol{\alpha}}$. Substitution into the state equations gives

$$\mathbf{V}\dot{\boldsymbol{\alpha}} = \mathbf{AV}\boldsymbol{\alpha} + \mathbf{Bu} \quad \text{or} \quad \dot{\boldsymbol{\alpha}} = \mathbf{V}^{-1}\mathbf{AV}\boldsymbol{\alpha} + \mathbf{V}^{-1}\mathbf{Bu} \quad \text{and} \quad \mathbf{y} = \mathbf{CV}\boldsymbol{\alpha} + \mathbf{Du}$$

By defining $\mathbf{A}' = \mathbf{V}^{-1}\mathbf{AV}$, $\mathbf{B}' = \mathbf{V}^{-1}\mathbf{B}$, $\mathbf{C}' = \mathbf{CV}$, and $\mathbf{x}' \equiv \boldsymbol{\alpha}$, it is seen that the change of basis vectors from the original set, which might have been the natural cartesian set, to $\{\mathbf{v}_i\}$ has created a different state variable model for the same system. In Chapter 3 various forms of the state variable models were derived. Is it possible that all the varieties which were presented can be related by a simple change of basis? That this is not always so is now demonstrated. Consider a system whose input output transfer function is $H(s) = (s+1)/(s^2 + 3s + 2)$. The controllable canonical form of the state equations are obtained from Figure 5.2 as

$$\dot{\mathbf{x}} = \begin{bmatrix} 0 & 1 \\ -2 & -3 \end{bmatrix}\mathbf{x} + \begin{bmatrix} 0 \\ 1 \end{bmatrix}u \qquad y = [1 \quad 1]\mathbf{x}$$

The observable canonical form is obtained from Figure 5.3 as

$$\dot{\mathbf{x}}' = \begin{bmatrix} -3 & 1 \\ -2 & 0 \end{bmatrix}\mathbf{x}' + \begin{bmatrix} 1 \\ 1 \end{bmatrix}u \qquad y = [1 \quad 0]\mathbf{x}'$$

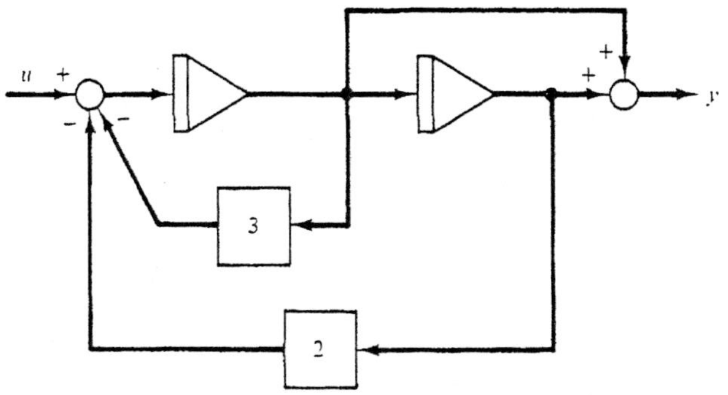

Figure 5.2

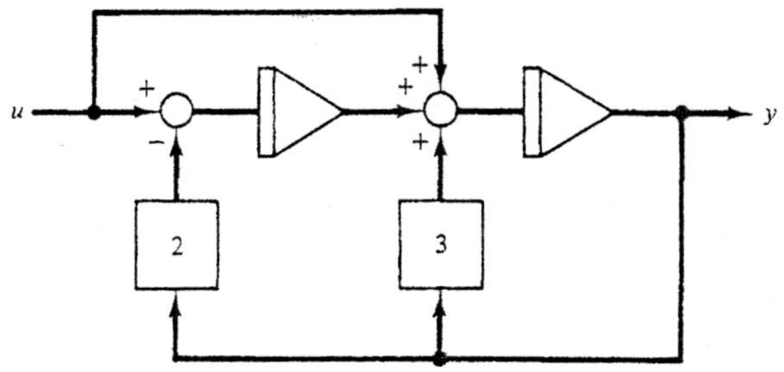

Figure 5.3

It is to be shown that no basis set (or equivalently, no nonsingular matrix $\mathbf{V}$) exists for which all three transformations from $\mathbf{A}, \mathbf{B}, \mathbf{C}$ to $\mathbf{A}', \mathbf{B}', \mathbf{C}'$ will be true. Consider first $\mathbf{V}^{-1}\mathbf{A}\mathbf{V} = \mathbf{A}'$, or $\begin{bmatrix} 0 & 1 \\ -2 & -3 \end{bmatrix}\mathbf{V} = \mathbf{V}\begin{bmatrix} -3 & 1 \\ -2 & 0 \end{bmatrix}$. Expanding $\mathbf{V}$ into components v_{ij} and writing out the matrix products shows that v_{11} must equal v_{22} but that $\mathbf{V}$ is otherwise unrestricted so far. From the expanded form of $\mathbf{C}\mathbf{V} = \mathbf{C}'$ it is found that equality requires in addition that $v_{12} = v_{21}$. Using both of these restrictions in $\mathbf{V}^{-1}\mathbf{B} = \mathbf{B}'$ converted to $\mathbf{B} = \mathbf{V}\mathbf{B}'$ gives

$$\begin{bmatrix} v_{11} & v_{12} \\ v_{12} & v_{11} \end{bmatrix}\begin{bmatrix} 1 \\ 1 \end{bmatrix} = \begin{bmatrix} 0 \\ 1 \end{bmatrix}$$

which is an impossible contradiction. These two particular state models are *not* related by a simple change of basis vectors.

Can some other type of transformation be found which relates the primed and unprimed state models? To answer this, differentiate both forms of the y equation, giving

$$\dot{y} = \mathbf{C}\dot{\mathbf{x}} = \mathbf{C}\mathbf{A}\mathbf{x} + \mathbf{C}\mathbf{B}u \quad \text{and} \quad \dot{y} = \mathbf{C}'\dot{\mathbf{x}}' = \mathbf{C}'\mathbf{A}'\mathbf{x}' + \mathbf{C}'\mathbf{B}'u$$

Grouping the differentiated and undifferentiated y equations together and combining all u terms gives

$$\begin{bmatrix} \mathbf{C}' \\ \mathbf{C}'\mathbf{A}' \end{bmatrix}\mathbf{x}' = \begin{bmatrix} \mathbf{C} \\ \mathbf{C}\mathbf{A} \end{bmatrix}\mathbf{x} + \begin{bmatrix} \mathbf{C}\mathbf{B} - \mathbf{C}'\mathbf{B}' \\ 0 \end{bmatrix}u$$

Call the 2×2 matrix coefficients of $\mathbf{x}$ and $\mathbf{x}'$, $\mathbf{Q}$ and $\mathbf{Q}'$, respectively, and let the column coefficient of u be $\mathbf{W}$. $\mathbf{Q}'$ is invertible in this case, so $\mathbf{x}' = [\mathbf{Q}']^{-1}\{\mathbf{Q}\mathbf{x} + \mathbf{W}u\}$. Comparing this with the transformation $\mathbf{x}' = \mathbf{V}\mathbf{x}$, which represents a change of basis, the presence of the $\mathbf{W}u$ term is an obvious difference. However, the important difference is that $[\mathbf{Q}']^{-1}\mathbf{Q}$ can never be represented by a nonsingular $\mathbf{V}$, since $\mathbf{Q}$ is singular for this system. Many variants of the state equations can be related by a change of basis vectors. The particular system models examined here cannot because of a failure in a basic property, to be examined in detail in Chapter 11.

Reciprocal Basis Vectors

When the basis set $\mathcal{B} = \{\mathbf{v}_i\}$ is not orthonormal, the preceding simple results no longer hold, but every vector $\mathbf{z} \in \mathcal{X}$ still has a unique expansion

$$\mathbf{z} = \sum_{i=1}^{n} a_i \mathbf{v}_i$$

Another set of n vectors, called the *reciprocal basis vectors* $\{\mathbf{r}_1, \mathbf{r}_2, \ldots, \mathbf{r}_n\}$, is introduced to facilitate finding the expansion coefficients. These reciprocal or dual basis vectors are defined by n^2 equations, each of the form

$$\langle \mathbf{r}_i, \mathbf{v}_j \rangle = \delta_{ij}$$

In matrix form this set of equations becomes

$$\mathbf{RB} = \mathbf{I}$$

where $\mathbf{B}$ is the $n \times n$ matrix whose columns are $\mathbf{v}_i$ and $\mathbf{R}$ is the $n \times n$ matrix whose *rows* are $\bar{\mathbf{r}}_i^T$. Thus $\mathbf{R} = \mathbf{B}^{-1}$, so the reciprocal basis vector $\mathbf{r}_i$ is the conjugate transpose of the ith row of $\mathbf{B}^{-1}$. With the reciprocal basis vectors available, it is apparent that the expansion coefficients are given by $a_i = \langle \mathbf{r}_i, \mathbf{z} \rangle$ so that

$$\mathbf{z} = \sum_{i=1}^{n} \langle \mathbf{r}_i, \mathbf{z} \rangle \mathbf{v}_i$$

EXAMPLE 5.10 Let $\mathbf{v}_1 = [1 \quad 0]^T$ and $\mathbf{v}_2 = [-1 \quad 1]^T$. Express the vector $\mathbf{z} = [3 \quad 3]^T$ in terms of this basis set.

First, the reciprocal basis set is found from

$$\mathbf{R} = \begin{bmatrix} 1 & -1 \\ 0 & 1 \end{bmatrix}^{-1} = \begin{bmatrix} 1 & 1 \\ 0 & 1 \end{bmatrix}$$

so that $\mathbf{r}_1 = [1 \quad 1]^T$ and $\mathbf{r}_2 = [0 \quad 1]^T$. The coefficients are

$$a_1 = \langle \mathbf{r}_1, \mathbf{z} \rangle = 6 \quad \text{and} \quad a_2 = \langle \mathbf{r}_2, \mathbf{z} \rangle = 3$$

so that $\mathbf{z} = 6\mathbf{v}_1 + 3\mathbf{v}_2$. A sketch of the $\mathbf{r}_i$ and $\mathbf{v}_i$ vectors may be informative. ■

The matrix $\mathbf{B}$ of basis vectors need not always be square. For example, the basis set for a two-dimensional subspace of a four-dimensional space would consist of two $\mathbf{b}_i$ vectors, each with four components. $\mathbf{B}$ is of dimension 4×2 and has no inverse in the usual sense. There will be two reciprocal basis vectors $\mathbf{r}_i$ also, and using their conjugate transposes as rows, $\mathbf{R}$ is of dimension 2×4. There are two ways for determining $\mathbf{R}$. One could augment the columns in $\mathbf{B}$ with two more columns $\mathbf{B}_a$ so that $[\mathbf{B} \mid \mathbf{B}_a]$ is square and invertible. If both columns in $\mathbf{B}_a$ are selected to be orthogonal to all columns in $\mathbf{B}$ (by forcing $\mathbf{B}^T\mathbf{B}_a = [0]$), then $[\mathbf{B} \quad \mathbf{B}_a]^{-1} = \begin{bmatrix} \mathbf{R} \\ \mathbf{R}_a \end{bmatrix}$. That is, the conjugate transposes of the desired reciprocal basis vectors are found in the first two rows of the augmented inverse. The second method of finding $\mathbf{R}$ is to notice that $\mathbf{R} = (\mathbf{B}^T\mathbf{B})^{-1}\mathbf{B}^T$ will have the desired orthogonality property $\mathbf{RB} = \mathbf{I}$. In fact, these two methods give the same result. The direct expression for $\mathbf{R}$ is called the *left pseudo-inverse* of $\mathbf{B}$ and appears in many applications involving projections, approximations, and least-squares solutions, as is seen in the next chapter.

EXAMPLE 5.11 Find the reciprocal basis vectors for the basis set $\mathbf{b}_1 = [1 \quad 1 \quad 0 \quad 1]^T$ and $b_2 = [2 \quad 1 \quad 1 \quad 0]^T$. Then use them to find the components of the vector $\mathbf{y} = [3 \quad 0 \quad 1 \quad 2]^T$ along

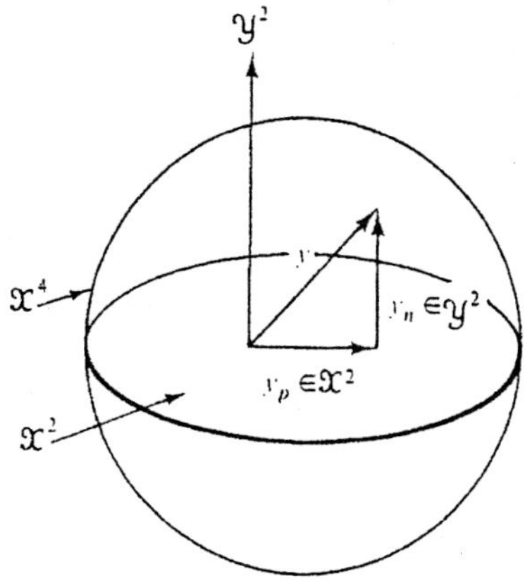

Figure 5.4

these two basis directions. That is, find the projection of $\mathbf{y}$ onto the two-dimensional space spanned by $\mathbf{b}_1$ and $\mathbf{b}_2$.

Using the augmentation approach first, we must find two independent solutions of

$$\begin{bmatrix} 1 & 1 & 0 & 1 \\ 2 & 1 & 1 & 0 \end{bmatrix} \begin{bmatrix} x_1 \\ x_2 \\ x_3 \\ x_4 \end{bmatrix} = \begin{bmatrix} 0 \\ 0 \end{bmatrix}$$

Two solutions are $\mathbf{b}_3 = [0.5 \quad 0 \quad -1 \quad 0.5]^T$ and $\mathbf{b}_4 = [0.5 \quad -1 \quad 0 \quad 0.5]^T$. Using these to form $\mathbf{B}_a$ and inverting gives

$$\begin{bmatrix} \mathbf{R} \\ \hline \mathbf{R}_a \end{bmatrix} = \begin{bmatrix} 0 & \frac{1}{3} & -\frac{1}{3} & \frac{2}{3} \\ \frac{1}{3} & 0 & \frac{1}{3} & -\frac{1}{3} \\ \hline \frac{1}{3} & 0 & -\frac{2}{3} & -\frac{1}{3} \\ \frac{1}{3} & -\frac{2}{3} & 0 & \frac{1}{3} \end{bmatrix}$$

Direct calculation of the second method shows that $(\mathbf{B}^T\mathbf{B})^{-1}\mathbf{B}^T$ gives the first two rows, namely, $\mathbf{R}$, so $\mathbf{r}_1 = [0 \quad \frac{1}{3} \quad -\frac{1}{3} \quad \frac{2}{3}]^T$ and $\mathbf{r}_2 = [\frac{1}{3} \quad 0 \quad \frac{1}{3} \quad -\frac{1}{3}]^T$. The expansion coefficients of the vector $\mathbf{y}$ along the basis vectors $\mathbf{b}_1$ and $\mathbf{b}_2$ are $\langle \mathbf{r}_1, \mathbf{y}\rangle = 1$ and $\langle \mathbf{r}_2, \mathbf{y}\rangle = \frac{2}{3}$. Note that both these calculations are given by the matrix product $\mathbf{Ry}$. The resulting vector, the projection of $\mathbf{y}$ onto the space of $\{\mathbf{b}_1, \mathbf{b}_2\}$, is $\mathbf{y}_p = [\frac{7}{3} \quad \frac{5}{3} \quad \frac{2}{3} \quad 1]^T$ when it is expressed in terms of the same basis vectors as the original $\mathbf{y}$. This vector could just as well be referred to in component form as $[1 \quad \frac{2}{3}]^T$, provided it is understood that the basis vectors being used are $\mathbf{b}_1$ and $\mathbf{b}_2$. The component of $\mathbf{y}$ which is normal to the space of $\{\mathbf{b}_1, \mathbf{b}_2\}$ is given by $\mathbf{y}_n = \langle \mathbf{r}_3, \mathbf{y}\rangle \mathbf{b}_3 + \langle r_4, \mathbf{y}\rangle \mathbf{b}_4 = -\frac{1}{3}\mathbf{b}_3 + \frac{5}{3}\mathbf{b}_4 = [\frac{2}{3} \quad -\frac{5}{3} \quad \frac{1}{3} \quad 1]^T$. The expansion coefficients can also be computed from $\mathbf{R}_a\,\mathbf{y} = [-\frac{1}{3} \quad \frac{5}{3}]^T$. The original vector has been decomposed into components $\mathbf{y} = \mathbf{y}_p + \mathbf{y}_n$, and it is easily verified that $\mathbf{y}_p$ and $\mathbf{y}_n$ are orthogonal. With a little imagination, Figure 5.4 represents the decomposition of a four-dimensional space $\mathcal{X}^4$ into two separate orthogonal two-dimensional spaces, $\mathcal{X}^2$ and $\mathcal{Y}^2$. The projection of $\mathbf{y}$ onto each subspace is also shown. These notions are formalized in the next section. ∎

9 LINEAR MANIFOLDS, SUBSPACES, AND PROJECTIONS

Let $\mathcal{X}$ be a linear vector space defined over the number field $\mathcal{F}$. A nonempty subset, $\mathcal{M}$, of $\mathcal{X}$ is called a *linear manifold* if for each vector $\mathbf{x}$ and $\mathbf{y}$ in $\mathcal{M}$, the combination $\alpha\mathbf{x} + \beta\mathbf{y}$ is also in $\mathcal{M}$ for arbitrary $\alpha, \beta \in \mathcal{F}$. The zero vector, of necessity, is included in every linear manifold.

A closed linear manifold is called a *subspace*. In finite dimensional spaces, there is no distinction between linear manifolds and subspaces, because every finite dimensional manifold is closed.

A subspace of an n-dimensional linear vector space $\mathcal{X}^n$ is itself a linear vector space contained within $\mathcal{X}^n$, but with dimension $m \le n$. A *proper* subspace has $m < n$.

EXAMPLE 5.12 The spaces $\mathcal{X}^2$ and $\mathcal{Y}^2$ of Example 5.11 are both two-dimensional subspaces of the four-dimensional space $\mathcal{X}^4$.

The space defined in Problem 5.11 is a three-dimensional subspace of $\mathcal{X}^5$, and the first space defined in Example 5.6 is a one-dimensional subspace of $\mathcal{X}^n$. In general, since $\mathcal{X}^n$ has n basis vectors, deleting any one of the basis vectors leaves a basis set for an $n-1$ dimensional subspace, deleting two allows the definition of an $n-2$ dimensional subspace, etc. Note that $\mathbf{0}$ must be an element of every subspace. If it is the only element, then that subspace is zero dimensional. ■

Starting with one vector space it is possible to define other spaces, called *subspaces,* by selecting subsets of the basis vectors. The process can also go the other way. Starting with two linear vector spaces $\mathcal{U}$ and $\mathcal{V}$ defined over the same number field $\mathcal{F}$, a new vector space $\mathcal{X}$ can be constructed from their sum:

$$\mathcal{X} = \mathcal{U} + \mathcal{V}$$

This means that every vector $\mathbf{x}$ in $\mathcal{X}$ can be written as

$$\mathbf{x} = \mathbf{u} + \mathbf{v}, \qquad \mathbf{u} \in \mathcal{U}, \qquad \mathbf{v} \in \mathcal{V}$$

If there is one and only one pair $\mathbf{u}, \mathbf{v}$ for each $\mathbf{x}$, then $\mathcal{X}$ is called the *direct sum* of $\mathcal{U}$ and $\mathcal{V}$, written

$$\mathcal{X} = \mathcal{U} \oplus \mathcal{V}$$

This implies that the only vector common to both $\mathcal{U}$ and $\mathcal{V}$ is $\mathbf{0}$. In this case,

$$\dim(\mathcal{X}) = \dim(\mathcal{U}) + \dim(\mathcal{V})$$

In Example 5.11 $\mathcal{X}^4 = \mathcal{X}^2 \oplus \mathcal{Y}^2$. This is a direct sum because the basis vectors $\mathbf{b}_3$ and $\mathbf{b}_4$ were constructed to be orthogonal to both $\mathbf{b}_1$ and $\mathbf{b}_2$, and therefore every $\mathbf{u} \in \mathcal{Y}^2$ is orthogonal to every $\mathbf{v} \in \mathcal{X}^2$. As a simple example of a sum (as opposed to a direct sum) of two spaces, define $\mathcal{U}$ as the linear space with basis $\{\mathbf{b}_1, \mathbf{b}_2, \mathbf{b}_3\}$ of Example 5.11. Then $\mathcal{X}^4 = \mathcal{U} + \mathcal{Y}^2$. The fact that both spaces have one basis vector in common prevents a unique decomposition of vectors and causes $\dim(\mathcal{X}^4) \neq \dim(\mathcal{U}) + \dim(\mathcal{Y}^2)$.

Regardless of whether a space $\mathcal{X}$ was constructed as a sum of two spaces $\mathcal{U}$ and $\mathcal{V}$ or if $\mathcal{U}$ and $\mathcal{V}$ were selected as subspaces of $\mathcal{X}$, each vector $\mathbf{x}$ can be written as $\mathbf{x} = \mathbf{u} + \mathbf{v}$. Then $\mathbf{u}$ is called the projection of $\mathbf{x}$ on $\mathcal{U}$ and $\mathbf{v}$ is the projection of $\mathbf{x}$ on $\mathcal{V}$.

The Projection Theorem

Let $\mathcal{X}^n$ be an n-dimensional vector space, and let $\mathcal{U}$ be a subspace of dimension $m < n$. Then for every $\mathbf{x} \in \mathcal{X}^n$ there exists a vector $\mathbf{u} \in \mathcal{U}$, called the projection of $\mathbf{x}$ on $\mathcal{U}$, which satisfies

$$\langle \mathbf{x} - \mathbf{u}, \mathbf{y} \rangle = 0$$

for every vector $\mathbf{y} \in \mathcal{U}$. This says that $\mathbf{w} = \mathbf{x} - \mathbf{u}$ is orthogonal to $\mathbf{y}$. In other words, $\mathbf{u}$ is the orthogonal projection of $\mathbf{x}$ on $\mathcal{U}$, and $\mathbf{w}$ is orthogonal to the subspace $\mathcal{U}$.

For proofs of the projection theorem, see References 1 and 2.

For a given $\mathbf{x}$ there is a unique projection $\mathbf{u}$, but there are infinitely many $\mathbf{x}$ vectors which have the same projection. The set of all vectors in $\mathcal{X}^n$ which are orthogonal to $\mathcal{U}$ forms an $n - m$ dimensional subspace of $\mathcal{X}^n$, called the *orthogonal complement* of $\mathcal{U}$, written $\mathcal{U}^\perp$. Every $\mathbf{w} \in \mathcal{U}^\perp$ is orthogonal to every $\mathbf{y} \in \mathcal{U}$. The set of all vectors which are orthogonal to $\mathcal{U}^\perp$ is the subspace $\mathcal{U}$, that is, $(\mathcal{U}^\perp)^\perp = \mathcal{U}$. The spaces $\mathcal{X}^2$ and $\mathcal{Y}^2$ of Example 5.11 are orthogonal complements of one another. By using any subspace $\mathcal{U}$ and its orthogonal complement, an n-dimensional space can be expressed as the direct sum $\mathcal{X}^n = \mathcal{U} \oplus \mathcal{U}^\perp$. Each vector $\mathbf{x} \in \mathcal{X}^n$ can be written uniquely as $\mathbf{x} = \mathbf{u} + \mathbf{v}$. It is easy to show that $\|\mathbf{x}\|^2 = \|\mathbf{u}\|^2 + \|\mathbf{v}\|^2$ because of the orthogonal nature of this decomposition.

The projection theorem and related concepts can be used to develop the theory of least squares estimation and the theory of *generalized* or *pseudo-inverses* of non-square or singular matrices. Some of these applications appear in the next chapter.

5.10 PRODUCT SPACES

Let $\mathcal{X}$ and $\mathcal{Y}$ be arbitrary linear vector spaces defined over the field $\mathcal{F}$. Let $\mathbf{x} \in \mathcal{X}$ and $\mathbf{y} \in \mathcal{Y}$. Then the *product space* $\mathcal{X} \times \mathcal{Y}$ is defined as all ordered pairs of vectors $(\mathbf{x}, \mathbf{y})$. It can be verified that the product space satisfies the conditions of Sec. 5.3 and is therefore a linear vector space. Let $\mathbf{z}_1 = (\mathbf{x}_1, \mathbf{y}_1)$ and $\mathbf{z}_2 = (\mathbf{x}_2, \mathbf{y}_2)$ belong to $\mathcal{X} \times \mathcal{Y}$. Then addition and scalar multiplication are defined by $\mathbf{z}_1 + \mathbf{z}_2 = (\mathbf{x}_1 + \mathbf{x}_2, \mathbf{y}_1 + \mathbf{y}_2) = \mathbf{z}_2 + \mathbf{z}_1$ and $\alpha\mathbf{z}_1 = (\alpha\mathbf{x}_1, \alpha\mathbf{y}_1)$. The zero vector in $\mathcal{X} \times \mathcal{Y}$ is the ordered pair of zero elements $\mathbf{0} \in \mathcal{X}$ and $\mathbf{0} \in \mathcal{Y}$.

Product spaces can be formed as the product of any number of spaces. The familiar Euclidean three-dimensional space is a product space formed from products of the real line $\mathcal{R}^1$, $\mathcal{R}^3 = \mathcal{R}^1 \times \mathcal{R}^1 \times \mathcal{R}^1$. Another common product space is formed from n products of the space of square integrable functions, $\mathcal{L}_2[a, b] \times \mathcal{L}_2[a, b] \times \cdots \times \mathcal{L}_2[a, b]$. Each element in this space is of the form $(f_1(t), f_2(t), \ldots, f_n(t))$, where $f_i(t) \in \mathcal{L}_2[a, b]$. Elements in this product space are usually written more simply as n component vectors $\mathbf{f}(t)$.

The spaces used in forming a product space need not be the same type of spaces. If $\mathcal{R}^1$ is considered as a vector space with elements t and if $\mathbf{x} \in \mathcal{X}^m$, $\mathbf{y} \in \mathcal{Y}$, then elements of $\mathcal{R}^1 \times \mathcal{X}^m \times \mathcal{Y}$ are $\mathbf{z} = (t, \mathbf{x}, \mathbf{y})$. Product spaces were used in Chapter 3 in the definitions of a dynamic system. It should now be clear that a number of diverse objects can be grouped together and treated as components of a single vector in a product space.

For example, two time points t_0 and t_1, each in some segment τ of the real line, an initial state vector $\mathbf{x}(t_0) \in \Sigma$ and a segment of input vector functions $\mathbf{u}_{[t0,t1]} \in \mathcal{U}$ can be used to define a point or vector $\mathbf{p} = (t_0, t_1, \mathbf{x}(t_0), \mathbf{u}_{[t0,t1]})$ which belongs to the product space $\tau \times \tau \times \Sigma \times \mathcal{U}$. One requirement of a dynamical system is that there exist a *unique* mapping $\mathbf{x}(t_1) = \mathbf{g}(\mathbf{p})$.

11 TRANSFORMATIONS OR MAPPINGS

The concepts of functions, which were introduced in Sec. 3.1, are now generalized to abstract vector spaces. Let $\mathcal{X}$ and $\mathcal{Y}$ be linear vector spaces (not necessarily distinct), which are defined over the same scalar number field $\mathcal{F}$. If for each vector $\mathbf{x} \in \mathcal{X}$ there is associated, according to some rule, a vector $\mathbf{y} \in \mathcal{Y}$, then that "rule" defines a mapping of $\mathbf{x}$ into $\mathbf{y}$. This mapping rule is referred to as a transformation (or an operator or a function). This relationship is expressed by

$$\mathcal{A} : \mathcal{X} \rightarrow \mathcal{Y}$$

The transformation is $\mathcal{A}$ and the mapping rule is $\mathcal{A}(\mathbf{x}) = \mathbf{y}$. The spaces $\mathcal{X}$ and $\mathcal{Y}$ are called the domain and codomain of $\mathcal{A}$, respectively. The domain is often written as $\mathcal{D}(\mathcal{A})$, and the range of $\mathcal{A}$ is $\mathcal{A}(\mathcal{X})$ or $\mathcal{R}(\mathcal{A})$. Obviously $\mathcal{R}(\mathcal{A})$ is contained within or equal to $\mathcal{Y}$. This is written as $\mathcal{R}(\mathcal{A}) \subseteq \mathcal{Y}$. In general, $\mathcal{A}$ maps $\mathcal{X}$ *into* $\mathcal{Y}$, but if the equality holds, it maps $\mathcal{X}$ *onto* $\mathcal{Y}$. Again, if $\mathcal{A}(\mathbf{x}) = \mathbf{y}$, $\mathbf{y}$ is called the image of $\mathbf{x}$ or $\mathbf{x}$ is the pre-image of $\mathbf{y}$. The transformation $\mathcal{A}$ is said to be one-to-one if

$$\mathbf{x}_1 \neq \mathbf{x}_2 \Rightarrow \mathcal{A}(\mathbf{x}_1) \neq \mathcal{A}(\mathbf{x}_2)$$

or equivalently, if

$$\mathcal{A}(\mathbf{x}_1) = \mathcal{A}(\mathbf{x}_2) \Rightarrow \mathbf{x}_1 = \mathbf{x}_2$$

If $\mathcal{A}$ is both one-to-one and onto, then for each $\mathbf{y} \in \mathcal{Y}$ there is a unique pre-image $\mathbf{x} \in \mathcal{X}$, and an inverse transformation $\mathcal{A}^{-1}$ maps $\mathbf{y}$ into $\mathbf{x}$. In this case, $\mathcal{A}(\mathbf{x}) = \mathbf{y}$ and $\mathcal{A}^{-1}(\mathbf{y}) = \mathbf{x}$, so $\mathcal{A}^{-1}(\mathcal{A}(\mathbf{x})) = \mathbf{x}$. Thus

$$\mathcal{A}^{-1}\mathcal{A} = \mathcal{I}$$

is the identity transformation which maps each vector in its domain into itself.

The *null space* $\mathcal{N}(\mathcal{A})$ of the transformation $\mathcal{A}$ is the set of all vectors $\mathbf{x} \in \mathcal{X}$, which are mapped into the zero vector in $\mathcal{Y}$:

$$\mathcal{N}(\mathcal{A}) \triangleq \{\mathbf{x} \in \mathcal{X} | \mathcal{A}(\mathbf{x}) = \mathbf{0}\}$$

Linear Transformations

A transformation $\mathcal{A} : \mathcal{X} \rightarrow \mathcal{Y}$ is said to be *linear* if the following two conditions are satisfied:

1. For any $\mathbf{x}_1$ and $\mathbf{x}_2 \in \mathcal{X}$, $\mathcal{A}(\mathbf{x}_1 + \mathbf{x}_2) = \mathcal{A}(\mathbf{x}_1) + \mathcal{A}(\mathbf{x}_2)$.
2. For any $\mathbf{x} \in \mathcal{X}$ and any scalar $\alpha \in \mathcal{F}$, $\mathcal{A}(\alpha\mathbf{x}) = \alpha\mathcal{A}(\mathbf{x})$.

Although nonlinear functions or transformations arise in connection with nonlinear control systems, major emphasis in this book is on linear systems or linear approximations to nonlinear ones. For this reason the rest of this chapter is devoted to linear transformations.

By far the most useful linear transformation for the purposes of this book is one whose domain and codomain are finite dimensional vector spaces. Every linear transformation of this type can be represented as a matrix, once suitable bases are selected. This is seen as follows.

Consider the linear transformation $\mathcal{A}: \mathcal{X}^n \to \mathcal{X}^m$ such that for $\mathbf{x} \in \mathcal{X}^n$ and $\mathbf{y} \in \mathcal{X}^m$, $\mathbf{y} = \mathcal{A}(\mathbf{x})$. Let $\{\mathbf{v}_i, i = 1, \ldots, n\}$ and $\{\mathbf{u}_i, i = 1, \ldots, m\}$ be basis sets for $\mathcal{X}^n$ and $\mathcal{X}^m$, respectively. Then $\mathbf{x} = \sum_{i=1}^{n} \alpha_i \mathbf{v}_i$ and the linearity properties of $\mathcal{A}$ give

$$\mathbf{y} = \sum_{i=1}^{n} \alpha_i \mathcal{A}(\mathbf{v}_i) = [\mathcal{A}(\mathbf{v}_1) \mid \mathcal{A}(\mathbf{v}_2) \mid \cdots \mid \mathcal{A}(\mathbf{v}_n)] \begin{bmatrix} \alpha_1 \\ \alpha_2 \\ \vdots \\ \alpha_n \end{bmatrix} \tag{5.3}$$

The vectors $\mathcal{A}(\mathbf{v}_i)$ are images of the basis vectors $\mathbf{v}_i$ under the transformation $\mathcal{A}$. Since $\mathbf{y}$ and each $\mathcal{A}(\mathbf{v}_i)$ belong to $\mathcal{X}^m$, they have unique expansions with respect to the basis set $\{\mathbf{u}_i\}$,

$$\mathbf{y} = \sum_{j=1}^{m} \beta_j \mathbf{u}_j \quad \text{and} \quad \mathcal{A}(\mathbf{v}_i) = \sum_{j=1}^{m} a_{ji} \mathbf{u}_j \tag{5.4}$$

Combining equations (5.3) and (5.4) gives

$$\mathbf{y} = \sum_{j=1}^{m} \beta_j \mathbf{u}_j = \sum_{i=1}^{n} \alpha_i \left[\sum_{j=1}^{m} a_{ji} \mathbf{u}_j \right]$$

Interchanging the order of summation and using the fact that the expansion coefficients β_j are unique lead to

$$\beta_j = \sum_{i=1}^{n} a_{ji} \alpha_i, \quad j = 1, 2, \ldots, m \tag{5.5}$$

Let $[\mathbf{x}]_{\mathbf{v}} \triangleq [\alpha_1 \ \ \alpha_2 \ \ \cdots \ \ \alpha_n]^T$ and $[\mathbf{y}]_{\mathbf{u}} \triangleq [\beta_1 \ \ \beta_2 \ \ \cdots \ \ \beta_m]^T$ be the coordinate representations of the vectors $\mathbf{x}$ and $\mathbf{y}$ with respect to the basis sets $\{\mathbf{v}_i\}$ and $\{\mathbf{u}_i\}$, respectively. (When the natural cartesian basis of Example 5.1 is used, this cumbersome notation is not necessary since then $\mathbf{x} = [x_1 \ \ x_2 \ \ \cdots \ \ x_n]^T$ and $\mathbf{y} = [y_1 \ \ y_2 \ \ \cdots \ \ y_m]^T$.) Regardless of which basis sets are selected, the transformation $\mathbf{y} = \mathcal{A}(\mathbf{x})$, or equivalently the set of Eqs. (5.5), can be represented by the matrix equation $[\mathbf{y}]_{\mathbf{u}} = \mathbf{A}[\mathbf{x}]_{\mathbf{v}}$. The matrix $\mathbf{A}$ is $m \times n$, and a typical element a_{ji} is seen to be the jth component (with respect to the basis $\{\mathbf{u}_i\}$) of the image of $\mathbf{v}_i$. The particular matrix representation $\mathbf{A}$ for $\mathcal{A}$ obviously depends on the choice of basis in both $\mathcal{X}^n$ and $\mathcal{X}^m$. Changing either basis set changes the resultant representation $\mathbf{A}$. However, many properties of $\mathcal{A}$ are independent of the particular representation $\mathbf{A}$. For example, the rank of $\mathcal{A}$ equals the rank of $\mathbf{A}$ regardless of which representation $\mathbf{A}$ is used. This is also the dimension of the range space of $\mathcal{A}$:

$$\text{rank}(\mathcal{A}) = r_A = \dim(\mathcal{R}(\mathcal{A}))$$

The range of $\mathcal{A}$ is frequently referred to as the column space of **A**.

Change of Basis

Consider the n-dimensional linear vector space $\mathcal{X}^n$. Let $\{\mathbf{v}_i, i = 1, \ldots, n\}$ and $\{\mathbf{v}_i', i = 1, \ldots, n\}$ be two basis sets. Each vector $\mathbf{x} \in \mathcal{X}^n$ can be expressed with respect to either basis; for example,

$$\mathbf{x} = \sum_{j=1}^{n} x_j \mathbf{v}_j = \sum_{i=1}^{n} x_i' \mathbf{v}_i'$$

where x_j and x_i' are scalar components. Since the basis vectors themsevles belong to $\mathcal{H}^n$, one set ca be expressed in terms of the other. For example,

$$\mathbf{v}_j = \sum_{i=1}^{n} b_{ij} \mathbf{v}_i'$$

Using this result to eliminate $\mathbf{v}_j$ in the expression for $\mathbf{x}$ gives

$$\sum_{j=1}^{n} x_j \sum_{i=1}^{n} b_{ij} \mathbf{v}_i' = \sum_{i=1}^{n} x_i' \mathbf{v}_i' \quad \text{or} \quad \sum_{i=1}^{n} \left(\sum_{j=1}^{n} b_{ij} x_j - x_i' \right) \mathbf{v}_i' = \mathbf{0}$$

Linear independence of the set $\{\mathbf{v}_i'\}$ requires that

$$\sum_{j=1}^{n} b_{ij} x_j = x_i'$$

The component vectors $[\mathbf{x}]_{\mathbf{v}}$ and $[\mathbf{x}]_{\mathbf{v}'}$ are thus related by a matrix multiplication:

$$[\mathbf{x}]_{\mathbf{v}'} = [\mathbf{B}][\mathbf{x}]_{\mathbf{v}}$$

A change of basis is seen to be equivalent to a matrix multiplication. The effect of a change of basis on the representation of a linear transformation is now considered. Let $\mathcal{A}$ map vectors in $\mathcal{X}^n$ into other vectors also in $\mathcal{X}^n : \mathcal{A} : \mathcal{X}^n \to \mathcal{X}^n$, where $\mathcal{A}(\mathbf{x}) = \mathbf{y}$. Let **A** be the representation of $\mathcal{A}$ when the basis $\{\mathbf{v}_i\}$ is used, and let $\mathbf{A}'$ be the representation when $\{\mathbf{v}_i'\}$ is used. The relation between **A** and $\mathbf{A}'$ is to be found. When using the unprimed basis set, the transformation is represented as

$$\mathbf{A}[\mathbf{x}]_{\mathbf{v}} = [\mathbf{y}]_{\mathbf{v}}$$

When using the primed basis set,

$$\mathbf{A}'[\mathbf{x}]_{\mathbf{v}'} = [\mathbf{y}]_{\mathbf{v}'}$$

But it was shown earlier that coordinate representations of any vector with respect to two sets of basis vectors are related by

$$[\mathbf{x}]_{\mathbf{v}'} = [\mathbf{B}][\mathbf{x}]_{\mathbf{v}} \qquad [\mathbf{y}]_{\mathbf{v}'} = [\mathbf{B}][\mathbf{y}]_{\mathbf{v}}$$

Thus

$$\mathbf{A}'[\mathbf{B}][\mathbf{x}]_{\mathbf{v}} = [\mathbf{B}][\mathbf{y}]_{\mathbf{v}}$$

The matrix **B** which represents the change of basis always has an inverse (see Problem 5.27), so

$$[\mathbf{B}^{-1}]\mathbf{A}'[\mathbf{B}][\mathbf{x}]_v = [\mathbf{y}]_v$$

This is the representation of the transformation in the unprimed system. The two representations for $\mathcal{A}$ are related by

$$\mathbf{B}^{-1}\mathbf{A}'\mathbf{B} = \mathbf{A}$$

This relationship between $\mathbf{A}'$ and $\mathbf{A}$ is called a *similarity transformation.* Any two matrices which are related by a similarity transformation are said to be *similar matrices.* In the present context similar matrices are representations of a linear transformation with respect to different basis vectors.

If both basis sets $\{\mathbf{v}_i\}$ and $\{\mathbf{v}_i'\}$ are orthonormal, then it can be shown (see Problem 5.28) that the matrix **B** is an orthogonal matrix. That is,

$$\mathbf{B}^{-1} = \mathbf{B}^T$$

In this case the two representations of $\mathcal{A}$ are related by an *orthogonal transformation,*

$$\mathbf{B}^T\mathbf{A}'\mathbf{B} = \mathbf{A}$$

Operations with Linear Transformations

Every linear transformation on finite dimensional spaces can be represented as a matrix. It is natural to expect that algebraic operations with linear transformations are governed by rules much like those of matrix algebra. Let $\mathcal{X}^n$, $\mathcal{X}^m$, and $\mathcal{X}^p$ be linear vector spaces defined over the same scalar number field $\mathcal{F}$. Then if

$$\mathcal{A}_1 : \mathcal{X}^n \to \mathcal{X}^m, \qquad \mathcal{A}_2 : \mathcal{X}^n \to \mathcal{X}^m \text{ and } \mathcal{A}_1(\mathbf{x}) = \mathbf{y}_1, \qquad \mathcal{A}_2(\mathbf{x}) = \mathbf{y}_2$$

then

$$(\mathcal{A}_1 + \mathcal{A}_2)(\mathbf{x}) = \mathcal{A}_1(\mathbf{x}) + \mathcal{A}_2(\mathbf{x}) = \mathbf{y}_1 + \mathbf{y}_2$$

If $\mathcal{A}_1 : \mathcal{X}^n \to \mathcal{X}^m$ and $\mathcal{A}_2 : \mathcal{X}^m \to \mathcal{X}^p$, then $\mathcal{A}_2\mathcal{A}_1 : \mathcal{X}^n \to \mathcal{X}^p$ and, in general, $\mathcal{A}_1\mathcal{A}_2$ is not defined. Thus linear transformations are distributive but not commutative.

A norm can be defined for linear transformations, as follows. If

$$\mathcal{A}(\mathbf{x}) = \mathbf{y}$$

then $\|\mathbf{y}\| = \|\mathcal{A}(\mathbf{x})\|$. If there exists a finite number K such that $\|\mathcal{A}(\mathbf{x})\| \le K\|\mathbf{x}\|$ for all $\mathbf{x}$, the linear transformation is said to be *bounded.* (Every linear transformation on finite dimensional spaces is bounded). Assuming that $\mathcal{A}$ is bounded, the norm of $\mathcal{A}$, written $\|\mathcal{A}\|$, is the smallest value of K which provides such a bound. Alternatively, $\|\mathcal{A}\|$ is the least upper bound (supremum or sup) of $\|\mathcal{A}(\mathbf{x})\|/\|\mathbf{x}\|$ for nonzero $\mathbf{x}$. Two equivalent formulas are

$$\|\mathcal{A}\| = \sup_{\mathbf{x} \neq 0} \frac{\|\mathcal{A}(\mathbf{x})\|}{\|\mathbf{x}\|} \quad \text{or} \quad \|\mathcal{A}\| = \sup_{\|\mathbf{x}\| = 1} \|\mathcal{A}(\mathbf{x})\| \tag{5.6}$$

There are various possible choices for the norm of the vector $\mathcal{A}(\mathbf{x})$. The most familiar is the Euclidean norm of Sec. 5.6, but see also Problems 5.33 and 5.34. Each particular choice induces a different form for $\|\mathcal{A}\|$. If $\mathbf{A}$ is the matrix representation for a finite dimensional transformation, $\mathcal{A}$, and if the quadratic vector norm is used, then

$$\|\mathcal{A}\|^2 = \max_{\|\mathbf{x}\|=1} \{\bar{\mathbf{x}}^T \bar{\mathbf{A}}^T \mathbf{A}\mathbf{x}\}$$

Some properties satisfied by the norm of a linear transformation are

$$\|\mathcal{A}(\mathbf{x})\| \leq \|\mathcal{A}\| \cdot \|\mathbf{x}\| \quad \text{for all } \mathbf{x}$$

$$\|\mathcal{A}_1 + \mathcal{A}_2\| \leq \|\mathcal{A}_1\| + \|\mathcal{A}_2\|$$

$$\|\mathcal{A}_1 \mathcal{A}_2\| \leq \|\mathcal{A}_1\| \cdot \|\mathcal{A}_2\|$$

$$\|\alpha\mathcal{A}\| = |\alpha| \cdot \|\mathcal{A}\|$$

As defined earlier, the norm of every linear transformation is a nonnegative number, and is zero only for a null transformation, i.e., a transformation which maps every vector into the zero vector.

A particular class of linear transformations is that which maps vectors into the one-dimensional vector space formed by the scalar number field. These transformations are called *linear functionals* [3].

12 ADJOINT TRANSFORMATIONS

Let $\mathcal{A}: \mathcal{X}_1 \to \mathcal{X}_2$ be a linear transformation, where $\mathcal{X}_1$ and $\mathcal{X}_2$ are inner product spaces, with inner products $\langle , \rangle_1$ and $\langle , \rangle_2$, respectively. For each $\mathbf{x} \in \mathcal{X}_1$, $\mathcal{A}(\mathbf{x}) = \mathbf{y} \in \mathcal{X}_2$. If $\mathbf{z}$ is an arbitrary vector in $\mathcal{X}_2$, then the inner product $\langle \mathbf{z}, \mathbf{y} \rangle_2 = \langle \mathbf{z}, \mathcal{A}(\mathbf{x}) \rangle_2$ is well defined and can be used to define the *adjoint transformation* $\mathcal{A}^*: \mathcal{X}_2 \to \mathcal{X}_1$, according to $\langle \mathbf{z}, \mathcal{A}(\mathbf{x}) \rangle_2 = \langle \mathcal{A}^*(\mathbf{z}), \mathbf{x} \rangle_1$. It can be shown for finite dimensional spaces that $\mathcal{A}^*$ is also a linear transformation, i.e., if $\mathcal{A}^*(\mathbf{z}_1) = \mathbf{w}_1$ and $\mathcal{A}^*(\mathbf{z}_2) = \mathbf{w}_2$, then $\mathcal{A}^*(\alpha_1 \mathbf{z}_1 + \alpha_2 \mathbf{z}_2) = \alpha_1 \mathbf{w}_1 + \alpha_2 \mathbf{w}_2$ for arbitrary scalars α_1 and $\alpha_2 \in \mathcal{F}$. This follows from the linearity properties of the inner product.

EXAMPLE 5.13 Let $\mathcal{A}$ be a transformation from an n-dimensional vector space to an m-dimensional space, with the usual definition of the complex inner products,

$$\langle \mathbf{x}_1, \mathbf{x}_2 \rangle \triangleq \bar{\mathbf{x}}_1^T \mathbf{x}_2, \qquad \langle \mathbf{y}_1, \mathbf{y}_2 \rangle \triangleq \bar{\mathbf{y}}_1^T \mathbf{y}_2$$

The operator $\mathcal{A}$ can be represented by an $m \times n$ matrix $\mathbf{A}$, so that

$$\langle \mathbf{z}, \mathcal{A}(\mathbf{x}) \rangle = \bar{\mathbf{z}}^T (\mathbf{A}\mathbf{x}) = \overline{(\bar{\mathbf{A}}^T \mathbf{z})}^T \mathbf{x} = \langle \bar{\mathbf{A}}^T \mathbf{z}, \mathbf{x} \rangle$$

For this example $\mathcal{A}^*$ is represented by the matrix $\bar{\mathbf{A}}^T$. ∎

The adjoint transformation defined by the inner product should not be confused with the adjoint matrix defined and used in Chapter 4. Adjoint transformations appear in several roles in modern control theory, and some of these will be developed later. Only a few properties of adjoint transformations are presented here.

If $\mathcal{A}:\mathcal{X}_1 \rightarrow \mathcal{X}_2$, then $\mathcal{A}^*:\mathcal{X}_2 \rightarrow \mathcal{X}_1$. Also, $\mathcal{A}^*\mathcal{A}:\mathcal{X}_1 \rightarrow \mathcal{X}_1$ and $\mathcal{A}\mathcal{A}^*:\mathcal{X}_2 \rightarrow \mathcal{X}_2$. If $\mathcal{X}_1 = \mathcal{X}_2$ and $\mathcal{A} = \mathcal{A}^*$, then $\mathcal{A}$ is said to be *self-adjoint*. In all cases, it can be shown that $\|\mathcal{A}\| = \|\mathcal{A}^*\|$, and that $(\mathcal{A}^*)^* = \mathcal{A}$. It is clear that $\mathcal{A}^*\mathcal{A}$ is generally not equal to $\mathcal{A}\mathcal{A}^*$. Those particular transformations for which $\mathcal{A}^*\mathcal{A} = \mathcal{A}\mathcal{A}^*$ are said to be *normal transformations.*

Let $\mathcal{A}:\mathcal{X}_1 \rightarrow \mathcal{X}_2$ be an arbitrary linear transformation. Then the linear vector spaces $\mathcal{X}_1$ and $\mathcal{X}_2$ can be written as direct sums

$$\mathcal{X}_1 = \mathcal{N}(\mathcal{A}) \oplus \overline{\mathcal{R}}(\mathcal{A}^*)$$

$$\mathcal{X}_2 = \mathcal{N}(\mathcal{A}^*) \oplus \overline{\mathcal{R}}(\mathcal{A}) \tag{5.7}$$

where $\mathcal{N}(\cdot)$ and $\mathcal{R}(\cdot)$ are the null space and range of the indicated tranformations. $\overline{\mathcal{R}}(\cdot)$ denotes the closure of the range $\mathcal{R}(\cdot)$, that is, $\mathcal{R}(\cdot)$ plus the limit of all convergent sequences of elements in $\mathcal{R}(\cdot)$. In finite dimensional spaces every subspace is closed, so that $\overline{\mathcal{R}}(\cdot) = \mathcal{R}(\cdot)$. Equation (5.7) constitutes an *orthogonal* decomposition of $\mathcal{X}_1$ into two linear subspaces. That is, for any vector $\mathbf{x} \in \mathcal{N}(\mathcal{A})$ and any vector $\mathbf{y} \in \mathcal{R}(\mathcal{A}^*)$, $\langle \mathbf{x}, \mathbf{y} \rangle = 0$. Equation (5.7) also provides an orthogonal decomposition for $\mathcal{X}_2$. Additional results for abstract transformations and their adjoints are found in the problems for this chapter. More concrete applications, where the operators are just matrices, are found in Sec. 5.13 and throughout the next chapter.

5.13 SOME FINITE-DIMENSIONAL TRANSFORMATIONS

Every linear transformation from one finite-dimensional space to another finite-dimensional space can be represented as a matrix. Within this general category, a few special transformations are now discussed.

Rotations

A particular transformation that frequently arises in control applications is a pure rotation. This can often be viewed in two ways. The result can be considered as a new vector obtained by rotating the original vector, or it can be considered as the same vector expressed in terms of a new coordinate system which is rotated with respect to the original coordinate system. The latter point of view is adopted for the time being, and the treatment is restricted to real, three-dimensional space, $\mathcal{R}^3$. Let $\{\mathbf{x}_1, \mathbf{x}_2, \mathbf{x}_3\}$ be an orthonormal basis set, and more specifically, let it define a right-handed cartesian coordinate system. Let $\{\mathbf{y}_1, \mathbf{y}_2, \mathbf{y}_3\}$ be another right-handed cartesian coordinate system.

The set $\{\mathbf{x}_i\}$ might represent an orthogonal triad fixed to an aerospace vehicle or a tracking antenna. The set $\{\mathbf{y}_i\}$ might represent an inertially fixed coordinate system. These two sets can be brought into coincidence by a sequence of rotations. The most familiar set of angles of rotation are the Euler angles [4], although the present discussion applies to any sequence of finite rotations such as those of Figure 5.5. A rotation θ about the $\mathbf{x}_2$ axis rotates $\mathbf{x}_1$ and $\mathbf{x}_3$ into $\mathbf{x}_1'$ and $\mathbf{x}_3'$, and leaves $\mathbf{x}_2' = \mathbf{x}_2$. A rotation ψ about $\mathbf{x}_1'$ gives $\mathbf{x}_1'' = \mathbf{x}_1'$ and $\mathbf{x}_2'', \mathbf{x}_3''$. The final rotation ϕ about $\mathbf{x}_3''$ gives $\mathbf{y}_1, \mathbf{y}_2$, and $\mathbf{y}_3 = \mathbf{x}_3''$.

Let $\mathbf{z}$ be an arbitrary vector and let $[\mathbf{z}]$, $[\mathbf{z}]'$, $[\mathbf{z}]''$, and $[\mathbf{z}]'''$ be its coordinate

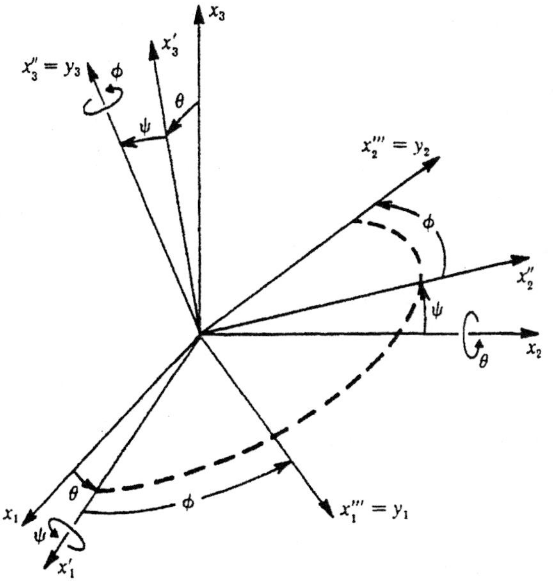

Figure 5.5

representations in the $\{\mathbf{x}_i\}$, $\{\mathbf{x}_i'\}$, $\{\mathbf{x}_i''\}$, and $\{\mathbf{y}_i\}$ coordinate systems, respectively. It is easily verified that

$$[\mathbf{z}]' = \begin{bmatrix} \cos\theta & 0 & -\sin\theta \\ 0 & 1 & 0 \\ \sin\theta & 0 & \cos\theta \end{bmatrix}[\mathbf{z}], \qquad [\mathbf{z}]'' = \begin{bmatrix} 1 & 0 & 0 \\ 0 & \cos\psi & \sin\psi \\ 0 & -\sin\psi & \cos\psi \end{bmatrix}[\mathbf{z}]'$$

$$[\mathbf{z}]''' = \begin{bmatrix} \cos\phi & \sin\phi & 0 \\ -\sin\phi & \cos\phi & 0 \\ 0 & 0 & 1 \end{bmatrix}[\mathbf{z}]'' \tag{5.8}$$

The overall transformation from $[\mathbf{z}]$ to $[\mathbf{z}]'''$ is given by the product of the three transformation matrices.

A symbolic method of representing coordinate rotations has been developed [5, 6]. These resolver-like diagrams, called Piograms, make it possible to write vector components in the new coordinate system without resorting to successive matrix multiplications (see Problem 5.37).

Reflections

The relation between a vector $\mathbf{x}$ and its reflection $\mathbf{x}_r$ at a plane surface defined by a unit normal $\mathbf{n}$ is $\mathbf{x}_r = \mathbf{x} - 2\langle\mathbf{n}, \mathbf{x}\rangle\mathbf{n}$. That is, $\mathbf{x}$ and $\mathbf{x}_r$ are equal except for a sign change in the component along $\mathbf{n}$. This can be rewritten as

$$\mathbf{x}_r = \mathbf{x} - 2\mathbf{n}\rangle\langle\mathbf{n}\mathbf{x} = [\mathbf{I} - 2\mathbf{n}\rangle\langle\mathbf{n}]\mathbf{x}$$

The matrix $\mathbf{A}_r = [\mathbf{I} - 2\mathbf{n}\rangle\langle\mathbf{n}]$ is the general representation of a reflection transformation. It is characterized by the fact that $|\mathbf{A}_r| = -1$, as verified by using the results of Problem 4.5.

Projections

A simple example of a transformation $\mathcal{A}(\mathbf{x})$ which maps $\mathbf{x}$ into its orthogonal projection on a hyperplane with a unit normal vector $\mathbf{n}$ is

$$\mathbf{A}_p = [\mathbf{I} - \mathbf{n}\rangle\langle\mathbf{n}]$$

This is fairly obvious since $\mathbf{A}_p\mathbf{x} = \mathbf{x} - \langle\mathbf{n}, \mathbf{x}\rangle\mathbf{n}$ has the effect of subtracting out the component of $\mathbf{x}$ along $\mathbf{n}$. It is easily verified that $\mathbf{A}_p\,\mathbf{A}_p = \mathbf{A}_p^2 = \mathbf{A}_p$. In general, any linear transformation which satisfies

$$\mathcal{A}^2 = \mathcal{A}$$

is a projection, although it need not be an orthogonal projection as in the above case. It is always possible to express a linear vector space as a direct sum $\mathcal{X} = \mathcal{U} \oplus \mathcal{V}$, where $\mathcal{U}$ and $\mathcal{V}$ are nonvoid subspaces of $\mathcal{X}$. This means that for each $\mathbf{x} \in \mathcal{X}$ there is one and only one way of writing

$$\mathbf{x} = \mathbf{u} + \mathbf{v}, \quad \text{where } \mathbf{u} \in \mathcal{U}, \mathbf{v} \in \mathcal{V}$$

A transformation $\mathcal{P}$ satisfying $\mathcal{P}(\mathbf{x}) = \mathbf{u}$ is said to be the projection on $\mathcal{U}$ along $\mathcal{V}$.

A Practical Application. Many control problems involve coordinate rotations. Some involve projections of vector quantities onto a sensor and others involve reflections. A typical kind of pointing and tracking example from geometrical optics is now given to demonstrate all three.

EXAMPLE 5.14 Suppose that an earth resource satellite consists of a steerable plane mirror and an imaging focal plane. The image of a right angle formed by the square corner of a Nebraska cornfield is to be captured on the focal plane. This image will be skewed or distorted—that is, the edges of the field will no longer appear orthogonal in general. Let $\mathbf{v}_1$ and $\mathbf{v}_2$ be unit vectors at the corner of the field, and let $\mathbf{v}_1'$ and $\mathbf{v}_2'$ be their images on the focal plane. Find expressions for these images and then evaluate their inner product to show nonorthogonality.

There are four coordinate systems involved in this problem, the ground-fixed system $\{x, y, z\}$, the satellite coordinate system, the mirror coordinates $\{x_m, y_m, z_m\}$, and the focal plane coordinates $\{x_f, y_f, z_f\}$. Figure 5.6 shows these and defines the satellite position with respect to the corner in terms of the azimuth angle ψ and zenith angle β and the slant range R. If $\mathbf{z}_m$ is the normal coordinate to the mirror, then the vector $\mathbf{z}_m$ can be written in terms of components in the $\{x, y, z\}$ system as

$$\mathbf{z}_m = \mathbf{T}_{GM}\begin{bmatrix}0\\0\\1\end{bmatrix} = \mathbf{T}_{GS}\,\mathbf{T}_{SM}\begin{bmatrix}0\\0\\1\end{bmatrix}$$

where $\mathbf{T}_{GM}$, $\mathbf{T}_{GS}$, and $\mathbf{T}_{SM}$ are 3×3 rotation matrices that transform vectors from mirror-to-gound, satellite-to-ground, and mirror-to-satellite, respectively. Note that $\mathbf{v}_1$ and $\mathbf{v}_2$ are assumed aligned with the ground x and y axes, respectively. The apparent reflections of $\mathbf{v}_1$ and $\mathbf{v}_2$ are given by

$$\mathbf{v}_1'' = [\mathbf{I} - 2\mathbf{z}_m\,\mathbf{z}_m^T]\mathbf{v}_1 = \mathbf{A}_r\,\mathbf{v}_1$$

$$\mathbf{v}_2'' = \mathbf{A}_r\,\mathbf{v}_2$$

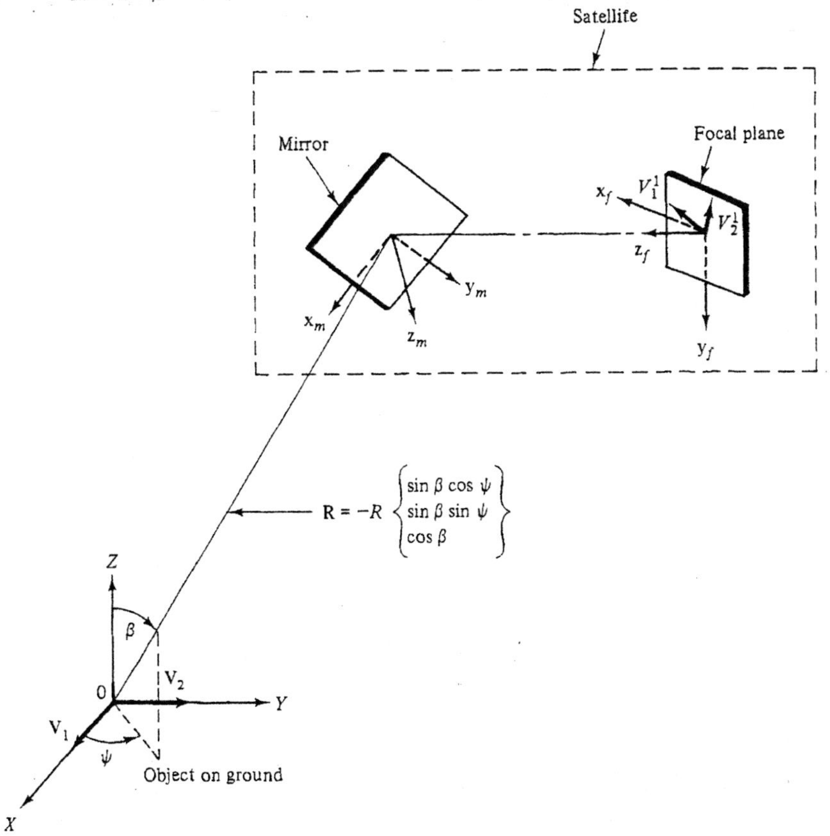

Figure 5.6

where $\mathbf{A}_r$ is the reflection matrix for the mirror. The reflected images are still expressed in ground coordinates. Let the normal to the focal plane be the vector $\mathbf{z}_f$ and assume that the x_f and y_f directions in the focal plane are suitably defined. When expressed in the focal plane coordinates, the reflected images of the two vectors are

$$\mathbf{v}_1''' = \mathbf{T}_{FG}\,\mathbf{A}_r\,\mathbf{v}_1 \quad \text{and} \quad \mathbf{v}_2 = \mathbf{T}_{FG}\,\mathbf{A}_r\,\mathbf{v}_2$$

where $\mathbf{T}_{FS}$ is the transformation from satellite to focal plane coordinates and where $\mathbf{T}_{FG} = \mathbf{T}_{FS}\,\mathbf{T}_{SG} = \mathbf{T}_{FS}\,\mathbf{T}_{GS}^T$. Note that because all the coordinate frames are orthogonal, the transformation matrices are orthogonal, so $\mathbf{T}_{SG} = \mathbf{T}_{GS}^{-1} = \mathbf{T}_{GS}^T$. The triply primed vectors are still three-dimensional. The focal plane images are the projection of these onto the focal plane:

$$\mathbf{v}_1' = [\mathbf{I} - \mathbf{n}\mathbf{n}^T]\mathbf{v}_1''' = A_p\,\mathbf{v}_1''' = \begin{bmatrix} 1 & 0 & 0 \\ 0 & 1 & 0 \\ 0 & 0 & 0 \end{bmatrix}\mathbf{v}_1'''$$

and $\mathbf{v}_2' = \mathbf{A}_p\,\mathbf{v}_2'''$, where $\mathbf{n} = [0 \quad 0 \quad 1]^T$. The projection matrix has been named $\mathbf{A}_p$. In general there would be a lens system between the mirror and the focal plane. It merely scales the vectors without changing their directions, so this complexity is neglected here. The true physical angle θ

on the ground is assumed to be 90° here, but in general it is given by $\cos(\theta) = \langle \mathbf{v}_1, \mathbf{v}_2 \rangle$. The apparent angle on the focal plane is found from $\cos(\theta_f) = \langle \mathbf{v}_1', \mathbf{v}_2' \rangle / \|\mathbf{v}_1'\| \|\mathbf{v}_2'\|$. ■

EXAMPLE 5.15 In order to use the relations of the previous example the various transformation matrices must be known. For simplicity the satellite coordinates are assumed aligned with the ground-fix coordinates so that $\mathbf{T}_{GS} = \mathbf{I}$. The optical focal plane is assumed fixed to the vehicle with an orientation which gives

$$\mathbf{T}_{FS} = \begin{bmatrix} -1 & 0 & 0 \\ 0 & 0 & -1 \\ 0 & -1 & 0 \end{bmatrix}$$

That leaves only the mirror orientation to be specified. However, it cannot be arbitrarily specified. The mirror must be steered so that the correct scene is reflected and projected upon the focal plane. Treat this as an open-loop control problem and determine $\mathbf{T}_{SM}$ so that the satellite-to-ground vector $\mathbf{R}$ is projected onto the focal plane origin.

Note that $\mathbf{T}_{SM}$ is needed to find the vector $\mathbf{z}_m$, which in turn is used to calculate $\mathbf{A}_r$. The desired $\mathbf{A}_r$ matrix will now be found directly. The $\mathbf{R}$ vector, after reflection, must be entirely along the normal to the focal plane, so

$$[0 \quad 0 \quad R]^T = \mathbf{T}_{FG}\, \mathbf{A}_r\, \mathbf{R} \tag{5.9}$$

is required. In order to determine the unknown mirror orientation matrix $\mathbf{A}_r$, two more independent equations are needed. One can be obtained by specifying how the x_f, y_f axes are rotated about the focal plane normal. One way of doing this is to force the unit vector $\mathbf{u}$, which is normal to both $\mathbf{R}$ and the ground x axis, to project along the $-y_f$ axis. $\mathbf{u} = \mathbf{R} \times \mathbf{x} / \|\mathbf{R} \times \mathbf{x}\|$ and then

$$[0 \quad -1 \quad 0]^T = \mathbf{T}_{FG}\, \mathbf{A}_r\, \mathbf{u} \tag{5.10}$$

A third independent equation is available from the cross product of Eq. (5.9) and (5.10):

$$[0 \quad 0 \quad R]^T \times [0 \quad -1 \quad 0]^T = \mathbf{T}_{FG}\, \mathbf{A}_r (\mathbf{R} \times \mathbf{u}) \tag{5.11}$$

Since $\mathbf{T}_{GS} = \mathbf{I}$, $\mathbf{T}_{FS} = \mathbf{T}_{FG}$. Thus Eqs. (5.9), (5.10) and (5.11) can be combined into one matrix equation and solved to give

$$\mathbf{A}_r = \begin{bmatrix} 0 & 0 & R \\ -R & 0 & 0 \\ 0 & 1 & 0 \end{bmatrix} [\mathbf{R} \mid \mathbf{u} \mid \mathbf{R} \times \mathbf{u}]^{-1}$$

Using the definition of $\mathbf{A}_r$, the orientation of the mirror normal vector can be found. The required gimbal angles for the mirror can then be calculated and used as the open-loop commands to the two axes of the mirror-drive servos. Closed-loop error-nulling controllers would normally be used in an actual system. The purpose here was to demonstrate that rotations, reflections, and projections are useful in real control problems. ■

EXAMPLE 5.16 For the system of the previous two examples, suppose the satellite is located with respect to the desired corner at $\beta = 30°$, $\psi = 45°$, and a slant range of 100 nautical miles. Compute the skew in the 90° corner.

With these values,

$$\mathbf{A}_r = \begin{bmatrix} -0.9524 & 0.13606 & 0.33328 \\ 0.30855 & 0.35998 & 0.88177 \\ 0.11783 & -0.94265 & 0.33673 \end{bmatrix} \quad \text{(rounded)}$$

The focal plane images are found to be $\mathbf{v}_1' = [0.95242 \quad -0.11783]^T$ and $\mathbf{v}_2' = [-0.13606 \quad 0.94265]^T$, so that the inner product gives $\theta_f = 105.266°$. The skew (distortion from the true angle) is 15.266°. As the angle β approaches zero (direct overhead viewing) the skew approaches zero for all ψ. The skew also decreases to zero if the corner is viewed from above either the x axis or the y axis, i.e., for ψ either 0° or 90°. Table 5.1 gives results for a few representative combinations.

TABLE 5.1

ψ	β	θ_f
45	30	105.266
45	20	96.903
45	10	91.741
45	0	90
0	30	90

■

5.14 SOME TRANSFORMATIONS ON INFINITE DIMENSIONAL SPACES

Most of the analysis of lumped-parameter systems in modern control theory can be considered in terms of a finite dimensional linear space, the state space. Consequently, the major emphasis is on finite dimensional transformations. However, transformations on infinite dimensional spaces do arise, and two of the more important ones are mentioned here.

It is recalled that the dimension of a space is equal to the number of elements in its basis set. The set of all periodic functions with period π is an example of an infinite dimensional space and its basis could be selected as the functions $\{\sin nt, n = 0, 1, \ldots\}$. The expansion with respect to this basis is the Fourier series. The set of all continuous functions, or of integrable functions, or of all square integrable functions are other examples of infinite dimensional spaces. A space is not necessarily infinite dimensional just because its elements are functions of time. For example, the space of all polynomials of degree 3 or less—i.e., $\{f(t)|f(t) = \alpha_0 + \alpha_1 t + \alpha_2 t^2 + \alpha_3 t^3, \alpha_i \in \mathcal{F}\}$—is a four-dimensional space.

Integral Relations

The integral form of a system's input-output equation was mentioned in Sec. 1.5 and a particular example is used in Problem 5.36. For a linear system the input and output can be related by

$$\mathbf{y}(T) = \int_{-\infty}^{T} \mathbf{W}(T, \tau)\mathbf{u}(\tau)\, d\tau$$

where $\mathbf{y}(T)$ is the $m \times 1$ output vector at time T, $\mathbf{u}(t)$ is an $r \times 1$ input vector for each value of t, and $\mathbf{W}(t, \tau)$ is the $m \times r$ weighting matrix. At each time T, $\mathbf{y}(T)$ is a vector in an m-dimensional space. A particular input function, $u(t), t \in (-\infty, T]$ can be considered as an element of the (infinite dimensional) input function space $\mathcal{U}$. The input-output integral represents a transformation $\mathcal{A}: \mathcal{U} \rightarrow \mathcal{Y}^m$, where $\mathcal{A}(u) = \mathbf{y}(T)$.

The equation for the Laplace transform

$$\mathbf{y}(s) = \int_{0}^{\infty} e^{-st} y(t)\, dt$$

provides another example of a linear transformation on infinite dimensional spaces. The domain of these transformations must be suitably restricted so that the indicated operations "make sense." In other words, nonintegrable functions cannot be integrated, and functions which cannot be bounded by some exponential function do not have Laplace transforms.

Differential Relations

A linear differential equation can be considered to be a transformation, but again infinite dimensional spaces (function spaces) are involved. The simplest case

$$\frac{dx}{dt} = u$$

maps a function $x(t)$ into another function $u(t)$. Of course, the domain of this transformation must be restricted to the class of functions which are differentiable. Other restrictions may be necessary as well. Perhaps only those functions for which $x(0) = 0$ are considered. This constitutes an initial condition. The relation

$$\left[\mathbf{I}\frac{d}{dt} - \mathbf{A}\right]\mathbf{x}(t) = \mathbf{B}\mathbf{u}(t)$$

is another example of a differential transformation which maps $\mathbf{x}(t)$ into $\mathbf{Bu}(t)$. Although this equation appears repeatedly in the modern formulation of control problems, it will not be necessary to consider it as an abstract mapping on function spaces. Rather, this brief section dealing with transformations on function spaces is intended only to hint at a direction for an abstract treatment of all linear transformations. If the function spaces are Hilbert spaces (i.e., complete inner product spaces), then the results parallel the finite dimensional results to a large degree [2]. In general, however, there will be some major differences. Every finite dimensional linear vector space is complete, and every transformation on finite dimensional spaces is bounded. These are not generally true in the infinite dimensional case. For example, the space of square integrable functions $\mathcal{L}_2[a, b]$ of Problem 5.22 is complete. But the space of all continuous functions $C[a, b)$ is not complete because a sequence of continuous functions may converge to a discontinuous function. Differential operators are examples of *unbounded* linear transformations. Some of the other differences arise because of the greater variety of definitions that can be given for norms and distance measures in function spaces. A complete treatment of these topics can be found in texts on functional analysis [1, 3, 7].

REFERENCES

1. Friedman, B.: *Principles and Techniques of Applied Mathematics,* John Wiley, New York, 1956.
2. Halmos, P. R.: *Finite Dimensional Vector Spaces,* 2d ed., D. Van Nostrand, Princeton, N.J., 1958.
3. Taylor, A. E.: *Functional Analysis,* John Wiley, New York, 1958.

4. Goldstein, H.: *Classical Mechanics*, Addison-Wesley, Reading, Mass., 1959.
5. Pio, R. L.: "Symbolic Representation of Coordinate Transformations," *IEEE Transactions on Aerospace and Navigational Electronics*, Vol. ANE-11, No. 2, June 1964, pp. 128–134.
6. Pio, R. L.: "Euler Angle Transformations," *IEEE Transactions on Automatic Control*, Vol. AC-11, No. 4, Oct. 1966, pp. 707–715.
7. Kolomogorov, A. N. and S. V. Fomin: *Elements of the Theory of Functional Analysis*, Vol. 1 (1957) and Vol. 2 (1961), Graylock Press, Albany, N.Y. (Translated from the Russian editions).
8. Strang, G.: *Linear Algebra and Its Applications*, Academic Press, New York, 1980.

ILLUSTRATIVE PROBLEMS

Vectors in Two and Three Dimensions

5.1 If two vectors $\mathbf{v}^T = [3 \quad -5 \quad 6]$ and $\mathbf{w}^T = [\alpha \quad 2 \quad 2]$ are known to be orthogonal, what is α?

Assuming that the given components are expressed with respect to a common coordinate system, orthogonality requires

$$\langle \mathbf{v}, \mathbf{w} \rangle = \mathbf{v}^T \mathbf{w} = 0 \quad \text{or} \quad 3\alpha - 10 + 12 = 0$$

so that $\alpha = -\frac{2}{3}$.

5.2 If $\mathbf{v}^T = [3 \quad -5 \quad 6]$ and $\mathbf{w}^T = [5 \quad 8]$, find $\langle \mathbf{v}, \mathbf{w} \rangle$ and the two outer products.

Since $\mathbf{v}$ and $\mathbf{w}$ have different numbers of components and hence belong to different dimensional spaces, their inner product is not defined. The outer products are, however,

$$\mathbf{v}\mathbf{w}^T = \begin{bmatrix} 3 \\ -5 \\ 6 \end{bmatrix} [5 \quad 8] = \begin{bmatrix} 15 & 24 \\ -25 & -40 \\ 30 & 48 \end{bmatrix} = (\mathbf{w}\mathbf{v}^T)^T$$

5.3 Consider the nonzero vector $\mathbf{v}$ with complex components $\mathbf{v} = \begin{bmatrix} j \\ 1 \end{bmatrix}$. Compute $\mathbf{v}^T\mathbf{v}$ and $\bar{\mathbf{v}}^T\mathbf{v}$.

Vector multiplication gives

$$\mathbf{v}^T\mathbf{v} = (j)(j) + 1 = 0$$

$$\bar{\mathbf{v}}^T\mathbf{v} = (-j)(j) + 1 = 2$$

Therefore, if the real form of the inner product $\langle \mathbf{v}, \mathbf{v} \rangle = \mathbf{v}^T\mathbf{v}$ is used to define length, nonzero vectors can have zero "length." When the complex form of the inner product $\langle \mathbf{v}, \mathbf{v} \rangle = \bar{\mathbf{v}}^T\mathbf{v}$ is used, this cannot happen.

5.4 Find the component of $\mathbf{v}^T = [2 \quad -3 \quad -4]$ in the direction of the vector $\mathbf{w}^T = [1 \quad 2 \quad 1]$.

First find the unit vector $\hat{\mathbf{w}}$ in the desired direction:

$$\|\mathbf{w}\| = \langle \mathbf{w}, \mathbf{w} \rangle^{1/2} = \sqrt{6}$$

$$\hat{\mathbf{w}} = \frac{1}{\sqrt{6}}\mathbf{w}$$

Then form the inner product:

$$\langle \mathbf{v}, \hat{\mathbf{w}} \rangle = \frac{1}{\sqrt{6}}(2 - 6 - 4) = \frac{-1}{\sqrt{6}}8$$

This result indicates that $\mathbf{v}$ has a component along the negative $\mathbf{w}$ direction and its magnitude is $8/\sqrt{6}$.

5.5 Find the cross products $\mathbf{v} \times \mathbf{w}$ and $\mathbf{w} \times \mathbf{v}$ if $\mathbf{v}^T = [v_1 \quad v_2 \quad v_3]$ and $\mathbf{w}^T = [w_1 \quad w_2 \quad w_3]$ are real vectors.

If we let $\mathbf{e}_1$, $\mathbf{e}_2$, and $\mathbf{e}_3$ be unit vectors along the three mutually orthogonal coordinate axes, the cross product can be computed using the following determinant:

$$\mathbf{v} \times \mathbf{w} = \left\| \begin{bmatrix} \mathbf{e}_1 & \mathbf{e}_2 & \mathbf{e}_3 \\ v_1 & v_2 & v_3 \\ w_1 & w_2 & w_3 \end{bmatrix} \right\|$$

Using Laplace expansion with respect to row one gives

$$\mathbf{v} \times \mathbf{w} = \mathbf{e}_1(v_2 w_3 - v_3 w_2) - \mathbf{e}_2(v_1 w_3 - v_3 w_1) + \mathbf{e}_3(v_1 w_2 - v_2 w_1)$$

Since $\mathbf{e}_1 = \begin{bmatrix} 1 \\ 0 \\ 0 \end{bmatrix}$, $\mathbf{e}_2 = \begin{bmatrix} 0 \\ 1 \\ 0 \end{bmatrix}$, $\mathbf{e}_3 = \begin{bmatrix} 0 \\ 0 \\ 1 \end{bmatrix}$, the column matrix representations are

$$\mathbf{v} \times \mathbf{w} = \begin{bmatrix} v_2 w_3 - v_3 w_2 \\ v_3 w_1 - v_1 w_3 \\ v_1 w_2 - v_2 w_1 \end{bmatrix}$$

$$\mathbf{w} \times \mathbf{v} = \begin{bmatrix} v_3 w_2 - v_2 w_3 \\ v_1 w_3 - v_3 w_1 \\ v_2 w_1 - v_1 w_2 \end{bmatrix}$$

5.6 Show that $\mathbf{v} \times \mathbf{w}$ of the previous problem can be written as the product of a skew-symmetric matrix and $\mathbf{w}$, or another skew-symmetric matrix and $\mathbf{v}$.

Direct matrix multiplication verifies that

$$\mathbf{v} \times \mathbf{w} = \begin{bmatrix} 0 & -v_3 & v_2 \\ v_3 & 0 & -v_1 \\ -v_2 & v_1 & 0 \end{bmatrix} \begin{bmatrix} w_1 \\ w_2 \\ w_3 \end{bmatrix} = \begin{bmatrix} 0 & w_3 & -w_2 \\ -w_3 & 0 & w_1 \\ w_2 & -w_1 & 0 \end{bmatrix} \begin{bmatrix} v_1 \\ v_2 \\ v_3 \end{bmatrix}$$

5.7 From the results of the preceding problem, what can be said about the product $\mathbf{x}^T \mathbf{A} \mathbf{x}$ if $\mathbf{x}$ is a real three-component vector and $\mathbf{A}$ is skew-symmetric?

Any 3×3 skew-symmetric matrix could be used to define a 3×1 vector, and then $\mathbf{Ax}$ would represent a cross product. Therefore, $\mathbf{Ax}$ is a vector perpendicular to $\mathbf{x}$. Therefore, $\langle \mathbf{x}, \mathbf{Ax} \rangle = \mathbf{x}^T \mathbf{Ax} = \mathbf{0}$ for every real $\mathbf{x}$. In fact, if $\mathbf{A}$ is skew-symmetric of arbitrary dimension, the result $\mathbf{x}^T \mathbf{Ax} = \mathbf{0}$ is true for any real, conformable vector $\mathbf{x}$.

Defining a Vector Space

5.8 Does the set of all vectors in the first and fourth quadrants of the plane form a vector space?

No. The first four conditions of Sec. 5.3, page 159, hold. However, if $\mathbf{x}$ is in the first or fourth quadrant, $-\mathbf{x}$ is in the second or third quadrant, so condition 5 is not satisfied and neither is 6 when negative scalars are considered.

5.9 Does the set of all three-dimensional vectors inside a sphere of finite radius constitute a linear vector space?

No. If $\mathbf{x}$ and $\mathbf{y}$ are inside the sphere, $\mathbf{x} + \mathbf{y}$ need not be. If a is sufficiently large, $a\mathbf{x}$ will also extend outside the sphere. Conditions 1 and 6 are not satisfied.

5.10 Consider all vectors defined by points in a plane passing through a three-dimensional space. Is this set a linear vector space?

This is a linear vector space if and only if the plane passes through the origin. If it does not, condition 4 is not satisfied.

5.11 What is the dimension of the space $\mathscr{X}$ defined as the set of all linear combinations of

$$\mathbf{x}_1^T = [1 \quad 2 \quad 3 \quad 4 \quad 5]$$
$$\mathbf{x}_2^T = [1 \quad 0 \quad 0 \quad 0 \quad 1]$$
$$\mathbf{x}_3^T = [0 \quad 1 \quad 1 \quad 0 \quad 0]$$

The manner in which $\mathscr{X}$ is defined guarantees that $\mathbf{x}_1$, $\mathbf{x}_2$, and $\mathbf{x}_3$ span this space. Since the Grammian gives $|\mathbf{G}| = 98$, the set is linearly independent and thus constitutes a basis set. Since there are three vectors in the basis set, the dimension of $\mathscr{X}$, written dim ($\mathscr{X}$), is three even though every vector in $\mathscr{X}$ has five components.

Linear Dependence, Independence, and Degeneracy

.12 Prove that the addition of the zero vector $\mathbf{0}$ to any set of linearly independent vectors yields a set of linearly dependent vectors.

Let $\mathscr{V} = \{\mathbf{x}_i, i = 1, n\}$ be a set of linearly independent vectors. This means that $a_1\mathbf{x}_1 + a_2\mathbf{x}_2 + \cdots + a_n\mathbf{x}_n = \mathbf{0}$ requires $a_i = 0, i = 1, n$. Select $a_{n+1} \neq 0$. Then

$$a_1\mathbf{x}_1 + a_2\mathbf{x}_2 + \cdots + a_n\mathbf{x}_n + a_{n+1}\mathbf{0} = \mathbf{0}$$

so the set of vectors $\{\mathbf{0}, \mathbf{x}_i, i = 1, n\}$ is a linearly dependent set.

.13 Use the Grammian to test the following vectors for linear dependence:

$$\mathbf{x}_1^T = [1 \quad 1 \quad 0 \quad 0], \qquad \mathbf{x}_2^T = [1 \quad 1 \quad 1 \quad 1], \qquad \mathbf{x}_3^T = [0 \quad 0 \quad 1 \quad 1]$$

The Grammian determinant is

$$|\mathbf{G}| = \left\| \begin{bmatrix} 2 & 2 & 0 \\ 2 & 4 & 2 \\ 0 & 2 & 2 \end{bmatrix} \right\| = 0$$

Hence the three vectors are linearly dependent. Note that the Grammian is symmetric for vectors with real components. In general, it is a Hermitian matrix.

.14 Consider two real vectors $\mathbf{x}_1$ and $\mathbf{x}_2$ expressed as 2×1 column vectors in terms of the natural coordinate vectors. Show that $|\mathbf{G}|$ is the square of the area of the parallelogram which has $\mathbf{x}_1$ and $\mathbf{x}_2$ as sides.

We have

$$|\mathbf{G}| = \begin{vmatrix} \mathbf{x}_1^T\mathbf{x}_1 & \mathbf{x}_1^T\mathbf{x}_2 \\ \mathbf{x}_2^T\mathbf{x}_1 & \mathbf{x}_2^T\mathbf{x}_2 \end{vmatrix} = \begin{vmatrix} x_1^2 & x_1x_2\cos\theta \\ x_1x_2\cos\theta & x_2^2 \end{vmatrix}$$

where x_1 and x_2 are the magnitudes of $\mathbf{x}_1$ and $\mathbf{x}_2$ and θ is the included angle. The definition of the inner product of Sec. 5.2 has been used in arriving at this result. Expanding gives

$$|\mathbf{G}| = x_1^2x_2^2(1 - \cos^2\theta) = x_1^2x_2^2\sin^2\theta$$

Considering x_1 as the base of the parallelogram, the height is $x_2 \sin\theta$, so the result is proven.

.15 What are the rank and degeneracy of the following matrices?

(a) $\mathbf{A} = \begin{bmatrix} 6 & 2 & 4 \\ 2 & 0 & 2 \\ 1 & -1 & 2 \end{bmatrix}$ (b) $\mathbf{B} = \begin{bmatrix} 4 & 3 & 7 & 1 \\ 2 & 6 & 2 & 10 \\ 8 & 6 & 14 & 2 \\ 1 & 3 & 1 & 5 \end{bmatrix}$

(c) $\mathbf{C} = \begin{bmatrix} 6 & -4 & -4 & -9 \\ 24 & 3 & 0 & -9 \\ -14 & 3 & 4 & 12 \\ 48 & 25 & 16 & 9 \end{bmatrix}$

(a) $|\mathbf{A}| = 0$ so that $r_A < 3$. Picking the submatrix $\mathbf{A}_1 = \begin{bmatrix} 6 & 2 \\ 2 & 0 \end{bmatrix}$ gives $|\mathbf{A}_1| = -4 \neq 0$, so $r_A = 2$. The degeneracy is $q_A = n - r_A$ or $q_A = 1$. The one linear dependency relation between columns can be written as $\mathbf{x}_2 = \mathbf{x}_1 - \mathbf{x}_3$.

(b) $|\mathbf{B}| = 0$ (row 2 is twice row 4). Therefore $r_B < 4$. Any 3×3 matrix $\mathbf{B}_1$ formed by crossing out a row and column also has $|\mathbf{B}_1| = 0$ since row 3 is twice row 1. Therefore $r_B < 3$. It is easy to find a nonzero 2×2 determinant, so $r_B = 2$ and $q_B = 4 - 2 = 2$.

(c) $|\mathbf{C}| = 0$ as does the determinant of every 3×3 submatrix. In fact, there are just two linearly independent column vectors

$$\mathbf{x}_1 = [1 \quad 3 \quad -2 \quad 5]^T \quad \text{and} \quad \mathbf{x}_2 = [-1 \quad 0 \quad 1 \quad 4]^T$$

which can be used to generate $\mathbf{C}$. Then column 1 of $\mathbf{C}$ is $\mathbf{c}_1 = 8\mathbf{x}_1 + 2\mathbf{x}_2$. Likewise, $\mathbf{c}_2 = 1\mathbf{x}_1 + 5\mathbf{x}_2$, $\mathbf{c}_3 = 4\mathbf{x}_2$, and $\mathbf{c}_4 = -3\mathbf{x}_1 + 6\mathbf{x}_2$. In this case $r_C = 2$ and $q_C = 2$.

Gram-Schmidt Process

5.16 Use the Gram-Schmidt process to construct a set of orthonormal vectors from

$$\mathbf{x}_1 = \begin{bmatrix} 1+j \\ 1-j \\ j \end{bmatrix}, \qquad \mathbf{x}_2 = \begin{bmatrix} 2j \\ 1-2j \\ 1+2j \end{bmatrix}, \qquad \mathbf{x}_3 = \begin{bmatrix} 1 \\ j \\ 5j \end{bmatrix}$$

The Grammian is first used to verify linear independence of the set $\{\mathbf{x}_i\}$. The complex form of the inner product must be used:

$$\mathbf{G} = [\langle \mathbf{x}_i, \mathbf{x}_j \rangle] = [\bar{\mathbf{x}}_i^T \mathbf{x}_j] = \begin{bmatrix} 5 & 7 & 5 \\ 7 & 14 & 8+4j \\ 5 & 8-4j & 27 \end{bmatrix}$$

The determinant is $|\mathbf{G}| = 377 \neq 0$; therefore, the $\mathbf{x}_i$ are linearly independent.

Step 1. Construct an orthogonal set of $\mathbf{v}_i$:

$$\mathbf{v}_1 = \mathbf{x}_1$$

$$\mathbf{v}_2 = \begin{bmatrix} 2j \\ 1-2j \\ 1+2j \end{bmatrix} - \frac{7}{5}\begin{bmatrix} 1+j \\ 1-j \\ j \end{bmatrix} = \frac{1}{5}\begin{bmatrix} -7+3j \\ -2-3j \\ 5+3j \end{bmatrix}$$

Anticipating their need in advance, the products $\langle \mathbf{v}_2, \mathbf{v}_2 \rangle = 21/5$ and $\langle \mathbf{v}_2, \mathbf{x}_3 \rangle = 1 + 4j$ are computed.

$$\mathbf{v}_3 = \begin{bmatrix} 1 \\ j \\ 5j \end{bmatrix} - \frac{5}{5}\begin{bmatrix} 1+j \\ 1-j \\ j \end{bmatrix} - \frac{5(1+4j)}{21(5)}\begin{bmatrix} -7+3j \\ -2-3j \\ 5+3j \end{bmatrix} = \frac{1}{21}\begin{bmatrix} 19+4j \\ -31+53j \\ 7+61j \end{bmatrix}$$

Step 2. Normalize to obtain

$$\hat{\mathbf{v}}_1 = \frac{1}{\sqrt{5}}\begin{bmatrix} 1+j \\ 1-j \\ j \end{bmatrix}, \qquad \hat{\mathbf{v}}_2 = \frac{1}{\sqrt{105}}\begin{bmatrix} -7+3j \\ -2-3j \\ 5+3j \end{bmatrix}$$

Using $\langle \mathbf{v}_3, \mathbf{v}_3 \rangle = \dfrac{7917}{441}$ gives

$$\hat{\mathbf{v}}_3 = \sqrt{\frac{441}{7917}}\,\mathbf{v}_3 = \frac{1}{\sqrt{7917}}\begin{bmatrix} 19+4j \\ -31+53j \\ 7+61j \end{bmatrix}$$

It is a good exercise in the use of the complex inner product to verify that these results satisfy $\langle \hat{\mathbf{v}}_i, \hat{\mathbf{v}}_j \rangle = \delta_{ij}$.

17 Given a set of independent, real vectors $\{\mathbf{y}_i\}$, show that the Gram-Schmidt process for generating the orthonormal set $\{\hat{\mathbf{v}}_i\}$ can be expressed as a recursive matrix calculation:

$$\mathbf{T}_1 = \mathbf{I} \quad \text{(the initial condition)}$$

$$\mathbf{v}_i = \mathbf{T}_i \mathbf{y}_i \quad \text{(removal of components along previously computed } \mathbf{v}_k\text{)}$$

$$\hat{\mathbf{v}}_i = \mathbf{v}_i / \|\mathbf{v}_i\| \quad \text{(normalization step)}$$

$$\mathbf{T}_{i+1} = \mathbf{T}_i - \hat{\mathbf{v}}_i \hat{\mathbf{v}}_i^T \quad \text{(the recursion step)}$$

The Gram-Schmidt formulas from Sec. 5.7 can be written as

$$\begin{aligned}
\mathbf{v}_{i+1} &= \mathbf{y}_{i+1} - \sum_{k=1}^{i} \langle \hat{\mathbf{y}}_k, \mathbf{y}_{i+1} \rangle \hat{\mathbf{u}}_k \\
&= \mathbf{y}_{i+1} - \sum_{k=1}^{i} \hat{\mathbf{u}}_k \rangle \langle \hat{\mathbf{y}}_k, \mathbf{y}_{i+1} \rangle \\
&= \left[\mathbf{I} - \sum_{k=1}^{i} \hat{\mathbf{u}}_k \rangle \langle \hat{\mathbf{y}}_k \right] \mathbf{y}_{i+1} \\
&= \left[\underbrace{\mathbf{I} - \sum_{k=1}^{i} \hat{\mathbf{u}}_k \rangle \langle \hat{\mathbf{y}}_k}_{} - \hat{\mathbf{v}}_i \rangle \langle \hat{\mathbf{v}}_i \right] \mathbf{y}_{i+1} \\
&= [\mathbf{T}_i \qquad\qquad - \hat{\mathbf{v}}_i \rangle \langle \hat{\mathbf{v}}_i] \mathbf{y}_{i+1} \\
&= \mathbf{T}_{i+1} \mathbf{y}_{i+1}
\end{aligned}$$

This proves the validity of the recursion from step i to step $i+1$. Since it is true for $i = 1$, this constitutes a proof by induction. Notice that each of the sequence of $\mathbf{T}_i$ operators is a projection operator, since $\mathbf{T}_i \mathbf{T}_i = \mathbf{T}_i$. This is also easily proven by induction. It is obviously true for $\mathbf{T}_1 = \mathbf{I}$. At a general step,

$$\begin{aligned}
\mathbf{T}_{i+1}\mathbf{T}_{i+1} &= [\mathbf{T}_i - \hat{\mathbf{v}}_i\rangle\langle\hat{\mathbf{v}}_i][\mathbf{T}_i - \hat{\mathbf{v}}_i\rangle\langle\hat{\mathbf{v}}_i] = \mathbf{T}_i\mathbf{T}_i - \hat{\mathbf{v}}_i\rangle\langle\hat{\mathbf{v}}_i\mathbf{T}_i - \hat{\mathbf{v}}_i\rangle\langle\hat{\mathbf{v}}_i\mathbf{T}_i - \hat{\mathbf{v}}_i\rangle\langle\mathbf{v}_i, \hat{\mathbf{v}}_i\rangle\langle\hat{\mathbf{v}}_i \\
&= \mathbf{T}_i\mathbf{T}_i - \hat{\mathbf{v}}_i\rangle\langle\hat{\mathbf{v}}_i = \mathbf{T}_i - \hat{\mathbf{v}}_i\rangle\langle\hat{\mathbf{v}}_i = \mathbf{T}_{i+1}
\end{aligned}$$

In this calculation the facts that $\langle \hat{\mathbf{v}}_i, \hat{\mathbf{v}}_i \rangle = 1$ and $\mathbf{T}_{i+1}\hat{\mathbf{v}}_i = 0$ were used.

.18 Give a matrix version of the modified Gram-Schmidt process and contrast it with the results of the previous problem.

Assume that $\mathbf{y}_1$ is selected as $\mathbf{v}_1$, as before. Then the modified Gram-Schmidt process immediately subtracts the components along $\mathbf{v}_1$ from all other $\mathbf{y}_i$ vectors, for $i = 2, \ldots, n$. This can be accomplished using the projection operator $\mathbf{P}_1 = \mathbf{I} - \hat{\mathbf{v}}_1\rangle\langle\hat{\mathbf{v}}_1$. That is, all the $\mathbf{y}_i$ vectors are replaced by $\mathbf{y}_i' = \mathbf{P}_1\mathbf{y}_i$, for $i = 2, \ldots, n$. These are the projections on a subspace normal to $\hat{\mathbf{v}}_1$. Then $\mathbf{y}_2'$ is selected as $\mathbf{v}_2$ and normalized to $\hat{\mathbf{v}}_2$, and the whole process is repeated. If at any step a vector is found with $\|\mathbf{y}_i'\| = 0$ (in practice, less than epsilon), then the original $\mathbf{y}_i$ is a linear combination of the previously calculated $\{\hat{\mathbf{v}}_j, j = 1, \ldots, i-1\}$. In that case, the ith vector is skipped and the process continues with the next $\mathbf{y}_{i+1}$ vector. The pseudocode for this calculation might look as follows:

```
Rank = 0
For i = 1 to m
  If ‖y_i‖ < ε increment i and test next vector
  If ‖y_i‖ ≥ ε then
     Rank = rank + 1
     If rank = n, quit. The entire set of n has been found.
```

$\hat{\mathbf{v}}_i = \mathbf{y}_i/\|\mathbf{y}_i\|$
$\mathbf{P} = \mathbf{I} - \hat{\mathbf{v}}_i\rangle\langle\hat{\mathbf{v}}_i$
For $j = i + 1$ to m
  $\mathbf{y}_j \leftarrow \mathbf{P}\mathbf{y}_j$ (Replace $\mathbf{y}_j$ by its projection)
Increment i and repeat

Upon completion, the number of orthonormal vectors will equal the rank of the matrix **A**, whose columns are the vectors $\mathbf{y}_i$. Often it is desired to find a full set of n orthonormal vectors, even though $n > m$ or rank $(\mathbf{A}) < n$ for some other reason. This can be done by selecting a sufficient number of extra column vectors with components chosen randomly and appending these to the matrix **A**.

The difference between the modified and unmodified Gram-Schmidt processes is that here each vector $\mathbf{y}_i$ is modified to $\mathbf{y}_i'$ repeatedly, but each vector $\hat{\mathbf{v}}_j$ is used only once in a projection. In the unmodified version, each vector $\mathbf{y}_i$ is adjusted only once, but the vectors $\hat{\mathbf{v}}_j$ are used repeatedly in the ever-more complicated $\mathbf{T}_i$ projection operators.

Geometry in n-Dimensional Spaces

5.19 The equation of a plane in n dimensions is $\langle \mathbf{c}, \mathbf{x}\rangle = a$, where **c** is the normal to the plane and a is a scalar constant. Find the point on the plane nearest the origin and find the distance to this point.

Any **x** can be decomposed into a component $\mathbf{x}_n$ normal to the plane plus $\mathbf{x}_p$ parallel to the plane. Then $\langle \mathbf{c}, \mathbf{x}\rangle = \langle \mathbf{c}, \mathbf{x}_n\rangle + \langle \mathbf{c}, \mathbf{x}_p\rangle = a$ for every **x** terminating on the plane. Since **c** and $\mathbf{x}_p$ are orthogonal, $\langle \mathbf{c}, \mathbf{x}_p\rangle = 0$. Since **c** and $\mathbf{x}_n$ are parallel, $\mathbf{x}_n = \pm\|\mathbf{x}_n\|\mathbf{c}/\|\mathbf{c}\|$. Then $\langle \mathbf{c}, \mathbf{x}_n\rangle = \pm\langle \mathbf{c}, \mathbf{c}\rangle\|\mathbf{x}_n\|/\|\mathbf{c}\| = a$. The + or − sign must be selected to agree with the sign of a. Solving gives the minimum distance as $\|\mathbf{x}_n\| = |a|/\|\mathbf{c}\|$ and the closest point is $\mathbf{x}_n = a\mathbf{c}/\langle \mathbf{c}, \mathbf{c}\rangle$.

5.20 Find the minimum distance from the origin to a point (x_1, x_2) on the line $6x_1 + 2x_2 = 4$, and find the coordinates of that point.

The normal to the line is $\mathbf{c} = [6 \quad 2]^T$, and so $\|\mathbf{c}\| = \sqrt{40}$. The results of Problem 5.19 apply, and the minimum distance is $\|\mathbf{x}_n\| = 4/\sqrt{40} = 2/\sqrt{10}$. The point nearest the origin is

$$\mathbf{x}_n = \frac{1}{5}\begin{bmatrix} 3 \\ 1 \end{bmatrix}$$

5.21 Generalize the concepts of lines, planes, spheres, cones, and convex sets to n-dimensional Euclidean spaces.

The generalization of a line is the set of all vectors satisfying

$$\mathbf{x} = a\mathbf{v} + \mathbf{k} \tag{1}$$

where **v** and **k** are constant vectors and a is a scalar.

The generalization of a plane is called a *hyperplane*. An $n - 1$ dimensional hyperplane consists of the set of n-dimensional vectors **x** satisfying

$$\langle \mathbf{c}, \mathbf{x}\rangle = a \tag{2}$$

where **c** and a are a constant vector and scalar, respectively. Since a subspace always contains the **0** vector, equation (*1*) represents a subspace only if **k** is zero. Equation (2) represents a subspace only if a is zero.

Points on or inside a hypersphere of radius R are defined by the set of all **x** satisfying

$$\langle \mathbf{x}, \mathbf{x}\rangle \le R^2$$

A right circular cone of semi-vertex angle θ, with its axis in the direction of a unit vector **n**, and with vertex at the origin, consists of the set of all **x** satisfying

$$\langle \mathbf{n}, \mathbf{x}\rangle/\|\mathbf{x}\| = \cos\theta$$

If $\theta = \pi/2$, the cone degenerates to a hyperplane containing the origin.

If the line segments connecting every two points in a set contain only points in the set, the set is *convex*. A convex set of vectors in n-dimensional space is a set for which the vector

$$\mathbf{z} = a\mathbf{x}_1 + (1 - a)\mathbf{x}_2$$

belongs to the set for every $\mathbf{x}_1, \mathbf{x}_2$ in the set and for every real scalar satisfying $0 \le a \le 1$.

Some Generalizations

.22 Let $\mathcal{L}_2[a, b]$ be the linear space consisting of all real square integrable functions of t, that is, all functions $f(t)$ satisfying $\int_a^b f(\tau)^2\, d\tau < \infty$. Define a suitable inner product, norm, and metric.

The three requirements for an inner product can be shown to be satisfied by

$$\langle f, g\rangle = \int_a^b f(\tau)g(\tau)\, d\tau \quad \text{where} \quad f, g \in \mathcal{L}_2[a, b]$$

As in other cases, a norm can always be defined as

$$\|f\| = \langle f, f\rangle^{1/2}$$

and the metric or distance measure between two functions can be defined in terms of the norm,

$$\rho(f, g) = \|f - g\|$$

The inner product space defined by this set of functions and the inner product definition is a complete infinite dimensional linear vector space. A space $\mathcal{X}$ is *complete* if every convergent (Cauchy) sequence of elements in $\mathcal{X}$ converges to a limit which is also in $\mathcal{X}$. Any infinite dimensional inner product space which is complete is called a *Hilbert space*.

.23 Use the results of the previous problem to prove that the mean value of a real function is always less than or equal to its root-mean-square (rms) value.

We are to prove that

$$\frac{1}{T}\int_0^T f(\tau)\, d\tau \le \left[\frac{1}{T}\int_0^T f^2(\tau)\, d\tau\right]^{1/2}$$

For any functions f and $g \in \mathcal{L}_2[0, T]$,

$$\|f - g\| \ge 0$$

or

$$\langle f - g, f - g\rangle \ge 0 \quad \text{or} \quad \langle f, f\rangle \ge 2\langle f, g\rangle - \langle g, g\rangle$$

Choosing the particular function $g = \text{constant} = \frac{1}{T}\int_0^T f(\tau)\, d\tau$ gives

$$\int_0^T f^2\, dt \ge \frac{1}{T}\left[\int_0^T f\, dt\right]^2$$

Dividing both sides by T and taking the square root gives the desired result.

.24 Is it always necessary to define the norm in terms of the inner product?

No. Linear spaces can be defined with a norm, but without any mention of an inner product. Such spaces are called *normed linear spaces*. If they are infinite dimensional spaces, and are complete, then they are usually referred to as *Banach spaces*.

Two examples of other valid norms for finite dimensional spaces whose elements $\mathbf{x}$ are ordered n-tuples of scalars belonging to $\mathcal{F}$ are

$$\|\mathbf{x}\| = \max_i \{|x_1|, |x_2|, \ldots, |x_n|\}$$

and

$$\|\mathbf{x}\|_p = [|x_1|^p + |x_2|^p + \cdots + |x_n|^p]^{1/p}$$

where p is real, $1 \le p \le \infty$. The quadratic norm is a special case with $p = 2$. All the required axioms for a norm are satisfied for these examples.

5.25 Let $\mathscr{X}$ be a linear space consisting of all $n \times n$ matrices defined over $\mathscr{F}$. (Let $n = 2$ for simplicity.) Show that
(a) $\langle \mathbf{A}, \mathbf{B} \rangle = \text{Tr}(\overline{\mathbf{A}}^T \mathbf{B})$ is a valid inner product; and
(b) an orthonormal basis for this space is the set of matrices

$$\left\{ \begin{bmatrix} 1 & 0 \\ 0 & 0 \end{bmatrix}, \begin{bmatrix} 0 & 1 \\ 0 & 0 \end{bmatrix}, \begin{bmatrix} 0 & 0 \\ 1 & 0 \end{bmatrix}, \begin{bmatrix} 0 & 0 \\ 0 & 1 \end{bmatrix} \right\}$$

(a) The inner product of two elements must yield a scalar. Obviously the trace gives a scalar. In addition, the three axioms must be satisfied:
(i) $\langle \mathbf{A}, \mathbf{B} \rangle = \overline{\langle \mathbf{B}, \mathbf{A} \rangle}$ but

$$\langle \mathbf{A}, \mathbf{B} \rangle = \text{Tr}(\overline{\mathbf{A}}^T \mathbf{B})$$

and

$$\overline{\langle \mathbf{B}, \mathbf{A} \rangle} = \overline{\text{Tr}(\overline{\mathbf{B}}^T \mathbf{A})} = \text{Tr}(\mathbf{B}^T \overline{\mathbf{A}}) = \text{Tr}(\mathbf{B}^T \overline{\mathbf{A}})^T = \text{Tr}(\overline{\mathbf{A}}^T \mathbf{B})$$

(ii) $\langle \mathbf{A}, \alpha\mathbf{B}_1 + \beta\mathbf{B}_2 \rangle = \alpha\langle \mathbf{A}, \mathbf{B}_1 \rangle + \beta\langle \mathbf{A}, \mathbf{B}_2 \rangle$ but

$$\langle \mathbf{A}, \alpha\mathbf{B}_1 + \beta\mathbf{B}_2 \rangle = \text{Tr}[\overline{\mathbf{A}}^T(\alpha\mathbf{B}_1 + \beta\mathbf{B}_2)] = \alpha \,\text{Tr}(\overline{\mathbf{A}}^T \mathbf{B}_1) + \beta \,\text{Tr}(\overline{\mathbf{A}}^T \mathbf{B}_2)$$
$$= \alpha\langle \mathbf{A}, \mathbf{B}_1 \rangle + \beta\langle \mathbf{A}, \mathbf{B}_2 \rangle$$

(iii) $\langle \mathbf{A}, \mathbf{A} \rangle \ge 0$ for all $\mathbf{A}$ and equals zero if and only if $\mathbf{A} = \mathbf{0}$. Let $\mathbf{A} = [a_{ij}]$. Then

$$\langle \mathbf{A}, \mathbf{A} \rangle = \text{Tr}(\overline{\mathbf{A}}^T \mathbf{A}) = \bar{a}_{11} a_{11} + \bar{a}_{21} a_{21} + \bar{a}_{12} a_{12} + \bar{a}_{22} a_{22}$$
$$= |a_{11}|^2 + |a_{21}|^2 + |a_{12}|^2 + |a_{22}|^2$$

This is obviously nonnegative and can vanish only if $a_{ij} = 0$ for all i and j.

(b) First show that the set is orthonormal. $\mathbf{G} = [\langle \mathbf{V}_i, \mathbf{V}_j \rangle]$, where $\mathbf{V}_i$ are the four indicated matrices. Simple calculation shows that $\mathbf{G}$ is the 4×4 unit matrix, and thus the set is orthonormal. They span the space, since every 2×2 matrix $[a_{ji}]$ can be written as

$$\mathbf{A} = a_{11}\mathbf{V}_1 + a_{12}\mathbf{V}_2 + a_{21}\mathbf{V}_3 + a_{22}\mathbf{V}_4$$

An orthonormal set which spans the space is an orthonormal basis set.

5.26 Let $\mathscr{X}$ be an n-dimensional linear vector space with $\mathbf{x} \in \mathscr{X}$. Discuss the transformation $\mathscr{A}(\mathbf{x}) = \|\mathbf{x}\|$.

This function maps $\mathscr{X}$ into the real line, $\mathscr{A}: \mathscr{X} \to \mathscr{R}^1$. In this case the range of $\mathscr{A}$ is the nonnegative half line, so $\mathscr{R}(\mathscr{A}) \ne \mathscr{R}^1$ and the mapping is not onto. It is not one-to-one either, because many different vectors can have the same norm. Thus, given a value for $\|\mathbf{x}\|$, it is not possible to determine which vector $\mathbf{x}$ is the pre-image, i.e., $\mathscr{A}^{-1}$ does not exist. This transformation is not a linear transformation since, in general,

$$\mathscr{A}(\mathbf{x}_1 + \mathbf{x}_2) = \|\mathbf{x}_1 + \mathbf{x}_2\| \ne \|\mathbf{x}_1\| + \|\mathbf{x}_2\| = \mathscr{A}(\mathbf{x}_1) + \mathscr{A}(\mathbf{x}_2)$$

The null space of this transformation consists of the single vector $\mathbf{x} = \mathbf{0}$, by the properties of the norm.

Change of Basis

5.27 A vector $\mathbf{x}$ is represented by the $n \times 1$ column matrix $[\mathbf{x}]_\mathbf{v}$ with respect to the basis $\{\mathbf{v}_i\}$ and by the $n \times 1$ column matrix $[\mathbf{x}]_{\mathbf{v}'}$ with respect to another basis $\{\mathbf{v}_i'\}$. Show that the matrix $\mathbf{B}$ satisfying $[\mathbf{x}]_{\mathbf{v}'} = \mathbf{B}[\mathbf{x}]_\mathbf{v}$ always has an inverse.

Expand a typical member of the first basis in terms of the second basis, $\mathbf{v}_j = \sum_{i=1}^{n} b_{ij}\mathbf{v}_i'$. Likewise, a typical $\mathbf{v}_j'$ can be expanded as $\mathbf{v}_j' = \sum_{i=1}^{n} c_{ij}\mathbf{v}_i$. Thus $\mathbf{v}_i' = \sum_{k=1}^{n} c_{ki}\mathbf{v}_k$. Eliminating $\mathbf{v}_i'$ from the first equation gives

$$\mathbf{v}_j = \sum_{i=1}^{n} b_{ij} \sum_{k=1}^{n} c_{ki}\mathbf{v}_k = \sum_{i=1}^{n}\sum_{k=1}^{n} b_{ij}c_{ki}\mathbf{v}_k.$$

Because the vectors $\mathbf{v}_j$ are linearly independent, this requires

$$\sum_{i=1}^{n} b_{ij}c_{ki} = \begin{cases} 1 & \text{if } k = j \\ 0 & \text{if } k \neq j \end{cases}$$

Letting $\mathbf{B} = [b_{ij}]$ and $\mathbf{C} = [c_{ij}]$, this can be written as $\mathbf{CB} = \mathbf{I}$. If the preceding procedure is modified slightly, by eliminating $\mathbf{v}_i$ from the second equation, we obtain $\mathbf{BC} = \mathbf{I}$. Taken together, the last two results imply that $\mathbf{C} = \mathbf{B}^{-1}$.

28 Show that if $\{\mathbf{v}_i\}$ and $\{\mathbf{v}_i'\}$ are both real, orthonormal basis sets, then $\mathbf{B}^{-1} = \mathbf{B}^T$.

Results of the previous problem give $\mathbf{B}^{-1} = \mathbf{C}$. But $c_{ij} = \langle \mathbf{r}_i, \mathbf{v}_j' \rangle$, where $\{\mathbf{r}_i\}$ are the reciprocal bases associated with $\{\mathbf{v}_i\}$. Since $\{\mathbf{v}_i\}$ is an orthonormal set, $\mathbf{r}_i = \mathbf{v}_i$ and $c_{ij} = \langle \mathbf{v}_i, \mathbf{v}_j' \rangle$. Similarly, $b_{ij} = \langle \mathbf{r}_i', \mathbf{v}_j \rangle$ in the general case, where $\{\mathbf{r}_i'\}$ is the set of reciprocal bases for $\{\mathbf{v}_i'\}$. For the orthonormal case, $b_{ij} = \langle \mathbf{v}_i', \mathbf{v}_j \rangle$. Interchanging subscripts gives $b_{ji} = \langle \mathbf{v}_j', \mathbf{v}_i \rangle$. For the case of real vectors this shows that $b_{ji} = c_{ij}$, or $\mathbf{B}^T = \mathbf{C} = \mathbf{B}^{-1}$.

Adjoint Transformations

.29 Consider the linear transformation $\mathcal{A}: \mathcal{X} \to \mathcal{Y}$, with the adjoint transformation $\mathcal{A}^*$. Prove that $\|\mathcal{A}\| = \|\mathcal{A}^*\|$.

The definition of the adjoint requires that $\langle \mathbf{y}, \mathcal{A}(\mathbf{x}) \rangle = \langle \mathcal{A}^*(\mathbf{y}), \mathbf{x} \rangle$ for all $\mathbf{x} \in \mathcal{X}, \mathbf{y} \in \mathcal{Y}$. The Cauchy-Schwarz inequality and the definition of $\|\mathcal{A}^*\|$ gives

$$|\langle \mathbf{y}, \mathcal{A}(\mathbf{x}) \rangle| = |\langle \mathcal{A}^*(\mathbf{y}), \mathbf{x} \rangle| \leq \|\mathcal{A}^*(\mathbf{y})\| \cdot \|\mathbf{x}\| \leq \|\mathcal{A}^*\| \cdot \|\mathbf{y}\| \cdot \|\mathbf{x}\|$$

This must be true for all $\mathbf{y}, \mathbf{x}$ including the particular pair related by $\mathcal{A}(\mathbf{x}) = \mathbf{y}$. Using this gives

$$\langle \mathcal{A}(x), \mathcal{A}(x) \rangle \leq \|\mathcal{A}^*\| \cdot \|\mathcal{A}(\mathbf{x})\| \cdot \|\mathbf{x}\| \quad \text{or} \quad \frac{\|\mathcal{A}(\mathbf{x})\|}{\|\mathbf{x}\|} \leq \|\mathcal{A}^*\|$$

for all $\mathbf{x} \neq \mathbf{0}$. Therefore,

$$\|\mathcal{A}^*\| \geq \|\mathcal{A}\| \tag{1}$$

Similarly, if particular $\mathbf{x}, \mathbf{y}$ pairs are chosen such that $\mathbf{x} = \mathcal{A}^*(\mathbf{y})$, then

$$|\langle \mathcal{A}^*(\mathbf{y}), \mathcal{A}^*(\mathbf{y}) \rangle| \leq \|\mathbf{y}\| \cdot \|\mathcal{A}(x)\| \leq \|\mathbf{y}\| \cdot \|\mathbf{x}\| \cdot \|\mathcal{A}\|$$

or

$$\|\mathcal{A}^*(y)\|^2 \leq \|\mathbf{y}\| \cdot \|\mathcal{A}^*(y)\| \cdot \|\mathcal{A}\| \quad \text{or} \quad \frac{\|\mathcal{A}^*(\mathbf{y})\|}{\|\mathbf{y}\|} \leq \|\mathcal{A}\| \quad \text{for all } \mathbf{y} \neq \mathbf{0}$$

Therefore, $\|\mathcal{A}\| \geq \|\mathcal{A}^*\|$. This, together with Eq. (*1*), gives $\|\mathcal{A}\| = \|\mathcal{A}^*\|$.

.30 Let $\mathcal{X}$ be the set of all n-component real functions defined over $[t_0, t_f]$ which have continuous first derivatives and let $\mathcal{Y}$ be the set of all continuous functions defined over the same interval. Find the adjoint of $\mathcal{A}: \mathcal{X} \to \mathcal{Y}$ if $\mathcal{A}(\mathbf{x}) = (d/dt)\mathbf{x} - \mathbf{A}\mathbf{x}$, where $\mathbf{A}$ is an $n \times n$ matrix.

The inner product for $\mathcal{X}$ and $\mathcal{Y}$ takes the form of equation (*1*). The adjoint is defined by equation (*2*):

$$\langle \mathbf{x}_1, \mathbf{x}_2 \rangle = \int_{t_0}^{t_f} \mathbf{x}_1^T(\tau)\mathbf{x}_2(\tau)\, d\tau \tag{1}$$

$$\langle \mathbf{y}, \mathscr{A}(\mathbf{x}) \rangle = \langle \mathscr{A}^*(\mathbf{y}), \mathbf{x} \rangle \tag{2}$$

Integrating the left-hand side of Eq. (2) gives

$$\int_{t_0}^{t_f} \mathbf{y}^T(\tau)\left[\frac{d\mathbf{x}}{dt} - \mathbf{A}\mathbf{x}(\tau)\right] d\tau = \mathbf{y}^T(\tau)\mathbf{x}(\tau)\Big|_{t_0}^{t_f} - \int_{t_0}^{t_f}\left[\frac{d\mathbf{y}^T}{dt} + \mathbf{y}(\tau)^T\mathbf{A}\right]\mathbf{x}(\tau)\, d\tau$$

The term involving the limits of integration, $\mathbf{y}^T(t_f)\mathbf{x}(t_f) - \mathbf{y}^T(t_0)\mathbf{x}(t_0)$, could be made to vanish by specifying appropriate boundary conditions. Ignoring this term, the remaining term is in the form of the right-hand side of Eq. (2). Therefore, $\mathscr{A}^*(\mathbf{y}) = -(d\mathbf{y}/dt) - \mathbf{A}^T\mathbf{y}(t)$. We conclude that the formal adjoint of $\dot{\mathbf{x}} = \mathbf{A}\mathbf{x}$ is $\dot{\mathbf{y}} = -\mathbf{A}^T\mathbf{y}$. This adjoint differential equation arises frequently in optimal control theory and in other applications.

5.31 Consider the equation $\dot{\mathbf{x}} = \mathbf{A}\mathbf{x}$ with $\mathbf{x}(t_0) = \mathbf{x}_0$. Show that $\mathbf{y}^T(t)\mathbf{x}(t)$ is a constant for all time, where $\mathbf{y}$ satisfies the adjoint equation $\dot{\mathbf{y}} = -\mathbf{A}^T\mathbf{y}$.

A necessary and sufficient condition that $\mathbf{y}^T\mathbf{x}$ be constant is that $(d/dt)(\mathbf{y}^T\mathbf{x}) = 0$. Using the chain rule for differentiating gives $(d/dt)(\mathbf{y}^T\mathbf{x}) = \dot{\mathbf{y}}^T\mathbf{x} + \mathbf{y}^T\dot{\mathbf{x}}$. Substituting in for $\dot{\mathbf{y}}$ and $\dot{\mathbf{x}}$ gives

$$(d/dt)(\mathbf{y}^T\mathbf{x}) = -\mathbf{y}^T\mathbf{A}\mathbf{x} + \mathbf{y}^T\mathbf{A}\mathbf{x} = 0$$

5.32 The input-output equation $\mathbf{y}(t) = \int_{t_0}^{\infty} \mathbf{W}(t, \tau)\mathbf{u}(\tau)\, d\tau$ defines a linear transformation from the space of r-component square integrable functions $\mathbf{u}(t) \in \mathscr{U}$ to the set of m-component continuous functions $\mathbf{y}(t) \in \mathscr{Y}$. The functions are defined over (t_0, ∞). Consider $\mathscr{A}: \mathscr{U} \to \mathscr{Y}$ and find $\mathscr{A}^*$, if the inner product for each space has the form

$$\langle \mathbf{x}_1, \mathbf{x}_2 \rangle = \int_{t_0}^{\infty} \mathbf{x}_1^T(t)\mathbf{x}_2(t)\, dt$$

The transformation $\mathscr{A}$ is $\mathscr{A}(\mathbf{u}) = \int_{t_0}^{\infty} \mathbf{W}(t, \tau)\mathbf{u}(\tau)\, d\tau$ and $\langle \mathbf{z}, \mathscr{A}(\mathbf{u}) \rangle = \langle \mathscr{A}^*(\mathbf{z}), \mathbf{u} \rangle$. Using the inner product definition gives

$$\langle \mathbf{z}, \mathscr{A}(\mathbf{u}) \rangle = \int_{t_0}^{\infty} \mathbf{z}^T(t) \int_{t_0}^{\infty} \mathbf{W}(t, \tau)\mathbf{u}(\tau)\, d\tau\, dt = \int_{t_0}^{\infty}\int_{t_0}^{\infty} \mathbf{z}^T(t)\mathbf{W}(t, \tau)\, dt\, \mathbf{u}(\tau)\, d\tau$$

Therefore, $\mathscr{A}^*(\mathbf{z}) = \int_{t_0}^{\infty} \mathbf{W}^T(t, \tau)\mathbf{z}(t)\, dt$. The weighting matrix for the adjoint equation is the transpose of $\mathbf{W}$ and the integration variable is t rather than τ as in the original transformation. This operator is self-adjoint if $\mathbf{W}(t, \tau) = \mathbf{W}^T(\tau, t)$.

Matrix Norms

5.33 Consider a linear transformation which maps vectors from one finite dimensional space to another. Let $\mathbf{A}$ be its matrix representation. If $\|\mathbf{x}\| = \max_i |x_i|$, find $\|\mathbf{A}\|$.

With this definition for the vector norm,

$$\|\mathbf{A}\mathbf{x}\| = \max_i \left|\sum_j a_{ij}x_j\right| \le \left\{\max_i \left|\sum_j a_{ij}\right|\right\}\left\{\max_j |x_j|\right\} \le \left\{\max_i \sum_j |a_{ij}|\right\}\|\mathbf{x}\|$$

The norm of $\mathbf{A}$ must satisfy $\|\mathbf{A}\mathbf{x}\| \le \|\mathbf{A}\|\cdot\|\mathbf{x}\|$. We see that $\max_i \sum_j |a_{ij}|$ qualifies as a bound for $\|\mathbf{A}\mathbf{x}\|/\|\mathbf{x}\|$, with $\|\mathbf{x}\| \neq 0$. To show that it is the *least* upper bound, and hence is equal to the norm, we must show that no smaller bound is possible. This can be done by demonstrating that the bound is actually attained for some $\mathbf{x}$.

Let i^* be the row i, which maximizes the above sum. Let

$$x_j = \begin{cases} 0 & \text{if } a_{i^*j} = 0 \\ \dfrac{\bar{a}_{i^*j}}{|a_{i^*j}|} & \text{otherwise} \end{cases}$$

For this choice, $\|\mathbf{x}\| = 1$ and

$$\max_i \left|\sum_j a_{ij}x_j\right| = \left|\sum_j \frac{a_{i^*j}\bar{a}_{i^*j}}{|a_{i^*j}|}\right| = \sum_j |a_{i^*j}|$$

Therefore, $\|\mathbf{A}\| = \max_i \sum_j |a_{ij}|$.

.34 Let the vector norm be defined by $\|\mathbf{x}\| = \sum_i |x_i|$ and find a bound for $\|\mathbf{A}\|$.

Beginning with the given definition for the norm, a series of manipulations gives

$$\|\mathbf{Ax}\| = \sum_i \left|\sum_j a_{ij}x_j\right| \le \sum_i \left|\sum_j a_{ij}\right| \cdot \left|\sum_j x_j\right| \le \sum_i \sum_j |a_{ij}| \sum_j |x_j| = \sum_i \sum_j |a_{ij}| \cdot \|\mathbf{x}\|$$

Since $\|\mathbf{Ax}\| \le \|\mathbf{x}\| \sum_i \sum_j |a_{ij}|$, the double summation term is an upper bound for $\|\mathbf{Ax}\|/\|\mathbf{x}\|$, $\|\mathbf{x}\| \ne 0$. Since $\|\mathbf{A}\|$ is the least such bound, $\|\mathbf{A}\| \le \sum_i \sum_j |a_{ij}|$.

Miscellaneous Applications

.35 Find the projection of $\mathbf{y} = [1 \quad -3 \quad 4 \quad 2 \quad 8]^T$ on the subspace spanned by

$$\mathbf{x}_1 = [1 \quad 2 \quad -3 \quad 1 \quad 0]^T \quad \text{and} \quad \mathbf{x}_2 = [0 \quad 1 \quad 3 \quad 3 \quad 1]^T$$

The dimension of the subspace is two, since $|\mathbf{G}| = 284 \ne 0$. An orthonormal basis is constructed:

$$\hat{\mathbf{v}}_1 = \frac{\mathbf{x}_1}{\|\mathbf{x}_1\|} = \frac{1}{\sqrt{15}}[1 \quad 2 \quad -3 \quad 1 \quad 0]^T$$

$$\hat{\mathbf{v}}_2 = \frac{\mathbf{x}_2 - \langle \mathbf{x}_2, \hat{\mathbf{v}}_1\rangle \hat{\mathbf{v}}_1}{\|\mathbf{x}_2 - \langle \mathbf{x}_2, \hat{\mathbf{v}}_1\rangle \hat{\mathbf{v}}_1\|} = \frac{1}{\sqrt{4260}}\begin{bmatrix} 4 \\ 23 \\ 33 \\ 49 \\ 15 \end{bmatrix}$$

The projection of $\mathbf{y}$ on this subspace is

$$\mathbf{y}_p = \langle \hat{\mathbf{v}}_1, \mathbf{y}\rangle \hat{\mathbf{v}}_1 + \langle \hat{\mathbf{v}}_2, \mathbf{y}\rangle \hat{\mathbf{v}}_2$$

$$= -\sqrt{15}\,\hat{\mathbf{v}}_1 + \frac{285}{\sqrt{4260}}\hat{\mathbf{v}}_2 = \begin{bmatrix} -1 \\ -2 \\ 3 \\ -1 \\ 0 \end{bmatrix} + \frac{285}{4260}\begin{bmatrix} 4 \\ 23 \\ 33 \\ 49 \\ 15 \end{bmatrix} = \frac{1}{284}\begin{bmatrix} -208 \\ -131 \\ 1479 \\ 647 \\ 285 \end{bmatrix}$$

The projection $\mathbf{y}_p$ is the closest vector in the subspace to $\mathbf{y}$, in the sense that

$$\|\mathbf{y} - \mathbf{y}_p\|^2 \le \|\mathbf{y} - \mathbf{z}\|^2$$

for every $\mathbf{z}$ in the subspace. These results are directly related to the problem of least squares approximations, considered in the next chapter.

.36 Consider the dc motor of Problem 2.5, with transfer function

$$\frac{\Omega(s)}{V(s)} = \frac{K'}{s + a}$$

If the motor is initially at rest, $\omega(0) = 0$, find the input $v(t)$ which gives an angular velocity $\omega(T) = 100$ at a fixed time T, while minimizing a measure of the input energy,

$$J = \int_0^T v^2(\tau)\, d\tau$$

The input-output relationship is written in terms of the *system weighting function*. Since $W(t, 0) = \mathcal{L}^{-1}\{K'/(s+a)\} = K'e^{-at}$, the weighting function is $W(t, \tau) = K'e^{-a(t-\tau)}$. Then

$$\omega(T) = \int_0^T W(T, \tau)v(\tau)\,d\tau$$

This is in the form of an inner product, $100 = \langle W(T, \tau), v(\tau)\rangle$, so the Cauchy-Schwarz inequality can be used to give

$$100 = |\langle W(T, \tau), v(\tau)\rangle| \le \|W(T, \tau)\|\cdot\|v(\tau)\|$$

The minimum value of $\|v(\tau)\|$ is obtained when the equality holds. Therefore,

$$\|v(\tau)\| = \frac{100}{\|W(T, \tau)\|} = \frac{100}{\left\{\int_0^T [K'e^{-a(T-\tau)}]^2\,d\tau\right\}^{1/2}} = \frac{100\sqrt{2a}}{K'[1-e^{-2aT}]^{1/2}}$$

The equality holds if and only if $v(t)$ and $W(T, t)$ are linearly dependent. This means $v(t) = kW(T, t)$ for some scalar k. Comparing gives $\|v\| = |k|\,\|W\| = 100/\|W\|$ and so

$$|k| = \frac{100}{\|W\|^2} \quad \text{and} \quad v_{\text{optimal}}(t) = \frac{100}{\|W\|^2}W(T, t) = \frac{200ae^{-a(T-t)}}{K'[1-e^{-2aT}]}$$

5.37 A satellite position vector has components (X, Y, Z) with respect to an inertially fixed coordinate system with origin at the earth's center. Determine the components of this position vector as measured by a tracking station at longitude L degrees east and latitude λ degrees north, if the angle between the X_I inertial axis and the 0° longitudinal meridian is ϕ degrees and the earth's radius is R_e. See Figure 5.7.

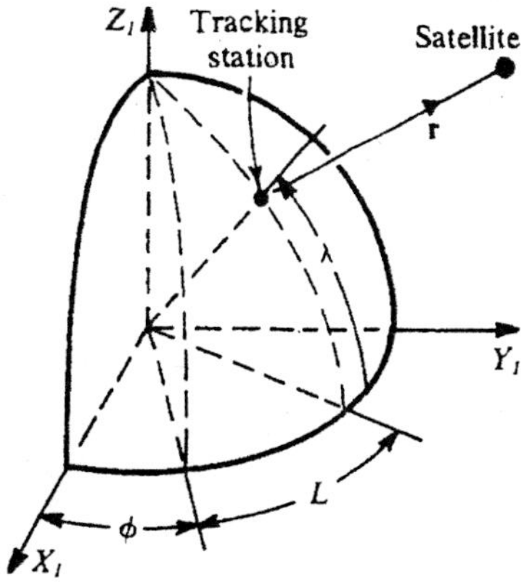

Figure 5.7

The coordinate transformations can be represented by the Piogram of Figure 5.8. The measured range vector has the following components in the Up-East-North coordinate system. (These components can be read directly from the Piogram by using a few standard conventions [5, 6].) The origin is shifted by subtracting R_e from the Up component:

$$\mathbf{r} = \begin{bmatrix} X[\cos(\phi+L)\cos\lambda] + Y\sin(\phi+L)\cos\lambda + Z\sin\lambda - R_e \\ -X\sin(\phi+L) + Y\cos(\phi+L) \\ -X\cos(\phi+L)\sin\lambda - Y\sin(\phi+L)\sin\lambda + Z\cos\lambda \end{bmatrix}$$

5.38 Demonstrate how the spatial attitude orientation of a vehicle (aircraft, satellite, etc.) can be determined by sighting two known stars.

The attitude of the vehicle will be characterized by the 3×3 transformation matrix $\mathbf{T}_{BE}$, which relates a set of orthonormal body fixed axes $\{x, y, z\}$ to a fixed orthonormal inertial

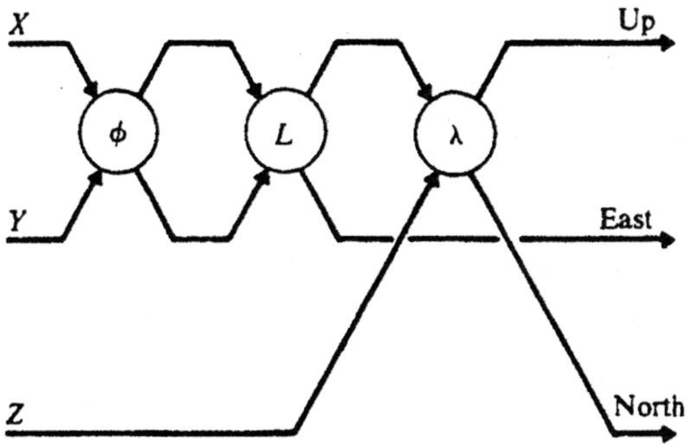

Figure 5.8

coordinate system $\{X, Y, Z\}$. This means a vector $\mathbf{v}$ expressed in the two coordinate systems satisfies

$$\begin{bmatrix} v_x \\ v_y \\ v_z \end{bmatrix} = \mathbf{T}_{BE} \begin{bmatrix} v_X \\ v_Y \\ v_Z \end{bmatrix}$$

$\mathbf{T}_{BE}$ is to be found. The unit vectors which point toward the two stars are assumed available from star catalogs, in inertial components. That is,

$$\hat{\mathbf{u}} = [u_X \quad u_Y \quad u_Z]$$

$$\hat{\mathbf{v}} = [v_X \quad v_Y \quad v_Z]$$

The pointing direction of a telescope mounted in the vehicle is described by two gimbal angles α and β as shown in Figure 5.9.

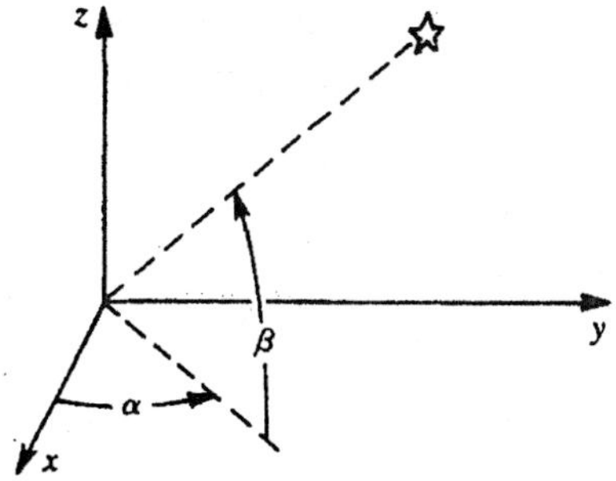

Figure 5.9

The unit vectors to the given stars are expressed in vehicle fixed coordinates as

$$\hat{\mathbf{u}}_B = \begin{bmatrix} \cos\alpha_1 \cos\beta_1 \\ \sin\alpha_1 \cos\beta_1 \\ \sin\beta_1 \end{bmatrix}, \qquad \hat{\mathbf{v}}_B = \begin{bmatrix} \cos\alpha_2 \cos\beta_2 \\ \sin\alpha_2 \cos\beta_2 \\ \sin\beta_2 \end{bmatrix}$$

where α_1, β_1 and α_2, β_2 are the pointing angles for the two stars. $\hat{\mathbf{u}}$ and $\hat{\mathbf{v}}$ are known, and $\hat{\mathbf{u}}_B$ and $\hat{\mathbf{v}}_B$ are available from measurements.

In order to determine $\mathbf{T}_{BE}$, one more relation is necessary. A vector normal to the plane defined by $\hat{\mathbf{u}}$ and $\hat{\mathbf{v}}$ is constructed using the cross product. Thus

$$\mathbf{w} \triangleq \hat{\mathbf{u}} \times \hat{\mathbf{v}} \quad \text{and} \quad \mathbf{w}_B \triangleq \hat{\mathbf{u}}_B \times \hat{\mathbf{v}}_B$$

Now $\hat{\mathbf{u}}_B = \mathbf{T}_{BE}\hat{\mathbf{u}}$, $\hat{\mathbf{v}}_B = \mathbf{T}_{BE}\hat{\mathbf{v}}$, and $\mathbf{w}_B = \mathbf{T}_{BE}\mathbf{w}$. These can be combined into

$$[\hat{\mathbf{u}}_B \quad \hat{\mathbf{v}}_B \quad \mathbf{w}_B] = \mathbf{T}_{BE}[\hat{\mathbf{u}} \quad \hat{\mathbf{v}} \quad \mathbf{w}]$$

If the vectors to the two stars are linearly independent, then the 3×3 matrix $[\hat{\mathbf{u}} \quad \hat{\mathbf{v}} \quad \mathbf{w}]$ is nonsingular and

$$\mathbf{T}_{BE} = [\hat{\mathbf{u}}_B \quad \hat{\mathbf{v}}_B \quad \mathbf{w}_B][\hat{\mathbf{u}} \quad \hat{\mathbf{v}} \quad \mathbf{w}]^{-1}$$

PROBLEMS

Linear Independence, Orthonormal Basis Vectors, and Reciprocal Basis Vectors

5.39 Consider $\mathbf{x}_1 = [1 \quad 2 \quad 3]^T$, $\mathbf{x}_2 = [1 \quad -2 \quad 3]^T$, $\mathbf{x}_3 = [0 \quad 1 \quad 1]^T$.
(a) Show that this set is linearly independent.
(b) Generate an orthonormal set using the Gram-Schmidt procedure.

5.40 Considering $\mathbf{x}_1$, $\mathbf{x}_2$, and $\mathbf{x}_3$ of Problem 5.39 as a basis set, find the reciprocal basis set.

5.41 Express the vector $\mathbf{z} = [6 \quad 4 \quad -3]^T$ in terms of the orthonormal basis set $\{\hat{\mathbf{v}}_i\}$ of Problem 5.39.

5.42 Express the vector $\mathbf{z} = [6 \quad 4 \quad -3]^T$ in terms of the original basis set $\{\mathbf{x}_i\}$ of Problem 5.39 by using the reciprocal basis vectors $\{\mathbf{r}_i\}$ found in Problem 5.40.

5.43 Find the reciprocal basis set if the basis vectors are $\mathbf{x}_1 = [4 \quad 2 \quad 1]^T$, $\mathbf{x}_2 = [2 \quad 6 \quad 3]^T$, $\mathbf{x}_3 = [1 \quad 3 \quad 5]^T$.

5.44 Given $\mathbf{x}_1 = [1 \quad 1 \quad 1]^T$, $\mathbf{x}_2 = [1 \quad -1 \quad 1]^T$, $\mathbf{x}_3 = [1 \quad 0 \quad 0]^T$. Use these as basis vectors and find reciprocal basis vectors. Also, express $\mathbf{z} = [6 \quad 3 \quad 1]^T$ in terms of the basis vectors.

5.45 Show that an orthonormal basis set and the corresponding set of reciprocal basis vectors are the same.

5.46 Verify that the following four $\mathbf{y}_i$ vectors are linearly independent by computing their Grammian. Then use them to construct an orthonormal set using the Gram-Schmidt process. Then use the orthonormal vectors to expand the vector $\mathbf{x} = [12.3 \quad 9.8 \quad -4.03 \quad 33.33]^T$.

The given **y** *vectors*

$$\mathbf{y}_1 = \begin{bmatrix} 4.4400002E+01 \\ 1.2800000E+01 \\ 1.5000000E+00 \\ -2.1000000E+01 \end{bmatrix}, \quad \mathbf{y}_2 = \begin{bmatrix} 7.7700000E+00 \\ 2.1500000E+01 \\ 1.0000000E+01 \\ 0.0000000E+00 \end{bmatrix},$$

$$\mathbf{y}_3 = \begin{bmatrix} -3.3329999E+00 \\ 4.1250000E+00 \\ 6.6670001E-01 \\ 1.0000000E+00 \end{bmatrix}, \quad \mathbf{y}_4 = \begin{bmatrix} 9.1250000E+00 \\ 2.1222000E+00 \\ -3.0500000E+00 \\ 4.4400001E+00 \end{bmatrix}$$

5.47 Compute the Grammian for the following vectors and draw conclusions about their linear independence.

The given **y** *vectors*

$$\mathbf{y}_1 = \begin{bmatrix} 1.0000000E+00 \\ 1.0000000E+00 \\ 1.0000000E+00 \\ 1.0000000E+00 \end{bmatrix}, \quad \mathbf{y}_2 = \begin{bmatrix} 1.0001000E+00 \\ 9.9989998E-01 \\ 1.0000000E+00 \\ 1.0000000E+00 \end{bmatrix}, \quad \mathbf{y}_3 = \begin{bmatrix} -2.0000000E+00 \\ -1.9999000E+00 \\ -2.0000000E+00 \\ -2.0000000E+00 \end{bmatrix}$$

.48 Find the orthogonal projection of the vectors

$$\mathbf{x}_1 = \begin{bmatrix} 1 \\ 1 \\ 1 \\ 1 \\ 1 \end{bmatrix} \quad \text{and} \quad \mathbf{x}_2 = \begin{bmatrix} -4 \\ 2 \\ -8 \\ 3 \\ 9 \end{bmatrix}$$

on the subspace spanned by the following three vectors.

The given **y** *vectors*

$$\mathbf{y}_1 = \begin{bmatrix} 1.0000000E+01 \\ 5.0000000E+00 \\ -5.0000000E+00 \\ 1.0000000E+00 \\ 6.0000000E+00 \end{bmatrix}, \qquad \mathbf{y}_2 = \begin{bmatrix} 1.1000000E+01 \\ 4.0000000E+00 \\ -1.1000000E+01 \\ -2.0000000E+00 \\ 6.0000000E+00 \end{bmatrix},$$

$$\mathbf{y}_3 = \begin{bmatrix} 1.0000000E+00 \\ 0.0000000E+00 \\ -1.0000000E+00 \\ -1.0000000E+00 \\ -1.0000000E+00 \end{bmatrix}$$

.49 Determine the dimension of the vector space spanned by $\mathbf{x}_1 = [1 \quad 2 \quad 2 \quad 1]^T$, $\mathbf{x}_2 = [1 \quad 0 \quad 0 \quad 1]^T$, $\mathbf{x}_3 = [3 \quad 4 \quad 4 \quad 3]^T$.

.50 Find the minimum distance from the origin to the plane $2x_1 + 3x_2 - x_3 = -5$ and find coordinates of the point on the plane nearest the origin.

.51 Let $\mathbf{c} = [1 \quad 2 \quad -1]^T$ and $\mathbf{y} = [2 \quad 5 \quad 3]^T$. Find the projection of $\mathbf{y}$ that is parallel to the family of planes defined by $\langle \mathbf{c}, \mathbf{x} \rangle = \text{constant}$.

.52 Show that the various Fourier series expansion formulas are special cases of the general expansion formula in an infinite dimensional linear inner product space:

$$\mathbf{x} = \sum_{i=1}^{\infty} \langle \mathbf{r}_i, \mathbf{x} \rangle \mathbf{v}_i$$

.53 Under what conditions does $\langle \mathbf{x}, \mathbf{y} \rangle = \mathbf{x}^T \mathbf{A} \mathbf{y}$ define a valid inner product for an n-dimension vector space defined over the real number field?

.54 If $\mathscr{X}^m$ and $\mathscr{X}^n$ are m- and n-dimensional linear vector spaces, respectively, then the product space $\mathscr{X}^m \times \mathscr{X}^n$ is itself a linear vector space consisting of all ordered pairs of $\mathbf{x} \in \mathscr{X}^m$, $\mathbf{y} \in \mathscr{X}^n$. That is,

$$\mathscr{X}^m \times \mathscr{X}^n = \{(\mathbf{x}, \mathbf{y}); \mathbf{x} \in \mathscr{X}^m, \ \mathbf{y} \in \mathscr{X}^n\}$$

If $\mathscr{X}^m$ has an inner product $\langle \mathbf{x}_1, \mathbf{x}_2 \rangle_m$ and $\mathscr{X}^n$ has an inner product $\langle \mathbf{y}_1, \mathbf{y}_2 \rangle_n$, show that the appropriate inner product for $\mathscr{X}^m \times \mathscr{X}^n$ is

$$\left\langle \begin{bmatrix} \mathbf{x}_1 \\ \mathbf{y}_1 \end{bmatrix}, \begin{bmatrix} \mathbf{x}_2 \\ \mathbf{y}_2 \end{bmatrix} \right\rangle = \langle \mathbf{x}_1, \mathbf{x}_2 \rangle_m + \langle \mathbf{y}_1, \mathbf{y}_2 \rangle_n$$

.55 Let $\mathscr{L}_2^n$ be the linear space consisting of all complex valued square integrable n component vector functions of a scalar variable $t \in [a, b]$, $\mathbf{f}(t) = [f_i(t)]$, where $i = 1, n$. Define an appropriate inner product and norm for this space.

.56 Prove the Cauchy-Schwarz inequality given on page 167.

.57 Let $\mathscr{R}^n$ be an n-dimensional Euclidean space with an orthonormal basis $\mathscr{B} = \{\mathbf{v}_i, i = 1, n\}$. Prove that for any $\mathbf{x} \in \mathscr{R}^n$,

$$\|\mathbf{x}\|^2 \geq \sum_{i=1}^{m} |\langle \mathbf{v}_i, \mathbf{x}\rangle|^2$$

where the summation is over any subset of m basis vectors. This is called Bessel's inequality. If $m = n$, the equality holds.

5.58 Consider the linear transformation $\mathcal{A}: \mathcal{R}^3 \to \mathcal{R}^3$. The basis vectors are selected as

$$\mathbf{v}_1 = [1 \quad 0 \quad 1]^T, \qquad \mathbf{v}_2 = [1 \quad 1 \quad 0]^T, \qquad \mathbf{v}_3 = [1 \quad 1 \quad 1]^T$$

The images of the basis vectors under the transformation $\mathcal{A}$ are

$$\mathbf{u}_1 = [2 \quad -1 \quad 3]^T, \qquad \mathbf{u}_2 = [-1 \quad -1 \quad 2]^T, \qquad \mathbf{u}_3 = [1 \quad 1 \quad 5]^T$$

with respect to the same basis. What is the matrix representation for $\mathcal{A}$?

5.59 The coordinate representations of a real vector $\mathbf{x}$ with respect to two different sets of orthonormal basis vectors are related by

$$[\mathbf{x}]' = \begin{bmatrix} \cos\alpha & \sin\alpha & 0 \\ -\sin\alpha & \cos\alpha & 0 \\ 0 & 0 & 1 \end{bmatrix} [\mathbf{x}]$$

Verify that $\sum_{i=1}^{3} x_i^2 = \sum_{i=1}^{3} (x_i')^2$. Would this be true for nonorthonormal bases?

5.60 If $\hat{\mathbf{u}}$ and $\hat{\mathbf{v}}$ are real unit vectors in three-dimensional space with an included angle α, and if $\mathbf{w} = \hat{\mathbf{u}} \times \hat{\mathbf{v}}$, find an expression for $\mathbf{A}^{-1} = [\hat{\mathbf{u}} \quad \hat{\mathbf{v}} \quad \mathbf{w}]^{-1}$.

5.61 Derive a matrix differential equation for the time rate of change $\dot{\mathbf{T}}_{BE}$ for the transformation of Problem 5.38.

5.62 A mirror lies in the plane defined by $-2x_1 + 3x_2 + x_3 = 0$. Find the reflected image of $\mathbf{y} = [4 \quad -2 \quad 3]^T$ and also find the orthogonal projection of $\mathbf{y}$ on the plane of the mirror.

5.63 Show that $\mathcal{A}^*(\mathcal{A}\mathcal{A}^*)^{-1}\mathcal{A}$ gives the orthogonal projection of $\mathcal{X}$ onto $\mathcal{R}(\mathcal{A}^*)$.

5.64 If $\mathbf{A}$ is the matrix representation of an operator which acts on $\mathbf{x} \in \mathcal{X}$, and if $\|\mathbf{x}\|^2 = \bar{\mathbf{x}}^T\mathbf{x}$, show that $\|\mathbf{A}\| \leq [\sum_i \sum_j |a_{ij}|^2]^{1/2}$.

5.65 The vector norm is defined by $\|\mathbf{x}\|^2 = \bar{\mathbf{x}}^T\mathbf{x}$ and $\mathbf{A}\mathbf{x} = \mathbf{y}$.
(a) Show that $\|\mathbf{A}\|^2 = \max\{\gamma_i\}$, where $\{\gamma_i\}$ is the set of eigenvalues for $\bar{\mathbf{A}}^T\mathbf{A}$.
(b) Show that if $\mathbf{A}$ is normal, $\|\mathbf{A}\| = \max_i |\lambda_i|$, where $\{\lambda_i\}$ is the set of eigenvalues for $\mathbf{A}$.

5.66 Let $\mathcal{X}$ and $\mathcal{Y}$ be linear vector spaces whose elements are summable n-component vector sequences, $\{\mathbf{x}(k), k = 0, 1, 2, \ldots\}$ and $\{\mathbf{y}(k), k = 0, 1, 2, \ldots\}$. A linear transformation $\mathcal{A}: \mathcal{X} \to \mathcal{Y}$ is defined by the first-order difference equation $\mathcal{A}(\mathbf{x}) = \mathbf{x}(k+1) - \mathbf{A}_k\mathbf{x}(k)$. Find $\mathcal{A}^*$ if the inner product is defined as $\langle \mathbf{x}_1, \mathbf{x}_2\rangle = \sum_{k=0}^{\infty} \mathbf{x}_1(k)^T\mathbf{x}_2(k)$.

6

Simultaneous Linear Equations

1 INTRODUCTION

The task of solving a set of simultaneous linear algebraic equations is frequently encountered by engineers and scientists in all fields. Many problems of estimation, control, system identification, pole-placement, and optimization depend on the solution of simultaneous equations. The properties of controllability and observability of linear systems are conditions which directly relate to the ability to solve a set of simultaneous equations. The stability and natural modes of a system are determined by the solution of an eigenvalue problem, which involves solution of simultaneous equations. This chapter uses the matrix theory and linear algebra of the last two chapters to study this class of problems. Several important applications are also introduced. The material in this chapter will be used in every chapter in the rest of this book.

.2 STATEMENT OF THE PROBLEM AND CONDITIONS FOR SOLUTIONS

Consider the set of simultaneous linear algebraic equations

$$\begin{aligned} a_{11}x_1 + a_{12}x_2 + \cdots + a_{1n}x_n &= y_1 \\ a_{21}x_1 + a_{22}x_2 + \cdots + a_{2n}x_n &= y_2 \\ &\vdots \\ a_{m1}x_1 + a_{m2}x_2 + \cdots + a_{mn}x_n &= y_m \end{aligned}$$

In matrix notation this is simply

$$\mathbf{Ax} = \mathbf{y} \tag{6.1}$$

where the elements of a_{ij} of the $m \times n$ matrix $\mathbf{A}$ are known, as are the scalar components y_i of the $m \times 1$ vector $\mathbf{y}$. The $n \times 1$ vector $\mathbf{x}$ contains the unknowns which are to be determined if possible. Any vector, say $\mathbf{x}_1$, which satisfies all m of these equations is called a *solution.* Not every set of simultaneous equations has a solution. The *augmented matrix,* defined by $\mathbf{W} = [\mathbf{A} \mid \mathbf{y}]$, indicates whether or not solutions exist. In fact,

1. If $r_W \neq r_A$, no solution exists. The equations are *inconsistent.*
2. If $r_W = r_A$, at least one solution exists.
   (a) If $r_W = r_A = n$, there is a *unique* solution for $\mathbf{x}$.
   (b) If $r_W = r_A < n$, then there is an infinite set of solution vectors.

It is clearly impossible for r_A to exceed n, so the only possibilities are that there are no solutions, or exactly one solution, or an infinity of solutions. In order to explain fully the basis for these results, two linear vector spaces and the mappings between them must be studied. Let $\mathcal{X}^n$ be the space of all n-dimensional $\mathbf{x}$ vectors and let $\mathcal{X}^m$ be the space of all m-dimensional $\mathbf{y}$ vectors. The matrix $\mathbf{A}$ can be considered as a concrete example of an operator which maps members of $\mathcal{X}^n$ into members of $\mathcal{X}^m$. As discussed in Chapter 5, there is another operator, the adjoint operator $\mathbf{A}^*$, which maps elements of $\mathcal{X}^m$ back into $\mathcal{X}^n$. In the present case the adjoint operator is just the conjugate transpose of $\mathbf{A}$, i.e., $\mathbf{A}^* = \overline{\mathbf{A}}^T$. It is very useful to know that the two spaces under discussion can each be written as an orthogonal sum of two subspaces. First $\mathcal{X}^m$ is considered. The primary subspace of interest is the range space of $\mathbf{A}$, $\mathcal{R}(\mathbf{A})$. This is the space of all $\mathbf{y}$ vectors which are images of some $\mathbf{x}$ vector. Since a column-partitioned version of Eq. (*6.1*) is

$$x_1\mathbf{a}_1 + x_2\mathbf{a}_2 + \cdots + x_n\mathbf{a}_n = \mathbf{y}$$

it is seen that $\mathcal{R}(\mathbf{A})$ is actually made up of all possible linear combinations of the columns $\mathbf{a}_j$ of $\mathbf{A}$. For this reason $\mathcal{R}(\mathbf{A})$ is also called the *column space* of $\mathbf{A}$, written as $L(\mathbf{a}_j)$. It should already be clear that for a particular $\mathbf{A}$ and $\mathbf{y}$ of Eq. (*6.1*), if $\mathbf{y} \notin L(\mathbf{a}_j)$, there is no $\mathbf{x}$ solution. Saying that $\mathbf{y} \in L(\mathbf{a}_j)$ is equivalent to saying that $\mathbf{y}$ is a linear combination of the columns of $\mathbf{A}$ and hence $\text{rank}(\mathbf{A}) = \text{rank}(\mathbf{W})$. This is the condition which is necessary for the existence of *at least one* solution $\mathbf{x}$. The columns $\mathbf{a}_j$ span the column space directly from the definition. If, moreover, these n columns form a *basis* for it, then the dimension of $L(\mathbf{a}_j)$ is n, meaning the following:

1. There is a *unique* set of n x_j coefficients for each $\mathbf{y}$ in $L(\mathbf{a}_j)$.
2. $L(\mathbf{a}_j) \equiv \mathcal{X}^n$, and hence $m = n$ is required.
3. $\text{Rank}(\mathbf{A}) = \text{rank}(\mathbf{W}) = n$.

The critical difference between *solutions* for Eq. (*6.1*) and *one unique solution* hinges upon whether the columns of $\mathbf{A}$ merely span the column space or form a basis set.

The orthogonal complement of a linear vector space such as $L(\mathbf{a}_j)$ is another linear vector space, denoted by $L(\mathbf{a}_j)^\perp$. $\mathcal{X}^m$ can be written as the direct sum $\mathcal{X}^m = L(\mathbf{a}_j) \oplus L(\mathbf{a}_j)^\perp$. *The space* $\mathcal{X}^m$ has been decomposed into two orthogonal subspaces as promised, but a more descriptive explanation of the orthogonal complement will be given shortly. First attention is directed to $\mathcal{X}^n$. From among all $\mathbf{x} \in \mathcal{X}^n$, those for which

$\mathbf{Ax} = \mathbf{0}$ form the *null space* of $\mathbf{A}$, written $\mathcal{N}(\mathbf{A})$. Using the orthogonal complement of $\mathcal{N}(\mathbf{A})$, we can write the direct sum $\mathscr{X}^n = \mathcal{N}(\mathbf{A}) \oplus \mathcal{N}(\mathbf{A})^\perp$. The two orthogonal complement spaces which have been introduced can be given more meaning by considering the adjoint operator $\overline{\mathbf{A}}^T$ mapping from $\mathscr{X}^m$ to $\mathscr{X}^n$. Let $\mathbf{c}_j$ be the columns of $\overline{\mathbf{A}}^T$, that is, the conjugates of the *rows* of $\mathbf{A}$. Then the adjoint mapping $\overline{\mathbf{A}}^T\mathbf{y} = \mathbf{x}$ can be written in column-partitioned form as

$$y_1\mathbf{c}_1 + y_2\mathbf{c}_2 + \cdots + y_m\mathbf{c}_m = \mathbf{x}$$

All $\mathbf{x}$ vectors in the range space of $\overline{\mathbf{A}}^T$ are linear combinations of the columns $\mathbf{c}_j$; hence $\mathcal{R}(\overline{\mathbf{A}}^T)$ is frequently called the *row-space* of $\mathbf{A}$, sometimes written $L(\mathbf{c}_j)$. Problem 6.21 shows this to be precisely the same space as the orthogonal complement of the null space of $\mathbf{A}$, so $\mathscr{X}^n$ is the direct sum of the null space of $\mathbf{A}$ and the row space of $\mathbf{A}$.

Returning to the space $\mathscr{X}^m$, those vectors $\mathbf{y}$ for which $\overline{\mathbf{A}}^T\mathbf{y} = \mathbf{0}$ form the null space of $\overline{\mathbf{A}}^T$. This space is frequently called the *left null space* of $\mathbf{A}$ because the conjugate transpose of the previous equation gives $\overline{\mathbf{y}}^T\mathbf{A} = \mathbf{0}$. Vectors $\mathbf{y}$ satisfying this relation are called the *left null vectors* of $\mathbf{A}$, and the left null space of $\mathbf{A}$ is the space of all left null vectors. This space is precisely the same space introduced earlier as the orthogonal complement of $L(\mathbf{a}_j)$. Therefore, $\mathscr{X}^m$ is the direct sum of the column space of $\mathbf{A}$ and the left null space of $\mathbf{A}$.

To summarize the results of this section, it has been found that every $m \times n$ matrix $\mathbf{A}$ has four important vector spaces associated with it. These are

The column space $L(\mathbf{a}_i) \equiv \mathcal{R}(\mathbf{A})$
The null space $\mathcal{N}(\mathbf{A})$
The row space $L(\mathbf{c}_i) \equiv \mathcal{R}(\overline{\mathbf{A}}^T)$
The left null space $\mathcal{N}(\overline{\mathbf{A}}^T)$

The primary vector spaces $\mathscr{X}^n$ and $\mathscr{X}^m$ can be written as direct sums, as suggested pictorially in Figure 6.1. For a particular matrix $\mathbf{A}$, some of these subspaces may be zero-dimensional—that is, contain only the zero vector.

6.3 THE ROW-REDUCED ECHELON FORM OF A MATRIX

The rank of certain matrices plays a vital role in the above discussion and in many other contexts in modern control theory. An efficient method of determining the rank of a matrix is to put the matrix into *row-reduced echelon* form [1]. This form is obtained

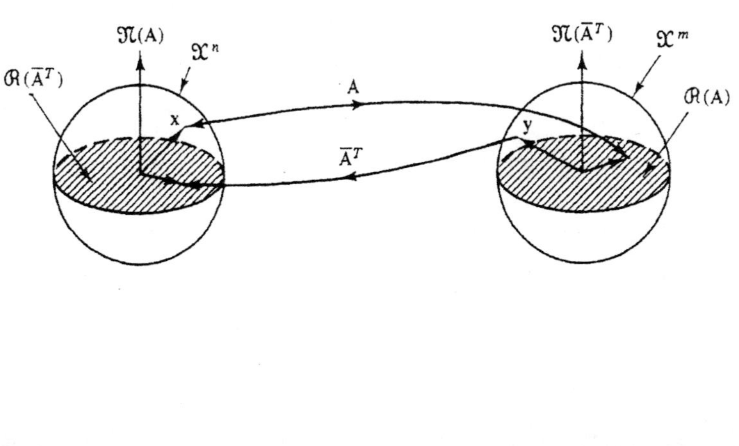

Figure 6.1

by using elementary row operations, just as described in Problem 4.18, page 147, for Gaussian elimination. Elementary row operations are performed until the first nonzero term in each row is unity and all terms below these ones are zero. This form of the matrix is the *echelon form,* but the matrix is not yet in row-reduced echelon form. Even in this intermediate form, the rank of the matrix is obvious by inspection. It is the number of nonzero rows in the matrix. Recall from Chapter 4 that the elementary matrices are nonsingular, and therefore multiplication of a matrix by them does not change the rank of that matrix.

Additional row operations are carried out until the leading ones in each row are the only nonzero terms in their respective columns. That is, any nonzero terms *above* the leading ones of the echelon form are now removed. The result is the desired row-reduced echelon form. Every matrix has a *unique* row-reduced echelon form. Some texts refer to this as the Hermite normal form of the matrix. The rank of a matrix is certainly obvious from its row-reduced echelon form, but its usefulness goes far beyond that. Any nonsingular **A** matrix just reduces to the unit matrix **I**. Therefore, aside from confirming the rank, it seems that all the information in **A** is "lost" by this reduction. The usefulness of the technique usually comes about by applying the reduction to some matrix other than just the coefficient matrix **A** in a set of simultaneous equations. If it is applied to the composite matrix $\mathbf{W} = [\mathbf{A} \,|\, \mathbf{y}]$ defined before, and if the resultant row-reduced echelon form is called $\mathbf{W}' = [\mathbf{A}' \,|\, \mathbf{y}']$, then

$$\text{rank}\,\mathbf{W} = \text{rank}\,\mathbf{W}' \quad \text{and} \quad \text{rank}\,\mathbf{A} = \text{rank}\,\mathbf{A}'$$

so that inspection of $\mathbf{W}'$ reveals instantly which of the previous categories applies. Further, when solutions do exist they are obtained directly from $\mathbf{W}'$ with little or no additional effort.

EXAMPLE 6.1 In the following nine situations a set of simultaneous equations of the form $\mathbf{Ax} = \mathbf{y}$ is being considered. In each case the **W** matrix containing **A** and **y** is displayed, followed by the row-reduced echelon form $\mathbf{W}'$. These are then used to draw conclusions about the original set of equations. Note that in cases 1, 2, and 3 the number of equations, m, is equal to the number of unknowns in **x**, n. In cases 4 and 5 m is less than n and in cases 6 through 9 m is greater than n.

1.

$$\mathbf{W} = \left[\begin{array}{ccc|c} 1 & 0 & 1 & 3 \\ 0 & 1 & 1 & -1 \\ 1 & 0 & 1 & 5 \end{array}\right] \qquad \mathbf{W}' = \left[\begin{array}{ccc|c} 1 & 0 & 1 & 0 \\ 0 & 1 & 1 & 0 \\ 0 & 0 & 0 & 1 \end{array}\right]$$

From the reduced form, $r_A = 2$ and $r_W = 3$, so no solutions exist.

2.

$$\mathbf{W} = \left[\begin{array}{ccc|c} 1 & 0 & 1 & 3 \\ 0 & 1 & 1 & -1 \\ 1 & 0 & 1 & 3 \end{array}\right] \qquad \mathbf{W}' = \left[\begin{array}{ccc|c} 1 & 0 & 1 & 3 \\ 0 & 1 & 1 & -1 \\ 0 & 0 & 0 & 0 \end{array}\right]$$

From this, $r_A = 2 = r_W < n = 3$. Therefore, an infinite number of solutions exist, and $\mathbf{W}'$ tells us that $\mathbf{x}_1 + \mathbf{x}_3 = 3$ and $\mathbf{x}_2 + \mathbf{x}_3 = -1$.

3.

$$\mathbf{W} = \left[\begin{array}{ccc|c} 1 & 2 & 3 & 4 \\ 2 & 1 & 2 & 7 \\ 3 & 2 & 1 & 1 \end{array}\right] \qquad \mathbf{W}' = \left[\begin{array}{ccc|c} 1 & 0 & 0 & 2.125 \\ 0 & 1 & 0 & -4.5 \\ 0 & 0 & 1 & 3.625 \end{array}\right]$$

Thus $r_A = r_W = n = 3$, so a unique solution exists and it is just $\mathbf{y}'$.

4.

$$\mathbf{W} = \left[\begin{array}{ccc|c} 1 & -1 & 2 & 8 \\ -1 & 2 & 0 & 2 \end{array}\right] \qquad \mathbf{W}' = \left[\begin{array}{ccc|c} 1 & 0 & 4 & 18 \\ 0 & 1 & 2 & 10 \end{array}\right]$$

Here $r_A = r_W = 2 < n$, so solutions exist but are not unique. They all must satisfy $\mathbf{x}_1 + 4\mathbf{x}_3 = 18$ and $\mathbf{x}_2 + 2\mathbf{x}_3 = 10$.

5.

$$\mathbf{W} = \left[\begin{array}{cccc|c} 1 & 2 & 3 & 1 & 1 \\ -4 & 5 & 1 & 9 & 2 \\ -2 & 8 & 6 & 10 & 3 \end{array}\right] \qquad \mathbf{W}' = \left[\begin{array}{cccc|c} 1 & 0 & 1 & -1 & 0 \\ 0 & 1 & 1 & 1 & 0 \\ 0 & 0 & 0 & 0 & 1 \end{array}\right]$$

This case has $r_A = 2, r_W = 3$, so no solutions exist.

6.

$$\mathbf{W} = \left[\begin{array}{cc|c} 1 & 2 & 2 \\ 3 & 4 & 3 \\ 5 & 6 & 4 \end{array}\right] \qquad \mathbf{W}' = \left[\begin{array}{cc|c} 1 & 0 & -1 \\ 0 & 1 & 1.5 \\ 0 & 0 & 0 \end{array}\right]$$

Since $r_A = r_W = 2 = n$, there is a unique solution given by the first two components of $\mathbf{y}'$, namely $\mathbf{x} = [-1 \quad 1.5]^T$.

7.

$$\mathbf{W} = \left[\begin{array}{cc|c} 1 & 2 & 2 \\ 3 & 4 & 3 \\ 5 & 6 & -4 \end{array}\right] \qquad \mathbf{W}' = \left[\begin{array}{cc|c} 1 & 0 & 0 \\ 0 & 1 & 0 \\ 0 & 0 & 1 \end{array}\right]$$

Note that this is the same as case 6 except for the sign of one component of $\mathbf{y}$. The results are quite different. Since $r_A = 2$ and $r_W = 3$, there is no solution.

8.

$$\mathbf{W} = \left[\begin{array}{ccc|c} 1 & 3 & 5 & 3 \\ 1 & 4 & 6 & 3.5 \\ -1 & 5 & 3 & 1 \\ -1 & 4 & 2 & 0.5 \\ 1 & 3 & 5 & 3 \end{array}\right] \qquad \mathbf{W}' = \left[\begin{array}{ccc|c} 1 & 0 & 2 & 1.5 \\ 0 & 1 & 1 & 0.5 \\ 0 & 0 & 0 & 0 \\ 0 & 0 & 0 & 0 \\ 0 & 0 & 0 & 0 \end{array}\right]$$

Here, even though there are more equations than unknowns, there are still solutions—in fact, an infinite number—all of which satisfy $\mathbf{x}_1 + 2\mathbf{x}_3 = 1.5$ and $\mathbf{x}_2 + \mathbf{x}_3 = 0.5$.

9. Changing only the $\mathbf{y}$ vector of the previous case to $[3 \quad 3 \quad 1 \quad 1 \quad 3]^T$ leads to

$$\mathbf{W}' = \left[\begin{array}{ccc|c} 1 & 0 & 2 & 0 \\ 0 & 1 & 1 & 0 \\ 0 & 0 & 0 & 1 \\ 0 & 0 & 0 & 0 \\ 0 & 0 & 0 & 0 \end{array}\right]$$

Since $r_A = 2$ and $r_W = 3$, there is no solution. In addition to demonstrating the various categories of simultaneous equations and illustrating row-reduced echelon forms, this example is intended to show that characterizations strictly in terms of numbers of equations and numbers of unknowns are clearly inadequate. ■

6.3.1 Applications to Polynomial Matrices

The previous examples of obtaining row-reduced echelon matrices by use of elementary operations all involved matrices defined over the real numbers. This can be applied to complex numbers as well. The notions of elementary matrices $\mathbf{E}_p(\alpha)$ and $\mathbf{E}_{p,q}(\alpha)$ in Sec. 4.10 are valid for α belonging to *any* scalar number field, including the rational polynomial functions. However, for many purposes, when dealing with polynomial matrices it is desirable to redefine slightly the elementary matrices and hence the allowed elementary operations. The purpose is to ensure that the results of our restricted elementary operations remain polynomial matrices (and not matrices of *ratios* of polynomials). We call the modified elementary matrices the *polynomial-restricted* elementary matrices, defined as follows:

1. The elementary matrix $\mathbf{E}_{p,q}$ to interchange rows or columns is not a function of α and requires no change.
2. The row (or column) multiplier $\mathbf{E}_p(\alpha)$ will be restricted to either real or complex α. Disallowing polynomial α ensures that $\mathbf{E}_p(\alpha)^{-1}$ is still an elementary matrix in the restricted sense. If α were a polynomial, the inverse would involve *ratios* of polynomials.
3. The matrix $\mathbf{E}_{p,q}(\alpha)$, which adds α times the pth row (column) to the qth row (column), does allow α to be a polynomial. Note that $|\mathbf{E}_{p,q}(\alpha)|$ is not a function of α, so that the inverse remains a polynomial-restricted elementary matrix.

The polynomial-restricted elementary operations (row or column) can be carried out by pre- or postmultiplication by the appropriate $\mathbf{E}$ matrix, as before. The notion of the row-reduced echelon form also must be modified in the case of polynomial matrices. The more general term Hermite form will be used to denote a matrix in which

1. The first nonzero entry in a row is a *monic* polynomial, that is, a polynomial in which the coefficient of the highest power is unity.
2. All terms in the same column and *below* this leading monic polynomial are zero.
3. All terms in this same column and *above* the leading term are polynomials of degree less than the degree of the leading monic polynomial.

Note that this reduces to the previous definition of the row-reduced echelon form when the leading monic polynomials are all of degree zero, i.e., just the scalar 1. It is understood that lower-degree polynomials are scalar zeros in this case. A more standard interpretation of a polynomial of degree less than zero (such as degree of -1) would lead us outside the realm of polynomials and into rational polynomial functions, which is what we are trying to avoid. Three examples of polynomial matrices in Hermite normal form are

$$\begin{bmatrix} 1 & 0 & 3 \\ 0 & 1 & 15 \\ 0 & 0 & 0 \end{bmatrix}, \quad \begin{bmatrix} s^2+2s+1 & 3 & s \\ 0 & s+1 & s^2+2s+1 \\ 0 & 0 & s^3+4s \end{bmatrix}, \quad \begin{bmatrix} s & 5s+1 \\ 0 & s^2+3s \\ 0 & 0 \end{bmatrix}$$

EXAMPLE 6.2 Use polynomial-restricted elementary operations to reduce $\mathbf{P}(s) = \begin{bmatrix} s+1 & s & 5 \\ s-1 & s^2+3s+2 & s \end{bmatrix}$ to Hermite normal form. A sequence of elementary row operations modifies $\mathbf{P}(s)$ successively to

$$\begin{bmatrix} s+1 & s & 5 \\ -2 & s^2+2s+2 & s-5 \end{bmatrix} \rightarrow \begin{bmatrix} 1 & -0.5(s^2+2s+2) & -0.5(s-5) \\ s+1 & s & 5 \end{bmatrix}$$

$$\rightarrow \begin{bmatrix} 1 & -0.5(s^2+2s+2) & -0.5(s-5) \\ 0 & s^3+3s^2+6s+2 & s^2-4s+5 \end{bmatrix}$$

Some obvious intermediate steps have not been given. The actual sequence of elementary operations used was

$$\mathbf{E}_2(2)\mathbf{E}_{1,2}(-s-1)\mathbf{E}_{1,2}\,\mathbf{E}_2(-0.5)\mathbf{E}_{1,2}(-1) = \begin{bmatrix} 0.5 & -0.5 \\ -s+1 & s+1 \end{bmatrix}$$

The order of application of the elementary matrices was right to left; that is, $\mathbf{E}_{1,2}(-1)$ was used first, then $\mathbf{E}_2(-0.5)$, and so on. ■

The elementary operations used in the reduction can be systematically applied [2]. The first nonzero term in each row is usually the diagonal term. On a column-by-column basis, the terms below the diagonal must be reduced to zero. For the general column j, assume that at some point in the reduction process polynomials $p_1(s)$ and $p_2(s)$ are in the jj and ij positions, with $i > j$. Row interchanges can be used to ensure that degree $(p_1) \leq$ degree(p_2). Standard long division gives $p_2(s)/p_1(s) = q(s) + r(s)/p_1(s)$, where q and r are quotient and remainder polynomials. Therefore, $p_2(s) - p_1(s)q(s) = r(s)$. Hence, premultiplication of the matrix by $\mathbf{E}_{j,i}(-q(s))$ will reduce the ij term from $p_2(s)$ to the lower-degree $r(s)$. This procedure can be applied to each nonzero term below the diagonal. A row interchange can then be used to bring the minimum degree nonzero remainder to the jj diagonal, and the whole column-reduction process can be repeated. If a constant remainder (polynomial of degree zero) is ever found, that term is placed on the diagonal, normalized to unity, and then used immediately to reduce all other terms in that column to zero. If all terms below the diagonal are reduced to zero while the diagonal term remains a finite-degree polynomial, then the terms above the diagonal need not be zero. However, they must be reduced to polynomials of a lower degree than the diagonal. This can be done using the same long-division procedure as before, leaving only the remainder terms above the diagonal. Although simple in concept, this reduction to Hermite normal form can be algebraically tedious. To demonstrate this, the reader should verify that

$$\mathbf{P}(s) = \begin{bmatrix} 3s+2 & 6 \\ s & s \\ s^2-1 & s+5 \end{bmatrix} \quad \text{can be reduced to} \quad \begin{bmatrix} 1 & 0 \\ 0 & 1 \\ 0 & 0 \end{bmatrix}$$

by a series of 11 elementary row operations.

6.3.2 Application to Matrix Fraction Description of Systems

The polynomial-restricted elementary matrices can be used systematically to reduce MFDs of transfer function matrices $\mathbf{H}(s)$. (See Eq. (*4.7*) and Problems 4.29 through 4.31.) The left-divisor form $\mathbf{H}(s) = \mathbf{P}_1^{-1}(s)\mathbf{N}_1(s)$ can be written as $\mathbf{P}_1(s)\mathbf{H}(s) = \mathbf{N}_1(s)$. Premultiplying by a sequence of elementary matrices leaves $\mathbf{H}(s)$ unchanged and therefore is equivalent to performing elementary row operations on $\mathbf{W} \equiv [\mathbf{P}_1(s) \mid \mathbf{N}_1(s)]$ to obtain $[\mathbf{T}(s)\mathbf{P}_1(s) \mid \mathbf{T}(s)\mathbf{N}_1(s)] = [\underline{\mathbf{P}}_1(s) \mid \underline{\mathbf{N}}_1(s)]$. The matrix $\mathbf{T}(s)$ is the product of elementary matrices and thus is invertible. Its inverse is still a polynomial matrix because of the restrictions placed on the polynomial-restricted elementary matrices. Therefore, $[\mathbf{P}_1(s) \mid \mathbf{N}_1(s)] = [\mathbf{T}^{-1}\underline{\mathbf{P}}_1 \mid \mathbf{T}^{-1}\underline{\mathbf{N}}_1]$. This shows that the matrix $\mathbf{T}^{-1}$ is a common factor of both $\mathbf{P}_1$ and $\mathbf{N}_1$, on the left. More commonly, $\mathbf{T}$ is called a *left common divisor* of $\mathbf{P}_1$ and $\mathbf{N}_1$. It is clear that the common divisor cancels, leaving a reduced MFD with the same original transfer function,

$$\mathbf{H}(s) = [\mathbf{T}^{-1}\underline{\mathbf{P}}_1]^{-1}[\mathbf{T}^{-1}\underline{\mathbf{N}}_1] = \underline{\mathbf{P}}_1^{-1}\underline{\mathbf{N}}_1$$

Similarly, starting with the right-hand divisor form of the MFD

$$\mathbf{H}(s) = \mathbf{N}_2(s)\mathbf{P}_2^{-1}(s)$$

it is clear that postmultiplication of $\mathbf{H}(s)\mathbf{P}_2(s) = \mathbf{N}_2(s)$ by elementary matrices leaves $\mathbf{H}(s)$ unchanged and is equivalent to elementary column operations on

$$\mathbf{Y} \equiv \begin{bmatrix} \mathbf{N}_2(s) \\ \hline \mathbf{P}_2(s) \end{bmatrix}$$

This leads to the notion of right common divisors for $\mathbf{N}_2$ and $\mathbf{P}_2$. In either case, these concepts are generalizations of the notion of pole-zero cancellations in scalar transfer functions. When the elementary row (column) operations are carried out on $\mathbf{W}$ (or $\mathbf{Y}$) to the limit—i.e., until the Hermite normal form is reached—the product $\mathbf{T}$ of the elementary matrices used will represent the *greatest* common left (or right) divisor. The resulting transfer function representations $\mathbf{H}(s) = \underline{\mathbf{P}}_1^{-1}(s)\underline{\mathbf{N}}_1(s) = \underline{\mathbf{N}}_2(s)\underline{\mathbf{P}}_2^{-1}(s)$ are maximally reduced in the sense that greatest common divisors have been removed and no further common polynomial factors can be canceled. This is a very important consideration in several instances when analyzing multiple-input-output systems within the transfer function and polynomial matrix domain. This will be useful in Chapter 13. The major emphasis in this book is on state variable representation of systems. Even here, the polynomial matrix representations often play a role in obtaining appropriate state models, as is shown in Chapter 12.

6.4 SOLUTION BY PARTITIONING

It is assumed in this section that $r_A = r_W$, so that one or more solutions exist. By definition, the $m \times n$ coefficient matrix $\mathbf{A}$ contains a nonsingular $r_A \times r_A$ matrix. The original equations $\mathbf{Ax} = \mathbf{y}$ can always be rearranged and partitioned into

$$\left[\begin{array}{c|c} \mathbf{A}_1 & \mathbf{A}_2 \\ \hline \mathbf{A}_3 & \mathbf{A}_4 \end{array}\right]\begin{bmatrix} \mathbf{x}_1 \\ \hline \mathbf{x}_2 \end{bmatrix} = \begin{bmatrix} \mathbf{y}_1 \\ \hline \mathbf{y}_2 \end{bmatrix}$$

where $\mathbf{A}_1$ is $r_A \times r_A$ and nonsingular. Depending on the relation between m, n, and r_A, some of the terms in the partitioned equation will not be required. For example, if $m = n = r_A$, then $\mathbf{A}_1 = \mathbf{A}$, $\mathbf{x}_1 = \mathbf{x}$, and $\mathbf{y}_1 = \mathbf{y}$. The general case is treated here, and then

$$\mathbf{A}_1\mathbf{x}_1 + \mathbf{A}_2\mathbf{x}_2 = \mathbf{y}_1 \quad \text{or} \quad \mathbf{x}_1 = \mathbf{A}_1^{-1}[\mathbf{y}_1 - \mathbf{A}_2\mathbf{x}_2]$$

The *degeneracy* of $\mathbf{A}$ is $q_A = n - r_A$. The values of the q_A components of $\mathbf{x}_2$ are completely arbitrary and generate the q_A parameter family of solutions for $\mathbf{x}$ mentioned on page 208, case 2(*b*). If $r_A = n$, as in case 2(*a*), then $\mathbf{A}_2$, $\mathbf{A}_4$, and $\mathbf{x}_2$ will not be present in the partitioned equation. In that case the unique solution is $\mathbf{x} = \mathbf{A}_1^{-1}\mathbf{y}_1$. If in addition $m = n$, then $\mathbf{A}_3$ and $\mathbf{y}_2$ will not be present and $\mathbf{x} = \mathbf{A}^{-1}\mathbf{y}$. This is the simple case mentioned in Chapter 4, and $\mathbf{x}$ could be computed by using Cramer's rule or various matrix inversion techniques. However, Gaussian elimination or similar reduction techniques are more efficient for large values of n.

EXAMPLE 6.3 Consider once more the situation of case 2 in Example 6.1, and find all solutions $\mathbf{x}$ for

$$\begin{bmatrix} 1 & 0 & 1 \\ 0 & 1 & 1 \\ 1 & 0 & 1 \end{bmatrix} \begin{bmatrix} x_1 \\ x_2 \\ x_3 \end{bmatrix} = \begin{bmatrix} 3 \\ -1 \\ 3 \end{bmatrix}$$

Here $r_A = 2$ (rows 1 and 3 are identical) and $r_W = 2$ also. An infinite set of solutions exists. Let

$$\mathbf{A}_1 = \begin{bmatrix} 1 & 0 \\ 0 & 1 \end{bmatrix} \qquad \mathbf{x}_1 = \begin{bmatrix} x_1 \\ x_2 \end{bmatrix} \qquad \mathbf{y}_1 = \begin{bmatrix} 3 \\ -1 \end{bmatrix}$$

$$\mathbf{A}_2 = \begin{bmatrix} 1 \\ 1 \end{bmatrix} \qquad \mathbf{x}_2 = x_3 \qquad \mathbf{y}_2 = 3$$

Then

$$\mathbf{x}_1 = \begin{bmatrix} 1 & 0 \\ 0 & 1 \end{bmatrix}^{-1} \left\{ \begin{bmatrix} 3 \\ -1 \end{bmatrix} - \begin{bmatrix} 1 \\ 1 \end{bmatrix} x_3 \right\} = \begin{bmatrix} 3 - x_3 \\ -1 - x_3 \end{bmatrix}$$

The one-parameter family of solutions is $\mathbf{x} = [3 - x_3 \quad -1 - x_3 \quad x_3]^T$ with x_3 arbitrary. ■

6.5 A GRAM-SCHMIDT EXPANSION METHOD OF SOLUTION

The set of m simultaneous equations in n unknowns $\mathbf{x}$ is again considered.

$$\mathbf{Ax} = \mathbf{y} \tag{6.1}$$

No special assumptions are made at the outset about r_A and r_W relative to each other or to m and n. By definition, there are r_A independent a_j columns in the matrix $\mathbf{A}$. These vectors can be used as a basis set for a linear vector space $L(a_j)$ called the *column space* of $\mathbf{A}$. The number of vectors in this basis set could be n or any smaller positive integer in a given case. In addition to these r_A vectors a_j, the $\mathbf{y}$ vector is considered, giving a set of $r_A + 1$ vectors. The Gram-Schmidt procedure is used on this set to form an orthonormal basis set $\{\hat{\mathbf{v}}_j\}$. The only possible exception is the last vector $\hat{\mathbf{v}}_{r_A+1}$. Since $\mathbf{y}$ may be linearly dependent on the columns $\mathbf{a}_j$, it might not be possible to form a nonzero vector from $\mathbf{y}$ which is orthogonal to all the $\mathbf{a}_j$ vectors. If $\mathbf{y}$ is linearly dependent on the $\mathbf{a}_j$ vectors, then the unnormalized vector $\hat{\mathbf{v}}_{r_A+1}$ will automatically come out zero during the Gram-Schmidt construction. Since the zero vector is orthogonal to every other vector, an orthogonal set of $\{\hat{\mathbf{v}}_j\}$ can thus be constructed in all cases. Each vector in the set is a unit vector with the possible exception of a zero vector as the last entry. Form the $m \times (r_A + 1)$ matrix $\mathbf{V}$ from the set. Premultiplying Eq. (*6.1*) by V^T is equivalent to premultiplying the previously defined $\mathbf{W}$ matrix. The result is

$$
\mathbf{V}^T\mathbf{W} = \underbrace{\left[\begin{array}{cccccc|ccc|c}
\langle\hat{\mathbf{v}}_1,\mathbf{a}_1\rangle & \langle\hat{\mathbf{v}}_1,\mathbf{a}_2\rangle & & \cdots & & \langle\hat{\mathbf{v}}_1,\mathbf{a}_r\rangle & \langle\hat{\mathbf{v}}_1,\mathbf{a}_{r+1}\rangle & \cdots & \langle\hat{\mathbf{v}}_1,\mathbf{a}_n\rangle & \langle\hat{\mathbf{v}}_1,\mathbf{y}\rangle \\
0 & \langle\hat{\mathbf{v}}_2,\mathbf{a}_2\rangle & & \cdots & & \langle\hat{\mathbf{v}}_2,\mathbf{a}_r\rangle & \langle\hat{\mathbf{v}}_2,\mathbf{a}_{r+1}\rangle & \cdots & \langle\hat{\mathbf{v}}_2,\mathbf{a}_n\rangle & \langle\hat{\mathbf{v}}_2,\mathbf{y}\rangle \\
0 & 0 & \langle\hat{\mathbf{v}}_3,\mathbf{a}_3\rangle & \cdots & & \langle\hat{\mathbf{v}}_3,\mathbf{a}_r\rangle & \langle\hat{\mathbf{v}}_3,\mathbf{a}_{r+1}\rangle & \cdots & \langle\hat{\mathbf{v}}_3,\mathbf{a}_n\rangle & \langle\hat{\mathbf{v}}_3,\mathbf{y}\rangle \\
\vdots & \vdots & \vdots & & & & & & & \\
0 & 0 & 0 & \cdots & & \langle\hat{\mathbf{v}}_r,\mathbf{a}_r\rangle & \langle\hat{\mathbf{v}}_r,\mathbf{a}_{r+1}\rangle & \cdots & \langle\hat{\mathbf{v}}_r,\mathbf{a}_n\rangle & \langle\hat{\mathbf{v}}_r,\mathbf{y}\rangle \\
\hline
0 & 0 & 0 & \cdots & & 0 & \cdots & & 0 & \langle\hat{\mathbf{v}}_{r+1},\mathbf{y}\rangle
\end{array}\right]}_{n+1 \text{ columns}}
\begin{array}{l} \left.\right\} r \text{ rows} \\ \left.\right\} r+1 \text{ rows} \end{array}
$$

r columns $\qquad$ $n - r$ columns

$n + 1$ columns

In writing this semitriangular form it is assumed that the first r columns of $\mathbf{A}$ are the r_A independent ones used in the Gram-Schmidt process. The entire last row of the above matrix will be zero with the possible exception of the very last term $\langle\hat{\mathbf{v}}_{r+1}, \mathbf{y}\rangle$. If $\mathbf{y}$ is dependent on the columns of $\mathbf{A}$, then this term will be zero, since then $\hat{\mathbf{v}}_{r+1}$ is exactly zero. This is the case for which there are solutions, since then $r_A = r_W$. When this last inner product is not zero, $r_W > r_A$, so no solutions exist. The last inner product is the component of $\mathbf{y}$ normal to the column space of $\mathbf{A}$. There is no $\mathbf{x}$ vector which will cause $\mathbf{Ax}$ to equal this part of $\mathbf{y}$. The vector $\mathbf{y}$ can be decomposed into a component $\mathbf{y}_p$ parallel to $L(\mathbf{a}_j)$ and a component $\mathbf{y}_e$ normal to $L(\mathbf{a}_j)$, $\mathbf{y} = \mathbf{y}_p + \mathbf{y}_e$. See Figure 6.2.

The best that can be done by choice of $\mathbf{x}$ is to force $\mathbf{Ax} = \mathbf{y}_p$. The unavoidable error committed in doing this is the residual $\mathbf{y} - \mathbf{Ax} = \mathbf{y}_p + \mathbf{y}_e - \mathbf{Ax} = \mathbf{y}_e$. The length or norm of this residual error is the lower corner element in $\mathbf{W}'$, namely $\|\mathbf{y}_e\| = \langle\hat{\mathbf{v}}_{r+1}, \mathbf{y}\rangle$. Thus, a glance at $\mathbf{W}'$ tells whether or not solutions exist, and if they do not, then the magnitude of the smallest possible error is also given.

The solution which satisfies $\mathbf{y}$, or $\mathbf{y}_p$ if necessary, is found from $\mathbf{W}'$ as in Section

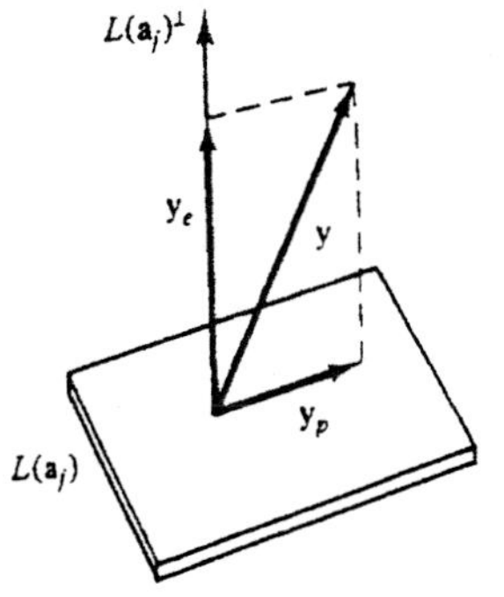

Figure 6.2

6.3, except the final row is ignored. This represents r_A equations in n unknowns. At least one solution always exists, and it will be unique if and only if $r_A = n$. When more than one solution exists, some additional criterion may be used to select one particular solution. These underdetermined problems are discussed further in Section 6.7. In cases where $\mathbf{y}$ has a nonzero component normal to $L(\mathbf{a}_j)$, the procedure given here leads to the least-squares solution (or solutions). This topic is pursued further in Section 6.8.

EXAMPLE 6.4 Analyze the following set of equations using the Gram-Schmidt expansion method (GSE).

$$\begin{bmatrix} 1 & 3 & 2 \\ 2 & 5 & 3 \\ 3 & 7 & 4 \\ 4 & 9 & 5 \end{bmatrix} \begin{bmatrix} x_1 \\ x_2 \\ x_3 \end{bmatrix} = \begin{bmatrix} 1 \\ 0 \\ -1 \\ 1 \end{bmatrix}$$

In this example it is easy to see that column 3 of $\mathbf{A}$ is the difference between columns 2 and 1, so $\mathbf{A}$ has rank 2. Columns 1 and 2 of $\mathbf{A}$ are used, along with $\mathbf{y}$, to form the following orthonormal set:

THE ORTHONORMAL BASIS SET

$$\mathbf{V} = \begin{bmatrix} 1.8257418\text{E}-01 & 8.1649667\text{E}-01 & 3.6514840\text{E}-01 \\ 3.6514837\text{E}-01 & 4.0824848\text{E}-01 & -1.8257420\text{E}-01 \\ 5.4772252\text{E}-01 & 5.8400383\text{E}-07 & -7.3029685\text{E}-01 \\ 7.3029673\text{E}-01 & -4.0824789\text{E}-01 & 5.4772240\text{E}-01 \end{bmatrix}$$

Then $\mathbf{V}^T\mathbf{W} = \mathbf{W}'$.

THE EXPANSION COEFFICIENT VECTOR(S)

$$\mathbf{W}' = \begin{bmatrix} 5.4772258\text{E}+00 & 1.2780193\text{E}+01 & 7.3029675\text{E}+00 & 3.6514840\text{E}-01 \\ 3.9339066\text{E}-06 & 8.1650591\text{E}-01 & 8.1650186\text{E}-01 & 4.0824819\text{E}-01 \\ -9.5367432\text{E}-07 & -2.3841858\text{E}-06 & -1.1920929\text{E}-06 & 1.6431677\text{E}+00 \end{bmatrix}$$

Since the first three entries in the last row and the first entry in row 2 are theoretically zero, a "machine zero" can be defined. Here any number of magnitude less than 4×10^{-6} is set to zero.

From this it is seen that

1. *There is no solution to the given set of equations, since $r_A = 2$ and $r_W = 3$.* The equations are inconsistent.

2. *The best that can be done is to satisfy the column space portion of the equations.* Let this be called a projected solution. If this is done the residual error will have the norm

$$\|\mathbf{Ax} - \mathbf{y}\| = 1.6431677$$

3. *There are an infinite number of projected solutions, all of which will give exactly this same residual error norm.* They all must satisfy (rounded)

$$0.8165(x_2 + x_3) = 0.40825$$

$$5.4772x_1 + 12.7802x_2 + 7.3030x_3 = 0.36515$$

If the row-reduced-echelon (RRE) method of Section 6.3 is applied to this problem instead, the resulting matrix W′ is

$$\mathbf{W}' = \begin{bmatrix} 1 & 0 & -1 & 0 \\ 0 & 1 & 1 & 0 \\ 0 & 0 & 0 & 1 \\ 0 & 0 & 0 & 0 \end{bmatrix}$$

This also indicates that the equations are inconsistent but gives no clue about how closely a solution can be approached. It also indicates, apparently, that if one row of inconsistent equations could be ignored, an infinite set of solutions would exist, and they would all satisfy $x_1 - x_3 = 0$ and $x_2 + x_3 = 0$. Although the RRE method has yielded somewhat less information than the GSE method, it has yielded the one-dimensional null space spanned by $\mathbf{e} = [1 \quad -1 \quad 1]^T$. It can be shown that any constant α times $\mathbf{e}$ can be added to any projected solution $\mathbf{x}_p$ found from the GSE method (or any other method) and the result will still be a projected solution. That is, $\mathbf{x}_p + \alpha\mathbf{e} = \mathbf{x}$ is also a projected solution. ∎

6.6 HOMOGENEOUS LINEAR EQUATIONS

The set of homogeneous equations $\mathbf{Ax} = \mathbf{0}$ always has at least one solution, $\mathbf{x} = \mathbf{0}$. This is true because $\mathbf{A}$ and $\mathbf{W}$ always have the same rank. However, $\mathbf{x} = \mathbf{0}$ is called the *trivial solution*. In order for *nontrivial solutions* to exist, it must be true that $r_A < n$. Of course, if one such nontrivial solution exists, there will be an infinite set of solutions with $n - r_A$ free parameters. The methods of the previous sections apply to the homogeneous case without modification.

It is pointed out that any set of nonhomogeneous equations $\mathbf{Ax} = \mathbf{y}$ can always be written as an equivalent set of homogeneous equations:

$$[\mathbf{A} \mid \mathbf{y}]\begin{bmatrix} \mathbf{x} \\ \hline -1 \end{bmatrix} = \mathbf{0} \quad \text{or} \quad \mathbf{W}\begin{bmatrix} \mathbf{x} \\ \hline -1 \end{bmatrix} = \mathbf{0}$$

EXAMPLE 6.5 Find all nontrivial solutions to the equations $\mathbf{Ax} = \mathbf{0}$, if $\mathbf{A} = \begin{bmatrix} 0 & 2 & 1 \\ 0 & 2 & 1 \\ 0 & -4 & -2 \end{bmatrix}$. The matrix W and its RRE form W′ are

THE MATRIX W

$$\begin{bmatrix} 0.0000000E+00 & 2.0000000E+00 & 1.0000000E+00 & 0.0000000E+00 \\ 0.0000000E+00 & 2.0000000E+00 & 1.0000000E+00 & 0.0000000E+00 \\ 0.0000000E+00 & -4.0000000E+00 & -2.0000000E+00 & 0.0000000E+00 \end{bmatrix}$$

RANK OF W IS 1: THE HERMITE FORM W′ FOLLOWS

$$\begin{bmatrix} 0.0000000E+00 & 1.0000000E-00 & 5.0000000E-01 & 0.0000000E+00 \\ 0.0000000E+00 & 0.0000000E-00 & 0.0000000E-00 & 0.0000000E+00 \\ 0.0000000E+00 & 0.0000000E-00 & 0.0000000E-00 & 0.0000000E+00 \end{bmatrix}$$

Therefore, all nontrivial solutions are linear combinations of the following two vectors, which constitute a basis set for the null space of **A**.

DEGENERACY OF A IS 2; NULL SPACE BASIS IS

$$\begin{bmatrix} -1.0000000E+00 \\ 0.0000000E+00 \\ 0.0000000E+00 \end{bmatrix} \begin{bmatrix} 0.0000000E+00 \\ 5.0000000E-01 \\ -1.0000000E+00 \end{bmatrix}$$

■

6.7 THE UNDERDETERMINED CASE

When the matrix **A** has $m < n$, there is no possibility of a unique solution **x**. The underdetermined case, which has an infinite number of solutions ($r_A = r_W$), is discussed here. The methods of Sections 6.3, 6.4, or 6.5 can be used to find the family of solutions. This section presents a method for singling out one particular solution, the solution with the minimum norm, $\|\mathbf{x}\|$. Problem 6.47 suggests that the procedure can be generalized to various other weighted norms.

The Minimum Norm Solution

Consider the equation $\mathbf{Ax} = \mathbf{y}$, where **A** is $m \times n$ with $m < n$ and with $r_A = r_W$. If $r_A < m$, some rows of **W** are linearly dependent. This means that some of the original equations are redundant and can be deleted without losing information. Assume that these deletions have been made and as a result $r_A = r_W = m$. The conjugate transpose of the m rows of any **A** can be used to define the n-component vectors $\{\mathbf{c}_i, i = 1, \ldots, m\}$. These vectors belong to $\mathcal{X}^n$. The space spanned by the set of $\mathbf{c}_i$ vectors is called the *row space* $L(\mathbf{c}_i)$ of **A**. In general, $L(\mathbf{c}_i)$ will be a subspace of $\mathcal{X}^n$, and here $r_A = m$ means that it is an m-dimensional subspace with the $\mathbf{c}_i$ vectors forming a basis. The space $\mathcal{X}^n$, which contains all possible **x** vectors, can be written as the direct sum

$$\mathcal{X}^n = L(\mathbf{c}_i) \oplus L(\mathbf{c}_i)^{\perp}$$

Every vector in $\mathcal{X}^n$ can be written

$$\mathbf{x} = \mathbf{x}_1 + \mathbf{x}_2 \quad \text{where } \mathbf{x}_1 \in L(\mathbf{c}_i), \mathbf{x}_2 \in L(\mathbf{c}_i)^{\perp}$$

Figure 6.3 illustrates this decomposition for $n = 3$ and $m = 2$.
The norm of **x** satisfies

$$\|\mathbf{x}\|^2 = \|\mathbf{x}_1\|^2 + \|\mathbf{x}_2\|^2$$

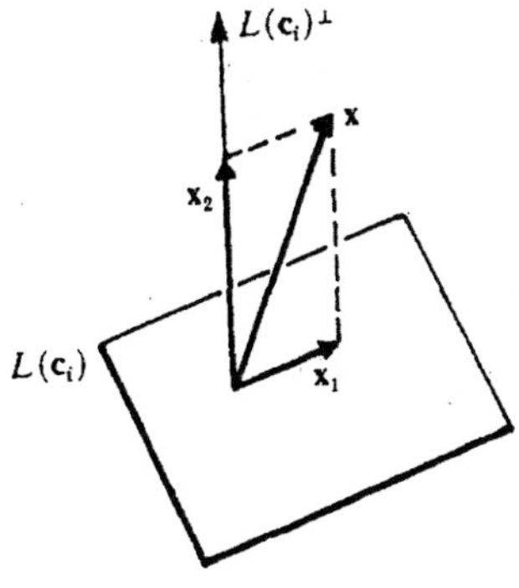

Figure 6.3

Since $\mathbf{x}_2 \in L(\mathbf{c}_i)^\perp$, $\langle \mathbf{c}_i, \mathbf{x}_2 \rangle = 0$ for each $\mathbf{c}_i$, so $\mathbf{A}\mathbf{x}_2 = \mathbf{0}$. Thus

$$\mathbf{A}\mathbf{x} = \mathbf{A}(\mathbf{x}_1 + \mathbf{x}_2) = \mathbf{A}\mathbf{x}_1 = \mathbf{y}$$

For every $\mathbf{x}_1 \in L(\mathbf{c}_i)$, $\mathbf{x}_1 = \sum_{i=1}^{m} \alpha_i \mathbf{c}_i = \overline{\mathbf{A}}^T \boldsymbol{\alpha}$. Using $\mathbf{A}\mathbf{x}_1 = \mathbf{y}$ gives

$$\mathbf{A}\overline{\mathbf{A}}^T \boldsymbol{\alpha} = \mathbf{y}$$

But $\mathbf{A}\overline{\mathbf{A}}^T$ is an $m \times m$ matrix with rank m and is therefore nonsingular. Solving for $\boldsymbol{\alpha}$ gives $\boldsymbol{\alpha} = (\mathbf{A}\overline{\mathbf{A}}^T)^{-1}\mathbf{y}$ and therefore $\mathbf{x}_1 = \overline{\mathbf{A}}^T\boldsymbol{\alpha} = \overline{\mathbf{A}}^T(\mathbf{A}\overline{\mathbf{A}}^T)^{-1}\mathbf{y}$. This $\mathbf{x}_1$ is the unique $\mathbf{x} \in L(\mathbf{c}_i)$, which satisfies $\mathbf{A}\mathbf{x} = \mathbf{y}$. From the norm relations, it is clear that $\mathbf{x}_1$ is the minimum norm solution, since any other solution must have a component in $L(\mathbf{c}_i)^\perp$ and this would increase the norm. The minimum norm solution then is

$$\mathbf{x} = \overline{\mathbf{A}}^T(\mathbf{A}\overline{\mathbf{A}}^T)^{-1}\mathbf{y}$$

This result can also be derived by using Lagrange multipliers [3] and straightforward minimization of $\|\mathbf{x}\|^2$ subject to the constraint $\mathbf{A}\mathbf{x} - \mathbf{y} = \mathbf{0}$. Although it is not pursued here, many other classes of problems amount to using the minimum (or maximum) of a cost function to single out one desirable solution from the infinite set available in the underdetermined case. If the cost function is linear, a linear programming problem results. The minimum norm problem is an example of quadratic programming. Other nonlinear programming problems can also be posed.

EXAMPLE 6.6 The minimum norm solution of $\begin{bmatrix} 1 & 1 & 0 \\ 0 & 0 & 1 \end{bmatrix}\begin{bmatrix} x_1 \\ x_2 \\ x_3 \end{bmatrix} = \begin{bmatrix} 1 \\ 4 \end{bmatrix}$ is

$$\mathbf{x}_1 = \begin{bmatrix} 1 & 0 \\ 1 & 0 \\ 0 & 1 \end{bmatrix}\begin{bmatrix} 2 & 0 \\ 0 & 1 \end{bmatrix}^{-1}\begin{bmatrix} 1 \\ 4 \end{bmatrix} = \frac{1}{2}\begin{bmatrix} 1 \\ 1 \\ 8 \end{bmatrix}$$ ∎

EXAMPLE 6.7 Find the minimum norm solution to the projected problem derived in Example 6.4. From the previous analysis the projected problem is

$$\begin{bmatrix} 5.4772 & 12.7802 & 7.3030 \\ 0 & 0.8165 & 0.8165 \end{bmatrix}\begin{bmatrix} x_1 \\ x_2 \\ x_3 \end{bmatrix} = \begin{bmatrix} 0.3651 \\ 0.4082 \end{bmatrix}$$

A direct calculation of $\mathbf{x} = \mathbf{A}^T[\mathbf{A}\mathbf{A}^T]^{-1}\mathbf{y}$ can be made in this simple case. Alternately one can form

$$\mathbf{A}\mathbf{A}^T = \begin{bmatrix} 246.6670 & 16.3979 \\ 16.3979 & 1.3333 \end{bmatrix}$$

and then solve $\mathbf{A}\mathbf{A}^T\mathbf{x}' = \mathbf{y}$ for $\mathbf{x}'$. Then, finally, $\mathbf{x} = \mathbf{A}^T\mathbf{x}'$. The advantage of doing this is that a matrix inversion is not directly needed, and a routine such as the RRE package can instead be used to solve for $\mathbf{x}'$. This is the approach used here.

$$\mathbf{W} = \left[\begin{array}{cc|c} 246.6670 & 16.3979 & 0.3651 \\ 16.3979 & 1.3333 & 0.4082 \end{array}\right] \qquad \mathbf{W}' = \left[\begin{array}{cc|c} 1 & 0 & -0.10346 \\ 0 & 1 & 1.57862 \end{array}\right]$$

Therefore, $\mathbf{x}' = \begin{bmatrix} -0.10346 \\ 1.57862 \end{bmatrix}$ and $\mathbf{x} = \begin{bmatrix} -0.5667 \\ -0.0334 \\ 0.5334 \end{bmatrix}$. ■

.8 THE OVERDETERMINED CASE

When there are more equations than unknowns, the $m \times n$ coefficient matrix $\mathbf{A}$ has $m > n$. If the equations are inconsistent, no solution exists. This situation often arises because of inaccuracies in measuring the components of the $\mathbf{y}$ vector, or because the relationship assumed to exist between $\mathbf{x}$ and $\mathbf{y}$, as expressed by $\mathbf{A}$, is oversimplified or wrong. *Approximate* solution vectors $\mathbf{x}$ are desired in this case. Three approaches are presented. The first method ignores some equations and places total reliance on those remaining. The second method (least squares) places equal reliance on all equations with the hope that the errors will average out. The third method (weighted least squares) uses all of the equations but weights some more heavily than others. An alternative computational procedure (recursive weighted least squares) is also given for obtaining the latter two approximations.

A considerable amount of information, which is useful for the overdetermined case, has already been given. The GSE method of Section 6.5 applies to this case, as already demonstrated. When the problem is overdetermined, $\langle \hat{\mathbf{v}}_{r+1}, \mathbf{y} \rangle \neq 0$. The so-called projected solution is then sought. If the projected solution is nonunique, the minimum norm solution is often singled out, as was done in Example 6.4 and continued in Example 6.7. This combination of GSE plus minimum norm solution always gives a solution to Eq. (*6.1*). It is true for the underdetermined, overdetermined, or uniquely determined cases. The only difficulty that might remain on a machine solution is the ability to recognize the difference between 0 and a very small number, or the difference between vectors that are linearly dependent or nearly linearly dependent. The notion of machine zero was introduced, and an example of how it can be determined on a given problem has been given. The GSE–minimum norm solution combination has many things in common with the method of *singular value decomposition,* but there are also some unique differences.

In this section a more traditional approach to the overdetermined problem is presented. It is assumed that $\mathbf{A}$ is of full rank n and that $m > n$.

Ignore Some Equations

If a subset of n equations is selected and the remaining $m - n$ are ignored, an approximate solution can be obtained. The basis for ignoring certain equations is a subjective matter. Perhaps certain equations are more reliable for one reason or another. Perhaps other results are obviously "wild points" and can be discarded. If $\mathbf{A}_1$ is a nonsingular $n \times n$ matrix formed by deleting rows from $\mathbf{A}$ and if $\mathbf{y}_1$ is the $n \times 1$ vector obtained from $\mathbf{y}$ by deleting the corresponding elements, then a result which satisfies n of the original equations is

$$\mathbf{x} = \mathbf{A}_1^{-1} \mathbf{y}_1$$

Least-Squares Approximate Solution

If all the equations are used correctly, errors may tend to average out, and a good approximation for $\mathbf{x}$ results. Since no one $\mathbf{x}$ can satisfy all the simultaneous equations, it is inappropriate to write the equality $\mathbf{Ax} = \mathbf{y}$. Rather, an $n \times 1$ error vector $\mathbf{e}$ is introduced:

$$\mathbf{e} = \mathbf{y} - \mathbf{Ax}$$

The least-squares approach yields the one $\mathbf{x}$ which minimizes the sum of the squares of the e_i components. That is, $\mathbf{x}$ is chosen to minimize

$$\|\mathbf{e}\|^2 = \mathbf{e}^T \mathbf{e} = (\mathbf{y} - \mathbf{Ax})^T(\mathbf{y} - \mathbf{Ax})$$

The vectors $\mathbf{e}$, $\mathbf{y}$, and $\mathbf{Ax}$ all belong to $\mathscr{X}^m$. But $\mathbf{Ax}$ belongs to the *column space* of $\mathbf{A}$, the space spanned by the columns $\mathbf{a}_j$ of $\mathbf{A}$, denoted by $L(\mathbf{a}_j)$. $\mathscr{X}^m$ is decomposed, as shown in Figure 6.1, into

$$\mathscr{X}^m = L(\mathbf{a}_j) \oplus L(\mathbf{a}_j)^{\perp}$$

The error has a unique decomposition:

$$\mathbf{e} = \mathbf{e}_1 + \mathbf{e}_2, \qquad \mathbf{e}_1 \in L(\mathbf{a}_j), \qquad \mathbf{e}_2 \in L(\mathbf{a}_j)^{\perp}$$

(The vector $\mathbf{e}_2$ is the $\mathbf{y}_e$ vector of Section 6.5). The norm of $\mathbf{e}$ satisfies $\|\mathbf{e}\|^2 = \|\mathbf{e}_1\|^2 + \|\mathbf{e}_2\|^2$.

Since $\mathbf{y}$ is given and since $\mathbf{Ax} \in L(\mathbf{a}_j)$, the choice of $\mathbf{x}$ cannot affect $\mathbf{e}_2$. The least-squares solution vector $\mathbf{x}$ is the one for which $\|\mathbf{e}_1\|^2 = 0$, so $\mathbf{e}_1 = \mathbf{0}$. This means that the projection of $\mathbf{y}$ on $L(\mathbf{a}_j)$, call it $\mathbf{y}_p$, must equal $\mathbf{Ax} = \sum_{j=1}^{n} x_j\, \mathbf{a}_j$. Since $r_A = n$, the columns $\mathbf{a}_j$ form a *basis* for $L(\mathbf{a}_j)$ so that $\mathbf{y}_p = \sum_{j=1}^{n} \alpha_j\, \mathbf{a}_j$. Because of the uniqueness of this expansion, $\alpha_j = x_j$, that is, $\mathbf{x} = \boldsymbol{\alpha}$. The set of n reciprocal basis vectors $\mathbf{r}_i$ is defined by $\langle \mathbf{r}_i, \mathbf{a}_j \rangle = \delta_{ij}$, or in matrix form

$$\underset{(n \times m)}{\mathbf{R}} \cdot \underset{(m \times n)}{\mathbf{A}} = \mathbf{I}$$

Since $\mathbf{A}$ is not square, it cannot be inverted to find $\mathbf{R}$. It is still true that $\alpha_j = \langle \mathbf{r}_j, \mathbf{y} \rangle = x_j$ so that

$$\boldsymbol{\alpha} = \mathbf{Ry} = \mathbf{x} \tag{6.2}$$

Therefore, $\mathbf{Ax} = \mathbf{ARy} = \mathbf{y}_p$. Using $\mathbf{e}_2 = \mathbf{y} - \mathbf{y}_p$ and the fact that $\langle \mathbf{a}_j, \mathbf{e}_2 \rangle = 0$ gives $\mathbf{A}^T[\mathbf{y} - \mathbf{ARy}] = \mathbf{0}$, or

$$\mathbf{Ry} = (\mathbf{A}^T \mathbf{A})^{-1} \mathbf{A}^T \mathbf{y} \tag{6.3}$$

Combining Eqs. (*6.2*) and (*6.3*) gives the least-squares solution

$$\mathbf{x} = (\mathbf{A}^T \mathbf{A})^{-1} \mathbf{A}^T \mathbf{y}$$

The amount of error in this approximate solution is indicated by

$$\|\mathbf{e}\|^2 = \|\mathbf{e}_2\|^2 = \mathbf{y}^T[\mathbf{I} - \mathbf{A}(\mathbf{A}^T \mathbf{A})^{-1} \mathbf{A}^T]\mathbf{y} = \|\mathbf{y} - \mathbf{Ax}\|^2$$

Recall that the square root of this quantity was given directly in the GSE method.

The matrix $\mathbf{R} = (\mathbf{A}^T\mathbf{A})^{-1}\mathbf{A}^T$ is a particular example of the *generalized* or *pseudo-inverse* [5] of $\mathbf{A}$, written $\mathbf{A}^{\dagger}$. If $\mathbf{A}^{-1}$ exists, then $\mathbf{A}^{\dagger} = \mathbf{A}^{-1}$ and $\|\mathbf{e}\|^2 = 0$. The minimum norm solution of Section 6.7 provides another example of the pseudo-inverse that was appropriate to those circumstances, namely, $\mathbf{A}^{\dagger} = \overline{\mathbf{A}}^T(\mathbf{A}\overline{\mathbf{A}}^T)^{-1}$. The general solutions to the $n \times n$ nonsingular case, the underdetermined minimum norm case, and the over-determined least-squares case can all be expressed in terms of the pseudo-inverse as $\mathbf{x} = \mathbf{A}^{\dagger}\mathbf{y}$.

Weighted Least-Squares Approximation to the Solution‡

Ignoring some equations or placing equal reliance on all equations represents two extremes. If some equations are more reliable than others, but all equations are to be retained, a weighted least-squares approximation can be used. That is, $\mathbf{x}$ should minimize $\mathbf{e}^T\mathbf{R}^{-1}\mathbf{e} = (\mathbf{y} - \mathbf{A}\mathbf{x})^T\mathbf{R}^{-1}(\mathbf{y} - \mathbf{A}\mathbf{x})$. $\mathbf{R}^{-1}$ is symmetric, $m \times m$, nonsingular, and often diagonal. Those familiar with random processes should know that $\mathbf{R}$ is generally selected as the covariance matrix for the noise on the vector $\mathbf{y}$. Smaller values of r_{ii} will cause e_i^2 to be smaller, and the ith equation is more nearly satisfied. If a norm $\|\mathbf{e}\|^2_{\mathbf{R}^{-1}} = \mathbf{e}^T\mathbf{R}^{-1}\mathbf{e}$ is defined, the method of orthogonal projections immediately leads to

$$\mathbf{A}^T\mathbf{R}^{-1}\mathbf{A}\mathbf{x} = \mathbf{A}^T\mathbf{R}^{-1}\mathbf{y}$$

If $r_A = n$, as assumed here, the $n \times n$ matrix $\mathbf{A}^T\mathbf{R}^{-1}\mathbf{A}$ is nonsingular and the weighted least-squares solution is

$$\mathbf{x} = (\mathbf{A}^T\mathbf{R}^{-1}\mathbf{A})^{-1}\mathbf{A}^T\mathbf{R}^{-1}\mathbf{y}$$

Notice that if $\mathbf{A}$ is not full rank, the required inverse will not exist, signaling that the least-squares solution is not unique.

The least-squares and weighted least-squares formulas can also be derived simply by setting $\partial\|\mathbf{e}\|^2/\partial x_i = 0$, using results of Sec. 4.12.

Recursive Weighted Least-Squares Solutions

The preceding sections dealt with what is commonly called "batch least squares," because all data equations are treated in one batch. A recursive method of using each new set of data as it is received is now presented.

Assume that a set of m equations

$$\mathbf{y}_k = \mathbf{A}\mathbf{x} + \mathbf{e}$$

has been used to obtain a weighted least-squares estimate for $\mathbf{x}$, denoted by $\mathbf{x}_k$:

$$\mathbf{x}_k = (\mathbf{A}^T\mathbf{R}^{-1}\mathbf{A})^{-1}\mathbf{A}^T\mathbf{R}^{-1}\mathbf{y}_k$$

As is often the case, assume that an additional set of relations

$$\mathbf{y}_{k+1} = \mathbf{H}_{k+1}\mathbf{x} + \mathbf{e}_{k+1}$$

‡The matrix $\mathbf{R}$ in this section is unrelated to the matrix of reciprocal basis vectors of previous sections.

then becomes available. It is desired to obtain a new estimate for $\mathbf{x}$, denoted as $\mathbf{x}_{k+1}$, which combines both sets of data and minimizes

$$J = [\mathbf{e}^T \quad \mathbf{e}_{k+1}^T]\left[\begin{array}{c|c} \mathbf{R}^{-1} & \mathbf{0} \\ \hline \mathbf{0} & \mathbf{R}_{k+1}^{-1} \end{array}\right]\left[\begin{array}{c} \mathbf{e} \\ \hline \mathbf{e}_{k+1} \end{array}\right]$$

$\mathbf{R}_{k+1}^{-1}$ is the weighting matrix, analogous to $\mathbf{R}^{-1}$ but applied to the new data $\mathbf{y}_{k+1}$. It is not necessary to reprocess the whole set of equations involving $[\mathbf{y}_k \mid \mathbf{y}_{k+1}]$ in order to determine $\mathbf{x}_{k+1}$. It is shown in Problem 6.12, using partitioned matrices and a matrix inversion identity, that

$$\mathbf{x}_{k+1} = \mathbf{x}_k + \mathbf{K}_k[\mathbf{y}_{k+1} - \mathbf{H}_{k+1}\mathbf{x}_k]$$

where $\mathbf{K}_k = \mathbf{P}_k\mathbf{H}_{k+1}^T[\mathbf{H}_{k+1}\mathbf{P}_k\mathbf{H}_{k+1}^T + \mathbf{R}_{k+1}]^{-1}$ and $\mathbf{P}_k \triangleq (\mathbf{A}^T\mathbf{R}^{-1}\mathbf{A})^{-1}$, which is available from the computation of $\mathbf{x}_k$. If still other sets of equations are to be incorporated, the above relations can be used recursively. A new matrix, $\mathbf{P}_{k+1}$, is then needed and is given by

$$\mathbf{P}_{k+1} = [\mathbf{P}_k^{-1} + \mathbf{H}_{k+1}^T\mathbf{R}_{k+1}^{-1}\mathbf{H}_{k+1}]^{-1} \tag{6.4}$$

Using the matrix inversion lemma, page 132, this can also be written as

$$\mathbf{P}_{k+1} = \mathbf{P}_k - \mathbf{P}_k\mathbf{H}_{k+1}^T[\mathbf{H}_{k+1}\mathbf{P}_k\mathbf{H}_{k+1}^T + \mathbf{R}_{k+1}]^{-1}\mathbf{H}_{k+1}\mathbf{P}_k \tag{6.5}$$

The latter form is often more convenient. For example, if $\mathbf{y}_{k+1}$ is a scalar, then matrix inversion is not required, just a scalar division.

Data Deweighting

A common occurrence is that data are received sequentially over time. As each new group of data is received, it is used to improve the estimate of $\mathbf{x}$. If this process is carried out over a sufficient number of steps, one will find that the $\mathbf{P}_{k+1}$ matrix has decreased to very small values due to the repeated addition of a nonnegative term to its inverse in Eq. (6.4). This in turn will cause the value of $\mathbf{K}_{k+1}$ to become small. This means that the corrections made to $\mathbf{x}_k$ in order to determine $\mathbf{x}_{k+1}$ get small, independent of what new or surprising information may be contained in the latest measurements. In order to prevent the recursive estimator from failing to respond adequately to new data (called going to sleep), some form of data deweighting is often used. Two types will be presented here, additive deweighting and multiplicative deweighting. In both cases the $\mathbf{P}$ matrix is prevented from getting too small. The easiest way to change the former algorithm is to introduce another matrix $\mathbf{M}_k$, given by

$$\mathbf{M}_k = \mathbf{P}_k/\beta \quad \text{with } \beta < 1\text{; this is multiplicative deweighting}$$

or

$$\mathbf{M}_k = \mathbf{P}_k + \mathbf{Q} \quad \text{with } \mathbf{Q} \text{ a positive definite matrix; this is additive deweighting}$$

The gain is now computed as

$$\mathbf{K}_k = \mathbf{M}_k\mathbf{H}_{k+1}^T[\mathbf{H}_{k+1}\mathbf{M}_k\mathbf{H}_{k+1}^T + \mathbf{R}]^{-1}$$

and the new $\mathbf{P}_{k+1}$ is given by either Eq. (6.4) or (6.5), but with $\mathbf{P}_k$ on the right-hand side replaced everywhere by $\mathbf{M}_k$. The formula for updating the estimate of $\mathbf{x}$ remains the same. If $\beta = 1$ or $\mathbf{Q} = 0$, both of these deweighting schemes revert to the original algorithm. Values used for the so-called forgetting factor β depend on how much deweighting is desired. A concept called asymptotic sample length, ASL, is a measure of how much past data are having a significant effect on the current estimate of $\mathbf{x}$. A relation between ASL and β is

$$\text{ASL} = 1/(1 - \beta)$$

Therefore, the commonly used values of β between 0.999 and 0.95 correspond to asymptotic sample lengths of 1000 past measurements down to 20. Although $\mathbf{Q}$ is often selected as a diagonal matrix with small diagonal elements, an idea of the appropriate magnitudes can be obtained by assuming that $\mathbf{Q} = \alpha\mathbf{P}$, with α a scalar. Then comparison of the two forms of deweighting shows that $\alpha = (1/\beta) - 1$. Therefore, to get an ASL of 1000, α would be 0.001 or $\mathbf{Q}$ should be about 0.1% of $\mathbf{P}$. Of course, since $\mathbf{P}$ is changing, this kind of comparison is not perfect. It does indicate roughly that a small $\mathbf{Q}$ matrix can be an effective deweighting scheme. Figure 6.4 shows the relative weight applied to past measurements when forming the current estimate of $\mathbf{x}$. The curve is normalized so that the current measurement weight is one. This is computed by using

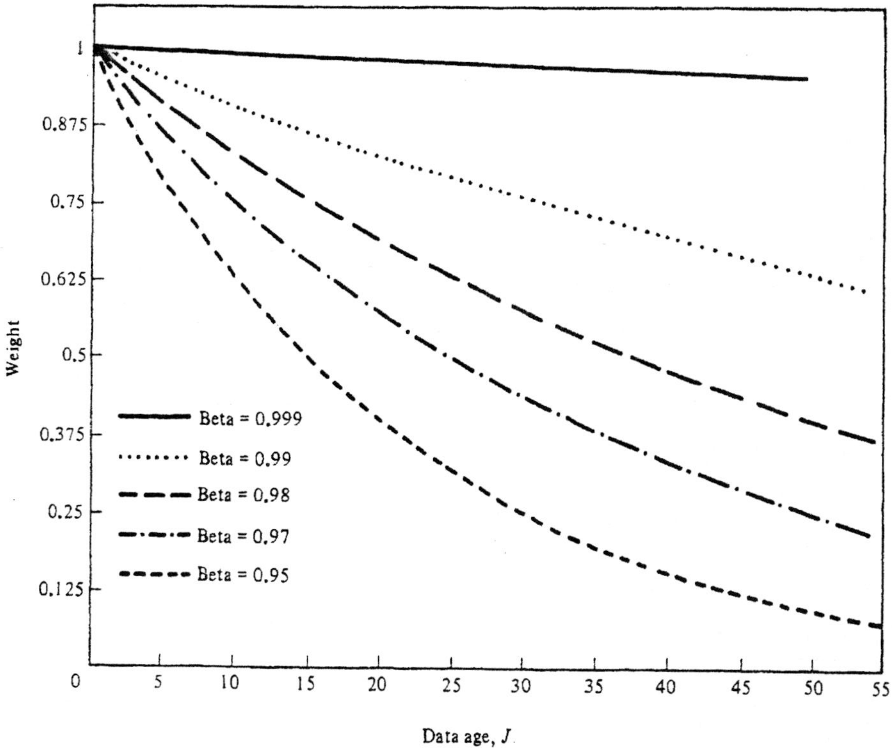

Figure 6.4 Deweighting for common forgetting factors

the recursive estimation algorithm backwards in time for a scalar **x**. Qualitatively similar deweighting occurs in the vector case, but it is harder to normalize and display.

If the matrices **P**, **Q**, and **R** are given the appropriate statistical interpretation as covariance matrices of certain signals, then the recursive equations constitute a simple example of the discrete *Kalman filtering* equations. The Kalman filter is used extensively in modern control theory in order to estimate the internal (state) variables of a linear system based on noisy measurements of the output variables [6, 7]. A deterministic state estimation procedure, the *observer*, is presented in Chapter 13. The least-squares approach finds many other applications in modern control theory as well.

6.9 TWO BASIC PROBLEMS IN CONTROL THEORY

Consider the discrete-time state equations of Chapter 3,

$$\mathbf{x}(k+1) = \mathbf{A}\mathbf{x}(k) + \mathbf{B}\mathbf{u}(k)$$

$$\mathbf{y}(k+1) = \mathbf{C}\mathbf{x}(k+1) + \mathbf{D}\mathbf{u}(k+1)$$

Let the initial conditions for the state vector be $\mathbf{x}(0)$. Then the states at succeeding time points are

$$\mathbf{x}(1) = \mathbf{A}\mathbf{x}(0) + \mathbf{B}\mathbf{u}(0)$$

$$\begin{aligned}\mathbf{x}(2) &= \mathbf{A}\mathbf{x}(1) + \mathbf{B}\mathbf{u}(1)\\ &= \mathbf{A}[\mathbf{A}\mathbf{x}(0) + \mathbf{B}\mathbf{u}(0)] + \mathbf{B}\mathbf{u}(1) = \mathbf{A}^2\mathbf{x}(0) + \mathbf{A}\mathbf{B}\mathbf{u}(0) + \mathbf{B}\mathbf{u}(1)\end{aligned}$$

Continuing in this fashion the state at a general time point k is found to be

$$\begin{aligned}\mathbf{x}(k) = \mathbf{A}^k\mathbf{x}(0) &+ \mathbf{B}\mathbf{u}(k-1) + \mathbf{A}\mathbf{B}\mathbf{u}(k-2) + \mathbf{A}^2\mathbf{B}\mathbf{u}(k-3) + \cdots\\ &+ \mathbf{A}^{k-2}\mathbf{B}\mathbf{u}(1) + \mathbf{A}^{k-1}\mathbf{B}\mathbf{u}(0)\end{aligned}$$

6.9.1 A CONTROL PROBLEM

One of the two basic control problems is the determination of a sequence of control inputs $\mathbf{u}(i)$ which will transfer a known initial state vector $\mathbf{x}(0)$ to the origin of state space at some finite time point k. It is convenient to stack up all the unknown control vectors into one composite vector $\mathbf{U}$ defined as

$$\mathbf{U} = \begin{bmatrix} \mathbf{u}(k-1) \\ \mathbf{u}(k-2) \\ \mathbf{u}(k-3) \\ \vdots \\ \mathbf{u}(1) \\ \mathbf{u}(0) \end{bmatrix}$$

Also define $\mathbf{P} = [\mathbf{B} \mid \mathbf{AB} \mid \mathbf{A}^2\mathbf{B} \mid \dots \mid \mathbf{A}^{k-1}\mathbf{B}]$. Then $\mathbf{x}(k) = \mathbf{A}^k\mathbf{x}(0) + \mathbf{PU}$. Setting the final state to zero gives a set of simultaneous linear equations for the unknown controls $\mathbf{U}$:

$$\mathbf{PU} = -\mathbf{A}^k\mathbf{x}(0)$$

Under what conditions will there be a control sequence $\mathbf{U}$ that will drive a *given* $\mathbf{x}(0)$ to zero? How about an *arbitrary* $\mathbf{x}(0)$? If there is a solution vector $\mathbf{U}$, is it unique? The composite partitioned matrix $\mathbf{P}$ plays the role of the general matrix $\mathbf{A}$, which was discussed throughout most of this chapter. If the state vector has n components and if each $\mathbf{u}(i)$ has r components, then $\mathbf{B}$ is $n \times r$, as are $\mathbf{AB}$ and any power of $\mathbf{A}$ times $\mathbf{B}$. There are k partitions of this type in $\mathbf{P}$, so $\mathbf{P}$ is an $n \times (kr)$ matrix; of course, $\mathbf{U}$ has kr unknown components. If the $n \times 1$ vector $-\mathbf{A}^k\mathbf{x}(0)$ happens to be a linear combination of the columns of $\mathbf{P}$, then a solution for $\mathbf{U}$ will exist. In order for this to be true for any arbitrary vector $-\mathbf{A}^k\mathbf{x}(0)$, and hence for any arbitrary $\mathbf{x}(0)$ initial condition, it is necessary that the range space of $\mathbf{P}$ must span the n-dimensional state space. This means that there are n linearly independent vectors among the columns of $\mathbf{P}$, and therefore the rank of $\mathbf{P}$ must be n. If $\mathbf{P}$ has full rank n, there will be solutions, but are they unique? And what about the final time index k? If $k = 1$, then $\mathbf{P} \equiv \mathbf{B}$ and $\text{rank}(\mathbf{P}) \leq r$. If the value of r is less than n, it is clear that no solutions can exist for arbitrary $\mathbf{x}(0)$, although they may for certain $\mathbf{x}(0)$. As more time is allowed (i.e., as k increases) the matrix $\mathbf{P}$ contains more columns, and its rank may increase. It is shown in Chapter 8 that the rank of $\mathbf{P}$ will never increase beyond the value achieved for $k = n$, the number of states, because further partitions will always be linear combinations of the first n. For this reason the test of whether controls can be found which will drive an arbitrary $\mathbf{x}(0)$ to zero in finite time is equivalent to testing whether $\text{rank}[\mathbf{B} \mid \mathbf{AB} \mid \mathbf{A}^2\mathbf{B} \mid \dots \mid \mathbf{A}^{n-1}\mathbf{B}] = n$. If a particular system passes this test, it is said to be *controllable*. This is pursued in more detail in Chapter 11. Assume that this controllability condition is met. In general the sequence of controls making up $\mathbf{U}$ is not unique for a given final time k. Many optional control sequences might all bring the final state to the origin. One desirable solution might be the minimum norm solution of Section 6.7. Let the subscript k indicate explicitly how many control cycles are being used in $\mathbf{U}$. Then

$$\mathbf{U}_k = -\mathbf{P}_k^T[\mathbf{P}_k\mathbf{P}_k^T]^{-1}\mathbf{A}^k\mathbf{x}(0)$$

Even though the rank of $\mathbf{P}_k$ will not increase for $k > n$, the amount of control effort required to drive the initial state to zero will change in general. The minimum norm squared is

$$\|\mathbf{U}_k\|^2 = \mathbf{x}(0)^T(\mathbf{A}^k)^T[\mathbf{P}_k\mathbf{P}_k^T]^{-1}\mathbf{A}^k\mathbf{x}(0)$$

The $\mathbf{A}^k$ terms could increase or decrease with k depending upon the system stability. The term inside the inverse is

$$\mathbf{PP}^T = \mathbf{BB}^T + \mathbf{ABB}^T\mathbf{A}^T + \cdots + \mathbf{A}^{k-1}\mathbf{BB}^T[\mathbf{A}^{k-1}]^T$$

It would be expected to grow with k as more terms are added to the sum. Thus the inverse itself will diminish. It seems intuitive that if the system is stable so that its unforced state is decaying toward zero, not much control effort will be required to help the state reach zero if k is large. This is borne out by the preceding results.

6.9.2 A State Estimation Problem

The same discrete-time system is considered, but now the inputs $\mathbf{u}(i)$ are all assumed known. The question to be addressed is whether the *unknown* initial state vector $\mathbf{x}(0)$ can be determined from knowledge of the sequence of output vectors $\mathbf{y}(j)$ for $j = 0, 1, \ldots, k$. If the first $k+1$ output vectors are stacked up into one composite vector $\mathbf{Y}_k = [\mathbf{y}(0)^T \;\; \mathbf{y}(1)^T \;\; \ldots \;\; \mathbf{y}(k)^T]^T$, the preceding results can be used to write

$$\mathbf{Y}_k = \begin{bmatrix} \mathbf{C} \\ \mathbf{CA} \\ \mathbf{CA}^2 \\ \vdots \\ \mathbf{CA}^k \end{bmatrix} \mathbf{x}(0) + \begin{bmatrix} \mathbf{D} & \mathbf{0} & \mathbf{0} & \mathbf{0} & \cdots & \mathbf{0} \\ \mathbf{CB} & \mathbf{D} & \mathbf{0} & \mathbf{0} & \cdots & \mathbf{0} \\ \mathbf{CAB} & \mathbf{CB} & \mathbf{D} & \mathbf{0} & \cdots & \mathbf{0} \\ \vdots & & & & & \\ \mathbf{CA}^{k-1}\mathbf{B} & \mathbf{CA}^{k-2}\mathbf{B} & & & \cdots & \mathbf{D} \end{bmatrix} \begin{bmatrix} \mathbf{u}(0) \\ \mathbf{u}(1) \\ \mathbf{u}(2) \\ \vdots \\ \mathbf{u}(k) \end{bmatrix}$$

Since $\mathbf{A}$, $\mathbf{B}$, $\mathbf{C}$, and $\mathbf{D}$ as well as all the input terms $\mathbf{u}(i)$ are assumed known, they can be brought to the left side of the equation to define a new vector $\mathbf{Y}_k'$ or all $\mathbf{u}(i)$ can be assumed zero without loss of generality. This leaves a set of simultaneous linear equations relating outputs in $\mathbf{Y}_k$ (or $\mathbf{Y}_k'$) and the unknown initial state $\mathbf{x}(0)$. Define the matrix $\mathbf{Q}_k^T = [\mathbf{C}^T \mid \mathbf{A}^T\mathbf{C}^T \mid (\mathbf{A}^2)^T\mathbf{C}^T \mid \ldots \mid (\mathbf{A}^k)^T\mathbf{C}^T]$. The simultaneous equations then become $\mathbf{Y}_k = \mathbf{Q}_k\,\mathbf{x}(0)$. If each $\mathbf{y}(i)$ vector has m components, the $\mathbf{Q}_k$ matrix will be of dimension $mk \times n$. The maximum rank is n. Just as with the previous matrix $\mathbf{P}$, it is shown in Chapter 8 that the rank of $\mathbf{Q}_k$ will not increase for values of k larger than $n-1$. If $\mathbf{Q}_k$ achieves its full rank n, then the $n \times n$ matrix $\mathbf{Q}_k^T\,\mathbf{Q}_k$ will be invertible. If this is true, then

$$\mathbf{x}(0|k) = [\mathbf{Q}_k^T\,\mathbf{Q}_k]^{-1}\mathbf{Q}_k^T\,\mathbf{Y}_k$$

Is this the unique solution for $\mathbf{x}(0)$, or is it merely a least-squares approximation for $\mathbf{x}(0)$? If the vector $\mathbf{Y}_k$ belongs to the n-dimensional range space of $\mathbf{Q}_k$, then this is indeed the solution. This is just another way of saying that even though there may be more equations than unknowns, they are *not* inconsistent. However, since the vector $\mathbf{Y}_k$ has mk components, it is very possible that $\mathbf{Y}_k$ might not lie in the n-dimensional range subspace, perhaps because of measurement errors or modeling errors in $\mathbf{A}$, $\mathbf{B}$, $\mathbf{C}$, or $\mathbf{D}$. In that case the equation for $\mathbf{x}(0|k)$ is a least-squares approximate solution for $\mathbf{x}(0)$ based on the k available data points. Note that the existence of a unique least-squares answer also requires that $\mathbf{Q}_k$ be of full rank n. The condition that

$$[\mathbf{C}^T \mid \mathbf{A}^T\mathbf{C}^T \mid (\mathbf{A}^2)^T\mathbf{C}^T \mid \ldots \mid (\mathbf{A}^{n-1})^T\mathbf{C}^T] \quad \text{has rank} \quad n$$

is necessary if $\mathbf{x}(0)$ is to be found from the observed data. A system which meets this criterion is said to be *observable*. Chapter 11 looks into this further. The intent here is to stress the importance of the theory of simultaneous linear equations. Problem 6.17 develops the equations for recursively updating the estimate of $\mathbf{x}(0)$ each time a new noisy measurement becomes available. This gives an improved estimate $\mathbf{x}(0|k+1)$ by using the newest measurement $\mathbf{y}(k+1)$ to add a correction to $\mathbf{x}(0|k)$. Problem 6.18

modifies these results so that estimates of the *current* state $\mathbf{x}(k)$ are recursively computed, rather than estimates of the initial state.

10 LYAPUNOV EQUATIONS

Another type of equation which occurs in control theory is

$$\mathbf{XA} + \mathbf{BX} = \mathbf{C} \tag{6.6}$$

where **A**, **B**, and **C** are known matrices and **X** is a matrix of unknowns. Assume that the dimensions of **A**, **B**, **C**, and **X** are $m \times m$, $n \times n$, $n \times m$, and $n \times m$, respectively. This is usually called the *Lyapunov equation*. It is linear in the unknowns, but it is of an entirely different type than those that have been treated previously. For small-dimensional problems it is not difficult to expand the equation into component form and thus obtain simultaneous equations in the unknown x_{ij} components. However, a more general procedure can also be developed using the concept of vectorized matrices, which was presented in Sec. 4.12. The columns of the unknown matrix **X** are stacked into a single column, referred to as (**X**). The two matrix products involving **X** can be epxressed in terms of the vectorized (**X**) by using the Kronecker product of Chapter 4. Specifically,

$$(\mathbf{XA}) = [\mathbf{A}^T \otimes \mathbf{I}_n](\mathbf{X})$$

$$(\mathbf{BX}) = [\mathbf{I}_m \otimes \mathbf{B}](\mathbf{X})$$

so that the total vectorized equation can be written

$$\{[\mathbf{A}^T \otimes \mathbf{I}_n] + [\mathbf{I}_m \otimes \mathbf{B}]\}(\mathbf{X}) = (\mathbf{C})$$

Of course, the vectors (**X**) and (**C**) are $nm \times 1$ columns, and the coefficient matrix **Q** created with the two Kronecker products is of size $nm \times nm$. While the size of the problem has seemingly been multiplied, what has been accomplished is the positioning of both unknown **X** terms on the same side of a known square matrix **Q.** If **Q** has an inverse, the unique solution for **X** is expressible, still in vectorized form, as

$$(\mathbf{X}) = \{[\mathbf{A}^T \otimes \mathbf{I}_n] + [\mathbf{I}_m \otimes \mathbf{B}]\}^{-1}(\mathbf{C}) = \mathbf{Q}^{-1}(\mathbf{C})$$

The matrix **X** is then recreated by undoing the vectorization process.

EXAMPLE 6.8 Let $\mathbf{A} = \begin{bmatrix} 0 & 1 \\ -2 & -3 \end{bmatrix}$, $\mathbf{B} = \begin{bmatrix} 0 & 1 & 0 \\ 0 & 0 & 1 \\ -18 & -27 & -10 \end{bmatrix}$, and $\mathbf{C} = \begin{bmatrix} 1 & 0 \\ 2 & 1 \\ 3 & 0 \end{bmatrix}$. Find the 3×2 **X** matrix which satisfies Eq. (*6.6*).

The vectorized form of the equations is

$$\begin{bmatrix} 0 & 1 & 0 & -2 & 0 & 0 \\ 0 & 0 & 1 & 0 & -2 & 0 \\ -18 & -27 & -10 & 0 & 0 & -2 \\ 1 & 0 & 0 & -3 & 1 & 0 \\ 0 & 1 & 0 & 0 & -3 & 1 \\ 0 & 0 & 1 & -18 & -27 & -13 \end{bmatrix} \begin{bmatrix} x_{11} \\ x_{21} \\ x_{31} \\ x_{12} \\ x_{22} \\ x_{32} \end{bmatrix} = \begin{bmatrix} 1 \\ 2 \\ 3 \\ 0 \\ 1 \\ 0 \end{bmatrix}$$

The 6×6 matrix $\mathbf{Q}$ has a determinant value of 6720. Solving and putting the x_{ij} components into their traditional rectangular array gives

$$\mathbf{X} \approx \begin{bmatrix} -1.407143 & -0.4785714 \\ 0.042857 & -0.0285714 \\ 1.942857 & 0.8714286 \end{bmatrix}$$

■

EXAMPLE 6.9 Repeat the previous example if the matrix $\mathbf{B}$ is changed to

$$\mathbf{B} = \begin{bmatrix} 0 & 1 & 0 \\ 0 & 0 & 1 \\ 18 & -9 & -8 \end{bmatrix}$$

Using the two Kronecker products forms a 6×6 coefficient matrix $\mathbf{Q}$, which is singular of rank 5. The preceding solution process is not possible, and no unique solution exists. Do nonunique solutions exist? The rank of $\mathbf{W} = [\mathbf{Q} \mid (\mathbf{C})]$ is 6, showing that the equations are now inconsistent, and *no* solution exists. ■

The conditions under which solutions to Eq. (*6.6*) exist are now stated without proof. The concept of ***eigenvalues*** must be anticipated from the next chapter. Let $\{\lambda_i, i = 1, \ldots, m\}$ be the eigenvalues of $\mathbf{A}$. Let $\{\mu_j, j = 1, \ldots, n\}$ be the eigenvalues of $\mathbf{B}$. Then a unique solution exists for Eq. (*6.6*) if and only if

$$\lambda_i + \mu_j \neq 0 \quad \text{for all } i, j \text{ pairs}$$

In the two preceding examples $\mathbf{A}$ has eigenvalues of -1 and -2. In Example 6.8 $\mathbf{B}$ has eigenvalues of -1, -3, and -6, so the conditions for a unique solution are satisfied. In Example 6.9 the modified matrix $\mathbf{B}$ has eigenvalues of 1, -3, and -6. Now $\lambda_1 + \mu_1 = 0$, so that the conditions for solutions are not satisfied. A special case of Eq. (*6.6*), which commonly occurs in stability, random processes, and optimal control problems, has $\mathbf{A}$ and $\mathbf{B}$ both $n \times n$ and transposes of each other. Then $\mathbf{X}$ and $\mathbf{C}$ must also be $n \times n$ matrices. Since $\mathbf{A}$ and $\mathbf{A}^T$ have the same eigenvalues, the existence conditions fail to be satisfied only when $\mathbf{A}$ has one or more pairs of eigenvalues positioned symmetrically with respect to the $j\omega$ axis, such as $-\alpha$ and $+\alpha$ for some scalar α.

If $\mathbf{C}$ is symmetric, then $\mathbf{X}$ will also be symmetric. In this case the vectorized equations will contain unnecessary redundancies in both $(\mathbf{C})$ and $(\mathbf{X})$, which could be removed. The corresponding rows of $\mathbf{Q}$ can then be removed, and the sum of the columns which multiply $\mathbf{X}_{ij}$ and $\mathbf{X}_{ji}$ is used. This gives a reduced set of equations for the $n(n+1)/2$ unknowns. This need not be done, however. The full set of n^2 equations is still solvable, and the resulting $\mathbf{X}$ will be symmetric (except possibly for rounding error).

EXAMPLE 6.10 Let $\mathbf{A} = \begin{bmatrix} 0 & -4 \\ 1 & -2 \end{bmatrix}$, $\mathbf{B} = \mathbf{A}^T$, and $\mathbf{C} = \begin{bmatrix} 0 & 0 \\ 0 & -5 \end{bmatrix}$. Then the solution for $\mathbf{X}$ is found as $\mathbf{X} = \text{diag}\{0.3125, \ \ 1.25\}$, which is symmetric, as promised. In fact, $\mathbf{X}$ is positive definite, a property which is defined in the next chapter. This is related to the fact that both eigenvalues of $\mathbf{A}$ have negative real parts and to the properties of $\mathbf{C}$. These issues are clarified in later chapters. ■

REFERENCES

1. Strang, G.: *Linear Algebra and Its Applications,* Academic Press, New York, 1980.
2. Kailath, T.: *Linear Systems,* Prentice Hall, Englewood Cliffs, N.J., 1980.
3. Taylor, A. E. and R. W. Mann: *Advanced Calculus,* Xerox, Boston, Mass., 1972.
4. Forsythe, G. E., M. A. Malcolm, and C. Moler: *Computer Methods for Mathematical Computations,* Prentice Hall, Englewood Cliffs, N.J., 1977.
5. Penrose, R.: "A Generalized Inverse for Matrices," *Proceedings of the Cambridge Philosophical Society,* Vol. 51, Part 3, 1955, pp. 406–413.
6. Kalman, R. E.: "A New Approach to Linear Filtering and Prediction Problems," *Trans. of the ASME, Journal of Basic Engineering,* Vol. 82, 1960, pp. 35–45.
7. Maybeck, P. S.: *Stochastic Models, Estimation and Control,* Vol. 1, Academic Press, New York, 1979.
8. Meditch, J. S., *Stochastic Optimal Linear Estimation and Control,* McGraw-Hill, New York, 1969.

ILLUSTRATIVE PROBLEMS

.1 Use arguments in $\mathscr{X}^n$ to draw conclusions about solutions to $\mathbf{Ax} = \mathbf{y}$.

Let the rows of $\mathbf{A}$ be considered as the conjugate transpose of n component vectors $\mathbf{c}_i$. Then the set of simultaneous equations is equivalent to m scalar equations of the form

$$\langle \mathbf{c}_i, \mathbf{x} \rangle = y_i \quad \text{where } i = 1, \ldots, m$$

Each of these equations defines an $n - 1$ dimensional hyperplane in $\mathscr{X}^n$, with a normal $\mathbf{c}_i$.

The existence of a solution means that there exists a vector $\mathbf{x}$ that simultaneously terminates in all m of the hyperplanes. If the set $\{\mathbf{c}_i\}$ is linearly independent, so that $r_A = m$, the intersection of m hyperplanes of dimension $n - 1$ defines an $n - m$ dimensional hyperplane. Every vector $\mathbf{x}$ terminating in this hyperplane is a solution. If $m = n$, the hyperplane is of zero dimension, i.e., a single point, and defines a unique solution. Obviously, whenever $r_A = m$, $r_W = m$ also.

If $r_A < m$, two or more of the $n - 1$ dimensional hyperplanes are either parallel or they coincide. If parallel but distinct, they never intersect and the equations are inconsistent. It can be shown that these geometrical conditions are equivalent to the algebraic conditions given in terms of the $\mathbf{W}$ matrix.

Homogeneous Equations

.2 Let $\mathbf{A}$ be an $n \times n$ matrix with $r_A = n - 1$. Show that a nontrivial solution to $\mathbf{Ax} = \mathbf{0}$ can be selected as any nonzero column of the matrix Adj $\mathbf{A}$.

For any $n \times n$ matrix, $\mathbf{A}[\text{Adj}(\mathbf{A})] = \mathbf{I}_n |\mathbf{A}|$. Any column j of this equation can be singled out and written as $\mathbf{A}[\text{Adj}(\mathbf{A})]_j = [\mathbf{I}_n]_j |\mathbf{A}|$. If $r_A < n$, the determinant is zero, and $\mathbf{A}$ times *any* column of Adj($\mathbf{A}$) then gives a zero vector. Selecting a particular column j which is not identically zero thus gives a nontrivial solution.

3 Use the results of Problem 6.2 to find a nontrivial solution for

$$\begin{bmatrix} 1 & 0 & 1 \\ 0 & 1 & 0 \\ 2 & 2 & 2 \end{bmatrix} \begin{bmatrix} x_1 \\ x_2 \\ x_3 \end{bmatrix} = \mathbf{0}$$

Since $r_A = 2$, the degeneracy is $n - r_A = 1$, so a one-parameter family of nontrivial solutions exists. It is found by computing $\text{Adj}\,\mathbf{A} = \begin{bmatrix} 2 & 2 & -1 \\ 0 & 0 & 0 \\ -2 & -2 & 1 \end{bmatrix}$. Thus $\mathbf{x} = k\begin{bmatrix} 1 \\ 0 \\ -1 \end{bmatrix}$, $k \neq 0$, generates the set of all nontrivial solutions.

6.4 If an $n \times n$ matrix has rank $r_A < n$, it can be shown that $\mathbf{Ax} = \mathbf{0}$ has $q = n - r_A$ linearly independent solutions. They may be chosen as linearly independent columns of

$$\frac{d^{q-1}}{d\epsilon^{q-1}}\{\text{Adj}[\mathbf{A} - \mathbf{I}\epsilon]\}\bigg|_{\epsilon=0}$$

Find nontrivial solutions for this problem when $\mathbf{A} = \begin{bmatrix} 1 & 1 & 1 \\ 2 & 2 & 2 \\ -2 & -2 & -2 \end{bmatrix}$.

The rank of $\mathbf{A}$ is 1, so $q = 2$ and

$$\lim_{\epsilon \to 0} \frac{d}{d\epsilon}\{\text{Adj}[\mathbf{A} - \mathbf{I}\epsilon]\} = \begin{bmatrix} 0 & 1 & 1 \\ 2 & 1 & 2 \\ -2 & -2 & -3 \end{bmatrix}$$

There are just two linearly independent columns, and nontrivial solutions are $\mathbf{x}_1 = [0 \quad 2 \quad -2]^T$, $\mathbf{x}_2 = [1 \quad 1 \quad -2]^T$ or any linear combination of these two. $\mathbf{x}_1$ and $\mathbf{x}_2$ form a basis for the two-dimensional subspace defined by $\langle \mathbf{c}, \mathbf{x} \rangle = 0$, where $\mathbf{c}$ is the transpose of any row of $\mathbf{A}$.

6.5 Find all nontrivial solutions to

$$\begin{bmatrix} 1 & 3 & 5 \\ 1 & 4 & 6 \\ -1 & 5 & 3 \\ -1 & 4 & 2 \\ 1 & 3 & 5 \end{bmatrix}\mathbf{x} = \begin{bmatrix} 0 \\ 0 \\ 0 \\ 0 \\ 0 \end{bmatrix}$$

This is the same $\mathbf{A}$ matrix as appeared in Example 6.1(8). The RRE form for $\mathbf{W}$ is the same as given in that example, except that the last column is all zeros. Therefore, all nontrivial solutions must satisfy $x_1 + 2x_3 = 0$ and $x_2 + x_3 = 0$. Thus $\mathbf{x} = a[-2 \quad -1 \quad 1]^T$, for any scalar a, constitutes the one parameter family of nontrivial solutions. From $\mathbf{W}'$ it can be seen that the rank of $\mathbf{A}$ is 2 and the number of unknowns is $n = 3$, so the degeneracy or nullity is $q = 3 - 2 = 1$. This is the dimension of the null space of $\mathbf{A}$.

6.6 Find all nontrivial solutions to

$$\begin{bmatrix} 4 & -2 & 3 \\ 1 & 3 & 1 \\ 1 & 3 & 1 \end{bmatrix}\mathbf{x} = \begin{bmatrix} 0 \\ 0 \\ 0 \end{bmatrix}$$

For this problem the RRE form of $\mathbf{W}$ is

$$\mathbf{W}' = \left[\begin{array}{ccc|c} 1 & 0 & 0.78571427 & 0 \\ 0 & 1 & 0.07142857 & 0 \\ 0 & 0 & 0 & 0 \end{array}\right]$$

Therefore, $x_1 + 0.78571427x_3 = 0$ and $x_2 + 0.07142857x_3 = 0$. All nontrivial solutions must be proportional to

$$\mathbf{x} = [11 \quad 1 \quad -14]^T$$

6.7 Do nontrivial solutions exist for the following?

$$\begin{bmatrix} 2 & -2 & 3 \\ 1 & 1 & 1 \\ 1 & 3 & -1 \end{bmatrix} \mathbf{x} = \begin{bmatrix} 0 \\ 0 \\ 0 \end{bmatrix}$$

Using the RRE method, or just computing its determinant, shows that the rank of **A** is 3. Its degeneracy is zero, it is nonsingular, and so the only solution to this problem is the trivial solution $\mathbf{x} = \mathbf{0}$.

Minimum Norm Solutions

6.8 Find the minimum norm solution for $[1 \quad 2]\begin{bmatrix} x_1 \\ x_2 \end{bmatrix} = 1$.

Identifying $\mathbf{A} = [1 \quad 2]$, the minimum norm solution is

$$\mathbf{x} = \begin{bmatrix} 1 \\ 2 \end{bmatrix} \left\{ [1 \quad 2] \begin{bmatrix} 1 \\ 2 \end{bmatrix} \right\}^{-1} (1) = \frac{1}{5} \begin{bmatrix} 1 \\ 2 \end{bmatrix}$$

6.9 A specified amount of constant current, i, must be delivered to the ground point of Figure 6.5. Specify v_1, v_2, and v_3 so that the total energy dissipated in the resistors is minimized.

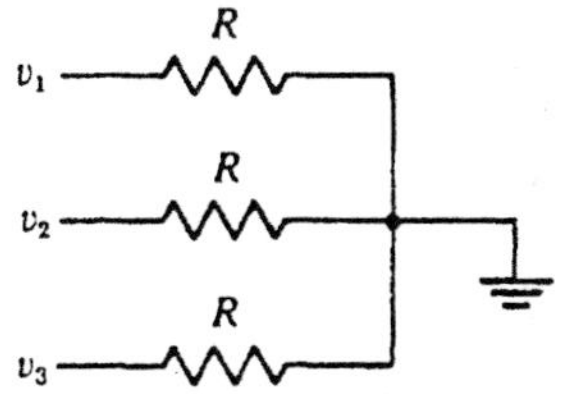

Figure 6.5

It is required that

$$v_1/R + v_2/R + v_3/R = i \quad \text{or} \quad \frac{1}{R}[1 \quad 1 \quad 1]\mathbf{v} = i$$

The total energy dissipated per unit time is

$$\dot{\varepsilon} = \frac{1}{R}[v_1^2 + v_2^2 + v_3^2] = \frac{1}{R}\|\mathbf{v}\|^2$$

The desired solution is thus the minimum norm solution.

$$\mathbf{v} = \frac{1}{R}\begin{bmatrix} 1 \\ 1 \\ 1 \end{bmatrix} \left\{ \frac{1}{R^2}[1 \quad 1 \quad 1] \begin{bmatrix} 1 \\ 1 \\ 1 \end{bmatrix} \right\}^{-1} i = \frac{R}{3}\begin{bmatrix} 1 \\ 1 \\ 1 \end{bmatrix} i$$

Frequently, the norm can be given a physical interpretation of energy or power. This is one reason why minimum norm solutions are often sought for underdetermined problems.

6.10 A small microcomputer has five terminals connected to it. The fraction of the time devoted to each terminal is x_i, so

$$x_1 + x_2 + x_3 + x_4 + x_5 = 1$$

The programmer at terminal 2 types four times as fast as the programmer at 1. They are both typing in the same code and must finish at the same time so $x_1 = 4x_2$. Both terminals 3 and 4 are sending mail files to 5, so $x_3 + x_4 = x_5$. Find the minimum norm solution for allocating CPU time.

In matrix form the constraints are

$$\begin{bmatrix} 1 & 1 & 1 & 1 & 1 \\ 1 & -4 & 0 & 0 & 0 \\ 0 & 0 & 1 & 1 & -1 \end{bmatrix} \mathbf{x} = \begin{bmatrix} 1 \\ 0 \\ 0 \end{bmatrix}$$

The minimum norm solution is

$$\mathbf{x} = \mathbf{A}^T[\mathbf{A}\mathbf{A}^T]^{-1}\mathbf{y} = \begin{bmatrix} 1 & 1 & 0 \\ 1 & -4 & 0 \\ 1 & 0 & 1 \\ 1 & 0 & 1 \\ 1 & 0 & -1 \end{bmatrix} \begin{bmatrix} 5 & -3 & 1 \\ -3 & 17 & 0 \\ 1 & 0 & 3 \end{bmatrix}^{-1} \begin{bmatrix} 1 \\ 0 \\ 0 \end{bmatrix} = \begin{bmatrix} 0.28436 \\ 0.07109 \\ 0.15602 \\ 0.15602 \\ 0.32739 \end{bmatrix}$$

6.11 Find the shortest four-dimensional vector from the origin to the four-dimensional hyperplane described by

$$5x_1 - 2x_2 + x_3 + 7x_4 = 12$$

This is the same as asking for the minimum norm solution to

$$[5 \quad -2 \quad 1 \quad 7]\mathbf{x} = 12$$

Therefore,

$$\mathbf{x} = \begin{bmatrix} 5 \\ -2 \\ 1 \\ 7 \end{bmatrix} [79]^{-1}(12) = \begin{bmatrix} 0.7595 \\ -0.3038 \\ 0.1519 \\ 1.0633 \end{bmatrix}$$

Least Squares, Weighted Least Squares, and Recursive Least Squares

6.12 Consider the set of simultaneous linear equations

$$\begin{bmatrix} \mathbf{y}_k \\ \hline \mathbf{y}_{k+1} \end{bmatrix} = \begin{bmatrix} \mathbf{A} \\ \hline \mathbf{H}_{k+1} \end{bmatrix} \mathbf{x} + \begin{bmatrix} \mathbf{e} \\ \hline \mathbf{e}_{k+1} \end{bmatrix}$$

Find the vector $\mathbf{x}$ which minimizes

$$J = [\mathbf{e}^T \quad \mathbf{e}_{k+1}^T] \left[\begin{array}{c|c} \mathbf{R}^{-1} & \mathbf{0} \\ \hline \mathbf{0} & \mathbf{R}_{k+1}^{-1} \end{array}\right] \begin{bmatrix} \mathbf{e} \\ \hline \mathbf{e}_{k+1} \end{bmatrix}$$

The weighted least-squares estimate is

$$\mathbf{x}_{k+1} = \left\{ [\mathbf{A}^T \mid \mathbf{H}_{k+1}^T] \left[\begin{array}{c|c} \mathbf{R}^{-1} & \mathbf{0} \\ \hline \mathbf{0} & \mathbf{R}_{k+1}^{-1} \end{array}\right] \begin{bmatrix} \mathbf{A} \\ \hline \mathbf{H}_{k+1} \end{bmatrix} \right\}^{-1} [\mathbf{A}^T \mid \mathbf{H}_{k+1}^T] \left[\begin{array}{c|c} \mathbf{R}^{-1} & \mathbf{0} \\ \hline \mathbf{0} & \mathbf{R}_{k+1}^{-1} \end{array}\right] \begin{bmatrix} \mathbf{y}_k \\ \hline \mathbf{y}_{k+1} \end{bmatrix}$$

$$= [\mathbf{A}^T\mathbf{R}^{-1}\mathbf{A} + \mathbf{H}_{k+1}^T\mathbf{R}_{k+1}^{-1}\mathbf{H}_{k+1}]^{-1}[\mathbf{A}^T\mathbf{R}^{-1}\mathbf{y}_k + \mathbf{H}_{k+1}^T\mathbf{R}_{k+1}^{-1}\mathbf{y}_{k+1}]$$

Defining $\mathbf{A}^T\mathbf{R}^{-1}\mathbf{A} = \mathbf{P}_k^{-1}$ and using the matrix inversion identity of Sec. 4.9 gives

$$\mathbf{x}_{k+1} = \{\mathbf{P}_k - \mathbf{P}_k\mathbf{H}_{k+1}^T[\mathbf{H}_{k+1}\mathbf{P}_k\mathbf{H}_{k+1}^T + \mathbf{R}_{k+1}]^{-1}\mathbf{H}_{k+1}\mathbf{P}_k\}\{\mathbf{A}^T\mathbf{R}^{-1}\mathbf{y}_k + \mathbf{H}_{k+1}^T\mathbf{R}_{k+1}^{-1}\mathbf{y}_{k+1}\}$$

Note that $\mathbf{P}_k\mathbf{A}^T\mathbf{R}^{-1}\mathbf{y}_k = \mathbf{x}_k$ is the weighted least-squares solution when only the first group of equations is used. Therefore,

$$\mathbf{x}_{k+1} = \mathbf{x}_k - \mathbf{P}_k\mathbf{H}_{k+1}^T[\mathbf{H}_{k+1}\mathbf{P}_k\mathbf{H}_{k+1}^T + \mathbf{R}_{k+1}]^{-1}\mathbf{H}_{k+1}\mathbf{x}_k$$
$$+ \mathbf{P}_k\mathbf{H}_{k+1}^T\{\mathbf{I} - [\mathbf{H}_{k+1}\mathbf{P}_k\mathbf{H}_{k+1}^T + \mathbf{R}_{k+1}]^{-1}\mathbf{H}_{k+1}\mathbf{P}_k\mathbf{H}_{k+1}^T\}\mathbf{R}_{k+1}^{-1}\mathbf{y}_{k+1}$$

The unit matrix in the last equation is written as

$$\mathbf{I} = [\mathbf{H}_{k+1}\mathbf{P}_k\mathbf{H}_{k+1}^T + \mathbf{R}_{k+1}]^{-1}[\mathbf{H}_{k+1}\mathbf{P}_k\mathbf{H}_{k+1}^T + \mathbf{R}_{k+1}]$$

This step is analogous to finding the common denominator in scalar algebra and leads to

$$\mathbf{x}_{k+1} = \mathbf{x}_k + \mathbf{P}_k \mathbf{H}_{k+1}^T[\mathbf{H}_{k+1}\mathbf{P}_k \mathbf{H}_{k+1}^T + \mathbf{R}_{k+1}]^{-1}\{\mathbf{y}_{k+1} - \mathbf{H}_{k+1}\mathbf{x}_k\}$$

If this recursive process is to be continued, then an expression for $\mathbf{P}_{k+1}$ is needed. If $\mathbf{A}$ is replaced by $\begin{bmatrix} \mathbf{A} \\ \hline \mathbf{H}_{k+1} \end{bmatrix}$ and if $\mathbf{R}^{-1}$ is replaced by $\left[\begin{array}{c|c} \mathbf{R}^{-1} & \mathbf{0} \\ \hline \mathbf{0} & \mathbf{R}_{k+1}^{-1} \end{array}\right]$, then the definition for $\mathbf{P}_k$ is modified to read

$$\mathbf{P}_{k+1} = \left\{[\mathbf{A}^T \,|\, \mathbf{H}_{k+1}^T]\left[\begin{array}{c|c} \mathbf{R}^{-1} & \mathbf{0} \\ \hline \mathbf{0} & \mathbf{R}_{k+1}^{-1} \end{array}\right]\begin{bmatrix} \mathbf{A} \\ \hline \mathbf{H}_{k+1} \end{bmatrix}\right\}^{-1}$$

$$= [\mathbf{A}^T\mathbf{R}^{-1}\mathbf{A} + \mathbf{H}_{k+1}^T\mathbf{R}_{k+1}^{-1}\mathbf{H}_{k+1}]^{-1} = [\mathbf{P}_k^{-1} + \mathbf{H}_{k+1}^T\mathbf{R}_{k+1}^{-1}\mathbf{H}_{k+1}]^{-1}$$

13 A tracking station measures $\dot{r}$, the time derivative of the range to a satellite, every second. The measurements are noisy. Find the least-squares fit to a straight line.

The measurements are assumed to fit the equation $\dot{r}(t) = a + bt + e(t)$, where a and b are to be determined. Measurement times are $t = 1, 2, \ldots, k$:

$$\begin{bmatrix} \dot{r}_1 \\ \dot{r}_2 \\ \vdots \\ \dot{r}_k \end{bmatrix} = \begin{bmatrix} 1 & 1 \\ 1 & 2 \\ \vdots & \vdots \\ 1 & k \end{bmatrix}\begin{bmatrix} a \\ b \end{bmatrix} + \begin{bmatrix} e_1 \\ e_2 \\ \vdots \\ e_k \end{bmatrix}$$

The least-squares solution gives

$$\begin{bmatrix} a \\ b \end{bmatrix} = \begin{bmatrix} k & \sum_{j=1}^{k} j \\ \sum_{j=1}^{k} j & \sum_{j=1}^{k} j^2 \end{bmatrix}^{-1}\begin{bmatrix} \sum_{j=1}^{k} \dot{r}_i \\ \sum_{j=1}^{k} j\dot{r}_j \end{bmatrix}$$

Using the identities $\sum_{j=1}^{k} j = \dfrac{k(k+1)}{2}$, $\sum_{j=1}^{k} j^2 = \dfrac{k(k+1)(2k+1)}{6}$, and carrying out the matrix inversion gives

$$a = \frac{2(2k+1)\sum_{j=1}^{k} \dot{r}_j - 6\sum_{j=1}^{k} j\dot{r}_j}{k(k-1)}, \quad b = \frac{-6(k+1)\sum_{j=1}^{k} \dot{r}_j + 12\sum_{j=1}^{k} j\dot{r}_j}{k(k^2-1)}$$

If $k = 1$, the results are indeterminate, indicating that an infinite number of lines can be passed through a single point.

14 Investigate the following equations:

$$\begin{bmatrix} 1 & 2 \\ 3 & 4 \\ 5 & 6 \end{bmatrix}\mathbf{x} = \begin{bmatrix} 2 \\ 3 \\ 14 \end{bmatrix}$$

The RRE form of $\mathbf{W}$ is $\mathbf{W}' = \left[\begin{array}{cc|c} 1 & 0 & 0 \\ 0 & 1 & 0 \\ 0 & 0 & 1 \end{array}\right]$.

Since the rank of $\mathbf{A}$ is 2 and the rank of $\mathbf{W}$ is 3, the equations are inconsistent. Least-squares solutions are investigated by several methods.

(a) The normal equation $\mathbf{A}^T\mathbf{A}\mathbf{x} = \mathbf{A}^T\mathbf{y}$ is in this case

$$\begin{bmatrix} 35 & 44 \\ 44 & 56 \end{bmatrix}\begin{bmatrix} x_1 \\ x_2 \end{bmatrix} = \begin{bmatrix} 81 \\ 100 \end{bmatrix}$$

Straightforward matrix inversion could be used, or the RRE method as follows:

$$\mathbf{W} = \left[\begin{array}{cc|c} 35 & 44 & 81 \\ 44 & 56 & 100 \end{array}\right] \longrightarrow \left[\begin{array}{cc|c} 1 & 0 & 5.6666 \\ 0 & 1 & -2.6666 \end{array}\right] \quad \text{so} \quad \begin{bmatrix} x_1 \\ x_2 \end{bmatrix} = \begin{bmatrix} 5.6666 \\ -2.6666 \end{bmatrix}$$

(b) Cholesky decomposition gives

$$\mathbf{A}^T\mathbf{A} = \begin{bmatrix} 5.91608 & 0 \\ 7.43736 & 0.82808 \end{bmatrix}\begin{bmatrix} 5.91608 & 7.43736 \\ 0 & 0.82808 \end{bmatrix} = \mathbf{S}^T\mathbf{S}$$

Solving $\mathbf{S}^T\mathbf{v} = \begin{bmatrix} 81 \\ 100 \end{bmatrix}$ gives $\mathbf{v} = \begin{bmatrix} 13.69150 \\ -2.20818 \end{bmatrix}$

Then solving $\mathbf{Sx} = \mathbf{v}$ gives the same answer for $\mathbf{x}$ as in part (a).

(c) Using the GSE method, the two columns of the original **A** matrix, plus the vector **y**, are used to generate an orthonormal basis set which is shown as columns of the matrix **V**.

$$\mathbf{V} = \begin{bmatrix} 1.6903085E-01 & 8.9708525E-01 & 4.0824756E-01 \\ 5.0709254E-01 & 2.7602646E-01 & -8.1649655E-01 \\ 8.4515423E-01 & -3.4503257E-01 & 4.0824878E-01 \end{bmatrix}$$

Then $\mathbf{V}^T\mathbf{Ax} = \mathbf{V}^T\mathbf{y}$ can be compactly written as $\mathbf{V}^T\mathbf{W} = \mathbf{W}'$. Rounding terms of order 10^{-6} to zero gives

$$\mathbf{W}' = \left[\begin{array}{cc|c} 5.91608 & 7.43736 & 13.69150 \\ 0 & 0.82808 & -2.20821 \\ 0 & 0 & 4.08249 \end{array}\right].$$

Compare this with the Cholesky results.

The last row indicates that the norm of the residual error is $\|\mathbf{y}_e\| = 4.08256$. Ignoring the last row and solving the first two obviously again give the same answer.

6.15 After seven semesters of college a student surmises that his cumulative grade point average (GPA) is a cubic function of the number of semesters completed. His record to date is

Semester, s	1	2	3	4	5	6	7	8
Cum. GPA	2.5	3.1	2.9	2.8	2.8	3.0	3.1	??

Find the coefficients for the least-squares fit to a cubic. Also determine the residual error norm. Then use the cubic to predict his GPA on graduation day after semester 8.

It is assumed that GPA $= a_0 + a_1 s + a_2 s^2 + a_3 s^3$. In matrix form this student's data and postulated model are

$$\begin{bmatrix} 1 & 1 & 1 & 1 \\ 1 & 2 & 4 & 8 \\ 1 & 3 & 9 & 27 \\ 1 & 4 & 16 & 64 \\ 1 & 5 & 25 & 125 \\ 1 & 6 & 36 & 216 \\ 1 & 7 & 49 & 343 \end{bmatrix}\begin{bmatrix} a_0 \\ a_1 \\ a_2 \\ a_3 \end{bmatrix} = \begin{bmatrix} 2.5 \\ 3.1 \\ 2.9 \\ 2.8 \\ 2.8 \\ 3.0 \\ 3.1 \end{bmatrix}$$

Using the GSE method leads to the following $\mathbf{W}'$ matrix, rounded off.

$$\mathbf{W}' = \left[\begin{array}{cccc|c} 2.646 & 10.583 & 52.915 & 296.324 & 7.635 \\ 0 & 5.292 & 42.322 & 291.033 & 0.283 \\ 0 & 0 & 9.165 & 109.982 & -0.033 \\ 0 & 0 & 0 & 14.700 & 0.327 \\ \hline 0 & 0 & 0 & 0 & 0.284 \end{array}\right]$$

From this the coefficients are determined as

$$\mathbf{a} = [1.828 \quad 0.993 \quad -0.270 \quad 0.022]^T \quad \text{and} \quad \|\mathbf{y}_e\| = 0.284$$

Using these coefficients the estimated GPA after eight semesters is

$$1.828 + 8(0.993) - 64(0.27) + 512(0.022) = 3.756.$$

16 Use the recursive least-squares algorithm to estimate $\mathbf{x}$ from Problem 6.14. Start with an initial estimate of $\mathbf{x} = \mathbf{0}$ and set $\mathbf{P}_0 = \text{diag}\,[10000, 10000]$. Use $\mathbf{R} = 10$ and $\mathbf{Q} = 0$ (no deweighting). Also add the following additional equations to be processed:

$$0 = 3x_1 + 6x_2, \quad 10 = 2x_1 + x_2$$

The recursive calculations are rounded off and tabulated in the order performed.

k	$\mathbf{M}_k$	$\mathbf{H}_k$	$\mathbf{y}_k$	$\mathbf{K}_k$	$\mathbf{y}_k - \mathbf{H}_k\mathbf{x}_{k-1}$	$\mathbf{x}_k$	$\mathbf{P}_k$
0	—	—	—	—	—	$\begin{pmatrix}0\\0\end{pmatrix}$	$\begin{pmatrix}10000 & 0\\0 & 10000\end{pmatrix}$
1	$\begin{pmatrix}10000 & 0\\0 & 10000\end{pmatrix}$	$(1 \quad 2)$	2	$\begin{pmatrix}0.2\\0.4\end{pmatrix}$	2	$\begin{pmatrix}0.4\\0.8\end{pmatrix}$	$\begin{pmatrix}8000 & -4000\\-4000 & 2000\end{pmatrix}$
2	$\begin{pmatrix}8000 & -4000\\-4000 & 2000\end{pmatrix}$	$(3 \quad 4)$	3	$\begin{pmatrix}0.993\\-0.495\end{pmatrix}$	−1.4	$\begin{pmatrix}-0.990\\1.493\end{pmatrix}$	$\begin{pmatrix}49.628 & -34.474\\-34.474 & 24.815\end{pmatrix}$
3	$\begin{pmatrix}49.628 & -34.474\\-34.474 & 24.815\end{pmatrix}$	$(5 \quad 6)$	14	$\begin{pmatrix}0.664\\-0.415\end{pmatrix}$	9.992	$\begin{pmatrix}5.648\\-2.652\end{pmatrix}$	$\begin{pmatrix}23.245 & -18.264\\-18.264 & 14.529\end{pmatrix}$
4	$\begin{pmatrix}23.245 & -18.264\\-18.264 & 14.529\end{pmatrix}$	$(3 \quad 6)$	0	$\begin{pmatrix}-0.470\\0.382\end{pmatrix}$	−1.033	$\begin{pmatrix}6.134\\-3.046\end{pmatrix}$	$\begin{pmatrix}4.507 & -3.037\\-3.037 & 2.155\end{pmatrix}$
5	$\begin{pmatrix}4.507 & -3.037\\-3.037 & 2.155\end{pmatrix}$	$(2 \quad 1)$	10	$\begin{pmatrix}0.331\\-0.217\end{pmatrix}$	0.779	$\begin{pmatrix}6.392\\-3.216\end{pmatrix}$	$\begin{pmatrix}2.526 & -1.739\\-1.739 & 1.304\end{pmatrix}$

Notice that the estimate of $\mathbf{x}$ after three measurements is not exactly the same as was found in Problem 6.14. The difference is due to the initial estimates used for $\mathbf{x}$ and $\mathbf{P}$.

Recursive Weighted Least Squares with Discrete-Time Systems

.17 A homogeneous linear discrete-time system is described by $\mathbf{x}(k+1) = \mathbf{A}(k)\mathbf{x}(k)$. Measured outputs are given by $\mathbf{y}(k) = \mathbf{C}(k)\mathbf{x}(k) + \mathbf{e}(k)$, where $\mathbf{e}(k)$ is an error vector due to imperfect measuring devices. The precise initial conditions for this system, $\mathbf{x}(0)$, are not known, although an estimate, $\hat{\mathbf{x}}(0)$, is available.‡ Use the recursive weighted least-squares technique to develop a scheme for improving on the estimate $\hat{\mathbf{x}}(0)$ each time a new measurement $\mathbf{y}(k)$ becomes available.

In order to place this problem in the framework developed in Sec. 6.8, page 223, all that is required is a few notational changes and the elimination of the difference equation. The solution to the difference equation is $\mathbf{x}(k) = \mathbf{\Phi}(k, 0)\mathbf{x}(0)$, so that the measurement equations can be written as

$$\mathbf{y}(k) = \mathbf{C}(k)\mathbf{\Phi}(k, 0)\mathbf{x}(0) + \mathbf{e}(k)$$

‡The circumflex ^ is used in this and the next problem to indicate an estimated quantity. It should not be confused with the notation for a unit vector used earlier.

This is in the form used in Sec. 6.8, where $\mathbf{C}(k)\mathbf{\Phi}(k, 0)$ corresponds to $\mathbf{H}_k$ used in that earlier section. In Sec. 6.8 the index k referred to the kth estimate of a constant vector $\mathbf{x}$. To avoid confusion here, $\hat{\mathbf{x}}(0|k)$ will be used to indicate the estimate of $\mathbf{x}(0)$ based on all measurements up to and including $\mathbf{y}(k)$. Assuming that the original estimate $\hat{\mathbf{x}}(0)$ is to be weighted by a nonsingular $n \times n$ matrix $\mathbf{P}_0^{-1}$ and that each succeeding measurement is weighted by the $m \times m$ nonsingular matrix $\mathbf{R}_k^{-1}$ the results of Sec. 6.8 give

$$\hat{\mathbf{x}}(0|1) = \hat{\mathbf{x}}(0) + \mathbf{K}_0[\mathbf{y}(1) - \mathbf{C}(1)\mathbf{\Phi}(1,0)\hat{\mathbf{x}}(0)]$$

The gain matrix $\mathbf{K}_0$ is given by

$$\mathbf{K}_0 = \mathbf{P}_0\mathbf{\Phi}^T(1,0)\mathbf{C}^T(1)[\mathbf{C}(1)\mathbf{\Phi}(1,0)\mathbf{P}_0\mathbf{\Phi}^T(1,0)\mathbf{C}^T(1) + \mathbf{R}_1]^{-1}$$

The estimate of $\mathbf{x}(0)$ may be further refined each time a new measurement $\mathbf{y}(k)$ is taken by using the recursive relations

$$\hat{\mathbf{x}}(0|k+1) = \hat{\mathbf{x}}(0|k) + \mathbf{K}_k[\mathbf{y}(k+1) - \mathbf{C}(k+1)\mathbf{\Phi}(k+1,0)\hat{\mathbf{x}}(0|k)]$$

where

$$K_k = \mathbf{P}_k\mathbf{\Phi}^T(k+1,0)\mathbf{C}^T(k+1)[\mathbf{C}(k+1)\mathbf{\Phi}(k+1,0)$$
$$\times \mathbf{P}_k\mathbf{\Phi}^T(k+1,0)\mathbf{C}^T(k+1) + \mathbf{R}_{k+1}]^{-1}$$

and where $\mathbf{P}_k$ is computed recursively using Eq. (6.5), page 224, which can be written as

$$\mathbf{P}_{k+1} = \mathbf{P}_k - \mathbf{K}_k\mathbf{C}(k+1)\mathbf{\Phi}(k+1,0)\mathbf{P}_k$$

If the error vector $\mathbf{e}(k)$ and the weighting matrices $\mathbf{P}_k$ and $\mathbf{R}_k$ are given the proper statistical interpretation, the above technique constitutes a simple example of the fixed-point smoothing algorithm [8]. A block diagram of the procedure is given in Figure 6.6.

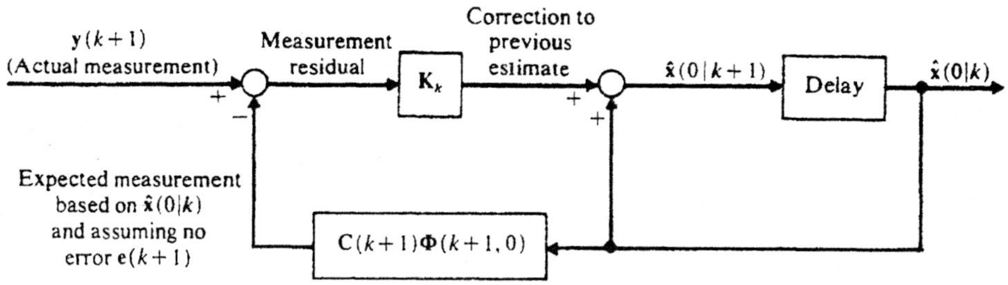

Figure 6.6

6.18 Consider the results of Problem 6.17. Instead of reestimating $\mathbf{x}(0)$ after each measurement, an estimate of the current state is now desired. Let $\hat{\mathbf{x}}(k|k)$ be the estimate of $\mathbf{x}(k)$ based on all measurements up to and including $\mathbf{y}(k)$. For this estimate, use the intuitively reasonable relation

$$\hat{\mathbf{x}}(k+1|k+1) = \mathbf{\Phi}(k+1,0)\hat{\mathbf{x}}(0|k+1)$$

Modify the previous block diagram so that $\hat{\mathbf{x}}(k|k)$ is the output.

If the transition matrix $\mathbf{\Phi}(k+1,0)$ is inserted into the diagram of Figure 6.6 before the delay, then the output will be $\hat{\mathbf{x}}(k|k)$ as shown in Figure 6.7. To maintain the correct relations in the rest of the diagram, the term $\hat{\mathbf{x}}(0|k) = \mathbf{\Phi}(0,k)\hat{\mathbf{x}}(k|k)$ is needed, so $\mathbf{\Phi}(0,k)$ is inserted in the feedback path as shown in Figure 6.7.

Using standard matrix block diagram manipulations, $\mathbf{\Phi}(k+1,0)$ is moved past the summing junction, into both paths. A new gain matrix $\mathbf{K}'_{k+1} = \mathbf{\Phi}(k+1,0)\mathbf{K}_k$ is defined. The other $\mathbf{\Phi}(k+1,0)$ term is shifted into the feedback path and combined with $\mathbf{\Phi}(0,k)$ to give $\mathbf{\Phi}(k+1,k) = \mathbf{\Phi}(k+1,0)\mathbf{\Phi}(0,k)$. This also removes the $\mathbf{\Phi}(k+1,0)$ term multiplying $\mathbf{C}(k+1)$.

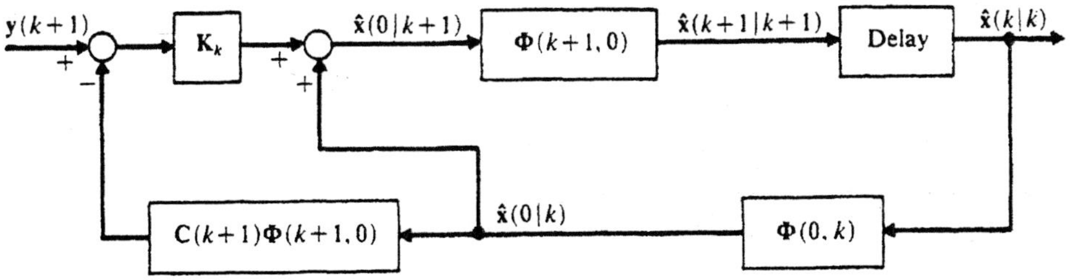

Figure 6.7

This results in the most commonly used form of the discrete Kalman filter [7, 8], shown in Figure 6.8.

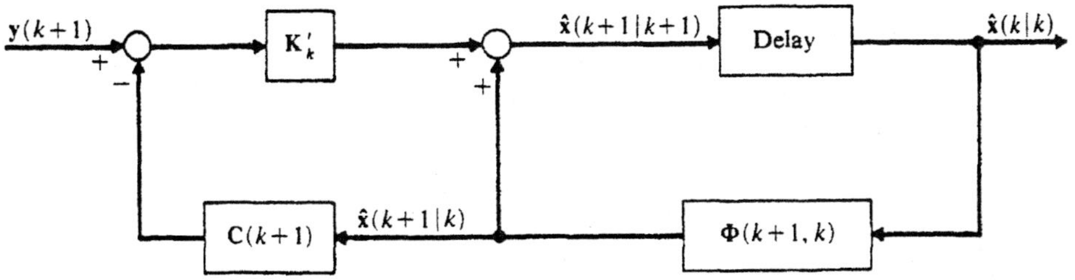

Figure 6.8

For easy reference the five equations which constitute the discrete Kalman filter are summarized below. An additive deweighting matrix $\mathbf{Q}_k$ (see Section 6.8) is included. To appreciate these results fully some knowledge of random processes is required [6]. Lacking this, the procedure can still be interpreted and used successfully as a recursive least-squares algorithm with ad hoc additive deweighting included.

To **find** the gain $\mathbf{K}'_k$, recursively compute

$$\mathbf{M}_k = \mathbf{\Phi}(k, k-1)\mathbf{P}_{k-1}\mathbf{\Phi}^T(k, k-1) + \mathbf{Q}_{k-1} \quad (1)$$

$$K'_k = \mathbf{M}_k\,\mathbf{C}^T(k)[\mathbf{C}(k)\mathbf{M}_k\,\mathbf{C}^T(k) + \mathbf{R}(k)]^{-1} \quad (2)$$

$$\mathbf{P}_k = [\mathbf{I} - \mathbf{K}'_k\,\mathbf{C}(k)]\mathbf{M}_k \quad (3)$$

To **use** the gain to estimate **x**, recursively compute

$$\hat{\mathbf{x}}(k+1|k) = \mathbf{\Phi}(k+1, k)\hat{\mathbf{x}}(k|k) \quad (4)$$

$$\hat{\mathbf{x}}(k+1|k+1) = \hat{\mathbf{x}}(k+1|k) + \mathbf{K}'_{k+1}[\mathbf{y}(k+1) - \mathbf{C}(k+1)\hat{\mathbf{x}}(k+1|k)] \quad (5)$$

To **initialize** the procedure, there are two possibilities:

(i) If $\hat{\mathbf{x}}(k+1|k)$ and $\mathbf{M}_{k+1}$ are given, then start by using equation (2) to find $\mathbf{K}'_{k+1}$, then use (5), along with the measurement $\mathbf{y}(k+1)$, to find $\hat{\mathbf{x}}(k+1|k+1)$. To get ready for the next cycle, use (3), (4), and (1).

(ii) If $\hat{\mathbf{x}}(k|k)$ and $\mathbf{P}_k$ are given, then start by using (1), then (2) to find $\mathbf{K}'_{k+1}$. Then use (4), followed by (5). To complete the first cycle and get ready for the next, (3) is then used.

Note that the above algorithm reduces to those at the end of Sec. 6.8 when $\mathbf{\Phi}(k+1, k) = \mathbf{I}$, that is, when $\mathbf{x}(k)$ is just a constant. Also be aware that the above algorithm can be written in several other forms which are *algebraically* equivalent, but which may have different numerical behavior on a finite word-length computer.

Properties of Linear Transformations

6.19 Prove that if $\mathcal{A}$ is any linear transformation there are only two possibilities regarding the null space: either *(1)* $\mathcal{N}(\mathcal{A})$ consists of only the zero vector or *(2)* $\mathcal{N}(\mathcal{A})$ contains an infinite number of vectors.

Consider $\mathcal{A}(\mathbf{x}) = \mathbf{y}$. Since $\mathcal{A}$ is linear, $\mathcal{A}(\mathbf{x}-\mathbf{x}) = \mathcal{A}(\mathbf{x}) - \mathcal{A}(\mathbf{x}) = \mathbf{y}-\mathbf{y}$ so $\mathbf{x} = \mathbf{0}$ is *always* a solution to $\mathcal{A}(\mathbf{x}) = \mathbf{0}$. Thus the null space always contains the zero vector. If $\mathbf{x}_1$ also belongs to the null space, $\mathcal{A}(\mathbf{x}_1) = \mathbf{0}$. Assume $\mathbf{x}_1 \neq \mathbf{0}$. Then the infinite set of vectors defined by $\mathbf{x} = \alpha\mathbf{x}_1$, with α any scalar, satisfies $\mathcal{A}(\mathbf{x}) = \mathcal{A}(\alpha\mathbf{x}_1) = \alpha\mathcal{A}(\mathbf{x}_1) = \mathbf{0}$. Thus if one nonzero vector belongs to the null space, then so does the infinite set of scalar multiples of it. These need not be the only vectors in $\mathcal{N}(\mathcal{A})$.

6.20 Consider the linear transformation $\mathcal{A}: \mathcal{X} \to \mathcal{Y}$. Prove that the null space of $\mathcal{A}$ is a subspace of $\mathcal{X}$.

The solution to this problem consists of combining the results of the previous problem with the definition of a subspace. A set of vectors is a subspace if for each $\mathbf{x}_1, \mathbf{x}_2$ in the set, $\alpha\mathbf{x}_1 + \beta\mathbf{x}_2$ is also in the set, for arbitrary scalars $\alpha, \beta \in \mathcal{F}$. Let $\mathbf{x}_1$ and $\mathbf{x}_2 \in \mathcal{N}(\mathcal{A})$. That is, $\mathcal{A}(\mathbf{x}_1) = \mathbf{0}$ and $\mathcal{A}(\mathbf{x}_2) = \mathbf{0}$. Then $\mathcal{A}(\alpha\mathbf{x}_1 + \beta\mathbf{x}_2) = \alpha\mathcal{A}(\mathbf{x}_1) + \beta\mathcal{A}(\mathbf{x}_2) = \mathbf{0}$. Thus $\alpha\mathbf{x}_1 + \beta\mathbf{x}_2 \in \mathcal{N}(\mathcal{A})$ and so the null space is a subspace. It is a zero dimensional subspace if its only element is the zero vector.

6.21 Let $\mathcal{A}: \mathcal{X} \to \mathcal{Y}$ be a linear transformation, with $\mathcal{X}$ and $\mathcal{Y}$ finite dimensional. Prove that the space $\mathcal{X}$ can be written as a direct sum $\mathcal{X} = \mathcal{N}(\mathcal{A}) \oplus \mathcal{R}(\mathcal{A}^*)$.

It was shown in Problem 6.20 that $\mathcal{N}(\mathcal{A})$ is a subspace of $\mathcal{X}$. Let $\mathcal{N}(\mathcal{A})^\perp$ be the orthogonal complement of $\mathcal{N}(\mathcal{A})$. Then from the results of Section 5.9, $\mathcal{X} = \mathcal{N}(\mathcal{A}) \oplus \mathcal{N}(\mathcal{A})^\perp$.

It remains to be shown that $\mathcal{R}(\mathcal{A}^*) = \mathcal{N}(\mathcal{A})^\perp$. Let $\mathbf{x}$ be an arbitrary vector in $\mathcal{N}(\mathcal{A})$ and let $\mathbf{z}$ be an arbitrary vector in $\mathcal{Y}$. Then $\mathcal{A}(\mathbf{x}) = \mathbf{0}$ so that $\langle \mathbf{z}, \mathcal{A}(\mathbf{x})\rangle = \langle \mathcal{A}^*(\mathbf{z}), \mathbf{x}\rangle = \mathbf{0}$. From this we see that $\mathbf{x}$ is orthogonal to $\mathcal{A}^*(\mathbf{z})$, which shows that $\mathcal{N}(\mathcal{A}) = \mathcal{R}(\mathcal{A}^*)^\perp$. The orthogonal complements are also equal, $\mathcal{N}(\mathcal{A})^\perp = (\mathcal{R}(\mathcal{A}^*)^\perp)^\perp$, but $(\mathcal{R}(\mathcal{A}^*)^\perp)^\perp = \mathcal{R}(\mathcal{A}^*)$. This completes the proof for $\mathcal{X}$ finite dimensional.

For the infinite dimensional case, the decomposition is valid if the closure of $\mathcal{R}(\mathcal{A}^*)$ is used in place of $\mathcal{R}(\mathcal{A}^*)$. Every finite dimensional space is closed.

6.22 Consider the linear transformation $\mathcal{A}(\mathbf{x}) = \mathbf{y}$, where $\mathcal{A}: \mathcal{X} \to \mathcal{Y}$. Show that there is no unique solution for $\mathbf{x}$ if $\mathcal{N}(\mathcal{A})$ contains vectors other than the zero vector.

Suppose $\mathbf{x}_1 \in \mathcal{N}(\mathcal{A})$ and $\mathbf{x}_1 \neq \mathbf{0}$. If $\mathbf{x}$ satisfies $\mathcal{A}(\mathbf{x}) = \mathbf{y}$, then $\mathbf{x} + \mathbf{x}_1$ is also a solution, since $\mathcal{A}(\mathbf{x}+\mathbf{x}_1) = \mathcal{A}(\mathbf{x}) + \mathcal{A}(\mathbf{x}_1) = \mathbf{y} + \mathbf{0} = \mathbf{y}$.

6.23 Prove that the linear transformation $\mathcal{A}: \mathcal{X} \to \mathcal{Y}$, with $\mathcal{A}(\mathbf{x}) = \mathbf{y}$, has a solution for every $\mathbf{y} \in \mathcal{Y}$ if and only if $\mathcal{N}(\mathcal{A}^*) = \{\mathbf{0}\}$. Assume that $\mathcal{Y}$ is finite dimensional.

The linear transformation $\mathcal{A}^*$ and its finite dimensional domain $\mathcal{Y}$ can be used in place of $\mathcal{A}$ and $\mathcal{X}$ in the result of Problem 6.21. That is, $\mathcal{Y} = \mathcal{N}(\mathcal{A}^*) \oplus \mathcal{R}((\mathcal{A}^*)^*) = \mathcal{N}(\mathcal{A}^*) \oplus \mathcal{R}(\mathcal{A})$. If $\mathcal{N}(\mathcal{A}^*) = \{\mathbf{0}\}$, then $\mathcal{Y} = \mathcal{R}(\mathcal{A})$, so that every $\mathbf{y} \in \mathcal{Y}$ is the image of at least one $\mathbf{x} \in \mathcal{X}$. Note that $\mathbf{y} \in \mathcal{R}(\mathcal{A})$ is equivalent to the requirement for existence of solutions given in Section 6.2: $\text{rank}[\mathbf{A}] = \text{rank}[\mathbf{A} \quad \mathbf{y}]$. If $\mathcal{N}(\mathcal{A}^*) \neq \{\mathbf{0}\}$, then $\mathcal{Y} \neq \mathcal{R}(\mathcal{A})$; so there exists some $\mathbf{y} \in \mathcal{Y}$ but $\mathbf{y} \notin \mathcal{R}(\mathcal{A})$. Such a $\mathbf{y}$ is not the image of any $\mathbf{x} \in \mathcal{X}$.

6.24 Prove: $\mathcal{A}(\mathbf{x}) = \mathbf{y}$ has a *unique* solution for every $\mathbf{y} \in \mathcal{Y}$ if $\mathcal{N}(\mathcal{A}^*) = \{\mathbf{0}\}$ and $\mathcal{N}(\mathcal{A}) = \{\mathbf{0}\}$.

The results of Problem 6.23 guarantee that at least one solution exists. Assume $\mathbf{x}_1$ and $\mathbf{x}_2$ are two solutions, that is, $\mathcal{A}(\mathbf{x}_1) = \mathbf{y}$ and $\mathcal{A}(\mathbf{x}_2) = \mathbf{y}$. Then $\mathcal{A}(\mathbf{x}_1) - \mathcal{A}(\mathbf{x}_2) = \mathbf{0}$ or $\mathcal{A}(\mathbf{x}_1 - \mathbf{x}_2) = \mathbf{0}$. But since $\mathcal{N}(\mathcal{A}) = \{\mathbf{0}\}$, this requires that $\mathbf{x}_1 - \mathbf{x}_2 = \mathbf{0}$ or $\mathbf{x}_1 = \mathbf{x}_2$ is the unique solution. Note that if the domain $\mathcal{X}$ is n-dimensional, $\mathcal{N}(\mathcal{A}) = \{\mathbf{0}\}$ implies that $\text{rank}(\mathcal{A}^*) = \dim \mathcal{R}(\mathcal{A}^*) = \dim \mathcal{X} = n$. Also $\mathcal{N}(\mathcal{A}^*) = \{\mathbf{0}\}$ implies that $\text{rank}(\mathcal{A}) = \dim \mathcal{R}(\mathcal{A}) = \dim \mathcal{Y}$. But $\text{rank}(\mathcal{A}) = \text{rank}(\mathcal{A}^*)$, so a unique solution requires $\text{rank}(\mathcal{A}) = n$ as shown in Section 6.2, using a matrix representation $\mathbf{A}$ for $\mathcal{A}$.

6.25 If the domain and codomain for $\mathcal{A}$ are restricted so that $\mathcal{A}: \mathcal{R}(\mathcal{A}^*) \to \mathcal{R}(\mathcal{A})$, show that $\mathcal{A}$ is one-to-one and onto, and thus possesses an inverse.

Since, in general, $\mathcal{X} = \mathcal{N}(\mathcal{A}) \oplus \mathcal{R}(\mathcal{A}^*)$ and $\mathcal{Y} = \mathcal{N}(\mathcal{A}^*) \oplus \mathcal{R}(\mathcal{A})$, restricting $\mathcal{A}$ as stated guarantees that the conditions for a unique solution, as stated in Problem 6.24, are satisfied, so the restriction of $\mathcal{A}$ is one-to-one and onto.

26 Let $\mathcal{A} : \mathcal{X} \to \mathcal{Y}$ be a linear transformation for which $\mathcal{N}(\mathcal{A}) \neq \{0\}$, but $\mathcal{N}(\mathcal{A}^*) = \{0\}$.

(a) Show that the most general solution to $\mathcal{A}(\mathbf{x}) = \mathbf{y}$ is given by $\mathbf{x} = \mathbf{x}_0 + \mathbf{x}_1$, where $\mathbf{x}_1 \in \mathcal{R}(\mathcal{A}^*)$, $\mathbf{x}_0 \in \mathcal{N}(\mathcal{A})$.

(b) Show that $\mathbf{x}_1$ is the minimum norm solution given by $\mathbf{x}_1 = \mathcal{A}^*(\mathcal{A}\mathcal{A}^*)^{-1}\mathbf{y}$.

(1) For every $\mathbf{x} \in \mathcal{X}$, a unique decomposition is possible, $\mathbf{x} = \mathbf{x}_0 + \mathbf{x}_1$ with $\mathbf{x}_1 \in \mathcal{R}(\mathcal{A}^*)$, $\mathbf{x}_0 \in \mathcal{N}(\mathcal{A})$.

(2) Using this decomposition gives $\mathcal{A}(\mathbf{x}) = \mathcal{A}(\mathbf{x}_0 + \mathbf{x}_1) = \mathcal{A}(\mathbf{x}_1) = \mathbf{y}$. Since $\mathbf{x}_1 \in \mathcal{R}(\mathcal{A}^*)$, there exists a $\mathbf{y}_1 \in \mathcal{Y}$ such that $\mathcal{A}^*(\mathbf{y}_1) = \mathbf{x}_1$, from which $\mathcal{A}(\mathbf{x}_1) = \mathbf{y} = \mathcal{A}\mathcal{A}^*(\mathbf{y}_1)$. Since $\mathcal{A}\mathcal{A}^*$ is a one-to-one linear transformation from $\mathcal{R}(\mathcal{A}) = \mathcal{Y}$ onto itself, and thus possesses an inverse, $\mathbf{y}_1 = (\mathcal{A}\mathcal{A}^*)^{-1}\mathbf{y}$. Using $\mathbf{x}_1 = \mathcal{A}^*(\mathbf{y}_1)$ gives the minimum norm solution $\mathbf{x}_1 = \mathcal{A}^*(\mathcal{A}\mathcal{A}^*)^{-1}\mathbf{y}$. Any vector $\mathbf{x}_0 \in \mathcal{N}(\mathcal{A})$ can be added to $\mathbf{x}_1$ and the result is still a solution, but with a larger norm.

27 If $\mathcal{N}(\mathcal{A}^*) \neq \{0\}$ and $\mathbf{y} \notin \mathcal{R}(\mathcal{A})$, no solution to $\mathcal{A}(\mathbf{x}) = \mathbf{y}$ exists. Give a geometrical interpretation of the least-squares approximate solution, $\mathbf{x} = (\mathcal{A}^*\mathcal{A})^{-1}\mathcal{A}^*\mathbf{y}$.

Figure 6.9 illustrates the decomposition of $\mathcal{X}$ and $\mathcal{Y}$. For any $\mathbf{y} \in \mathcal{Y}$, $\mathbf{y} = \mathbf{z} + \mathbf{w}$ with $\mathbf{z} \in \mathcal{R}(\mathcal{A})$ and $\mathbf{w} \in \mathcal{N}(\mathcal{A}^*)$. Therefore

$$\mathcal{A}^*(\mathbf{y}) = \mathcal{A}^*(\mathbf{z} + \mathbf{w}) = \mathcal{A}^*(\mathbf{z})$$

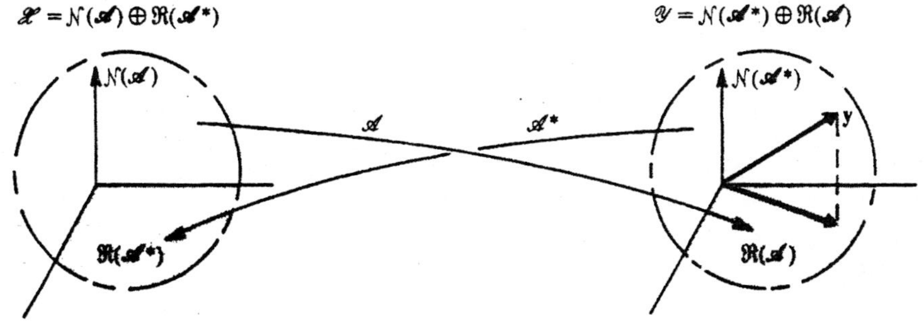

Figure 6.9

Since $\mathbf{z} \in \mathcal{R}(\mathcal{A})$, there exists some $\mathbf{x} \in \mathcal{X}$ such that $\mathcal{A}(\mathbf{x}) = \mathbf{z}$. Combining these results gives

$$\mathcal{A}^*(\mathbf{z}) = \mathcal{A}^*\mathcal{A}(\mathbf{x}) = \mathcal{A}^*(\mathbf{y})$$

If $\mathcal{N}(\mathcal{A}) \neq \{0\}$ there will be many vectors $\mathbf{x}$ satisfying $\mathcal{A}(\mathbf{x}) = \mathbf{z}$. The one such vector with minimum norm is selected, i.e. the one belonging to $\mathcal{R}(\mathcal{A}^*)$. With this restriction, $\mathcal{A}^*\mathcal{A}$ is a one-to-one transformation from $\mathcal{R}(\mathcal{A}^*)$ onto itself and thus possesses an inverse. The least-squares solution is $\mathbf{x} = (\mathcal{A}^*\mathcal{A})^{-1}\mathcal{A}^*(\mathbf{y})$. This solution is the pre-image of the projection of $\mathbf{y}$ on $\mathcal{R}(\mathcal{A})$.

28 Prove that every finite dimensional linear transformation is bounded.

Let $\mathcal{A} : \mathcal{X}^n \to \mathcal{X}^m$, with $\mathbf{x} \in \mathcal{X}^n$. Let $\{\mathbf{v}_i, i = 1, 2, \ldots, n\}$ be an orthonormal basis for $\mathcal{X}^n$. Then $\mathbf{x} = \sum_{i=1}^{n} \alpha_i \mathbf{v}_i$ and $\mathcal{A}(\mathbf{x}) = \sum_{i=1}^{n} \alpha_i \mathcal{A}(\mathbf{v}_i)$. Therefore,

$$\|\mathcal{A}(\mathbf{x})\| = \left\|\sum_{i=1}^{n} \alpha_i \mathcal{A}(\mathbf{v}_i)\right\| \le \sum_{i=1}^{n} |\alpha_i| \|\mathcal{A}(\mathbf{v}_i)\|$$

Since the vectors $\mathbf{y}_i \triangleq \mathcal{A}(\mathbf{v}_i)$ belong to $\mathcal{X}^m$, they have a finite norm. Define $K = \max_{i=1,\ldots,n} \|\mathbf{y}_i\|$. Then

$$\|\mathcal{A}(\mathbf{x})\| \le \sum_{i=1}^{n} |\alpha_i| \|\mathbf{y}_i\| \le K \sum_{i=1}^{n} |\alpha_i|$$

But $\alpha_i = \langle \mathbf{v}_i, \mathbf{x}\rangle$ so that the Cauchy-Schwarz inequality gives $|\alpha_i| \le \|\mathbf{v}_i\|\cdot\|\mathbf{x}\| = \|\mathbf{x}\|$ since $\|\mathbf{v}_i\| = 1$. Using this gives $\|\mathcal{A}(\mathbf{x})\| \le Kn\|\mathbf{x}\|$, so that $\mathcal{A}$ is clearly bounded [Eq. (5.6)]. Notice the dependence on the finite dimension n.

6.29 Let $\mathcal{A}: \mathcal{X} \to \mathcal{Y}$ be a linear transformation such that $(\mathcal{A}^*\mathcal{A})^{-1}$ exists. Prove that $\mathcal{A}(\mathcal{A}^*\mathcal{A})^{-1}\mathcal{A}^*$ is the orthogonal projection of $\mathcal{Y}$ onto $\mathcal{R}(\mathcal{A})$.

The indicated transformation is a projection if and only if

$$[\mathcal{A}(\mathcal{A}^*\mathcal{A})^{-1}\mathcal{A}^*][\mathcal{A}(\mathcal{A}^*\mathcal{A})^{-1}\mathcal{A}^*] = \mathcal{A}(\mathcal{A}^*\mathcal{A})^{-1}\mathcal{A}^*$$

This condition is obviously satisfied. Decompose $\mathcal{Y} = \mathcal{N}(\mathcal{A}^*) \oplus \mathcal{R}(\mathcal{A})$ and let $\mathbf{y} = \mathbf{y}_0 + \mathbf{y}_1$, where $\mathbf{y}_0 \in \mathcal{N}(\mathcal{A}^*)$, $\mathbf{y}_1 \in \mathcal{R}(\mathcal{A})$. The orthogonal projection of $\mathbf{y}$ onto $\mathcal{R}(\mathcal{A})$ is $\mathbf{y}_1$ by definition.

We must show that $\mathcal{A}(\mathcal{A}^*\mathcal{A})^{-1}\mathcal{A}^*\mathbf{y} = \mathbf{y}_1$. First note that

$$\mathcal{A}^*(\mathbf{y}) = \cancelto{0}{\mathcal{A}^*(\mathbf{y}_0)} + \mathcal{A}^*(\mathbf{y}_1)$$

Both $\mathbf{y}_1$ and $\mathcal{A}(\mathcal{A}^*\mathcal{A})^{-1}\mathcal{A}^*\mathbf{y}_1 \in \mathcal{R}(\mathcal{A}) = \mathcal{N}(\mathcal{A}^*)^\perp$ so their difference, $\mathbf{e}$, does also. Applying the adjoint transformation $\mathcal{A}^*$ to their difference gives

$$\mathcal{A}^*[\mathcal{A}(\mathcal{A}^*\mathcal{A})^{-1}\mathcal{A}^*\mathbf{y}_1 - \mathbf{y}_1] = \mathbf{0}$$

Since $\mathbf{e} \in \mathcal{N}(\mathcal{A}^*)^\perp$ and $\mathcal{A}^*(\mathbf{e}) = \mathbf{0}$, the conclusion is that $\mathbf{e} = \mathbf{0}$. This leads to $\mathbf{y}_1 = \mathcal{A}(\mathcal{A}^*\mathcal{A})^{-1}\mathcal{A}^*\mathbf{y}$.

PROBLEMS

Miscellaneous

6.30 Solve for $\mathbf{x}$ if $\begin{bmatrix} 8 & 2 & 1 \\ 1 & 1 & 3 \\ 2 & 5 & 4 \end{bmatrix}\mathbf{x} = \begin{bmatrix} 10 \\ 5 \\ 1 \end{bmatrix}$.

6.31 Assume that a dynamic system can be described by a vector of time-varying parameters $\mathbf{x}(t)$, with initial conditions $\mathbf{x}(0)$. The relation between $\mathbf{x}(t)$ and $\mathbf{x}(0)$ is $\mathbf{x}(t) = \mathbf{\Phi}(t)\mathbf{x}(0)$, where $\mathbf{\Phi}(t)$ is an $n \times n$ matrix. Let $\mathbf{x}$ be partitioned into $\begin{bmatrix} \mathbf{x}_1(t) \\ \hline \mathbf{x}_2(t) \end{bmatrix}$. If $\mathbf{x}_1(0)$ and $\mathbf{x}_2(T)$ are known, find $\mathbf{x}_2(0)$.

Homogeneous Equations

6.32 If $\mathbf{Ax} = \mathbf{0}$ has q linearly independent solutions $\mathbf{x}_i$ and $\mathbf{Ax} = \mathbf{y}$ has $\mathbf{x}_0$ as a solution, show that

(a) $\mathbf{x}_c = \sum_{i=1}^{q} \alpha_i \mathbf{x}_i$ is also a solution of $\mathbf{Ax} = \mathbf{0}$,

(b) $\mathbf{x} = \mathbf{x}_0 + \sum_{i=1}^{q} \alpha_i \mathbf{x}_i$ is a solution of $\mathbf{Ax} = \mathbf{y}$.

6.33 Find the nontrivial solutions for $\begin{bmatrix} 1 & 0 & 4 \\ 2 & 3 & 8 \end{bmatrix}\begin{bmatrix} x_1 \\ x_2 \\ x_3 \end{bmatrix} = \mathbf{0}$.

6.34 Find all nontrivial solutions for $\begin{bmatrix} 1 & 1 & 1 \\ 2 & 2 & 2 \end{bmatrix}\begin{bmatrix} x_1 \\ x_2 \\ x_3 \end{bmatrix} = \mathbf{0}$.

35 Determine whether nontrivial solutions exist for $\begin{bmatrix} 1 & 2 \\ 2 & 4 \\ 0 & 0 \\ -1 & -2 \end{bmatrix}\begin{bmatrix} x_1 \\ x_2 \end{bmatrix} = \mathbf{0}.$

36 Find all nontrivial solutions of $\mathbf{Ax} = \mathbf{0}$, i.e, the null space, of

$$\mathbf{A} = \begin{bmatrix} 26 & 17 & 8 & 39 & 35 \\ 17 & 13 & 9 & 29 & 28 \\ 8 & 9 & 10 & 19 & 21 \\ 39 & 29 & 19 & 65 & 62 \\ 35 & 28 & 21 & 62 & 61 \end{bmatrix}$$

Least Squares

37 Solve for x_1 and x_2 if
(a) $2x_1 - x_2 = 5, x_1 + 2x_2 = 3$,
(b) in addition to the equations in **a**, a third equation is $-x_1 + x_2 = -1$. Use least squares.

38 Given that $\begin{bmatrix} y_1 \\ y_2 \end{bmatrix} = \begin{bmatrix} 2 \\ 1 \end{bmatrix} x + \begin{bmatrix} e_1 \\ e_1 \end{bmatrix}$. Measurements give $[y_1 \quad y_2] = [3 \quad 4]$. Find the least-squares estimate for x. Use a sketch in the y_1, y_2 plane to indicate the geometrical interpretation.

39 Verify the result of Problem 5.35, page 201 by determining the least-squares solution for $\mathbf{x}$:

$$\begin{bmatrix} 1 & 0 \\ 2 & 1 \\ -3 & 3 \\ 1 & 3 \\ 0 & 1 \end{bmatrix}\begin{bmatrix} x_1 \\ x_2 \end{bmatrix} = \begin{bmatrix} 1 \\ -3 \\ 4 \\ 2 \\ 8 \end{bmatrix}$$

and then use the fact that the orthogonal projection of $\mathbf{y}$ on the column space of $\mathbf{A}$ is $\mathbf{y}_p = \mathbf{Ax}$.

40 A physical device is shown in Figure 6.10. It is believed that the output y is linearly related to the input u. That is, $y = au + b$. What are the values of a and b if the following data are taken?

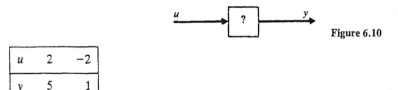

Figure 6.10

u	2	−2
y	5	1

41 The same device as in Problem 6.40 is considered. One more set of readings is taken as

$$u = 5, \qquad y = 7$$

Find a least-squares estimate of a and b. Also find the minimum mean-squared error in this straight line fit to the three points.

42 Consider the data of Problem 6.41, but assume that the first two equations are much more reliable. Use

$$\mathbf{R}^{-1} = \begin{bmatrix} 10 & 0 & 0 \\ 0 & 10 & 0 \\ 0 & 0 & 1 \end{bmatrix}$$

and show that the resultant weighted least-squares estimate is much closer to the values obtained in Problem 6.40.

6.43 Estimate the initial current $i(0)$ in the circuit of Figure 6.11, if $R = 10\ \Omega$, $L = 3.56$ H, and the following voltmeter readings are taken:

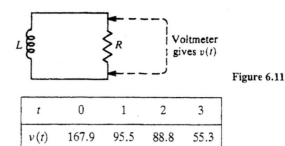

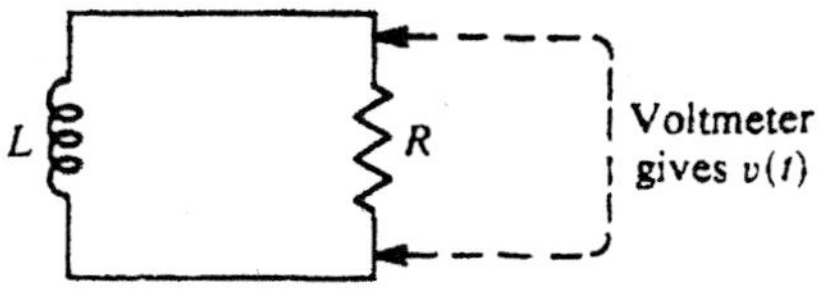

Figure 6.11

t	0	1	2	3
$v(t)$	167.9	95.5	88.8	55.3

6.44 Least-squares fit a quadratic function to the data of Problem 6.15. Determine the coefficients and the norm of the residual error. Then predict the GPA after semester eight.

6.45 An empirical theory used by many distance runners states that the time T_i required to race a distance D_i can be expressed as $T_i = C(D_i)^\alpha$, where C and α are constants for a given person, determined by lung capacity, body build, etc. Obtain a least-squares fit to the following data for one middle-aged jogger. (Convert to a linear equation in the unknowns C and α by taking the logarithm of the above expression.) Predict the time for one mile.

Time	185 min	79.6 min	60 min	37.9 min	11.5 min
Distance	26.2 mi	12.4 mi	9.5 mi	6.2 mi	2 mi

6.46 Apply the recursive least squares algorithm to the data of Examples 6.4 and 6.7. Try different starting assumptions to determine their effect on the estimate. Recall that **A** is not full rank and thus a unique least-squares solution does not exist. Add a fifth equation,

$$4x_1 - x_2 + 6x_3 = 2$$

so that the enlarged **A** matrix is of full rank. How do your results compare with Example 6.7?

6.47 How should the results for the minimum norm and least-squares solutions of Problems 6.26 and 6.27 be modified if a weighted norm or weighted least-squares solution is desired?

Lyapunov Equations

6.48 Solve $\mathbf{XA} + \mathbf{BX} = \mathbf{C}$ for $\mathbf{X}$ if

$$\mathbf{A} = \begin{bmatrix} 1 & 3 \\ 0 & 5 \end{bmatrix}, \quad \mathbf{B} = \mathbf{A}^T, \quad \text{and} \quad \mathbf{C} = \begin{bmatrix} -1 & 0 \\ 0 & -1 \end{bmatrix}$$

6.49 Solve $\mathbf{XA} + \mathbf{BX} = \mathbf{C}$ for $\mathbf{X}$ if

$$\mathbf{A} = \begin{bmatrix} 0 & 1 \\ -2 & -3 \end{bmatrix}, \quad \mathbf{B} = \begin{bmatrix} 1 & 0 \\ 3 & 5 \end{bmatrix} \quad \text{and} \quad \mathbf{C} = \begin{bmatrix} -1 & 0 \\ 0 & -1 \end{bmatrix}$$

7

Eigenvalues and Eigenvectors

7.1 INTRODUCTION

This chapter defines eigenvalues and eigenvectors (also referred to as proper, or characteristic, values and vectors). Methods of determining eigenvalues and eigenvectors are presented as well as some of their more important properties and uses. For a linear continuous-time system—i.e., Eqs. (*3.11*), (*3.12*)—or for a linear discrete-time system—i.e., Eqs. (*3.13*), (*3.14*)—the eigenvalues of the $\mathbf{A}$ matrix completely determine system stability. The eigenvectors of $\mathbf{A}$ form a very convenient choice for basis vectors in state space. When a full set of eigenvectors can be found, it will be shown that the nth-order system can be transformed into an *uncoupled* set of n first-order equations. Each equation describes one natural mode of the system. The uncoupled form allows for easier analysis as well as providing greater insight into the system's structural properties. Unfortunately, not all matrices have a full set of eigenvectors. This can happen only when the matrix has repeated eigenvalues, plus an additional condition to be described in detail later. The most annoying complications that arise in the eigenvalue-eigenvector problem are due to this degenerate case, where less than a full set of eigenvectors exist. It has sometimes been argued (erroneously) that the repeated eigenvalue case is purely academic because small computational differences will always exist between any two eigenvalues. In fact, the degenerate case cannot be avoided so easily. Additional vectors, called generalized eigenvectors, will be defined and used to supplement the eigenvectors when necessary. Doing this will lead to a system which is as close to being uncoupled as possible.

7.2 DEFINITION OF THE EIGENVALUE-EIGENVECTOR PROBLEM

Let $\mathcal{A}$ be any linear transformation with domain $\mathfrak{D}(\mathcal{A})$ and range $\mathfrak{R}(\mathcal{A})$, both contained within the same linear vector space $\mathcal{X}$. Let elements of $\mathcal{X}$ be denoted as $\mathbf{x}_i$. Those particular elements $\mathbf{x}_i \neq \mathbf{0}$ and the particular scalars $\lambda_i \in \mathcal{F}$ which satisfy

$$\mathcal{A}(\mathbf{x}_i) = \lambda_i \mathbf{x}_i \tag{7.1}$$

are called *eigenvectors* and *eigenvalues,* respectively. Note that the trivial case $\mathbf{x} = \mathbf{0}$ is explicitly excluded. Thus λ_i is an eigenvalue if and only if the transformation $\mathcal{A} - \mathcal{I}\lambda_i$ has no inverse. The set of all scalars λ for which this is true is called the *spectrum of* $\mathcal{A}$.

The eigenvector problem of Eq. (*7.1*) applies to a more general class of operators than is needed here [1]. With the exception of the material on singular value decomposition, the transformations considered in this chapter map elements in $\mathcal{X}^n$ into other elements in $\mathcal{X}^n$, so $\mathcal{A}$ can be represented by an $n \times n$ matrix $\mathbf{A}$. The identity transformation is represented by the unit matrix $\mathbf{I}$. The matrix representation of Eq. (*7.1*) is

$$(\mathbf{A} - \mathbf{I}\lambda_i)\mathbf{x}_i = \mathbf{0} \tag{7.2}$$

and the determination of eigenvectors is a matter of finding nontrivial solutions to a set of n homogeneous equations. If scalar eigenvalues λ_i are known, then any of the techniques of the previous chapter for solving simultaneous linear homogeneous equations can be used to find the corresponding eigenvectors $\mathbf{x}_i$. The various row-reduced-echelon- and Gram-Schmidt-based methods are easily adaptable to machine computation for this purpose. However, the eigenvalue-eigenvector problem is actually more difficult than those considered in Sec. 6.6 because the scalar λ_i is also unknown. This leads to a *nonlinear* problem because the product of the unknowns λ_i and $\mathbf{x}_i$ enters into the equations. As a starting point for this discussion, the determination of the eigenvalues is isolated and solved first. Once this is done, the remaining problem of determining the eigenvectors is linear, exactly of the type treated in Sec. 6.6. While splitting the problem this way is a customary method of discussion, it is not necessarily the best computational approach. A direct computational attack on the simultaneous determination of eigenvalues and eigenvectors is more efficient for many problems.

7.3 EIGENVALUES

It was shown in Chapter 6 that a necessary condition for the existence of nontrivial solutions to the set of n homogeneous equations (7.2) is that $\text{rank}(\mathbf{A} - \mathbf{I}\lambda_i) < n$. This is equivalent to requiring $|\mathbf{A} - \mathbf{I}\lambda_i| = 0$. When the determinant is expanded, it yields an nth-degree polynomial in the scalar λ_i—that is,

$$|\mathbf{A} - \mathbf{I}\lambda| = (-\lambda)^n + c_{n-1}\lambda^{n-1} + c_{n-2}\lambda^{n-2} + \cdots + c_1\lambda + c_0 = \Delta(\lambda) \tag{7.3}$$

The roots of this algebraic equation are the eigenvalues λ_i. A fundamental result in algebra states that an nth degree polynomial has exactly n roots, so every $n \times n$ matrix $\mathbf{A}$ has exactly n eigenvalues. The nth-degree polynomial in λ is called the *characteristic polynomial* and the characteristic equation is $\Delta(\lambda) = 0$. In factored form,

$$\Delta(\lambda) = (-1)^n(\lambda - \lambda_1)(\lambda - \lambda_2)\cdots(\lambda - \lambda_n) = 0$$

and the roots are $\lambda_1, \lambda_2, \ldots, \lambda_n$. In general, some of these roots may be equal. If there are $p < n$ distinct roots, $\Delta(\lambda)$ takes the form

$$\Delta(\lambda) = (-1)^n(\lambda - \lambda_1)^{m_1}(\lambda - \lambda_2)^{m_2}\cdots(\lambda - \lambda_p)^{m_p}$$

This indicates that $\lambda = \lambda_1$ is an m_1-order root, $\lambda = \lambda_2$ is an m_2-order root, etc. The integer m_i is called the *algebraic multiplicity* of the eigenvalue λ_i. Of course, $m_1 + m_2 + \cdots + m_p = n$.

The problem of determining eigenvalues for $\mathscr{A}$ amounts to factoring an nth degree polynomial. For large n this is not an easy computational problem, although it is conceptually simple and will not be discussed here. Section 7.6 presents a direct iterative method of determining eigenvalues and eigenvectors, which, when applicable, avoids factoring the characteristic polynomial.

Many relationships exist between $\mathbf{A}$, c_i, and λ_i, three of which are

1. If the scalars c_i in Eq. (7.3) are real (the usual case), then if λ_i is a complex eigenvalue, so is $\bar{\lambda}_i$.
2. $\text{Tr}(\mathbf{A}) = \lambda_1 + \lambda_2 + \cdots + \lambda_n = (-1)^{n+1} c_{n-1}$
3. $|\mathbf{A}| = \lambda_1 \lambda_2 \cdots \lambda_n = c_0$.

The characteristic polynomial is often defined by $|\mathbf{I}\lambda_i - \mathbf{A}|$ rather than $|\mathbf{A} - \mathbf{I}\lambda_i|$. This leaves the roots unaltered but changes Eq. (7.3) by a factor $(-1)^n$ (see Problem 7.46).

7.4 DETERMINATION OF EIGENVECTORS

The procedure for determining eigenvectors can be divided into two possible cases, depending on the results of the eigenvalue calculations.

Case I: All the eigenvalues are distinct.

Case II: Some eigenvalues are multiple roots of the characteristic equation.

Case I: Distinct Eigenvalues

When each of the eigenvalues has algebraic multiplicity of one (i.e., they are all simple, distinct roots), then rank$(\mathbf{A} - \mathbf{I}\lambda)$ will be $n - 1$. This means that there is only one independent nontrivial solution to the homogeneous equation

$$(\mathbf{A} - \mathbf{I}\lambda_i)\mathbf{x}_i = \mathbf{0}$$

There are any number of methods of solving this equation for the eigenvector $\mathbf{x}_i$. One which is easy to use for hand computations on small matrices is to compute the adjoint matrix Adj$(\mathbf{A} - \mathbf{I}_n \lambda)$, leaving λ as a parameter. Then, successively substituting the value for each λ_i and selecting any nonzero column will give each $\mathbf{x}_i$ eigenvector in turn. The overhead of computing the adjoint matrix is done only once, and the result gives all simple eigenvectors. Other methods involve reducing $\mathbf{A} - \mathbf{I}_n \lambda_j$ to a purely numeric matrix for each eigenvalue. Then row-reduced-echelon methods, Gram-Schmidt decomposition methods, singular-value decomposition (SVD) methods (see Problems 7.29 through 7.34), or other numerical methods of solution can be applied. The eigenvectors are not unique. If $\mathbf{x}_i$ is an eigenvector, then so is $\alpha\mathbf{x}_i$ for any nonzero scalar α. This fact is often used to normalize the eigenvectors, perhaps so that $\|\mathbf{x}_j\| = 1$. Another common normalization forces the largest component of $\mathbf{x}_j$ to be unity. Regardless of method, a full set of n linearly independent eigenvectors can always be found for this case. They satisfy $\mathbf{Ax}_1 = \lambda_1 \mathbf{x}_1, \mathbf{Ax}_2 = \lambda_2 \mathbf{x}_2, \ldots, \mathbf{Ax}_n = \lambda_n \mathbf{x}_n$. By defining an $n \times n$ *modal matrix* $\mathbf{M} = [\mathbf{x}_1\ \mathbf{x}_2\ \ldots\ \mathbf{x}_n]$ and an $n \times n$ diagonal matrix $\mathbf{\Lambda}$ with the ith eigenvalue in the i, i position, all n eigenvalue-eigenvector equations can be combined into one matrix equation, $\mathbf{AM} = \mathbf{M\Lambda}$. Since the eigenvectors form a linearly independent set, the rank of $\mathbf{M}$ is n, and $\mathbf{M}^{-1}$ exists. Therefore, $\mathbf{\Lambda} = \mathbf{M}^{-1}\mathbf{AM}$. This shows that a matrix $\mathbf{A}$, which has distinct eigenvalues, can always be transformed to a diagonal matrix $\mathbf{\Lambda} = \text{diag}[\lambda_1\ \lambda_2\ \lambda_3\ \ldots\ \lambda_n]$ by a *similarity transformation.* A similarity transformation is a relationship between two square matrices $\mathbf{A}$ and $\mathbf{B}$ of the form $\mathbf{B} = \mathbf{Q}^{-1}\mathbf{AQ}$ for any nonsingular matrix $\mathbf{Q}$. In the particular case before where $\mathbf{B}$ was the diagonal matrix, the modal matrix played the role of $\mathbf{Q}$. In some cases (see Problems 7.25 and 7.27) the eigenvectors are mutually orthogonal and, when normalized, constitute an orthonormal set. In this case $\mathbf{M}$ is an orthogonal matrix, i.e., $\mathbf{M}^{-1} = \mathbf{M}^T$. The similarity transformation simplifies to an *orthogonal* transformation $\mathbf{\Lambda} = \mathbf{M}^T\mathbf{AM}$ in that case.

EXAMPLE 7.1 Consider the unforced portion of the state variable model

$$\dot{\mathbf{x}} = \begin{bmatrix} 0 & 1 & 0 \\ 0 & 0 & 1 \\ -18 & -27 & -10 \end{bmatrix} \mathbf{x}$$

(These equations are from Problem 3.3.) Find the eigenvalues and eigenvectors of the 3×3 matrix $\mathbf{A}$. Then form the modal matrix.

First find the eigenvalues. The characteristic equation is

$$|\mathbf{A} - \mathbf{I}\lambda| = -\lambda^3 - 10\lambda^2 - 27\lambda - 18 = 0$$

Note the correspondence between the coefficients of the characteristic polynomial and the entries in the last row of $\mathbf{A}$. This is not a coincidence and always occurs when $\mathbf{A}$ is expressed in companion form, as it is here. To eliminate the minus signs, the characteristic equation can obviously be multiplied by -1 without altering its roots. This is equivalent to writing $|\mathbf{I}\lambda - \mathbf{A}| = 0$. The cubic polynomial has three roots, $\lambda_1 = -1$, $\lambda_2 = -3$, and $\lambda_3 = -6$. Note that these three distinct roots have the same values as the poles of the transfer function from which the state equations were derived (Problem 3.3). This is also not a coincidence. Next find the eigenvectors. Four different methods are demonstrated.

1. The adjoint matrix is

$$\text{Adj}(\mathbf{A} - \mathbf{I}\lambda) = \begin{bmatrix} \lambda^2 + 10\lambda + 27 & \lambda + 10 & 1 \\ -18 & \lambda^2 + 10\lambda & \lambda \\ -18\lambda & -27\lambda - 18 & \lambda^2 \end{bmatrix}$$

Any nonzero column can be selected to form the eigenvectors. For example, column 1 with $\lambda = -1$ gives $\mathbf{x}_1 = [18\quad -18\quad 18]^T$, column 2 with $\lambda = -3$ gives $\mathbf{x}_2 = [7\quad -21\quad 63]^T$, and so on. However, column 3 is clearly the easiest to use for all three eigenvectors. It is seen that $\mathbf{x}_i = [1\quad \lambda_i\quad \lambda_i^2]^T$, and the modal matrix becomes

$$\mathbf{M} = \begin{bmatrix} 1 & 1 & 1 \\ \lambda_1 & \lambda_2 & \lambda_3 \\ \lambda_1^2 & \lambda_2^2 & \lambda_3^2 \end{bmatrix} = \begin{bmatrix} 1 & 1 & 1 \\ -1 & -3 & -6 \\ 1 & 9 & 36 \end{bmatrix}$$

This matrix has a very special form, due to the special companion form of the matrix **A**, and is called a *Vandermonde matrix*. Notice that the numerical values selected before from the other columns are just scalar multiples of the same vectors.

2. The row-reduced echelon form of $\mathbf{A} - \mathbf{I}(-3)$ is easily found to be $\begin{bmatrix} 1 & 0 & -\frac{1}{9} \\ 0 & 1 & \frac{1}{3} \\ 0 & 0 & 0 \end{bmatrix}$. From this it is clear that the rank is 2, as it must be for a nonrepeated root, and that the eigenvector for $\lambda = -3$ is $\mathbf{x} = [1 \quad -3 \quad 9]^T$, as before.

3. The Gram-Schmidt-based **QR** decomposition of $\mathbf{A} - \mathbf{I}(-1)$ is found by the computer to be

$$\begin{bmatrix} 0.05547 & -0.44597 & -0.89332 \\ 0 & 0.89470 & -0.44666 \\ -0.99846 & -0.024776 & 0.09629 \end{bmatrix} \begin{bmatrix} 18.0277 & 27.0139 & 8.9861 \\ 0 & 1.11769 & 1.11769 \\ 0 & 0 & 0 \end{bmatrix} = [\mathbf{Q}][\mathbf{R}]$$

Since **Q** is nonsingular, $\mathbf{QRx} = \mathbf{0}$ is equivalent to just $\mathbf{Rx} = \mathbf{0}$. This triangular set of equations has all the same advantages as the RRE form and leads to the eigenvector $\mathbf{x} = [1 \quad -1 \quad 1]^T$ for $\lambda = -1$.

4. The singular-value decomposition of $\mathbf{A} - \mathbf{I}(-6)$ is $\mathbf{U\Sigma V}^T$, where

$$\mathbf{U} = \begin{bmatrix} -0.12408 & 0.79904 & -0.58835 \\ -0.15213 & -0.60122 & -0.78446 \\ 0.98055 & 0.00777 & -0.196116 \end{bmatrix}, \qquad \mathbf{\Sigma} = \begin{bmatrix} 33.3441 & 0 & 0 \\ 0 & 5.5832 & 0 \\ 0 & 0 & 0 \end{bmatrix}$$

and

$$\mathbf{V} = \begin{bmatrix} -0.55164 & 0.83363 & 0.0273896 \\ -0.82508 & -0.54058 & -0.164337 \\ -0.12219 & -0.11325 & 0.986024 \end{bmatrix}$$

Solving $\mathbf{U\Sigma V}^T\mathbf{x} = \mathbf{0}$ is simplified to solving $\mathbf{\Sigma V}^T\mathbf{x} = \mathbf{0}$ because **U** is orthogonal. Let $\mathbf{V}^T\mathbf{x} = \mathbf{w}$ temporarily. Then, because $\Sigma_{3,3} = 0$, $\mathbf{\Sigma w} = \mathbf{0}$ implies that $\mathbf{w} = [0 \quad 0 \quad 1]^T$ or some scalar multiple. **V** is also orthogonal, so that $\mathbf{x} = \mathbf{Vw}$, that is, the eigenvector **x** is that column (or those columns) of **V** which correspond to the zero elements in $\mathbf{\Sigma}$. In this case, column 3 of **V** is the eigenvector, and it is proportional to $[1 \quad -6 \quad 36]^T$, as found earlier. The extra complications of the **QR** and SVD decompositions would normally rule out these methods for hand calculations. However, they form the basis for reliable machine computations. Section 7.6 shows how the **QR** decomposition can be used to find the eigenvalues as well as the eigenvectors. ■

EXAMPLE 7.2 Use the modal matrix found in the previous example to decouple the modes of the state variable system.

Define a new vector by $\mathbf{x} = \mathbf{Mw}$. Actually this is the same state vector expressed with respect to a new basis set consisting of the eigenvectors. That is, **x** can be written in expanded form as

$$\mathbf{x} = w_1\mathbf{x}_1 + w_2\mathbf{x}_2 + \cdots + w_n\mathbf{x}_n$$

The original state equations become $\mathbf{M\dot{w}} = \mathbf{AMw}$, and upon premultiplying by $\mathbf{M}^{-1}$, the result is totally uncoupled, $\mathbf{\dot{w}} = \mathbf{\Lambda w}$. The three components of **w** satisfy $\dot{w}_1 = -w_1$, $\dot{w}_2 = -3w_2$, and $\dot{w}_3 = -6w_3$. These uncoupled scalar equations each have an exponential solution of the form $w_i(t) = \exp(\lambda_i t)w_i(0)$. This shows that if any eigenvalue has a positive real part, the corresponding component of $\mathbf{w}$ will grow without bound, thus forcing the entire vector $\mathbf{w}$ to infinity. The lesson is that the eigenvalues of the matrix $\mathbf{A}$ completely determine the stability of a linear, constant coefficient system. This should come as no surprise, since it was shown earlier that transfer function poles are also eigenvalues. Pole locations are at the center of stability discussions in classical control theory, as reviewed in Chapter 2. ■

Case II: Repeated Eigenvalues

When one or more eigenvalues are repeated roots of the characteristic equation, a full set of eigenvectors may or may not exist, and a deeper analysis is required. The question of whether two roots such as 4.000001 and 3.99999 are really numerical approximations of the same root or if they are distinct is postponed temporarily. The clean idealistic case with a binary yes or no answer to the repeated-root question is addressed first. The number of linearly independent eigenvectors associated with an eigenvalue λ_i repeated with an algebraic multiplicity m_i is equal to the dimension of the null space of $\mathbf{A} - \mathbf{I}\lambda_i$. This dimension is given by

$$q_i = n - \text{rank}(\mathbf{A} - \mathbf{I}\lambda_i) \qquad \text{(see Problem 7.18)}$$

and is called the degeneracy of $\mathbf{A} - \mathbf{I}\lambda_i$. The degeneracy is also called the *geometric multiplicity* of λ_i because it is the dimension of the subspace spanned by the eigenvectors. The distinction between the algebraic multiplicity m_i and the geometric multiplicity q_i of a repeated eigenvalue is crucial to finding the associated eigenvectors and, if needed, generalized eigenvectors. First, notice that the range of possible values for the integer q_i is given by $1 \le q_i \le m_i$. For example, a given matrix $\mathbf{A}$ might have λ_i as a triple root of $\Delta(\lambda) = 0$ so that $m_i = 3$. Yet there may only be one eigenvector ($q_i = 1$) or perhaps two eigenvectors ($q_i = 2$) or even a full set of three. An important point is restated for emphasis. *Every $n \times n$ matrix $\mathbf{A}$ always has a full set of n eigenvalues, but it might not have a full set of n independent eigenvectors.* It is convenient to consider three subclassifications for Case II.

***Case II$_1$*: The fully degenerate case, $q_i = m_i$.** The fully degenerate case has a full set of m_i eigenvectors associated with the repeated root λ_i. They can be found by the same types of methods described for Case I, with only minor modifications. In numerical methods such as the row-reduced-echelon, Gram-Schmidt **QR**, or SVD methods, the modification is fairly obvious: There will be q_i independent solutions to $(\mathbf{A} - \mathbf{I}\lambda_i)\mathbf{x}_i = \mathbf{0}$ instead of just one, as in Case I and Example 7.1. In the adjoint matrix method the modification is not quite so obvious, but it is based upon Problem 6.4. The m_i independent eigenvectors associated with the repeated root can be selected as independent columns of the differentiated adjoint matrix

$$\frac{1}{(m_i - 1)!}\left\{\frac{d^{m_i - 1}}{d\lambda^{m_i - 1}}[\text{Adj}(\mathbf{A} - \mathbf{I}\lambda)]\right\}_{\lambda = \lambda_i}$$

A further complication here as compared with Case I is that if $\mathbf{x}_1, \mathbf{x}_2, \ldots, \mathbf{x}_m$ are eigenvectors of $\mathbf{A}$, then so is *every* vector $\mathbf{y}$ in the subspace spanned by the $\mathbf{x}_i$ vectors. That is, any $\mathbf{y} = \Sigma\alpha_i \mathbf{x}_i$ also satisfies $(\mathbf{A} - \mathbf{I}\lambda_i)\mathbf{y} = \mathbf{0}$. This makes it more difficult to recognize the equivalence of the eigenvectors found by different solution methods, as is demonstrated in Example 7.3. Notice that Case I is really a subset of Case II_1 with $m_i = 1$. The category must be determined separately for each eigenvalue. The only time a matrix will have a full set of eigenvectors is when each of its eigenvalues is in either Case I or Case II_1. In this situation the modal matrix **M** is formed as before, and the similarity transformation $\mathbf{M}^{-1}\mathbf{AM}$ will again give a diagonal matrix $\mathbf{\Lambda}$ with λ_i as its diagonal elements. The state equations associated with a matrix **A** with these properties can be fully decoupled, just as in Example 7.2. It is known in advance that real, symmetric matrices and Hermitian matrices will always meet these conditions and thus have a full set of eigenvectors. This property extends to the entire class of *normal* transformations defined in Sec. 5.12 and revisited in Problems 7.26 through 7.28. Many physical matrices, including impedance and admittance matrices of circuit analysis, fall into this group. This might lead to the erroneous conclusion that in practical problems a full set of eigenvectors will always exist. In control theory, the controllable canonical form of the state equations gives a matrix **A** in companion form. A companion form matrix will *always* have just one eigenvector for each eigenvalue, regardless of the multiplicity of the eigenvalues. (See Problem 7.36.)

EXAMPLE 7.3 Find the eigenvalues, eigenvectors, modal matrix, and diagonal form of

$$\mathbf{A} = \begin{bmatrix} \frac{10}{3} & 1 & -1 & -\frac{1}{3} \\ 0 & 4 & 0 & 0 \\ -\frac{2}{3} & 1 & 3 & -\frac{1}{3} \\ -\frac{2}{3} & 1 & -1 & \frac{11}{3} \end{bmatrix}$$

It is impossible to represent this matrix exactly with a finite number of digits. Keeping six-digit input accuracy, a computer routine gave the characteristic equation

$$\Delta(\lambda) = \lambda^4 - 14\lambda^3 + 72\lambda^2 - 160\lambda + 128 = 0$$

and the (approximate) roots were found to be

$$\lambda_1 = 1.999999$$

$$\lambda_2 = 3.999980$$

$$\lambda_3 = 3.999982$$

$$\lambda_4 = 4.000017$$

Is this a case of repeated roots? Here it is known that the *exact* eigenvalues are $\{2, 4, 4, 4\}$, but in a general computer solution how is the question answered? The answer directly relates to the numerical determination of the rank of a matrix. Several solutions to this problem will be presented to demonstrate this point. Actually, $\lambda_1 = 2$ is a simple root, so Case I applies. Also $\lambda_2 = 4$ is a triple root ($m_2 = 3$). In order to determine the degeneracy q_2, rank$(\mathbf{A} - 4\mathbf{I})$ must be found. Three of the four solution procedures will indicate rank automatically as part of the solution process. First the simple root is treated.

$$(2\mathbf{I} - \mathbf{A})\mathbf{x} = \begin{bmatrix} 1.33333 & -1 & 1 & 0.33333 \\ 0 & -2 & 0 & 0 \\ 0.66666 & -1 & -1 & 0.33333 \\ 0.66666 & -1 & 1 & -1.66666 \end{bmatrix} \mathbf{x} = \mathbf{0}$$

gives the solution

$$\mathbf{x}_1 = [-0.99999 \quad 0 \quad -1 \quad -0.999999]^T \approx [-1 \quad 0 \quad -1 \quad -1]^T$$

The repeated root case gives

$$(4\mathbf{I} - \mathbf{A})\mathbf{x} = \begin{bmatrix} \frac{2}{3} & -1 & 1 & \frac{1}{3} \\ 0 & 0 & 0 & 0 \\ \frac{2}{3} & -1 & 1 & \frac{1}{3} \\ \frac{2}{3} & -1 & 1 & \frac{1}{3} \end{bmatrix} \mathbf{x} = \mathbf{0}$$

Solution Method 1: Row-reduced echelon solution shows that the rank is 1, so $q_2 = 3$; this is the fully degenerate case. There is just one independent equation for the four components of $\mathbf{x}$, and as a result three independent solutions can be found, all of which satisfy $[\frac{2}{3} \quad -1 \quad 1 \quad \frac{1}{3}]\mathbf{x} = \mathbf{0}$. Among the infinite number of possibilities, the three used here are $\mathbf{x}_2 = [1 \quad 0 \quad 0 \quad -2]^T$, $\mathbf{x}_3 = [0 \quad 1 \quad 1 \quad 0]^T$, and $\mathbf{x}_4 = [0 \quad 1 \quad 0 \quad 3]^T$.

Solution Method 2: The modified Gram-Schmidt process was used to find the **QR** decomposition

$$4\mathbf{I} - \mathbf{A} \approx \begin{bmatrix} 0.5774 & 0.2123 & 0.2634 & 0.7431 \\ 0 & 0.6501 & 0.6384 & -0.4121 \\ 0.5774 & 0.3987 & -0.6258 & -0.3407 \\ 0.5774 & -0.6110 & 0.3624 & -0.4024 \end{bmatrix} \begin{bmatrix} 1.1547 & -1.732 & 1.732 & 0.5773 \\ 0 & 10^{-16} & 10^{-16} & 10^{-6} \\ 0 & 0 & 10^{-16} & 10^{-6} \\ 0 & 0 & 0 & 10^{-6} \end{bmatrix}$$

For the small numbers only the power-of-ten magnitude is shown. The "almost" upper-triangular **R** part shows the type of judgment necessary to determine rank. The input data were accurate only to about six decimal places, so it is reasonable to conclude that the last three rows of **R** are actually zero, giving a rank of 1 to **R**, and hence to $4\mathbf{I} - \mathbf{A}$, since **Q** has full rank. Thus $q_2 = 3$, and there are three independent solutions of

$$[1.1547 \quad -1.732 \quad 1.732 \quad 0.5773]\mathbf{x} = \mathbf{0}$$

This is proportional to the equation found using the row-reduced-echelon form, so the same eigenvectors are again valid. It is very likely that a computer would give three different members of the eigenspace, however. This will be evident in the SVD solution.

Solution Method 3: The SVD decomposition of $4\mathbf{I} - \mathbf{A}$ gave $\Sigma = \text{diag}(2.7688 \quad 10^{-8} \quad 10^{-7} \quad 0)$. The last three singular values are zero to within the accuracy of the input data. The last three colunns of **V** are, therefore, eigenvectors for $\lambda = 4$. These columns are

$$\mathbf{x}_2 = [-0.90767 \quad -0.29505 \quad 0.29505 \quad 0.04505]^T$$

$$\mathbf{x}_3 = [0.04715 \quad -0.14711 \quad 0.14711 \quad -0.97698]^T$$

$$\mathbf{x}_4 = [0 \quad -0.7071 \quad -0.7071 \quad 0]^T$$

The last vector is clearly recognizable as being a normalized form of the previously found $\mathbf{x}_3$. The other two are linear combinations of the vectors found using the previous methods, but this is not obvious.

Solution Method 4: The adjoint matrix, as obtained by computer [2], is

$$\text{Adj}(\lambda\mathbf{I} - \mathbf{A}) = \lambda^3\mathbf{I}_4 + \lambda^2\mathbf{F} + \lambda\mathbf{G} + \mathbf{H}$$

where

$$\mathbf{F} = \begin{bmatrix} -10.66667 & 1 & -1 & -0.33333 \\ 0 & -10 & 0 & 0 \\ -0.66667 & 1 & -11 & -0.33333 \\ -0.66667 & 1 & -1 & -10.33333 \end{bmatrix}, \quad \mathbf{G} = \begin{bmatrix} 37.3333 & -8 & 8 & 2.6667 \\ 0 & 32 & 0 & 0 \\ 5.3333 & -8 & 40 & 2.6667 \\ 5.3333 & -8 & 8 & 34.6667 \end{bmatrix},$$

and

$$\mathbf{H} = \begin{bmatrix} -42.6667 & 16 & -16 & -5.3333 \\ 0 & -32 & 0 & 0 \\ -10.6667 & 16 & -48 & -5.3333 \\ -10.6667 & 16 & -16 & -5.3333 \end{bmatrix}$$

When $\lambda = 2$ is substituted,

$$\text{Adj}(2\mathbf{I} - \mathbf{A}) = \begin{bmatrix} -2.66667 & 4 & -4 & -1.33333 \\ 0 & 0 & 0 & 0 \\ -2.66667 & 4 & -4 & -1.33333 \\ -2.66667 & 4 & -4 & -1.33333 \end{bmatrix}$$

All columns are proportional to the previously found $\mathbf{x}_1$. Then, with $\lambda = 4$, $\text{Adj}(4\mathbf{I} - \mathbf{A}) = [0]$, and $d/d\lambda[\text{Adj}(\lambda\mathbf{I} - \mathbf{A})] = 3\lambda^2\mathbf{I} + 2\lambda\mathbf{F} + \mathbf{G}$. When $\lambda = 4$ is substituted, this again gives the matrix $[0]$. Another derivative gives $\frac{1}{2}d^2/d\lambda^2[\text{Adj}(\lambda\mathbf{I} - \mathbf{A})] = 3\lambda\mathbf{I} + \mathbf{F}$. With $\lambda = 4$, this gives the following 4×4 matrix, but only three columns are independent:

$$\begin{bmatrix} 1.3333 & 1 & -1 & -0.3333 \\ 0 & 2 & 0 & 0 \\ -0.6667 & 1 & 1 & -0.3333 \\ -0.6667 & 1 & -1 & 1.6667 \end{bmatrix}$$

Any three of these four columns could be selected as eigenvectors, or any combination of them. For example, the sum of columns 2 and 3 is $[0 \quad 2 \quad 2 \quad 0]^T$, which has appeared as an eigenvector in the other solution methods.

Any valid set of eigenvectors can be used to form $\mathbf{M} = [\mathbf{x}_1 \quad \mathbf{x}_2 \quad \mathbf{x}_3 \quad \mathbf{x}_4]$, and then $\mathbf{M}^{-1}\mathbf{AM} \approx \text{Diag}[2 \quad 4 \quad 4 \quad 4]$ to within the 10^{-6} accuracy established by the input. ■

***Case II_2*: Simple Degeneracy, $q_i = 1$.** For this case there is just one eigenvector for each eigenvalue, regardless of the algebraic multiplicity. It can be found by using any of the methods mentioned in Case I. The more interesting question here is how to fill in for the missing eigenvectors. If the purpose is to construct a basis set, then the eigenvectors must be augmented with additional linearly independent vectors. This can be done in various ways. The additional vectors could be constructed to be orthogonal to all of the eigenvectors by using a Gram-Schmidt process. Assume this is done and the resulting set of vectors is used to form columns of the $n \times n$ matrix $\mathbf{T}$. The result of the similarity transformation $\mathbf{T}^{-1}\mathbf{AT}$ will not be diagonal, although it will be upper triangular and perhaps close to diagonal, depending upon how many non-eigenvectors are included in $\mathbf{T}$. This means that state equations with this $\mathbf{A}$ matrix cannot be fully decoupled. In fact, no similarity transformation exists which will diagonalize $\mathbf{A}$ in Case II_2 or Case II_3 to follow. If a special class of augmenting vectors, called *generalized eigenvectors,* is used instead of constructing some arbitrary orthogonal set, the diagonal matrix (i.e., the possibility of decoupling) is more nearly achieved. From here forward it is assumed that generalized eigenvectors will be used to fill in where needed. The matrix formed from the set of n eigenvectors and generalized eigenvectors will again be referred to as the modal matrix $\mathbf{M}$ rather than the matrix $\mathbf{T}$ just used. The claim is that $\mathbf{M}^{-1}\mathbf{AM} = \mathbf{J}$ will be as nearly diagonal as possible. The matrix $\mathbf{J}$ is called the *Jordan form.*

The determination of generalized eigenvectors and the Jordan form is now discussed in detail. Suppose that λ_i has an algebraic multiplicity m_i. Since $q_i = 1$ by assumption in Case II_2, there is one eigenvector $\mathbf{x}_1$ and $m_i - 1$ generalized eigenvectors are required. They are defined by the string or chain of equations

$$\mathbf{A}\mathbf{x}_1 = \lambda_i \mathbf{x}_1 \qquad \text{(the usual eigenvalue equation)}$$

$$\mathbf{A}\mathbf{x}_2 = \lambda_i \mathbf{x}_2 + \mathbf{x}_1, \qquad \mathbf{A}\mathbf{x}_3 = \lambda_i \mathbf{x}_3 + \mathbf{x}_2, \ldots, \mathbf{A}\mathbf{x}_{mi} = \lambda_i \mathbf{x}_{mi} + \mathbf{x}_{mi-1}$$

Each equation in the chain except the first is coupled to the preceding equation. Assume for the moment that there are no other eigenvalues, that is $m_i = n$. The preceding chain of equations can be written as one matrix equation:

$$\mathbf{A}[\mathbf{x}_1 \quad \mathbf{x}_2 \ldots \mathbf{x}_{n-1} \quad \mathbf{x}_n] = [\mathbf{x}_1 \quad \mathbf{x}_2 \ldots \mathbf{x}_{n-1} \quad \mathbf{x}_n] \begin{bmatrix} \lambda_i & 1 & 0 & \cdots & & 0 \\ 0 & \lambda_i & 1 & \cdots & & 0 \\ 0 & 0 & \lambda_i & \cdots & & 0 \\ \vdots & & & & & \\ 0 & 0 & 0 & \cdots & \lambda_i & 1 \\ 0 & 0 & 0 & \cdots & 0 & \lambda_i \end{bmatrix}$$

This explicitly shows one example of the Jordan form matrix $\mathbf{J}$. It has the same repeated eigenvalue in every diagonal position and a 1 in every position above the main diagonal. In the more general case where there are other eigenvalues in addition to the m_i multiple root, the Jordan form will be a block diagonal matrix

$$\mathbf{J} = \text{Diag}[\mathbf{J}_1, \mathbf{J}_2, \ldots, \mathbf{J}_p]$$

with each of the $\mathbf{J}_i$ submatrices, called *Jordan blocks*, having the structure just shown explicitly. There will be one $m_i \times m_i$ Jordan block associated with each eigenvalue of multiplicity m_i that satisfies the conditions of Case II_2. Repeated eigenvalues satisfying Case II_1 will have m_i separate 1×1 Jordan blocks $\mathbf{J}_i = [\lambda_i]$, and the nonrepeated Case I will also have separate 1×1 blocks along the diagonal. That is, the diagonal matrix $\mathbf{\Lambda}$ is included in the definition of the Jordan form $\mathbf{J}$ as a special case. From what has been presented so far, it may be surmised that the Jordan form of a matrix will have as many separate Jordan blocks as there are eigenvectors and as many ones just above the main diagonal as there are generalized eigenvectors. This observation is true in general, even for Case II_3, which is discussed shortly.

EXAMPLE 7.4 Find the eigenvalues, eigenvectors, generalized eigenvectors, if needed, and the Jordan form for the companion form matrix

$$\mathbf{A} = \begin{bmatrix} 0 & 1 & 0 & 0 \\ 0 & 0 & 1 & 0 \\ 0 & 0 & 0 & 1 \\ -8 & -20 & -18 & -7 \end{bmatrix}$$

The characteristic equation $\lambda^4 + 7\lambda^3 + 18\lambda^2 + 20\lambda + 8 = 0$ has roots $\lambda_i = \{-1, -2, -2, -2\}$. The simple root $\lambda_1 = -1$ belongs to Case I, and the corresponding eigenvector is easily found to be

$$\mathbf{x}_1 = [-1 \quad 1 \quad -1 \quad 1]^T$$

For $\lambda_2 = -2$, $m_2 = 3$, and $q_2 = 1$, since the row-reduced-echelon form of

$$\mathbf{A}-\mathbf{I}\lambda_2=\begin{bmatrix}2&1&0&0\\0&2&1&0\\0&0&2&1\\-8&-20&-18&-5\end{bmatrix}\quad\text{is}\quad\begin{bmatrix}1&0&0&0.125\\0&1&0&-0.25\\0&0&1&0.5\\0&0&0&0\end{bmatrix}$$

From this, the only eigenvector for λ_2 is $\mathbf{x}_2=[0.125\quad -0.25\quad 0.5\quad -1]^T$. The generalized eigenvector $\mathbf{x}_3$ must satisfy $(\mathbf{A}-\mathbf{I}\lambda_2)\mathbf{x}_3=\mathbf{x}_2$, which gives $\mathbf{x}_3=[0.1875\quad -0.25\quad 0.25\quad 0]^T$. Then $(\mathbf{A}-\mathbf{I}\lambda_2)\mathbf{x}_4=\mathbf{x}_3$, giving $\mathbf{x}_4=[0.1875\quad -0.1875\quad 0.125\quad 0]^T$. There are many other valid answers. Using the four $\mathbf{x}_i$ vectors as columns in $\mathbf{M}$ gives

$$\mathbf{M}^{-1}\mathbf{A}\mathbf{M}=\left[\begin{array}{c|ccc}-1&0&0&0\\\hline 0&-2&1&0\\0&0&-2&1\\0&0&0&-2\end{array}\right]=\mathbf{J}=\text{Diag}[-1,\mathbf{J}_2]$$

where $\mathbf{J}_2$ is a 3×3 Jordan block for $\lambda_2=-2$. ■

Case II$_3$. The general case for an eigenvalue of algebraic multiplicity m_i and degeneracy q_i satisfying $1\le q_i\le m_i$ still has q_i eigenvectors associated with λ_i. There will be one Jordan block for each eigenvector; that is, λ_i will have q_i blocks associated with it. Case II$_3$ is really just a combination of the previous two cases, but knowledge of m_i and q_i still leaves some ambiguity. Assume λ_1 is a fourth-order root of the characteristic equation and assume $q_1=2$. Then it is known that there are two eigenvectors and two generalized eigenvectors. The eigenvectors satisfy $\mathbf{A}\mathbf{x}_a=\lambda_1\mathbf{x}_a$ and $\mathbf{A}\mathbf{x}_b=\lambda_1\mathbf{x}_b$, but it is still uncertain whether the generalized eigenvectors are both associated with $\mathbf{x}_a$ or both with $\mathbf{x}_b$ or one with each. That is, the two Jordan blocks could take one of the following forms:

$$\mathbf{J}_1=\begin{bmatrix}\lambda_1&1&0\\0&\lambda_1&1\\0&0&\lambda_1\end{bmatrix},\qquad \mathbf{J}_2=[\lambda_1]\quad\text{or}\quad \mathbf{J}_1=\begin{bmatrix}\lambda_1&1\\0&\lambda_1\end{bmatrix},\qquad \mathbf{J}_2=\begin{bmatrix}\lambda_1&1\\0&\lambda_1\end{bmatrix}$$

The first pair corresponds to the equations

$$\mathbf{A}\mathbf{x}_1=\lambda_1\mathbf{x}_1,\qquad \mathbf{A}\mathbf{x}_2=\lambda_1\mathbf{x}_2+\mathbf{x}_1,\qquad \mathbf{A}\mathbf{x}_3=\lambda_1\mathbf{x}_3+\mathbf{x}_2,\qquad \mathbf{A}\mathbf{x}_4=\lambda_1\mathbf{x}_4$$

The second pair corresponds to

$$\mathbf{A}\mathbf{x}_1=\lambda_1\mathbf{x}_1,\qquad \mathbf{A}\mathbf{x}_2=\lambda_1\mathbf{x}_2+\mathbf{x}_1,\qquad \mathbf{A}\mathbf{x}_3=\lambda_1\mathbf{x}_3,\qquad \mathbf{A}\mathbf{x}_4=\lambda_1\mathbf{x}_4+\mathbf{x}_3$$

Ambiguities such as this can be resolved by a trial-and-error process [3] or they can be avoided by using a systematic method from Section 7.5. Combinations of the preceding cases may be required for a given $n\times n$ matrix $\mathbf{A}$. The applicable case can be different for each eigenvalue. For example, if the eigenvalues for some 9×9 matrix were $\{2,3,3,5,5,5,6,6,6\}$, $\lambda=2$ is of necessity an example of Case I. $\lambda=3$ might have two eigenvectors, i.e., Case II$_1$; $\lambda=5$ might have only one eigenvector, i.e., Case II$_2$; and $\lambda=6$ might have two eigenvectors and one generalized eigenvector, i.e., Case II$_3$. Finding a total of n vectors, m_i for each eigenvalue with multiplicity m_i, allows the nonsingular modal matrix $\mathbf{M}$ to be formed. The similarity transformation $\mathbf{M}^{-1}\mathbf{A}\mathbf{M}$ again gives the Jordan form $\mathbf{J}$. The diagonal matrix $\mathbf{\Lambda}$ of Case I is considered a special case of the Jordan form with all of its Jordan blocks being 1×1.

Summary. Every $n \times n$ matrix has n eigenvalues and n linearly independent vectors, either eigenvectors or generalized eigenvectors. The eigenvalues are roots of an nth degree polynomial. For each repeated eigenvalue the degeneracy q_i should be found. There will be q_i eigenvectors and Jordan blocks associated with λ_i. If $q_i < m_i$, then generalized eigenvectors will be required. This type of analysis makes it clear how many eigenvectors, generalized eigenvectors, and Jordan blocks there are, as demonstrated in Example 7.5. The actual determination of the generalized eigenvectors and the removal of any remaining ambiguities are discussed in more detail in Section 7.5.

EXAMPLE 7.5 Let $\mathbf{A}$ be an 8×8 matrix and assume that the eigenvalues have been found as $\lambda_1 = \lambda_2 = 2$, $\lambda_3 = \lambda_4 = \lambda_5 = \lambda_6 = -3$, $\lambda_7 = \lambda_8 = 4$. If $\text{rank}[\mathbf{A} - 2\mathbf{I}] = 7$, $\text{rank}[\mathbf{A} + 3\mathbf{I}] = 6$, and $\text{rank}[\mathbf{A} - 4\mathbf{I}] = 6$, find the degeneracies and determine how many eigenvectors and generalized eigenvectors there are. Also, write down the Jordan form.

For $\lambda_1 = 2$, $q_1 = 8 - 7 = 1$. This is the simple degeneracy Case II$_2$, so $\mathbf{x}_1$ is one eigenvector and $\mathbf{x}_2$ must be a generalized eigenvector. For $\lambda_3 = -3$, $q_3 = 8 - 6 = 2$. This falls into Case II$_3$ and there are two eigenvectors (and Jordan blocks) and two generalized eigenvectors must be associated with this root. For $\lambda_7 = 4$, $q_7 = 8 - 6 = 2$. This is Case II$_1$, since $q_7 = m_7 = 2$. There are two eigenvectors and no generalized eigenvectors associated with this eigenvalue. There are a total of five eigenvectors (and Jordan blocks), and three generalized eigenvectors. The Jordan form is either

$$\mathbf{J} = \begin{bmatrix} 2 & 1 & & & & & & \\ 0 & 2 & & & & & \mathbf{0} & \\ & & -3 & 1 & & & & \\ & & 0 & -3 & & & & \\ & & & & -3 & 1 & & \\ & \mathbf{0} & & & 0 & -3 & & \\ & & & & & & 4 & \\ & & & & & & & 4 \end{bmatrix} \quad \text{or} \quad \mathbf{J} = \begin{bmatrix} 2 & 1 & & & & & & \\ 0 & 2 & & & & & & \mathbf{0} \\ & & -3 & 1 & 0 & & & \\ & & 0 & -3 & 1 & & & \\ & & 0 & 0 & -3 & & & \\ & & & & & -3 & & \\ & \mathbf{0} & & & & & 4 & \\ & & & & & & & 4 \end{bmatrix}$$ ■

7.5 DETERMINATION OF GENERALIZED EIGENVECTORS

It is assumed throughout this section that one or more multiple eigenvalues exist for a matrix $\mathbf{A}$ and that a need for generalized eigenvectors has already been established. Three alternative methods are presented for finding generalized eigenvectors.

1. *The first method is a bottom-up method in that the eigenvectors are found first and then a chain of one or more generalized eigenvectors is built up from these.* That is, first find all solutions of the homogeneous equation

$$(\mathbf{A} - \mathbf{I}\lambda_i)\mathbf{x}_i = \mathbf{0}$$

for the repeated eigenvalue λ_i. For each $\mathbf{x}_i$ thus determined, try to construct a generalized eigenvector using

$$(\mathbf{A} - \mathbf{I}\lambda_i)\mathbf{x}_{i+1} = \mathbf{x}_i$$

If the resultant vector $\mathbf{x}_{i+1}$ is linearly independent of all vectors already found, it is a valid generalized eigenvector. If still more generalized eigenvectors are needed for λ_i, then solve

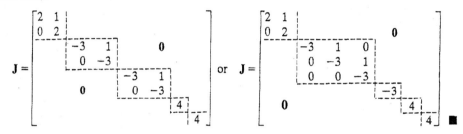

$$(\mathbf{A} - \mathbf{I}\lambda_i)\mathbf{x}_{i+2} = \mathbf{x}_{i+1}$$

and so on until all needed vectors are found. This method can be efficiently used in the simply degenerate cases such as Example 7.4 because there is only a single eigenvector and a single chain of generalized eigenvector equations. Because of possible ambiguities about how the chains of equations are connected in the general case, a more systematic method is desirable.

2. A second method is to use the adjoint matrix $\text{Adj}(\mathbf{I}\lambda - \mathbf{A})$, which is also called the *resolvent* matrix of $\mathbf{A}$, and its various derivatives. Effective algorithms for computing the resolvent matrix are available [2]. This is a bottom-up method also, since eigenvectors are found first. This is done by selecting linearly independent columns of $\text{Adj}(\mathbf{I}\lambda - \mathbf{A})$ with a particular eigenvalue λ_i. Judgment is withheld about the final set of vectors to be retained, since repetitions often will occur in the process to follow. If λ_i is an m_i-repeated eigenvalue, it is possible that fewer than the required m_i vectors will be found on the first step. Some or even all columns of the resolvent matrix may be zero, and others may be linearly dependent. The derivative of the resolvent matrix is then evaluated at the repeated eigenvalue. Some columns may still be zero on this second step. If a given column j is not zero on step 2, then (1) it is an eigenvector if column j was zero on the previous step and (2) it is a generalized eigenvector if column j was not zero on the previous step. Step 2 may still not yield the required m_i independent vectors, so the second derivative of the resolvent matrix is taken. Again, any column j which is nonzero is either an eigenvector or a generalized eigenvector, depending on whether column j was zero or not on the previous step. These relationships require the retention of the factorial divisor, which appeared to be an irrelevant scale factor in Case II_1. The vectors obtained by this process cannot be arbitrarily rescaled on a given step, since they are all tied together in an interdependent chain. Also note that the same eigenvector can appear more than once. That is, column j on one step may yield the same eigenvector as column k on some other step. For this reason the final selection of eigenvectors should be made only after seeing all the columns from all of the steps.

EXAMPLE 7.6 Consider the matrix $\mathbf{A} = \begin{bmatrix} 1 & 2 & 3 \\ 0 & 1 & 4 \\ 0 & 0 & 1 \end{bmatrix}$. Clearly $\lambda_i = 1$ has algebraic multiplicity $m = 3$, and the rank of $\mathbf{A} - \mathbf{I}\lambda$ is 2, so $q = 1$. There is just one eigenvector, so two generalized eigenvectors are required (Case II_2). The adjoint matrix is

$$\text{Adj}(\mathbf{A} - \mathbf{I}\lambda) = \begin{bmatrix} (1-\lambda)^2 & -2(1-\lambda) & 8 - 3(1-\lambda) \\ 0 & (1-\lambda)^2 & -4(1-\lambda) \\ 0 & 0 & (1-\lambda)^2 \end{bmatrix} = \lambda^2\mathbf{I} + \lambda\mathbf{F} + \mathbf{G}$$

where

$$\mathbf{F} = \begin{bmatrix} -2 & 2 & 3 \\ 0 & -2 & 4 \\ 0 & 0 & -2 \end{bmatrix} \quad \text{and} \quad \mathbf{G} = \begin{bmatrix} 1 & -2 & 5 \\ 0 & 1 & -4 \\ 0 & 0 & 1 \end{bmatrix}$$

With $\lambda = 1$, this gives $\begin{bmatrix} 0 & 0 & 8 \\ 0 & 0 & 0 \\ 0 & 0 & 0 \end{bmatrix}$, and one eigenvector is evident in column 3. The derivative gives $d[\text{Adj}(\mathbf{A}-\mathbf{I}\lambda)/d\lambda = 2\lambda\mathbf{I}+\mathbf{F}$. When evaluated at $\lambda = 1$, this gives $\begin{bmatrix} 0 & 2 & 3 \\ 0 & 0 & 4 \\ 0 & 0 & 0 \end{bmatrix}$. Column 2 is nonzero for the first time and hence is an eigenvector. However, it is just a rescaled copy of the one found on the first step in column 3. On this second step column 3 is a generalized eigenvector. The final derivative (the m_ith for this Case II_2) is $\frac{1}{2}d^2[\text{Adj}(\mathbf{A}-\mathbf{I}\lambda)]/d\lambda^2 = \mathbf{I}$. Column 1 is nonzero for the first time and hence is an eigenvector. But, except for scaling, it is the same one found twice before. Column 2 is a generalized eigenvector if column 2 of step 2 is used as the eigenvector. This would still leave us one short of the needed three vectors, and further derivatives will only give zero columns. Therefore, column 2 must be rejected. From column 3 the final set is $\mathbf{x}_1 = [8 \quad 0 \quad 0]^T$, $\mathbf{x}_2 = [3 \quad 4 \quad 0]^T$, and $\mathbf{x}_3 = [0 \quad 0 \quad 1]^T$. ■

The adjoint (or resolvent matrix) method is workable for small hand calculations and can be adapted to machine calculation as well. The next method may be better suited to machine implementation for larger problems and can also be used for hand calculation with small problems.

3. The third method of finding eigenvectors and generalized eigenvectors is a *top-down* method. Rather than finding all the eigenvectors first and then building the necessary chains of generalized eigenvectors on them, we find the maximum number m_i of linearly independent vector solutions to a modified problem $(\mathbf{A}-\mathbf{I}\lambda_i)^k\mathbf{x} = \mathbf{0}$. All the eigenvectors and generalized eigenvectors associated with λ_i belong to the m_i-dimensional space spanned by the m_i solution vectors. The eigenvectors belong, because for any integer $j > 1$, $(\mathbf{A}-\mathbf{I}\lambda_i)^j\mathbf{x} = \mathbf{0}$ if it is true for $j = 1$. A jth-order generalized eigenvector must satisfy $(\mathbf{A}-\mathbf{I}\lambda_i)^j\mathbf{x} = \mathbf{0}$ and $(\mathbf{A}-\mathbf{I}\lambda_i)^{j-1}\mathbf{x} = \mathbf{x}_c \neq \mathbf{0}$, for $j = k, k-1, \ldots, 2$. This is consistent with the bottom-up construction equations for generalized eigenvectors:

$$
\begin{aligned}
&(\mathbf{A}-\mathbf{I}\lambda_i)\mathbf{x}_1 = \mathbf{0} \\
&(\mathbf{A}-\mathbf{I}\lambda_i)\mathbf{x}_2 = \mathbf{x}_1 \Rightarrow (\mathbf{A}-\mathbf{I}\lambda_i)^2\mathbf{x}_2 = (\mathbf{A}-\mathbf{I}\lambda_i)\mathbf{x}_1 = \mathbf{0} \\
&(\mathbf{A}-\mathbf{I}\lambda_i)\mathbf{x}_3 = \mathbf{x}_2 \Rightarrow (\mathbf{A}-\mathbf{I}\lambda_i)^2\mathbf{x}_3 = (\mathbf{A}-\mathbf{I}\lambda_i)\mathbf{x}_2 = \mathbf{x}_1 \neq \mathbf{0} \\
&\qquad\qquad\qquad\qquad (\mathbf{A}-\mathbf{I}\lambda_i)^3\mathbf{x}_3 = (\mathbf{A}-\mathbf{I}\lambda_i)\mathbf{x}_1 = \mathbf{0} \\
&(\mathbf{A}-\mathbf{I}\lambda_i)\mathbf{x}_4 = \mathbf{x}_3 \Rightarrow (\mathbf{A}-\mathbf{I}\lambda_i)^2\mathbf{x}_4 = (\mathbf{A}-\mathbf{I}\lambda_i)\mathbf{x}_3 = \mathbf{x}_2 \neq \mathbf{0} \\
&\qquad\qquad\qquad\qquad (\mathbf{A}-\mathbf{I}\lambda_i)^3\mathbf{x}_4 = (\mathbf{A}-\mathbf{I}\lambda_i)\mathbf{x}_2 = \mathbf{x}_1 \neq \mathbf{0} \\
&\qquad\qquad\qquad\qquad (\mathbf{A}-\mathbf{I}\lambda_i)^4\mathbf{x}_4 = (\mathbf{A}-\mathbf{I}\lambda_i)\mathbf{x}_1 = \mathbf{0}
\end{aligned}
\tag{7.4}
$$

This pattern continues up to some maximum integer k_i, the index of λ_i. The key to the top-down method is finding the correct integer k. It is the *index* of the eigenvalue and is the smallest integer such that

$$\text{rank}(\mathbf{A}-\mathbf{I}\lambda_i)^k = n - m_i$$

The index k_i indicates the length of the longest chain of eigenvectors-generalized eigenvectors for λ_i. It also is the size of the largest Jordan block for λ_i in the Jordan form.

After finding the index and the m_i independent solution vectors, a simple testing procedure, consisting of matrix multiplications from the left side of Eq. (7.4), indicates whether each vector is a generalized eigenvector or an eigenvector. The same matrix multiplications also give each successive vector in the chain until the final member—i.e., the eigenvector—is found. The procedure is complete if it is explicitly ensured that the eigenvectors are included in the set of m_i vectors found at the top.

Finding the index is the key to avoiding the ambiguities about how the various chains should be formed. If in Example 7.5 the index for $\lambda_i = -3$ were found to be 2, then the first form for **J** would be correct. If the index were 3, then the second Jordan form would be correct.

EXAMPLE 7.7 Find the eigenvalues, eigenvectors, generalized eigenvectors, and Jordan form for

$$\mathbf{A} = \begin{bmatrix} -5 & \frac{1}{6} & -\frac{1}{6} & 0 \\ -\frac{1}{2} & -\frac{16}{3} & \frac{1}{3} & \frac{1}{2} \\ -\frac{1}{2} & -\frac{1}{3} & -\frac{14}{3} & \frac{1}{2} \\ 0 & -\frac{1}{6} & \frac{1}{6} & -5 \end{bmatrix}$$

The characteristic equation is $\Delta(\lambda) = (\lambda + 5)^4 = 0$, so $\lambda = -5$ has algebraic multiplicity $m = 4$. Since $n = 4$, the index k must be found for which $\text{rank}(\mathbf{A} - \mathbf{I}\lambda)^k = 0$. Singular-value decomposition, Gram-Schmidt **QR** decomposition, or RRE methods can be applied to find that $\text{rank}(\mathbf{A} - \mathbf{I}\lambda) = 2$. To be specific, the SVD decomposition gives $\mathbf{\Sigma} = \text{diag}(1.2176, 0.27395, 0, 0)$, and the last two columns of **V** give eigenvectors $\mathbf{x}_a = [0 \;\; 1 \;\; 1 \;\; 0]^T$ and $\mathbf{x}_b = [1 \;\; 0 \;\; 0 \;\; 1]^T$. These are the only two independent eigenvectors, but so far it is not clear if the two generalized eigenvectors are connected in a single chain to one eigenvector (giving a 3×3 Jordan block and a 1×1 Jordan block) or if they form two separate chains (giving two 2×2 Jordan blocks). Forming

$$(\mathbf{A} - \mathbf{I}\lambda)^2 = \begin{bmatrix} 0 & 0 & 0 & 0 \\ 0 & -\frac{1}{6} & \frac{1}{6} & 0 \\ 0 & -\frac{1}{6} & \frac{1}{6} & 0 \\ 0 & 0 & 0 & 0 \end{bmatrix}$$

reveals that its rank is 1. So far it has been found that the index k is neither 1 nor 2. Forming $(\mathbf{A} - \mathbf{I}\lambda)^3 = [\mathbf{0}]$ (on the computer it was zero to within order 10^{-7}) shows that $k = 3$. It also indicates that one chain is of length 3. The other will be just an isolated eigenvector, and **J** will have 1×1 and 3×3 Jordan blocks. All ambiguity has been removed, except for the unimportant order of the Jordan blocks within the Jordan form. This depends only on the order in which the isolated eigenvector and the chain of three are placed in the modal matrix **M**. The actual finding of these vectors is now demonstrated using the top-down method. Any nonzero vector is a nontrivial solution of $(\mathbf{A} - \mathbf{I}\lambda)^3\mathbf{x} = \mathbf{0}$ in this case. Choices consisting of the two known eigenvectors plus any two additional independent vectors will suffice. (Actually, if the eigenvectors are not explicitly included, one will be found automatically at the end of the chain of length 3. The eigenvector equation can then be used to find the other one.) Four convenient linearly independent vectors for this problem are

$$\mathbf{x}_a = [1 \;\; 0 \;\; 0 \;\; 1]^T, \qquad \mathbf{x}_b = [0 \;\; 1 \;\; 1 \;\; 0]^T$$

$$\mathbf{x}_c = [0 \;\; 1 \;\; 0 \;\; 0]^T, \qquad \mathbf{x}_d = [0 \;\; 0 \;\; 0 \;\; 1]^T$$

It is not clear yet how these vectors chain together, so the testing procedure is invoked. Let $\mathbf{C} = (\mathbf{A} - \mathbf{I}\lambda_i)$, with $\lambda_i = -5$. Then $\mathbf{C}^2\mathbf{x}_a$ and $\mathbf{C}^2\mathbf{x}_b$ are zero, confirming that $\mathbf{x}_a$ and $\mathbf{x}_b$ are not generalized eigenvectors. It is also found that $\mathbf{C}^2\mathbf{x}_d = \mathbf{0}$, so $\mathbf{x}_d$ is also not a generalized eigenvector. Only $\mathbf{C}^2\mathbf{x}_c$ is nonzero, so $\mathbf{x}_c$ is the generalized eigenvector which starts the chain of three. Call it $\mathbf{x}_4$, thus indicating its ultimate column position in the modal matrix. Then $(\mathbf{A} - \mathbf{I}\lambda)\mathbf{x}_4 = \mathbf{x}_3 = [0.16667 \quad -0.33333 \quad -0.33333 \quad -0.166667]^T$. Next in the chain is $(\mathbf{A} - \mathbf{I}\lambda)\mathbf{x}_3 = \mathbf{x}_2 = [0 \quad -0.16667 \quad -0.16667 \quad 0]^T$. This is a multiple of $\mathbf{x}_b$ found earlier and is an eigenvector (not a generalized eigenvector), as seen at the next step, $(\mathbf{A} - \mathbf{I}\lambda)\mathbf{x}_2 = \mathbf{0}$. The zero vector always signals the end of the chain. The fourth vector is a stand-alone eigenvector $\mathbf{x}_a = [1 \quad 0 \quad 0 \quad 1]^T$, renamed $\mathbf{x}_1$. Using these four vectors as columns in $\mathbf{M}$ gives

$$\mathbf{M}^{-1}\mathbf{A}\mathbf{M} = \left[\begin{array}{c|ccc} -5 & 0 & 0 & 0 \\ \hline 0 & -5 & 1 & 0 \\ 0 & 0 & -5 & 1 \\ 0 & 0 & 0 & -5 \end{array}\right]$$

where the zero elements all had magnitude on the order of 10^{-7} or less. ■

EXAMPLE 7.8 Let

$$\mathbf{A} = \begin{bmatrix} 0 & 0 & 1 & 0 \\ 0 & 0 & 0 & 1 \\ 0 & 0 & 0 & 0 \\ 0 & 0 & 0 & 0 \end{bmatrix}$$

The characteristic equation is $\lambda^4 = 0$, so $\lambda_1 = 0$ with $m_1 = 4$. Since $\text{rank}(\mathbf{A} - \mathbf{I}\lambda_1) = 2$, there are $q = 2$ eigenvectors and also 2 generalized eigenvectors. Since $n - m = 0$, and since $\text{rank}[\mathbf{A} - \mathbf{I}\lambda_1]^2 = 0$, $k_1 = 2$. The largest Jordan block is 2×2. Since there are just two blocks, they both must be 2×2. To find the eigenvectors and generalized eigenvectors, consider $(\mathbf{A} - \mathbf{I}\lambda_1)^2\mathbf{x} = \mathbf{0}$. Any vector satisfies this equation, but there are at most four linearly independent solutions. Select $\mathbf{x}_a = [1 \quad 0 \quad 0 \quad 0]^T$. This is *not* a generalized eigenvector since $(\mathbf{A} - \mathbf{I}\lambda_1)\mathbf{x}_a = \mathbf{0}$. Similarly $\mathbf{x}_b = [0 \quad 1 \quad 0 \quad 0]^T$ is not a generalized eigenvector. Select $\mathbf{x}_c = [0 \quad 0 \quad 1 \quad 0]^T$. Then since $[\mathbf{A} - \mathbf{I}\lambda_1]\mathbf{x}_c = [1 \quad 0 \quad 0 \quad 0]^T \neq \mathbf{0}$, $\mathbf{x}_c$ is a generalized eigenvector associated with the eigenvector $[1 \quad 0 \quad 0 \quad 0]^T$. Finally $\mathbf{x}_d = [0 \quad 0 \quad 0 \quad 1]^T$ is a generalized eigenvector and $[\mathbf{A} - \mathbf{I}\lambda_1]\mathbf{x}_d = [0 \quad 1 \quad 0 \quad 0]^T$ is the associated eigenvector. Thus the modal matrix and the Jordan form are

$$\mathbf{M} = \begin{bmatrix} 1 & 0 & 0 & 0 \\ 0 & 0 & 1 & 0 \\ 0 & 1 & 0 & 0 \\ 0 & 0 & 0 & 1 \end{bmatrix} \quad \text{and} \quad \mathbf{J} = \left[\begin{array}{cc|cc} 0 & 1 & 0 & 0 \\ 0 & 0 & 0 & 0 \\ \hline 0 & 0 & 0 & 1 \\ 0 & 0 & 0 & 0 \end{array}\right]$$

See page 259 of Reference 3 for a trial-and-error solution to the same problem. ■

7.6 ITERATIVE COMPUTER METHODS FOR DETERMINING EIGENVALUES AND EIGENVECTORS

In all discussions to this point the eigenvalue-eigenvector problem has been split into two parts. First the roots of the characteristic equation—i.e., the eigenvalues—were found by some sort of polynomial root-finding routine such as Newton-Raphson. Only then was the eigenvector problem considered. Because of the difficulty in accurately factoring high-degree polynomials, other iterative computer algorithms are often used to determine the eigenvalues more directly. In some cases, such as real, symmetric matrices, the eigenvectors are found simultaneously with the eigenvectors. In other procedures the determination of the eigenvectors is still a separate calculation. Two general methods are presented in this section. The first method is restricted to real, symmetric matrices. Thus it is known at the outset that (1) all λ_i will be real, (2) even if some eigenvalue is repeated, a full set of eigenvectors will always exist, and (3) the eigenvectors form an orthogonal set. The simple version of the first algorithm to be presented here assumes that no two eigenvalues have the same magnitude, that is, $|\lambda_i| \neq |\lambda_j|$ if $i \neq j$. Any vector $\mathbf{z}_0$ can be written as

$$\mathbf{z}_0 = \alpha_1 \mathbf{x}_1 + \alpha_2 \mathbf{x}_2 + \cdots + \alpha_n \mathbf{x}_n$$

Therefore,

$$\mathbf{A}\mathbf{z}_0 = \alpha_1 \lambda_1 \mathbf{x}_1 + \alpha_2 \lambda_2 \mathbf{x}_2 + \cdots + \alpha_n \lambda_n \mathbf{x}_n \triangleq \mathbf{z}_1$$

$$\mathbf{A}\mathbf{z}_1 = \alpha_1 \lambda_1^2 \mathbf{x}_1 + \alpha_2 \lambda_2^2 \mathbf{x}_2 + \cdots + \alpha_n \lambda_n^2 \mathbf{x}_n \triangleq \mathbf{z}_2$$

$$\vdots$$

$$\mathbf{A}\mathbf{z}_k = \alpha_1 \lambda_1^{k+1} \mathbf{x}_1 + \alpha_2 \lambda_2^{k+1} \mathbf{x}_2 + \cdots + \alpha_n \lambda_n^{k+1} \mathbf{x}_n \triangleq \mathbf{z}_{k+1}$$

If λ_1 is the eigenvalue with the largest absolute value, then for k sufficiently large,

$$\mathbf{z}_{k+1} \cong \alpha_1 \lambda_1^{k+1} \mathbf{x}_1 = \beta \mathbf{x}_1 \quad \text{and} \quad \mathbf{A}\mathbf{z}_{k+1} \cong \lambda_1 \mathbf{z}_{k+1} = \lambda_1 \beta \mathbf{x}_1$$

Hence starting with an arbitrary vector $\mathbf{z}_0$ and repeatedly calculating $\mathbf{z}_{\text{new}} = \mathbf{A}\mathbf{z}_{\text{old}}$ until $\mathbf{z}_{\text{new}}$ is proportional to $\mathbf{z}_{\text{old}}$ leads to the maximum magnitude eigenvalue (the constant of proportionality) and the corresponding eigenvector. At each step of the iterative calculations, the vectors $\mathbf{z}$ can be normalized in any number of ways. In the subsequent discussion it is assumed that the final vector is normalized to a unit vector.

The next largest eigenvalue and its eigenvector can be found by constraining all vectors in the iteration process to be orthogonal to the first eigenvector. From Chapter 5, the projection operator $\mathbf{P}_1 = \mathbf{I} - \mathbf{x}_1 \mathbf{x}_1^T$ takes any arbitrary vector $\mathbf{z}$ into the subspace orthogonal to $\mathbf{x}_1$. Thus an arbitrary $\mathbf{z}_{\text{free}}$ gets mapped into $\mathbf{P}_1 \mathbf{z}_{\text{free}} = \mathbf{z}_{\text{constrained}}$ and $\mathbf{A}\mathbf{z}_{\text{constrained}}$ is equivalent to defining $\mathbf{A}_1 = \mathbf{A}\mathbf{P}_1$. Then the same iteration process is performed with $\mathbf{A}_1$ and freely selected $\mathbf{z}$ vectors, $\mathbf{A}_1 \mathbf{z}_k = \mathbf{z}_{k+1}$. The end results will be the next largest $|\lambda_2|$ and its eigenvector $\mathbf{x}_2$. For the eigenvalue with the third largest magnitude, iteration vectors $\mathbf{z}$ are restricted to be orthogonal to both $\mathbf{x}_1$ and $\mathbf{x}_2$. This can be done by defining a new projection matrix $\mathbf{P}_2 = \mathbf{I} - \mathbf{x}_1 \mathbf{x}_1^T - \mathbf{x}_2 \mathbf{x}_2^T$ and then using $\mathbf{A}_2 = \mathbf{A}\mathbf{P}_2$ in place of $\mathbf{A}$. The sequence of $\mathbf{P}_i$ matrices are often called *sweep matrices*. The similarity with the vector version of the Gram-Schmidt construction process of Problem 5.17 is noted. The process continues in an obvious way until all eigenvalues and eigenvectors are found. This is a very rapid and effective calculation method for the limited class of matrices to which it applies. Ordinarily the eigenvalues are not known at the outset, so it is not clear whether this method applies or not. Actually, the method sometimes gives the correct answers even when two eigenvalues have the same magnitude. For a trivial example, consider $\mathbf{A} = \text{Diag}(2, 2)$. If the starting vector is

$\mathbf{z}_0 = [1 \quad 0]^T$, then one iteration gives $\mathbf{z}_1 = \mathbf{z}_0$ and $\lambda_1 = 2$. Then $\mathbf{P}_1$ is found to be Diag(0, 1) and $\mathbf{A}_1 = \text{Diag}(0, 2)$. If the same $\mathbf{z}_0 = [1 \quad 0]^T$ is used to start the search for $\mathbf{x}_2$ and λ_2, one iteration gives the *wrong* final answer $\lambda_2 = 0$. If $\mathbf{z}_0 = [1 \quad 1]^T$ is used to start the second stage, two iterations give the *correct* final answer $\mathbf{x}_2 = [0 \quad 1]^T$ and $\lambda_2 = 2$. This type of dangerous behavior can be minimized but not avoided by starting with randomly selected $\mathbf{z}_0$ vectors. Another type of failure, which is less dangerous because it is recognized as a failure, is illustrated by

$$\mathbf{A} = \begin{bmatrix} 0 & 2 \\ 2 & 0 \end{bmatrix}$$

This matrix has $\lambda = \pm 2$. For *any* initial vector, the two components flip-flop back and forth each iteration, and convergence never occurs.

Another commonly used method of finding the eigenvalues by iteration is the so-called **QR** method. Only the rudiments of the method are given here. It, like many other methods, depends on transforming the original matrix to a form in which the eigenvalues are obvious. For example, if a similarity transformation could be found such that $\mathbf{T}^{-1}\mathbf{A}\mathbf{T} = \mathbf{D}$ is diagonal, then those diagonal elements are the eigenvalues of **A**. **D** and **A** have the same eigenvalues, since

$$|\mathbf{D} - \mathbf{I}\lambda| = |\mathbf{T}^{-1}\mathbf{A}\mathbf{T} - \mathbf{T}^{-1}\mathbf{T}\lambda| = |\mathbf{T}^{-1}||\mathbf{A} - \mathbf{I}\lambda||\mathbf{T}| = |\mathbf{A} - \mathbf{I}\lambda|$$

The eigenvectors of **D** are not the same as the eigenvectors of **A**, however. If the decomposition $\mathbf{A} = \mathbf{QR}$ is found and then a new matrix $\mathbf{A}_1 = \mathbf{RQ}$ is formed, $\mathbf{A}_1$ and the original **A** are related by a similarity transformation. Since **Q** is orthogonal, it can be inverted to give $\mathbf{R} = \mathbf{Q}^{-1}\mathbf{A}$. Using this in the reversed-order product gives $\mathbf{A}_1 = \mathbf{Q}^{-1}\mathbf{A}\mathbf{Q}$. If $\mathbf{A}_1$ is now decomposed into $\mathbf{A}_1 = \mathbf{Q}_1\mathbf{R}_1$ and then $\mathbf{A}_2 = \mathbf{R}_1\mathbf{Q}_1$ is formed, it can be seen that $\mathbf{A}_2 = \mathbf{Q}_1^{-1}\mathbf{Q}^{-1}\mathbf{A}\mathbf{Q}\mathbf{Q}_1 = (\mathbf{Q}\mathbf{Q}_1)^{-1}\mathbf{A}(\mathbf{Q}\mathbf{Q}_1)$. Thus $\mathbf{A}_2$ is also related to **A** by a similarity transformation, and they have the same eigenvalues. This remains true for any number of steps. It is an interesting fact that because of the upper triangular nature of $\mathbf{R}_i$ at each step, this procedure will (usually) converge to a matrix with either a 1×1 or a 2×2 block **E** in the lower-right corner, $\mathbf{A}_k = \begin{bmatrix} \mathbf{F} & \mathbf{G} \\ \mathbf{0} & \mathbf{E} \end{bmatrix}$. The eigenvalues of this block-triangular structure are the eigenvalues of **F** and of **E**. The eigenvalues of **E** are just **E** itself if it is 1×1. If **E** is 2×2, a simple quadratic equation can be solved to find its eigenvalues. Complex conjugate pairs of eigenvalues can be found using only real arithmetic by this method. In either case, the **E** portion can be stripped out, and the $\mathbf{QR} \rightarrow \mathbf{RQ}$ procedure can be continued on just the **F** portion. As just described, the convergence would be very slow. Good **QR** eigenvalue procedures have various refinements, including initial conditioning on **A** to speed convergence [4, 5]. Another modification which speeds convergence is to do the $\mathbf{Q}_{k+1}\mathbf{R}_{k+1}$ decomposition on $\mathbf{R}_k\mathbf{Q}_k - \mathbf{I}\alpha$ rather than on $\mathbf{R}_k\mathbf{Q}_k$, for some judicious choice of α. A common choice for α is the current lower-right corner element in $\mathbf{R}_k\mathbf{Q}_k$. Once all the eigenvalues are found, one more **QR** decomposition of $\mathbf{A} - \mathbf{I}\lambda$ can be carried out to find each eigenvector. Since **Q** is nonsingular, $(\mathbf{A} - \mathbf{I}\lambda)\mathbf{x} = \mathbf{0} \Rightarrow \mathbf{Rx} = \mathbf{0}$, and this is easily solved due to the triangular nature of **R**.

7 SPECTRAL DECOMPOSITION AND INVARIANCE PROPERTIES

Definition 7.1. Let $\mathscr{X}$ be a linear vector space defined over the complex number field. Let $\mathscr{A}:\mathscr{X}\to\mathscr{X}$ be a linear transformation and let $\mathscr{X}_1$ be a subspace of $\mathscr{X}$. Then $\mathscr{X}_1$ is said to be *invariant* under the transformation $\mathscr{A}$ if for every $\mathbf{x}\in\mathscr{X}_1$, $\mathscr{A}(\mathbf{x})$ also belongs to $\mathscr{X}_1$.

Definition 7.2. The set of all vectors $\mathbf{x}_i$ satisfying

$$\mathscr{A}(\mathbf{x}_i) = \lambda_i\,\mathbf{x}_i$$

for a particular λ_i is called the *eigenspace* of λ_i.

It consists of all the eigenvectors associated with that particular eigenvalue λ_i, plus the zero vector. The eigenspace of λ_i is a subspace of $\mathscr{X}$, and may alternatively be characterized as the null space of the transformation $(\mathscr{A}-\mathscr{I}\lambda_i)$, denoted as $\mathscr{N}_i$ for brevity.

Theorem 7.1. $\mathscr{N}_i$ is a q_i dimensional subspace of $\mathscr{X}$ which is invariant under $\mathscr{A}$, where q_i is the degeneracy, $q_i = n - \operatorname{rank}(\mathscr{A}-\mathscr{I}\lambda_i)$.

Definition 7.3. If the linear transformation $\mathscr{A}$ has a complete set of n linearly independent eigenvectors (i.e., Case I or Case II_1), then $\mathscr{A}$ is said to be a *simple* linear transformation.

Theorem 7.2. If $\mathscr{A}$ is simple, then

$$\mathscr{X} = \mathscr{N}_1 \oplus \mathscr{N}_2 \oplus \cdots \oplus \mathscr{N}_p$$

where the direct sum is taken over the p distinct eigenvalues. (Note that $p<n$ for Case II_1.)

Theorem 7.3. If $\mathscr{A}$ is normal, $\mathscr{N}_i$ and $\mathscr{N}_j$ are orthogonal to each other, for all $i\neq j$. Note that normal transformations are a subset of simple transformations.

Let $\mathscr{A}:\mathscr{X}\to\mathscr{X}$ be a simple linear transformation, with the matrix representation $\mathbf{A}$. Then the eigenvectors of $\mathbf{A}$, $\{\mathbf{x}_i\}$, form a basis for $\mathscr{X}$. Let $\{\mathbf{r}_i\}$ be the reciprocal basis vectors. Then, for every $\mathbf{z}\in\mathscr{X}$,

$$\mathbf{z} = \sum_{i=1}^{n} \langle \mathbf{r}_i, \mathbf{z}\rangle \mathbf{x}_i \quad\text{and}\quad \mathbf{A}\mathbf{z} = \sum_{i=1}^{n} \langle \mathbf{r}_i, \mathbf{z}\rangle \mathbf{A}\mathbf{x}_i = \sum_{i=1}^{n} \lambda_i \langle \mathbf{r}_i, \mathbf{z}\rangle \mathbf{x}_i$$

This allows $\mathbf{A}$ to be written as

$$\mathbf{A} = \sum_{i=1}^{n} \lambda_i\, \mathbf{x}_i\rangle\langle\mathbf{r}_i \tag{7.5}$$

This is called the *spectral representation* of $\mathbf{A}$. If $\mathscr{A}$ is normal, then its eigenvectors are mutually orthogonal (see Problem 7.27) so that the reciprocal basis vector $\mathbf{r}_i$ can be made equal to $\mathbf{x}_i$ by normalizing the eigenvectors. Then

$$\mathbf{A} = \sum_{i=1}^{n} \lambda_i \, \mathbf{x}_i \rangle\langle \mathbf{x}_i \tag{7.6}$$

When a linear transformation is not simple, its eigenvectors do not form a basis for $\mathscr{X}$ (Cases II_2 and II_3). It is still possible to construct a basis by adding generalized eigenvectors, as discussed in Section 7.5. These vectors, along with the eigenvectors, belong to $\mathcal{N}_i^{k_i}$, the null space of $(\mathscr{A} - \mathscr{I}\lambda_i)^{k_i}$ (see Problem 7.35). The power k_i is called the *index* of the eigenvalue λ_i and is one for simple transformations. Using this generalization, it is again possible to write $\mathscr{X}$ as a direct sum of invariant subspaces of $\mathscr{A}$.

$$\mathscr{X} = \mathcal{N}_1^{k_1} \oplus \mathcal{N}_2^{k_2} \oplus \cdots \oplus \mathcal{N}_p^{k_p} \tag{7.7}$$

Alternatively, the space $\mathscr{X}$ can be decomposed into

$$\mathscr{X} = \mathcal{N}_i^{k_i} \oplus \mathscr{R}_i^{k_i \perp}$$

Since all m_i eigenvectors and generalized eigenvectors associated with λ_i belong to $\mathcal{N}_i^{k_i}$, $\dim(\mathcal{N}_i^{k}) = m_i$. Thus, $\text{rank}(\mathbf{A} - \mathbf{I}\lambda_i)^{k_i} = n - m_i$ and the index k_i is the smallest integer for which this is true.

Theorem 7.4. If $\mathscr{A} : \mathscr{X} \to \mathscr{X}$, with $\dim(\mathscr{X}) = n$, and if $\mathscr{X}$ can be expressed as the direct sum of p invariant subspaces, as in Eq. (7.7), then $\mathscr{A}$ can be represented by a block diagonal matrix, with p blocks, each of dimension k_i, provided a suitable basis is selected.

The block diagonal representation for $\mathscr{A}$ is the Jordan form, and "the suitable basis" consists of eigenvectors, and if necessary, generalized eigenvectors. This provides the simplest possible representation for a linear transformation, and will be most useful in analyzing systems in later chapters.

7.8 BILINEAR AND QUADRATIC FORMS

The expression $\langle \mathbf{y}, \mathbf{Ax} \rangle$ is called a *bilinear form,* since if $\mathbf{y}$ is held fixed, it is linear in $\mathbf{x}$: and if $\mathbf{x}$ is held fixed, it is linear in $\mathbf{y}$. When $\mathbf{x} = \mathbf{y}$, the result is the *quadratic form,* $Q(\mathbf{x}) = \langle \mathbf{x}, \mathbf{Ax} \rangle$. Every matrix $\mathbf{A}$ can be written as the sum of a Hermitian matrix and a skew-Hermitian matrix. It will be assumed that all quadratic forms are defined in terms of a Hermitian matrix $\mathbf{A}$. For real quadratic forms, there is no loss of generality since $\langle \mathbf{x}, \mathbf{Ax} \rangle = 0$ for all $\mathbf{x}$ if $\mathbf{A}$ is skew-symmetric.

Quadratic forms arise in connection with performance criteria in optimal control problems, in consideration of system stability, and in other applications. Here the several types of quadratic forms are defined and means for establishing the type of a given quadratic form are summarized.

Definitions.

1. Q (or the defining matrix $\mathbf{A}$) is said to be *positive definite* if and only if $\langle \mathbf{x}, \mathbf{Ax} \rangle > 0$ for all $\mathbf{x} \neq 0$.

2. Q is *positive semidefinite* if $\langle \mathbf{x}, \mathbf{Ax} \rangle \geq 0$ for all $\mathbf{x}$. That is, $Q = 0$ is possible for some $\mathbf{x} \neq \mathbf{0}$.
3. Q is *negative definite* if and only if $\langle \mathbf{x}, \mathbf{Ax} \rangle < 0$ for all $\mathbf{x} \neq \mathbf{0}$.
4. Q is *negative semidefinite* if $\langle \mathbf{x}, \mathbf{Ax} \rangle \leq 0$ for all $\mathbf{x}$.
5. Q is said to be *indefinite* if $\langle \mathbf{x}, \mathbf{Ax} \rangle > 0$ for some $\mathbf{x}$ and $\langle \mathbf{x}, \mathbf{Ax} \rangle < 0$ for other $\mathbf{x}$.

Tests for Definiteness. Let $\mathbf{A}$ be an $n \times n$ real symmetric matrix with eigenvalues λ_i. Define

$$\Delta_1 = a_{11}, \qquad \Delta_2 = \begin{vmatrix} a_{11} & a_{12} \\ a_{21} & a_{22} \end{vmatrix}, \qquad \Delta_3 = \begin{vmatrix} a_{11} & a_{12} & a_{13} \\ a_{21} & a_{22} & a_{23} \\ a_{31} & a_{32} & a_{33} \end{vmatrix}, \ldots, \Delta_n = |\mathbf{A}|$$

The Δ_i are called the *principal minors* of $\mathbf{A}$. Two possible methods of determining the definiteness of a Hermitian matrix $\mathbf{A}$ are given in Table 7.1.

TABLE 7.1

	Tests Using:	
Class	Eigenvalues of $\mathbf{A}$	Principal Minors of $\mathbf{A}$ (for real symmetric $\mathbf{A}$)
1. Positive definite	All $\lambda_i > 0$	$\Delta_1 > 0, \Delta_2 > 0, \ldots, \Delta_n > 0$
2. Positive semidefinite	All $\lambda_i \geq 0$	$\Delta_1 \geq 0, \Delta_2 \geq 0, \ldots, \Delta_n \geq 0$
3. Negative definite	All $\lambda_i < 0$	$\Delta_1 < 0, \Delta_2 > 0, \Delta_3 < 0, \ldots$ (note alternating signs)
4. Negative semidefinite	All $\lambda_i \leq 0$	$\Delta_1 \leq 0, \Delta_2 \geq 0, \Delta_3 \leq 0, \ldots$
5. Indefinite	Some $\lambda_i > 0$, some $\lambda_j < 0$	None of the above

.9 MISCELLANEOUS USES OF EIGENVALUES AND EIGENVECTORS

Eigenvalues and eigenvectors are useful in many contexts. Four of the more important uses in modern control theory are mentioned.

1. *Existence of solutions for sets of linear equations:* In Chapter 6, the existence of nontrivial solutions for homogeneous equations and of a unique solution for nonhomogeneous equations was seen to depend upon whether or not the coefficient matrix $\mathbf{A}$ had a zero determinant. Stated differently, the existence of unique solutions depended upon whether or not the null space of a linear transformation contained nonzero vectors. Both of these conditions are related to the question of whether or not zero is an eigenvalue. Even when the transformation maps vectors from a space of one dimension to a space of another dimension (and thus cannot define an eigenvalue problem), conditions can be expressed in terms of the eigenvalues of transformations $\mathcal{A}\mathcal{A}^*$ and/or $\mathcal{A}^*\mathcal{A}$. This will be done in Chapter 11 when discussing controllability and observability.

2. *Stability of linear differential and difference equations:* It is not difficult to show that the characteristic equation of the companion matrix (Problem 7.36) is the same as the denominator of the input-output transfer function for the nth order differential equation. Thus the system poles are the same as the eigenvalues of the matrix. The influence of pole locations on system stability has been discussed in Chapter 2. It will be seen in Chapter 10 that the eigenvalues of a system matrix, not necessarily the companion matrix, determine the stability of linear systems described either by differential or difference equations.

3. *Eigenvectors are convenient basis vectors:* When eigenvectors and, if needed, generalized eigenvectors are used as basis vectors, a linear transformation assumes its simplest possible form. In this simple form independent modes of system behavior become apparent. As a simple example, consider the equation

$$\mathbf{y} = \mathbf{A}\mathbf{x}$$

where $\mathbf{A}$ is an $n \times n$ matrix, and $\mathbf{y}$ might be a set of time derivatives or any other $n \times 1$ vector. If a change of basis is used, $\mathbf{x} = \mathbf{M}\mathbf{z}$, $\mathbf{y} = \mathbf{M}\mathbf{w}$, where $\mathbf{M}$ is the modal matrix, then

$$\mathbf{M}\mathbf{w} = \mathbf{A}\mathbf{M}\mathbf{z} \quad \text{or} \quad \mathbf{w} = \mathbf{M}^{-1}\mathbf{A}\mathbf{M}\mathbf{z} = \mathbf{J}\mathbf{z}$$

$\mathbf{J}$ is the Jordan form in general, but will simply be the diagonal matrix $\mathbf{\Lambda}$ in many cases. The simultaneous equations are now as nearly uncoupled as is possible.

When considering the real quadratic form $Q = \langle \mathbf{x}, \mathbf{A}\mathbf{x} \rangle$, with $\mathbf{A}$ symmetric, all second-order products of the components of $\mathbf{x}$ are usually present. If a change of basis $\mathbf{x} = \mathbf{M}\mathbf{z}$ is used, then $Q = \langle \mathbf{M}\mathbf{z}, \mathbf{A}\mathbf{M}\mathbf{z} \rangle = \langle \mathbf{z}, \mathbf{M}^T\mathbf{A}\mathbf{M}\mathbf{z} \rangle$. Since $\mathbf{A}$ is real and symmetric, it can always be diagonalized by an orthogonal transformation, so $Q = \langle \mathbf{z}, \mathbf{\Lambda}\mathbf{z} \rangle$ reduces to the sum of the squares of the z_i components, weighted by the eigenvalues λ_i. This makes the relationships between eigenvalues and the various kinds of definiteness rather transparent.

4. *Sufficient conditions for relative maximum or minimum:* When considering a smooth function of a single variable on an open interval, the necessary condition for a relative maximum or minimum is that the first derivative vanish. To determine whether a maximum, minimum, or saddle point exists, the sign of the second derivative must be determined. In multidimensional cases it is again necessary that the first derivatives (all of them) vanish at a point of relative maximum or minimum. The test of the sign of the second derivative in the scalar case is replaced by a test for positive or negative definiteness of a matrix of second derivative terms. Eigenvalue-eigenvector theory plays an important part in the investigation of these and many other questions.

Additional material related to this chapter may be found in References 3 through 7.

REFERENCES

1. Brogan, W. L.: "Optimal Control Theory Applied to Systems Described by Partial Differential Equations," *Advances in Control Systems,* Vol. 6, C. T. Leondes, Ed., Academic Press, New York, 1968.
2. Melsa, J. L. and S. K. Jones: *Computer Programs for Computational Assistance in the Study of Linear Control Theory,* 2nd ed., McGraw-Hill, New York, 1973.

3. De Russo, P. M., R. J. Roy, and C. M. Close: *State Variables for Engineers,* John Wiley, New York, 1965.
4. Strang, G.: *Linear Algebra and Its Applications,* Academic Press, New York, 1980.
5. Golub, G. H. and C. F. Van Loan: *Matrix Computations,* Johns Hopkins University Press, Baltimore, 1983.
6. Forsythe, G. F., M. A. Malcolm, and C. Moler: *Computer Methods for Mathematical Computations,* Prentice Hall, Englewood Cliffs, N.J., 1977.
7. Courant, R. and D. Hilbert: *Methods of Mathematical Physics,* Vol. 1, Interscience: John Wiley, New York, 1953.

ILLUSTRATIVE PROBLEMS

Determination of Eigenvalues, Eigenvectors, and the Jordan Form

1 Find the eigenvalues and eigenvectors and then use a similarity transformation to diagonalize $\mathbf{A} = \begin{bmatrix} 0 & 1 \\ -3 & -4 \end{bmatrix}$.

The characteristic equation is $|\mathbf{A} - \mathbf{I}\lambda| = \lambda^2 + 4\lambda + 3 = 0$. Therefore $\lambda_1 = -1, \lambda_2 = -3$. For simple roots (Case I) compute

$$\operatorname{Adj}[\mathbf{A} - \mathbf{I}\lambda] = \begin{bmatrix} -4-\lambda & -1 \\ 3 & -\lambda \end{bmatrix}$$

Substituting in $\lambda = -1$ gives $\mathbf{x}_1 = [-1 \quad 1]^T$ or any vector proportional to this. Using $\lambda = -3$ gives $\mathbf{x}_2 = [-1 \quad 3]^T$:

$$\Lambda = \begin{bmatrix} -1 & -1 \\ 1 & 3 \end{bmatrix}^{-1} \mathbf{A} \begin{bmatrix} -1 & -1 \\ 1 & 3 \end{bmatrix} = \begin{bmatrix} -1 & 0 \\ 0 & -3 \end{bmatrix}$$

2 Consider the eigenvalue-eigenvector problem for $\mathbf{A} = \begin{bmatrix} 1 & 2 \\ -2 & -3 \end{bmatrix}$.

The characteristic equation is $|\mathbf{A} - \mathbf{I}\lambda| = \lambda^2 + 2\lambda + 1 = (\lambda + 1)^2$. Therefore, $\lambda = -1$ with algebraic multiplicity 2. $\operatorname{Rank}[\mathbf{A} - \mathbf{I}\lambda]|_{\lambda=-1} = 1$ and degeneracy $q = 2 - 1 = 1$. This is an example of simple degeneracy, so there is one eigenvector $\mathbf{x}_1$ and one generalized eigenvector $\mathbf{x}_2$:

$$\operatorname{Adj}[\mathbf{A} - \mathbf{I}\lambda]\Big|_{\lambda=-1} = \begin{bmatrix} -2 & -2 \\ 2 & 2 \end{bmatrix}$$

Select $\mathbf{x}_1 = \begin{bmatrix} -1 \\ 1 \end{bmatrix}$.

Generalized eigenvector, method 1:

Set $\mathbf{A}\mathbf{x}_2 = -\mathbf{x}_2 + \mathbf{x}_1$ and let $\mathbf{x}_2 = [a \quad b]^T$. Then $a + 2b = -a - 1$ or $2a + 2b = -1$. We could set $a = 1$, then $b = -\frac{3}{2}$ and $\mathbf{x}_2 = [1 \quad -\frac{3}{2}]^T$, or if $a = -1, b = \frac{1}{2}$. If $\mathbf{x}_1$ was selected as $\mathbf{x}_1 = [2 \quad -2]^T$, then $\mathbf{x}_2 = [1 \quad 0]^T$. Either of these choices is valid and each gives $\mathbf{J} = \mathbf{M}^{-1}\mathbf{A}\mathbf{M} = \begin{bmatrix} -1 & 1 \\ 0 & -1 \end{bmatrix}$.

Generalized eigenvector, method 2:

If $\mathbf{x}_1$ is selected as column 1 of $\operatorname{Adj}[\mathbf{A} + \mathbf{I}]$, i.e., $\mathbf{x}_1 = [-2 \quad 2]^T$, then $\mathbf{x}_2$ is column 1 of the differentiated adjoint matrix

$$\mathbf{x}_2 = \frac{d}{d\lambda}\begin{bmatrix} -3-\lambda \\ 2 \end{bmatrix}\Big|_{\lambda=-1} = \begin{bmatrix} -1 \\ 0 \end{bmatrix}$$

If $\mathbf{x}_1 = [-2 \quad 2]^T$ (column 2 of Adj[$\mathbf{A} - \mathbf{I}$]), then $\mathbf{x}_2$ is column 2 of the differentiated adjoint matrix

$$\mathbf{x}_2 = \frac{d}{d\lambda}\begin{bmatrix} -2 \\ 1-\lambda \end{bmatrix}_{\lambda=-1} = \begin{bmatrix} 0 \\ -1 \end{bmatrix}$$

Eigenvector and generalized eigenvector, method 3:

Since $n = m = 2$, we seek the smallest k such that $\text{rank}[\mathbf{A} - \mathbf{I}\lambda]^k|_{\lambda_1=-1} = 0$. The index k is 2. Then $(\mathbf{A} - \mathbf{I}\lambda_1)^2\mathbf{x} = \mathbf{0}$ has two independent solutions, one of which is $\mathbf{x}_2 = [1 \quad 0]^T$. Since $[\mathbf{A} - \mathbf{I}\lambda_1]\mathbf{x}_2 = [2 \quad -2]^T \neq \mathbf{0}$, $\mathbf{x}_2$ is a generalized eigenvector and $\mathbf{x}_1$ is the eigenvector. Alternatively, one could select $\mathbf{x}_2 = [0 \quad 1]^T$ and this also leads to $\mathbf{x}_1 = [2 \quad -2]^T$. All of these methods give the same Jordan form.

7.3 Find the eigenvalues-eigenvectors and the Jordan form for $\mathbf{A} = \begin{bmatrix} 1 & 1 \\ 1 & 1 \end{bmatrix}$.

The characteristic equation is $|\mathbf{A} - \mathbf{I}\lambda| = \lambda^2 - 2\lambda = \lambda(\lambda - 2)$. Therefore, $\lambda_1 = 0, \lambda_2 = 2$. Since $\lambda_1 \neq \lambda_2$, Case I applies:

$$\text{Adj}[\mathbf{A} - \mathbf{I}\lambda] = \begin{bmatrix} 1-\lambda & -1 \\ -1 & 1-\lambda \end{bmatrix}$$

Using the first column with $\lambda = 0$ gives $\mathbf{x}_1 = [1 \quad -1]^T$. Using the first column with $\lambda = 2$ gives $\mathbf{x}_2 = [-1 \quad -1]^T$ or $[1 \quad 1]^T$:

$$\mathbf{M} = \begin{bmatrix} 1 & 1 \\ -1 & 1 \end{bmatrix}, \qquad \mathbf{M}^{-1} = \frac{1}{2}\begin{bmatrix} 1 & -1 \\ 1 & 1 \end{bmatrix}, \qquad \mathbf{J} = \mathbf{M}^{-1}\mathbf{A}\mathbf{M} = \begin{bmatrix} 0 & 0 \\ 0 & 2 \end{bmatrix}$$

7.4 Find the eigenvalues, eigenvectors, and Jordan form for

$$\mathbf{A} = \begin{bmatrix} 1 & 0 & 0 & -3 \\ 0 & 1 & -3 & 0 \\ -0.5 & -3 & 1 & 0.5 \\ -3 & 0 & 0 & 1 \end{bmatrix}$$

The characteristic equation is $\lambda^4 - 4\lambda^3 - 12\lambda^2 + 32\lambda + 64 = 0$. The roots are found to be $\lambda_i = -2, -2, 4, 4$. With $\lambda = -2$

$$\mathbf{A} - \lambda\mathbf{I} = \mathbf{A} + 2\mathbf{I} = \begin{bmatrix} 3 & 0 & 0 & -3 \\ 0 & 3 & -3 & 0 \\ -0.5 & -3 & 3 & 0.5 \\ -3 & 0 & 0 & 3 \end{bmatrix}$$

This matrix has rank 2, so $q = n - r = 2$. This shows that there are two eigenvectors associated with $\lambda = -2$, and since the multiplicity of that root is also 2, no generalized eigenvectors are needed for this eigenvalue. Two linearly independent solutions of $[\mathbf{A} + 2\mathbf{I}]\mathbf{x}_i = \mathbf{0}$ are $\mathbf{x}_1 = [0 \quad 1 \quad 1 \quad 0]^T$ and $\mathbf{x}_2 = [1 \quad 0 \quad 0 \quad 1]^T$. With $\lambda = 4$, $\mathbf{A} - \lambda\mathbf{I} = \begin{bmatrix} -3 & 0 & 0 & -3 \\ 0 & -3 & -3 & 0 \\ -0.5 & -3 & -3 & 0.5 \\ -3 & 0 & 0 & -3 \end{bmatrix}$. The rank is 3 and $q = 1$. There is only one eigenvector, and since the algebraic multiplicity $m = 2$, a generalized eigenvector is needed. Solving $[\mathbf{A} - 4\mathbf{I}]\mathbf{x}_i = 0$ gives only one independent solution, $\mathbf{x}_3 = [0 \quad 1 \quad -1 \quad 0]^T$. Therefore, a generalized eigenvector is needed and it can be found by solving

$$[\mathbf{A} - 4\mathbf{I}]\mathbf{x}_4 = \mathbf{x}_3$$

The result is $\mathbf{x}_4 = [2 \quad -\frac{1}{3} \quad 0 \quad -2]^T$. The modal matrix is thus

$$\mathbf{M} = [\mathbf{x}_1 \quad \mathbf{x}_2 \quad \mathbf{x}_3 \quad \mathbf{x}_4]$$

and the Jordan form is

$$\mathbf{J} = \begin{bmatrix} -2 & 0 & 0 & 0 \\ 0 & -2 & 0 & 0 \\ 0 & 0 & 4 & 1 \\ 0 & 0 & 0 & 4 \end{bmatrix}$$

5 **A** is a 5×5 matrix for which the following information has been found:

$$\lambda_1 = \lambda_2 = 2, \qquad \text{Rank}[\mathbf{A} - 2\mathbf{I}] = 4$$

$$\lambda_3 = \lambda_4 = \lambda_5 = -2, \qquad \text{Rank}[\mathbf{A} + 2\mathbf{I}] = 3$$

Determine the Jordan form for **A**.

For $\lambda = 2$, the degeneracy is $q_1 = 1$, so there is a single Jordan block $\mathbf{J}_1 = \begin{bmatrix} 2 & 1 \\ 0 & 2 \end{bmatrix}$. For $\lambda = -2$, the degeneracy is $q_3 = 2$. Thus there are two Jordan blocks $\mathbf{J}_2 = \begin{bmatrix} -2 & 1 \\ 0 & -2 \end{bmatrix}$ and $\mathbf{J}_3 = [-2]$. The arrangement of these blocks within the Jordan form depends upon the ordering of the eigenvectors and generalized eigenvectors within **M**. Assuming that $\mathbf{x}_1$ and $\mathbf{x}_2$ are an eigenvector and generalized eigenvector for $\lambda = 2$, $\mathbf{x}_3$ and $\mathbf{x}_4$ are an eigenvector and generalized eigenvector for $\lambda = -2$, and $\mathbf{x}_5$ is the second eigenvector associated with $\lambda = -2$, then

$$\text{diag}\,[\mathbf{J}_1, \mathbf{J}_2, \mathbf{J}_3] = \left[\begin{array}{cc|cc|c} 2 & 1 & & & \\ 0 & 2 & & \mathbf{0} & \\ \hline & & -2 & 1 & \\ & \mathbf{0} & 0 & -2 & \\ \hline & & & & -2 \end{array}\right] = \mathbf{J}$$

6 Let $\mathbf{A} = \begin{bmatrix} 3 & 1 & 0 & 0 & 0 & 0 & 0 \\ 0 & 3 & 0 & 0 & 0 & 0 & 0 \\ 0 & 0 & 3 & 0 & 0 & 0 & 0 \\ 0 & 0 & 0 & 4 & 1 & 0 & 0 \\ 0 & 0 & 0 & 0 & 4 & 0 & 0 \\ 0 & 0 & 0 & 0 & 0 & 4 & 1 \\ 0 & 0 & 0 & 0 & 0 & 0 & 4 \end{bmatrix}$.

(a) What are the eigenvalues?
(b) How many linearly independent eigenvectors does **A** have?
(c) How many generalized eigenvectors?

a. **A** is upper triangular and so is $\mathbf{A} - \mathbf{I}\lambda$. Thus the eigenvalues are the diagonal elements of **A**, $\lambda = 3, 3, 3, 4, 4, 4, 4$. b. The matrix **A** is already in Jordan form with four Jordan blocks. There are four eigenvectors. c. There are three generalized eigenvectors. The number of "ones" above the main diagonal is always equal to the number of generalized eigenvectors.

7 Find the eigenvalues, eigenvectors, and, if needed, the generalized eigenvectors for $\mathbf{A} = \begin{bmatrix} 4 & 2 & 1 \\ 0 & 6 & 1 \\ 0 & -4 & 2 \end{bmatrix}$. Also find the Jordan form.

The characteristic equation is $|\mathbf{A} - \mathbf{I}\lambda| = (4 - \lambda)^3$. Then $\lambda_1 = \lambda_2 = \lambda_3 = 4$ with algebraic multiplicity 3.

$[\mathbf{A} - 4\mathbf{I}] = \begin{bmatrix} 0 & 2 & 1 \\ 0 & 2 & 1 \\ 0 & -4 & -2 \end{bmatrix}$ has rank $r = 1$. The degeneracy is $q = 2$. This is an example of the general Case II_3 with two eigenvectors and one generalized eigenvector.

Method 1:

$$\text{Adj}[\mathbf{A} - \mathbf{I}\lambda] = \begin{bmatrix} (6-\lambda)(2-\lambda)+4 & 2\lambda - 8 & \lambda - 4 \\ 0 & (4-\lambda)(2-\lambda) & \lambda - 4 \\ 0 & 16 - 4\lambda & (4-\lambda)(6-\lambda) \end{bmatrix}$$

With $\lambda = 4$, this reduces to the null matrix.

$$\frac{d}{d\lambda}\{\text{Adj}[\mathbf{A} - \mathbf{I}\lambda]\}\bigg|_{\lambda=4} = \begin{bmatrix} 2\lambda - 8 & 2 & 1 \\ 0 & 2\lambda - 6 & 1 \\ 0 & -4 & 2\lambda - 10 \end{bmatrix}\Bigg|_{\lambda=4} = \begin{bmatrix} 0 & 2 & 1 \\ 0 & 2 & 1 \\ 0 & -4 & -2 \end{bmatrix}$$

One eigenvector can be selected as $\mathbf{x}_2 = [1 \quad 1 \quad -2]^T$.

$$\frac{1}{2}\frac{d^2}{d\lambda^2}\{\text{Adj}[\mathbf{A} - \mathbf{I}\lambda]\}\bigg|_{\lambda=4} = \frac{1}{2}\begin{bmatrix} 2 & 0 & 0 \\ 0 & 2 & 0 \\ 0 & 0 & 2 \end{bmatrix}$$

The first column can be selected as another eigenvector, call it $\mathbf{x}_1 = [1 \quad 0 \quad 0]^T$. A generalized eigenvector is given by the third column $\mathbf{x}_3 = [0 \quad 0 \quad 1]^T$. Note that $\mathbf{x}_1$ and $\mathbf{x}_2$ are selected as columns that are nonzero for the first time as the adjoint matrix is repeatedly differentiated. Note also that $\mathbf{x}_3$ is selected from the same column that gave $\mathbf{x}_2$, but with one more differentiation.

Method 2:

We require two independent solutions of $[\mathbf{A} - 4\mathbf{I}]\mathbf{x} = \mathbf{0}$. Let $\mathbf{x} = [a \quad b \quad c]^T$. Then $2b + c = 0$ is the only restriction placed on a, b and c. Since a is arbitrary, set $a = 1$ and $b = c = 0$, or $\mathbf{x}_1 = [1 \quad 0 \quad 0]^T$. Another solution is $a = 1, b = 1, c = -2$, or $\mathbf{x}_2 = [1 \quad 1 \quad -2]^T$.

A generalized eigenvector is needed, and it must satisfy $(\mathbf{A} - 4\mathbf{I})\mathbf{x}_3 = \mathbf{x}_2$ or $(\mathbf{A} - 4\mathbf{I})^2\mathbf{x}_3 = (\mathbf{A} - 4\mathbf{I})\mathbf{x}_2 = \mathbf{0}$. This reduces to $[0]\mathbf{x}_3 = \mathbf{0}$, so $\mathbf{x}_3$ is arbitrary, except it must be nonzero and linearly independent of $\mathbf{x}_1$ and $\mathbf{x}_2$. $\mathbf{x}_3 = [0 \quad 0 \quad 1]^T$ is one such vector.

Setting $\mathbf{M} = \begin{bmatrix} 1 & 1 & 0 \\ 0 & 1 & 0 \\ 0 & -2 & 1 \end{bmatrix}$ gives $\mathbf{M}^{-1} = \begin{bmatrix} 1 & -1 & 0 \\ 0 & 1 & 0 \\ 0 & 2 & 1 \end{bmatrix}$, so that $\mathbf{J} = \mathbf{M}^{-1}\mathbf{A}\mathbf{M} = \left[\begin{array}{c|cc} 4 & 0 & 0 \\ \hline 0 & 4 & 1 \\ 0 & 0 & 4 \end{array}\right]$.

Method 3:

Alternatively, the index of $\lambda = 4$ is $k = 2$, since $\text{rank}[\mathbf{A} - 4\mathbf{I}]^2 = n - m = \mathbf{0}$. A generalized eigenvector satisfying $[\mathbf{A} - 4\mathbf{I}]^2\mathbf{x}_3 = \mathbf{0}$, $[\mathbf{A} - 4\mathbf{I}]\mathbf{x}_3 \neq \mathbf{0}$ is $\mathbf{x}_3 = [0 \quad 0 \quad 1]^T$. Then $\mathbf{x}_2 = [\mathbf{A} - 4\mathbf{I}]\mathbf{x}_3 = [1 \quad 1 \quad -2]^T$. This is one eigenvector, with the other one obviously being $\mathbf{x}_1 = [1 \quad 0 \quad 0]^T$.

7.8 Rework Example 7.6 using the top-down method of finding the eigenvectors or generalized eigenvectors.

Since $\lambda = 1$ has algebraic multiplicity $m = 3$ and since the matrix $\mathbf{A}$ has $n = 3$, the index k must give $\text{rank}\{[\mathbf{A} - \mathbf{I}]^k\} = 0$. It is easily verified that $k = 3$ and that $[\mathbf{A} - \mathbf{I}]^3 = [0]$. Therefore, any vector will satisfy the kth-order generalized eigenvector equation. Select $\mathbf{x}_a$, $\mathbf{x}_b$, and $\mathbf{x}_c$ as columns of the unit matrix. Define $\mathbf{C} = [\mathbf{A} - \mathbf{I}]$. The testing procedure shows that $\mathbf{C}\mathbf{x}_a = \mathbf{0}$, indicating that $\mathbf{x}_a$ is not the sought-for generalized eigenvector. It is an eigenvector, but we do not select it at this point because of scaling considerations. Then $\mathbf{C}\mathbf{x}_b = [2 \quad 0 \quad 0]^T \neq \mathbf{0}$ is found, indicating that $\mathbf{x}_b$ is a generalized eigenvector. It is not selected yet either. Continuing with the testing shows $\mathbf{C}\mathbf{x}_c = [3 \quad 4 \quad 0]^T = \mathbf{x}_d \neq \mathbf{0}$, so $\mathbf{x}_c$ is also a generalized eigenvector. Since $\mathbf{C}\mathbf{x}_b$ gives an eigenvector (and not a generalized eigenvector), it cannot be used to start our chain of three vectors. Testing at the next level shows that $\mathbf{C}\mathbf{x}_d = [8 \quad 0 \quad 0]^T$, another copy of the eigenvector. The final selection can now be made as $\mathbf{x}_3 = \mathbf{x}_c$, $\mathbf{x}_2 = \mathbf{x}_d$, and $\mathbf{x}_1 = 8\mathbf{x}_a$. This is the same set found in Example 7.6.

7.9 Find the eigenvalues, eigenvectors, and if needed, generalized eigenvectors, and the Jordan form for

$$\mathbf{A} = \begin{bmatrix} -4.5 & -1 & 0.5 & 0.5 \\ 0.333333 & -5.33333 & 0.666667 & 0 \\ -0.083333 & 0.33333 & -4.916667 & -0.25 \\ 0.25 & 0 & 0.75 & -5.25 \end{bmatrix}$$

Computer solution gives

$$\lambda_i = \{-4.99959, -5.00063, -4.99958, -5.00035\}$$

It seems likely but not certain that this is a case of repeated roots, but with rounding and truncation errors. The matrix $\mathbf{C} = 5\mathbf{I} - \mathbf{A}$ was formed and subjected to both SVD and **QR** decomposition. The rank was determined to be two, since the singular values Σ_{ii} were 1.5759, 0.8989, 10^{-5}, and 10^{-5}. This indicates that we have at least two repeated roots. Next $\mathbf{C}^2$ was computed and found to be [0] to within 10^{-6}. This indicates that the second-order generalized eigenvector problem has four independent solutions for $\lambda \approx -5$. The conclusion is that $m = 4$, $n = 4$, $q = 2$, and $k = 2$. There are two eigenvectors and two generalized eigenvectors and the Jordan form will have two 2×2 Jordan blocks $\mathbf{J} = \text{Diag}[\mathbf{J}_1, \mathbf{J}_2]$ with $\mathbf{J}_1 = \mathbf{J}_2 = \begin{bmatrix} -5 & 1 \\ 0 & -5 \end{bmatrix}$.

There are many ways to find the eigenvectors. The last two columns of the SVD $\mathbf{V}$ matrix could be used. Here we note that $\mathbf{C}^2 = [\mathbf{0}]$, so any column of $\mathbf{C}$ could be selected as an eigenvector. Rank($\mathbf{C}$) = 2, so only two independent columns exist. We select $\mathbf{x}_a = [0.5 \quad 0.33333 \quad -0.08333 \quad 0.25]^T$ and $\mathbf{x}_b = [0.5 \quad 0.66667 \quad 0.08333 \quad 0.75]^T$. Two other independent vectors $\mathbf{x}_c = [1 \quad 0 \quad 0 \quad 0]^T$ and $\mathbf{x}_d = [0 \quad 0 \quad 1 \quad 0]^T$ are selected. They all satisfy the 2nd-order generalized eigenvector problem. The top-down testing procedure shows that $\mathbf{C}\mathbf{x}_c = \mathbf{x}_a$ and that $\mathbf{C}\mathbf{x}_d = \mathbf{x}_b$. The columns of the modal matrix are thus selected as $\mathbf{x}_1 = \mathbf{x}_a$, $\mathbf{x}_2 = \mathbf{x}_c$, $\mathbf{x}_3 = \mathbf{x}_b$, and $\mathbf{x}_4 = \mathbf{x}_d$.

.10 Noting that $\mathbf{AM} - \mathbf{MJ} = [0]$ is in the class of problems considered in Section 6.10, the solution for the vectorized version of $\mathbf{M}$ can be found by solving the $n^2 \times n^2$ problem $[\mathbf{I}_n \otimes \mathbf{A} - \mathbf{J}^T \otimes \mathbf{I}_n](\mathbf{M}) = (0)$.

(a) Apply this approach to Problems 7.1, 7.2, and 7.13 assuming $\mathbf{J}$ is known.

(b) Use the same approach on Problem 7.2, but this time try using $\mathbf{J} = \text{Diag}[-1, -1]$.

(a) Computer solution of the four simultaneous equations obtained for Problem 7.1 gives two independent nontrivial solutions, $(\mathbf{M})_1 = [1 \quad -1 \quad 0 \quad 0]^T$ and $(\mathbf{M})_2 = [0 \quad 0 \quad \frac{1}{3} \quad -1]^T$. Neither of these provides a valid nonsingular matrix $\mathbf{M}$, but any linear combination of $(\mathbf{M})_1$ and $(\mathbf{M})_2$ is also a solution. Using $(\mathbf{M}) = -(\mathbf{M})_1 - 3(\mathbf{M})_2$ gives the previously obtained modal matrix. For Problem 7.2 there are also two solutions to the 4×4 problem, $(\mathbf{M})_1 = [1 \quad -1 \quad 0.5 \quad 0]^T$ and $(\mathbf{M})_2 = [0 \quad 0 \quad 1 \quad -1]^T$. The first solution gives an acceptable modal matrix $\mathbf{M} = \begin{bmatrix} 1 & 0.5 \\ -1 & 0 \end{bmatrix}$. For Problem 7.3 a similar result is obtained. Two vectorized solutions $(\mathbf{M})_1 = [1 \quad -1 \quad 0 \quad 0]^T$ and $(\mathbf{M})_2 = [0 \quad 0 \quad -1 \quad -1]^T$ are obtained. The former answer is obtained from $(\mathbf{M}) = (\mathbf{M})_1 - (\mathbf{M})_2$.

(b) When the wrong $\mathbf{J}$ matrix is used, two solutions are obtained; $(\mathbf{M})_1$ is the same as when the correct $\mathbf{J}$ was used in (a) and $(\mathbf{M})_2$ is $[0 \quad 0 \quad 1 \quad -1]^T$. No linear combination of these will give a nonsingular matrix $\mathbf{M}$ because the two nonzero columns are linearly dependent. This problem suggests another approach to determining eigenvectors and generalized eigenvectors, based on assuming $\mathbf{J}$ and testing the resulting $\mathbf{M}$. Although it works, the dimension of the vectorized problems quickly get out of hand. Using the method on Problem 7.4 requires solution of a 16×16 set of equations. When this was done, six linearly independent $(\mathbf{M})_i$ vectors were found, and the previous answer was a linear combination of these six.

.11 If $\lambda_i = \sigma + j\omega$ is a complex eigenvalue for $\mathbf{A}$, with the associated complex eigenvector $\mathbf{x}_i = \mathbf{xr}_i + \mathbf{xc}_i j$, then the complex conjugate of λ_i is also an eigenvalue associated with an eigenvector $\mathbf{x}_{i+1} = \mathbf{xr}_i - \mathbf{xc}_i j$. The methods already presented for solving the eigenvalue-eigenvector problem apply on the complex number field. Most examples have been restricted to real eigenvalues to maintain simplicity. In the complex case, purely real arithmetic can again be used on a double-sized problem. By combining $\mathbf{A}\mathbf{x}_i = \lambda_i \mathbf{x}_i$ and $\mathbf{A}\bar{\mathbf{x}}_i = \bar{\lambda}_i \bar{\mathbf{x}}_i$ and equating real parts and imaginary parts, the following a set of equations are obtained:

$$\mathbf{A}[\mathbf{xr}_i \quad \mathbf{xc}_i] = [\mathbf{xr}_i \quad \mathbf{xc}_i]\mathbf{E}$$

where

$$\mathbf{E} = \begin{bmatrix} \sigma & \omega \\ -\omega & \sigma \end{bmatrix}$$

This is the type of equation dealt with in Sec. 6.10; it can be written in vectorized form as a set of six simultaneous equations:

$$[(\mathbf{I}_2 \otimes \mathbf{A}) - (\mathbf{E}^T \otimes \mathbf{I}_2)]\begin{bmatrix} \mathbf{xr}_i \\ \mathbf{xc}_i \end{bmatrix} = (0)$$

Find the eigenvalues, eigenvectors, and Jordan form for

$$\mathbf{A} = \begin{bmatrix} -4 & -2 & 1 \\ -1 & -2 & 1 \\ -1 & -4 & -6 \end{bmatrix}$$

The iterative **QR** method and direct solution for the characteristic equation roots both give eigenvalues as $\lambda = \{-1.56516, -5.21742 \pm 1.85843j\}$. The eigenvector associated with the real root is found by solving $(\mathbf{A} - \mathbf{I}\lambda_1)\mathbf{x}_1 = \mathbf{0}$ and is $\mathbf{x}_1 = [-1 \quad 0.916739 \quad -0.60136]^T$. With $\mathbf{E} = \begin{bmatrix} -5.21742 & -1.85643 \\ 1.85643 & -5.21742 \end{bmatrix}$ the 6×6 coefficient matrix $(\mathbf{I}_2 \otimes \mathbf{A}) - (\mathbf{E}^T \otimes \mathbf{I}_3)$ is

Row 1	1.217420E + 00	−2.000000E + 00	1.000000E + 00	−1.856430E + 00	0.000000E + 00	0.000000E + 00
Row 2	−1.000000E + 00	3.217420E + 00	1.000000E + 00	0.000000E + 00	−1.856430E + 00	0.000000E + 00
Row 3	−1.000000E + 00	−4.000000E + 00	−7.825799E − 01	0.000000E + 00	0.000000E + 00	−1.856430E + 00
Row 4	1.856430E + 00	0.000000E + 00	0.000000E + 00	1.217420E + 00	−2.000000E + 00	1.000000E + 00
Row 5	0.000000E + 00	1.856430E + 00	0.000000E + 00	−1.000000E + 00	3.217420E + 00	1.000000E + 00
Row 6	0.000000E + 00	0.000000E + 00	1.856430E + 00	−1.000000E + 00	−4.000000E + 00	−7.825799E − 01

Its rank is 4, yielding two independent solutions for the stacked eigenvector,

Row 1	1.041709E − 01	6.530732E − 01
Row 2	1.696022E − 01	3.008392E − 01
Row 3	−1.000000E + 00	0.000000E + 00
Row 4	−6.530732E − 01	1.041709E − 01
Row 5	−3.008392E − 01	1.696022E − 01
Row 6	0.000000E + 00	−1.000000E + 00

From this, $\mathbf{x}_2 = [0.10417 - 0.65307j \quad 0.169602 - 0.30084j \quad -1]^T$, and $\mathbf{x}_3 = \bar{\mathbf{x}}_2$. The complex modal matrix is formed from these columns, and it yields

$$\mathbf{J} = \mathbf{M}^{-1}\mathbf{AM} = \begin{bmatrix} -1.56516 & 0 & 0 \\ 0 & -5.21742 - 1.85643j & 0 \\ 0 & 0 & -5.21742 + 1.85643j \end{bmatrix}$$

Note that if columns of $\mathbf{T} = [\mathbf{x}_1 \quad \mathbf{xr}_2 \quad \mathbf{xc}_2]$ are used as basis vectors instead of $\mathbf{M}$, a block diagonal real matrix is obtained in place of $\mathbf{J}$, namely, $\mathbf{T}^{-1}\mathbf{AT} = \text{Diag}[\lambda_1, \mathbf{E}]$.

Similar Matrices

12 Prove that two similar matrices have the same eigenvalues.

A and B are similar matrices if they are related by $\mathbf{A} = \mathbf{Q}^{-1}\mathbf{BQ}$ for some nonsingular matrix Q. The characteristic equation for A is

$$|\mathbf{A} - \mathbf{I}\lambda| = |\mathbf{Q}^{-1}\mathbf{BQ} - \mathbf{Q}^{-1}\mathbf{Q}\lambda| = 0$$

or

$$|\mathbf{Q}^{-1}[\mathbf{B} - \mathbf{I}\lambda]\mathbf{Q}| = |\mathbf{Q}^{-1}|\cdot|\mathbf{Q}||\mathbf{B} - \mathbf{I}\lambda| = |\mathbf{B} - \mathbf{I}\lambda| = 0$$

The characteristic equations for A and B are the same so they have the same eigenvalues.

13 Are the following matrices similar?

$$\begin{bmatrix} 2 & 0 & 0 & 0 \\ 0 & 2 & 0 & 0 \\ 0 & 0 & 2 & 0 \\ 0 & 0 & 0 & 2 \end{bmatrix}, \quad \begin{bmatrix} 2 & 1 & 0 & 0 \\ 0 & 2 & 0 & 0 \\ 0 & 0 & 2 & 1 \\ 0 & 0 & 0 & 2 \end{bmatrix}, \quad \begin{bmatrix} 2 & 0 & 0 & 0 \\ 0 & 2 & 1 & 0 \\ 0 & 0 & 2 & 1 \\ 0 & 0 & 0 & 2 \end{bmatrix}, \quad \begin{bmatrix} 2 & 1 & 0 & 0 \\ 0 & 2 & 1 & 0 \\ 0 & 0 & 2 & 1 \\ 0 & 0 & 0 & 2 \end{bmatrix}$$

All four matrices have the same characteristic equation $(2 - \lambda)^4 = 0$, so $\lambda = 2$ is the eigenvalue with algebraic multiplicity $m = 4$. The given matrices are expressed in Jordan form. Since similar matrices must have the same Jordan form, the answer is no. Note that the index of the eigenvalue is 1, 2, 3, and 4, respectively, for these matrices.

Miscellaneous Properties

14 Prove that $|\mathbf{A}| = \lambda_1\lambda_2\cdots\lambda_n$.

Any $n \times n$ matrix A can be reduced to the Jordan form $\mathbf{J} = \mathbf{M}^{-1}\mathbf{AM}$; so $\mathbf{A} = \mathbf{MJM}^{-1}$. From this $|\mathbf{A}| = |\mathbf{MJM}^{-1}| = |\mathbf{M}||\mathbf{J}||\mathbf{M}^{-1}| = |\mathbf{J}|$. Since J is upper triangular, with the eigenvalues on the main diagonal, $|\mathbf{A}| = |\mathbf{J}| = \lambda_1\lambda_2\cdots\lambda_n$. Thus $|\mathbf{A}| = 0 \Leftrightarrow$ at least one $\lambda_i = 0$.

15 Prove $\text{Tr}(\mathbf{A}) = \lambda_i + \lambda_2 + \cdots + \lambda_n$.

Since $\mathbf{A} = \mathbf{MJM}^{-1}$, $\text{Tr}(\mathbf{A}) = \text{Tr}(\mathbf{MJM}^{-1})$.

But $\text{Tr}(\mathbf{AB}) = \text{Tr}(\mathbf{BA})$, so $\text{Tr}(\mathbf{A}) = \text{Tr}(\mathbf{JM}^{-1}\mathbf{M})$ or $\text{Tr}(\mathbf{A}) = \text{Tr}(\mathbf{J}) = \lambda_1 + \lambda_2 + \cdots + \lambda_n$.

16 Prove that if A is nonsingular with eigenvalues λ_i, then $1/\lambda_i$ are the eigenvalues of $\mathbf{A}^{-1}$.

The n roots λ_i are defined by $|\mathbf{A} - \mathbf{I}\lambda| = 0$. But $|\mathbf{A} - \mathbf{I}\lambda| = |\mathbf{A}[\mathbf{I} - \mathbf{A}^{-1}\lambda]| = |\mathbf{A}|\cdot|\mathbf{I} - \mathbf{A}^{-1}\lambda| = 0$. Since A is nonsingular, $|\mathbf{A}|$ can be divided out leaving $|\mathbf{I} - \mathbf{A}^{-1}\lambda| = |\mathbf{I}(1/\lambda) - \mathbf{A}^{-1}|\lambda^n = 0$. Since $|\mathbf{A}|$, and hence λ, are not zero, the characteristic equation for A leads to $|\mathbf{A}^{-1} - \mathbf{I}(1/\lambda)| = 0$. Thus if λ_i is an eigenvalue of A, then $1/\lambda_i$ is an eigenvalue of $\mathbf{A}^{-1}$.

17 Let A be an $n \times n$ matrix with n distinct eigenvalues. Prove that the set of n eigenvectors $\mathbf{x}_i$ are linearly independent.

Let

$$a_1\mathbf{x}_1 + a_2\mathbf{x}_2 + \cdots + a_n\mathbf{x}_n = \mathbf{0} \qquad (1)$$

If it can be shown that this implies that $a_1 = a_2 = \cdots = a_n = 0$, then the set $\{\mathbf{x}_i\}$ is linearly independent. Define $\mathbf{T}_i = \mathbf{A} - \mathbf{I}\lambda_i$ and note that $\mathbf{T}_i\mathbf{x}_i = \mathbf{0}$. $\mathbf{T}_i\mathbf{x}_j = (\lambda_j - \lambda_i)\mathbf{x}_j$ if $i \neq j$. Multiplying equation (1) by $\mathbf{T}_1$ gives

$$a_2(\lambda_2 - \lambda_1)\mathbf{x}_2 + a_3(\lambda_3 - \lambda_1)\mathbf{x}_3 + \cdots + a_n(\lambda_n - \lambda_1)\mathbf{x}_n = \mathbf{0}$$

Multiplying this in turn by $\mathbf{T}_2$, then $\mathbf{T}_3, \ldots, \mathbf{T}_{n-1}$ gives

$$a_3(\lambda_3 - \lambda_1)(\lambda_3 - \lambda_2)\mathbf{x}_3 + \cdots + a_n(\lambda_n - \lambda_1)(\lambda_n - \lambda_2)\mathbf{x}_n = \mathbf{0}$$

$$\vdots$$

$$a_{n-1}(\lambda_{n-1} - \lambda_1)(\lambda_{n-1} - \lambda_2)\cdots(\lambda_{n-1} - \lambda_{n-2})\mathbf{x}_{n-1} + a_n(\lambda_n - \lambda_1)(\lambda_n - \lambda_2)\cdots(\lambda_n - \lambda_{n-2})\mathbf{x}_n = \mathbf{0} \quad (2)$$

$$a_n(\lambda_n - \lambda_1)(\lambda_n - \lambda_2)\cdots(\lambda_n - \lambda_{n-2})(\lambda_n - \lambda_{n-1})\mathbf{x}_n = \mathbf{0} \quad (3)$$

Since $\mathbf{x}_n \neq \mathbf{0}$, and $\lambda_n \neq \lambda_i$ for $i \neq n$, equation (3) requires that $a_n = 0$. This plus Eq. (2) requires that $a_{n-1} = 0$. Continuing this reasoning shows that Eq. (1) requires $a_i = 0$ for $i = 1, 2, \ldots, n$, so the eigenvectors are linearly independent.

7.18 Let $\mathbf{T}_i : \mathscr{X} \to \mathscr{X}$ be defined by $\mathbf{T}_i = \mathbf{A} - \mathbf{I}\lambda_i$, where $\mathscr{X}$ is an n-dimensional space. Prove that there are always $q_i = n - \text{rank}(\mathbf{T}_i)$ linearly independent eigenvectors associated with eigenvalue λ_i.

The space $\mathscr{X}$ can be written as the direct sum $\mathscr{X} = \mathcal{R}(\mathbf{T}_i^*) \oplus \mathcal{N}(\mathbf{T}_i)$ and $\dim(\mathscr{X}) = \dim(\mathcal{R}(\mathbf{T}_i^*)) + \dim(\mathcal{N}(\mathbf{T}_i))$ or $n = \text{rank}(\mathbf{T}_i^*) + \dim(\mathcal{N}(\mathbf{T}_i))$. But since rank $(\mathbf{T}_i) = \text{rank}(\mathbf{T}_i^*)$, $n - \text{rank}(\mathbf{T}_i) = q_i = \dim(\mathcal{N}(\mathbf{T}_i))$. The null space of $\mathbf{T}_i$ is of dimension q_i and, therefore, it contains q_i linearly independent vectors, all of which are eigenvectors.

7.19 Let $\mathbf{A}$ be an arbitrary $n \times r$ matrix and let $\mathbf{B}$ be an arbitrary $r \times n$ matrix, so that $\mathbf{AB}$ and $\mathbf{BA}$ are $n \times n$ and $r \times r$ matrices respectively. Assume that $n \geq r$ and prove:

(a) The scalar λ is a nonzero eigenvalue of $\mathbf{AB}$ if and only if it is a nonzero eigenvalue of $\mathbf{BA}$.

(b) If $\mathbf{x}_i$ is an eigenvector (or generalized eigenvector) of $\mathbf{AB}$ associated with a nonzero eigenvalue, then $\zeta_i \triangleq \mathbf{Bx}_i$ is an eigenvector (or generalized eigenvector) or $\mathbf{BA}$.

(c) $\mathbf{AB}$ has at least $n - r$ zero eigenvalues.

Assume $\mathbf{ABx}_i = \lambda\mathbf{x}_i$ with $\lambda \neq 0, \mathbf{x}_i \neq \mathbf{0}$. Then multiplying by $\mathbf{B}$ gives $\mathbf{BA}(\mathbf{Bx}_i) = \lambda\mathbf{Bx}_i$ or $\mathbf{BA}\zeta_i = \lambda\zeta_i$. Thus λ and ζ_i are an eigenvalue and eigenvector of $\mathbf{BA}$ provided $\zeta_i \neq \mathbf{0}$. But since $\lambda\mathbf{x}_i \neq \mathbf{0}$, $\mathbf{Bx}_i \neq \mathbf{0}$; otherwise $\mathbf{ABx}_i = \mathbf{0}$. Therefore, λ and ζ_i are an eigenvalue and eigenvector of $\mathbf{BA}$, provided $\lambda \neq 0$ and $\mathbf{x}_i$ are an eigenvalue and eigenvector of $\mathbf{AB}$. Now assume $\mathbf{BA}\zeta_i = \lambda\zeta_i$, $\lambda \neq 0$, and $\zeta_i \neq \mathbf{0}$. Using the same kind of arguments as above show that λ and $\mathbf{x}_i = \mathbf{A}\zeta_i$ are an eigenvalue and eigenvector of $\mathbf{AB}$. These results can be generalized for the case of generalized eigenvectors. This proves a and b. To prove c, it is only necessary to note that $\mathbf{AB}$ has n eigenvalues and each nonzero eigenvalue is simultaneously an eigenvalue of $\mathbf{BA}$. Since $\mathbf{BA}$ has r eigenvalues, $\mathbf{AB}$ has at most r nonzero eigenvalues and, therefore, at least $n - r$ zero eigenvalues.

7.20 Let $\mathbf{A}$ and $\mathbf{B}$ be defined as in Problem 7.19. Define $\mathbf{N} = \mathbf{I}_n + \mathbf{AB}$ and $\mathbf{R} = \mathbf{I}_r + \mathbf{BA}$. Prove:

(a) $\mathbf{x}_i$ is an eigenvector of $\mathbf{AB}$ if and only if it is an eigenvector of $\mathbf{N}$ and λ is an eigenvalue of $\mathbf{AB}$ if and only if $1 + \lambda$ is an eigenvalue of $\mathbf{N}$.

(b) ζ_i is an eigenvector of $\mathbf{BA}$ if and only if it is an eigenvector of $\mathbf{R}$, and λ is an eigenvalue of $\mathbf{BA}$ if and only if $1 + \lambda$ is an eigenvalue of $\mathbf{R}$.

(a) The proof requires showing that

$$(\mathbf{AB} - \lambda_i \mathbf{I}_n)\mathbf{x}_i = \mathbf{0} \Leftrightarrow (\mathbf{N} - (\lambda_i + 1)\mathbf{I}_n)\mathbf{x}_i = \mathbf{0}$$

But $\mathbf{N} - (\lambda_i + 1)\mathbf{I}_n = \mathbf{AB} + \mathbf{I}_n - \lambda_i \mathbf{I}_n - \mathbf{I}_n = \mathbf{AB} - \lambda_i \mathbf{I}_n$.

(b) The proof is a simple matter of applying the definitions of the two eigenvalue problems in question, just as in part a.

7.21 Show that the r eigenvalues of $\mathbf{R}$ defined in Problem 7.20 are also eigenvalues of $\mathbf{N}$. The remaining $n - r$ eigenvalues of $\mathbf{N}$ are all equal to one.

If λ is an eigenvalue of $\mathbf{BA}$, then it is also an eigenvalue of $\mathbf{AB}$. Since the eigenvalues of $\mathbf{N}$ and $\mathbf{R}$ are shifted by one from these eigenvalues, the result is proven for the r eigenvalues of $\mathbf{BA}$. It was shown in Problem 7.19 that the remaining $n - r$ eigenvalues of $\mathbf{AB}$ must be zero, so the corresponding eigenvalues of $\mathbf{N}$ are one.

22 Let $\mathbf{A}$ and $\mathbf{B}$ be as defined in Problem 7.19. Prove the determinant identity of Problem 4.5, page 142, i.e., prove $|\mathbf{I}_n \pm \mathbf{AB}| = |\mathbf{I}_r \pm \mathbf{BA}|$.

Since the determinant is equal to the product of the eigenvalues (see Problem 7.14) and since $\mathbf{N} = \mathbf{I}_n + \mathbf{AB}$ and $\mathbf{R} = \mathbf{I}_r + \mathbf{BA}$ have been shown to have r eigenvalues in common with the rest of the $n - r$ eigenvalues equal to one, the identity is proven for the plus sign. The identity also holds for the minus sign, since we can always define $\mathbf{B}_1 = -\mathbf{B}$ or $\mathbf{A}_1 = -\mathbf{A}$. The previous results placed no restrictions on $\mathbf{A}$ and $\mathbf{B}$ other than their dimensions. Another form of the same identity is

$$|\mathbf{A}_1\mathbf{B} \pm \lambda\mathbf{I}_n| = (\pm\lambda)^{n-r}|\mathbf{BA}_1 \pm \lambda\mathbf{I}_r|$$

which is established by defining $\lambda\mathbf{A} = \mathbf{A}_1$ and using the rule for multiplying a determinant by a scalar.

23 Let $\mathbf{P}$ be a nonsingular $n \times n$ matrix whose determinant and inverse are known. Let $\mathbf{C}$ and $\mathbf{D}$ be arbitrary $n \times r$ and $r \times n$ matrices, respectively. Show that $|\mathbf{P} + \mathbf{CD}| = |\mathbf{P}|\cdot|\mathbf{I}_r + \mathbf{DP}^{-1}\mathbf{C}|$.

Simple manipulations show $|\mathbf{P} + \mathbf{CD}| = |\mathbf{P}[\mathbf{I}_n + \mathbf{P}^{-1}\mathbf{CD}]| = |\mathbf{P}|\cdot|\mathbf{I}_n + \mathbf{P}^{-1}\mathbf{CD}|$. Using the result of Problem 7.22 allows the interchange of factors $(\mathbf{P}^{-1}\mathbf{C})$ and $(\mathbf{D})$ and the corresponding change from an $n \times n$ determinant to an $r \times r$ determinant.

24 Let $\mathbf{N}$ be an $n \times n$ matrix given by $\mathbf{N} = \mathbf{I}_n + \mathbf{AB}$, where $\mathbf{A}$ and $\mathbf{B}$ are $n \times 1$ and $1 \times n$ matrices. Show that the $n \times n$ determinant can be expressed in terms of the easily evaluated trace: $|\mathbf{N}| = \mathrm{Tr}(\mathbf{N}) + 1 - n$.

Since $|\mathbf{N}| = |\mathbf{I}_n + \mathbf{AB}| = |\mathbf{I}_r + \mathbf{BA}|$ and since in this case the dimension r of $\mathbf{BA}$ is one, $|\mathbf{N}| = 1 + \mathbf{BA} = 1 + \mathrm{Tr}(\mathbf{BA})$. But $\mathrm{Tr}(\mathbf{BA}) = \mathrm{Tr}(\mathbf{AB})$ and $\mathrm{Tr}(\mathbf{N}) = \mathrm{Tr}(\mathbf{AB}) + \mathrm{Tr}(\mathbf{I}_n)$ or $\mathrm{Tr}(\mathbf{AB}) = \mathrm{Tr}(\mathbf{N}) - n$, so $|\mathbf{N}| = 1 + \mathrm{Tr}(\mathbf{N}) - n$.

Self-Adjoint Transformation

25 If $\mathscr{A}$ is a self-adjoint transformation (see Section 5.12), show that all of its eigenvalues are real and that the eigenvectors associated with two different eigenvalues are orthogonal.

Consider $\mathscr{A}(\mathbf{x}_i) = \lambda_i\mathbf{x}_i$ and form the inner product $\langle\mathbf{x}_i, \mathscr{A}(\mathbf{x}_i)\rangle = \langle\mathbf{x}_i, \lambda_i\mathbf{x}_i\rangle = \lambda_i\langle\mathbf{x}_i, \mathbf{x}_i\rangle$. The definition of $\mathscr{A}^*$ ensures that $\langle\mathbf{x}_i, \mathscr{A}(\mathbf{x}_i)\rangle = \langle\mathscr{A}^*(\mathbf{x}_i), \mathbf{x}_i\rangle$ and if $\mathscr{A} = \mathscr{A}^*$, this gives $\langle\lambda_i\mathbf{x}_i, \mathbf{x}_i\rangle = \bar{\lambda}_i\langle\mathbf{x}_i, \mathbf{x}_i\rangle$. Subtracting gives $0 = (\bar{\lambda}_i - \lambda_i)\langle\mathbf{x}_i, \mathbf{x}_i\rangle$. Since $\mathbf{x}_i$ is an eigenvector $\|\mathbf{x}_i\|^2 \neq 0$, so $\bar{\lambda}_i = \lambda_i$ and all eigenvalues are real.

Now consider $\mathscr{A}(\mathbf{x}_i) = \lambda_i\mathbf{x}_i$, $\mathscr{A}(\mathbf{x}_j) = \lambda_j\mathbf{x}_j$ with $\lambda_i \neq \lambda_j$. Then $\langle\mathbf{x}_j, \mathscr{A}(\mathbf{x}_i)\rangle = \lambda_i\langle\mathbf{x}_j, \mathbf{x}_i\rangle$. Also $\langle\mathbf{x}_j, \mathscr{A}(\mathbf{x}_i)\rangle = \langle\mathscr{A}^*(\mathbf{x}_j), \mathbf{x}_i\rangle = \langle\mathscr{A}(\mathbf{x}_j), \mathbf{x}_i\rangle = \bar{\lambda}_j\langle\mathbf{x}_j, \mathbf{x}_i\rangle$. But $\bar{\lambda}_j = \lambda_j$, so substracting gives $0 = (\lambda_i - \lambda_j)\langle\mathbf{x}_j, \mathbf{x}_i\rangle$. Since $\lambda_i \neq \lambda_j$, we have $\langle\mathbf{x}_j, \mathbf{x}_i\rangle = 0$ and $\mathbf{x}_j$ is orthogonal to $\mathbf{x}_i$.

Normal Transformation

.26 Let $\mathscr{A}$ be a normal transformation (see Section 5.12). Prove that $\mathscr{A}(\mathbf{x}_i) = \lambda_i\mathbf{x}_i$ if and only if $\mathscr{A}^*(\mathbf{x}_i) = \bar{\lambda}_i\mathbf{x}_i$.

This is equivalent to showing $(\mathscr{A} - \mathscr{I}\lambda_i)\mathbf{x}_i = \mathbf{0} \Leftrightarrow (\mathscr{A}^* - \mathscr{I}\bar{\lambda}_i)\mathbf{x}_i = \mathbf{0}$.

$$\begin{aligned}\langle(\mathscr{A} - \mathscr{I}\lambda_i)\mathbf{x}_i, (\mathscr{A} - \mathscr{I}\lambda_i)\mathbf{x}_i\rangle &= \langle\mathscr{A}(\mathbf{x}_i), \mathscr{A}(\mathbf{x}_i)\rangle - \langle\lambda_i\mathbf{x}_i, \mathscr{A}(\mathbf{x}_i)\rangle \\ &\quad - \langle\mathscr{A}(\mathbf{x}_i), \lambda_i\mathbf{x}_i\rangle + \langle\lambda_i\mathbf{x}_i, \lambda_i\mathbf{x}_i\rangle \\ &= \langle\mathscr{A}^*\mathscr{A}(\mathbf{x}_i), \mathbf{x}_i\rangle - \bar{\lambda}_i\langle\mathscr{A}^*(\mathbf{x}_i), \mathbf{x}_i\rangle \\ &\quad - \lambda_i\langle\mathbf{x}_i, \mathscr{A}^*(\mathbf{x}_i)\rangle - \bar{\lambda}_i\lambda_i\langle\mathbf{x}_i, \mathbf{x}_i\rangle\end{aligned}$$

Using $\mathscr{A}^*\mathscr{A} = \mathscr{A}\mathscr{A}^*$ allows this to be rewritten as

$$\begin{aligned}&\langle\mathscr{A}^*(\mathbf{x}_i), \mathscr{A}^*(\mathbf{x}_i)\rangle - \langle\mathscr{A}^*(\mathbf{x}_i), \bar{\lambda}_i\mathbf{x}_i\rangle - \langle\bar{\lambda}_i\mathbf{x}_i, \mathscr{A}^*(\mathbf{x}_i)\rangle + \langle\bar{\lambda}_i\mathbf{x}_i, \bar{\lambda}_i\mathbf{x}_i\rangle \\ &\qquad = \langle(\mathscr{A}^* - \mathscr{I}\bar{\lambda}_i)\mathbf{x}_i, (\mathscr{A}^* - \mathscr{I}\bar{\lambda}_i)\mathbf{x}_i\rangle\end{aligned}$$

or

$$\|(\mathcal{A} - \mathcal{I}\lambda_i)\mathbf{x}_i\|^2 = \|(\mathcal{A}^* - \mathcal{I}\bar{\lambda}_i)\mathbf{x}_i\|^2$$

The desired result follows.

7.27 Prove that the eigenvectors $\mathbf{x}_i$ and $\mathbf{x}_j$ associated with eigenvalues λ_i and λ_j are orthogonal for any normal transformation, provided $\lambda_i \neq \lambda_j$.

A normal transformation satisfies $\mathcal{A}^*\mathcal{A} = \mathcal{A}\mathcal{A}^*$, and therefore the class of normal transformations includes self-adjoint transformations as a subclass. Consider $\mathcal{A}(\mathbf{x}_i) = \lambda_i \mathbf{x}_i$ and $\mathcal{A}(\mathbf{x}_j) = \lambda_j \mathbf{x}_j$ with $\lambda_i \neq \lambda_j$. Then

$$\langle \mathbf{x}_j, \mathcal{A}(\mathbf{x}_i)\rangle = \lambda_i\langle \mathbf{x}_j, \mathbf{x}_i\rangle = \langle \mathcal{A}^*(\mathbf{x}_j), \mathbf{x}_i\rangle \tag{1}$$

From the previous problem $\mathcal{A}^*(\mathbf{x}_j) = \bar{\lambda}_j \mathbf{x}_j$, so

$$\langle \mathcal{A}^*(\mathbf{x}_j), \mathbf{x}_i\rangle = \langle \bar{\lambda}_j \mathbf{x}_j, \mathbf{x}_i\rangle = \lambda_j\langle \mathbf{x}_j, \mathbf{x}_i\rangle \tag{2}$$

Subtracting Eq. (2) from Eq. (1) gives $0 = (\lambda_i - \lambda_j)\langle \mathbf{x}_j, \mathbf{x}_i\rangle$. Since $\lambda_i \neq \lambda_j$, it follows that $\langle \mathbf{x}_j, \mathbf{x}_i\rangle =$ 0 and, therefore, $\mathbf{x}_i$ and $\mathbf{x}_j$ are orthogonal.

7.28 Prove that if $\mathbf{A}$ is the $n \times n$ matrix representation of a normal transformation, then $\mathbf{A}$ has a full set of n linearly independent eigenvectors, regardless of the multiplicity of the eigenvalues.

Let $\mathbf{T}_i \triangleq \mathbf{A} - \mathbf{I}\lambda_i$. The eigenvectors satisfy $\mathbf{T}_i \mathbf{x}_i = \mathbf{0}$, and a generalized eigenvector $\mathbf{x}_{i+1}$ must satisfy $\mathbf{T}_i \mathbf{x}_{i+1} = \mathbf{x}_i$. The required proof consists of showing that if $\mathbf{A}$ is normal, the condition on $\mathbf{x}_{i+1}$ leads to a contradiction and thus cannot be satisfied. For $\mathbf{A}$ normal, $\mathbf{A}^*\mathbf{x}_i = \bar{\lambda}_i \mathbf{x}_i$ for each eigenvector $\mathbf{x}_i$; that is, $\mathbf{T}_i^* \mathbf{x}_i = \mathbf{0}$. Then $\mathbf{T}_i^* \mathbf{T}_i \mathbf{x}_{i+1} = \mathbf{T}_i^* \mathbf{x}_i = \mathbf{0}$. This means that

$$\langle \mathbf{x}_{i+1}, \mathbf{T}_i^* \mathbf{T}_i \mathbf{x}_{i+1}\rangle = \langle \mathbf{T}_i \mathbf{x}_{i+1}, \mathbf{T}_i \mathbf{x}_{i+1}\rangle = 0 \quad \text{or} \quad \|\mathbf{T}_i \mathbf{x}_{i+1}\|^2 = 0$$

This requires that $\mathbf{T}_i \mathbf{x}_{i+1} = \mathbf{0}$, but this contradicts the original assumption, since $\mathbf{x}_i \neq \mathbf{0}$. When $\mathbf{A}$ is normal, it cannot have generalized eigenvectors and, therefore, must have a full set of n linearly independent eigenvectors.

Singular Value Decomposition SVD [6]

7.29 Consider an $m \times n$ matrix $\mathbf{A}$ with $m \geq n$, and with $\text{rank}(\mathbf{A}) = r$. Show that $\mathbf{A}$ can be written as $\mathbf{A} = \mathbf{U}\mathbf{\Sigma}\mathbf{V}^T$, where $\mathbf{U}$ and $\mathbf{V}$ are $m \times m$ and $n \times n$ orthogonal matrices respectively and where $\mathbf{\Sigma}$ is $m \times n$ and "diagonal." (A nonsquare matrix is diagonal if all i, j entries are zero for $i \neq j$.)

Since $\mathbf{A}$ is not square, it cannot be used directly in an eigenvalue problem. However, two related problems are pertinent. Consider $\mathbf{A}\mathbf{A}^T\xi_1 = \sigma_i^2 \xi_i$ and $\mathbf{A}^T\mathbf{A}\eta_i = \lambda_i^2 \eta_i$. $\mathbf{A}\mathbf{A}^T$ is $m \times m$, symmetric, and hence normal. It is also positive semidefinite, and thus the σ_i^2 notation for the eigenvalue is justified. From Problem 7.28 there is a full set of m eigenvectors. From Problem 7.25 or 7.26 these eigenvectors are mutually orthogonal, at least for two different eigenvalues. They still can be selected as orthogonal even if there are repeated eigenvalues. For a multiplicity k we are assured there are k independent eigenvectors, and Gram-Schmidt can be used to construct k orthogonal vectors from them. The new vectors are still eigenvectors. By proper normalization, all ξ_i are also unit vectors, i.e., they are orthonormal. These vectors form the columns of an $m \times m$ orthogonal matrix $\mathbf{U}$.

$\mathbf{A}^T\mathbf{A}$ is $n \times n$ symmetric and at least positive semidefinite. Thus it also has a full set of n orthonormal eigenvectors η_i and nonnegative eigenvalues λ_i^2. Use the η_i vectors to form columns of an $n \times n$ orthogonal matrix $\mathbf{V}$. From Problem 7.19 it is known that $\mathbf{A}^T\xi_i \triangleq \zeta_i$ will be an eigenvector of $\mathbf{A}^T\mathbf{A}$, at least in the case where $\sigma_i \neq 0$. This is still true even for $\sigma_i = 0$, as will be seen when the length of ζ_i is computed below. From that same problem the nonzero values of σ_i and λ_i are the same. Thus $\mathbf{A}\mathbf{A}^T\xi_i = \sigma_i^2 \xi_i$ becomes $\mathbf{A}\zeta_i = \sigma_i^2 \xi_i$. But ζ_i is generally not a unit vector. In fact its length is found from $\|\zeta_i\|^2 = \langle \zeta_i, \zeta_i\rangle = \langle \mathbf{A}^T\xi_i, \mathbf{A}^T\xi_i\rangle = \langle \mathbf{A}\mathbf{A}^T\xi_i, \xi_i\rangle = \sigma_i^2\langle \xi_i, \xi_i\rangle = \sigma_i^2$. Therefore, dividing the earlier equation by σ_i gives $\mathbf{A}\eta_i = \sigma_i \xi_i$. The entire set of such equations is

$$A[\eta_1\,\eta_2\ldots\eta_n] = [\sigma_1\,\xi_1\ldots\sigma_m\,\xi_m] = [\xi_1\,\xi_2\ldots\xi_m]\begin{bmatrix} \sigma_1 & & & \\ & \sigma_2 & & 0 \\ & & \ddots & \\ 0 & & & \sigma_n \\ \hline & & 0 & \end{bmatrix}\begin{matrix} \left.\begin{matrix} \\ \\ \\ \\ \end{matrix}\right\} n\times n \text{ diagonal, rank } r \\ \}\,(m-n)\times n \text{ zero} \end{matrix}$$

or $AV = U\Sigma$. Using the orthogonality of V gives $V^{-1} = V^T$. The final result is $A = U\Sigma V^T$. For exposition purposes it was assumed that $m > n$. This was not at all essential. For example, if $m < n$, define $B = A^T$ and then all the above applies to B.

The positive square roots of the eigenvalues of $A^T A$, namely the $\sigma_i = \lambda_i$, are called the singular values of A. The eigenvectors of AA^T, namely the ξ_i, are called the left singular vectors of A and the η_i are called the right singular vectors of A.

30 Show that any $m \times n$ matrix A of rank r can be written as

$$A = U'\Sigma'V'^T$$

where U' and V' are $m \times r$ and $r \times n$ matrices, respectively, with orthonormal columns, and where Σ' is an $r \times r$ full rank diagonal matrix.

Starting with the previous problem results, all the zero columns of Σ can be deleted as long as the corresponding rows of V^T are also deleted to maintain conformability. The values in the resulting matrix product are unchanged. Likewise, all the zero rows of Σ can be deleted without changing the answer, so long as the corresponding columns of U are deleted to maintain a conformable product. Actually this last set of deletions can be done in every case, whether A is full rank n or not. The first set of deletions only applies when A is of less than full rank, say r, because in the full rank case there are no zero columns in Σ. The row-deleted version of V^T is V'^T and the column-deleted version of U is U'. Likewise for Σ and Σ'. The primed matrices form what has been called the economy-sized version of singular value decomposition. It can save a lot of computer storage. Even though U' and V' are no longer square, it is still true that $U'^T U' = I$ and $V'^T V' = I$. In both this form of the singular value decomposition and the previous full-sized form, the rank of A is the number of nonzero singular values in Σ or Σ'.

.31 Show how SVD can be used to solve simultaneous linear equations $Ax = y$.

Using the SVD form for A gives $U\Sigma V^T x = y$. Using the orthogonality of U and defining $U^T y \triangleq w$ and $V^T x \triangleq v$ gives $\Sigma v = w$. Because of the diagonal nature of Σ, these are easily solved for v in most cases. Then a simple matrix product gives $x = Vv$. An expanded form of the crucial equation is

$$\Sigma v = w \Rightarrow \left[\begin{array}{cccc|c} \sigma_1 & & & & \\ & \cdot & & & \\ & & \cdot & & 0 \\ & & & \cdot & \\ & & & \sigma_r & \\ \hline & & 0 & & 0 \end{array}\right]\begin{bmatrix} v_1 \\ \cdot \\ \cdot \\ v_r \\ \hline v_{r+1} \\ \cdot \\ \cdot \\ \cdot \\ v_n \end{bmatrix} = \begin{bmatrix} w_1 \\ \cdot \\ \cdot \\ w_r \\ \hline w_{r+1} \\ \cdot \\ \cdot \\ w_m \end{bmatrix}$$

From this it can be seen that the original equations are inconsistent and have no solution if A and Σ have rank $r < n$ unless w also has these last $m - r$ rows zero. A least-squares solution is still possible. The solution (or least-squares solution) for v will have some arbitrary components whenever there are zero columns in Σ. Setting these components of v to zero will give the minimum norm solution (or least-squares solution if required) for v. Since x and v are related by an orthogonal matrix, they have the same norm, so x is also minimum norm in that case.

7.32 Show that the SVD provides the means for extending the spectral representation of Eq. (7.5) to nonsquare matrices.

Starting with $\mathbf{A} = \mathbf{U}\boldsymbol{\Sigma}\mathbf{V}^T$ as defined in Problem 7.29,

$$\mathbf{A} = [\boldsymbol{\xi}_1\, \boldsymbol{\xi}_2 \ldots \boldsymbol{\xi}_m][\boldsymbol{\Sigma}]\begin{bmatrix} \boldsymbol{\eta}_1^T \\ \vdots \\ \boldsymbol{\eta}_n^T \end{bmatrix} = [\sigma_1 \boldsymbol{\xi}_1 \quad \sigma_2 \boldsymbol{\xi}_2 \quad \cdots \quad \sigma_n \boldsymbol{\xi}_n]\begin{bmatrix} \boldsymbol{\eta}_1^T \\ \vdots \\ \boldsymbol{\eta}_n^T \end{bmatrix}$$

$$= \sum_{i=1}^{n} \sigma_i \boldsymbol{\xi}_i \boldsymbol{\eta}_i^T = \sum_{i=1}^{n} \sigma_i \boldsymbol{\xi}_i\rangle\langle\boldsymbol{\eta}_i$$

This is of the form of Eq. (7.5). It has similar uses. For example, in approximation theory this series can be truncated prior to including all n terms if some values of σ_i are considered to be sufficiently small to be neglected.

7.33 Find the singular value decomposition for the matrix $\mathbf{A}$ of Problem 6.14.

$$\mathbf{A} = \begin{bmatrix} 1 & 2 \\ 3 & 4 \\ 5 & 6 \end{bmatrix} \quad \text{so} \quad \mathbf{A}^T\mathbf{A} = \begin{bmatrix} 35 & 44 \\ 44 & 56 \end{bmatrix} \quad \text{and} \quad \mathbf{A}\mathbf{A}^T = \begin{bmatrix} 5 & 11 & 17 \\ 11 & 25 & 39 \\ 17 & 39 & 61 \end{bmatrix}$$

The nonzero eigenvalues are approximately $\sigma_1^2 = 90.7355$ and $\sigma_2^2 = 0.2645$. The square roots of these form the diagonal terms in $\boldsymbol{\Sigma}$ below, and the two sets of normalized eigenvectors are shown as columns of $\mathbf{U}$ and rows of $\mathbf{V}^T$ next.

$$\mathbf{A} = \overbrace{\begin{bmatrix} 0.2298 & -0.8835 & 0.4082 \\ 0.5247 & -0.2408 & -0.8165 \\ 0.8196 & 0.4019 & 0.4082 \end{bmatrix}}^{\mathbf{U}} \overbrace{\begin{bmatrix} 9.5255 & 0 \\ 0 & 0.5143 \\ \hdashline 0 & 0 \end{bmatrix}}^{\boldsymbol{\Sigma}} \overbrace{\begin{bmatrix} 0.6196 & 0.7849 \\ 0.7849 & -0.6196 \end{bmatrix}}^{\mathbf{V}^T}$$

The efficient computational determination of the SVD form is crucial if it is to be useful. The indicated eigenvectors can be determined by the means presented in this chapter. This easily leads to the SVD form in simple cases like this one. However, Reference 6 should be consulted for a superior algorithm for use in more realistic cases. The real value of the discussion of this and the four previous problems is in understanding the concepts of the method, not in developing a general-purpose algorithm.

7.34 Resolve the equations of Problem 6.14 using the SVD results of Problem 7.33 and the method of Problem 7.31.

The simultaneous equations are

$$\begin{bmatrix} 1 & 2 \\ 3 & 4 \\ 5 & 6 \end{bmatrix}\mathbf{x} = \begin{bmatrix} 2 \\ 3 \\ 14 \end{bmatrix}$$

Using $\mathbf{U}$ from Problem 7.33 gives $\mathbf{w} \triangleq \mathbf{U}^T\mathbf{y} = \begin{bmatrix} 13.5095 \\ 3.1372 \end{bmatrix}$.

Then $\mathbf{v}_1 = 13.5095/\sigma_1 = 1.4182$ an $\mathbf{v}_2 = 3.1372/\sigma_2 = 6.0999$. A matrix product then gives $\mathbf{x} = \mathbf{V}\mathbf{v} = \begin{bmatrix} 5.666 \\ -2.666 \end{bmatrix}$.

Independence of Generalized Eigenvectors

7.35 If $\mathbf{x}_1$ is an eigenvector and $\mathbf{x}_2, \mathbf{x}_3, \ldots, \mathbf{x}_k$ are generalized eigenvectors, all associated with the same eigenvalue λ_1, show that:

(a) All of these vectors belong to the null space of $(\mathbf{A}-\mathbf{I}\lambda_1)^k$.
(b) This set of vectors is linearly independent.
(a) The defining equations for the set of vectors are

$$\mathbf{A}\mathbf{x}_1 = \lambda_1\mathbf{x}_1, \quad \mathbf{x}_1 \neq \mathbf{0}, \quad \text{or} \quad (\mathbf{A}-\mathbf{I}\lambda_1)\mathbf{x}_1 = \mathbf{0} \tag{1}$$

$$\mathbf{A}\mathbf{x}_2 = \lambda_1\mathbf{x}_2 + \mathbf{x}_1 \qquad (\mathbf{A}-\mathbf{I}\lambda_1)\mathbf{x}_2 = \mathbf{x}_1 \tag{2}$$

$$\mathbf{A}\mathbf{x}_3 = \lambda_1\mathbf{x}_3 + \mathbf{x}_2 \qquad (\mathbf{A}-\mathbf{I}\lambda_1)\mathbf{x}_3 = \mathbf{x}_2 \tag{3}$$

$$\vdots \qquad\qquad \vdots$$

$$\mathbf{A}\mathbf{x}_k = \lambda_1\mathbf{x}_k + \mathbf{x}_{k-1} \qquad (\mathbf{A}-\mathbf{I}\lambda_1)\mathbf{x}_k = \mathbf{x}_{k-1}$$

From Eqs. (1) and (2), $(\mathbf{A}-\mathbf{I}\lambda_1)^2\mathbf{x}_2 = (\mathbf{A}-\mathbf{I}\lambda_1)\mathbf{x}_1 = \mathbf{0}$. Multiplying Eq. (3) by $(\mathbf{A}-\mathbf{I}\lambda_1)^2$ gives $(\mathbf{A}-\mathbf{I}\lambda_1)^3\mathbf{x}_3 = (\mathbf{A}-\mathbf{I}\lambda_1)^2\mathbf{x}_2 = \mathbf{0}$. In general, it can be seen that $(\mathbf{A}-\mathbf{I}\lambda_1)^p\mathbf{x}_p = \mathbf{0}$, and $(\mathbf{A}-\mathbf{I}\lambda_1)^{p-1}\mathbf{x}_p = \mathbf{x}_1$. Since $(\mathbf{A}-\mathbf{I}\lambda_1)^k = (\mathbf{A}-\mathbf{I}\lambda_1)^{k-p}(\mathbf{A}-\mathbf{I}\lambda_1)^p$, $(\mathbf{A}-\mathbf{I}\lambda_1)^k\mathbf{x}_p = \mathbf{0}$ for $p = 1, 2, \ldots, k$ and part a is proven.

(b) Let

$$a_1\mathbf{x}_1 + a_2\mathbf{x}_2 + \cdots + a_k\mathbf{x}_k = \mathbf{0} \tag{4}$$

and show that this implies that each $a_i = 0$. Multiplying Eq. (4) by $(\mathbf{A}-\mathbf{I}\lambda_1)^{k-1}$ gives $a_k(\mathbf{A}-\mathbf{I}\lambda_1)^{k-1}\mathbf{x}_k = \mathbf{0}$. Since $(\mathbf{A}-\mathbf{I}\lambda_1)^{k-1}\mathbf{x}_k = \mathbf{x}_1 \neq \mathbf{0}$, a_k must be zero. Using this fact and then multiplying equation (4) by $(\mathbf{A}-\mathbf{I}\lambda_1)^{k-2}$ shows that $a_{k-1} = 0$. Continuing this process shows that if $\sum_{i=1}^{k} a_i\mathbf{x}_i = \mathbf{0}$, then $a_i = 0$ for $i = 1, 2, \ldots, k$. This means the set $\{\mathbf{x}_i\}$ is linearly independent.

Companion Matrix

.36 When considering nth-order linear differential equations of the type

$$\frac{d^n x}{dt^n} + a_{n-1}\frac{d^{n-1}x}{dt^{n-1}} + a_{n-2}\frac{d^{n-2}x}{dt^{n-2}} + \cdots + a_1\frac{dx}{dt} + a_0 x = u(t)$$

the $n \times n$ matrix $\mathbf{A} = \begin{bmatrix} 0 & 1 & 0 & 0 & \cdots & 0 \\ 0 & 0 & 1 & 0 & & 0 \\ 0 & 0 & 0 & 1 & & 0 \\ \vdots & & & & & \vdots \\ -a_0 & -a_1 & -a_2 & -a_3 & \cdots & -a_{n-1} \end{bmatrix}$

will often arise. $\mathbf{A}$ is called the companion matrix. Show that the companion matrix always has just one eigenvector for each eigenvalue, regardless of its algebraic multiplicity.

The matrix $\mathbf{A}-\mathbf{I}\lambda$ always has rank r, which satisfies $r \geq n-1$. To see this, delete the first column and the last row, leaving a lower triangular $n-1 \times n-1$ matrix with ones on the diagonal. If λ is an eigenvalue, $\text{rank}(\mathbf{A}-\mathbf{I}\lambda) < n$. Together, these results imply $\text{rank}(\mathbf{A}-\mathbf{I}\lambda_i) = n-1$, so the degeneracy is $q = n - r = 1$. The case of simple degeneracy always applies, so there is exactly one eigenvector for each eigenvalue.

Quadratic Form

.37 Show that the eigenvalue problem for a real, symmetric matrix can be characterized as one of maximizing or minimizing a quadratic form subject to the constraint that $\mathbf{x}$ be a unit vector.

Consider $Q = \langle\mathbf{x}, \mathbf{A}\mathbf{x}\rangle$. If the change of basis $\mathbf{x} = \mathbf{M}\mathbf{z}$ is used, $Q = \sum_{i=1}^{n} \lambda_i z_i^2$. Because of the orthogonal transformation, $\mathbf{z}$ is also a unit vector. Since all $z_i^2 \leq 1$, this suggests that the maximum value of Q will be attained if all $z_i = 0$ except $z_k^2 = 1$, where $\lambda_k = \lambda_{max}$. Then $Q = \lambda_{max}$. Alternatively, adjoining the constraint $\|\mathbf{x}\|^2 = 1$ to Q by means of the Lagrange multiplier λ shows this directly. That is, maximizing $\mathbf{x}^T\mathbf{A}\mathbf{x} - \lambda(\mathbf{x}^T\mathbf{x} - 1)$ requires that the derivative with respect to each component x_i must vanish. This gives the set $\mathbf{A}\mathbf{x} - \lambda\mathbf{x} = \mathbf{0}$, which is the eigenvalue-eigenvector equation. If $\mathbf{x}$ satisfies this condition, then $Q = \mathbf{x}^T\mathbf{A}\mathbf{x} = \mathbf{x}^T\mathbf{x}\lambda = \lambda$, so $Q_{max} = \lambda_{max}$. Also $Q_{min} = \lambda_{min}$.

If $\mathbf{x}_1$ is the eigenvector associated with λ_{max}, then selecting $\mathbf{x}$ to maximize Q subject to $\langle \mathbf{x}_1, \mathbf{x}\rangle = 0$ and $\|\mathbf{x}\| = 1$ will lead to the second largest eigenvalue and its associated eigenvector. The remaining eigenvalues-eigenvectors are found in a similar way by requiring orthogonality with all previously found eigenvectors.

7.38 Reduce the quadratic form $Q = \frac{1}{3}[16y_1^2 + 10y_2^2 + 16y_3^2 - 4y_1y_2 + 16y_1y_3 + 4y_2y_3]$ to a sum of squared terms only by selecting a suitable change of coordinates.

This quadratic form can be expressed in matrix form as $Q = \mathbf{y}^T\mathbf{A}\mathbf{y}$, where $\mathbf{y} = [y_1 \;\; y_2 \;\; y_3]^T$ and $\mathbf{A} = \frac{1}{3}\begin{bmatrix} 16 & -2 & 8 \\ -2 & 10 & 2 \\ 8 & 2 & 16 \end{bmatrix}$. The eigenvalues of $\mathbf{A}$ are $\lambda_1 = 8, \lambda_2 = 4, \lambda_3 = 2$, and since $\mathbf{A}$ is real and symmetric, a set of orthonormal eigenvectors can be found. They are used as columns of the modal matrix

$$\mathbf{M} = \begin{bmatrix} 1/\sqrt{2} & -1/\sqrt{6} & -1/\sqrt{3} \\ 0 & 2/\sqrt{6} & -1/\sqrt{3} \\ 1/\sqrt{2} & 1/\sqrt{6} & 1/\sqrt{3} \end{bmatrix}$$

Since $\mathbf{M}$ is orthogonal, $\mathbf{M}^{-1} = \mathbf{M}^T$ and $\mathbf{A}$ is diagonalized by the orthogonal transformation $\mathbf{M}^T\mathbf{A}\mathbf{M} = \text{diag}[8, 4, 2]$. If the change of variables $\mathbf{y} = \mathbf{M}\mathbf{z}$ is used, then

$$Q = \mathbf{z}^T\mathbf{M}^T\mathbf{A}\mathbf{M}\mathbf{z} = 8z_1^2 + 4z_2^2 + 2z_3^2$$

PROBLEMS

7.39 Find the eigenvalues, eigenvectors, and Jordan form for $\mathbf{A} = \begin{bmatrix} 2 & -2 & 3 \\ 1 & 1 & 1 \\ 1 & 3 & -1 \end{bmatrix}$.

7.40 Find the eigenvectors of $\mathbf{A} = \begin{bmatrix} 2 & 0 \\ 0 & 2 \end{bmatrix}$.

7.41 Analyze the eigenvalue-eigenvector problem for $\mathbf{A} = \begin{bmatrix} 2 & 0 & 1 & 0 \\ 0 & 0 & 0 & 1 \\ 0 & 0 & 0 & 0 \\ 0 & 0 & 0 & 0 \end{bmatrix}$.

7.42 Compute the eigenvalues, eigenvectors, and Jordan form for $\mathbf{A} = \begin{bmatrix} 4 & -2 & 0 \\ 1 & 2 & 0 \\ 0 & 0 & 6 \end{bmatrix}$.

7.43 Are $\mathbf{A} = \frac{1}{2}\begin{bmatrix} 3 & 1 \\ -1 & 5 \end{bmatrix}$ and $\mathbf{B} = \begin{bmatrix} 2 & 0 \\ 0 & 2 \end{bmatrix}$ similar matrices?

7.44 Use the iterative technique of Sec. 7.6 to find approximate eigenvalues and eigenvectors for $\mathbf{A} = \begin{bmatrix} 8 & 2 & -5 \\ 2 & 11 & -2 \\ -5 & -2 & 8 \end{bmatrix}$.

45 Find an approximate set of eigenvalues and eigenvectors for $\mathbf{A} = \begin{bmatrix} 3 & 2 & 1 \\ 2 & 2 & 1 \\ 1 & 1 & 1 \end{bmatrix}$ using iteration.

46 If $\Delta(\lambda)$ is defined as in equation (7.3) and if

$$\Delta'(\lambda) \triangleq |\mathbf{I}\lambda - \mathbf{A}| = \lambda^n + c'_{n-1}\lambda^{n-1} + c'_{n-2}\lambda^{n-2} + \cdots + c'_1\lambda + c'_0$$

show that
(a) $c_0 = |\mathbf{A}|;\ c'_0 = |-\mathbf{A}| = (-1)^n|\mathbf{A}|$.
(b) $(-1)^{n-1}c_{n-1} = \mathrm{Tr}(\mathbf{A});\ c'_{n-1} = -\mathrm{Tr}(\mathbf{A})$.

47 Find the eigenvalues, eigenvectors, and Jordan form for

$$\mathbf{A} = \begin{bmatrix} 1 & 2 & 3 \\ -2 & 3 & -4 \\ 1 & 1 & -4 \end{bmatrix}$$

48 Find the eigenvalues, eigenvectors, and Jordan form for

$$\mathbf{A} = \begin{bmatrix} 4 & 2 & 1 \\ 1 & 2 & 1 \\ -1 & -4 & 8 \end{bmatrix}$$

.49 Draw conclusions about the sign definiteness of

(a) $\mathbf{A} = \begin{bmatrix} -6 & 2 \\ 2 & -1 \end{bmatrix}$,

(b) $\mathbf{A} = \begin{bmatrix} 13 & 4 & -13 \\ 4 & 22 & -4 \\ -13 & -4 & 13 \end{bmatrix}$,

(c) $\mathbf{A} = \begin{bmatrix} -1 & 3 & 0 & 0 \\ 3 & -9 & 0 & 0 \\ 0 & 0 & -6 & 2 \\ 0 & 0 & 2 & -1 \end{bmatrix}$,

(d) $\mathbf{A} = \begin{bmatrix} 8 & 2 & -5 \\ 2 & 11 & -2 \\ -5 & -2 & 8 \end{bmatrix}$,

(e) $\mathbf{A} = \begin{bmatrix} -3 & 2 & 0 & 1 & 7 \\ 2 & 1 & -2 & 1 & 0 \\ 0 & -2 & 6 & 3 & 8 \\ 1 & 1 & 3 & 2 & 4 \\ 7 & 0 & 8 & 4 & 5 \end{bmatrix}$.

.50 Let $\mathbf{A} = \begin{bmatrix} 3 & -1 \\ -1 & 3 \end{bmatrix}$ and let $\mathbf{x} = [x_1 \ \ x_2]^T$ be any *unit* vector. Consider $Q = \langle \mathbf{x}, \mathbf{A}\mathbf{x} \rangle$ as a scalar function of $\mathbf{x}$. From among all unit vectors find the one which gives Q its maximum value. Also determine $Q_{\max}$.

.51 Show that any simple linear transformation can be represented as $\mathbf{A} = \sum_{i=1}^{n} \lambda_i \mathbf{E}_i$, where $\mathbf{E}_i$ is a projection onto $\mathcal{N}(\mathbf{A} - \mathbf{I}\lambda_i)$.

.52 Show that any normal linear transformation can be written as a sum of orthogonal projection transformations.

.53 Analyze the simultaneous equations of Example 6.4 using the SVD method.

8

Functions of Square Matrices and the Cayley-Hamilton Theorem

8.1 INTRODUCTION

Functions of square matrices arise in connection with the solution of vector-matrix differential and difference equations. Some scalar-valued functions of matrices have already been considered, namely, $|\mathbf{A}|$, $\text{Tr}(\mathbf{A})$, $\|\mathbf{A}\|$, etc. In this chapter matrix-valued functions $f(\mathbf{A})$ of square matrices $\mathbf{A}$ are considered. These functions are themselves matrices of the same size as $\mathbf{A}$, and their element values depend upon the particular function as well as on the values of $\mathbf{A}$. This chapter is devoted to explaining when these functions can be defined, what these functions are, and how to compute them. Before beginning, it may be helpful to state clearly what they are *not*. If $\mathbf{A} = [a_{ij}]$, a matrix function $f(\mathbf{A})$ *is not* just the matrix made up of the elements $f(a_{ij})$ except in special cases.

Specific attention is given to the $n \times n$ matrix exponential function $f(\mathbf{A}) = e^{\mathbf{A}t}$ and to the matrix power function $f(\mathbf{A}) = \mathbf{A}^k$. Solutions of continuous-time state variable equations depend upon the matrix exponential. Solutions of discrete-time state equations depend in a similar way on powers of the $\mathbf{A}$ matrix. Methods of evaluating these two functions are stressed because of their importance in the analysis of control systems expressed in state variable format.

8.2 POWERS OF A MATRIX AND MATRIX POLYNOMIALS

A matrix is conformable with itself only if it is square. Only square, $n \times n$ matrices are considered in this chapter. The product $\mathbf{AA}$ will be referred to as $\mathbf{A}^2$ for obvious reasons. The product of k such factors, $\mathbf{AA}\cdots\mathbf{A}$, is defined as $\mathbf{A}^k$. By definition, $\mathbf{A}^0 = \mathbf{I}$. With these definitions all the usual rules of exponents apply. That is,

$$\mathbf{A}^m\mathbf{A}^n = \underbrace{(\mathbf{AA}\cdots\mathbf{A})}_{m \text{ factors}}\underbrace{(\mathbf{AA}\cdots\mathbf{A})}_{n \text{ factors}} = A^{m+n}$$

$$(\mathbf{A}^m)^n = (\underbrace{\mathbf{AA}\cdots\mathbf{A}}_{m \text{ factors}})^n = (\underbrace{\mathbf{A}^n\mathbf{A}^n\cdots\mathbf{A}^n}_{m \text{ factors}})$$

$$= (\underbrace{\mathbf{AA}\cdots\mathbf{A}}_{n \text{ factors}})(\underbrace{\mathbf{AA}\cdots\mathbf{A}}_{n \text{ factors}})\cdots(\underbrace{\mathbf{AA}\cdots\mathbf{A}}_{n \text{ factors}}) = \mathbf{A}^{mn}$$

If $\mathbf{A}$ is nonsingular, then $(\mathbf{A}^{-1})^n = \mathbf{A}^{-n}$ and $\mathbf{A}^n\mathbf{A}^{-n} = \mathbf{A}^n(\mathbf{A}^{-1})^n = (\mathbf{AA}^{-1})^n = \mathbf{I}^n = \mathbf{I}$ or $\mathbf{A}^{n-n} = \mathbf{A}^0 = \mathbf{I}$ as agreed upon earlier.

The notion of matrix powers can be used to define matrix polynomials in a natural way. For example, if $P(x) = c_m x^m + c_{m-1}x^{m-1} + \cdots + c_1 x + c_0$ is an mth degree polynomial in the scalar variable x, then a corresponding matrix polynomial can be defined as

$$P(\mathbf{A}) = c_m\mathbf{A}^m + c_{m-1}\mathbf{A}^{m-1} + \cdots + c_1\mathbf{A} + c_0\mathbf{I}$$

If the scalar polynomial can be written in factored form as

$$P(x) = c(x - a_1)(x - a_2)\cdots(x - a_m)$$

then this is also true for the matrix polynomial

$$P(\mathbf{A}) = c(\mathbf{A} - \mathbf{I}a_1)(\mathbf{A} - \mathbf{I}a_2)\cdots(\mathbf{A} - \mathbf{I}a_m)$$

Thus a matrix polynomial of an $n \times n$ matrix $\mathbf{A}$ is just another $n \times n$ matrix whose elements depend on $\mathbf{A}$ as well as on the coefficients of the polynomial. A clear distinction should be made between the matrix polynomial function defined here as a combination of powers of a matrix $\mathbf{A}$, and the polynomial matrices (discussed in Chapter 4 and in Section 6.3.1), which are matrices whose elements are polynomials in some variable such as the Laplace transform variable s.

.3 INFINITE SERIES AND ANALYTIC FUNCTIONS OF MATRICES

Let $\mathbf{A}$ be an $n \times n$ matrix with eigenvalues $\lambda_1, \lambda_2, \ldots, \lambda_n$. Consider the infinite series in a scalar variable x,

$$\sigma(x) = a_0 + a_1 x + a_2 x^2 + \cdots + a_k x^k + \cdots$$

It is well known that a given infinite series may converge or diverge depending on the value of x. For example, the geometric series

$$1 + x + x^2 + x^3 + \cdots + x^k + \cdots$$

converges for $|x| < 1$ and diverges otherwise. Some infinite series are convergent for all values of x, such as

$$1 + x + \frac{x^2}{2!} + \frac{x^3}{3!} + \cdots + \frac{x^k}{k!} + \cdots$$

Because this infinite series is so widely useful, it is given a special name, e^x. The various test for convergence of series will not be considered here. The following theorem is of major importance.

Theorem 8.1. Let A be an $n \times n$ matrix with eigenvalues λ_i. If the infinite series $\sigma(x) = a_0 + a_1 x + a_2 x^2 + \cdots$ is convergent for each of the n values $x = \lambda_i$, then the corresponding matrix infinite series

$$\sigma(\mathbf{A}) = a_0 \mathbf{I} + a_1 \mathbf{A} + a_2 \mathbf{A}^2 + \cdots + a_k \mathbf{A}^k + \cdots = \sum_{k=0}^{\infty} a_k \mathbf{A}^k$$

converges.

Definition 8.1 [1]. A single-valued function $f(z)$, with z a complex scalar, is said to be *analytic* at a point z_0 if and only if its derivative exists at every point in some neighborhood of z_0. Points at which the function is not analytic are called *singular points*. For example, $f(z) = 1/z$ has $z = 0$ as its only singular point and is analytic at every other point.

The result, which makes Theorem 8.1 useful for the purposes of this book, is Theorem 8.2.

Theorem 8.2. If a function $f(z)$ is analytic (contains no singularities) at every point in some circle Ω in the complex plane, then $f(z)$ can be represented as a convergent power series (the Taylor series) at every point z inside Ω [1].

Taken together, Theorems 8.1 and 8.2 give Theorem 8.3.

Theorem 8.3. If $f(z)$ is any function which is analytic within a circle in the complex plane which contains all eigenvalues λ_i of **A**, then a corresponding matrix function $f(\mathbf{A})$ can be defined by a convergent power series.

EXAMPLE 8.1 The function $e^{\alpha x}$ is analytic for all values of x. Thus it has a convergent series representation

$$e^{\alpha x} = 1 + \alpha x + \frac{\alpha^2 x^2}{2!} + \frac{\alpha^3 x^3}{3!} + \cdots + \frac{\alpha^k x^k}{k!} + \cdots$$

The corresponding matrix function is defined as

$$e^{\alpha \mathbf{A}} = \mathbf{I} + \alpha \mathbf{A} + \frac{\alpha^2 \mathbf{A}^2}{2!} + \frac{\alpha^3 \mathbf{A}^3}{3!} + \cdots + \frac{\alpha^k \mathbf{A}^k}{k!} + \cdots$$ ■

At any point within the circles of convergence, the power series representations for two functions $f(z)$ and $g(z)$ can be added, differentiated, or integrated term by term. Also, the product of the two series gives another series which converges to $f(z)g(z)$ [1]. Because of Theorems 8.1 and 8.2 the same properties are valid for analytic functions of the matrix **A**. The usual cautions with matrix algebra must be observed, however. In particular, matrix multiplication is not commutative.

EXAMPLE 8.2 $e^{\mathbf{A}t} e^{\mathbf{B}t} = e^{(\mathbf{A}+\mathbf{B})t}$ if and only if $\mathbf{AB} = \mathbf{BA}$. This is easily verified by multiplying out the first few terms for $e^{\mathbf{A}t}$ and $e^{\mathbf{B}t}$. ■

EXAMPLE 8.3 Find $\frac{d}{dt}[e^{\mathbf{A}t}]$. The series representation is

$$e^{\mathbf{A}t} = \mathbf{I} + \mathbf{A}t + \frac{\mathbf{A}^2 t^2}{2!} + \frac{\mathbf{A}^3 t^3}{3!} + \cdots$$

Term-by-term differentiation gives

$$\frac{de^{\mathbf{A}t}}{dt} = \mathbf{A} + \frac{2\mathbf{A}^2 t}{2!} + \frac{3\mathbf{A}^3 t^2}{3!} + \cdots$$

Since $\mathbf{A}$ can be factored out on either the left or the right,

$$\frac{de^{\mathbf{A}t}}{dt} = \mathbf{A}\left[\mathbf{I} + \mathbf{A}t + \frac{\mathbf{A}^2 t^2}{2!} + \cdots\right] = \left[\mathbf{I} + \mathbf{A}t + \frac{\mathbf{A}^2 t^2}{2!} + \cdots\right]\mathbf{A} = \mathbf{A}e^{\mathbf{A}t} = e^{\mathbf{A}t}\mathbf{A} \qquad \blacksquare$$

EXAMPLE 8.4 Compute $\int_0^t e^{\mathbf{A}\tau}\, d\tau$. Using the series representation, term-by-term integration gives

$$\int_0^t e^{\mathbf{A}\tau}\, d\tau = \int_0^t \mathbf{I}\, d\tau + \mathbf{A}\int_0^t \tau\, d\tau + \frac{\mathbf{A}^2}{2!}\int_0^t \tau^2\, d\tau + \cdots = \mathbf{I}t + \frac{\mathbf{A}t^2}{2} + \frac{\mathbf{A}^2 t^3}{3!} + \cdots$$

Therefore, $\mathbf{A}\int_0^t e^{\mathbf{A}\tau}\, d\tau + \mathbf{I} = e^{\mathbf{A}t}$ or, if $\mathbf{A}^{-1}$ exists,

$$\int_0^t e^{\mathbf{A}\tau}\, d\tau = \mathbf{A}^{-1}[e^{\mathbf{A}t} - \mathbf{I}] = [e^{\mathbf{A}t} - \mathbf{I}]\mathbf{A}^{-1} \qquad \blacksquare$$

Since the exponential function is analytic for all finite arguments, it is possible to define

$$e^{-\alpha\mathbf{A}} = \mathbf{I} - \alpha\mathbf{A} + \frac{\alpha^2\mathbf{A}^2}{2} - \frac{\alpha^3\mathbf{A}^3}{3!} + \cdots - \cdots$$

Since $\mathbf{A}$ commutes with itself,

$$e^{-\alpha\mathbf{A}} e^{\alpha\mathbf{A}} = e^{\alpha(\mathbf{A}-\mathbf{A})} = e^{\alpha[0]} = \mathbf{I}$$

which is analogous to the scalar result.

Although the exponential function of a matrix is the one which will be most useful in this text, many other functions can be defined, for example:

$$\sin\mathbf{A} = \mathbf{A} - \frac{\mathbf{A}^3}{3!} + \frac{\mathbf{A}^5}{5!} - \cdots \qquad \sinh\mathbf{A} = \mathbf{A} + \frac{\mathbf{A}^3}{3!} + \frac{\mathbf{A}^5}{5!} + \cdots$$

$$\cos\mathbf{A} = \mathbf{I} - \frac{\mathbf{A}^2}{2!} + \frac{\mathbf{A}^4}{4!} - \cdots \qquad \cosh\mathbf{A} = \mathbf{I} + \frac{\mathbf{A}^2}{2!} + \frac{\mathbf{A}^4}{4!} + \cdots$$

Using these definitions, it can be verifed that relations analogous to the scalar results hold. For example,

$$\sin^2\mathbf{A} + \cos^2\mathbf{A} = \mathbf{I}$$

$$\sin\mathbf{A} = \frac{e^{j\mathbf{A}} - e^{-j\mathbf{A}}}{2j}, \qquad \cos\mathbf{A} = \frac{e^{j\mathbf{A}} + e^{-j\mathbf{A}}}{2}$$

$$\cosh^2 \mathbf{A} - \sinh^2 \mathbf{A} = \mathbf{I}$$

$$\sinh \mathbf{A} = \frac{e^{\mathbf{A}} - e^{-\mathbf{A}}}{2}, \qquad \cosh \mathbf{A} = \frac{e^{\mathbf{A}} + e^{-\mathbf{A}}}{2}$$

Although all of the preceding matrix functions are defined in terms of their power series representations, the series converge to $n \times n$ matrices. It is known from Chapter 7 that every square matrix has a unique Jordan form $\mathbf{J} = \mathbf{M}^{-1}\mathbf{AM}$, from which $\mathbf{A} = \mathbf{MJM}^{-1}$. Thus (see Problem 8.5) for any integer k, $\mathbf{A}^k = \mathbf{MJ}^k\mathbf{M}^{-1}$. Therefore, $f(\mathbf{A}) = \mathbf{M}f(\mathbf{J})\mathbf{M}^{-1}$ for f any finite-degree polynomial or infinite series. Any analytic function satisfying the conditions of Theorem 8.2 can be expressed in terms of the modal matrix and the Jordan form. In those cases for which $\mathbf{A}$ is diagonalizable,

$$\mathbf{J} = \mathbf{\Lambda} = \text{Diag}[\lambda_1, \ldots, \lambda_n] \quad \text{and} \quad f(\mathbf{\Lambda}) = \text{Diag}[f(\lambda_1), \ldots, f(\lambda_n)]$$

The Jordan form decomposition of $\mathbf{A}$ may not be the most efficient method of computing $f(\mathbf{A})$ because it requires determination of the modal matrix. Section 8.5 develops methods of determining the closed form expressions for analytic functions of square matrices. However, the Jordan form–modal matrix approach does show several useful results in a simple fashion. For example, if $\{\lambda_i\}$ are the eigenvalues of $\mathbf{A}$, then $f(\lambda_i)$ are the eigenvalues of $f(\mathbf{A})$. This result is called Frobenius' theorem. As applied to the state equations, this indicates that if all eigenvalues of $\mathbf{A}$ are in the left-half complex plane (stable poles) then all eigenvalues of $e^{\mathbf{A}t}$ will have magnitudes less than one, that is, they are inside the unit circle. This is a result that could have been anticipated from the Chapter 2 discussion of Z-transforms and the stability regions for continuous and discrete systems.

8.4 THE CHARACTERISTIC POLYNOMIAL AND CAYLEY-HAMILTON THEOREM

Although arbitrary matrix polynomials have been discussed, one very special polynomial is the characteristic polynomial. If the characteristic polynomial for the matrix $\mathbf{A}$ is written as

$$|\mathbf{A} - \mathbf{I}\lambda| = (-\lambda)^n + c_{n-1}\lambda^{n-1} + c_{n-2}\lambda^{n-2} + \cdots + c_1\lambda + c_0 = \Delta(\lambda)$$

then the corresponding matrix polynomial is

$$\Delta(\mathbf{A}) = (-1)^n\mathbf{A}^n + c_{n-1}\mathbf{A}^{n-1} + c_{n-2}\mathbf{A}^{n-2} + \cdots + c_1\mathbf{A} + c_0\mathbf{I}$$

Cayley-Hamilton Theorem. Every matrix satisfies its own characteristic equation; that is, $\Delta(\mathbf{A}) = [\mathbf{0}]$.

Proof. (Valid when $\mathbf{A}$ is similar to a diagonal matrix. For the general case, see Problems 8.5 and 8.6.) A similarity transformation reduces $\mathbf{A}$ to the diagonal matrix $\mathbf{\Lambda}$, so

$$\mathbf{A} = \mathbf{M\Lambda M}^{-1}, \qquad \mathbf{A}^2 = \mathbf{M\Lambda}^2\mathbf{M}^{-1}, \ldots, \mathbf{A}^k = \mathbf{M\Lambda}^k\mathbf{M}^{-1}$$

Therefore,

$$\Delta(\mathbf{A}) = \mathbf{M}[(-1)^n \mathbf{\Lambda}^n + c_{n-1}\mathbf{\Lambda}^{n-1} + c_{n-2}\mathbf{\Lambda}^{n-2} + \cdots + c_1 \mathbf{\Lambda} + c_0 \mathbf{I}]\mathbf{M}^{-1}$$

Each term inside the brackets is a diagonal matrix. The sum of a typical i, i element is

$$(-\lambda_i)^n + c_{n-1}\lambda_i^{n-1} + \cdots + c_1\lambda_i + c_0$$

which is zero because λ_i is a root of the characteristic equation. Therefore,

$$\Delta(\mathbf{A}) = \mathbf{M}[\mathbf{0}]\mathbf{M}^{-1} = [\mathbf{0}]$$

EXAMPLE 8.5 Let $\mathbf{A} = \begin{bmatrix} 3 & 1 \\ 1 & 2 \end{bmatrix}$, $|\mathbf{A} - \mathbf{I}\lambda| = (3-\lambda)(2-\lambda) - 1 = \lambda^2 - 5\lambda + 5$. Then

$$\Delta(\mathbf{A}) = \mathbf{A}^2 - 5\mathbf{A} + 5\mathbf{I} = \begin{bmatrix} 10 & 5 \\ 5 & 5 \end{bmatrix} - 5\begin{bmatrix} 3 & 1 \\ 1 & 2 \end{bmatrix} + 5\begin{bmatrix} 1 & 0 \\ 0 & 1 \end{bmatrix} = \begin{bmatrix} 0 & 0 \\ 0 & 0 \end{bmatrix}$$ ■

Definition 8.2 [2]. The minimum polynomial of a square matrix **A** is the lowest-degree monic polynomial (that is, the coefficient of the highest power is normalized to one) which satisfies

$$m(\mathbf{A}) = [\mathbf{0}]$$

It would be slightly more efficient to use the minimum polynomial rather than the characteristic polynomial in some cases (in Section 8.5, for example). The minimum polynomial $m(\mathbf{A})$ and the characteristic polynomial $\Delta'(\mathbf{A})$ are often the same (if all λ_i are distinct, or in the simple degeneracy case). In factored form the only possible differences between $m(\mathbf{A})$ and $\Delta'(\mathbf{A})$ are the powers of the terms involving repeated roots. Since for the definitions given in Chapter 7 $\Delta(\lambda) = (-1)^n \Delta'(\lambda)$, then

$$\Delta(\lambda) = (-1)^n(\lambda - \lambda_1)^{m_1}(\lambda - \lambda_2)^{m_2}\cdots(\lambda - \lambda_p)^{m_p}$$

$$m(\lambda) = (\lambda - \lambda_1)^{k_1}(\lambda - \lambda_2)^{k_2}\cdots(\lambda - \lambda_p)^{k_p}$$

In all cases $k_i \le m_i$. In this text the characteristic polynomial will be used in most cases rather than the minimum polynomial because it is more familiar, and only occasionally requires a small amount of extra calculations (see Problem 8.18).

8.5 SOME USE OF THE CAYLEY-HAMILTON THEOREM

Matrix Inversion

Let **A** be an $n \times n$ matrix with the characteristic equation $\Delta(\lambda) = (-\lambda)^n + c_{n-1}\lambda^{n-1} + \cdots + c_1\lambda + c_0 = 0$. Recall that the constant $c_0 = \lambda_1\lambda_2\cdots\lambda_n = |\mathbf{A}|$ and is zero if and only if **A** is singular. Using $\Delta(\mathbf{A}) = (-1)^n\mathbf{A}^n + c_{n-1}\mathbf{A}^{n-1} + \cdots + c_1\mathbf{A} + c_0\mathbf{I} = \mathbf{0}$, and assuming $\mathbf{A}^{-1}$ exists, multiplication by $\mathbf{A}^{-1}$ gives

$$(-1)^n\mathbf{A}^{n-1} + c_{n-1}\mathbf{A}^{n-2} + \cdots + c_1\mathbf{I} + c_0\mathbf{A}^{-1} = \mathbf{0}$$

or

$$\mathbf{A}^{-1} = \frac{-1}{c_0}[(-1)^n\mathbf{A}^{n-1} + c_{n-1}\mathbf{A}^{n-2} + \cdots + c_1\mathbf{I}]$$

EXAMPLE 8.6 Let $\mathbf{A} = \begin{bmatrix} 3 & 1 \\ 1 & 2 \end{bmatrix}$, $\Delta(\lambda) = \lambda^2 - 5\lambda + 5$. Then

$$\Delta(\mathbf{A}) = \mathbf{A}^2 - 5\mathbf{A} + 5\mathbf{I} = \mathbf{0}, \qquad \mathbf{A}^{-1} = -\tfrac{1}{5}[\mathbf{A} - 5\mathbf{I}] = -\frac{1}{5}\begin{bmatrix} -2 & 1 \\ 1 & -3 \end{bmatrix}$$ ∎

Reduction of a Polynomial in A to One of Degree n − 1 or Less

Let $P(x)$ be a scalar polynomial of degree m. Let $P_1(x)$ be another polynomial of degree n, where $n < m$. Then $P(x)$ can always be written $P(x) = Q(x)P_1(x) + R(x)$, where $Q(x)$ is a polynomial of degree $m - n$ and $R(x)$ is a remainder polynomial of degree $n - 1$ or less. For this scalar case, $Q(x)$ and $R(x)$ could be found by formally dividing $P(x)$ by $P_1(x)$, since this gives $P(x)/P_1(x) = Q(x) + R(x)/P_1(x)$.

EXAMPLE 8.7 Let $P(x) = 3x^4 + 2x^2 + x + 1$ and $P_1(x) = x^2 - 3$. Then it is easily verified that

$$P(x) = (3x^2 + 11)(x^2 - 3) + (x + 34)$$

so that $Q(x) = 3x^2 + 11$ and $R(x) = x + 34$. ∎

Similarly, the matrix polynomial $P(\mathbf{A})$ can be written

$$P(\mathbf{A}) = Q(\mathbf{A})P_1(\mathbf{A}) + R(\mathbf{A})$$

since it is always defined in the same manner as its scalar counterpart. If the arbitrary polynomial P_1 used above is selected as the characteristic polynomial of $\mathbf{A}$, then the scalar version of P is

$$P(x) = Q(x)\Delta(x) + R(x)$$

Note that $\Delta(x) \neq 0$ except for those specific values $x = \lambda_i$, the eigenvalues. The matrix version of P is

$$P(\mathbf{A}) = Q(\mathbf{A})\Delta(\mathbf{A}) + R(\mathbf{A})$$

By the Cayley-Hamilton theorem, $\Delta(\mathbf{A}) = \mathbf{0}$, so $P(\mathbf{A}) = R(\mathbf{A})$. The coefficients of the matrix remainder polynomial R can be found by long division of the corresponding scalar polynomials. Alternatively, the Cayley-Hamilton theorem can be used to reduce each individual term in $P(\mathbf{A})$ to one of degree $n - 1$ or less.

EXAMPLE 8.8 Let $\mathbf{A} = \begin{bmatrix} 3 & 1 \\ 1 & 2 \end{bmatrix}$, $\Delta(\lambda) = \lambda^2 - 5\lambda + 5$. Compute $P(\mathbf{A}) = \mathbf{A}^4 + 3\mathbf{A}^3 + 2\mathbf{A}^2 + \mathbf{A} + \mathbf{I}$.

Method 1: By long division,

$$\frac{P(x)}{\Delta(x)} = x^2 + 8x + 37 + \frac{146x - 184}{x^2 - 5x + 5} \quad \text{or} \quad P(x) = (x^2 + 8x + 37)\Delta(x) + (146x - 184)$$

Therefore, $R(x) = 146x - 184$ and the Cayley-Hamilton theorem guarantees that $P(\mathbf{A}) = R(\mathbf{A}) = 146\mathbf{A} - 184\mathbf{I}$.

Method 2: From the Cayley-Hamilton theorem,

$$\mathbf{A}^2 - 5\mathbf{A} + 5\mathbf{I} = [\mathbf{0}] \quad \text{or} \quad \mathbf{A}^2 = 5(\mathbf{A} - \mathbf{I})$$

Hence

$$\mathbf{A}^4 = \mathbf{A}^2\mathbf{A}^2 = 25(\mathbf{A}-\mathbf{I})(\mathbf{A}-\mathbf{I}) = 25(\mathbf{A}^2 - 2\mathbf{A} + \mathbf{I}) = 25[5(\mathbf{A}-\mathbf{I}) - 2\mathbf{A} + \mathbf{I}] = 25[3\mathbf{A} - 4\mathbf{I}]$$

$$\mathbf{A}^3 = \mathbf{A}(\mathbf{A}^2) = 5(\mathbf{A}^2 - \mathbf{A}) = 5[5(\mathbf{A}-\mathbf{I}) - \mathbf{A}] = 5[4\mathbf{A} - 5\mathbf{I}]$$

Thus

$$P(\mathbf{A}) = 25[3\mathbf{A} - 4\mathbf{I}] + 15[4\mathbf{A} - 5\mathbf{I}] + 10(\mathbf{A}-\mathbf{I}) + \mathbf{A} + \mathbf{I} = 146\mathbf{A} - 184\mathbf{I}$$ ■

Closed Form Solution for Analytic Functions of Matrices

Let $f(x)$ be a function which is analytic in a region Ω of the complex plane and let $\mathbf{A}$ be an $n \times n$ matrix whose eigenvalues $\lambda_i \in \Omega$. Then $f(x)$ has a power series representation

$$f(x) = \sum_{k=0}^{\infty} \alpha_k x^k$$

It is possible to regroup the infinite series for $f(x)$ so that

$$f(x) = \Delta(x) \sum_{k=0}^{\infty} \beta_k x^k + R(x)$$

The remainder R will have degree less than or equal to $n - 1$. The analytic function of the square matrix $\mathbf{A}$ is defined by the same series as its scalar counterpart, but with $\mathbf{A}$ replacing x. Therefore, $f(\mathbf{A}) = R(\mathbf{A})$, since $\Delta(\mathbf{A})$ is always the null matrix.

Although the form of $R(x)$ is known to be

$$R(x) = \alpha_0 + \alpha_1 x + \alpha_2 x^2 + \cdots + \alpha_{n-1} x^{n-1}$$

it is clearly impossible to find the coefficients α_i by long division as in Example 8.8. However, if the n eigenvalues λ_i are distinct, n equations for determining the n α_i terms are available. Since $\Delta(\lambda_i) = 0$, setting $x = \lambda_i$ gives $f(\lambda_i) = R(\lambda_i)$, $i = 1, 2, \ldots, n$.

EXAMPLE 8.9 Find the closed form expression for $\sin \mathbf{A}$ if $\mathbf{A} = \begin{bmatrix} -3 & 1 \\ 0 & -2 \end{bmatrix}$.

$\Delta(\lambda) = (-3-\lambda)(-2-\lambda)$, so $\lambda_1 = -3$, $\lambda_2 = -2$. Since $\mathbf{A}$ is a 2×2 matrix, it is known that R is of degree one (or less):

$$R(x) = \alpha_0 + \alpha_1 x$$

Also, using $x = \lambda_1$ and $x = \lambda_2$ gives

$$\sin \lambda_1 = R(\lambda_1) = \alpha_0 + \alpha_1 \lambda_1$$

$$\sin \lambda_2 = R(\lambda_2) = \alpha_0 + \alpha_1 \lambda_2$$

Solving gives

$$\alpha_0 = \frac{\lambda_1 \sin \lambda_2 - \lambda_2 \sin \lambda_1}{\lambda_1 - \lambda_2}, \qquad \alpha_1 = \frac{\sin \lambda_1 - \sin \lambda_2}{\lambda_1 - \lambda_2}$$

or $\alpha_0 = 3\sin(-2) - 2\sin(-3)$, $\alpha_1 = \sin(-2) - \sin(-3)$. Using these in $f(\mathbf{A}) = R(\mathbf{A})$ gives

$$\sin \mathbf{A} = \alpha_0 \mathbf{I} + \alpha_1 \mathbf{A} = \begin{bmatrix} \alpha_0 - 3\alpha_1 & \alpha_1 \\ 0 & \alpha_0 - 2\alpha_1 \end{bmatrix} = \begin{bmatrix} \sin(-3) & \sin(-2) - \sin(-3) \\ 0 & \sin(-2) \end{bmatrix}$$ ■

When λ_i is a repeated root, this procedure must be modified. Some of the equations $f(\lambda_i) = R(\lambda_i)$ will be repeated, so they do not form a set of n linearly independent equations. However, for λ_i a repeated root, $\left.\dfrac{d\Delta(\lambda)}{d\lambda}\right|_{\lambda=\lambda_i} = 0$ also, and so

$$\left.\frac{df(\lambda)}{d\lambda}\right|_{\lambda=\lambda_i} = \left.\frac{d\Delta}{d\lambda}\sum_{k=0}^{\infty} \beta_k \lambda_i^k + \Delta(\lambda_i)\frac{d}{d\lambda}[\Sigma\beta_k\lambda^k]\right|_{\lambda=\lambda_i} + \left.\frac{dR}{d\lambda}\right|_{\lambda=\lambda_i} = \left.\frac{dR}{d\lambda}\right|_{\lambda=\lambda_i}$$

For an eigenvalue with algebraic multiplicity m_i, the first $m_i - 1$ derivatives of Δ all vanish and thus

$$f(\lambda_i) = R(\lambda_i), \qquad \left.\frac{df}{d\lambda}\right|_{\lambda_i} = \left.\frac{dR}{d\lambda}\right|_{\lambda_i}, \qquad \left.\frac{d^2 f}{d\lambda^2}\right|_{\lambda_i} = \left.\frac{d^2 R}{d\lambda^2}\right|_{\lambda_i}, \ldots,$$

$$\left.\frac{d^{m_i-1} f}{d\lambda^{m_i-1}}\right|_{\lambda_i} = \left.\frac{d^{m_i-1} R}{d\lambda^{m_i-1}}\right|_{\lambda_i}$$

form a set of m_i linearly independent equations. Thus a full set of n equations is always available for finding the α_i coefficients of the remainder term R.

EXAMPLE 8.10 Find the closed form expression for $e^{\mathbf{A}t}$ if $\mathbf{A} = \begin{bmatrix} 0 & 1 & 0 \\ 0 & 0 & 1 \\ 27 & -27 & 9 \end{bmatrix}$.

We have

$$|\mathbf{A} - \mathbf{I}\lambda| = \Delta(\lambda) = -\lambda^3 + 9\lambda^2 - 27\lambda + 27 = (3 - \lambda)^3$$

Therefore, $\lambda_1 = \lambda_2 = \lambda_3 = 3$, and

$$e^{\mathbf{A}t} = R(\mathbf{A}) = \alpha_0 \mathbf{I} + \alpha_1 \mathbf{A} + \alpha_2 \mathbf{A}^2$$

where

$$e^{3t} = \alpha_0 + 3\alpha_1 + 9\alpha_2$$

$$\left.\frac{de^{\lambda t}}{d\lambda}\right|_{\lambda=3} = \left.\frac{d}{d\lambda}[\alpha_0 + \lambda\alpha_1 + \lambda^2\alpha_2]\right|_{\lambda=3} \quad \text{or} \quad te^{3t} = \alpha_1 + 6\alpha_2$$

$$\left.\frac{d^2 e^{\lambda t}}{d\lambda^2}\right|_{\lambda=3} = \left.\frac{d^2}{d\lambda^2}[\alpha_0 + \lambda\alpha_1 + \lambda^2\alpha_2]\right|_{\lambda=3} \quad \text{or} \quad t^2 e^{3t} = 2\alpha_2$$

Solving for α_0, α_1, and α_2 gives

$$\alpha_2 = \tfrac{1}{2}t^2 e^{3t}, \qquad \alpha_1 = te^{3t} - 6\alpha_2 = (t - 3t^2)e^{3t},$$

$$\alpha_0 = e^{3t} - 3\alpha_1 - 9\alpha_2 = (1 - 3t + \tfrac{9}{2}t^2)e^{3t}$$

Using these coefficients in $R(\mathbf{A})$ gives

$$e^{\mathbf{A}t} = \begin{bmatrix} 1 - 3t + \frac{9}{2}t^2 & t - 3t^2 & \frac{1}{2}t^2 \\ \frac{27}{2}t^2 & 1 - 3t - 9t^2 & t + \frac{3}{2}t^2 \\ 27t + \frac{81}{2}t^2 & -27t - 27t^2 & 1 + 6t + \frac{9}{2}t^2 \end{bmatrix} e^{3t}$$

■

When some eigenvalues are repeated and others are simple roots, a full set of n independent equations are still available for computing the α_i coefficients.

The method presented above for computing functions of a matrix requires a knowledge of the eigenvalues. If the eigenvectors are also known, an alternative method can be used (see Problem 8.21). For the particular function $f(\mathbf{A}) = e^{At}$, a third alternative is available (see Problems 8.19 and 8.20). Finally, an aproximation can be obtained by truncating the infinite series after a finite number of terms.

6 SOLUTION OF THE UNFORCED STATE EQUATIONS

Complete solutions of the state equations are considered in the next chapter. In order to emphasize the importance of the matrix exponential and power functions, the initial condition response of the state equations are considered briefly here. Only the constant coefficient case is considered, that is the **A** matrix is not a function of time.

The Continuous-Time Case

When the control input $\mathbf{u}(t)$ is zero, the state vector $\mathbf{x}(t)$ evolves in time according to solutions of

$$\dot{\mathbf{x}}(t) = \mathbf{A}\mathbf{x}(t) \tag{8.1}$$

starting with initial conditions $\mathbf{x}(0)$. One classic method of solving differential equations is to guess a solution form, perhaps with adjustable parameters, and then see if it can be made to satisfy (a) the initial condition and (b) the differential equation. Here the "guess" is $\mathbf{x}(t) = e^{At}\mathbf{x}(0)$, and we merely verify that this is correct. In Example 8.4 $e^{[0]} = \mathbf{I}$ was introduced, so when $t = 0$ is substituted, the assumed solution reduces to $\mathbf{I}\mathbf{x}(0)$ and the initial conditions are satisfied. From Example 8.3 $d[e^{At}]/dt = \mathbf{A}e^{At}$, so that substitution of the assumed solution into Eq. (*8.1*) gives the self-consistent result $\dot{\mathbf{x}}(t) = \mathbf{A}e^{At}\mathbf{x}(0) = \mathbf{A}\mathbf{x}(t)$.

In Chapter 4 it was stated that the Laplace transform of a matrix of time functions can be calculated term by term on each element of the matrix. This provides an alternative approach to solving Eq. (*8.1*). The state vector $\mathbf{x}(t)$ is an example of a time-variable column matrix. Let $\mathcal{L}\{\mathbf{x}(t)\} = \mathbf{X}(s)$. Then $\mathcal{L}\{\dot{\mathbf{x}}(t)\} = s\mathbf{X}(s) - \mathbf{x}(0)$, so that Eq. (*8.1*) transforms to $s\mathbf{X}(s) - \mathbf{x}(0) = \mathbf{A}\mathbf{X}(s)$. Solving for $\mathbf{X}(s)$ gives $\mathbf{X}(s) = [s\mathbf{I} - \mathbf{A}]^{-1}\mathbf{x}(0)$. The inverse Laplace transform then gives $\mathbf{x}(t) = \mathcal{L}^{-1}\{[s\mathbf{I} - \mathbf{A}]^{-1}\}\mathbf{x}(0)$. Not only is this the solution, but comparing it with the previous solution shows that

$$e^{\mathbf{A}t} = \mathcal{L}^{-1}\{[s\mathbf{I} - \mathbf{A}]^{-1}\}$$

(See also Problem 8.19.) One final method of solving Eq. (*8.1*) utilizes modal decoupling. The matrix **A** is assumed diagonalizable to keep the discussion simple. Equation (*8.1*) can be written as $\dot{\mathbf{x}} = \mathbf{M}\mathbf{\Lambda}\mathbf{M}^{-1}\mathbf{x}$. If a change of variables (change of basis) $\mathbf{M}^{-1}\mathbf{x} = \mathbf{w}$ is used, after premultiplying by **M** this becomes $\dot{\mathbf{w}} = \mathbf{\Lambda}\mathbf{w}$. Because $\mathbf{\Lambda}$ is diagonal this represents n scalar equations $\dot{w}_i = \lambda_i w_i$. Each of these equations has a solution $w_i(t) = e^{\lambda_i t} w_i(0)$. The entire set of solution components can be written as $\mathbf{w}(t) = \text{Diag}[e^{\lambda_1 t}, e^{\lambda_2 t}, \ldots, e^{\lambda_n t}]\mathbf{w}(0)$. The solution for **x** and not **w** is desired, but $\mathbf{x}(t) = \mathbf{M}\mathbf{w}(t)$. The initial conditions are presumed given for $\mathbf{x}(0)$ and not $\mathbf{w}(0)$, but $\mathbf{w}(0) = \mathbf{M}^{-1}\mathbf{x}(0)$, so the final solution is $\mathbf{x}(t) = \mathbf{M}\,\text{Diag}[e^{\lambda_1 t}, e^{\lambda_2 t}, \ldots, e^{\lambda_n t}]\mathbf{M}^{-1}\mathbf{x}(0)$. Comparison with the previous solutions shows that $e^{\mathbf{A}t} = \mathbf{M}e^{\Lambda t}\mathbf{M}^{-1}$. This is an explicit verification of the result given at the end of Sec. 8.3 for a general function $f(\mathbf{A})$.

The Discrete-Time Case

The initial condition response of an unforced constant coefficient discrete-time system evolves according to

$$\mathbf{x}(k+1) = \mathbf{A}\mathbf{x}(k) \tag{8.2}$$

The state at any general time point t_k is known to be (see Sec. 6.9) $\mathbf{x}(k) = \mathbf{A}^k\mathbf{x}(0)$. The Z-transform of any time-variable matrix, such as the column matrix $\mathbf{x}(k)$, is obtained by transforming each scalar element in the matrix. If $\mathbf{Z}\{\ \}$ represents the Z-transform operator and if the column of transformed elements is defined as $\mathbf{X}(z) = Z\{\mathbf{x}(k)\}$, then $Z\{\mathbf{x}(k+1)\} = z\mathbf{X}(z) - z\mathbf{x}(0)$, and the transformed Eq. (8.2) can be written as $(z\mathbf{I} - \mathbf{A})\mathbf{X}(z) = z\mathbf{x}(0)$. Solving for $\mathbf{X}(z)$ gives $\mathbf{X}(z) = [z\mathbf{I} - \mathbf{A}]^{-1}z\mathbf{x}(0)$. Using $Z^{-1}\{\ \}$ to represent the inverse Z-transform, the time domain solution is

$$\mathbf{x}(k) = Z^{-1}\{[z\mathbf{I} - \mathbf{A}]^{-1}z\}\mathbf{x}(0)$$

Comparison with the previous solution shows that the kth power of a matrix can be computed from $\mathbf{A}^k = Z^{-1}\{[z\mathbf{I} - \mathbf{A}]^{-1}z\}$, which is similar to the Laplace transform result for the continuous-time system. Writing $\mathbf{A} = \mathbf{M}\Lambda\mathbf{M}^{-1}$ and repeating the steps used in decoupling the continuous-time system shows that $\mathbf{A}^k = \mathbf{M}\Lambda^k\mathbf{M}^{-1}$ is an alternative way of computing the power of a matrix, as was already known from Sec. 8.3.

REFERENCES

1. Churchill, R. V.: *Introduction to Complex Variables and Applications,* 2nd ed., McGraw-Hill, New York, 1960.
2. Zadeh, L. A. and C. A. Desoer: *Linear System Theory, The State Space Approach,* McGraw-Hill, New York, 1963.
3. Moler, C. B. and C. F. Van Loan: "Nineteen Dubious Ways to Compute the Exponential of a Matrix," *SIAM Review,* Vol. 20, 1978, pp. 801–836.
4. Franklin, G. F. and J. D. Powell: *Digital Control of Dynamic Systems,* Addison-Wesley, Reading, Mass., 1980.
5. Brogan, W. L.: "Optimal Control Theory Applied to Systems Described by Partial Differential Equations," *Advances in Control Systems,* Vol. 6, C. T. Leondes, ed., Academic Press, New York, 1968.

ILLUSTRATIVE PROBLEMS

Inversion of Matrices and Reduction of Polynomials

8.1 If $\mathbf{A} = \begin{bmatrix} 1 & -1 \\ 1 & 1 \end{bmatrix}$, use the Cayley-Hamilton theorem to compute a. $\mathbf{A}^{-1}$ and b. $P(\mathbf{A}) = \mathbf{A}^5 + 16\mathbf{A}^4 + 32\mathbf{A}^3 + 16\mathbf{A}^2 + 4\mathbf{A} + \mathbf{I}$.

(a) $\Delta(\lambda) = |\mathbf{A} - \mathbf{I}\lambda| = \lambda^2 - 2\lambda + 2$. Therefore,

$$\mathbf{A}^2 - 2\mathbf{A} + 2\mathbf{I} = [\mathbf{0}] \quad \text{or} \quad \mathbf{A}^{-1} = -\tfrac{1}{2}[\mathbf{A} - 2\mathbf{I}] = \frac{1}{2}\begin{bmatrix} 1 & 1 \\ -1 & 1 \end{bmatrix}$$

(b) $P(\mathbf{A}) = R(\mathbf{A})$, where R is the remainder term in

$$\frac{P(x)}{\Delta(x)} = x^3 + 18x^2 + 66x + 112 + \frac{96x - 223}{\Delta(x)}$$

Therefore, $P(\mathbf{A}) = 96\mathbf{A} - 223\mathbf{I} = \begin{bmatrix} -127 & -96 \\ 96 & -127 \end{bmatrix}$.

8.2 $\mathbf{A} = \begin{bmatrix} 1 & 2 \\ 3 & 4 \end{bmatrix}$. Find a. $\mathbf{A}^{-1}$ and b. $P(\mathbf{A}) = \mathbf{A}^5 + \mathbf{A}^3 + \mathbf{A} + \mathbf{I}$.

(a) $\Delta(\lambda) = \lambda^2 - 5\lambda - 2$. Therefore, $\Delta(\mathbf{A}) = \mathbf{A}^2 - 5\mathbf{A} - 2\mathbf{I} = [\mathbf{0}]$ by the Cayley-Hamilton theorem and $\mathbf{A} - 5\mathbf{I} - 2\mathbf{A}^{-1} = [\mathbf{0}]$ or $\mathbf{A}^{-1} = \frac{1}{2}[\mathbf{A} - 5\mathbf{I}] = \begin{bmatrix} -2 & 1 \\ \frac{3}{2} & -\frac{1}{2} \end{bmatrix}$.

(b) $P(x)/\Delta(x) = x^3 + 5x^2 + 28x + 150 + [(807x + 301)/\Delta(x)]$. Therefore,

$$P(x) = \Delta(x)[x^3 + 5x^2 + 28x + 150] + \underbrace{807x + 301}_{R(x)}$$

and $P(\mathbf{A}) = R(\mathbf{A}) = 807\mathbf{A} + 301\mathbf{I} = \begin{bmatrix} 1108 & 1614 \\ 2421 & 3529 \end{bmatrix}$.

Functions with Singularities

8.3 Comment on the following functions in view of Theorems 8.1, 8.2, and 8.3:
(a) $f_a(x) = 1/(1 - x)$, and (b) $f_b(x) = \tan x$.

(a) $f_a(x)$ has a singularity (pole) at $x = 1$, but is analytic elsewhere. In particular, it is analytic at all points x in the complex plane inside the circle $|x| < 1$, and for these points

$$f_a(x) = 1 + x + x^2 + x^3 + \cdots$$

If all the eigenvalues of $\mathbf{A}$ satisfy $|\lambda_i| < 1$, then $f_a(\mathbf{A}) = (\mathbf{I} - \mathbf{A})^{-1}$ exists and can be written as a convergent series

$$f_a(\mathbf{A}) = \mathbf{I} + \mathbf{A} + \mathbf{A}^2 + \mathbf{A}^3 + \cdots$$

Note that $(\mathbf{I} - \mathbf{A})f_a(\mathbf{A}) = \mathbf{I} + (\mathbf{A} - \mathbf{A}) + (\mathbf{A}^2 - \mathbf{A}^2) + \cdots = \mathbf{I}$ as required. When $\mathbf{A}$ has $\lambda = 1$ as an eigenvalue, then $(\mathbf{I} - \mathbf{A})$ is singular and the inverse does not exist. If $\lambda = 1$ is not an eigenvalue, but at least one eigenvalue has a magnitude larger than unity, then $(\mathbf{I} - \mathbf{A})^{-1}$ exists but cannot be represented by the above infinite series.

(b) $f_b(x) = \sin x/\cos x$ has a singularity (pole) at each zero of $\cos x$, that is, at $x = \pm\pi/2$, $\pm 3\pi/2, \ldots$, but is analytic elsewhere. If all eigenvalues of $\mathbf{A}$ satisfy $|\lambda| < \pi/2$, then we could define $\tan \mathbf{A} = \mathbf{A} + \mathbf{A}^3/3 + 2\mathbf{A}^5/15 + \cdots$ and be assured that this series is convergent (see Problem 8.11).

Powers of a Jordan Block

8.4 Let $\mathbf{J}_1$ be an $m \times m$ Jordan block. Show that for any integer $k > 0$,

$$\mathbf{J}_1^k = \begin{bmatrix} \lambda^k & k\lambda^{k-1} & \frac{1}{2!}k(k-1)\lambda^{k-2} & \frac{1}{3!}k(k-1)(k-2)\lambda^{k-3} & \cdots & \dfrac{k!\lambda^{k-m+1}}{(k-m+1)!(m-1)!} \\ 0 & \lambda^k & k\lambda^{k-1} & \frac{1}{2!}k(k-1)\lambda^{k-2} & & \\ 0 & 0 & \lambda^k & k\lambda^{k-1} & & \vdots \\ 0 & 0 & 0 & \lambda^k & & \\ 0 & 0 & 0 & & \ddots & \\ \vdots & & & & & k\lambda^{k-1} \\ 0 & 0 & 0 & & & \lambda^k \end{bmatrix}$$

With $k = 1$, $\mathbf{J}_1$ satisfies the preceding equation. To see this, it must be recalled that $(k - m + 1)! = \infty$ if $k - m + 1 < 0$. Multiplying gives $\mathbf{J}_1$ squared:

$$\mathbf{J}_1 = \begin{bmatrix} \lambda & 1 & 0 & \cdots & 0 \\ 0 & \lambda & 1 & & \\ 0 & 0 & \lambda & & \vdots \\ 0 & 0 & 0 & & \\ \vdots & & & & 1 \\ 0 & 0 & 0 & & \lambda \end{bmatrix}, \qquad \mathbf{J}_1^2 = \begin{bmatrix} \lambda^2 & 2\lambda & 1 & 0 & \cdots & 0 \\ 0 & \lambda^2 & 2\lambda & 1 & \cdot & \vdots \\ 0 & 0 & \lambda^2 & 2\lambda & \cdot & \\ \vdots & & & & \cdot & 1 \\ & & & & & 2\lambda \\ 0 & 0 & 0 & 0 & & \lambda \end{bmatrix}$$

Use induction, assuming the stated form holds for k, and show that it holds for $k + 1$ by computing

$$\begin{bmatrix} \lambda & 1 & 0 & \cdots & 0 \\ 0 & \lambda & 1 & & \\ 0 & 0 & \lambda & & \\ 0 & 0 & 0 & & \vdots \\ \vdots & & & & \\ 0 & 0 & 0 & & \lambda \end{bmatrix} \begin{bmatrix} \lambda^k & k\lambda^{k-1} & \frac{1}{2!}k(k-1)\lambda^{k-2} & \cdots & \\ 0 & \lambda^k & k\lambda^{k-1} & & \vdots \\ 0 & 0 & \lambda^k & & \\ 0 & 0 & 0 & \cdot & \\ \vdots & \vdots & & \cdot & k\lambda^{k-1} \\ 0 & 0 & & & \lambda^k \end{bmatrix} =$$

$$\begin{bmatrix} \lambda^{k+1} & (k+1)\lambda^k & \frac{1}{2!}k(k-1)\lambda^{k-1} + k\lambda^{k-1} & \frac{1}{3!}k(k-1)(k-2)\lambda^{k-2} + \frac{1}{2!}k(k-1)\lambda^{k-2} & \cdots \\ 0 & \lambda^{k+1} & (k+1)\lambda^k & & \\ 0 & 0 & \lambda^{k+1} & & \vdots \\ 0 & 0 & 0 & \vdots & \\ \vdots & & & & \\ 0 & 0 & 0 & & \lambda^{k+1} \end{bmatrix}$$

But

$$\left[\frac{1}{2!}k(k-1) + k\right]\lambda^{k-1} = \frac{k}{2!}(k - 1 + 2)\lambda^{k-1} = \frac{(k+1)k}{2!}\lambda^{k-1}$$

$$\left[\frac{1}{3!}k(k-1)(k-2) + \frac{1}{2!}k(k-1)\right]\lambda^{k-2} = \frac{1}{3!}k(k-1)(k+1)\lambda^{k-2}$$

etc.

so the stated result holds.

Proof of Cayley-Hamilton Theorem

8.5 Prove the Cayley-Hamilton theorem when $\mathbf{A}$ is not diagonalizable.

Any square $\mathbf{A}$ can be reduced to Jordan canonical form $\mathbf{J} = \mathbf{M}^{-1}\mathbf{A}\mathbf{M}$, so that $\mathbf{A} = \mathbf{MJM}^{-1}$, $\mathbf{A}^2 = \mathbf{MJM}^{-1}\mathbf{MJM}^{-1} = \mathbf{MJ}^2\mathbf{M}^{-1}$, $\mathbf{A}^k = \mathbf{MJ}^k\mathbf{M}^{-1}$. If the characteristic polynomial is $\Delta(\lambda) = c_n\lambda^n + c_{n-1}\lambda^{n-1} + \cdots + c_1\lambda + c_0$, then

$$\Delta(\mathbf{A}) = c_n\mathbf{A}^n + c_{n-1}\mathbf{A}^{n-1} + \cdots + c_1\mathbf{A} + c_0\mathbf{I}$$
$$= \mathbf{M}[c_n\mathbf{J}^n + c_{n-1}\mathbf{J}^{n-1} + \cdots + c_1\mathbf{J} + c_0\mathbf{I}]\mathbf{M}^{-1}$$

It is to be shown that the matrix polynomial inside the brackets sums to the zero matrix. The Jordan form is $\mathbf{J} = \text{diag}[\mathbf{J}_1, \mathbf{J}_2, \ldots, \mathbf{J}_p]$, where $\mathbf{J}_i$ are Jordan blocks. Also, $\mathbf{J}^k = \text{diag}[\mathbf{J}_1^k, \mathbf{J}_2^k, \ldots, \mathbf{J}_p^k]$. Therefore, to prove the theorem it is only necessary to show that $c_n\mathbf{J}_i^n + c_{n-1}\mathbf{J}_i^{n-1} + \cdots + c_1\mathbf{J}_i + c_0\mathbf{I} = [\mathbf{0}]$ for a typical block $\mathbf{J}_i$. Using the result of the previous problem, all terms below the main diagonal are zero. The sum of the terms in a typical main diagonal position is $c_n \lambda_i^n + c_{n-1}\lambda_i^{n-1} + \cdots + c_1\lambda_i + c_0$ and equals zero because this is just the characteristic polynomial evaluated with a root λ_i. A typical term of the sum in the diagonal just above the main diagonal is $nc_n\lambda_i^{n-1} + (n-1)c_{n-1}\lambda_i^{n-2} + \cdots + c_2\lambda_i + c_1$. This sum is zero since it equals $\left.\dfrac{d\Delta(\lambda)}{d\lambda}\right|_{\lambda=\lambda_i}$. The root λ_i is a multiple root, so

$$\Delta(\lambda) = c(\lambda - \lambda_i)^m(\lambda - \lambda_j)(\lambda - \lambda_k)\ldots, \quad \text{and} \quad \left.\frac{d\Delta}{d\lambda}\right|_{\lambda=\lambda_i} = 0$$

If $\mathbf{J}_i$ is an $r \times r$ block, λ_i is at least an rth-order root, so

$$\left.\frac{1}{2}\frac{d^2\Delta(\lambda)}{d\lambda^2}\right|_{\lambda=\lambda_i} = 0, \qquad \left.\frac{1}{3!}\frac{d^3\Delta(\lambda)}{d\lambda^3}\right|_{\lambda=\lambda_i} = 0, \ldots, \qquad \left.\frac{1}{(r-1)!}\frac{d^{r-1}\Delta(\lambda)}{d\lambda^{r-1}}\right|_{\lambda=\lambda_i} = 0$$

The successive diagonals above the main diagonal give terms which sum to these derivatives of the characteristic equation. Hence

$$c_n\mathbf{J}_i^n + \cdots + c_1\mathbf{J}_i + c_0\mathbf{I} = [\mathbf{0}]$$

and so

$$c_n\mathbf{A}^n + c_{n-1}\mathbf{A}^{n-1} + \cdots + c_1\mathbf{A} + c_0\mathbf{I} = [\mathbf{0}]$$

.6 Give a general proof of the Cayley-Hamilton theorem without using the Jordan form.

A is an $n \times n$ matrix with n eigenvalues λ_i, some of which may be equal. The characteristic polynomial is

$$\Delta(\lambda) = |\mathbf{A} - \mathbf{I}\lambda| = (-\lambda)^n + c_{n-1}\lambda^{n-1} + c_{n-2}\lambda^{n-2} + \cdots + c_1\lambda + c_0 \tag{1}$$

We are to show that

$$\Delta(\mathbf{A}) = (-1)^n\mathbf{A}^n + c_{n-1}\mathbf{A}^{n-1} + c_{n-2}\mathbf{A}^{n-2} + \cdots + c_1\mathbf{A} + c_0\mathbf{I} = [\mathbf{0}] \tag{2}$$

Consider Adj$[\mathbf{A} - \mathbf{I}\lambda]$. Its elements are formed from $n-1 \times n-1$ determinants obtained by deleting a row and a column of $\mathbf{A} - \mathbf{I}\lambda$. Therefore, the highest power of λ that can be in any element of Adj$[\mathbf{A} - \mathbf{I}\lambda]$ is λ^{n-1}. This means it is possible to write

$$\text{Adj}[\mathbf{A} - \mathbf{I}\lambda] = \mathbf{B}_{n-1}\lambda^{n-1} + \mathbf{B}_{n-2}\lambda^{n-2} + \cdots + \mathbf{B}_1\lambda + \mathbf{B}_0 \tag{3}$$

where the $\mathbf{B}_i$ terms are $n \times n$ matrices not containing λ, but are otherwise unknown. We use the known result

$$[\mathbf{A} - \mathbf{I}\lambda]\ \text{Adj}[\mathbf{A} - \mathbf{I}\lambda] = |\mathbf{A} - \mathbf{I}\lambda|\mathbf{I} \tag{4}$$

Substituting Eq. (*3*) into the left side of Eq. (*4*) gives

$$[\mathbf{A} - \mathbf{I}\lambda]\ \text{Adj}[\mathbf{A} - \mathbf{I}\lambda] = -\mathbf{B}_{n-1}\lambda^n + (\mathbf{AB}_{n-1} - \mathbf{B}_{n-2})\lambda^{n-1} + (\mathbf{AB}_{n-2} - \mathbf{B}_{n-3})\lambda^{n-2}$$
$$+ \cdots + (\mathbf{AB}_2 - \mathbf{B}_1)\lambda^2 + (\mathbf{AB}_1 - \mathbf{B}_0)\lambda + \mathbf{AB}_0$$

Using Eq. (*1*) on the right side of Eq. (*4*) gives

$$\Delta(\lambda)\mathbf{I} = (-\lambda)^n\mathbf{I} + c_{n-1}\lambda^{n-1}\mathbf{I} + c_{n-2}\lambda^{n-2}\mathbf{I} + \cdots + c_1\lambda\mathbf{I} + c_0\mathbf{I}$$

The left side equals the right side, and the coefficients of like powers of λ on the two sides must be equal. This leads to the following set of equations:

$$-\mathbf{B}_{n-1} = (-1)^n\mathbf{I}$$

$$\mathbf{AB}_{n-1} - \mathbf{B}_{n-2} = c_{n-1}\mathbf{I}$$

$$\mathbf{AB}_{n-2} - \mathbf{B}_{n-3} = c_{n-2}\mathbf{I}$$

$$\vdots$$

$$\mathbf{AB}_2 - \mathbf{B}_1 = c_2\mathbf{I}$$

$$\mathbf{AB}_1 - \mathbf{B}_0 = c_1\mathbf{I}$$

$$\mathbf{AB}_0 = c_0\mathbf{I}$$

If the first of these equations is premultiplied by $\mathbf{A}^n$, the second by $\mathbf{A}^{n-1}$, etc., the sum of the right-side terms is $\Delta(\mathbf{A})$. The left-side sum must also equal $\Delta(\mathbf{A})$, and takes the form

$$(-\mathbf{A}^n\mathbf{B}_{n-1} + \mathbf{A}^n\mathbf{B}_{n-1}) + (-\mathbf{A}^{n-1}\mathbf{B}_{n-2} + \mathbf{A}^{n-1}\mathbf{B}_{n-2}) + (-\mathbf{A}^{n-2}\mathbf{B}_{n-3} + \mathbf{A}^{n-2}\mathbf{B}_{n-3})$$
$$+ \cdots + (-\mathbf{A}^2\mathbf{B}_1 + \mathbf{A}^2\mathbf{B}_1) + (-\mathbf{AB}_0 + \mathbf{AB}_0) \equiv [\mathbf{0}]$$

Therefore, $\Delta(\mathbf{A}) = [\mathbf{0}]$.

Functions of a Jordan Block

8.7 If $\mathbf{J}_1$ is an $m \times m$ Jordan block with eigenvalue λ_1, find a general expression for the coefficients α_i for $e^{\mathbf{J}_1 t}$.

The characteristic equation is $\Delta(\lambda) = (\lambda - \lambda_1)^m$ and

$$e^{\mathbf{J}_1 t} = \alpha_0\mathbf{I} + \alpha_1\mathbf{J}_1 + \alpha_2\mathbf{J}_1^2 + \cdots + \alpha_{m-1}\mathbf{J}_1^{m-1}$$

where

$$e^{\lambda_1 t} = \alpha_0 + \alpha_1\lambda_1 + \alpha_2\lambda_1^2 + \cdots + \alpha_{m-1}\lambda_1^{m-1}$$

$$\left.\frac{de^{\lambda t}}{d\lambda}\right|_{\lambda_1} = te^{\lambda_1 t} = \alpha_1 + 2\alpha_2\lambda_1 + 3\alpha_3\lambda_1^2 + \cdots + (m-1)\alpha_{m-1}\lambda_1^{m-2}$$

More symmetry is achieved if the kth derivative term is divided by $1/k!$:

$$\tfrac{1}{2}t^2 e^{\lambda_1 t} = \alpha_2 + 3\alpha_3\lambda_1 + \cdots + \tfrac{1}{2}(m-1)(m-2)\alpha_{m-1}\lambda_1^{m-3}$$

$$\frac{1}{3!}t^3 e^{\lambda_1 t} = \alpha_3 + 4\lambda_1 + \cdots + \frac{1}{3!}(m-1)(m-2)(m-3)\alpha_{m-1}\lambda_1^{m-4}$$

$$\vdots$$

$$\frac{1}{(m-1)!}t^{m-1}e^{\lambda_1 t} = \alpha_{m-1}$$

or

$$\begin{bmatrix} 1 \\ t \\ \frac{1}{2}t^2 \\ \frac{1}{3!}t^3 \\ \vdots \\ \frac{1}{(m-1)!}t^{m-1} \end{bmatrix} e^{\lambda_1 t} = \begin{bmatrix} 1 & \lambda_1 & \lambda_1^2 & \lambda_1^3 & \cdots & \lambda_1^{m-1} \\ 0 & 1 & 2\lambda_1 & 3\lambda_1^2 & & (m-1)\lambda_1^{m-2} \\ 0 & 0 & 1 & 3\lambda_1 & & \frac{1}{2}(m-1)(m-2)\lambda_1^{m-3} \\ \vdots & & & & & \vdots \\ 0 & 0 & 0 & 0 & & 1 \end{bmatrix} \begin{bmatrix} \alpha_0 \\ \alpha_1 \\ \alpha_2 \\ \vdots \\ \alpha_{m-1} \end{bmatrix} \triangleq \mathbf{F}\boldsymbol{\alpha}$$

$$\alpha = [\mathbf{F}]^{-1}\begin{bmatrix} 1 \\ t \\ \frac{1}{2}t^2 \\ \frac{1}{3!}t^3 \\ \vdots \\ \frac{1}{(m-1)!}t^{m-1} \end{bmatrix} e^{\lambda_1 t}$$

But using Gaussian elimination, for example, gives

$$\mathbf{F}^{-1} = \begin{bmatrix} 1 & -\lambda_1 & \lambda_1^2 & -\lambda_1^3 & \lambda_1^4 & \cdots & (-\lambda_1)^{m-1} \\ 0 & 1 & -2\lambda_1 & 3\lambda_1^2 & -4\lambda_1^3 & & (m-1)(-\lambda_1)^{m-2} \\ 0 & 0 & 1 & -3\lambda_1 & 6\lambda_1^2 & & \frac{1}{2}(m-1)(m-2)(-\lambda_1)^{m-3} \\ 0 & 0 & 0 & 1 & -4\lambda_1 & & \frac{1}{3!}(m-1)(m-2)(m-3)(-\lambda_1)^{m-4} \\ \vdots & & & & & \cdot & \vdots \\ & & & & & \cdot & -(m-1)\lambda_1 \\ 0 & 0 & 0 & 0 & 0 & \cdots & 1 \end{bmatrix}$$

so

$$\alpha_0 = e^{\lambda_1 t}\left[1 - \lambda_1 t + \frac{\lambda_1^2 t^2}{2!} - \frac{\lambda_1^3 t^3}{3!} + \frac{\lambda_1^4 t^4}{4!} - \cdots + \frac{(-\lambda_1)^{m-1} t^{m-1}}{(m-1)!}\right]$$

$$\alpha_1 = e^{\lambda_1 t}\left[t - \lambda_1 t^2 + \frac{\lambda_1^2 t^3}{2} - \frac{\lambda_1^3 t^4}{3!} + \cdots + \frac{t^{m-1}(-\lambda_1)^{m-2}}{(m-2)!}\right]$$

$$\alpha_2 = e^{\lambda_1 t}\left[\frac{1}{2}t^2 - \frac{\lambda_1 t^3}{2} + \frac{\lambda_1^2 t^4}{4} + \cdots + \frac{t^{m-1}(-\lambda)^{m-3}}{2(m-3)!}\right]$$

$$\vdots$$

$$\alpha_{m-2} = e^{\lambda_1 t}\left[\frac{1}{(m-2)!}t^{m-2} - \frac{1}{(m-2)!}\lambda_1 t^{m-1}\right]$$

$$\alpha_{m-1} = e^{\lambda_1 t}\frac{t^{m-1}}{(m-1)!}$$

8.8 Use the results of the previous problem to find $e^{\mathbf{J}_1 t}$ if a. $\mathbf{J}_1$ is a 3×3 Jordan block, and b. $\mathbf{J}_1$ is a 4×4 Jordan block.

(a) $$\mathbf{J}_1 = \begin{bmatrix} \lambda_1 & 1 & 0 \\ 0 & \lambda_1 & 1 \\ 0 & 0 & \lambda_1 \end{bmatrix}, \qquad \mathbf{J}_1^2 = \begin{bmatrix} \lambda_1^2 & 2\lambda_1 & 1 \\ 0 & \lambda_1^2 & 2\lambda_1 \\ 0 & 0 & \lambda_1^2 \end{bmatrix}$$

$$e^{\mathbf{J}_1 t} = \alpha_0 \mathbf{I} + \alpha_1 \mathbf{J}_1 + \alpha_2 \mathbf{J}_1^2$$

$$= \begin{bmatrix} \alpha_0 + \alpha_1\lambda_1 + \alpha_2\lambda_1^2 & \alpha_1 + 2\lambda_1\alpha_2 & \alpha_2 \\ 0 & \alpha_0 + \alpha_1\lambda_1 + \alpha_2\lambda_1^2 & \alpha_1 + 2\lambda_1\alpha_2 \\ 0 & 0 & \alpha_0 + \alpha_1\lambda_1 + \alpha_2\lambda_1^2 \end{bmatrix}$$

$$= e^{\lambda_1 t}\begin{bmatrix} 1 & t & \frac{1}{2}t^2 \\ 0 & 1 & t \\ 0 & 0 & 1 \end{bmatrix}$$

(b) $$\mathbf{J}_1 = \begin{bmatrix} \lambda_1 & 1 & 0 & 0 \\ 0 & \lambda_1 & 1 & 0 \\ 0 & 0 & \lambda_1 & 1 \\ 0 & 0 & 0 & \lambda_1 \end{bmatrix} \qquad \mathbf{J}_1^2 = \begin{bmatrix} \lambda_1^2 & 2\lambda_1 & 1 & 0 \\ 0 & \lambda_1^2 & 2\lambda_1 & 1 \\ 0 & 0 & \lambda_1^2 & 2\lambda_1 \\ 0 & 0 & 0 & \lambda_1^2 \end{bmatrix},$$

$$\mathbf{J}_1^3 = \begin{bmatrix} \lambda_1^3 & 3\lambda_1^2 & 3\lambda_1 & 1 \\ 0 & \lambda_1^3 & 3\lambda_1^2 & 3\lambda_1 \\ 0 & 0 & \lambda_1^3 & 3\lambda_1^2 \\ 0 & 0 & 0 & \lambda_1^3 \end{bmatrix}$$

so

$$e^{\mathbf{J}_1 t} = \alpha_0 \mathbf{I} + \alpha_1 \mathbf{J}_1 + \alpha_2 \mathbf{J}_1^2 + \alpha_3 \mathbf{J}_1^3$$
$$= e^{\lambda_1 t} \begin{bmatrix} 1 & t & \frac{1}{2}t^2 & \frac{1}{3!}t^3 \\ 0 & 1 & t & \frac{1}{2}t^2 \\ 0 & 0 & 1 & t \\ 0 & 0 & 0 & 1 \end{bmatrix}$$

The pattern illustrated by these two cases continues for any $m \times m$ Jordan block.

Some Matrix Identities

8.9 (a) If $\mathbf{A} = \begin{bmatrix} 1 & 2 \\ -2 & -3 \end{bmatrix}$, compute the 2×2 matrices $\sin \mathbf{A}$ and $\cos \mathbf{A}$.

(b) Verify that $\sin^2 \mathbf{A} + \cos^2 \mathbf{A} = \mathbf{I}$.

(a) From Problem 7.2, $\Delta(\lambda) = \lambda^2 + 2\lambda + 1$ and $\lambda_1 = \lambda_2 = -1$. Now $\sin \mathbf{A} = \alpha_0 \mathbf{I} + \alpha_1 \mathbf{A}$ and $\cos \mathbf{A} = \alpha_2 \mathbf{I} + \alpha_3 \mathbf{A}$, where

$$\left.\begin{aligned} \sin \lambda_1 &= \alpha_0 + \alpha_1 \lambda_1 \\ \frac{d}{d\lambda}(\sin \lambda)\Big|_{\lambda=\lambda_1} &= \cos \lambda_1 = \alpha_1 \end{aligned}\right\} \Rightarrow \left\{\begin{aligned} \alpha_1 &= \cos(-1) = \cos(1) \\ \alpha_0 &= \sin \lambda_1 + \alpha_1 = \cos(1) - \sin(1) \end{aligned}\right.$$

and

$$\left.\begin{aligned} \cos \lambda_1 &= \alpha_2 + \lambda_1 \alpha_3 \\ \frac{d}{d\lambda}(\cos \lambda)\Big|_{\lambda=\lambda_1} &= -\sin \lambda_1 = \alpha_3 \end{aligned}\right\} \Rightarrow \left\{\begin{aligned} \alpha_3 &= \sin(1) \\ \alpha_2 &= \cos(1) + \alpha_3 = \cos(1) + \sin(1) \end{aligned}\right.$$

Therefore,

$$\sin \mathbf{A} = \begin{bmatrix} 2\cos(1) - \sin(1) & 2\cos(1) \\ -2\cos(1) & -2\cos(1) - \sin(1) \end{bmatrix}$$

and

$$\cos \mathbf{A} = \begin{bmatrix} \cos(1) + 2\sin(1) & 2\sin(1) \\ -2\sin(1) & \cos(1) - 2\sin(1) \end{bmatrix}$$

(b) Multiplication gives

$$\sin^2 \mathbf{A} = \begin{bmatrix} \sin^2(1) - 4\cos(1)\sin(1) & -4\cos(1)\sin(1) \\ 4\cos(1)\sin(1) & \sin^2(1) + 4\cos(1)\sin(1) \end{bmatrix}$$

and

$$\cos^2 \mathbf{A} = \begin{bmatrix} \cos^2(1) + 4\cos(1)\sin(1) & 4\cos(1)\sin(1) \\ -4\cos(1)\sin(1) & \cos^2(1) - 4\cos(1)\sin(1) \end{bmatrix}$$

so $\sin^2 \mathbf{A} + \cos^2 \mathbf{A} = \begin{bmatrix} \sin^2(1) + \cos^2(1) & 0 \\ 0 & \sin^2(1) + \cos^2(1) \end{bmatrix} = \mathbf{I}.$

10 (a) Compute $e^{\mathbf{A}t}$, $e^{-\mathbf{A}t}$, $\sinh \mathbf{A}t$, and $\cosh \mathbf{A}t$ for $\mathbf{A} = \begin{bmatrix} 1 & 1 \\ 1 & 1 \end{bmatrix}$.

(b) Verify that $\sinh \mathbf{A}t = (e^{\mathbf{A}t} - e^{-\mathbf{A}t})/2$, $\cosh \mathbf{A}t = (e^{\mathbf{A}t} + e^{-\mathbf{A}t})/2$, and $\cosh^2 \mathbf{A}t - \sinh^2 \mathbf{A}t = \mathbf{I}$.

(a) The eigenvalues are $\lambda_1 = 0$ and $\lambda_2 = 2$, so $f(\mathbf{A}) = \alpha_0 \mathbf{I} + \alpha_1 \mathbf{A} = \begin{bmatrix} \alpha_0 + \alpha_1 & \alpha_1 \\ \alpha_1 & \alpha_0 + \alpha_1 \end{bmatrix}$. The coefficients α_0 and α_1 are found from $f(\lambda_1) = \alpha_0$ and $f(\lambda_2) = \alpha_0 + 2\alpha_1$. Therefore, $\alpha_1 = [f(\lambda_2) - f(\lambda_1)]/2$ and $\alpha_0 + \alpha_1 = [f(\lambda_2) + f(\lambda_1)]/2$ for any analytic $f(\lambda)$. Letting $f(\lambda)$ be each of the four required functions gives, in turn,

$$e^{\mathbf{A}t} = \begin{bmatrix} \dfrac{e^{2t}+1}{2} & \dfrac{e^{2t}-1}{2} \\ \dfrac{e^{2t}-1}{2} & \dfrac{e^{2t}+1}{2} \end{bmatrix}, \qquad e^{-\mathbf{A}t} = \begin{bmatrix} \dfrac{e^{-2t}+1}{2} & \dfrac{e^{-2t}-1}{2} \\ \dfrac{e^{-2t}-1}{2} & \dfrac{e^{-2t+1}}{2} \end{bmatrix}$$

$$\sinh \mathbf{A}t = \begin{bmatrix} \dfrac{\sinh 2t}{2} & \dfrac{\sinh 2t}{2} \\ \dfrac{\sinh 2t}{2} & \dfrac{\sinh 2t}{2} \end{bmatrix}, \qquad \cosh \mathbf{A}t = \begin{bmatrix} \dfrac{\cosh 2t + 1}{2} & \dfrac{\cosh 2t - 1}{2} \\ \dfrac{\cosh 2t - 1}{2} & \dfrac{\cosh 2t + 1}{2} \end{bmatrix}$$

(b) Since $(e^{2t} - e^{-2t})/2 = \sinh 2t$ and $(e^{2t} + e^{-2t})/2 = \cosh 2t$, it is clear that $(e^{\mathbf{A}t} + e^{-\mathbf{A}t})/2 = \cosh \mathbf{A}t$ and $(e^{\mathbf{A}t} - e^{-\mathbf{A}t})/2 = \sinh \mathbf{A}t$. Computing

$$\cosh^2 \mathbf{A}t = \begin{bmatrix} \dfrac{\cosh^2 2t + 1}{2} & \dfrac{\cosh^2 2t - 1}{2} \\ \dfrac{\cosh^2 2t - 1}{2} & \dfrac{\cosh^2 2t + 1}{2} \end{bmatrix}$$

$$\sinh^2 \mathbf{A}t = \begin{bmatrix} \dfrac{\sinh^2 2t}{2} & \dfrac{\sinh^2 2t}{2} \\ \dfrac{\sinh^2 2t}{2} & \dfrac{\sinh^2 2t}{2} \end{bmatrix}$$

and using $\cosh^2 2t - \sinh^2 2t = 1$ gives $\cosh^2 \mathbf{A}t - \sinh^2 \mathbf{A}t = \mathbf{I}$.

11 Let $\mathbf{A} = \begin{bmatrix} -1 & 1 \\ 1 & 1 \end{bmatrix}$.

(a) Find $\sin \mathbf{A}$, $\cos \mathbf{A}$, and $\tan \mathbf{A}$.
(b) Show that $\tan \mathbf{A} = (\sin \mathbf{A})(\cos \mathbf{A})^{-1}$.
(a) The eigenvalues are $\lambda_1 = \sqrt{2}$ and $\lambda_2 = -\sqrt{2}$. Since both eigenvalues satisfy $|\lambda_i| < \pi/2$, use of the results for $\tan \mathbf{A}$ of Problem 8.3 is justified. For all three functions, $f(\mathbf{A}) = \alpha_0 \mathbf{I} + \alpha_1 \mathbf{A}$, where $\alpha_0 = (1/2)[f(\sqrt{2}) + f(-\sqrt{2})]$ and $\alpha_1 = (1/\sqrt{2})[\alpha_0 - f(-\sqrt{2})]$, so

$$\sin \mathbf{A} = \frac{\sin\sqrt{2}}{\sqrt{2}} \begin{bmatrix} -1 & 1 \\ 1 & 1 \end{bmatrix}, \qquad \cos \mathbf{A} = (\cos\sqrt{2}) \begin{bmatrix} 1 & 0 \\ 0 & 1 \end{bmatrix},$$

$$\tan \mathbf{A} = \frac{\tan\sqrt{2}}{\sqrt{2}} \begin{bmatrix} -1 & 1 \\ 1 & 1 \end{bmatrix}$$

(b) Using $(\cos \mathbf{A})^{-1} = (1/\cos\sqrt{2})\mathbf{I}$ shows that $\tan \mathbf{A} = (\sin \mathbf{A})(\cos \mathbf{A})^{-1}$.

Closed Form for Functions of a Matrix

8.12 If $\mathbf{A} = \begin{bmatrix} -2 & 2 \\ 1 & -3 \end{bmatrix}$, what is $\sin \mathbf{A}t$?

Since $\Delta(\lambda) = \lambda^2 + 5\lambda + 4 = (\lambda + 4)(\lambda + 1)$, the eigenvalues are $\lambda_1 = -1$ and $\lambda_2 = -4$. The general form of the solution is

$$\sin \mathbf{A}t = \alpha_0 \mathbf{I} + \alpha_1 \mathbf{A}$$

where

$$\left.\begin{aligned} \sin(-4t) &= \alpha_0 - 4\alpha_1 \\ \sin(-t) &= \alpha_0 - \alpha_1 \end{aligned}\right\} \Rightarrow \left\{\begin{aligned} \alpha_1 &= -\tfrac{1}{3}[\sin(-4t) - \sin(-t)] \\ \alpha_0 &= -\tfrac{1}{3}[\sin(-4t) - 4\sin(-t)] \end{aligned}\right.$$

Then $\sin \mathbf{A}t = \dfrac{1}{3}\begin{bmatrix} \sin(-4t) + 2\sin(-t) & -2\sin(-4t) + 2\sin(-t) \\ -\sin(-4t) + \sin(-t) & 2\sin(-4t) + \sin(-t) \end{bmatrix}$.

8.13 If $\mathbf{A} = \begin{bmatrix} 2 & 0 & 0 \\ 0 & -2 & 2 \\ 0 & 1 & -3 \end{bmatrix}$, find $e^{\mathbf{A}t}$.

Note that $\mathbf{A}$ is block diagonal and the lower block is the same matrix as in the previous problem. Since the algebra involved in finding the α's is the same when finding any analytic function, the answer can be written down by replacing $\sin \lambda_i t$ by $e^{\lambda_i t}$, so

$$e^{\mathbf{A}t} = \left[\begin{array}{c|cc} e^{2t} & 0 & 0 \\ \hline 0 & \dfrac{(e^{-4t} + 2e^{-t})}{3} & \dfrac{2(-e^{-4t} + e^{-t})}{3} \\ 0 & \dfrac{(-e^{-4t} + e^{-t})}{3} & \dfrac{(2e^{-4t} + e^{-t})}{3} \end{array}\right]$$

A useful partial check is that $e^{\mathbf{A}t}$ must always equal $\mathbf{I}$ when $t = 0$.

8.14 Compute $\begin{bmatrix} 1 & -1 & 1 \\ 0 & 1 & 1 \\ 0 & 0 & 1 \end{bmatrix}^k$ for any arbitrary integer k.

Computing $\Delta(\lambda) = (1 - \lambda)^3$ gives $\lambda_1 = \lambda_2 = \lambda_3 = 1$. Since $\mathbf{A}$ is 3×3, $\mathbf{A}^k$ can always be written as a polynomial of degree 2 (or less):

$$\mathbf{A}^k = \alpha_0 \mathbf{I} + \alpha_1 \mathbf{A} + \alpha_2 \mathbf{A}^2 = \begin{bmatrix} \alpha_0 + \alpha_1 + \alpha_2 & -\alpha_1 - 2\alpha_2 & \alpha_1 + \alpha_2 \\ 0 & \alpha_0 + \alpha_1 + \alpha_2 & \alpha_1 + 2\alpha_2 \\ 0 & 0 & \alpha_0 + \alpha_1 + \alpha_2 \end{bmatrix}$$

where

$$(\lambda_1)^k = 1 = \alpha_0 + \alpha_1 + \alpha_2$$

$$\left.\frac{d(\lambda)^k}{d\lambda}\right|_{\lambda=\lambda_1} = k\lambda_1^{k-1} = k = \alpha_1 + 2\alpha_2$$

$$\left.\frac{d^2(\lambda)^k}{d\lambda^2}\right|_{\lambda=\lambda_1} = k(k-1)\lambda_1^{k-2} = k(k-1) = 2\alpha_2$$

Note that in this case all the individual α's need not be found. Combining the last two equations gives $\alpha_1 + \alpha_2 = k - k(k-1)/2 = k(3-k)/2$. All the combinations of α_i that are required in $\mathbf{A}^k$ are now available, and

$$\mathbf{A}^k = \begin{bmatrix} 1 & -k & k(3-k)/2 \\ 0 & 1 & k \\ 0 & 0 & 1 \end{bmatrix}$$

15 Compute $e^{\mathbf{A}t}$ if $\mathbf{A} = \begin{bmatrix} -1 & 1 \\ 1 & 1 \end{bmatrix}$.

We know that $e^{\mathbf{A}t} = \alpha_0 \mathbf{I} + \alpha_1 \mathbf{A} = \begin{bmatrix} \alpha_0 - \alpha_1 & \alpha_1 \\ \alpha_1 & \alpha_0 + \alpha_1 \end{bmatrix}$. Since this is the same A matrix as in Problem 8.11, the coefficients are

$$\alpha_0 = \tfrac{1}{2}[e^{\sqrt{2}t} + e^{-\sqrt{2}t}] = \cosh \sqrt{2}t, \quad \alpha_1 = \frac{e^{\sqrt{2}t} - e^{-\sqrt{2}t}}{2\sqrt{2}} = \frac{1}{\sqrt{2}} \sinh \sqrt{2}t$$

$$\text{so } e^{\mathbf{A}t} = \begin{bmatrix} \cosh \sqrt{2}t - \frac{1}{\sqrt{2}} \sinh \sqrt{2}t & \frac{1}{\sqrt{2}} \sinh \sqrt{2}t \\ \frac{1}{\sqrt{2}} \sinh \sqrt{2}t & \cosh \sqrt{2}t + \frac{1}{\sqrt{2}} \sinh \sqrt{2}t \end{bmatrix}.$$

16 Given $\mathbf{A} = \begin{bmatrix} 0 & -3 & 0 \\ 3 & 0 & 0 \\ 0 & 0 & -1 \end{bmatrix}$. a. Find $\mathbf{A}^{-1}$ using the Cayley-Hamilton theorem. b. Compute $e^{\mathbf{A}t}$.

(a) The characteristic polynomial is $\Delta(\lambda) = (1+\lambda)(\lambda^2+9) = \lambda^3 + \lambda^2 + 9\lambda + 9$. Therefore $\Delta(\mathbf{A}) = [0] = \mathbf{A}^3 + \mathbf{A}^2 + 9\mathbf{A} + 9\mathbf{I}$ so

$$\mathbf{A}^{-1} = -\frac{1}{9}[\mathbf{A}^2 + \mathbf{A} + 9\mathbf{I}] = \begin{bmatrix} 0 & \frac{1}{3} & 0 \\ -\frac{1}{3} & 0 & 0 \\ 0 & 0 & -1 \end{bmatrix}$$

(b) The eigenvalues are roots of $\Delta(\lambda) = 0$, so $\lambda_1 = -1, \lambda_2 = 3j, \lambda_3 = -3j$. Since A is block diagonal,

$$e^{\mathbf{A}t} = \left[\begin{array}{c|c} e^{\begin{bmatrix} 0 & -3 \\ 3 & 0 \end{bmatrix} t} & \begin{matrix} 0 \\ 0 \end{matrix} \\ \hline \begin{matrix} 0 & 0 \end{matrix} & e^{-t} \end{array}\right]$$

and

$$e^{\begin{bmatrix} 0 & -3 \\ 3 & 0 \end{bmatrix} t} = \alpha_0 \mathbf{I} + \alpha_1 \begin{bmatrix} 0 & -3 \\ 3 & 0 \end{bmatrix} = \begin{bmatrix} \alpha_0 & -3\alpha_1 \\ 3\alpha_1 & \alpha_0 \end{bmatrix}$$

where $e^{3jt} = \alpha_0 + \alpha_1 3j$ and $e^{-3jt} = \alpha_0 - \alpha_1 3j$. From this,

$$\alpha_0 = \tfrac{1}{2}[e^{3jt} + e^{-3jt}] = \cos 3t$$

$$\alpha_1 = \frac{1}{3}\left[\frac{e^{3jt} - e^{-3jt}}{2j}\right] = \tfrac{1}{3} \sin 3t$$

$$\text{and so } e^{\mathbf{A}t} = \begin{bmatrix} \cos 3t & -\sin 3t & 0 \\ \sin 3t & \cos 3t & 0 \\ 0 & 0 & e^{-t} \end{bmatrix}.$$

17 Find $e^{\mathbf{A}t}$ if $\mathbf{A} = \left[\begin{array}{cc|cc} 0 & -\Omega & a & 0 \\ \Omega & 0 & 0 & a \\ \hline 0 & 0 & 0 & -\Omega \\ 0 & 0 & \Omega & 0 \end{array}\right]$

The algebra can be simplified in this case if it is noted that

$$\mathbf{A} = \left[\begin{array}{c|c} \mathbf{B}_1 & \mathbf{0} \\ \hline \mathbf{0} & \mathbf{B}_1 \end{array}\right] + \left[\begin{array}{c|c} \mathbf{0} & \mathbf{B}_2 \\ \hline \mathbf{0} & \mathbf{0} \end{array}\right] \triangleq \mathbf{A}_1 + \mathbf{A}_2 \quad \text{and} \quad \mathbf{A}_1\mathbf{A}_2 = \mathbf{A}_2\mathbf{A}_1 = \left[\begin{array}{c|c} \mathbf{0} & a\mathbf{B}_1 \\ \hline \mathbf{0} & \mathbf{0} \end{array}\right]$$

Since $\mathbf{A}_1$ and $\mathbf{A}_2$ commute, $e^{(\mathbf{A}_1+\mathbf{A}_2)t} = e^{\mathbf{A}_1 t}e^{\mathbf{A}_2 t}$. But $\mathbf{A}_1$ is block diagonal, so

$$e^{\mathbf{A}_1 t} = \left[\begin{array}{c|c} e^{\mathbf{B}_1 t} & \mathbf{0} \\ \hline \mathbf{0} & e^{\mathbf{B}_1 t} \end{array}\right]$$

$e^{\mathbf{B}_1 t} = \alpha_0 \mathbf{I} + \alpha_1 \mathbf{B}_1$ and eigenvalues of $\mathbf{B}_1$ are $\pm j\Omega$:

$$\left.\begin{array}{l} e^{j\Omega t} = \alpha_0 + j\Omega\alpha_1 \\ e^{-j\Omega t} = \alpha_0 - j\Omega\alpha_1 \end{array}\right\} \Rightarrow e^{\mathbf{B}_1 t} = \begin{bmatrix} \cos\Omega t & -\sin\Omega t \\ \sin\Omega t & \cos\Omega t \end{bmatrix}$$

Consider $\mathbf{A}_2$: $|\mathbf{A}_2 - \mathbf{I}\lambda| = \lambda^4 = 0$. $\mathbf{A}_2$ has $\lambda_i = 0$ with algebraic multiplicity of four:

$$e^{\mathbf{A}_2 t} = \beta_0 \mathbf{I} + \beta_1 \mathbf{A}_2 + \beta_2 \mathbf{A}_2^2 + \beta_3 \mathbf{A}_2^3$$

Since $\mathbf{A}_2^2 = \mathbf{A}_2^3 = [\mathbf{0}]$, coefficients β_2 and β_3 are not needed. The remaining coefficients are found to be $e^{0t} = 1 = \beta_0$ and $\left.\dfrac{de^{\lambda t}}{d\lambda}\right|_{\lambda=0} = t = \beta_1$. Thus

$$e^{\mathbf{A}_2 t} = \left[\begin{array}{cc|cc} 1 & 0 & at & 0 \\ 0 & 1 & 0 & at \\ \hline 0 & 0 & 1 & 0 \\ 0 & 0 & 0 & 1 \end{array}\right]$$

$$\text{Combining gives } e^{\mathbf{A}t} = e^{\mathbf{A}_1 t}e^{\mathbf{A}_2 t} = \left[\begin{array}{cc|cc} \cos\Omega t & -\sin\Omega t & at\cos\Omega t & -at\sin\Omega t \\ \sin\Omega t & \cos\Omega t & at\sin\Omega t & at\cos\Omega t \\ \hline 0 & 0 & \cos\Omega t & -\sin\Omega t \\ 0 & 0 & \sin\Omega t & \cos\Omega t \end{array}\right].$$

Minimum Polynomial

8.18 In computing $e^{\mathbf{A}_2 t}$ in the previous problem, it was found that only first-order terms in $\mathbf{A}_2$ were needed even though $\mathbf{A}_2$ was a 4×4 matrix. This is a case where the minimum polynomial is of lower order than the characteristic polynomial. Find the minimal polynomial of $\mathbf{A}_2$.

The minimal polynomial will have the same factors as the characteristic polynomial, but perhaps raised to smaller powers. (There could also be a sign difference, depending on how the characteristic polynomial is defined.)

Here $\Delta(\lambda) = \lambda^4$, so $m(\lambda) = \lambda^k$, where k is the smallest integer for which $m(\mathbf{A}) = [\mathbf{0}]$. From the results of the previous problem, $k = 2$ and therefore $m(\lambda) = \lambda^2$. If this had been known in advance, then it would have been known that $e^{\mathbf{A}_2 t} = \beta_0 \mathbf{I} + \beta_1 \mathbf{A}_2$ and a slight amount of matrix algebra would be avoided. This savings is usually offset by the effort required to find $m(\lambda)$, and in this text the minimum polynomial is seldom used.

Alternative Methods

8.19 It was shown in the text that $de^{\mathbf{A}t}/dt = \mathbf{A}e^{\mathbf{A}t}$ if $\mathbf{A}$ is a constant matrix. Use this result, plus the fact that $e^{\mathbf{A}\cdot 0} = \mathbf{I}$, to derive an alternative method of computing $e^{\mathbf{A}t}$.

Let $e^{\mathbf{A}t} = \mathbf{F}(t)$. Then $\mathbf{F}$ satisfies the matrix differential equation $\dot{\mathbf{F}} = \mathbf{A}\mathbf{F}$ with initial conditions $\mathbf{F}(0) = \mathbf{I}$. Laplace transforms can be used to solve this equation:

$$\mathscr{L}\{\dot{\mathbf{F}}(t)\} = s\mathbf{F}(s) - \mathbf{F}(0) = s\mathbf{F}(s) - \mathbf{I}$$

$$\mathscr{L}\{\mathbf{A}\mathbf{F}(t)\} = \mathbf{A}\mathscr{L}\{\mathbf{F}(t)\} = \mathbf{A}\mathbf{F}(s)$$

Therefore, $[s\mathbf{I}-\mathbf{A}]\mathbf{F}(s)=\mathbf{I}$ or $\mathbf{F}(s)=[s\mathbf{I}-\mathbf{A}]^{-1}$, and

$$e^{\mathbf{A}t}=\mathbf{F}(t)=\mathcal{L}^{-1}\{[s\mathbf{I}-\mathbf{A}]^{-1}\}$$

20 Use the method of the previous problem to compute $e^{\mathbf{A}t}$ for $\mathbf{A}=\begin{bmatrix}-2 & -2 & 0\\ 0 & 0 & 1\\ 0 & -3 & -4\end{bmatrix}$.

Form $s\mathbf{I}-\mathbf{A}=\begin{bmatrix}s+2 & 2 & 0\\ 0 & s & -1\\ 0 & 3 & s+4\end{bmatrix}$ and $|s\mathbf{I}-\mathbf{A}|=(s+2)(s^2+4s+3)$

$=(s+1)(s+2)(s+3)$. Then

$$\mathbf{F}(s)=[s\mathbf{I}-\mathbf{A}]^{-1}$$
$$=\frac{1}{(s+1)(s+2)(s+3)}\begin{bmatrix}(s+1)(s+3) & -2(s+4) & -2\\ 0 & (s+2)(s+4) & s+2\\ 0 & -3(s+2) & s(s+2)\end{bmatrix}$$

The inverse Laplace transforms are computed term by term:

$$F_{11}(t)=\mathcal{L}^{-1}\left\{\frac{1}{s+2}\right\}=e^{-2t};\qquad F_{21}=F_{31}=0$$

$$F_{12}(t)=\mathcal{L}^{-1}\left\{\frac{-2(s+4)}{(s+1)(s+2)(s+3)}\right\}=-3e^{-t}+4e^{-2t}-e^{-3t}$$

$$F_{13}(t)=\mathcal{L}^{-1}\left\{\frac{-2}{(s+1)(s+2)(s+3)}\right\}=-e^{-t}+2e^{-2t}-e^{-3t}$$

$$F_{22}(t)=\mathcal{L}^{-1}\left\{\frac{s+4}{(s+1)(s+3)}\right\}=\tfrac{3}{2}e^{-t}-\tfrac{1}{2}e^{-3t}$$

$$F_{32}(t)=\mathcal{L}^{-1}\left\{\frac{-3}{(s+1)(s+3)}\right\}=-\tfrac{3}{2}e^{-t}+\tfrac{3}{2}e^{-3t}$$

$$F_{23}(t)=\mathcal{L}^{-1}\left\{\frac{1}{(s+1)(s+3)}\right\}=\tfrac{1}{2}e^{-t}-\tfrac{1}{2}e^{-3t}$$

$$F_{33}(t)=\mathcal{L}^{-1}\left\{\frac{s}{(s+1)(s+3)}\right\}=-\tfrac{1}{2}e^{-t}+\tfrac{3}{2}e^{-3t}$$

.21 Find the closed-form expression for $e^{\mathbf{A}t}$, where $\mathbf{A}=\begin{bmatrix}0 & 1 & 0 & 0\\ 0 & 0 & 1 & 0\\ 0 & 0 & 0 & 1\\ -27 & 54 & -36 & 10\end{bmatrix}$.

Writing $|\mathbf{A}-\mathbf{I}\lambda|=\Delta(\lambda)=\lambda^4-10\lambda^3+36\lambda^2-54\lambda+27=(\lambda-1)(\lambda-3)^3$ shows that $\lambda_1=1$, $\lambda_2=\lambda_3=\lambda_4=3$. For the repeated eigenvalue, $\text{rank}[\mathbf{A}-\mathbf{I}\lambda_2]=3$ and $q_2=1$. Thus the Jordan

form is $\mathbf{J}=\left[\begin{array}{c|ccc}1 & 0 & 0 & 0\\ \hline 0 & 3 & 1 & 0\\ 0 & 0 & 3 & 1\\ 0 & 0 & 0 & 3\end{array}\right]$. Using the results of Problem 8.8,

$$e^{\mathbf{J}t} = \left[\begin{array}{c|ccc} e^t & 0 & 0 & 0 \\ \hline 0 & & & \\ 0 & & e^{\mathbf{J}_2 t} & \\ 0 & & & \end{array}\right] = \begin{bmatrix} e^t & 0 & 0 & 0 \\ 0 & e^{3t} & te^{3t} & \frac{1}{2}t^2e^{3t} \\ 0 & 0 & e^{3t} & te^{3t} \\ 0 & 0 & 0 & e^{3t} \end{bmatrix}$$

and $e^{\mathbf{A}t} = \mathbf{M}e^{\mathbf{J}t}\mathbf{M}^{-1}$, where $\mathbf{M}$ is the modal matrix consisting of two eigenvectors and two generalized eigenvectors. The first column of $\text{Adj}[\mathbf{A} - \mathbf{I}\lambda]$ is

$$\begin{bmatrix} \lambda^2(10-\lambda) - 36\lambda + 54 \\ 27 \\ 27\lambda \\ 27\lambda^2 \end{bmatrix}, \quad \text{so with } \lambda = 1, \quad \mathbf{x} = \begin{bmatrix} 1 \\ 1 \\ 1 \\ 1 \end{bmatrix}$$

and with $\lambda = 3$, $\mathbf{x}_2 = [1 \quad 3 \quad 9 \quad 27]^T$ is an eigenvector. Generalized eigenvectors are solutions of $\mathbf{A}\mathbf{x}_3 = 3\mathbf{x}_3 + \mathbf{x}_2$ and $\mathbf{A}\mathbf{x}_4 = 3\mathbf{x}_4 + \mathbf{x}_3$. Solutions are $\mathbf{x}_3 = [1 \quad 4 \quad 15 \quad 54]^T$ and $\mathbf{x}_4 = [0 \quad 1 \quad 7 \quad 36]^T$, so

$$\mathbf{M} = \begin{bmatrix} 1 & 1 & 1 & 0 \\ 1 & 3 & 4 & 1 \\ 1 & 9 & 15 & 7 \\ 1 & 27 & 54 & 36 \end{bmatrix} \quad \text{and} \quad \mathbf{M}^{-1} = \frac{1}{8}\begin{bmatrix} 27 & -27 & 9 & -1 \\ -85 & 133 & -55 & 7 \\ 66 & -106 & 46 & -6 \\ -36 & 60 & -28 & 4 \end{bmatrix}$$

Thus

$$e^{\mathbf{A}t} = \mathbf{M}e^{\mathbf{J}t}\mathbf{M}^{-1} =$$

$$\frac{1}{8}\left[\begin{array}{ll} 27e^t + (-19 + 30t - 18t^2)e^{3t} & -27e^t + (27 - 46t + 30t^2)e^{3t} \\ 27e^t + (-27 + 54t - 54t^2)e^{3t} & -27e^t + (35 - 78t + 90t^2)e^{3t} \\ 27e^t + (-27 + 54t - 162t^2)e^{3t} & -27e^t + (27 - 54t + 270t^2)e^{3t} \\ 27e^t + (-27 - 162t - 486t^2)e^{3t} & -27e^t + (27 + 378t + 810t^2)e^{3t} \end{array}\right.$$

$$\left.\begin{array}{ll} 9e^t + (-9 + 18t - 14t^2)e^{3t} & -e^t + (1 - 2t + 2t^2)e^{3t} \\ 9e^t + (-9 + 26t - 42t^2)e^{3t} & -e^t + (1 - 2t + 6t^2)e^{3t} \\ 9e^t + (-1 - 6t - 126t^2)e^{3t} & -e^t + (1 + 6t + 18t^2)e^{3t} \\ 9e^t + (-9 - 270t - 378t^2)e^{3t} & -e^t + (9 + 54t + 54t^2)e^{3t} \end{array}\right]$$

8.22 Another method of computing an analytic function $f(\mathbf{A})$ of an $n \times n$ matrix $\mathbf{A}$ with distinct eigenvalues is given by

$$f(\mathbf{A}) = \sum_{i=1}^{n} f(\lambda_i)\mathbf{Z}_i(\lambda)$$

where λ_i are the eigenvalues of $\mathbf{A}$ and the $n \times n$ $\mathbf{Z}_i$ matrices are given by

$$\mathbf{Z}_i(\lambda) = \frac{\prod_{\substack{j=1 \\ j \neq i}}^{n} (\mathbf{A} - \lambda_j\mathbf{I})}{\prod_{\substack{j=1 \\ j \neq i}}^{n} (\lambda_i - \lambda_j)}$$

This method is sometimes called Sylvester's expansion and sometimes it is referred to as Lagrange interpolation [2]. This is a restricted form, but it can be extended to the case of repeated eigenvalues. Use it to compute $e^{\mathbf{A}t}$ for

$$\mathbf{A} = \begin{bmatrix} 2 & -2 & 3 \\ 1 & 1 & 1 \\ 1 & 3 & -1 \end{bmatrix}$$

The eigenvalues are $\lambda_i = 1, -2, 3$ (see Problem 7.39).

$$\mathbf{Z}_1 = \frac{(\mathbf{A}+2\mathbf{I})(\mathbf{A}-3\mathbf{I})}{(1+2)(1-3)} = -\frac{1}{6}\begin{bmatrix} 4 & -2 & 3 \\ 1 & 3 & 1 \\ 1 & 3 & 1 \end{bmatrix}\begin{bmatrix} -1 & -2 & 3 \\ 1 & -2 & 1 \\ 1 & 3 & -4 \end{bmatrix} = -\frac{1}{6}\begin{bmatrix} -3 & 5 & -2 \\ 3 & -5 & 2 \\ 3 & -5 & 2 \end{bmatrix}$$

$$\mathbf{Z}_2 = \frac{(\mathbf{A}-\mathbf{I})(\mathbf{A}-3\mathbf{I})}{(-2-1)(-2-3)} = \frac{1}{15}\begin{bmatrix} 0 & 11 & -11 \\ 0 & 1 & -1 \\ 0 & -14 & 14 \end{bmatrix}$$

$$\mathbf{Z}_3 = \frac{(\mathbf{A}-\mathbf{I})(\mathbf{A}+2\mathbf{I})}{(3-1)(3+2)} = \frac{1}{10}\begin{bmatrix} 5 & 1 & 4 \\ 5 & 1 & 4 \\ 5 & 1 & 4 \end{bmatrix}$$

Therefore,

$$e^{\mathbf{A}t} = \frac{1}{6}e^{t}\begin{bmatrix} 3 & -5 & 2 \\ -3 & 5 & -2 \\ -3 & 5 & -2 \end{bmatrix} + \frac{1}{15}e^{-2t}\begin{bmatrix} 0 & 11 & -11 \\ 0 & 1 & -1 \\ 0 & -14 & 14 \end{bmatrix} + \frac{1}{10}e^{3t}\begin{bmatrix} 5 & 1 & 4 \\ 5 & 1 & 4 \\ 5 & 1 & 4 \end{bmatrix}$$

23 Find $\mathbf{A}^k$ for the $\mathbf{A}$ matrix of the previous problem.

The expansion matrices $\mathbf{Z}_i$ depend only on $\mathbf{A}$, not on the particular function of $\mathbf{A}$ that is being computed. They are the same as in the previous problem, so with $f(\mathbf{A}) = \mathbf{A}^k$, the result is

$$\mathbf{A}^k = \frac{1}{6}\begin{bmatrix} 3 & -5 & 2 \\ -3 & 5 & -2 \\ -3 & 5 & -2 \end{bmatrix} + \frac{(-2)^k}{15}\begin{bmatrix} 0 & 11 & -11 \\ 0 & 1 & -1 \\ 0 & -14 & 14 \end{bmatrix} + \frac{(3)^k}{10}\begin{bmatrix} 5 & 1 & 4 \\ 5 & 1 & 4 \\ 5 & 1 & 4 \end{bmatrix}$$

24 Find the closed form expression for $e^{\mathbf{A}t}$ if $\mathbf{A} = \begin{bmatrix} -13 & -15 \\ -15 & -13 \end{bmatrix}$. Investigate using the truncated infinite series to get a numerical approximation, with $t = 1$.

The eigenvalues of $\mathbf{A}$ are $\lambda = 2$ and -28. The closed form answer is

$$e^{\mathbf{A}t} = e^{2t}\begin{bmatrix} 0.5 & -0.5 \\ -0.5 & 0.5 \end{bmatrix} + e^{-28t}\begin{bmatrix} 0.5 & 0.5 \\ 0.5 & 0.5 \end{bmatrix}$$

With $t = 1$ this becomes

$$e^{\mathbf{A}} \approx 0.5\begin{bmatrix} e^2 & -e^2 \\ -e^2 & e^2 \end{bmatrix} \approx \begin{bmatrix} 3.6945 & -3.6945 \\ -3.6945 & 3.6946 \end{bmatrix}$$

Using the first 30 terms in the infinite series gives the *bad* answer $e^{\mathbf{A}} = \begin{bmatrix} 2.435 & 2.435 \\ 2.435 & 2.435 \end{bmatrix} \times 10^{10}$. Keeping more terms does not help. The problem is that successive terms in the series depend upon $(\lambda_i t)^j/j!$. The alternating signs caused by $(-28)^j/j!$ cause loss of all significance, due to small differences of large numbers. A way to avoid this problem is to note that $e^{\mathbf{A}t} = [e^{\mathbf{A}t/m}]^m$. By picking some integer m such that $|\lambda|t/m$ is sufficiently small, the series for $e^{\mathbf{A}t/m}$ will converge rapidly. In fact the number of terms that need to be retained is K, where $[|\lambda|_{\max} t/m]^K/K! < \epsilon$. Note specifically that this depends on the largest-magnitude eigenvalue and on t/m. Once this truncated series is found, raising the answer to the mth power gives $e^{\mathbf{A}t}$. By selecting m as a power of 2, several successive doublings of $e^{\mathbf{A}t/m}$ efficiently give $e^{\mathbf{A}t}$. In this example it is found that $t/m = 0.25$ (i.e., $m = 4$) gives the correct answer, with about 20 terms retained in the series. Two successive doublings of $e^{\mathbf{A}t/m}$ are required at the end. If $t/m = 0.125$ ($m = 8$) is used, only 7 or 8 terms are needed in the series. This savings more than makes up for the extra matrix product (doubling) required to compute $e^{\mathbf{A}} = [e^{\mathbf{A}/8}]^8 = \{[(e^{\mathbf{A}/8})^2]^2\}^2$. (See Reference 3.)

PROBLEMS

8.25 Find the inverse of $\mathbf{A} = \begin{bmatrix} -1 & 2 & 0 \\ 1 & 1 & 0 \\ 2 & -1 & 2 \end{bmatrix}$ using the Cayley-Hamilton theorem.

8.26 Find $e^{\mathbf{A}t}$ if $\mathbf{A} = \begin{bmatrix} -1 & \frac{1}{2} \\ 0 & 1 \end{bmatrix}$.

8.27 Find $e^{\mathbf{A}t}$ with $\mathbf{A} = \begin{bmatrix} -3 & 2 \\ 0 & -3 \end{bmatrix}$.

8.28 Show that $e^{\begin{bmatrix} 0 & 1 \\ -3 & -4 \end{bmatrix}t} = \begin{bmatrix} \frac{1}{2}(3e^{-t} - e^{-3t}) & \frac{1}{2}(e^{-t} - e^{-3t}) \\ -\frac{3}{2}(e^{-t} - e^{-3t}) & \frac{1}{2}(3e^{-3t} - e^{-t}) \end{bmatrix}$.

8.29 Use $\mathbf{A} = \begin{bmatrix} 3 & 5 \\ \frac{16}{5} & 3 \end{bmatrix}$ to compute $e^{\mathbf{A}t}$.

8.30 Compute $e^{\mathbf{A}t}$ for $\mathbf{A} = \begin{bmatrix} -3 & 1 \\ 2 & -2 \end{bmatrix}$.

8.31 Find $\begin{bmatrix} \frac{1}{2} & -\frac{1}{2} & 1 \\ 0 & \frac{1}{2} & 2 \\ 0 & 0 & \frac{1}{2} \end{bmatrix}^k$.

8.32 Compute $e^{\mathbf{A}t}$ for $\mathbf{A} = \begin{bmatrix} -5 & -6 & 0 \\ 2 & 2 & 0 \\ 0 & 0 & -3 \end{bmatrix}$.

8.33 Find $e^{\mathbf{A}t}$ with $\mathbf{A} = \begin{bmatrix} 0 & 1 & 0 \\ 0 & 0 & 1 \\ 1 & -3 & 3 \end{bmatrix}$.

8.34 Find $e^{\mathbf{A}t}$ for $\mathbf{A} = \begin{bmatrix} -10 & 0 & -10 & 0 \\ 0 & -0.7 & 9 & 0 \\ 0 & -1 & -0.7 & 0 \\ 1 & 0 & 0 & 0 \end{bmatrix}$.

8.35 Find a closed-form expression for

$$\mathbf{A}^k = \begin{bmatrix} 0 & 1 & 0 \\ 0 & 0 & 1 \\ 0 & -0.5 & 1.5 \end{bmatrix}^k$$

8.36 Show that $\begin{bmatrix} 0 & 1 & 0 \\ 0 & 0 & 1 \\ 0.1653 & -0.9425 & 1.7085 \end{bmatrix}^k = (0.367955)^k \mathrm{E} + (0.665229)^k \mathrm{F} + (0.675317)^k \mathrm{G}$, where

$$\mathrm{E} = \begin{bmatrix} 4.916679^*\mathrm{E}+00 & -1.467152^*\mathrm{E}+01 & 1.094444^*\mathrm{E}+01 \\ 1.809116^*\mathrm{E}+00 & -5.398457^*\mathrm{E}+00 & 4.027060^*\mathrm{E}+00 \\ 6.656732^*\mathrm{E}-01 & -1.986389^*\mathrm{E}+00 & 1.481777^*\mathrm{E}+00 \end{bmatrix}$$

$$\mathbf{F} = \begin{bmatrix} -8.285841{*}E+01 & 3.478820{*}E+02 & -3.334531{*}E+02 \\ -5.511979{*}E+01 & 2.314211{*}E+02 & -2.218225{*}E+02 \\ -3.666726{*}E+01 & 1.539479{*}E+02 & -1.475626{*}E+02 \end{bmatrix}$$

$$\mathbf{G} = \begin{bmatrix} 7.894173{*}E+01 & -3.332105{*}E+02 & 3.225086{*}E+02 \\ 5.331067{*}E+01 & -2.250226{*}E+02 & 2.177954{*}E+02 \\ 3.600158{*}E+01 & -1.519615{*}E+02 & 1.470809{*}E+02 \end{bmatrix}$$

37 Show that $\begin{bmatrix} \frac{1}{2} & \frac{1}{2} & 0 \\ 0 & 1 & 0 \\ \frac{5}{6} & -\frac{13}{6} & -\frac{1}{3} \end{bmatrix}^k = \begin{bmatrix} (\frac{1}{2})^k & 1-(\frac{1}{2})^k & 0 \\ 0 & 1 & 0 \\ (\frac{1}{2})^k - (-\frac{1}{3})^k & 2(-\frac{1}{3})^k - (\frac{1}{2})^k - 1 & (-\frac{1}{3})^k \end{bmatrix}$.

38 The matrix $\mathbf{A} = \begin{bmatrix} 4 & -2 & 0 \\ 1 & 2 & 0 \\ 0 & 0 & 6 \end{bmatrix}$ is known to have eigenvalues $\lambda = \{6, 3+j, 3-j\}$ (see Problem 7.42). Find $\mathbf{A}_1 = e^{\mathbf{A}T}$ for $T = 0.2$, and then verify that the eigenvalues of $\mathbf{A}_1$ satisfy Frobenius' theorem.

39 Prove that $[e^{\mathbf{A}t}]^T = e^{\mathbf{F}t}$, where $\mathbf{F} = \mathbf{A}^T$.

40 The state of the unforced continuous-time system $\dot{\mathbf{x}} = \mathbf{A}\mathbf{x}$ is observed only at the periodic instants $t = 0, T, 2T, \ldots, kT, \ldots$. Use the exponent power law $[e^{\mathbf{A}T}]^k = e^{\mathbf{A}kT}$ to show that the initial-condition response can be described by either $\mathbf{x}(t_k) = e^{\mathbf{A}kT}\mathbf{x}(0)$ or $\mathbf{x}(k+1) = [\mathbf{A}_1]^k\mathbf{x}(0)$.

41 Use the modal decomposition of Sec. 8.6 to relate the approximate settling time T_s of a stable continuous-time system like the one in the previous problem to the eigenvalues of **A**. *Settling* is defined here as being within 2% of the final value.

42 If **A** has complex eigenvalues in the previous problem, the imaginary parts ω determine the modal frequencies of oscillation. This system is to be approximated by its states at discrete periodic times t_k (perhaps because of sampling instrumentation or digital computer control). What sampling time $T = t_{k+1} - t_k$ would you recommend? You may wish to reread Problem 2.21 at this point.

9

Analysis of Continuous- and Discrete-Time Linear State Equations

9.1 INTRODUCTION

The description of a physical system by a mathematical model was discussed in Chapter 1. The model often takes the form of a set of coupled differential equations of various orders. In other cases the original model takes the form of a set of discrete-time difference equations, as was the case when fitting empirical data with an ARMA model in Chapter 1. In Chapter 2, linear models were described by input-output transfer functions. The continuous-time case used Laplace transfer functions and the discrete-time case used Z-transforms. It was pointed out in Chapter 2 that the discrete-time model may represent an approximation of a continuous-time system or it may be necessitated because of sampling sensors or digital controllers. Chapter 3 developed state variable models for these same classes of physical systems, starting from either the coupled differential equation (difference equation) model or the Laplace transform (Z-transform) transfer function description. Figure 9.1 presents the modeling paradigm under discussion. It emphasizes that there are two distinct routes to the determination of a discrete-time approximate state model for a continuous-time system. This approximation problem is revisited in Sec. 9.8 because it is needed so frequently in control applications.

This chapter is devoted to the *solution* of the resulting vector-matrix state variable equations. It will be assumed here that the control input variable is known. In later chapters the typical control system design problem, the determination of the control inputs $\mathbf{u}(t)$ which will cause the states $\mathbf{x}(t)$ and/or outputs $\mathbf{y}(t)$ to behave as desired, will be discussed. The determination of the control for linear systems will be considered from two different points of view. In Chapter 13 the controller is designed to give specified closed-loop poles. This is the pole-placement problem. In Chapter 14 the controller is designed in order to minimize a quadratic cost function. This is a widely used subclass of optimal control theory. Before considering the controller design problem, the behavior of the system response for a known input should be

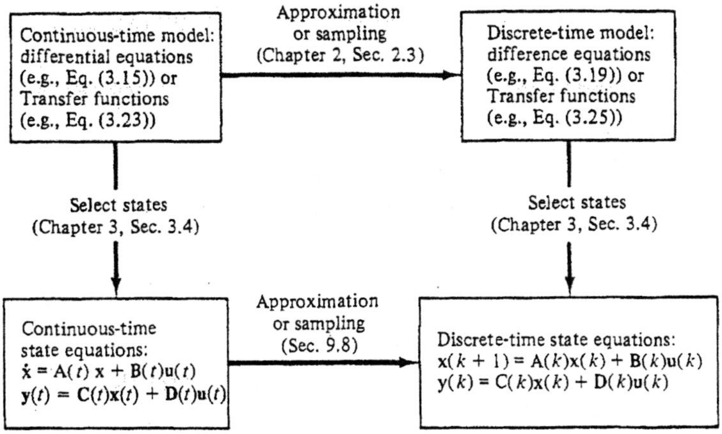

Figure 9.1 State variable modeling paradigm.

understood. The major effort is spent in solving the differential or difference equations for $\mathbf{x}(t)$ or $\mathbf{x}(k)$ in terms of $\mathbf{u}$. Then the output $\mathbf{y}$ is related to the state $\mathbf{x}$ and the input $\mathbf{u}$ in a simple algebraic fashion.

2 FIRST-ORDER SCALAR DIFFERENTIAL EQUATIONS

The familiar scalar differential equation

$$\dot{x} = a(t)x(t) + b(t)u(t) \tag{9.1}$$

is reviewed before considering the nth-order matrix case. When the input $u(t)$ is zero, the differential equation for x is said to be homogeneous. In this case,

$$\frac{dx}{dt} = a(t)x(t) \quad \text{or} \quad \frac{dx}{x} = a(t)\,dt$$

In the latter form the dependent variable x and the independent variable t are separated so that both sides of the equation represent an exact differential and can be integrated:

$$\int_{x(t_0)}^{x(t)} \frac{dx}{x} = \ln(x)\Big|_{x(t_0)}^{x(t)} = \int_{t_0}^{t} a(\tau)\,d\tau$$

or

$$\ln x(t) - \ln x(t_0) = \int_{t_0}^{t} a(\tau)\,d\tau$$

Using $\ln x(t) - \ln x(t_0) = \ln[x(t)/x(t_0)]$ and the fact that $e^{\ln x} = x$ gives

$$x(t) = x(t_0)e^{\int_{t_0}^{t} a(\tau)\,d\tau} \tag{9.2a}$$

In the particular case where $a(t) = a$ is constant, this reduces to

$$x(t) = x(t_0)e^{(t-t_0)a} \tag{9.2b}$$

As expected, the initial condition $x(t_0)$ must be specified before a unique solution $x(t)$ can be determined.

When the nonhomogeneous Eq. (*9.1*) is considered, a solution can still be obtained by reducing the equation to a form which can be easily integrated. One extra step is required first. Consider

$$\frac{d}{dt}[k(t)x(t)] = k(t)\dot{x}(t) + \dot{k}(t)x(t)$$

If Eq. (*9.1*) is multiplied by $k(t)$ and rearranged, the result is

$$k(t)\dot{x}(t) - k(t)a(t)x(t) = k(t)b(t)u(t)$$

The left-hand side can be made an exact differential provided a function $k(t)$ is selected that satisfies $\dot{k}(t) = -k(t)a(t)$. This requirement on $k(t)$ represents a first-order homogeneous equation of the type just considered. Its solution is $k(t) = k(t_0)e^{-\int_{t_0}^{t} a(\tau)\,d\tau}$. Using this, the nonhomogeneous equation for $x(t)$ can be written in terms of exact differentials,

$$d[k(t)x(t)] = k(t)b(t)u(t)\,dt$$

Carrying out the integration of both sides and solving for $x(t)$ gives

$$x(t) = \left[\frac{k(t_0)}{k(t)}\right]x(t_0) + \int_{t_0}^{t} \frac{k(\tau)}{k(t)}\, b(\tau)u(\tau)\,d\tau$$

Using the agreed-upon form for $k(t)$, the general solution becomes

$$x(t) = \{e^{\int_{t_0}^{t} a(\tau)\,d\tau}\}x(t_0) + \int_{t_0}^{t} e^{\int_{\tau}^{t} a(\zeta)\,d\zeta}\, b(\tau)u(\tau)\,d\tau \qquad (9.3a)$$

If the coefficient a is constant, this reduces to

$$x(t) = e^{(t-t_0)a}x(t_0) + \int_{t_0}^{t} e^{(t-\tau)a}\, b(\tau)u(\tau)\,d\tau \qquad (9.3b)$$

The last result can be derived directly using Laplace transforms and the convolution theorem (see Problem 9.1). A direct verification that this represents a solution of the differential equation is given in Problem 9.2.

When b is constant as well as a, Eq. (*9.3b*) indicates that $x(t)$ depends only on the time difference $t - t_0$ and so the starting time t_0 is often replaced by 0 for simplicity. When both a and b are constant, the system is said to be time-invariant because the response due to a given input is always the same regardless of the label attached to the starting time.

9.3 THE CONSTANT COEFFICIENT MATRIX CASE

The homogeneous set of n state equations

$$\dot{\mathbf{x}} = \mathbf{A}\mathbf{x}, \quad \mathbf{x}(t_0) \text{ given, } \mathbf{A} \text{ constant} \qquad (9.4)$$

has a solution which is completely analogous to the scalar result of equation (*9.2b*):

$$\mathbf{x}(t) = e^{(t-t_0)\mathbf{A}}\mathbf{x}(t_0) \tag{9.5}$$

There are several methods of verifying that this is a solution to the state equations (see Problems 9.3 and 9.4). First, note that the initial conditions are satisfied. That is,

$$\mathbf{x}(t_0) = e^{(t_0-t_0)\mathbf{A}}\mathbf{x}(t_0) = e^{[\mathbf{0}]}\mathbf{x}(t_0) = \mathbf{x}(t_0)$$

Differentiating both sides of Eq. (*9.5*) and using the result of Example 8.3, it is easily verified that $\dot{\mathbf{x}} = \mathbf{A}\mathbf{x}(t)$. Since Eq. (*9.5*) satisfies the initial conditions and the differential equation, it represents a unique solution (see page 20 of Reference 1) of Eq. (*9.4*).

The nonhomogeneous set of state equations is now considered. The system matrix $\mathbf{A}$ is still constant, but $\mathbf{B}(t)$ may be time-varying. Components of $\mathbf{B}(t)\mathbf{u}(t)$ are assumed to be piecewise continuous to guarantee a unique solution (see page 74 of Reference 1):

$$\dot{\mathbf{x}} = \mathbf{A}\mathbf{x} + \mathbf{B}(t)\mathbf{u}(t), \quad \mathbf{x}(t_0) \text{ given} \tag{9.6}$$

The technique used in solving the scalar equation is repeated with only minor dimensional modifications. Let $\mathbf{K}(t)$ be an $n \times n$ matrix. Premultiplying Eq. (*9.6*) by $\mathbf{K}(t)$ and rearranging gives

$$\mathbf{K}(t)\dot{\mathbf{x}}(t) - \mathbf{K}(t)\mathbf{A}\mathbf{x}(t) = \mathbf{K}(t)\mathbf{B}(t)\mathbf{u}(t)$$

Since $d[\mathbf{K}(t)\mathbf{x}(t)]/dt = \mathbf{K}\dot{\mathbf{x}} + \dot{\mathbf{K}}\mathbf{x}$, the left-hand side can be written as an exact (vector) differential provided $\dot{\mathbf{K}} = -\mathbf{K}(t)\mathbf{A}$. One such matrix is $\mathbf{K}(t) = e^{-(t-t_0)\mathbf{A}}$. Agreeing that this is the $\mathbf{K}$ matrix to be used, the differential equation can be written

$$d[\mathbf{K}(t)\mathbf{x}(t)] = \mathbf{K}(t)\mathbf{B}(t)\mathbf{u}(t)\,dt$$

Integration gives

$$\mathbf{K}(t)\mathbf{x}(t) - \mathbf{K}(t_0)\mathbf{x}(t_0) = \int_{t_0}^{t} \mathbf{K}(\tau)\mathbf{B}(\tau)\mathbf{u}(\tau)\,d\tau$$

The selected form for $\mathbf{K}$ always has an inverse, so

$$\mathbf{x}(t) = \mathbf{K}^{-1}(t)\mathbf{K}(t_0)\mathbf{x}(t_0) + \int_{t_0}^{t} \mathbf{K}^{-1}(t)\mathbf{K}(\tau)\mathbf{B}(\tau)\mathbf{u}(\tau)\,d\tau$$

or

$$\mathbf{x}(t) = e^{(t-t_0)\mathbf{A}}\mathbf{x}(t_0) + \int_{t_0}^{t} e^{(t-\tau)\mathbf{A}}\mathbf{B}(\tau)\mathbf{u}(\tau)\,d\tau \tag{9.7}$$

This represents the solution for any system equation in the form of Eq. (*9.6*). Note that it is composed of a term depending only on the initial state and a convolution integral involving the input but not the initial state. These two terms are known by various names such as the homogeneous solution and the particular integral, the force-free response and the forced response, the zero input response and the zero state response, etc.

9.4 SYSTEM MODES AND MODAL DECOMPOSITION [2, 3]

Equation (9.6) is considered again. It is emphasized that the matrix $\mathbf{A}$ is constant, but $\mathbf{B}(t)$ may be time-varying. Assume that the n eigenvalues λ_i and n independent vectors, either eigenvectors or generalized eigenvectors, have been found for the matrix $\mathbf{A}$. These vectors are denoted by ξ_i to avoid confusion with the state vector $\mathbf{x}$. Since the set $\{\xi_i\}$ is linearly independent, it can be used as a basis for the state space Σ. Thus at any given time t, the state $\mathbf{x}(t)$ can be expressed as

$$\mathbf{x}(t) = q_1(t)\xi_1 + q_2(t)\xi_2 + \cdots + q_n(t)\xi_n \tag{9.8}$$

The time variation of $\mathbf{x}$ is contained in the expansion coefficients q_i since $\mathbf{A}$, and hence the ξ_i, are constant. At any given time t, the vector $\mathbf{B}(t)\mathbf{u}(t) \in \Sigma$ and therefore it, too, can be expanded as

$$\mathbf{B}(t)\mathbf{u}(t) = \beta_1(t)\xi_1 + \beta_2(t)\xi_2 + \cdots + \beta_n(t)\xi_n$$

In fact, $\beta_i(t) = \langle \mathbf{r}_i, \mathbf{B}(t)\mathbf{u}(t)\rangle$, where $\{\mathbf{r}_i\}$ is the set of reciprocal basis vectors.

Using the above expansion, Eq. (9.6) becomes

$$\dot{q}_1\xi_1 + \dot{q}_2\xi_2 + \cdots + \dot{q}_n\xi_n = q_1\mathbf{A}\xi_1 + q_2\mathbf{A}\xi_2 + \cdots + q_n\mathbf{A}\xi_n + \beta_1\xi_1 + \beta_2\xi_2 + \cdots + \beta_n\xi_n$$

Assume for the moment that $\mathbf{A}$ is normal (see Sec. 5.12, page 184, and Problem 7.28, page 276) so that all ξ_i are eigenvectors rather than generalized eigenvectors. Then $\mathbf{A}\xi_i = \lambda_i\xi_i$, so that

$$(\dot{q}_1 - \lambda_1 q_1 - \beta_1)\xi_1 + (\dot{q}_2 - \lambda_2 q_2 - \beta_2)\xi_2 + \cdots + (\dot{q}_n - \lambda_n q_n - \beta_n)\xi_n = \mathbf{0}$$

Since the set $\{\xi_i\}$ is linearly independent, this requires that

$$\dot{q}_i = \lambda_i q_i + \beta_i \quad \text{for } i = 1, 2, \ldots, n$$

This demonstrates that when $\mathbf{A}$ is constant and has a full set of eigenvectors, the system is completely described by a set of n uncoupled scalar equations whose solutions are of the form

$$q_i(t) = e^{(t-t_0)\lambda_i}q_i(t_0) + \int_{t_0}^{t} e^{(t-\tau)\lambda_i}\beta_i(\tau)\,d\tau$$

Of course, $q_i(t_0) = \langle \mathbf{r}_i, \mathbf{x}(t_0)\rangle$. The state vector is given by

$$\mathbf{x}(t) = q_1(t)\xi_1 + q_2(t)\xi_2 + \cdots + q_n(t)\xi_n$$

The terms in this sum are called the system modes. The general response of a complicated system can be broken down into the sum of n simple modal responses.

It should be recognized that Eq. (9.8) can be written in terms of the modal matrix $\mathbf{M} = [\xi_1 \ldots \xi_n]$ as $\mathbf{x} = \mathbf{M}\mathbf{q}$, and as such, represents a change of basis. Using this notation, Eq. (9.6) is considered again:

$\dot{\mathbf{x}}$ becomes $\mathbf{M}\dot{\mathbf{q}}$, since $\mathbf{M}$ is constant

$\mathbf{A}\mathbf{x}$ becomes $\mathbf{A}\mathbf{M}\mathbf{q}$ and $\mathbf{B}\mathbf{u}$ remains unchanged

Therefore,

$$\mathbf{M}\dot{\mathbf{q}} = \mathbf{AMq} + \mathbf{Bu}$$

or

$$\dot{\mathbf{q}} = \mathbf{M}^{-1}\mathbf{AMq} + \mathbf{M}^{-1}\mathbf{Bu} = \mathbf{Jq} + \mathbf{B}_n\mathbf{u}$$

where $\mathbf{B}_n \triangleq \mathbf{M}^{-1}\mathbf{B}$ and the assumption regarding a full set of eigenvectors is dropped. $\mathbf{J}$ is the Jordan canonical form (or the diagonal matrix $\mathbf{\Lambda}$ in many cases). If the same change of basis is used in the output equation, then the system is described by the pair of *normal form* equations

$$\dot{\mathbf{q}} = \mathbf{Jq} + \mathbf{B}_n\mathbf{u} \tag{9.9}$$

$$\mathbf{y} = \mathbf{C}_n\mathbf{q} + \mathbf{Du} \tag{9.10}$$

where $\mathbf{C}_n \triangleq \mathbf{CM}$. One advantage of the normal form is that the state equations are as nearly uncoupled as possible. Each component of $\mathbf{q}$ is coupled to at most one other component because of the nature of the Jordan form matrix $\mathbf{J}$. The solution for Eq. (9.9) can be written as

$$\mathbf{q}(t) = e^{(t-t_0)\mathbf{J}}\mathbf{q}(t_0) + \int_{t_0}^{t} e^{(t-\tau)\mathbf{J}}\mathbf{B}_n(\tau)\mathbf{u}(\tau)\,d\tau$$

Relating this to the original state vector gives

$$\mathbf{x}(t) = \mathbf{Mq}(t) = \mathbf{M}e^{(t-t_0)\mathbf{J}}\mathbf{M}^{-1}\mathbf{x}(t_0) + \int_{t_0}^{t} \mathbf{M}e^{(t-\tau)\mathbf{J}}\mathbf{M}^{-1}\mathbf{B}(\tau)\mathbf{u}(\tau)\,d\tau \tag{9.11}$$

The preceding equation used the fact that $\mathbf{M}^{-1}$ is the matrix of transposed reciprocal basis vectors, which means that

$$\mathbf{q}(t_0) = \mathbf{M}^{-1}\mathbf{x}(t_0)$$

Comparing Eqs. (9.11) and (9.7) shows that

$$e^{(t-t_0)\mathbf{A}} = \mathbf{M}e^{(t-t_0)\mathbf{J}}\mathbf{M}^{-1}$$

a result given earlier in Chapter 8.

Modal decomposition is useful because of the insight it gives regarding the intrinsic properties of the system. The properties of controllability, observability, stabilizability, and detectability (Chapter 11) are more easily understood and evaluated. The stability properties (Chapter 10) of the system are also more clearly revealed. Modal decomposition provides a simple geometrical picture for the motion of the state vector versus time. By retaining only the dominant modes, a high-order system can be approximated by a lower-order system.

It should be kept in mind that the modal decomposition technique is useful only if $\mathbf{A}$, and thus ξ_i, λ_i are constant. It is the invariance of the vector parts of $\mathbf{x}(t)$, that is, the ξ_i terms, that gives value to the method. If the modal matrix were time-varying and had to be continually reevaluated, most of the advantages of modal decomposition would be lost.

EXAMPLE 9.1 A system is described by

$$\begin{bmatrix}\dot{x}_1\\ \dot{x}_2\end{bmatrix}=\begin{bmatrix}0 & 1\\ 8 & -2\end{bmatrix}\begin{bmatrix}x_1\\ x_2\end{bmatrix}+\begin{bmatrix}1\\ 1\end{bmatrix}u$$

$$y(t)=[4 \quad 1]\mathbf{x}(t)$$

The initial conditions are $\mathbf{x}(0)=[1 \quad -4]^T$. Assume that $u(t)=0$ and analyze this system.

With $\mathbf{A}=\begin{bmatrix}0 & 1\\ 8 & -2\end{bmatrix}$, the eigenvalues are $\lambda_1=-4, \lambda_2=2$.

The eigenvectors are $\xi_1=[1 \quad -4],^T \xi_2=[1 \quad 2]^T$.

The modal matrix and its inverse are $\mathbf{M}=\begin{bmatrix}1 & 1\\ -4 & 2\end{bmatrix}, \mathbf{M}^{-1}=\frac{1}{6}\begin{bmatrix}2 & -1\\ 4 & 1\end{bmatrix}$.

Any one of several methods gives

$$e^{\mathbf{A}t}=\frac{1}{6}\begin{bmatrix}2e^{-4t}+4e^{2t} & -e^{-4t}+e^{2t}\\ -8e^{-4t}+8e^{2t} & 4e^{-4t}+2e^{2t}\end{bmatrix}$$

so the homogeneous solution is $\mathbf{x}(t)=e^{\mathbf{A}t}\mathbf{x}(0)=[e^{-4t} \quad -4e^{-4t}]^T$, and the output is $y(t)=4e^{-4t}-4e^{-4t}=0$ for all t. ■

EXAMPLE 9.2 Modal decomposition is now applied in an attempt to gain insight into the unusual result of Example 9.1.

Since the eigenvalues are distinct, $\mathbf{M}^{-1}\mathbf{AM}=\mathbf{\Lambda}$ for this system, and Eq. (9.9) becomes

$$\begin{bmatrix}\dot{q}_1\\ \dot{q}_2\end{bmatrix}=\begin{bmatrix}-4 & 0\\ 0 & 2\end{bmatrix}\begin{bmatrix}q_1\\ q_2\end{bmatrix}+\begin{bmatrix}\frac{1}{6}\\ \frac{5}{6}\end{bmatrix}u$$

The initial conditions are $\mathbf{q}(0)=\mathbf{M}^{-1}\mathbf{x}(0)=[1 \quad 0]^T$. Equation (9.10) becomes $y=[0 \quad 6]\mathbf{q}$. The state vector $\mathbf{x}(t)$ can be written as the sum of two modes,

$$\mathbf{x}(t)=q_1(0)e^{-4t}\begin{bmatrix}1\\ -4\end{bmatrix}+q_2(0)e^{2t}\begin{bmatrix}1\\ 2\end{bmatrix}$$

The particular initial condition selected here has no component along the direction of mode 2, as evidenced by $q_2(0)=0$. Thus the second mode is not excited, since the input $u(t)$ has been assumed zero. The output of this system consists only of the second mode contribution, as evidenced by $\mathbf{C}_n=[0 \quad 6]$. Mode 1 contributes nothing to the output and mode 2 is not excited, so the output remains identically zero. ■

9.5 THE TIME-VARYING MATRIX CASE

The time-varying homogeneous state equations

$$\dot{\mathbf{x}}=\mathbf{A}(t)\mathbf{x} \tag{9.12}$$

are considered first. In order that this qualify as a valid state equation, it is required that there be a *unique* solution for every $\mathbf{x}(t_0)\in\Sigma$. This places some restriction on the kind of time variation allowed on the matrix $\mathbf{A}$. A *sufficient* condition for the existence of unique solutions is to require that all elements $a_{ij}(t)$ of $\mathbf{A}(t)$ be continuous. Weaker conditions may be found in textbooks on differential equations [1, 4].

Since $\dim(\Sigma) = n$, n linearly independent initial vectors $\mathbf{x}_i(t_0)$ can be found, and each one defines a unique solution of Eq. (*9.12*), called $\mathbf{x}_i(t)$, $t \geq t_0$. Define an $n \times n$ matrix $\mathbf{U}(t_0)$ with columns formed by the independent initial condition vectors $\mathbf{x}_i(t_0)$. (A particular set $\mathbf{U}(t_0) = I_n$ is sometimes used, but that restriction is unnecessary.) The n solutions corresponding to these initial conditions are used as the columns in forming an $n \times n$ matrix $\mathbf{U}(t) = [\mathbf{x}_1(t)]\ \ \mathbf{x}_2(t)\ \ \ldots\ \ \mathbf{x}_n(t)]$. Any matrix $\mathbf{U}(t)$ satisfying

$$\dot{\mathbf{U}}(t) = \mathbf{A}(t)\mathbf{U}(t) \tag{9.13}$$

is called a *fundamental solution matrix*, provided that $|\mathbf{U}(t_0)| \neq 0$. Assuming that the fundamental solution matrix is available, the solution to Eq. (*9.12*) with an arbitrary initial condition vector $\mathbf{x}(t_0)$ is

$$\mathbf{x}(t) = \mathbf{U}(t)\mathbf{U}^{-1}(t_0)\mathbf{x}(t_0) \tag{9.14}$$

This is easily verified. Checking initial conditions,

$$\mathbf{x}(t_0) = \mathbf{U}(t_0)\mathbf{U}^{-1}(t_0)\mathbf{x}(t_0) = \mathbf{I}_n\,\mathbf{x}(t_0) = \mathbf{x}(t_0)$$

Checking to see that this solution satisfies the differential equation,

$$\dot{\mathbf{x}}(t) = \dot{\mathbf{U}}(t)\mathbf{U}^{-1}(t_0)\mathbf{x}(t_0) = \mathbf{A}(t)\mathbf{U}(t)\mathbf{U}^{-1}(t_0)\mathbf{x}(t_0) = \mathbf{A}(t)\mathbf{x}(t)$$

Both the initial conditions and the differential equation are satisfied, so this represents the unique solution to the homogeneous problem.

The nonhomogeneous time-varying state equation is solved in an analogous manner to the scalar and constant matrix cases. That is, the equation is reduced to exact differentials so that it can be integrated. Preliminary to this, it is noted that $\mathbf{U}^{-1}(t)$ can be shown to exist for all $t \geq t_0$ and that

$$\mathbf{U}(t)\mathbf{U}^{-1}(t) = \mathbf{I}_n \quad \text{so that} \quad \frac{d}{dt}(\mathbf{U}(t)\mathbf{U}^{-1}(t)) = [\mathbf{0}]$$

or

$$\frac{d\mathbf{U}}{dt}\mathbf{U}^{-1} + \mathbf{U}\frac{d\mathbf{U}^{-1}}{dt} = [\mathbf{0}] \quad \text{or} \quad \frac{d\mathbf{U}^{-1}}{dt} = -\mathbf{U}^{-1}\frac{d\mathbf{U}}{dt}\mathbf{U}^{-1}$$

Therefore,

$$\frac{d\mathbf{U}^{-1}}{dt} = -\mathbf{U}^{-1}(t)\mathbf{A}(t) \tag{9.15}$$

Note that the matrix $\mathbf{K}(t)$ of Sec. 9.3 is an example of $\mathbf{U}^{-1}(t)$. Premultiplying the time-varying version of Eq. (*9.6*) by $\mathbf{U}^{-1}(t)$, postmultiplying Eq. (*9.15*) by $\mathbf{x}(t)$, and adding the results gives

$$\mathbf{U}^{-1}(t)\dot{\mathbf{x}} + \frac{d\mathbf{U}^{-1}}{dt}\mathbf{x}(t) = \mathbf{U}^{-1}(t)\mathbf{B}(t)\mathbf{u}(t)$$

or

$$\frac{d}{dt}[\mathbf{U}^{-1}(t)\mathbf{x}(t)] = \mathbf{U}^{-1}(t)\mathbf{B}(t)\mathbf{u}(t)$$

The nonhomogeneous solution is obtained by integrating both sides from t_0 to t, that is,

$$\mathbf{U}^{-1}(t)\mathbf{x}(t) - \mathbf{U}^{-1}(t_0)\mathbf{x}(t_0) = \int_{t_0}^{t} \mathbf{U}^{-1}(\tau)\mathbf{B}(\tau)\mathbf{u}(\tau)\, d\tau$$

or

$$\mathbf{x}(t) = \mathbf{U}(t)\mathbf{U}^{-1}(t_0)\mathbf{x}(t_0) + \int_{t_0}^{t} \mathbf{U}(t)\mathbf{U}^{-1}(\tau)\mathbf{B}(\tau)\mathbf{u}(\tau)\, d\tau \tag{9.16}$$

The result again takes the form of a term depending on the initial state and a convolution integral involving the input function. In fact, the first term is the same homogeneous solution given by Eq. (*9.14*). This result shows the *form* of the solution, but it may not be immediately useful. It assumes knowledge of the fundamental solution matrix $\mathbf{U}(t)$, and actually finding $\mathbf{U}$ has not yet been addressed.

9.6 THE TRANSITION MATRIX

The preceding results prompt the definition of an important matrix that can be associated with any linear system, namely, the *transition matrix:*

$$\mathbf{\Phi}(t, \tau) \triangleq \mathbf{U}(t)\mathbf{U}^{-1}(\tau) \tag{9.17}$$

This $n \times n$ matrix is a linear transformation or mapping of Σ onto itself. That is, in the absence of any input $\mathbf{u}(t)$, given the state $\mathbf{x}(\tau)$ at any time τ, the state at any other time t is given by the mapping

$$\mathbf{x}(t) = \mathbf{\Phi}(t, \tau)\mathbf{x}(\tau)$$

The mapping of $\mathbf{x}(\tau)$ into itself requires that

$$\mathbf{\Phi}(\tau, \tau) = \mathbf{I}_n \quad \text{for any } \tau \tag{9.18}$$

This is obviously true from Eq. (*9.17*). Differentiating $\mathbf{\Phi}(t, \tau)$ with respect to its first argument t gives

$$\frac{d\mathbf{\Phi}(t, \tau)}{dt} = \frac{d\mathbf{U}(t)}{dt}\mathbf{U}^{-1}(\tau) = \mathbf{A}(t)\mathbf{U}(t)\mathbf{U}^{-1}(\tau)$$

so

$$\frac{d\mathbf{\Phi}(t, \tau)}{dt} = \mathbf{A}(t)\mathbf{\Phi}(t, \tau) \tag{9.19}$$

The set of differential equations (*9.19*), along with the initial condition, Eq. (*9.18*), is often considered as the definition for $\mathbf{\Phi}(t, \tau)$.

Two other important properties of the transition matrix are the semigroup property, mentioned in Chapter 3 while defining state,

$$\Phi(t_2, t_0) = \Phi(t_2, t_1)\Phi(t_1, t_0) \quad \text{for any } t_0, t_1, t_2$$

and the relationship between Φ^{-1} and Φ:

$$\Phi^{-1}(t, t_0) = \Phi(t_0, t) \quad \text{for any } t_0, t$$

Both of these propeties are immediately obvious if the definition of Eq. (*9.17*) is considered.

Methods of Computing the Transition Matrix

If the matrix **A** is constant, then

$$\Phi(t, \tau) = e^{(t-\tau)\mathbf{A}} \qquad \text{(Compare Eqs. (9.7) and (9.16).)}$$

Therefore, all the methods of Chapter 8 are applicable for finding Φ, including

1. $\Phi(t, 0) = \mathcal{L}^{-1}\{[\mathbf{I}s - \mathbf{A}]^{-1}\}$. $\Phi(t, \tau)$ is then found by replacing t by $t - \tau$, since $\Phi(t, \tau) = \Phi(t - \tau, 0)$ when **A** is constant.
2. $\Phi(t, \tau) = \alpha_0\mathbf{I} + \alpha_1\mathbf{A} + \cdots + \alpha_{n-1}\mathbf{A}^{n-1}$, where $e^{\lambda_i(t-\tau)} = \alpha_0 + \alpha_1\lambda_i + \cdots + \alpha_{n-1}\lambda_i^{n-1}$ and, if some eigenvalues are repeated, derivatives of the above expression with respect to λ must be used.
3. $\Phi(t, \tau) = \mathbf{M}e^{\mathbf{J}(t-\tau)}\mathbf{M}^{-1}$, where **J** is the Jordan form (or the diagonal matrix Λ), and **M** is the modal matrix.
4. $\Phi(t, \tau) = \sum_{i=1}^{n} e^{\lambda_i(t-\tau)}\mathbf{Z}_i(\lambda)$, where the $n \times n$ matrices $\mathbf{Z}_i$ are defined in Problem 8.22.
5. $\Phi(t, \tau) \cong \mathbf{I} + \mathbf{A}(t-\tau) + \frac{1}{2}\mathbf{A}^2(t-\tau)^2 + \frac{1}{3!}\mathbf{A}^3(t-\tau)^3 + \cdots$. This infinite series can be truncated after a finite number of terms to obtain an approximation for the transition matrix. See Problem 9.10 for a more efficient computational form of this series.

A modification of method 1, using signal flow graphs to avoid the matrix inversion, can also be used. Since $\phi_{ij}(s) \triangleq \mathcal{L}\{\phi_{ij}(t, 0)\}$ is the transfer function from the input to the jth integrator to the output of the ith integrator, that is, the ith state variable x_i, Mason's gain rule [5] can be used to write the components $\phi_{ij}(s)$ directly. Inverse Laplace transformations then give the elements of $\Phi(t, 0)$.

When $\mathbf{A}(t)$ is time-varying, the choices for finding $\Phi(t, \tau)$ are more restricted:

1. *Computer solution of* $\dot{\Phi} = \mathbf{A}(t)\Phi$ *with* $\Phi(\tau, \tau) = \mathbf{I}$. This is expensive in terms of computer time if the transition matrix is required for all t and τ. It means solving the matrix differential equation many times, using a large set of different τ values as initial times.

2. *Let* $\mathbf{B}(t, \tau) = \int_\tau^t \mathbf{A}(\zeta)\,d\zeta$. Unlike the time-varying scalar case, $\Phi(t, \tau) \neq e^{\mathbf{B}(t,\tau)}$ unless $\mathbf{B}(t, \tau)$ and $\mathbf{A}(t)$ commute. Unfortunately, they generally do not commute, but two cases for which they do are when **A** is constant and when **A** is diagonal. Whenever $\mathbf{BA} = \mathbf{AB}$, any method may be used for computing $\Phi(t, \tau) = e^{\mathbf{B}(t,\tau)}$.

3. *Successive approximations may be used to obtain an approximate transition matrix, as derived in Problem 9.5:*

$$\Phi(t, t_0) = \mathbf{I}_n + \int_{t_0}^{t} \mathbf{A}(\tau_0)\,d\tau_0 + \int_{t_0}^{t} \mathbf{A}(\tau_0) \int_{t_0}^{\tau_0} \mathbf{A}(\tau_1)\,d\tau_1\,d\tau_0$$

$$+ \int_{t_0}^{t} \mathbf{A}(\tau_0) \int_{t_0}^{\tau_0} \mathbf{A}(\tau_1) \int_{t_0}^{\tau_1} \mathbf{A}(\tau_2)\,d\tau_2\,d\tau_1\,d\tau_0 + \cdots$$

4. *In some special cases closed form solutions to the equations may be possible.*

In many cases it is necessary or desirable to select a set of discrete time points, t_k, such that $\mathbf{A}(t)$ can be approximated by a constant matrix over each interval $[t_k, t_{k+1}]$. Then a set of difference equations can be used to describe the state of the system at these discrete times. The approximating difference equation is derived in Sec. 9.8 and Problem 9.10. Solutions of this type of equation are discussed in Sec. 9.9.

9.7 SUMMARY OF CONTINUOUS-TIME LINEAR SYSTEM SOLUTIONS

The most general state space description of a linear system is given by

$$\dot{\mathbf{x}}(t) = \mathbf{A}(t)\mathbf{x}(t) + \mathbf{B}(t)\mathbf{u}(t)$$

$$\mathbf{y}(t) = \mathbf{C}(t)\mathbf{x}(t) + \mathbf{D}(t)\mathbf{u}(t)$$

The form of the solution for $\mathbf{x}(t)$ has been shown to be

$$\mathbf{x}(t) = \Phi(t, t_0)\mathbf{x}(t_0) + \int_{t_0}^{t} \Phi(t, \tau)\mathbf{B}(\tau)\mathbf{u}(\tau)\,d\tau \tag{9.20}$$

Equation (*9.20*) is the explicit form of the (linear system) transformation

$$\mathbf{x}(t) = \mathbf{g}(\mathbf{x}(t_0), \mathbf{u}(t), t_0, t)$$

introduced in Chapter 3 when defining state. When the system matrix $\mathbf{A}$ is constant, the transition matrix can always be found in closed form, although it may be tedious to do so for high-order systems. In the time-varying case numerical solutions or approximations must be relied upon. When considering certain questions, it is valuable to know that a solution exists in the stated form, even if it cannot be easily computed.

When the solution for $\mathbf{x}(t)$ is used, the expression for the output becomes

$$\mathbf{y}(t) = \mathbf{C}(t)\Phi(t, t_0)\mathbf{x}(t_0) + \int_{t_0}^{t} \mathbf{C}(t)\Phi(t, \tau)\mathbf{B}(\tau)\mathbf{u}(\tau)\,d\tau + \mathbf{D}(t)\mathbf{u}(t)$$

or

$$\mathbf{y}(t) = \mathbf{C}(t)\Phi(t, t_0)\mathbf{x}(t_0) + \int_{t_0}^{t} [\mathbf{C}(t)\Phi(t, \tau)\mathbf{B}(\tau) + \delta(t-\tau)\mathbf{D}(\tau)]\mathbf{u}(\tau)\,d\tau$$

The term inside the integral is an explicit expression for the weighting matrix $\mathbf{W}(t, \tau)$ used in the integral form of the input-output description for the system:

$$\mathbf{y}(t) = \int_{t_0}^{t} \mathbf{W}(t, \tau)\mathbf{u}(\tau)\,d\tau$$

It is seen that this input-output description is only valid when $\mathbf{x}(t_0) = \mathbf{0}$, the so-called zero state response case. This difficulty can be overcome by considering the initial state part of $\mathbf{y}(t)$ as having arisen because of some input between $t = -\infty$ and $t = t_0$. Then

$$\mathbf{y}(t) = \int_{-\infty}^{t} \mathbf{W}(t, \tau)\mathbf{u}(\tau)\, d\tau$$

8 DISCRETE-TIME MODELS OF CONTINUOUS-TIME SYSTEMS

A multivariable continuous-time system with r inputs $u_i(t)$ and m outputs $y_j(t)$ is considered. The mathematical model for such a system may originally be given in various forms, including transfer functions, coupled differential equations in y_j variables, or state variable format. These options correspond to the left side of Figure 9.1. There are at least three possible reasons for being interested in a discrete-time model of such a system, as suggested by Figure 9.2.

1. *Sampled outputs:* Sampling or time-shared sensors may provide output data only at discrete time points t_k. A scanning radar gives measurements to a target only once per scan cycle as the transmitted beam sweeps across the object being tracked. A digital voltmeter may be monitoring several signals via a multiplexed A/D input channel. No information is available between the sample times.

2. *Sampled inputs:* A digital controller calculates new values for the control inputs only once per control cycle. A zero-order hold converts the digital commands into a sequence of piecewise constant analog levels. These levels change only at the discrete time points t_k.

3. *Digital simulation:* Even though all the actual system input and output signals are continuous, a digital simulation may be desired to study the time response. This inherently involves discrete approximations of all the signals, and hence it is equivalent to a combination of sampled outputs and inputs. The goal is to pick a stepsize which is sufficiently small so that the continuous signals $\mathbf{u}$, $\mathbf{y}$, and $\mathbf{x}$ can be represented by piecewise constant approximations within an acceptable error.

Regardless of the reason for using the discrete model, the goal is to make the sampled values of system variables at times $t_0, t_1, \ldots, t_k, \ldots$ be an acceptably accurate representation of the corresponding continuous signals. Several methods of obtaining a discrete Z-transfer function from a continuous Laplace transfer function have been discussed in Chapter 2. See Problems 2.17 through 2.21 or the references on sampled-data control systems. The approximation of differential equations by difference

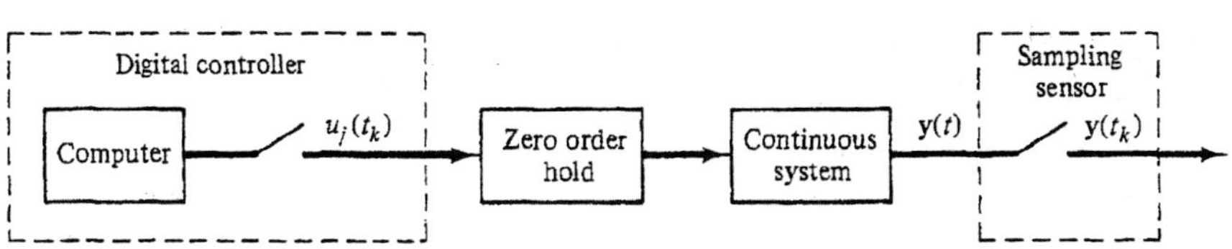

Figure 9.2

equations—or, equivalently, the various numerical integration techniques—is a standard topic in texts on numerical methods. Thus it is assumed that the top-level left-to-right transition in our modeling paradigm is understood. Chapter 3 gave details of finding state variable models from transfer functions and differential or difference equations. Thus both the continuous and sampled versions of the vertical transitions in Figure 9.1 have been explained. The last link in Figure 9.1, the lower-level horizontal transition from continuous-time to discrete-time state equations, is now discussed. Consider the system

$$\dot{\mathbf{x}} = \mathbf{A}(t)\mathbf{x}(t) + \mathbf{B}(t)\mathbf{u}(t) \quad \text{with } \mathbf{x}(t_0) \text{ known}$$

Assume that $\{t_0, t_1, \ldots, t_k, \ldots\}$ is a set of discrete time points sufficiently close together so that during any interval $[t_k, t_{k+1}]$ the input vector $\mathbf{u}(t)$ can be approximated by $\mathbf{u}(t_k)$. Note that if the inputs are processed through a zero-order hold as part of a digital controller, then they are automatically constant over the sampling interval. Equation (*9.20*) can be used to write the solution at t_{k+1} by treating $\mathbf{x}(t_k)$ as the initial condition.

$$\mathbf{x}(t_{k+1}) = \mathbf{\Phi}(t_{k+1}, t_k)\mathbf{x}(t_k) + \int_{t_k}^{t_{k+1}} \mathbf{\Phi}(t_{k+1}, \tau)\mathbf{B}(\tau)\, d\tau\, \mathbf{u}(t_k) \tag{9.21}$$

Equation (*9.21*) is an approximating difference equation for the states. Note that the input $\mathbf{u}(\tau)$ has been replaced by $\mathbf{u}(t_k)$ and taken out of the integral sign because of its piecewise constant behavior. If $\mathbf{A}(t)$ and/or $\mathbf{B}(t)$ are also approximately constant over $[t_k, t_{k+1}]$, further simplifications can be made. For example, if $\mathbf{A}$ is (approximately) constant, then $\mathbf{\Phi}(t_{k+1}, t_k) = e^{\mathbf{A}T}$, where $T = t_{k+1} - t_k$ and where $\mathbf{A}$ is the value of $\mathbf{A}(t_k)$. If $\mathbf{B}$ is (approximately) constant over $[t_{k+1}, t_k]$, then it can be removed from the integral sign. This leads to a commonly used approximation for the discrete state equations,

$$\mathbf{x}(k+1) = \mathbf{A}_1\mathbf{x}(k) + \mathbf{B}_1\mathbf{u}(k) \tag{9.22}$$

where $\mathbf{A}_1$ and $\mathbf{B}_1$ are used in the discrete model to distinguish them from the continuous model matrices $\mathbf{A}$ and $\mathbf{B}$. The relationships are

$$\mathbf{A}_1 = e^{\mathbf{A}T} \quad \text{and} \quad \mathbf{B}_1 = \int e^{\mathbf{A}(t_{k+1} - \tau)}\, d\tau \mathbf{B}$$

Even though these results have been referred to as discrete *approximations*, they are exact for constant-coefficient systems whose inputs pass through a zero-order hold, as is common in digital controllers. A further analytical simplification can be made in the special case where $\mathbf{A}(t_k)^{-1}$ exists by using results from Chapter 8 for integrating the exponential matrix,

$$\mathbf{B}_1 = [\mathbf{A}_1(t_k) - \mathbf{I}]\mathbf{A}^{-1}(t_k)\mathbf{B}(t_k)$$

Although the sample times t_k are usually equally spaced, this is not required by Eq. (*9.21*). Reevaluation of $\mathbf{A}_1$ and $\mathbf{B}_1$ would be necessary for each cycle of Eq. (*9.22*) in the variable sample-rate case. Problem 9.10 gives an efficient algorithm for evaluating $\mathbf{A}_1$ and $\mathbf{B}_1$ using a truncated infinite series approximation.

EXAMPLE 9.3 Consider the second-order system $\ddot{y} + 3\dot{y} + 2y = u(t)$, which has the transfer function $y(s)/u(s) = 1/[s^2 + 3s + 2] = 1/[(s+1)(s+2)]$. Use this model as the starting point in the upper left corner of the paradigm of Figure 9.1. Obtain approximate discrete state models by going around both transition paths in Figure 9.1. Use $T = t_{k+1} - t_k = 0.2$ seconds. A continuous-state model is selected first. Recall from Chapter 3 that there are many different methods for picking states, and each method will give a different model. The controllable canonical form of the state equations is

$$\begin{bmatrix} \dot{x}_1 \\ \dot{x}_2 \end{bmatrix} = \begin{bmatrix} 0 & 1 \\ -2 & -3 \end{bmatrix} \begin{bmatrix} x_1 \\ x_2 \end{bmatrix} + \begin{bmatrix} 0 \\ 1 \end{bmatrix} u(t)$$

with $y(t) = x_1(t)$. Since $\mathbf{A}$ and $\mathbf{B}$ are constant and it is assumed that $u(t)$ will be piecewise constant over each sample period T, the discrete matrices $\mathbf{A}_1$ and $\mathbf{B}_1$ can be calculated as shown before. A truncated infinite series is used in the numerical evaluation (see Problem 9.10) and gives

$$\mathbf{x}(k+1) = \begin{bmatrix} 0.967141 & 0.148411 \\ -0.296821 & 0.521909 \end{bmatrix} \mathbf{x}(k) + \begin{bmatrix} 0.016429 \\ 0.148411 \end{bmatrix} u(k) \tag{9.23}$$

The sampled output equation is

$$y(k) = [1 \quad 0]\mathbf{x}(k)$$

To find the second form of the discrete model, a discrete Z-transfer function is found first. As discussed in Chapter 2, there are several ways of performing this step, such as approximating derivatives by forward or backward differences. In this example the exact conversion of the zero-order hold–continuous system combination gives

$$\begin{aligned} G(z) &= (1 - z^{-1})Z\{G(s)/s\} \\ &= (0.01643z + 0.013452)/(z^2 - 1.48905z + 0.548811) \end{aligned} \tag{9.24}$$

The denominator of this transfer function factors into $(z - e^{-T})(z - e^{-2T})$, where $T = 0.2$ has been used. There are many ways of picking a state model from this transfer function, as discussed in Chapter 3. Here the observable canonical form is used because then the discrete state x_1 will be the same physical variable as was the continuous state x_1. To see this, the simulation diagrams for the two models should be drawn. If this system model represents an armature-controlled dc motor, for example, x_1 is the motor shaft angle in both the preceding continuous- and discrete-state models and in the model from the current approach,

$$\begin{bmatrix} x_1(k+1) \\ x_2(k+1) \end{bmatrix} = \begin{bmatrix} 1.48905 & 1 \\ -0.548811 & 0 \end{bmatrix} \begin{bmatrix} x_1(k) \\ x_2(k) \end{bmatrix} + \begin{bmatrix} 0.01643 \\ 0.013452 \end{bmatrix} u(k)$$

$$y(k) = [1 \quad 0]\mathbf{x}(k) \tag{9.25}$$

Equations (*9.23*) and (*9.25*) are both valid discrete models of the same system, with the same sampling rate and same assumptions, yet they are obviously different. Both models have the same input-output characteristics—namely, those described by the transfer function of Eq. (*9.24*). Their internal descriptions differ because different state variables were selected. Many other forms of the model could also be found. If a close correspondence with the continuous physical variables is desired, then the first procedure would be preferred. That is, if the continuous variable $x_2(t)$ represents the angular velocity of a motor shaft, then $x_2(k)$ of Eq. (*9.23*) is an approximation of that same variable. In Eq. (*9.25*) the meaning of x_2 is totally different and its time behavior, say for a step input, is completely different. ■

The sample rate used in developing a discrete model is often fixed by the sensor or controller cycle time. However, when a choice is still possible, such as early in the system design or in the case of digital simulation, a rough order of magnitude guide is useful. The piecewise constant approximations used before for **A** and **B** may be adequate if 6 to 10 samples occur per period of the highest-system modal frequency or per fastest time constant. The input frequency content also must be considered, and again the factor of 6 to 10 is a suggested starting range. Note that the theoretical lower limit on sample rate, the Nyquist rate of Sec. 2.5, is only two samples per period. Such slow sampling is never adequate in real systems. For a finer-tuned answer regarding sample rate, each case should be analyzed separately. The meaning of "acceptable" accuracy will be problem- and system-dependent. Furthermore, accuracy versus computer burden is a common design trade-off.

9.9 ANALYSIS OF CONSTANT COEFFICIENT DISCRETE-TIME STATE EQUATIONS

In this and subsequent sections the subscripts on the matrices $\mathbf{A}_1$ and $\mathbf{B}_1$ of the discrete state models will be dropped for convenience. It is assumed here that $\mathbf{A}(k)$ is constant, so the index k can be omitted. The homogeneous case is first considered:

$$\mathbf{x}(k+1) = \mathbf{A}\mathbf{x}(k)$$

The initial conditions $\mathbf{x}(0)$ are assumed known, so that $\mathbf{x}(1) = \mathbf{A}\mathbf{x}(0)$. Using this in the difference equation gives $\mathbf{x}(2) = \mathbf{A}\mathbf{x}(1) = \mathbf{A}^2\mathbf{x}(0)$. Continuing this process, the solution at a general time t_k is expressed in terms of $\mathbf{x}(0)$ as

$$\mathbf{x}(k) = \mathbf{A}^k\mathbf{x}(0) \tag{9.26}$$

The methods of Chapter 8 can be used to determine $\mathbf{A}^k$ as a general function of k, so that repeated matrix multiplications are unnecessary.

The nonhomogeneous case is now considered. A sequence of input vectors $\mathbf{u}(0)$, $\mathbf{u}(1)$, $\mathbf{u}(2)$, ... is given, as well as the initial conditions $\mathbf{x}(0)$. Then

$$\mathbf{x}(1) = \mathbf{A}\mathbf{x}(0) + \mathbf{B}(0)\mathbf{u}(0)$$

$$\mathbf{x}(2) = \mathbf{A}\mathbf{x}(1) + \mathbf{B}(1)\mathbf{u}(1) = \mathbf{A}^2\mathbf{x}(0) + \mathbf{A}\mathbf{B}(0)\mathbf{u}(0) + \mathbf{B}(1)\mathbf{u}(1)$$

$$\mathbf{x}(3) = \mathbf{A}\mathbf{x}(2) + \mathbf{B}(2)\mathbf{u}(2) = \mathbf{A}^3\mathbf{x}(0) + \mathbf{A}^2\mathbf{B}(0)\mathbf{u}(0) + \mathbf{A}\mathbf{B}(1)\mathbf{u}(1) + \mathbf{B}(2)\mathbf{u}(2)$$

At a general time t_k, this leads to

$$\mathbf{x}(k) = \mathbf{A}^k\mathbf{x}(0) + \sum_{j=0}^{k-1} \mathbf{A}^{k-1-j}\mathbf{B}(j)\mathbf{u}(j) \tag{9.27}$$

A change in the dummy summation index allows this result to be written in the alternative form

$$\mathbf{x}(k) = \mathbf{A}^k\mathbf{x}(0) + \sum_{j=1}^{k} \mathbf{A}^{k-j}\mathbf{B}(j-1)\mathbf{u}(j-1) \tag{9.28}$$

Either of these forms may be used. The close analogy with the continuous-time system results is made more apparent by using the definition for the discrete system transition matrix. Whenever **A** is constant, the discrete transition matrix is given by

$$\mathbf{\Phi}(k, j) = \mathbf{A}^{k-j}$$

Then

$$\mathbf{x}(k) = \mathbf{\Phi}(k, 0)\mathbf{x}(0) + \sum_{j=1}^{k} \mathbf{\Phi}(k, j)\mathbf{B}(j-1)\mathbf{u}(j-1) \tag{9.29}$$

and the only difference from the continuous result is the replacement of the convolution integration by a discrete summation.

10 MODAL DECOMPOSITION

The system matrix $\mathbf{A}$ is still considered constant, and its eigenvalues and eigenvectors (or generalized eigenvectors) are λ_i and ξ_i, respectively. Then, if the change of basis $\mathbf{x}(k) = \mathbf{M}\mathbf{q}(k)$ is used, where $\mathbf{M} = [\xi_1 \quad \xi_2 \quad \cdots \quad \xi_n]$, the state equations reduce to

$$\mathbf{M}\mathbf{q}(k+1) = \mathbf{A}\mathbf{M}\mathbf{q}(k) + \mathbf{B}(k)\mathbf{u}(k)$$

or

$$\mathbf{q}(k+1) = \mathbf{J}\mathbf{q}(k) + \mathbf{B}_n(k)\mathbf{u}(k) \tag{9.30}$$

and

$$\mathbf{y}(k) = \mathbf{C}(k)\mathbf{M}\mathbf{q}(k) + \mathbf{D}(k)\mathbf{u}(k)$$

or

$$\mathbf{y}(k) = \mathbf{C}_n(k)\mathbf{q}(k) + \mathbf{D}(k)\mathbf{u}(k) \tag{9.31}$$

where $\mathbf{J} = \mathbf{M}^{-1}\mathbf{A}\mathbf{M}$, $\mathbf{B}_n(k) = \mathbf{M}^{-1}\mathbf{B}(k)$, and $\mathbf{C}_n(k) = \mathbf{C}(k)\mathbf{M}$. Just as in the continuous case, the equations in $\mathbf{q}$ are as nearly uncoupled as possible and provide the same advantages. When $\mathbf{A}$ has a full set of eigenvectors, then $\mathbf{J}$ will be the diagonal matrix $\mathbf{\Lambda}$. A typical equation for the q_i components then takes the form

$$q_i(k+1) = \lambda_i q_i(k) + \langle \mathbf{r}_i, \mathbf{B}(k)\mathbf{u}(k) \rangle$$

The solution is

$$q_i(k) = \lambda_i^k q_i(0) + \sum_{j=1}^{k} \lambda_i^{k-j} \langle \mathbf{r}_i, \mathbf{B}(j-1)\mathbf{u}(j-1) \rangle$$

so that

$$\mathbf{q}(k) = \mathbf{\Lambda}^k \mathbf{q}(0) + \sum_{j=1}^{k} \mathbf{\Lambda}^{k-j}\mathbf{M}^{-1}\mathbf{B}(j-1)\mathbf{u}(j-1)$$

Using $\mathbf{x}(k) = \mathbf{M}\mathbf{q}(k)$ and $\mathbf{q}(0) = \mathbf{M}^{-1}\mathbf{x}(0)$ gives

$$\mathbf{x}(k) = \mathbf{M}\mathbf{\Lambda}^k\mathbf{M}^{-1}\mathbf{x}(0) + \sum_{j=1}^{k} \mathbf{M}\mathbf{\Lambda}^{k-j}\mathbf{M}^{-1}\mathbf{B}(j-1)\mathbf{u}(j-1)$$

This demonstrates again that $\mathbf{A}^k = \mathbf{M}\mathbf{\Lambda}^k\mathbf{M}^{-1}$ and provides a means of computing the transiton matrix, provided $\mathbf{A}$ has a full set of eigenvectors.

The modal decomposition technique provides geometrical insight into the system's structure. The behavior of $\mathbf{x}(k)$ versus the time index k can be represented as the vector sum of the eigenvectors ξ_i multiplied by the easily evaluated time-variable coefficients $q_i(k)$. That is,

$$\mathbf{x}(k) = \mathbf{M}\mathbf{q}(k) = \xi_1 q_1(k) + \xi_2 q_2(k) + \cdots + \xi_n q_n(k)$$

9.11 TIME-VARIABLE COEFFICIENTS

When $\mathbf{A}(k)$ is a time-variable matrix, then the solution technique of Sec. 9.9 must be modified slightly. Rather than the powers of $\mathbf{A}$, products of $\mathbf{A}$ evaluated at successive time points k are obtained. That is, the solution for $\mathbf{x}(k)$ at a general time t_k becomes

$$\mathbf{x}(k) = \mathbf{A}(k-1)\mathbf{A}(k-2)\cdots\mathbf{A}(0)\mathbf{x}(0) + \sum_{j=1}^{k}\left[\prod_{p=j}^{k-1}\mathbf{A}(p)\right]\mathbf{B}(j-1)\mathbf{u}(j-1) \qquad (9.32)$$

In Eq. (9.32) the notation $\prod_{p=j}^{k-1}\mathbf{A}(p)$ indicates that the product $\mathbf{A}(k-1)\mathbf{A}(k-2)\ldots$ $\mathbf{A}(j+1)\mathbf{A}(j)$. It is understood that if $j = k-1$, the product is just $\mathbf{A}(k-1)$ and if $j = k$, then $\prod_{p=k}^{k-1}\mathbf{A}(p) \triangleq \mathbf{I}_n$. The transition matrix for the time-varying case is given by

$$\mathbf{\Phi}(k, j) = \prod_{p=j}^{k-1}\mathbf{A}(p) \qquad (9.33)$$

When this definition is used, the solution for the time-variable case, Eq. (9.32), is exactly that given in Eq. (9.29). Evaluation of the transition matrix is much more cumbersome for the time-variable case, however.

9.12 THE DISCRETE-TIME TRANSITION MATRIX

The discrete-time transition matrix has been defined and used in the previous sections. The principal properties of this important matrix are summarized here. For the most part, the same properties hold for both the continuous-time and discrete-time transition matrices. In particular, the transition matrix $\mathbf{\Phi}(k, j)$ represents the mapping of the state at time t_j into the state at time t_k provided the input sequence $\mathbf{u}$ is zero in that interval. It completely describes the unforced behavior of the state vector.

The semigroup property applies, that is, $\mathbf{\Phi}(k, m)\mathbf{\Phi}(m, j) = \mathbf{\Phi}(k, j)$ for any k, m, j satisfying $j \le m \le k$. The identity property holds, that is, $\mathbf{\Phi}(k, k) = \mathbf{I}_n$ for any time index k.

One major difference for the discrete-time transition matrix is that its inverse *need not* exist. When the inverse does exist, then the reversed time property holds:

$$\mathbf{\Phi}^{-1}(k, j) = \mathbf{\Phi}(j, k)$$

The inverse will exist if the discrete system is correctly derived as an approximation to a continuous system, since then $\mathbf{A}(k) = \mathbf{\Phi}(t_{k+1}, t_k)$ and $\mathbf{\Phi}(k, j) = \mathbf{\Phi}(t_k, t_j)$ and the continuous system transition matrix is always nonsingular.

In Problem 5.66, page 206, the formal adjoint for the discrete-time system operator was shown to be

$$\mathbf{w}(k-1) = \mathbf{A}^T(k)\mathbf{w}(k)$$

Notice the backward time indexing. If the transition matrix for this adjoint system is defined as $\boldsymbol{\Theta}(k, j)$, then many (but not all) of the relationships existing between $\boldsymbol{\Phi}$ and $\boldsymbol{\Theta}$ in the continuous case (Problems 9.16, 9.17, 9.19) will also be true in the discrete case. These properties are less useful in the discrete case and are not presented.

13 SUMMARY OF DISCRETE-TIME LINEAR SYSTEM SOLUTIONS

The most general solution for the linear discrete-time state Eq. (*9.22*) is given by Eq. (*9.29*), repeated here:

$$\mathbf{x}(k) = \boldsymbol{\Phi}(k, 0)\mathbf{x}(0) + \sum_{j=1}^{k} \boldsymbol{\Phi}(k, j)\mathbf{B}(j-1)\mathbf{u}(j-1) \tag{9.29}$$

The output is given by

$$\mathbf{y}(k) = \mathbf{C}(k)\boldsymbol{\Phi}(k, 0)\mathbf{x}(0) + \sum_{j=1}^{k} \mathbf{C}(k)\boldsymbol{\Phi}(k, j)\mathbf{B}(j-1)\mathbf{u}(j-1) + \mathbf{D}(k)\mathbf{u}(k)$$

When the discrete-time system matrix $\mathbf{A}$ is constant, then the transition matrix $\boldsymbol{\Phi}(k, j) = \mathbf{A}^{k-j}$ can be computed by any one of the several methods presented in Chapter 8. When $\mathbf{A}(k)$ is time-varying, no simple method exists for evaluating $\boldsymbol{\Phi}$ other than the direct calculation of the products indicated in Eq. (*9.33*). One hopes a digital computer would be available for this task.

REFERENCES

1. Coddington, E. A. and N. Levinson: *Theory of Ordinary Differential Equations*, McGraw-Hill, New York, 1955.
2. De Russo, P. M., R. J. Roy, and C. M. Close: *State Variables for Engineers*, John Wiley, New York, 1965.
3. Zadeh, L. A. and C. A. Desoer: *Linear System Theory, The State Space Approach*, McGraw-Hill, New York, 1963.
4. MacCamy, R. C. and V. J. Mizel: *Linear Analysis and Differential Equations*, MacMillan, New York, 1969.
5. Dorf, R. C.: *Modern Control Systems*, 5th ed., Addison-Wesley, Reading, Mass., 1989.

ILLUSTRATIVE PROBLEMS

Derivation and Verification of Solutions

.1 Use Laplace transforms to solve $\dot{x} = ax(t) + b(t)u(t)$, with the initial condition $x(0)$, and a is constant.

Transforming gives

$$sx(s) - x(0) = ax(s) + \mathcal{L}\{b(t)u(t)\} \quad \text{or} \quad x(s) = \frac{x(0)}{s-a} + \frac{\mathcal{L}\{b(t)u(t)\}}{s-a}$$

The inverse transform gives

$$x(t) = \mathcal{L}^{-1}\{x(s)\} = x(0)e^{at} + \mathcal{L}^{-1}\left\{\frac{\mathcal{L}\{b(t)u(t)\}}{s-a}\right\}$$

Using the convolution theorem $\mathcal{L}^{-1}\{g_1(s)g_2(s)\} = \int_0^t g_1(t-\tau)g_2(\tau)\,d\tau$ on the last term gives

$$\mathcal{L}^{-1}\left\{\frac{\mathcal{L}\{b(t)u(t)\}}{s-a}\right\} = \int_0^t e^{a(t-\tau)}b(\tau)u(\tau)\,d\tau$$

so that

$$x(t) = e^{at}x(0) + \int_0^t e^{a(t-\tau)}b(\tau)u(\tau)\,d\tau$$

If b is also constant, the system is time-invariant. The solution due to any other initial condition $x(t_0)$ at time t_0 is

$$x(t) = e^{a(t-t_0)}x(t_0) + \int_{t_0}^t e^{a(t-\tau)}b(\tau)u(\tau)\,d\tau$$

9.2 Verify that Eq. (*9.3a*) is the solution of Eq. (*9.1*).

Verification requires showing that the postulated solution $x(t)$ satisfies the initial condition and the differential equation.

Initial condition check: With $t = t_0$,

$$x(t_0) = \{e^{\int_{t_0}^{t_0} a(\tau)d\tau}\}x(t_0) + \int_{t_0}^{t_0} e^{\int_\tau^{t_0} a(\zeta)d\zeta}b(\tau)u(\tau)\,d\tau$$

Since $a(t)$ is continuous, $\int_{t_0}^{t_0} a(\tau)\,d\tau = 0$ so $e^{\int_{t_0}^{t_0} a(\tau)\,d\tau} = 1$.

Likewise, $\int_{t_0}^{t_0} e^{\int a(\zeta)d\zeta}b(\tau)u(\tau)\,d\tau = 0$ provided that $b(t)$ and $u(t)$ remain finite. Therefore, $x(t_0) = x(t_0)$.

Differential equation check: Differentiating the postulated solution gives

$$\dot{x}(t) = \frac{d}{dt}\left[\int_{t_0}^t a(\tau)\,d\tau\right]e^{\int_{t_0}^t a(\tau)\,d\tau}x(t_0) + \frac{d}{dt}\left[\int_{t_0}^t e^{\int_\tau^t a(\zeta)\,d\zeta}b(\tau)u(\tau)\,d\tau\right]$$

Using the general formula for differentiating an integral term,

$$\frac{d}{dt}\left[\int_{f(t)}^{g(t)} h(t,\tau)\,d\tau\right] = \int_{f(t)}^{g(t)} \frac{\partial h(t,\tau)}{\partial t}\,d\tau + h(t,g(t))\frac{dg}{dt} - h(t,f(t))\frac{df}{dt}$$

gives

$$\frac{d}{dt}\left[\int_{t_0}^t a(\tau)\,d\tau\right] = a(t)$$

$$\frac{d}{dt}\left[\int_{t_0}^t e^{\int_\tau^t a(\zeta)\,d\zeta}b(\tau)u(\tau)\,d\tau\right] = e^{\int_t^t a(\zeta)\,d\zeta}b(t)u(t) + \int_{t_0}^t \frac{\partial}{\partial t}[e^{\int_\tau^t a(\zeta)\,d\zeta}]b(\tau)u(\tau)\,d\tau$$

$$= b(t)u(t) + a(t)\int_{t_0}^t e^{\int_\tau^t a(\zeta)\,d\zeta}b(\tau)u(\tau)\,d\tau$$

so that

$$\dot{x}(t) = a(t)\left[e^{\int_{t_0}^t a(\tau)\,d\tau}x(t_0) + \int_{t_0}^t e^{\int_\tau^t a(\zeta)\,d\zeta}b(\tau)u(\tau)\,d\tau\right] + b(t)u(t)$$

The term in brackets is the postulated solution for $x(t)$, so $\dot{x} = a(t)x(t) + b(t)u(t)$ and the equation is satisfied.

.3 Solve $\dot{\mathbf{x}} = \mathbf{Ax} + \mathbf{B}(t)\mathbf{u}(t)$ using Laplace transforms.

Let the vector $\mathbf{B}(t)\mathbf{u}(t) = \mathbf{f}(t)$ for convenience. $s\mathbf{x}(s) - \mathbf{x}(0) = \mathbf{Ax}(s) + \mathbf{f}(s)$ or $[s\mathbf{I} - \mathbf{A}]\mathbf{x}(s) = \mathbf{x}(0) + \mathbf{f}(s)$ so that $\mathbf{x}(s) = [s\mathbf{I} - \mathbf{A}]^{-1}\mathbf{x}(0) + [s\mathbf{I} - \mathbf{A}]^{-1}\mathbf{f}(s)$. Taking the inverse transform gives $\mathbf{x}(t) = \mathcal{L}^{-1}\{[s\mathbf{I} - \mathbf{A}]^{-1}\}\mathbf{x}(0) + \mathcal{L}^{-1}\{[s\mathbf{I} - \mathbf{A}]^{-1}\} * \mathbf{f}(t)$, where $g(t) * f(t)$ is used to indicate convolution. Since $\mathcal{L}^{-1}\{[s\mathbf{I} - \mathbf{A}]^{-1}\} = e^{\mathbf{A}t}$,

$$\mathbf{x}(t) = e^{\mathbf{A}t}\mathbf{x}(0) + \int_0^t e^{\mathbf{A}(t-\tau)}\mathbf{f}(\tau)\,d\tau = e^{\mathbf{A}t}\mathbf{x}(0) + \int_0^t e^{\mathbf{A}(t-\tau)}\mathbf{B}(\tau)\mathbf{u}(\tau)\,d\tau$$

.4 Verify that Eq. (*9.7*) is the solution of Eq. (*9.6*).

Setting $t = t_0$ gives $\mathbf{x}(t_0) = e^{[0]}\mathbf{x}(t_0) = \mathbf{x}(t_0)$, assuming that $\mathbf{B}(t)\mathbf{u}(t)$ remains finite, i.e., contains no impulse functions.

Differentiating Eq. (*9.7*) gives

$$\dot{\mathbf{x}} = \mathbf{A}e^{\mathbf{A}(t-t_0)}\mathbf{x}(t_0) + \mathbf{B}(t)\mathbf{u}(t) + \mathbf{A}\int_{t_0}^t e^{\mathbf{A}(t-\tau)}\mathbf{B}(\tau)\mathbf{u}(\tau)\,d\tau$$

Using Eq. (*9.7*) reduces this to $\dot{\mathbf{x}} = \mathbf{Ax}(t) + \mathbf{B}(t)\mathbf{u}(t)$, indicating that Eq. (*9.7*) does satisfy the differential equation.

.5 Use a sequence of approximations for the solution of $\dot{\mathbf{x}} = \mathbf{A}(t)\mathbf{x}(t)$, $\mathbf{x}(t_0)$ given, and derive an approximation for the transition matrix $\mathbf{\Phi}(t, t_0)$.

As the zeroth approximation, let $\mathbf{x}^{(0)}(t) = \mathbf{x}(t_0)$. Then use the differential equation to find the next approximation $\mathbf{x}^{(1)}(t)$ by solving $\dot{\mathbf{x}}^{(1)}(t) = \mathbf{A}(t)\mathbf{x}^{(0)}(t)$. The solution is

$$\mathbf{x}^{(1)}(t) = \mathbf{x}(t_0) + \int_{t_0}^t \dot{\mathbf{x}}^{(1)}(\tau)\,d\tau = \left[\mathbf{I} + \int_{t_0}^t \mathbf{A}(\tau_0)\,d\tau_0\right]\mathbf{x}(t_0)$$

Let $\dot{\mathbf{x}}^{(2)} = \mathbf{A}(t)\mathbf{x}^{(1)}(t)$. Then

$$\begin{aligned}
\mathbf{x}^{(2)}(t) &= \mathbf{x}(t_0) + \int_{t_0}^t \dot{\mathbf{x}}^{(2)}(\tau)\,d\tau \\
&= \mathbf{x}(t_0) + \left[\int_{t_0}^t \mathbf{A}(\tau_0)\,d\tau_0 + \int_{t_0}^t \mathbf{A}(\tau_0)\int_{t_0}^{\tau_0} \mathbf{A}(\tau_1)\,d\tau_1\,d\tau_0\right]\mathbf{x}(t_0) \\
&= \left[\mathbf{I} + \int_{t_0}^t \mathbf{A}(\tau_0)\,d\tau_0 + \int_{t_0}^t \mathbf{A}(\tau_0)\int_{t_0}^{\tau_0} \mathbf{A}(\tau_1)\,d\tau_1\,d\tau_0\right]\mathbf{x}(t_0)
\end{aligned}$$

Continuing this procedure with $\dot{\mathbf{x}}^{(k+1)}(t) = \mathbf{A}(t)\mathbf{x}^{(k)}(t)$ leads to

$$\begin{aligned}
\mathbf{x}(t) \cong \Bigg[\mathbf{I} &+ \int_{t_0}^t \mathbf{A}(\tau_0)\,d\tau_0 + \int_{t_0}^t \mathbf{A}(\tau_0)\int_{t_0}^{\tau_0} \mathbf{A}(\tau_1)\,d\tau_1\,d\tau_0 \\
&+ \int_{t_0}^t \mathbf{A}(\tau_0)\int_{t_0}^{\tau_0} \mathbf{A}(\tau_1)\int_{t_0}^{\tau_1} \mathbf{A}(\tau_2)\,d\tau_2\,d\tau_1\,d\tau_0 + \cdots\Bigg]\mathbf{x}(t_0)
\end{aligned}$$

Truncating the series in the brackets after a finite number of terms gives an approximation for $\mathbf{\Phi}(t, t_0)$.

Miscellaneous Applications

.6 The satellite of Problem 3.16 is considered. If the two input torques are programmed to give $u_1(t) = (1/J_y)T_y(t) = C\sin\alpha t$, and $u_2(t) = (1/J_y)T_z(t) = C\cos\alpha t$, find the resultant time history of the state $\mathbf{x}(t) = [\omega_y \quad \omega_z]^T$. Use arbitrary initial conditions at time $t = 0$.

The state equations are $\dot{\mathbf{x}} = \begin{bmatrix} 0 & -\Omega \\ \Omega & 0 \end{bmatrix}\mathbf{x} + \begin{bmatrix} u_1(t) \\ u_2(t) \end{bmatrix}$. From Problem 8.17, the transition matrix is $\mathbf{\Phi}(t, 0) = \begin{bmatrix} \cos \Omega t & -\sin \Omega t \\ \sin \Omega t & \cos \Omega t \end{bmatrix}$ and

$$\mathbf{x}(t) = \mathbf{\Phi}(t, 0)\mathbf{x}(0) + \int_0^t \mathbf{\Phi}(t, \tau)\mathbf{u}(\tau)\, d\tau$$

The transition matrix properties can be used to write $\mathbf{\Phi}(t, \tau) = \mathbf{\Phi}(t, 0)\mathbf{\Phi}(0, \tau)$ and $\mathbf{\Phi}(0, \tau) = \mathbf{\Phi}(-\tau, 0)$, so

$$\mathbf{x}(t) = \mathbf{\Phi}(t, 0)\left\{\mathbf{x}(0) + C\int_0^t \begin{bmatrix} \cos \Omega\tau \sin \alpha\tau + \sin \Omega\tau \cos \alpha\tau \\ -\sin \Omega\tau \sin \alpha\tau + \cos \Omega\tau \cos \alpha\tau \end{bmatrix} d\tau\right\}$$

The trigonometric identities $\cos a \sin b + \sin a \cos b = \sin(a + b)$ and $\cos a \cos b - \sin a \sin b = \cos(a + b)$ are used inside the integral to give

$$\mathbf{x}(t) = \mathbf{\Phi}(t, 0)\left\{\mathbf{x}(0) + \frac{C}{\Omega + \alpha}\begin{bmatrix} 1 - \cos(\Omega + \alpha)t \\ \sin(\Omega + \alpha)t \end{bmatrix}\right\}$$

9.7 The input to the circuit of Figure 9.3 is an ideal current source $u(t)$. The output (and also the state) is the voltage across the capacitor $x(t)$. If

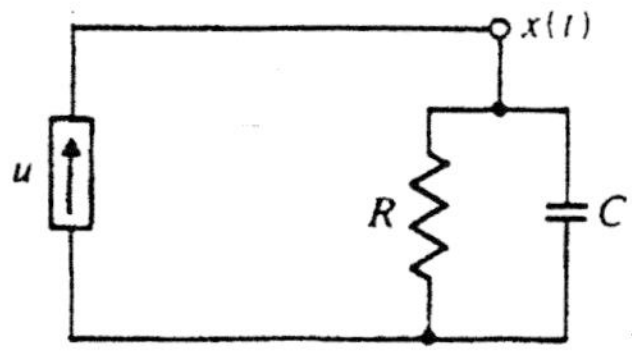

Figure 9.3

$$u(t) = e^{t/RC}\frac{10 - e^{-t_f/RC}x_0}{R \sinh(t_f/RC)}$$

and if $x(0) = x_0$, find the output $x(t_f)$ at some final time $t = t_f$.

The state equation is $\dot{x} = -x/RC + u/C$. The solution is

$$x(t) = e^{-t/RC}x_0 + \frac{1}{C}\int_0^t e^{-(t-\tau)/RC}u(\tau)\, d\tau$$

Letting $u(t) = Ke^{t/RC}$ for simplicity gives

$$x(t) = e^{-t/RC}x_0 + \frac{K}{C}e^{-t/RC}\int_0^t e^{2\tau/RC}\, d\tau$$

$$= e^{-t/RC}x_0 + \frac{K}{C}e^{-t/RC}\left\{\frac{RC}{2}[e^{2t/RC} - 1]\right\} = e^{-t/RC}x_0 + KR \sinh\left(\frac{t}{RC}\right)$$

Using $K = [10 - e^{-t_f/RC}x_0]/[R \sinh(t_f/RC)]$ and evaluating at $t = t_f$ gives $x(t_f) = 10$.

Although not proven here, the specified input $u(t)$ is the one which charges the capacitor from $x(0) = x_0$ to $x(t_f) = 10$ while minimizing the energy dissipated in R.

9.8 A system is described by $\begin{bmatrix} \dot{x}_1 \\ \dot{x}_2 \\ \dot{x}_3 \end{bmatrix} = \begin{bmatrix} -2 & -2 & 0 \\ 0 & 0 & 1 \\ 0 & -3 & -4 \end{bmatrix}\begin{bmatrix} x_1 \\ x_2 \\ x_3 \end{bmatrix} + \begin{bmatrix} 1 & 0 \\ 0 & 1 \\ 1 & 1 \end{bmatrix}\begin{bmatrix} u_1(t) \\ u_2(t) \end{bmatrix}$

(a) Find the change of variables $\mathbf{x} = \mathbf{M}\mathbf{q}$ which uncouples this system.

(b) If $\mathbf{x}(0) = [10 \quad 5 \quad 2]^T$ and if $\mathbf{u}(t) = [t \quad 1]^T$, find $\mathbf{x}(t)$.

(a) The modal matrix $\mathbf{M}$ must be found. $|\mathbf{A} - \mathbf{I}\lambda| = -(\lambda + 1)(\lambda + 2)(\lambda + 3)$, so the eigenvalues are $\lambda_i = -1, -2, -3$:

$$\operatorname{Adj}[\mathbf{A}-\mathbf{I}\lambda]=\begin{bmatrix}\lambda^2+4\lambda+3 & -2(\lambda+4) & -2\\ 0 & (\lambda+2)(\lambda+4) & 2+\lambda\\ 0 & -3(2+\lambda) & \lambda(2+\lambda)\end{bmatrix}$$

From this, the eigenvectors are $\xi_1=[-2\quad 1\quad -1]^T$, $\xi_2=[1\quad 0\quad 0]^T$, and $\xi_3=[-2\quad -1\quad 3]^T$,

so that $\mathbf{x}=\begin{bmatrix}-2 & 1 & -2\\ 1 & 0 & -1\\ -1 & 0 & 3\end{bmatrix}\mathbf{q}$ is the decoupling transformation. Using this substitution along

with $\mathbf{M}^{-1}=\begin{bmatrix}0 & \frac{3}{2} & \frac{1}{2}\\ 1 & 4 & 2\\ 0 & \frac{1}{2} & \frac{1}{2}\end{bmatrix}$ leads to

$$\dot{\mathbf{q}}=\begin{bmatrix}-1 & 0 & 0\\ 0 & -2 & 0\\ 0 & 0 & -3\end{bmatrix}\mathbf{q}+\begin{bmatrix}\frac{1}{2} & 2\\ 3 & 6\\ \frac{1}{2} & 1\end{bmatrix}\begin{bmatrix}u_1\\ u_2\end{bmatrix}$$

(b) The three uncoupled equations and their solutions are

$$\dot q_1=-q_1+\tfrac{1}{2}t+2\Rightarrow q_1(t)=e^{-t}q_1(0)+\tfrac{1}{2}t+\tfrac{3}{2}(1-e^{-t})$$

$$\dot q_2=-2q_2+3t+6\Rightarrow q_2(t)=e^{-2t}q_2(0)+\tfrac{3}{2}t+\tfrac{9}{4}(1-e^{-2t})$$

$$\dot q_3=-3q_3+\tfrac{1}{2}t+1\Rightarrow q_3(t)=e^{-3t}q_3(0)+\tfrac{1}{6}t+\tfrac{5}{18}(1-e^{-3t})$$

Since $\mathbf{q}(0)=\mathbf{M}^{-1}\mathbf{x}(0)=[\frac{17}{2}\quad 34\quad \frac{7}{2}]^T$, and since $\mathbf{x}(t)=\mathbf{M}\mathbf{q}(t)$, the solution is

$$\mathbf{x}(t)=\begin{bmatrix}-14e^{-t}+(\frac{127}{4})e^{-2t}-(\frac{58}{9})e^{-3t}+(\frac{1}{6})t-\frac{47}{36}\\ 7e^{-t}-(\frac{29}{9})e^{-3t}+(\frac{1}{3})t+\frac{11}{9}\\ -7e^{-t}+(\frac{29}{3})e^{-3t}-\frac{2}{3}\end{bmatrix}$$

.9 The motor-generator system of Problem 3.13, page 114, has been driving the load at a constant speed $\Omega=100$ rad/sec for some time. At time $t=0$ the input voltage $e_f(t)$ is suddenly removed, that is, $e_f(t)=0$ for $t\geq 0$. Find the resulting motion of the system. Assume the linear relation $e_g=K_g i_f$ and use the parameter values $b/J=1$, $K_m/J=2$, $K_m/(L_g+L_m)=2.5$, $(R_g+R_m)/(L_g+L_m)=7$, $K_g/(L_g+L_m)=4$, $R_f/L_f=5$, $L_f=1$.

The state equations are $\begin{bmatrix}\dot x_1\\ \dot x_2\\ \dot x_3\end{bmatrix}=\begin{bmatrix}-1 & 2 & 0\\ -2.5 & -7 & 4\\ 0 & 0 & -5\end{bmatrix}\begin{bmatrix}x_1\\ x_2\\ x_3\end{bmatrix}+\begin{bmatrix}0\\ 0\\ 1\end{bmatrix}u(t)$.

The desired solution is $\mathbf{x}(t)=\mathbf{\Phi}(t,0)\mathbf{x}(0)=e^{\mathbf{A}t}\mathbf{x}(0)$. The initial value of $\Omega=x_1(0)$ is 100. The initial values of the other state variables can be determined from the fact that the system was initially in steady-state, $\dot\Omega=0$ and $T(0)=b\Omega(0)=K_m i_m(0)$. From this, $x_2(0)=i_m(0)=(b/K_m)\Omega(0)=(b/J)(J/K_m)\Omega(0)=50$. Also $di_m/dt=0$ for $t\leq 0$, so $e_g-e_m=(R_g+R_m)i_m$ at $t=0$. Therefore, $e_g(0)=e_m(0)+(R_g+R_m)i_m(0)$ and $i_f(0)=x_3(0)=e_g(0)/K_g$. $x_3(0)=[K_m\Omega(0)+(R_g+R_m)i_m(0)]/K_g=2.5(100)/4+7(50)/4=150$ or $\mathbf{x}(0)=[100\quad 50\quad 150]^T$.

To find $\mathbf{\Phi}(t,0)$, the eigenvalues of $\mathbf{A}$ are found.

$$|\mathbf{A}-\mathbf{I}\lambda|=(-\lambda-5)\begin{vmatrix}-\lambda-1 & 2\\ -2.5 & -\lambda-7\end{vmatrix}=(-\lambda-5)(\lambda+2)(\lambda+6);$$

$$\lambda_1=-2,\ \lambda_2=-5,\ \lambda_3=-6$$

Using the Cayley-Hamilton remainder technique,

$$e^{\mathbf{A}t}=\alpha_0\mathbf{I}+\alpha_1\mathbf{A}+\alpha_2\mathbf{A}^2=\begin{bmatrix}\alpha_0-\alpha_1-4\alpha_2 & 2\alpha_1-16\alpha_2 & 8\alpha_2\\ -2.5\alpha_1+20\alpha_2 & \alpha_0-7\alpha_1+44\alpha_2 & 4\alpha_1-48\alpha_2\\ 0 & 0 & \alpha_0-5\alpha_1+25\alpha_2\end{bmatrix}$$

where

$$\left.\begin{aligned} e^{-2t} &= \alpha_0 - 2\alpha_1 + 4\alpha_2 \\ e^{-5t} &= \alpha_0 - 5\alpha_1 + 25\alpha_2 \\ e^{-6t} &= \alpha_0 - 6\alpha_1 + 36\alpha_2 \end{aligned}\right\} \Rightarrow \left\{\begin{aligned} \alpha_0 &= (5/2)e^{-2t} - 4e^{-5t} + (5/2)e^{-6t} \\ \alpha_1 &= (11/12)e^{-2t} - (8/3)e^{-5t} + (7/4)e^{-6t} \\ \alpha_2 &= (1/12)e^{-2t} - (1/3)e^{-5t} + (1/4)e^{-6t} \end{aligned}\right.$$

Using these gives

$$\Phi(t,0) = \begin{bmatrix} \frac{5}{4}e^{-2t} - \frac{1}{4}e^{-6t} & \frac{1}{2}e^{-2t} - \frac{1}{2}e^{-6t} & \frac{2}{3}e^{-2t} - \frac{8}{3}e^{-5t} + 2e^{-6t} \\ -\frac{5}{8}e^{-2t} + \frac{5}{8}e^{-6t} & -\frac{1}{4}e^{-2t} + \frac{5}{4}e^{-6t} & -\frac{1}{3}e^{-2t} + \frac{16}{3}e^{-5t} - 5e^{-6t} \\ 0 & 0 & e^{-5t} \end{bmatrix}$$

Then $\mathbf{x}(t) = \Phi(t,0)\mathbf{x}(0)$; and since $\Omega(t) = y = [1 \quad 0 \quad 0]\mathbf{x}(t)$, $\Omega(t) = 250e^{-2t} - 400e^{-5t} + 250e^{-6t}$.

9.10 A system is described by $\dot{\mathbf{x}} = \mathbf{Ax} + \mathbf{Bu}$, with $\mathbf{A}$ and $\mathbf{B}$ constant. Develop an efficient computational procedure for finding $\mathbf{x}(T)$, assuming $\mathbf{u}(t)$ is constant over $[0, T]$.

Setting $t_{k+1} = T$ and $t_k = 0$ in Eq. (*9.21*) gives the form of $\mathbf{x}(T)$

$$\mathbf{x}(t) = \Phi(T, 0)\mathbf{x}(0) + \int_0^T \Phi(T, \tau)\, d\tau\, \mathbf{Bu}$$

It is known that $\Phi(T, 0) = e^{\mathbf{A}T}$ and the series form is

$$\begin{aligned} \Phi(T, 0) &= \mathbf{I} + \mathbf{A}T + (\mathbf{A}T)^2/2 + (\mathbf{A}T)^3/3! + (\mathbf{A}T)^4/4! + \cdots \\ &= \mathbf{I} + \mathbf{A}T\{\mathbf{I} + \mathbf{A}T/2 + (\mathbf{A}T)^2/3! + (\mathbf{A}T)^3/4! + \cdots\} \\ &\;\vdots \\ &= \mathbf{I} + \mathbf{A}T\{\mathbf{I} + \mathbf{A}T/2[\mathbf{I} + \mathbf{A}T/3(\mathbf{I} + \mathbf{A}T/4(\mathbf{I} + \cdots (\mathbf{I} + \mathbf{A}T/N)))]\} \end{aligned}$$

This nested form for $\Phi(T, 0)$ does not require the direct computation of increasingly high powers of $\mathbf{A}T$ and therefore avoids many overflow and underflow problems. How many terms need to be retained depends upon $|\lambda_{max}|T$, where $|\lambda_{max}|$ is the largest magnitude eigenvalue of $\mathbf{A}$. This test is not normally used, however. On the Nth step in the nested sequence, the first neglected term would be $(\mathbf{A}T)^2/[(N)(N+1)]$, and this should be acceptably small compared with $\mathbf{A}T/N$.

The previous problem gave a result for $\mathbf{B}_1$ which depends on the existence of $\mathbf{A}^{-1}$. That is too restrictive in many cases, so another form which is better suited to machine computation is sought. Clearly,

$$\mathbf{B}_1 = \int_0^T e^{\mathbf{A}(T-\tau)}\, d\tau\, \mathbf{B} = -\int_T^0 e^{\mathbf{A}\xi}\, d\xi\, \mathbf{B} = \int_0^T e^{\mathbf{A}\xi}\, d\xi\, \mathbf{B}$$

Direct term-by-term integration of the exponential matrix gives

$$\begin{aligned} \mathbf{B}_1 &= \{\mathbf{I}T + \mathbf{A}T^2/2 + \mathbf{A}^2\mathbf{T}^3/3! + \cdots\}\mathbf{B} \\ &= T\{\mathbf{I} + \mathbf{A}T/2 + (\mathbf{A}T)^2/3! + \cdots\}\mathbf{B} \end{aligned}$$

Note that the series inside { } is the same as the one which appeared in the calculation of $\Phi(T, 0)$. Therefore, the same nested form is possible. Define this part of the solution as Ψ. That is,

$$\Psi = \mathbf{I} + \mathbf{A}T/2[\mathbf{I} + \mathbf{A}T/3(\mathbf{I} + \mathbf{A}T/4(\mathbf{I} + \ldots (\mathbf{I} + \mathbf{A}T/N)))]$$

This partial result is then used to obtain the desired approximations

$$\mathbf{B}_1 = T\Psi\mathbf{B}$$

$$\Phi = \mathbf{I} + \mathbf{A}T\Psi$$

These are widely used in obtaining discrete approximations to continuous-time systems.

1 A second-order system is described by $\ddot{x} + 2\dot{x} + 4x = u(t)$. Using $x = x_1$ and $x_2 = \dot{x}$ as states, find the state equations and evaluate the exact transition matrix $\mathbf{\Phi}(T, 0)$ and input matrix $\mathbf{B}_1$ using the results of Sec. 9.8 and Chapter 8 with $T = 0.2$. Then use results of Problem 9.10 to obtain approximate numerical results. Compare these.

The state equation is

$$\dot{\mathbf{x}} = \begin{bmatrix} 0 & 1 \\ -4 & -2 \end{bmatrix}\mathbf{x} + \begin{bmatrix} 0 \\ 1 \end{bmatrix}u$$

The exact state transition matrix is found to be

$$\mathbf{\Phi}(T, 0) = \begin{bmatrix} C+S & S \\ -4S & C-S \end{bmatrix}, \quad \text{where } C = e^{-T}\cos(\sqrt{3}T) \text{ and } S = e^{-T}\sin(\sqrt{3}T)/\sqrt{3}$$

Using $T = 0.2$ gives $\mathbf{\Phi} = \begin{bmatrix} 0.9306 & 0.1605 \\ -0.6420 & 0.6096 \end{bmatrix}$. Since $\mathbf{A}$ is nonsingular,

$$\mathbf{B}_1 = \mathbf{A}^{-1}[\mathbf{\Phi}(T, 0) - \mathbf{I}]\mathbf{B} = \begin{bmatrix} 0.017 \\ 0.160 \end{bmatrix}$$

The truncated series defined as $\mathbf{\Psi}$ in Problem 9.10 is now used. Note that for an Nth order approximation in T for $\mathbf{\Phi}$ and $\mathbf{B}_1$, an $(N-1)$st order approximation in $\mathbf{\Psi}$ is used.

Highest power of T	$\mathbf{\Psi}$	$\mathbf{\Phi}(T, 0)$	$\mathbf{B}_1$
1	$\begin{bmatrix} 1 & 0 \\ 0 & 1 \end{bmatrix}$	$\begin{bmatrix} 1 & 0.2 \\ -0.8 & 0.6 \end{bmatrix}$	$\begin{bmatrix} 0 \\ 0.2 \end{bmatrix}$
2	$\begin{bmatrix} 1 & 0.1 \\ -0.4 & 0.8 \end{bmatrix}$	$\begin{bmatrix} 0.92 & 0.16 \\ -0.64 & 0.60 \end{bmatrix}$	$\begin{bmatrix} 0.02 \\ 0.16 \end{bmatrix}$
3	$\begin{bmatrix} 0.9733 & 0.0867 \\ -0.3467 & 0.8 \end{bmatrix}$	$\begin{bmatrix} 0.931 & 0.160 \\ -0.64 & 0.611 \end{bmatrix}$	$\begin{bmatrix} 0.017 \\ 0.160 \end{bmatrix}$

Depending on the application, the second- or third-order approximation may suffice. The first-order approximation probably would not, because very large differences between $(1)^k$ and $(0.93)^k$ will quickly appear in $\mathbf{\Phi}(k, 0)$ as the approximate difference equations are solved over k time steps.

.12 Find a discrete-time approximate model for the system of Figure 9.4. Use $t_{k+1} - t_k = \Delta t = 1$ and approximate u_1 and u_2 as piecewise constant functions.

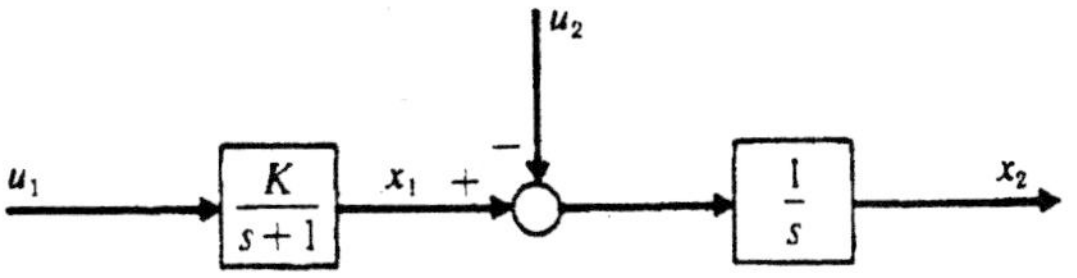

Figure 9.4

The continuous-time state equations are

$$\begin{bmatrix} \dot{x}_1 \\ \dot{x}_2 \end{bmatrix} = \begin{bmatrix} -1 & 0 \\ 1 & 0 \end{bmatrix}\begin{bmatrix} x_1 \\ x_2 \end{bmatrix} + \begin{bmatrix} K & 0 \\ 0 & -1 \end{bmatrix}\begin{bmatrix} u_1 \\ u_2 \end{bmatrix}$$

and the transition matrix is

$$\mathbf{\Phi}(t, 0) = e^{\mathbf{A}t} = \mathcal{L}^{-1}\{[s\mathbf{I} - \mathbf{A}]^{-1}\}, \qquad \mathbf{\Phi}(s) = \frac{\begin{bmatrix} s & 0 \\ 1 & s+1 \end{bmatrix}}{s(s+1)}$$

or

$$\Phi(t, 0) = \begin{bmatrix} e^{-t} & 0 \\ (1 - e^{-t}) & 1 \end{bmatrix}$$

The state at time $t_{k+1} = t_k + \Delta t$ can be written as

$$\mathbf{x}(t_{k+1}) = \Phi(t_{k+1}, t_k)\mathbf{x}(t_k) + \int_{t_k}^{t_{k+1}} \Phi(t_{k+1}, \tau)\, d\tau \begin{bmatrix} K & 0 \\ 0 & -1 \end{bmatrix} \begin{bmatrix} u_1(t_k) \\ u_2(t_k) \end{bmatrix}$$

But

$$\Phi(t_{k+1}, t_k) = \Phi(t_{k+1} - t_k, 0) = \begin{bmatrix} e^{-1} & 0 \\ 1 - e^{-1} & 1 \end{bmatrix} = \begin{bmatrix} 0.368 & 0 \\ 0.632 & 1 \end{bmatrix}$$

and

$$\int_{t_k}^{t_{k+1}} \Phi(t_{k+1}, \tau)\, d\tau = \begin{bmatrix} 1 - e^{-1} & 0 \\ e^{-1} & 1 \end{bmatrix} = \begin{bmatrix} 0.632 & 0 \\ 0.368 & 1 \end{bmatrix}$$

The approximating difference equation is

$$\begin{bmatrix} x_1(t_{k+1}) \\ x_2(t_{k+1}) \end{bmatrix} = \begin{bmatrix} 0.368 & 0 \\ 0.632 & 1 \end{bmatrix} \begin{bmatrix} x_1(t_k) \\ x_2(t_k) \end{bmatrix} + \begin{bmatrix} 0.632K & 0 \\ 0.368K & -1 \end{bmatrix} \begin{bmatrix} u_1(t_k) \\ u_2(t_k) \end{bmatrix}$$

9.13 The system of Figure 9.4 represents a simple model of a production and inventory control system. The input $u_1(t)$ represents the scheduled production rate, $x_1(t)$ represents the actual production rate, $u_2(t)$ represents the sales rate, and $x_2(t)$ represents the current inventory level. Suppose that the production schedule is selected as $u_1(t) = c - x_2(t)$, where c is the desired inventory level. This is a feedback control policy. The system is originally in equilibrium with $x_1(0)$ equal to the sales rate and $x_2(0) = c$. At time $t = 0$ the sales rate suddenly increases by 10%. That is, $u_2(t) = 1.1x_1(0)$ for $t \geq 0$. Find the resulting system response. Use $K = \frac{3}{16}$.

The simulation diagram for the feedback system is shown in Figure 9.5.

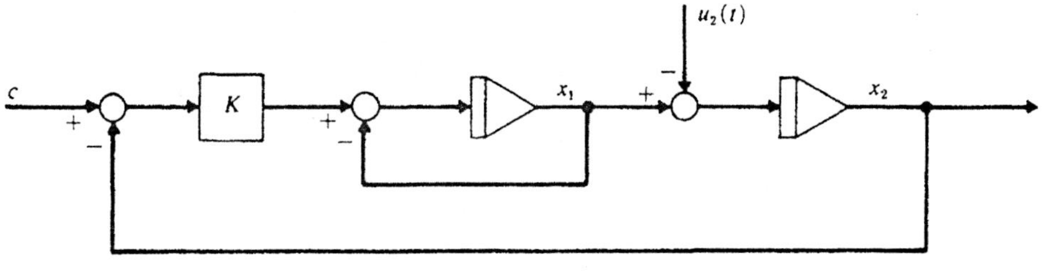

Figure 9.5

The state equations are $\dot{\mathbf{x}} = \begin{bmatrix} -1 & -K \\ 1 & 0 \end{bmatrix}\mathbf{x} + \begin{bmatrix} K & 0 \\ 0 & -1 \end{bmatrix}\begin{bmatrix} c \\ u_2 \end{bmatrix}$. The eigenvalues are determined from $|\mathbf{A} - \mathbf{I}\lambda| = \lambda^2 + \lambda + K = 0$ so that, with $K = \frac{3}{16}$, $\lambda_1 = -\frac{1}{4}$ and $\lambda_2 = -\frac{3}{4}$.

The transition matrix is

$$\Phi(t, 0) = e^{\mathbf{A}t} = \alpha_0 \mathbf{I} + \alpha_1 \mathbf{A} = \begin{bmatrix} \alpha_0 - \alpha_1 & -3\alpha_1/16 \\ \alpha_1 & \alpha_0 \end{bmatrix}$$

Using the eigenvalues to solve for α_0 and α_1 gives

$$\Phi(t, 0) = \begin{bmatrix} -\frac{1}{2}e^{-t/4} + \frac{3}{2}e^{-3t/4} & -\frac{3}{8}(e^{-t/4} - e^{-3t/4}) \\ 2(e^{-t/4} - e^{-3t/4}) & \frac{3}{2}e^{-t/4} - \frac{1}{2}e^{-3t/4} \end{bmatrix}$$

The solution is

$$\mathbf{x}(t) = \mathbf{\Phi}(t, 0)\begin{bmatrix} x_1(0) \\ c \end{bmatrix} + \int_0^t \mathbf{\Phi}(t, \tau)\, d\tau\, \mathbf{B}\begin{bmatrix} c \\ 1.1x_1(0) \end{bmatrix}$$

Using $\mathbf{\Phi}(t, \tau) = \mathbf{\Phi}(t - \tau, 0)$, carrying out the integration, and simplifying give

$$x_1(t) = x_1(0)\{1.1 - (4.3/2)e^{-t/4} + (4.1/2)e^{-3t/4}\}$$

$$x_2(t) = c + x_1(0)\{-17.6/3 + 8.6e^{-t/4} - (8.2/3)e^{-3t/4}\}$$

Additional Properties of the Transition Matrix

4 Assume that the eigenvectors of the constant system matrix $\mathbf{A}$ form a basis and show that $\mathbf{\Phi}(t, t_0) = \sum_{i=1}^{n} e^{\lambda_i(t - t_0)}\, \boldsymbol{\xi}_i\rangle\langle \mathbf{r}_i$, where λ_i, $\boldsymbol{\xi}_i$, and $\mathbf{r}_i$ are eigenvalues, eigenvectors, and reciprocal basis vectors of $\mathbf{A}$, respectively.

The modal decomposition developed in Sec. 9.4 led to the expression

$$\mathbf{x}(t) = q_1(t)\boldsymbol{\xi}_1 + q_2(t)\boldsymbol{\xi}_2 + \cdots + q_n(t)\boldsymbol{\xi}_n = \sum_{i=1}^{n} q_i(t)\boldsymbol{\xi}_i$$

For the homogeneous system $\dot{\mathbf{x}} = \mathbf{A}\mathbf{x}$,

$$q_i(t) = e^{\lambda_i(t - t_0)}\, q_i(t_0) = e^{\lambda_i(t - t_0)}\langle \mathbf{r}_i, \mathbf{x}(t_0)\rangle$$

so that

$$\mathbf{x}(t) = \sum_{i=1}^{n} e^{\lambda_i(t - t_0)}\langle \mathbf{r}_i, \mathbf{x}(t_0)\rangle\boldsymbol{\xi}_i = \left[\sum_{i=1}^{n} e^{\lambda_i(t - t_0)}\, \boldsymbol{\xi}_i\rangle\langle \mathbf{r}_i\right]\mathbf{x}(t_0)$$

Comparing this with the known solution $\mathbf{x}(t) = \mathbf{\Phi}(t, t_0)\mathbf{x}(t_0)$ gives the desired result.

15 For fixed times t_0 and t, the transition matrix is a transformation of the state space Σ onto itself. A linear transformation which possesses a full set of n linearly independent eigenvectors has a spectral representation

$$\mathbf{\Phi}(t, t_0) = \sum_{i=1}^{n} \gamma_i\, \boldsymbol{\eta}_i\rangle\langle \mathbf{v}_i$$

where γ_i, $\boldsymbol{\eta}_i$ are the eigenvalues and eigenvectors of $\mathbf{\Phi}(t, t_0)$ and $\mathbf{v}_i$ are reciprocal to $\boldsymbol{\eta}_i$. This is the result of Eq. (7.5), page 263, with notational changes. Comparing this with the previous problem, draw conclusions about the relationships between eigenvalues and eigenvectors of $\mathbf{A}$ and of $\mathbf{\Phi}(t, t_0)$.

The indicated comparison suggests the following relationship between eigenvalues, a result known as Frobenius' theorem. If $\lambda_1, \lambda_2, \ldots, \lambda_n$ are eigenvalues of the $n \times n$ matrix $\mathbf{A}$, and if $f(x)$ is a function which is analytic inside a circle in the complex plane which contains all the λ_i, then $f(\lambda_1), f(\lambda_2), \ldots, f(\lambda_n)$ are the eigenvalues of the matrix function $f(\mathbf{A})$. In the present case the eigenvalues of $\mathbf{\Phi}(t, t_0) = e^{\mathbf{A}(t - t_0)}$ are $\gamma_i = e^{\lambda_i(t - t_0)}$. Furthermore, it can be verified that the eigenvectors of $\mathbf{A}$ and of $\mathbf{\Phi}(t, t_0)$ are the same, that is, $\boldsymbol{\xi}_i = \boldsymbol{\eta}_i$.

The Adjoint Equations

16 The formal adjoint of the differential equation $\dot{\mathbf{x}} = \mathbf{A}\mathbf{x}$ is $\dot{\mathbf{y}} = -\mathbf{A}^T\mathbf{y}$ (see Problem 5.30). Let the transition matrix for the adjoint equation be $\mathbf{\Theta}(t, \tau)$. Show that $\mathbf{\Theta}(t, \tau) = \mathbf{\Phi}^T(\tau, t)$, where $\mathbf{\Phi}(t, \tau)$ is the transition matrix for the equation in $\mathbf{x}$.

Since $\Theta(t, \tau)$ is the transition matrix, it satisfies

$$\frac{d}{dt}[\Theta(t, \tau)] = -\mathbf{A}^T\Theta(t, \tau), \qquad \Theta(\tau, \tau) = \mathbf{I}$$

The desired result is established by showing that $\Phi^T(\tau, t)$ satisfies the same differential equation and initial conditions, since these equations define a unique solution. Since $\Phi(\tau, t) = \Phi^{-1}(t, \tau)$,

$$\frac{d}{dt}[\Phi(\tau, t)] = \frac{d}{dt}[\Phi^{-1}(t, \tau)] = -\Phi^{-1}(t, \tau)\frac{d}{dt}[\Phi(t, \tau)]\Phi^{-1}(t, \tau)$$

But $\frac{d}{dt}[\Phi(t, \tau)] = \mathbf{A}\Phi(t, \tau)$. Therefore, $\frac{d}{dt}[\Phi(\tau, t)] = -\Phi^{-1}(t, \tau)\mathbf{A} = -\Phi(\tau, t)\mathbf{A}$. Transposing shows that $\frac{d}{dt}[\Phi(\tau, t)]^T = -\mathbf{A}^T[\Phi(\tau, t)]^T$. This, plus the fact that $\Phi(\tau, \tau) = \mathbf{I}$, establishes that

$$\Theta(t, \tau) = \Phi^T(\tau, t)$$

It follows that the adjoint transition matrix can be expressed in terms of the fundamental matrix $\mathbf{U}(t)$ as $\Theta^T(t, \tau) = \mathbf{U}(\tau)\mathbf{U}^{-1}(t)$.

9.17 Suppose a simulation of the system $\dot{\mathbf{x}} = \mathbf{A}\mathbf{x}$ is available, as well as a simulation of the adjoint system $\dot{\mathbf{y}} = -\mathbf{A}^T\mathbf{y}$. Interpret the meaning of the vector functions $\mathbf{x}(t)$ and $\mathbf{y}(t)$ which are obtained if the initial conditions used are $\mathbf{x}(t_0) = [1 \quad 0 \quad 0 \quad \cdots \quad 0]^T$ and $\mathbf{y}(t_0) = [1 \quad 0 \quad 0 \quad \cdots \quad 0]^T$.

Since the solutions are $\mathbf{x}(t) = \Phi(t, t_0)\mathbf{x}(t_0)$ and $\mathbf{y}(t) = \Theta(t, t_0)\mathbf{y}(t_0)$, the first simulation generates the first column of $\Phi(t, t_0)$ for all $t \geq t_0$ and the second generates the first column of $\Theta(t, t_0)$ for all $t \geq t_0$. But the first column of $\Theta(t, t_0)$ equals the first *row* of $\Phi(t_0, t)$.

The adjoint system simulation provides a means of reversing the roles of t_0 and t. In a sense, a reversed time impulse response for the original system can be generated. The complete matrix $\Phi(t_0, t)$ can be obtained, one row at a time, by modifying the initial conditions for the adjoint simulation. This property has several uses (see pages 379–394 of Reference 2).

State Equations as Linear Transformations

9.18 Consider the operator $\mathcal{A}(\mathbf{x}) \triangleq [\mathbf{I}(d/dt) - \mathbf{A}]\mathbf{x}$ as a transformation on infinite dimensional function spaces. Discuss the form of the solution for $\mathcal{A}(\mathbf{x}) = \mathbf{B}\mathbf{u}(t)$ given by Eq. (*9.20*) in terms of the results of Problem 6.22, page 240.

Problem 6.22 indicates that if $\mathbf{x}_1$ is a nonzero solution of the homogeneous equation $\mathcal{A}(\mathbf{x}) = \mathbf{0}$, then the most general solution of $\mathcal{A}(\mathbf{x}) = \mathbf{B}\mathbf{u}$ takes the form $\mathbf{x} + \mathbf{x}_1$, where $\mathbf{x}$ is a solution to the nonhomogeneous equation. This is precisely the form of Eq. (*9.20*), with $\mathbf{x}_1 = \Phi(t, t_0)\mathbf{x}(t_0)$ being the homogeneous solution. A unique solution is not possible without specifying initial conditions.

9.19 Discuss the implications of Problem 6.23, page 240, in the context of linear state equations.

Problem 6.23 indicates that $\mathcal{A}(\mathbf{x}) = \mathbf{B}\mathbf{u}$ will have a solution for all $\mathbf{B}\mathbf{u}(t)$ if and only if the only solution to $\mathcal{A}^*(\mathbf{y}) = \mathbf{0}$ is the trivial solution. This poses an *apparent* contradiction, since nontrivial solutions to the adjoint equation have been discussed in Problem 9.17 and since solutions to $\mathcal{A}(\mathbf{x}) = \mathbf{B}\mathbf{u}(t)$ have been explicitly displayed in Eq. (*9.20*). The difficulty arises because of the differences between the adjoint transformation of Chapters 5 and 6 and the *formal* adjoint as used in this chapter.

A heuristic reconciliation is provided in a nonrigorous, formal manner. In Problem 5.30, page 199, it is shown that

$$\langle \mathbf{y}(t), \mathcal{A}(x(t)) \rangle = \mathbf{y}^T(t_f)\mathbf{x}(t_f) - \mathbf{y}^T(t_0)\mathbf{x}(t_0) - \left\langle \frac{d\mathbf{y}}{dt} + \mathbf{A}^T\mathbf{y}, \mathbf{x}(t) \right\rangle$$

The *formal* adjoint was defined by dropping the two boundary terms. These two terms automatically cancel if $\mathbf{u}(t) = \mathbf{0}$ (see Problem 5.31). Generally, they are nonzero and can be included as follows:

$$\mathbf{y}^T(t_f)\mathbf{x}(t_f) - \mathbf{y}^T(t_0)\mathbf{x}(t_0) - \int_{t_0}^{t_f}\left[\frac{d\mathbf{y}^T(\tau)}{dt} + \mathbf{y}^T(\tau)\mathbf{A}\right]\mathbf{x}(\tau)\,d\tau$$
$$= \int_{t_0}^{t_f}\left\{-\frac{d\mathbf{y}^T(\tau)}{dt} - \mathbf{y}^T(\tau)\mathbf{A} + \mathbf{y}^T(t_f)\delta(t_f-\tau) - \mathbf{y}^T(t_0)\delta(\tau-t_0)\right\}\mathbf{x}(\tau)\,d\tau$$

The term in brackets is the transpose of the adjoint transformation, that is,

$$\mathcal{A}^*(\mathbf{y}) = -\frac{d\mathbf{y}}{dt} - \mathbf{A}^T\mathbf{y}(t) + \mathbf{y}(t_f)\delta(t_f-t) - \mathbf{y}(t_0)\delta(t-t_0) \tag{1}$$

Then treating the two impulse terms as forcing terms, the solution of Eq. (*1*) is

$$\mathbf{y}(t) = \mathbf{\Theta}(t, t_0)\mathbf{y}(t_0) + \int_{t_0}^{t}\mathbf{\Theta}(t,\tau)[\mathbf{y}(t_f)\delta(t_f-\tau) - \mathbf{y}(t_0)\delta(\tau-t_0)]\,d\tau$$

But

$$\int_{t_0}^{t}\mathbf{\Theta}(t,\tau)\mathbf{y}(t_f)\delta(t_f-\tau)\,d\tau = \mathbf{0} \qquad \text{for all } t < t_f$$

and

$$\int_{t_0}^{t}\mathbf{\Theta}(t,\tau)\mathbf{y}(t_0)\delta(\tau-t_0)\,d\tau = \mathbf{\Theta}(t,t_0)\mathbf{y}(t_0)$$

because of the sifting property of the impulse function. Therefore, the solution is $\mathbf{y}(t) = \mathbf{0}$ for all $t < t_f$. At the final time an identity is obtained, $\mathbf{y}(t_f) = \mathbf{y}(t_f)$. The solution $\mathbf{y}(t)$ is therefore zero for all t except possibly at the single time $t = t_f$. Such a function will be considered $\mathbf{0}$, since its norm is zero (the Hilbert inner product norm, for example). Thus the only solution to $\mathcal{A}^*(\mathbf{y}) = \mathbf{0}$ is the trivial solution, and the results of Problem 6.23 are still true and do not lead to a contradiction.

Solution of Linear Discrete-Time State Equations

20 Solve for $\mathbf{x}(k)$ if

$$x_1(k+1) = \tfrac{1}{2}x_1(k) - \tfrac{1}{2}x_2(k) + x_3(k) \qquad x_3(k+1) = \tfrac{1}{2}x_3(k)$$
$$x_2(k+1) = \tfrac{1}{2}x_2(k) + 2x_3(k) \qquad x(0) = [2 \quad 4 \quad 6]^T$$

When put in matrix form $\mathbf{x}(k+1) = \mathbf{A}\mathbf{x}(k)$, the system matrix is $\mathbf{A} = \begin{bmatrix} \frac{1}{2} & -\frac{1}{2} & 1 \\ 0 & \frac{1}{2} & 2 \\ 0 & 0 & \frac{1}{2} \end{bmatrix}$. The solution for this homogeneous system is $\mathbf{x}(k) = \mathbf{A}^k\mathbf{x}(0)$. The matrix $\mathbf{A}^k$ was found in Problem 8.31, page 306. Using that result,

$$x_1(k) = 2(\tfrac{1}{2})^k - 4k(\tfrac{1}{2})^k + 6k(2-k)(\tfrac{1}{2})^{k-1}$$
$$x_2(k) = 4(\tfrac{1}{2})^k + 6k(\tfrac{1}{2})^{k-2}$$
$$x_3(k) = 6(\tfrac{1}{2})^k$$

21 Write the solution for the homogeneous discrete-time system

$$\begin{bmatrix} x_1(k+1) \\ x_2(k+1) \end{bmatrix} = \frac{1}{12}\begin{bmatrix} 5 & 1 \\ 1 & 5 \end{bmatrix}\begin{bmatrix} x_1(k) \\ x_2(k) \end{bmatrix}, \qquad \mathbf{x}(0) = \begin{bmatrix} 2 \\ 1 \end{bmatrix}$$

in the modal expansion form $\mathbf{x}(k) = \sum_{i=1}^{2} \langle \mathbf{r}_i, \mathbf{x}(0)\rangle \lambda_i^k\, \boldsymbol{\xi}_i$.

The matrix $\mathbf{A} = \frac{1}{12}\begin{bmatrix} 5 & 1 \\ 1 & 5 \end{bmatrix}$ has as its eigenvalues $\lambda_1 = \frac{1}{2}, \lambda_2 = \frac{1}{3}$. The eigenvectors are $\xi_1 = [1 \quad 1]^T$ and $\xi_2 = [-1 \quad 1]^T$. Thus $\mathbf{M} = \begin{bmatrix} 1 & -1 \\ 1 & 1 \end{bmatrix}$ and $\mathbf{M}^{-1} = \frac{1}{2}\begin{bmatrix} 1 & 1 \\ -1 & 1 \end{bmatrix}$.

The rows of $\mathbf{M}^{-1}$ give the reciprocal basis vectors $\mathbf{r}_1 = [\frac{1}{2} \quad \frac{1}{2}]^T$, $\mathbf{r}_2 = [-\frac{1}{2} \quad \frac{1}{2}]^T$. Since $\langle \mathbf{r}_1, \mathbf{x}(0)\rangle = \frac{3}{2}, \langle \mathbf{r}_2, \mathbf{x}(0)\rangle = -\frac{1}{2}$, the solution is

$$\mathbf{x}(k) = \frac{3}{2}\left(\frac{1}{2}\right)^k \begin{bmatrix} 1 \\ 1 \end{bmatrix} - \frac{1}{2}\left(\frac{1}{3}\right)^k \begin{bmatrix} -1 \\ 1 \end{bmatrix}$$

9.22 Consider the discrete-time system

$$\begin{bmatrix} x_1(k+1) \\ x_2(k+1) \end{bmatrix} = \begin{bmatrix} \frac{1}{2} & \frac{1}{8} \\ \frac{1}{8} & \frac{1}{2} \end{bmatrix} \begin{bmatrix} x_1(k) \\ x_2(k) \end{bmatrix} + \begin{bmatrix} 1 & 0 \\ 0 & 1 \end{bmatrix} \begin{bmatrix} u_1(k) \\ u_2(k) \end{bmatrix}$$

$$y(k) = x_1(k) + 2x_2(k)$$

Find $y(k)$ if $x_1(0) = -1, x_2(0) = 3$. The input $u_1(k)$ is obtained by sampling the ramp function t at times $t_0 = 0, t_1 = 1, \ldots, t_k = k$, and $u_2(k)$ is obtained by sampling e^{-t} at the same set of discrete times.

The transition matrix is first found:

$$\mathbf{A}^k = \alpha_0 \mathbf{I} + \alpha_1 \mathbf{A} = \begin{bmatrix} \alpha_0 + \frac{1}{2}\alpha_1 & \frac{1}{8}\alpha_1 \\ \frac{1}{8}\alpha_1 & \alpha_0 + \frac{1}{2}\alpha_1 \end{bmatrix}$$

The eigenvalues are required. $|\mathbf{A} - \mathbf{I}\lambda| = \lambda^2 - \lambda + \frac{15}{64}$ so $\lambda_1 = \frac{3}{8}, \lambda_2 = \frac{5}{8}$. Solving for α_0 and α_1,

$$\left.\begin{array}{l} (\frac{3}{8})^k = \alpha_0 + (\frac{3}{8})\alpha_1 \\ (\frac{5}{8})^k = \alpha_0 + (\frac{5}{8})\alpha_1 \end{array}\right\} \Rightarrow \left\{\begin{array}{l} 4[(\frac{5}{8})^k - (\frac{3}{8})^k] = \alpha_1 \\ (\frac{5}{2})(\frac{3}{8})^k - (\frac{3}{2})(\frac{5}{8})^k = \alpha_0 \end{array}\right.$$

Thus

$$\mathbf{A}^k = \mathbf{\Phi}(k, 0) = \begin{bmatrix} \frac{1}{2}[(\frac{5}{8})^k + (\frac{3}{8})^k] & \frac{1}{2}[(\frac{5}{8})^k - (\frac{3}{8})^k] \\ \frac{1}{2}[(\frac{5}{8})^k - (\frac{3}{8})^k] & \frac{1}{2}[(\frac{5}{8})^k + (\frac{3}{8})^k] \end{bmatrix}$$

Using this and the fact that $\mathbf{\Phi}(k, j) = \mathbf{A}^{k-j}$ gives

$$\mathbf{x}(k) = \begin{bmatrix} (\frac{5}{8})^k - 2(\frac{3}{8})^k \\ (\frac{5}{8})^k + 2(\frac{3}{8})^k \end{bmatrix} + \sum_{j=1}^{k} \begin{bmatrix} \frac{1}{2}[(\frac{5}{8})^{k-j} + (\frac{3}{8})^{k-j}] & \frac{1}{2}[(\frac{5}{8})^{k-j} - (\frac{3}{8})^{k-j}] \\ \frac{1}{2}[(\frac{5}{8})^{k-j} - (\frac{3}{8})^{k-j}] & \frac{1}{2}[(\frac{5}{8})^{k-j} + (\frac{3}{8})^{k-j}] \end{bmatrix} \begin{bmatrix} j-1 \\ e^{1-j} \end{bmatrix}$$

The output is $y(k) = [1 \quad 2]\mathbf{x}(k)$.

9.23 Express the following discrete-time state equations in normal form using the modal matrix in a change of basis:

$$\begin{bmatrix} x_1(k+1) \\ x_2(k+1) \\ x_3(k+1) \end{bmatrix} = \begin{bmatrix} \frac{1}{2} & \frac{1}{2} & 0 \\ 0 & 1 & 0 \\ \frac{5}{6} & -\frac{13}{6} & -\frac{1}{3} \end{bmatrix} \begin{bmatrix} x_1(k) \\ x_2(k) \\ x_3(k) \end{bmatrix} + \begin{bmatrix} 3 & 1 \\ 2 & 0 \\ -1 & 1 \end{bmatrix} \begin{bmatrix} u_1(k) \\ u_2(k) \end{bmatrix}$$

$$\begin{bmatrix} y_1(k) \\ y_2(k) \end{bmatrix} = \begin{bmatrix} -1 & 3 & 1 \\ 0 & 1 & 1 \end{bmatrix} \begin{bmatrix} x_1(k) \\ x_2(k) \\ x_3(k) \end{bmatrix}$$

When the equations are expressed as in Eqs. (*9.30*) and (*9.31*), they are said to be in normal form. To put the equations in this form, the eigenvalues and eigenvectors must be determined:

$$|\mathbf{A} - \mathbf{I}\lambda| = (1 - \lambda)(\tfrac{1}{2} - \lambda)(-\tfrac{1}{3} - \lambda)$$

The eigenvectors are $\xi_1 = [1 \quad 1 \quad -1]^T$, $\xi_2 = [1 \quad 0 \quad 1]^T$, $\xi_3 = [0 \quad 0 \quad 1]^T$, so that

$$\mathbf{M} = \begin{bmatrix} 1 & 1 & 0 \\ 1 & 0 & 0 \\ -1 & 1 & 1 \end{bmatrix} \quad \text{and} \quad \mathbf{M}^{-1} = \begin{bmatrix} 0 & 1 & 0 \\ 1 & -1 & 0 \\ -1 & 2 & 1 \end{bmatrix}$$

$$\mathbf{\Lambda} = \mathbf{M}^{-1}\mathbf{A}\mathbf{M} = \begin{bmatrix} 1 & 0 & 0 \\ 0 & \frac{1}{2} & 0 \\ 0 & 0 & -\frac{1}{3} \end{bmatrix}, \qquad \mathbf{B}_n = \mathbf{M}^{-1}\mathbf{B} = \begin{bmatrix} 2 & 0 \\ 1 & 1 \\ 0 & 0 \end{bmatrix},$$

$$\mathbf{C}_n = \mathbf{C}\mathbf{M} = \begin{bmatrix} 1 & 0 & 1 \\ 0 & 1 & 1 \end{bmatrix}$$

Collecting and using these results in $\mathbf{q}(k+1) = \mathbf{\Lambda}\mathbf{q}(k) + \mathbf{B}_n\mathbf{u}(k)$ and $\mathbf{y}(k) = \mathbf{C}_n\mathbf{q}(k)$ gives the normal form equations.

24 Find an expression for $\mathbf{y}(k)$, valid for all time t_k, for the system of Problem 9.23. Use $\mathbf{x}(0) = [1 \quad 2 \quad 1]^T$, $\mathbf{u}(k) = [k \quad -1-k]^T$.

Using the normal form equations,

$$\mathbf{q}(0) = \mathbf{M}^{-1}\mathbf{x}(0) = [2 \quad -1 \quad 4]^T$$

$$q_1(k) = q_1(0) + \sum_{j=1}^{k} 2u_1(j-1) = 2 + k(k-1)$$

$$q_2(k) = (\tfrac{1}{2})^k q_2(0) + \sum_{j=1}^{k} (\tfrac{1}{2})^{k-j}[u_1(j-1) + u_2(j-1)] = -(\tfrac{1}{2})^k\left[1 + \sum_{j=1}^{k} 2^j\right]$$

$$q_3(k) = (-\tfrac{1}{3})^k q_3(0) = 4(-\tfrac{1}{3})^k$$

Using the output equation $\mathbf{y}(k) = \mathbf{C}_n\mathbf{q}(k)$ yields

$$y_1(k) = q_1(k) + q_3(k) = 2 + k(k-1) + 4(-\tfrac{1}{3})^k$$

$$y_2(k) = q_2(k) + q_3(k) = (-\tfrac{1}{2})^k\left[1 + \sum_{j=1}^{k} 2^j\right] + 4(-\tfrac{1}{3})^k$$

25 Consider a homogeneous discrete-time system described by $\mathbf{x}(k+1) = \mathbf{A}\mathbf{x}(k)$.

(a) Show that if a nontrivial steady-state (constant) solution is to exist, the matrix $\mathbf{A}$ must have unity as an eigenvalue.

(b) Construct a 2×2 nondiagonal, symmetric matrix with this property and find the steady-state solution.

(a) A constant steady-state solution implies that for k sufficiently large, $\mathbf{x}(k+1) = \mathbf{x}(k)$. Call this solution $\mathbf{x}_e$. Then the difference equation requires that $\mathbf{x}_e = \mathbf{A}\mathbf{x}_e$. But this is just the eigenvalue equation $\mathbf{A}\mathbf{x}_e = \lambda\mathbf{x}_e$, with $\lambda = 1$.

(b) Let $\mathbf{A} = \begin{bmatrix} a_{11} & a_{12} \\ a_{12} & a_{22} \end{bmatrix}$. Then

$$|\mathbf{A} - \mathbf{I}\lambda| = \lambda^2 - (a_{11} + a_{22})\lambda + (a_{11}a_{22} - a_{12}^2) = 0$$

The roots are

$$\lambda_{1,2} = \tfrac{1}{2}(a_{11} + a_{22}) \pm \sqrt{[\tfrac{1}{2}(a_{11} + a_{22})]^2 - (a_{11}a_{22} - a_{12}^2)}$$

There are many possible solutions. One is obtained by arbitrarily setting $a_{11} = a_{22} = 2$. Then $\lambda_{1,2} = 2 \pm \sqrt{4 - 4 + a_{12}^2}$. If a root is to be $\lambda = 1$, then $a_{12} = 1$. Using $\mathbf{A} = \begin{bmatrix} 2 & 1 \\ 1 & 2 \end{bmatrix}$, the steady-state solution $\mathbf{x}_e$ is just the eigenvector associated with the root $\lambda = 1$:

$$\text{Adj}[\mathbf{A} - \mathbf{I}\lambda]|_{\lambda=1} = \begin{bmatrix} 2-\lambda & -1 \\ -1 & 2-\lambda \end{bmatrix}\Bigg|_{\lambda=1} = \begin{bmatrix} 1 & -1 \\ -1 & 1 \end{bmatrix}$$

The steady-state solution will be proportional to $\mathbf{x}_e = \begin{bmatrix} 1 \\ -1 \end{bmatrix}$.

9.26 Assume that a system is described by Eqs. (*3.13*) and (*3.14*) with **A, B, C,** and **D** constant. Use Z-transforms to find the transforms of $\mathbf{x}(k)$ and $\mathbf{y}(k)$. Then use the inverse Z-transform to find $\mathbf{x}(k)$. The initial condition $\mathbf{x}(0)$ is known and the input $\mathbf{u}(k)$ is zero for $k < 0$.

The Z-transform introduced in Chapters 1 and 2 applies to vector-matrix equations on an obvious component-by-component basis. If $Z\{\mathbf{x}(k)\} \triangleq \mathbf{X}(z)$, then $Z\{\mathbf{x}(k+1)\} = z\mathbf{X}(z) - z\mathbf{x}(0)$. Note the z multiplier on $\mathbf{x}(0)$, which deviates from the analogy expected from the Laplace transform of a derivative term. With this result and the linearity of the Z-transform operator, Eq. (*3.13*) gives

$$z\mathbf{X}(z) - z\mathbf{x}(0) = \mathbf{AX}(z) + \mathbf{BU}(z)$$

or

$$\mathbf{X}(z) = [z\mathbf{I} - \mathbf{A}]^{-1}\{\mathbf{BU}(z) + z\mathbf{x}(0)\} \tag{1}$$

Since the form of the initial condition response is known from earlier time-domain analysis, it is noted that

$$\mathbf{\Phi}(k, 0) \triangleq Z^{-1}\{[z\mathbf{I} - \mathbf{A}]^{-1} z\}$$

Using this definition allows the forcing function term to be written as

$$[z\mathbf{I} - \mathbf{A}]^{-1}\mathbf{BU}(z) = [z\mathbf{I} - \mathbf{A}]^{-1} zz^{-1}\mathbf{BU}(z) = Z\{\mathbf{\Phi}(k, 0)\}Z\{\mathbf{Bu}(k-1)\}$$

Then the convolution theorem of Z-transforms gives the inverse as

$$Z^{-1}\{[z\mathbf{I} - \mathbf{A}]^{-1}\mathbf{BU}(z)\} = \sum_{j=0}^{k} \mathbf{\Phi}(k, j)\mathbf{Bu}(j-1)$$

Since $\mathbf{u}(j-1) = \mathbf{0}$ for $j \le 0$, the lower summation limit is changed to $j = 1$. The solution for $\mathbf{x}(k)$ is then exactly Eq. (*9.29*). Note that the assumption that **B** was constant is unnecessary. From the transform of Eq. (*3.14*),

$$\mathbf{Y}(z) = \mathbf{CX}(z) + \mathbf{DU}(z) \tag{2}$$

Combined with (*1*) this gives

$$\mathbf{Y}(z) = \{\mathbf{C}[z\mathbf{I} - \mathbf{A}]^{-1}\mathbf{B} + \mathbf{D}\}\mathbf{U}(z) + [z\mathbf{I} - \mathbf{A}]^{-1} z\mathbf{x}(0) \tag{3}$$

Note that a useful formula for computing input-output transfer functions has been found, namely,

$$T(z) = \mathbf{C}[z\mathbf{I} - \mathbf{A}]^{-1}\mathbf{B} + \mathbf{D}$$

Finally, the sequence $\mathbf{y}(k)$ could be computed by inverse transforming (*3*) or by first finding $\mathbf{x}(k)$ as the inverse transform of (*1*) and then using that result in Eq. (*3.14*).

Approximation of a Continuous-Time System

9.27 A simple scalar system is described by $\dot{x} = -x + u$, $x(0) = 10$.

(a) Solve for $x(t)$ if $u(t) = e^t$.

(b) Derive a discrete approximation for the above system, using $t_{k+1} - t_k = 1$. Solve this discrete system and compare the results with the continuous solution.

(a) By inspection, $\phi(t, \tau) = e^{-(t-\tau)}$, so

$$x(t) = 10e^{-t} + \int_0^t e^{-(t-\tau)} e^{\tau}\, d\tau = 10e^{-t} + \sinh t$$

(b) Using the scalar transition matrix, the relation between $x(k+1)$ and $x(k)$ is

$$x(k+1) = e^{-1}x(k) + \int_{t_k}^{t_{k+1}} e^{-(t_{k+1}-\tau)}u(\tau)\,d\tau$$

Assuming $u(\tau)$ is constant over the interval $[t_k, t_{k+1}]$ and carrying out the integration gives

$$x(k+1) = e^{-1}x(k) + [1 - e^{-1}]u(k) \cong 0.368x(k) + 0.632u(k)$$

The solution for the discrete system is

$$x(k) = 10(0.368)^k + \sum_{j=1}^{k}(0.368)^{k-j}[0.632u(j-1)]$$

The initial condition term is the same for both the continuous and the discrete cases. A comparison of the forced response for the first five sampling periods is given in Table 9.1. The two separate approximations of Figure 9.6 are used for $u(k)$.

TABLE 9.1

		$t=k=0$	$t=k=1$	$t=k=2$	$t=k=3$	$t=k=4$
Continuous result: $\sinh t$		0	1.1752	3.6269	10.018	27.290
Discrete result: $\sum_{j=1}^{k}(0.368)^{k-j}[0.632u(j-1)]$	input (*a*)	0	0.632	1.951	5.388	14.677
	input (*b*)	0	1.175	3.626	10.016	27.286

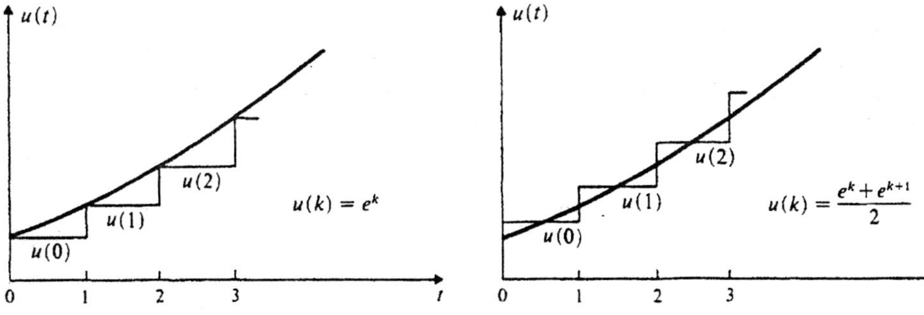

Approximation (*a*) Approximation (*b*)

Figure 9.6

The accuracy of the discrete approximation depends on how closely the piecewise constant discrete input matches the continuous input. The accuracy can be good or poor as demonstrated in the example, but improves as the discrete step-size is decreased.

28 Apply the method of Problem 9.10 to the observable canonical form of the continuous-time system in Example 3.8 to obtain discrete state equations.

To facilitate comparisons, $T = 0.2$ is again used. Keeping fifth-order terms leads to

$$\Phi(T, 0) = \begin{bmatrix} 0.0056 & 0.0793 & 0.1100 \\ -2.1224 & 0.7194 & 0.1779 \\ -1.5861 & -0.2191 & 0.9823 \end{bmatrix}, \qquad \mathbf{B}_1 = \begin{bmatrix} 0.0136 \\ 0.2347 \\ 0.5797 \end{bmatrix}$$

The output matrix is still $\mathbf{C} = [1 \quad 0 \quad 0]$, and $\mathbf{D} = 0$.

PROBLEMS

9.29 A system has two inputs $\mathbf{u} = [u_1 \quad u_2]^T$, and two outputs $\mathbf{y} = [y_1 \quad y_2]^T$. The input-output equations are $\dot{y}_1 + 3(y_1 + y_2) = u_1$ and $\ddot{y}_2 + 4\dot{y}_2 + 3y_2 = u_2$. Find $\mathbf{y}(t)$ if $y_1(0) = 1$, $y_2(0) = 2$, $\dot{y}_2(0) = 1$, and $\mathbf{u}(t) = \mathbf{0}$.

9.30 A system is described by the coupled input-output equations $\dot{y}_1 + 2(y_1 + y_2) = u_1$ and $\ddot{y}_2 + 4\dot{y}_2 + 3y_2 = u_2$. Find the output $\mathbf{y}(t) = [y_1(t) \quad y_2(t)]^T$ if $y_1(0) = 1, y_2(0) = 2, \dot{y}_2(0) = 0, u_1(t) = 0$, $u_2(t) = \delta(t)$ (i.e., an impulse at $t = 0$).

9.31 A system is described by $\begin{bmatrix} \dot{x}_1 \\ \dot{x}_2 \end{bmatrix} = \begin{bmatrix} -2 & 1 \\ -1 & 0 \end{bmatrix}\begin{bmatrix} x_1 \\ x_2 \end{bmatrix} + \begin{bmatrix} 3 \\ 1 \end{bmatrix} u(t)$. If $\mathbf{x}(0) = [10 \quad 1]^T$ and if $u(t) = 0$, find $\mathbf{x}(t)$.

9.32 If the input to the system of the previous problem is $u(t) = e^{2t}$, what is $\mathbf{x}(t)$?

9.33 The wobbling satellite of Problems 3.16, page 117, and 9.6, page 327, has the initial state $\mathbf{x}(0) = [\omega_y(0) \quad \omega_z(0)]^T$. If the input torques are programmed as

$$u_1(t) = -\frac{1}{t_f}[\omega_y(0) \cos \Omega t - \omega_z(0) \sin \Omega t]$$

and

$$u_2(t) = -\frac{1}{t_f}[\omega_y(0) \sin \Omega t + \omega_z(0) \cos \Omega t]$$

find the state (wobble) $\mathbf{x}$ at time $t = t_f$.

9.34 Show that the approximate numerical solution of

$$\dot{\mathbf{x}} = \mathbf{A}\mathbf{x} + \mathbf{B}\mathbf{u} \qquad (1)$$

at time T, expressed as

$$\mathbf{x}(T) = \mathbf{x}(0) + \dot{\mathbf{x}}(0)T + \ddot{\mathbf{x}}(0)T^2/2 + \dddot{\mathbf{x}}(0)T^3/3! + \cdots$$

leads to exactly the same series representation for $\Phi(T, 0)$ and $\mathbf{B}_1$ as found in Problem 9.10. *Hint:* Repeatedly use Eq. (*1*) and its derivatives to express all derivatives of $\mathbf{x}$ in terms of $\mathbf{x}$ and $\mathbf{u}$. Treat $\mathbf{A}$, $\mathbf{B}$, and $\mathbf{u}$ as constants. Notice that the first-order approximation is just rectangular integration of $\dot{\mathbf{x}}$, the second-order approximation is trapezoidal integration of $\dot{\mathbf{x}}$, etc.

9.35 Find the transition matrix $\Phi(t, 0)$ for the feedback system of Problem 9.13 if K is increased to 2.5

9.36 Let $\mathbf{A}$ be a constant $n \times n$ matrix with n linearly independent eigenvectors. Use the Cayley-Hamilton remainder form for $\Phi(t, t_0) = e^{\mathbf{A}(t - t_0)}$ to verify the results stated in Problem 9.15. That is, show that $\Phi(t, t_0)\xi_i = e^{\lambda_i(t - t_0)}\xi_i$, where λ_i and ξ_i are eigenvalues and eigenvectors of $\mathbf{A}$.

9.37 Solve the following homogeneous difference equations:

$$x_1(k + 1) = x_1(k) - x_2(k) + x_3(k)$$

$$x_2(k + 1) = x_2(k) + x_3(k)$$

$$x_3(k + 1) = x_3(k)$$

with $x_1(0) = 2, x_2(0) = 5, x_3(0) = 10$.

38 Find the time response of the discrete model developed in Problem 9.12 if $x_1(0) = (0), x_2(0) = 10$, $u_1(k) = 1/K$, and $u_2(k) = 1$.

39 A single-input, single-output system is described by

$$\mathbf{x}(k+1) = \begin{bmatrix} 1 & 0 \\ -\frac{1}{2} & \frac{1}{2} \end{bmatrix} \mathbf{x}(k) + \begin{bmatrix} 1 \\ -1 \end{bmatrix} u(k) \quad \text{and} \quad y(k) = [5 \quad 1]\mathbf{x}(k)$$

Use a change of basis to determine the normal form equations.

40 If a system is described by $\mathbf{x}(k+1) = \begin{bmatrix} 3 & 2 & 3 \\ 2 & 1 & 1 \\ 1 & 1 & 2 \end{bmatrix} \mathbf{x}(k)$, is it true that $\mathbf{\Phi}(j, k) = \mathbf{\Phi}^{-1}(k, j)$?

41 A simplified model of a motor is given by the transfer function $\theta(s)/u(s) = K/[s(\tau s + 1)]$. Let $x_1 = \theta, x_2 = \dot{\theta}$ and develop the continuous state equations. Then determine the approximate discrete-time state equations, using time points separated by $t_{k+1} - t_k = \Delta t$.

42 The motor of Problem 9.41 is used in a sampled-data feedback system as shown in Figure 9.7. The signal $u(k)$ is $e(t_k) = r(t_k) - \theta(k)$. Write the discrete state equations, using $r(t_k)$ as the input and $\theta(t_k)$ as the output.

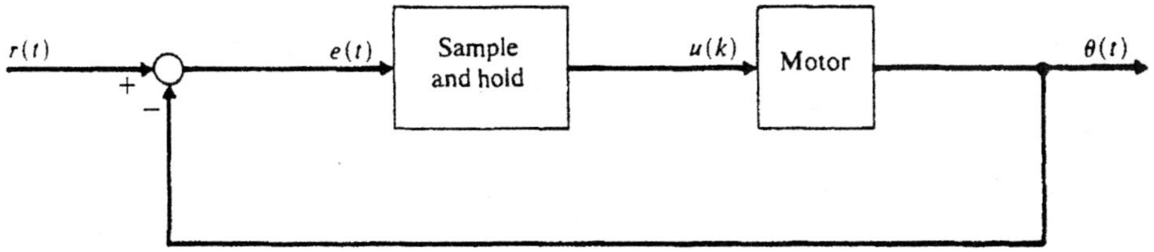

Figure 9.7

43 Apply the method of Problem 9.10 to the cascade realization of the continuous-time system in Example 3.8. Use $T = 0.2$ and keep terms through fifth order. After finding the approximate discrete models for **A** and **B**, find the transfer function by using

$$T(z) = \mathbf{C}[z\mathbf{I} - \mathbf{A}]^{-1}\mathbf{B}$$

44 Find a discrete-time state variable model for a system with transfer function

$$T(z) = \frac{0.006745(z + 0.0672)(z + 1.2416)}{(z - 0.04979)(z - 0.22313)(z - 0.60653)}$$

10 Stability

10.1 INTRODUCTION

Stability of single-input, single-output linear time-invariant systems was discussed from the transfer function point of view in Chapter 2. There the conditions for stability were given in terms of pole locations. The left half of the complex s-plane was found to be the stable region for continuous-time systems. The interior of the unit circle, centered at the origin of the Z-plane, was the stable region for discrete-time systems. Classical methods of stability analysis, including those of Nyquist, Bode, and root-locus, were presented in Chapter 2.

The goal of this chapter is to extend the previous stability concepts to multivariable systems described by state variable models. Although this chapter is primarily concerned with linear system stability, much of the machinery needed for nonlinear systems is also established here. Chapter 15 is devoted to several aspects of nonlinear control system analysis, including additional applications of stability theory.

In earlier discussions a system was either said to be stable or unstable, with perhaps some uncertainty about how to label systems which fall on the dividing line. Actually there are many different definitions of stability. A few of the more common ones are given here, along with methods of investigating them. Furthermore, a given system can exhibit behavior that is considered stable in some region of state space and unstable in other regions. Thus the question of stability should properly be addressed to the various *equilibrium points* (sometimes called critical points) of a system rather than to the system itself. This distinction is largely unnecessary for linear systems, as will be seen, but it is stressed here in preparation for nonlinear systems.

A sampling of the many treatments of stability from various points of view may be found in References 1 through 6.

2 EQUILIBRIUM POINTS AND STABILITY CONCEPTS

A heuristic discussion of stability is first given to help make the later mathematical treatment more intuitive. Consider the ball which is free to roll on the surface shown in Figure 10.1. The ball could be made to rest at points A, E, F, and G and anywhere between points B and D, such as at C. Each of these points is an equilibrium point of the system.

In state space, an equilibrium point for a continuous-time system is a point at which $\dot{\mathbf{x}}$ is zero in the absence of all inputs and disruptive disturbances. Thus if the system is placed in that state, it will remain there. For discrete-time systems, an equilibrium point is one for which $\mathbf{x}(k+1) = \mathbf{x}(k)$ in the absence of all control inputs or disturbances.

An infinitesimal perturbation away from points A or F will cause the ball to diverge from these points. This behavior intuitively justifies labeling A and F as *unstable* equilibrium points. After small perturbations away from E or G, the ball will eventually return to rest at these points. Thus E and G are labeled as stable equilibrium points. If the ball is displaced slightly from point C, in the absence of an initial velocity it will stay at the new position. Points like C are sometimes said to be *neutrally* stable.

Assume that the shape of the surface in Figure 10.1 changes with time. Specifically, assume that point E moves vertically so that the slope at that point is always zero, but the surface is sometimes concave upward (as shown) and sometimes concave downward. Point E is still an equilibrium point, but whether it is stable or not now depends upon time.

Thus far only *local* stability has been considered, since the perturbations were assumed to be small. If the ball were displaced sufficiently far from point G, it would not return to that point. Stability therefore depends on the size of the original perturbation and on the nature of any disturbances which may be acting. These intuitive notions are now developed for dynamical systems.

A particular point $\mathbf{x}_e \in \Sigma$ is an equilibrium point of a dynamical system if the system's state at t_0 is $\mathbf{x}_e$ and $\mathbf{x}(t) = \mathbf{x}_e$ for all $t \geq t_0$ in the absence of inputs or disturbances. For the continuous-time system $\dot{\mathbf{x}} = \mathbf{f}(\mathbf{x}(t), \mathbf{u}(t), t)$, this means that $\mathbf{f}(\mathbf{x}_e, \mathbf{0}, t) = \mathbf{0}$ for $t \geq t_0$. For the discrete-time system $\mathbf{x}(k+1) = \mathbf{f}(\mathbf{x}(k), \mathbf{u}(k), k)$, this means that $\mathbf{f}(\mathbf{x}_e, \mathbf{0}, k) = \mathbf{x}_e$ for all $k > 0$.

The origin of the state space is always an equilibrium point for linear systems, although it need not be the only one. In the continuous-time case, if the system matrix

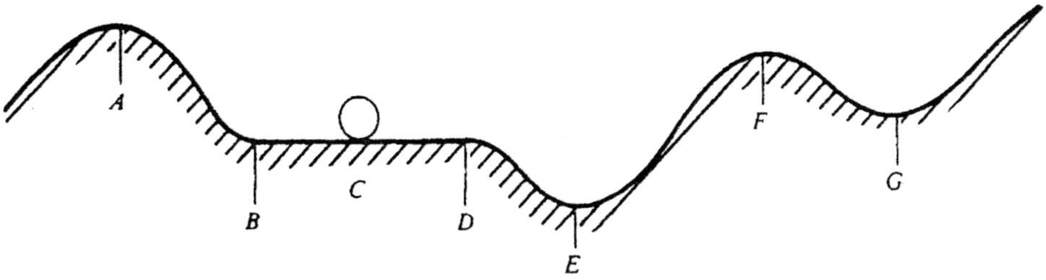

Figure 10.1

$\mathbf{A}$ has a zero eigenvalue, then there is an infinity of vectors (eigenvectors) satisfying $\mathbf{Ax}_e = \mathbf{0}$. In the discrete-time case a unity eigenvalue of $\mathbf{A}$ means there is an infinity of vectors satisfying $\mathbf{Ax}_e = \mathbf{x}_e$. These points loosely correspond to points between B and D of Figure 10.1. Only *isolated equilibrium points* will be considered in this text, and for linear systems the only isolated equilibrium point is the origin.

Any isolated singular point can be transferred to the origin by a change of variables, $\mathbf{x}' = \mathbf{x} - \mathbf{x}_e$. For this reason it is often assumed in the sequel that $\mathbf{x}_e = \mathbf{0}$.

Stability deals with the following questions. If at time t_0 the state is perturbed from its equilibrium point, does the state return to $\mathbf{x}_e$, or remain close to $\mathbf{x}_e$, or diverge from it? Similar questions could be raised if system inputs or disturbances are allowed. Another class of stability questions deals with the state trajectories of an unperturbed system and of a perturbed system. Let the solution to $\dot{\mathbf{x}}_1 = \mathbf{f}(\mathbf{x}_1(t), \mathbf{u}(t), t)$, with $\mathbf{x}_1(t_0)$ given, define the unperturbed trajectory $\mathbf{x}_1(t)$. Let the perturbed trajectory $\mathbf{x}_2(t)$ be defined by $\dot{\mathbf{x}}_2 = \mathbf{f}(\mathbf{x}_2(t), \mathbf{u}(t) + \mathbf{v}(t), t)$, where $\mathbf{x}_2(t_0) = \mathbf{x}_1(t_0) + \mathbf{e}(t_0)$. The initial state and control perturbations are $\mathbf{e}(t_0)$ and $\mathbf{v}(t)$, respectively. Does $\mathbf{x}_2(t)$ return to $\mathbf{x}_1(t)$, or remain close to it, or diverge from it? These questions can be studied by considering the difference $\mathbf{e}(t) = \mathbf{x}_2(t) - \mathbf{x}_1(t)$, which satisfies $\dot{\mathbf{e}} = \mathbf{f}(\mathbf{x}_1(t) + \mathbf{e}(t),\ \mathbf{u}(t) + \mathbf{v}(t), t) - \mathbf{f}(\mathbf{x}_1(t), \mathbf{u}(t), t)$ or simply $\dot{\mathbf{e}} = \mathbf{f}'(\mathbf{e}(t), \mathbf{v}(t), t)$ with $\mathbf{e}(t_0)$, $\mathbf{x}_1(t)$, and $\mathbf{u}(t)$ given. Now $\mathbf{e} = \mathbf{0}$ is an equilibrium point and the questions regarding the perturbed motion can be studied in terms of perturbations about the origin, as before.

Whether an equilibrium point is stable or not depends upon what is meant by remaining close, the magnitude of state or input disturbances, and their time of application. These qualifying conditions are the reasons for the existence of a variety of stability definitions.

10.3 STABILITY DEFINITIONS

Consider the continuous-time system with the input set to zero,

$$\dot{\mathbf{x}} = \mathbf{f}(\mathbf{x}, \mathbf{0}, t), \qquad \mathbf{x}(t_0) = \mathbf{x}_0 \tag{10.1}$$

As was pointed out in Sec. 3.3, page 78, there will *exist* a *unique* solution to these differential equations provided that the function $\mathbf{f}(\mathbf{x}, 0, t)$ satisfies a Lipschitz condition [7] with respect to $\mathbf{x}$ and is at least piecewise continuous with respect to t throughout some region of the product space $\Sigma \times \tau$, which contains $\mathbf{x}_0, t_0$. Furthermore, the solution depends on its arguments in a continuous fashion. The solution is frequently written as $\boldsymbol{\Phi}(t; x_0, t_0)$ to show its arguments explicitly, but we often refer to it simply as $\mathbf{x}(t)$. Note that in the linear case discussed in Chapter 9, $\boldsymbol{\Phi}(t; x_0, t_0) = \boldsymbol{\Phi}(t, t_0)\mathbf{x}_0$.

It is assumed that an equilibrium point for system (*10.1*) is at or has been transferred to the origin. Then the following definitions apply. For the continuous-time case with zero input

$$\dot{\mathbf{x}} = \mathbf{f}(\mathbf{x}, \mathbf{0}, t), \qquad \mathbf{x}(t_0) = \mathbf{x}_0$$

with the origin an equilibrium point, the following apply.

Definition 10.1. The origin is a *stable* equilibrium point if for any given value $\epsilon > 0$ there exists a number $\delta(\epsilon, t_0) > 0$ such that if $\|\mathbf{x}(t_0)\| < \delta$, then the resultant motion $\mathbf{x}(t)$ satisfies $\|\mathbf{x}(t)\| < \epsilon$ for all $t > t_0$.

This definition of stability is sometimes called *stability in the sense of Lyapunov,* abbreviated as stable i.s.L. If a system possesses this type of stability, then it is ensured that the state can be kept within ϵ, in norm, of the origin by restricting the initial perturbation to be less than δ, in norm. Note that it is necessarily true that $\delta \leq \epsilon$.

Definition 10.2 The origin is an *asymptotically stable* equilibrium point if (a) it is stable, and if in addition, (b) there exists a number $\delta'(t_0) > 0$ such that whenever $\|\mathbf{x}(t_0)\| < \delta'(t_0)$ the resultant motion satisfies $\lim_{t\to\infty} \|\mathbf{x}(t)\| = 0$.

Figure 10.2a illustrates these definitions for the two-dimensional state space Σ_2. The same notions conceptually apply in higher dimensions. Two examples of possible trajectories are shown in Figure 10.2a, one for a system which is stable i.s.L. and the other for an asymptotically stable system. A projection onto the state space Σ_2 is shown in Figure 10.2b. An *unstable* trajectory—i.e., one which is not stable—has been added to the original two.

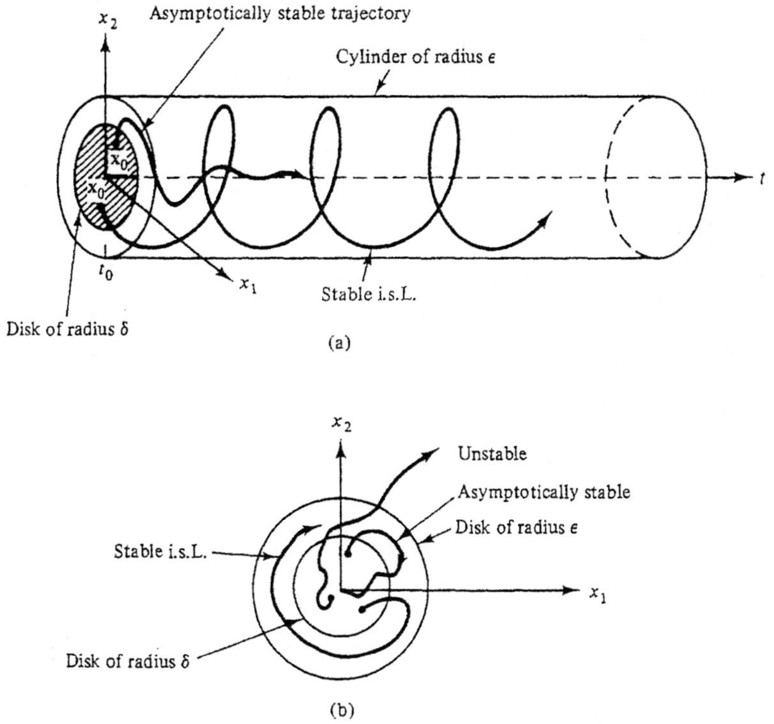

Figure 10.2a Illustrations of stable trajectories in $\Sigma_2 \times \tau$.
Figure 10.2b Illustrations of possible trajectories in Σ_2.

The two most basic definitions of stability have been given for unforced continuous-time systems. Variations upon these are defined by adding additional qualifying adjectives [2]. If δ and δ' are not functions of t_0, then the origin is said to be *uniformly stable* and *uniformly asymptotically stable*, respectively. If $\delta'(t_0)$ in Definition 10.2 can be made arbitrarily large—i.e., if all $\mathbf{x}(t_0)$ converge to $\mathbf{0}$—then the origin is said to be *globally* asymptotically stable or asymptotically stable *in the large*.

Stability definitions for the discrete-time system with zero input

$$\mathbf{x}(k+1) = \mathbf{f}(\mathbf{x}(k), \mathbf{0}, k), \qquad \mathbf{x}(0) = \mathbf{x}_0 \tag{10.2}$$

are identical to those given earlier, provided the discrete-time index k is used in place of t. As before, it is assumed that the coordinates have been chosen so that the origin is an equilibrium state.

When nonzero inputs $\mathbf{u}(t)$ or $\mathbf{u}(k)$ are considered, two additional types of stability are often used.

Definition 10.3. (Bounded input, bounded state stability.) If there is a fixed, finite constant K such that $\|\mathbf{u}\| \le K$ for every t (or k), then the input is said to be bounded. If for every bounded input, and for arbitrary initial conditions $\mathbf{x}(t_0)$, there exists a scalar $0 < \delta(K, t_0, \mathbf{x}(t_0))$ such that the resultant state satisfies $\|\mathbf{x}\| \le \delta$, then the system is *bounded input, bounded state* stable, abbreviated as BIBS stable.

All the previous definitions of stability deal with the behavior of the state vector relative to an equilibrium state. Frequently, the main interest is in the system output behavior. This motivates the final stability definition.

Definition 10.4. (Bounded input, bounded output stability.) Let $\mathbf{u}$ be a bounded input with K_m as the least upper bound. If there exists a scalar α such that for every t (or k), the output satisfies $\|\mathbf{y}\| \le \alpha K_m$, then the system is *bounded input, bounded ouptut* stable, abbreviated as BIBO stable.

10.4 LINEAR SYSTEM STABILITY

The following linear continuous-time system is considered:

$$\begin{aligned} \dot{\mathbf{x}} &= \mathbf{A}(t)\mathbf{x}(t) + \mathbf{B}(t)\mathbf{u}(t) \\ \mathbf{y}(t) &= \mathbf{C}(t)\mathbf{x}(t) + \mathbf{D}(t)\mathbf{u}(t) \end{aligned} \tag{10.3}$$

The unforced case is treated first. With $\mathbf{u}(t) = \mathbf{0}$, the state vector is given by

$$\mathbf{x}(t) = \mathbf{\Phi}(t, t_0)\mathbf{x}(t_0) \tag{10.4}$$

The norm of $\mathbf{x}(t)$ is a measure of the distance of the state from the origin:

$$\|\mathbf{x}(t)\| = \|\mathbf{\Phi}(t, t_0)\mathbf{x}(t_0)\| \le \|\mathbf{\Phi}(t, t_0)\| \, \|\mathbf{x}(t_0)\| \tag{10.5}$$

Suppose there exists a number $N(t_0)$, possibly depending on t_0, such that

$$\|\mathbf{\Phi}(t, t_0)\| \le N(t_0) \quad \text{for all } t \ge t_0 \tag{10.6}$$

Then the conditions of Definition 10.1 can be satisfied for any $\epsilon > 0$ by letting $\delta(t_0, \epsilon) = \epsilon/N(t_0)$. It follows from Eq. (*10.5*) that Eq. (*10.6*) is *sufficient* to ensure that the origin is stable in the sense of Lyapunov. It is easy to show that this condition is also *necessary*. The origin is asymptotically stable if and only if Eq. (*10.6*) holds, and if in addition, $\|\boldsymbol{\Phi}(t, t_0)\| \to 0$ for $t \to \infty$. Note that for a linear system, asymptotic stability does not depend on $\mathbf{x}(t_0)$. If a linear system is asymptotically stable, it is globally asymptotically stable.

The stability types which depend upon the input $\mathbf{u}(t)$ are now considered. For the linear continuous-time system, the state vector is given by

$$\mathbf{x}(t) = \boldsymbol{\Phi}(t, t_0)\mathbf{x}(t_0) + \int_{t_0}^{t} \boldsymbol{\Phi}(t, \tau)\mathbf{B}(\tau)\mathbf{u}(\tau)\, d\tau \tag{10.7}$$

BIBS stability requires that $\mathbf{x}(t)$ remain bounded for all bounded inputs. Since $\mathbf{u}(t) = \mathbf{0}$ is bounded, it is clear that stability i.s.L. is a necessary condition for BIBS stability. By taking the norm of both sides of Eq. (*10.7*) and using well-known properties of the norm, it is found that $\|\mathbf{x}(t)\|$ remains bounded, and thus the origin is BIBS stable, if Eq. (*10.6*) holds and if in addition there exists a number $N_1(t_0)$ such that

$$\int_{t_0}^{t} \|\boldsymbol{\Phi}(t, \tau)\mathbf{B}(\tau)\|\, d\tau \le N_1(t_0) \quad \text{for all } t \ge t_0 \tag{10.8}$$

Similar arguments show that a linear discrete-time system is BIBS stable if the discrete transition matrix satisfies Eq. (*10.6*) and if $\sum_{k=0}^{k_1} \|\boldsymbol{\Phi}(k_1, k)\mathbf{B}(k-1)\| \le N_1$.

BIBO stability is investigated by considering the output of a linear system

$$\mathbf{y}(t) = \mathbf{C}(t)\mathbf{x}(t) + \mathbf{D}(t)\mathbf{u}(t) \tag{10.9}$$

Substitution of Eq. (*10.7*) into Eq. (*10.9*) and viewing the initial state $\mathbf{x}(t_0)$ as having arisen because of a bounded input over the interval $(-\infty, t_0)$ gives

$$\mathbf{y}(t) = \int_{-\infty}^{t} \mathbf{W}(t, \tau)\mathbf{u}(\tau)\, d\tau \tag{10.10}$$

Only bounded inputs are considered, that is,

$$\|\mathbf{u}(\tau)\| \le K \quad \text{for all } \tau \tag{10.11}$$

The output remains bounded in norm if there exists a constant $M > 0$ such that the impulse response or weighting matrix $\mathbf{W}(t, \tau)$ satisfies

$$\int_{-\infty}^{t} \|\mathbf{W}(t, \tau)\|\, d\tau \le M \quad \text{for all } t \tag{10.12}$$

Equation (*10.12*) is the necessary and sufficient condition for BIBO stability of continuous-time systems. The analogous result, with a summation replacing the integration, holds for discrete-time systems.

The matrix norm $\|\boldsymbol{\Phi}(t, t_0)\|$ plays a central role in the stability conditions. This norm can be defined in various ways, including the inner product definition

$$\|\boldsymbol{\Phi}(t, t_0)\|^2 = \max_{\mathbf{x}} \{\langle \boldsymbol{\Phi}(t, t_0)\mathbf{x}, \boldsymbol{\Phi}(t, t_0)\mathbf{x}\rangle \mid \langle \mathbf{x}, \mathbf{x}\rangle = 1\} \tag{10.13}$$

By introducing the adjoint of $\Phi(t, t_0)$, it is found that Eq. (10.13) leads to

$$\|\Phi(t, t_0)\|^2 = \text{max eigenvalue of } \overline{\Phi}^T(t, t_0)\Phi(t, t_0) \tag{10.14}$$

If $\Phi(t, t_0)$ is normal, then $\overline{\Phi}^T(t, t_0)\Phi(t, t_0) = \Phi(t, t_0)\overline{\Phi}^T(t, t_0)$ and then

$$\|\Phi(t, t_0)\| = \max_i |\alpha_i| \tag{10.15}$$

where α_i is an eigenvalue of $\Phi(t, t_0)$. In all cases a useful lower bound on the norm is given by

$$\|\Phi(t, t_0)\|^2 \geq |\alpha_i|^2 \quad \text{for any eigenvalue } \alpha_i \text{ of } \Phi(t, t_0) \tag{10.16}$$

10.5 LINEAR CONSTANT SYSTEMS

Whenever the system under consideration has a constant system matrix **A**, the following results hold:

$$\Phi(t, t_0) = e^{\mathbf{A}(t - t_0)} \quad \text{(continuous-time)} \tag{10.17}$$

$$\Phi(k, 0) = \mathbf{A}^k \quad \text{(discrete-time)} \tag{10.18}$$

By virtue of the Cayley-Hamilton theorem, Chapter 8, both of these results can be expressed as polynomials in **A**. Then by Frobenius' theorem (Problem 9.15, page 333) the eigenvalues α_i of Φ are related to the eigenvalues λ_i of **A** by

$$\alpha_i = e^{\lambda_i(t - t_0)} \quad \text{or} \quad \alpha_i = \lambda_i^k$$

for the continuous-time and discrete-time cases, respectively. It is relatively simple to express the previous stability conditions in terms of the eigenvalues of the system matrix **A**. Letting these eigenvalues be $\lambda_i = \beta_i \pm j\omega_i$, the resulting conditions are summarized in Table 10.1.

The results of Table 10.1 again show the left-half plane and the interior of the unit circle as stability regions, this time in terms of locations of the eigenvalues of the **A** matrix instead of transfer function poles. The consistency between the continuous-

TABLE 10-1 STABILITY CRITERIA FOR LINEAR CONSTANT SYSTEMS

(Eigenvalues of **A** are $\lambda_i = \beta_i \pm j\omega_i$)

	Continuous Time $\dot{\mathbf{x}} = \mathbf{A}\mathbf{x}$	Discrete Time $\mathbf{x}(k+1) = \mathbf{A}\mathbf{x}(k)$
Unstable	If $\beta_i > 0$ for any simple root or $\beta_i \geq 0$ for any repeated root	If $\lvert\lambda_i\rvert > 1$ for any simple root or $\lvert\lambda_i\rvert \geq 1$ for any repeated root
Stable i.s.L.	If $\beta_i \leq 0$ for all simple roots and $\beta_i < 0$ for all repeated roots	If $\lvert\lambda_i\rvert \leq 1$ for all simple roots and $\lvert\lambda_i\rvert < 1$ for all repeated roots
Asymptotically Stable	If $\beta_i < 0$ for all roots	$\lvert\lambda_i\rvert < 1$ for all roots

time and discrete-time columns of Table 10.1 is noted. Recall from Problem 2.21 that any s-plane pole with a negative real part maps into a complex number in the Z-plane with a magnitude less than unity.

Classification of Equilibrium Points. Equilibrium points can be classified into several types, beyond just labeling them stable or unstable. For this purpose it is useful to expand the initial condition response, using the eigenvectors ξ_i of $\mathbf{A}$ as basis vectors:

$$\mathbf{x}(t) = w_1(0)\exp\{\lambda_1 t\}\xi_1 + w_2(0)\exp\{\lambda_2 t\}\xi_2 + \cdots + w_n(0)\exp\{\lambda_n t\}\xi_n \tag{10.19}$$

where $\mathbf{w}(0) = \mathbf{M}^{-1}\mathbf{x}_0$ and $\mathbf{M}$ is the modal matrix.

Equation (*10.19*) is used to discuss the initial condition response of a second-order system. The two eigenvalues could both be real or a complex conjugate pair. In the real case, if both eigenvalues are stable, the equilibrium point is called a *stable node,* and the phase portrait for various initial conditions is shown in Figure 10.3*a.* If one eigenvalue is stable and the other is unstable, the behavior is as shown in Figure 10.3*b,* and this is called a *saddle point.* When both eigenvalues are unstable, trajectories such as those in Figure 10.3*c* result, and this is called an *unstable node.* When the eigenvalues are complex, Eq. (*10.19*) can be rewritten as

$$\mathbf{x}(t) = \exp\{\beta t\}\{[a\cos(\omega t) + b\sin(\omega t)]\xi_R + [b\cos(\omega t) - a\sin(\omega t)]\xi_I\}$$

where ξ_R and ξ_I are the real and imaginary parts of ξ and a, b are real constants dependent upon the initial conditions. If the real part of the eigenvalue, β, is negative the phase portraits of Figure 10.4*a* result, and this is called a *stable focus.* If $\beta > 0$, the *unstable focus* of Figure 10.4*b* results. If $\beta = 0$, the equilibrium point is called a *center,* and the phase portraits are as shown in Figure 10.4*c.* Generally, graphical display of trajectories becomes difficult or impossible in higher dimensions. However, Eq. (*10.19*) is still useful for conceptualizing the response. For example, if a third-order system has two stable complex eigenvalues and one real, unstable eigenvalue with eigenvector ξ_3, a three-dimensional phase trajectory would look somewhat like Figure 10.4*a* but with the center stretched out along the vector ξ_3, as sketched in Figure 10.5.

Phase-space methods have proven very useful in analyzing linear and nonlinear second-order systems. A method of extending many of phase-plane advantages to higher-order systems is called the *center manifold theory* [4]. This approach essentially selects the two modes that are providing the dominant action, those with eigenvalues nearest the stability boundary. The full development of classical phase plane methods and their more recent extensions are left for the references.

10.6 THE DIRECT METHOD OF LYAPUNOV

The general stability results for linear systems have been presented in Sec. 10.4 in terms of the properties of the transition matrix $\Phi(t, t_0)$. In effect this means that the solution of the system's differential or difference equations needs to be known before stability conclusions can be drawn. In the case of linear *constant* systems, this does not

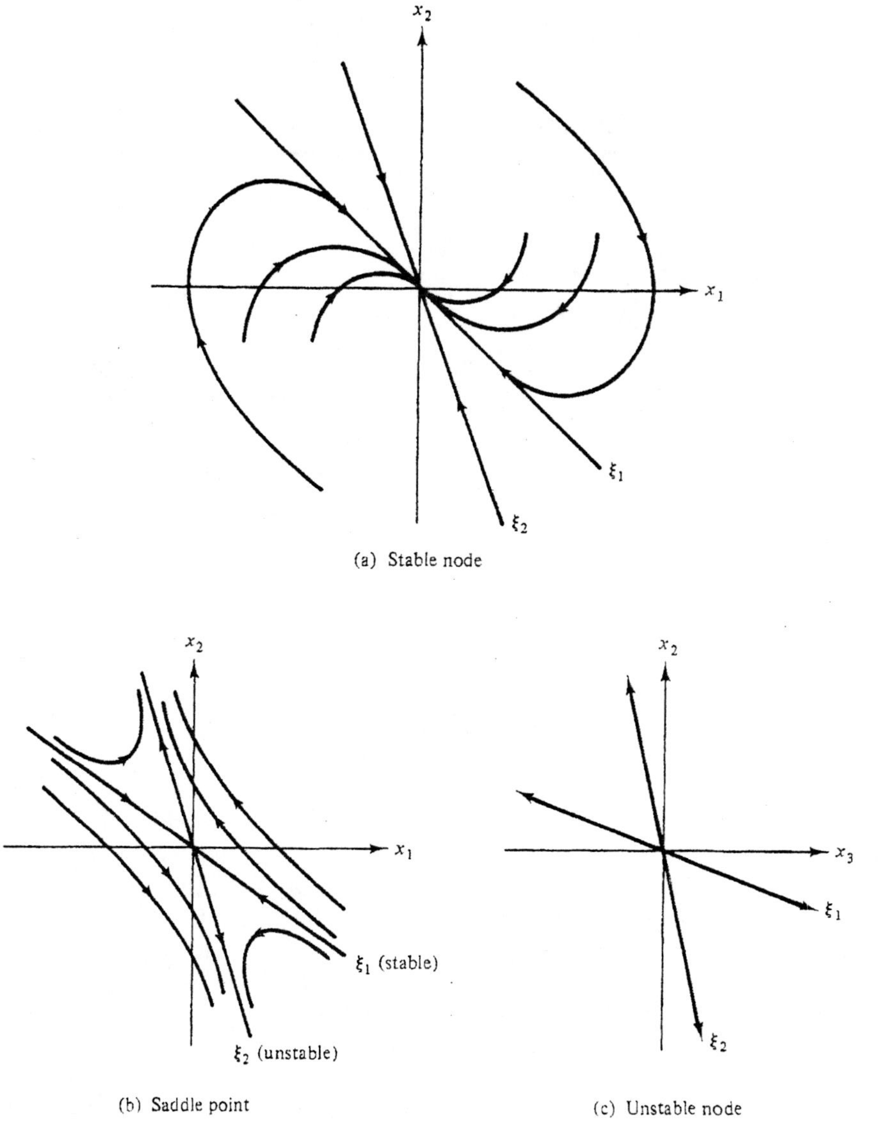

Figure 10.3

appear to be so because Sec. 10.5 expresses the stability conditions in terms of the eigenvalues of the system matrix **A.** Late in the nineteenth century the Russian mathematician A. M. Lyapunov developed an approach to stability analysis, now known as the direct method (or second method) of Lyapunov. The unique thing about this method is that only the form of the differential or difference equations need be known, not their solutions. Lyapunov's direct method is now widely used for stability analysis of linear and nonlinear systems, both time-invariant and time-varying. Although this chapter is primarily concerned with linear systems, Lyapunov's method is presented in

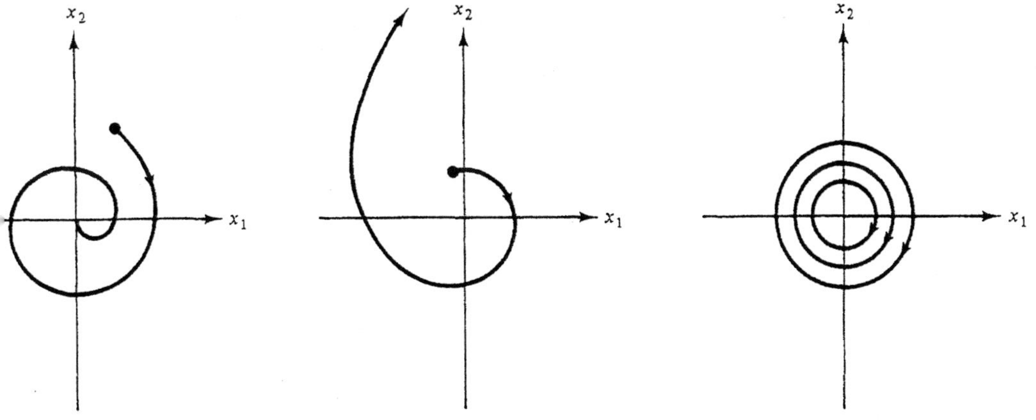

(a) Stable focus (b) Unstable focus (c) Center

Figure 10.4

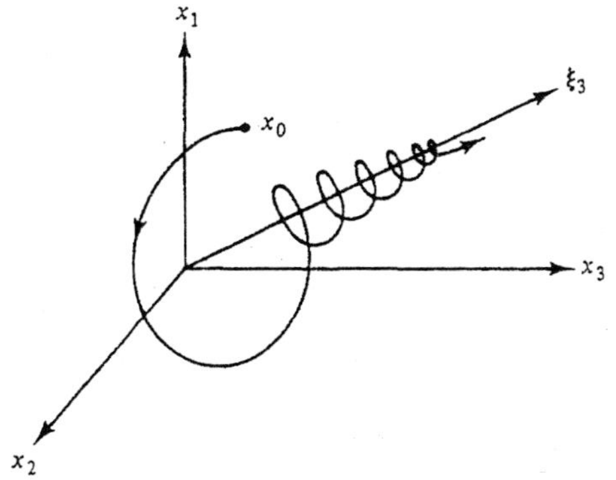

Figure 10.5

sufficient detail so that it can be later applied to nonlinear systems as well. The approach will be to present an intuitive discussion of the method first. Then a few of the main and most useful theorems will be presented without proof. Uses of the theorems are illustrated in the problems and in Chapter 15. The mathematical overhead may seem excessive for dealing with linear systems. However, for *time-varying* linear systems, reliable answers to stability questions, which might otherwise be difficult to obtain, can be found. Even for constant linear systems, the Lyapunov method provides an informative alternative approach.

Energy concepts are widely used and easily understood by engineers. Lyapunov's direct method can be viewed as a generalized energy method. Consider a second-order system, such as the unforced *LC* circuit of Figure 10.6*a* or the mass-spring system of Figure 10.6*b*.

In the first case the capacitor voltage v and inductor current i can be used as state variables $\mathbf{x}$, and the total energy (magnetic plus electric) in the system at any time is $Li^2/2 + Cv^2/2$. In the second case the position y (measured from the free length of the spring) and the velocity $\dot{y}$ can be used as state variables $\mathbf{x}$, and the total energy (kinetic

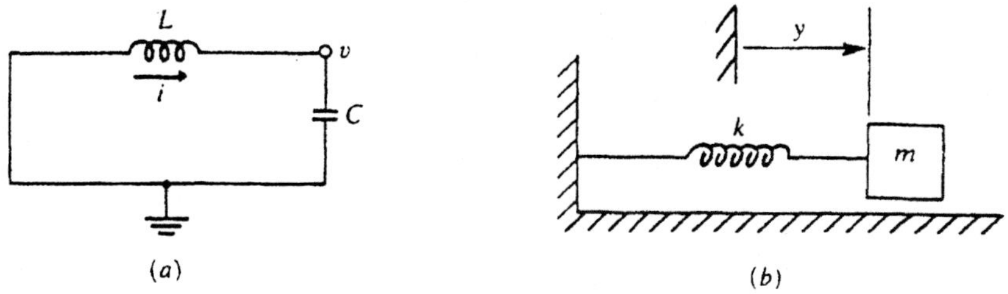

Figure 10.6

plus potential) at any time is $m\dot{y}^2/2 + ky^2/2$. In both cases the energy $\mathscr{E}$ is a quadratic function of the state variables. Thus $\mathscr{E}(\mathbf{x}) > 0$ if $\mathbf{x} \neq \mathbf{0}$, and $\mathscr{E} = 0$ if and only if $\mathbf{x} = \mathbf{0}$. If the time rate of change of the energy $\dot{\mathscr{E}}$ is always negative, except at $\mathbf{x} = \mathbf{0}$, then $\mathscr{E}$ will be continually decreasing and will eventually approach zero. Because of the nature of the energy function, $\mathscr{E} = 0$ implies $\mathbf{x} = \mathbf{0}$. Therefore, if $\dot{\mathscr{E}} < 0$ for all t, except when $\mathbf{x} = \mathbf{0}$, it is concluded that $\mathbf{x}(t) \to \mathbf{0}$ for sufficiently large t. The close relationship of this conclusion to asymptotic stability, Definition 10.2, is obvious.

If the time rate of change of energy is never positive, that is, $\dot{\mathscr{E}} \leq 0$, then $\mathscr{E}$ can never increase, but it need not approach zero either. It can then be concluded that $\mathscr{E}$, and hence $\mathbf{x}$, remain bounded in some sense. This situation is related in an obvious way to stability i.s.L., Definition 10.1.

For both systems of Figure 10.6, the energy can be expressed in the form $\mathscr{E}(\mathbf{x}) = \frac{1}{2}a_1 x_1^2 + \frac{1}{2}a_2 x_2^2$. Assuming for the present that the coefficients a_1 and a_2 are constant, the time rate of change is given by

$$\dot{\mathscr{E}} = a_1 x_1 \dot{x}_1 + a_2 x_2 \dot{x}_2 \tag{10.20}$$

Knowledge of the *form* of the differential equations $\dot{x}_1 = f_1(\mathbf{x})$ and $\dot{x}_2 = f_2(\mathbf{x})$ allows both $\mathscr{E}$ and $\dot{\mathscr{E}}$ to be expressed as a function of the state $\mathbf{x}$. No knowledge of the *solutions* of the differential equations is required in order to draw conclusions regarding stability. Lyapunov's direct method is a generalization of these ideas.

Notice that in Figure 10.1 those points E and G that were described as stable in the earlier discussion are points of relative minimum total energy. If the ball is at rest at one of these points, its kinetic energy is zero and its potential energy is at a relative minimum. A small displacement would increase the potential energy slightly. In the absence of all surface and air friction or any other dissipative phenomenon, the total energy would remain constant after release, being passed back and forth between potential and kinetic energy. The ball would oscillate forever with constant amplitude. This is an example of stability i.s.L. In the real-world case there will always be some energy dissipation, and the ball would eventually return to the point of (relative) minimum total energy. This is an example of asymptotic stability.

EXAMPLE 10.1 For the system of Figure 10.6*a*, with $x_1 = i$ and $x_2 = v$, $\dot{x}_1 = -x_2/L$, $\dot{x}_2 = x_1/C$, so that $\dot{\mathscr{E}} = Lx_1(-x_2/L) + Cx_2(x_1/C) = 0$. Thus $\dot{\mathscr{E}} \equiv 0$ for all t and $\mathscr{E}$ is constant. This conservative system is stable i.s.L. but not asymptotically stable. Of course, this is a well-known result for this undamped oscillator. If the system is modified to include a positive resistance, then the system is dissipative and $\dot{\mathscr{E}}$ would be always negative, except when $\mathbf{x} = \mathbf{0}$. The system is then asymptotically stable. ■

EXAMPLE 10.2 The system of Figure 10.6*b* is assumed to have a nonlinear frictional force d acting between the mass and the supporting surface. This does not change the definition of the energy $\mathscr{E}$. Letting $x_1 = y$ and $x_2 = \dot{y}$, the differential equations are $\dot{x}_1 = x_2$, $\dot{x}_2 = (-kx_1 + d)/m$. Then

$$\dot{\mathscr{E}} = kx_1\dot{x}_1 + mx_2\dot{x}_2 = kx_1x_2 - kx_1x_2 + x_2 d = x_2 d$$

$\dot{\mathscr{E}}$ is nonpositive if the friction force d is always opposing the direction of the velocity x_2. In this case $\mathscr{E}$ can never increase, and the system is clearly stable i.s.L. It may, in fact, be asymptotically stable, as will become clear later. On the other hand, if d and x_2 have the same sign, then $\dot{\mathscr{E}} > 0$ and $\mathscr{E}$ may increase. Neither stability nor instability can be concluded without further analysis. ■

Lyapunov's direct method makes use of a *Lyapunov function* $V(\mathbf{x})$. This scalar function of the state may be thought of as a generalized energy. In many problems the energy function can serve as a Lyapunov function. In cases where a system model is described mathematically, it may not be clear what "energy" means. The conditions which $V(\mathbf{x})$ must satisfy in order to be a Lyapunov function are therefore based on mathematical rather than physical considerations.

A single-valued function $V(\mathbf{x})$ which is continuous and has continuous partial derivatives is said to be *positive definite* in some region Ω about the origin of the state space if (1) $V(\mathbf{0}) = 0$ and (2) $V(\mathbf{x}) > 0$ for all nonzero $\mathbf{x}$ in Ω. A special case of a positive definite function was the quadratic form discussed in Sec. 7.8, page 264. If condition (2) is relaxed to $V(\mathbf{x}) \geq 0$ for all $\mathbf{x} \in \Omega$, then $V(\mathbf{x})$ is said to be *positive semidefinite*. Reversing the inequalities leads to corresponding definitions of *negative definite* and *negative semidefinite* functions.

Consider the autonomous (i.e., unforced, no explicit time dependence) system

$$\dot{\mathbf{x}} = \mathbf{f}(\mathbf{x}) \tag{10.21}$$

The origin is assumed to be an equilibrium point, that is, $\mathbf{f}(\mathbf{0}) = \mathbf{0}$. The stability of this equilibrium point can be investigated by means of the following theorems.

Theorem 10.1. If a positive definite function $V(\mathbf{x})$ can be determined such that $\dot{V}(\mathbf{x}) \leq 0$ (negative semidefinite), then the origin is stable i.s.L.

A function $V(\mathbf{x})$ satisfying these requirements is called a Lyapunov function. The Lyapunov function is not unique; rather, many different Lyapunov functions may be found for a given system. Likewise, the inability to find a satisfactory Lyapunov function does not mean that the system is unstable.

Theorem 10.2. If a positive definite function $V(\mathbf{x})$ can be found such that $\dot{V}(\mathbf{x})$ is negative definite, then the origin is asymptotically stable.

Both of the previous theorems relate to local stability in the neighborhood Ω of the origin. Global asymptotic stability is considered next.

Theorem 10.3. The origin is a globally asymptotically stable equilibrium point for the system of Eq. (*10.21*) if a Lyapunov function $V(\mathbf{x})$ can be found such that (1) $V(\mathbf{x}) > 0$ for all $\mathbf{x} \neq \mathbf{0}$ and $V(\mathbf{0}) = 0$, (2) $\dot{V}(\mathbf{x}) < 0$ for all $\mathbf{x} \neq 0$, and (3) $V(\mathbf{x}) \to \infty$ as $\|\mathbf{x}\| \to \infty$.

In both the preceding theorems the negative definite requirement on $\dot{V}$ can be relaxed to negative semidefinite, and asymptotic stability can still be concluded, if it can be shown that $\dot{V}$ is never zero along a solution trajectory.

As an aid in understanding these three theorems and their differences, a family of contours of $V(\mathbf{x}) =$ constant is shown in Figure 10.7 for a two-dimensional state space. These contours can never intersect because $V(\mathbf{x})$ is single-valued. They are smooth curves because $V(\mathbf{x})$ and all its partial derivatives are required to be continuous.

For the situation shown, condition (3), Theorem 10.3 is not met. It is possible for $\|\mathbf{x}\| \to \infty$ while $V(\mathbf{x})$ remains finite; for example, along the contour $V(\mathbf{x}) = C_4$. It could happen that $V(\mathbf{x})$ continually decreases, causing $V(\mathbf{x})$ to decrease from C_6 through C_5 and asymptotically approach C_4, while $\|\mathbf{x}(t)\|$ continues to grow without bound. The trajectory starting at $\mathbf{x}_3(t_0)$ illustrates this. Condition (3) of theorem 10.3 is necessary to rule out this possibility.

Referring to Figure 10.7, if $\dot{V}(\mathbf{x}) < 0$, then the origin is asymptotically stable for any initial state sufficiently close to the origin, specifically for any $\mathbf{x}(t_0)$ inside contour C_3. Any initial state inside contour C_3, such as $\mathbf{x}_2(t_0)$, must eventually approach the origin if $\dot{V}(\mathbf{x})$ is always negative for $\mathbf{x} \neq \mathbf{0}$. This is the essence of Theorem 10.2. If the restriction is relaxed so that $\dot{V}(\mathbf{x}) \leq 0$, as in Theorem 10.1, then $V(\mathbf{x})$ can never

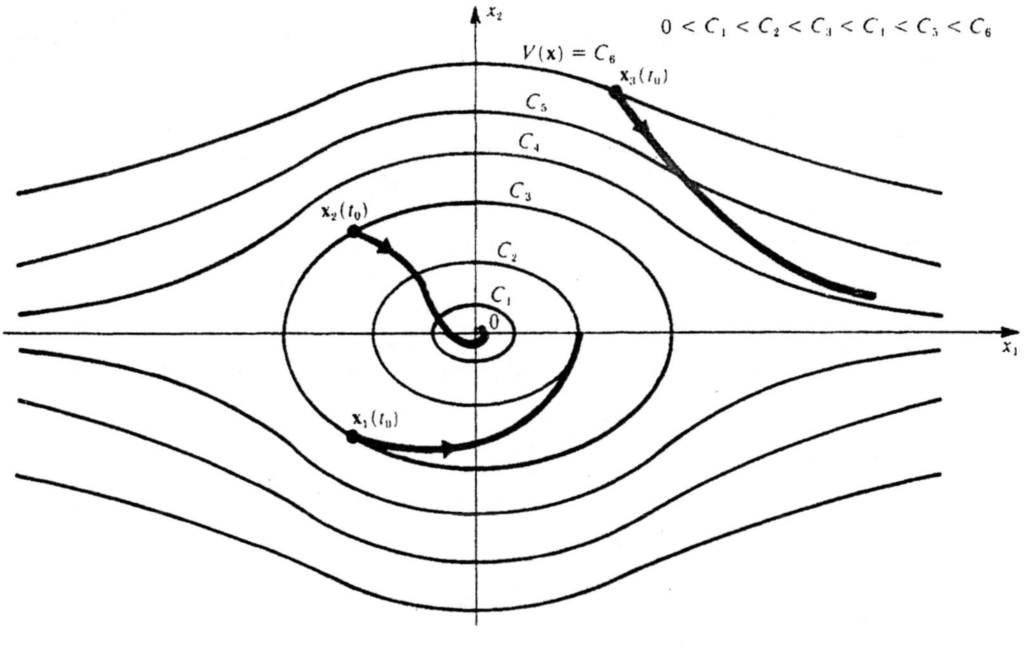

Figure 10.7

increase, but might approach a nonzero constant. If this is true and if the initial state is inside contour C_3, the state will remain bounded but need not approach zero. Initial state $\mathbf{x}_1(t_0)$ illustrates this case and indicates that the origin is stable i.s.L. but not necessarily asymptotically stable. If $\dot{V}$ is only negative semidefinite but is never zero on a solution trajectory (except at the origin), then V will always be decreasing. This rules out the behavior shown with $\mathbf{x}_1(t_0)$, and asymptotic stability is again the conclusion.

Unfortunately, the Lyapunov theorems give no indication of how a Lyapunov function might be found. There is no one unique Lyapunov function for a given system. Some are better than others. It might happen that a $V_1(\mathbf{x})$ can be found which indicates stability i.s.L., $V_2(\mathbf{x})$ might indicate asymptotic stability for initial states quite close to the origin, and $V_3(\mathbf{x})$ might indicate asymptotic stability for a much larger region, or even global asymptotic stability. The inability to find a suitable Lyapunov function does not prove instability. If a system is stable in one of the senses mentioned, it is ensured that an appropriate Lyapunov function does exist. Much ingenuity may be required to find it, however.

There is no universally best method of searching for Lyapunov functions. A form for $V(\mathbf{x})$ can be assumed, either as a pure guess or tempered by physical insight and energy-like considerations. Then $\dot{V}(\mathbf{x})$ can be tested, bringing in the system equations $\dot{\mathbf{x}} = \mathbf{f}(\mathbf{x})$ in the process. Another approach is to assume a form for the derivatives of $V(\mathbf{x})$, either $\dot{V}(\mathbf{x})$ or $\nabla_{\mathbf{x}} V(\mathbf{x})$. Then $V(\mathbf{x})$ can be determined by integration and tested to see if it meets the required conditions. Examples of these techniques are found in the problems for this chapter and, for nonlinear systems, in Chapter 15.

The task of finding suitable Lyapunov functions for linear systems is much more routine. A quadratic form can be used,

$$V(\mathbf{x}) = \mathbf{x}^T \mathbf{P} \mathbf{x}$$

where $\mathbf{P}$ is a real, symmetric, positive definite matrix. Then the time rate-of-change is $\dot{V}(\mathbf{x}) = \dot{\mathbf{x}}^T \mathbf{P}\mathbf{x} + \mathbf{x}^T \mathbf{P}\dot{\mathbf{x}} + \mathbf{x}^T \dot{\mathbf{P}}\mathbf{x}$. Using the linear system state equation $\dot{\mathbf{x}} = \mathbf{A}\mathbf{x}$, this becomes

$$\dot{V}(\mathbf{x}) = \mathbf{x}^T[\mathbf{A}^T \mathbf{P} + \mathbf{P}\mathbf{A} + \dot{\mathbf{P}}]\mathbf{x}$$

If the system is asymptotically stable, $\dot{V}$ must be negative definite, and this requires that the matrix equation

$$\mathbf{A}^T \mathbf{P} + \mathbf{P}\mathbf{A} + \dot{\mathbf{P}} = -\mathbf{Q}$$

must be true for some positive definite matrix $\mathbf{Q}$. If $\mathbf{A}$ is constant, then a constant $\mathbf{P}$ will suffice, so $\dot{\mathbf{P}} = \mathbf{0}$. The remaining equation is often called the Lyapunov equation,

$$\mathbf{A}^T \mathbf{P} + \mathbf{P}\mathbf{A} = -\mathbf{Q} \tag{10.22}$$

Numerical solution of equations of this type were discussed at length in Sec. 6.10. From those results it is recalled that a unique solution $\mathbf{P}$ will exist for each $\mathbf{Q}$ provided that no two eigenvalues of $\mathbf{A}$ satisfy $\lambda_i + \lambda_j = 0$. Therefore, for any arbitrary $\mathbf{Q}$, $\mathbf{P}$ can be found by solving n^2 linear equations (because $\mathbf{P} = \mathbf{P}^T$, only $(n^2 + n)/2$ are really needed):

$$[\mathbf{I}_n \otimes \mathbf{A}^T + \mathbf{A}^T \otimes \mathbf{I}_n](\mathbf{P}) = -(\mathbf{Q}) \tag{10.23}$$

where (P) and (Q) are the column vectorized versions of **P** and **Q**. One approach to using Eq. (*10.22*) would be to guess a trial positive definite symmetric matrix **P** and then check to see whether the calculated **Q** is positive definite. If it is, asymptotic stability is established. If **Q** is only positive semidefinite, then stability i.s.L. is established. If **Q** is negative definite, instability is concluded. If **Q** is indefinite, *nothing* can be concluded. A more powerful result is obtained by working in the reverse direction, that is, by starting with a positive definite **Q** matrix and using Eq. (*10.22*) to determine **P**. It turns out that any symmetric positive definite **Q** will suffice, and typically $\mathbf{Q} = \mathbf{I}$ is used. The result is formalized as follows.

Theorem 10.4. The constant matrix **A** is asymptotically stable—i.e., has all its eigenvalues strictly in the left-half plane—if and only if the solution **P** of Eq. (*10.22*) is positive definite when **Q** is positive definite.

Note that this is not just a special case of Theorem 10.2 because both necessary and sufficient conditions are given here. In view of the one-to-one relationship between **P** and **Q** established by Eq. (*10.22*), it may seem strange that starting with an arbitrary positive definite **Q** gives a definite answer, whereas starting with an arbitrary positive definite **P** may not. The Venn diagram of Figure 10.8 provides the explanation.

EXAMPLE 10.3 A linear system with $\mathbf{A} = \begin{bmatrix} 0 & 1 \\ -2 & -3 \end{bmatrix}$ is known to be asymptotically stable. If $\mathbf{P} = \mathbf{I}_2$ is arbitrarily selected, then Eq. (*10.22*) gives $\mathbf{Q} = -\begin{bmatrix} 0 & -1 \\ -1 & -6 \end{bmatrix}$. **Q** is *not* positive definite. If, instead, $\mathbf{Q} = \mathbf{I}_2$ is selected, then solving Eq. (*10.22*) for **P** gives $\mathbf{P} = \begin{bmatrix} 1.25 & 0.25 \\ 0.25 & 0.25 \end{bmatrix}$, which *is* positive definite. ■

EXAMPLE 10.4 A third-order system expressed in controllable canonical form is

$$\dot{\mathbf{x}} = \begin{bmatrix} 0 & 1 & 0 \\ 0 & 0 & 1 \\ -a_0 & -a_1 & -a_2 \end{bmatrix} \mathbf{x}$$

The Lyapunov equation can be reduced to

$$\begin{bmatrix} -2a_0P_{13} & P_{11} - a_0P_{23} - a_1P_{13} & P_{12} - a_0P_{33} - a_2P_{13} \\ & 2P_{12} - 2a_1P_{23} & P_{22} + P_{13} - a_1P_{33} - a_2P_{23} \\ & & 2P_{23} - 2a_2P_{33} \end{bmatrix} = -\mathbf{I}_3$$

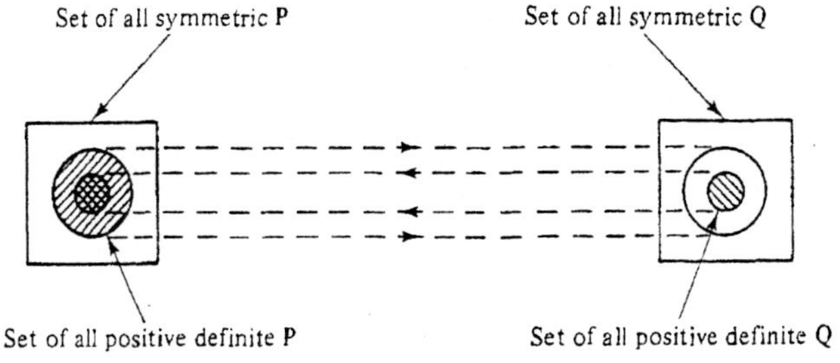

Figure 10.8 The mapping of Eq. (*10.22*) with stable **A**.

(Only the upper triangle of these symmetric equations has been shown, and the symmetry of **P** has been used explicitly.) From the 1, 1 element it is immediate that $P_{13} = 1/(2a_0)$. The five remaining unknowns are related by

$$\begin{bmatrix} 1 & 0 & 0 & -a_0 & 0 \\ 0 & 1 & 0 & 0 & -a_0 \\ 0 & 0 & 1 & -a_2 & -a_1 \\ 0 & 2 & 0 & -2a_1 & 0 \\ 0 & 0 & 0 & 2 & -2a_2 \end{bmatrix} \begin{bmatrix} P_{11} \\ P_{12} \\ P_{22} \\ P_{23} \\ P_{33} \end{bmatrix} = \begin{bmatrix} a_1/(2a_0) \\ a_2/(2a_0) \\ -1/(2a_0) \\ -1 \\ -1 \end{bmatrix}$$

These are solved using Gaussian elimination to give

$$P_{33} = [a_0 a_1 + a_0 + a_2]/[2a_0 a_1 a_2 - 2a_0^2]$$

$$P_{23} = 1/(2a_1) + a_2/(2a_0 a_1) + a_0 P_{33}/a_1$$

$$P_{22} = -1/(2a_0) + a_1 P_{33} + a_2 P_{23}$$

$$P_{12} = a_2/(2a_0) + a_0 P_{33}$$

$$P_{11} = a_1/(2a_0) + a_0 P_{23}$$

The principal minors of **P** are

$$\Delta_1 = P_{11}, \qquad \Delta_2 = P_{22}\Delta_1 - P_{12}^2$$

$$\Delta_3 = P_{33}\Delta_2 - P_{23}(P_{11}P_{23} - P_{12}P_{13}) + P_{13}(P_{12}P_{23} - P_{22}P_{13})$$

The requirement for asymptotic stability is that each of these be positive. This gives expressions somewhat like those of Routh's criterion for checking stability. This example demonstrates that the procedure can become quite messy if carried out analytically. Numerical solution, for a specific system, is quite straightforward. This is demonstrated in the problems. ∎

A more general, time-varying system is now considered:

$$\dot{\mathbf{x}} = \mathbf{f}(\mathbf{x}, t) \qquad (10.24)$$

It is assumed that the origin is an equilibrium point, $\mathbf{f}(\mathbf{0}, t) = \mathbf{0}$ for all t. The previous stability theorems are still basically what is needed to conclude stability. However, now the Lyapunov function and its time derivative may be explicit functions of time as well as the state. The definitions of positive and negative definite must be modified to reflect this added generality. Only Theorem 10.3 is generalized.

Theorem 10.5. If a single-valued scalar function $V(\mathbf{x}, t)$ exists, which is continuous and has continuous first partial derivatives and for which

(1) $V(\mathbf{0}, t) = 0$ for all t;

(2) $V(\mathbf{x}, t) \geq \sigma(\|\mathbf{x}\|) > 0$ for all $\mathbf{x} \neq \mathbf{0}$ and for all t, where $\sigma(\cdot)$ is a continuous, nondecreasing scalar function with $\sigma(0) = 0$;

(3) $\dot{V}(\mathbf{x}, t) \leq -\kappa(\|\mathbf{x}\|) < 0$ for all $\mathbf{x} \neq \mathbf{0}$ and for all t, where $\kappa(\cdot)$ is a continuous nondecreasing scalar function with $\kappa(0) = 0$;

(4) $V(\mathbf{x}, t) \leq \nu(\|\mathbf{x}\|)$ for all $\mathbf{x}$ and t, where $\nu(\cdot)$ is a continuous nondecreasing scalar function with $\nu(0) = 0$;

(5) $\sigma(\|\mathbf{x}\|) \to \infty$ as $\|\mathbf{x}\| \to \infty$;

then $\mathbf{x} = \mathbf{0}$ is uniformly globally asymptotically stable.

The functions σ, κ, and ν are all positive definite in the earlier sense, where time did not explicitly appear. $V(\mathbf{x}, t)$ is said to be positive definite if $V(\mathbf{0}, t) = 0$ and if $V(\mathbf{x}, t)$ is always greater than or equal to a time-invariant positive definite function such as σ. Conditions (1) and (2) simply require $V(\mathbf{x}, t)$ to be positive definite. Similarly, condition (3) requires $\dot{V}(\mathbf{x}, t)$ to be negative definite. Condition (5) requires $V(\mathbf{x}, t)$ to become infinite as $\|\mathbf{x}\| \to \infty$, as in Theorem 10.3, and condition (4) prevents $V(\mathbf{x}, t)$ from becoming infinite when $\|\mathbf{x}\|$ is finite. The "uniformly" in the conclusion indicates that the asymptotic stability does not depend on any particular initial time t_0.

There are a number of instability theorems [1, 8] which are useful in avoiding fruitless searches for Lyapunov functions for unstable systems. One typical theorem of this sort follows.

Theorem 10.6. If a scalar function $V(\mathbf{x}, t)$ is continuous, single valued, and has continuous first partial derivatives, and if

(1) $\dot{V}(\mathbf{x}, t)$ is positive definite in some region Ω containing the origin;
(2) $V(\mathbf{0}, t) = 0$ for all t;
(3) $V(\mathbf{x}, t) > 0$ at some point in Ω arbitrarily near the origin;

then the origin is an *unstable* equilibrium point of the system of Eq. *(10.24)*.

Lyapunov's stability theorems can also be used to investigate the stability of discrete-time systems:

$$\mathbf{x}(k+1) = \mathbf{f}(\mathbf{x}(k)) \tag{10.25}$$

The origin is assumed to be an equilibrium point, so that $\mathbf{f}(\mathbf{0}) = \mathbf{0}$. The preceding theorems for stability, asymptotic stability, and global asymptotic stability apply provided the time derivative $\dot{V}(\mathbf{x})$ is replaced by the first difference, $\Delta V = V(\mathbf{x}(k+1)) - V(\mathbf{x}(k))$.

Lyapunov's direct method of stability analysis is a general method of approach, but its successful use requires considerable ingenuity. Beginning with the very general Lyapunov philosophy, various methods of generating Lyapunov functions and easy-to-use results have been developed by restricting the system under consideration in various ways. The methods of Zubov, Lure, Popov, and others fall into this category [8, 9]. The circle criterion is one popular result of this type, which is in essence a generalized frequency domain criterion similar to the Nyquist criterion for linear systems [10]. The circle criterion is discussed in Chapter 15 in conjunction with nonlinear systems.

10.7 A CAUTIONARY NOTE ON TIME-VARYING SYSTEMS

Many control systems can be modeled as linear systems with time-varying coefficients. Examples include aircraft or spacecraft whose mass decreases as fuel is burned; machines for processing paper, wire, or other materials which cause the driven inertia to change as material is wound on the spool; robotic manipulators with unknown or varying loads; and linearized approximations to nonlinear systems whose coefficients change as they are evaluated at different points along the nominal trajectory.

A commonly used stability analysis technique is the so-called frozen coefficient method, in which all time-varying coefficients are frozen and then the system stability is analyzed as if it were a constant coefficient system. It is almost a folk theorem that if the eigenvalues (poles) are safely within the stability region at all time points (or at least a representative sampling of them) and if the coefficients and eigenvalues are not changing "too rapidly," then the time-varying system can be presumed stable. Empirical evidence suggests that when used with caution, this approach will *usually* give correct results. However, the theoretical basis for the folk theorem and a precise definition of "too rapidly" are not well understood. Total dependence on eigenvalue or pole location can be misleading for time-varying systems. The direct method of Lyapunov can be used to advantage in making stability decisions in these cases. Two carefully selected examples are given to reinforce these words of caution and to give examples of using the Lyapunov techniques.

EXAMPLE 10.5 [6] Consider the unforced time-varying linear system

$$\dot{\mathbf{x}}(t) = \begin{bmatrix} -1 + \alpha\cos^2(t) & 1 - \alpha\sin(t)\cos(t) \\ -1 - \alpha\sin(t)\cos(t) & -1 + \alpha\sin^2(t) \end{bmatrix}\mathbf{x}(t)$$

Find the eigenvalues of the 2×2 $\mathbf{A}(t)$ matrix and then use Lyapunov's direct method to draw conclusions about system stability as a function of the scalar parameter α. The characteristic equation is $|\lambda\mathbf{I} - \mathbf{A}(t)| = 0$ or $\lambda^2 + (2-\alpha)\lambda + (2-\alpha) = 0$, from which the eigenvalues are

$$\lambda_i = -(2-\alpha)/2 \pm \sqrt{(2-\alpha)^2/4 - (2-\alpha)}$$

If α is any real number less than 2, the system has strictly left-half-plane eigenvalues. The eigenvalues are not functions of time, so the question of whether they are moving too rapidly does not apply here. It is not always possible to find the transition matrix for time-varying systems, but in this case it is known to be [6]

$$\boldsymbol{\Phi}(t, 0) = \begin{bmatrix} e^{(\alpha-1)t}\cos(t) & e^{-t}\sin(t) \\ -e^{(\alpha-1)t}\sin(t) & e^{-t}\cos(t) \end{bmatrix}$$

This can be verified by showing that both $\boldsymbol{\Phi}(0,0) = \mathbf{I}$ and $\dot{\boldsymbol{\Phi}} = \mathbf{A}(t)\boldsymbol{\Phi}$ are satisfied. From knowledge of $\boldsymbol{\Phi}$, it is clear that for any real value of $\alpha > 1$, exponential growth occurs and the system is unstable by any definition, even though both eigenvalues are in the left-half plane for $\alpha < 2$.

A quadratic Lyapunov function is adequate for analyzing linear systems, and in this case a very simple one suffices. let $V = \frac{1}{2}\mathbf{x}^T\mathbf{x}$. Then $\dot{V} = \frac{1}{2}\mathbf{x}^T[\mathbf{A}^T(t) + \mathbf{A}(t)]\mathbf{x}$. The quadratic form $\dot{V}$ is negative definite if the matrix $\mathbf{A}_s = \frac{1}{2}[\mathbf{A}^T(t) + \mathbf{A}(t)]$ is negative definite. Since this is a symmetric matrix, its sign definiteness can be checked by using its principal minors or its eigenvalues, as presented in Sec. 7.8. It is worth emphasizing that the system stability can be assured if *all* eigenvalues of $\mathbf{A}_s$—i.e., the symmetric part of $\mathbf{A}(t)$—are in the left-half plane. (The factor $\frac{1}{2}$ is superfluous. It does not change the sign of the eigenvalues.) For this example, the eigenvalues of the symmetric part of $\mathbf{A}(t)$ are $\lambda = -2$ and $-2(1-\alpha)$. Therefore $\dot{V}$ is negative definite for all real $\alpha < 1$. Lyapunov's second method correctly indicates that the system is stable for $\alpha < 1$. If the transition matrix were not known, the question of whether the system might still be stable for some values of α larger than 1 could legitimately be raised because Lyapunov's method does not always give a tight bound. To answer this question one could search for a better Lyapunov function or attempt to use the instability Theorem 10.5. ■

It is not true that the system must be unstable just because $\mathbf{A}$ (or its symmetric part, $\mathbf{A}_s$) has an eigenvalue in the right-half plane. It is true that if *all* eigenvalues of $\mathbf{A}_s$ are in the right-half plane, the system is unstable. These results are proven as follows. From Problem 7.37 it is seen that

$$\lambda_{\min}\|\mathbf{x}(t)\|^2 \le \mathbf{x}^T\mathbf{A}_s\mathbf{x} \equiv \dot{V} \le \lambda_{\max}\|\mathbf{x}(t)\|^2$$

and this remains true for all t if the eigenvalues are time-varying. Dividing by $V \equiv \|\mathbf{x}(t)\|^2$, this inequality can be written

$$\lambda_{\min}\,dt \le dV/V \le \lambda_{\max}\,dt$$

Integrating gives $\int \lambda_{\min}\,dt \le \ln\{V(t)/V(t_0)\} \le \int \lambda_{\max}\,dt$. When raised to the exponential, this inequality gives

$$\exp\{\textstyle\int \lambda_{\min}\,dt\} \le V(t)/V(t_0) \le \exp\{\textstyle\int \lambda_{\max}\,dt\}$$

or

$$\|\mathbf{x}(t_0)\|^2 \exp\{\textstyle\int \lambda_{\min}\,dt\} \le \|\mathbf{x}(t)\|^2 \le \|\mathbf{x}(t_0)\|^2 \exp\{\textstyle\int \lambda_{\max}\,dt\}$$

Since $\mathbf{A}_s$ is symmetric, its eigenvalues are all real (Problem 7.25). If $\lambda_{\max}$ of $\mathbf{A}_s$ is negative (then they all are), the integral in the exponent will approach $-\infty$ as $t \to \infty$. Thus $\|\mathbf{x}(t)\| \to 0$, and asymptotic stability is assured. Likewise if $\lambda_{\min}$ is positive (then so are all of the eigenvalues of $\mathbf{A}_s$), the leftmost inequality shows that $\|\mathbf{x}(t)\| \to \infty$, and the system is unstable. Although it is the sign of the time integral of the eigenvalues and not the eigenvalues themselves that is important here, constant bounds on the eigenvalue excursions are easier to work with. These kinds of results are occasionally useful, but they are usually ultraconservative. A system is now demonstrated whose matrix $\mathbf{A}$ has a right-half plane eigenvalue for all time, and yet the system is asymptotically stable for a certain range of the parameter ω.

EXAMPLE 10.6 Consider the linear, time-varying system

$$\dot{\mathbf{x}} = \begin{bmatrix} -4 - \sqrt{50}\sin(\omega t) & 1 \\ 25\cos(2\omega t) & -4 + \sqrt{50}\sin(\omega t) \end{bmatrix}\mathbf{x}$$

The eigenvalues of $\mathbf{A}(t)$ are $\lambda_1 = 1$ and $\lambda_2 = -9$ for all t. For this system we have

$$\mathbf{A}_s = \begin{bmatrix} -8 + 2\sqrt{50}\sin(\omega t) & 1 + 25\cos(2\omega t) \\ 1 + 25\cos(2\omega t) & -8 + 2\sqrt{50}\sin(\omega t) \end{bmatrix}$$

The eigenvalues of $\mathbf{A}_s$ are given by

$$\lambda = -8 \pm \{200\sin^2(\omega t) + [1 + 25\cos(2\omega t)]^2\}^{1/2}$$

The minimum eigenvalue is clearly always negative. The maximum eigenvalue is approximately bounded by $7.179 < \lambda_{\max} < 19.856$. Yet this system is asymptotically stable for $\omega > 6.14$. This is only an approximate bound, found by numerical integration. Simulation shows that if ω is reduced to 5.9, for instance, the system is definitely unstable—but not in the simple exponential way the eigenvalue of $\mathbf{A}$ at $+1$ might suggest. When ω is further decreased to about 1, then the response looks more like the expected exponential growth. Even there, the rate of growth does not appear as simple as e^t. When ω is increased to about 6.2, the oscillations are clearly seen to be decaying. Figures 10.9 and 10.10 show the initial condition response of the two state components of **x** for $\omega = 5$ and $\omega = 6.8$, respectively. ■

Even though the previous system has a right-half plane eigenvalue, it is still stable if the oscillations are fast enough. Note that the folk theorem states that the frozen coefficient analysis based on eigenvalue locations will usually give the correct stability result if the system parameters are not changing too rapidly. If the oscillations in this system are slowed down by reducing ω, then the eigenvalue locations (one unstable) do correctly predict an unstable system. This example gives a quantitative meaning to *too rapidly* for this problem. Also notice that the simple bound using the integral of the maximum eigenvalue was *not* evaluated explicitly. Numerical integration indicates that the bounds will not be satisfied and therefore cannot be used to establish the stability regime. The simple use of the constant upper limit for λ_{max} is even more obviously of no help here. The existence of a suitable Lyapunov function is guaranteed. Finding it will require looking beyond the simple unit diagonal form used earlier.

10.8 USE OF LYAPUNOV'S METHOD IN FEEDBACK DESIGN

Stability is an overriding requirement in almost all control system designs. It is possible to use the Lyapunov method directly in the design of the feedback controller in some situations. Since response time or settling time is frequently involved in the system-design specifications, it is first noted that the Lyapunov function and its derivative can be used to estimate the system's speed of response, i.e., its dominant time constant [2, 3, 9]. Assume that the origin is a stable equilibrium point. Let $V(\mathbf{x}, t)$ be a Lyapu-

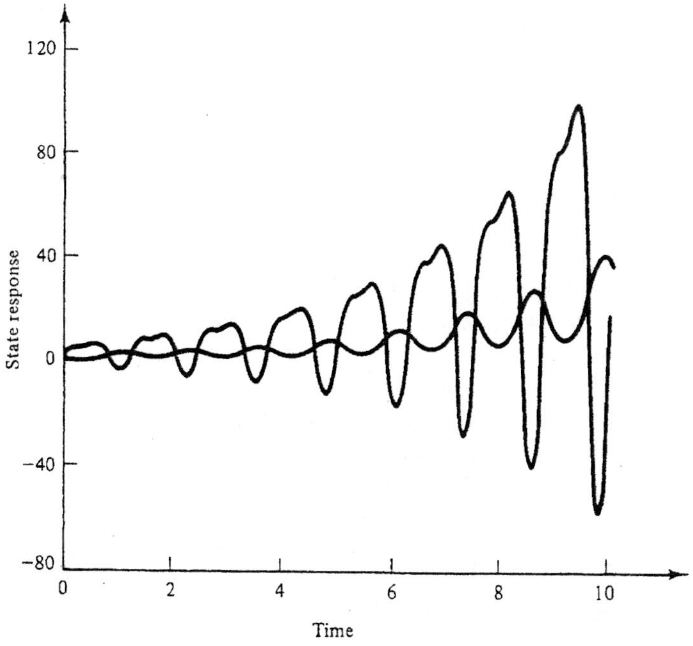

Figure 10.9 Response with $\omega = 5$.

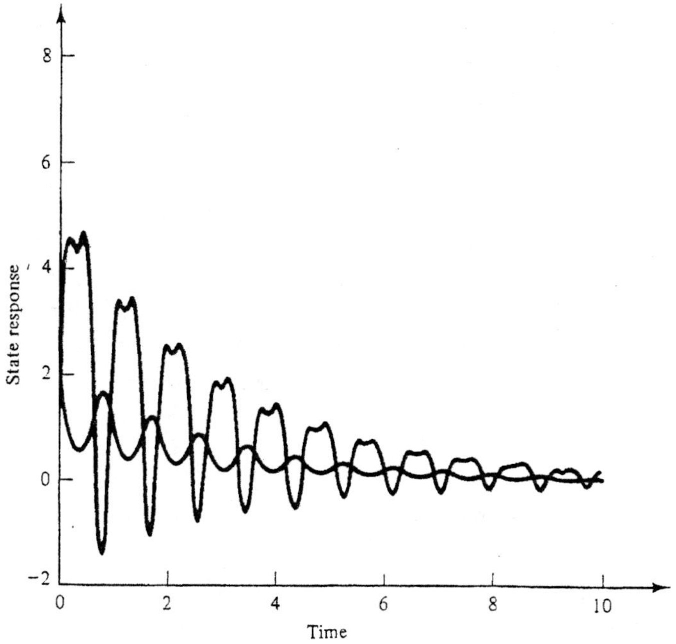

Figure 10.10 Response with $\omega = 6.8$.

nov function, and define $\eta = -\dot{V}(\mathbf{x}, t)/V(\mathbf{x}, t)$. Integrating both sides of this definition gives

$$\int_{t_0}^{t} \eta\, dt = -\int_{V(\mathbf{x}(t_0), t_0)}^{V(\mathbf{x}(t), t)} \frac{dV}{V}$$

from which

$$\ln\left[\frac{V(\mathbf{x}(t), t)}{V(\mathbf{x}(t_0), t_0)}\right] = -\int_{t_0}^{t} \eta\, dt \quad \text{or} \quad V(\mathbf{x}(t), t) = V(\mathbf{x}(t_0), t_0)e^{-\int_{t_0}^{t} \eta\, dt}$$

Although η is not generally constant, if $\eta_{\min}$ is defined as the minimum value of η, then

$$V(\mathbf{x}(t), t) \le V(\mathbf{x}(t_0), t_0)e^{-\eta_{\min}(t - t_0)}$$

For an asymptotically stable system, $\mathbf{x}(t) \to \mathbf{0}$ as $V(\mathbf{x}, t) \to 0$. It is seen that $1/\eta_{\min}$ is an approximate bound on the energy decay rate time constant. This is sometimes taken as a bound on the system's dominant time constant. However, since energy (and frequently the generalized energy V) is quadratic in $\mathbf{x}$, $V \approx V_0 e^{-\eta t}$ suggests that $\mathbf{x}(t) \approx \mathbf{x}(0)e^{-\eta t/2}$ might be a better bound on the state behavior. From this it is suggested that $2/\eta_{\min}$ be used as an approximation of the system's dominant time constant. This will be used later in designing a controller with a settling time specification. With or without the factor of 2, when several candidate Lyapunov functions are under consideration, it is seen that the one with the larger ratio $-\dot{V}/V$ will give the faster time response.

Since $V(\mathbf{x}, t)$ does not depend on $\dot{\mathbf{x}}$ and hence on $\mathbf{u}$—but $\dot{V}$ does—the following controller design procedure can sometimes be used to advantage. First select a Lyapunov function for the homogeneous system equations. This V should be positive definite, and $\dot{V}$ should be at least negative semidefinite. This assumes that the uncontrolled system is at least stable i.s.L. Then allow the control variable **u** to be nonzero and add these control-dependent terms to form a modified $\dot{V}$ term. V itself will not change. With V known from the unforced analysis, the only unknowns in the modified $\dot{V}$ are terms containing **u**. These can be selected to force $\dot{V}$ to be as negative as possible within available control limits. This would yield a control system which approximates a minimum time response system. It typically leads to a nonlinear control law, even when the original system is linear.

EXAMPLE 10.7 Consider the system

$$\dot{\mathbf{x}} = \begin{bmatrix} -3 & 2 \\ -1 & -1 \end{bmatrix}\mathbf{x} + \begin{bmatrix} 1 & -1 \\ 0 & 1 \end{bmatrix}\mathbf{u}(t)$$

The input control vector is limited by $\mathbf{u}^T(t)\mathbf{u}(t) \le 1$ for all time t. Use the quadratic Lyapunov function found for this system in Problem 10.10. Specify the feedback control law for computing $\mathbf{u}(t)$ as a function of $\mathbf{x}(t)$ which will make the modified $\dot{V}$ as negative as possible. This controller will rapidly drive the state back to $\mathbf{x}(t) = \mathbf{0}$ if it is perturbed by initial conditions or disturbances.

Using the Lyapunov function of Problem 10.10, $V(\mathbf{x}) = \mathbf{x}^T\mathbf{P}\mathbf{x}$ and then $\dot{V}(\mathbf{x}) = \dot{\mathbf{x}}^T\mathbf{P}\mathbf{x} + \mathbf{x}^T\mathbf{P}\dot{\mathbf{x}} = -\mathbf{x}^T\mathbf{x} + 2\mathbf{u}^T\mathbf{B}^T\mathbf{P}\mathbf{x}$. The control $\mathbf{u}(t)$ is selected to minimize $\mathbf{u}^T\mathbf{B}^T\mathbf{P}\mathbf{x}$ subject to the restriction placed upon $\mathbf{u}(t)$. Obviously $\mathbf{u}(t)$ should be parallel to $\mathbf{B}^T\mathbf{P}\mathbf{x}$, but with the opposite sign, and have its largest possible magnitude. That is,

$$\mathbf{u}(t) = \frac{-\mathbf{B}^T\mathbf{P}\mathbf{x}(t)}{\|\mathbf{B}^T\mathbf{P}\mathbf{x}(t)\|}$$

Using the given form for the matrices **B** and **P**, this gives

$$\mathbf{u}(t) = \frac{\begin{bmatrix} -7x_1(t) + x_2(t) \\ 8x_1(t) - 19x_2(t) \end{bmatrix}}{(113x_1^2 - 318x_1x_2 + 362x_2^2)^{1/2}}$$ ■

Other control system designs might require a specified response time rather than the maximal effort, minimum time response just demonstrated. One such example follows.

EXAMPLE 10.8 Consider the linearized equations of motion for the spin-stabilized satellite of Problem 3.16. Assume that the value of Ω is 8 rad/min. Design a linear controller which will drive any wobbling errors to zero with a time constant of 0.5 min.

The equations of motion are rewritten here as

$$\dot{\mathbf{x}} = \begin{bmatrix} 0 & -\Omega \\ \Omega & 0 \end{bmatrix}\mathbf{x} + \mathbf{u}(t)$$

A quadratic Lyapunov function $V = \mathbf{x}^T\mathbf{P}\mathbf{x}$ is selected, and $\mathbf{P} = \mathbf{I}_2$ is adequate for present purposes. Then $\dot{V} = \mathbf{x}^T[\mathbf{A}^T + \mathbf{A}]\mathbf{x} + \mathbf{u}^T\mathbf{x} + \mathbf{x}^T\mathbf{u}$. With $\mathbf{u} = \mathbf{0}$, $\dot{V}$ is identically zero, due to the skew-symmetric matrix **A**. The conclusion to be drawn from this particular Lyapunov function is that the uncontrolled system is stable i.s.L. but not necessarily asymptotically stable. Actually, this was known at the outset, since no energy dissipation has been included in the model. If the linear controller $\mathbf{u} = -k\mathbf{x}$ is selected, then the **u**-dependent terms give the modified $\dot{V} = -2k\mathbf{x}^T\mathbf{x}$. This is negative definite for any positive real scalar k. The ratio $\eta = -\dot{V}/V = 2k$ can be used to select k in order to meet the settling time-constant specification $\tau = 0.5 \approx 2/(2k)$. Therefore, $k = 2$ is selected, meaning that the feedback control law is $\mathbf{u}(t) = -2\mathbf{x}(t)$. Figure 10.11 shows the time response due to a unit initial condition error. The $\pm e^{-2t}$ decay envelope is also included.

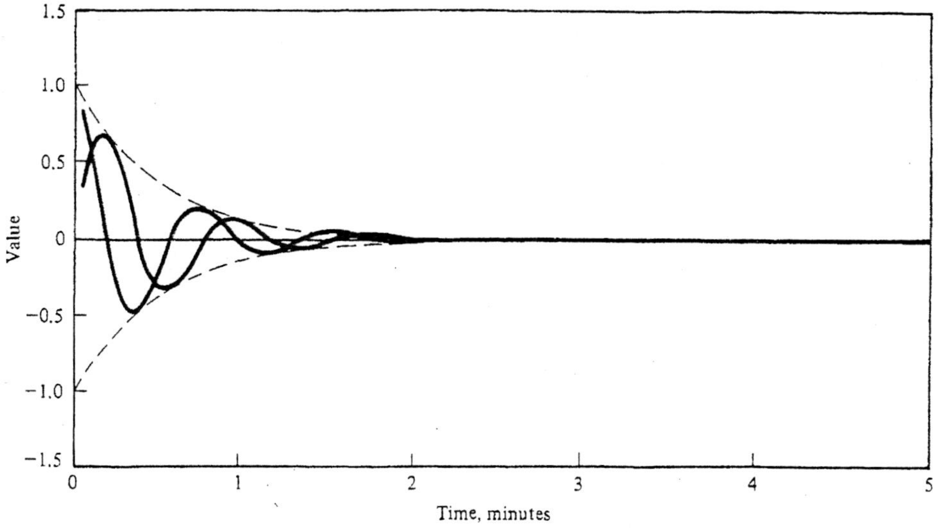

Figure 10.11 Satellite wobble control example, Example 10.7, Linear Feedback $u = -2x$.

REFERENCES

1. LaSalle, J. and S. Lefschetz: *Stability by Liapunov's Direct Method, with Applications,* Academic Press, New York, 1961.
2. Kalman, R. E. and J. E. Bertram: "Control System Analysis and Design Via the Second Method of Lyapunov I. Continuous-Time Systems," *Trans. ASME Journal of Basic Engineering,* Vol. 82D, June 1960, pp. 371–393.
3. Kalman, R. E. and J. E. Bertram: "Control System Analysis and Design Via the Second Method of Lyapunov II. Discrete-Time Systems," *Trans. ASME Journal of Basic Engineering,* Vol. 82D, June 1960, pp. 394–400.
4. Guckenheimer, J. and P. Holmes: *Nonlinear Oscillations, Dynamical Systems, and Bifurcations of Vector Fields,* Springer-Verlag, New York, 1983.
5. Schultz, D. G. and J. E. Gibson: "The Variable Gradient Method for Generating Liapunov Functions," *AIEE Trans.,* Part II, Vol. 81, Sept. 1962, pp. 203–210.
6. Desoer, C. A.: *Notes for a Second Course on Linear Systems,* Van Nostrand Reinhold, New York, 1970.
7. Coddington, E. A. and N. Levinson: *Theory of Ordinary Differential Equations,* McGraw-Hill, New York, 1955.
8. Willems, J. L.: *Stability Theory of Dynamical Systems,* John Wiley, New York, 1970.
9. Ogata, K.: *State Space Analysis of Control Systems,* Prentice Hall, Englewood Cliffs, N.J., 1967.
10. Atherton, D. P.: *Nonlinear Control Engineering,* Van Nostrand Reinhold, New York, 1975.
11. Reiss, R. and G. Geiss: "The Construction of Liapunov Functions," *IEEE Transactions on Automatic Control,* Vol. AC-8, No. 4, Oct. 1963, pp. 382–383.
12. Puri, N. N. and C. N. Weygandt: "Second Method of Liapunov and Routh's Canonical Form," *Journal of The Franklin Institute,* Vol. 276, No. 5, 1963, pp. 365–384.
13. Ku, Y. H. and N. N. Puri: "On Liapunov Functions of High Order Nonlinear Systems," *Journal of The Franklin Institute,* Vol. 276, No. 5, 1963, pp. 349–364.

ILLUSTRATIVE PROBLEMS

Constant, Linear System Stability

.1 Comment on the stability of the four constant coefficient systems of Problem 3.1, page 104.

(a) The eigenvalue is $\lambda = -\alpha$ and the system is asymptotically stable if $\lambda < 0$. If $\lambda = 0$, the system is stable in the sense of Lyapunov. If $\lambda > 0$, it is unstable.

(b) The eigenvalue is again $\lambda = -\alpha$ and the results of part (a) still apply.

(c) The eigenvalues of the 2×2 **A** matrix satisfy $\lambda^2 + 2\zeta\omega\lambda + \omega^2 = 0$. They are $\lambda = -\zeta\omega \pm \omega\sqrt{\zeta^2 - 1}$. Both eigenvalues are negative for any positive value of ζ and then the system is asymptotically stable. This assumes that ω is a positive constant. If $\zeta = 0$, then $\lambda = \pm j\omega$ and the system is stable i.s.L. For $\zeta < 0$, the system is unstable.

(d) The results of part (c) again apply.

These examples illustrate that the poles of a transfer function are eigenvalues of the matrix **A** in the state space representation. This will always be true, but the converse need not be true.

.2 Is the system of Problem 3.2, page 105, stable? Is it asymptotically stable?

Evaluating $|\mathbf{A} - \mathbf{I}\lambda| = 0$ gives $\lambda^4 + 3\lambda^3 + 3\lambda^2 + \lambda = 0$. Thus one eigenvalue is $\lambda = 0$. The remaining cubic $\lambda^3 + 3\lambda^2 + 3\lambda + 1 = 0$ can be investigated using Routh's criterion. This indicates that there are no right-half-plane eigenvalues. Thus the system is stable, but not asymptotically stable because of the root $\lambda = 0$.

.3 **(a)** Is the system of Problem 3.3, page 106, asymptotically stable?

(b) Is the system of Problem 3.4, page 107, asymptotically stable?

(a) All three of the forms given for **A** have the same eigenvalues, $\lambda = -3, -1$, and -6. (These are also poles of the original transfer function.) This system is asymptotically stable.

(b) The 6×6 **A** matrix is given in Jordan form, from which it is apparent that the eigenvalues are $0, 0, -3, -3, -3, -1$. This system is unstable because of the double root $\lambda = 0$. With $\mathbf{u} = \mathbf{0}$, $\dot{x}_2 = 0$ or $x_2 =$ constant. Since $\dot{x}_1 = x_2$, $x_1(t)$ is a linear function of time. Thus unless $x_2(0) = 0$, $x_1(t) \rightarrow \pm\infty$ and as a result $\|\mathbf{x}(t)\| \rightarrow \infty$.

.4 Investigate the stability of the continuous-time system and its discrete-time approximation, as described in Problem 9.12, page 331.

The continuous-time system matrix **A** has eigenvalues $\lambda = 0, -1$. Therefore, the system is stable in the sense of Lyapunov but not asymptotically stable. The discrete-time system matrix **A** has eigenvalues $\lambda_1 = 0.368$ and $\lambda_2 = 1$. Since $|\lambda_1| < 1$, but $|\lambda_2| = 1$, the discrete system is stable in the sense of Lyapunov, but not asymptotically stable.

.5 Comment on the stability of the linear discrete-time systems described in

(a) Problem 9.20, page 335;

(b) Problem 9.21, page 335;

(c) Problem 9.23, page 336.

(a) Since the **A** matrix is triangular, its eigenvalues are found by inspection as $\lambda_1 = \lambda_2 = \lambda_3 = \frac{1}{2}$. Since all $|\lambda_i| < 1$, the system is asymptotically stable. This is verified by the explicit solution found in Problem 9.20, which shows that $\mathbf{x}(k) \rightarrow \mathbf{0}$ as $k \rightarrow \infty$.

(b) The eigenvalues were given earlier as $\lambda_1 = \frac{1}{2}, \lambda_2 = \frac{1}{3}$. Since $|\lambda_i| < 1$, the system is asymptotically stable.

(c) The eigenvalues are $\lambda_1 = 1, \lambda_2 = \frac{1}{2}, \lambda_3 = -\frac{1}{3}$. This system is stable i.s.L., but not asymptotically stable, since $\lambda_1 = 1$.

.6 Show that if a continuous-time, linear constant system is asymptotically stable, then its adjoint is unstable.

Asymptotic stability of $\dot{\mathbf{x}} = \mathbf{A}\mathbf{x}$ implies that all eigenvalues of **A** have negative real parts. The adjoint equation is $\dot{\mathbf{z}} = -\mathbf{A}^T\mathbf{z}$. The eigenvalues of $-\mathbf{A}^T$ are the roots of $|-\mathbf{A}^T - \mathbf{I}\gamma| = 0$ or $|\mathbf{A}^T - \mathbf{I}(-\gamma)| = \mathbf{0}$. Since $\mathbf{A}^T$ and **A** have the same eigenvalues, the roots γ are the negative of the eigenvalues of **A**. Then **A** having all left half plane eigenvalues implies that $-\mathbf{A}^T$ has all right half plane eigenvalues, and the adjoint system is unstable.

10.7 Show that if a continuous-time, linear constant system is asymptotically stable, it is also BIBS stable.

The conditions for BIBS stability are given by Eqs. (*10.6*) and (*10.8*). If the system is asymptotically stable, then $\|\mathbf{\Phi}(t, t_0)\| \le N(t_0)$ for all $t \ge t_0$. Furthermore, using norm inequalities gives

$$\int_{t_0}^{t} \|\mathbf{\Phi}(t, \tau)\mathbf{B}\|\, d\tau \le \int_{t_0}^{t} \|\mathbf{\Phi}(t, \tau)\| \cdot \|\mathbf{B}\|\, d\tau$$

Since **B** and hence $\|\mathbf{B}\|$ are constant, this term can be taken outside the integral. Also, expressing $\mathbf{\Phi}$ in terms of the Jordan form and modal matrix of **A** (see page 317) leads to

$$\int_{t_0}^{t} \|\mathbf{\Phi}(t, \tau)\mathbf{B}\|\, d\tau \le \|\mathbf{M}^{-1}\| \cdot \|\mathbf{M}\| \int_{t_0}^{t} \|e^{\mathbf{J}(t-\tau)}\|\, d\tau \|\mathbf{B}\|$$

If all the eigenvalues of **A** are distinct, then $\|e^{\mathbf{J}(t-\tau)}\| = e^{\beta_i(t-\tau)}$, where β_i is the largest real part of the eigenvalues of **A**. But β_i is negative since the system is asymptotically stable. In this case the integral is bounded and thus BIBS stability follows. If **A** has repeated eigenvalues, then $\|e^{\mathbf{J}(t-\tau)}\|$ can be bounded by $\sqrt{p((t)}\, e^{\beta_i(t-\tau)}$, where $p(t)$ is a polynomial in t. Asymptotic stability ensures that $\beta_i < 0$, and again the integral is bounded. Therefore, for linear constant systems, asymptotic stability implies BIBS stability.

10.8 Show that if a continuous-time, linear constant system is asymptotically stable, it is also BIBO stable.

The system $\dot{\mathbf{x}} = \mathbf{Ax} + \mathbf{Bu}$, $\mathbf{y} = \mathbf{Cx} + \mathbf{Du}$ has the output solution

$$\mathbf{y}(t) = \mathbf{C\Phi}(t, t_0)\mathbf{x}(t_0) + \int_{t_0}^{t} \mathbf{C\Phi}(t, \tau)\mathbf{Bu}(\tau)\, d\tau + \mathbf{Du}(t)$$

Since the norm of any constant matrix is bounded, straightforward application of norm inequalities leads to

$$\|\mathbf{y}(t)\| \le \|\mathbf{C}\|\, \|\mathbf{\Phi}(t, t_0)\|\, \|\mathbf{x}(t_0)\| + \|\mathbf{C}\| \cdot \|\mathbf{B}\| K \int_{t_0}^{t} \|\mathbf{\Phi}(t, \tau)\|\, d\tau + \|\mathbf{D}\| K$$

where $\|\mathbf{u}(t)\| \le K$ for all t. Asymptotic stability ensures that $\|\mathbf{\Phi}(t, t_0)\|$ is bounded by a decaying exponential and thus $\int_{t_0}^{t} \|\mathbf{\Phi}(t, \tau)\|\, d\tau$ is also bounded for all $t \ge t_0$. It follows that the output is bounded in norm also.

It is also known [8] that *if the system is completely controllable and completely observable*, then the implication can also be reversed. In that case asymptotic stability $\Leftrightarrow$ BIBO stability.

Lyapunov Methods

10.9 Verify the stability of the third-order system of Problem 3.3. This is of the same form as the problem treated in Example 10.3 with parameter values $a_0 = 18$, $a_1 = 27$, and $a_2 = 10$. Numerical solution of the Lyapunov equation (*10.22*) with $\mathbf{Q} = \mathbf{I}_3$, using the full 9×9 Kronecker product form of Eq. (*10.23*), yields

$$\mathbf{P} = \begin{bmatrix} 1.94841 & 1.297619 & 0.027777 \\ 1.297619 & 2.167769 & 0.066578 \\ 0.027777 & 0.066578 & 0.056657 \end{bmatrix}$$

From this the principal minors are $\Delta_1 = 1.94841$, $\Delta_2 = 2.53989$, and $\Delta_3 = 0.138395$. Since they are all positive, **P** is positive definite and the system is asymptotically stable. This verifies a known result, since the eigenvalues of **A** were given in Problem 3.3 as $\lambda = -1, -3$, and -6.

10.10 Investigate the stability of the system described by

$$\dot{\mathbf{x}} = \begin{bmatrix} -3 & 2 \\ -1 & -1 \end{bmatrix} \mathbf{x} \qquad (1)$$

Let $\mathbf{P} = [P_{ij}]$ with $P_{12} = P_{21}$. Then

$$\mathbf{A}^T\mathbf{P} + \mathbf{PA} = \begin{bmatrix} -6P_{11} - 2P_{12} & -4P_{12} - P_{22} + 2P_{11} \\ -4P_{12} - P_{22} + 2P_{11} & 4P_{12} - 2P_{22} \end{bmatrix}$$

Using $\mathbf{Q} = \mathbf{I}_2$, the unit matrix, and solving $\mathbf{A}^T\mathbf{P} + \mathbf{PA} = -\mathbf{Q}$ gives $\mathbf{P} = \begin{bmatrix} \frac{7}{40} & -\frac{1}{40} \\ -\frac{1}{40} & \frac{18}{40} \end{bmatrix}$. The principal minors of $\mathbf{P}$ are $\Delta_1 = \frac{7}{40} > 0$ and $\Delta_2 = |\mathbf{P}| = \frac{5}{64} > 0$. Therefore, $\mathbf{P}$ is positive definite and the system is asymptotically stable.

11 Derive conditions for asymptotic stability of the origin for the following systems. Use Lyapunov's direct method and proceed by assuming a suitable form for $\dot{V}(\mathbf{x})$. $V(\mathbf{x})$ is then found by integration.

(a) $\dot{x} = ax$

(b) $\dot{x}_1 = x_2, \dot{x}_2 = -ax_1 - bx_2$

(c) $\dot{x}_1 = x_2, \dot{x}_2 = x_3, \dot{x}_3 = -ax_1 - bx_2 - cx_3$

A form for $\dot{V}$ is assumed, which is at least negative semidefinite. A choice which is often successful is $\dot{V} = -x_n^2$ [11]. Then $V(\mathbf{x}(t)) - V(\mathbf{x}(t_1)) = \int_{t_1}^{t} \dot{V}(\mathbf{x})\, dt$. The lower limit is selected so that $\mathbf{x}(t_1) = \mathbf{0}$ and $V(\mathbf{0}) = 0$. If the $V(\mathbf{x})$ found in this manner is positive definite, then the method is successful.

(a) Try $\dot{V} = -x^2$. Assuming $a \neq 0$, this gives $\dot{V} = -x\dot{x}/a$. Then

$$V(x) = -\frac{1}{a}\int_{t_1}^{t} x\dot{x}\, dt = -\frac{1}{a}\int_0^x x\, dx = -\frac{x^2}{2a}$$

$V(x)$ is positive definite if $a < 0$. Since $\dot{V}$ is negative definite, asymptotic stability results if $a < 0$.

(b) Try $\dot{V} = -x_2^2 = (-x_2)x_2$. If $a \neq 0$, then $-x_2 = (\dot{x}_2 + ax_1)/b$ so that $\dot{V} = \dot{x}_2 x_2/b + ax_1x_2/b$. Using $x_2 = \dot{x}_1$ gives

$$V(\mathbf{x}) = (a/b)\int_0^{x_1} x_1\, dx_1 + (1/b)\int_0^{x_2} x_2\, dx_2 = \frac{ax_1^2}{2b} + \frac{x_2^2}{2b}$$

If $a > 0$ and $b > 0$, $V(\mathbf{x})$ is positive definite. $\dot{V}$ is negative semidefinite, but is never zero on any trajectory of this system except at $\mathbf{x} = \mathbf{0}$. Therefore, $a > 0$ and $b > 0$ ensure asymptotic stability.

(c) Try $\dot{V} = -x_3^2$. Then $V(\mathbf{x}) = -\int_{t_1}^{t} x_3^2\, dt = -\int_{t_1}^{t} x_3\dot{x}_2\, dt$. Using integration by parts, $V(\mathbf{x}) = -x_3x_2 + \int x_2\dot{x}_3\, dt$. Using the differential equation to replace $\dot{x}_3$,

$$V(\mathbf{x}) = -x_3x_2 - \int x_2(ax_1 + bx_2 + cx_3)\, dt = -x_3x_2 - \frac{ax_1^2}{2} - b\int x_2^2\, dt - \frac{cx_2^2}{2}$$

The integral term is evaluated, using $\dot{x}_1 = x_2$ and integration by parts:

$$b\int x_2^2\, dt = bx_1x_2 - b\int x_1\dot{x}_2\, dt = bx_1x_2 - b\int x_1x_3\, dt$$

From the differential equation, $-x_1 = (1/a)(\dot{x}_3 + bx_2 + cx_3)$. Hence

$$b\int x_2^2\, dt = bx_1x_2 + \frac{b}{2a}x_3^2 + \frac{b^2}{2a}x_2^2 + (bc/a)\int_{t_2}^{t} x_3^2\, dt$$

which gives

$$V(\mathbf{x}) = -\frac{a}{2}\left(x_1 + \frac{b}{a}x_2\right)^2 - \frac{b}{2a}\left(x_3 + \frac{a}{b}x_2\right)^2 - \frac{c - a/b}{2}x_2^2 - \frac{bc}{a}\int x_3^2\, dt$$

This expression is *not* positive definite. In fact, it can be made negative definite. A new trial function is selected as $V'(\mathbf{x}) = -V(\mathbf{x}) - (bc/a)\int_{t_1}^{t} x_3^2\, dt$. Then

$$\dot{V}'(\mathbf{x}) = -\dot{V}(x) - (bc/a)x_3^2 = -(bc/a - 1)x_3^2$$

$\dot{V}'(\mathbf{x})$ is negative semidefinite if $bc/a > 1$. $V'(\mathbf{x})$ is positive definite if, in addition, $a > 0, b > 0, c > 0$. Since $\dot{V}'(\mathbf{x})$ can never vanish on a trajectory of this system, the conditions for asymptotic stability are $a > 0$, $b > 0$, $c > 0$, and, $bc - a > 0$.

In all three cases the well-known results of the Routh stability criterion have been determined. The procedure illustrated by these examples can be extended to nonlinear problems [11].

10.12 Assume that a system is described by a matrix $\mathbf{A}(t)$ that is composed of a sign-definite matrix $\mathbf{A}_1(t)$ plus a skew-symmetric term $\mathbf{S}(t)$. Show that the system is asymptotically stable if $\mathbf{A}_1$ is negative definite and unstable if $\mathbf{A}_1$ is positive definite.

Use the quadratic Lyapunov function $V = \mathbf{x}^T\mathbf{P}\mathbf{x}$, with $\mathbf{P} = \mathbf{I}$. Then $\dot{V} = \mathbf{x}^T[\mathbf{A}^T + \mathbf{A}]\mathbf{x} \equiv \mathbf{x}^T[\mathbf{A}_1^T + \mathbf{A}_1]\mathbf{x}$, since $\mathbf{S}^T + \mathbf{S} = \mathbf{0}$. Thus by Theorem 10.2 (or Theorem 10.4), the system is asymptotically stable if $\mathbf{A}_1$ is negative definite. By Theorem 10.6 the system is unstable if $\mathbf{A}_1$ is positive definite. This result also shows that many (but not all) of the surprises regarding stability and eigenvalue locations in the linear time-varying case of Sec. 10.7 can be resolved by examining the eigenvalues of $\mathbf{A}_s = \mathbf{A}^T + \mathbf{A}$ rather than the eigenvalues of $\mathbf{A}$.

10.13 Find a constant feedback gain matrix which will ensure asymptotic stability for the system of Example 10.6, regardless of the frequency ω.

If a feedback term $\mathbf{u} = -K\mathbf{x} = -k\mathbf{I}_2\mathbf{x}$ is added to the system equation in Example 10.6, the $\mathbf{A}_s$ matrix is modified to

$$\mathbf{A}_s = \begin{bmatrix} -8 - 2k + 2\sqrt{50}\sin(\omega t) & 1 + 25\cos(2\omega t) \\ 1 + 25\cos(2\omega t) & -8 - 2k + 2\sqrt{50}\sin(\omega t) \end{bmatrix}$$

This means that the eigenvalues are both shifted to the left by $2k$. From the bound previously given for $\lambda_{\max}$, it is clear that both eigenvalues will always remain in the left-hand plane for $k > 19.856/2 = 9.928 \approx 10$. This will ensure asymptotic stability.

10.14 Describe a method of generating Lyapunov functions, beginning with an assumed form for the gradient $\nabla_x V$. This method is known as the variable gradient method [5, 8, 9].

Consider a general class of nonlinear, autonomous systems, $\dot{\mathbf{x}} = \mathbf{f}(\mathbf{x})$, for which $\mathbf{f}(\mathbf{0}) = \mathbf{0}$. Then for any candidate Lyapunov function $V(\mathbf{x})$,

$$\frac{dV(\mathbf{x})}{dt} = [\nabla_{\mathbf{x}} V]^T\dot{\mathbf{x}} = [\nabla_{\mathbf{x}} V]^T\mathbf{f}(\mathbf{x}) \tag{1}$$

A general form for the gradient is assumed, such as

$$\nabla_{\mathbf{x}} V = \begin{bmatrix} a_{11}x_1 + a_{12}x_2 + \cdots + a_{1n}x_n \\ a_{21}x_1 + a_{22}x_2 + \cdots + a_{2n}x_n \\ \vdots \\ a_{n1}x_1 + a_{n2}x_2 + \cdots + a_{nn}x_n \end{bmatrix} \tag{2}$$

The coefficients a_{ij} need not be constants, but may be functions of the components of $\mathbf{x}$. The function $V(\mathbf{x})$ can be determined by evaluating the line integral in state space from $\mathbf{0}$ to a general point $\mathbf{x}$, $V(\mathbf{x}) = \int_0^{\mathbf{x}} (\nabla_{\mathbf{x}} V)^T\,d\mathbf{x}$. The line integral can be made independent of the path of integration if a set of generalized curl conditions is imposed.

Let the ith component of $\nabla_{\mathbf{x}} V$ be called ∇V_i. Then the curl requirements are that $\partial\nabla V_i/\partial x_j = \partial\nabla V_j/\partial x_i$ for $i, j = 1, 2, \ldots, n$. That is, the matrix of second partials $\nabla_{\mathbf{x}}(\nabla_{\mathbf{x}} V)^T = [\partial^2 V/\partial x_i\,\partial x_j]$ must be symmetric. Satisfying this requirement allows the line integral to be written as the sum of n simple scalar integrals:

$$V(\mathbf{x}) = \int_0^{x_1} \nabla V_1\,dx_1\bigg|_{x_2 = x_3 = \cdots = x_n = 0} + \int_0^{x_2} \nabla V_2\,dx_2\bigg|_{\substack{x_1 = x_1 \\ x_3 = x_4 = \cdots = x_n = 0}} + \cdots$$

$$+ \int_0^{x_n} \nabla V_n\,dx_n\bigg|_{\substack{x_1 = x_1 \\ x_2 = x_2 \\ \cdots\cdots}} \tag{3}$$

The procedure then consists of assuming $\nabla_{\mathbf{x}} V$ as in Eq. (2), specifying the coefficients a_{ij} such that (1) $\dot{V}(\mathbf{x})$, Eq. (1), is at least negative semidefinite and (2) the generalized curl equations are satisfied. Then $V(\mathbf{x})$ is found as in Eq. (3) and checked for positive definiteness. Usually it will be possible to let many a_{ij} terms equal zero. Generally $a_{nn} = 1$ is chosen at the outset to ensure that $V(\mathbf{x})$ is quadratic in x_n. This technique is illustrated in Chapter 15 on nonlinear systems. It can also be used on linear systems, either constant or time-varying.

.15 Use the variable gradient method to determine sufficient conditions for asymptotic stability of the time-varying second-order system

$$\dot{x}_1 = x_2, \qquad \dot{x}_2 = a_1(t)x_1 + a_2(t)x_2 \tag{1}$$

Assume

$$\nabla V = \begin{bmatrix} \alpha_{11}x_1 + \alpha_{12}x_2 \\ \alpha_{21}x_1 + \alpha_{22}x_2 \end{bmatrix} \tag{2}$$

Select $\alpha_{12} = \alpha_{21}$ so that the curl equations will be satisfied. This allows determination of V by a path-independent line integral of ∇V from $\mathbf{0}$ to $\mathbf{x}$. That is,

$$\begin{aligned} V(\mathbf{x}, t) &= \int_0^{x_1} (\alpha_{11}x_1 + \alpha_{12}x_2)\, dx_1 \Big|_{x_2 = 0} \\ &\quad + \int_0^{x_2} (\alpha_{12}x_1 + \alpha_{22}x_2)\, dx_2 \Big|_{x_1 \text{ fixed}} \\ &= \tfrac{1}{2}\mathbf{x}^T \begin{bmatrix} \alpha_{11} & \alpha_{12} \\ \alpha_{12} & \alpha_{22} \end{bmatrix} \mathbf{x} \end{aligned} \tag{3}$$

In order to make V positive definite, it is required that $\alpha_{11}(t) > 0$ and $\alpha_{11}\alpha_{22} - \alpha_{12}^2 > 0$ for all t. The time rate-of-change is given by

$$\dot{V} = (\nabla V)^T \dot{\mathbf{x}} + \partial V/\partial t = \mathbf{x}^T \begin{bmatrix} \alpha_{12}a_1 + \dot{\alpha}_{11}/2 & \alpha_{11} + \alpha_{12}a_2 + \dot{\alpha}_{12}/2 \\ \alpha_{22}a_1 + \dot{\alpha}_{12}/2 & \alpha_{12} + \alpha_{22}a_2 + \dot{\alpha}_{22}/2 \end{bmatrix} \mathbf{x} \tag{4}$$

This must be forced to be negative definite for all t by proper choice of α_{ij}. It is suggested [5] that the coefficient of the highest state component can often be set to 1, so $\alpha_{22} = 1$ is selected. Also, some of the α_{ij} terms can usually be set to zero. Here we tentatively try $\alpha_{12} = 0$. Then Eq. (4) reduces to

$$\dot{V} = \mathbf{x}^T \begin{bmatrix} \dot{\alpha}_{11}/2 & \alpha_{11} \\ a_1 & a_2 \end{bmatrix} \mathbf{x}$$

and will be negative definite if $\dot{\alpha}_{11} < 0$ and $\dot{\alpha}_{11}a_2 - 2\alpha_{11}a_1 > 0$. By selecting $\alpha_{11} = -a_1$, these requirements become $\dot{a}_1 > 0$ and $-\dot{a}_1 a_2 + 2a_1^2 > 0$. V is assured of being positive definite if $a_1 < 0$. In order to satisfy all requirements, it is sufficient that

$$a_1 < 0, \qquad a_2 < 0, \quad \text{and} \quad \dot{a}_1 > 0 \quad \text{for all } t \tag{5}$$

(Actually a_2 need only be less than $2a_1^2/\dot{a}_1$ for $\dot{a}_1$ strictly positive). This same problem has been considered [9] using a slightly different Lyapunov function, and a different set of sufficient conditions were found:

$$a_1 < 0, \qquad a_2 < 0, \quad \text{and} \quad \dot{a}_1 < 2a_1 a_2 \quad \text{for all } t. \tag{6}$$

The major difference is in the $\dot{a}_1$ condition. Equation (5) allows all positive $\dot{a}_1$ and Eq. (6) allows all negative $\dot{a}_1$ (plus some positive). Taken together, it is clear that asymptotic stability is assured for *any* $\dot{a}_1$ rate as long as a_1 and a_2 remain negative for all t. Note that these conditions are the same as obtained from applying Routh's criterion to the constant-coefficient case.

PROBLEMS

10.16 Comment on the stability of the systems described in
(a) Example 9.1, page 314;
(b) Problem 9.8, page 328;
(c) Figure 9.5, page 332, with $K = \frac{3}{16}$.

10.17 Investigate the stability of the systems described in conjunction with controllability and observability in
(a) Problem 11.2, page 388;
(b) Problem 11.21, page 397;
(c) Problem 11.26, page 401.

10.18 If the input of Example 10.7 has constraints on the individual components, $|u_i(t)| \leq M_i$, what should the control law be if it is desired to rapidly correct for initial state perturbations?

10.19 Verify by simulation that the time-varying system

$$\dot{\mathbf{x}} = \begin{bmatrix} -2 & g(t) \\ -g(t) & -3 \end{bmatrix} \mathbf{x}$$

is asymptotically stable for any reasonable $g(t)$ function. Note that numerical overflow may occur on the computer if a $g(t)$ which grows without bound is selected. One suggestion is $g(t) = \alpha e^{bt} \sin(\gamma t)$, with $b \leq 0$.

10.20 Modify the matrix $\mathbf{A}(t)$ used in Problem 10.19 by changing the sign of A_{21}. Analyze the stability of the new system.

10.21 Modify the matrix $\mathbf{A}(t)$ used in Problem 10.19 by changing the amplitude coefficient on the A_{21} term. Do not change the A_{12} term. Analyze the stability of the new system.

10.22 Consider the second-order system of Problem 10.15. Form the symmetric matrix $\mathbf{A}_s = \mathbf{A} + \mathbf{A}^T$. Show that the stability results of Sec. 10.7, which are expressed in terms of the eigenvalues of $\mathbf{A}_s$, are not particularly useful because λ_{max} will always be positive or zero.

10.23 The system

$$\ddot{x} + (\gamma - \sin \omega t)\dot{x} + (2 + \beta e^{-\alpha t})x = 0$$

is assured of being asymptotically stable by the results of Problem 10.15 if $\beta > 0$, $\alpha > 0$, and $\gamma > 1$. Use $\alpha = 0.01$, $\beta = 1$, $\gamma = 2$ and $\omega = 15$ and verify by simulation that the origin is a stable focus, as shown in Figure 10.12. Note that $\dot{a}_1 = \beta\alpha e^{-\alpha t}$ is positive.

10.24 Repeat Problem 10.23, but let $\beta = -1$ so that $\dot{a}_1 = \beta\alpha e^{-\alpha t}$ is now negative, thus satisfying Eq. (6) of Problem 10.15.

10.25 Demonstrate by simulation that the sufficient conditions of Eq. (6) in Problem 10.15 are *not* necessary conditions. This can be demonstrated using the system of Problem 10.23 with $\gamma < 1$, so that a_2 is positive part of the time. With $\alpha = 0.01$, $\beta = 1$, $\gamma = 0.25$, and $\omega = 10$, the phase portrait of a stable focus, shown in Figure 10.13, should be obtained.

10.26 Repeat the variable gradient steps used in Problem 10.15 to develop sufficient conditions for asymptotic stability of a general third-order time-varying system of the form

$$\dddot{y} + a_2(t)\ddot{y} + a_1(t)\dot{y} + a_0(t)y = 0$$

10.27 Consider the motion of the mass-spring-damper system of Figure 10.14 acting under the influence of gravity, g.
(a) Measure the position y of the mass from the unloaded free length of the spring and write the equation of motion

$$m\ddot{y} + b\dot{y} + ky = mg \tag{1}$$

Show that the equilibrium point is $y_e = mg/k$, $\dot{y}_e = 0$. Then, by referencing the position of the

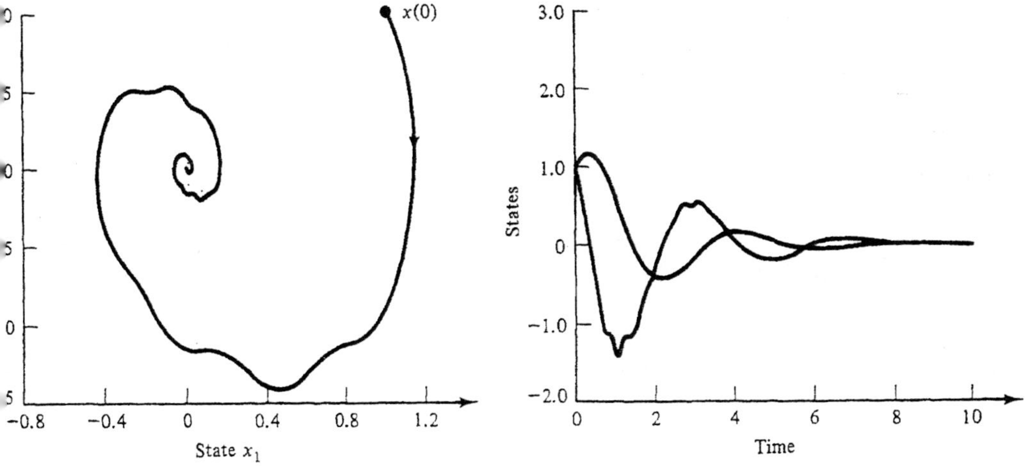

Figure 10.12

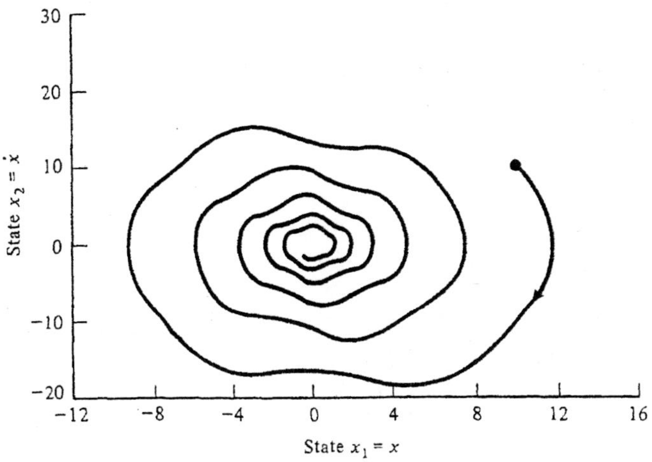

Figure 10.13

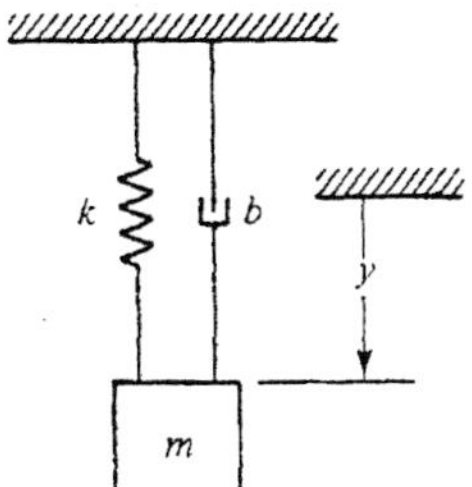

Figure 10.14

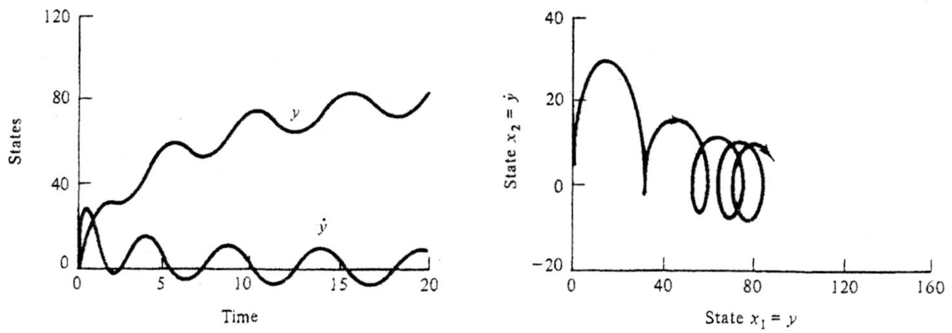

Figure 10.15

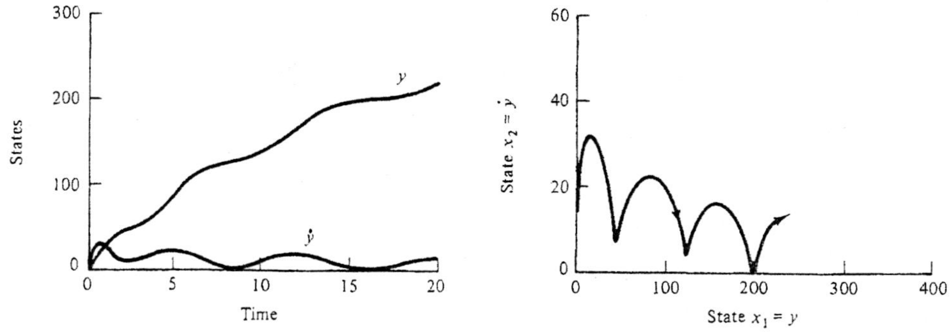

Figure 10.16

mass with respect to the loaded free length of the spring, transfer the equilibrium point to the origin of the new coordinates, so that

$$m\ddot{y}_1 + b\dot{y}_1 + ky_1 = 0 \tag{2}$$

(b) Assume that the damping coefficient b is constant, but that the spring coefficient k is time variable. Show that the frozen coefficient approach mentioned in Sec. 10.7 indicates asymptotic stability for all $k > 0$ and $b > 0$, no matter how small.

(c) Let $b = 1$ lb/ft/s, and $mg = 100$ lb, and assume that the spring becomes softer with time according to

$$k(t) = 50/(1 + 10t)$$

Investigate the behavior of y and $\dot{y}$ in Eq. (*1*). Is the system stable? Response curves for $\mathbf{x}(0) = \mathbf{0}$ are given in Figure 10.15.

(d) Repeat part (c) with $b = 0.2$ and verify the response curves of Figure 10.16.

11

Controllability and Observability for Linear Systems

1 INTRODUCTION

In Chapter 10 system stability was discussed. In this chapter two additional equally important system properties, controllability and observability, are defined. These were both mentioned briefly in Sec. 6.9 while discussing solution of simultaneous equations. Several criteria are presented here for determining whether a linear system possesses these properties.

Similarity transformations can be used to decompose a linear system into various canonical forms. When the modal matrix is used in this transformation, the decoupled or nearly decoupled Jordan canonical form of the system equations result. From these, controllability and observability properties of individual system modes are rather obvious. In this context, two weaker system properties can be defined which connect controllability and observability with modal stability. These properties, stabilizability and detectability, are also defined and discussed.

An alternate similarity transformation is presented, which transforms the state equations into Kalman's controllable form and/or Kalman's observable form. These transformations use orthogonal vectors obtained from a **QR** decomposition, rather than eigenvectors. This approach also provides the basis for one way of determining minimal-order state variable models from transfer function matrices or from arbitrary nonminimal realizations. This is pursued in Chapter 12.

2 DEFINITIONS

It has been shown previously that the description of a linear system, either continuous-time or discrete-time, depends upon four matrices **A**, **B**, **C**, and **D**. Depending on the choice of state variables, or alternatively on the choice of the basis for the state space

Σ, different matrices can be used to describe the same system. A particular set $\{\mathbf{A}, \mathbf{B}, \mathbf{C}, \mathbf{D}\}$ is called a *system representation* or *realization*. In some cases these matrices will be constant. In other cases they will depend on time, in either a continuous fashion $\{\mathbf{A}(t), \mathbf{B}(t), \mathbf{C}(t), \mathbf{D}(t)\}$ or in a discrete fashion, $\{\mathbf{A}(k), \mathbf{B}(k), \mathbf{C}(k), \mathbf{D}(k)\}$. Both the continuous-time and discrete-time cases are considered simultaneously using the notation of Chapter 3. The times of interest will be referred to as the set of scalars $\mathcal{T}$, where $\mathcal{T}$ can be a continuous interval $[t_0, t_f]$, or a set of discrete points $[t_0, t_1, \ldots, t_N]$. At any particular time $t \in \mathcal{T}$, the four system matrices are representations of transformations on the n-dimensional state space Σ, the r-dimensional input space $\mathcal{U}^r$, and the m-dimensional output space $\mathcal{Y}^m$. That is,

$$\mathbf{A}: \Sigma \to \Sigma$$

$$\mathbf{B}: \mathcal{U}^r \to \Sigma$$

$$\mathbf{C}: \Sigma \to \mathcal{Y}^m$$

$$\mathbf{D}: \mathcal{U}^r \to \mathcal{Y}^m$$

Figure 11.1 symbolizes these relationships.

11.21. Controllability

Controllability is a property of the coupling between the input and the state, and thus involves the matrices **A** and **B**.

Definition 11.1. A linear system is said to be *controllable* at t_0 if it is possible to find some input function (or sequence in the discrete case) $\mathbf{u}(t)$, defined over $t \in \mathcal{T}$, which will transfer the initial state $\mathbf{x}(t_0)$ to the origin at some finite time $t_1 \in \mathcal{T}, t_1 > t_0$. That is, there exists some input $\mathbf{u}_{[t_0, t_1]}$, which gives $\mathbf{x}(t_1) = 0$ at a finite $t_1 \in \mathcal{T}$. If this is true for all initial times t_0 and all initial states $\mathbf{x}(t_0)$, the system is *completely controllable*.

Some authors define another kind of controllability involving the output $\mathbf{y}(t)$ [1]. The definition given above is referred to as state controllability. It is the most common definition, and is the only type used in this text, so the adjective "state" is omitted. Complete controllability is obviously a very important property. If a system is not completely controllable, then for some initial states no input exists which can drive the

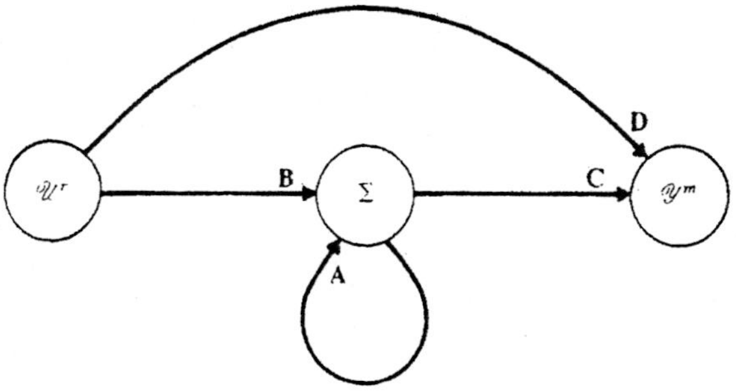

Figure 11.1

system to the zero state. It would be meaningless to search for an *optimal control* in this case. A trivial example of an uncontrollable system arises when the matrix **B** is zero, because then the input is disconnected from the state.

The full significance of controllability is realized in Chapters 13 and 14. It is seen there that if a linear system is controllable, it is possible to design a linear state feedback control law that will give arbitrarily specified closed-loop eigenvalues (poles). Thus an unstable system can be stabilized, a slow system can be speeded up, the natural frequencies can be changed, and so on, if the system is controllable. The existence of solutions to certain optimal control problems can be assured if the system is controllable.

11.2.2 Observability

Observability is a property of the coupling between the state and the output and thus involves the matrices **A** and **C**.

Definition 11.2. A linear system is said to be *observable* at t_0 if $\mathbf{x}(t_0)$ can be determined from the output function $\mathbf{y}_{[t_0,t_1]}$ (or output sequence) for $t_0 \in \mathcal{T}$ and $t_0 \leq t_1$, where t_1 is some *finite* time belonging to $\mathcal{T}$. If this is true for all t_0 and $\mathbf{x}(t_0)$, the system is said to be *completely observable.*

Clearly the observability of a system will be a major requirement in filtering and state estimation or reconstruction problems. In many feedback control problems, the controller must use output variables **y** rather than the state vector **x** in forming the feedback signals. If the system is observable, then **y** contains sufficient information about the internal states so that most of the power of state feedback can still be realized. A more complicated controller is needed to achieve these results. This is discussed fully in Chapter 13.

11.2.3 Dependence Upon the Model

Both controllability and observability are defined in terms of the state of the system. For a given physical system there are many ways of selecting state variables, as discussed in Chapter 3. Two such model forms for single-input–single-output systems were called the *controllable canonical form* and the *observable canonical form.* As might be expected, the state variable model will always be controllable in the first case and always observable in the second case. (See Problems 11.6 and 11.7.) It is therefore possible that a given physical system will have one state model which is controllable but not observable and another state model which is observable but not controllable. These properties are characteristics of the model $\{\mathbf{A}, \mathbf{B}, \mathbf{C}, \mathbf{D}\}$ rather than the physical system per se. However, if one nth-order state variable model is both controllable and observable, then *all* possible state variable models of order n will have these properties. If either property is lacking in a given nth-order state variable model, then *every* state variable model of that order will fail to have either one or the other property.

Ideally it would be preferred that the system model be both controllable and observable. If this is not true, then weaker conditions of stabilizability and detectability (defined in Sec. 11.8) *may* allow the control system design to proceed in an acceptable fashion. It is noted that a control system designer frequently can dictate whether the model is controllable or observable. If observability is originally lacking, this might be changed by adding additional sensors. If controllability is originally lacking, this might signal the need for additional control actuators.

If a system is not completely observable, then the initial state $\mathbf{x}(t_0)$ cannot be determined from the output, no matter how long the output is observed. The system of Example 9.1 gave an output which was identically zero for all time. That system is obviously not completely observable.

11.3 TIME-INVARIANT SYSTEMS WITH DISTINCT EIGENVALUES

Controllability and observability for time-invariant systems depend only on the constant matrices $\{\mathbf{A}, \mathbf{B}, \mathbf{C}\}$. $\mathbf{D}$ is not included because it is a direct mapping from input space to output space, without affecting the internal states. No reference need be made to a particular interval $[t_0, t_1]$. In this section the n eigenvalues of $\mathbf{A}$ are assumed to be distinct. Then the Jordan form representation for continuous-time systems is

$$\dot{\mathbf{q}} = \mathbf{\Lambda q} + \mathbf{B}_n \mathbf{u}(t) \tag{11.1}$$

$$\mathbf{y}(t) = \mathbf{C}_n \mathbf{q}(t) + \mathbf{Du}(t) \tag{11.2}$$

For discrete-time systems,

$$\mathbf{q}(k+1) = \mathbf{\Lambda q}(k) + \mathbf{B}_n \mathbf{u}(k) \tag{11.3}$$

$$\mathbf{y}(k) = \mathbf{C}_n \mathbf{q}(k) + \mathbf{Du}(k) \tag{11.4}$$

The following definitions from earlier chapters are recalled. The modal matrix is $\mathbf{M}$, and $\mathbf{\Lambda} = \mathbf{M}^{-1}\mathbf{AM}$, $\mathbf{B}_n = \mathbf{M}^{-1}\mathbf{B}$, $\mathbf{C}_n = \mathbf{CM}$. Clearly, if any one row of $\mathbf{B}_n$ contains only zero elements, then the corresponding mode q_i is unaffected by the input. Then $\dot{q}_i = \lambda_i q_i$ or $q_i(k+1) = \lambda_i q_i(k)$. In this case the homogeneous solution for q_i may eventually approach zero as t (or k) $\to \infty$, but there is no *finite* time at which this component of $\mathbf{q}$ will be zero. Thus there is no finite time at which $\mathbf{q}$, and consequently $\mathbf{x}$, can be driven to zero.

Controllability Criterion 1

The constant coefficient system, for which $\mathbf{A}$ has distinct eigenvalues, is completely controllable if and only if there are no zero rows of $\mathbf{B}_n = \mathbf{M}^{-1}\mathbf{B}$.

EXAMPLE 11.1 The system of Problem 9.8, page 328, has $\mathbf{B}_n = \begin{bmatrix} \frac{1}{2} & 2 \\ 3 & 6 \\ \frac{1}{2} & 1 \end{bmatrix}$. Since $\mathbf{A}$ is constant and its eigenvalues are distinct, the above criterion applies. This system is completely controllable because no row of $\mathbf{B}_n$ is all zero. ■

EXAMPLE 11.2 The system of Problem 9.23, page 336, has $\mathbf{B}_n = \begin{bmatrix} 2 & 0 \\ 1 & 1 \\ 0 & 0 \end{bmatrix}$. Since $\mathbf{A}$ is constant and has distinct eigenvalues, the above criterion indicates that this system is not completely controllable. Since the third row of $\mathbf{B}_n$ contains only zeros, no control can affect the third mode $q_3(k)$ of this system. Since $q_3(k) = (-\frac{1}{3})^k q_3(0)$, this component approaches zero only as $k \to \infty$. ■

Equations (*11.2*) and (*11.4*) indicate that the output $\mathbf{y}$ will not be influenced by the ith system mode q_i if column i of $\mathbf{C}_n$ contains only zero elements. If this is true, then $q_i(t_0)$ could take on any arbitrary value without influencing $\mathbf{y}$. There is no possibility of determining $\mathbf{q}(t_0)$ or $\mathbf{x}(t_0)$ in this case.

Observability Criterion 1

The constant coefficient system, for which $\mathbf{A}$ has distinct eigenvalues, is completely observable if and only if there are no zero columns of $\mathbf{C}_n = \mathbf{CM}$.

EXAMPLE 11.3 The system of Example 9.2 has $\mathbf{C}_n = [0 \quad 6]$. Since $\mathbf{A}$ is constant, with distinct eigenvalues, the above criterion applies. Column one of $\mathbf{C}_n$ is zero, so this system is not completely observable.

The discrete-time system of Problem 9.23, page 336, has $\mathbf{C}_n = \begin{bmatrix} 1 & 0 & 1 \\ 0 & 1 & 1 \end{bmatrix}$. Since no columns are all zero, this system is completely observable. ■

The criteria for complete controllability and observability given in this section (see also Problem 11.13) are useful because of the geometrical insight they give. They also make it possible to speak of individual system modes as being controllable or uncontrollable, observable or unobservable. However, these criteria are not the most useful because they are restricted to the distinct eigenvalue case (and not merely a diagonal Jordan form). These results are generalized in Problems 11.16 and 11.17 for the case of repeated eigenvalues. There it is shown that each mode must have a direct connection to the input (nonzero row of $\mathbf{B}_n$) *or* be coupled to another mode which has such a direct control connection. This means that zero rows can be tolerated in $\mathbf{B}_n$, provided they are not the *last* row associated with a given Jordan block. A similar result (Problem 11.17) states that the system is observable provided that the *first* column associated with each Jordan block is not identically zero.

The application of the preceding criteria requires determination of the Jordan normal form, which requires finding the modal matrix $\mathbf{M}$ and its inverse. For most applications it is easier to use the results of the following section, which are stated directly in terms of $\{\mathbf{A}, \mathbf{B}, \mathbf{C}\}$ in any arbitrary form.

11.4 TIME-INVARIANT SYSTEMS WITH ARBITRARY EIGENVALUES

Controllability Criterion 2

A constant coefficient linear system with the representation $\{\mathbf{A}, \mathbf{B}, \mathbf{C}, \mathbf{D}\}$ is completely controllable if and only if the $n \times rn$ matrix of

$$\mathbf{P} \triangleq [\mathbf{B} \mid \mathbf{AB} \mid \mathbf{A}^2\mathbf{B} \mid \cdots \mid \mathbf{A}^{n-1}\mathbf{B}] \tag{11.5}$$

has rank n. See Problem 11.14. The form of this condition is exactly the same for both continuous-time and discrete-time systems. The sufficiency of this condition was proven in Sec. 6.9 for the discrete-time case. With the standard definition of controllability—i.e., the ability to drive any initial state to the origin in finite time—this condition is slightly stronger than necessary. If $\mathbf{A}$ has a zero eigenvalue, then certain initial conditions (those aligned with the corresponding eigenvector) can be driven to zero in one time step without $\mathbf{P}$ having full rank. When a modified definition of controllability—i.e., the ability to drive any initial state to *any* other state in finite time—is used, then the full rank of $\mathbf{P}$ is both necessary and sufficient. Problem 11.12 shows that both definitions of controllability are equivalent for continuous-time systems. Problem 11.14 proves that criterion 2 is a necessary condition for controllability of a continuous-time system. It is also a sufficient condition.

EXAMPLE 11.4 For the continuous-time system of Problem 9.8,

$$\mathbf{A} = \begin{bmatrix} -2 & -2 & 0 \\ 0 & 0 & 1 \\ 0 & -3 & -4 \end{bmatrix}, \qquad \mathbf{B} = \begin{bmatrix} 1 & 0 \\ 0 & 1 \\ 1 & 1 \end{bmatrix}$$

and

$$\mathbf{P} = \left[\begin{array}{cc|cc|cc} 1 & 0 & -2 & -2 & 2 & 2 \\ 0 & 1 & 1 & 1 & -4 & -7 \\ 1 & 1 & -4 & -7 & 13 & 25 \end{array}\right]$$

The rank of $\mathbf{P}$ is 3, since the determinant of the first three columns is nonzero ($= -3$). Therefore, this system is completely controllable. ■

EXAMPLE 11.5 For the discrete-time system considered in Problem 9.23,

$$\mathbf{P} = \left[\begin{array}{cc|cc|cc} 3 & 1 & \frac{5}{2} & \frac{1}{2} & \frac{9}{4} & \frac{1}{4} \\ 2 & 0 & 2 & 0 & 2 & 0 \\ -1 & 1 & -\frac{3}{2} & \frac{1}{2} & -\frac{7}{4} & \frac{1}{4} \end{array}\right]$$

The rank of $\mathbf{P}$ is not 3, since subtracting row 3 from row 1 gives twice row 2. Since rank $\mathbf{P} \neq n$, this system is not completely controllable. ■

Observability Criterion 2

A constant coefficient linear system is completely observable if and only if the $n \times mn$ matrix of Eq. (*11.6*) has rank n:

$$\mathbf{Q} \triangleq [\overline{\mathbf{C}}^T \mid \overline{\mathbf{A}}^T\overline{\mathbf{C}}^T \mid \overline{\mathbf{A}}^{2T}\overline{\mathbf{C}}^T \mid \cdots \mid \overline{\mathbf{A}}^{n-1T}\overline{\mathbf{C}}^T] \tag{11.6}$$

The form of observability criterion 2 is exactly the same for both continuous-time and discrete-time systems. The proof of sufficiency was given in Sec. 6.9 for the discrete-time case with real $\mathbf{A}$ and $\mathbf{C}$.

EXAMPLE 11.6 The continuous-time system of Example 9.1 has $\mathbf{A} = \begin{bmatrix} 0 & 1 \\ 8 & -2 \end{bmatrix}$, $\mathbf{C} = [4 \quad 1]$. These are both real, so the complex conjugate in $\mathbf{Q}$ is unnecessary for this example, and $\mathbf{Q} = \left[\begin{array}{c|c} 4 & 8 \\ 1 & 2 \end{array}\right]$. The second column is twice the first, so $r_Q = 1$. Since $r_Q \neq 2$, this system is not completely observable. Similar manipulations show that the system of Problem 9.23 is completely observable. ■

The forms of Eqs. (*11.5*) and (*11.6*) are the same. Therefore, a computer algorithm for forming P and checking its rank will also form $\mathbf{Q}$ and check its rank. In forming $\mathbf{Q}$, $\overline{\mathbf{A}}^T$ replaces $\mathbf{A}$ and $\overline{\mathbf{C}}^T$ replaces $\mathbf{B}$. In the most common case where $\mathbf{A}$ and $\mathbf{B}$ are real, the complex conjugates are superfluous. In developing such a computer algorithm, the crucial consideration is accurate determination of rank. If a row-reduced echelon technique is used, the notion of "machine zero" must be used to distinguish between legitimate nonzero but small divisors and divisors which would have been zero were it not for round-off error. The discussion of the GSE method in Sec. 6.5, the **QR** decomposition based on the modified Gram-Schmidt process, and the SVD method in Problems 7.29 through 7.34 are of significance in this regard. These same considerations arise many other places as well; e.g., do nontrivial solutions exist for a set of homogeneous equations? See Reference 2 for a fuller discussion of the implications of the fact that rank of a matrix is a discontinuous function. The wrong rank can easily be computed due to very small computer errors, unless due precautions are taken.

.5 YET ANOTHER CONTROLLABILITY-OBSERVABILITY CONDITION

The controllability and observability of linear, constant systems can be characterized in still another way. This characterization is not especially convenient in checking whether a given system has these properties or not. However, it will be useful in solving the pole-placement problem of Chapter 13.

The nth-order system realization $\{\mathbf{A}, \mathbf{B}, \mathbf{C}\}$ is controllable if and only if $[s\mathbf{I} - \mathbf{A} \mid \mathbf{B}]$ has rank n *for all* values of s. The same system is observable if and only if $[s\mathbf{I} - \mathbf{A}^T \mid \mathbf{C}^T]$ has rank n for all s. Since it has already been established that $\text{rank}(\mathbf{P}) = n$ is necessary and sufficient for controllability, it is next established that $\{\text{rank}(\mathbf{P}) = n\} \Rightarrow \{\text{rank}[s\mathbf{I} - \mathbf{A} \mid \mathbf{B}] = n\}$ by proving that $\{\text{rank}[s\mathbf{I} - \mathbf{A} \mid \mathbf{B}] \neq n \text{ for some } s\} \Rightarrow \{\text{rank}(\mathbf{P}) \neq n\}$. If $\text{rank}[s\mathbf{I} - \mathbf{A} \mid \mathbf{B}] \neq n$ for some s, then there exists a nonzero vector $\boldsymbol{\eta}$ such that $\boldsymbol{\eta}^T[s\mathbf{I} - \mathbf{A} \mid \mathbf{B}] = 0$, so that $\boldsymbol{\eta}^T\mathbf{B} = 0$ and $\boldsymbol{\eta}^T s = \boldsymbol{\eta}^T\mathbf{A}$. Postmultiplying the last equation by $\mathbf{A}$ gives $\boldsymbol{\eta}^T\mathbf{A}s = \boldsymbol{\eta}^T\mathbf{A}^2 = \boldsymbol{\eta}^T s^2$. Likewise, $\boldsymbol{\eta}^T\mathbf{A}^3 = \boldsymbol{\eta}^T s^3, \ldots$. Therefore,

$$\begin{aligned}\boldsymbol{\eta}^T\mathbf{P} &= [\boldsymbol{\eta}^T\mathbf{B} \quad \boldsymbol{\eta}^T\mathbf{A}\mathbf{B} \quad \cdots \quad \boldsymbol{\eta}^T\mathbf{A}^{n-1}\mathbf{B}] \\ &= [\boldsymbol{\eta}^T\mathbf{B} \quad s\boldsymbol{\eta}^T\mathbf{B} \quad s^2\boldsymbol{\eta}^T\mathbf{B} \quad \cdots \quad s^{n-1}\boldsymbol{\eta}^T\mathbf{B}] \\ &= [0]\end{aligned}$$

Thus $\mathbf{P}$ cannot have rank n if $[s\mathbf{I} - \mathbf{A} \mid \mathbf{B}]$ does not have rank n. Minor notational changes are all that are needed to prove the corresponding observability result.

If s is not an eigenvalue of $\mathbf{A}$, then it is clear that $s\mathbf{I} - \mathbf{A}$ has rank n all by itself without any assistance from $\mathbf{B}$, regardless of whether the system is controllable or not. In a controllable system, $\mathbf{A}$ and $\mathbf{B}$ must work together so that when $s\mathbf{I} - \mathbf{A}$ becomes rank-deficient, $\mathbf{B}$ fills in the deficiency. The condition must be true for any scalar s, real or complex, including the special case $s = 0$. Thus, a simple rank test on just $[\mathbf{B} \mid \mathbf{A}]$ is sometimes useful. Note that {rank[**B** **A**] $\neq n$} $\Rightarrow$ {system is not controllable}. The reverse implication is *not* true.

11.6 TIME-VARYING LINEAR SYSTEMS

11.6.1 Controllability of Continuous-Time Systems

A continuous-time system with the representation $\{\mathbf{A}(t), \mathbf{B}(t), \mathbf{C}(t), \mathbf{D}(t)\}$ is considered. For a given input function $\mathbf{u}(t)$, the solution for the state at a fixed time t_1 is

$$\mathbf{x}(t_1) = \mathbf{\Phi}(t_1, t_0)\mathbf{x}(t_0) + \int_{t_0}^{t_1} \mathbf{\Phi}(t_1, \tau)\mathbf{B}(\tau)\mathbf{u}(\tau)\, d\tau$$

The vector defined by $\mathbf{x}_1 = \mathbf{x}(t_1) - \mathbf{\Phi}(t_1, t_0)\mathbf{x}(t_0)$ is a constant vector in Σ for any fixed time t_1. The notation of Chapters 5 and 6 is used to define the linear transformation

$$\mathcal{A}_c(\mathbf{u}) \triangleq \int_{t_0}^{t_1} \mathbf{\Phi}(t_1, \tau)\mathbf{B}(\tau)\mathbf{u}(\tau)\, d\tau$$

The transformation $\mathcal{A}_c$ maps functions in $\mathcal{U}$ into vectors in Σ. The question of complete controllability on $[t_0, t_1]$ reduces to asking whether $\mathcal{A}_c(\mathbf{u}) = \mathbf{x}_1$ has a solution $\mathbf{u}(t)$ for arbitrary $\mathbf{x}_1 \in \Sigma$. It was shown in Problem 6.23, page 240, that a necessary and sufficient condition for the existence of such a solution is that the null space of $\mathcal{A}_c^*$ contain only the zero element $\mathcal{N}(\mathcal{A}_c^*) = \{0\}$. This is the requirement for complete controllability on $[t_0, t_1]$, but it can be put into a more useful form. Since $\mathcal{A}_c^*: \Sigma \to \mathcal{U}$, the range of $\mathcal{A}_c^*$ is an infinite dimensional function space. The following lemma allows the use of a finite dimensional transformation.

Lemma 11.1. The null space of $\mathcal{A}_c^*$ is the same as the null space of $\mathcal{A}_c\mathcal{A}_c^*$. That is, $\mathcal{N}(\mathcal{A}_c^*) = \mathcal{N}(\mathcal{A}_c\mathcal{A}_c^*)$.

Proof. Let $\mathbf{v} \in \mathcal{N}(\mathcal{A}_c^*)$. Then $\mathcal{A}_c^*(\mathbf{v}) = \mathbf{0}$. Therefore, $\mathcal{A}_c\mathcal{A}_c^*(\mathbf{v}) = \mathcal{A}_c(\mathbf{0}) = \mathbf{0}$, so $\mathbf{v} \in \mathcal{N}(\mathcal{A}_c\mathcal{A}_c^*)$ also. Now assume that $\mathbf{v} \in \mathcal{N}(\mathcal{A}_c\mathcal{A}_c^*)$. Then $\mathcal{A}_c\mathcal{A}_c^*(\mathbf{v}) = \mathbf{0}$. Therefore, $\langle \mathcal{A}_c\mathcal{A}_c^*(\mathbf{v}), \mathbf{v}\rangle = 0$ or $\langle \mathcal{A}_c^*(\mathbf{v}), \mathcal{A}_c^*(\mathbf{v})\rangle = 0$. But this indicates that $\|\mathcal{A}_c^*(\mathbf{v})\|^2 = 0$, so that $\mathcal{A}_c^*(\mathbf{v}) = \mathbf{0}$. Thus for every $\mathbf{v} \in \mathcal{N}(\mathcal{A}_c^*)$, $\mathbf{v}$ also belongs to $\mathcal{N}(\mathcal{A}_c\mathcal{A}_c^*)$ and conversely. The two null spaces are therefore equal.

To use the lemma in developing the criterion for complete controllability, an expression for the transformation $\mathcal{A}_c\mathcal{A}_c^*$ must be found:

$$\mathcal{A}_c(\mathbf{u}) = \int_{t_0}^{t_1} \mathbf{\Phi}(t_1, \tau)\mathbf{B}(\tau)\mathbf{u}(\tau)\, d\tau$$

Therefore,

$$\langle \mathbf{v}, \mathcal{A}_c(\mathbf{u})\rangle = \langle \mathcal{A}_c^*(\mathbf{v}), \mathbf{u}\rangle = \int_{t_0}^{t_1} \overline{\mathbf{v}}^T \mathbf{\Phi}(t_1, \tau)\mathbf{B}(\tau)\mathbf{u}(\tau)\, d\tau$$

so that $\mathcal{A}_c^*(\mathbf{v}) = \overline{\mathbf{B}}^T(t)\overline{\mathbf{\Phi}}^T(t_1, t)\mathbf{v}$. Then

$$\mathcal{A}_c\mathcal{A}_c^*(\mathbf{v}) = \int_{t_0}^{t_1} \mathbf{\Phi}(t_1, \tau)\mathbf{B}(\tau)\overline{\mathbf{B}}^T(\tau)\overline{\mathbf{\Phi}}^T(t_1, \tau)\, d\tau\mathbf{v}$$

The transformation $\mathcal{A}_c\mathcal{A}_c^*$ is just an $n \times n$ matrix, redefined as $\mathbf{G}(t_1, t_0)$,

$$\mathbf{G}(t_1, t_0) \triangleq \int_{t_0}^{t_1} \mathbf{\Phi}(t_1, \tau)\mathbf{B}(\tau)\overline{\mathbf{B}}^T(\tau)\overline{\mathbf{\Phi}}^T(t_1, \tau)\, d\tau \tag{11.7}$$

The null space of $\mathcal{A}_c\mathcal{A}_c^*$ will contain only the zero element if and only if $\mathbf{G}(t_1, t_0)$ does not have zero as an eigenvalue.

Controllability Criterion 3

The system described by $\dot{\mathbf{x}} = \mathbf{A}(t)\mathbf{x} + \mathbf{B}(t)\mathbf{u}(t)$ is completely controllable on the interval $[t_0, t_1]$ if any of the following equivalent conditions is satisfied:

(a) The matrix $\mathbf{G}(t_1, t_0)$ is positive definite.
(b) Zero is not an eigenvalue of $\mathbf{G}(t_1, t_0)$.
(c) $|\mathbf{G}(t_1, t_0)| \neq 0$.

The time-varying controllability criterion 3 can be shown to reduce to criterion 2 in the special case where **A** and **B** are constant. Because of the quadratic form of the integrand of **G**, it is clear that **G** is always at least positive semidefinite. Therefore, **G** having full rank and being positive definite are equivalent conditions. First it will be proven that if **G** has full rank, then **P** must also have full rank. The Cayley-Hamilton theorem gives

$$\mathbf{\Phi}(t_1, \tau)\mathbf{B} = \alpha_0(\tau)\mathbf{B} + \alpha_1(\tau)\mathbf{AB} + \cdots + \alpha_{n-1}(\tau)\mathbf{A}^{n-1}\mathbf{B}$$

$$= [\mathbf{B} \quad \mathbf{AB} \quad \mathbf{A}^2\mathbf{B} \quad \cdots \quad \mathbf{A}^{n-1}\mathbf{B}]\begin{bmatrix} \alpha_0 \mathbf{I}_r \\ \alpha_1 \mathbf{I}_r \\ \vdots \\ \alpha_{n-1}\mathbf{I}_r \end{bmatrix} = \mathbf{PS}(\tau)$$

Because **A**, **B**, and (hence) **P** are constant,

$$\mathbf{G}(t_1, t_0) = \mathbf{P}\int_{t_0}^{t_1} \mathbf{S}(\tau)\overline{\mathbf{S}}^T(\tau)\, d\tau\overline{\mathbf{P}}^T = \mathbf{PR}\overline{\mathbf{P}}^T$$

Sylvester's law says that $\text{rank}(\mathbf{G}) \leq \min\{\text{rank}(\mathbf{P}), \text{rank}(\mathbf{R})\}$. Therefore, {**G** is positive definite}$\Rightarrow\{\text{rank}(\mathbf{G}) = n\}\Rightarrow\{\text{rank}(\mathbf{P}) = n\}$. The reverse implication is also true. Assume $\text{rank}(\mathbf{P}) = n$. Then $\mathbf{P}^T$ has n independent columns. Since, by the Cayley-Hamilton theorem, for any $m \geq n$, $\mathbf{A}^m$ can be expressed as a linear combination of lower powers $\mathbf{A}^j, j < n$, the matrix

$$\begin{bmatrix} \overline{\mathbf{B}}^T \\ \overline{\mathbf{B}}^T\overline{\mathbf{A}}^T \\ \overline{\mathbf{B}}^T(\overline{\mathbf{A}}^T)^2 \\ \vdots \\ \overline{\mathbf{B}}^T(\overline{\mathbf{A}}^T)^m \\ \vdots \end{bmatrix}$$

still has n linearly independent columns, even as $m \rightarrow \infty$. This remains true when groups of rows are weighted by various powers of $(t_1 - t_0)$ and summed, so $\overline{\mathbf{B}}^T\overline{\mathbf{\Phi}}^T(t_1, t_0)$

also has n independent columns and has full rank n. This is the operator $\mathcal{A}_c^*$. Therefore $\mathcal{A}_c^* \mathbf{w} = \mathbf{0}$ if and only if $\mathbf{w} \equiv \mathbf{0}$, i.e., $\mathcal{N}(\mathcal{A}_c^*) = \{0\}$. By Lemma 11.1,

$$\mathcal{N}(\mathcal{A}_c \mathcal{A}_c^*) \equiv \mathcal{N}(\mathcal{A}_c^*)$$

The conclusion is that $\mathbf{G}(t_1, t_0) = \mathcal{A}_c \mathcal{A}_c^*$ has only the zero vector in its null space, which means $\text{rank}(\mathbf{G}) = n$ if $\text{rank}(\mathbf{P}) = n$.

11.6.2 Observability of Continuous-Time Systems

The general form for the output $\mathbf{y}(t)$ is

$$\mathbf{y}(t) = \mathbf{C}(t)\mathbf{\Phi}(t, t_0)\mathbf{x}(t_0) + \int_{t_0}^{t_1} \mathbf{C}(t)\mathbf{\Phi}(t, \tau)\mathbf{B}(\tau)\mathbf{u}(\tau)\, d\tau + \mathbf{D}(t)\mathbf{u}(t)$$

Since the input $\mathbf{u}(t)$ is assumed known, the two terms containing the input could be combined with the output function $\mathbf{y}(t)$ to give a modified function $\mathbf{y}_1(t)$. Alternatively, only the unforced solution could be considered. In either case complete observability requires that a knowledge of $\mathbf{y}(t)$ (or $\mathbf{y}_1(t)$) be sufficient for the determination of $\mathbf{x}(t_0)$. Defining the linear transformation $\mathcal{A}_0(\mathbf{x}(t_0)) = \mathbf{C}(t)\mathbf{\Phi}(t, t_0)\mathbf{x}(t_0)$, the requirement for complete observability is that a unique $\mathbf{x}(t_0)$ can be associated with each output function $\mathbf{y}(t)$. This requires that $\mathcal{N}(\mathcal{A}_0) = \{0\}$ (see Problem 6.22, page 240). Using only minor changes in the previous lemma, it can be shown that $\mathcal{N}(\mathcal{A}_0^* \mathcal{A}_0) = \mathcal{N}(\mathcal{A}_0)$. To find the adjoint transformation, consider

$$\langle \mathbf{w}(t), \mathcal{A}_0(\mathbf{x}(t_0))\rangle = \langle \mathcal{A}_0^* \mathbf{w}(t), \mathbf{x}(t_0)\rangle = \int_{t_0}^{t_1} \bar{\mathbf{w}}^T(\tau)\mathbf{C}(\tau)\mathbf{\Phi}(\tau, t_0)\, d\tau\, \mathbf{x}(t_0)$$

Thus

$$\mathcal{A}_0^*(\mathbf{w}) = \int_{t_0}^{t_1} \bar{\mathbf{\Phi}}^T(\tau, t_0)\bar{\mathbf{C}}^T(\tau)\mathbf{w}(\tau)\, d\tau$$

and

$$\mathcal{A}_0^* \mathcal{A}_0(\mathbf{x}(t_0)) = \int_{t_0}^{t_1} \bar{\mathbf{\Phi}}^T(\tau, t_0)\bar{\mathbf{C}}^T(\tau)\mathbf{C}(\tau)\mathbf{\Phi}(\tau, t_0)\, d\tau\, \mathbf{x}(t_0)$$

The transformation $\mathcal{A}_0^* \mathcal{A}_0 \colon \Sigma \rightarrow \Sigma$ is just an $n \times n$ matrix, redefined as $\mathbf{H}(t_1, t_0)$,

$$\mathbf{H}(t_1, t_0) \triangleq \int_{t_0}^{t_1} \bar{\mathbf{\Phi}}^T(\tau, t_0)\bar{\mathbf{C}}^T(\tau)\mathbf{C}(\tau)\mathbf{\Phi}(\tau, t_0)\, d\tau \tag{11.8}$$

Observability Criterion 3

The system

$$\dot{\mathbf{x}} = \mathbf{A}(t)\mathbf{x}(t) + \mathbf{B}(t)\mathbf{u}(t)$$

$$\mathbf{y}(t) = \mathbf{C}(t)\mathbf{x}(t) + \mathbf{D}(t)\mathbf{u}(t)$$

is completely observable at t_0 if there exists some finite time t_1 for which any one of the following equivalent conditions holds:

(a) The matrix $\mathbf{H}(t_1, t_0)$ is positive definite.
(b) Zero is not a an eigenvalue of $\mathbf{H}(t_1, t_0)$.
(c) $|\mathbf{H}(t_1, t_0)| \neq 0$.

The proof that $\{\text{rank}[\mathbf{H}(t_1, t_0)] = n\} \rightleftarrows \{\text{rank}(\mathbf{Q}) = n\}$ is the same as the proof that $\{\text{rank}[\mathbf{G}(t_1, t_0)] = n\} \rightleftarrows \{\text{rank}(\mathbf{P}) = n\}$ given in the last section, with notational changes.

11.6.3 Discrete-Time Systems

The corresponding forms of the controllability and observability criteria 3 for discrete systems are derived in the same way. However, since the input and output spaces have sequences, rather than functions, as their elements, the appropriate inner product is a summation rather than an integral:

$$\langle \mathbf{w}(k), \mathbf{y}(k) \rangle = \sum_{k=0}^{N} \overline{\mathbf{w}}^T(k)\mathbf{y}(k)$$

The criteria may be stated as follows.

Controllability and Observability Criteria 3, Discrete Systems

The system

$$\mathbf{x}(k+1) = \mathbf{A}(k)\mathbf{x}(k) + \mathbf{B}(k)\mathbf{u}(k)$$
$$\mathbf{y}(k) = \mathbf{C}(k)\mathbf{x}(k) + \mathbf{D}(k)\mathbf{u}(k)$$

is completely controllable at $k = 0$ if and only if for some finite time index N, the $n \times n$ matrix

$$\sum_{k=0}^{N} \mathbf{\Phi}(N, k)\mathbf{B}(k)\overline{\mathbf{B}}^T(k)\overline{\mathbf{\Phi}}^T(N, k)$$

is positive definite (or does not have zero as an eigenvalue, or has a nonzero determinant). This system is completely observable at $k = 0$ if and only if there exists some finite index N such that the $n \times n$ matrix

$$\sum_{k=0}^{N} \overline{\mathbf{\Phi}}^T(k, 0)\overline{\mathbf{C}}^T(k)\mathbf{C}(k)\mathbf{\Phi}(k, 0)$$

is positive definite (or does not have zero as an eigenvalue, or has a nonzero determinant).

.7 KALMAN CANONICAL FORMS

It can be seen from Problem 11.12 that any vector $\mathbf{x}(t_0)$ that belongs to the subspace spanned by the columns of $\mathbf{P}$ can be driven to zero, that is, these states are controllable. If the columns of $\mathbf{P}$ span the entire n-dimensional state space, then the system is controllable. When rank(**P**) $= r_P < n$, state space can be decomposed into two orthogonal subspaces, $\Sigma = \Sigma_1 \oplus \Sigma_2$, with Σ_1 being the subspace spanned by the columns of **P**. Σ_1 is called the controllable subspace for obvious reasons. Select a set of basis vector for Σ consisting of r_P orthogonal vectors belonging to Σ_1 and the remaining $n - r_P$ vectors orthogonal to these. Let these vectors form the columns of a transformation matrix $\mathbf{T} = [\mathbf{T}_1 \mid \mathbf{T}_2]$. The original state equations are transformed to the Kalman controllable canonical form by letting $\mathbf{x} = \mathbf{Tw}$. Then, the orthogonal basis set gives $\mathbf{T}^{-1} = \mathbf{T}^T$, so that

$$\dot{\mathbf{w}} = \mathbf{T}^T\mathbf{ATw} + \mathbf{T}^T\mathbf{Bu} \quad \text{and} \quad \mathbf{y} = \mathbf{CTw} + \mathbf{Du}$$

Partitioning these equations according to the dimensions of Σ_1 and Σ_2 gives

$$\begin{bmatrix} \dot{\mathbf{w}}_1 \\ \dot{\mathbf{w}}_2 \end{bmatrix} = \left[\begin{array}{c|c} \mathbf{T}_1^T\mathbf{AT}_1 & \mathbf{T}_1^T\mathbf{AT}_2 \\ \hline \mathbf{T}_2^T\mathbf{AT}_1 & \mathbf{T}_2^T\mathbf{AT}_2 \end{array}\right] \begin{bmatrix} \mathbf{w}_1 \\ \mathbf{w}_2 \end{bmatrix} + \left[\begin{array}{c} \mathbf{T}_1^T\mathbf{B} \\ \hline \mathbf{T}_2^T\mathbf{B} \end{array}\right]\mathbf{u}$$

and

$$\mathbf{y} = [\mathbf{CT}_1 \mid \mathbf{CT}_2]\mathbf{w} + \mathbf{Du}$$

Since columns of $\mathbf{T}_2$ were selected orthogonal to all columns in **P** (this includes the columns in **B**), it is clear that $\mathbf{T}_2^T\mathbf{B} = [0]$. The control variables have no direct input to the states $\mathbf{w}_2$. Also, the term $\mathbf{T}_2^T\mathbf{AT}_1 = [0]$ (proven in Problem 11.22a), so that the states $\mathbf{w}_2$ are not coupled to states $\mathbf{w}_1$. The Kalman controllable canonical decomposition is thus

$$\begin{bmatrix} \dot{\mathbf{w}}_1 \\ \dot{\mathbf{w}}_2 \end{bmatrix} = \left[\begin{array}{c|c} \mathbf{T}_1^T\mathbf{AT}_1 & \mathbf{T}_1^T\mathbf{AT}_2 \\ \hline [0] & \mathbf{T}_2^T\mathbf{AT}_2 \end{array}\right] \begin{bmatrix} \mathbf{w}_1 \\ \mathbf{w}_2 \end{bmatrix} + \left[\begin{array}{c} \mathbf{T}_1^T\mathbf{B} \\ \hline [0] \end{array}\right]\mathbf{u}$$

and

$$\mathbf{y} = [\mathbf{CT}_1 \mid \mathbf{CT}_2]\mathbf{w} + \mathbf{Du} \tag{11.9}$$

State variables $\mathbf{w}_2$ are not connected to the input, neither directly nor indirectly through $\mathbf{w}_1$ coupling. Thus they are uncontrollable.

A straightforward method of finding the required orthogonal basis vectors in $\mathbf{T}_1$ and $\mathbf{T}_2$ is to perform a **QR** decomposition on **P**, (using the modified Gram-Schmidt procedure because it has better numerical properties than the standard Gram-Schmidt procedure.) The **QR** decomposition is a reliable method of determining the rank of **P**, so it will often be carried out anyway in the test of controllability criterion 2. If rank(**P**) $= r_P < n$, then a random-number generator is used to select components of additional vectors to augment the columns of **P** until a full set of n orthogonal basis vectors is found. The randomly selected augmenting vectors are transformed into columns of $\mathbf{T}_2$ as the modified Gram-Schmidt process is carried out.

EXAMPLE 11.7 A state variable model has

$$\mathbf{A} = \begin{bmatrix} 3 & 6 & 4 \\ 9 & 6 & 10 \\ -7 & -7 & -9 \end{bmatrix}, \quad \mathbf{B} = \begin{bmatrix} -0.666667 & 0.333333 \\ 0.333333 & -0.666667 \\ 0.333333 & 0.333333 \end{bmatrix}, \quad \mathbf{C}^T = \begin{bmatrix} 2 & 2 \\ 3 & 1 \\ 4 & 3 \end{bmatrix}$$

Evaluate the controllability matrix **P** and then use it to find the Kalman controllable canonical form.

Using Eq. (*11.5*) gives

$$\mathbf{P} = \left[\begin{array}{cc|cc|cc} -0.666667 & 0.333333 & 1.333333 & -1.666667 & -2.666667 & 6.333333 \\ 0.333333 & -0.666667 & -0.666667 & 2.333333 & 1.333333 & -7.666667 \\ 0.333333 & 0.333333 & -0.666667 & -0.666667 & 1.333333 & 1.333333 \end{array}\right]$$

The **QR** decomposition of this matrix is $\mathbf{P} = \mathbf{T}_1\mathbf{R}$, with

$$\mathbf{T}_1 = \begin{bmatrix} -0.816497 & 0 \\ 0.408248 & -0.707107 \\ 0.408248 & 0.707107 \end{bmatrix}$$

$$\mathbf{R} = \begin{bmatrix} 0.816496 & -0.408248 & -1.63299 & 2.04124 & 3.265986 & -7.75671 \\ 0 & 0.707107 & 0 & -2.12132 & 0 & 6.36396 \end{bmatrix}$$

From this, rank($\mathbf{P}$) = 2, so the system is uncontrollable. The two columns of $\mathbf{T}_1$ are an orthogonal basis for the controllable subspace. A third orthogonal basis vector is found and forms the column of $\mathbf{T}_2 = [0.57735 \quad 0.57735 \quad 0.57735]^T$. If $\mathbf{P}$ is augmented with an additional independent column, then $\mathbf{T}_2$ is found automatically via the modified Gram-Schmidt **QR** decomposition. The **R** factor will be found to have an additional all-zero row. Using $\mathbf{T} = [\mathbf{T}_1 \mid \mathbf{T}_2]$ gives the Kalman controllable canonical form

$$\dot{\mathbf{w}} = \left[\begin{array}{cc|c} -2 & 1.73204 & -5.65684 \\ 0 & -3 & -19.5959 \\ \hline 0 & 0 & 5 \end{array}\right]\mathbf{w} + \left[\begin{array}{cc} 0.816495 & -0.408247 \\ 0 & 0.707107 \\ \hline 0 & 0 \end{array}\right]\mathbf{u}$$

and

$$\mathbf{y} = \left[\begin{array}{cc|c} 1.224756 & 0.707107 & 5.19615 \\ 0 & 1.41421 & 3.46410 \end{array}\right]\mathbf{w}$$

The third component of $\mathbf{w}$ is uncontrollable. ■

An exactly analogous Kalman observable canonical form can be found by selecting an orthonormal basis set from columns of $\mathbf{Q}$ (these form columns of a matrix $\mathbf{V}_1$) and, if needed, augmenting vectors which form the columns of $\mathbf{V}_2$. Then the transformation $\mathbf{x} = \mathbf{V}\mathbf{v}$ leads to the desired form

$$\begin{bmatrix} \dot{\mathbf{v}}_1 \\ \dot{\mathbf{v}}_2 \end{bmatrix} = \left[\begin{array}{c|c} \mathbf{V}_1^T\mathbf{A}\mathbf{V}_1 & [\mathbf{0}] \\ \hline \mathbf{V}_2^T\mathbf{A}\mathbf{V}_1 & \mathbf{V}_2^T\mathbf{A}\mathbf{V}_2 \end{array}\right]\begin{bmatrix} \mathbf{v}_1 \\ \mathbf{v}_2 \end{bmatrix} + \left[\begin{array}{c} \mathbf{V}_1^T\mathbf{B} \\ \hline \mathbf{V}_2^T\mathbf{B} \end{array}\right]\mathbf{u}$$

and

$$\mathbf{y} = [\mathbf{C}\mathbf{V}_1 \mid [\mathbf{0}]]\begin{bmatrix} \mathbf{v}_1 \\ \mathbf{v}_2 \end{bmatrix} + \mathbf{D}\mathbf{u} \tag{11.10}$$

The facts that $\mathbf{V}_2^T\mathbf{C}^T = [\mathbf{0}]$ by construction and that $\mathbf{V}_1^T\mathbf{A}\mathbf{V}_2 = [\mathbf{0}]$ (see Problem 11.22b) have been used in arriving at this so-called Kalman observable canonical form of the state equations. Note that states $\mathbf{v}_2$ are not directly contributing to the output $\mathbf{y}$. Information about $\mathbf{v}_2$ is also not available indirectly in $\mathbf{y}$ through the $\mathbf{v}_1$ variables because $\mathbf{v}_2$ has no effect upon the $\mathbf{v}_1$ states.

The same **QR** decomposition method, this time applied to the matrix $\mathbf{Q}$ of Eq. (*11.6*), can be used to find the required transformation matrix $\mathbf{V}$.

EXAMPLE 11.8 A state-variable model has the same matrix **A** as in Example 11.7, and

$$\mathbf{B} = \begin{bmatrix} 0.333333 & 1.333333 \\ 1.333333 & 0.333333 \\ -0.666667 & 0.333333 \end{bmatrix}, \qquad \mathbf{C} = \begin{bmatrix} 1 & 2 & 3 \\ 3 & 3 & 6 \end{bmatrix}$$

Calculate the observability matrix **Q** and then use it to evaluate system observability and to find the Kalman observable canonical form.

From Eq. (11.6),

$$\mathbf{Q} = \left[\begin{array}{cc|cc|cc} 1 & 3 & 0 & -6 & -6 & 12 \\ 2 & 3 & -3 & -6 & 3 & 12 \\ 3 & 6 & -3 & -12 & -3 & 24 \end{array}\right] = \mathbf{V}_1\mathbf{R}$$

where

$$\mathbf{V}_1 = \begin{bmatrix} 0.26726 & 0.771517 \\ 0.53452 & -0.617213 \\ 0.801784 & 0.154303 \end{bmatrix}$$

and

$$\mathbf{R} = \begin{bmatrix} 3.741657 & 7.21605 & -4.0089 & -14.43211 & -2.40535 & 28.86421 \\ 0 & 1.38873 & 1.38873 & -2.77746 & -6.94365 & 5.55492 \end{bmatrix}$$

Since there are only two columns in $\mathbf{V}_1$, rank(**Q**) is 2 and the system is not observable. Using these results, augmented by $\mathbf{V}_2 = [-0.57735 \quad -0.57735 \quad 0.57735]^T$, Eq. (11.9) gives

$$\dot{\mathbf{v}} = \left[\begin{array}{cc|c} -1.07143 & 0.371154 & 0 \\ -4.82499 & -3.92857 & 0 \\ \hline -19.4422 & -3.74171 & 5 \end{array}\right]\mathbf{v} + \left[\begin{array}{cc} 0.26726 & 0.80178 \\ -0.66865 & 0.87439 \\ \hline -1.3472 & -0.76980 \end{array}\right]\mathbf{u}$$

$$\mathbf{y} = \left[\begin{array}{cc|c} 3.74165 & 0 & 0 \\ 7.21604 & 1.38873 & 0 \end{array}\right]\mathbf{v}$$

The third component of **v** is unobservable. ■

11.8 STABILIZABILITY AND DETECTABILITY

In general the properties of controllability and observability, discussed in this chapter, and stability, discussed in Chapter 10, are independent. None of these properties implies or is implied by any of the others. In this section two additional system properties are defined, which are useful whenever a system fails to be either completely controllable or completely observable.

Definition 11.3. A linear system is said to be *stabilizable* if all its unstable modes, if any, are controllable.

If a system is stable it is stabilizable. If a system is completely controllable, it is stabilizable. In the general case, the subsystem defined by the modes or states in $\mathbf{w}_2$ must be stable in order that the system be stabilizable. For a linear, constant system the requirement is that all eigenvalues of $\mathbf{T}_2^T\mathbf{A}\mathbf{T}_2$ must lie in the stable region, that is, the left-half of the s-plane (for continuous-time systems). The system of Example 11.7 is not stabilizable because the uncontrollable mode $\mathbf{w}_3$ has an eigenvalue of 5. The same definitions, decompositions, and analysis can be applied to discrete-time systems. Everything remains the same except that the stable region is the interior of the unit circle of the complex Z-plane in that case. The significance of stabilizability is that even though certain modes cannot be controlled by choice of input or feedback, if they are stable (better yet asymptotically stable), these modes will stay bounded (or better yet decay to zero). This modal behavior can often be tolerated in the overall control system.

Definition 11.4. A linear system is said to be *detectable* if all of its unstable modes, if any, are observable.

If the system is stable, it is detectable. If it is observable, it is also detectable. In general, the condition is met if the subsystem described by modes $\mathbf{v}_2$ are stable. In the linear constant case, the requirement is that all eigenvalues of $\mathbf{V}_2^T \mathbf{A} \mathbf{V}_2$ fall in the stable region of the complex plane (s or Z). the system of Example 11.8 is not detectable because the unobservable state v_3 is unstable (eigenvalue of 5). The significance of the detectability property is that if certain modes are unstable and hence subject to growth without bound, at least this undesirable behavior will be obvious from the output signals $\mathbf{y}$. No "hidden modes" such as those contained in $\mathbf{v}_2$ can be allowed to grow secretly in an unstable fashion.

REFERENCES

1. Chen, C. T.: *Introduction to Linear System Theory,* Holt, Rinehart and Winston, New York, 1970.
2. Forsythe, G. E., M. A. Malcolm, and C. Moler: *Computer Methods for Mathematical Computations,* Prentice Hall, Englewood Cliffs, N.J., 1977.
3. Chen, C. T. and C. A. Desoer: "A Proof of Controllability of Jordan Form State Equations," *IEEE Transactions on Automatic Control,* Vol. AC-13, No. 2, April 1968, pp. 195–196.
4. Kalman, R. E.: "Mathematical Description of Linear Dynamical Systems," *Jour. Soc. Ind. Appl. Math-Control Series,* Series A, Vol. 1, No. 2, 1963, pp. 152–192.
5. Elgerd, O. I.: *Control Systems Theory,* McGraw-Hill, New York, 1967.
6. Friedland, B.: *Control System Design,* McGraw-Hill, New York, 1986.
7. Alag, G. and H. Kaufman: "An Implementable Digital Adaptive Flight Controller Designed Using Stabilized Single-Stage Algorithms," *IEEE Transactions on Automatic Control,* Vol. AC-22, No. 5, October 1977, pp. 780–788.

ILLUSTRATIVE PROBLEMS

Application of the Criteria

.1 Is the following system completely controllable and completely observable?

$$\dot{\mathbf{x}} = \begin{bmatrix} -\frac{3}{4} & -\frac{1}{4} \\ -\frac{1}{2} & -\frac{1}{2} \end{bmatrix}\mathbf{x}(t) + \begin{bmatrix} 1 \\ 1 \end{bmatrix}u(t), \qquad y(t) = [4 \quad 2]x(t)$$

Using criteria 2, $\mathbf{P} = [\mathbf{B} \mid \mathbf{AB}] = \begin{bmatrix} 1 & -1 \\ 1 & -1 \end{bmatrix}$ has rank 1 and $\mathbf{Q} = [\mathbf{C}^T \mid \mathbf{A}^T\mathbf{C}^T] = \begin{bmatrix} 4 & -4 \\ 2 & -2 \end{bmatrix}$ has rank 1. Therefore, the system is neither completely controllable nor completely observable.

11.2 Is the following discrete-time system completely controllable and completely observable?

$$\mathbf{x}(k+1) = \begin{bmatrix} 1 & 0 \\ -\frac{1}{2} & \frac{1}{2} \end{bmatrix}\mathbf{x}(k) + \begin{bmatrix} 1 \\ -1 \end{bmatrix}u(k), \qquad y(k) = [5 \quad 1]\mathbf{x}(k)$$

Using criteria 2, $\mathbf{P} = \begin{bmatrix} 1 & 1 \\ -1 & -1 \end{bmatrix}$, rank $\mathbf{P} = 1$ but $n = 2$. Therefore, the system is *not* completely controllable. $\mathbf{Q} = \begin{bmatrix} 5 & \frac{9}{2} \\ 1 & \frac{1}{2} \end{bmatrix}$ has rank 2. The system is completely observable.

11.3 Investigate the controllability and observability of the systems in Figure 11.2(*a*) and (*b*) individually and when connected in series as in (*c*).

For system (*a*), $\dot{y}_1 + \beta y_1 = \dot{u}_1 + \alpha u_1$. Letting $x_1 = y_1 - u_1$ gives the state equation $\dot{x}_1 = -\beta x_1 + (\alpha - \beta)u_1$. This system is completely controllable if $\alpha \neq \beta$. It is always completely observable.

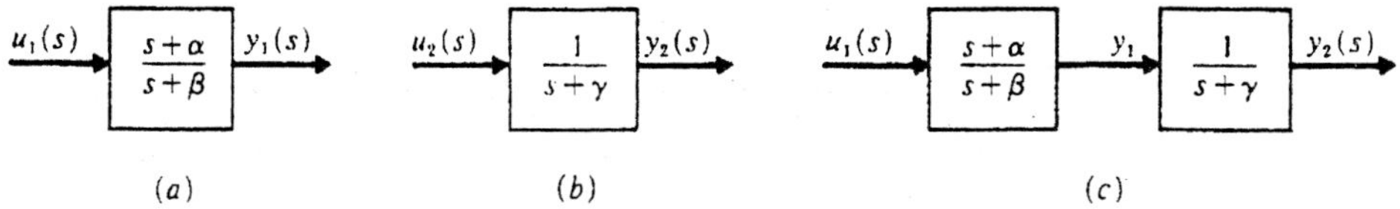

Figure 11.2

For system (*b*), $\dot{y}_2 + \gamma y_2 = u_2$. Letting $x_2 = y_2$ gives the state equation $\dot{x}_2 = -\gamma x_2 + u_2$. This system is completely controllable and observable.

Using the same definition for x_1 and x_2 in system (*c*), and noting that $y_1 = x_1 + u_1$ replaces u_2, the state equations are

$$\begin{bmatrix} \dot{x}_1 \\ \dot{x}_2 \end{bmatrix} = \begin{bmatrix} -\beta & 0 \\ 1 & -\gamma \end{bmatrix}\begin{bmatrix} x_1 \\ x_2 \end{bmatrix} + \begin{bmatrix} \alpha - \beta \\ 1 \end{bmatrix}u_1$$

The controllability matrix is

$$\mathbf{P} = \begin{bmatrix} \alpha - \beta & -\beta(\alpha - \beta) \\ 1 & \alpha - \beta - \gamma \end{bmatrix} \quad \text{and} \quad |\mathbf{P}| = (\alpha - \beta)(\alpha - \gamma)$$

The rank of $\mathbf{P}$ is 2 and system (*c*) is completely controllable, unless $\alpha = \beta$ or $\alpha = \gamma$. If either of these conditions is satisfied, the pole-zero cancellation leads to an uncontrollable system. The observability matrix is $\mathbf{Q} = \begin{bmatrix} 0 & 1 \\ 1 & -\gamma \end{bmatrix}$; and since its rank is 2, system (*c*) is completely observable. With other choices of states, this system is controllable but not observable.

11.4 Investigate the controllability and observability of the two systems shown in Figure 11.3.

System (*a*) is described by $\dot{x}_1 = -\alpha x_1 + Ku$, $y = x_1$, and is completely controllable and observable.

System (*b*) can be described by

$$\begin{bmatrix} \dot{x}_1 \\ \dot{x}_2 \end{bmatrix} = \begin{bmatrix} -\alpha & 0 \\ 0 & -\beta \end{bmatrix}\begin{bmatrix} x_1 \\ x_2 \end{bmatrix} + \begin{bmatrix} K_1 \\ K_2 \end{bmatrix}u, \qquad y = [C_1 \quad C_2]\mathbf{x}$$

The controllability and observability matrices of criterion 2 are

$$\mathbf{P} = \begin{bmatrix} K_1 & -\alpha K_1 \\ K_2 & -\beta K_2 \end{bmatrix} \quad \text{and} \quad \mathbf{Q} = \begin{bmatrix} C_1 & -\alpha C_1 \\ C_2 & -\beta C_2 \end{bmatrix}$$

System (b) is completely controllable and observable except when $\alpha = \beta$.

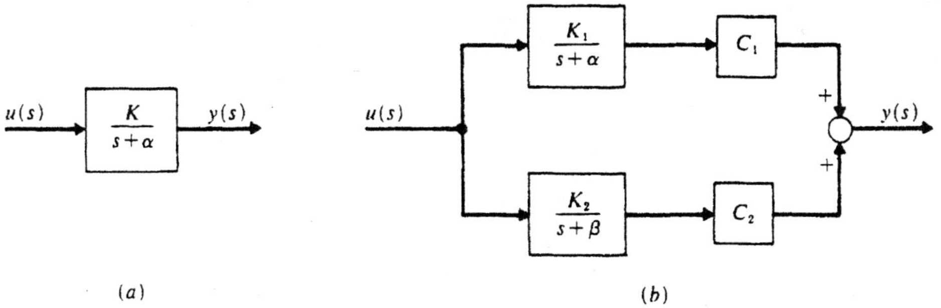

Figure 11.3

1.5 Use the system of Problem 9.6 to draw conclusions regarding the observability of a discrete-time system obtained by sampling a completely observable continuous-time system. Assume that the output is $y = x_1$.

To consider observability, it is only necessary to consider the unforced system $\dot{\mathbf{x}} = \begin{bmatrix} 0 & -\Omega \\ \Omega & 0 \end{bmatrix}\mathbf{x}$, $\mathbf{y} = [1 \quad 0]\mathbf{x}$. This system is completly observable since $\mathbf{Q} = \begin{bmatrix} 1 & 0 \\ 0 & -\Omega \end{bmatrix}$ has rank 2.

Using the state transition matrix $\mathbf{\Phi}(t_{k+1}, t_k) = \mathbf{\Phi}(\Delta t, 0)$, with $\Delta t \triangleq t_{k+1} - t_k$, the discrete-time equations are

$$\mathbf{x}(k+1) = \begin{bmatrix} \cos\Omega\,\Delta t & -\sin\Omega\,\Delta t \\ \sin\Omega\,\Delta t & \cos\Omega\,\Delta t \end{bmatrix}\mathbf{x}(k), \qquad \mathbf{y}(k) = [1 \quad 0]\mathbf{x}(k)$$

The discrete observability matrix is $\mathbf{Q} = \begin{bmatrix} 1 & \cos\Omega\,\Delta t \\ 0 & -\sin\Omega\,\Delta t \end{bmatrix}$. The rank is 2 unless the sampling period Δt is an integer multiple of π/Ω. The property of complete observability is lost if an oscillatory system is sampled at its natural frequency.

1.6 Prove that the name attached to the *controllable* canonical form of the state equations for a single input system is justified.

The controllable canonical form of the single-input state equations were given in Chapter 3 and always have

$$\mathbf{A} = \begin{bmatrix} 0 & 1 & 0 & \cdots & 0 \\ 0 & 0 & 1 & \cdots & 0 \\ \vdots & & & & \\ 0 & 0 & 0 & \cdots & 1 \\ a & b & c & \cdots & d \end{bmatrix} \qquad \mathbf{B} = \begin{bmatrix} 0 \\ 0 \\ 0 \\ 0 \\ 1 \end{bmatrix}$$

where $a, b, c, \ldots, d$ are arbitrary coefficients. Direct application of controllability criterion 2 shows that

$$\mathbf{AB} = \begin{bmatrix} 0 \\ 0 \\ \vdots \\ 1 \\ d \end{bmatrix} \qquad \mathbf{A}^2\mathbf{B} = \begin{bmatrix} 0 \\ 0 \\ \vdots \\ 1 \\ d \\ d^2 \end{bmatrix} \quad \mathbf{A}^3\mathbf{B} = \begin{bmatrix} 0 \\ 0 \\ 1 \\ d \\ d^2 \\ d^3 \end{bmatrix} \qquad \mathbf{A}^{n-1}\mathbf{B} = \begin{bmatrix} 1 \\ d \\ d^2 \\ d^3 \\ \vdots \\ d^{n-1} \end{bmatrix}$$

so that for any $n > 0$, the $n \times n$ matrix $\mathbf{P}$ has a nonzero determinant independent of the system coefficient values $a, b, c, \ldots, d$. Therefore, this form of the state equations is always controllable, and the name is aptly chosen.

11.7 Prove that the name attached to the *observable* canonical form of the state equations for a single output system is justified.

The observable canonical form for the single-input, single-output state equations were given in Chapter 3 and always have

$$\mathbf{A} = \begin{bmatrix} a & 1 & 0 & \cdots & 0 \\ b & 0 & 1 & \cdots & 0 \\ \vdots & & & & \\ c & 0 & 0 & \cdots & 1 \\ d & 0 & 0 & \cdots & 0 \end{bmatrix} \qquad \mathbf{C}^T = \begin{bmatrix} 0 \\ 0 \\ 0 \\ 0 \\ 1 \end{bmatrix} \qquad \mathbf{B} = [\text{not important}]$$

where $a, b, c, \ldots, d$ are arbitrary coefficients. Direct application of observability criterion 2 shows that the $n \times n$ matrix $\mathbf{Q}$ always has a nonzero determinant, and therefore this form of the state equations is always observable.

11.8 Is the following time-variable system completely controllable?

$$\dot{\mathbf{x}} = \frac{1}{12}\begin{bmatrix} 5 & 1 \\ 1 & 5 \end{bmatrix}\mathbf{x} + e^{t/2}\begin{bmatrix} 1 \\ 1 \end{bmatrix}u(t)$$

Since $\mathbf{B}(t)$ is time-varying, criterion 3 is used. The controllability matrix of Eq. (*11.7*) can be written as

$$\mathbf{G}(t_1, t_0) = \mathbf{\Phi}(t_1, 0)\int_{t_0}^{t_1} \mathbf{\Phi}^{-1}(\tau, 0)\mathbf{B}(\tau)\mathbf{B}^T(\tau)[\mathbf{\Phi}^{-1}(\tau, 0)]^T\, d\tau\, \mathbf{\Phi}^T(t_0, 0)$$

The transition matrix $\mathbf{\Phi}(t, 0)$ can be found by any of the methods of Chapter 8, and then

$$\mathbf{\Phi}^{-1}(\tau, 0) = \mathbf{\Phi}(-\tau, 0) = \frac{1}{2}\begin{bmatrix} e^{-\tau/2} + e^{-\tau/3} & e^{-\tau/2} - e^{-\tau/3} \\ e^{-\tau/2} - e^{-\tau/3} & e^{-\tau/2} + e^{-\tau/3} \end{bmatrix}$$

Therefore, $\mathbf{\Phi}^{-1}(\tau, 0)\mathbf{B}(\tau) = \begin{bmatrix} 1 \\ 1 \end{bmatrix}$ so that

$$|\mathbf{G}(t_1, t_0)| = |\mathbf{\Phi}(t_1, 0)|\left|\begin{bmatrix} t_1 - t_0 & t_1 - t_0 \\ t_1 - t_0 & t_1 - t_0 \end{bmatrix}\right||\mathbf{\Phi}^T(t_1, 0)| = 0$$

This is true for all t_0, t_1. The system is not completely controllable.

11.9 An approximate linear model of the lateral dynamics of an aircraft, for a particular set of flight conditions, has [7] the state and control vectors in the perturbation quantities

$$\mathbf{x} = [p \quad r \quad \beta \quad \phi]^T \quad \text{and} \quad \mathbf{u} = [\delta_a \quad \delta_r]^T$$

where p and r are incremental roll and yaw rates, β is an incremental sideslip angle, and ϕ is an incremental roll angle. The control inputs are the incremental changes in the aileron angle δ_a and in the rudder angle δ_r, respectively. These variables are shown in Figure 11.4. In a consistent set of units this linearized model has

$$\mathbf{A} = \begin{bmatrix} -10 & 0 & -10 & 0 \\ 0 & -0.7 & 9 & 0 \\ 0 & -1 & -0.7 & 0 \\ 1 & 0 & 0 & 0 \end{bmatrix} \qquad \mathbf{B} = \begin{bmatrix} 20 & 2.8 \\ 0 & -3.13 \\ 0 & 0 \\ 0 & 0 \end{bmatrix}$$

Suppose a malfunction prevents manipulation of the input δ_r. Is it possible to control the aircraft using only δ_a? Is the aircraft controllable with just δ_r? Verify that it is controllable with both inputs operable.

When δ_a is the only input, just the first column of the $\mathbf{B}$ matrix must be used in checking for controllability. The $\mathbf{P}$ matrix is determined to be

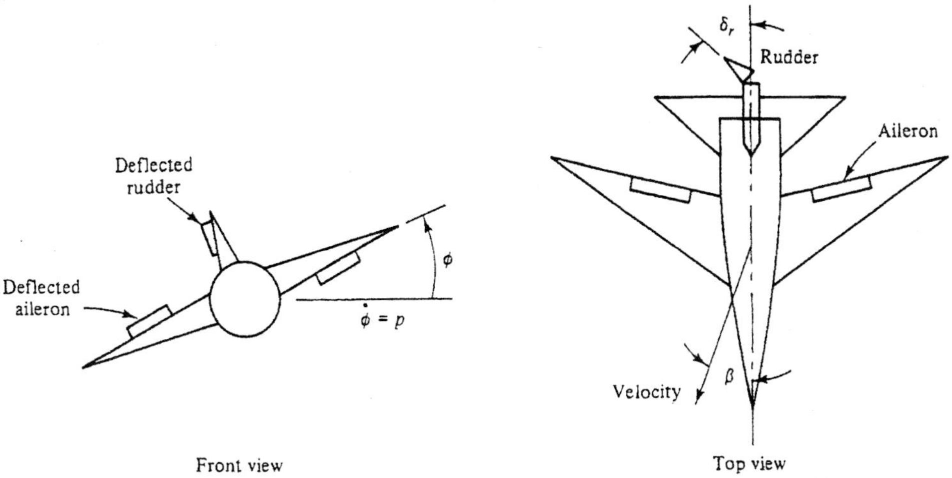

Figure 11.4

$$\mathbf{P} = \left[\begin{array}{c|c|c|c} 20 & -200 & 2000 & -2\times 10^4 \\ 0 & 0 & 0 & 0 \\ 0 & 0 & 0 & 0 \\ 0 & 20 & -200 & 2\times 10^3 \end{array}\right]$$

and its rank is $2 < n$. Thus it is not controllable. The aircraft can be made to roll using only the ailerons, but it cannot be made to turn, at least insofar as this linearized model is concerned.

With δ_r as the only input, column 2 of **B** is used to compute another **P** matrix:

$$\mathbf{P} = \left[\begin{array}{c|c|c|c} 2.8 & -28 & 248.7 & -2443.18 \\ -3.13 & 2.191 & 26.636 & -58.08 \\ 0 & 3.13 & -4.382 & -23.57 \\ 0 & 2.8 & -28 & 248.7 \end{array}\right]$$

The rank is now 4, and the system is controllable. The controllability index (the number of partitions in **P** that are required before full rank is achieved) is 4. The maneuverability of the aircraft would be greatly degraded, and a very sloppy flight profile would be expected under such conditions. Controllability does not guarantee a high-quality control, it just guarantees that all the states can be manipulated to zero in some fashion in some finite time. Adding the second input only adds more columns to **P** so its rank is also 4. Now, however, a rank 4 matrix can be formed from just the first two partitions of **P**. The controllability index is 2, indicating a stronger degree of controllability in some sense.

.10 If the only output is a measurement of the roll rate p (provided by a rate gyro) in the previous problem, is the system observable?

The output matrix is $\mathbf{C} = [1 \quad 0 \quad 0 \quad 0]$. Using this, the matrix **Q** is

$$\mathbf{Q} = \left[\begin{array}{c|c|c|c} 1 & -10 & 100 & -1000 \\ 0 & 0 & 10 & -114 \\ 0 & -10 & 107 & -984.9 \\ 0 & 0 & 0 & 0 \end{array}\right]$$

The rank is $3 < 4$, so the system is not observable. Measurements of roll rate allow the *change* in roll angle to be monitored, but will never allow determination of the roll angle itself because its initial value is unknown. In fact, if any one or all of the states except ϕ are measured outputs, this system remains unobservable. A bank indicator or some other means of measuring ϕ is required in order to obtain an observable system. If ϕ is the only measurement, then $\mathbf{C} = [0 \quad 0 \quad 0 \quad 1]$, which leads to

$$\mathbf{Q} = \begin{bmatrix} 0 & 1 & -10 & 100 \\ 0 & 0 & 0 & 10 \\ 0 & 0 & -10 & 107 \\ 1 & 0 & 0 & 0 \end{bmatrix}$$

This has rank 4, so the system is observable. The observability index is 4, meaning that all four partitions in $\mathbf{Q}$ are required to give a rank 4 result. If other states can also be measured and combined with ϕ to form a vector rather than scalar output, the observability index will improve (decrease).

11.11 (a) Show that a system governed by

$$\mathbf{x}(k+1) = \mathbf{x}(k)$$
$$\mathbf{y}(k) = \mathbf{C}\mathbf{x}(k)$$

with $\mathbf{C}$ constant, is never observable unless $\text{rank}(\mathbf{C}) = n$.

(b) Show that if $\mathbf{C}(k)$ is time-varying, observability criterion 3 leads to the condition that the normal equations of least squares must eventually become invertible for some finite time N if the system is to be observable.

(a) With $\mathbf{A} = \mathbf{I}$, criterion 2 gives

$$\mathbf{Q} = [\mathbf{C}^T \mid \mathbf{C}^T \mid \cdots \mid \mathbf{C}^T]$$

Therefore, $\text{rank}(\mathbf{Q}) = \text{rank}(\mathbf{C}^T) = \text{rank}(\mathbf{C})$. Observability requires that $\text{rank}(\mathbf{C}) = n$, where n is the number of components in $\mathbf{x}$.

(b) Again, $\mathbf{A} = \mathbf{I}$, so the observability matrix of criterion 3 becomes

$$\mathbf{Q}' = \sum_{k=0}^{N} \mathbf{C}^T(k)\mathbf{C}(k) = [\mathbf{C}^T(0) \quad \mathbf{C}^T(1) \quad \cdots \quad \mathbf{C}^T(N)] \begin{bmatrix} \mathbf{C}(0) \\ \mathbf{C}(1) \\ \vdots \\ \mathbf{C}(N) \end{bmatrix}$$

Define the stacked-up measurement vector and measurement matrix as

$$\mathbf{Y} = \begin{bmatrix} \mathbf{y}(0) \\ \mathbf{y}(1) \\ \vdots \\ \mathbf{y}(N) \end{bmatrix}, \qquad \mathcal{H} = \begin{bmatrix} \mathbf{C}(0) \\ \mathbf{C}(1) \\ \vdots \\ \mathbf{C}(N) \end{bmatrix}$$

Then the entire group of measurements, as it would be processed in batch least squares, is $\mathbf{Y} = \mathcal{H}\mathbf{x}(0)$, and $\mathbf{Q}' = \mathcal{H}^T\mathcal{H}$. The normal equation is $\mathcal{H}^T\mathcal{H}\mathbf{x}(0) = \mathcal{H}^T\mathbf{Y}$, and is invertible if and only if $\mathcal{H}$ has rank n. This is the same as requiring that $\mathbf{Q}'$ have rank n. Thus, observability is seen to be the same as having a *unique* least-squares solution in this case of a constant state vector.

Extensions and Proofs

11.12 Show that if a continuous-time linear system is completely controllable at t_0, then any initial state $\mathbf{x}(t_0)$ can be transferred to any other state $\mathbf{x}(t_1)$ at some finite time t_1.

Complete controllability means that any $\mathbf{x}(t_0)$ can be transferred to the origin $\mathbf{x}(t_1) = \mathbf{0}$. The solution for a given input is of the form

$$\mathbf{x}(t_1) = \mathbf{\Phi}(t_1, t_0)\mathbf{x}(t_0) + \int_{t_0}^{t_1} \mathbf{\Phi}(t_1, \tau)\mathbf{B}(\tau)\mathbf{u}(\tau)\, d\tau \tag{1}$$

This could be written as

$$\mathbf{0} = \mathbf{\Phi}(t_1, t_0)[\mathbf{x}(t_0) - \mathbf{\Phi}(t_0, t_1)\mathbf{x}(t_1)] + \int_{t_0}^{t_1} \mathbf{\Phi}(t_1, \tau)\mathbf{B}(\tau)\mathbf{u}(\tau)\, d\tau \tag{2}$$

Since $\mathbf{x}' \triangleq \mathbf{x}(t_0) - \mathbf{\Phi}(t_0, t_1)\mathbf{x}(t_1)$ belongs to Σ, it is a possible initial state. Complete controllability at t_0 guarantees that $\mathbf{x}'$ can be driven to the origin (Eq. (*2*)), which means that $\mathbf{x}(t_0)$ can be driven to any arbitrary $\mathbf{x}(t_1)$ (Eq. (*1*)).

.13 The arguments in Sec. 11.3 establish the necessity of controllability criterion 1. Show that this criterion is sufficient by assuming no zero rows of $\mathbf{B}_n$ and deriving the control which drives an arbitrary initial state to the origin.

The normal form description of the system is $\dot{\mathbf{q}} = \mathbf{\Lambda q} + \mathbf{B}_n \mathbf{u}$ and the solution at t_1 is

$$\mathbf{q}(t_1) = e^{\mathbf{\Lambda} t_1}\mathbf{q}(0) + \int_0^{t_1} e^{\mathbf{\Lambda} t_1} e^{-\mathbf{\Lambda}\tau}\mathbf{B}_n \mathbf{u}(\tau)\, d\tau$$

Let the ith row of $\mathbf{B}_n$ define the *row* vector $\mathbf{b}_i$, and let $\mathbf{u}(t) = \sum_{j=1}^{n} \beta_j e^{-\bar{\lambda}_j t}\bar{\mathbf{b}}_j^T$. The coefficients β_j are unknown constants. It is to be shown that these constants can be selected in such a way that $\mathbf{q}(t_1) = \mathbf{0}$ if none of the rows $\mathbf{b}_i$ are identically zero. Using the assumed form for $\mathbf{u}(t)$, a typical component of $\mathbf{q}(t_1)$ is

$$q_i(t_1) = e^{\lambda_i t_1} q_i(0) + \sum_{j=1}^{n} e^{\lambda_i t_1} \int_0^{t_1} e^{-\lambda_i \tau}\mathbf{b}_i \bar{\mathbf{b}}_j^T e^{-\bar{\lambda}_j \tau}\, d\tau\, \beta_j$$

or

$$e^{-\lambda_i t_1} q_i(t_1) - q_i(0) = \sum_{j=1}^{n} \langle \boldsymbol{\theta}_i(\tau), \boldsymbol{\theta}_j(\tau)\rangle \beta_j$$

The integral inner product (Problem 5.22, page 197) of the functions $\boldsymbol{\theta}_j(\tau) = e^{-\bar{\lambda}_j \tau}\bar{\mathbf{b}}_j^T$ is used. The unknown coefficients can be obtained by solving n simultaneous equations, and are given by

$$\begin{bmatrix} \beta_1 \\ \beta_2 \\ \vdots \\ \beta_n \end{bmatrix} = [\langle \boldsymbol{\theta}_i(\tau), \boldsymbol{\theta}_j(\tau)\rangle]^{-1}[e^{-\mathbf{\Lambda} t_1}\mathbf{q}(t_1) - \mathbf{q}(0)]$$

The indicated inverse is guaranteed to exist if $\mathbf{b}_i \neq \mathbf{0}$ for all i and if all the λ_i are distinct. This is true because under these conditions the set of functions $\{\boldsymbol{\theta}_j(\tau)\}$ is linearly independent over every finite interval $[0, t_1]$. The matrix $[\langle \boldsymbol{\theta}_i(\tau), \boldsymbol{\theta}_j(\tau)\rangle]$, which can also be written as $\int_0^{t_1} e^{-\mathbf{\Lambda}\tau}\mathbf{B}_n \bar{\mathbf{B}}_n^T e^{-\bar{\mathbf{\Lambda}}\tau}\, d\tau$, is the Grammian matrix and is nonsingular. The conditions of controllability criterion 1 are sufficient to guarantee that any $\mathbf{q}(0)$ can be driven to any $\mathbf{q}(t_1)$, including $\mathbf{q}(t_1) = \mathbf{0}$. An input function which drives $\mathbf{q}_1(0)$ to the origin at t_1 is

$$\mathbf{u}(t) = \bar{\mathbf{B}}_n^T e^{\bar{\mathbf{\Lambda}} t}\boldsymbol{\beta} = -\bar{\mathbf{B}}_n^T e^{-\bar{\mathbf{\Lambda}} t}\left[\int_0^{t_1} e^{-\mathbf{\Lambda}\tau}\mathbf{B}_n \bar{\mathbf{B}}_n^T e^{-\bar{\mathbf{\Lambda}}\tau}\, d\tau\right]^{-1}\mathbf{q}(0)$$

.14 Assume that the time-invariant system $\dot{\mathbf{x}} = \mathbf{Ax} + \mathbf{Bu}$ is completely controllable. Prove that the controllability criterion 2 is a necessary condition.

Complete controllability means that for every $\mathbf{x}_0$ there is some finite time t_1 and some input function $\mathbf{u}(t)$ such that

$$\mathbf{0} = e^{\mathbf{A} t_1}\mathbf{x}_0 + \int_0^{t_1} e^{\mathbf{A}(t_1 - \tau)}\mathbf{Bu}(\tau)\, d\tau \quad \text{or} \quad -\mathbf{x}_0 = \int_0^{t_1} e^{-\mathbf{A}\tau}\mathbf{Bu}(\tau)\, d\tau$$

Using the remainder form from the matrix exponential,

$$e^{-\mathbf{A}\tau} = \alpha_0(\tau)\mathbf{I} + \alpha_1(\tau)\mathbf{A} + \alpha_2(\tau)\mathbf{A}^2 + \cdots + \alpha_{n-1}(\tau)\mathbf{A}^{n-1}$$

gives

$$-\mathbf{x}_0 = \sum_{j=0}^{n-1} \mathbf{A}^j \mathbf{B} \int_0^{t_1} \alpha_j(\tau)\mathbf{u}(\tau)\, d\tau$$

Each integral term is an $r \times 1$ constant vector, defined as

$$\mathbf{v}_j = \int_0^{t_1} \alpha_j(\tau)\mathbf{u}(\tau)\, d\tau$$

Then

$$-\mathbf{x}_0 = [\mathbf{B} \mid \mathbf{A}\mathbf{B} \mid \mathbf{A}^2\mathbf{B} \mid \cdots \mid \mathbf{A}^{n-1}\mathbf{B}]\begin{bmatrix} \mathbf{v}_0 \\ \hline \mathbf{v}_1 \\ \hline \vdots \\ \hline \mathbf{v}_{n-1} \end{bmatrix}$$

This result states that every *vector* $-\mathbf{x}_0$ can be expressed as some linear combination of the columns of $\mathbf{P} = [\mathbf{B} \mid \mathbf{AB} \mid \cdots \mid \mathbf{A}^{n-1}\mathbf{B}]$. These columns must span the n-dimensional state space Σ, that is, it is necessary that rank $\mathbf{P} = n$. The necessity of the observability condition 2 can be established in a similar manner.

11.15 Assume that the following system is completely controllable and completely observable over the interval $[t_0, t_1]$:

$$\dot{\mathbf{x}} = \mathbf{A}(t)\mathbf{x}(t) + \mathbf{B}(t)\mathbf{u}(t), \qquad \mathbf{y}(t) = \mathbf{C}(t)\mathbf{x}(t) + \mathbf{D}(t)\mathbf{u}(t)$$

(a) Derive an explicit expression for an input which transfers the state from $\mathbf{x}(t_0)$ to $\mathbf{x}(t_1)$.
(b) If the input is zero, find an explicit expression for $\mathbf{x}(t_0)$ in terms of the output function $\mathbf{y}(t)$, $t_0 \le t \le t_1$.

(a) The solution for the state at t_1 can be written in terms of the transformation $\mathcal{A}_c: \mathcal{U} \rightarrow \Sigma$,

$$\mathbf{x}(t_1) - \mathbf{\Phi}(t_1, t_0)\mathbf{x}(t_0) = \mathcal{A}_c(\mathbf{u})$$

Let $\mathbf{u}(t) = \mathcal{A}_c^*(\mathbf{w})$, where $\mathbf{w}$ is an unknown vector in Σ. Then $\mathbf{x}(t_1) - \mathbf{\Phi}(t_1, t_0)\mathbf{x}(t_0) = \mathcal{A}_c\mathcal{A}_c^*(\mathbf{w})$. The condition for complete controllability is that the $n \times n$ matrix $\mathcal{A}_c\mathcal{A}_c^*$ has an inverse. Inverting this matrix to solve for $\mathbf{w}$ leads to

$$\mathbf{u}(t) = \mathcal{A}_c^*(\mathcal{A}_c\mathcal{A}_c^*)^{-1}[\mathbf{x}(t_1) - \mathbf{\Phi}(t_1, t_0)\mathbf{x}(t_0)]$$

$$= \overline{\mathbf{B}}^T(t)\overline{\mathbf{\Phi}}^T(t_1, t)\left[\int_{t_0}^{t_1} \mathbf{\Phi}(t_1, \tau)\mathbf{B}(\tau)\overline{\mathbf{B}}^T(\tau)\overline{\mathbf{\Phi}}^T(t_1, \tau)\, d\tau\right]^{-1}[\mathbf{x}(t_1) - \mathbf{\Phi}(t_1, t_0)\mathbf{x}(t_0)]$$

(b) In terms of the transformation $\mathcal{A}_0$, the output of the unforced system is $\mathbf{y}(t) = \mathcal{A}_0(\mathbf{x}(t_0))$. Operating on both sides with the adjoint transformation $\mathcal{A}_0^*$ gives $\mathcal{A}_0^*(\mathbf{y}(t)) = \mathcal{A}_0^*\mathcal{A}_0(\mathbf{x}(t_0))$. The criterion for complete observability ensures that the matrix $\mathcal{A}_0^*\mathcal{A}_0$ has an inverse, so

$$\mathbf{x}(t_0) = (\mathcal{A}_0^*\mathcal{A}_0)^{-1}\mathcal{A}_0^*(\mathbf{y}(t))$$

$$= \left[\int_{t_0}^{t_1} \overline{\mathbf{\Phi}}^T(\tau, t_0)\overline{\mathbf{C}}^T(\tau)\mathbf{C}(\tau)\mathbf{\Phi}(\tau, t_0)\, d\tau\right]^{-1}\int_{t_0}^{t_1} \overline{\mathbf{\Phi}}(\tau, t_0)\overline{\mathbf{C}}^T(\tau)\mathbf{y}(\tau)\, d\tau$$

11.16 A system with n state variables and r inputs is expressed in Jordan form

$$\dot{\mathbf{x}} = \begin{bmatrix} \mathbf{J}_1 & & & \\ & \mathbf{J}_2 & & \\ & & \ddots & \\ & & & \mathbf{J}_p \end{bmatrix}\mathbf{x} + \begin{bmatrix} \mathbf{B}_1 \\ \hline \mathbf{B}_2 \\ \hline \vdots \\ \hline \mathbf{B}_p \end{bmatrix}\mathbf{u} \tag{1}$$

Show that the controllability of this system is determined entirely by the last rows $\mathbf{b}_{il}^T$ of each $\mathbf{B}_i$ submatrix. (The subscript l signifies the last row in a given block and is not a fixed integer.) In particular, show that the system is completely controllable if and only if

1. $\{\mathbf{b}_{il}, \mathbf{b}_{jl}, \dots, \mathbf{b}_{kl}\}$ is a linearly independent set if $\mathbf{J}_i, \mathbf{J}_j, \dots, \mathbf{J}_k$ are Jordan blocks *with the same eigenvalue* λ_i, and
2. $\mathbf{b}_{pl} \neq \mathbf{0}$ if $\mathbf{J}_p$ is the *only* Jordan block with eigenvalue λ_p.

Note that if all $\mathbf{J}_i$ blocks are 1×1 blocks so that $\mathbf{A}$ is diagonal, the controllability criterion 1 of Sec. 11.3 requires that all $\mathbf{b}^T \neq \mathbf{0}$ for all rows of $\mathbf{B}$. Controllability criterion 2 is used to investigate the more general case. For simplicity, assume there are just three blocks,

$$\mathbf{J}_1 = \begin{bmatrix} \lambda_1 & 1 & 0 \\ 0 & \lambda_1 & 1 \\ 0 & 0 & \lambda_1 \end{bmatrix}, \qquad \mathbf{J}_2 = \begin{bmatrix} \lambda_1 & 1 \\ 0 & \lambda_1 \end{bmatrix}, \qquad \mathbf{J}_3 = \begin{bmatrix} \lambda_3 & 1 \\ 0 & \lambda_3 \end{bmatrix}$$

with $\mathbf{B}^T = [\mathbf{b}_{11} \;\; \mathbf{b}_{12} \;\; \mathbf{b}_{13} \mid \mathbf{b}_{21} \;\; \mathbf{b}_{22} \mid \mathbf{b}_{31} \;\; \mathbf{b}_{32}]$. The controllability matrix is

$$\mathbf{P} = [\mathbf{B} \mid \mathbf{AB} \mid \mathbf{A}^2\mathbf{B} \mid \mathbf{A}^3\mathbf{B} \mid \mathbf{A}^4\mathbf{B} \mid \mathbf{A}^5\mathbf{B} \mid \mathbf{A}^6\mathbf{B}]$$

$$= \begin{bmatrix} \mathbf{B}_1 & \mathbf{J}_1\mathbf{B}_1 & \mathbf{J}_1^2\mathbf{B}_1 & \mathbf{J}_1^3\mathbf{B}_1 & \mathbf{J}_1^4\mathbf{B}_1 & \mathbf{J}_1^5\mathbf{B}_1 & \mathbf{J}_1^6\mathbf{B}_1 \\ \mathbf{B}_2 & \mathbf{J}_2\mathbf{B}_2 & \mathbf{J}_2^2\mathbf{B}_2 & \mathbf{J}_2^3\mathbf{B}_2 & \mathbf{J}_2^4\mathbf{B}_2 & \mathbf{J}_2^5\mathbf{B}_2 & \mathbf{J}_2^6\mathbf{B}_2 \\ \mathbf{B}_3 & \mathbf{J}_3\mathbf{B}_3 & \mathbf{J}_3^2\mathbf{B}_3 & \mathbf{J}_3^3\mathbf{B}_3 & \mathbf{J}_3^4\mathbf{B}_3 & \mathbf{J}_3^5\mathbf{B}_3 & \mathbf{J}_3^6\mathbf{B}_3 \end{bmatrix}$$

The results of Problem 8.4, page 293, are used for the various powers $\mathbf{J}_i^k$. Then

$$\mathbf{J}_1^k\mathbf{B}_1 = \begin{bmatrix} \lambda_1^k\mathbf{b}_{11}^T + k\lambda_1^{k-1}\mathbf{b}_{12}^T + \tfrac{1}{2}k(k-1)\lambda_1^{k-2}\mathbf{b}_{13}^T \\ \lambda_1^k\mathbf{b}_{12}^T + k\lambda_1^{k-1}\mathbf{b}_{13}^T \\ \lambda_1^k\mathbf{b}_{13}^T \end{bmatrix}$$

$$\mathbf{J}_2^k\mathbf{B}_2 = \begin{bmatrix} \lambda_1^k\mathbf{b}_{21}^T + k\lambda_1^{k-1}\mathbf{b}_{22}^T \\ \lambda_1^k\mathbf{b}_{22}^T \end{bmatrix}$$

$$\mathbf{J}_3^k\mathbf{B}_3 = \begin{bmatrix} \lambda_3^k\mathbf{b}_{31}^T + k\lambda_3^{k-1}\mathbf{b}_{32}^T \\ \lambda_3^k\mathbf{b}_{32}^T \end{bmatrix}$$

At this point the necessity of condition (2) is obvious since, for example, if $\mathbf{b}_{32} = \mathbf{0}$, the entire seventh row of $\mathbf{P}$ would be zero and rank $\mathbf{P} < n$. To see the necessity of condition (1), let $\mathbf{b}_{22} = \alpha\mathbf{b}_{13}$. Then an elementary row operation ($-\alpha$ times row 3 added to row 5) would make row 5 zero. This would again give rank $\mathbf{P} < n$, so the system would be uncontrollable.

It is tedious but trivial to show that a sequence of elementary column operations can be used to reduce $\mathbf{P}$ to $\mathbf{P}'$. Specifically, subtract λ_1 times each of the first r columns from the corresponding column in the second group of r columns. Then subtract λ_1^2 times column 1 and $2\lambda_1$ times the modified $(r+1)$st column from the $(2r+1)$st column and so on. Continuing this process leads to $\mathbf{P}'$:

$$\mathbf{P}' = \begin{bmatrix} \mathbf{b}_{11}^T & \mathbf{b}_{12}^T & \mathbf{b}_{13}^T & \cdots \\ \mathbf{b}_{12}^T & \mathbf{b}_{13}^T & \mathbf{0} & \cdots \\ \mathbf{b}_{13}^T & \mathbf{0} & \mathbf{0} & \\ \mathbf{b}_{21}^T & \mathbf{b}_{22}^T & \mathbf{0} & \cdots \\ \mathbf{b}_{22}^T & \mathbf{0} & \mathbf{0} & \\ \mathbf{b}_{31}^T & (\lambda_3-\lambda_1)\mathbf{b}_{31}^T + \mathbf{b}_{32}^T & (\lambda_3^2-\lambda_1^2)\mathbf{b}_{31}^T + 2(\lambda_3-\lambda_1)\mathbf{b}_{32}^T - 2\lambda_1(\lambda_3-\lambda_1)\mathbf{b}_{31}^T & \cdots \\ \mathbf{b}_{32}^T & (\lambda_3-\lambda_1)\mathbf{b}_{32}^T & (\lambda_3^2-\lambda_1^2)\mathbf{b}_{32}^T - 2\lambda_1(\lambda_3-\lambda_1)\mathbf{b}_{32}^T & \end{bmatrix}$$

The last two rows of $\mathbf{P}'$ should be recognized as consisting of $\mathbf{B}_3, (\mathbf{J}_3 - \mathbf{I}\lambda_1)\mathbf{B}_3, (\mathbf{J}_3 - \mathbf{I}\lambda_1)^2\mathbf{B}_3, (\mathbf{J}_3 - \mathbf{I}\lambda_1)^3\mathbf{B}_3, \dots$. Finally, a series of elementary row operations (row interchanges) gives

$$\mathbf{P}'' = \left[\begin{array}{c|c|c|c|c|c}
\mathbf{b}_{13}^T & & & & & \\
\mathbf{b}_{22}^T & & & \mathbf{0} & & \\
\mathbf{b}_{12}^T & \mathbf{b}_{13}^T & & & & \\
\mathbf{b}_{21}^T & \mathbf{b}_{22}^T & & & & \\
\mathbf{b}_{11}^T & \mathbf{b}_{12}^T & \mathbf{b}_{13}^T & & & \\
\hline
\mathbf{B}_3 & (\mathbf{J}_2 - \mathbf{I}\lambda_1)\mathbf{B}_3 & (\mathbf{J}_3 - \mathbf{I}\lambda_1)^2\mathbf{B}_3 & (\mathbf{J}_3 - \mathbf{I}\lambda_1)^3\mathbf{B}_3 & (\mathbf{J}_3 - \mathbf{I}\lambda_1)^4\mathbf{B}_3 & \cdots
\end{array}\right]$$

By definition, the rank of $\begin{bmatrix} \mathbf{b}_{13}^T \\ \mathbf{b}_{22}^T \end{bmatrix}$ is 2 if and only if $\mathbf{b}_{13}$ and $\mathbf{b}_{22}$ are linearly independent. If they are, then a 2×2 nonzero determinant can be formed by deleting columns. Linear independence requires $\mathbf{b}_{13} \neq \mathbf{0}$. It is easy to show that if any one element of $\mathbf{b}_{32}$ is nonzero, then a nonzero 2×2 determinant can be obtained from $[(\mathbf{J}_3 - \mathbf{I}\lambda_1)^3\mathbf{B}_3 \quad (\mathbf{J}_3 - \mathbf{I}\lambda_1)^4\mathbf{B}_3]$. Thus an $n \times n$ lower block triangular matrix can be formed from $\mathbf{P}''$, whose determinant is nonzero. Rank $\mathbf{P}'' =$ rank $\mathbf{P} = n$ implies complete controllability. The procedure used for this example quickly becomes unwieldy, but the result generalizes to any number of Jordan blocks [3].

11.17 If the output equation for the system in Problem 11.16 is

$$\mathbf{y} = [\mathbf{C}_1 \mid \mathbf{C}_2 \mid \cdots \mid \mathbf{C}_p]\mathbf{x} + \mathbf{D}\mathbf{u}$$

show that observability depends only on the first columns $\mathbf{c}_{i1}$ of each $\mathbf{C}_i$ block. In particular, show that the system is completely observable if and only if

1. $\{\mathbf{c}_{i1}, \mathbf{c}_{j1}, \ldots, \mathbf{c}_{k1}\}$ is a linearly independent set if $\mathbf{J}_i, \mathbf{J}_j, \ldots, \mathbf{J}_k$ are Jordan blocks *with the same eigenvalue* λ_i, and
2. $\mathbf{c}_{p1} \neq \mathbf{0}$ if $\mathbf{J}_p$ is the *only* Jordan block with eigenvalue λ_p.

Employ the same seventh-order system of Problem 11.16 for simplicity, and use $\mathbf{C} = [\mathbf{c}_{11} \quad \mathbf{c}_{12} \quad \mathbf{c}_{13} \mid \mathbf{c}_{21} \quad \mathbf{c}_{22} \mid \mathbf{c}_{31} \quad \mathbf{c}_{32}]$. The observability matrix is

$$\mathbf{Q} = [\overline{\mathbf{C}}^T \mid \overline{\mathbf{A}}^T\overline{\mathbf{C}}^T \mid (\overline{\mathbf{A}}^T)^2\overline{\mathbf{C}}^T \mid \cdots \mid (\overline{\mathbf{A}}^T)^6\overline{\mathbf{C}}^T]$$

Using the same kind of elementary column and row operations as in Problem 11.16, $\mathbf{Q}$ can be reduced to $\mathbf{Q}''$, a matrix of a similar form to $\mathbf{P}''$. Because now $\overline{\mathbf{A}}^T$ is used instead of $\mathbf{A}$, the *first* rows of $\overline{\mathbf{C}}_i^T$ play the role of the *last* rows of $\mathbf{B}_i$. Thus the conclusions regarding the leading columns of $\mathbf{C}_i$ follow directly from the results of Problem 11.16.

Decomposition

11.18 Show that at any time $t_1 > t_0$, the state space Σ can be expressed as the direct sum $\Sigma = \mathscr{X}_1 \oplus \mathscr{X}_2$, where $\mathscr{X}_1$ is the subspace which contains all the controllable initial states and $\mathscr{X}_2$ is the null space of $\mathbf{G}'(t_1, t_0)$ of Problem 11.33.

Let $\mathbf{x}(t_0)$ be a controllable initial state. Then by definition, there exists an input such that

$$-\mathbf{x}(t_0) = \int_{t_0}^{t_1} \mathbf{\Phi}(t_0, \tau)\mathbf{B}(\tau)\mathbf{u}(\tau)\, d\tau \triangleq \mathscr{A}(\mathbf{u})$$

All controllable initial states belong to $\mathscr{R}(\mathscr{A}) \triangleq \mathscr{X}_1$. Since $\mathscr{A}: \mathscr{U} \to \Sigma$, the results of Problem 6.21 give $\Sigma = \mathscr{R}(\mathscr{A}) \oplus \mathscr{N}(\mathscr{A}^*)$. From Sec. 11.6, $\mathscr{N}(\mathscr{A}^*) = \mathscr{N}(\mathscr{A}\mathscr{A}^*)$. For this example, $\mathscr{A}^* = \overline{\mathbf{B}}^T(t)\overline{\mathbf{\Phi}}^T(t_0, t)$ so that $\mathscr{A}\mathscr{A}^* = \mathbf{G}'(t_1, t_0)$. Therefore, $\mathscr{N}(\mathscr{A}^*) = \mathscr{N}(\mathbf{G}'(t_1, t_0))$. Note that this result implies that for each $\mathbf{x}(t_0) \in \Sigma$, $\mathbf{x}(t_0) = \mathbf{x}_1 + \mathbf{x}_2$ with $\mathbf{x}_1 \in \mathscr{X}_1$ (controllable) and $\mathbf{x}_2 \in \mathscr{X}_2$ (not controllable). However, it does not imply that the set of all uncontrollable states is the subspace $\mathscr{X}_2$.

Let $\mathbf{x}_a(t_0) = \mathbf{x}_1 + \mathbf{x}_2$ and $\mathbf{x}_b(t_0) = \mathbf{x}_1 - \mathbf{x}_2$. Since $\mathbf{x}_a(t_0)$ and $\mathbf{x}_b(t_0)$ do not belong to $\mathscr{X}_1$ they are not controllable. They do not belong to $\mathscr{X}_2$ either. Still, $\mathbf{x}_a(t_0) + \mathbf{x}_b(t_0) = 2\mathbf{x}_1$ belongs to $\mathscr{X}_1$ and is therefore controllable. The set of all uncontrollable initial states is *not* a subspace.

19 Let $\mathcal{A}_0 : \Sigma \to \mathcal{Y}$ be the output transformation defined in Sec. 11.6. Then the results of Problem 6.21 allow the decomposition $\Sigma = \mathcal{N}(\mathcal{A}_0) \oplus \mathcal{R}(\mathcal{A}_0^*)$. It has been shown that $\mathcal{N}(\mathcal{A}_0) = \mathcal{N}(\mathcal{A}_0^* \mathcal{A}_0) = \mathcal{N}(\mathbf{H}(t_1, t_0))$. Define this null space as $\mathcal{X}_3$. Then every $\mathbf{x}(t_0) \in \mathcal{X}_3$ contributes nothing to the output $\mathbf{y}(t)$, and these are referred to as unobservable states. Use this and the results of Problem 11.18 to show that for all $\mathbf{x}(t_0) \in \Sigma$, $\mathbf{x}(t_0) = \mathbf{x}_a + \mathbf{x}_b + \mathbf{x}_c + \mathbf{x}_d$, where $\mathbf{x}_a$ is controllable but unobservable, $\mathbf{x}_b$ is controllable and observable, $\mathbf{x}_c$ is uncontrollable but observable, and $\mathbf{x}_d$ is uncontrollable and unobservable.

Define $\mathcal{X}_4 = \mathcal{X}_3^\perp = \mathcal{R}(\mathcal{A}_0^*)$. Each $\mathbf{x}(t_0) \in \mathcal{X}_4$ is observable in the sense that a unique $\mathbf{x}(t_0) \in \mathcal{X}_4$ can be associated with a given unforced output record $\mathbf{y}(t)$. (Of course, $\mathbf{x}'(t_0) = \mathbf{x}(t_0) + \mathbf{x}_3$ will give the same $\mathbf{y}(t)$ if $\mathbf{x}_3 \in \mathcal{X}_3$, so it is not possible to determine whether $\mathbf{x}(t_0)$ or $\mathbf{x}'(t_0)$ is the actual initial state.)

Every $\mathbf{x}(t_0)$ can be written as $\mathbf{x}(t_0) = \mathbf{x}_1 + \mathbf{x}_2$ with $\mathbf{x}_1 \in \mathcal{X}_1, \mathbf{x}_2 \in \mathcal{X}_2$. The orthogonal projection of $\mathbf{x}_1$ into $\mathcal{X}_3$ gives $\mathbf{x}_a$. The projection into $\mathcal{X}_4$ gives $\mathbf{x}_b$. Similarly, projecting $\mathbf{x}_2$ into $\mathcal{X}_4$ gives $\mathbf{x}_c$ and projecting $\mathbf{x}_2$ into $\mathcal{X}_3$ gives $\mathbf{x}_d$.

20 Indicate how a time-invariant linear system with distinct eigenvalues can be decomposed into four possible subsystems with the respective properties (1) controllable but unobservable, (2) controllable and observable, (3) uncontrollable but observable, and (4) uncontrollable and unobservable.

The system can be put into normal form, giving $\mathbf{x}(t_0) = q_1(t_0)\xi_1 + q_2(t_0)\xi_2 + \cdots + q_n(t_0)\xi_n$. For this class of systems, controllability and observability criteria 1 apply. The controllability and observability can be ascertained for each mode individually. The modes are each assigned to one of the four categories. The resulting decomposition is illustrated in Figure 11.5.

Notice that there is no signal path from the input to an uncontrollable subsystem, either directly or through other subsystems. Also, there is no signal path from an unobservable subsystem to the output. The decomposition of Figure 11.5 can be accomplished for any linear system, but the process is not always this simple [4].

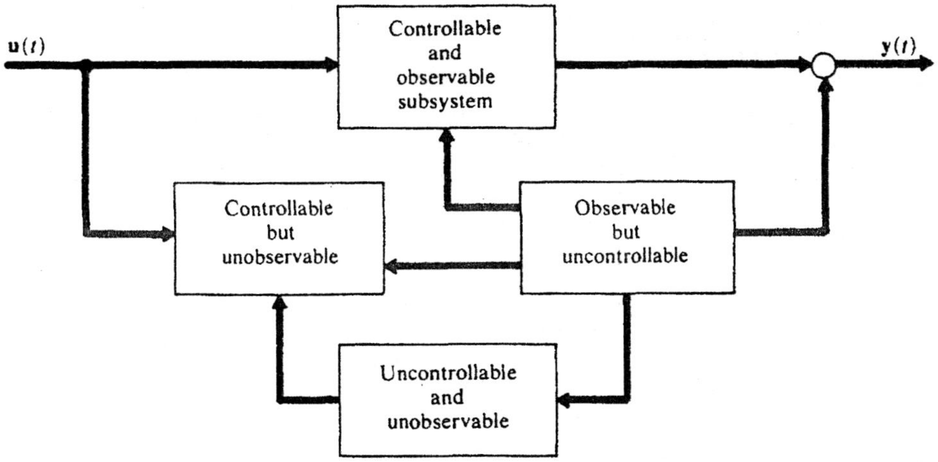

Figure 11.5

21 Subdivide the following system into subsystems as discussed in Problem 11.20:

$$\dot{\mathbf{x}} = \begin{bmatrix} -7 & -2 & 6 \\ 2 & -3 & -2 \\ -2 & -2 & 1 \end{bmatrix}\mathbf{x} + \begin{bmatrix} 1 & 1 \\ 1 & -1 \\ 1 & 0 \end{bmatrix}\mathbf{u}, \qquad \mathbf{y} = \begin{bmatrix} -1 & -1 & 2 \\ 1 & 1 & -1 \end{bmatrix}\mathbf{x}$$

The eigenvalues of **A** are $\lambda_i = -1, -3,$ and -5. The Jordan normal form will be used, since the controllability and observability criteria 1 apply. The modal matrix containing the eigenvectors is

$$\mathbf{M} = \begin{bmatrix} 1 & 1 & 1 \\ 0 & 1 & -1 \\ 1 & 1 & 0 \end{bmatrix} \quad \text{and} \quad \mathbf{M}^{-1} = \begin{bmatrix} -1 & -1 & 2 \\ 1 & 1 & -1 \\ 1 & 0 & -1 \end{bmatrix}$$

so that

$$\dot{\mathbf{q}} = \begin{bmatrix} -1 & 0 & 0 \\ 0 & -3 & 0 \\ 0 & 0 & -5 \end{bmatrix}\mathbf{q} + \begin{bmatrix} 0 & 0 \\ 1 & 0 \\ 0 & 1 \end{bmatrix}\mathbf{u}, \qquad \mathbf{y} = \begin{bmatrix} 1 & 0 & 0 \\ 0 & 1 & 0 \end{bmatrix}\mathbf{q}$$

The first mode is uncontrollable and the third mode is unobservable. The second mode is both controllable and observable. There is no mode which is both uncontrollable and unobservable. Figure 11.6 illustrates the three subsystems.

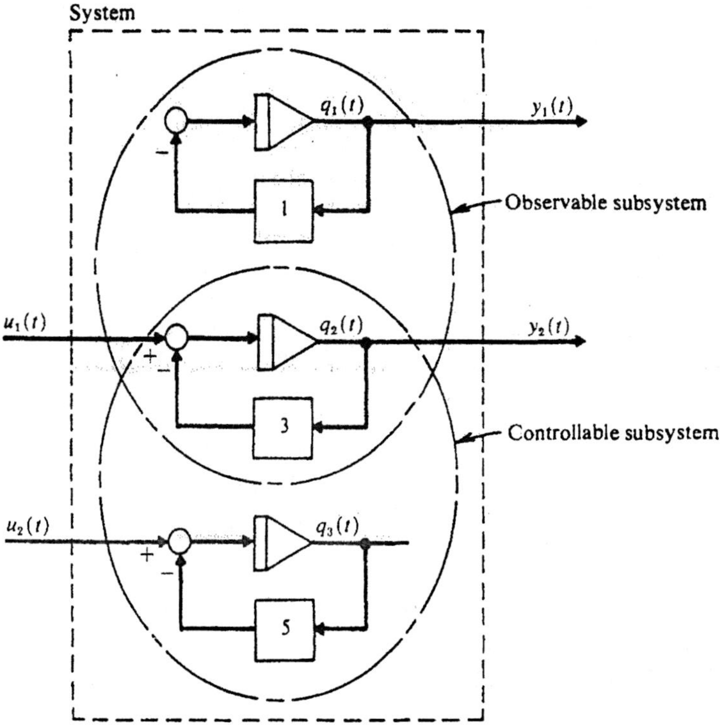

Figure 11.6

11.22 (a) Let matrix $\mathbf{T}_2$ have columns which are orthogonal to the subspace spanned by the columns of **P**, and let columns of $\mathbf{T}_1$ form an orthogonal basis for this same subspace. Show that $\mathbf{T}_2^T\mathbf{A}\mathbf{T}_1 = [\mathbf{0}]$.

(b) Let matrix $\mathbf{V}_2$ have columns which are orthogonal to the subspace spanned by the columns of **Q**, and let columns of $\mathbf{V}_1$ form an orthogonal basis for this same subspace. Show that $\mathbf{V}_1^T\mathbf{A}\mathbf{V}_2 = [\mathbf{0}]$.

(a) The controllability matrix can be written as $\mathbf{P} = [\mathbf{B} \mid \mathbf{AB} \mid \mathbf{A}^2\mathbf{B} \mid \cdots \mid \mathbf{A}^{n-1}\mathbf{B}] = \mathbf{T}_1\mathbf{R}$ by using a **QR** decomposition. This implies that $\mathbf{B} = \mathbf{T}_1\mathbf{R}_1$, $\mathbf{AB} = \mathbf{T}_1\mathbf{R}_2$, $\mathbf{A}^2\mathbf{B} = \mathbf{T}_1\mathbf{R}_3, \ldots,$ $\mathbf{A}^{n-1}\mathbf{B} = \mathbf{T}_1\mathbf{R}_n$, where the $\mathbf{R}_i$ terms are partitions of **R**. By construction, $\mathbf{T}_2^T\mathbf{P} = [\mathbf{0}]$. By looking at individual partitions of this equation and using the previous results for the $\mathbf{A}^i\mathbf{B}$ terms, it is found that

$$\mathbf{T}_2^T\mathbf{B} = [0] \Rightarrow \mathbf{T}_2^T\mathbf{T}_1\mathbf{R}_1 = [0]$$

$$\mathbf{T}_2^T\mathbf{A}\mathbf{B} = [0] \Rightarrow \mathbf{T}_2^T\mathbf{A}\mathbf{T}_1\mathbf{R}_1 = [0]$$

$$\mathbf{T}_2^T\mathbf{A}^2\mathbf{B} = [0] \Rightarrow \mathbf{T}_2^T\mathbf{A}(\mathbf{A}\mathbf{B}) = \mathbf{T}_2^T\mathbf{A}\mathbf{T}_1\mathbf{R}_2 = [0]$$

and so on. The first of these expressions just gives the lower partition of the new matrix $\mathbf{B}$ in the Kalman canonical form. The second expression gives the desired result *if* $\mathbf{R}_1$ is nonsingular. If $\mathbf{R}_2$ is nonsingular, the third equation proves the desired result, and so on. Each $\mathbf{R}_i$ partition is square, of dimension equal to the rank of $\mathbf{P}$, so it is assured that there is a nonsingular matrix that can be formed from one $\mathbf{R}_i$ or from a combination of $\mathbf{R}_i$ columns.

(b) Since $\mathbf{Q} = [\mathbf{C}^T \mid \mathbf{A}^T\mathbf{C}^T \mid (\mathbf{A}^2)^T\mathbf{C}^T \mid \cdots \mid (\mathbf{A}^{n-1})^T\mathbf{C}^T]$, repeating this procedure with notational changes proves that $\mathbf{V}_2^T\mathbf{A}^T\mathbf{V}_1 = [0]$. The transpose of this gives the desired result.

23 **(a)** Find the Kalman controllable canonical form for the system of Problem 11.21.
(b) Find the Kalman observable canonical form for the system of Problem 11.21.

(a) The controllability matrix of criterion 2 is

$$\mathbf{P} = \left[\begin{array}{cc|cc|cc} 1 & 1 & -3 & -5 & 9 & 25 \\ 1 & -1 & -3 & 5 & 9 & -25 \\ 1 & 0 & -3 & 0 & 9 & 0 \end{array}\right]$$

Using $\mathbf{QR}$ decomposition, this matrix can be expressed as

$$\mathbf{P} = \begin{bmatrix} 0.57735 & 0.707107 \\ 0.57735 & -0.707107 \\ 0.57735 & 0 \end{bmatrix} \begin{bmatrix} 1.732 & 0 & -5.19615 & 0 & 15.588 & 0 \\ 0 & 1.4142 & 0 & -7.07107 & 0 & 35.355 \end{bmatrix}$$

This shows that $\text{rank}(\mathbf{P}) = 2$, and the system is uncontrollable. The two-dimensional controllable subspace is spanned by the columns of $\mathbf{T}_1$, the first factor in the preceding decomposition. A third orthogonal vector makes up $\mathbf{T}_2$ and is found to be $[-0.40825 \quad -0.40825 \quad 0.816496]^T$. The orthogonal transformation $\mathbf{T} = [\mathbf{T}_1 \mid \mathbf{T}_2]$ then gives

$$\dot{\mathbf{w}} = \mathbf{T}^T\mathbf{A}\mathbf{T}\mathbf{w} + \mathbf{T}^T\mathbf{B}\mathbf{u}$$

$$= \left[\begin{array}{cc|c} -3 & 0 & 5.6569 \\ 0 & -5 & 6.9282 \\ \hline 0 & 0 & -1 \end{array}\right]\mathbf{w} + \left[\begin{array}{cc} 1.73205 & 0 \\ 0 & 1.4142 \\ \hline 0 & 0 \end{array}\right]\mathbf{u}$$

and

$$\mathbf{y} = \mathbf{C}\mathbf{T}\mathbf{w} = \left[\begin{array}{cc|c} 0 & 0 & 2.4495 \\ 0.57735 & 0 & -1.6330 \end{array}\right]\mathbf{w}$$

This form verifies that the system is not controllable. It is stabilizable, since the uncontrollable mode has its eigenvalue located at -1, and thus is stable.

(b) The observability matrix of criterion 2 is

$$\mathbf{Q} = \left[\begin{array}{cc|cc|cc} -1 & 1 & 1 & -3 & -1 & 9 \\ -1 & 1 & 1 & -3 & -1 & 9 \\ 2 & -1 & -2 & 3 & 2 & -9 \end{array}\right] = \mathbf{V}_1\mathbf{R}_1$$

where

$$\mathbf{V}_1 = \begin{bmatrix} -0.40825 & 0.57735 \\ -0.40825 & 0.57735 \\ 0.81650 & 0.57735 \end{bmatrix}$$

and

$$\mathbf{R}_1 = \begin{bmatrix} 2.4495 & -1.6330 & -2.4495 & 4.89898 & 2.4495 & -14.6969 \\ 0 & 0.57735 & 0 & -1.73205 & 0 & 5.19615 \end{bmatrix}$$

This shows that rank(**Q**) = 2, and the system is unobservable. The two-dimensional controllable subspace is spanned by the columns of $\mathbf{V}_1$. A third orthogonal vector makes up $\mathbf{V}_2 = [-0.707107 \quad 0.707107 \quad 0]^T$. The orthogonal transformation $\mathbf{V} = [\mathbf{V}_1 \mid \mathbf{V}_2]$ then gives

$$\dot{\mathbf{v}} = \mathbf{V}^T\mathbf{A}\mathbf{V}\mathbf{v} + \mathbf{V}^T\mathbf{B}\mathbf{u}$$

$$= \left[\begin{array}{cc|c} -1 & 0 & 0 \\ 5.6569 & -3 & 0 \\ \hline -6.9282 & 0 & -5 \end{array}\right]\mathbf{v} + \left[\begin{array}{cc} 0 & 0 \\ 1.73205 & 0 \\ \hline 0 & -1.4142 \end{array}\right]\mathbf{u}$$

and

$$\mathbf{y} = \mathbf{C}\mathbf{V}\mathbf{v} = \left[\begin{array}{cc|c} 2.44949 & 0 & 0 \\ -1.6330 & 0.57735 & 0 \end{array}\right]\mathbf{v}$$

This form verifies that the system is not observable. It is detectable, since the unobservable mode has its eigenvalue located at -5 and thus is stable.

11.24 Consider a system which has the **B** and **C** matrices of Example 11.7 and has

$$\mathbf{A} = \begin{bmatrix} -6 & -3 & -5 \\ 0 & -3 & 1 \\ 2 & 2 & 0 \end{bmatrix}$$

Find the Kalman controllable canonical form. Is this system controllable? Is it stabilizable?

The controllability matrix is

$$\mathbf{P} = \left[\begin{array}{cc|cc|cc} -0.66667 & 0.33333 & 1.33333 & -1.66667 & -2.66667 & 6.33333 \\ 0.33333 & -0.66667 & -0.66667 & 2.33333 & 1.33333 & -7.66667 \\ 0.33333 & 0.33333 & -0.66667 & -0.66667 & 1.33333 & 1.33333 \end{array}\right]$$

This can be decomposed into the product of

$$\mathbf{T}_1 = \begin{bmatrix} -0.81650 & 0 \\ 0.40825 & -0.707107 \\ 0.40825 & 0.707107 \end{bmatrix}$$

and

$$\mathbf{R}_1 = \begin{bmatrix} 0.81649 & -0.40824 & -1.6330 & 2.04124 & 3.2660 & -7.75671 \\ 0 & 0.70711 & 0 & -2.12132 & 0 & 6.36396 \end{bmatrix}$$

The rank of **P** is 2, so the system is not controllable. A third orthogonal basis vector is $[0.57735 \quad 0.57735 \quad 0.57735]^T$, and this is used for $\mathbf{T}_2$ in the orthogonal matrix $\mathbf{T} = [\mathbf{T}_1 \mid \mathbf{T}_2]$. Using this, the Kalman controllable canonical form is

$$\dot{\mathbf{w}} = \left[\begin{array}{cc|c} -2 & 1.73205 & 0.707107 \\ 0 & -3 & 2.4495 \\ \hline 0 & 0 & -4 \end{array}\right]\mathbf{w} + \left[\begin{array}{cc} 0.81650 & -0.40825 \\ 0 & 0.707107 \\ \hline 0 & 0 \end{array}\right]\mathbf{u}$$

and

$$\mathbf{y} = \begin{bmatrix} 1.2248 & 0.707107 & 5.1962 \\ 0 & 1.4142 & 3.4641 \end{bmatrix}\mathbf{w}$$

Since the eigenvalue of the uncontrollable mode is -4, this system is stabilizable.

25 The state variable system matrices are

$$\mathbf{A} = \begin{bmatrix} 1 & 1 & -1 \\ 2 & 1 & -3 \\ 1 & 2 & 0 \end{bmatrix} \qquad \mathbf{B} = \begin{bmatrix} 2 & 1 \\ 3 & 1 \\ 3 & 2 \end{bmatrix} \qquad \mathbf{C} = \begin{bmatrix} 1 & 1 & -1 \\ 3 & 3 & -3 \end{bmatrix}$$

Find the Kalman observable canonical form.

The observability matrix is

$$\mathbf{Q} = \left[\begin{array}{rr|rr|rr} 1 & 3 & 2 & 6 & -2 & -6 \\ 1 & 3 & 0 & 0 & -6 & -18 \\ -1 & -3 & -4 & -12 & -2 & -6 \end{array}\right] = \mathbf{V}_1 \mathbf{R}_1$$

where

$$\mathbf{V}_1 = \begin{bmatrix} 0.57735 & 0 \\ 0.57735 & -0.707107 \\ -0.57735 & -0.707107 \end{bmatrix}$$

and

$$\mathbf{R}_1 = \begin{bmatrix} 1.723 & 5.196 & 3.464 & 10.392 & -3.464 & -10.392 \\ 0 & 0 & 2.828 & 8.485 & 5.657 & 16.971 \end{bmatrix}$$

Since $\mathbf{Q}$ has rank 2, the system is not observable. By augmenting $\mathbf{V}_1$ with a third orthonormal column, $[0.816496 \quad -0.408248 \quad 0.408248]^T$, the requested canonical form is found to be

$$\dot{\mathbf{v}} = \left[\begin{array}{rr|r} 2 & 1.6330 & 0 \\ -3.6742 & 0 & 0 \\ \hline 0.7011 & -1.1547 & 0 \end{array}\right] \mathbf{v} + \left[\begin{array}{rr} 1.1547 & 0 \\ -4.2426 & -2.1213 \\ \hline 1.6330 & 1.2247 \end{array}\right] \mathbf{u}$$

$$\mathbf{y} = \left[\begin{array}{rr|r} 1.7321 & 0 & 0 \\ 5.1962 & 0 & 0 \end{array}\right] \mathbf{v}$$

The eigenvalue of the unobservable mode is at zero. Note that even though the second mode, $\mathbf{v}_2$, does not directly affect $\mathbf{y}$, it does so indirectly by virtue of its coupling into $\mathbf{v}_1$, and it is observable.

PROBLEMS

1.26 A continuous-time system is represented by $\mathbf{A} = \begin{bmatrix} 2 & -5 \\ -4 & 0 \end{bmatrix}$, $\mathbf{B} = \begin{bmatrix} 1 \\ -1 \end{bmatrix}$, $\mathbf{C} = [1 \quad 1]$. Is this system completely controllable and completely observable?

1.27 Investigate the controllability properties of time-invariant systems $\dot{\mathbf{x}} = \mathbf{A}\mathbf{x} + \mathbf{B}u$ if $u(t)$ is a scalar, and

(a) $\mathbf{A} = \begin{bmatrix} -5 & 1 \\ 0 & 4 \end{bmatrix}$, $\quad \mathbf{B} = \begin{bmatrix} 1 \\ 1 \end{bmatrix}$; $\quad$ (b) $\mathbf{A} = \begin{bmatrix} 0 & 1 & 0 \\ 0 & 0 & 1 \\ 0 & 0 & 0 \end{bmatrix}$, $\quad \mathbf{B} = \begin{bmatrix} 0 \\ 0 \\ 1 \end{bmatrix}$;

(c) $\mathbf{A} = \begin{bmatrix} 3 & 3 & 6 \\ 1 & 1 & 2 \\ 2 & 2 & 4 \end{bmatrix}$, $\quad \mathbf{B} = \begin{bmatrix} 0 \\ 0 \\ 1 \end{bmatrix}$.

1.28 Investigate the controllability and observability of the following systems. Note the results are unaffected by whether or not the system has a nonzero $\mathbf{D}$ matrix.

(a) $\mathbf{A} = \begin{bmatrix} -1 & 0 & 0 & 0 & 0 & 0 \\ 0 & -2 & 1 & 0 & 0 & 0 \\ 0 & 0 & -1 & 0 & 0 & 0 \\ 0 & 0 & 0 & -3 & 0 & 0 \\ 0 & 0 & 0 & 0 & -3 & 1 \\ 0 & 0 & 0 & 0 & 0 & -1 \end{bmatrix}, \quad \mathbf{B} = \begin{bmatrix} 1 & 0 \\ 0 & 0 \\ 0 & 1 \\ 0 & 1 \\ 0 & 0 \\ 1 & 0 \end{bmatrix}, \quad \mathbf{C}^T = \begin{bmatrix} 1 & 0 \\ 2 & 0 \\ 0 & 0 \\ 0 & 1 \\ 0 & 1 \\ 0 & 0 \end{bmatrix}$

(b) $\mathbf{A} = \begin{bmatrix} -6 & 1 & 0 \\ -11 & 0 & 1 \\ -6 & 0 & 0 \end{bmatrix}, \quad \mathbf{B} = \begin{bmatrix} 1 \\ 6 \\ 5 \end{bmatrix}, \quad \mathbf{C} = [1 \quad 0 \quad 0]$

(c) $\mathbf{A} = \begin{bmatrix} -1 & 3 & 0 & 0 \\ -3 & -1 & 0 & 0 \\ 0 & 0 & -5 & 0 \\ 0 & 0 & 0 & -5 \end{bmatrix}, \quad \mathbf{B} = \begin{bmatrix} 1 & 0 \\ 0 & 0 \\ 1 & 0 \\ 0 & -1 \end{bmatrix}, \quad \mathbf{C} = \frac{1}{25}\begin{bmatrix} -3 & -4 & 3 & 250 \\ 4 & -3 & -4 & 100 \end{bmatrix}$

11.29 A factor in determining useful life of a flexible structure, such as a ship, a tall building, or a large airplane, is the possibility of fatigue failures due to structural vibrations. Each vibration mode is described by an equation of the form $m\ddot{x} + kx = u(t)$, where $u(t)$ is the input force. Is it possible to find an input which will drive both the deflection $x(t)$ and the velocity $\dot{x}(t)$ to zero in finite time for arbitrary initial conditions?

11.30 Investigate the controllability and observability of the mechanical system of Figure 11.7. Use x_1 and x_2 as state variables, $u(t)$ as the input force, and $y(t) = x_1(t)$ as the output. Assume the masses m_1 and m_2 are negligible [5].

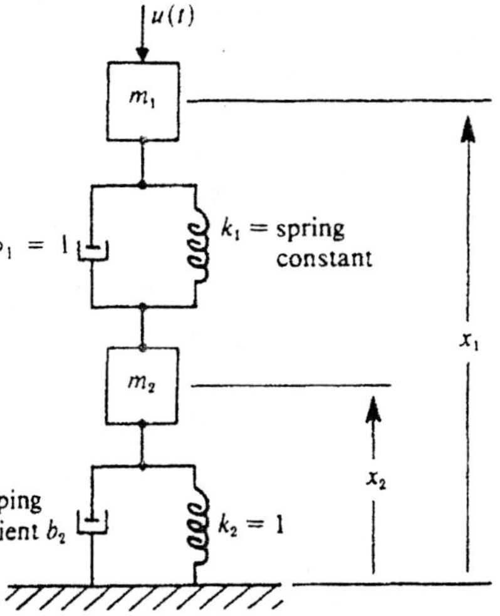

Figure 11.7

11.31 Is the system of Figure 11.8 completely controllable and completely observable?

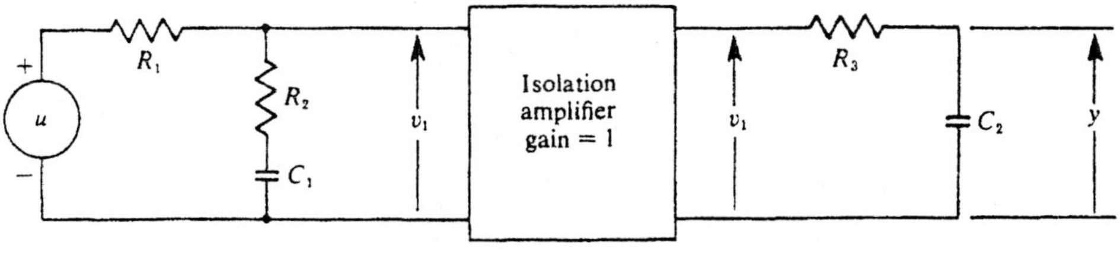

Figure 11.8

32 Determine whether the circuits of Figure 11.9 are completely controllable and completely observable.

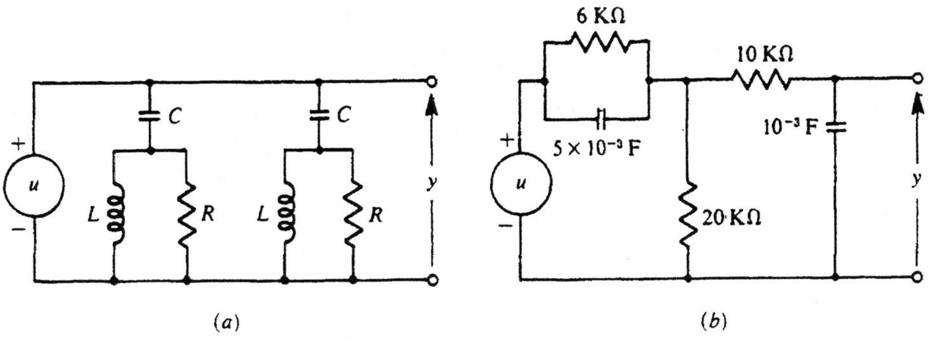

Figure 11.9

33 The criterion for complete controllability is often stated as the requirement that

$$\mathbf{G}'(t_1, t_0) = \int_{t_0}^{t_1} \mathbf{\Phi}(t_0, t)\mathbf{B}(t)\mathbf{B}^T(t)\mathbf{\Phi}^T(t_0, t)\, dt$$

be positive definite for some finite $t_1 > t_0$. Show that this is equivalent to the controllability criterion 3, provided that the system matrices **A** and **B** are real.

34 Is the system of Problem 11.25 controllable?

35 Change the output matrix of Problem 11.25 to $\mathbf{C} = \begin{bmatrix} 1 & 0 & 1 \\ 0 & 1 & 1 \end{bmatrix}$ and find the new Kalman observable canonical form. Is the system observable?

36 Consider a system which has the matrix **A** of Problem 11.24 along with the matrices **B** and **C** of Example 11.8.

(a) Find the Kalman controllable canonical form.

(b) Find the Kalman observable canonical form.

12

The Relationship Between State Variable and Transfer Function Description of Systems

12.1 INTRODUCTION

Classical control theory makes extensive use of input-output transfer function models to describe physical systems. When such a system has multiple inputs and/or multiple outputs, a matrix array of transfer function elements H_{ij} can be used to relate the jth input to the ith output. Since transfer functions are largely restricted to linear constant coefficient systems, this chapter is similarly restricted.

The objective of this chapter is to explore the relationships between the transfer function matrix and the state variable model for the same system. Treatment of continuous-time and discrete-time systems proceed with only minor notational differences.

12.2 TRANSFER FUNCTION MATRICES FROM STATE EQUATIONS

For a given state variable model there is one unique transfer function matrix. The most general state space description of a linear, constant system with r inputs $\mathbf{u}(t)$, m outputs $\mathbf{y}(t)$, and n state variables $\mathbf{x}(t)$ is given by

$$\dot{\mathbf{x}}(t) = \mathbf{A}\mathbf{x}(t) + \mathbf{B}\mathbf{u}(t) \tag{12.1}$$

$$\mathbf{y}(t) = \mathbf{C}\mathbf{x}(t) + \mathbf{D}\mathbf{u}(t) \tag{12.2}$$

where $\mathbf{A}$, $\mathbf{B}$, $\mathbf{C}$, and $\mathbf{D}$ are constant matrices of dimensions $n \times n$, $n \times r$, $m \times n$, and $m \times r$, respectively. The Laplace transforms of Eqs. (*12.1*) and (*12.2*) are

$$\begin{aligned} s\mathbf{x}(s) - \mathbf{x}(t=0) &= \mathbf{A}\mathbf{x}(s) + \mathbf{B}\mathbf{u}(s) \\ \mathbf{y}(s) &= \mathbf{C}\mathbf{x}(s) + \mathbf{D}\mathbf{u}(s) \end{aligned} \tag{12.3}$$

As usual when dealing with transfer functions, the initial conditions $\mathbf{x}(t = 0)$ will be ignored. Solving for $\mathbf{x}(s)$ gives

$$\mathbf{x}(s) = [s\mathbf{I}_n - \mathbf{A}]^{-1}\mathbf{B}\mathbf{u}(s)$$

Using this result leads to Eq. (*12.4*), the input-output relationship for the transformed variables:

$$\mathbf{y}(s) = \{\mathbf{C}[s\mathbf{I}_n - \mathbf{A}]^{-1}\mathbf{B} + \mathbf{D}\}\mathbf{u}(s) \tag{12.4}$$

The $m \times r$ matrix which premultiplies $\mathbf{u}(s)$ is the *transfer matrix* $\mathbf{H}(s)$,

$$\mathbf{H}(s) = \mathbf{C}[s\mathbf{I}_n - \mathbf{A}]^{-1}\mathbf{B} + \mathbf{D} \tag{12.5}$$

This result was also given as Eq. (*4.5*). The discrete-time equivalent of this result was derived in Problem 9.26. A typical element $H_{ij}(s)$ of $\mathbf{H}(s)$ is the transfer function relating the jth input component u_j to the ith output component y_i, $H_{ij}(s) = y_i(s)/u_j(s)$, with all inputs equal to zero except u_j and all initial conditions being zero.

It is convenient to define $\Delta(s) = |s\mathbf{I}_n - \mathbf{A}|$.‡ Then $[s\mathbf{I}_n - \mathbf{A}]^{-1} = \text{Adj}[s\mathbf{I}_n - \mathbf{A}]/\Delta(s)$ and the transfer matrix can be rewritten as

$$\mathbf{H}(s) = \frac{\mathbf{C}\ \text{Adj}[s\mathbf{I}_n - \mathbf{A}]\mathbf{B} + \mathbf{D}\Delta(s)}{\Delta(s)} \tag{12.6}$$

Each element of the adjoint matrix is a polynomial in s of degree less than or equal to $n - 1$. Since $\Delta(s)$ is an nth degree polynomial in s, each element $H_{ij}(s)$ is a ratio of polynomials in s, with the degree of the denominator at least as great as the degree of the numerator. Such an $\mathbf{H}$ matrix is called a *proper* rational matrix (loosely, at least as many poles as zeros). If $\mathbf{D} = [\mathbf{0}]$, then the numerator of every element in $\mathbf{H}(s)$ will be of degree less than the denominator. In this case, $\mathbf{H}(s)$ is a *strictly proper* rational matrix (loosely, more poles than zeros). Clearly, from Eq. (*12.6*),

$$\mathbf{D} = \lim_{s \to \infty} \mathbf{H}(s) \tag{12.7}$$

Equation (*12.5*) indicates that a proper transfer function matrix can be written as the sum of a strictly proper transfer function matrix plus a constant matrix $\mathbf{D}$.

When a state variable description of a system is given, the unique corresponding transfer function matrix is given by Eq. (*12.5*) or (*12.6*). The reverse process is *not* unique. If the matrix $\mathbf{H}(s)$ (or $\mathbf{H}(z)$) is given, the matrix $\mathbf{D}$ is immediately given by Eq. (*12.7*). However, the determination of $\{\mathbf{A}, \mathbf{B}, \mathbf{C}\}$ from a knowledge of $\mathbf{H}$ is not a unique process. Many different state variable realizations $\{\mathbf{A}, \mathbf{B}, \mathbf{C}, \mathbf{D}\}$ yield the same transfer function. The determination of state variable models from a given transfer function is treated in the following sections.

EXAMPLE 12.1 A single-input, single-output system is described in state variable form with

$$\mathbf{A} = \begin{bmatrix} -5 & 1 & 0 \\ 0 & -2 & 1 \\ 0 & 0 & -2 \end{bmatrix}, \qquad \mathbf{B} = \begin{bmatrix} 0 \\ 1 \\ 1 \end{bmatrix}, \qquad \mathbf{C} = [1 \quad 0 \quad 0], \qquad D = 0$$

‡ Recall that in Chapter 7, page 246, $\Delta(\lambda) \triangleq |\mathbf{A} - \mathbf{I}\lambda|$. The above definition is more convenient for this chapter and the next. They differ by the inconsequential factor $(-1)^n$.

Then

$$[s\mathbf{I}-\mathbf{A}]^{-1} = \frac{\begin{bmatrix} (s+2)^2 & s+2 & 1 \\ 0 & (s+2)(s+5) & (s+5) \\ 0 & 0 & (s+2)(s+5) \end{bmatrix}}{(s+5)(s+2)^2}$$

From Eq. (12.5),

$$H(s) = \frac{s+3}{(s+2)^2(s+5)}$$

This agrees with the results of Example 3.8, where the same problem was worked in reverse order. Note that $\lim_{s\to\infty} H(s) = 0$ as expected, since $D = 0$ in this case. ■

12.3 STATE EQUATIONS FROM TRANSFER MATRICES: REALIZATIONS

Several methods of selecting state variables were presented in Sec. 3.4. These methods are entirely satisfactory for single-input, single-output systems but were also applied to multiple input-output systems without dwelling on the consequences. Interconnections of such systems were discussed in Sec. 3.5. Multivariable systems are now discussed in detail, and the consequences of previous methods will be explored.

If a pair of equations (*12.1*) and (*12.2*) can be found which has a specified transfer matrix $\mathbf{H}(s)$, then those equations are called a *realization* of $\mathbf{H}(s)$. For brevity, it is common to refer to the matrices $\{\mathbf{A}, \mathbf{B}, \mathbf{C}, \mathbf{D}\}$ as the realization of $\mathbf{H}(s)$. Specifying $\{\mathbf{A}, \mathbf{B}, \mathbf{C}, \mathbf{D}\}$ is equivalent to giving a prescription for synthesizing a system with a given transfer matrix. Figure 3.3, for example, can be mechanized with physical devices such as operational amplifiers and other standard analog computer equipment. The discrete equivalent in Figure 3.4 would most often use digital hardware.

Perhaps the simplest method of realizing a given transfer matrix $\mathbf{H}(s)$ is illustrated in Figure 12.1. Each scalar component $H_{ij}(s)$ is considered individually, and the methods discussed for scalar transfer function in Sec. 3.4 can be used on each. This method is directly related to the techniques given earlier (Sec. 3.5) for dealing with composite systems.

EXAMPLE 12.2 A system with two inputs and two outputs has the transfer matrix

$$\mathbf{H}(s) = \begin{bmatrix} 1/(s+1) & 2/[(s+1)(s+2)] \\ 1/[(s+1)(s+3)] & 1/(s+3) \end{bmatrix}$$

Using the approach suggested in Figure 12.1, four separate scalar transfer functions are simulated as shown in Figure 12.2.

Using the state variables x_1 through x_6 as defined in Figure 12.2, a realization of $\mathbf{H}(s)$ is given by

$$\mathbf{A} = \begin{bmatrix} -1 & 0 & 0 & 0 & 0 & 0 \\ 0 & -2 & 1 & 0 & 0 & 0 \\ 0 & 0 & -1 & 0 & 0 & 0 \\ 0 & 0 & 0 & -3 & 0 & 0 \\ 0 & 0 & 0 & 0 & -3 & 1 \\ 0 & 0 & 0 & 0 & 0 & -1 \end{bmatrix}, \quad \mathbf{B} = \begin{bmatrix} 1 & 0 \\ 0 & 0 \\ 0 & 1 \\ 0 & 1 \\ 0 & 0 \\ 1 & 0 \end{bmatrix}, \quad \mathbf{C} = \begin{bmatrix} 1 & 0 \\ 2 & 0 \\ 0 & 0 \\ 0 & 1 \\ 0 & 1 \\ 0 & 0 \end{bmatrix}^T, \quad \mathbf{D} = \begin{bmatrix} 0 & 0 \\ 0 & 0 \end{bmatrix}$$

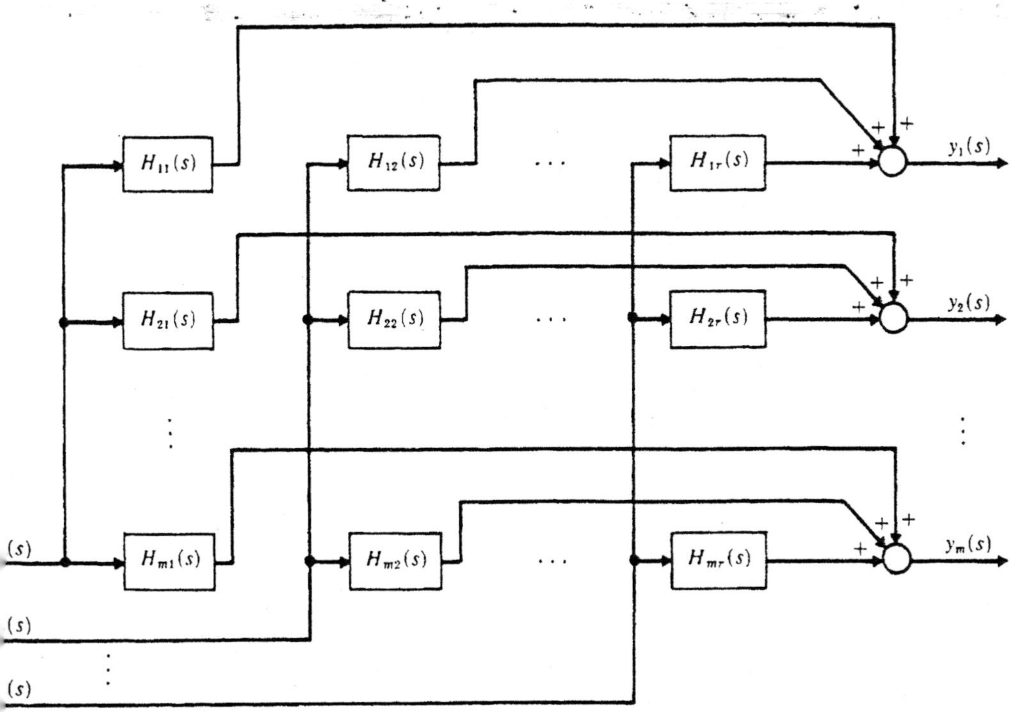

Figure 12.1 Block diagram of $H(s)$ without internal coupling.

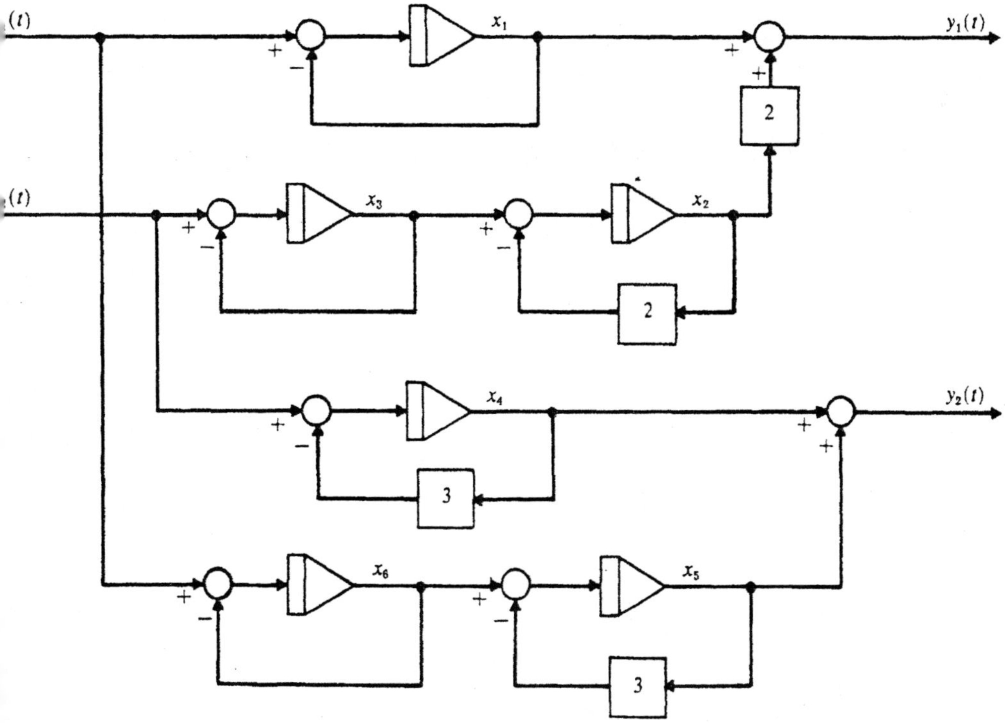

Figure 12.2 Realization for Example 12.2.

Application of Eq. (*12.5*) shows that this sixth-order state representation does have the specified input-output transfer matrix. ■

The method just presented yields a realization for any transfer matrix. The realization of $\mathbf{H}(s)$ is not unique. If one nth-order realization of $\mathbf{H}(s)$ can be found, then there are an infinite number of nth-order realizations.

EXAMPLE 12.3 Let $\{\mathbf{A}_1, \mathbf{B}_1, \mathbf{C}_1, \mathbf{D}_1\}$ be a realization of $\mathbf{H}(s)$, where $\mathbf{A}_1$ is an $n \times n$ matrix. Let $\mathbf{M}$ be any constant nonsingular $n \times n$ matrix and define a new state vector $\mathbf{x}_2$ by the transformation (change of basis in the state space Σ) $\mathbf{x}_1 = \mathbf{M}\mathbf{x}_2$. Now

$$\dot{\mathbf{x}}_1 = \mathbf{A}_1\mathbf{x}_1 + \mathbf{B}_1\mathbf{u},\ \mathbf{y} = \mathbf{C}_1\mathbf{x}_1 + \mathbf{D}_1\mathbf{u}$$

become

$$\mathbf{M}\dot{\mathbf{x}}_2 = \mathbf{A}_1\mathbf{M}\mathbf{x}_2 + \mathbf{B}_1\mathbf{u},\ \mathbf{y} = \mathbf{C}_1\mathbf{M}\mathbf{x}_2 + \mathbf{D}_1\mathbf{u}.$$

Defining $\mathbf{A}_2 = \mathbf{M}^{-1}\mathbf{A}_1\mathbf{M}$, $\mathbf{B}_2 = \mathbf{M}^{-1}\mathbf{B}_1$, $\mathbf{C}_2 = \mathbf{C}_1\mathbf{M}$, and $\mathbf{D}_2 = \mathbf{D}_1$ gives

$$\dot{\mathbf{x}}_2 = \mathbf{A}_2\mathbf{x}_2 + \mathbf{B}_2\mathbf{u},\ \mathbf{y} = \mathbf{C}_2\mathbf{x}_2 + \mathbf{D}_2\mathbf{u}.$$

Then

$$\mathbf{C}_2[s\mathbf{I} - \mathbf{A}_2]^{-1}\mathbf{B}_2 + \mathbf{D}_2 = \mathbf{C}_1\mathbf{M}[s\mathbf{I} - \mathbf{M}^{-1}\mathbf{A}_1\mathbf{M}]^{-1}\mathbf{M}^{-1}\mathbf{B}_1 + \mathbf{D}_1 = \mathbf{C}_1[s\mathbf{I} - \mathbf{A}_1]^{-1}\mathbf{B}_1 + \mathbf{D}_1$$

Thus $\{\mathbf{A}_1, \mathbf{B}_1, \mathbf{C}_1, \mathbf{D}_1\}$ and $\{\mathbf{A}_2, \mathbf{B}_2, \mathbf{C}_2, \mathbf{D}_2\}$ are two different realizations. Eq. (*12.7*) indicates that $\mathbf{D}$ is determined only by $\mathbf{H}(s)$. The choice of basis vectors for Σ does not influence $\mathbf{D}$. Hence $\mathbf{D}_1 = \mathbf{D}_2$. ■

This establishes the nonuniqueness of system representations. A more interesting result is that $\dim(\Sigma)$ is not uniquely defined by $\mathbf{H}(s)$.

12.4 DEFINITION AND IMPLICATION OF IRREDUCIBLE REALIZATIONS

The methods of Sec. 12.3 will produce state variable models for either continuous-time or discrete-time systems from a transfer function matrix. However, this method ignores any commonalities that may exist among the various elements H_{ij}. Such commonalities *may* allow a sharing of integrators or delay elements by more than one element H_{ij}. When these opportunities are ignored, the resulting state equations will be of an unnecessarily high order. The state space is thus of unnecessarily high dimension.

Definition 12.1. Of all the possible realizations of $\mathbf{H}(s)$, $\{\mathbf{A}, \mathbf{B}, \mathbf{C}, \mathbf{D}\}$ is said to be an *irreducible* (or minimum) *realization* if the associated state space has the smallest possible $\dim(\Sigma)$.

Important Fact

A minimal realization is both completely controllable and completely observable.

In the case of a scalar transfer function, the minimum dimension required is equal to the order of the denominator of the transfer function after all common pole-zero cancellations are made. The corresponding result for transfer matrices is not so obvious and is discussed in Sec. 12.5.

Whenever a pole-zero pair is cancelled from a transfer function, the system mode associated with the cancelled pole will not be evident in the state equations. Yet, in order to achieve an irreducible realization, these cancellations must be made. What is the implication of irreducible realizations in view of this apparent loss of information about the system?

An irreducible realization is a system, of minimal dimension, which is capable of reproducing the measurable relationships between inputs and outputs. This assumes that the system is originally relaxed, i.e., the initial state vector is zero. For this reason a system and its irreducible realization are said to be *zero state equivalent.* Often, the only knowledge about a system is the information which is obtainable from measurements of inputs and outputs. Transfer functions and matrices can be experimentally determined from these measurements. The irreducible realization does not cause loss of information in this case, because nothing was known about the system's internal structure in the first place. It is not claimed that an irreducible realization is the best description of the internal structure of a system.

If the internal structure of a system is known (for example, a circuit diagram), then an irreducible realization may not be the appropriate one to use. State equations can always be written directly from the system's linear graph, as described in Chapter 3. The realization $\{\mathbf{A}, \mathbf{B}, \mathbf{C}, \mathbf{D}\}$ obtained in this manner is the most complete system description, whether it is irreducible or not. Let this realization have $\dim(\Sigma) = n$. The transfer matrix $\mathbf{H}(s)$ can be found using Eq. (*12.5*). Starting with $\mathbf{H}(s)$, an irreducible realization $\{\mathbf{A}_1, \mathbf{B}_1, \mathbf{C}_1, \mathbf{D}\}$ can be found. If $\mathbf{A}_1$ is $n_1 \times n_1$, with $n_1 < n$, then information about $n - n_1$ modes would be lost by using the irreducible realization (see Problems 12.4 and 12.6). In Example 12.5, the irreducible realization gives no indication that the system is actually unstable. If $n_1 = n$, the system is said to be *completely characterized* by $\mathbf{H}(s)$. In general, transfer matrices, and irreducible realizations of them, describe only that subsystem which is both completely controllable and completely observable (see Problems 12.12 and 12.13). The incompleteness of a transfer matrix description is another reason for preferring state space techniques.

EXAMPLE 12.4 The input-output equations for a system are

$$\dot{y}_1 + 2(y_1 - y_2) = 4u_1 - u_2$$
$$\dot{y}_2 + 3(y_2 - y_1) = 4u_1 - u_2 \tag{12.8}$$

Using the simulation diagram of Figure 12.3, the state equations (*12.9*) and (*12.10*) are obtained:

$$\begin{bmatrix} \dot{x}_1 \\ \dot{x}_2 \end{bmatrix} = \begin{bmatrix} -2 & 2 \\ 3 & -3 \end{bmatrix} \begin{bmatrix} x_1 \\ x_2 \end{bmatrix} + \begin{bmatrix} 4 & -1 \\ 4 & -1 \end{bmatrix} \begin{bmatrix} u_1 \\ u_2 \end{bmatrix} \tag{12.9}$$

$$\begin{bmatrix} y_1 \\ y_2 \end{bmatrix} = \begin{bmatrix} 1 & 0 \\ 0 & 1 \end{bmatrix} \begin{bmatrix} x_1 \\ x_2 \end{bmatrix} \tag{12.10}$$

The state vector has dimension 2. The transfer matrix can be derived directly from Eq. (*12.8*) by Laplace transforming and a matrix inversion,

$$\mathbf{H}(s) = \begin{bmatrix} 4/s & -1/s \\ 4/s & -1/s \end{bmatrix}$$

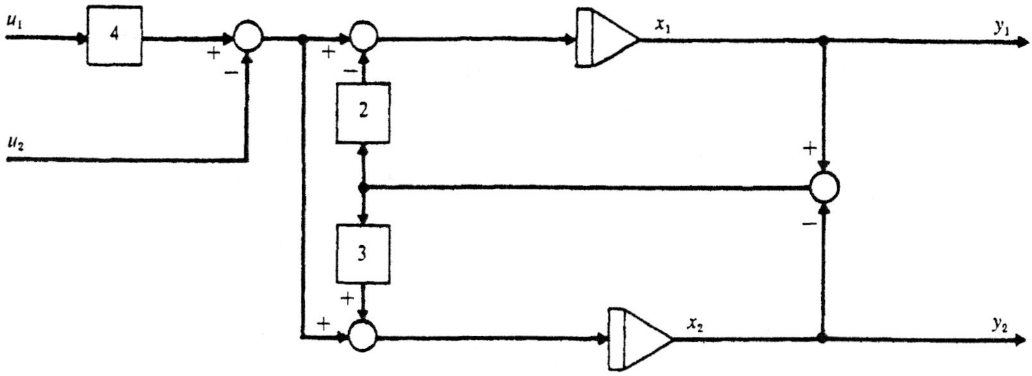

Figure 12.3

Assume now that $\mathbf{H}(s)$ is the only information known about the system. The method of Sec. 12.3 leads to the fourth-order system realization shown in Figure 12.4.

Two other realizations for $\mathbf{H}(s)$ are shown in Figures 12.5 and 12.6.

None of these three realizations resembles the original system. The one shown in Figure 12.5 agrees insofar as $\dim(\Sigma) = 2$, but the mode corresponding to the pole $(s + 5)$ is not present. The minimum realization, shown in Figure 12.6 has $n = 1$. The original second-order model was not of minimal order and therefore fails to have either one or both of the properties of controllability and observability. Examination shows that the original system is observable but not cotrollable. If the original system is decomposed into subsystems as in Problem 11.21, page 397, this is verified. The only mode that is completely controllable and completely observable is the one associated with the eigenvalue (pole) $s = 0$. ■

Consider a scalar transfer function, given in the form of Eq. (*12.6*). The eigenvalues of the matrix $\mathbf{A}$ are roots of $\Delta(\lambda) = |\lambda\mathbf{I} - \mathbf{A}| = 0$. The poles of $\mathbf{H}(s)$ will be the roots of $\Delta(s) = |s\mathbf{I} - \mathbf{A}| = 0$, *unless* some of the factors of $\Delta(s)$ are cancelled by terms in the numerator of $\mathbf{H}(s)$. Thus the poles of a transfer function will always be eigenvalues of the system matrix $\mathbf{A}$. Eigenvalues of $\mathbf{A}$ need not always be poles of the transfer function because cancellations may occur. If a system is completely characterized by its

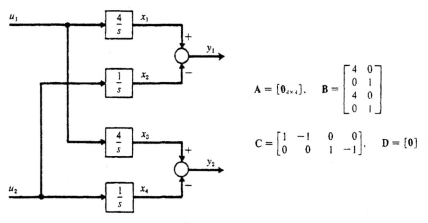

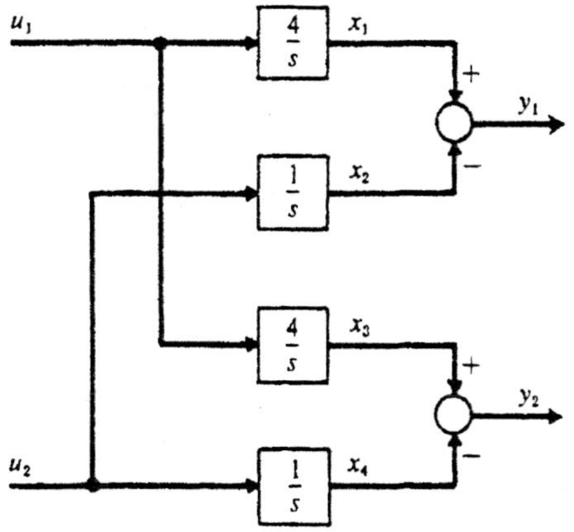

$$\mathbf{A} = [\mathbf{0}_{4\times 4}], \quad \mathbf{B} = \begin{bmatrix} 4 & 0 \\ 0 & 1 \\ 4 & 0 \\ 0 & 1 \end{bmatrix}$$

$$\mathbf{C} = \begin{bmatrix} 1 & -1 & 0 & 0 \\ 0 & 0 & 1 & -1 \end{bmatrix}, \quad \mathbf{D} = [\mathbf{0}]$$

Figure 12.4

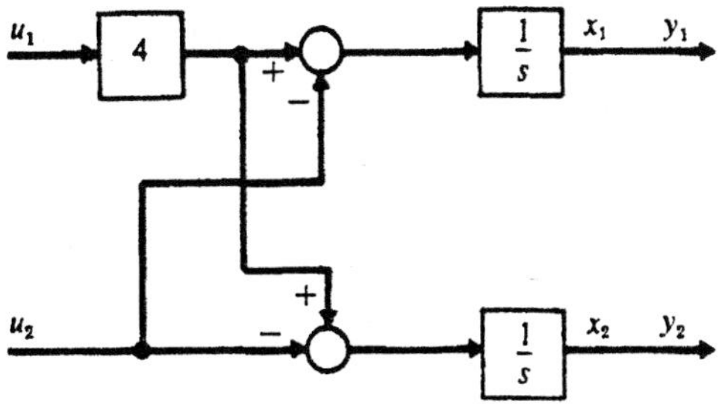

$$\mathbf{A} = [\mathbf{0}_{2\times 2}], \quad \mathbf{B} = \begin{bmatrix} 4 & -1 \\ 4 & -1 \end{bmatrix}$$

$$\mathbf{C} = \begin{bmatrix} 1 & 0 \\ 0 & 1 \end{bmatrix}, \quad \mathbf{D} = [\mathbf{0}]$$

Figure 12.5

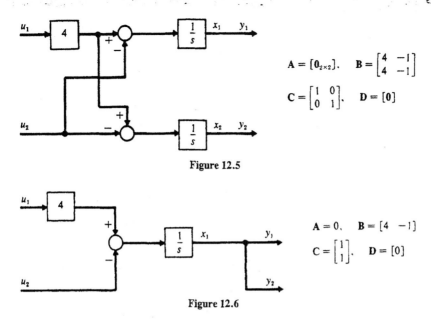

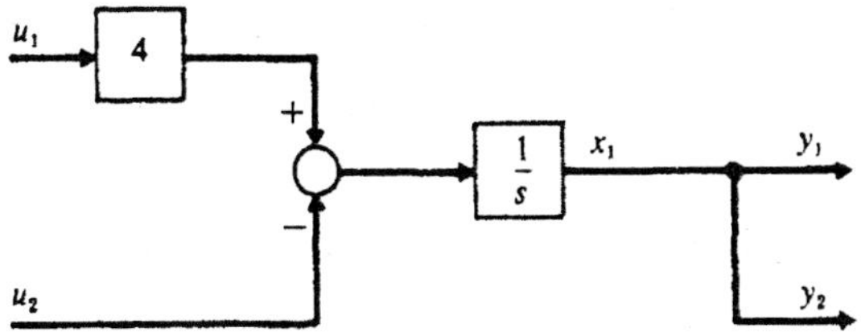

$$\mathbf{A} = 0, \quad \mathbf{B} = [4 \quad -1]$$

$$\mathbf{C} = \begin{bmatrix} 1 \\ 1 \end{bmatrix}, \quad \mathbf{D} = [0]$$

Figure 12.6

transfer function (equivalently if **A** is associated with an irreducible realization of **H**(s)), then the poles and the eigenvalues are the same. This will be true if the system is completely controllable and completely observable. A similar relationship exists between the eigenvalues of **A** and the roots of the "denominator" of a transfer matrix. This assumes a suitable definition for a matrix denominator (see Sec. 12.6).

.5 THE DETERMINATION OF IRREDUCIBLE REALIZATIONS

There are two generic approaches to finding a minimal realization for a given matrix **H**(s). The first consists of finding a state realization by any available means and then reducing it, if necessary, in the state space domain. Methods of Chapter 3 or Sec. 12.4 may be used to find the original realization. The reduction process consists of using a suitable decomposition in order to determine the controllable and observable subsystem. This section presents Jordan form and Kalman canonical form approaches to the decomposition and reduction process.

The second generic approach consists of appropriate reduction of **H**(s) (or **H**(z)) first and then finding a state variable realization of the reduced transfer function. In the single-input, single-output case, reduction of **H** simply consists of canceling any common pole-zero pairs. Reduction of a matrix **H**(s) is not so obvious. Sec. 12.6 gives a procedure which uses elementary (polynomial) operations on the matrix fraction description (MFD) in order to achieve the required reduction.

12.5.1. Jordan Canonical Form Approach

Any state variable model can be decomposed into four subsystems which are (1) completely controllable and completely observable, (2) completely controllable but not observable, (3) completely observable but not controllable, and (4) neither controllable nor observable (see Problems 11.18 through 11.21). The subsystem which is completely controllable and completely observable constitutes an irreducible realization of $\mathbf{H}(s)$ [1, 2].

EXAMPLE 12.5 Suppose that a system realization has been obtained, and it has been transformed into the Jordan canonical form

$$\begin{bmatrix} \dot{q}_1 \\ \dot{q}_2 \\ \dot{q}_3 \end{bmatrix} = \begin{bmatrix} -5 & 0 & 0 \\ 0 & +1 & 0 \\ 0 & 0 & -3 \end{bmatrix} \begin{bmatrix} q_1 \\ q_2 \\ q_3 \end{bmatrix} + \begin{bmatrix} 1 & 0 \\ 0 & 1 \\ 0 & 0 \end{bmatrix} \begin{bmatrix} u_1 \\ u_2 \end{bmatrix}$$

$$\begin{bmatrix} y_1 \\ y_2 \end{bmatrix} = \begin{bmatrix} 1 & 0 & 0 \\ 0 & 0 & 1 \end{bmatrix} \begin{bmatrix} q_1 \\ q_2 \\ q_3 \end{bmatrix} + \begin{bmatrix} 0 & 0 \\ 0 & 1 \end{bmatrix} \begin{bmatrix} u_1 \\ u_2 \end{bmatrix}$$

Because this Jordan form system has distinct eigenvalues, controllability criterion 1 of Sec. 11.3 applies. Since row 3 of $\mathbf{B}_n$ is all zero, mode q_3 is uncontrollable. The second column of $\mathbf{C}_n$ is all zero, so q_2 is an unobservable mode. Modes q_2 and q_3 have no effect on the input-output behavior and can be dropped. An irreducible realization is

$$\dot{q}_1 = -5q_1 + [1 \quad 0]\mathbf{u}$$

$$\mathbf{y} = \begin{bmatrix} 1 \\ 0 \end{bmatrix} q_1 + \begin{bmatrix} 0 & 0 \\ 0 & 1 \end{bmatrix} \mathbf{u}$$

The transfer matrix is $\mathbf{H}(s) = \begin{bmatrix} 1/(s+5) & 0 \\ 0 & 1 \end{bmatrix}$ for both forms of the state equations. ■

EXAMPLE 12.6 A system realization is found to be

$$\begin{bmatrix} \dot{q}_1 \\ \dot{q}_2 \\ \dot{q}_3 \end{bmatrix} = \begin{bmatrix} -a & 0 & 0 \\ 0 & -a & 0 \\ 0 & 0 & -a \end{bmatrix} \begin{bmatrix} q_1 \\ q_2 \\ q_3 \end{bmatrix} + \begin{bmatrix} 1 & 0 \\ 0 & 1 \\ 1 & 0 \end{bmatrix} \begin{bmatrix} u_1 \\ u_2 \end{bmatrix}$$

$$\mathbf{y} = \begin{bmatrix} 0 & 1 & 1 \\ 1 & 0 & 1 \end{bmatrix} \begin{bmatrix} q_1 \\ q_2 \\ q_3 \end{bmatrix} + [\mathbf{D}]\mathbf{u}$$

Even though the coefficient matrix $\mathbf{A}$ is diagonal, controllability and observability criteria 1 of Sec. 11.3 cannot be used because of the repeated eigenvalues of $\mathbf{A}$. This realization is not controllable because rows 1 and 3 of $\mathbf{B}$ are not independent (see Problem 11.16). It is not observable because column 3 of $\mathbf{C}$ is a linear combination of columns 1 and 2 (see Problem 11.17). In this simple case direct manipulations show that two new states, which are both controllable and observable, can be formed from the original $q_1 + q_3$ and $q_2 + q_3$.

$$\mathbf{y} = \begin{bmatrix} 0 \\ 1 \end{bmatrix} q_1 + \begin{bmatrix} 1 \\ 0 \end{bmatrix} q_2 + \begin{bmatrix} 1 \\ 1 \end{bmatrix} q_3 + \mathbf{Du} = \begin{bmatrix} 0 \\ 1 \end{bmatrix} (q_1 + q_3) + \begin{bmatrix} 1 \\ 0 \end{bmatrix} (q_2 + q_3) + \mathbf{Du}$$

Because of the linear dependence of the columns of $\mathbf{C}_n$, it is possible to define two new state variables $\bar{q}_1 = q_1 + q_3$ and $\bar{q}_2 = q_2 + q_3$. Then $\dot{\bar{q}}_1 = \dot{q}_1 + \dot{q}_3 = -a(\bar{q}_1) + 2u_1$ and $\dot{\bar{q}}_2 = -a\bar{q}_2 + u_1 + u_2$, or

$$\dot{\bar{\mathbf{q}}} = \begin{bmatrix} -a & 0 \\ 0 & -a \end{bmatrix} \bar{\mathbf{q}} + \begin{bmatrix} 2 & 0 \\ 1 & 1 \end{bmatrix} \mathbf{u}$$

$$\mathbf{y} = \begin{bmatrix} 0 & 1 \\ 1 & 0 \end{bmatrix}\bar{\mathbf{q}} + \mathbf{Du}$$

This reduced system is completely controllable and observable, hence irreducible. The transfer matrix for it and for the original third-order system is

$$\mathbf{H}(s) = \begin{bmatrix} 1/(s+a) & 1/(s+a) \\ 2/(s+a) & 0 \end{bmatrix} + \mathbf{D} \qquad \blacksquare$$

Rather than first finding some arbitrary realization and then using the modal matrix in a similarity transformation (see Secs. 9.4 and 9.10), it is possible to obtain a Jordan form realization directly from $\mathbf{H}(s)$ or $\mathbf{H}(z)$. This is done by expanding each $H_{ij}(s)$ element, using partial fractions. Regroupng terms gives a matrix version of the partial fraction expansion of $\mathbf{H}(s)$. A completely controllable realization can be written directly from this.

EXAMPLE 12.7 The transfer matrix of Example 12.2 is reconsidered. Using partial fractions, it can be written as

$$\mathbf{H}(s) = \begin{bmatrix} \dfrac{1}{s+1} & \dfrac{2}{s+1} - \dfrac{2}{s+2} \\ \dfrac{1/2}{s+1} - \dfrac{1/2}{s+3} & \dfrac{1}{s+3} \end{bmatrix} = \frac{\begin{bmatrix} 1 & 2 \\ \frac{1}{2} & 0 \end{bmatrix}}{s+1} + \frac{\begin{bmatrix} 0 & -2 \\ 0 & 0 \end{bmatrix}}{s+2} + \frac{\begin{bmatrix} 0 & 0 \\ -\frac{1}{2} & 1 \end{bmatrix}}{s+3}$$

$$= \frac{\begin{bmatrix} 1 \\ 0 \end{bmatrix}[1 \;\; 2] + \begin{bmatrix} 0 \\ 1 \end{bmatrix}[\frac{1}{2} \;\; 0]}{s+1} + \frac{\begin{bmatrix} 1 \\ 0 \end{bmatrix}[0 \;\; -2]}{s+2} + \frac{\begin{bmatrix} 0 \\ 1 \end{bmatrix}[-\frac{1}{2} \;\; 1]}{s+3}$$

In this case $\lim_{s\to\infty} \mathbf{H}(s) = [\mathbf{0}] = \mathbf{D}$, so that, from Eq. (*12.5*), $\mathbf{H}(s) = \mathbf{C}[s\mathbf{I} - \mathbf{A}]^{-1}\mathbf{B}$. Comparing this with the expanded form gives

$$\mathbf{H}(s) = \left[\begin{array}{cc|c|c} 1 & 0 & 1 & 0 \\ 0 & 1 & 0 & 1 \end{array}\right] \begin{bmatrix} \dfrac{1}{s+1} & & & \mathbf{0} \\ & \dfrac{1}{s+1} & & \\ & & \dfrac{1}{s+2} & \\ \mathbf{0} & & & \dfrac{1}{s+3} \end{bmatrix} \left[\begin{array}{cc} 1 & 2 \\ \frac{1}{2} & 0 \\ \hline 0 & -2 \\ \hline -\frac{1}{2} & 1 \end{array}\right]$$

In this form the matrices $\mathbf{C}$, $[s\mathbf{I} - \mathbf{A}]^{-1}$, and $\mathbf{B}$ are clearly evident. Since $[s\mathbf{I} - \mathbf{A}]^{-1}$ is diagonal, $\mathbf{A}$ is also diagonal and a realization of $\mathbf{H}(s)$ is given by

$$\mathbf{A} = \text{diag}[-1, -1, -2, -3], \qquad \mathbf{B} = \begin{bmatrix} 1 & 2 \\ \frac{1}{2} & 0 \\ 0 & -2 \\ -\frac{1}{2} & 1 \end{bmatrix}, \qquad \mathbf{C} = \begin{bmatrix} 1 & 0 & 1 & 0 \\ 0 & 1 & 0 & 1 \end{bmatrix}, \qquad \mathbf{D} = [\mathbf{0}]$$

This is an irreducible realization, since it is completely controllable and observable. This realization is completely controllable because the rows of $\mathbf{B}$ associated with the simple eigenvalues -2 and -3 are both nonzero and the two rows of $\mathbf{B}$ associated with the two 1×1 blocks for the eigenvalue -1 are independent. The controllability is ensured by the construction process. The observability is not guaranteed by the process in cases where multiple poles (terms such as $1/(s+p_i)^m$) occur. In this example the realization is observable, since the columns of $\mathbf{C}$ satisfy the independence requirements of Problem 11.17. Therefore this realization is irreducible. The required state space satisfies $\dim(\Sigma) = 4$, and not 6 as in Example 12.2. ■

The method illustrated by Example 12.7 consists of

1. Writing $\mathbf{H}(s)$ in a matrix form of partial fraction expansion:

$$\mathbf{H}(s) = \mathbf{D} + \sum_{i=1}^{k} \frac{\mathbf{N}_i}{s+p_i}$$

(note that terms like $1/(s+p_i)^m$ are *not* considered here, but will be later);

2. Determining the rank r_i of each of the $\mathbf{N}_i$ matrices;
3. Factoring each $\mathbf{N}_i$ into the sum of r_i outer products $\mathbf{N}_i = \sum_{j=1}^{r_i} \mathbf{c}_{ij}\rangle\langle\mathbf{b}_{ij}$;

(The double subscripts on the column vectors $\mathbf{c}$ and the row vectors $\mathbf{b}^T$ are used to distinguish which pole they are associated with. This distinction is not always necessary and a single subscript then suffices.)

4. Setting

$$\mathbf{A} = \operatorname{diag}[-p_1\mathbf{I}_{r_1}, -p_2\mathbf{I}_{r_2}, \ldots, -p_k\mathbf{I}_{r_k}]$$

$$\mathbf{B} = \begin{bmatrix} \mathbf{b}_{11}^T \\ \vdots \\ \mathbf{b}_{1r_1}^T \\ \hline \mathbf{b}_{21}^T \\ \vdots \\ \mathbf{b}_{2r_2}^T \\ \hline \vdots \\ \hline \mathbf{b}_{k1}^T \\ \vdots \\ \mathbf{b}_{kr_k}^T \end{bmatrix}, \qquad \mathbf{C} = [\mathbf{c}_{11} \cdots \mathbf{c}_{1r_1} \mid \mathbf{c}_{21} \cdots \mathbf{c}_{2r_2} \mid \cdots \mid \mathbf{c}_{k1} \cdots \mathbf{c}_{kr_k}]$$

The dimension of $\mathbf{A}$ is $n \times n$, where $n = r_1 + r_2 + \cdots + r_k$. Since $\mathbf{A}$ is diagonal, it is in Jordan form. There is just a single 1×1 Jordan block associated with those poles p_i for which $r_i = 1$. The controllability and observability of these modes is ensured by criteria 1 of Sec. 11.3, since the rows $\mathbf{b}_{i1}^T$ and columns $\mathbf{c}_{i1}$ are not zero. In general, there will be r_i 1×1 Jordan blocks associated with a given pole p_i. These modes will be both controllable and observable because rank $\mathbf{N}_i = r_i$ implies that the r_i rows $\{\mathbf{b}_{i1}^T, \mathbf{b}_{i2}^T, \ldots, \mathbf{b}_{ir_i}^T\}$ are linearly independent and so are the r_i columns $\{\mathbf{c}_{i1}, \mathbf{c}_{i2}, \ldots, \mathbf{c}_{ir_i}\}$ (see Problems 11.16 and 11.17). Since every mode of this realization is both controllable and observable, it is irreducible.

Multiple Poles. If any one element $H_{ij}(s)$ has a multiple pole, say $1/(s+p_1)^m$, then the matrix partial fraction expansion for $\mathbf{H}(s)$ will also contain this term, plus terms to the $m-1, m-2, \ldots, 1$ powers. This requires a modification of the method, which is best explained by example. The essence of the modification amounts to obeying the admonition of Sec. 3.4 against using *unnecessary* integrators when using simulation diagrams. This results in realizations with nondiagonal **A** matrices.

EXAMPLE 12.8 Find an irreducible realization of

$$\mathbf{H}(s) = \begin{bmatrix} \dfrac{1}{(s+2)^3(s+5)} & \dfrac{1}{s+5} \\ \dfrac{1}{s+2} & 0 \end{bmatrix}$$

Using partial fraction expansion gives

$$\mathbf{H}(s) = \frac{\begin{bmatrix} \frac{1}{3} & 0 \\ 0 & 0 \end{bmatrix}}{(s+2)^3} + \frac{\begin{bmatrix} -\frac{1}{9} & 0 \\ 0 & 0 \end{bmatrix}}{(s+2)^2} + \frac{\begin{bmatrix} \frac{1}{27} & 0 \\ 1 & 0 \end{bmatrix}}{s+2} + \frac{\begin{bmatrix} -\frac{1}{27} & 1 \\ 0 & 0 \end{bmatrix}}{s+5}$$

$$= \frac{\begin{bmatrix} \frac{1}{3} \\ 0 \end{bmatrix}[1 \quad 0]}{(s+2)^3} + \frac{\begin{bmatrix} -\frac{1}{9} \\ 0 \end{bmatrix}[1 \quad 0]}{(s+2)^2} + \frac{\begin{bmatrix} \frac{1}{27} \\ 1 \end{bmatrix}[1 \quad 0]}{s+2} + \frac{\begin{bmatrix} 1 \\ 0 \end{bmatrix}[-\frac{1}{27} \quad 1]}{s+5}$$

There is only one **b** vector associated with the three terms involving $s+2$, namely, $\mathbf{b}_1^T = [1 \quad 0]$. This means that these three terms can be simulated from a series connection of $1/(s+2)$ terms with a single input $\mathbf{b}_1^T\mathbf{u}$. These terms will also form a single Jordan block. Figure 12.7 gives a simulation diagram from which state equations can be written. Note that

$$x_1(s) = \frac{1}{(s+2)^3}[1 \quad 0]\mathbf{u}(s) \qquad x_2(s) = \frac{1}{(s+2)^2}[1 \quad 0]\mathbf{u}(s)$$

$$x_3(s) = \frac{1}{(s+2)}[1 \quad 0]\mathbf{u}(s)$$

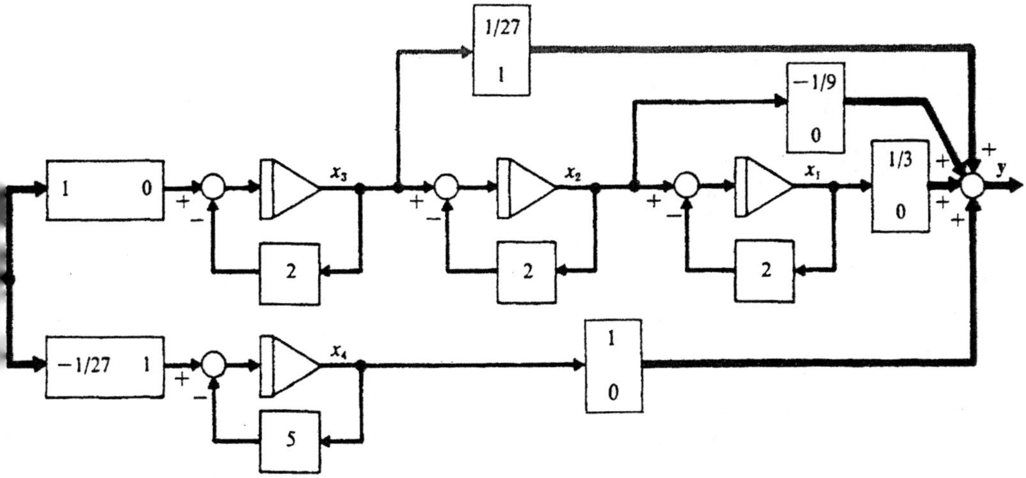

Figure 12.7

From Figure 12.7,

$$\begin{bmatrix}\dot{x}_1\\ \dot{x}_2\\ \dot{x}_3\\ \dot{x}_4\end{bmatrix} = \begin{bmatrix}-2 & 1 & 0 & 0\\ 0 & -2 & 1 & 0\\ 0 & 0 & -2 & 0\\ 0 & 0 & 0 & -5\end{bmatrix}\begin{bmatrix}x_1\\ x_2\\ x_3\\ x_4\end{bmatrix} + \begin{bmatrix}0 & 0\\ 0 & 0\\ 1 & 0\\ -\frac{1}{27} & 1\end{bmatrix}\mathbf{u}$$

and

$$\mathbf{y} = \begin{bmatrix}\frac{1}{3}\\ 0\end{bmatrix}x_1 + \begin{bmatrix}-\frac{1}{9}\\ 0\end{bmatrix}x_2 + \begin{bmatrix}\frac{1}{27}\\ 1\end{bmatrix}x_3 + \begin{bmatrix}1\\ 0\end{bmatrix}x_4 = \begin{bmatrix}\frac{1}{3} & -\frac{1}{9} & \frac{1}{27} & 1\\ 0 & 0 & 1 & 0\end{bmatrix}\mathbf{x}$$

This realization is both completely controllable and completely observable and is therefore irreducible. ■

For each repeated pole the previous method will yield a number of Jordan blocks equal to the number of *linearly independent* $\mathbf{b}_i$ vectors associated with that pole. The realization obtained in this manner will always be completely controllable because one of these $\mathbf{b}_i^T$ vectors will form the last row for each block in $\mathbf{B}$ associated with a Jordan block. (See Problem 11.16.) Observability cannot be ensured in advance. Depending on the example, observability may or may not result with the first realization. If it does not, then less than systematic modifications, i.e., eliminating states, will be required until observability is achieved. In general, the value of the Jordan form approach is greatly diminished in the multiple-pole case. The next two examples illustrate what can happen and show a more or less trial-and-error path to a minimal realization. Problem 12.7 presents another example, which uses a somewhat more systematic—but still tedious—approach to the reduction process. The Kalman canonical form approach of Sec. 12.5.2, by way of contrast, provides an algorithm which is easily implementable on a computer.

EXAMPLE 12.9 Suppose

$$\mathbf{H}(s) = \frac{\begin{bmatrix}1\\1\\1\end{bmatrix}[1 \quad 0 \quad 0]}{(s+p)^3} + \frac{\begin{bmatrix}0\\1\\0\end{bmatrix}[0 \quad 1 \quad 0]}{(s+p)^3} + \frac{\begin{bmatrix}1\\0\\1\end{bmatrix}[1 \quad 0 \quad 0]}{(s+p)^2} + \frac{\begin{bmatrix}2\\5\\2\end{bmatrix}[0 \quad 0 \quad 1]}{(s+p)^2} + \frac{\begin{bmatrix}2\\2\\3\end{bmatrix}[-1 \quad -1 \quad 0]}{s+p}$$

There are three independent $\mathbf{b}_i^T$ vectors, labeled $\mathbf{b}_1 = \begin{bmatrix}1\\0\\0\end{bmatrix}$, $\mathbf{b}_2 = \begin{bmatrix}0\\1\\0\end{bmatrix}$, and $\mathbf{b}_3 = \begin{bmatrix}0\\0\\1\end{bmatrix}$. The last $\mathbf{b}^T$ vector can be written as $\begin{bmatrix}-1\\-1\\0\end{bmatrix} = -\mathbf{b}_1 - \mathbf{b}_2$.

Defining $\mathbf{c}_1 = \begin{bmatrix}1\\1\\1\end{bmatrix}$, $\mathbf{c}_2 = \begin{bmatrix}0\\1\\0\end{bmatrix}$, $\mathbf{c}_3 = \begin{bmatrix}1\\0\\1\end{bmatrix}$, $\mathbf{c}_4 = \begin{bmatrix}2\\5\\2\end{bmatrix}$, and $\mathbf{c}_5 = \begin{bmatrix}2\\2\\3\end{bmatrix}$ allows $\mathbf{H}(s)$ to be written as

$$\mathbf{H}(s)=\frac{\mathbf{c}_1\rangle\langle\mathbf{b}_1}{(s+p)^3}+\frac{\mathbf{c}_3\rangle\langle\mathbf{b}_1}{(s+p)^2}-\frac{\mathbf{c}_5\rangle\langle\mathbf{b}_1}{s+p}+\frac{\mathbf{c}_2\rangle\langle\mathbf{b}_2}{(s+p^3)}-\frac{\mathbf{c}_5\rangle\langle\mathbf{b}_2}{s+p}+\frac{\mathbf{c}_4\rangle\langle\mathbf{b}_3}{(s+p)^2}$$

A simulation diagram for this equation is given in Figure 12.8. The three terms with input $\mathbf{b}_1^T\mathbf{u}$ are linked together in what is called a *Jordan chain*. The three associated state variables will be described by a 3×3 Jordan block. The two terms with input $\mathbf{b}_2^T\mathbf{u}$ form another Jordan chain and hence a Jordan block. Likewise, the term with input $\mathbf{b}_3^T\mathbf{u}$ leads to a third Jordan block. The dimension of a Jordan block is determined by the number of integrators required to simulate the Jordan chain.

The completely controllable realization is given by

$$\mathbf{A}=\left[\begin{array}{ccc|ccc|cc} -p & 1 & 0 & & & & & \\ 0 & -p & 1 & & & & \mathbf{0} & \\ 0 & 0 & -p & & & & & \\ \hline & & & -p & 1 & 0 & & \\ & & & 0 & -p & 1 & & \\ & \mathbf{0} & & 0 & 0 & -p & & \\ \hline & & & & & & -p & 1 \\ & & & & & & 0 & -p \end{array}\right],\quad \mathbf{B}=\left[\begin{array}{c} \mathbf{0}\\ \mathbf{0}\\ \mathbf{b}_1^T\\ \hline \mathbf{0}\\ \mathbf{0}\\ \mathbf{b}_2^T\\ \hline \mathbf{0}\\ \mathbf{b}_3^T \end{array}\right],$$

$$\mathbf{C}=\left[\begin{array}{ccc|ccc|cc}\mathbf{c}_1 & \mathbf{c}_3 & -\mathbf{c}_5 & \mathbf{c}_2 & \mathbf{0} & -\mathbf{c}_5 & \mathbf{c}_4 & \mathbf{0}\end{array}\right]$$

It is *not* completely observable because $\{\mathbf{c}_1, \mathbf{c}_2, \mathbf{c}_4\}$ is not a linearly independent set. (See Problem 11.17.) In fact, $\mathbf{c}_4 = 2\mathbf{c}_1 + 3\mathbf{c}_2$. Therefore, this realization can be reduced. ■

EXAMPLE 12.10 Find an *irreducible* realization for $\mathbf{H}(s)$ of the previous example.

In order to reduce the order of the realization by one, one integrator must be eliminated from Figure 12.8 without altering the input-output characteristics, that is, $\mathbf{H}(s)$. Using $\mathbf{c}_4 = 2\mathbf{c}_1 + 3\mathbf{c}_2$ allows $\mathbf{H}(s)$ to be rewritten as

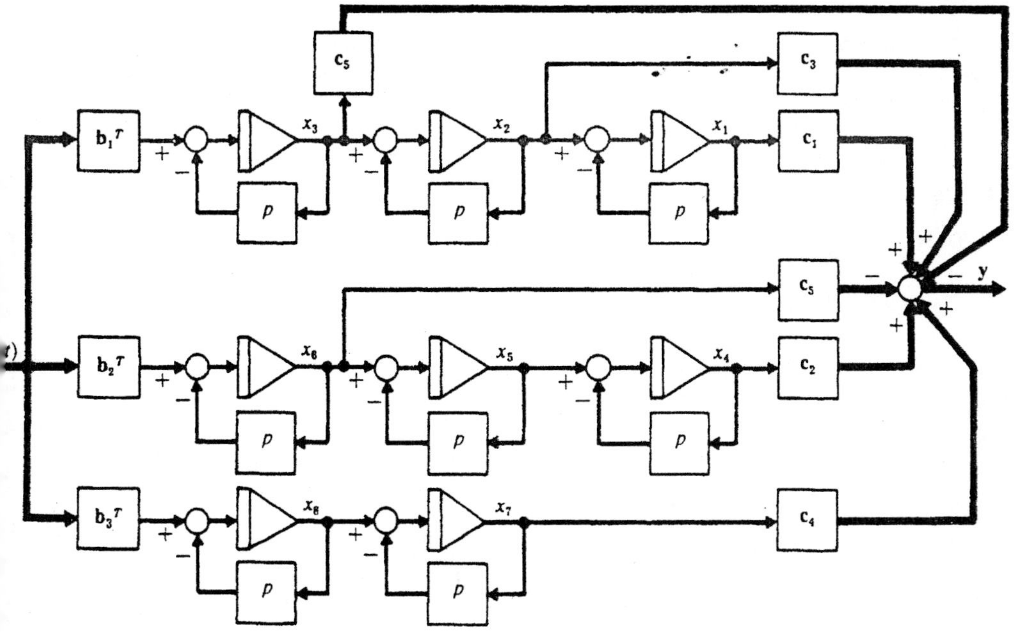

Figure 12.8

$$\mathbf{H}(s) = \frac{\mathbf{c}_1\rangle\langle[\mathbf{b}_1 + 2(s+p)\mathbf{b}_3]}{(s+p)^3} + \frac{\mathbf{c}_3\rangle\langle\mathbf{b}_1}{(s+p)^2} - \frac{\mathbf{c}_5\rangle\langle\mathbf{b}_1}{s+p}$$
$$+ \frac{\mathbf{c}_2\rangle\langle[\mathbf{b}_2 + 3(s+p)\mathbf{b}_3]}{(s+p)^3} - \frac{\mathbf{c}_5\rangle\langle\mathbf{b}_2}{s+p} \qquad (12.11)$$

This form indicates that if additional inputs to the first and second Jordan blocks are used, the second-order term in the third Jordan block can be eliminated. However, a new first-order term will apparently be required in the third Jordan block to subtract out all additional unwanted outputs from blocks one and two due to their added inputs. This is made clear by rewriting Eq. (*12.11*) as

$$\mathbf{H}(s) = \frac{\mathbf{c}_1\rangle\langle[\mathbf{b}_1 + 2(s+p)\mathbf{b}_3]}{(s+p)^3} + \frac{\mathbf{c}_3\rangle\langle[\mathbf{b}_1 + 2(s+p)\mathbf{b}_3]}{(s+p)^2} - \frac{\mathbf{c}_5\rangle\langle\mathbf{b}_1}{s+p}$$
$$+ \frac{\mathbf{c}_2\rangle\langle[\mathbf{b}_2 + 3(s+p)\mathbf{b}_3]}{(s+p)^3} - \frac{\mathbf{c}_5\rangle\langle\mathbf{b}_2}{s+p} - \frac{2\mathbf{c}_3\rangle\langle\mathbf{b}_3}{s+p} \qquad (12.12)$$

Figure 12.9 gives the reduced simulation diagram. The new realization is given by

$$\mathbf{A} = \left[\begin{array}{ccc|ccc|c} -p & 1 & 0 & & & & \\ 0 & -p & 1 & & \mathbf{0} & & \\ 0 & 0 & -p & & & & \\ \hline & & & -p & 1 & 0 & \\ & \mathbf{0} & & 0 & -p & 1 & \\ & & & 0 & 0 & -p & \\ \hline & & & & & & -p \end{array}\right], \qquad \mathbf{B} = \left[\begin{array}{c} \mathbf{0} \\ 2\mathbf{b}_3^T \\ \mathbf{b}_1^T \\ \hline \mathbf{0} \\ 3\mathbf{b}_3^T \\ \mathbf{b}_2^T \\ \hline \mathbf{b}_3^T \end{array}\right],$$

$$\mathbf{C} = [\mathbf{c}_1 \quad \mathbf{c}_3 \quad -\mathbf{c}_5 \mid \mathbf{c}_2 \quad \mathbf{0} \quad -\mathbf{c}_5 \mid -2\mathbf{c}_3]$$

This realization is still not observable, and can be further reduced since $-2\mathbf{c}_1 + 2\mathbf{c}_2 = -2\mathbf{c}_3$. This result could have been recognized from Eq. (*12.12*) and the reduction could have been completed in one step.

Rewriting Eq. (*12.12*) gives

$$\mathbf{H}(s) = \frac{\mathbf{c}_1\rangle\langle\mathbf{b}_1 + 2(s+p)\mathbf{b}_3 - 2(s+p)^2\mathbf{b}_3]}{(s+p)^3} + \frac{\mathbf{c}_3\rangle\langle[\mathbf{b}_1 + 2(s+p)\mathbf{b}_3]}{(s+p)^2} - \frac{\mathbf{c}_5\rangle\langle\mathbf{b}_1}{s+p}$$
$$+ \frac{\mathbf{c}_2\rangle\langle[\mathbf{b}_2 + 3(s+p)\mathbf{b}_3 + 2(s+p)^2\mathbf{b}_3]}{(s+p)^3} - \frac{\mathbf{c}_5\rangle\langle\mathbf{b}_2}{s+p}$$

The simulation diagram of Figure 12.9 can be modified to obtain an irreducible sixth-order realization:

$$\mathbf{A} = \left[\begin{array}{ccc|ccc} -p & 1 & 0 & & & \\ 0 & -p & 1 & & \mathbf{0} & \\ 0 & 0 & -p & & & \\ \hline & & & -p & 1 & 0 \\ & \mathbf{0} & & 0 & -p & 1 \\ & & & 0 & 0 & -p \end{array}\right], \qquad \mathbf{B} = \left[\begin{array}{c} -2\mathbf{b}_3^T \\ 2\mathbf{b}_3^T \\ \mathbf{b}_1^T \\ \hline 2\mathbf{b}_3^T \\ 3\mathbf{b}_3^T \\ \mathbf{b}_2^T \end{array}\right],$$

$$\mathbf{C} = [\mathbf{c}_1 \quad \mathbf{c}_3 \quad -\mathbf{c}_5 \mid \mathbf{c}_2 \quad \mathbf{0} \quad -\mathbf{c}_5]$$

■

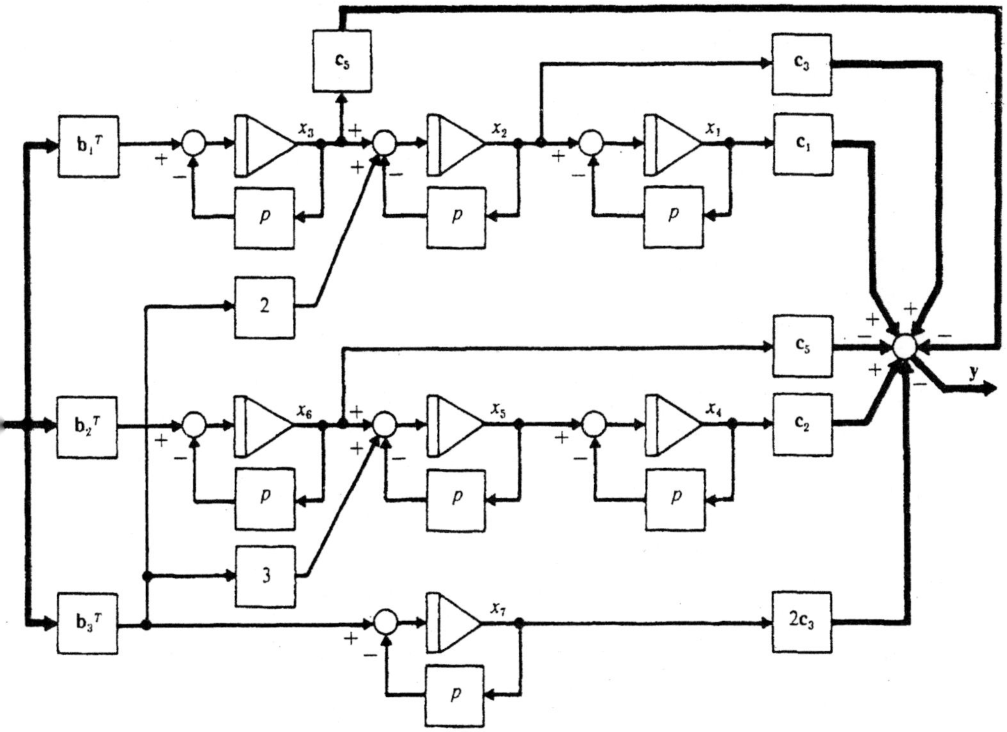

Figure 12.9

12.5.2 Kalman Canonical Form Approach to Minimal Realizations

In Sec. 11.7 it was shown that the dimension of the controllable subspace, here called Σ_c, for a given state variable realization is equal to the rank of the controllability matrix **P**. In fact, the independent columns of **P** form a basis for Σ_c. An orthogonal decomposition of the state space, $\Sigma = \Sigma_c \oplus \Sigma_c^\perp$, was obtained, with the orthonormal basis vectors of these two subspaces forming columns of the matrices $\mathbf{T}_1$ and $\mathbf{T}_2$. The modified Gram-Schmidt process was used to obtain the **QR** decomposition of **P**, and this gave $\mathbf{T}_1$ directly. For an alternate method using singular value decomposition, see Reference 3.

Likewise, the observable subspace, call it Σ_o, was found by applying the **QR** decomposition algorithm to the observability matrix **Q**. Σ_o has an orthonormal basis set, which forms the columns of the matrix $\mathbf{V}_1$. An alternative orthonormal decomposition of the state space, $\Sigma = \Sigma_o \oplus \Sigma_o^\perp$, was obtained by completing the set of basis vectors with $\mathbf{V}_2$.

It is possible to use these notions to decompose the state variable model into four subsystems [1]:

1. The intersection $\Sigma_c \cap \Sigma_o$ contains all the states which are both controllable and observable. This is the only part of the system that is described by the transfer function, and its model is a minimal realization.

2. The portion of Σ_c that is spanned by columns of $\mathbf{T}_1$ which are *not* included in (1) contains states which are controllable but not observable.
3. The portion of Σ_o that is spanned by columns of $\mathbf{V}_1$ which are *not* included in (1) contains the states which are observable but not controllable.
4. The remainder of Σ contains those states which are neither controllable nor observable.

The full decomposition just described is unnecessary for the determination of a minimal realization. If only $\mathbf{T}_1$ is found from $\mathbf{P}$, then

$$\dot{\mathbf{w}}_1 = \mathbf{T}_1^T\mathbf{A}\mathbf{T}_1\mathbf{w}_1 + \mathbf{T}_1^T\mathbf{B}\mathbf{u} \quad \text{and} \quad \mathbf{y} = \mathbf{C}\mathbf{T}_1\mathbf{w}_1 + \mathbf{D}\mathbf{u} \tag{12.13}$$

is a realization of the controllable subsystem. Comparing this with Eq. (*11.8*) shows that the $\mathbf{w}_2$ states have been deleted. Even though the $\mathbf{w}_2$ states may be observable in the output $\mathbf{y}$, and even though the $\mathbf{w}_2$ states may *in general* be nonzero due to initial conditions, recall that the transfer function describes only the zero state response. Since $\mathbf{w}_2$ is not controllable, if it is initially zero, it stays zero. Thus in the case of zero initial conditions, $\mathbf{w}_2$ plays no role and can be dropped.

Starting with the reduced order controllable system of Eq. (*12.13*), a new observability matrix $\mathbf{Q}$ can be calculated. If this system is observable, then Eq. (*12.13*) constitutes a minimal realization. If it is not, then the **QR** decomposition process can be applied to $\mathbf{Q}$ to find $\mathbf{V}_1$, an orthonormal basis for the observable subspace of the reduced system. A similarity transformation (actually an orthogonal transformation because of the orthonormality of the basis set) on Eq. (*12.13*) gives

$$\dot{\mathbf{q}} = \mathbf{V}_1^T\mathbf{T}_1^T\mathbf{A}\mathbf{T}_1\mathbf{V}_1\mathbf{q} + \mathbf{V}_1^T\mathbf{T}_1^T\mathbf{B}\mathbf{u} \quad \text{and} \quad \mathbf{y} = \mathbf{C}\mathbf{T}_1\mathbf{V}_1\mathbf{q} + \mathbf{D}\mathbf{u} \tag{12.14}$$

Equation (*12.14*) models a system that is both controllable and observable and hence is a minimal realization. That is, Eq. (*12.14*) and the original system $\{\mathbf{A}, \mathbf{B}, \mathbf{C}, \mathbf{D}\}$ have exactly the same transfer function.

Clearly the preceding sequence of the two orthogonal transformations could have been reversed. That is, an observable realization could have been constructed first from the original $\mathbf{Q}$ to yield a reduced system

$$\dot{\mathbf{v}}_1 = \mathbf{V}_1^T\mathbf{A}\mathbf{V}_1\mathbf{v}_1 + \mathbf{V}_1^T\mathbf{B}\mathbf{u} \quad \text{and} \quad \mathbf{y} = \mathbf{C}\mathbf{V}_1\mathbf{v}_1 + \mathbf{D}\mathbf{u} \tag{12.15}$$

This model describes the observable part of Eq. (*11.9*). Then a new matrix $\mathbf{P}$ could be constructed for this system. If it is full rank, then Eq. (*12.15*) constitutes a minimal realization. If it is not, then a second **QR** decomposition would give a matrix $\mathbf{T}_1$ to be used in the final orthogonal transformation, yielding

$$\dot{\mathbf{q}} = \mathbf{T}_1^T\mathbf{V}_1^T\mathbf{A}\mathbf{V}_1\mathbf{T}_1\mathbf{q} + \mathbf{T}_1^T\mathbf{V}_1^T\mathbf{B}\mathbf{u} \quad \text{and} \quad \mathbf{y} = \mathbf{C}\mathbf{V}_1\mathbf{T}_1\mathbf{q} + \mathbf{D}\mathbf{u} \tag{12.16}$$

In general, $\mathbf{q}$, $\mathbf{T}_1$, and $\mathbf{V}_1$ in Eq. (*12.14*) and Eq. (*12.16*) will not be the same. These constitute two alternative and equally valid forms of minimum realizations. By checking the ranks of both $\mathbf{P}$ and $\mathbf{Q}$ first, using the one with the smaller rank first, *might* possibly eliminate the need for the second orthogonal transformation.

EXAMPLE 12.11 The second-order realization of Example 12.4 is reconsidered, using Kalman canonical forms. Since $\mathbf{C} = \mathbf{I}_2$, the observability matrix $\mathbf{Q}$ has full rank 2, so the system is observable. It is important to realize that any similarity transformation on an observable (or controllable) system will give a new system which is also observable (or controllable), so that the observability of any reduced order system derived from this example will not need to be rechecked. The controllability matrix is $\mathbf{P} = \begin{bmatrix} 4 & -1 & 0 & 0 \\ 4 & -1 & 0 & 0 \end{bmatrix}$ and has rank 1. A single basis vector spans the controllable subspace. It is selected as $\mathbf{T}_1 = [0.7071 \quad 0.7071]^T$. Then the first-order controllable subsystem is given by

$$\dot{w}_1 = \mathbf{T}_1^T \mathbf{A} \mathbf{T}_1 w_1 + \mathbf{T}_1^T \mathbf{B} u = \mathbf{0} w_1 + [5.65685 \quad -1.4142]\mathbf{u}$$

and $\mathbf{y} = \mathbf{I}_2 \mathbf{T}_1 w_1 = [0.707 \quad 0.707]^T w_1$. This is a minimal realization and agrees with the minimal realization presented earlier in Figure 12.6 (except for an inconsequential scaling up of **B** and scaling down of **C** by a factor of 1.414). ■

EXAMPLE 12.12 Start with the sixth-order realization of Example 12.2 and find a minimal realization using the Kalman canonical form technique. Note that this same system was reduced to a fourth-order minimal realization in Example 12.7 using the Jordan form method. The 6×12 controllability matrix is

$$\mathbf{P} = \left[\begin{array}{cc|cc|cc|cc|cc|cc} 1 & 0 & -1 & 0 & 1 & 0 & -1 & 0 & 1 & 0 & -1 & 0 \\ 0 & 0 & 0 & 1 & 0 & -3 & 0 & 7 & 0 & -15 & 0 & 31 \\ 0 & 1 & 0 & -1 & 0 & 1 & 0 & -1 & 0 & 1 & 0 & -1 \\ 0 & 1 & 0 & -3 & 0 & 9 & 0 & -27 & 0 & 81 & 0 & -243 \\ 0 & 0 & 1 & 0 & -4 & 0 & 13 & 0 & -40 & 0 & 121 & 0 \\ 1 & 0 & -1 & 0 & 1 & 0 & -1 & 0 & 1 & 0 & -1 & 0 \end{array}\right]$$

The rank of **P** is 5. The observability matrix is

$$\mathbf{Q} = \left[\begin{array}{cc|cc|cc|cc|cc|cc} 1 & 0 & -1 & 0 & 1 & 0 & -1 & 0 & 1 & 0 & -1 & 0 \\ 2 & 0 & -4 & 0 & 8 & 0 & -16 & 0 & 32 & 0 & -64 & 0 \\ 0 & 0 & 2 & 0 & -6 & 0 & 14 & 0 & -30 & 0 & 62 & 0 \\ 0 & 1 & 0 & -3 & 0 & 9 & 0 & -27 & 0 & 81 & 0 & -243 \\ 0 & 1 & 0 & -3 & 0 & 9 & 0 & -27 & 0 & 81 & 0 & -243 \\ 0 & 0 & 0 & 1 & 0 & -4 & 0 & 13 & 0 & -40 & 0 & 121 \end{array}\right]$$

Its rank is 4. A set of four orthonormal six-component vectors is selected from **Q** and used to form the columns of

$$\mathbf{V}_1 = \begin{bmatrix} 0.4472136 & 0 & 0.3651484 & 0 \\ 0.8944272 & 0 & -0.1825742 & 0 \\ 0 & 0 & 0.9128709 & 0 \\ 0 & 0.7071068 & 0 & 0 \\ 0 & 0.7071068 & 0 & 0 \\ 0 & 0 & 0 & 1 \end{bmatrix}$$

The fourth-order observable subsystem is found to be (rounded)

$$\dot{\mathbf{q}} = \begin{bmatrix} -1.8 & 0 & 0.9797959 & 0 \\ 0 & -3 & 0 & 0.707107 \\ 0.163299 & 0 & -1.2 & 0 \\ 0 & 0 & 0 & -1 \end{bmatrix}\mathbf{q} + \begin{bmatrix} 0.447214 & 0 \\ 0 & 0.707107 \\ 0.365148 & 0.912871 \\ 1 & 0 \end{bmatrix}\mathbf{u}$$

$$\mathbf{y} = \begin{bmatrix} 2.23607 & 0 & 0 & 0 \\ 0 & 1.41421 & 0 & 0 \end{bmatrix}\mathbf{q}$$

The new 4×8 matrix $\mathbf{P}$ is of full rank, so that this realization is both controllable and observable and hence irreducible. Note that if the original matrix $\mathbf{P}$ had been used to construct five orthonormal basis vectors, a fifth-order controllable realization would be found, but it would not be observable (rank($\mathbf{Q}$) = 4). It would then be necessary to construct a set of four orthonormal basis vectors from the new $\mathbf{Q}$ and carry out a second orthogonal transformation. It should not be inferred that just because the original realization had a controllable subspace Σ_c of dimension 5, which is larger than the dimension of the original Σ_o of 4, that $\Sigma_o \subset \Sigma_c$, as was the case in this example. Problem 12.10 illustrates a case where $\dim(\Sigma) = 8$, $\dim(\Sigma_c) = 7$, and $\dim(\Sigma_o) = 6$. Selecting a sixth-order observable realization and then testing its new controllability matrix shows that rank($\mathbf{P}$) = 5 = order of the minimal realization. The following sketch helps understand this behavior.

$|\longleftarrow \dim(\Sigma) = 8 \longrightarrow|$

$|\longleftarrow \dim(\Sigma_c) = 7 \longrightarrow|$

$|\longleftarrow \dim(\Sigma_c \cap \Sigma_o) = 5 \longrightarrow|$

$|\longleftarrow \dim(\Sigma_o) = 6 \longrightarrow|$

■

A bounding inequality on the minimum order $n_{\min}$ is

$$\dim(\Sigma_c) + \dim(\Sigma_o) - n \le n_{\min} \le \min\{\dim(\Sigma_c), \dim(\Sigma_o)\}.$$

12.6 MINIMAL REALIZATIONS FROM MATRIX FRACTION DESCRIPTION

For a scalar transfer function, the order of the irreducible realization is the degree of the denominator after all common pole/zero pairs are canceled. The poles of the reduced denominator will be eigenvalues of the matrix $\mathbf{A}$ of the minimal realization. A similar result is true for matrix transfer functions. It hinges on what is meant by the denominator. Chen [4] defines the denominator of a rational transfer function matrix as the lowest common denominator of all minors of all orders, after all possible cancellations are made in each minor. This applies to sampled and continuous transfer functions.

EXAMPLE 12.13 A Z-transform transfer function matrix is given by

$$\mathbf{H}(z) = \begin{bmatrix} 0.2z/[(z-1)(z-0.5)] & 3z(z+0.3)/[(z-1)(z-0.8)] \\ (z+0.8)/[(z-0.5)^2(z-0.7)] & z/(z-0.5) \end{bmatrix}$$

The four first-order minors are just the entries of $\mathbf{H}(z)$, and the lowest common denominator is $(z-1)(z-0.5)^2(z-0.7)(z-0.8)$. The only second-order minor is the determinant

$$H_{11}H_{22} - H_{12}H_{21} = \{0.2z^2(z-0.7)(z-0.8) - 3z(z+0.3)(z+0.8)\}$$
$$/[(z-1)(z-0.5)^2(z-0.7)(z-0.8)]$$

In this case the lowest common denominator of all minors is the same as the lowest common denominator of the entries in $\mathbf{H}$. This will not usually be the case. Here the denominator is of degree 5 and has poles at $z = 1, 0.5, 0.5, 0.7$, and 0.8. ■

The order of the minimum or irreducible state variable realization is equal to the degree of the denominator in the sense just defined. For the example, a completely controllable and completely observable fifth-order realization of $\mathbf{H}(z)$ is possible, and its matrix A will have eigenvalues at 1, 0.5, 0.5, 0.7, and 0.8. The *determination* of a minimal realization by means of Jordan form or Kalman canonical form techniques was described previously. The matrix fraction description MFD of the transfer function can also be used. (See Chapter 4 and Problems 4.30 through 4.32.) The matrix equivalent of canceling all common pole-zero pairs is accomplished by using a series of polynomial-restricted elementary operations. (See Sec. 6.3.1 and 6.3.2.) Recall that a given transfer matrix can be written as either a left or right MFD,

$$\mathbf{H}(s) = \mathbf{P}_1^{-1}(s)\mathbf{N}_1(s) = \mathbf{N}_2(s)\mathbf{P}_2^{-1}(s)$$

Thus a series of elementary row operations on $\mathbf{P}_1(s)\mathbf{H}(s) = \mathbf{N}_1(s)$ or a series of elementary column operations on $\mathbf{H}(s)\mathbf{P}_2(s) = \mathbf{N}_2(s)$ will leave $\mathbf{H}(s)$ unaltered. The order of the state variable realization required for a given MFD description is the degree of the determinant of the "denominator matrix" $\mathbf{P}_1$ or $\mathbf{P}_2$. Do not confuse the denominator matrix or its determinant degree with Chen's denominator, which is always a scalar and whose degree is uniquely determined by **H.** Here the determinant degree of the denominator matrix can be changed by choice of the elementary operations that are used. A systematic sequence of elementary row operations can be carried out on $[\mathbf{P}_1 \mid \mathbf{N}_1]$ (or elementary column operations on $\begin{bmatrix}\mathbf{P}_2 \\ \hline \mathbf{N}_2\end{bmatrix}$) until the determinant of the modified $\mathbf{P}_i$ portion takes on the a priori known minimal degree. It is a fact that the same minimal order applies, whether the left or right MFD version is being used. The systematic reduction procedure can be carried out until the Hermite (row or column) form is reached, but this is more than necessary. When the minimal order is reached, the factors $\mathbf{N}_i$ and $\mathbf{P}_i$ are said to be *coprime* (left coprime in the case $i = 1$ and right coprime in the case $i = 2$). Being coprime is the matrix equivalent of being maximally reduced—that is, having all common pole-zero factors canceled in the scalar case. For a much more thorough treatment of these topics and the related proofs, see Kailath [5]. Once the reduced form of the MFD is obtained, the state equations can be obtained using one of the methods of Chapter 3. For example, if the left MFD is being used then the corresponding coupled differential (or difference) equations are obtained from

$$\mathbf{P}_1\mathbf{Y} = \mathbf{N}_1\mathbf{U}$$

Then the nested integrator approach of Sec. 3.4 is used to obtain the matrix version of the observable canonical form.

If the right MFD is being used, the intermediate variables g_i are first simulated from $\mathbf{P}_2\mathbf{g} = \mathbf{U}$ and then the outputs are constructed from $\mathbf{Y} = \mathbf{N}_2\mathbf{g}$. This is the matrix version of the controllable canonical form approach in Figure 3.9. In both cases the state variables are the outputs of integrators, or delay elements, and will be of the minimum possible number. The realizations thus obtained will be both controllable and observable and hence irreducible.

EXAMPLE 12.14 Find a minimal realization for the system of Example 12.13.

An initial MFD can always be obtained by finding the common denominator $d(z)$ of all the first-order minors and then writing either $\mathbf{H}(z) = [\mathbf{I}d(z)]^{-1}\mathbf{N}(z)$ or $\mathbf{N}(z)[\mathbf{I}d(z)]^{-1}$. Here the left MFD will be used, and the starting form of $[\mathbf{P}_1(z) \mid \mathbf{N}_1(z)]$ is of determinant degree 10:

$$\left[\begin{matrix}(z-1)(z-0.5)^2(z-0.7)(z-0.8) & 0 \\ 0 & (z-1)(z-0.5)^2(z-0.7)(z-0.8)\end{matrix}\right.$$

$$\left.\begin{matrix}0.2z(z-0.5)(z-0.7)(z-0.8) & 3z(z+0.3)(z-0.5)^2(z-0.7) \\ (z+0.8)(z-1)(z-0.8) & z(z-0.5)(z-0.7)(z-0.8)(z-1)\end{matrix}\right]$$

When using the so-called polynomial restricted elementary row operations, the division operation is allowed only if it leaves no remainder. In row 1 a factor $(z-0.5)(z-0.7)$ can be canceled from each term. In row 2 a factor $(z-1)(z-0.8)$ can be canceled, leaving a determinant degree of 6:

$$\left[\begin{array}{cc|cc}(z-1)(z-0.5)(z-0.8) & 0 & 0.2z(z-0.8) & 3z(z+0.3)(z-0.5) \\ 0 & (z-0.5)^2(z-0.7) & (z+0.8) & z(z-0.5)(z-0.7)\end{array}\right]$$

No more easy common factors remain, but it is known that the degree must be reduced by one more factor, and the factor to be removed must be $(z-0.5)$, since currently there are three of these and the final result must only have two. If $\alpha = 0.03/1.3$ is multiplied times row 2 and if the result is added to row 1, then every term in the altered row 1 will have at least one factor $(z-0.5)$. Doing this and canceling leaves the final form with determinant degree 5:

$$\left[\begin{array}{cc|cc}(z-1)(z-0.8) & \alpha(z-0.5)(z-0.7) & 0.2z+(\alpha-0.06) & z[3(z+0.3)+\alpha(z-0.7)] \\ 0 & (z-0.5)^2(z-0.7) & (z+0.8) & z(z-0.5)(z-0.7)\end{array}\right]$$

From this, a pair of coupled difference equations are written:

$$y_1(k+2) - 1.8y_1(k+1) + 0.8y_1(k) + \alpha y_2(k+2) - 1.2\alpha y_2(k+1) + 0.35\alpha y_2(k)$$
$$= 0.2u_1(k+1) + (\alpha - 0.06)u_1(k) + (\alpha+3)u_2(k+2) + (0.9 - 0.7\alpha)u_2(k+1)$$
$$y_2(k+3) - 1.7y_2(k+2) + 0.95y_2(k+1) - 0.175y_2(k)$$
$$= u_1(k+1) + 0.8u_1(k) + u_2(k+3) - 1.2u_2(k+2) + 0.35u_2(k+1)$$

By delaying the first equation twice and the second equation three times, a nested-delay form of these equations is obtained:

$$y_1(k) = -\alpha y_2(k) + (\alpha+3)u_2(k) + \mathfrak{D}\{1.8y_1(k) + 1.2\alpha y_2(k) + 0.2u_1(k)$$
$$+ (0.9-0.7\alpha)u_2(k) + \mathfrak{D}[-0.8y_1(k) - 0.35\alpha y_2(k) + (\alpha-0.06)u_1(k)]\}$$
$$y_2(k) = u_2(k) + \mathfrak{D}\{1.7y_2(k) - 1.2u_2(k) + \mathfrak{D}[-0.95y_2(k) + u_1(k) + 0.35u_2(k)$$
$$+ \mathfrak{D}\langle 0.175y_2(k) + 0.8u_1(k)\rangle]\}$$

where $\mathfrak{D}$ is a unit delay.

These can be represented as an interconnection of five delay elements. Picking the delay outputs as states leads to the minimal realization

$$\mathbf{A} = \begin{bmatrix} 1.8 & 0 & -0.6\alpha & 0 & 0 \\ -0.8 & 0 & 0.45\alpha & 0 & 0 \\ 0 & 0 & 1.7 & 1 & 0 \\ 0 & 0 & -0.95 & 0 & 1 \\ 0 & 0 & 0.175 & 0 & 0 \end{bmatrix}, \qquad \mathbf{B} = \begin{bmatrix} 0.2 & (6.3+0.5\alpha) \\ (\alpha-0.06) & -(2.4+0.35\alpha) \\ 0 & 0.5 \\ 1 & -0.6 \\ 0.8 & 0.175 \end{bmatrix},$$

$$\mathbf{C} = \begin{bmatrix} 1 & 0 & -\alpha & 0 & 0 \\ 0 & 0 & 1 & 0 & 0 \end{bmatrix}, \quad \text{and} \quad \mathbf{D} = \begin{bmatrix} 0 & 3 \\ 0 & 1 \end{bmatrix} \quad \text{with} \quad \alpha = 0.0230769$$

■

7 CONCLUDING COMMENTS

Linear constant coefficient systems can be described by either state variable techniques or by transfer functions. This chapter has explored the relationships between these alternatives. If the state equations are given, the transfer function is given uniquely by Eq. (*12.5*). The reverse process is not unique. For a given transfer function there exist many sets of corresponding state equations. Several methods of determining state equations, called realizations, have been presented, starting with Chapter 3. In this chapter a major emphasis has been placed on finding state variable realizations with the minimum possible number of states. Three different approaches to the determination of minimal realizations have been presented. Each has advantages and disadvantages, but the Kalman canonical form technique seems best matched to the developments in this book. It is also well suited to machine implementation.

Minimal realizations are both completely controllable and observable. These properties are important in the following chapters, since they guarantee the existence of solutions to various control design problems.

An alternative method for determining irreducible realizations was presented in the fundamental contributions by B. L. Ho [6, 7]. His work is also discussed in Reference 4. Raven [8] presents a method for determining *approximate* realizations when only approximations of the transfer functions are known. Additional related material may be found in References 9, 10, and 11.

REFERENCES

1. Kalman, R. E.: "Mathematical Description of Linear Dynamical Systems," *Journ. Soc. Ind. Appl. Math-Control Series,* Series A, Vol. 1, No. 2, 1963, pp. 152–192.
2. Kalman, R. E., "Irreducible Realizations and the Degree of a Rational Matrix," *Journ. Soc. Ind. Appl. Math-Control Series,* Series A, Vol. 13, 1965, pp. 520–544.
3. DeCarlo, R. A.: *Linear Systems,* Prentice Hall, Englewood Cliffs, N.J., 1989.
4. Chen, C. T.: *Introduction to Linear Systems Theory,* Holt, Rinehart and Winston, New York, 1970.
5. Kailath, T.: *Linear Systems,* Prentice Hall, Englewood Cliffs, N.J., 1980.
6. Ho, B. L. and R. E. Kalman: "Effective Construction of Linear State-Variable Models From Input/Output Functions," *Proc. Third Allerton Conf.,* 1965, pp. 449–459.
7. Kalman, R. E. and N. DeClaris: *Aspects of Network and System Theory,* Holt, Rinehart and Winston, New York, 1971, pp. 385–407.
8. Raven, E. A.: "A Minimum Realization Method," *IEEE Control System Magazine,* Vol. 1, No. 3, Sept. 1981, pp. 14–20.
9. Chen, C. T. and D. P. Mital: "A Simplified Irreducible Algorithm," *IEEE Transactions on Automatic Control,* Vol. AC-17, No. 4, Aug. 1972, pp. 535–537.
10. Desoer, C. A.: *Notes for a Second Course on Linear Systems,* Van Nostrand Reinhold, New York, 1970.
11. Leondes, C. T. and L. M. Novak: "Optimal Minimal-Order Observers for Discrete-Time Systems—A Unified Theory," *Automatica,* Vol. 8, No. 4, 1972, pp. 379–387.

ILLUSTRATIVE PROBLEMS

Transfer Functions and State Equations

12.1 A single-input, single-output system has the transfer function

$$H(s) = \frac{(s+5)(s+\alpha)}{(s+2)(s+3)(s+\alpha)}$$

(a) Cancel the common pole, zero pair and write the input-output differential equation for this system. Use a simulation diagram to select state variables.
(b) Repeat part a without canceling the pole, zero pair.
(c) Compare the controllability and observability of the two realizations obtained in (a) and (b).

(a) The transfer function $y(s)/u(s) = (s+5)/[(s+2)(s+3)]$ implies the differential equation $\ddot{y} + 5\dot{y} + 6y = 5u + \dot{u}$. A simulation diagram is given in Figure 12.10. Then

$$\begin{bmatrix}\dot{x}_1\\ \dot{x}_2\end{bmatrix} = \begin{bmatrix}0 & 1\\ -6 & -5\end{bmatrix}\begin{bmatrix}x_1\\ x_2\end{bmatrix} + \begin{bmatrix}1\\ 0\end{bmatrix}u, \qquad y = [1 \quad 0]\mathbf{x}$$

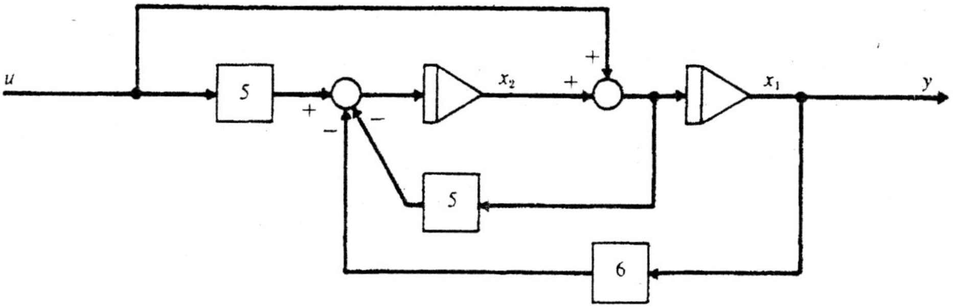

Figure 12.10

(b) The differential equation now is $\dddot{y} + (5+\alpha)\ddot{y} + (6+5\alpha)\dot{y} + 6\alpha y = \ddot{u} + (5+\alpha)\dot{u} + 5\alpha u$. From this,

$$y = \int\left\{[u - (5+\alpha)y] + \int\left[(5+\alpha)u - (6+5\alpha)y + \int(5\alpha u - 6\alpha y)\,dt\right]dt\right\}dt$$

A simulation diagram is given in Figure 12.11. Then

$$\begin{bmatrix}\dot{x}_1\\ \dot{x}_2\\ \dot{x}_3\end{bmatrix} = \begin{bmatrix}-(5+\alpha) & 1 & 0\\ -(6+5\alpha) & 0 & 1\\ -6\alpha & 0 & 0\end{bmatrix}\begin{bmatrix}x_1\\ x_2\\ x_3\end{bmatrix} + \begin{bmatrix}1\\ 5+\alpha\\ 5\alpha\end{bmatrix}u, \qquad y = [1 \quad 0 \quad 0]\mathbf{x}$$

(c) For system a, $\mathbf{P} = \begin{bmatrix}1 & 0\\ 0 & -6\end{bmatrix}$ has rank 2. $\mathbf{Q}\begin{bmatrix}1 & 0\\ 0 & 1\end{bmatrix}$ has rank 2. System a is completely controllable and observable. For system b, $\mathbf{P} = \begin{bmatrix}1 & 0 & -6\\ 5+\alpha & -6 & -6\alpha\\ 5\alpha & -6\alpha & 0\end{bmatrix}$ has rank 2. $\mathbf{Q} = \begin{bmatrix}1 & -5-\alpha & \alpha^2+5\alpha+19\\ 0 & 1 & -5-\alpha\\ 0 & 0 & 1\end{bmatrix}$ has rank 3. System b is completely observable but not controllable. Other simulation diagrams can be drawn for system b, which result in a realization which is completely controllable but not observable. It is not possible to obtain a third-order realization which is both completely controllable and observable.

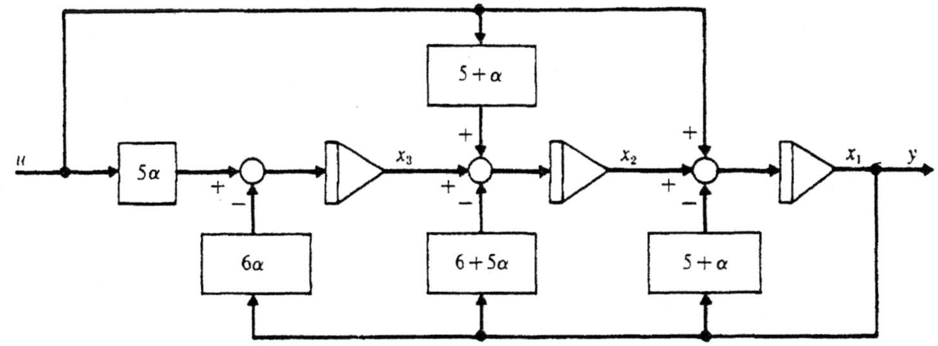

Figure 12.11

2 Find state equations and the input-output transfer matrix for the system of Figure 12.12.

The fact that there are multiple inputs and outputs does not change the procedure for selecting state variables from the linear graph, described in Sec. 3.4.5. The tree to be used is shown in heavy lines. Use the capacitor voltage x_1 and inductor current x_2 as state variables. Then $\dot{x}_1 = x_2/C + u_1/C$, $\dot{x}_2 = -x_1/L - R_2x_2/L + u_2/L$, and $y_1 = x_1$, $y_2 = x_1 + R_2x_2$. Hence

$$\mathbf{A} = \begin{bmatrix} 0 & 1/C \\ -1/L & -R_2/L \end{bmatrix}, \qquad \mathbf{B} = \begin{bmatrix} 1/C & 0 \\ 0 & 1/L \end{bmatrix}, \qquad \mathbf{C} = \begin{bmatrix} 1 & 0 \\ 1 & R_2 \end{bmatrix}, \qquad D = 0$$

Then

$$\mathbf{H}(s) = \mathbf{C}[s\mathbf{I} - \mathbf{A}]^{-1}\mathbf{B} = \frac{\begin{bmatrix} (s + R_2/L)/C & 1/LC \\ s/C & (R_2s + 1/C)/L \end{bmatrix}}{s^2 + R_2s/L + 1/LC}$$

3 A system with three inputs and two outputs is described by

$$\ddot{y}_1 + 6\dot{y}_1 - \dot{y}_2 + 10y_1 - 6y_2 = \frac{\dot{u}_1}{2} - 2u_1 + \frac{3\dot{u}_2}{2} + 4u_2 - 5\dot{u}_3$$

$$\ddot{y}_2 + 8\dot{y}_2 - 2\dot{y}_1 + 12y_2 - 4y_1 = \dot{u}_1 + 2u_1 + \dot{u}_2 + 2u_2 + \dot{u}_3 + 2u_3$$

Find the transfer matrix $\mathbf{H}(s)$.

A simulation diagram could be drawn, utilizing four integrators. Then Eq. (*12.5*) could be used to obtain $\mathbf{H}(s)$. Alternatively, the Laplace transforms of the original equations are used

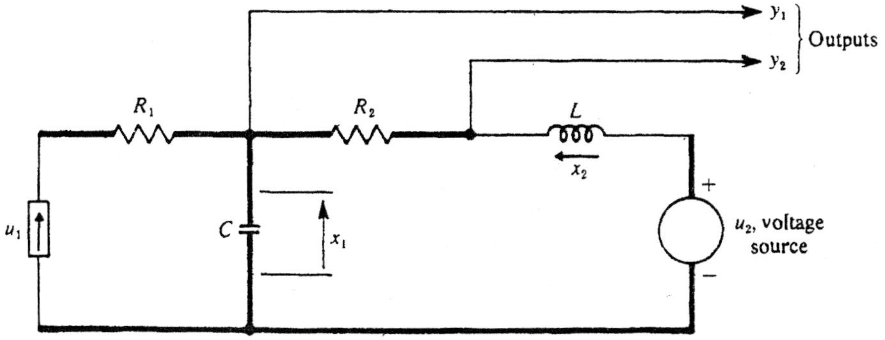

Figure 12.12

here, giving

$$\mathbf{y}(s) = \begin{bmatrix} s^2+6s+10 & -s-6 \\ -2s-4 & s^2+8s+12 \end{bmatrix}^{-1} \begin{bmatrix} s/2-2 & 3s/2+4 & -5 \\ s+2 & s+2 & s+2 \end{bmatrix} \mathbf{u}(s)$$

Therefore,

$$\mathbf{H}(s) = \frac{\begin{bmatrix} s/2-1 & 3s/2+5 & -4 \\ s+1 & s+3 & s \end{bmatrix}}{(s+2)(s+4)}$$

12.4 Determine the input-output transfer matrix for a system described by

$$\dot{\mathbf{x}} = \begin{bmatrix} -1 & 1 & 0 \\ 0 & -1 & 0 \\ 0 & 0 & -1 \end{bmatrix} \mathbf{x} + \begin{bmatrix} 2 & 0 \\ 1 & -1 \\ -1 & 1 \end{bmatrix} \mathbf{u}, \qquad \mathbf{y} = \begin{bmatrix} 2 & -1 & 3 \\ 0 & 1 & 0 \\ 3 & -1 & 6 \end{bmatrix} \mathbf{x} + \begin{bmatrix} 1 & 1 \\ 0 & 0 \\ 0 & 0 \end{bmatrix} \mathbf{u}$$

Using Eq. (*12.5*), the transfer matrix is $\mathbf{H}(s) = \mathbf{C}[s\mathbf{I} - \mathbf{A}]^{-1}\mathbf{B} + \mathbf{D}$. Substituting for **A**, **B**, **C**, and **D** and carrying out the multiplication gives

$$\mathbf{H}(s) = \begin{bmatrix} \dfrac{s^2+2s+3}{(s+1)^2} & \dfrac{s^2+6s+3}{(s+1)^2} \\[2ex] \dfrac{1}{s+1} & \dfrac{-1}{s+1} \\[2ex] \dfrac{-s+2}{(s+1)^2} & \dfrac{7s+4}{(s+1)^2} \end{bmatrix}$$

12.5 Use the linearized state variable model of the aircraft in Problem 11.9. Find the transfer function matrix which relates the aircraft aileron and rudder deflections to the state variables p, r, β, and ϕ. Draw a block diagram of the relationships contained in the state equations and verify the entries in the transfer function matrix. Use the block diagram to explain in physical terms the mathematical results on controllability (Problem 11.9) and observability (Problem 11.10).

Since the transfer function to *all* the states is desired, set $\mathbf{C} = \mathbf{I}$. Then

$$\mathbf{H}(s) = \mathbf{C}[s\mathbf{I} - \mathbf{A}]^{-1}\mathbf{B} = \begin{bmatrix} s+10 & 0 & 10 & 0 \\ 0 & s+0.7 & -9 & 0 \\ 0 & 1 & s+0.7 & 0 \\ 1 & 0 & 0 & s \end{bmatrix}^{-1} \begin{bmatrix} 20 & 2.8 \\ 0 & -3.13 \\ 0 & 0 \\ 0 & 0 \end{bmatrix}$$

$$= \begin{bmatrix} \dfrac{20}{s+10} & \dfrac{2.8(s-0.776)(s+2.176)}{(s+10)(s^2+1.4s+9.49)} \\[2ex] 0 & \dfrac{-3.13(s+0.7)}{s^2+1.4s+9.49} \\[2ex] 0 & \dfrac{3.13}{s^2+1.4s+9.49} \\[2ex] \dfrac{20}{s(s+10)} & \dfrac{2.8(s-0.776)(s+2.176)}{s(s+10)(s^2+1.4s+9.49)} \end{bmatrix}$$

The block diagram of Figure 12.13 is drawn directly from the state equations. Using elementary block diagram reduction technques, or Mason's gain formula for a one-step reduction, the transfer functions H_{ij} from each input j to each of the four outputs i can be found. The eight results are of course the entries in the previous matrix.

Figure 12.13 makes it clear why the system is uncontrollable with just δ_a. There is no signal path from this input to either r or β. The other input does feed into all four state variables, either

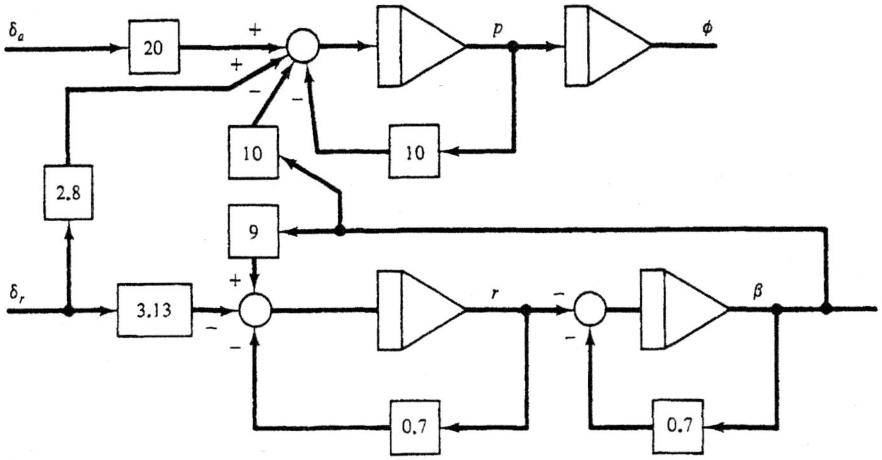

Figure 12.13

directly or indirectly. The observability results of Problem 11.10 are equally obvious. If ϕ is not an output, it can never be determined, since there is no path by which ϕ affects any other state or output. Conversely, if ϕ can be monitored, it is intuitive that its rate, i.e., p, can be deduced. Likewise, the rate of change of p can be deduced, and it contains information about β. This in turn is strongly related to r.

Jordan Form Irreducible Realizations

2.6 Find an irreducible realization of $\mathbf{H}(s)$ from Problem 12.4.

Expanding each H_{ij} element yields

$$\mathbf{H}(s) = \begin{bmatrix} 1+\dfrac{2}{(s+1)^2} & 1-\dfrac{2}{(s+1)^2}+\dfrac{4}{s+1} \\ \dfrac{1}{s+1} & -\dfrac{1}{s+1} \\ \dfrac{3}{(s+1)^2}-\dfrac{1}{s+1} & \dfrac{-3}{(s+1)^2}+\dfrac{7}{s+1} \end{bmatrix}$$

$$= \frac{\begin{bmatrix} 2 & -2 \\ 0 & 0 \\ 3 & -3 \end{bmatrix}}{(s+1)^2} + \frac{\begin{bmatrix} 0 & 4 \\ 1 & -1 \\ -1 & 7 \end{bmatrix}}{s+1} + \begin{bmatrix} 1 & 1 \\ 0 & 0 \\ 0 & 0 \end{bmatrix}$$

This form indicates that $\mathbf{D} = \begin{bmatrix} 1 & 1 \\ 0 & 0 \\ 0 & 0 \end{bmatrix}$. The remaining two matrices are of rank 1 and 2, respectively. Writing them as outer products gives

$$\mathbf{H}(s) = \frac{\begin{bmatrix} 2 \\ 0 \\ 3 \end{bmatrix}[1 \quad -1]}{(s+1)^2} + \frac{\begin{bmatrix} 0 \\ 1 \\ -1 \end{bmatrix}[1 \quad -1]}{s+1} + \frac{\begin{bmatrix} 4 \\ 0 \\ 6 \end{bmatrix}[0 \quad 1]}{s+1} + \mathbf{D}$$

The vectors $\mathbf{b}_1^T = [1 \quad -1]$, $\mathbf{b}_3^T = [0 \quad 1]$ are linearly independent, and could be used to form the last rows associated with two Jordan blocks $\mathbf{B} = \begin{bmatrix} 0 & 0 \\ 1 & -1 \\ \hdashline 0 & 1 \end{bmatrix}$. However, the corresponding output matrix $\mathbf{C} = \left[\begin{array}{cc:c} 2 & 0 & 4 \\ 0 & 1 & 0 \\ 3 & -1 & 6 \end{array}\right]$ has its first and third columns linearly dependent. This would lead to a realization which is completely controllable, but not observable and hence reducible. Since $\mathbf{c}_3 = 2\mathbf{c}_1$, the first and third terms can be combined:

$$\mathbf{H}(s) = \frac{\mathbf{c}_1\rangle\langle[\mathbf{b}_1 + 2\mathbf{b}_3(s+1)]}{(s+1)^2} + \frac{\mathbf{c}_2\rangle\langle\mathbf{b}_1}{s+1} + \mathbf{D}$$

The corresponding simulation diagram is given in Figure 12.14. The irreducible realization is

$$\begin{bmatrix} \dot{x}_1 \\ \dot{x}_2 \end{bmatrix} = \begin{bmatrix} -1 & 1 \\ 0 & -1 \end{bmatrix}\begin{bmatrix} x_1 \\ x_2 \end{bmatrix} + \begin{bmatrix} 0 & 2 \\ 1 & -1 \end{bmatrix}\begin{bmatrix} u_1 \\ u_2 \end{bmatrix},$$

$$\begin{bmatrix} y_1 \\ y_2 \\ y_3 \end{bmatrix} = \begin{bmatrix} 2 & 0 \\ 0 & 1 \\ 3 & -1 \end{bmatrix}\begin{bmatrix} x_1 \\ x_2 \end{bmatrix} + \begin{bmatrix} 1 & 1 \\ 0 & 0 \\ 0 & 0 \end{bmatrix}\begin{bmatrix} u_1 \\ u_2 \end{bmatrix}$$

This realization is of lesser dimension than the actual system from which the transfer function was obtained. The original system was not completely controllable.

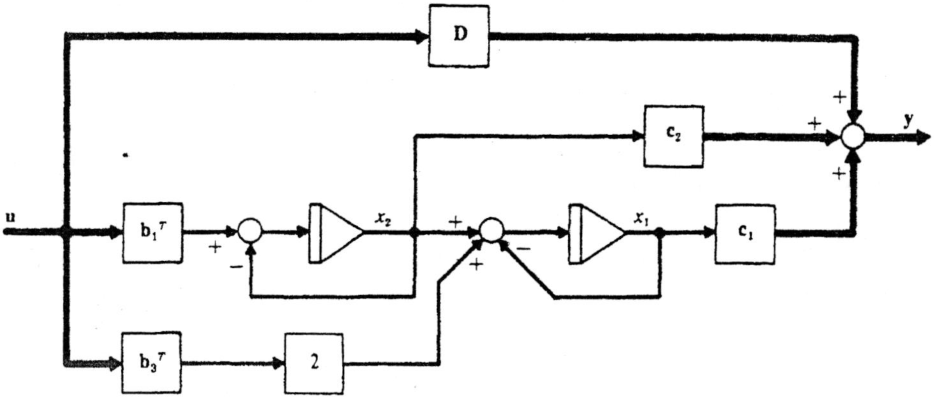

Figure 12.14

12.7 Find a minimal order state variable realization for the discrete system described by

$$\mathbf{T}(z) = \begin{bmatrix} \dfrac{0.2z}{(z-1)(z-0.5)} & \dfrac{3z(z+0.3)}{(z-1)(z-0.8)} \\[2ex] \dfrac{(z+0.8)}{(z-0.5)^2(z-0.7)} & \dfrac{z}{(z-0.5)} \end{bmatrix}$$

A partial fraction expanded form for $\mathbf{T}(z)$ is

$$\mathbf{T}(z) = \frac{1}{(z-0.5)}\begin{bmatrix} -0.2 & 0 \\ -37.5 & 0.5 \end{bmatrix} + \frac{1}{(z-0.5)^2}\begin{bmatrix} 0 & 0 \\ -6.5 & 0 \end{bmatrix}$$

$$+ \frac{1}{(z-1)}\begin{bmatrix} 0.4 & 19.5 \\ 0 & 0 \end{bmatrix} + \frac{1}{(z-0.8)}\begin{bmatrix} 0 & -13.2 \\ 0 & 0 \end{bmatrix}$$

$$+ \frac{1}{(z-0.7)}\begin{bmatrix} 0 & 0 \\ 37.5 & 0 \end{bmatrix} + \begin{bmatrix} 0 & 3 \\ 0 & 1 \end{bmatrix}$$

All of the 2 × 2 expansion matrices are of rank one except the first, which has rank 2. Therefore, a seventh-order state variable model could be written down immediately as

$$\mathbf{A} = \text{diag}\left[0.5, 0.5, \begin{bmatrix} 0.5 & 1 \\ 0 & 0.5 \end{bmatrix}, 1, 0.8, 0.7\right]$$

$$\mathbf{B} = \begin{bmatrix} 1 & 0 \\ 0 & 1 \\ \hline 1 & 0 \\ 0 & 0 \\ \hline 0.4 & 19.5 \\ 0 & 1 \\ 1 & 0 \end{bmatrix} \qquad \mathbf{C} = \left[\begin{array}{cc|cc|ccc} -0.2 & 0 & 0 & 0 & 1 & -13.2 & 0 \\ -37.5 & 0.5 & 0 & -6.5 & 0 & 0 & 37.5 \end{array}\right]$$

$$\mathbf{D} = \begin{bmatrix} 0 & 3 \\ 0 & 1 \end{bmatrix}$$

Note that a 2 × 2 Jordan block is included for the $(z - 0.5)^2$ factor. The first state in this segment has a direct connection to the input but not the output. The reverse is true for the second state in this subsegment. These two states are related, as shown in Figure 12.15. It can be shown (easily with the right computational aids, otherwise more laboriously) that the above seventh-order system is neither controllable nor observable. The ranks of the controllability and observability matrices are both 5. This indicates that two unneeded states can be deleted. The two states shown in Figure 12.15, plus the other two uncoupled states associated with $z = 0.5$, must be reduced to just two states. A systematic way of accomplishing this is to assume the segment shown in Figure 12.16 and select the eight scalar parameters $a, \ldots, h$ so that this segment realization yields the same output as the original fourth-order segment. That is,

$$\mathbf{w} = \left\{ \frac{\begin{bmatrix} e \\ f \end{bmatrix}[a \quad b] + \begin{bmatrix} g \\ h \end{bmatrix}[c \quad d]}{(z - 0.5)} + \frac{\begin{bmatrix} g \\ h \end{bmatrix}[a \quad b]}{(z - 0.5)^2} \right\} \mathbf{u}$$

$$= \left\{ \frac{\begin{bmatrix} -0.2 & 0 \\ -37.5 & 0.5 \end{bmatrix}}{(z - 0.5)} + \frac{\begin{bmatrix} 0 & 0 \\ -6.5 & 0 \end{bmatrix}}{(z - 0.5)^2} \right\} \mathbf{u}$$

This requires that $ae + cg = -0.2, be + dg = 0, af + ch = -37.5, bf + dh = 0.5, ag = 0, bg = 0, ah = -6.5, bh = 0$. Of the many possibilities, pick $b = c = g = 0$ and $a = 1$. Then solve for $e = -0.2$, $h = -6.5$, $d = -0.076923$, and $f = -37.5$. The 2 × 2 segment associated with $z = 0.5$

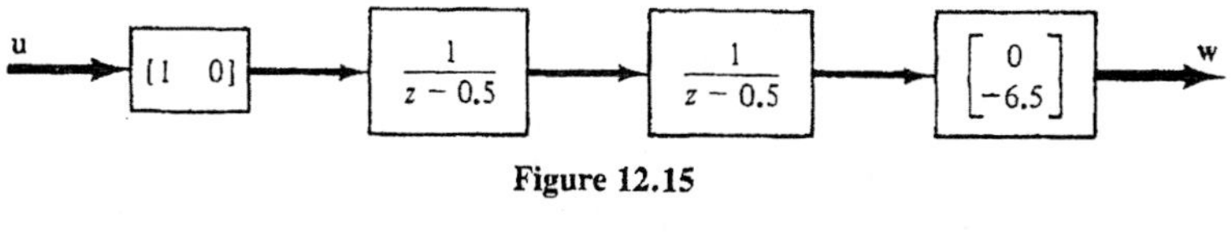

Figure 12.15

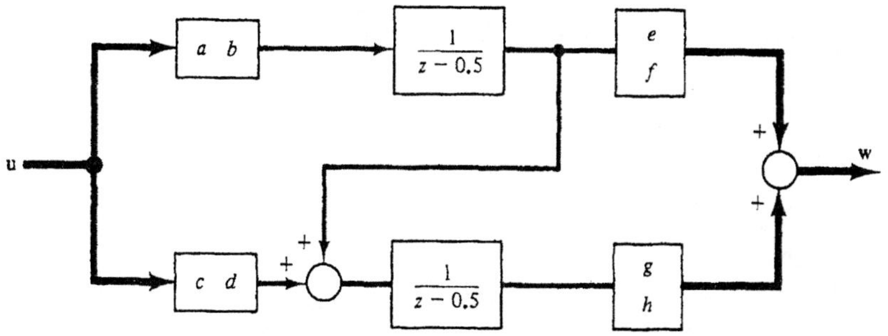

Figure 12.16

thus has the state equations

$$\begin{bmatrix} x_1(k+1) \\ x_2(k+1) \end{bmatrix} = \begin{bmatrix} 0.5 & 1 \\ 0 & 0.5 \end{bmatrix} \begin{bmatrix} x_1(k) \\ x_2(k) \end{bmatrix} + \begin{bmatrix} 0 & -0.076923 \\ 1 & 0 \end{bmatrix} \mathbf{u}(k)$$

$$\mathbf{w}(k) = \begin{bmatrix} 0 & -0.2 \\ -6.5 & -37.5 \end{bmatrix} \begin{bmatrix} x_1(k) \\ x_2(k) \end{bmatrix}$$

By substituting this second-order segment for the original fourth-order segment, the final fifth-order system realization, which is controllable with index 3 and observable with index 3, is

$$\mathbf{A} = \begin{bmatrix} 0.5 & 1 & 0 & 0 & 0 \\ 0 & 0.5 & 0 & 0 & 0 \\ 0 & 0 & 1 & 0 & 0 \\ 0 & 0 & 0 & 0.8 & 0 \\ 0 & 0 & 0 & 0 & 0.7 \end{bmatrix} \qquad \mathbf{B} = \begin{bmatrix} 0 & -0.07693 \\ 1 & 0 \\ 0.4 & 19.5 \\ 0 & 1 \\ 1 & 0 \end{bmatrix}$$

$$\mathbf{C} = \begin{bmatrix} 0 & -0.2 & 1 & -13.2 & 0 \\ -6.5 & -37.5 & 0 & 0 & 37.5 \end{bmatrix} \qquad \mathbf{D} = \begin{bmatrix} 0 & 3 \\ 0 & 1 \end{bmatrix}$$

12.8 Find an irreducible realization for the transfer matrix

$$\mathbf{H}(s) = \begin{bmatrix} \dfrac{3}{(s^2+2s+10)(s+5)} & \dfrac{2s}{s+5} \\ \dfrac{s+1}{(s^2+2s+10)(s+5)} & \dfrac{s+1}{s+5} \end{bmatrix}$$

Partial fraction expansion gives

$$H_{11}(s) = \frac{\alpha}{s+1+3j} + \frac{\bar{\alpha}}{s+1-3j} + \frac{3/25}{s+5}, \quad H_{12}(s) = 2 - \frac{10}{s+5}$$

$$H_{21}(s) = \frac{-j\alpha}{s+1+3j} + \frac{-j\bar{\alpha}}{s+1-3j} - \frac{4/25}{s+5}, \quad H_{22}(s) = 1 - \frac{4}{s+5}$$

where $\alpha = -3/50 + 2j/25$. Let $p_1 = 1 + 3j$. Then

$$\mathbf{H}(s) = \frac{\begin{bmatrix} \alpha & 0 \\ -j\alpha & 0 \end{bmatrix}}{s+p_1} + \frac{\begin{bmatrix} \bar{\alpha} & 0 \\ -j\bar{\alpha} & 0 \end{bmatrix}}{s+\bar{p}_1} + \frac{\begin{bmatrix} \frac{3}{25} & -10 \\ -\frac{4}{25} & -4 \end{bmatrix}}{s+5} + \begin{bmatrix} 0 & 2 \\ 0 & 1 \end{bmatrix}$$

$$= \frac{\begin{bmatrix} \alpha \\ -j\alpha \end{bmatrix}[1 \;\; 0]}{s+p_1} + \frac{\begin{bmatrix} \bar{\alpha} \\ -j\bar{\alpha} \end{bmatrix}[1 \;\; 0]}{s+\bar{p}_1} + \frac{\begin{bmatrix} \frac{3}{25} \\ -\frac{4}{25} \end{bmatrix}[1 \;\; 0]}{s+5} + \frac{\begin{bmatrix} 10 \\ 4 \end{bmatrix}[0 \;\; -1]}{s+5} + \mathbf{D}$$

This transfer matrix can be realized with four 1×1 Jordan blocks. Since $\mathbf{c}_3$ and $\mathbf{c}_4$ associated with the pole at -5 are linearly independent, an irreducible realization is given by

$$\mathbf{A} = \text{diag}[-p_1, -\bar{p}_1, -5, -5], \qquad \mathbf{B} = \begin{bmatrix} 1 & 0 \\ 1 & 0 \\ 1 & 0 \\ 0 & -1 \end{bmatrix}$$

$$\mathbf{C} = \begin{bmatrix} \alpha & \bar{\alpha} & \frac{2}{25} & 10 \\ -j\alpha & -j\bar{\alpha} & -\frac{4}{25} & 4 \end{bmatrix}, \qquad \mathbf{D} = \begin{bmatrix} 0 & 2 \\ 0 & 1 \end{bmatrix}$$

Because of the complex numbers in **A** and **C**, this realization cannot be directly mechanized. The next problem illustrates a transformation which eliminates the complex numbers.

.9 Find a suitable transformation (change of basis in Σ) which transforms the realization of Problem 12.8 into one with all real coefficients.

The original state vector is $\mathbf{x}$. A new state vector $\mathbf{x}'$ is sought, satisfying $\mathbf{x} = \mathbf{T}\mathbf{x}'$. Then

$$\dot{\mathbf{x}}' = \mathbf{T}^{-1}\mathbf{A}\mathbf{T}\mathbf{x}' + \mathbf{T}^{-1}\mathbf{B}\mathbf{u}, \quad \mathbf{y} = \mathbf{C}\mathbf{T}\mathbf{x}' + \mathbf{D}\mathbf{u}$$

and it is required that $\mathbf{A}' = \mathbf{T}^{-1}\mathbf{A}\mathbf{T}$, $\mathbf{B}' = \mathbf{T}^{-1}\mathbf{B}$, and $\mathbf{C}' = \mathbf{C}\mathbf{T}$ all be real.

A tentative transformation is $\mathbf{T} = \left[\begin{array}{cc|c} 1 & j & \mathbf{0} \\ 1 & -j & \\ \hline \multicolumn{2}{c|}{\mathbf{0}} & \mathbf{I}_2 \end{array}\right]$. Then

$$\mathbf{C}' = \begin{bmatrix} \alpha + \bar{\alpha} & j\alpha - j\bar{\alpha} & \frac{3}{25} & 10 \\ -j\alpha + \overline{j\alpha} & \alpha + \bar{\alpha} & -\frac{4}{25} & 4 \end{bmatrix} \quad \text{or} \quad \mathbf{C}' = \begin{bmatrix} -\frac{3}{25} & -\frac{4}{25} & \frac{3}{25} & 10 \\ \frac{4}{25} & -\frac{3}{25} & -\frac{4}{25} & 4 \end{bmatrix}$$

The selected transformation does yield a real $\mathbf{C}'$. To evaluate $\mathbf{A}'$ and $\mathbf{B}'$,

$$\mathbf{T}^{-1} = \left[\begin{array}{cc|c} 1/2 & 1/2 & \mathbf{0} \\ -j/2 & j/2 & \\ \hline \multicolumn{2}{c|}{\mathbf{0}} & \mathbf{I}_2 \end{array}\right]$$

is required. Matrix multiplication then gives

$$\mathbf{A}' = \left[\begin{array}{cc|cc} -1 & 3 & 0 & 0 \\ -3 & -1 & 0 & 0 \\ \hline 0 & 0 & -5 & 0 \\ 0 & 0 & 0 & -5 \end{array}\right] \quad \text{and} \quad \mathbf{B}' = \left[\begin{array}{cc} 1 & 0 \\ 0 & 0 \\ \hline 1 & 0 \\ 0 & -1 \end{array}\right]$$

The $\mathbf{D}$ matrix is unaffected by this change of basis, so a real, irreducible realization is $\{\mathbf{A}', \mathbf{B}', \mathbf{C}', \mathbf{D}\}$.

Irreducible Realization Using Kalman Canonical Form

10 Find another minimal realization for the transfer function of Examples 12.13 and 12.14 and Problem 12.7 by first finding an eighth-order realization with no internal coupling (as in Sec. 3.5) and then reducing it using the Kalman canonical decomposition method.

By using a controllable canonical form realization for each of the four scalar transfer function elements, two second-order subsystems, one third-order, and one first-order are obtained. The total eighth-order realization has

$$\mathbf{A} = \begin{bmatrix} 0 & 1 & 0 & 0 & 0 & 0 & 0 & 0 \\ -0.5 & 1.5 & 0 & 0 & 0 & 0 & 0 & 0 \\ 0 & 0 & 0 & 1 & 0 & 0 & 0 & 0 \\ 0 & 0 & 0 & 0 & 1 & 0 & 0 & 0 \\ 0 & 0 & 0.175 & -0.95 & 1.7 & 0 & 0 & 0 \\ 0 & 0 & 0 & 0 & 0 & 0 & 1 & 0 \\ 0 & 0 & 0 & 0 & 0 & -0.8 & 1.8 & 0 \\ 0 & 0 & 0 & 0 & 0 & 0 & 0 & 0.5 \end{bmatrix}$$

$$\mathbf{B} = \begin{bmatrix} 0 & 0 \\ 1 & 0 \\ 0 & 0 \\ 0 & 0 \\ 1 & 0 \\ 0 & 0 \\ 0 & 1 \\ 0 & 1 \end{bmatrix} \quad \mathbf{C}^T = \begin{bmatrix} 0 & 0 \\ 0.2 & 0 \\ 0 & 0.8 \\ 0 & 1 \\ 0 & 0 \\ -2.4 & 0 \\ 6.3 & 0 \\ 0 & 0.5 \end{bmatrix} \quad \mathbf{D} = \begin{bmatrix} 0 & 3 \\ 0 & 1 \end{bmatrix}$$

Computer evaluation of the 8×16 controllability matrix $\mathbf{P}$ shows that it has rank 7, so the eighth-order system is not controllable. The 8×16 observability matrix is found to have rank 6. A set of six orthonormal basis vectors are selected from $\mathbf{Q}$ and used to find a sixth-order observable realization having

$$\mathbf{A} = \begin{bmatrix} 1.50534 & 0 & 0.226920 & 0 & 0 & 0 \\ 0 & 0.48942 & 0 & 0.81318 & 0 & 0 \\ -1.57135 & 0 & 0.29339 & 0 & 0.027686 & 0 \\ 0 & -0.78092 & 0 & 1.22054 & 0 & 1.61228 \\ -0.50807 & 0 & -0.00469 & 0 & 0.50127 & 0 \\ 0 & 0.24190 & 0 & 0.16458 & 0 & -0.009961 \end{bmatrix}$$

$$\mathbf{B} = \begin{bmatrix} 0.02965 & 0.93407 \\ 0 & 0.36370 \\ -0.00070 & -0.35521 \\ 0.89450 & 0.004733 \\ -0.70741 & 0.00335 \\ 0.43832 & 0.17404 \end{bmatrix} \quad \mathbf{C}^T = \begin{bmatrix} 6.7446 & 0 \\ 0 & 1.37477 \\ 0 & 0 \\ 0 & 0 \\ 0 & 0 \\ 0 & 0 \end{bmatrix} \quad \mathbf{D} = \begin{bmatrix} 0 & 3 \\ 0 & 1 \end{bmatrix}$$

The new 6×12 controllability matrix has rank 5, so the reduced system is observable but not controllable. A new matrix $\mathbf{T}_1$ is selected from columns of $\mathbf{P}$ and used in an orthonormal transformation to give the final fifth-order minimal realization

$$\mathbf{A} = \begin{bmatrix} 1.28861 & -0.018929 & -0.33158 & -0.02789 & -0.023257 \\ 0.17151 & 1.60667 & 0.03270 & -0.46532 & -0.78351 \\ 0.96580 & -0.22085 & 0.25438 & 0.07253 & 0.04190 \\ 0 & 1.20235 & -0.48397 & 0.09238 & -0.45202 \\ 0 & 0 & 0.68778 & 0.13848 & 0.25796 \end{bmatrix}$$

$$\mathbf{B} = \begin{bmatrix} 1.2221 & 0.08681 \\ 0 & 1.07412 \\ 0 & 0 \\ 0 & 0 \\ 0 & 0 \end{bmatrix} \quad \mathbf{C}^T = \begin{bmatrix} 0.16365 & 0 \\ 5.85202 & 0.46550 \\ -1.00342 & 0.76456 \\ -1.28731 & -0.40300 \\ -2.92306 & 0.86074 \end{bmatrix} \quad \mathbf{D} = \begin{bmatrix} 0 & 3 \\ 0 & 1 \end{bmatrix}$$

There are an infinite number of other valid fifth-order realizations. To check the validity of other answers, use Eq. (*12.5*) to see if the original $\mathbf{H}(s)$ is reconstructed.

12.11 Use the method of Sec. 12.5.2 to find a minimal realization for the state variable model of Problem 12.4.

The observability matrix $\mathbf{Q}$ is

$$\mathbf{Q} = \begin{bmatrix} 2 & 0 & 3 & -2 & 0 & -3 & 2 & 0 & 3 \\ -1 & 1 & -1 & 3 & -1 & 4 & -5 & 1 & -7 \\ 3 & 0 & 6 & -3 & 0 & -6 & 3 & 0 & 6 \end{bmatrix} \quad \text{and} \quad \text{rank}(\mathbf{Q}) = 3$$

The controllability matrix is given by $\mathbf{P} = \begin{bmatrix} 2 & 0 & -1 & -1 & 0 & 2 \\ 1 & -1 & -1 & 1 & 1 & -1 \\ -1 & 1 & 1 & -1 & -1 & 1 \end{bmatrix}$. Using **QR** decomposition, this becomes

$$\begin{bmatrix} 0.81650 & 0.57735 \\ 0.40825 & -0.57735 \\ -0.40825 & 0.57735 \end{bmatrix} \begin{bmatrix} 2.4495 & -0.8165 & -1.6330 & 0 & 0.81650 & 0.81650 \\ 0 & 1.1547 & 0.57735 & -1.7321 & -1.1547 & 2.3094 \end{bmatrix}$$

From this it is obvious that rank($\mathbf{P}$) = 2 and that an orthonormal basis for the two-dimensional controllable subspace is given by the first two columns above. Using these for $\mathbf{T}_1$ in Eq. (*12.13*) gives the minimal realization

$$\mathbf{A} = \begin{bmatrix} -0.6667 & -0.4714 \\ 0.2357 & -1.3333 \end{bmatrix}, \qquad \mathbf{B} = \begin{bmatrix} 2.4495 & -0.8165 \\ 0 & 1.1547 \end{bmatrix}$$

$$\mathbf{C} = \begin{bmatrix} 0 & 3.4641 & 1.41421 \\ 0.40825 & -0.57735 & 0.707107 \\ -0.40825 & 0.57735 & 3.53553 \end{bmatrix}$$

Of course, $\mathbf{D}$ remains unchanged.

12 Find a minimal realization for the state variable model which has

$$\mathbf{A} = \begin{bmatrix} -\frac{8}{3} & -1 & -\frac{5}{3} \\ -\frac{2}{3} & -2 & -\frac{2}{3} \\ -\frac{4}{3} & 1 & -\frac{7}{3} \end{bmatrix}, \qquad \mathbf{B} = \begin{bmatrix} 1 & -1 \\ 0 & -1 \\ -1 & 1 \end{bmatrix}, \qquad \mathbf{C} = \begin{bmatrix} 2 & -1 & 1 \\ 1 & 1 & 1 \end{bmatrix}$$

The controllability matrix is

$$\mathbf{P} = \begin{bmatrix} 1 & -1 & -1 & 2 & 1 & -4 \\ 0 & -1 & 0 & 2 & 0 & -4 \\ -1 & 1 & 1 & -2 & -1 & 4 \end{bmatrix} = \begin{bmatrix} 0.707107 & 0 \\ 0 & 1 \\ -0.707106 & 0 \end{bmatrix} \mathbf{R} = \mathbf{T}_1\mathbf{R}$$

where $\mathbf{R}$ is a 2×6 upper triangular matrix. Thus $\mathbf{P}$ has rank 2, and the system is not controllable. The observability matrix $\mathbf{Q}$ is of rank three, so the original system is observable. Using $\mathbf{T}_1$ leads to the Kalman controllable canonical form of Eq. (*12.13*),

$$\dot{\mathbf{w}} = \left[\begin{array}{cc|c} -1 & 1.41421 & -0.333333 \\ 0 & -2 & 0.942809 \\ \hline 0 & 0 & -4 \end{array}\right]\mathbf{w} + \left[\begin{array}{cc} 1.41421 & -1.41421 \\ 0 & 1 \\ \hline 0 & 0 \end{array}\right]\mathbf{u}$$

$$\mathbf{y} = \left[\begin{array}{cc|c} 0.707106 & 1 & 2.12132 \\ 0 & -1 & 1.41421 \end{array}\right]\mathbf{w}$$

The minimal realization is thus the second-order system with

$$\mathbf{A} = \begin{bmatrix} -1 & 1.41421 \\ 0 & -2 \end{bmatrix}, \qquad \mathbf{B} = \begin{bmatrix} 1.41421 & -1.41421 \\ 0 & 1 \end{bmatrix}, \qquad \mathbf{C} = \begin{bmatrix} 0.707106 & 1 \\ 0 & -1 \end{bmatrix}$$

Both this system and the original system can be verified to have the same transfer function matrix $\mathbf{H}(s) = \text{Diag}[1/(s+1) \quad 1/(s+2)]$.

Irreducible Realizations Using MFD

13 Use the methods of Sec. 12.6 to find a minimal realization for [5]

$$\mathbf{H}(s) = \frac{\begin{bmatrix} s+6 & s+3 \\ 4 & s+3 \end{bmatrix}}{(s+2)(s+3)}$$

The order of the minimal realization is first found. The common denominator of all first-order minors is $(s+2)(s+3)$. The only second-order minor is the determinant

$$(s+6)/[(s+3)(s+2)^2] - 4/[(s+3)(s+2)^2] = (s+2)/[(s+3)(s+2)^2]$$

After canceling the common factor, this also has $(s+2)(s+3)$ as the denominator, so the minimal order is 2. The left MFD form is used, starting with $\mathbf{D}(s) = [(s+2)(s+3)\mathbf{I}]$ and

$\mathbf{N}(s) = \begin{bmatrix} s+6 & s+3 \\ 4 & s+3 \end{bmatrix}$. Therefore, $\mathbf{D}(s)\mathbf{H}(s) = \mathbf{N}(s)$ can be reduced by using elementary row operations shown next. Subtract row 2 from row 1.

$$\left[\begin{array}{cc|cc} s^2+5s+6 & 0 & s+6 & s+3 \\ 0 & s^2+5s+6 & 4 & s+3 \end{array}\right] \rightarrow \left[\begin{array}{cc|cc} s^2+5s+6 & -(s^2+5s+6) & s+2 & 0 \\ 0 & s^2+5s+6 & 4 & s+3 \end{array}\right]$$

Now a factor of $s+2$ can be canceled from each term in row 1. Subtract 4 times row 1 from row 2 to get

$$\left[\begin{array}{cc|cc} s+3 & -(s+3) & 1 & 0 \\ -4(s+3) & s^2+9s+16 & 0 & s+3 \end{array}\right] \rightarrow \left[\begin{array}{cc|cc} s+3 & -(s+3) & 1 & 0 \\ -4 & s+6 & 0 & 1 \end{array}\right]$$

The reduction process could stop here (the determinant of the reduced $\mathbf{D}(s)$ has the minimal degree 2), but the state realization is slightly easier if row 2 is first added to row 1. This leaves $\begin{bmatrix} s-1 & s \\ -4 & s+6 \end{bmatrix}\begin{bmatrix} y_1 \\ y_2 \end{bmatrix} = \begin{bmatrix} 1 & 1 \\ 0 & 1 \end{bmatrix}\begin{bmatrix} u_1 \\ u_2 \end{bmatrix}$. From this, the simulation diagram is drawn. Then the minimal-order state equations are immediately written.

$$\dot{\mathbf{x}} = \begin{bmatrix} 1 & -3 \\ 4 & -6 \end{bmatrix}\mathbf{x} + \begin{bmatrix} 1 & 1 \\ 0 & 1 \end{bmatrix}\mathbf{u} \quad \text{and} \quad \mathbf{y} = \mathbf{I}\mathbf{x}$$

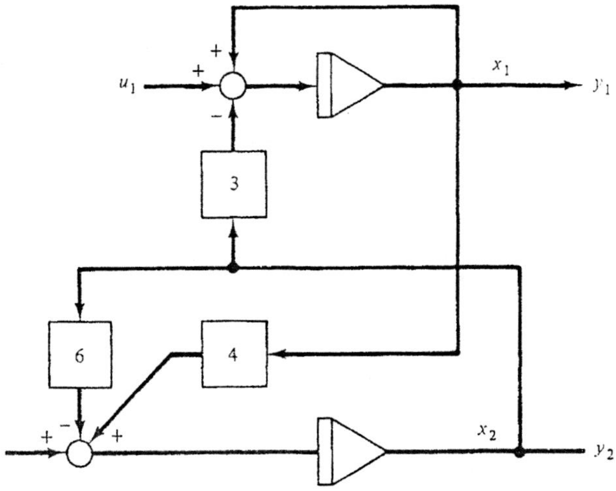

12.14 The transfer function for a system with two inputs and three outputs is

$$\mathbf{H}(s) = \begin{bmatrix} 2/(s+1)^2 & (4s+2)/(s+1)^2 \\ 1/(s+1) & -1/(s+1) \\ (-s+2)/(s+1)^2 & (7s+4)/(s+1)^2 \end{bmatrix}$$

Find a minimal realization using elementary operations on a right MFD. Note that this is the same $\mathbf{H}(s)$ as in Problem 12.4 but with $\mathbf{D}$ subtracted out.

The minimal order is first determined. The six first-order minors are just the six entries of $\mathbf{H}$, and the lowest common denominator of these is $(s+1)^2$. There are three second-order minors:

$$H_{11}H_{22} - H_{12}H_{21} = -2(s+3)(s+1)/(s+1)^3$$

$$H_{11}H_{32} - H_{12}H_{31} = (8s+6)(s+1)^2/(s+1)^4$$

$$H_{21}H_{32} - H_{22}H_{32} = 6(s+1)/(s+1)^3$$

After canceling common factors, the lowest common denominator of these nine terms is $(s+1)^2$. Therefore, this system can be realized by a second-order model. From a right MFD we have

$$\begin{bmatrix}\mathbf{N}(s)\\ \hline \mathbf{P}(s)\end{bmatrix} = \begin{bmatrix} 2 & 4s+2 \\ s+1 & -(s+1) \\ -s+2 & 7s+4 \\ \hline (s+1)^2 & 0 \\ 0 & (s+1)^2 \end{bmatrix} \to \begin{bmatrix} 2 & 4 \\ s+1 & 0 \\ -s+2 & 6 \\ \hline (s+1)^2 & (s+1) \\ 0 & (s+1) \end{bmatrix}$$

Column 1 was added to column 2 and then a factor of $(s+1)$ was canceled from column 2. This leaves a determinant degree of 3. Subtracting 0.5 times column 2 from column 1 leaves a factor of $(s+1)$ in every nonzero member of column 1. Canceling gives a form in which $\mathbf{P}(s)$ has the desired determinant degree of 2, as follows. However, one more operation, subtracting column 1 from column 2, reduces the degree of P_{12}, and this will simplify the construction of a simulation diagram.

$$\begin{bmatrix}\mathbf{N}(s)\\ \hline \mathbf{P}(s)\end{bmatrix} = \begin{bmatrix} 0 & 4 \\ s+1 & 0 \\ -s-1 & 6 \\ \hline (s+1)(s+0.5) & (s+1) \\ -0.5(s+1) & (s+1) \end{bmatrix} \to \begin{bmatrix} 0 & 4 \\ 1 & 0 \\ -1 & 6 \\ \hline (s+0.5) & (s+1) \\ -0.5 & (s+1) \end{bmatrix} \to \begin{bmatrix} 0 & 4 \\ 1 & -1 \\ -1 & 7 \\ \hline s+0.5 & 0.5 \\ -0.5 & s+1.5 \end{bmatrix}$$

Thus $\mathbf{H}(s) = \begin{bmatrix} 0 & 4 \\ 1 & -1 \\ -1 & 7 \end{bmatrix}\begin{bmatrix} s+0.5 & 0.5 \\ -0.5 & s+1.5 \end{bmatrix}^{-1}$. Then $\mathbf{Pg} = \mathbf{u}$ and $\mathbf{Ng} = \mathbf{y}$ can be simulated as shown. The second-order state model is

$$\dot{\mathbf{x}} = \begin{bmatrix} -0.5 & -0.5 \\ 0.5 & -1.5 \end{bmatrix}\mathbf{x} + \begin{bmatrix} 1 & 0 \\ 0 & 1 \end{bmatrix}\mathbf{u} \quad \text{and} \quad \mathbf{y} = \begin{bmatrix} 0 & 4 \\ 1 & -1 \\ -1 & 7 \end{bmatrix}\mathbf{x}$$

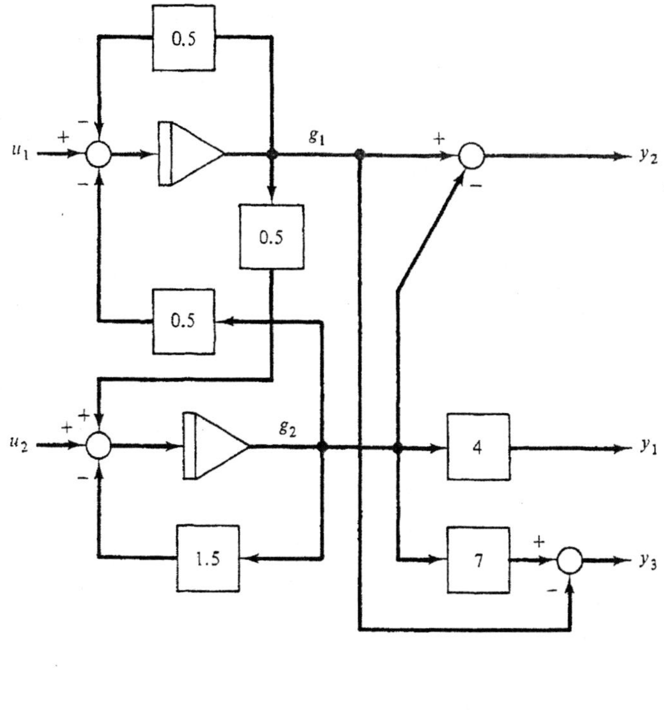

12.15 In Problems 4.30 and 4.31 the transfer function $\mathbf{H}(s) = \begin{bmatrix} 1/(s+1) & 2/[(s+1)(s+2)] \\ 1/[(s+1)(s+3)] & 1/(s+3) \end{bmatrix}$ was expressed in left and right MFD form, respectively. In both cases elementary operations were used to reduce this to a form with a determinant degree 4.

(a) Use Chen's definition of the denominator of **H** to show that the minimal realization will be of fourth order.

(b) Use the left MFD to find a fourth-order realization and verify that its matrix **A** has eigenvalues which agree with the poles of Chen's denominator.

(a) The common denominator of the four first-order minors is $(s+1)(s+2)(s+3)$. This was used in Problem 4.30 as an initial MFD with determinant degree 6, due to the $\mathbf{I}_2$ factor. The second-order minor of **H** is the determinant; its denominator is $(s+1)^2(s+2)(s+3)$, so this is the common denominator of all minors. The degree is four, so the minimal realization will be fourth order. Its eigenvalues will be at -1, -1, -2, and -3.

(b) From the final result in Problem 4.30,

$$\begin{bmatrix} (s+1)(s+2) & 0 \\ 0 & (s+1)(s+3) \end{bmatrix} \begin{bmatrix} y_1 \\ y_2 \end{bmatrix} = \begin{bmatrix} s+2 & 2 \\ 1 & s+1 \end{bmatrix} \begin{bmatrix} u_1 \\ u_2 \end{bmatrix}$$

A nested integrator form of the simulation diagram is as follows. From the diagram the state equations are

$$\dot{\mathbf{x}} = \begin{bmatrix} -3 & 1 & 0 & 0 \\ -2 & 0 & 0 & 0 \\ 0 & 0 & -4 & 1 \\ 0 & 0 & -3 & 0 \end{bmatrix} \mathbf{x} + \begin{bmatrix} 1 & 0 \\ 2 & 2 \\ 0 & 1 \\ 1 & 1 \end{bmatrix} \mathbf{u}$$

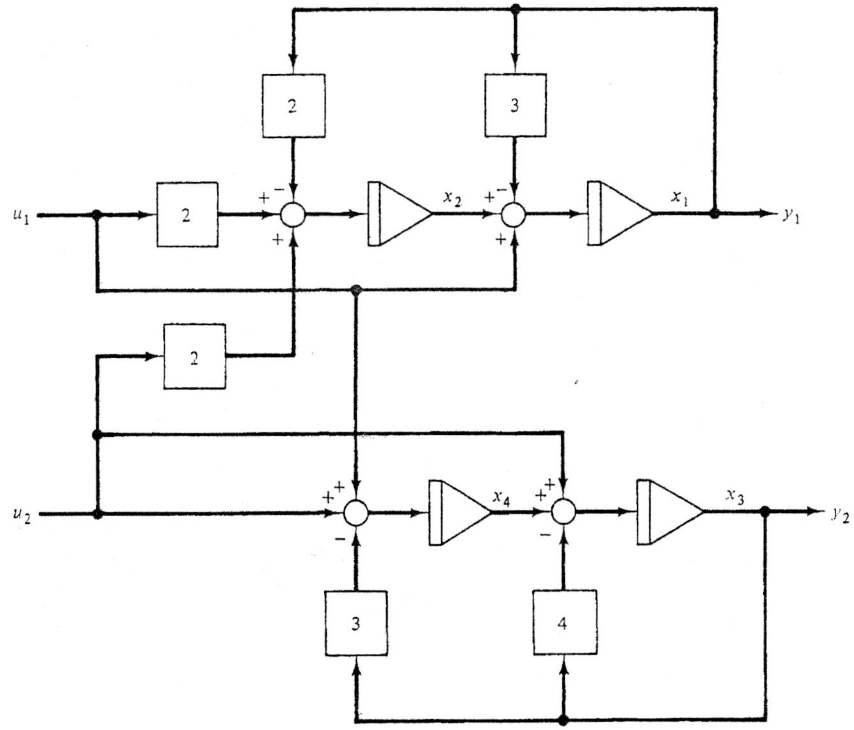

and

$$\mathbf{y} = \begin{bmatrix} 1 & 0 & 0 & 0 \\ 0 & 0 & 1 & 0 \end{bmatrix} \mathbf{x}$$

Because the form of the denominator polynomial $\mathbf{P}(s)$ was block diagonal, so is $\mathbf{A}$. Thus it is easy to verify that its eigenvalues are at -1, -1, -2, and -3.

16 Use Chen's definition of the denominator of $\mathbf{H}(s)$ to verify that the minimal order realization of the following transfer function is $n = 3$.

$$\mathbf{H}(s) = \begin{bmatrix} \dfrac{1}{s+1} & \dfrac{2}{(s+2)} \\ \dfrac{1}{(s+1)(s+3)} & \dfrac{1}{s+3} \end{bmatrix}.$$ The first-order minors are just the H_{ij} elements, and the lowest common denominator of these four terms is $(s+1)(s+2)(s+3)$. There is just one second-order minor, namely,

$$|\mathbf{H}(s)| = \frac{1}{(s+1)(s+3)} - \frac{2}{(s+1)(s+2)(s+3)} = \frac{s}{(s+1)(s+2)(s+3)}$$

The lowest common denominator of all first- and second-order minors is $(s+1) \times (s+2)(s+3)$. It has the order $n = 3$. Furthermore, the poles at $s = -1$, -2, and -3 will be the eigenvalues of the 3×3 irreducible $\mathbf{A}$ matrix.

Relation Between Irreducibility and Controllability, Observability

17 Show that an irreducible realization of $\mathbf{H}(s)$ must be completely controllable and observable. Show that a completely controllable and observable realization cannot be reduced. Assume that $\mathbf{D} = [\mathbf{0}]$.

Every $m \times r$ transfer matrix can be expanded into an infinite series $\mathbf{H}(s) = \mathbf{H}_1/s + \mathbf{H}_2/s^2 + \mathbf{H}_3/s^3 + \cdots$, where $\mathbf{H}_i$ are $m \times r$ constant matrices. Every realization satisfies $\mathbf{H}(s) = \mathbf{C}\mathbf{\Phi}(s)\mathbf{B}$, and $\mathbf{\Phi}(s) = \mathcal{L}\{e^{\mathbf{A}t}\} = \mathbf{I}/s + \mathbf{A}/s^2 + \mathbf{A}^2/s^3 + \cdots$. Therefore, for every realization, $\mathbf{H}_i = \mathbf{C}\mathbf{A}^{i-1}\mathbf{B}$.

Let n be the smallest integer such that $\mathbf{H}_j$ can be written as a combination of $\mathbf{H}_1$ through $\mathbf{H}_n$, for $j > n$. Then $\mathbf{H}(s) = \beta_1(s)\mathbf{H}_1 + \beta_2(s)\mathbf{H}_2 + \cdots + \beta_n(s)\mathbf{H}_n$. If $\dim(\Sigma) = n$ so that $\mathbf{A}$ is $n \times n$, the Cayley-Hamilton theorem gives $\mathbf{\Phi}(s) = \alpha_0\mathbf{I} + \alpha_1\mathbf{A} + \cdots + \alpha_{n-1}\mathbf{A}^{n-1}$. $\mathbf{A}$ cannot be less than $n \times n$, otherwise $\mathbf{A}^{n-1}$ could be expressed as a linear combination of lower powers of $\mathbf{A}$. This implies that $\mathbf{H}_n$ can be expressed as a combination of $\mathbf{H}_1$ through $\mathbf{H}_{n-1}$ and contradicts the manner in which n was chosen. Thus minimum $\dim(\Sigma) = n$.

Suppose $\mathbf{A}$ is $p \times p$, with $p > n$. Then

$$\mathbf{H}(s) = \beta_1\mathbf{CB} + \beta_2\mathbf{CAB} + \cdots + \beta_n\mathbf{CA}^{n-1}\mathbf{B}$$

must equal

$$\mathbf{H}(s) = \mathbf{C}[\alpha_0\mathbf{I} + \alpha_1\mathbf{A} + \cdots + \alpha_{p-1}\mathbf{A}^{p-1}]\mathbf{B}$$

This requires that for $i \geq n$, either $\mathbf{A}^i\mathbf{B}$ is a linear combination of $\mathbf{B}, \mathbf{AB}, \ldots, \mathbf{A}^{n-1}\mathbf{B}$, or $\mathbf{CA}^i$ is a linear combination of $\mathbf{C}, \mathbf{CA}, \ldots, \mathbf{CA}^{n-1}$. The first possibility means the system is uncontrollable. The second means the system is unobservable. A realization can have $\dim(\Sigma) > n$ only if it is either not controllable or not observable. Conversely, if $\mathbf{A}$ is $n \times n$ and both controllable and observable, then $\mathbf{H}(s)$ will contain terms up to and including $\mathbf{H}_n$, and thus cannot be reduced.

18 Give a geometrical interpretation of the relationship between irreducible realizations of transfer matrices and controllability and observability.

Assume that $\mathbf{D} = [\mathbf{0}]$ since $\mathbf{D}$ is not influenced by the dimension of the state space Σ. Then

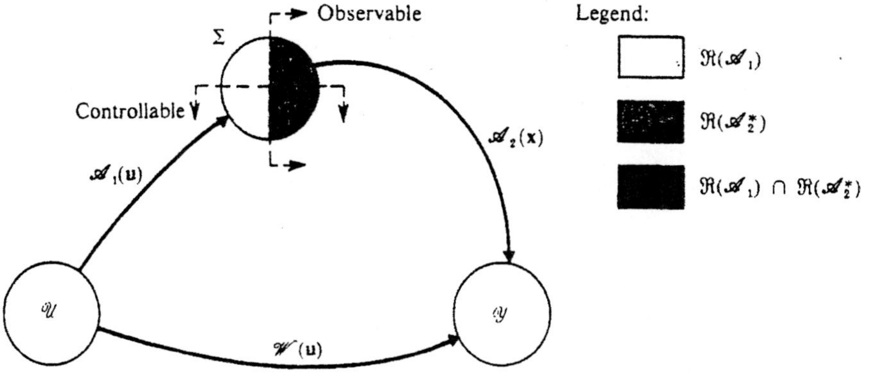

Figure 12.17

$\mathbf{H}(s) = \mathbf{C}\boldsymbol{\Phi}(s)\mathbf{B}$. In the time domain the mapping from the input space $\mathscr{U}$ to the output space $\mathscr{Y}$ is given by

$$\mathbf{y}(t) = \mathbf{C}\int_0^t e^{\mathbf{A}(t-\tau)}\mathbf{B}\mathbf{u}(\tau)\, d\tau$$

Define $\mathscr{A}_1 : \mathscr{U} \to \Sigma$ by $\mathbf{x} = \int_0^t e^{-\mathbf{A}\tau}\mathbf{B}\mathbf{u}(\tau)\, d\tau$ and $\mathscr{A}_2 : \Sigma \to \mathscr{Y}$ by $\mathbf{y} = \mathbf{C}e^{\mathbf{A}t}\mathbf{x}$. Then $\mathbf{y} = \mathscr{A}_2(\mathscr{A}_1(\mathbf{u}))$. The state space can be decomposed into $\Sigma = \mathscr{R}(\mathscr{A}_1) \oplus \mathscr{N}(\mathscr{A}_1^*)$ or alternatively into $\Sigma = \mathscr{R}(\mathscr{A}_2^*) \oplus \mathscr{N}(\mathscr{A}_2)$. Thus $\dim(\Sigma) = \dim(\mathscr{R}(\mathscr{A}_1) + \dim(\mathscr{N}(\mathscr{A}_1^*) = \dim(\mathscr{R}(\mathscr{A}_2^*)) + \dim(\mathscr{N}(\mathscr{A}_2))$. No matter which input $\mathbf{u}(t)$ is used, $\mathscr{A}_1(\mathbf{u}) \in \mathscr{R}(\mathscr{A}_1)$, so the component of $\mathbf{x} \in \mathscr{N}(\mathscr{A}_1^*)$ will be zero. Since $\mathscr{A}_2(\mathbf{0}) = \mathbf{0}$, no part of the output $\mathbf{y}$ due to any input $\mathbf{u}$ will depend upon states in $\mathscr{N}(\mathscr{A}_1^*)$. Thus $\dim(\Sigma)$ can be reduced without affecting the input-output relationship by requiring $\dim(\mathscr{N}(\mathscr{A}_1^*)) = 0$, that is, $\mathscr{N}(\mathscr{A}_1^*) = \{\mathbf{0}\}$. This is the condition for complete controllability. Likewise, any state $\mathbf{x} \in \mathscr{N}(\mathscr{A}_2)$ contributes nothing to the output $\mathbf{y}$.

Making $\dim(\mathscr{N}(\mathscr{A}_2) = 0$, that is, $\mathscr{N}(\mathscr{A}_2) = \{\mathbf{0}\}$, reduces the dimension of Σ without affecting the input-output relationship. But $\mathscr{N}(\mathscr{A}_2) = \{\mathbf{0}\}$ is the condition for complete observability. If the realization is completely controllable and observable, the $\dim(\Sigma) = \dim(\mathscr{R}(\mathscr{A}_1)) = \dim(\mathscr{R}(\mathscr{A}_2^*)) = n$, the dimension of the irreducible realization. A suggestive sketch is given in Figure 12.17.

PROBLEMS

12.19 Find the state equations and the input-output transfer function for the circuit of Figure 12.18.

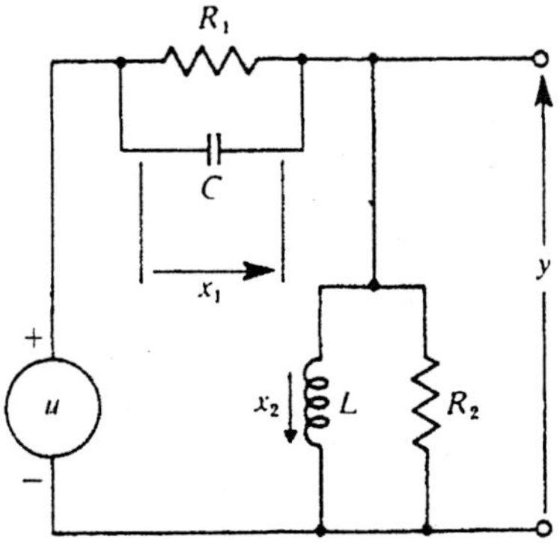

Figure 12.18

20 Find the state equations for the circuit of Figure 12.19.

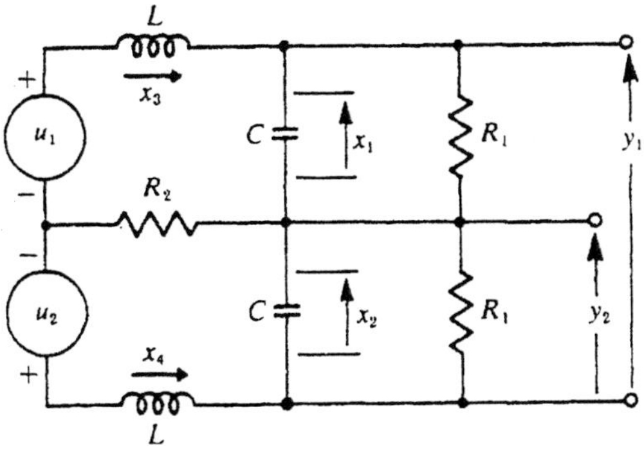

Figure 12.19

.21 Find the transfer matrix for the system given in Problem 11.21, page 397.

.22 Interchange the current source and voltage source of Figure 12.12, and repeat Problem 12.2.

.23 Find an irreducible realization for $\mathbf{H}(s) = \begin{bmatrix} \dfrac{1}{(s+1)} & \dfrac{1}{(s+1)(s+2)} & \dfrac{s}{(s+1)} \\ 0 & \dfrac{1}{(s+1)(s+3)} & \dfrac{(s+1)}{(s+3)} \end{bmatrix}$.

.24 Which of the following system realizations $\dot{\mathbf{x}} = \mathbf{Ax} + \mathbf{Bu}, \mathbf{y} = \mathbf{Cx} + \mathbf{Du}$ are irreducible?

(a) $\mathbf{A} = \begin{bmatrix} -6 & 1 & 0 & 0 \\ 0 & -6 & 0 & 0 \\ 0 & 0 & -6 & 0 \\ 0 & 0 & 0 & 6 \end{bmatrix}$, $\mathbf{B} = \begin{bmatrix} 1 & 1 & 1 \\ 1 & 1 & 1 \\ 2 & 2 & 2 \\ 0 & 1 & 0 \end{bmatrix}$

$\mathbf{C} = \begin{bmatrix} 3 & 1 & 4 & 0 \\ 0 & 1 & 1 & 1 \end{bmatrix}$, $\mathbf{D} = [\mathbf{0}]$

(b) Same $\mathbf{A}$ as in (a), $\mathbf{B} = \begin{bmatrix} 0 & 0 & 0 \\ 1 & 1 & 1 \\ 2 & 0 & 2 \\ 0 & 1 & 0 \end{bmatrix}$, $\mathbf{C} = \begin{bmatrix} 3 & 1 & 4 & 0 \\ 0 & 1 & 0 & 1 \end{bmatrix}$, $\mathbf{D} = [\mathbf{0}]$.

(c) Same $\mathbf{A}$ as in (a), same $\mathbf{B}$ as in (b), same $\mathbf{C}$ as in (a), $\mathbf{D} = [\mathbf{0}]$.

.25 Find an irreducible realization of $\mathbf{H}(s) = \begin{bmatrix} \dfrac{s}{s+1} & \dfrac{1}{(s+1)(s+2)} & \dfrac{1}{s+3} \\ \dfrac{-1}{s+1} & \dfrac{1}{(s+1)(s+2)} & \dfrac{1}{s} \end{bmatrix}$. See page 220 of Reference 4.

.26 Find an irreducible realization of $\mathbf{H}(s) = \begin{bmatrix} \dfrac{-s}{(s+1)^2} & \dfrac{1}{s+1} \\ \dfrac{2s+1}{s(s+1)} & \dfrac{1}{s+1} \end{bmatrix}$.

.27 Find an irreducible realization of

$$\mathbf{H}(s) = \frac{1}{s^4}\begin{bmatrix} s^3 - s^2 + 1 & 1 & -s^3 + s^2 - 2 \\ 1.5s + 1 & s + 1 & -1.5s - 2 \\ s^3 - 9s^2 - s + 1 & -s^2 + 1 & s^3 - s - 2 \end{bmatrix}$$

See pages 244 and 249 of Reference 4.

12.28 A discrete-time system's matrices are

$$\mathbf{A} = \begin{bmatrix} 1 & 0 & 0 \\ 0 & 0.5 & 0.8 \\ 0 & -0.8 & 0.5 \end{bmatrix}, \qquad \mathbf{B} = \begin{bmatrix} 1 & 0 \\ 0 & 1 \\ 1 & 1 \end{bmatrix}$$

$$\mathbf{C} = \begin{bmatrix} 0.5 & -1 & 0 \\ 0.5 & 1 & 1 \end{bmatrix}, \qquad \mathbf{D} = [0]$$

Find the input-output transfer function matrix $\mathbf{T}(z)$.

12.29 Find a minimal realization for the following system. Be sure that the final system matrices **A**, **B**, **C**, and **D** are all real so that they can be synthesized with real hardware.

$$\mathbf{T}(z) = \begin{bmatrix} \dfrac{z}{(z^2 - z + 0.5)(z-1)(z-0.3679)} \\[2ex] \dfrac{(z+0.2)}{(z-1)(z-0.3679)} \end{bmatrix}$$

Hint: Use partial fraction expansion on all real poles, but leave the complex conjugate pair in the form of a second-order segment.

12.30 A system has the Z-transform transfer function

$$\mathbf{H}(z) = \begin{bmatrix} 0.2z/[(z-1)(z-0.5)] & 3z(z+0.3)/[(z-1)(z-0.5)] \\ (z+0.8)/[(z-0.5)^2(z-0.7)] & z/(z-0.5) \end{bmatrix}$$

Use a left MFD, reduce it, and then find a minimal state variable realization. Note that $\mathbf{H}(z)$ is identical to that of Example 12.14 except for one denominator factor.

12.31 Consider the same system as in Problem 12.30. Construct an eighth-order state variable realization with no internal coupling. Then use the Kalman canonical form method to obtain a fifth-order (minimal) realization.

12.32 Rework Problem 12.8 using a right MFD approach. Subtract out the constant **D** matrix and then work with $\mathbf{H}'(s) = \mathbf{H}(s) - \mathbf{D}$.

12.33 Rework Problem 12.8 by first finding a reducible realization with no internal coupling and then reduce to minimal order using the Kalman canonical form procedure.

12.34 Find a minimal realization for

$$\mathbf{H}(s) = \begin{bmatrix} (s+2)/[(s+1)^2(s+3)] & s/[(s+1)(s+5)] \\ 2/(s+1) & (s+1)/[(s+2)(s+5)] \end{bmatrix}$$

using a left MFD.

12.35 Use a left MFD representation for the system in Example 12.8. Reduce it and thus find another minimal realization.

13

Design of Linear Feedback Control Systems

13.1 INTRODUCTION

A discussion of open-loop versus closed-loop control was presented in Chapter 1. Important advantages of feedback were discussed at that time. Chapter 2 presented a review of the analysis and design of single-input, single-output feedback systems using classical control techniques. It will be recalled that a fundamental method of classical design consists of forcing the dominant closed-loop poles to be suitably located in the s-plane or the Z-plane. Just what constitutes a suitable location depends upon the design specifications regarding relative stability, response times, accuracy, and so on.

This chapter considers the design of feedback compensators for linear, constant coefficient multivariable systems. One of the fundamental design objectives is, again, the achievement of suitable pole locations in order to ensure satisfactory transient response. This problem is analyzed, first under the assumption that *all* state variables can be used in forming feedback signals. Output feedback, i.e., incomplete state feedback, is also considered.

An additional design objective, which cannot arise in single-input, single-output systems, is the achievement of a decoupled or noninteracting system. This means that each input component affects just one output component, or possibly some prescribed subset of output components.

13.2 STATE FEEDBACK AND OUTPUT FEEDBACK

The state equations were introduced in Chapter 3 and analyzed extensively in the intervening chapters. It is assumed here that the open-loop system, often called the plant, is described in state variable form. Most of what is to be discussed in this chapter applies equally well to continuous-time systems,

$$\dot{\mathbf{x}} = \mathbf{Ax} + \mathbf{Bu}$$

$$\mathbf{y} = \mathbf{Cx} + \mathbf{Du} \tag{13.1}$$

or to discrete-time systems,

$$\mathbf{x}(k+1) = \mathbf{Ax}(k) + \mathbf{Bu}(k)$$

$$\mathbf{y}(k) = \mathbf{Cx}(k) + \mathbf{Du}(k) \tag{13.2}$$

The system matrices $\{\mathbf{A}, \mathbf{B}, \mathbf{C}, \mathbf{D}\}$ have different meanings in the two cases, and of course the locations of good poles will differ between the s-plane and the Z-plane. However, if s_1 is a good pole location in the s-plane, then its image $z_1 = \exp(s_1 T)$ will inherit the same good features in the Z-plane. In any event, the methods and procedures to be developed will look the same for both types of systems in terms of the four system matrices.

It is assumed that plant—and hence these four matrices—are specified and cannot be altered by the designer to improve performance. It will be consistently assumed that the state vector $\mathbf{x}$ is $n \times 1$, the input vector $\mathbf{u}$ is $r \times 1$ and the output or measurement vector $\mathbf{y}$ is $m \times 1$. Thus $\mathbf{A}$ is $n \times n$, $\mathbf{B}$ is $n \times r$, $\mathbf{C}$ is $m \times n$, and $\mathbf{D}$ is $m \times r$. If the system performance is to be altered, it must be accomplished by some form of signal manipulation outside of the given open-loop system. Two commonly used possibilities are shown in Figures 13.1 and 13.2 for continuous-time systems. These arrangements are referred to as state variable feedback and output feedback, respectively.

The feedback gain matrices $\mathbf{K}$ and $\mathbf{K}'$ are $r \times n$ and $r \times m$, respectively, and are assumed constant. The external inputs $\mathbf{v}$ and $\mathbf{v}'$ are assumed to be $p \times 1$ vectors for generality, although usually $\mathbf{v}$ will have the same number of components as $\mathbf{u}$, so that $r = p$. The feed-forward matrices $\mathbf{F}$ and $\mathbf{F}'$ in the most general case could be of dimension $r \times p$ but are most often square $r \times r$ constant matrices.

It could be justifiably argued that state variable feedback is only of academic interest because, by definition, the outputs are the only signals which are accessible. State variable feedback seems to violate our dictum about using only signals external

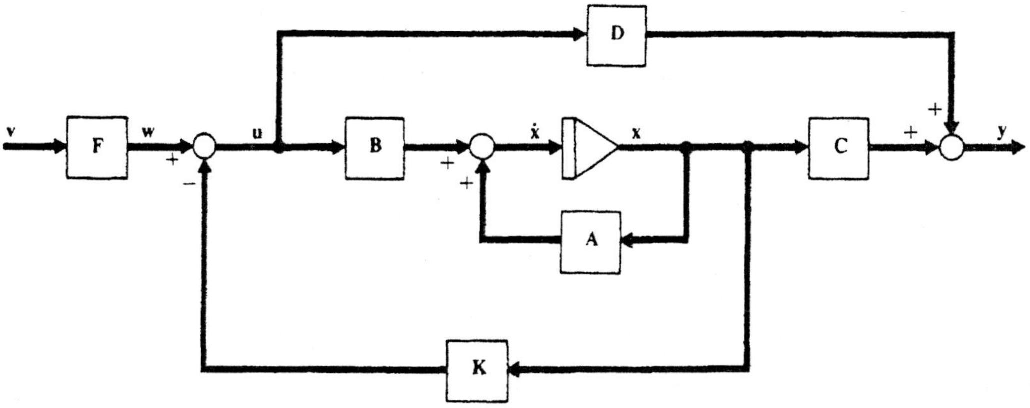

Figure 13.1 State variable feedback system.

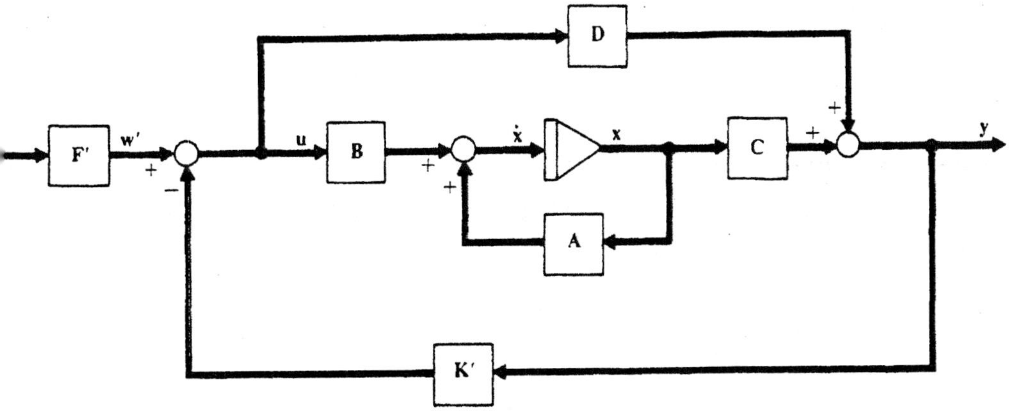

Figure 13.2 Output feedback system.

to the given open-loop system. In spite of this objection, state feedback is considered for the following reasons:

1. The state **x** contains all pertinent information about the system. It is of interest to determine what can be accomplished by using feedback in this ideal limiting case.
2. There are instances for which the state variables are all measurable, i.e., outputs. This will be the case if $\mathbf{C} = \mathbf{I}_n$ and $\mathbf{D} = [\mathbf{0}]$.
3. Several optimal control laws (see Chapter 14) take the form of a state feedback control. Anticipating this result, it is worthwhile to have an understanding of the effects of state feedback.
4. There are effective means available for estimating or reconstructing the state variables from the available inputs and outputs (see Sec. 13.6).

The equations which describe the state feedback problem are Eq. (*13.1*) or (*13.2*) plus the relation

$$\mathbf{u}(t) = \mathbf{F}\mathbf{v}(t) - \mathbf{K}\mathbf{x}(t) \quad \text{or} \quad \mathbf{u}(k) = \mathbf{F}\mathbf{v}(k) - \mathbf{K}\mathbf{x}(k) \tag{13.3}$$

Combining gives

$$\dot{\mathbf{x}} = [\mathbf{A} - \mathbf{B}\mathbf{K}]\mathbf{x} + [\mathbf{B}\mathbf{F}]\mathbf{v} \quad \text{or} \quad \mathbf{x}(k+1) = [\mathbf{A} - \mathbf{B}\mathbf{K}]\mathbf{x}(k) + [\mathbf{B}\mathbf{F}]\mathbf{v}(k) \tag{13.4}$$

and

$$\mathbf{y} = [\mathbf{C} - \mathbf{D}\mathbf{K}]\mathbf{x} + [\mathbf{D}\mathbf{F}]\mathbf{v} \quad \text{or} \quad \mathbf{y}(k) = [\mathbf{C} - \mathbf{D}\mathbf{K}]\mathbf{x}(k) + [\mathbf{D}\mathbf{F}]\mathbf{v}(k) \tag{13.5}$$

Equations (*13.4*) and (*13.5*) are of the same form as Eqs. (*13.1*) and (*13.2*). Considering $\{\mathbf{A}, \mathbf{B}, \mathbf{C}, \mathbf{D}\}$ as fixed system elements, the question is, "What changes in overall system characteristics can be achieved by choice of **K** and **F**?" *Stability* of the state feedback system depends on the eigenvalues of $[\mathbf{A} - \mathbf{B}\mathbf{K}]$. *Controllability* depends on the pair $\{[\mathbf{A} - \mathbf{B}\mathbf{K}], \mathbf{B}\mathbf{F}\}$. *Observability* depends on the pair $\{[\mathbf{A} - \mathbf{B}\mathbf{K}], [\mathbf{C} - \mathbf{D}\mathbf{K}]\}$. The effect of feedback on these properties is investigated in Sec. 13.3.

The output feedback system is described by Eq. (*13.1*) or (*13.2*) plus $\mathbf{u}(t) =$

$\mathbf{F'v' - K'y}$. Therefore, for continuous-time systems

$$\mathbf{y}(t) = \mathbf{Cx} + \mathbf{DF'v'} - \mathbf{DK'y}$$

or

$$\mathbf{y}(t) = [\mathbf{I}_m + \mathbf{DK'}]^{-1}\{\mathbf{Cx} + \mathbf{DF'v'}\} \tag{13.6}$$

Using this gives $\dot{\mathbf{x}} = \{\mathbf{A} - \mathbf{BK'}[\mathbf{I}_m + \mathbf{DK'}]^{-1}\mathbf{C}\}\mathbf{x} + \mathbf{B}\{\mathbf{F'} - \mathbf{K'}[\mathbf{I}_m + \mathbf{DK'}]^{-1}\mathbf{DF'}\}\mathbf{v'}$. The matrix inversion identity $\mathbf{I}_r - \mathbf{K'}[\mathbf{I}_m + \mathbf{DK'}]^{-1}\mathbf{D} \equiv [\mathbf{I}_r + \mathbf{K'D}]^{-1}$ simplifies this to

$$\dot{\mathbf{x}} = \{\mathbf{A} - \mathbf{BK'}[\mathbf{I}_m + \mathbf{DK'}]^{-1}\mathbf{C}\}\mathbf{x} + \mathbf{B}[\mathbf{I}_r + \mathbf{K'D}]^{-1}\mathbf{F'v'} \tag{13.7}$$

Again, Eqs. (*13.7*) and (*13.6*) are of the same form as Eq. (*13.1*). The properties of stability, controllability, and observability are now determined by $\{\mathbf{A} - \mathbf{BK'}[\mathbf{I}_m + \mathbf{DK'}]^{-1}\mathbf{C}\}$, $\mathbf{B}[\mathbf{I}_r + \mathbf{K'D}]^{-1}\mathbf{F'}$, and $[\mathbf{I}_m + \mathbf{DK'}]^{-1}\mathbf{C}$. Exactly the same kind of results applies to the discrete-time system.

13.3 THE EFFECT OF FEEDBACK ON SYSTEM PROPERTIES

The closed-loop systems, obtained by using either state feedback or output feedback, are described by four new system matrices. The reason for adding feedback is to improve the system characteristics in some sense. The effect of feedback on the properties of controllability, observability, and stability should be understood.

Controllability

Let $\mathbf{P}$ be the controllability matrix of Chapter 11. The open-loop system has $\mathbf{P} = [\mathbf{B} \mid \mathbf{AB} \mid \mathbf{A}^2\mathbf{B} \mid \cdots \mid \mathbf{A}^{n-1}\mathbf{B}]$. With state feedback the controllability matrix becomes

$$\bar{\mathbf{P}} = [\mathbf{BF} \mid (\mathbf{A} - \mathbf{BK})\mathbf{BF} \mid (\mathbf{A} - \mathbf{BK})^2\mathbf{BF} \mid \cdots \mid (\mathbf{A} - \mathbf{BK})^{n-1}\mathbf{BF}]$$

If the feed-forward matrix $\mathbf{F}$ satisfies rank$(\mathbf{F}) = r$, then $\mathbf{F}$ will not affect the rank of $\bar{\mathbf{P}}$. Physically, this means that there are as many independent input components $\mathbf{v}$ after adding feedback as there were in the input $\mathbf{u}$ without feedback. Assuming this is true, and letting $\mathbf{F} = \mathbf{I}_r$ for convenience, a series of elementary column operations can be used to reduce $\bar{\mathbf{P}}$ to $\mathbf{P}$. For example, the columns of $\mathbf{BKB}$ are linear combinations of the columns of $\mathbf{B}$. Elementary operations can reduce these, as well as all other extra terms in $\bar{\mathbf{P}}$, to $\mathbf{0}$. Therefore, rank $(\bar{\mathbf{P}}) = $ rank$(\mathbf{P})$ for any gain matrix $\mathbf{K}$. Thus state feedback does not alter the controllability of the open-loop system.

Since $\bar{\mathbf{P}}$ and $\mathbf{P}$ have the same rank for *any* $\mathbf{K}$, including the special case $\mathbf{K} = \mathbf{K'}[\mathbf{I}_m + \mathbf{DK'}]^{-1}\mathbf{C}$, system controllability is also unaltered when output feedback is used. This assumes that $\mathbf{I}_r + \mathbf{K'D}$ is nonsingular and rank $(\mathbf{F'}) = r$.

Observability

The open-loop observability matrix is $\mathbf{Q} = [\overline{\mathbf{C}}^T \mid \overline{\mathbf{A}}^T\overline{\mathbf{C}}^T \mid \cdots \mid (\overline{\mathbf{A}}^{n-1})^T\overline{\mathbf{C}}^T]$. From Eq. (*13.5*), it is obvious that when state feedback is used, observability is lost if $\mathbf{C} = \mathbf{DK}$. This set of simultaneous linear equations has a solution $\mathbf{K}$ if rank $[\mathbf{D}] = $ rank $[\mathbf{D} \mid \mathbf{C}]$.

This is just one illustration of the general result: state feedback *can* cause a loss of observability. When output feedback is used, observability of the open- and closed-loop systems is the same, as shown next.

Assume, without loss of generality, that $\mathbf{D} = [\mathbf{0}]$. Then the output feedback observability matrix is

$$\tilde{\mathbf{Q}} = [\overline{\mathbf{C}}^T \mid (\overline{\mathbf{A}} - \overline{\mathbf{BK'C}})^T \overline{\mathbf{C}}^T \mid \cdots \mid (\overline{\mathbf{A}} - \overline{\mathbf{BK'C}})^{n-1T} \overline{\mathbf{C}}^T]$$

A series of elementary column operations, precisely like those used on $\tilde{\mathbf{P}}$, can be used to reduce $\tilde{\mathbf{Q}}$ to $\mathbf{Q}$. This proves that $\text{rank}(\tilde{\mathbf{Q}}) = \text{rank}(\mathbf{Q})$. Observability is preserved when output feedback is used, regardless of $\mathbf{K}'$. The reason why system observability is invariant for output feedback and not for state feedback is the presence of the matrix $\mathbf{C}$ in the $\mathbf{A} - \mathbf{BK'C}$ terms. That is, columns of $\overline{\mathbf{C}}^T \overline{\mathbf{K}}'^T \overline{\mathbf{B}}^T \overline{\mathbf{C}}^T$ can be shown to be linearly related to columns of $\overline{\mathbf{C}}^T$, whereas columns of $\overline{\mathbf{K}}^T \overline{\mathbf{B}}^T \overline{\mathbf{C}}^T$ need not be.

Stability

Stability of linear, constant systems depends entirely on the location of the eigenvalues in the complex plane. Both state and output feedback yield closed-loop eigenvalues which differ from the open-loop eigenvalues. This means that the stability of the closed-loop system is not necessarily the same as that of the open-loop system. The degree of freedom we have in specifying closed-loop eigenvalue locations by choice of $\mathbf{K}$ or $\mathbf{K}'$ is the crux of the pole-assignment problem. This topic is the central theme of this chapter.

Poles and Zeros

Poles and zeros are primarily transfer function concepts. The poles are the values of the complex variable s (or z in discrete-time cases) for which one or more elements of the transfer function matrix have unbounded magnitude. In the single-input, single-output case, the transfer function is a scalar function, a ratio of polynomials. Those values of s (or z) for which the denominator is zero are poles. In the case of matrix transfer functions, nothing surprising happens to the definition of poles. This is because if the common denominator vanishes, all terms in the transfer function matrix increase without bound unless the offending denominator root happens to cancel. From Chapter 12 it is seen that the transfer function denominator is the characteristic determinant. Thus the poles are the same as the system eigenvalues except in the unusual case of cancellation. Actually, the poles will always be eigenvalues, but all eigenvalues may not appear as poles because of cancellation. Loosely speaking, the poles and the eigenvalues are the same. This chapter is predominantly devoted to methods of designing feedback systems which give "good" closed loop poles.

In the single-input, single-output case, zeros are those values of s (or z) for which the numerator of the transfer function is zero. With such a value of s (or z), nonzero inputs u will cause zero output from the system. When multiple-input, multiple-output systems are considered, the generalization of the concept of zero is somewhat more involved. The easiest case to generalize is the one where the numbers of inputs and outputs are equal. The transfer function matrix is thus square. The question of transmission zeros reduces to the question of whether a nonzero input vector $\mathbf{u}$ can cause a zero output $\mathbf{y}$. Since $\mathbf{y}(s) = \mathbf{H}(s)\mathbf{u}(s)$, it is known from Chapters 4 and 6 that nonzero inputs $\mathbf{u}(s)$ will allow for zero outputs if $\mathbf{H}(s)$ is singular. The zeros in this square transfer function matrix case are the values of s (or z) for which the determinant of the transfer function is zero. This definition leads to situations where a transfer function matrix has *no* finite zeros even though the numerators of the individual matrix elements are functions of s and can thus go to zero individually. Individual terms vanishing is not sufficient for causing the total output vector $\mathbf{y}$ to vanish.

It is shown in Problem 13.20 that state feedback does not alter the transfer function zeros. If s_1 is a zero of the open-loop system, it is still a zero of the closed-loop system.

For nonsquare transfer functions, zeros are defined as those particular values of s or z which cause the rank of the transfer function matrix to drop below its usual value. Further discussion and alternate definitions may be found in the references.

13.4 POLE ASSIGNMENT USING STATE FEEDBACK [1]

Equation (*13.4*) indicates that the eigenvalues of the closed-loop state feedback system are roots of

$$\Delta'(\lambda) \triangleq |\lambda\mathbf{I} - \mathbf{A} + \mathbf{BK}| = 0 \qquad (13.8)$$

It has been proven [2, 3] that if (and only if) the open-loop system $(\mathbf{A}, \mathbf{B})$ is completely controllable, then any set of desired closed-loop eigenvalues $\Gamma = \{\lambda_1, \lambda_2, \ldots, \lambda_n\}$ can be achieved using a constant state feedback matrix $\mathbf{K}$. In order to synthesize the system with real hardware, all elements of $\mathbf{K}$ must be real. This will be the case if, for each complex $\lambda_i \in \Gamma$, $\bar{\lambda}_i$ is also assigned to Γ.

One direct and simple method of finding the values of the unknown gain matrix $\mathbf{K}$ which gives specified eigenvalues is to expand Eq. (*13.8*) and equate it to the desired characteristic polynomial. Then equating like powers of λ gives expressions or, in simple cases, the values directly for the elements of $\mathbf{K}$.

EXAMPLE 13.1 Let $\mathbf{A} = \begin{bmatrix} 0 & 2 \\ 0 & 3 \end{bmatrix}$, $\mathbf{B} = \begin{bmatrix} 0 \\ 1 \end{bmatrix}$, and assume that closed-loop eigenvalues $\lambda = -3$ and -4 are desired. This means that the desired characteristic polynomial is

$$(\lambda + 3)(\lambda + 4) = \lambda^2 + 7\lambda + 12$$

Expanding the determinant yields

$$\begin{vmatrix} \lambda & -2 \\ K_1 & \lambda - 3 + K_2 \end{vmatrix} = \lambda^2 + (K_2 - 3)\lambda + 2K_1$$

Equating the constant terms gives $2K_1 = 12$. Equating the first-order terms gives $K_2 - 3 = 7$, so that $\mathbf{K} = [6 \quad 10]$. ■

The explicit method just demonstrated works well for low-order problems. It generalizes easily to single-input, nth-order systems whose state equations are in the controllable canonical form, introduced in Chapter 3:

$$\dot{\mathbf{x}} = \begin{bmatrix} 0 & 1 & 0 & \cdots & 0 \\ 0 & 0 & 1 & \cdots & 0 \\ \vdots & & & & \\ -a_0 & -a_1 & -a_2 & \cdots & -a_{n-1} \end{bmatrix} \mathbf{x} + \begin{bmatrix} 0 \\ 0 \\ \vdots \\ 1 \end{bmatrix} u$$

For this special form of the matrices **A** and **B**, the determinant of Eq. (*13.8*) becomes

$$\begin{vmatrix} \lambda & -1 & 0 & 0 & \cdots & 0 \\ 0 & \lambda & -1 & 0 & \cdots & 0 \\ 0 & 0 & \lambda & -1 & \cdots & 0 \\ \vdots & & & & & -1 \\ K_1 + a_0 & K_2 + a_1 & K_3 + a_2 & \cdots & & \{\lambda + K_n + a_{n-1}\} \end{vmatrix}$$

$$= \lambda^n + (K_n + a_{n-1})\lambda^{n-1} + \cdots + (K_2 + a_1)\lambda + (K_1 + a_0)$$

Equating like powers of this expression and the desired characteristic polynomial gives one equation for each unknown gain component, namely,

$$K_i = c_{i-1} - a_{i-1} \tag{13.9}$$

where c_i is the coefficient of λ^i in the desired closed-loop characteristic polynomial. This expanded form makes it obvious that the more the closed-loop polynomial differs from the open-loop polynomial, the larger are the required feedback gains. Note that this procedure applies to the controllable canonical form. The system of Example 13.1 is *not* in this canonical form, so Eq. (*13.9*) cannot be applied directly. However, for any single-input controllable system there exists a nonsingular transformation **T**, which maps the state vector **x** into a new state vector **x**′ in such a way as to give the **x**′ state equations in controllable canonical form. For multiple inputs the simple notion of the canonical forms become more complicated. There is not universal agreement on what the controllable canonical form is in the multiple input case. In this paragraph we consider only the single-input case. Let $\mathbf{x}' = \mathbf{T}\mathbf{x}$, and then the corresponding matrices $\mathbf{A}'$ and $\mathbf{B}'$ for these new states are $\mathbf{A}' = \mathbf{T}\mathbf{A}\mathbf{T}^{-1}$ and $\mathbf{B}' = \mathbf{T}\mathbf{B}$. Equation (*13.9*) can be used to find the feedback gains $\mathbf{K}'$ that are appropriate for the new state vector. Then equating the feedback signals $\mathbf{K}'\mathbf{x}' = \mathbf{K}\mathbf{x}$ and using $\mathbf{x}' = \mathbf{T}\mathbf{x}$ gives $\mathbf{K} = \mathbf{K}'\mathbf{T}$.

EXAMPLE 13.2 The system of Example 13.1 is transformed to controllable canonical form by using $\mathbf{T} = \begin{bmatrix} \frac{1}{2} & 0 \\ 0 & 1 \end{bmatrix}$, giving $\mathbf{A}' = \begin{bmatrix} 0 & 1 \\ 0 & 3 \end{bmatrix}$ and $\mathbf{B}' = \begin{bmatrix} 0 \\ 1 \end{bmatrix}$. From the last row of $\mathbf{A}'$, $a_0 = 0$ and $a_1 = -3$. From Eq. (*13.9*), $\mathbf{K}_1' = 12$ and $\mathbf{K}_2' = 7 + 3 = 10$. The gain **K** to be used with the original state variables is, therefore, $\mathbf{K} = \mathbf{K}'\mathbf{T} = [6 \quad 10]$. This is the same unique answer found in Example 13.1. ■

When there are multiple inputs or when the state equations are not in controllable canonical form, the preceding simple methods may not provide the best approach to the determination of feedback gain matrices. An approach which works for any order, any number of inputs, and any arbitrary form of the state equations is now developed. The problem is to determine a gain matrix K such that Eq. (*13.8*) is satisfied for each of n specified values $\lambda_i \in \Gamma$.

If Eq. (*13.8*) is true, then there exists at least one nonzero vector $\boldsymbol{\psi}_i$ such that

$$(\lambda_i \mathbf{I} - \mathbf{A} + \mathbf{BK})\boldsymbol{\psi}_i = \mathbf{0} \tag{13.10}$$

Rearranged, this says that

$$(\mathbf{A} - \mathbf{BK})\boldsymbol{\psi}_i = \lambda_i \boldsymbol{\psi}_i \tag{13.11}$$

This makes it clear that $\boldsymbol{\psi}_i$ is an eigenvector of the closed-loop system matrix $(\mathbf{A} - \mathbf{BK})$ associated with the closed-loop eigenvalue (pole) λ_i. Rewriting Eq. (*13.10*) in yet another way gives

$$(\lambda_i \mathbf{I} - \mathbf{A})\boldsymbol{\psi}_i = -\mathbf{BK}\boldsymbol{\psi}_i$$

or

$$[(\lambda_i \mathbf{I} - \mathbf{A}) \mid \mathbf{B}]\begin{bmatrix} \boldsymbol{\psi}_i \\ \mathbf{K}\boldsymbol{\psi}_i \end{bmatrix} = [0] \tag{13.12}$$

At this point the vector $\boldsymbol{\psi}_i$ and the matrix $\mathbf{K}$ are both unknown. Therefore, nothing is lost, and notational simplicity is gained by defining the $(n + r) \times 1$ unknown vector as

$$\boldsymbol{\xi}_i = \begin{bmatrix} \boldsymbol{\psi}_i \\ \hline \mathbf{K}\boldsymbol{\psi}_i \end{bmatrix} \tag{13.13}$$

The determination of $\mathbf{K}$ consists of two general steps. First, a sufficient number of solution vectors $\boldsymbol{\xi}_i$ is found. Then the internal structure among the components of these vectors, as expressed by Eq. (*13.13*), is used to find $\mathbf{K}$.

If the open-loop system described by the pair $\{\mathbf{A}, \mathbf{B}\}$ is controllable, it is known from Sec. 11.5 that the $n \times (n + r)$ coefficient matrix in the homogeneous equation (*13.12*) has full rank n for any value of λ_i. The solution of homogeneous equations has been discussed in Chapters 6 and 7. From this work it is known that there will thus be r independent solution vectors $\boldsymbol{\xi}_i$ for each λ_i.

EXAMPLE 13.3 Let the system of Eq. (*13.1*) have $\mathbf{A} = \begin{bmatrix} 0 & 2 \\ 0 & 3 \end{bmatrix}$, $\mathbf{B} = \begin{bmatrix} 0 \\ 1 \end{bmatrix}$. Find the solution vectors $\boldsymbol{\xi}_i$ for $\lambda_i = -3$ and $\lambda_i = -4$.

The controllability of this system is easily verified. Since $r = 1$, in this case there will be only one independent $\boldsymbol{\xi}_i$ for each λ_i. With $\lambda_1 = -3$, Eq. (*13.12*) gives

$$\begin{bmatrix} -3 & -2 & 0 \\ 0 & -6 & 1 \end{bmatrix}\boldsymbol{\xi}_1 = \mathbf{0}$$

Arbitrarily selecting $\xi_2 = 1$ gives $\xi_1 = -\frac{2}{3}$ and $\xi_3 = 6$. Recalling from Eq. (*13.13*) how $\boldsymbol{\xi}_i$ was defined, it becomes obvious that

$$\boldsymbol{\xi}_1 = [-\tfrac{2}{3} \quad 1 \quad 6]^T \quad \text{is equivalent to} \quad \mathbf{K}\begin{bmatrix} -\frac{2}{3} \\ 1 \end{bmatrix} = 6 \tag{13.14}$$

Similarly, with $\lambda_2 = -4$ the homogeneous equation (*13.12*) has a solution

$$\boldsymbol{\xi}_2 = [-\tfrac{1}{2} \quad 1 \quad 7]^T, \quad \text{which implies} \quad \mathbf{K}\begin{bmatrix} -\frac{1}{2} \\ 1 \end{bmatrix} = 7 \tag{13.15}$$

Taken together, these two equations are

$$\mathbf{K}\begin{bmatrix} -\frac{2}{3} & -\frac{1}{2} \\ 1 & 1 \end{bmatrix} = [6 \quad 7]$$

From Chapter 7 recall that eigenvectors for two different eigenvalues must be independent. This can be observed explicitly here. The 2×2 matrix that multiplies $\mathbf{K}$ has independent columns and hence is nonsingular. An inversion thus allows the unknown $\mathbf{K}$ to be found.

$$\mathbf{K} = \frac{[6 \quad 7]\begin{bmatrix} 1 & \frac{1}{2} \\ -1 & -\frac{2}{3} \end{bmatrix}}{(-\frac{1}{6})} = [6 \quad 10]$$

Note that although the ξ_i vectors are not unique, all other solutions for each λ_i will be multiples of those given earlier. The arbitrary nonzero constant multipliers will cancel from Eqs. (*13.14*) and (*13.15*), thus giving a *unique* solution for **K**. This is *not* the case when there is more than one input, i.e., $r > 1$. This issue is explored and the preceding method of finding the feedback matrix **K** is generalized next. ■

Let the maximal set of r linearly independent solution vectors of Eq. (*13.12*) for a given eigenvalue form the columns of an $(n + r) \times r$ matrix $\mathbf{U}(\lambda_i)$. These entire $(n + r)$ component columns constitute a basis for the null space of $[(\lambda_i \mathbf{I} - \mathbf{A} \mid \mathbf{B}]$. From Eq. (*13.13*), the top n components of each column form a closed-loop eigenvector and the remaining bottom r components are that same vector multiplied by an as yet unknown gain matrix **K**. The matrix **U** is partitioned accordingly as

$$\mathbf{U}(\lambda_i) = \begin{bmatrix} \boldsymbol{\psi}_1 & \boldsymbol{\psi}_2 & \cdots & \boldsymbol{\psi}_r \\ \mathbf{f}_1 & \mathbf{f}_2 & \cdots & \mathbf{f}_r \end{bmatrix} = \begin{bmatrix} \boldsymbol{\Psi}(\lambda_i) \\ \hline \mathcal{F}(\lambda_i) \end{bmatrix}$$

where the substitution $\mathbf{f}_i = \mathbf{K}\boldsymbol{\psi}_i$ has been made. Collectively, all these relations for the n eigenvalues can be written

$$\mathbf{K}[\boldsymbol{\Psi}(\lambda_1) \quad \boldsymbol{\Psi}(\lambda_2) \quad \cdots \quad \boldsymbol{\Psi}(\lambda_n)] = [\mathcal{F}(\lambda_1) \quad \mathcal{F}(\lambda_2) \quad \cdots \quad \mathcal{F}(\lambda_n)] \tag{13.16}$$

Equation (*13.16*) cannot be solved directly for **K** because if $r > 1$, it is overdetermined and thus represents inconsistent equations. However, if the system is controllable, a nonsingular $n \times n$ matrix of $\boldsymbol{\psi}_j(\lambda_i)$'s can be found by selecting n linearly independent columns from both sides of Eq. (*13.16*) and deleting the remaining columns. In this selection process, one column must be selected for each specified eigenvalue λ_i. Let the selected n columns from the left-hand side form a matrix **G** and the corresponding columns from the right side form a matrix $\mathcal{J}$. If this is done, the result $\mathbf{KG} = \mathcal{J}$ can then be solved for the feedback gain matrix **K**,

$$\mathbf{K} = \mathcal{J}\mathbf{G}^{-1} \tag{13.17}$$

The freedom in arbitrarily selecting a subset of n columns, subject only to the requirement that one column is chosen for each desired λ_i and that $\mathbf{G}^{-1}$ exists, is what leads to the multiplicity of possible feedback gain matrices in the multiple-input case. Actually, any linear combination of columns can be selected from the r columns found for each λ_i. That is, any vector belonging to the null space can be used.

EXAMPLE 13.4 Consider the same matrix A as in the previous problem; however, now there are two inputs, with $\mathbf{B} = \mathbf{I}_2$. Find a feedback gain matrix which yields closed-loop eigenvalues at $\lambda = -3$ and -5.

With $\lambda_1 = -3$, Eq. (*13.12*) gives

$$\begin{bmatrix} -3 & -2 & 1 & 0 \\ 0 & -6 & 0 & 1 \end{bmatrix} \xi = \begin{bmatrix} 0 \\ 0 \end{bmatrix}$$

Letting $\xi_1 = \alpha$ and $\xi_2 = \beta$ leads to a solution $\xi_1 = [\alpha \quad \beta \quad 3\alpha + 2\beta \quad 6\beta]^T$. It is possible to find two and only two independent vectors by specifying various values for α and β. What has been found so far can be written as

$$\mathbf{K}\begin{bmatrix} \alpha & \cdot \\ \beta & \end{bmatrix} = \begin{bmatrix} 3\alpha + 2\beta & \cdot \\ 6\beta & \end{bmatrix}$$

where the extra space has been left intentionally as a reminder that another equation will be found and filled in for $\lambda_2 = -5$. With $\lambda_2 = -5$, the same procedure gives

$$\begin{bmatrix} -5 & -2 & 1 & 0 \\ 0 & -8 & 0 & 1 \end{bmatrix} \xi = \mathbf{0}$$

Letting $\xi_1 = \gamma$ and $\xi_2 = \delta$ gives $\xi_2 = [\gamma \quad \delta \quad 5\gamma + 2\delta \quad 8\delta]^T$, which means that

$$\mathbf{K}\begin{bmatrix} \cdot & \gamma \\ & \delta \end{bmatrix} = \begin{bmatrix} \cdot & 5\gamma + 2\delta \\ & 8\delta \end{bmatrix}$$

Taken together the equations for determining $\mathbf{K}$ are

$$\mathbf{K}\begin{bmatrix} \alpha & \gamma \\ \beta & \delta \end{bmatrix} = \begin{bmatrix} 3\alpha + 2\beta & 5\gamma + 2\delta \\ 6\beta & 8\delta \end{bmatrix} \tag{13.18}$$

Any values of α, β, γ, and δ will give a valid gain matrix as long as the required inverse exists. The preceding degree of generality is not normally required and will be dispensed with in most of the examples to follow. To simplify calculations but also to give an orthogonal set of closed-loop eigenvectors, we select $\alpha = \delta = 1$ and $\beta = \gamma = 0$. Then, by inspection, the solution of Eq. (*13.18*) is $\mathbf{K} = \begin{bmatrix} 3 & 2 \\ 0 & 8 \end{bmatrix}$. ■

Since multiple-input systems allow an infinite number of choices for the feedback gain matrix, the designer must make choices. This freedom of choice can be exercised in various ways, which are discussed throughout this and the next chapter. One related factor can be simply stated now. Knowing that ψ_i are closed-loop eigenvectors may help guide the choice of which columns to retain. Actually, any linear combination of columns from a given partition of Ψ can be used as long as the same linear combination of $\mathbf{f}_i$ columns is used. It is often worthwhile to try to make $\mathbf{G}$ be as nearly orthogonal as possible because this improves its invertibility robustness and tends to give less interaction between modes of the closed-loop system. This orthogonality usually improves system sensitivity to parameter variations.

One final complication must be settled before a completely general method of pole assignment can be claimed. Suppose it is desired that λ_i be a pth-order root of the characteristic equation. If r, the dimension of the null space or, equivalently, the rank of Ψ, is greater than or equal to p, then the preceding procedure will suffice. If $r < p$, then it is not possible to specify p eigenvectors. It is still possible to achieve the desired p eigenvalues, however. The excess vectors for such a repeated eigenvalue will be generalized eigenvectors, as demonstrated next. Let ψ_i be an eigenvector which has already been found using this procedure. Let ψ_g denote an associated generalized eigenvector. Recall from Chapter 7 that it satisfies

$$(\mathbf{A} - \mathbf{BK})\psi_g = \lambda\psi_g + \psi_i$$

This can be rearranged to read

$$(\lambda\mathbf{I} - \mathbf{A} + \mathbf{BK})\psi_g = -\psi_i$$

or

$$[(\lambda\mathbf{I} - \mathbf{A}) \mid \mathbf{B}]\begin{bmatrix} \psi_g \\ \mathbf{K}\psi_g \end{bmatrix} = -\psi_i \tag{13.19}$$

Equation (*13.19*) replaces Eq. (*13.12*) for the first required generalized eigenvector. If more than one generalized eigenvector is required, the second and subsequent choices may satisfy Eq. (*13.19*) but with a different eigenvector on the right-hand side, *or* the additional generalized eigenvectors may be chained to the previous ones. Which alternative prevails is dictated only by the requirement that a full set of *independent* vectors be found so that the matrix **G** is invertible. (See Illustrative Problems 13.4 and 13.5.) Generalized eigenvectors are brought into the pole-placement discussion solely because we know that they provide a needed set of independent vectors. As was shown in Chapter 7, there are several ways of finding generalized eigenvectors.

EXAMPLE 13.5 Consider the system of Example 13.1. Find a feedback gain matrix that gives both closed-loop eigenvalues $\lambda = -1$. In this case Eq. (*13.12*) specializes to

$$\begin{bmatrix} -1 & -2 & 0 \\ 0 & -4 & 1 \end{bmatrix}\xi = \mathbf{0}$$

All nontrivial solutions are proportional to

$$\xi = [-2 \quad 1 \quad 4]^T$$

Obviously only one eigenvector $\psi = [-2 \quad 1]$ can be selected from this one-dimensional space, so a generalized eigenvector is needed. It is found from Eq. (*13.19*), that is,

$$\begin{bmatrix} -1 & -2 & 0 \\ 0 & -4 & 1 \end{bmatrix}\xi_g = \begin{bmatrix} 2 \\ -1 \end{bmatrix}$$

Selecting $\xi_2 = 1$ gives $\xi_1 = -4$ and $\xi_3 = 3$, so that a solution is $\xi_g = [-4 \quad 1 \quad 3]^T$. Therefore, the two equations for finding **K** are

$$\mathbf{K}\begin{bmatrix} 2 & -4 \\ -1 & 1 \end{bmatrix} = [4 \quad 3]$$

Matrix inversion or its equivalent gives $\mathbf{K} = [0.5 \quad 5]$. ■

Summary of the Pole Placement Algorithm (Eigenvalue-Eigenvector Assignment)

Given **A**, **B**, and the desired set of eigenvalues, carry out the following steps.

I. For each λ_i:

1. Form $[(\lambda_i\mathbf{I} - \mathbf{A} \mid \mathbf{B}]$.

2. Find the null space basis set U by finding all independent solutions of Eq. (*13.12*).
3. Partition U, using the top n rows as the $n \times r$ matrix $\boldsymbol{\Psi}(\lambda_i)$.
4. Use the remaining r rows of U as the $r \times r$ matrix $\mathcal{F}(\lambda_i)$.

II. Form the composite matrices $\boldsymbol{\Omega} = [\boldsymbol{\Psi}(\lambda_1) \quad \boldsymbol{\Psi}(\lambda_2) \quad \cdots \quad \boldsymbol{\Psi}(\lambda_n)]$; $n \times nr$, rank n if controllable and $\boldsymbol{\Lambda} = [\mathcal{F}(\lambda_1) \quad \mathcal{F}(\lambda_2) \quad \cdots \quad \mathcal{F}(\lambda_n)]$; $r \times nr$. Equation (*13.16*) can now be compactly written as $\mathbf{K}\boldsymbol{\Omega} = \boldsymbol{\Lambda}$.

III. Select n linearly independent columns of $\boldsymbol{\Omega}$ to form the $n \times n$ matrix **G**. One column (or any linear combinations of columns) must be selected from each $\boldsymbol{\Psi}(\lambda_i)$ partition. The selected columns will be closed-loop eigenvectors. As a preliminary screening out of linearly dependent columns, the inner products should be checked. This can be done by forming $\boldsymbol{\Omega}^T\boldsymbol{\Omega}$. The i, j element, normalized by the square root of the i, i and j, j elements, is the cosine of the angle between columns i and j. If this generalized cosine is 1 for any pair of selected columns, the columns are linearly dependent and the selection must be modified accordingly.

IV. Use the same column numbers selected in step III to form the $r \times n$ matrix $\mathcal{I}$ from $\boldsymbol{\Lambda}$.

V. Solve $\mathbf{KG} = \mathcal{I}$ for the $r \times n$ gain matrix **K**. It may be more convenient with some software packages to solve $\mathbf{G}^T\mathbf{K}^T = \mathcal{I}^T$ instead. Note that passing the pairwise inner product test in step III is necessary but not sufficient for the existence of $\mathbf{G}^{-1}$. Therefore, it may be necessary to return to step III and modify the eigenvector column selections. Also note that two columns being nearly collinear will lead to a nearly singular matrix to be inverted. This situation can be expected to give a poor, or non-robust, solution. It can be avoided by selecting columns which are as nearly orthogonal as possible.

VI. General Comments:

1. The process of selecting orthogonal columns could be automated, perhaps by using a modified Gram-Schmidt orthogonalization process or a singular value decomposition. However, allowing the designer the discretion of interactively making the selection allows judgments to be made regarding the desirability of certain closed-loop eigenvectors.
2. If a certain eigenvalue is specified to be a pth-order repeated root, with $p > r$, then $p - r$ generalized eigenvectors will need to be found, using Eq. (*13.19*) in place of Eq. (*13.12*) in step I.2.
3. It is never necessary to invert a complex matrix in Eq. (*13.17*), even when complex-valued eigenvalues and eigenvectors are selected. The result in Problem 4.23 allows the use of purely real arithmetic, as demonstrated in Problem 13.3.
4. The single-input problem ($r = 1$) can be solved by simpler methods that deal directly with the coefficients of the characteristic polynomial. A program which uses the approach of Example 13.2 is STVARFDBK [4].

EXAMPLE 13.6 A model [5] of the lateral dynamics of an F-8 aircraft at a particular set of flight condition was given in Problem 11.9. A discrete approximation for this system is obtained using the method of Problem 9.10, with $T = 0.2$ and retaining terms through thirtieth order.

The Approximate Discrete Transition Matrix

$$\begin{bmatrix} 1.3533533E-01 & 9.8391518E-02 & -7.2121400E-01 & 0.0000000E+00 \\ 0.0000000E+00 & 7.1751231E-01 & 1.4726298E+00 & 0.0000000E+00 \\ 0.0000000E+00 & -1.6362552E-01 & 7.1751231E-01 & 0.0000000E+00 \\ 8.6466469E-02 & 7.8584068E-03 & -1.0389241E-01 & 1.0000000E+00 \end{bmatrix}$$

The Approximate Input Matrix **B**

$$\begin{bmatrix} 1.7293293E+00 & 2.1750930E-01 \\ 0.0000000E+00 & -5.5092329E-01 \\ 0.0000000E+00 & 5.5393357E-02 \\ 2.2706707E-01 & 3.0423844E-02 \end{bmatrix}$$

Design a state feedback controller which will provide closed-loop Z-plane poles corresponding to s-plane poles at $s = -2, -5, -8$, and -10. Using $z = e^{Ts}$, this means that the desired Z-plane poles are

$$\lambda_i = 0.6703, 0.3679, 0.2019, \text{ and } 0.1353$$

Solving Eq. (*13.12*) for each λ_i in turn gives

$$U(\lambda_1) = \begin{bmatrix} -1.0000000E+00 & 0.0000000E+00 \\ 0.0000000E+00 & 4.4247019E-01 \\ 0.0000000E+00 & 3.5991812E-01 \\ 4.7545883E-01 & 1.0787997E-02 \\ \hdashline 3.0937374E-01 & 8.4780902E-04 \\ 0.0000000E-00 & -1.0000000E+00 \end{bmatrix}$$

$$U(\lambda_2) = \begin{bmatrix} -1.0000000E+00 & 0.0000000E+00 \\ 0.0000000E+00 & 7.5502163E-01 \\ 0.0000000E+00 & 1.9485569E-01 \\ 1.8517032E-01 & 5.8909208E-03 \\ \hdashline 1.3450530E-01 & 8.7471344E-02 \\ 0.0000000E+00 & -1.0000000E+00 \end{bmatrix}$$

$$U(\lambda_3) = \begin{bmatrix} -1.0000000E+00 & 0.0000000E+00 \\ 0.0000000E+00 & 7.2149646E-01 \\ 0.0000000E+00 & 1.2148339E-01 \\ 1.1934122E-01 & 3.6517035E-03 \\ \hdashline 3.8512699E-02 & 1.1616344E-01 \\ 0.0000000E+00 & -1.0000000E+00 \end{bmatrix}$$

$$U(\lambda_4) = \begin{bmatrix} -1.0000000E+00 & 0.0000000E+00 \\ 0.0000000E+00 & 6.9379878E-01 \\ 0.0000000E+00 & 9.9803299E-02 \\ 1.0003469E-01 & 2.9885261E-03 \\ \hdashline 0.0000000E+00 & 1.2362902E-01 \\ 0.0000000E+00 & -1.0000000E+00 \end{bmatrix}$$

The portions above the partition lines are $\Psi(\lambda_i)$ and the lower portions are $\mathcal{F}(\lambda_i)$. In order to find a suitable gain, some combination of columns (1 or 2) and (3 or 4) and (5 or 6) and (7 or 8) must be selected. For example, if columns 1, 4, 6, and 7 are used, Eqs. (*13.16*) and (*13.18*) give

The Feedback Gain Matrix

$$\mathbf{K} = \begin{bmatrix} 8.2435042\text{E}-02 & 2.4582298\text{E}-01 & -5.2851838\text{E}-01 & 8.2406455\text{E}-01 \\ 0.0000000\text{E}+00 & -1.5015275\text{E}+00 & 6.8607545\text{E}-01 & 0.0000000\text{E}+00 \end{bmatrix}$$

As a check of the result, the closed-loop system matrix $\mathbf{A}-\mathbf{BK}$ is formed. The eigenvalues of this closed-loop system matrix are given below, and verify that the desired pole locations have been achieved.

The resulting closed-loop eigenvalues are

Real Part	Imaginary Part
1.3529983E − 01	0.0000000E + 00
2.0190017E − 01	0.0000000E + 00
3.6790001E − 01	0.0000000E + 00
6.7029989E − 01	0.0000000E + 00

If columns 2, 4, 5, and 7 are selected, the following alternative gain matrix is obtained. Notice that the two sets of results have totally different **K** matrices and closed-loop eigenvectors, although the closed-loop eigenvalues will be the same.

The Feedback Gain Matrix

$$\mathbf{K} = \begin{bmatrix} 1.9954935\text{E}-01 & 1.6860582\text{E}-01 & -2.6471341\text{E}-01 & 1.9948014\text{E}+00 \\ 0.0000000\text{E}+00 & -8.8968951\text{E}-01 & -1.6846579\text{E}+00 & 0.0000000\text{E}+00 \end{bmatrix}$$ ■

Before concluding the discussion of state variable feedback, assume that the gain matrix $\mathbf{K}$ is factored into $\mathbf{K} = \mathbf{F}_1\mathbf{K}_1$, and that $\mathbf{K}_1$ is left in the feedback path and $\mathbf{F}_1$ is placed in the forward path, as shown in Figure 13.3. There is no change in the characteristic equation and hence the poles. The numerator of the closed-loop transfer function can differ greatly, however. If $\mathbf{F}_1$ is just a scalar, then only the gains of the transfer function elements change. If $\mathbf{F}_1$ is a full $r \times r$ matrix, the transfer function can take on a greatly modified appearance. Even though the functions of s (or z) that appear in the numerator elements may look totally different, the zeros of the transfer function remain unchanged. As in Sec. 13.3, it is assumed here that $\mathbf{H}(s)$ or $\mathbf{H}(z)$ is square, so that zeros are easily defined as the values of s or z that give a zero determinant of $\mathbf{H}$.

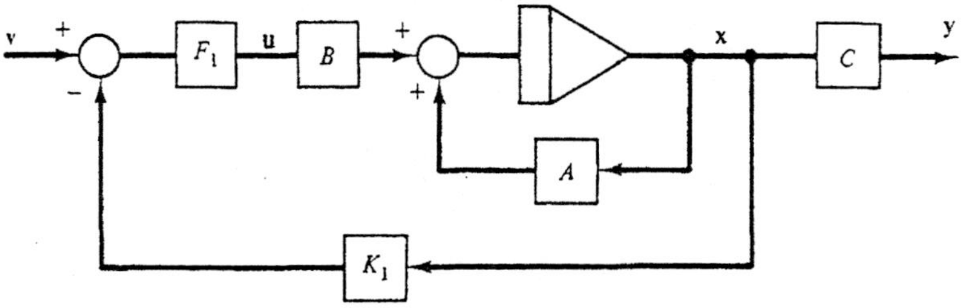

Figure 13.3

EXAMPLE 13.7 Show that if

$$\mathbf{K} = \begin{bmatrix} 2 & 0 \\ -4.5 & 9 \end{bmatrix}$$

is used in a state feedback controller for the second-order system of Example 12.6, the closed-loop poles are at -5 and -10. Then consider two optional implementations,

$$\mathbf{K} = \begin{bmatrix} 2 & 0 \\ 1 & 1 \end{bmatrix}\begin{bmatrix} 1 & 0 \\ -5.5 & 9 \end{bmatrix} = \begin{bmatrix} 1 & 1 \\ -1 & 1 \end{bmatrix}\begin{bmatrix} 3.25 & -4.5 \\ -1.25 & 4.5 \end{bmatrix}$$

and find the closed-loop transfer functions.

If the control law is $\mathbf{u} = -\mathbf{Kx} + \mathbf{F}_1\mathbf{v}$, the expression for the closed-loop transfer function is

$$\mathbf{H}(s) = \mathbf{C}[s\mathbf{I} - \mathbf{A} + \mathbf{BK}]^{-1}\mathbf{BF}_1 \quad \text{and} \quad \mathbf{A} - \mathbf{BK} = \begin{bmatrix} -5 & 0 \\ 2.5 & -10 \end{bmatrix}$$

For the three values of $\mathbf{F}_1$, namely $\mathbf{I}$, $\begin{bmatrix} 2 & 0 \\ 1 & 1 \end{bmatrix}$, and $\begin{bmatrix} 1 & 1 \\ -1 & 1 \end{bmatrix}$, the resultant transfer functions are

$$\mathbf{H}(s) = \begin{bmatrix} \dfrac{1}{s+5} & \dfrac{1}{s+10} \\ \dfrac{2}{s+5} & 0 \end{bmatrix}, \quad \begin{bmatrix} \dfrac{3(s+\frac{25}{3})}{(s+5)(s+10)} & \dfrac{1}{s+10} \\ \dfrac{4}{s+5} & 0 \end{bmatrix},$$

$$\text{and} \quad \begin{bmatrix} \dfrac{5}{(s+5)(s+10)} & \dfrac{2(s+\frac{15}{2})}{(s+5)(s+10)} \\ \dfrac{2}{s+5} & \dfrac{2}{s+5} \end{bmatrix}$$

The determinants of these three transfer functions are

$$-2/[(s+5)(s+10)], \qquad -4/[(s+5)(s+10)], \quad \text{and} \quad -4(s+5)/[(s+5)^2(s+10)]$$

respectively. In each case there is no finite value of s that gives a zero value to the determinant. Hence there are no zeros. ■

.5 PARTIAL POLE ASSIGNMENT USING STATIC OUTPUT FEEDBACK [6]

The feedback signal considered in this section is formed by premultiplying the output $\mathbf{y}$ by a constant $r \times m$ gain matrix called $\mathbf{K}'$. If $\mathbf{D}$ is not zero, a modified output vector $\mathbf{y}' = \mathbf{y} - \mathbf{Du}$ could be formed. Then a modified gain matrix $\mathbf{K}_*$ could be used with $\mathbf{y}'$. It is not difficult to show that $\mathbf{K}_*$ and $\mathbf{K}'$ are related by

$$\mathbf{K}_* = \mathbf{K}'[\mathbf{I} + \mathbf{DK}']^{-1} \quad \text{or} \quad \mathbf{K}' = [\mathbf{I} - \mathbf{K}_*\mathbf{D}]^{-1}\mathbf{K}_* \tag{13.20}$$

If $\mathbf{D} = \mathbf{0}$, these two gains are the same. For nonzero $\mathbf{D}$ it is easiest first to determine $\mathbf{K}_*$ (which is equivalent to assuming $\mathbf{D} = \mathbf{0}$) and then calculate $\mathbf{K}'$ using Eq. (*13.20*). This is the approach to be followed next. By introducing $\mathbf{K}_*$ into Eq. (*13.7*), it is apparent that when static output feedback is used, the closed-loop eigenvalues are the roots of

$$|\lambda\mathbf{I} - \mathbf{A} + \mathbf{BK}_*\mathbf{C}| = 0 \tag{13.21}$$

Following the same arguments as in Sec. 13.4, this implies the existence of one or more nonzero vectors $\boldsymbol{\psi}$ that satisfy

$$[\lambda\mathbf{I} - \mathbf{A} \,\vdots\, \mathbf{B}]\begin{bmatrix} \boldsymbol{\psi} \\ \mathbf{K}_* \mathbf{C}\boldsymbol{\psi} \end{bmatrix} = \mathbf{0} \tag{13.22}$$

Except for the presence of **C**, this is the same as Eq. (*13.12*). A vector

$$\boldsymbol{\xi} = \begin{bmatrix} \boldsymbol{\psi} \\ \mathbf{K}_* \mathbf{C}\boldsymbol{\psi} \end{bmatrix}$$

is defined. All independent nontrivial solutions for $\boldsymbol{\xi}$ must be found for each λ_i. Then $\mathbf{U}(\lambda_i)$, $\boldsymbol{\Psi}(\lambda_i)$, and $\boldsymbol{\Omega}$ are formed as in the previous algorithm, without change. The upper n components of each column $\boldsymbol{\xi}_i$, now in $\mathbf{U}(\lambda_i)$, are closed-loop eigenvectors. Likewise, the lower r components form the bottom partition of $\mathbf{U}(\lambda_i)$. These are used to form $\mathcal{F}'(\lambda_i)$. The meaning of these lower-partition columns is now different because of the presence of **C**. Specifically, $\mathbf{f}_i = \mathbf{K}_* \mathbf{C}\boldsymbol{\psi}_i$. These are used to form $\boldsymbol{\Lambda}'$ essentially as before. One extra operation is now required because of the presence of the matrix **C**. Define $\boldsymbol{\Omega}' = \mathbf{C}\boldsymbol{\Omega}$. Then the counterpart to Eq. (*13.16*) is $\mathbf{K}_* \boldsymbol{\Omega}' = \boldsymbol{\Lambda}'$. Now $\boldsymbol{\Omega}'$ is an $m \times nr$ matrix whose rank is m if **C** is full rank. Therefore, at most m linearly independent columns of $\boldsymbol{\Omega}'$ can be selected in step III of the algorithm to form $\mathbf{G}'$. The corresponding columns of $\boldsymbol{\Lambda}'$ are used to form $\mathcal{J}'$. It is clear that at most m poles can be arbitrarily placed by using static output feedback. The solution for $\mathbf{K}_*$ is found as in step V of Sec. 13.4: $\mathbf{K}_* = \mathcal{J}(\mathbf{G}')^{-1}$.

Finally, if **D** is not zero, the feedback signal is

$$\mathbf{u} = -\mathbf{K}_*(\mathbf{y} - \mathbf{D}\mathbf{u}) + \mathbf{F}'\mathbf{v}'$$

Combining the two **u** terms leads to

$$\mathbf{u} = -[\mathbf{I} - \mathbf{K}_* \mathbf{D}]^{-1}\mathbf{K}_* \mathbf{y} + [\mathbf{I} - \mathbf{K}_* \mathbf{D}]^{-1}\mathbf{F}'\mathbf{v}'$$

The output feedback gain matrix, as stated in Eq. (*13.20*), is

$$\mathbf{K}' = [\mathbf{I} - \mathbf{K}_* \mathbf{D}]^{-1}\mathbf{K}_*$$

Note that the multiplier of the external input $\mathbf{v}'$ is also modified by the presence of a nonzero **D**. The term common to both the **v** and the **y** multipliers can be placed in the forward loop, as shown in Figure 13.4*a*. Figure 13.4*b* shows the same system using the $\mathbf{y}'$ signal fed back through the gain $\mathbf{K}_*$. Matrix block diagram manipulations such as those in Problems 4.2 through 4.4 can be used to reduce Figure 13.4*b* to Figure 13.4*a*, thus giving an alternate method of establishing Eq. (*13.20*).

EXAMPLE 13.8 Specify an output feedback matrix $\mathbf{K}'$ so that the controllable and observable system

$$\mathbf{A} = \begin{bmatrix} 0 & 1 \\ -3 & -4 \end{bmatrix}, \quad \mathbf{B} = \begin{bmatrix} 1 & 0 \\ 0 & 1 \end{bmatrix}, \quad \mathbf{C} = [1 \quad 1], \quad \mathbf{D} = [0]$$

will have $\lambda_1 = -5$ as a closed-loop eigenvalue.

Setting $\lambda = -5$ in Eq. (*13.22*) gives

$$\begin{bmatrix} -5 & -1 & 1 & 0 \\ 3 & -9 & 0 & 1 \end{bmatrix}\boldsymbol{\xi} = \mathbf{0}$$

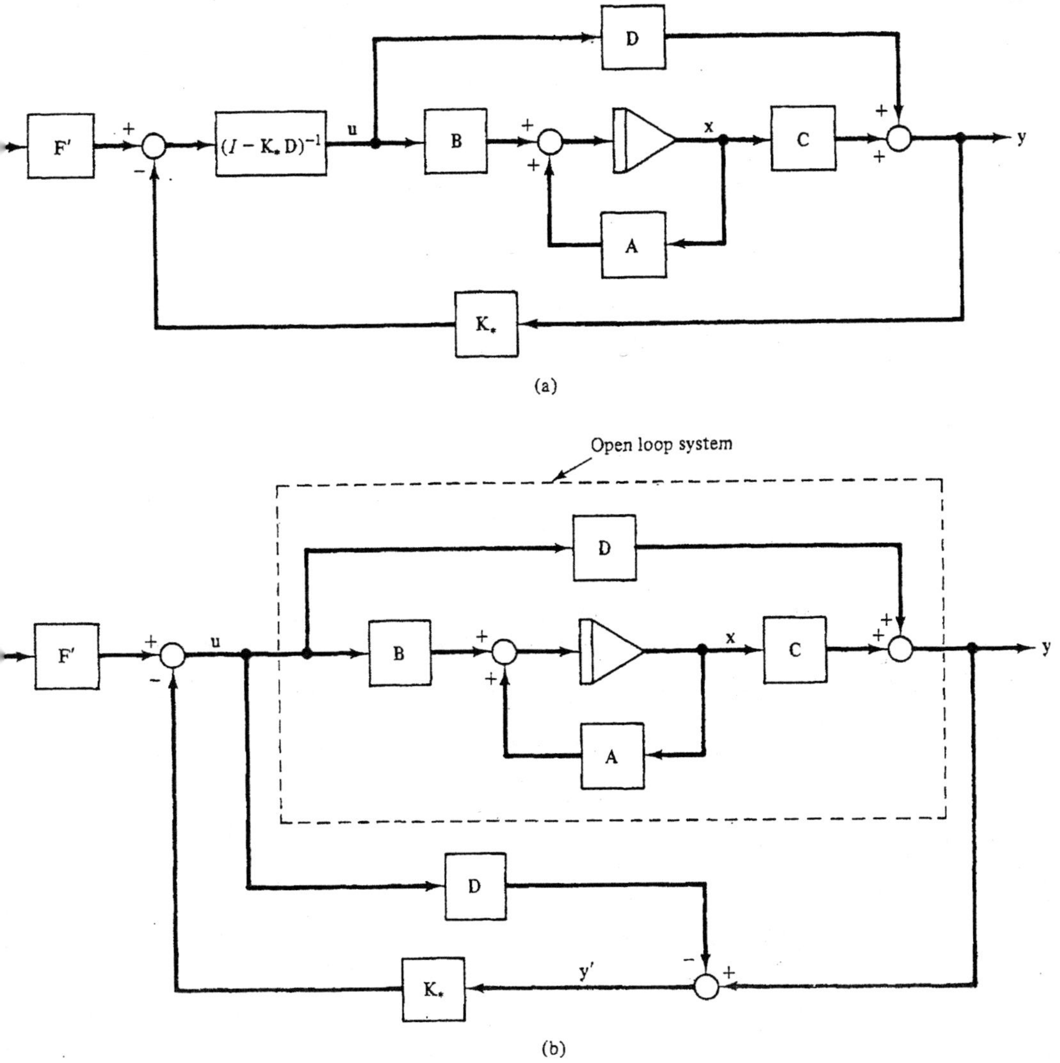

Figure 13.4

Letting $\xi_1 = \alpha$ and $\xi_2 = \beta$ be arbitrary constants allows the last two components to be found as $\xi_3 = 5\alpha + \beta$ and $\xi_4 = -3\alpha + \beta$. This means that

$$\mathbf{K}_* \mathbf{C} \begin{bmatrix} \alpha \\ \beta \end{bmatrix} = \begin{bmatrix} 5\alpha + \beta \\ -3\alpha + \beta \end{bmatrix} \tag{13.23}$$

Two such independent equations can be found. For example, $\alpha = 1, \beta = 0$ gives one and $\alpha = 0$, $\beta = 1$ gives another. From these we may be tempted to write

$$\mathbf{K}_* \mathbf{C} = \begin{bmatrix} 5 & 1 \\ -3 & 1 \end{bmatrix}$$

but this is not particularly helpful, since **C** cannot be inverted to find $\mathbf{K}_*$. What must be done is to return to Eq. (*13.23*) and explicitly combine the known **C** with the now-known values of α and β.

That is, $\mathbf{K}_*\{\alpha+\beta\} = \begin{bmatrix} 5\alpha+\beta \\ -3\alpha+\beta \end{bmatrix}$. Now the *scalar* $\alpha+\beta$ can be divided out to give

$$\mathbf{K}_* = \frac{\begin{bmatrix} 5\alpha+\beta \\ -3\alpha+\beta \end{bmatrix}}{\alpha+\beta}$$

An infinite number of valid choices for α and β are possible, such as $\alpha = \frac{1}{4}, \beta = \frac{3}{4}$, giving $\mathbf{K}_* = \mathbf{K}' = [2 \quad 0]^T$ or $\alpha = \frac{1}{2}, \beta = \frac{1}{2}$, giving $\mathbf{K}_* = \mathbf{K}' = [\frac{7}{2} \quad -1]^T$. We leave α and β general and proceed to check the closed-loop eigenvalues. The question arises whether the choice of α and β can be used to affect the eigenvalues. The matrix

$$\mathbf{A} - \mathbf{B}\mathbf{K}_*\mathbf{C} = \frac{\begin{bmatrix} -(5\alpha+\beta) & -4\alpha \\ -4\beta & -(\alpha+5\beta) \end{bmatrix}}{\alpha+\beta}$$

can be shown to have eigenvalues at -5 (the one requested) and at -1 (the one over which we have no choice) regardless of the α, β values. The original propostion that only m eigenvalues can be aribtrarily assigned is reaffirmed. ■

Static output feedback is not able to reposition all n eigenvalues, as just demonstrated. Even worse, those eigenvalues that are not reassignable may be unstable and hence unacceptable.

EXAMPLE 13.9 Let $\mathbf{A} = \begin{bmatrix} -2 & 1 & 0 \\ 0 & -2 & 0 \\ 0 & 0 & 4 \end{bmatrix}$, $\mathbf{B} = \begin{bmatrix} 0 & 0 \\ 0 & 1 \\ 1 & 0 \end{bmatrix}$, $\mathbf{C} = \begin{bmatrix} 0 & 0 & 1 \\ 1 & 0 & 0 \end{bmatrix}$, and $\mathbf{D} = \begin{bmatrix} 1 & 0 \\ 0 & 0 \end{bmatrix}$. Use static output feedback to achieve eigenvalues at $\lambda = -5$ and -6.

Note that the open-loop system is unstable due to the eigenvalue at $\lambda = 4$. With $\lambda_1 = -5$, Eq. (*13.22*) becomes

$$\begin{bmatrix} -3 & -1 & 0 & 0 & 0 \\ 0 & -3 & 0 & 0 & 1 \\ 0 & 0 & -9 & 1 & 0 \end{bmatrix} \boldsymbol{\xi}_1 = \mathbf{0}$$

from which we find solutions of the form $\boldsymbol{\xi}_1 = [\alpha \quad -3\alpha \quad \beta \quad 9\beta \quad -9\alpha]^T$. Similarly, with $\lambda_2 = -6$, Eq. (*13.22*) reduces to

$$\begin{bmatrix} -4 & -1 & 0 & 0 & 0 \\ 0 & -4 & 0 & 0 & 1 \\ 0 & 0 & -10 & 1 & 0 \end{bmatrix} \boldsymbol{\xi}_2 = \mathbf{0}$$

from which $\boldsymbol{\xi}_2 = [\delta \quad -4\delta \quad \gamma \quad 10\gamma \quad -16\delta]^T$. Once again, the first three elements of the $\boldsymbol{\xi}$ vectors are closed-loop eigenvectors $\boldsymbol{\psi}$. The fourth and fifth components are the results of multiplying $\boldsymbol{\psi}$ by $\mathbf{K}_*\mathbf{C}$. Therefore, explicitly multiplying out the $\mathbf{C}\boldsymbol{\psi}$ products gives

$$\mathbf{K}_* \begin{bmatrix} \beta & \gamma \\ \alpha & \delta \end{bmatrix} = \begin{bmatrix} 9\beta & 10\gamma \\ -9\alpha & -16\delta \end{bmatrix}$$

One convenient choice, which avoids a matrix inversion, is $\alpha = \gamma = 0$ and $\beta = \delta = 1$. Then $\mathbf{K}_* = \begin{bmatrix} 9 & 0 \\ 0 & -16 \end{bmatrix}$, from which $\mathbf{K}' = \begin{bmatrix} -\frac{9}{8} & 0 \\ 0 & -16 \end{bmatrix}$. This solution does give closed-loop eigenvalues at

$\lambda = -5$ and -6 as requested, but the third eigenvalue is at $\lambda = +2$, and hence the closed-loop system is still unstable. Note that this system is stabilizable (see Sec. 11.8) when full-state feedback can be used. ■

13.6 OBSERVERS—RECONSTRUCTING THE STATE FROM AVAILABLE OUTPUTS

In many systems all components of the state vector are not directly available as output signals. For example, a radar may be tracking the position (states) of a vehicle and it is desired to know the velocities, accelerations, or other states. Sometimes a knowledge of the state values is desired as an end in itself, simply for performance evaluation. In many system-control problems the reason for wanting knowledge of the states is for forming feedback signals. As was shown earlier, if all states can be used for feedback, complete control over all the eigenvalues is possible, assuming the system is controllable. Several approaches to the problem of unmeasured states can be considered. First, it might be possible to add additional sensors to provide the measurements. This is generally the most expensive option. Second, some sort of ad hoc differentiation of measured states may provide an estimate of unmeasured states. This option may not give sufficiently accurate performance, especially in the case of noisy data. The third option is to use full knowledge of the mathematical models of the system in a systematic way in an attempt to estimate or reconstruct the states. This third approach is developed now. The resulting estimator algorithm is called an *observer* [7]. There are continuous-time observers and discrete-time observers. *Full state observers,* sometimes called *identity observers,* produce estimates of all state components, even those that are measured directly. This redundancy can be removed with a *reduced state observer,* which estimates only the states that are not measured and uses raw measurement data for those that are measured. When measurements are noisy, the beneficial smoothing effects which are provided by the full state observer may be more important than the elimination of redundancy.

13.6.1 Continuous-Time Full State Observers

The continuous-time system described by Eq. (*13.1*) is considered first. For simplicity it is now assumed that $\mathbf{D} = [0]$. If this is not true, then an equivalent output $\mathbf{y}' = \mathbf{y} - \mathbf{D}\mathbf{u}$ can be used in place of $\mathbf{y}$ in what follows, as was demonstrated in the previous section. It is desired to obtain a good estimate of the state $\mathbf{x}(t)$, given a knowledge of the output $\mathbf{y}(t)$, the input $\mathbf{u}(t)$, and the system matrices $\mathbf{A}, \mathbf{B}, \mathbf{C}, \mathbf{D}$. This problem is referred to as the *state reconstruction problem.* If $\mathbf{C}$ is square and nonsingular, then $\mathbf{x}(t) = \mathbf{C}^{-1}\mathbf{y}(t)$. In general, this trivial result will not be applicable.

Another dynamic system, called an *observer* [7], is to be constructed. Its input will depend on $\mathbf{y}$ and $\mathbf{u}$ and its state (output) should be a good approximation to $\mathbf{x}(t)$. The form of the observer is selected as

$$\dot{\hat{\mathbf{x}}} = \mathbf{A}_c\hat{\mathbf{x}} + \mathbf{L}\mathbf{y} + \mathbf{z} \qquad (13.24)$$

where $\hat{\mathbf{x}}$ is the $n \times 1$ vector approximation to $\mathbf{x}$.‡ $\mathbf{A}_c$ and $\mathbf{L}$ are $n \times n$ and $n \times m$ matrices and $\mathbf{z}$ is an $n \times 1$ vector, to be determined. $\mathbf{A}_c$, $\mathbf{L}$, and $\mathbf{z}$ will be selected next in such a way that the postulated observer system meets the goal of giving good estimates for the state variables. Defining $\mathbf{e} = \mathbf{x} - \hat{\mathbf{x}}$ and using Eq. (*13.1*) gives

$$\dot{\mathbf{e}} = \mathbf{A}\mathbf{x} - \mathbf{A}_c\hat{\mathbf{x}} - \mathbf{L}\mathbf{y} + \mathbf{B}\mathbf{u} - \mathbf{z} \tag{13.25}$$

By selecting $\mathbf{z} = \mathbf{B}\mathbf{u}$ and using $\mathbf{y} = \mathbf{C}\mathbf{x}$, this reduces to

$$\dot{\mathbf{e}} = (\mathbf{A} - \mathbf{L}\mathbf{C})\mathbf{x} - \mathbf{A}_c\hat{\mathbf{x}}$$

Selecting $\mathbf{A}_c = \mathbf{A} - \mathbf{L}\mathbf{C}$ gives

$$\dot{\mathbf{e}} = \mathbf{A}_c\mathbf{e} \tag{13.26}$$

If the eigenvalues of $\mathbf{A}_c$ all have negative real parts, then an asymptotically stable error equation results. This indicates that $\mathbf{e}(t) \to \mathbf{0}$, or $\hat{\mathbf{x}}(t) \to \mathbf{x}(t)$ as $t \to \infty$. The term $\mathbf{L}$ of the observer is still unspecified. If the original system (*13.1*) is completely observable, then it is always possible to find an $\mathbf{L}$ which will yield any set of desired eigenvalues for $\mathbf{A}_c$. Thus it is possible to control the rate at which $\hat{\mathbf{x}} \to \mathbf{x}$.

Compare the $n \times n$ matrix $\mathbf{A} - \mathbf{L}\mathbf{C}$ of Eq. (*13.26*) with the $n \times n$ matrix $\mathbf{A} - \mathbf{B}\mathbf{K}$ of Eq. (*13.4*). Both have a known $n \times n$ system matrix $\mathbf{A}$, modified by a term that is partially selectable by the designer. The order of the known terms $\mathbf{B}$ and $\mathbf{C}$ and the selectable gains $\mathbf{K}$ and $\mathbf{L}$ are reversed, so the comparison is not exact. In earlier sections methods of determining $\mathbf{K}$ were developed which give $\mathbf{A} - \mathbf{B}\mathbf{K}$ a specified set of eigenvalues. A square matrix and its transpose have the same determinant and hence the same eigenvalues. Therefore, if $\mathbf{L}$ can be selected to force the eigenvalues of

$$\mathbf{A}_c^T = \mathbf{A}^T - \mathbf{C}^T\mathbf{L}^T \tag{13.27}$$

to have specified values, then $\mathbf{A}_c$ will have those same desirable values. The methods presented in Sec. 13.4 for pole placement apply without change to the observer problem, which is its dual. Replace $\mathbf{A}$ by $\mathbf{A}^T$ and $\mathbf{B}$ by $\mathbf{C}^T$, and find $\mathbf{L}^T$ (instead of $\mathbf{K}$) by any of the previous pole placement algorithms. The requirement in Sec. 13.4 that the system $\{\mathbf{A}, \mathbf{B}\}$ be controllable now becomes an observability requirement for $\{\mathbf{A}, \mathbf{C}\}$. To see this, note that a substitution of $\mathbf{A}^T$ and $\mathbf{C}^T$ for $\mathbf{A}$ and $\mathbf{B}$ in Eq. (*11.5*) for the controllability test automatically gives the observability test of Eq. (*11.6*) for $\{\mathbf{A}, \mathbf{C}\}$. The full-state observer design problem is the dual of the pole placement problem using full state feedback. The discussion and algorithm of Sec. 13.4 are totally applicable. For reference, the fundamental Eq. (*13.12*) is repeated here with notational changes appropriate for the observer problem.

$$[\lambda\mathbf{I} - \mathbf{A}^T \mid \mathbf{C}^T]\boldsymbol{\xi} = \mathbf{0} \tag{13.28}$$

Equation (*13.28*) is used to determine the observer gain matrix $\mathbf{L}$. As before, there will be times when generalized eigenvectors will be required in order to obtain multiple observer poles. Figure 13.5 shows the observer mechanization. Part (*a*) is directly from

‡ The circumflex ˆ is used in this chapter to indicate an estimate for a quantity, e.g., $\hat{\mathbf{x}}$ is an estimate of $\mathbf{x}$. This should not be confused with the notation for a unit vector introduced in Chapter 5.

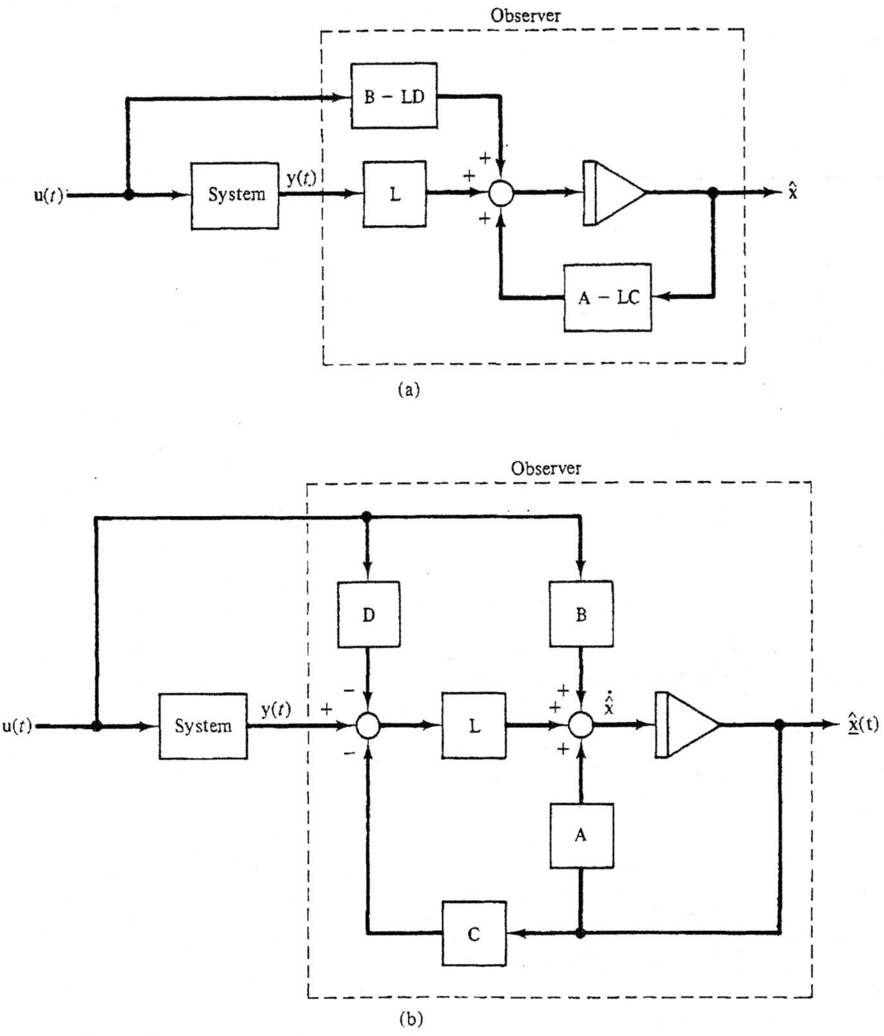

Figure 13.5

Eq. (*13.24*) and part (*b*) is obtained by matrix block diagram manipulations. This form better shows that a key factor in determining $\dot{\hat{\mathbf{x}}}$ is the difference between the actual measurement y and the "expected measurement" $\mathbf{C}\hat{\mathbf{x}}$ weighted by the observer gain **L.**

EXAMPLE 13.10 Design an observer for the system of Example 13.1 if only x_1 can be measured. Put both observer eigenvalues at $\lambda = -8$. This will cause the error $\mathbf{e}(t)$ to approach zero rapidly regardless of any error in the initial estimate $\hat{\mathbf{x}}$.

Using

$$\mathbf{A} = \begin{bmatrix} 0 & 2 \\ 0 & 3 \end{bmatrix} \quad \text{and} \quad \mathbf{C} = [1 \quad 0]$$

in $[-8\mathbf{I} - \mathbf{A}^T \mid \mathbf{C}^T]\boldsymbol{\xi} = \mathbf{0}$ gives component equations $8\xi_1 = \xi_3$ and $2\xi_1 = -11\xi_2$. Arbitrarily setting

$\xi_1 = 1$ gives $\xi_2 = -\frac{2}{11}$ and $\xi_3 = 8$. The first two components of ξ form the eigenvector ψ. A generalized eigenvector is needed to find a second independent column. Thus

$$[-8\mathbf{I} - \mathbf{A}^T \mid \mathbf{C}^T]\xi_g = -\psi = \begin{bmatrix} -1 \\ \frac{2}{11} \end{bmatrix}$$

is solved to find $\xi_g = [1 \quad -\frac{24}{121} \quad 7]^T$. The columns ξ and ξ_g form the matrix $\mathbf{U}(-8)$. Partitioning it gives

$$\mathbf{L}^T\begin{bmatrix} 1 & 1 \\ -\frac{2}{11} & -\frac{24}{121} \end{bmatrix} = [8 \quad 7]$$

and the solution is $\mathbf{L}^T = [19 \quad 60.5]$. As a check, the matrix $\mathbf{A}_c = \begin{bmatrix} -19 & 2 \\ -60.5 & 3 \end{bmatrix}$ is found to have $(\lambda + 8)^2$ as its characteristic polynomial. The complete observer description is

$$\dot{\hat{\mathbf{x}}} = \mathbf{A}_c\hat{\mathbf{x}} + \mathbf{L}\mathbf{y} + \mathbf{z}$$

or

$$\dot{\hat{\mathbf{x}}} = \begin{bmatrix} -19 & 2 \\ -60.5 & 3 \end{bmatrix}\hat{\mathbf{x}} + \begin{bmatrix} 19 \\ 60.5 \end{bmatrix}y + \begin{bmatrix} 0 \\ 1 \end{bmatrix}u$$

Figure 13.6 gives the observer cascaded with the original system. ■

13.6.2 Discrete-Time Full-State Observers

For the discrete-time system of Eq. (*13.2*) one possible form for the observer is postulated as

$$\hat{\mathbf{x}}(k+1) = \mathbf{A}_c\hat{\mathbf{x}}(k) + \mathbf{L}\mathbf{y}(k) + \mathbf{z}(k) \tag{13.29}$$

Notice that this implies that the estimate being calculated at time-step $k+1$ is based upon measured data one sample old, namely, $\mathbf{y}(k)$. This time delay may be forced upon the system designer to allow time for the computational lag of the algorithm. If the sample time between k and $k+1$ is large, this delay may be undesirable. An alternate form of the observer, which replaces $\mathbf{y}(k)$ by $\mathbf{y}(k+1)$, will be given later.

The preceding observer's estimation error $\mathbf{e}(k) = \mathbf{x}(k) - \hat{\mathbf{x}}(k)$ will satisfy the homogeneous difference equation

$$\mathbf{e}(k+1) = (\mathbf{A} - \mathbf{L}\mathbf{C})\mathbf{e}(k) = \mathbf{A}_c\mathbf{e}(k) \tag{13.30}$$

provided that $\mathbf{z}(k)$ is selected to cancel all the $\mathbf{u}$ terms that come into the error equation through $\mathbf{L}\mathbf{y}$. This requirement is met if $\mathbf{z} = (\mathbf{B} - \mathbf{L}\mathbf{D})\mathbf{u}(k)$. Just as in the continuous-time case, the error $\mathbf{e}(k)$ will decay toward zero if $\mathbf{A}_c$ is stable. This means that all its eigenvalues must be inside the unit circle. The decay rate depends on the location of these eigenvalues. If the original system of Eq. (*13.2*) is observable, it is possible to force the n eigenvalues to any desired locations by proper choice of $\mathbf{L}$. The mechanics of doing this are exactly the same as in the continuous-time case.

Figure 13.7*a* shows the observer as described in Eq. (*13.29*). Figure 13.7*b* is a rearrangement which shows how the gain-weighted difference between the actual $\mathbf{y}$ and a predicted $\mathbf{y}$, given by $\mathbf{C}\hat{\mathbf{x}} + \mathbf{D}\mathbf{u}$, drives the observer.

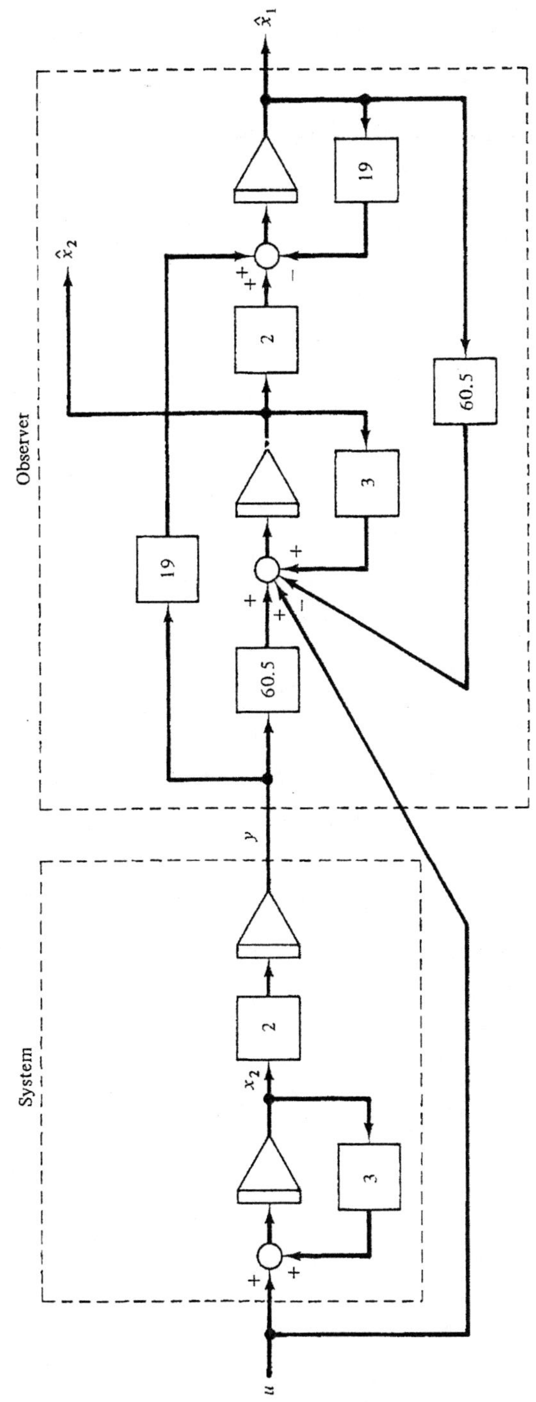

Figure 13.6

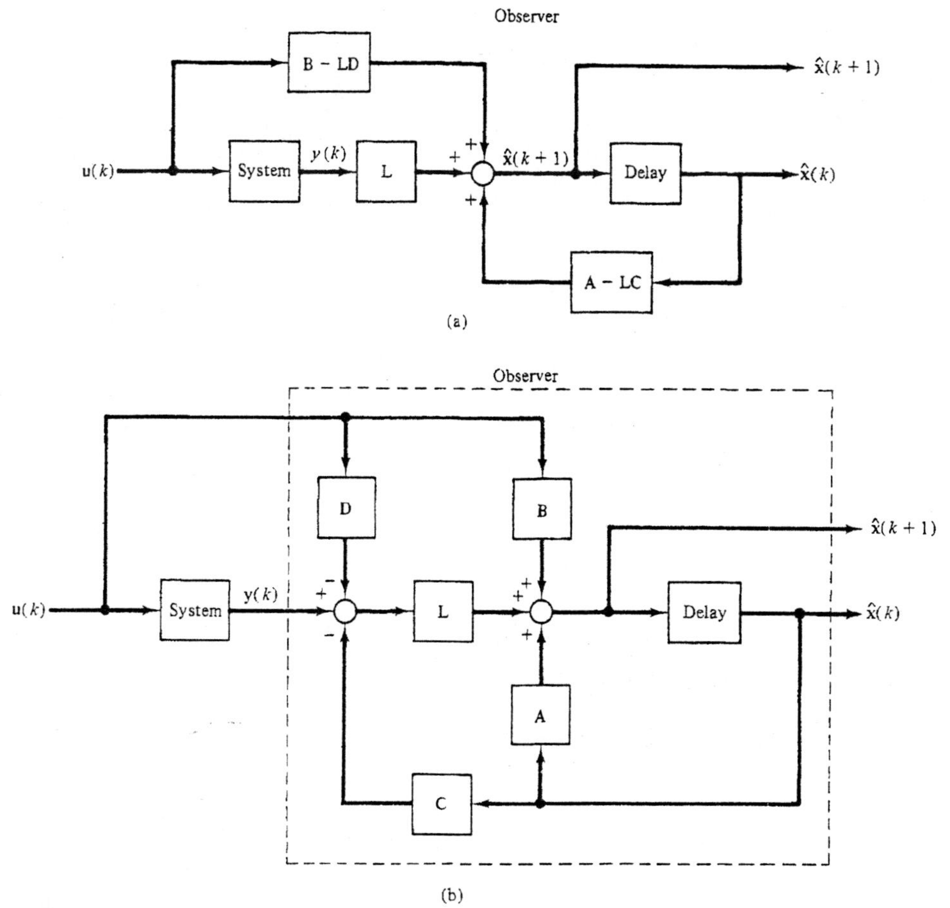

Figure 13.7

EXAMPLE 13.11 Design a full-state observer with eigenvalues at $z = 0.1 \pm 0.2j$ for the discrete system with $\mathbf{A} = \begin{bmatrix} 0 & 1 \\ 0.2 & 0.5 \end{bmatrix}$, $\mathbf{B} = \begin{bmatrix} 0 \\ 1 \end{bmatrix}$, $\mathbf{D} = [0]$, and $\mathbf{C} = [1 \quad 0]$.

For this case Eq. (*13.28*) becomes, with $z = 0.1 + 0.2j$,

$$\begin{bmatrix} 0.1+0.2j & -0.2 & 1 \\ -1 & -0.4+0.2j & 0 \end{bmatrix} \boldsymbol{\xi} = \mathbf{0}$$

Selecting $\xi_2 = 1$ gives $\xi_1 = -0.4 + 0.2j$ and $\xi_3 = 0.28 + 0.06j$. Using the conjugate of z gives the conjugate of $\boldsymbol{\xi}$. Therefore,

$$\mathbf{L}^T \begin{bmatrix} -0.4+0.2j & -0.4-0.2j \\ 1 & 1 \end{bmatrix} = [0.28+0.06j \quad 0.28-0.06j]$$

Using Problem 4.23, this can be expressed with purely real values: $\mathbf{L}^T \begin{bmatrix} -0.4 & 0.2 \\ 1 & 0 \end{bmatrix} =$ $[0.28 \quad 0.06]$, from which $\mathbf{L}^T = [0.3 \quad 0.4]$. The observer system matrix is $\mathbf{A}_c = \begin{bmatrix} -0.3 & 1 \\ -0.2 & 0.5 \end{bmatrix}$. The final observer is shown in Figure 13.8. ■

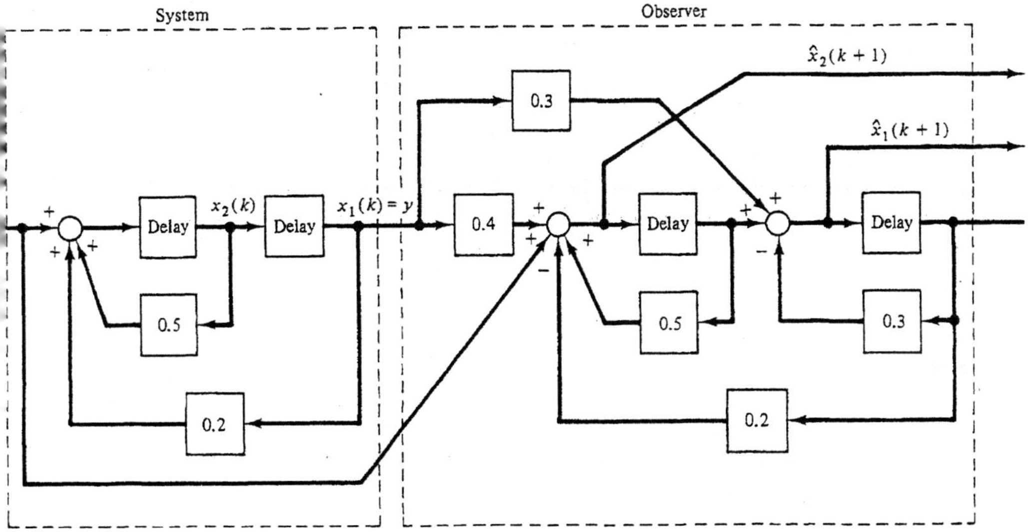

Figure 13.8

A second form of the discrete-time full state observer can be postulated as

$$\hat{\mathbf{x}}(k+1) = \mathbf{A}_c\hat{\mathbf{x}}(k) + \mathbf{L}\mathbf{y}(k+1) + \mathbf{z}(k) \tag{13.31}$$

Using the state equations to eliminate the $\mathbf{y}(k+1)$ term gives

$$\begin{aligned}\hat{\mathbf{x}}(k+1) &= \mathbf{A}_c\hat{\mathbf{x}}(k) + \mathbf{L}\{\mathbf{C}\mathbf{x}(k+1) + \mathbf{D}\mathbf{u}(k+1)\} + \mathbf{z}(k)\\ &= \mathbf{A}_c\hat{\mathbf{x}}(k) + \mathbf{LCAx}(k) + \mathbf{LCBu}(k) + \mathbf{LDu}(k+1) + \mathbf{z}(k)\end{aligned}$$

Forming the difference from the true state equation, the error equation is

$$\begin{aligned}\mathbf{e}(k+1) = (\mathbf{A} - \mathbf{LCA})\mathbf{x}(k) - \mathbf{A}_c\hat{\mathbf{x}}(k) + (\mathbf{B} - \mathbf{LCB})\mathbf{u}(k)\\ - \mathbf{LDu}(k+1) - \mathbf{z}(k)\end{aligned} \tag{13.32}$$

As before, the observer input $\mathbf{z}(k)$ is selected to cancel all the $\mathbf{u}$ terms, so $\mathbf{z}(k) = (\mathbf{B} - \mathbf{LCB})\mathbf{u}(k) - \mathbf{LDu}(k+1)$. The observer matrix is selected as

$$\mathbf{A}_c = (\mathbf{A} - \mathbf{LCA}) = (\mathbf{I} - \mathbf{LC})\mathbf{A} \tag{13.33}$$

so that the error equation is once again a simple homogeneous equation. Its response will decay to zero if the eigenvalues of $\mathbf{A}_c$ are all stable. Note, however, that in forcing the eigenvalues of $\mathbf{A}_c$ to prespecified stable locations by choice of the gain $\mathbf{L}$, a different kind of problem is being posed. The question now is whether the original system matrix $\mathbf{A}$ can be *multiplied* by a factor $(\mathbf{I} - \mathbf{LC})$ to force the result, $\mathbf{A}_c$, to have arbitrary eigenvalues. The answer is affirmative if the system $\mathbf{A}$ is nonsingular, as can be seen by defining a new "given" matrix $\mathbf{C}' = \mathbf{CA}$. This allows Eq. (*13.33*) to be written in a form which hides the product problem and forces the equation to appear as the one treated in previous sections, $\mathbf{A}_c = \mathbf{A} - \mathbf{LC}'$. Remember that the ability to arbitrarily prescribe eigenvalues to $\mathbf{A}_c$ required observability. Assuming $\{\mathbf{A}, \mathbf{C}\}$ is observable, is $\{\mathbf{A}, \mathbf{C}'\}$ also observable? Call the observability test matrix of Eq. (*11.6*) $\mathbf{Q}$ when $\mathbf{C}$ is used and $\mathbf{Q}'$

when $\mathbf{C}'$ is used. Clearly $\mathbf{Q}' = \mathbf{A}^T\mathbf{Q}$, so that if $\mathbf{A}$ is nonsingular, $\text{rank}(\mathbf{Q}') = \text{rank}(\mathbf{Q})$. Any discrete system which is derived by sampling a continuous system described by Eq. (*13.1*) will have a nonsingular $\mathbf{A}$ matrix, since $\mathbf{A}$ is then just the transition matrix $\mathbf{\Phi}$ over one sample period and $\mathbf{\Phi}^{-1}$ always exists. Other discrete systems, such as those having pure time delays, need not have $\mathbf{A}$ full rank. When $\mathbf{A}$ is not full rank, complete control over the eigenvalues of $\mathbf{A}_c$ is not possible. By forcing $n - \text{rank}(\mathbf{A})$ eigenvalues to be at the origin, the procedure can still be carried out. See Reference 8 for a further discussion of this point. Figure 13.9 shows two arrangements for this observer. Note

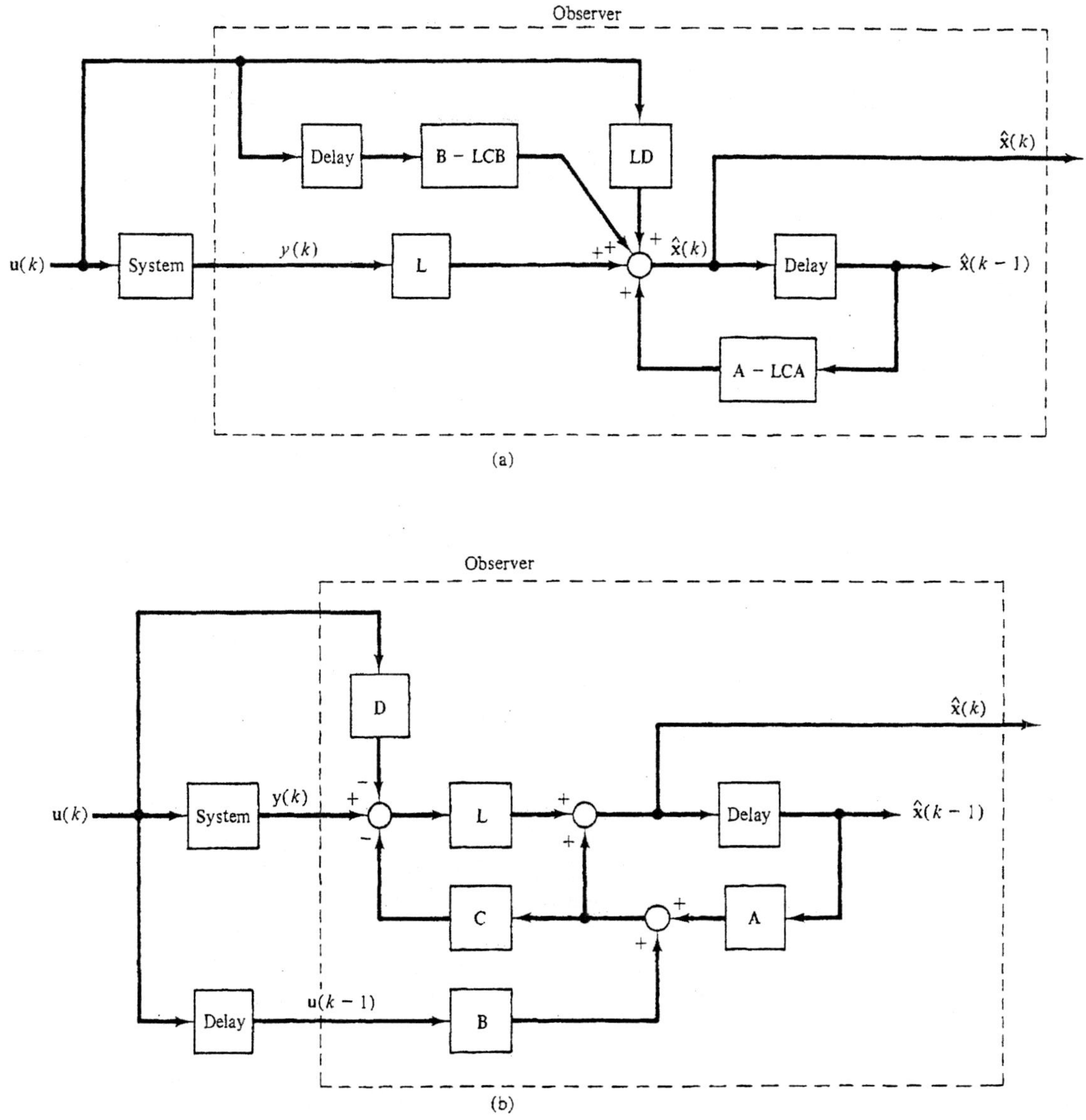

Figure 13.9

that whenever **D** is not zero, the input signals **u** at two successive time steps are required.

EXAMPLE 13.12 Use the same system and eigenvalue specifications as Example 13.11, but use $\mathbf{y}(k+1)$ as the input rather than $\mathbf{y}(k)$.

First calculate $\mathbf{C}' = \mathbf{CA} = [0 \quad 1]$ and then solve

$$\begin{bmatrix} 0.1+0.2j & -0.2 & 0 \\ -1 & -0.4+0.2j & 1 \end{bmatrix} \boldsymbol{\xi} = \mathbf{0}$$

The two conjugate solutions are used to write

$$\mathbf{L}^T \begin{bmatrix} 1 & 1 \\ 0.5+j & 0.5-j \end{bmatrix} = [1.4+0.3j \quad 1.4-0.3j]$$

or $\mathbf{L}^T \begin{bmatrix} 1 & 0 \\ 0.5 & 1 \end{bmatrix} = [1.4 \quad 0.3]$, giving $\mathbf{L}^T = [1.25 \quad 0.3]$. This version of the observer has

$$\mathbf{A}_c = \begin{bmatrix} 0 & -0.25 \\ 0.2 & 0.2 \end{bmatrix}, \qquad \mathbf{B} - \mathbf{LCB} = \begin{bmatrix} -1.25 \\ 0.7 \end{bmatrix}$$

The implementation is given in Figure 13.10. The time argument k on the input $\mathbf{u}(k)$ may seem inconsistent with Figure 13.9. Actually when $\mathbf{D} \neq [\mathbf{0}]$, both $\mathbf{u}(k)$ and $\mathbf{u}(k+1)$ directly affect $\mathbf{y}(k+1)$, so either notational convention can be used provided the correct representation inside the system block is used. ■

EXAMPLE 13.13 Suppose the previous model is modified to $\mathbf{A} = \begin{bmatrix} 0 & 1 \\ 0 & 0.5 \end{bmatrix}$, with **B** and **C** unchanged. Note that **A** is singular and that with $\mathbf{C}' = \mathbf{CA} = [0 \quad 1]$, the system is not observable. Investigate the implications.

Eq. (*13.28*) gives $\begin{bmatrix} \lambda & 0 & 0 \\ -1 & \lambda - 0.5 & 1 \end{bmatrix} \boldsymbol{\xi} = \mathbf{0}$, from which the requirements are $\lambda \xi_1 = 0$ and

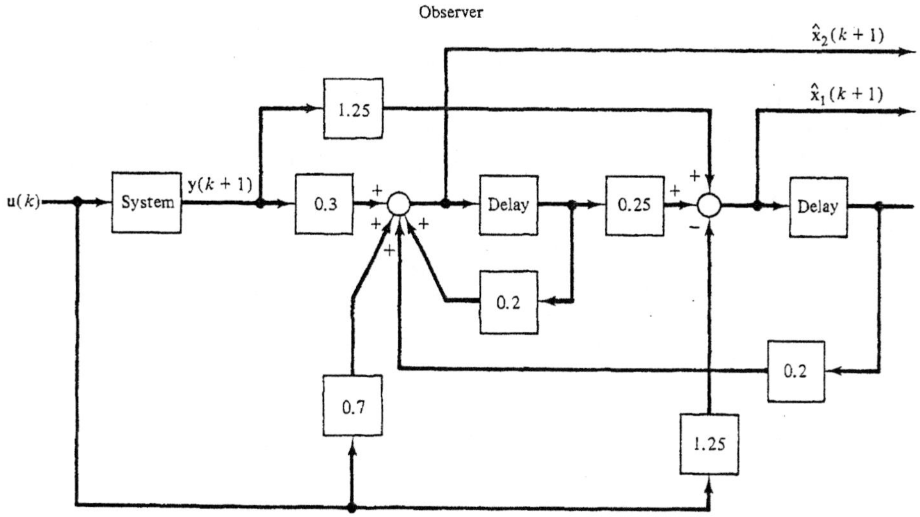

Figure 13.10 Observer for Example 13.12.

$-\xi_1 + (\lambda - 0.5)\xi_2 + \xi_3 = 0$. If any nonzero eigenvalue is specified, then ξ_1 must be zero. Assume this is so and then find that ξ_2 can be set arbitrarily to 1 and $\xi_3 = 0.5 - \lambda$. One of the equations for $\mathbf{L}^T$ is

$$\mathbf{L}^T\begin{bmatrix} 0 & \\ 1 & \end{bmatrix} = [0.5 - \lambda \quad] \tag{13.34}$$

Any choice other than 0 for the second eigenvalue will require another zero in the first row of Eq. (*13.34*), resulting in a singular matrix, so $\mathbf{L}^T$ cannot be determined. The only choice is to put an eigenvalue at $z = 0$ (which may have been a good value anyway, but there is no choice in the matter here.) With $\lambda = 0$ it is possible to set $\xi_1 = 1$ and $\xi_2 = 1$ and find $\xi_3 = 1.5$. With these selections, the observer gain matrix is $\mathbf{L}^T = [z_1 + 1 \quad 0.5 - z_1]$, a valid solution for any real value of observer eigenvalue $\lambda = z_1$. The other observer eigenvalue is at $\lambda = 0$. ■

13.6.3 Continuous-Time Reduced Order Observers

Assume that the state equations have been selected in such a way that the m outputs constitute the first m states, perhaps modified by a $\mathbf{Du}$ term. If this is not true, a transformation to a new set of states can make it true. (See Problem 13.12.) Then partitioning the state equations the obvious way gives

$$\mathbf{x} = \begin{bmatrix} \mathbf{x}_1 \\ \mathbf{x}_2 \end{bmatrix}, \qquad \mathbf{A} = \begin{bmatrix} \mathbf{A}_{11} & \mathbf{A}_{12} \\ \mathbf{A}_{21} & \mathbf{A}_{22} \end{bmatrix}, \qquad \mathbf{B} = \begin{bmatrix} \mathbf{B}_1 \\ \mathbf{B}_2 \end{bmatrix}, \qquad \mathbf{C} = [\mathbf{I}_m \quad \mathbf{0}]$$

The subset of states $\mathbf{x}_1$ are treated as knowns because they are given by $\mathbf{x}_1(t) = \mathbf{y}(t) - \mathbf{Du}(t)$. The lower half of the partitioned state differential equations are

$$\dot{\mathbf{x}}_2 = \mathbf{A}_{22}\mathbf{x}_2 + \{\mathbf{A}_{21}\mathbf{x}_1 + \mathbf{B}_2\mathbf{u}\} \tag{13.35}$$

The top half of these partitioned equations can be written as

$$\{\dot{\mathbf{x}}_1 - \mathbf{A}_{11}\mathbf{x}_1 - \mathbf{B}_1\mathbf{u}\} = \mathbf{A}_{12}\mathbf{x}_2 \tag{13.36}$$

In both Eqs. (*13.35*) and (*13.36*) all terms inside the braces can be treated as known. Only the $\dot{\mathbf{x}}_1$ may cause some concern, since differentiation seems to be required. This will be dealt with later, in the same way that $\dot{\mathbf{u}}$ terms were avoided in Chapter 3. By way of analogy with the full-state observer, Eq. (*13.35*) represents the dynamics and Eq. (*13.36*) represents the measurements of the reduced observer. Table 13.1 lists the corresponding terms in the two cases.

TABLE 13.1 SYSTEM ANALOGIES FOR REDUCED ORDER OBSERVER

Term	Full State Observer	Reduced Order Observer
State	$\mathbf{x}$ $(n \times 1)$	$\mathbf{x}_2$ $(n - m) \times 1$
Input	$\mathbf{Bu}$	$\{\mathbf{A}_{21}\mathbf{x}_1 + \mathbf{B}_2\mathbf{u}\}$
System matrix	$\mathbf{A}$ $(n \times n)$	$\mathbf{A}_{22}$ $(n - m) \times (n - m)$
Output matrix	$\mathbf{C}$	$\mathbf{A}_{12}$
Output signal	$\mathbf{y}$	$\{\dot{\mathbf{x}}_1 - \mathbf{A}_{11}\mathbf{x}_1 - \mathbf{B}_1\mathbf{u}\}$
Direct signal	$\mathbf{Du}$	None

Using these set-up analogies, the reduced order observer equations for the two portions of the partitioned state vector are:

$$\hat{\mathbf{x}}_1 = \mathbf{x}_1 = \mathbf{y} - \mathbf{D}\mathbf{u} \qquad \text{(just use the given data)}$$

$$\dot{\hat{\mathbf{x}}}_2 = \mathbf{A}_r\hat{\mathbf{x}}_2 + \mathbf{L}_r\mathbf{y}_r + \mathbf{z}_r \tag{13.37}$$

where

$$\mathbf{A}_r = \mathbf{A}_{22} - \mathbf{L}_r\mathbf{A}_{12}$$

$$\mathbf{y}_r = \dot{\mathbf{x}}_1 - \mathbf{A}_{11}\mathbf{x}_1 - \mathbf{B}_1\mathbf{u}$$

$$\mathbf{z}_r = \mathbf{A}_{21}\mathbf{x}_1 + \mathbf{B}_2\mathbf{u}$$

The reduced order observer gain matrix $\mathbf{L}_r$ can be calculated with the same algorithm as used in the full state observer. It can be selected to give any specified set of observer eigenvalues, provided the system $\{\mathbf{A}_{22}, \mathbf{A}_{12}\}$ is observable. This is automatically true if the original full system is observable. The mechanics of the calculation are exactly as in the full state case. The implementation is, of course, a little different. Two versions are shown in Figure 13.11. Figure 13.11*a* shows the $\dot{\mathbf{x}}_1$ term as it appears in the preceding equations. The need for actually differentiating the data is avoided by moving that signal to the output side of the integrator, as shown in Figure 13.11*b*.

EXAMPLE 13.14 Consider the system

$$\mathbf{A} = \begin{bmatrix} 0 & 1 & 0 \\ 0 & 0 & 1 \\ -1 & -2 & -3 \end{bmatrix}, \quad \mathbf{B} = \begin{bmatrix} 0 & 1 \\ 1 & 0 \\ 0 & 1 \end{bmatrix}, \quad \mathbf{C} = \begin{bmatrix} 1 & 0 & 0 \\ 0 & 1 & 0 \end{bmatrix}, \quad \mathbf{D} = [0]$$

Design a first-order observer with an eigenvalue at $\lambda = -5$ to estimate the unmeasured third state component. The partitioned matrix terms for this example are $\mathbf{A}_{22} = -3, \mathbf{A}_{12} = [0 \quad 1]^T$. Therefore, $[s\mathbf{I} - \mathbf{A}_{22}^T \mid \mathbf{A}_{12}^T]\boldsymbol{\xi} = 0$ reduces to a scalar equation $[-2 \quad 0 \quad 1]\boldsymbol{\xi} = 0$. Clearly, ξ_2 is arbitrary, and it is set to 0. Also, $\xi_1 = 1$ is selected and then $\xi_3 = 2$ is determined. Thus $\mathbf{L}^T[1] = \begin{bmatrix} 0 \\ 2 \end{bmatrix}$, or $\mathbf{L} = [0 \quad 2]$.

The general arrangement of Figure 13.11*b* specializes in this case to the first-order observer of Figure 13.12. Note that there are four inputs and just one nontrivial output, $\hat{\mathbf{x}}_3$. By using Mason's gain formula on Figure 13.12 or the Laplace transform of Eq. (*13.37*), it is found that

$$\hat{\mathbf{x}}_3(s) = [1/(s+5)][-x_1(s) + 2(s-1)x_2(s) - 2u_1(s) + u_2(s)]$$

which shows that the pole at $s = -5$ has been attained.

13.6.4 Discrete-Time Reduced Order Observers

The discrete-time state equations (*13.2*) are considered, but again it is assumed that $\mathbf{C} = [\mathbf{I}_m \quad 0]$. The derivation in this section is almost identical to the continuous case of Sec. 13.6.3. Both forms of full state discrete observers can be developed in the reduced state case, but given here is only the one using outputs that are one sample period old.

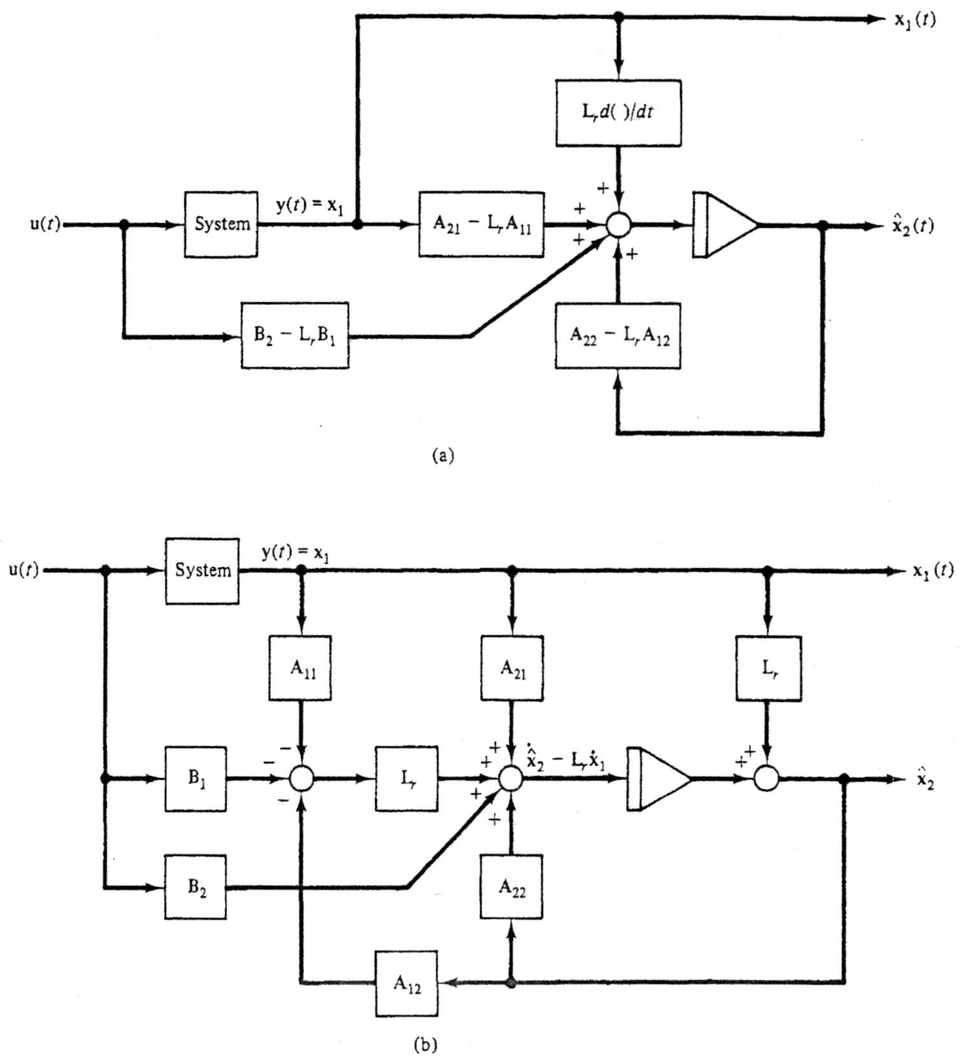

Figure 13.11

Partition the state vector into the m measured components and the n-m remaining components. Then the state equations become

$$\mathbf{x}_1(k+1) = \mathbf{A}_{11}\mathbf{x}_1(k) + \mathbf{A}_{12}\mathbf{x}_2(k) + \mathbf{B}_1\mathbf{u}(k)$$

$$\mathbf{x}_2(k+1) = \mathbf{A}_{21}\mathbf{x}_1(k) + \mathbf{A}_{22}\mathbf{x}_2(k) + \mathbf{B}_2\mathbf{u}(k)$$

and

$$\mathbf{y}(k) = \mathbf{x}_1(k) + \mathbf{D}\mathbf{u}(k)$$

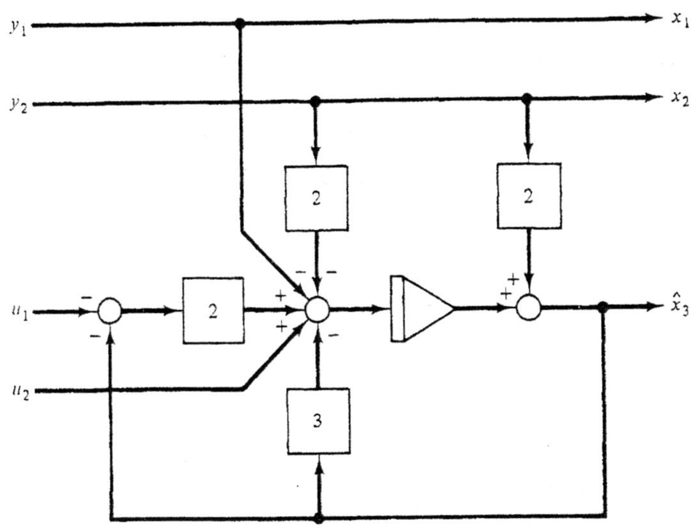

Figure 13.12

Since only the noise-free case is being discussed, $\mathbf{y}$ and $\mathbf{u}$ are assumed known exactly at all times. The third equation plays only the role of giving $\mathbf{x}_1$ if $\mathbf{D}$ is nonzero. Therefore $\mathbf{x}_1$ can be treated as known quantity at all times. Define two more "known" quantities

$$\mathbf{y}_r(k) \triangleq \mathbf{x}_1(k+1) - \mathbf{A}_{11}\mathbf{x}_1(k) - \mathbf{B}_1\mathbf{u}(k)$$
$$\mathbf{v}_r(k) \triangleq \mathbf{A}_{21}\mathbf{x}_1(k) + \mathbf{B}_2\mathbf{u}(k) \tag{13.38}$$

Then the new equivalent dynamic equation is

$$\mathbf{x}_2(k+1) = \mathbf{A}_{22}\mathbf{x}_2(k) + \mathbf{v}_r(k)$$

and the new equivalent measurement equation is

$$\mathbf{y}_r(k) = \mathbf{A}_{12}\mathbf{x}_2(k)$$

A linear observer form is postulated as

$$\hat{\mathbf{x}}_2(k+1) = \mathbf{A}_r\hat{\mathbf{x}}_2(k) + \mathbf{L}_r\mathbf{y}_r(k) + \mathbf{z}_r(k)$$

The error $\mathbf{e}(k) \triangleq \mathbf{x}_2(k) - \hat{\mathbf{x}}_2(k)$ satisfies

$$\mathbf{e}(k+1) = \mathbf{A}_{22}\mathbf{x}_2(k) + \mathbf{v}_r(k) - \mathbf{A}_r\hat{\mathbf{x}}_2 - \mathbf{L}_r\mathbf{y}_r(k) - \mathbf{z}_r(k)$$

If the selections $\mathbf{z}_r(k) = \mathbf{v}_r(k)$ and $\mathbf{A}_r = \mathbf{A}_{22} - \mathbf{L}_r\mathbf{A}_{12}$ are made, then

$$\mathbf{e}(k+1) = (\mathbf{A}_{22} - \mathbf{L}_r\mathbf{A}_{12})\mathbf{e}(k)$$

The convergence rate of $\mathbf{e}(k)$ to zero can be controlled by proper choice of $\mathbf{L}_r$, just as in the case of the full state observer. Figure 13.13 gives the reduced order observer in block diagram form.

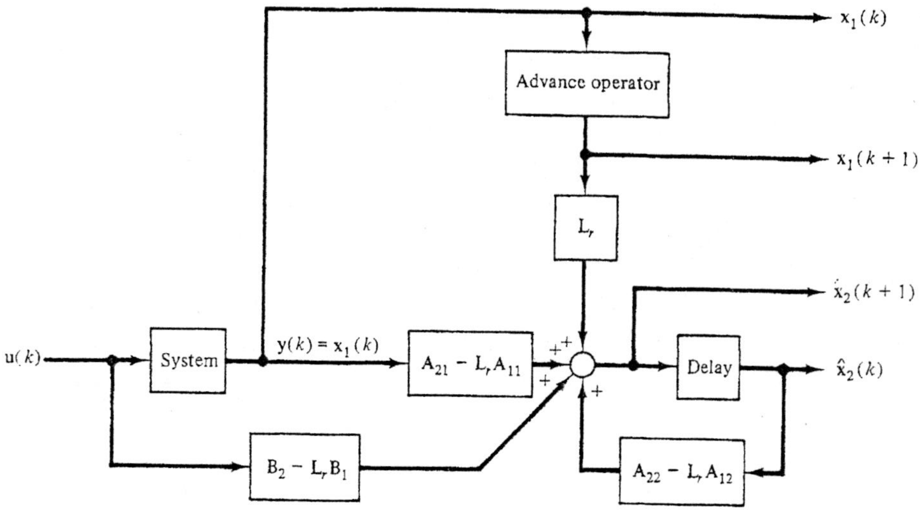

(a)

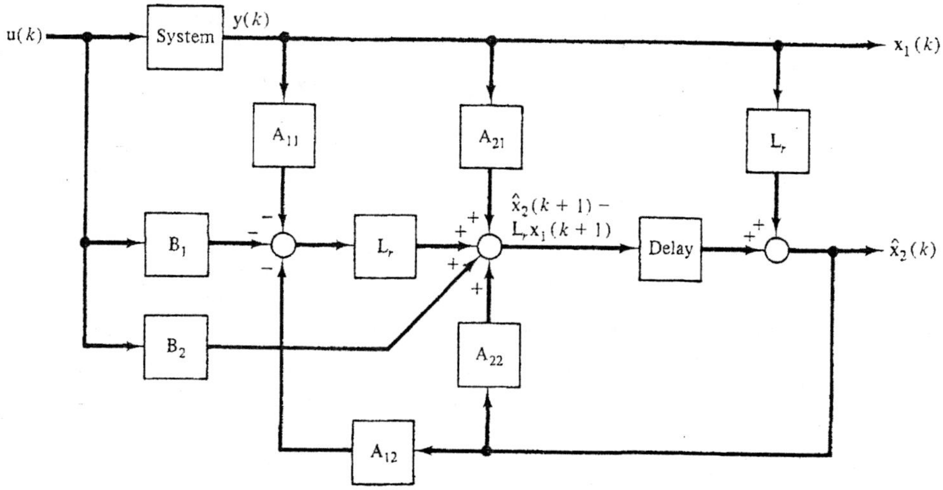

(b)

Figure 13.13 (a) Nonrealizable advanced used (b) No advance operator required.

13.7 A SEPARATION PRINCIPLE FOR FEEDBACK CONTROLLERS

One of the major uses of the state reconstruction observers of the previous section is state feedback control system design.

The observers described previously are used to estimate $\mathbf{x}$. If a constant-state feedback matrix $\mathbf{K}$ is then used, with $\hat{\mathbf{x}}$ as input instead of $\mathbf{x}$, the system shown in Figure 13.14 is obtained. The composite system is of order $2n$. By proper selection of $\mathbf{K}$, n of

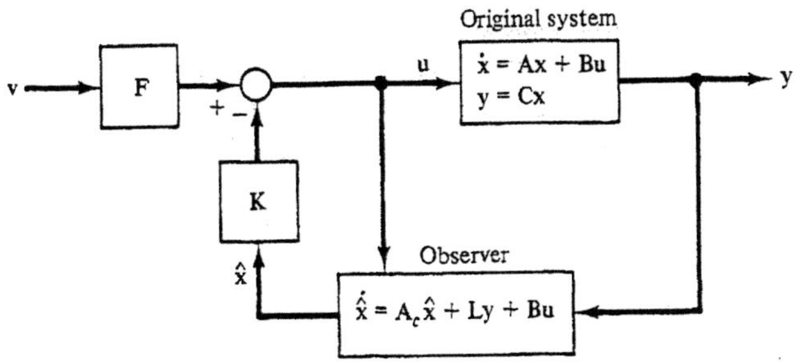

Figure 13.14

the closed-loop eigenvalues can be specified as in Sec. 13.4. By proper selection of **L**, the remaining n eigenvalues of the observer can be specified. This represents a *separation principle.* That is, a feedback system with the desired poles can be designed, proceeding as if all states were measurable. Then a separate design of the observer can be used to provide the desired observer poles. To see this, note that

$$\dot{\mathbf{x}} = \mathbf{Ax} + \mathbf{Bu}, \qquad \dot{\hat{\mathbf{x}}} = \mathbf{A}_c\hat{\mathbf{x}} + \mathbf{LCx} + \mathbf{Bu}, \qquad \mathbf{u} = \mathbf{Fv} - \mathbf{K}\hat{\mathbf{x}}$$

can be combined into

$$\begin{bmatrix} \dot{\mathbf{x}} \\ \dot{\hat{\mathbf{x}}} \end{bmatrix} = \begin{bmatrix} \mathbf{A} & -\mathbf{BK} \\ \mathbf{LC} & \mathbf{A}_c - \mathbf{BK} \end{bmatrix} \begin{bmatrix} \mathbf{x} \\ \hat{\mathbf{x}} \end{bmatrix} + \begin{bmatrix} \mathbf{BF} \\ \mathbf{BF} \end{bmatrix} \mathbf{v}$$

The $2n$ closed-loop eigenvalues are roots of

$$\Delta_T(\lambda) = \begin{vmatrix} \mathbf{I}_n\lambda - \mathbf{A} & \mathbf{BK} \\ -\mathbf{LC} & \mathbf{I}_n\lambda - \mathbf{A}_c + \mathbf{BK} \end{vmatrix} = 0$$

This can be reduced to a block triangular form by a series of elementary operations. Subtracting rows $i(1 \le i \le n)$ from rows $n+i$, and then adding columns $n+j$ $(1 \le j \le n)$ to columns j gives

$$\Delta_T(\lambda) = \begin{vmatrix} \mathbf{I}_n\lambda - \mathbf{A} & \mathbf{BK} \\ -\mathbf{I}_n\lambda + \mathbf{A} - \mathbf{LC} & \mathbf{I}_n\lambda - \mathbf{A}_c \end{vmatrix} = \begin{vmatrix} \mathbf{I}_n\lambda - \mathbf{A} + \mathbf{BK} & \mathbf{BK} \\ \mathbf{0} & \mathbf{I}_n\lambda - \mathbf{A}_c \end{vmatrix}$$

$$= |\mathbf{I}_n\lambda - \mathbf{A} + \mathbf{BK}| \cdot |\mathbf{I}_n\lambda - \mathbf{A}_c| = \Delta'(\lambda) \cdot \Delta_c(\lambda)$$

This verifies that the two sets of n eigenvalues can be specified separately, provided $(\mathbf{A}, \mathbf{B}, \mathbf{C})$ is completely controllable and completely observable. Experience indicates that a good design usually results if the continuous-time observer poles are selected to be a little farther to the left in the s-plane than the desired closed-loop state feedback poles. The discrete-time observer poles should be somewhat nearer the Z-plane origin so that the transients die out faster than the dominant system modes. Sec. 14.7 points out that there are other factors involved in the selection of observer pole locations.

13.8 TRANSFER FUNCTION VERSION OF POLE PLACEMENT–OBSERVER DESIGN

In this section the combined pole placement–observer design problem is reconsidered using the transfer function point of view. For the single-input, single-output case, a complete and very satisfying set of results is obtained. Only a brief introduction to the multiple-input, multiple-output problem is presented. The full treatment is left for the references because the main focus of this book is on state variable methods.

Consider the single-input, single-output nth-order system

$$\dot{\mathbf{x}} = \mathbf{Ax} + \mathbf{B}u$$
$$y = \mathbf{Cx} + \mathbf{D}u$$

The Laplace transform input-output relation is

$$y(s) = \{\mathbf{C}(s\mathbf{I} - \mathbf{A})^{-1}\mathbf{B} + D\}u(s)$$

Throughout this section the input-output transfer function will be written as the ratio of two polynomials $a(s)$ and $b(s)$. That is,

$$\mathbf{C}(s\mathbf{I} - \mathbf{A})^{-1}\mathbf{B} + D = \frac{b(s)}{a(s)}$$

In this case, the degree of the denominator polynomial is n, the number of state variables. The degree of the numerator polynomial $b(s)$ will clearly be n or less for any transfer function derived from state equations. If $D = 0$, it will always have a degree less than n. Any general transfer function $b(s)/a(s)$, not necessarily derived from state equations, is said to be proper if

$$\text{degree}[b(s)] \le \text{degree}[a(s)]$$

If the strict inequality holds, the transfer function is said to be *strictly proper*. Therefore, assuming that $b(s)/a(s)$ is strictly proper is equivalent to assuming that $D = 0$. For simplicity, this assumption is made throughout most of this section. The treatment of nonzero D is similar to that presented in Sec. 13.5 and 13.6. That is, a modified output $y' = y - Du$ can be used in the implementation.

From the treatment of Sec. 13.6, the form of the full state observer is

$$\dot{\hat{\mathbf{x}}} = \mathbf{A}_c\hat{\mathbf{x}} + \mathbf{L}y + \mathbf{B}u$$

The observer output $\hat{\mathbf{x}}$ is then multiplied by the pole placement–control feedback gain matrix $\mathbf{K}$, a $1 \times n$ row. The final quantity being fed back is a scalar $w = \mathbf{K}\hat{\mathbf{x}}$. Figure 13.15$a$ shows the standard state variable form of the system, the observer, and the control feedback gain. In this section it is convenient to write the transform of the feedback signal $w(s)$ as

$$\begin{aligned} w(s) &= \mathbf{K}(s\mathbf{I} - \mathbf{A}_c)^{-1}[\mathbf{L}y(s) + \mathbf{B}u(s)] \\ &= [q(s)/d(s)]y(s) + [e(s)/d(s)]u(s) \end{aligned}$$

Clearly the denominator of both transfer functions from y to w and from u to w is $d(s) = \det\{s\mathbf{I} - \mathbf{A}_c\}$. Figure 13.15$b$ gives the same pole placement–observer system in transfer function form. Figure 13.15b,c, and d give several possible rearrangements of this basic diagram. In each case the final closed-loop transfer function relation between the external reference signal $v(s)$ and the output $y(s)$ is found to be

$$\frac{y(s)}{v(s)} = \frac{b(s)d(s)}{a(s)[d(s) + e(s)] + b(s)q(s)} \tag{13.39}$$

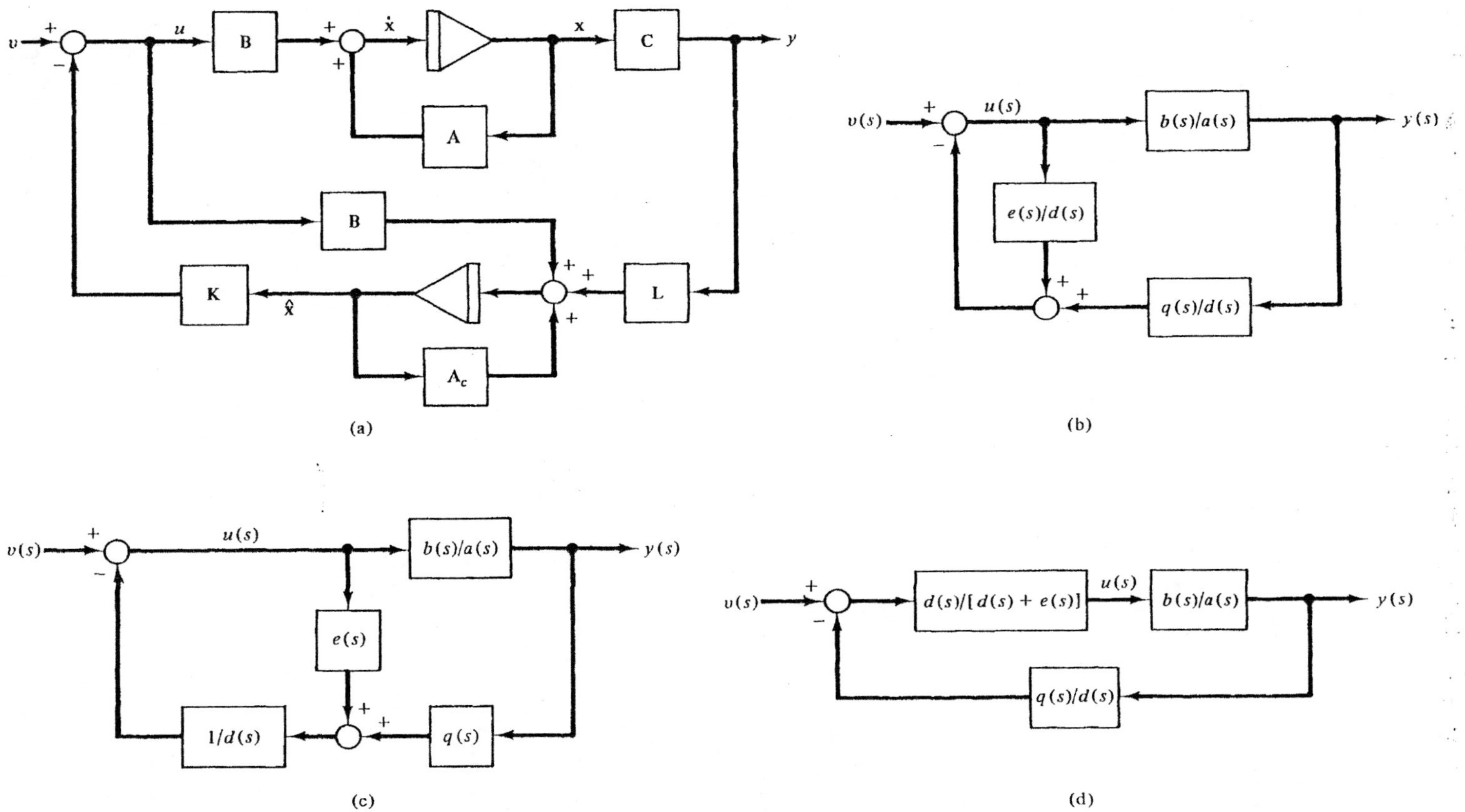

Figure 13.15 Various rearrangements of the pole placement–observer system.

Obviously, if the controller gain **K**, the observer gain **L**, and the observer system matrix $\mathbf{A}_c$ were known, say from methods presented in Secs. 13.4 and 13.6, then the complete input-output transfer function of Eq. (*13.39*) would be known. The central purpose of this section is to reverse the process. Given a system described by $b(s)/a(s)$, is it possible to determine the polynomials $d(s)$, $e(s)$, and $q(s)$ so that the closed-loop system has arbitrarily specified pole locations? From previous sections the answer is known to be yes if the open-loop system is completely controllable and completely observable. In the present transfer function context, the question can be stated entirely in terms of polynomial algebra. Let $d(s)+e(s)$ be renamed $p(s)$. Let the desired closed-loop characteristic polynomial be $c(s)$. For specified $a(s)$, $b(s)$, and $c(s)$, is it possible to find polynomials $p(s)$ and $q(s)$ which satisfy the Diophantine equation

$$a(s)p(s)+b(s)q(s)=c(s) \tag{13.40}$$

Several things are obvious directly from Eq. (*13.40*). First, if $a(s)$ and $b(s)$ have one or more common factors, then $c(s)$ cannot be specified arbitrarily. Of necessity, $c(s)$ must possess the same common factors as a and b. It was shown in Chapter 11 that system pole-zero cancellations indicate a loss either of controllability or observability or both. Common factors do not occur in a completely controllable and completely observable system. The absence of common factors in $a(s)$ and $b(s)$ will be shown to be both necessary and sufficient in the following development, provided that certain meaningful polynomial degree restrictions are enforced. For example, it is obvious from Eq. (*13.40*) that if $a(s)$ is of degree 5 and $b(s)$ is of degree 3, then a $c(s)$ with degree 2 cannot be specified.

13.8.1 The Full State Observer

With the hindsight of the state variable origin of Eq. (*13.40*), it is known that a full state observer $d(s)$ will have the same degree n as $a(s)$. We assume strictly proper transfer functions (for now), so that $b(s)$, $q(s)$, and $e(s)$ are assumed to have degree $n-1$. Some of their coefficients could be zero, leading to a lower degree. Therefore, $p(s)$ is of degree n also. A meaningfully restricted version of the Diophantine equation then is

Given arbitrary polynomials $a(s)$ of degree n,
$b(s)$ of degree $n-1$
$c(s)$ of degree $2n$

Find solution polynomials $p(s)$ of degree n
$q(s)$ of degree $n-1$

That satisfy Eq. (*13.40*)

After determining $p(s)$, a further specification of the observer characteristic equation $d(s)$ allows determination of

$$e(s)=p(s)-d(s)$$

In the state variable treatment of Sec. 13.7, the separation theorem results showed that the n desired control poles ($r(s)$) and the n more rapidly decaying observer poles ($d(s)$)

could be specified separately by choice of **K** and **L**. Then the composite system characteristic polynomial is the product $c(s) = d(s)r(s)$. With this (usual) choice, the final closed-loop system transfer function of Eq. (*13.39*) becomes

$$\frac{b(s)d(s)}{r(s)d(s)} = \frac{b(s)}{r(s)}$$

The cancellation leaves the n desired control poles and the original open-loop zeros. The observer modes completely cancel. They are unobservable hidden modes. In the present transfer function treatment, somewhat more freedom of choice exists. It is possible, for example, to select $c(s) = r(s)b(s)$ in order to cancel the open-loop zeros, if stable, and replace them by freely selected poles of the observer, $d(s)$. Other choices for $c(s)$ might cancel only certain factors of $b(s)$ (see also Sec. 14.7 regarding choice of observer poles). The difference between the approach of this section and the earlier sections is that here $c(s)$ and $d(s)$ may be selected arbitrarily, subject only to degree constraints. In the state variable treatment, $d(s)$ and $r(s)$ were selected.

Solution of the Diophantine equation. Consider the known polynomials

$$\begin{aligned} a(s) &= s^n + a_{n-1}s^{n-1} + a_{n-2}s^{n-2} + \cdots + a_1 s + a_0 \\ b(s) &= b_{n-1}s^{n-1} + b_{n-2}s^{n-2} + \cdots + b_1 s + b_0 \\ c(s) &= s^{2n} + c_{2n-1}s^{2n-1} + \cdots + c_1 s + c_0 \end{aligned} \tag{13.41}$$

Determine the coefficients of the unknown polynomials of Eq. (*13.40*),

$$\begin{aligned} p(s) &= s^n + p_{n-1}s^{n-1} + p_{n-2}s^{n-2} + \cdots + p_1 s + p_0 \\ q(s) &= q_{n-1}s^{n-1} + q_{n-2}s^{n-2} + \cdots + q_1 s + q_0 \end{aligned} \tag{13.42}$$

Note that $a(s)$ and $c(s)$ are assumed to be monic (i.e., highest coefficient normalized to unity), and thus $p(s)$ must also be monic. As a result, there are n unknowns in p and n more unknowns in q. By substituting the polynomial forms into the Diophantine equation and then equating coefficients of like powers of s, $2n$ equations are obtained. These will be referred to as the *matrix form of the Diophantine equation.*

$$\begin{bmatrix} 1 & 0 & 0 & \cdots & 0 & 0 & 0 & 0 & \cdots & 0 \\ a_{n-1} & 1 & 0 & \cdots & 0 & b_{n-1} & 0 & 0 & \cdots & 0 \\ a_{n-2} & a_{n-1} & 1 & \cdots & 0 & b_{n-2} & b_{n-1} & 0 & \cdots & 0 \\ a_{n-3} & a_{n-2} & a_{n-1} & \cdots & 0 & b_{n-3} & b_{n-2} & b_{n-1} & \cdots & 0 \\ \vdots & & & & & \vdots & & & & \vdots \\ a_1 & a_2 & a_3 & \cdots & 1 & b_1 & b_2 & b_3 & \cdots & 0 \\ a_0 & a_1 & a_2 & \cdots & a_{n-1} & b_0 & b_1 & b_2 & \cdots & b_{n-1} \\ 0 & a_0 & a_1 & \cdots & a_{n-2} & 0 & b_0 & b_1 & \cdots & b_{n-2} \\ 0 & 0 & a_0 & \cdots & a_{n-3} & 0 & 0 & b_0 & \cdots & b_{n-3} \\ & \vdots & & & \vdots & & & & & \vdots \\ 0 & 0 & 0 & & a_1 & 0 & 0 & 0 & \cdots & b_1 \\ 0 & 0 & 0 & & a_0 & 0 & 0 & 0 & \cdots & b_0 \end{bmatrix} \begin{bmatrix} p_{n-1} \\ p_{n-2} \\ p_{n-3} \\ p_{n-4} \\ \vdots \\ p_0 \\ q_{n-1} \\ q_{n-2} \\ q_{n-3} \\ \vdots \\ q_1 \\ q_0 \end{bmatrix} = \begin{bmatrix} c_{2n-1} - a_{n-1} \\ c_{2n-2} - a_{n-2} \\ c_{2n-3} - a_{n-3} \\ c_{2n-4} - a_{n-4} \\ \vdots \\ c_n - a_0 \\ c_{n-1} \\ c_{n-2} \\ c_{n-3} \\ \vdots \\ c_1 \\ c_0 \end{bmatrix} \tag{13.43}$$

It is known from Chapters 5 and 6 that unique solutions for the $2n$ unknowns exist if and only if the $2n \times 2n$ coefficient matrix is nonsingular. It can be shown that the matrix is nonsingular if and only if the polynomials $a(s)$ and $b(s)$ have no common factors. A direct algebraic proof in the general case is tedious. A demonstration for a second-order system is now given. The general proof, using other methods, may be found in Reference 9.

Consider the second-order transfer function

$$\frac{b(s)}{a(s)} = \frac{b_1 s + b_0}{s^2 + a_1 s + a_0}$$

In this case there are $2n = 4$ simultaneous equations, given by

$$\begin{bmatrix} 1 & 0 & 0 & 0 \\ a_1 & 1 & b_1 & 0 \\ a_0 & a_1 & b_0 & b_1 \\ 0 & a_0 & 0 & b_0 \end{bmatrix} \begin{bmatrix} p_1 \\ p_0 \\ q_1 \\ q_0 \end{bmatrix} = \begin{bmatrix} c_3 - a_1 \\ c_2 - a_0 \\ c_1 \\ c_0 \end{bmatrix}$$

A unique solution exists if the 4×4 matrix has a nonzero determinant. Laplace expansion with respect to row 1 reduces the problem to a 3×3 determinant, which when set to zero gives

$$b_0^2 - a_1 b_1 b_0 + b_1^2 a_0 = 0$$

If $b_1 = 0$, then b_0 must also be zero in order to yield a singular matrix. Assume this degenerate case does not apply and solve for the ratio $b_0/b_1 = a_1/2 \pm [(a_1/2)^2 - a_0]^{1/2}$. This ratio for b_0/b_1, which forces the 4×4 determinant to zero, also makes $s + b_0/b_1$ a factor of the quadratic $a(s)$. Aside from the degenerate case where both b_1 and b_0 are zero, the determinant can vanish only if the numerator term is a factor in the denominator.

EXAMPLE 13.15 A second-order system has the transfer function

$$\frac{y(s)}{u(s)} = \frac{s + 1}{s^2 + 3s + 2}$$

Can the preceding scheme for designing a pole placement–observer system be used to give an arbitrary fourth-order closed-loop characteristic equation $c(s)$?

The 4×4 coefficient matrix is

$$\begin{bmatrix} 1 & 0 & 0 & 0 \\ 3 & 1 & 1 & 0 \\ 2 & 3 & 1 & 1 \\ 0 & 2 & 0 & 1 \end{bmatrix}$$

Its determinant is easily seen to be zero, and its rank is 3. This indicates that no unique solution exists for arbitrary $c(s)$ polynomials. However, nonunique solutions will exist for *certain* $c(s)$ polynomials (see Chapter 6). The zero determinant has been caused by the factor $s + 1$, which is common to both $a(s)$ and $b(s)$. ■

EXAMPLE 13.16 Assume that the system of the previous example is changed so that $b(s) = s + 5$. Design a full state observer-controller which gives closed-loop poles at $s = -3 \pm 4j$, $s = -5$, and $s = -10$.

The desired polynomial $c(s)$ is

$$c(s) = (s^2 + 6s + 25)(s + 5)(s + 10)$$
$$= s^4 + 21s^3 + 165s^2 + 675s + 1250$$

The matrix form of the appropriate Diophantine equation is

$$\begin{bmatrix} 1 & 0 & 0 & 0 \\ 3 & 1 & 1 & 0 \\ 2 & 3 & 5 & 1 \\ 0 & 2 & 0 & 5 \end{bmatrix} \begin{bmatrix} p_1 \\ p_0 \\ q_1 \\ q_0 \end{bmatrix} = \begin{bmatrix} 21 - 3 \\ 165 - 2 \\ 675 \\ 1250 \end{bmatrix}$$

Generally a machine solution would be in order, but for low-order problems such as this the solution can easily be obtained manually. It is found that $p(s) = s^2 + 18s + 65$ and $q(s) = 44s + 224$. Knowing $p(s)$, the selection of $d(s)$ will give $e(s)$. Suppose the observer characteristic polynomial $d(s)$ is selected as $(s + 10)(s + 6) = s^2 + 16s + 60$. Note specifically that the factors of $d(s)$ were *not* both selected to be factors of $c(s)$. This will cancel the open-loop zero at $s = -5$ and replace it by a zero at $s = -6$. Finally, $e(s) = p(s) - d(s) = 2s + 5$. By using the $q(s)/d(s)$ and $e(s)/d(s)$ transfer functions in various ways, the final compensated system can be represented in the different forms of Figure 13.15. Note, however, that the hidden modes $(s + 5)$ and $(s + 10)$, while not observable in the output, still exist internally. If parameter errors have caused inexact cancellation, some small contribution from these modes may be observed. In this case, these modes are sufficiently stable so that their contribution due to initial conditions or disturbances would decay rapidly. ■

EXAMPLE 13.17 Consider the system used earlier in Examples 13.1 and 13.10. For comparison purposes, use the transfer function methods to design a closed-loop system with poles at $s = -3$ and -4. Put the two full state observer poles at $s = -8$, as in Example 13.10.

For the given **A**, **B**, and **C**, it is easy to show that the open-loop transfer function is $\mathbf{C}[s\mathbf{I} - \mathbf{A}]^{-1}\mathbf{B} = 2/[s(s - 3)] = b(s)/a(s)$. The desired closed-loop characteristic equation is

$$c(s) = (s + 3)(s + 4)(s + 8)^2$$
$$= s^4 + 23s^3 + 188s^2 + 640s + 768$$

The matrix form of the Diophantine equation is

$$\begin{bmatrix} 1 & 0 & 0 & 0 \\ -3 & 1 & 0 & 0 \\ 0 & -3 & 2 & 0 \\ 0 & 0 & 0 & 2 \end{bmatrix} \begin{bmatrix} p_1 \\ p_0 \\ q_1 \\ q_0 \end{bmatrix} = \begin{bmatrix} 26 \\ 188 \\ 640 \\ 768 \end{bmatrix}$$

The solution is especially easy here because both b_1 and a_0 are zero; it is given by

$$p(s) = s^2 + 26s + 266$$
$$q(s) = 719s + 384$$

Both observer poles are specified to be at $s = -8$, so $d(s) = (s + 8)^2$. This gives $e(s) = 10s + 202$.

The feedback signal is

$$w(s) = \mathbf{K}\hat{\mathbf{x}}(s) = \frac{(719s + 384)y(s) + (10s + 202)u(s)}{s^2 + 16s + 64}$$

and the closed-loop transfer function, after cancellations, is

$$\frac{y(s)}{v(s)} = \frac{2}{s^2 + 7s + 12}$$

For comparison purposes, the observer of Example 13.9 gave

$$\dot{\hat{\mathbf{x}}} = \begin{bmatrix} -19 & 2 \\ -\frac{121}{2} & 3 \end{bmatrix}\hat{\mathbf{x}} + \begin{bmatrix} 19 \\ \frac{121}{2} \end{bmatrix} y + \begin{bmatrix} 0 \\ 1 \end{bmatrix} u$$

The pole placement gain of Example 13.1 was $\mathbf{K} = [6 \quad 10]$. By taking Laplace transforms, it is found that

$$\mathbf{K}\hat{\mathbf{x}} = \mathbf{K}[s\mathbf{I} - \mathbf{A}_c]^{-1}\left\{\begin{bmatrix} 19 \\ \frac{121}{2} \end{bmatrix} y(s) + \begin{bmatrix} 0 \\ 1 \end{bmatrix} u(s)\right\}$$

When expanded, this agrees precisely with the preceding expression for $w(s) = q(s)/d(s)y(s) + e(s)/d(s)u(s)$. ■

EXAMPLE 13.18 Can the desired closed-loop poles of the previous example be achieved using only dynamic output feedback?

Dynamic output feedback means using only the $y(s)$ signal, passed through a strictly proper transfer function $q(s)/d(s)$ but without an $e(s)/d(s)u(s)$ component. This means that $e(s) = 0$ is required; hence $d(s) = p(s)$. If this is the selection, the resulting closed-loop transfer function is found to be

$$\frac{y(s)}{v(s)} = \frac{d(s)b(s)}{c(s)}$$

$$= \frac{2(s^2 + 26s + 266)}{(s^2 + 7s + 12)(s^2 + 16s + 64)}$$

The conclusion is that the desired control poles can be achieved, as well as the other two poles at $s = -8$. The open-loop $b(s)$ is still a numerator factor. However, the feedback path denominator poles (which are no longer associated with an observer) must be accepted as whatever they come out, i.e., $p(s)$. The observer poles are replaced by the dynamic compensator poles, and these are no longer factors in the overall closed-loop denominator. Hence the $d(s) = p(s)$ factor in the numerator no longer cancels. The separation principle property has been lost by not using the extra feedback path from $u(s)$ to $w(s)$. ■

13.8.2 The Reduced Order Observer

From the state variable treatment of the pole placement–observer problem, it is known that it is necessary to estimate only $n - m$ states, where m is rank of the output matrix $\mathbf{C}$. If $\mathbf{C}$ is full rank, this is the same as the number of outputs. The reduced order observer requires a dynamical system of order $n - m$ rather than n. In the single output case, this means that $d(s)$ and $p(s)$ need be only of degree $n - 1$, and hence $c(s)$ need be only of degree $2n - 1$. The previous results are only slightly modified by these changes in degree. Thus the modified problem becomes:

Given $a(s) = n$th degree monic polynomial, as before
$b(s) = (n - 1)$st degree polynomial, as before
$c(s) = (2n - 1)$st degree monic polynomial

Find $p(s) = (n-1)$st degree monic polynomial
$q(s) = (n-1)$st degree polynomial

Such that the modified Diophantine equation

$$a(s)p(s) + b(s)q(s) = c(s)$$

is satisfied. The polynomial coefficients are numbered as before; that is, the subscript agrees with the power of s it multiplies so the constant term has the 0 subscript. Substituting the polynomials into the Diophantine equation, expanding, and equating like powers of s gives the following $2n-1$ equations for the $2n-1$ unknowns (n coefficients q_i and $n-1$ coefficients p_i).

$$\begin{bmatrix}
1 & 0 & 0 & \cdots & 0 & 0 & b_{n-1} & 0 & 0 & \cdots & 0 \\
a_{n-1} & 1 & 0 & \cdots & 0 & 0 & b_{n-2} & b_{n-1} & 0 & \cdots & 0 \\
a_{n-2} & a_{n-1} & 1 & \cdots & 0 & 0 & b_{n-3} & b_{n-2} & b_{n-1} & \cdots & 0 \\
\cdot & & & \cdots & 1 & 0 & \vdots & & \vdots & & \vdots \\
\cdot & & & & a_{n-1} & 1 & & & & & 0 \\
\cdot & & & & & a_{n-1} & b_0 & & & & b_{n-1} \\
a_0 & a_1 & a_2 & \cdots & & a_{n-2} & & & & & \vdots \\
\vdots & & & & & \vdots & & & b_1 & b_2 & b_3 \\
0 & 0 & 0 & \cdots & a_1 & a_2 & 0 & \cdots & b_0 & b_1 & b_2 \\
0 & 0 & 0 & \cdots & a_0 & a_1 & 0 & \cdots & 0 & b_0 & b_1 \\
0 & 0 & 0 & \cdots & 0 & a_0 & 0 & \cdots & 0 & 0 & b_0
\end{bmatrix}
\begin{bmatrix}
p_{n-2} \\ p_{n-3} \\ p_{n-4} \\ \vdots \\ p_0 \\ q_{n-1} \\ q_{n-2} \\ \vdots \\ q_2 \\ q_1 \\ q_0
\end{bmatrix}$$

$$= \left[c_{2n-2} - a_{n-1} \;\; c_{2n-3} - a_{n-2} \;\; c_{2n-4} - a_{n-3} \;\; \cdots \;\; c_n - a_1 \;\; c_{n-1} - a_0 \;\; c_{n-2} \;\; \cdots \;\; c_2 \;\; c_1 \;\; c_0\right]^T \tag{13.44}$$

As in the full state observer case, it can be shown that the $(2n-1) \times (2n-1)$ coefficient matrix is nonsingular if and only if $a(s)$ and $b(s)$ have no common factors. If this is true, which it will be for all open-loop systems which are completely controllable and completely observable, a unique solution for the coefficients of $p(s)$ and $q(s)$ can be found. As in the full state case, after finding $p(s)$, a specification of the observer denominator $d(s)$ gives $e(s)$ through the polynomial relation $p(s) = d(s) + e(s)$. The block diagram representations of Figure 13.15 remain valid.

EXAMPLE 13.19 Design a reduced-order observer-pole placement controller for a system with the open-loop transfer function

$$\frac{b(s)}{a(s)} = \frac{1}{s(s+1)}$$

such that the desired closed-loop poles are at $s = -1 \pm j$ and the observer pole is at $s = -2$. This same example is given in References 9 and 10.

The given polynomials have coefficients $a_0 = 0$, $a_1 = 1$, $b_0 = 1$, and $b_1 = 0$. The total system order will be 3, (second-order system plus first-order observer) and the desired $c(s)$ coefficients are $c_0 = 4$, $c_1 = 6$, and $c_2 = 4$. The three simultaneous equations are

$$\begin{bmatrix} 1 & 0 & 0 \\ 1 & 1 & 0 \\ 0 & 1 & 1 \end{bmatrix}\begin{bmatrix} p_0 \\ q_1 \\ q_0 \end{bmatrix} = \begin{bmatrix} 4-1 \\ 6-0 \\ 4 \end{bmatrix}$$

and the solution gives $p(s) = s + 3, q(s) = 3s + 4$. Since it has been specified that the observer should have a pole at $s = -2$, $d(s) = s + 2$, so $e(s) = 1$. The two transfer functions $q(s)/d(s)$ and $e(s)/d(s)$ can be used to manipulate the composite system into various forms, as before. Figure 13.16 presents the solution in a form which shows a simple lead compensator in the feedback path and another lead compensator in the forward path. After cancellations, the closed-loop transfer function is $y(s)/v(s) = 1/[(s + 1 + j)(s + 1 - j)]$, as desired. ■

13.8.3 The Discrete-Time Pole Placement–Observer Problem

If a discrete-time system is described by a strictly proper Z-transfer function $y(z)/u(z) = b(z)/a(z)$, then all the previous material for continuous-time Laplace-transformed systems carries over exactly. The substitution of the complex variable z for s is the only change required. This is true in both the full state and reduced-order observer cases which have been considered.

EXAMPLE 13.20 *Discrete-time full state observer:* Consider the third-order system of Problem 13.13 but modified to satisfy the current restriction of a single input and output,

$$\mathbf{B} = [-1 \quad 1 \quad 3]^T$$

The open-loop input-output transfer function is easily found to be

$$\frac{y(z)}{u(z)} = \frac{-z^2 - 0.5z + 0.61}{z^3 - 0.3z^2 - 0.51z + 0.194}$$

Using the transfer function approach, design a system with control poles at $z = 0.5 + 0.4j$, $0.5 - 0.4j$, and 0.3. Put the full-state observer poles at $z = 0, 0.1$, and 0.2.

Expanding the desired factors gives the sixth-order characteristic polynomial

$$c(z) = z^6 - 1.6z^5 + 1.12z^4 - 0.362z^3 + 0.0511z^2 - 0.00246z$$

The matrix Diophantine equation to be solved for $p(z)$ and $q(z)$ is

$$\begin{bmatrix} 1 & 0 & 0 & 0 & 0 & 0 \\ -0.3 & 1 & 0 & -1 & 0 & 0 \\ -0.51 & -0.3 & 1 & -0.5 & -1 & 0 \\ 0.194 & -0.51 & -0.3 & 0.61 & -0.5 & -1 \\ 0 & 0.194 & -0.51 & 0 & 0.61 & -0.5 \\ 0 & 0 & 0.194 & 0 & 0 & 0.61 \end{bmatrix} \begin{bmatrix} p_2 \\ p_1 \\ p_0 \\ q_2 \\ q_1 \\ q_0 \end{bmatrix} = \begin{bmatrix} -1.6 + 0.3 \\ 1.12 + 0.51 \\ -.362 - 0.194 \\ 0.0511 \\ -0.00246 \\ 0 \end{bmatrix}$$

The solution gives $q(z) = 1.1559z^2 - 1.6987z + 0.5155$ and $p(z) = z^3 - 1.3z^2 + 2.3959z - 1.621$. Selecting $d(z) = z^3 - 0.3z^2 + 0.02z$ gives $e(z) = -z^2 + 2.3759z - 1.621$, so that

$$\mathbf{K}\hat{\mathbf{x}} = \frac{-z^2 + 2.3759z - 1.621}{z^3 - 0.3z^2 + 0.02z} u(z) + \frac{1.1559z^2 - 1.6987z + 0.5155}{z^3 - 0.3z^2 + 0.02z} y(z)$$

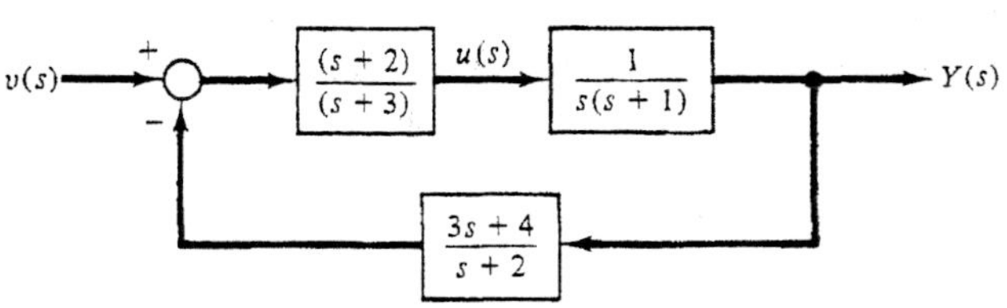

Figure 13.16

As a check on this result, the state variable method of Sec. 13.4 shows that state feedback gains of

$$\mathbf{K} = [0.51923 \quad -3.3423 \quad 0.953846]$$

will give the desired poles at $z = 0.5 \pm 0.4j$ and $z = 0.3$. The same algorithm applied to the dual problem gives observer gains

$$\mathbf{L} = [0 \quad -0.765734 \quad -1.471328]^T$$

to achieve observer poles at $z = 0, 0.1$, and 0.2. Using these in the observer equation

$$\hat{\mathbf{x}}(k+1) = (\mathbf{A} - \mathbf{LC})\hat{\mathbf{x}}(k) + \mathbf{L}y(k) + \mathbf{B}u(k)$$

allows verification that $\mathbf{K}(z\mathbf{I} - \mathbf{A} + \mathbf{LC})^{-1}\mathbf{L} = \mathbf{K}\hat{\mathbf{x}}(z)/y(z) = q(z)/d(z)$ and $\mathbf{K}(z\mathbf{I} - \mathbf{A} + \mathbf{LC})^{-1}\mathbf{B} = \mathbf{K}\hat{\mathbf{x}}/u = e(z)/d(z)$ agree with the transfer function results. The closed-loop transfer function for the composite system is given by

$$\frac{y(z)}{v(z)} = \frac{b(z)}{r(z)} = \frac{-z^2 - 0.5z + 0.61}{(z - 0.5 + 0.4j)(z - 0.5 - 0.4j)(z - 0.3)} \qquad (13.45)$$

Although this is what was asked for in the design specifications, it gives a less-than-desirable step response. Its time response due to a step input has excessive overshoot and oscillations. ∎

EXAMPLE 13.21 *Discrete-time reduced order observer:* Repeat the previous example, but use a reduced order observer with poles at $z = 0.1$ and 0.2.

Using the same desired control poles with the two observer poles gives $c(z) = z^5 - 1.6z^4 + 1.12z^3 - 0.362z^2 + 0.0511z - 0.00246$. The matrix version of the Diophantine equation gives five simultaneous equations:

$$\begin{bmatrix} 1 & 0 & -1 & 0 & 0 \\ -0.3 & 1 & -0.5 & -1 & 0 \\ -0.51 & -0.3 & 0.61 & -0.5 & -1 \\ 0.194 & -0.51 & 0 & 0.61 & -0.5 \\ 0 & 0.194 & 0 & 0 & 0.61 \end{bmatrix} \begin{bmatrix} p_1 \\ p_0 \\ q_2 \\ q_1 \\ q_0 \end{bmatrix} = \begin{bmatrix} -1.6 + 0.3 \\ 1.12 + 0.51 \\ -0.362 - 0.194 \\ 0.0511 \\ -0.00246 \end{bmatrix}$$

Numerical solution yields $p(z) = z^2 - 3.9573z + 1.0672$ and $q(z) = -2.6573z^2 + 1.9531z - 0.34345$. Using the specified observer poles gives the observer characteristic polynomial $d(z) = z^2 - 0.3z + 0.02$. This, along with $p(z)$, gives $e(z) = -3.6573z + 1.0472$. As usual, the controller-observer system is described by the two compensator transfer functions,

$$w(z) = \mathbf{K}\hat{\mathbf{x}}(z) = [e(z)/d(z)]u(z) + [q(z)/d(z)]y(z) \qquad (13.46)$$

The closed-loop transfer function is the same as in Example 13.20, Eq. (*13.45*). ∎

EXAMPLE 13.22 *Reduced order observer, state variable method:* Repeat the controller-observer design, using state variable methods, to verify the results of Example 13.21.

The control gains are unchanged, that is, $\mathbf{K} = [0.51923 \quad -3.3423 \quad 0.953846]$. Let $\mathbf{x}_2 = [x_2 \quad x_3]^T$ denote the states to be estimated by the reduced order observer. Using results from Problem 13.13, but now using only one input $u_1 = u$ (and the corresponding single column $\mathbf{B} = [-1 \quad 1 \quad 3]^T$), the observer equations are

$$\hat{\mathbf{x}}_2(k+1) = \begin{bmatrix} -0.504196 & 0.239161 \\ -1.7902 & 0.804196 \end{bmatrix} \hat{\mathbf{x}}_2(k)$$

$$+ \begin{bmatrix} 0.391608 \\ -1.95804 \end{bmatrix} [y(k+1) + u(k)] + \begin{bmatrix} -0.1 \\ 0.8 \end{bmatrix} y(k) + \begin{bmatrix} 1 \\ 3 \end{bmatrix} u(k)$$

Bringing $x_1(k) = y(k)$ back in, the entire three-component state reconstruction process is described by

$$\hat{\mathbf{x}}(k+1) = \begin{bmatrix} 0 & 0 & 0 \\ 0 & -0.504196 & 0.239161 \\ 0 & -1.77902 & 0.804196 \end{bmatrix} \hat{\mathbf{x}}(k) + \begin{bmatrix} 1 \\ 0.391608 \\ -1.95804 \end{bmatrix} y(k+1)$$

$$+ \begin{bmatrix} 0 \\ -0.1 \\ 0.8 \end{bmatrix} y(k) + \begin{bmatrix} 0 \\ 1.391688 \\ 1.04196 \end{bmatrix} u(k)$$

Taking Z-transforms, solving for $\hat{\mathbf{x}}(z)$, and then premultiplying by the gain row matrix $\mathbf{K}$ gives

$$\mathbf{K}\hat{\mathbf{x}}(z) = \mathbf{K}[z\mathbf{I} - \mathbf{A}_c]^{-1} \left\{ \begin{bmatrix} z \\ 0.391608z - 0.1 \\ -1.95804z + 0.8 \end{bmatrix} y(z) + \begin{bmatrix} 0 \\ 1.391608 \\ 1.04196 \end{bmatrix} u(z) \right\}$$

where $\mathbf{A}_c$ has been used to indicate the 3×3 matrix. Substituting values for $\mathbf{K}$ and simplifying leads to

$$\mathbf{K}\hat{\mathbf{x}}(z) = \frac{(-2.6573z^3 + 1.9531z^2 - 0.34345z)}{(z^2 - 0.3z + 0.02)z} y(z)$$

$$+ \frac{(-3.65756z + 1.0473)z}{(z^2 - 0.3z + 0.02)z} u(z)$$

After canceling a factor of z, this is the same result as obtained in Example 13.21, Eq. (*13.46*). ■

13.9 DESIGN OF DECOUPLED OR NONINTERACTING SYSTEMS [11,12]

A system with an equal number m of inputs and outputs is considered in this section. If the $m \times m$ transfer matrix $\mathbf{H}(s)$ is diagonal and nonsingular, the system is said to be *decoupled* because each input affects one and only one output. When the number of inputs and outputs are not equal, $\mathbf{H}(s)$ is not square, and thus cannot be diagonal. Various other kinds of decoupling and partial decoupling have been defined where $\mathbf{H}(s)$ is triangular or block diagonal and so on [13, 14].

The problem of reducing a system with m inputs and m outputs to decoupled form, using a state feedback control law $\mathbf{u} = -\mathbf{K}_d\mathbf{x} + \mathbf{F}_d\mathbf{v}$, is considered. Assuming that $\mathbf{D} = [\mathbf{0}]$, the transfer matrix for the state feedback system of Eqs. (*13.4*) and (*13.5*) is

$$\mathbf{H}(s) = \mathbf{C}[s\mathbf{I} - \mathbf{A} + \mathbf{B}\mathbf{K}_d]^{-1}\mathbf{B}\mathbf{F}_d \tag{13.47}$$

The decoupling problem is that of selecting matrices $\mathbf{F}_d(m \times m)$ and $\mathbf{K}_d(m \times n)$ so that $\mathbf{H}(s)$ is diagonal and nonsingular. Consider the inverse transform of equation (*13.47*) and the Cayley-Hamilton theorem applied to $e^{[\mathbf{A} - \mathbf{B}\mathbf{K}_d]t}$. This leads to an alternative statement of decoupling. The matrices

$$\mathbf{C}[\mathbf{A} - \mathbf{B}\mathbf{K}_d]^j\mathbf{B}\mathbf{F}_d, \qquad j = 0, 1, \ldots, n-1 \tag{13.48}$$

must all be diagonal if the system is decoupled. Let the ith row of $\mathbf{C}$ be $\mathbf{c}_i$ and define a set of m integers by

$$d_i = \min_j \{j | \mathbf{c}_i \mathbf{A}^j \mathbf{B} \neq \mathbf{0}, j = 0, 1, \ldots, n-1\} \tag{13.49}$$

or

$$d_i = n - 1 \qquad \text{if } \mathbf{c}_i \mathbf{A}^j \mathbf{B} = \mathbf{0} \text{ for all } j$$

The original system can be decoupled using state feedback [11] if and only if the following $m \times m$ matrix is nonsingular:

$$\mathbf{N} = \begin{bmatrix} \mathbf{c}_1 \mathbf{A}^{d_1} \mathbf{B} \\ \mathbf{c}_2 \mathbf{A}^{d_2} \mathbf{B} \\ \vdots \\ \mathbf{c}_m \mathbf{A}^{d_m} \mathbf{B} \end{bmatrix}$$

In particular, one set of decoupling matrices is

$$\mathbf{F}_d = \mathbf{N}^{-1} \quad \text{and} \quad \mathbf{K}_d = \mathbf{N}^{-1} \begin{bmatrix} \mathbf{c}_1 \mathbf{A}^{d_1+1} \\ \vdots \\ \mathbf{c}_m \mathbf{A}^{d_m+1} \end{bmatrix} \tag{13.50}$$

EXAMPLE 13.23 Determine whether the open-loop system of Example 13.9 can be decoupled using state feedback. Assume now that $\mathbf{D} = [\mathbf{0}]$.

Since $\mathbf{c}_1 \mathbf{A}^0 \mathbf{B} = [1 \quad 0] \neq \mathbf{0}$, set $d_1 = 0$. Also, $\mathbf{c}_2 \mathbf{A}^0 \mathbf{B} = \mathbf{0}$, but $\mathbf{c}_2 \mathbf{A} \mathbf{B} = [0 \quad 1] \neq \mathbf{0}$, so that $d_2 = 1$. Therefore, $\mathbf{N} = \begin{bmatrix} \mathbf{c}_1 \mathbf{B} \\ \mathbf{c}_2 \mathbf{A}\mathbf{B} \end{bmatrix} = \begin{bmatrix} 1 & 0 \\ 0 & 1 \end{bmatrix}$ is nonsingular and decoupling is possible. The decoupling matrices are $\mathbf{F}_d = \mathbf{N}^{-1} = \mathbf{I}$ and $\mathbf{K}_d = \mathbf{N}^{-1} \begin{bmatrix} \mathbf{c}_1 \mathbf{A} \\ \mathbf{c}_2 \mathbf{A}^2 \end{bmatrix} = \begin{bmatrix} 0 & 0 & 4 \\ 4 & -4 & 0 \end{bmatrix}$. Using these, the decoupled transfer matrix is obtained from Eq. (*13.47*):

$$\mathbf{H}(s) = \begin{bmatrix} 1/s & 0 \\ 0 & 1/s^2 \end{bmatrix}$$

■

Results of the previous example are typical in that all the poles of the decoupled system are at the origin. This is always the result if Eq. (*13.50*) is used. In fact, the general result is $\mathbf{H}(s) = \text{diag}[s^{-d_1-1} \quad s^{-d_2-1} \quad \cdots \quad s^{-d_m-1}]$. This system is said to be *integrator decoupled*. The peformance of such a decoupled system would usually not be acceptable. Two questions naturally arise. First, can decoupling be accomplished by using only the available outputs? The answer to this is yes, provided dynamic feedback compensators (observers for example) of sufficiently high dimension are allowed [15]. Second, can other state feedback matrices or output feedback matrices be used to prespecify closed-loop pole locations while preserving the decoupled nature of the system? The answer to this question is yes in some cases, no in others. At least $m + \sum_{i=1}^{m} d_i$ poles can be prespecified, but not necessarily all of the poles [11].

EXAMPLE 13.24 Consider the decoupled system found in Example 13.23. Find a constant feedback matrix which moves the closed-loop poles from zero to $\{-5, -5, -10\}$. Is the resultant system still decoupled?

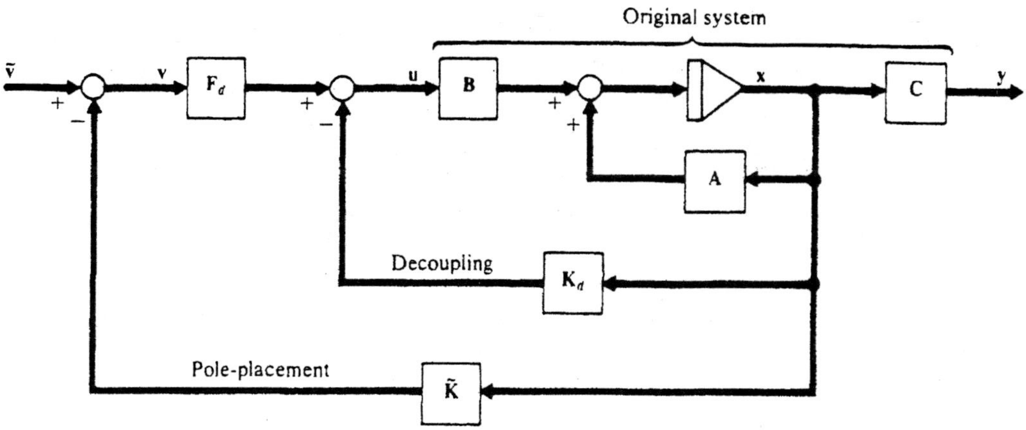

Figure 13.17

The decoupled system of Example 13.23 is treated as the open-loop system. That is,

$$\bar{\mathbf{A}} = \mathbf{A} - \mathbf{B}\mathbf{K}_d = \begin{bmatrix} -2 & 1 & 0 \\ -4 & 2 & 0 \\ 0 & 0 & 0 \end{bmatrix}, \quad \bar{\mathbf{B}} = \mathbf{B}\mathbf{F}_d = \begin{bmatrix} 0 & 0 \\ 0 & 1 \\ 1 & 0 \end{bmatrix}, \quad \bar{\mathbf{C}} = \mathbf{C} = \begin{bmatrix} 0 & 0 & 1 \\ 1 & 0 & 0 \end{bmatrix}$$

Since $\mathbf{F}_d$ is nonsingular, this system is still completely controllable (Sec. 13.3). Thus the specified eigenvalues can be achieved using a constant state feedback matrix $\tilde{\mathbf{K}}$. The method of Sec. 13.4 gives $\tilde{\mathbf{K}} = \begin{bmatrix} 0 & 0 & 5 \\ 20 & 15 & 0 \end{bmatrix}$.

The combined system, using $\mathbf{K}_d$ and $\mathbf{F}_d$ to achieve decoupling and $\tilde{\mathbf{K}}$ to achieve pole placement, is shown in Figure 13.17. In this case the resulting system is still decoupled and has closed-loop poles, as specified, at $-5, -5, -10$. The closed-loop transfer function is

$$\mathbf{H}(s) = \mathbf{C}[s\mathbf{I} - \mathbf{A} + \mathbf{B}\mathbf{K}_d + \mathbf{B}\mathbf{F}_d\tilde{\mathbf{K}}]^{-1}\mathbf{B}\mathbf{F}_d = \begin{bmatrix} 1/(s+5) & 0 \\ 0 & 1/[(s+5)(s+10)] \end{bmatrix}$$ ■

REFERENCES

1. Brogan, W. L.: "Applications of a Determinant Identity to Pole-Placement and Observer Problems," *IEEE Transactions on Automatic Control,* Vol. AC-19, Oct. 1974, pp. 612–614.
2. Davison, E. J.: "On Pole Assignment in Multivariable Linear Systems," *IEEE Transactions on Automatic Control,* Vol. AC-13, No. 6, Dec. 1968, pp. 747–748.
3. Wonham, W. M.: "On Pole Assignment in Multi-input, Controllable Linear Systems," *IEEE Transactions on Automatic Control,* Vol. AC-12, No. 6, Dec. 1967, pp. 660–665.
4. Melsa, J. L. and S. K. Jones: *Computer Programs for Computational Assistance in the Study of Linear Control Theory,* 2d ed., McGraw-Hill, New York, 1973.
5. Alag, G. and H. Kaufman: "An Implementable Digital Adaptive Flight Controller Designed Using Stabilized Single-Stage Algorithms," *IEEE Transactions on Automatic Control,* Vol. AC-22, No. 5, Oct. 1977, pp. 780–788.
6. Davison, E. J.: "On Pole Assignment in Linear Systems with Incomplete State Feedback," *IEEE Transactions on Automatic Control,* Vol. AC-15, No. 3, June 1970, pp. 348–351.

7. Luenberger, D. G.: "An Introduction to Observers," *IEEE Transactions on Automatic Control,* Vol. AC-16, No. 6, Dec. 1971, pp. 596–602.
8. Franklin, G. F. and J. D. Powell: *Digital Control of Dynamic Systems,* Addison-Wesley, Reading, Mass., 1980.
9. Kailath, T.: *Linear Systems,* Prentice Hall, Englewood Cliffs, N.J., 1980.
10. Chen, C. T.: *Introduction to Linear Systems Theory,* Holt, Rinehart and Winston, New York, 1970.
11. Falb, P. L. and W. A. Wolovich: "Decoupling in the Design and Synthesis of Multivariable Control Systems," *IEEE Transactions on Automatic Control,* Vol. AC-12, No. 6, Dec. 1967, pp. 651–659.
12. Morse, A. S. and W. M. Wonham: "Status of Noninteracting Control," *IEEE Transactions on Automatic Control,* Vol. AC-16, No. 6, Dec. 1971, pp. 568–581.
13. Morse, A. S. and W. M. Wonham: "Triangular Decoupling of Linear Multivariable Systems," *IEEE Transactions on Automatic Control,* Vol. AC-15, No. 4, Aug. 1970, pp. 447–449.
14. Sato, S. M. and P. V. Lopresti: "On the Generalization of State Variable Decoupling Theory," *IEEE Transactions on Automatic Control,* Vol. AC-16, No. 2, Apr. 1971, pp. 133–139.
15. Howze, J. W. and J. B. Pearson: "Decoupling and Arbitrary Pole Placement in Linear Systems Using Output Feedback," *IEEE Transactions on Automatic Control,* Vol. AC-15, No. 6, Dec. 1970, pp. 660–663.
16. Melsa, J. L. and D. G. Schultz: *Linear Control Systems,* McGraw-Hill, New York, 1969.
17. MacFarlane, A. G. J.: "A Survey of Some Results in Linear Multivariable Feedback Theory," *Automatic,* Vol. 8, No. 4, July 1972, pp. 455–492.
18. Perkins, W. R. and J. B. Cruz, Jr.: "Feedback Properties of Linear Regulators," *IEEE Transactions on Automatic Control,* Vol. AC-16, No. 6, Dec. 1971, pp. 659–664.

ILLUSTRATIVE PROBLEMS

State Feedback

1 A system is described by $\mathbf{A} = \begin{bmatrix} -2 & 1 & 0 \\ 0 & -2 & 0 \\ 0 & 0 & 4 \end{bmatrix}$, $\mathbf{B} = \begin{bmatrix} 0 & 0 \\ 0 & 1 \\ 1 & 0 \end{bmatrix}$. Find a constant state feedback matrix $\mathbf{K}$ which yields closed-loop poles $\Gamma = \{-2, -3, -4\}$.

Define $\mathbf{X}(\lambda) = [\mathbf{I}\lambda - \mathbf{A} \mid \mathbf{B}]$ and $\boldsymbol{\xi}^T = [\boldsymbol{\psi}^T \quad (\mathbf{K}\boldsymbol{\psi})^T]$, as in Eqs. (*13.12*) and (*13.13*). Then

$$\mathbf{X}(\lambda) = \begin{bmatrix} \lambda+2 & -1 & 0 & 0 & 0 \\ 0 & \lambda+2 & 0 & 0 & 1 \\ 0 & 0 & \lambda-4 & 1 & 0 \end{bmatrix}$$

One solution to $\mathbf{X}\boldsymbol{\xi} = \mathbf{0}$ is needed for each of the three desired eigenvalues, such that the resulting three $\boldsymbol{\psi}$ columns form a nonsingular matrix. With $\lambda = -2$, $\mathbf{X}\boldsymbol{\xi} = \mathbf{0}$ expands into the component equations $-\xi_2 = 0$, $\xi_5 = 0$, and $-6\xi_3 + \xi_4 = 0$. Clearly, ξ_1 is arbitrary, and it is selected as 0. Furthermore, $\xi_4 = -1$ is selected, giving $\boldsymbol{\xi} = [0 \quad 0 \quad -\frac{1}{6} \quad -1 \quad 0]^T$. A multitude of other choices could have been made. With $\lambda = -3$, the corresponding component equations are $-\xi_1 - \xi_2 = 0$, $-\xi_2 + \xi_5 = 0$, and $-7\xi_3 + \xi_4 = 0$. One valid solution is $\boldsymbol{\xi} = [1 \quad -1 \quad 0 \quad 0 \quad -1]^T$. With $\lambda = -4$, $\mathbf{X}\boldsymbol{\xi} = \mathbf{0}$ expands into $-2\xi_1 - \xi_2 = 0$, $-2\xi_2 + \xi_5 = 0$ and $-8\xi_3 + \xi_4 = 0$. One solution is $\boldsymbol{\xi} = [\frac{1}{4} \quad -\frac{1}{2} \quad 0 \quad 0 \quad -1]^T$. These three columns are a subset of the possible columns which form the matrix $\mathbf{U}(\lambda)$, and the top 3×3 partition $\mathbf{G} = \begin{bmatrix} 0 & 1 & \frac{1}{4} \\ 0 & -1 & -\frac{1}{2} \\ -\frac{1}{6} & 0 & 0 \end{bmatrix}$ is nonsingular. The desired gain is given by the bottom 2×3 partition $\mathcal{F}$, postmultiplied by $\mathbf{G}^{-1}$, or $\mathbf{K} = \begin{bmatrix} -1 & 0 & 0 \\ 0 & -1 & -1 \end{bmatrix} \mathbf{G}^{-1} = \begin{bmatrix} 0 & 0 & 6 \\ 2 & 3 & 0 \end{bmatrix}$, as specified in Eq. *(13.17)*.

13.2 Repeat Problem 13.1 if $\Gamma = \{-2, -2, -20\}$.

Since $\lambda = -2$ is requested as a double eigenvalue, two independent $\boldsymbol{\xi}$ solutions must be found. One is available from Problem 13.1. Another is obtained from the same component equations by selecting $\xi_1 = -1$ and $\xi_3 = 0$, giving $\boldsymbol{\xi} = [-1 \quad 0 \quad 0 \quad 0 \quad 0]^T$. Substituting $\lambda = -20$ into $\mathbf{X}(\lambda)$ leads to the third column. One valid choice is found by setting $\xi_3 = 0$ and $\xi_5 = -1$. Solving for the remaining components gives $\boldsymbol{\xi} = [1/(18)^2 \quad -1/(18) \quad 0 \quad 0 \quad -1]^T$. Then Eq. *(13.17)* gives the desired gain as $\mathbf{K} = \begin{bmatrix} 0 & 0 & 6 \\ 0 & 18 & 0 \end{bmatrix}$.

13.3 Repeat Problem 13.1 if $\Gamma = \{-3, -2 + j, -2 - j\}$.

$\boldsymbol{\xi} = [1 \quad -1 \quad 0 \quad 0 \quad -1]^T$ was selected for $\lambda = -3$ in Problem 13.1. With $\lambda = -2 + j$, the component equations are $j\xi_1 - \xi_2 = 0$, $j\xi_2 + \xi_5 = 0$, and $(-6 + j)\xi_3 + \xi_4 = 0$. One easy choice is $\xi_3 = 0, \xi_1 = 1$. Then $\boldsymbol{\xi} = [1 \quad j \quad 0 \quad 0 \quad 1]^T$. With $\lambda = -2 - j$, the solution $\boldsymbol{\xi}$ is the complex conjugate, so all three solutions have been found. Note, however, that row 3 of the 5×3 matrix is all zeros, so $\mathbf{G}$ is not invertible. An alternate solution for $\lambda = -3$ is easily found to be $[0 \quad 0 \quad -\frac{1}{7} \quad -1 \quad 0]$. Then

$$\mathbf{K} = \begin{bmatrix} -1 & 0 & 0 \\ 0 & 1 & 1 \end{bmatrix} \begin{bmatrix} 0 & 1 & 1 \\ 0 & j & -j \\ -\frac{1}{7} & 0 & 0 \end{bmatrix}^{-1}$$

In general it is easier to avoid the complex matrix inverse, and by the results of Problem 4.23

$$\mathbf{K} = \begin{bmatrix} -1 & 0 & 0 \\ 0 & 1 & 0 \end{bmatrix} \begin{bmatrix} 0 & 1 & 0 \\ 0 & 0 & 1 \\ -\frac{1}{7} & 0 & 0 \end{bmatrix}^{-1} = \begin{bmatrix} 0 & 0 & 7 \\ 1 & 0 & 0 \end{bmatrix}$$

13.4 Repeat Problem 13.1 if $\Gamma = \{-3, -3, -3\}$.

Since the rank of $\mathbf{X}(-3)$ is three, there are only two independent solutions to $\mathbf{X}(-3)\boldsymbol{\xi} = \mathbf{0}$. One was given in Problem 13.1 and another in Problem 13.3. A triple eigenvalue is requested, so the third column vector must be a generalized eigenvector, as in Eq. *(13.19)*. It will satisfy $\mathbf{X}(-3)\boldsymbol{\xi}_g = -\boldsymbol{\psi}$, where $\boldsymbol{\psi} = [1 \quad -1 \quad 0]^T$ or $[0 \quad 0 \quad -\frac{1}{7}]$—i.e., the top partitions of the two $\boldsymbol{\xi}$ solutions. The second choice must be ruled out because it does not give an independent solution. Using the first choice gives $\boldsymbol{\xi}_g = [2 \quad -1 \quad 0 \quad 0 \quad 0]^T$. The gain is then given by Eq. *(13.17)* as

$$\mathbf{K} = -\begin{bmatrix} 0 & 1 & 0 \\ 1 & 0 & 0 \end{bmatrix} \begin{bmatrix} 1 & 0 & 2 \\ -1 & 0 & -1 \\ 0 & -\frac{1}{7} & 0 \end{bmatrix}^{-1} = \begin{bmatrix} 0 & 0 & 7 \\ 1 & 2 & 0 \end{bmatrix}$$

13.5 A system is described by Eq. *(13.1)* with $\mathbf{A} = \begin{bmatrix} -2 & 1 & 0 \\ 0 & -2 & 1 \\ 0 & 0 & -2 \end{bmatrix}$, $\mathbf{B} = \begin{bmatrix} 0 \\ 0 \\ 1 \end{bmatrix}$. Find a constant state feedback matrix $\mathbf{K}$ which gives closed-loop eigenvalues $\lambda_1 = \lambda_2 = \lambda_3 = -1$.

Form $\mathbf{X}(-1) = [\lambda\mathbf{I} - \mathbf{A} \mid \mathbf{B}] = \begin{bmatrix} 1 & -1 & 0 & 0 \\ 0 & 1 & -1 & 0 \\ 0 & 0 & 1 & 1 \end{bmatrix}$. Its rank is 3, so there is only one independent solution to $\mathbf{X}\boldsymbol{\xi} = \mathbf{0}$, $\boldsymbol{\xi} = [1 \;\; 1 \;\; 1 \;\; -1]^T$. A chain of two generalized eigenvectors can be found. $\mathbf{X}\boldsymbol{\xi}_{g_1} = -[1 \;\; 1 \;\; 1]^T$ gives the first generalized vector as $\boldsymbol{\xi}_{g_1} = -[3 \;\; 2 \;\; 1 \;\; 0]^T$. Using this in the second generalized equation $\mathbf{X}\boldsymbol{\xi}_{g_2} = [3 \;\; 2 \;\; 1]^T$ gives the solution $\boldsymbol{\xi}_{g_2} = [6 \;\; 3 \;\; 1 \;\; 0]^T$. The top 3×3 partition forms the matrix $\mathbf{G}$, and row four forms the matrix $\mathcal{J}$. Then Eq. (*13.17*) gives the gain matrix as

$$\mathbf{K} = [-1 \;\; 0 \;\; 0]\begin{bmatrix} 1 & -3 & 6 \\ 1 & -2 & 3 \\ 1 & -1 & 1 \end{bmatrix}^{-1} = [-1 \;\; 3 \;\; -3]$$

Output Feedback

6 Specify a constant output feedback matrix so that a system with $\mathbf{A} = \begin{bmatrix} -1 & 0 \\ 0 & -4 \end{bmatrix}$, $\mathbf{B} = \begin{bmatrix} 0 & 1 \\ 1 & 0 \end{bmatrix}$, $\mathbf{C} = [0 \;\; 1]$, and $\mathbf{D} = [0 \;\; 0]$ will have a closed-loop eigenvalue at -10.

Form $[\lambda\mathbf{I} - \mathbf{A} \mid \mathbf{B}] = \begin{bmatrix} -9 & 0 & 0 & 1 \\ 0 & -6 & 1 & 0 \end{bmatrix}$ and find $\boldsymbol{\xi} = [0 \;\; 1 \;\; 6 \;\; 0]^T$ as a nontrivial solution.

The meaning of the vector $\boldsymbol{\xi}$ in the case of output feedback is $\boldsymbol{\xi} = \begin{bmatrix} \boldsymbol{\psi} \\ \mathbf{K}'\mathbf{C}\boldsymbol{\psi} \end{bmatrix}$, as shown in Eq. (*13.22*) (with $\mathbf{D} = 0$). Thus $\mathbf{K}'\mathbf{C}\begin{bmatrix} 0 \\ 1 \end{bmatrix} = \begin{bmatrix} 6 \\ 0 \end{bmatrix}$. Using $\mathbf{C} = [0 \;\; 1]$ gives $\mathbf{K}' = \begin{bmatrix} 6 \\ 0 \end{bmatrix}$.

7 Another output measurement is added to the system of Problem 13.6, so that now $\mathbf{C} = \begin{bmatrix} 1 & 1 \\ 0 & 1 \end{bmatrix}$. Is it possible to use constant output feedback so that both λ_1 and λ_2 equal -10 for the closed-loop system? If yes, find the gain matrix $\mathbf{K}'$.

Since rank $\mathbf{C} = 2 = n$, both eigenvalues can be prescribed, and two independent solutions of $\mathbf{X}\boldsymbol{\xi} = \mathbf{0}$ are needed. A second solution for $\boldsymbol{\xi}$ in Problem 13.6 is $\boldsymbol{\xi} = [1 \;\; 0 \;\; 0 \;\; 9]^T$. This solution could not have been used there to solve for $\mathbf{K}'$ (Why?), but with the modified $\mathbf{C}$ we can write $\mathbf{K}'\begin{bmatrix} 1 & 1 \\ 0 & 1 \end{bmatrix}\begin{bmatrix} 1 & 0 \\ 0 & 1 \end{bmatrix} = \begin{bmatrix} 0 & 6 \\ 9 & 0 \end{bmatrix}$, from which $\mathbf{K}' = \begin{bmatrix} 0 & 6 \\ 9 & -9 \end{bmatrix}$.

8 A system [6] is described by Eq. (*13.1*) with $\mathbf{A} = \begin{bmatrix} 0 & 1 & 0 \\ 0 & 0 & 1 \\ 1 & 0 & 0 \end{bmatrix}$, $\mathbf{B} = \begin{bmatrix} 0 \\ 1 \\ 0 \end{bmatrix}$, $\mathbf{C} = \begin{bmatrix} 1 & 0 & 0 \\ 1 & 1 & 0 \end{bmatrix}$, $\mathbf{D} = \begin{bmatrix} 0 \\ 0 \end{bmatrix}$. Since this system is completely controllable and rank $\mathbf{C} = m = 2$, two closed-loop eigenvalues can be made arbitrarily close to any specified values by using output feedback. Find $\mathbf{K}'$ so that $\lambda_1 = 1$, $\lambda_2 = \epsilon$, where ϵ is arbitrarily close to zero.

Nontrivial solutions to $[\lambda\mathbf{I} - \mathbf{A} \mid \mathbf{B}]\boldsymbol{\xi} = \mathbf{0}$ must be found. For $\lambda = 1$, $\boldsymbol{\xi} = [-1 \;\; -1 \;\; -1 \;\; 0]^T$. With $\lambda = \epsilon$, the component equations are $\epsilon\xi_1 - \xi_2 = 0$, $\epsilon\xi_2 - \xi_3 + \xi_4 = 0$, and $-\xi_1 + \epsilon\xi_3 = 0$. Arbitrarily selecting $\xi_1 = 1$ gives $\xi_2 = \epsilon$, $\xi_3 = 1/\epsilon$, and $\xi_4 = 1/\epsilon - \epsilon^2$. The gain can be determined from

$$\mathbf{K}'\mathbf{C}\begin{bmatrix} -1 & 1 \\ -1 & \epsilon \\ -1 & \frac{1}{\epsilon} \end{bmatrix} = \begin{bmatrix} 0 & \frac{1}{\epsilon} - \epsilon^2 \end{bmatrix} \quad \text{or} \quad \mathbf{K}'\begin{bmatrix} -1 & 1 \\ -2 & 1+\epsilon \end{bmatrix} = \begin{bmatrix} 0 & \frac{1}{\epsilon} - \epsilon^2 \end{bmatrix}$$

Matrix inversion gives $\mathbf{K}' = [2/(1/\epsilon - \epsilon^2) \quad (\epsilon^2 - 1/\epsilon)]/(1-\epsilon)$, which for very small ϵ approaches $\mathbf{K}' = [2/\epsilon \quad -1/\epsilon]$. This result indicates that $\lambda_2 = 0$ can be achieved only by using infinite feedback gains. This is analogous to the well-known result for single-input, single-output systems. An infinite gain is required to cause a closed-loop pole to coincide with an open-loop zero.

Observers

13.9 A system described by Eq. (*13.1*) has $\mathbf{A} = \begin{bmatrix} -2 & -2 & 0 \\ 0 & 0 & 1 \\ 0 & -3 & -4 \end{bmatrix}$, $\mathbf{B} = \begin{bmatrix} 1 & 0 \\ 0 & 0 \\ 0 & 1 \end{bmatrix}$, $\mathbf{C} = \begin{bmatrix} 1 & 0 & 1 \\ 0 & 1 & 0 \end{bmatrix}$, $\mathbf{D} = [\mathbf{0}]$. Design an observer whose output settles to the actual state vector within 1 s.

It should be verified that the system is completly observable. The determination of the observer gain is carried out by using the duality with the pole placement problem. Using $\mathbf{A}^T$ instead of $\mathbf{A}$ and $\mathbf{C}^T$ instead of $\mathbf{B}$ will give $\mathbf{L}^T$ in place of $\mathbf{K}$. Proceeding, $\mathbf{X}(\lambda) = \begin{bmatrix} \lambda+2 & 0 & 0 & 1 & 0 \\ 2 & \lambda & 3 & 0 & 1 \\ 0 & -1 & \lambda+4 & 1 & 0 \end{bmatrix}$. With $\lambda = -5$, its rank is 3, so it is possible to find two independent solutions to $\mathbf{X}(-5)\boldsymbol{\xi} = \mathbf{0}$. They are found numerically as $\xi_1 = [0.33333 \;\; 0.45833 \;\; 0.541667 \;\; 1 \;\; 0]^T$ and $[0 \;\; -0.125 \;\; 0.125 \;\; 0 \;\; -1]^T$. With $\lambda = -6$, a single solution is needed, and one is $\boldsymbol{\xi} = [0 \;\; -0.13333 \;\; 0.066667 \;\; 0 \;\; -1]^T$, leading to

$$\mathbf{L}^T = \begin{bmatrix} 1 & 0 & 0 \\ 0 & -1 & -1 \end{bmatrix} \begin{bmatrix} 0.33333 & 0 & 0 \\ 0.45833 & -0.125 & -0.133333 \\ 0.541667 & 0.125 & 0.066667 \end{bmatrix}^{-1}$$

$$= \begin{bmatrix} 3 & 0 & 0 \\ -8 & 7 & -1 \end{bmatrix}$$

and then $\mathbf{A}_c = \begin{bmatrix} -5 & 6 & -3 \\ 0 & -7 & 1 \\ 0 & -2 & -4 \end{bmatrix}$. The observer is defined by $\dot{\hat{\mathbf{x}}} = \mathbf{A}_c\hat{\mathbf{x}} + \mathbf{L}\mathbf{y} + \mathbf{B}\mathbf{u}$, where $\mathbf{u}$ and $\mathbf{y}$ are the input and output of the system being observed.

13.10 Consider the open-loop system described by the matrices **A**, **B**, **C**, and **D** of Example 13.9. Design an observer with eigenvalues of -8, -8, and -10.

Since the output is $\mathbf{y} = \mathbf{C}\mathbf{x} + \mathbf{D}\mathbf{u}$, the observer equations are modified to take into account the nonzero **D** matrix. Let $\dot{\hat{\mathbf{x}}} = \mathbf{A}_c\hat{\mathbf{x}} + \mathbf{L}\mathbf{y} + \mathbf{z} = \mathbf{A}_c\hat{\mathbf{x}} + \mathbf{L}\mathbf{C}\mathbf{x} + \mathbf{L}\mathbf{D}\mathbf{u} + \mathbf{z}$. It is seen that setting $\mathbf{z} = (\mathbf{B} - \mathbf{L}\mathbf{D})\mathbf{u}$ again leads to the error Eq. (*13.30*). Therefore, the observer gain matrix **L** is designed using the dual pole-placement procedure—that is, $\mathbf{A}^T$ and $\mathbf{C}^T$ are used in place of **A** and **B**. Solving Eq. (*13.12*) for each desired λ leads to $\mathbf{L} = \begin{bmatrix} 0 & 14 \\ 0 & 48 \\ 12 & 0 \end{bmatrix}$. Then using $\mathbf{A}_c = \mathbf{A} - \mathbf{L}\mathbf{C}$ and $\mathbf{z} = (\mathbf{B} - \mathbf{L}\mathbf{D})\mathbf{u}$ gives the observer shown in Figure 13.18.

13.11 Consider the open-loop unstable system of Example 13.9. Design a closed-loop system with dynamic feedback, which has eigenvalues -5, -6, and -6. The eigenvalues added by the feedback compensation should have real parts more negative than -7.

It is first assumed that all state variables are available for feedback. A state feedback matrix **K** is selected to satisfy the specification $\lambda_i \in \{-5, -6, -6\}$. This is found from Eqs. (*13.16*) and (*13.17*) using

$$\mathbf{G} = [\boldsymbol{\psi}(-5) \;\; \boldsymbol{\psi}_1(-6) \;\; \boldsymbol{\psi}_2(-6)] = \begin{bmatrix} 1 & 0 & 1 \\ -3 & 0 & -4 \\ 0 & -1 & 0 \end{bmatrix}$$

and

$$\mathcal{F} = [\mathbf{f}(-5) \;\; \mathbf{f}_1(-6) \;\; \mathbf{f}_2(-6)] = \begin{bmatrix} 0 & -1 & 0 \\ -9 & 0 & -16 \end{bmatrix}$$

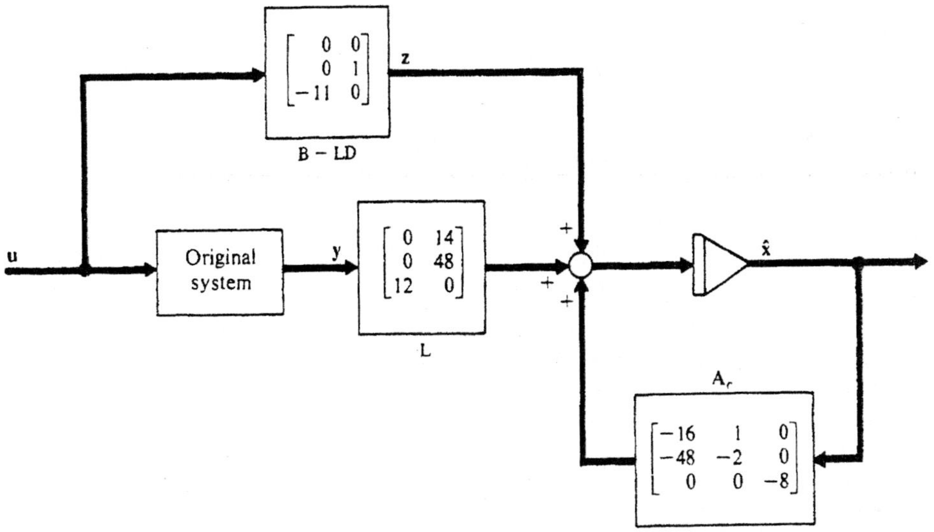

Figure 13.18

Then $\mathbf{K} = \mathcal{J}\mathbf{G}^{-1} = \begin{bmatrix} 0 & 0 & 10 \\ 12 & 7 & 0 \end{bmatrix}$. This result cannot be used directly since all state variables are not available. However, an observer was designed for this system in Problem 13.10. Since that observer meets the specification, it can be used to provide an estimate of **x**. Then **K** is applied to that estimate, as shown in Figure 13.19.

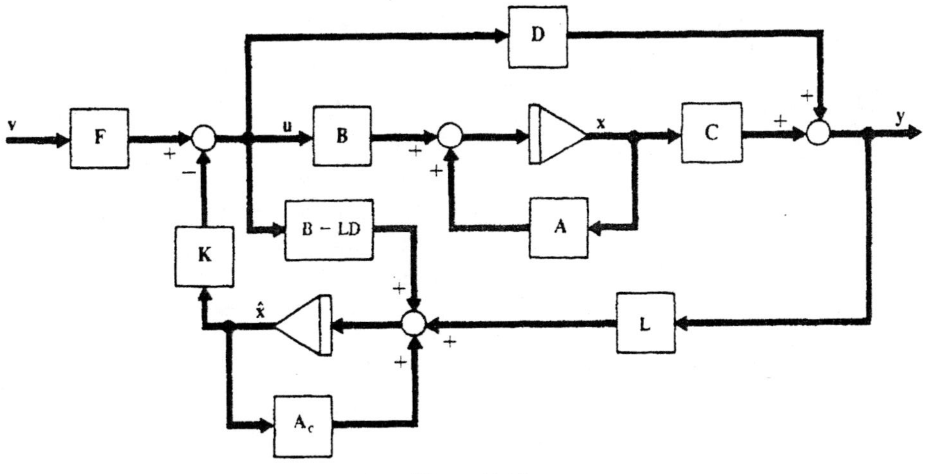

Figure 13.19

12 When can a system of the general form of Eq. (*13.1*) or (*13.2*) be transformed so that $\mathbf{C} = [\mathbf{I} \mid \mathbf{0}]$?

From Sec. 4.10 it is clear that the first requirement is that **C** must have full rank m. Assume that this is so and that the first m columns of **C** are linearly independent. Then by a series of row operations (equivalent to premultiplication by a nonsingular transformation matrix $\mathbf{T}_1$) **C** can be brought into row-reduced-echelon form. That is

$$\mathbf{T}_1\mathbf{y} = \mathbf{T}_1\mathbf{C}\mathbf{x} + \mathbf{T}_1\mathbf{D}\mathbf{u}$$

or

$$\mathbf{y}' = [\mathbf{I} \mid \mathbf{C}']\mathbf{x} + \mathbf{D}'\mathbf{u}$$

To further reduce **C** (i.e., to eliminate **C**′), column operations are required. These are accomplished by postmultiplying by a nonsingular transformation matrix $\mathbf{T}_2$. The only way this can be accomplished is to redefine the state vector according to

$$\mathbf{x} = \mathbf{T}_2\mathbf{x}'$$

Then $\mathbf{y}' = [\mathbf{I} \mid \mathbf{0}]\mathbf{x}' + \mathbf{D}'\mathbf{u}$. The dynamics equation must similarly be transformed into $\mathbf{x}'$ variables

$$\mathbf{x}'(k+1) = \mathbf{T}_2^{-1}\mathbf{A}\mathbf{T}_2\mathbf{x}'(k) + \mathbf{T}_2^{-1}\mathbf{B}\mathbf{u}(k)$$
$$= \mathbf{A}'\mathbf{x}'(k) + \mathbf{B}'\mathbf{u}(k)$$

13.13 A third-order discrete-time system with two inputs and one output is described by

$$\mathbf{A} = \begin{bmatrix} 0 & -0.5 & -0.1 \\ -0.1 & -0.7 & 0.2 \\ 0.8 & -0.8 & 1 \end{bmatrix} \quad \mathbf{B} = \begin{bmatrix} 1 & -1 \\ -1 & 1 \\ 0 & 3 \end{bmatrix} \quad \mathbf{C} = [1 \quad 0 \quad 0]$$

Design a reduced-order observer to estimate x_2 and x_3. Place its poles at $z = 0.1$ and 0.2.

Using the algorithm of Sec. 13.4 with the equivalent **A** and **B** matrices selected as

$$\mathbf{A}_{22}^T = \begin{bmatrix} -0.7 & -0.8 \\ 0.2 & 1 \end{bmatrix} \quad \text{and} \quad \mathbf{A}_{12}^T = \begin{bmatrix} -0.5 \\ -0.1 \end{bmatrix}$$

gives a gain matrix **K**, which is

$$\mathbf{L}_r^T = [0.391608 \quad -1.95804]$$

Using this to form $\mathbf{A}_r = \mathbf{A}_{22} - \mathbf{L}_r\mathbf{A}_{12}$ gives the observer the result shown in Figure 13.20.

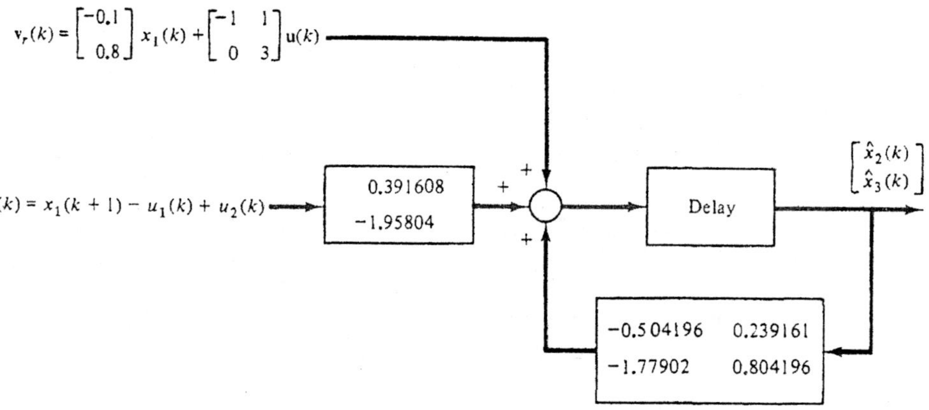

Figure 13.20

Decoupling

13.14 Use state feedback to decouple the system [11] described by

$$\mathbf{A} = \begin{bmatrix} 1 & 1 & 0 \\ 0 & 1 & 0 \\ 0 & 0 & 1 \end{bmatrix}, \quad \mathbf{B} = \begin{bmatrix} 0 & 1 \\ 1 & 0 \\ 1 & 0 \end{bmatrix}, \quad \mathbf{C} = \begin{bmatrix} 1 & 1 & -1 \\ 0 & 1 & 0 \end{bmatrix}, \quad \mathbf{D} = [0]$$

Using Eq. (*13.49*), $\mathbf{c}_1\mathbf{B} = [0 \quad 1] \neq \mathbf{0}$, so $d_1 = 0$. $\mathbf{c}_2\mathbf{B} = [1 \quad 0] \neq \mathbf{0}$, so $d_2 = \mathbf{0}$. Then $\mathbf{N} = \begin{bmatrix} 0 & 1 \\ 1 & 0 \end{bmatrix}$ is nonsingular and the system can be decoupled. $\mathbf{F}_d = \mathbf{N}^{-1} = \begin{bmatrix} 0 & 1 \\ 1 & 0 \end{bmatrix}$ and $\mathbf{K}_d = \begin{bmatrix} 0 & 1 \\ 1 & 0 \end{bmatrix}\begin{bmatrix} \mathbf{c}_1\mathbf{A} \\ \mathbf{c}_2\mathbf{A} \end{bmatrix} = \begin{bmatrix} 0 & 1 & 0 \\ 1 & 2 & -1 \end{bmatrix}$. Using these results in Eq. (*13.47*) gives the integrator decoupled result, $\mathbf{H}(s) = \begin{bmatrix} 1/s & 0 \\ 0 & 1/s \end{bmatrix}$.

Note that the original system is not completely controllable, and thus the decoupled system is not controllable either. Any attempt to specify all three eigenvalues using the method of Sec. 13.4 will fail.

15 Can the system of Example 12.8 be decoupled using state feedback?

The irreducible realization found in Example 12.8 is used. The product $\mathbf{c}_1\mathbf{B} = [0 \quad 1]$ is not zero, so $d_1 = 0$. Likewise, $\mathbf{c}_2\mathbf{B} = [1 \quad 0]$, so $d_2 = 0$. Then $\mathbf{N} = \begin{bmatrix} 0 & 1 \\ 1 & 0 \end{bmatrix}$ is nonsingular, so the system can be decoupled.

The integrator decoupled system is obtained by using $\mathbf{F}_d = \mathbf{N}^{-1} = \begin{bmatrix} 0 & 1 \\ 1 & 0 \end{bmatrix}$, and $\mathbf{K}_d = \mathbf{N}^{-1}\begin{bmatrix} \mathbf{c}_1\mathbf{A} \\ \mathbf{c}_2\mathbf{A} \end{bmatrix} = \begin{bmatrix} 0 & 0 & -2 & 0 \\ -\frac{2}{3} & \frac{5}{9} & -\frac{5}{27} & -5 \end{bmatrix}$. Equation (*13.47*) gives the decoupled closed-loop transfer function $\mathbf{H}(s) = \begin{bmatrix} 1/s & 0 \\ 0 & 1/s \end{bmatrix}$.

16 Consider the open-loop, decoupled system described by $\bar{\mathbf{A}}$, $\bar{\mathbf{B}}$, and $\bar{\mathbf{C}}$ of Example 13.24 and with $\mathbf{D} = \begin{bmatrix} 1 & 0 \\ 0 & 0 \end{bmatrix}$. Find an output feedback matrix which places closed-loop poles at $\lambda_1 = -5$, $\lambda_2 = -10$. Is the resultant system still decoupled? Is it stable?

We first find the gain $\mathbf{K}_*$ which will multiply $\mathbf{y}' = \mathbf{y} - \mathbf{Du}$. The desired values of λ are -5 and -10. One solution of $[\lambda\mathbf{I} - \mathbf{A} \mid \mathbf{B}]\boldsymbol{\xi} = \mathbf{0}$ is found for each λ. Then the $\boldsymbol{\xi}$ vectors are partitioned to give $\mathbf{K}_*\mathbf{C}[\boldsymbol{\psi}(-5) \quad \boldsymbol{\psi}(-10)] = [\boldsymbol{\psi}(-5) \quad \boldsymbol{\psi}(-10)]$. The numerical values are $\mathbf{K}'\mathbf{C}\begin{bmatrix} 0 & 0.01 \\ 0 & 0.08 \\ -0.2 & 0 \end{bmatrix} = \begin{bmatrix} -1 & 0 \\ 0 & -1 \end{bmatrix}$. Using $\mathbf{C}$ and taking the inverse of a 2×2 matrix gives $\mathbf{K}_* = \begin{bmatrix} 5 & 0 \\ 0 & -100 \end{bmatrix}$. The gain which multiplies the output $\mathbf{y}$ is $\mathbf{K}' = [\mathbf{I} - \mathbf{K}_*\mathbf{D}]^{-1}\mathbf{K}_* = \begin{bmatrix} -\frac{5}{4} & 0 \\ 0 & -100 \end{bmatrix}$.
When this feedback is used, the characteristic equation (*13.21*) is

$$\Delta_0'(\lambda) = |\lambda\mathbf{I}_3 - \bar{\mathbf{A}} + \bar{\mathbf{B}}\mathbf{K}'[\mathbf{I}_2 + \bar{\mathbf{D}}\mathbf{K}']^{-1}\bar{\mathbf{C}} = \left|\begin{bmatrix} \lambda+2 & -1 & 0 \\ -96 & \lambda-2 & 0 \\ 0 & 0 & \lambda+5 \end{bmatrix}\right|$$

$$= (\lambda+5)(\lambda^2 - 100) = 0$$

Therefore, the eigenvalues are at $-5, -10$, and $+10$ and the system is unstable. Using Eq. (*13.6*) and (*13.7*), and the fact that $\mathbf{F}' = \mathbf{I}_2$ leads to

$$\mathbf{H}(s) = [\mathbf{I}_2 + \mathbf{DK}']^{-1}\{\bar{\mathbf{C}}\{s\mathbf{I}_3 - \bar{\mathbf{A}} + \bar{\mathbf{B}}\mathbf{K}'[\mathbf{I}_2 + \mathbf{DK}']^{-1}\bar{\mathbf{C}}\}^{-1}\bar{\mathbf{B}}[\mathbf{I}_2 + \mathbf{K}'\mathbf{D}]^{-1} + \mathbf{D}\}$$

$$= \begin{bmatrix} -4(s+1)/(s+5) & 0 \\ 0 & 1/(s^2-100) \end{bmatrix}$$

The system is still uncoupled. If $\mathbf{D} = [0]$, as in Example 13.23, then $\mathbf{K}' = \begin{bmatrix} 5 & 0 \\ 0 & -100 \end{bmatrix}$ and $\mathbf{H}(s) = \begin{bmatrix} 1/(s+5) & 0 \\ 0 & 1/(s^2-100) \end{bmatrix}$.

13.17 Assume that a particular system can be decoupled by use of state feedback. Assume that the system is also completely observable. Show that the system can be decoupled by using an observer to estimate the state and then by using the estimated state with the decoupling matrices.

Let the system be described by $\dot{\mathbf{x}} = \mathbf{Ax} + \mathbf{Bu}$, $\mathbf{y} = \mathbf{Cx}$, and the decoupling state feedback is $\mathbf{u} = \mathbf{F}_d\mathbf{v} - \mathbf{K}_d\mathbf{x}$. Let the observer be described by $\dot{\hat{\mathbf{x}}} = \mathbf{A}_c\hat{\mathbf{x}} + \mathbf{Ly} + \mathbf{Bu}$. It is to be shown that $\mathbf{x}$ can be replaced by $\hat{\mathbf{x}}$ in the control law without altering the decoupling. Writing $\hat{\mathbf{x}} = \mathbf{x} - \mathbf{e}$ and using $\mathbf{u} = \mathbf{F}_d\mathbf{v} - \mathbf{K}_d\hat{\mathbf{x}} = \mathbf{F}_d\mathbf{v} - \mathbf{K}_d\mathbf{x} + \mathbf{K}_d\mathbf{e}$ gives

$$\begin{bmatrix} \dot{\mathbf{x}} \\ \dot{\mathbf{e}} \end{bmatrix} = \begin{bmatrix} \mathbf{A} - \mathbf{BK}_d & \mathbf{BK}_d \\ \mathbf{0} & \mathbf{A}_c \end{bmatrix} \begin{bmatrix} \mathbf{x} \\ \mathbf{e} \end{bmatrix} + \begin{bmatrix} \mathbf{BF}_d \\ \mathbf{0} \end{bmatrix} \mathbf{v}$$

Using Laplace transforms gives

$$\begin{bmatrix} \mathbf{x}(s) \\ \mathbf{e}(s) \end{bmatrix} = \begin{bmatrix} s\mathbf{I} - \mathbf{A} + \mathbf{BK}_d & -\mathbf{BK}_d \\ \mathbf{0} & s\mathbf{I} - \mathbf{A}_c \end{bmatrix}^{-1} \begin{bmatrix} \mathbf{BF}_d \\ \mathbf{0} \end{bmatrix} \mathbf{v}(s)$$

Only the upper-left block of the inverse is needed in order to solve for $\mathbf{x}(s)$. Using results of Sec. 4.9, this reduces to $\mathbf{x}(s) = [s\mathbf{I} - \mathbf{A} + \mathbf{BK}_d]^{-1}\mathbf{BF}_d\mathbf{v}(s)$ and the output is $\mathbf{y} = \mathbf{C}[s\mathbf{I} - \mathbf{A} + \mathbf{BK}_d]^{-1}$ $\mathbf{BF}_d\mathbf{v}(s)$. Comparing this with Eq. (*13.47*) shows that the system is still decoupled.

13.18 Figure 13.21 shows a system described by $\{\mathbf{A}, \mathbf{B}, \mathbf{C}\}$, an observer described by $\{\mathbf{A}_c, \mathbf{L}\}$, decoupling matrices $\{\mathbf{K}_d, \mathbf{F}_d\}$, and a pole placement feedback matrix $\bar{\mathbf{K}}$. Let the system be the one in Example 13.9, but with $\mathbf{D} = [\mathbf{0}]$. The decoupling matrices are calculated in Example 13.23, and the pole placement matrix is the one found in Example 13.24. Show that using the output and an observer yields the same decoupled system transfer matrix as was obtained in Example 13.24.

The dynamic equations are $\dot{\mathbf{x}} = \mathbf{Ax} + \mathbf{Bu}$ and $\dot{\hat{\mathbf{x}}} = \mathbf{A}_c\hat{\mathbf{x}} + \mathbf{LCx} + \mathbf{Bu}$. The control law is $\mathbf{u} = \mathbf{F}_d\bar{\mathbf{v}} - (\mathbf{F}_d\bar{\mathbf{K}} + \mathbf{K}_d)\hat{\mathbf{x}}$. Using $\mathbf{e} = \mathbf{x} - \hat{\mathbf{x}}$ to eliminate $\hat{\mathbf{x}}$ leads to

$$\begin{bmatrix} \dot{\mathbf{x}} \\ \dot{\mathbf{e}} \end{bmatrix} = \begin{bmatrix} \mathbf{A} - \mathbf{B}(\mathbf{F}_d\bar{\mathbf{K}} + \mathbf{K}_d) & \mathbf{B}(\mathbf{F}_d\bar{\mathbf{K}} + \mathbf{K}_d) \\ \mathbf{0} & \mathbf{A}_c \end{bmatrix} \begin{bmatrix} \mathbf{x} \\ \mathbf{e} \end{bmatrix} + \begin{bmatrix} \mathbf{BF}_d \\ \mathbf{0} \end{bmatrix} \bar{\mathbf{v}}$$

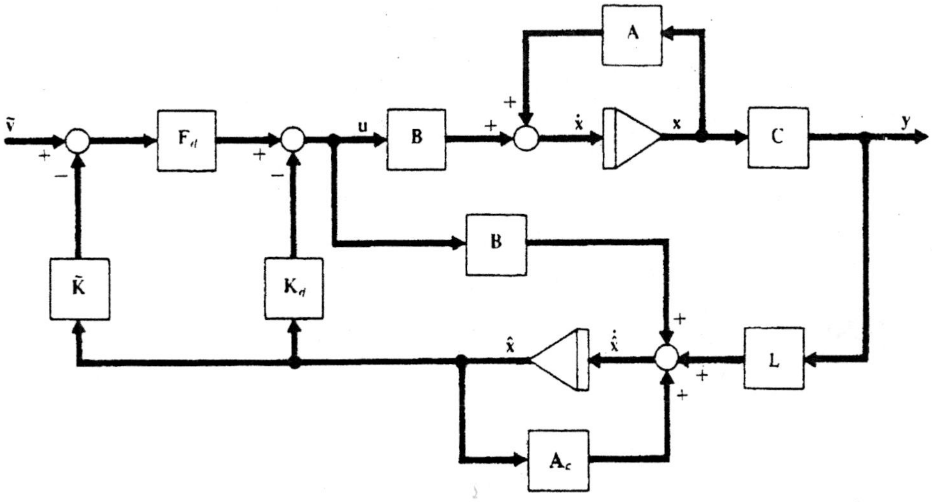

Figure 13.21

Taking the Laplace transform and solving for $\mathbf{x}(s)$ gives

$$\mathbf{y}(s) = \mathbf{C}\mathbf{x}(s) = \mathbf{C}[s\mathbf{I} - \mathbf{A} + \mathbf{B}(\mathbf{F}_d\tilde{\mathbf{K}} + \mathbf{K}_d)]^{-1}\mathbf{B}\mathbf{F}_d\bar{\mathbf{v}}$$

This is the same result as in Example 13.24 and gives

$$\mathbf{H}(s) = \begin{bmatrix} 1/(s+5) & 0 \\ 0 & 1/[(s+5)(s+10)] \end{bmatrix}$$

19 Design a controller/observer for the discrete-time system whose open-loop transfer function is

$$\frac{y(z)}{u(z)} = \frac{z + 0.866}{(z-1)(z^2 - 0.8z + 0.32)}$$

Put the closed-loop system poles at $z = 0$ and $0.2 \pm 0.6j$ and put the observer poles at $z = 0$ and $\pm 0.2j$.

The numerator and denominator polynomials are

$$b(z) = z + 0.866$$

$$a(z) = z^3 - 1.8z^2 + 1.12z - 0.32$$

Let the desired closed-loop poles define the polynomials $r(z)$ and $d(z)$, respectively. Then the desired closed-loop characteristic equation is

$$c(z) = (z^3 - 0.4z^2 + 0.4z)(z^3 + 0.04z) = r(z)d(z)$$
$$= z^6 - 0.4z^5 + 0.44z^4 - 0.016z^3 + 0.016z^2$$

The coefficients of these polynomials are used in Eq. (*13.43*) in order to determine the unknown polynomials $p(z)$ and $q(z)$,

$$\begin{bmatrix} 1 & 0 & 0 & 0 & 0 & 0 \\ -1.18 & 1 & 0 & 0 & 0 & 0 \\ 1.12 & -1.8 & 1 & 1 & 0 & 0 \\ -0.32 & 1.12 & -1.8 & 0.866 & 1 & 0 \\ 0 & -0.32 & 1.12 & 0 & 0.866 & 1 \\ 0 & 0 & -0.32 & 0 & 0 & 0.866 \end{bmatrix} \begin{bmatrix} p_2 \\ p_1 \\ p_0 \\ q_2 \\ q_1 \\ q_0 \end{bmatrix} = \begin{bmatrix} 1.4 \\ -0.68 \\ 0.304 \\ 0.016 \\ 0 \\ 0 \end{bmatrix}$$

Numerical solution gives the coefficients for the polynomials $p(z) = z^3 + 1.4z^2 + 1.84z + 0.92346$ and $q(z) = 1.1245z^2 - 0.90843z + 0.34123$. Using the polynomial $d(z)$ of the desired observer poles gives $e(z) = p(z) - d(z) = 1.4z^2 + 1.8z + 0.92346$. All needed terms are available, and the system can be expressed in any of the optional configurations of Figure 13.15*b*, *c*, or *d*. (For this discrete system, the polynomials are functions of z instead of s). After simplification and cancellation, the closed-loop transfer function is $y(z)/v(z) = b(z)/r(z) = (z + 0.866)/(z^3 - 0.4z^2 + 0.4z)$

20 Show that state feedback does not alter the transfer function zeros.

It is assumed that there are an equal number of inputs and outputs, so the zeros of the square transfer function matrix can be defined as the values of s (or z) which give a zero value to the transfer function determinant. The open-loop transfer function is $\mathbf{H}_0 = \mathbf{C}(s\mathbf{I} - \mathbf{A})^{-1}\mathbf{B}$ and the closed-loop transfer function under state feedback is $\mathbf{H}_c = \mathbf{C}(s\mathbf{I} - \mathbf{A} + \mathbf{B}\mathbf{K})^{-1}\mathbf{B}$. This can be rearranged as follows:

$$\mathbf{H}_c = \mathbf{C}\{(s\mathbf{I} - \mathbf{A})[\mathbf{I} + (s\mathbf{I} - \mathbf{A})^{-1}\mathbf{B}\mathbf{K}]\}^{-1}\mathbf{B}$$
$$= \mathbf{C}[\mathbf{I} + (s\mathbf{I} - \mathbf{A})^{-1}\mathbf{B}\mathbf{K}]^{-1}(s\mathbf{I} - \mathbf{A})^{-1}\mathbf{B}$$
$$= \mathbf{C}(s\mathbf{I} - \mathbf{A})^{-1}\mathbf{B}[\mathbf{I} + \mathbf{K}(s\mathbf{I} - \mathbf{A})^{-1}\mathbf{B}]^{-1}$$

The last step made use of a rearrangement identity proven in Problem 4.4. Taking the determinant gives

$$|\mathbf{H}_c(s)| = |\mathbf{C}(s\mathbf{I}-\mathbf{A})^{-1}\mathbf{B}|\,|[\mathbf{I}+\mathbf{K}(s\mathbf{I}-\mathbf{A})^{-1}\mathbf{B}]^{-1}|$$
$$= |\mathbf{H}_0(s)|/|[\mathbf{I}+\mathbf{K}(s\mathbf{I}-\mathbf{A})^{-1}\mathbf{B}]|$$

In this form it is obvious that the open- and closed-loop transfer function determinants vanish for the same values of s, so the zeros are the same. Note the similarity with the single-input, single-output results of Chapter 2. There it was found that the closed-loop zeros are the union of the forward path (open-loop) transfer function zeros and the poles of the feedback path transfer function. When only constants occur in the feedback path, the open- and closed-loop zeros are the same.

13.21 Discuss the design techniques of this chapter as contrasted to classical techniques.

In the classical methods, reviewed in Chapter 2, there are usually just one input and one output. Feedback is always output feedback, but perhaps it involves dynamic compensation (lead networks, lead-lag networks, etc.). The design is often developed (at the introductory level) as a trial and error process. The goal is to obtain satisfactory locations for the dominant poles, to ensure acceptable overshoot, settling time, natural frequency, etc. In addition, sensitivity to parameter variations and disturbances is to be reduced and steady-state accuracy must be acceptable. See Reference 16 for a treatment of classical problems using the state variable approach. See also Reference 17.

In this chapter a systematic means of locating *all* poles is presented, provided the system is completely controllable and observable. Multiple inputs and outputs can be treated, but a computer will be required in analyzing most realistic applications. When the feedback is restricted to output signals, observers (or reduced dimensional observers or other dynamic compensators) will be required. This is analogous to using dynamic network compensators in classical design. Sensitivity reduction is directly related to the concept of return difference and the return difference matrix [18].

The design procedures in this chapter assume that the system is controllable and observable. The decomposition techniques of Chapters 11 and 12 allow any linear system to be decomposed into a controllable and an uncontrollable subsystem. The poles of the controllable part can be relocated as desired. If all unstable open-loop poles are associated with the controllable part, the system can be stabilized by use of state feedback. The poles associated with the uncontrollable part are not affected by state feedback. This is what motivated the definition of the property of stabilizability in Chapter 11. Likewise, unobservable states cannot be reconstructed by observers. The consideration of similar decompositions into observable and unobservable subsystems is what prompted the definition of the detectibility property of Chapter 11. These issues did not arise in Chapter 2 because the classical transfer function description of a system, after canceling any common factors, describes the controllable and observable portion of the system, as discussed in Chapter 12.

PROBLEMS

13.22 Assuming that all state variables can be measured, find a state feedback matrix **K** for the system of Problem 13.9. The closed-loop poles are to be placed at $\lambda_i = -3, -3, -4$.

13.23 The open-loop system of Problem 13.9 is compensated by adding feedback consisting of the observer designed in Problem 13.9 followed by the constant matrix **K** of Problem 13.22. Verify that the sixth-order closed-loop system has eigenvalues at $\{-3, -3, -4, -5, -5, -6\}$.

13.24 What should the constant output feedback matrix **K**′ be if the system in Example 13.7 is to have a closed-loop eigenvalue at -20?

13.25 If Problem 13.6 is reconsidered, with $\mathbf{C} = \begin{bmatrix} 1 & 0 \\ 0 & 1 \end{bmatrix}$, find **K**′ so that $\lambda_1 = \lambda_2 = -10$.

26 Repeat Problem 13.8 with the desired eigenvalues $\lambda_1 = \epsilon, \lambda_2 = -\epsilon$.

27 Let $\mathbf{A} = \begin{bmatrix} 0 & 1 & 0 \\ 0 & 0 & 1 \\ 1 & 0 & 0 \end{bmatrix}, \mathbf{B} = \begin{bmatrix} 0 & 1 \\ 1 & 0 \\ 0 & 0 \end{bmatrix}, \mathbf{C} = \begin{bmatrix} 1 & 0 & 0 \\ 0 & 0 & 1 \end{bmatrix}$. Can this system be decoupled?

28 Can the system described by $\mathbf{A} = \begin{bmatrix} 1 & 0 & 0 \\ 0 & 0 & 1 \\ 0 & 1 & 0 \end{bmatrix}, \mathbf{B} = \begin{bmatrix} 1 & 0 \\ 1 & 0 \\ 1 & 1 \end{bmatrix}, \mathbf{C} = \begin{bmatrix} 1 & 0 & 0 \\ 0 & 1 & 0 \end{bmatrix}$ be decoupled?

29 Find the matrices $\mathbf{F}_d$ and $\mathbf{K}_d$ such that state feedback reduces the system of Problem 13.9 to integrator decoupled form. Find the decoupled transfer matrix also.

30 Find a state feedback gain matrix which will give the following discrete-time system closed-loop poles at $z = 0$ and 0.5.

$$\mathbf{x}(k+1) = \begin{bmatrix} 0 & 1 \\ 1 & 1 \end{bmatrix} \mathbf{x}(k) + \begin{bmatrix} 0 \\ 1 \end{bmatrix} u(k)$$

.31 Repeat Problem 13.30 but force the poles to be at $z = 0.3 \pm j0.5$.

.32 A discrete-time system has

$$\mathbf{A} = \begin{bmatrix} 0.65 & -0.15 \\ -0.15 & 0.65 \end{bmatrix} \quad \text{and} \quad \mathbf{B} = \begin{bmatrix} 1 & 1 \\ -1 & 1 \end{bmatrix}$$

Find two different feedback gain matrices which will both give closed-loop poles at $z = 0$ and 0.2.

.33 A system is described by

$$\dot{y}_1 + 3y_1 - y_2 = u_1$$

$$\ddot{y}_2 + 2(\dot{y}_1 + \dot{y}_2 - \dot{y}_3) + 4(y_2 - y_1) = u_2 + 5u_1$$

$$\ddot{y}_3 + 6\dot{y}_3 - 2\dot{y}_1 + y_3 = u_2$$

Put the system into state variable form, with the first three states being y_1, y_2, and y_3 respectively. Design a state feedback controller which will give closed-loop poles at $-3, -4, -4, -5$, and -5. Simulate the system for various test inputs, assuming all states are available for use. Then design a full state observer, with poles at $-6, -7, -8, -9$, and -10. Again on a simulation, compare the transient response of the cascaded observer, state feedback system with full state feedback results.

.34 Design a reduced-order observer to estimate only the nonmeasured states of Problem 13.33. Use this observer, along with the previous state feedback controller, and compare transient behavior with that obtained in the previous problem.

.35 Using the continuous system model of Problem 11.9, show that the closed-loop poles will be at $\lambda_i = -2, -5, -8$, and -10 if any of the following state feedback matrices are used. In each case the columns selected for use in Eq. (*13.6*) are shown. Compare these results with the discrete equivalent of Example 13.6.

The Feedback Gain Matrix Using Columns 1, 3, 6, 8

$$\mathbf{K} = \begin{bmatrix} -1.5000001\text{E}-01 & 7.4249178\text{E}-01 & -3.1340556\text{E}+00 & 5.0000000\text{E}-01 \\ 0.0000000\text{E}+90 & -5.3035126\text{E}+00 & 1.8814680\text{E}+01 & 0.0000000\text{E}+00 \end{bmatrix}$$

The Feedback Gain Matrix Using Columns 2, 4, 5, 7

$$\mathbf{K} = \begin{bmatrix} 4.0000007\text{E}-01 & 2.5047928\text{E}-01 & -3.4747612\text{E}-01 & 4.0000005\text{E}+00 \\ 0.0000000\text{E}+00 & -1.7891376\text{E}+00 & -1.0894562\text{E}+00 & 0.0000000\text{E}+00 \end{bmatrix}$$

The Feedback Gain Matrix Using Columns 1, 4, 6, 7

$$\mathbf{K} = \begin{bmatrix} 9.9999994E-02 & 5.1884985E-01 & -1.5014696E+00 & 1.0000000E+00 \\ 0.0000000E+00 & -3.7060702E+00 & 7.1533542E+00 & 0.0000000E+00 \end{bmatrix}$$

13.36 An open-loop system has the transfer function

$$\frac{y(z)}{u(z)} = \frac{z}{(z-1)(z-0.5)}$$

Design a controller and full state observer such that the closed-loop transfer function has poles at $z = 0.6 \pm 0.3j$. Put the observer poles at $z = 0.1$ and -0.1.

13.37 Repeat the previous problem, but use a reduced-order observer. Put the observer pole at $z = -0.1$.

14

An Introduction to Optimal Control Theory

1 INTRODUCTION

The word *optimal* intuitively means doing a job in the best possible way. Before beginning a search for such an optimal solution, the *job* must be defined, a mathematical scale must be established for quantifying what *best* means, and the *possible* alternatives must be spelled out. Unless there is agreement on these qualifiers, a claim that a system is optimal is really meaningless. A crude, inaccurate system might be considered optimal because it is inexpensive, is easy to fabricate, and gives adequate performance. Conversely, a very precise and elegant system could be rejected as nonoptimal because it is too expensive or is too heavy or would take too long to develop.

An introductory account of a dynamic programming [1] approach to optimal control problems is given in this chapter. The emphasis is on linear system, quadratic cost problems, both discrete-time and continuous-time. Other optimization techniques are discussed in many references, including References 2, 3, 4, and 5. A brief look at one of thse techniques, the minimum principle, is included in Section 14.5 and related problems. An elementary example of a minimum norm problem was given in Problem 5.36. Methods of generalizing this approach are suggested by Problem 14.1.

.2 STATEMENT OF THE OPTIMAL CONTROL PROBLEM

A mathematical statement of the optimal control problem consists of

1. A description of the system to be controlled.
2. A description of system constraints and possible alternatives.
3. A description of the task to be accomplished.
4. A statement of the criterion for judging optimal performance.

The dynamic systems to be considered are described in state variable form by one of the forms in Eq. (*14.1*). It is assumed that all states are available as output measurements, i.e., $\mathbf{y}(t) = \mathbf{x}(t)$. If this is not the case, the state of an *observable* linear system can be estimated by using an observer or a Kalman filter.

$$\dot{\mathbf{x}} = \mathbf{f}(\mathbf{x}(t), \mathbf{u}(t), t) \quad \text{or} \quad \mathbf{x}(k+1) = \mathbf{f}(\mathbf{x}(k), \mathbf{u}(k)) \tag{14.1}$$

Constraints will sometimes exist on allowable values of the state variables. However, in this chapter only control variable constraints are considered. The set of admissible controls U is a subset of the r-dimensional input space $\mathcal{U}^r$ of Chapter 3, $U \subseteq \mathcal{U}^r$. For example, U could be defined as the set of all piecewise continuous vectors $\mathbf{u}(t) \in \mathcal{U}^r$ satisfying $|\mathbf{u}_i(t)| \leq M$ or $\|\mathbf{u}(t)\| \leq M$ for all t, and for some positive constant M. If there are no constraints, $U = \mathcal{U}^r$.

The task to be performed often takes the form of additional boundary conditions on Eq. (*14.1*). An example might be to transfer the state from a known initial state $\mathbf{x}(t_0)$ to a specified final state $\mathbf{x}(t_f) = \mathbf{x}_d$ at a specified time t_f, or the minimum possible t_f. The task might be to transfer the state to a specified region of state space, called a *target set*, rather than to a specified point. Often, the task to be performed is implicitly specified by the performance criterion.

The most general continuous or discrete general performance criteria to be considered are

$$J = S(\mathbf{x}(t_f), t_f) + \int_{t_0}^{t_f} L(\mathbf{x}(t), \mathbf{u}(t), t)\, dt$$

or

$$J = S(\mathbf{x}(N)) + \sum_{k=0}^{N-1} L(\mathbf{x}(k), \mathbf{u}(k)) \tag{14.2}$$

S and L are real, scalar-valued functions of the indicated arguments. S is the cost or penalty associated with the error in the stopping or terminal state at time t_f (or N). L is the cost or loss function associated with the transient state errors and control effort. The functions S and L must be selected by the system designer to put more or less emphasis on terminal accuracy, transient behavior, and the expended control effort in the total cost function J.

EXAMPLE 14.1 Set $S = 0, L = 1$. Then $J = \int_{t_0}^{t_f} dt = t_f - t_0$. This is the minimum time problem. ■

EXAMPLE 14.2 Set $S = 0, L = \mathbf{u}^T\mathbf{u}$. Then $J = \int_{t_0}^{t_f} \mathbf{u}^T\mathbf{u}\, dt$ is a measure of the control effort expended. In many cases this term can be interpreted as control energy. This is called the least-effort problem. ■

EXAMPLE 14.3 Set $S = [\mathbf{x}(t_f) - \mathbf{x}_d]^T[\mathbf{x}(t_f) - x_d]$ and $L = 0$. Then minimizing J is equivalent to minimizing the square of the norm of the error between the final state $\mathbf{x}(t_f)$ and a desired final state $\mathbf{x}_d$. This is the minimum terminal error problem. ■

EXAMPLE 14.4 Set $S = 0, L = [\mathbf{x}(t) - \eta(t)]^T[\mathbf{x}(t) - \eta(t)]$. Then minimizing J is equivalent to minimizing the integral of the norm squared of the transient error between the actual state trajectory $\mathbf{x}(t)$ and a desired trajectory $\eta(t)$. ■

EXAMPLE 14.5 A general quadratic criterion which gives a weighted trade-off between the previous three criteria uses $S = [\mathbf{x}(t_f) - \mathbf{x}_d]^T \mathbf{M}[x(t_f) - \mathbf{x}_d]$ and $L = [\mathbf{x}(t) - \eta(t)]^T \mathbf{Q}[\mathbf{x}(t) - \eta(t)] + \mathbf{u}(t)^T \mathbf{R}\mathbf{u}(t)$. The $n \times n$ weighting matrices $\mathbf{M}$ and $\mathbf{Q}$ are assumed to be positive semidefinite to ensure a well-defined finite minimum for J. The $r \times r$ matrix $\mathbf{R}$ is assumed to be positive definite because $\mathbf{R}^{-1}$ will be required in future manipulations. Values for $\mathbf{M}$, $\mathbf{Q}$, and $\mathbf{R}$ should be selected to give the desired trade-offs among terminal error, transient error, and control effort. ■

The discrete-time versions of the above examples require obvious minor modifications. Other performance criteria would be appropriate in specific cases.

The *optimal control problem* is now stated as:

From among all admissible control functions (or sequences) $\mathbf{u} \in U$, find that one which minimizes J of Eq. (*14.2*) subject to the dynamic system constraints of Eq. (*14.1*) and all initial and terminal boundary conditions that may be specified.

If the control is determined as a function of the initial state and other given system parameters, the control is said to be open-loop. If the control is determined as a function of the current state, then it is a closed-loop or feedback control law. Examples of both types will be given in the sequel.

The importance of the property of controllability should be evident. If the system is completely controllable, there is at least one control which will transfer any initial state to any desired final state. If the system is not controllable, it is not meaningful to search for the optimal control. However, controllability does not guarantee that a solution exists for every optimal control problem. Whenever the admissible controls are restricted to the set U, certain final states may not be attainable. Even though the system is completely controllable, the required control may not belong to U (see Problem 14.42).

.3 DYNAMIC PROGRAMMING

14.3.1 General Introduction to the Principle of Optimality

Dynamic programming provides an efficient means for sequential decision-making. Its basis is R. Bellman's principle of optimality, "An optimal policy has the property that whatever the initial state and the initial decision are, the remaining decisions must constitute an optimal policy with regard to the state resulting from the first decision." As used here, a "decision" is a choice of control at a particular time and the "policy" is the entire control sequence (or function).

Consider the nodes in Figure 14.1 as states, in a general sense. A decision is the choice of alternative paths leaving a given node. The goal is to move from state a to state l with minimum cost. A cost is associated with each segment of the line graph. Define J_{ab} as the cost between a and b. J_{bd} is the cost between b and d, etc. For path a, b, d, l the total cost is $J = J_{ab} + J_{bd} + J_{dl}$, and the optimal path (policy) is defined by

$$\min J = \min[J_{ab} + J_{bd} + J_{dl}, J_{ab} + J_{be} + J_{el}, J_{ac} + J_{ch} + J_{hl}, J_{ac} + J_{ck} + J_{kl}] \tag{14.3}$$

If the initial state is a and if the initial decision is to go to b, then the path from b to l

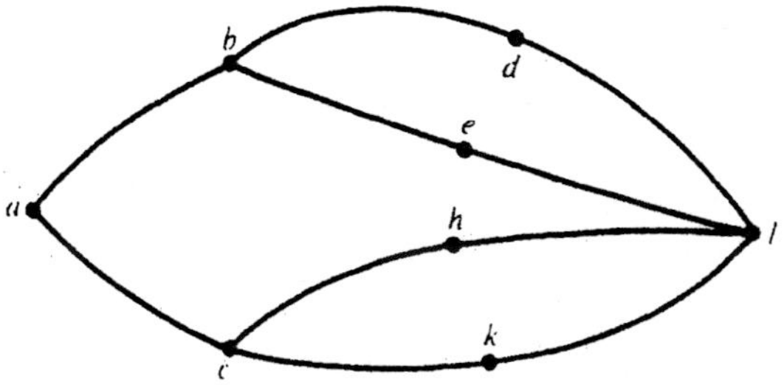

Figure 14.1

must certainly be selected optimally if the overall path from a to l is to be optimum. If the final decision is to go to c, then the path from c to l must then be selected optimally.

Let g_b and g_c be the minimum costs from b and c, respectively, to l. Then $g_b = \min[J_{bd} + J_{dl}, J_{be} + J_{el}]$ and $g_c = \min[J_{ch} + J_{hl}, J_{ck} + J_{kl}]$. The principle of optimality allows equation (*14.3*) to be written as

$$g_a \triangleq \min J = \min[J_{ab} + g_b, J_{ac} + g_c] \tag{14.4}$$

The key feature is that the quantity to be minimized consists of two parts:

1. The part directly attributable to the current decision, such as costs J_{ab} and J_{ac}.
2. The part representing the minimum value of all future costs, starting with the state which results from the first decision.

The principle of optimality replaces a choice between all alternatives (Eq. (*14.3*)) by a sequence of decisions between fewer alternatives (find g_b, g_c, and then g_a from Eq. (*14.4*)). Dynamic programming allows us to concentrate on a sequence of current decisions rather than being concerned about all decisions simultaneously.

Division of cost into the two parts, current and future, is typical, but these parts do not necessarily appear as a sum. A simple example illustrates the sequential nature of the method.

EXAMPLE 14.6 Given N numbers $x_1, x_2, \ldots, x_N$, find the smallest one.

Rather than consider all N numbers simultaneously, define g_k as the minimum of x_k through x_N. Then $g_N = x_N$, $g_{N-1} = \min\{x_{N-1}, g_N\}$, $g_{N-2} = \min\{x_{N-2}, g_{N-1}\}, \ldots, g_k = \min\{x_k, g_{k+1}\}$. Continuing to choose between two alternatives eventually leads to $g_1 = \min\{x_1, g_2\} = \min\{x_1, x_2, \ldots, x_N\}$.

The desired result g_1 need not be unique, since more than one number may have the same smallest value. The recursive nature of the formula $g_k = \min\{x_k, g_{k+1}\}$ is typical of all discrete dynamic programming solutions. ■

EXAMPLE 14.7 Use dynamic programming to find the minimum cost route between node a and node l of Figure 14.2. The line segments can only be traversed from left to right, and the travel costs are shown beside each segment.

Define g_k as the minimum cost from a general node k to l. Obviously, $g_l = 0$. Since there is only one admissible path from h to l, $g_h = 4$. Similarly, $g_j = 5$, $g_d = 2 + g_h = 6$, and $g_f = 6 + g_j = 11$. The first point where a decision must be made is at node e. $g_e = \min\{3 + g_h, 4 + g_j\} = \min\{3 + 4, 4 + 5\} = 7$. The best route from e to l passes through h. Continuing, it is found that $g_b = \min\{8 + g_d, 5 + g_e\} = 12$, $g_c = \min\{4 + g_e, 6 + g_f\} = 11$, and $g_a = \min\{3 + g_b, 5 + g_c\} = 15$. The minimum cost is 15, and the best path is a, b, e, h, l.

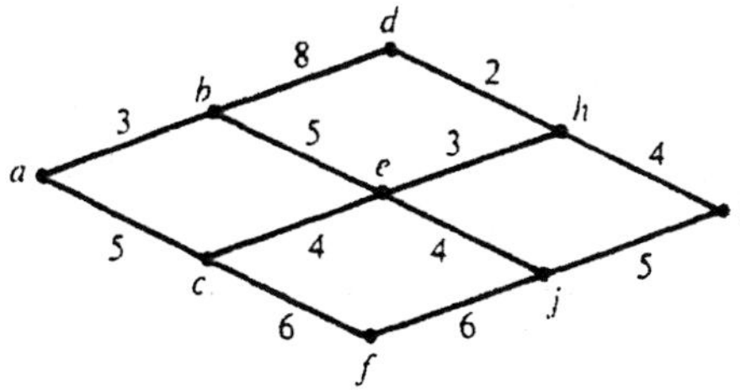

Figure 14.2 Simple routine problem.

The preceding procedure has actually answered the sequence of questions: If past decisions cause the route to be at point k, what is the best decision to make, starting at that point? The answers to these questions must be stored for future use and are shown by the arrows in Figure 14.3. The minimum cost path from any node to l is now obvious. For example, if the path starts at c, then c, e, h, l is cheapest. From f, the cheapest path is f, j, l. ■

14.3.2 Application to Discrete-Time Optimal Control

To make the transition from the simple graph to the more general control problems, the following analogies are made. The graph of Figure 14.2 or 14.3 is a plot of possible states (nodes) at discrete-time points t_k. Point a is at t_0, b and c at $t_1, \ldots, l$ at t_4. The choice of possible directions, say up or down from e, is analogous to the set of admissible controls. Selecting a control $\mathbf{u}(k)$ is analogous to selecting a direction of departure from a given node $\mathbf{x}(k)$. The line segments connecting nodes play the same role as the difference Equation (*14.1*), since both determine the next node $\mathbf{x}(k+1)$ to be encountered.

EXAMPLE 14.8 Consider the scalar system $x(k+1) = x(k) + u(k)$ with boundary conditions $x(0) = 0$ and $x(3) = 3$. Find the controls $u(0)$, $u(1)$, and $u(2)$ which minimize $J = \sum_{k=0}^{2} \{u(k)^2 + \Delta t_k^2\}$. This performance criterion is the sum of the squares of three hypotenuses in the tx plane. This is a form of a minimum distance problem and the optimal sequence of points $x(k)$ will lie on a straight line in the tx plane. Verifying this obvious result will serve to illustrate the dynamic programming procedure.

Let $g(x(k))$ be the minimum cost from $x(k)$ to the terminal point. Since $x(3)$ is the terminal point, $g(x(3)) = 0$. Then

$$g(x(2)) = \min_{u(2)} \{\text{cost from } x(2) \text{ to the terminal point}\} = \min_{u(2)} \{u(2)^2 + \Delta t_2^2 + g(x(3))\}$$

This minimization is not without restriction since $x(3)$ must equal 3. Using $x(3) = x(2) + u(2)$, it is found that $u(2) = 3 - x(2)$. For any $x(2)$, there is a uniquely required $u(2)$. This is analogous to the decisions at points h and j of Example 14.7.

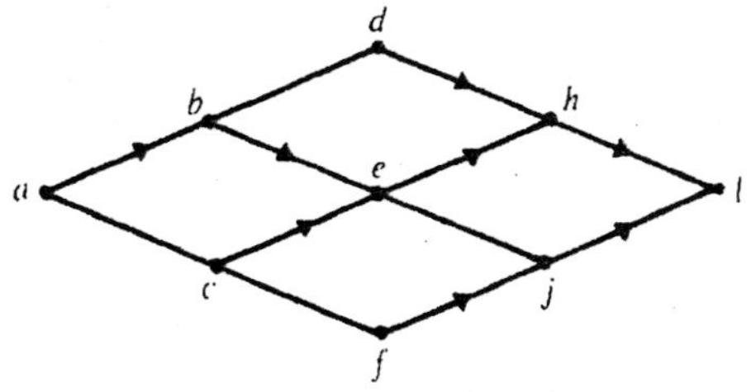

Figure 14.3 The optimal paths.

For convenience, let $\Delta t_k = 1$. Then $g(x(2)) = [3 - x(2)]^2 + 1$. Moving back one stage, $g(x(1)) = \min_{u(1)}\{u(1)^2 + 1 + g(x(2))\}$. Using $x(2) = x(1) + u(1)$ leads to

$$g(x(1)) = \min_{u(1)}\{u(1)^2 + 2 + [3 - x(1) - u(1)]^2\}$$

There are no restrictions on $u(1)$, so the minimization can be accomplished by setting $\frac{\partial}{\partial u(1)}\{u(1)^2 + 2 + [3 - x(1) - u(1)]^2\} = 0$. This gives $u(1) = [3 - x(1)]/2$, and then

$$g(x(1)) = \left[3 - x(1) - \frac{3 - x(1)}{2}\right]^2 + 2 + \left[\frac{3 - x(1)}{2}\right]^2$$

This is used along with $x(1) = x(0) + u(0)$ in $g(x(0)) = \min_{u(0)}\{u(0)^2 + 1 + g(x(1))\}$. Minimizing again gives $u(0) = [3 - x(0)]/3$. Since $x(0) = 0$ is given, $u(0) = 1$. The difference equation gives $x(1) = 1$. This is used in the previously computed expression to give $u(1) = 1$. Then $x(2) = x(1) + u(1) = 2$ and $u(2) = 3 - x(2) = 1$. Finally, $x(3) = x(2) + u(2) = 3$ as required. Note that the sequence $x(0), x(1), x(2), x(3)$ lies on a straight line in tx space, as expected. ∎

Discrete dynamic programming solutions of optimal control problems usually consist of two stage-by-stage passes through the time stages. First, a *backward* pass answers the questions, "What is the minimum cost if the problem is started at time t_k with state $\mathbf{x}(k)$, and what is the optimal control as a function of $\mathbf{x}(k)$?" If $\mathbf{u}(k)$ is unrestricted and the criterion function is simple enough, the optimal $\mathbf{u}(k)$ can be obtained in explicit equation form in terms of $\mathbf{x}(k)$. This can be done by setting the gradient with respect to $\mathbf{u}(k)$ equal to zero as in Example 14.8. When this is not possible, a discrete set of grid points for the state and control variables would be used. A computer search routine would be used to find the optimal $\mathbf{u}(k)$, in tabular form, for each discrete value of $\mathbf{x}(k)$. This is a generalization of the method in Example 14.7.

The backward pass is completed when time t_0 is reached. Since $\mathbf{x}(0)$ is known, $\mathbf{u}(0)$ can be found in terms of that specific state. The second pass is in the *forward* direction. The system difference equation uses $\mathbf{x}(0)$ and $\mathbf{u}(0)$ to obtain $\mathbf{x}(1)$. The previously computed function or table is used to determine the value of $\mathbf{u}(1)$ associated with $\mathbf{x}(1)$. Then, in turn, $\mathbf{u}(1)$ gives $\mathbf{x}(2)$ and $\mathbf{x}(2)$ gives $\mathbf{u}(2)$, and so on.

The preceding verbal description is now expressed in equation form for the discrete versions of Eqs. (*14.1*) and (*14.2*). Define $g(\mathbf{x}(k))$ as the minimum cost of the process, starting at $t_k, \mathbf{x}(k)$. Obviously, $g(\mathbf{x}(N)) = S(\mathbf{x}(N))$. Then the principle of optimality gives

$$g(\mathbf{x}(N-1)) = \min_{\mathbf{u}(N-1)}\{L(\mathbf{x}(N-1), \mathbf{u}(N-1)) + g(\mathbf{x}(N))\}$$

Equation (*14.1*) is used to eliminate $\mathbf{x}(N)$. The minimization with respect to $\mathbf{u}(N-1)$ is carried out by setting the gradient with respect to $\mathbf{u}(N-1)$ equal to zero (*if* there are no restrictions on $\mathbf{u}$), or by a computer search routine. In either case the optimal $\mathbf{u}(N-1)$ must be stored for each possible $\mathbf{x}(N-1)$. Also, $g(\mathbf{x}(N-1))$ must be stored for use in the next stage. This continues, stage-by-stage, with a typical step being

$$g(\mathbf{x}(k)) = \min_{\mathbf{u}(k)}\{L(\mathbf{x}(k), \mathbf{u}(k)) + g(\mathbf{x}(k+1))\}$$

$$= \min_{\mathbf{u}(k)} \{L(\mathbf{x}(k), \mathbf{u}(k)) + g(\mathbf{f}(x(k), \mathbf{u}(k)))\} \quad (14.5)$$

Equation (*14.5*) is an extremely powerful and general result. It is the key to the solution of many discrete-time optimal control problems. This nonlinear difference equation has the boundary condition $g(\mathbf{x}(N)) = S(\mathbf{x}(N))$. The solution of Eq. (*14.5*) yields the optimal control $\mathbf{u}^*(k)$‡ at each time step, and also the optimal trajectory $\mathbf{x}^*(k)$.

The most general case of Eq. (*14.5*) can be solved (in principle at least) by using a tabular computational approach. That is, a discretized grid of possible $\mathbf{x}(k)$ and $\mathbf{u}(k)$ values is determined at each time point. The results of the backward-in-time pass through this grid will consist of a table of optimal $\mathbf{u}^*(k)$ values for each possible $\mathbf{x}(k)$ value. The storage requirements quickly become excessive for all but the lowest order systems. The emphasis here will be on the case of linear systems with quadratic cost functions.

14.3.3 The Discrete-Time Linear Quadratic Problem

The linear system, quadratic cost function (LQ) optimal control problem has received special attention in the literature and in applications. This is because it can be solved analytically, and the resulting optimal controller is expressed in easy-to-implement state feedback form.

Consider the system $\mathbf{x}(k+1) = \mathbf{A}\mathbf{x}(k) + \mathbf{B}\mathbf{u}(k)$ with $\mathbf{x}(0)$ known. The goal is to find the control sequence $\mathbf{u}(k)$ which minimizes the quadratic cost function for the finite-time regulator problem,

$$J = \tfrac{1}{2}\mathbf{x}(N)^T\mathbf{M}\mathbf{x}(N) + \tfrac{1}{2}\sum_{k=0}^{N-1} \{\mathbf{x}^T(k)\mathbf{Q}\mathbf{x}(k) + \mathbf{u}^T(k)\mathbf{R}\mathbf{u}(k)\} \quad (14.6)$$

$\mathbf{M}$ and $\mathbf{Q}$ are symmetric positive semidefinite $n \times n$ matrices and $\mathbf{R}$ is a symmetric positive definite $r \times r$ matrix. No restrictions are placed on $\mathbf{u}(k)$.

Let $g[\mathbf{x}(k)]$ = min cost from k, $\mathbf{x}(k)$ to N, $\mathbf{x}(N)$. Equation (*14.5*) gives

$$g[\mathbf{x}(k)] = \min_{\mathbf{u}(k)} \{\tfrac{1}{2}\mathbf{x}^T(k)\mathbf{Q}\mathbf{x}(k) + \tfrac{1}{2}\mathbf{u}^T(k)\mathbf{R}\mathbf{u}(k) + g[\mathbf{x}(k+1)]\} \quad (14.7)$$

This is a difference equation and the boundary condition is $g[\mathbf{x}(N)] = \tfrac{1}{2}\mathbf{x}(N)^T\mathbf{M}\mathbf{x}(N)$. This equation is solved by assuming a solution

$$g[\mathbf{x}(k)] = \tfrac{1}{2}\mathbf{x}(k)^T\mathbf{W}(N-k)\mathbf{x}(k) + \mathbf{x}(k)^T\mathbf{V}(N-k) + Z(N-k) \quad (14.8)$$

where $\mathbf{W}$, $\mathbf{V}$, and Z are an $n \times n$ matrix, an $n \times 1$ vector, and a scalar, respectively. They will be selected so as to force Eq. (*14.8*) to satisfy Eq. (*14.7*). Equation (*14.7*) becomes

$$\begin{aligned}&\tfrac{1}{2}\mathbf{x}(k)^T\mathbf{W}(N-k)\mathbf{x}(k) + \mathbf{x}(k)^T\mathbf{V}(N-k) + Z(N-k)\\ &= \min_{\mathbf{u}(k)} \{\tfrac{1}{2}\mathbf{x}^T(k)\mathbf{Q}\mathbf{x}(k) + \tfrac{1}{2}\mathbf{u}^T(k)\mathbf{R}\mathbf{u}(k) + \tfrac{1}{2}\mathbf{x}(k+1)^T\mathbf{W}(N-k-1)\mathbf{x}(k+1) \quad (14.9)\\ &+ \mathbf{x}(k+1)^T\mathbf{V}(N-k-1) + Z(N-k-1)\}\end{aligned}$$

‡The * on a vector, such as $\mathbf{u}^*(k)$, indicates the optimal vector. It should not be confused with the notation for an adjoint transformation, such as $\mathcal{A}^*$.

Substituting $\mathbf{x}(k+1) = \mathbf{A}\mathbf{x}(k) + \mathbf{B}\mathbf{u}(k)$ and regrouping, the right-hand side of Eq. (*14.9*) becomes

$$\begin{aligned}\text{R.H.S.} = \min_{\mathbf{u}(k)} \{&\tfrac{1}{2}\mathbf{x}(k)^T[\mathbf{Q} + \mathbf{A}^T\mathbf{W}(N-k-1)\mathbf{A}]\mathbf{x}(k) \\ &+ \tfrac{1}{2}\mathbf{u}(k)^T[\mathbf{R} + \mathbf{B}^T\mathbf{W}(N-k-1)\mathbf{B}]\mathbf{u}(k) \\ &+ \mathbf{u}(k)^T[\mathbf{B}^T\mathbf{V}(N-k-1) + \mathbf{B}^T\mathbf{W}(N-k-1)\mathbf{A}\mathbf{x}(k)] \\ &+ \mathbf{x}(k)^T\mathbf{A}^T\mathbf{V}(N-k-1) + Z(N-k-1)\}\end{aligned} \tag{14.10}$$

Since there are no restrictions on $\mathbf{u}(k)$, the minimizing $\mathbf{u}(k)$ is found by setting $\partial\{\ \}/\partial\mathbf{u}(k) = 0$. This yields

$$\begin{aligned}\mathbf{u}^*(k) = &-[\mathbf{R} + \mathbf{B}^T\mathbf{W}(N-k-1)\mathbf{B}]^{-1}\mathbf{B}^T[\mathbf{V}(N-k-1) \\ &+ \mathbf{W}(N-k-1)\mathbf{A}\mathbf{x}(k)] = \mathbf{F}_c(k)\mathbf{V}(N-k-1) - \mathbf{G}(k)\mathbf{x}(k)\end{aligned} \tag{14.11}$$

The problem remains to determine the feed-forward and feedback gain matrices $\mathbf{F}_c(k)$ and $\mathbf{G}(k)$ and the external input $\mathbf{V}(k)$. Rearranging and simplifying Eq. (*14.10*) gives

$$\begin{aligned}&\text{R.H.S.} \\ &= \tfrac{1}{2}\mathbf{x}(k)^T[\mathbf{Q} + \mathbf{A}^T\mathbf{W}(N-k-1)\mathbf{A} - \mathbf{A}^T\mathbf{W}(N-k-1)\mathbf{B}\mathbf{U}\mathbf{B}^T\mathbf{W}(N-k-1)\mathbf{A}]\mathbf{x}(k) \\ &\quad + \mathbf{x}(k)^T[\mathbf{A}^T\mathbf{V}(N-k-1) - \mathbf{A}^T\mathbf{W}(N-k-1)\mathbf{B}\mathbf{U}\mathbf{B}^T\mathbf{V}(N-k-1)] \\ &\quad + [Z(N-k-1) - \tfrac{1}{2}\mathbf{V}(N-k-1)^T\mathbf{B}\mathbf{U}\mathbf{B}^T\mathbf{V}(N-k-1)]\end{aligned}$$

where $\mathbf{U} \triangleq [\mathbf{R} + \mathbf{B}^T\mathbf{W}(N-k-1)\mathbf{B}]^{-1}$. Equating the left-hand and right-hand sides, the assumed form for $g[\mathbf{x}(k)]$ can be forced to be a solution *for all* $\mathbf{x}(k)$ by requiring that the quadratic terms, the linear terms, and the terms not involving $\mathbf{x}$ all balance individually. This requires

$$\begin{aligned}\mathbf{W}(N-k) = \mathbf{Q} &+ \mathbf{A}^T\mathbf{W}(N-k-1)\mathbf{A} \\ &- \mathbf{A}^T\mathbf{W}(N-k-1)\mathbf{B}\mathbf{U}\mathbf{B}^T\mathbf{W}(N-k-1)\mathbf{A}\end{aligned} \tag{14.12}$$

$$\mathbf{V}(N-k) = \mathbf{A}^T\mathbf{V}(N-k-1) - \mathbf{A}^T\mathbf{W}(N-k-1)\mathbf{B}\mathbf{U}\mathbf{B}^T\mathbf{V}(N-k-1) \tag{14.13}$$

$$Z(N-k) = Z(N-k-1) - \tfrac{1}{2}\mathbf{V}(N-k-1)^T\mathbf{B}\mathbf{U}\mathbf{B}^T\mathbf{V}(N-k-1) \tag{14.14}$$

The boundary conditions are $\mathbf{W}(N-N) = \mathbf{M}$, $\mathbf{V}(N-N) = 0$, and $Z(N-N) = 0$. A computer solution, backward in time, easily gives $\mathbf{W}(N-k)$. Normally (*14.12*) is solved first. Then its solution $\mathbf{W}(N-k)$ is used as a known coefficient matrix while solving (*14.13*) for $\mathbf{V}(N-k)$. Then $\mathbf{V}$ acts as a known forcing function in (*14.14*). Actually, (*14.14*) never needs to be solved if the only interest is in finding the optimal control. $Z(N-k)$ is only needed if $J_{\min}$ must be calculated. For the cost function considered here (the so-called regulator problem), (*14.13*) is a homogeneous equation with zero initial conditions, so $\mathbf{V}(N-k)$ is zero for all stages. Therefore, (*14.13*) is not needed either.

Equation (*14.12*) remains as the principal result. It is interesting to rewrite (*14.12*) as follows: Let $k' = N - k$ be a backward running time index and let

$\mathbf{W}(N-k) = \mathbf{M}(k')$. Introducing two new intermediate variables $\mathbf{K}$ and $\mathbf{P}$ allows Eq. (*14.12*) to be replaced by

$$\mathbf{M}(k') = \mathbf{A}^T\mathbf{P}(k'-1)\mathbf{A} + \mathbf{Q} \tag{14.15}$$

$$\mathbf{K}(k') = \mathbf{M}(k')\mathbf{B}[\mathbf{B}^T\mathbf{M}(k')\mathbf{B} + \mathbf{R}]^{-1} \tag{14.16}$$

$$\mathbf{P}(k') = [\mathbf{I} - \mathbf{K}(k')\mathbf{B}^T]\mathbf{M}(k') \tag{14.17}$$

These are exactly the same as (*1*), (*2*), and (*3*) of Problem 6.18 (the Kalman filter algorithm), except that $\mathbf{B}^T$ replaces $\mathbf{C}$, $\mathbf{A}^T$ replaces $\mathbf{\Phi}$, and k' replaces k. The optimal mean square estimator problem and the optimal regulator problem are duals of each other. The initial condition on (*14.15*) is $\mathbf{M}(0) = \mathbf{M}$. The optimal feedback control is given by

$$\mathbf{u}^*(k) = -\mathbf{K}^T(k'-1)\mathbf{A}\mathbf{x}(k)$$

so that the feedback gain matrix is $\mathbf{G}(k) = \mathbf{K}^T(k'-1)\mathbf{A}$. This duality has a practical significance. If a computer program is available to compute the Kalman gain matrix $\mathbf{K}(k)$ (Eq. (2) of Problem 6.18), then the same algorithm can be used to find the control gains $\mathbf{G}(k)$. Just make the interchanges with $\mathbf{B}$, $\mathbf{C}$, $\mathbf{A}$, and $\mathbf{\Phi}$ as mentioned above and remember that time is reversed.

The optimal control sequence can next be found for the same linear system, but with the more general cost function for the tracking problem,

$$J = \tfrac{1}{2}[\mathbf{x}(N) - \mathbf{x}_d]^T\mathbf{M}[\mathbf{x}(N) - \mathbf{x}_d] + \tfrac{1}{2}\sum_{k=0}^{N-1}\{[\mathbf{x}(k) - \boldsymbol{\eta}(k)]^T\mathbf{Q}[\mathbf{x}(k) - \boldsymbol{\eta}(k)] + \mathbf{u}(k)^T\mathbf{R}\mathbf{u}(k)\} \tag{14.18}$$

This cost function attempts to make $\mathbf{x}(k)$ follow the specified sequence $\boldsymbol{\eta}(k)$.

Most of the solution details are the same as for the regulator problem and will not be repeated. Expanding the quadratic terms in J shows that there are four additional terms to deal with, due to $\mathbf{x}_d(k)$ and $\boldsymbol{\eta}(k)$. Two of these terms become forcing functions on Eqs. (*14.13*) and (*14.14*), modified here as

$$\mathbf{V}(N-k) = \mathbf{A}^T\mathbf{V}(N-k-1) - \mathbf{A}^T\mathbf{W}(N-k-1)\mathbf{B}\mathbf{U}\mathbf{B}^T\mathbf{V}(N-k-1) - \mathbf{Q}\boldsymbol{\eta}(k) \tag{14.19}$$

$$Z(N-k) = Z(N-k-1) - \tfrac{1}{2}\mathbf{V}^T(N-k-1)\mathbf{B}\mathbf{U}\mathbf{B}^T\mathbf{V}(N-k-1) + \tfrac{1}{2}\boldsymbol{\eta}(k)^T\mathbf{Q}\boldsymbol{\eta}(k) \tag{14.20}$$

Equation (*14.12*)—or its expanded counterparts (*14.15*), (*14.16*) and (*14.17*)—remains unchanged. The other two additional terms come into the boundary conditions, which are

$$\mathbf{W}(N-N) = \mathbf{M} \quad \text{(unchanged)}$$

$$\mathbf{V}(N-N) = -\mathbf{M}\mathbf{x}_d$$

$$Z(N-N) = \tfrac{1}{2}\mathbf{x}_d^T\mathbf{M}\mathbf{x}_d$$

A form of Eq. (*14.19*) that is consistent with the notation introduced in (*14.15*), (*14.16*), and (*14.17*) is

$$\mathbf{V}(k') = \mathbf{A}^T[\mathbf{I} - \mathbf{K}^T(k'-1)\mathbf{B}^T]\mathbf{V}(k'-1) - \mathbf{Q}\eta(k) \tag{14.21}$$

The control law is

$$\mathbf{u}^*(k) = -\mathbf{K}^T(k'-1)\mathbf{A}\mathbf{x}(k) - \mathbf{U}(k'-1)\mathbf{B}^T\mathbf{V}(k'-1) \tag{14.22}$$

In both the regulator and tracking problems, the feedback gain

$$\mathbf{G}(k) = [\mathbf{R} + \mathbf{B}^T\mathbf{W}(N-k-1)\mathbf{B}]^{-1}\mathbf{B}^T\mathbf{W}(N-k-1)\mathbf{A}$$

and the feed-forward gain

$$\mathbf{F}_c(k) = -[\mathbf{R} + \mathbf{B}^T\mathbf{W}(N-k-1)\mathbf{B}]^{-1}\mathbf{B}^T$$

depend on the positive definite symmetric matrix $\mathbf{W}$, which must be obtained by solving Eq. (*14.12*) or the equivalent triple, Eqs. (*14.15*) through (*14.17*). This is called the discrete-time Riccati equation. In terms of the backward running time index k', this discrete Riccati equation is

$$\begin{aligned}\mathbf{W}(k') = \mathbf{Q} + \mathbf{A}^T\mathbf{W}(k'-1)\mathbf{A} \\ - \mathbf{A}^T\mathbf{W}(k'-1)\mathbf{B}[\mathbf{R} + \mathbf{B}^T\mathbf{W}(k'-1)\mathbf{B}]^{-1}\mathbf{B}^T\mathbf{W}(k'-1)\mathbf{A}\end{aligned} \tag{14.23}$$

Assume that $\mathbf{A}$ is nonsingular. Then this nonlinear difference equation can be reduced to a pair of coupled linear equations by replacing the $n \times n$ nonsingular matrix $\mathbf{W}$ by $\mathbf{W}(k') = \mathbf{E}(k')\mathbf{F}(k')^{-1}$. Making this substitution in Eq. (*14.23*) and then postmultiplying by $\mathbf{F}(k')$ and premultiplying by $\mathbf{A}^{-T}$ (the transpose of $\mathbf{A}^{-1}$) gives

$$\begin{aligned}\mathbf{A}^{-T}\mathbf{E}(k') = \mathbf{A}^{-T}\mathbf{Q}\mathbf{F}(k') + \mathbf{W}(k'-1)\mathbf{A}\mathbf{F}(k') \\ - \mathbf{W}(k'-1)\mathbf{B}[\mathbf{R} + \mathbf{B}^T\mathbf{W}(k'-1)\mathbf{B}]^{-1}\mathbf{B}^T\mathbf{W}(k'-1)\mathbf{A}\mathbf{F}(k')\end{aligned} \tag{14.24}$$

If the judicious choice

$$\mathbf{E}(k'-1) = \mathbf{A}^{-T}\mathbf{E}(k') - \mathbf{A}^{-T}\mathbf{Q}\mathbf{F}(k') \tag{14.25}$$

is made, then Eq. (*14.24*) becomes

$$\begin{aligned}\mathbf{E}(k'-1) = \mathbf{W}(k'-1)\mathbf{A}\mathbf{F}(k') \\ - \mathbf{W}(k'-1)\mathbf{B}[\mathbf{R} + \mathbf{B}^T\mathbf{W}(k'-1)\mathbf{B}]^{-1}\mathbf{B}^T\mathbf{W}(k'-1)\mathbf{A}\mathbf{F}(k')\end{aligned}$$

Premultiplication by $\mathbf{W}(k'-1)^{-1} = \{\mathbf{E}(k'-1)\mathbf{F}(k'-1)^{-1}\}^{-1}$ gives

$$\mathbf{F}(k'-1) = \mathbf{A}\mathbf{F}(k') - \mathbf{B}[\mathbf{R} + \mathbf{B}^T\mathbf{W}(k'-1)\mathbf{B}]^{-1}\mathbf{B}^T\mathbf{W}(k'-1)\mathbf{A}\mathbf{F}(k')$$

The matrix inversion lemma of Section 4.9, $[\mathbf{R} + \mathbf{B}^T\mathbf{W}\mathbf{B}]^{-1} = \mathbf{R}^{-1} - \mathbf{R}^{-1}\mathbf{B}^T[\mathbf{B}\mathbf{R}^{-1}\mathbf{B}^T + \mathbf{W}^{-1}]^{-1}\mathbf{B}\mathbf{R}^{-1}$, is used to rewrite this as

$$\begin{aligned}\mathbf{F}(k'-1) = \mathbf{A}\mathbf{F}(k') \\ - \{\mathbf{I} - \mathbf{B}\mathbf{R}^{-1}\mathbf{B}^T[\mathbf{B}\mathbf{R}^{-1}\mathbf{B}^T + \mathbf{W}(k'-1)^{-1}]^{-1}\}\mathbf{B}\mathbf{R}^{-1}\mathbf{B}^T\mathbf{W}(k'-1)\mathbf{A}\mathbf{F}(k')\end{aligned}$$

When the unit matrix is written

$$\mathbf{I} = [\mathbf{BR}^{-1}\mathbf{B}^T + \mathbf{W}(k'-1)^{-1}][\mathbf{BR}^{-1}\mathbf{B}^T + \mathbf{W}(k'-1)^{-1}]^{-1}$$

the preceding equation can be rearranged to

$$\mathbf{F}(k'-1) - \mathbf{AF}(k') = -\mathbf{W}(k'-1)^{-1}[\mathbf{BR}^{-1}\mathbf{B}^T + \mathbf{W}(k'-1)^{-1}]^{-1}\mathbf{BR}^{-1}\mathbf{B}^T\mathbf{W}(k'-1)\mathbf{AF}(k')$$

Using $\mathbf{W}^{-1}[\mathbf{BR}^{-1}\mathbf{B}^T + \mathbf{W}^{-1}]^{-1} = [\mathbf{BR}^{-1}\mathbf{B}^T\mathbf{W} + \mathbf{I}]^{-1}$ and then premultiplying by $[\mathbf{BR}^{-1}\mathbf{B}^T\mathbf{W} + \mathbf{I}]$ and replacing the remaining $\mathbf{W}$ terms by $\mathbf{EF}^{-1}$ allows the linear equation to be obtained:

$$\mathbf{F}(k'-1) = \mathbf{AF}(k') - \mathbf{BR}^{-1}\mathbf{B}^T\mathbf{E}(k'-1)$$

Using Eq. (*14.25*) to eliminate $\mathbf{E}(k'-1)$ gives the final form

$$\mathbf{F}(k'-1) = [\mathbf{A} + \mathbf{BR}^{-1}\mathbf{B}^T\mathbf{A}^{-T}\mathbf{Q}]\mathbf{F}(k') - \mathbf{BR}^{-1}\mathbf{B}^T\mathbf{A}^{-T}\mathbf{E}(k') \tag{14.26}$$

Equations (*14.25*) and (*14.26*) can be combined into

$$\begin{bmatrix}\mathbf{F}(k'-1)\\ \mathbf{E}(k'-1)\end{bmatrix} = \begin{bmatrix}\mathbf{A} + \mathbf{BR}^{-1}\mathbf{B}^T\mathbf{A}^{-T}\mathbf{Q} & -\mathbf{BR}^{-1}\mathbf{B}^T\mathbf{A}^{-T}\\ -\mathbf{A}^{-T}\mathbf{Q} & \mathbf{A}^{-T}\end{bmatrix}\begin{bmatrix}\mathbf{F}(k')\\ \mathbf{E}(k')\end{bmatrix} \tag{14.27}$$

The $2n \times 2n$ coefficient matrix on the right side of Eq. (*14.27*) is often called the Hamiltonian matrix **H**. When **A**, **B**, **Q**, and **R** (and hence **H**) are all constant, the solution to Eq. (*14.27*) can be written as

$$\begin{bmatrix}\mathbf{F}(k')\\ \mathbf{E}(k')\end{bmatrix} = \mathbf{H}^{-k'}\begin{bmatrix}\mathbf{I}\\ \mathbf{M}\end{bmatrix} \tag{14.28}$$

where the initial condition $\mathbf{W}(k'=0) = \mathbf{M}$ was used to select $\mathbf{E}(0) = \mathbf{M}$ and $\mathbf{F}(0) = \mathbf{I}$.

In order to bring Eq. (*14.27*) into the standard form, it was premultiplied by $\mathbf{H}^{-1}$ before solving. This is what causes the negative power on the exponent of **H** in Eq. (*14.28*). It is known [2] that if λ is an eigenvalue of **H**, then so is $1/\lambda$. That is, n of the eigenvalues are stable (inside the unit circle) and n are unstable (outside the unit circle). The Jordan form of **H** is $\mathbf{J} = \text{Diag}[\mathbf{J}_s \quad \mathbf{J}_u]$, where $\mathbf{J}_s$ and $\mathbf{J}_u$ are the $n \times n$ blocks associated with the stable and unstable eigenvalues, respectively. In the case of distinct eigenvalues, these will be diagonal blocks and $\mathbf{J}_u = \mathbf{J}_s^{-1}$ because the eigenvalues occur in reciprocal pairs. Let the modal matrix of eigenvectors for **H** be written as $\mathbf{T} = \begin{bmatrix}\mathbf{T}_{11} & \mathbf{T}_{12}\\ \mathbf{T}_{21} & \mathbf{T}_{22}\end{bmatrix}$ and let $\mathbf{T}^{-1} = \begin{bmatrix}\mathbf{V}_{11} & \mathbf{V}_{12}\\ \mathbf{V}_{21} & \mathbf{V}_{22}\end{bmatrix}$. The columns of $\begin{bmatrix}\mathbf{T}_{11}\\ \mathbf{T}_{21}\end{bmatrix}$ represent the eigenvectors associated with the stable eigenvalues. Then

$$\mathbf{H}^{-k'} = \mathbf{TJ}^{-k'}\mathbf{T}^{-1} = \begin{bmatrix}\mathbf{T}_{11}\mathbf{J}_s^{-k'}\mathbf{V}_{11} + \mathbf{T}_{12}\mathbf{J}_u^{-k'}\mathbf{V}_{21} & \mathbf{T}_{11}\mathbf{J}_s^{-k'}\mathbf{V}_{12} + \mathbf{T}_{12}\mathbf{J}_u^{-k'}\mathbf{V}_{22}\\ \mathbf{T}_{21}\mathbf{J}_s^{-k'}\mathbf{V}_{11} + \mathbf{T}_{22}\mathbf{J}_u^{-k'}\mathbf{V}_{21} & \mathbf{T}_{21}\mathbf{J}_s^{-k'}\mathbf{V}_{12} + \mathbf{T}_{22}\mathbf{J}_u^{-k'}\mathbf{V}_{22}\end{bmatrix}$$

Using this result in Eq. (*14.28*) and then using $\mathbf{W} = \mathbf{EF}^{-1}$ gives

$$\begin{aligned}\mathbf{W}(k') = {} & [\mathbf{T}_{21}\mathbf{J}_s^{-k'}\mathbf{V}_{11} + \mathbf{T}_{22}\mathbf{J}_u^{-k'}\mathbf{V}_{21} + (\mathbf{T}_{21}\mathbf{J}_s^{-k'}\mathbf{V}_{12} + \mathbf{T}_{22}\mathbf{J}_u^{-k'}\mathbf{V}_{22})M] * \\ & [\mathbf{T}_{11}\mathbf{J}_s^{-k'}\mathbf{V}_{11} + \mathbf{T}_{12}\mathbf{J}_u^{-k'}\mathbf{V}_{21} + (\mathbf{T}_{11}\mathbf{J}_s^{-k'}\mathbf{V}_{12} + \mathbf{T}_{12}\mathbf{J}_u^{-k'}\mathbf{V}_{22})M]^{-1}\end{aligned} \tag{14.29}$$

The solution for $\mathbf{W}(k') = \mathbf{W}(N - k - 1)$ can then be used in the equations given previously for the feedback matrix $\mathbf{G}(k)$ and the feed-forward matrix $\mathbf{F}_c(k)$, leading to the implementations shown in Figure 14.4. Figure 14.4*b* is obtained by placing the factor common to both $\mathbf{G}$ and $\mathbf{F}_c$ in the forward path.

14.3.4 The Infinite Horizon, Constant Gain Solution

For constant coefficient systems whose operating time is very long compared with the system time constants, it is often justifiable to assume that the terminal time is infinitely far in the future. This so-called infinite horizon case leads to a constant feedback gain matrix, with attendant implementation advantages. This approximation may cause little or no degradation in optimality because the optimal time-varying gains approach constant values in a few time constants (backward from the final time). Thus the optimal gains are constant for most of the operating period. Furthermore, in the regulator problem the states are driven to zero in a few time constants after the initial time t_0. Therefore, the control $\mathbf{u} = -\mathbf{G}\mathbf{x}$ will be essentially zero during the final part of the operation, regardless of whether a constant or time-varying $\mathbf{G}$ is used.

As the time remaining until the end of the problem approaches infinity, $k' \to \infty$, the general solution for the time-varying matrix $\mathbf{W}(k')$ simplifies to a constant. One approach to finding the infinite-time-to-go solution is to set $\mathbf{W}(k') = \mathbf{W}(k' - 1)$ in Eq. (*14.23*) and solve the so-called discrete algebraic Riccati equation (DARE). This solution is easily obtained from the general solution of Sec. 14.3.3, Eq. (*14.29*). Note that $\mathbf{J}_u^{-k'} \to \mathbf{0}$ and $\mathbf{J}_s^{-k'} \to \infty$ as $k' \to \infty$. The $\mathbf{J}_u$ terms are dropped and the $\mathbf{J}_s$ terms are retained (although they too will be found to drop out), giving

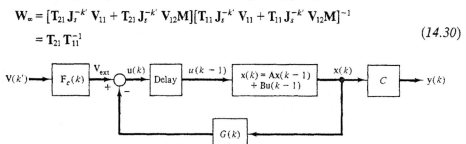

$$\begin{aligned}\mathbf{W}_\infty &= [\mathbf{T}_{21}\mathbf{J}_s^{-k'}\mathbf{V}_{11} + \mathbf{T}_{21}\mathbf{J}_s^{-k'}\mathbf{V}_{12}\mathbf{M}][\mathbf{T}_{11}\mathbf{J}_s^{-k'}\mathbf{V}_{11} + \mathbf{T}_{11}\mathbf{J}_s^{-k'}\mathbf{V}_{12}\mathbf{M}]^{-1} \\ &= \mathbf{T}_{21}\mathbf{T}_{11}^{-1}\end{aligned} \tag{14.30}$$

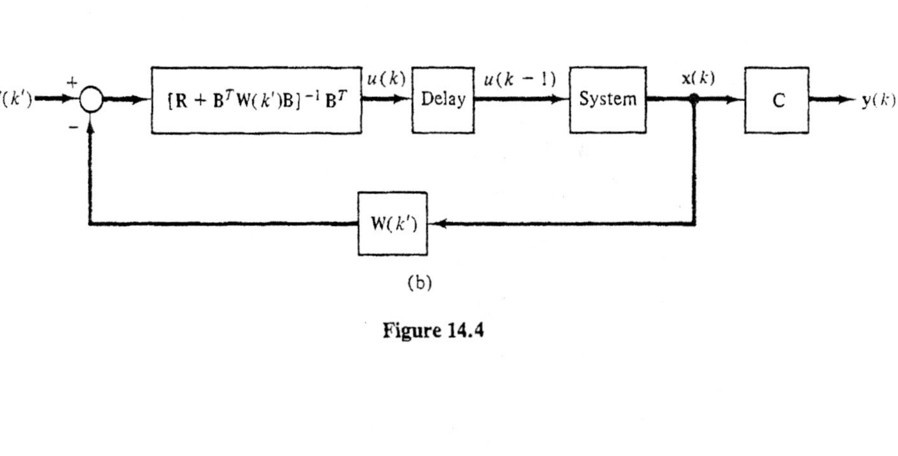

Figure 14.4

Thus the desired answer is found by determining the eigenvectors associated with the stable eigenvalues of $\mathbf{H}$ and partitioning them into the two $n \times n$ blocks required in Eq. (*14.30*). Another obvious way of calculating $\mathbf{W}_\infty$ without solving an eigenvalue problem is to cycle through Eq. (*14.23*) or (*14.27*) until an unchanging result is obtained. In either case, the constant feedback gain matrix is then given by

$$\mathbf{G}_\infty = [\mathbf{B}^T\mathbf{W}_\infty\mathbf{B} + \mathbf{R}]^{-1}\mathbf{B}^T\mathbf{W}_\infty\mathbf{A}$$

and the control law is

$$\mathbf{u}(k) = -\mathbf{G}_\infty\mathbf{x}(k) + \mathbf{F}_c\mathbf{V}(k)$$

In the regulator problem the input term $\mathbf{V}(k)$ is zero. In tracking problems this extra term must be evaluated from Eq. (*14.13*) or (*14.21*). When needed, the feed-forward control matrix $\mathbf{F}_c$ is given by

$$\mathbf{F}_c = -[\mathbf{B}^T\mathbf{W}_\infty\mathbf{B} + \mathbf{R}]^{-1}\mathbf{B}^T$$

EXAMPLE 14.9 Consider the second-order continuous-time plant shown in Figure 14.5. A sample and zero-order hold is placed in each input path, so that a discrete-time controller can be implemented. The continuous-time state equations are selected as

$$\dot{\mathbf{x}} = \begin{bmatrix} 0 & 1 \\ 0 & -0.5 \end{bmatrix}\mathbf{x} + \begin{bmatrix} 0 & 1 \\ 1 & 0 \end{bmatrix}\mathbf{u} \quad \text{and} \quad \mathbf{y} = [1 \quad 0]\mathbf{x}$$

The results of Sec. 9.8 are used to derive discrete-time state equations $\mathbf{x}(k+1) = \mathbf{A}_1\mathbf{x}(k) + \mathbf{B}_1\mathbf{u}(k)$ for three different sampling times $T = 1, \frac{1}{3}$, and $\frac{1}{10}$. Note that the system time constant is $\tau = 2$, so these will provide 2, 6, and 20 samples per time constant, respectively.

T	$\mathbf{A}_1$	$\mathbf{B}_1$
1	$\begin{bmatrix} 1 & 0.786939 \\ 0 & 0.606531 \end{bmatrix}$	$\begin{bmatrix} 0.426123 & 1 \\ 0.786939 & 0 \end{bmatrix}$
$\frac{1}{3}$	$\begin{bmatrix} 1 & 0.307036 \\ 0 & 0.846482 \end{bmatrix}$	$\begin{bmatrix} 0.052593 & 0.333333 \\ 0.307036 & 0 \end{bmatrix}$
$\frac{1}{10}$	$\begin{bmatrix} 1 & 0.097541 \\ 0 & 0.951229 \end{bmatrix}$	$\begin{bmatrix} 0.004918 & 0.1 \\ 0.097541 & 0 \end{bmatrix}$

To simplify notation, the subscripts on $\mathbf{A}_1$ and $\mathbf{B}_1$ will be omitted in the following. Figure 14.6 shows the transient behavior of the gain components derived from the discrete Riccati equation plotted versus time remaining. For this example the parameters were $T = \frac{1}{3}$, $\mathbf{Q} = \mathbf{R} = \mathbf{I}$, with the boundary condition for $\mathbf{W}(k' = 0) = \mathbf{M} = [\mathbf{0}]$. Note that the final constant values are essentially achieved by $t_r = 4$—that is, within just two system time constants. Similar results are found for

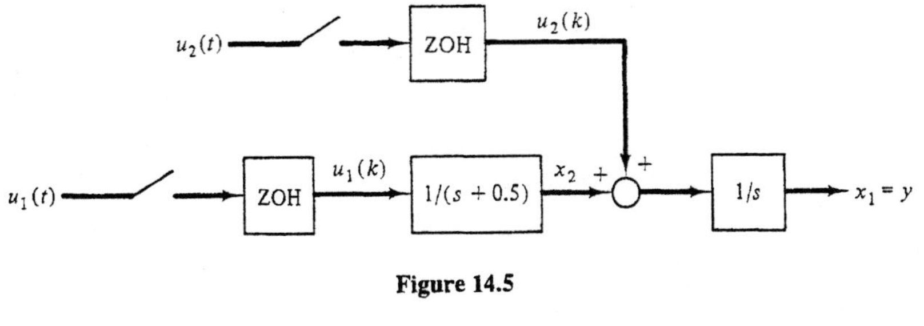

Figure 14.5

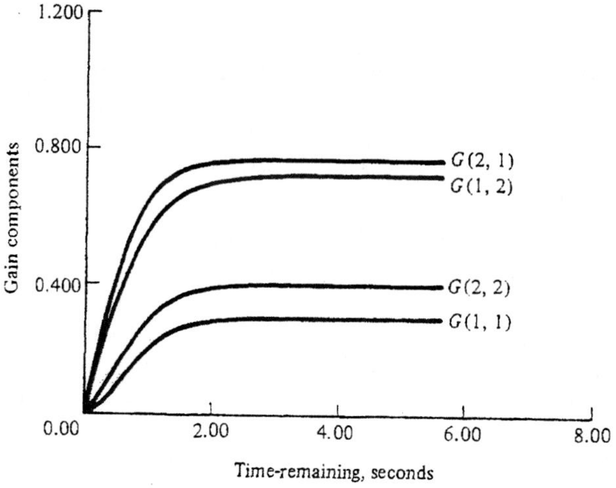

Figure 14.6 Transient gain behavior: $T = 0.3333$, $Q = R = I$.

the other sampling times and for other choices of **Q**, **R**, and **M**. For most problems four times the dominant time constant is a good conservative estimate of the time to settle. Note, however, that the number of discrete *cycles* of Eq. (*14.23*) required to reach these constant levels is a function of T.

The constant gain versions of the *LQ* state feedback controller are now used. To show the effects of the weights **Q** and **R**, three choices for **Q** are used for each sample time T. $\mathbf{R} = I$ is used in all nine cases. For the comparisons, a simple tracking problem is defined with $\boldsymbol{\eta} = [1 \quad 0]^T$, which asks for a unit step in x_1 while maintaining x_2 at or near zero. This requires solution for the infinite time-remaining version of $\mathbf{V}(k')$ and $\mathbf{F}_c$. By setting $\mathbf{V}(k') = \mathbf{V}(k'-1) = \mathbf{V}_\infty$ in Eq. (*14.21*), it is found that

$$\begin{aligned} \mathbf{V}_\infty &= -\{\mathbf{I} - \mathbf{A}^T + \mathbf{A}^T\mathbf{W}_\infty\mathbf{B}[\mathbf{R} + \mathbf{B}^T\mathbf{W}_\infty\mathbf{B}]^{-1}\mathbf{B}^T\}^{-1}\mathbf{Q}\boldsymbol{\eta} \\ &= -\{\mathbf{I} - \mathbf{A}^T[\mathbf{I} + \mathbf{W}_\infty\mathbf{B}\mathbf{F}_{c\infty}]\}^{-1}\mathbf{Q}\boldsymbol{\eta} \end{aligned} \tag{14.31}$$

The external input vector $\mathbf{v}_{ext}$ of Figure 14.4*a* is given by $\mathbf{F}_{c\infty}\mathbf{V}_\infty$. Table 14.1 gives the constant feedback gains, the required external input, and the resulting closed-loop eigenvalues for each case.

Note that in each case $\mathbf{v}_{ext}$ is just column one of $\mathbf{G}_\infty$, that is, $\mathbf{G}_\infty\boldsymbol{\eta}$. This is true here because the selected $\boldsymbol{\eta}$ happens to be an equilibrium point of the system; that is, it can be maintained with zero control, as seen from $\boldsymbol{\eta} = \mathbf{A}\boldsymbol{\eta}$. In this special case, Figure 14.4 shows that $\mathbf{F}_{c\infty}\mathbf{V}_\infty = \mathbf{G}_\infty\boldsymbol{\eta}$ if $\mathbf{u}$ is to be zero. To see that this is not generally true, the reader should rework a case from Table 14.1 using $\boldsymbol{\eta} = [0 \quad 1]^T$. With $T = 1$ and $\mathbf{Q} = \mathbf{R} = \mathbf{I}$, it will be found that $\mathbf{G}_\infty$ is unchanged but that $\mathbf{v}_{ext} = [0.4405 \quad -0.27403]^T$, the state vector approaches $[0 \quad 0.4444]^T$ rather than $\boldsymbol{\eta}$, and the control does not go to zero but rather to $\mathbf{u} = [0.2222 \quad -0.4444]^T$.

The transient responses for the cases of Table 14.1 are given in Figure 14.7*a*, *b*, and *c*. Notice that as **Q** increases, which is equivalent to making **R** relatively smaller, the response gets faster. This is also borne out by the magnitude of the eigenvalues becoming smaller in the tabulated data. At the same time the gains generally get larger. This is most evident in the off-diagonal components. The corresponding transient control signals would also get larger as **Q** increases. Regarding variations in sampling time, it is seen that all three cases give about the same results, except that the slower sampling rate gives a less smooth response due to the larger but less frequent changes at sampling times. As a general rule of thumb, sampling rates in the range of six to ten per dominant time constant are frequently used. ■

TABLE 14.1

T	$\mathbf{Q}$	$\mathbf{G}_\infty$	$\mathbf{v}_{ext}$	λ_{cl}
1	$10\mathbf{I}$	$\begin{bmatrix} 0.053625 & 0.69693 \\ 0.89339 & 0.45096 \end{bmatrix}$	$\begin{bmatrix} 0.053625 \\ 0.89339 \end{bmatrix}$	$0.07091 \pm j0.03848$
	$\mathbf{I}$	$\begin{bmatrix} 0.17640 & 0.49114 \\ 0.53973 & 0.38343 \end{bmatrix}$	$\begin{bmatrix} 0.17640 \\ 0.53973 \end{bmatrix}$	$0.30257 \pm j0.14195$
	$0.1\mathbf{I}$	$\begin{bmatrix} 0.16088 & 0.28932 \\ 0.18717 & 0.20982 \end{bmatrix}$	$\begin{bmatrix} 0.16088 \\ 0.18717 \end{bmatrix}$	$0.56156 \pm j0.15516$
$\frac{1}{3}$	$10\mathbf{I}$	$\begin{bmatrix} 0.19094 & 1.6476 \\ 1.8740 & 0.47914 \end{bmatrix}$	$\begin{bmatrix} 0.19094 \\ 1.8740 \end{bmatrix}$	$0.35295 \pm j0.05835$
	$\mathbf{I}$	$\begin{bmatrix} 0.29835 & 0.72404 \\ 0.76431 & 0.39803 \end{bmatrix}$	$\begin{bmatrix} 0.29835 \\ 0.76431 \end{bmatrix}$	$0.67685 \pm j0.09853$
	$0.1\mathbf{I}$	$\begin{bmatrix} 0.19409 & 0.33294 \\ 0.21669 & 0.21348 \end{bmatrix}$	$\begin{bmatrix} 0.19409 \\ 0.21669 \end{bmatrix}$	$0.83091 \pm j0.07419$
$\frac{1}{10}$	$10\mathbf{I}$	$\begin{bmatrix} 0.36152 & 2.3921 \\ 2.6645 & 0.48467 \end{bmatrix}$	$\begin{bmatrix} 0.36152 \\ 2.6645 \end{bmatrix}$	$0.72484 \pm j0.03560$
	$\mathbf{I}$	$\begin{bmatrix} 0.36586 & 0.82744 \\ 0.86783 & 0.39982 \end{bmatrix}$	$\begin{bmatrix} 0.36586 \\ 0.86783 \end{bmatrix}$	$0.89097 \pm j0.03861$
	$0.1\mathbf{I}$	$\begin{bmatrix} 0.20775 & 0.34882 \\ 0.22792 & 0.21390 \end{bmatrix}$	$\begin{bmatrix} 0.20775 \\ 0.22792 \end{bmatrix}$	$0.94670 \pm j0.02527$

4 DYNAMIC PROGRAMMING APPROACH TO CONTINUOUS-TIME OPTIMAL CONTROL

Dynamic programming applies in a similar way to continuous-time systems. The cost of operating the system from a general time and state t, $\mathbf{x}(t)$ to the terminal time and state $t_f, \mathbf{x}(t_f)$ is defined as

$$g(\mathbf{x}(t), t_f - t) \triangleq \min_{\mathbf{u}(t)} \left\{ S(\mathbf{x}(t_f), t_f) + \int_t^{t_f} L(\mathbf{x}(t), \mathbf{u}(t), t)\, dt \right\}$$

Breaking the integral into two segments gives

$$\begin{aligned} g(\mathbf{x}(t), t_f - t) = \min_{\mathbf{u}(t)} \Bigg\{ & S(\mathbf{x}(t_f), t_f) + \int_{t+\delta t}^{t_f} L(\mathbf{x}(t), \mathbf{u}(t), t)\, dt \\ & + \int_t^{t+\delta t} L(\mathbf{x}(t), \mathbf{u}(t), t)\, dt \Bigg\} \end{aligned} \tag{14.32}$$

The principle of optimality states that if the total cost is to be minimum, then the cost

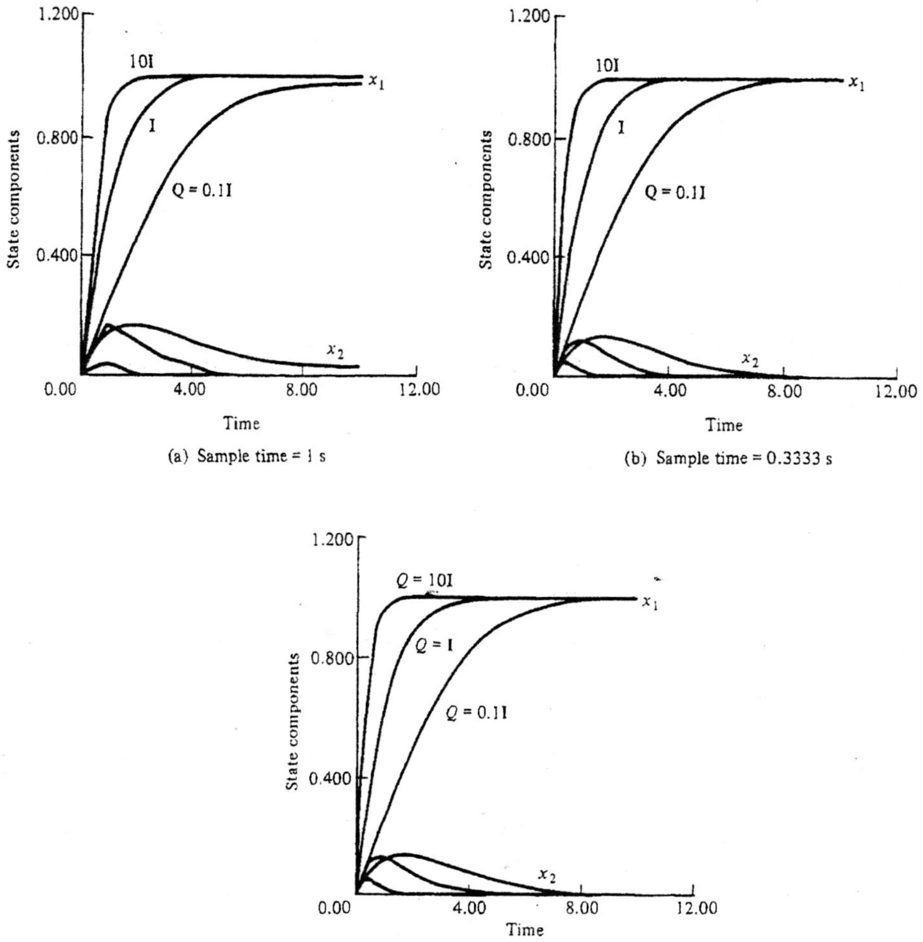

(a) Sample time = 1 s

(b) Sample time = 0.3333 s

(c) Sample time = 0.1 s

Figure 14.7 Transient response, constant LQ gains

from $t + \delta t$, $\mathbf{x}(t + \delta t)$, must also be minimum. The first two terms on the right side of Eq. (*14.32*) must therefore be equal to $g(\mathbf{x}(t + \delta t), t_f - t - \delta t)$, so that

$$g(\mathbf{x}(t), t_f - t) = \min_{\mathbf{u}(t)} \left\{ g(\mathbf{x}(t + \delta t), t_f - t - \delta t) + \int_t^{t+\delta t} L(\mathbf{x}(t), \mathbf{u}(t), t)\, dt \right\} \tag{14.33}$$

For δt sufficiently small,

$$\int_t^{t+\delta t} L(\mathbf{x}(t), \mathbf{u}(t), t)\, dt \cong L(\mathbf{x}(t), \mathbf{u}(t), t)\, \delta t \quad \text{and} \quad \mathbf{x}(t + \delta t) \cong \mathbf{x}(t) + \dot{\mathbf{x}}\delta t$$

Taylor series expansion gives

$$g(\mathbf{x}(t + \delta t), t_f - t - \delta t) \cong g(\mathbf{x}(t), t_f - t) + [\nabla_{\mathbf{x}} g]^T \dot{\mathbf{x}}\, \delta t - \frac{\partial g}{\partial t_r} \delta t$$

where $t_r \triangleq t_f - t$ is the time remaining. Using these in Eq. (*14.33*) gives

$$g(\mathbf{x}(t), t_r) = \min_{\mathbf{u}(t)} \left\{ g(\mathbf{x}(t), t_r) + [\nabla_{\mathbf{x}} g]^T \dot{\mathbf{x}}\, \delta t - \frac{\partial g}{\partial t_r} \delta t + L\, \delta t \right\} \tag{14.34}$$

By definition, $g(\mathbf{x}(t), t_r)$ is a function only of the current state and the time remaining (and not a function of $\mathbf{u}(t)$). Therefore, Eq. (*14.34*) leads to the *Hamilton-Jacobi-Bellman* (H.J.B.) partial differential equation [3]:

$$\frac{\partial g}{\partial t_r} = \min_{\mathbf{u}(t)} \{L(\mathbf{x}(t), \mathbf{u}(t), t) + [\nabla_{\mathbf{x}} g]^T \dot{\mathbf{x}}\} \tag{14.35}$$

The boundary condition is $g(\mathbf{x}(t), t_r)|_{t_r=0} = S(\mathbf{x}(t_f), t_f)$. Equation (*14.35*) is the continuous-time counterpart of Eq. (*14.5*). It is the major key to solving continuous-time optimal control problems. Whether it can be solved or not, and with what difficulty level, depends upon the class of systems, cost functions, admissible controls, and state constraints. The optimal control is the one which minimizes the right side of Eq. (*14.35*). If there are no restrictions on $\mathbf{u}(t)$, i.e., the admissible control set $\mathbf{U}$ is the entire space $\mathcal{U}^r$, then the minimum can be found by differentiating with respect to $\mathbf{u}(t)$ and setting the resultant gradient vector to zero. This gives the necessary condition for optimality.

$$\frac{\partial L}{\partial \mathbf{u}} + \frac{\partial \mathbf{f}}{\partial \mathbf{u}} \nabla_{\mathbf{x}} g = 0$$

If the system is linear, $\dot{\mathbf{x}} = \mathbf{A}\mathbf{x} + \mathbf{B}\mathbf{u}$ and $\partial \mathbf{f}/\partial \mathbf{u} = \mathbf{B}^T(t)$. If the loss function L is quadratic, i.e., L of Example 14.5, then $\partial L/\partial \mathbf{u} = 2\mathbf{R}\mathbf{u}(t)$ so that the optimal $\mathbf{u}(t)$ for the continuous version of the *LQ* problem is given by

$$\mathbf{u}^*(t) = -\tfrac{1}{2}\mathbf{R}^{-1}\mathbf{B}^T \nabla_{\mathbf{x}} g(\mathbf{x}(t), t_r) \tag{14.36}$$

The *LQ* problem is examined in detail next.

14.4.1 Linear-Quadratic (LQ) Problem; the Continuous Riccati Equation

Consider the linear system $\dot{\mathbf{x}} = \mathbf{A}\mathbf{x} + \mathbf{B}\mathbf{u}$. The optimal control is sought to minimize the quadratic performance criterion of Example 14.5, i.e., the tracking problem. There are no restrictions on $\mathbf{u}(t)$, and t_f is fixed.

Equation (*14.35*) specializes to

$$\frac{\partial g}{\partial t_r} = \min_{\mathbf{u}(t)} \{[\mathbf{x}(t) - \boldsymbol{\eta}(t)]^T \mathbf{Q}[\mathbf{x}(t) - \boldsymbol{\eta}(t)] + \mathbf{u}(t)^T \mathbf{R}\mathbf{u}(t) + (\nabla_{\mathbf{x}} g)^T [\mathbf{A}\mathbf{x} + \mathbf{B}\mathbf{u}]\}$$

with boundary conditions

$$g[\mathbf{x}(t), t_r]|_{t_r=0} = [\mathbf{x}(t_f) - \mathbf{x}_d]^T \mathbf{M}[\mathbf{x}(t_f) - \mathbf{x}_d]$$

Since $\mathbf{u}(t)$ is unrestricted, taking the derivative with respect to $\mathbf{u}(t)$ and setting it equal to zero gives $\mathbf{u}^*(t) = -\frac{1}{2}\mathbf{R}^{-1}\mathbf{B}^T \nabla_{\mathbf{x}} g[\mathbf{x}(t), t_r]$. The H.J.B. equation then reduces to

$$\frac{\partial g}{\partial t_r} = [\mathbf{x}(t) - \boldsymbol{\eta}(t)]^T \mathbf{Q}[\mathbf{x}(t) - \boldsymbol{\eta}(t)] + (\nabla_{\mathbf{x}} g)^T \mathbf{A}\mathbf{x}(t) - \tfrac{1}{4}(\nabla_{\mathbf{x}} g)^T \mathbf{B}\mathbf{R}^{-1}\mathbf{B}^T \nabla_{\mathbf{x}} g$$

This nonlinear partial differential equation can be solved by assuming a solution $g[\mathbf{x}(t), t_r] = \mathbf{x}^T(t)\mathbf{W}(t_r)\mathbf{x}(t) + \mathbf{x}^T(t)\mathbf{V}(t_r) + Z(t_r)$, where $\mathbf{W}$ is an unknown symmetric $n \times n$ matrix, $\mathbf{V}$ is an unknown $n \times 1$ vector, and Z is an unknown scalar. Differentiating the assumed answer, treating $\mathbf{x}(t)$ and t as independent variables, gives

$$\frac{\partial g}{\partial t_r} = \mathbf{x}^T(t)\frac{d\mathbf{W}}{dt_r}\mathbf{x}(t) + \mathbf{x}^T(t)\frac{d\mathbf{V}}{dt_r} + \frac{dZ}{dt_r} \quad \text{and} \quad \nabla_{\mathbf{x}} g = 2\mathbf{W}(t_r)\mathbf{x}(t) + \mathbf{V}(t_r)$$

Using these, the H.J.B. equation becomes

$$\begin{aligned}\mathbf{x}^T(t)\frac{dW}{dt_r}\mathbf{x}(t) + \mathbf{x}^T(t)\frac{d\mathbf{V}}{dt_r} + \frac{dZ}{dt_r} &= \mathbf{x}^T\{\mathbf{Q} + 2\mathbf{WA} - \mathbf{WBR}^{-1}\mathbf{B}^T\mathbf{W}\}\mathbf{x} \\ &+ \mathbf{x}^T\{-2\mathbf{Q}\boldsymbol{\eta} + \mathbf{A}^T\mathbf{V} - \mathbf{WBR}^{-1}\mathbf{B}^T\mathbf{V}\} \\ &+ \{\boldsymbol{\eta}^T\mathbf{Q}\boldsymbol{\eta} - \tfrac{1}{4}\mathbf{V}^T\mathbf{BR}^{-1}\mathbf{B}^T\mathbf{V}\}\end{aligned}$$

In order for the assumed form to actually be a solution for all $\mathbf{x}(t)$, the quadratic terms in $\mathbf{x}$, the linear terms, and the terms not involving $\mathbf{x}$ must balance individually. Therefore,

$$\frac{dZ}{dt_r} = \boldsymbol{\eta}(t)^T\mathbf{Q}\boldsymbol{\eta}(t) - \tfrac{1}{4}\mathbf{V}^T(t_r)\mathbf{BR}^{-1}\mathbf{B}^T\mathbf{V}(t_r) \qquad \text{(terms not involving } \mathbf{x}(t)) \tag{14.37}$$

The linear terms in $\mathbf{x}(t)$ require

$$\frac{d\mathbf{V}}{dt_r} = -2\mathbf{Q}\boldsymbol{\eta}(t) + \mathbf{A}^T\mathbf{V}(t_r) - \mathbf{W}(t_r)\mathbf{BR}^{-1}\mathbf{B}^T\mathbf{V}(t_r) \tag{14.38}$$

Each matrix involved in the quadratic terms is symmetric except $\mathbf{x}^T\{2\mathbf{WA}\}\mathbf{x}$, which is rewritten as $\mathbf{x}^T\{\mathbf{WA} + \mathbf{A}^T\mathbf{W}\}\mathbf{x} + \mathbf{x}^T\{\mathbf{WA} - \mathbf{A}^T\mathbf{W}\}\mathbf{x}$. The second term is always zero since the matrix is skew-symmetric. All quadratic terms in the H.J.B. equation can be combined into the form $\mathbf{x}^T\mathbf{Px} = 0$. Since every matrix term in $\mathbf{P}$ is now symmetric, it can be concluded that $\mathbf{P} = \mathbf{0}$, or

$$\frac{d\mathbf{W}}{dt_r} = \mathbf{Q} + \mathbf{WA} + \mathbf{A}^T\mathbf{W} - \mathbf{WBR}^{-1}\mathbf{B}^T\mathbf{W} \tag{14.39}$$

The boundary conditions for the three sets of differential equations are $\mathbf{W}(t_r = 0) = \mathbf{M}$, $\mathbf{V}(t_r = 0) = -2\mathbf{Mx}_d$, and $Z(t_r = 0) = \mathbf{x}_d^T\mathbf{Mx}_d$. Equation (*14.39*) is known as the matrix Riccati differential equation. It can be solved first, and the result can then be used in solving Eq. (*14.38*), after which Eq. (*14.37*) can be integrated. The optimal feedback control system can be represented by either Figure 14.8*a* or *b*. The equivalence is established by using the relations $\mathbf{G}(t_r) = \mathbf{R}^{-1}\mathbf{B}^T\mathbf{W}(t_r)$ and $\mathbf{F}_c(t_r) = -\tfrac{1}{2}\mathbf{R}^{-1}\mathbf{B}^T$. Since the major effort involves solving the Riccati equation, it is considered in detail next. The form of the matrix Riccati equation that is found most frequently in the controls literature is

$$-\dot{\mathbf{W}}(t) = \mathbf{W}(t)\mathbf{A} + \mathbf{A}^T\mathbf{W}(t) - \mathbf{W}(t)\mathbf{BR}^{-1}\mathbf{B}^T\mathbf{W}(t) + \mathbf{Q} \tag{14.40}$$

The alternate form expressed in terms of time remaining, t_r, has a sign change on the derivative term, since $d\{\ \}/dt_r = -d\{\ \}/dt$. This nonlinear differential equation can be

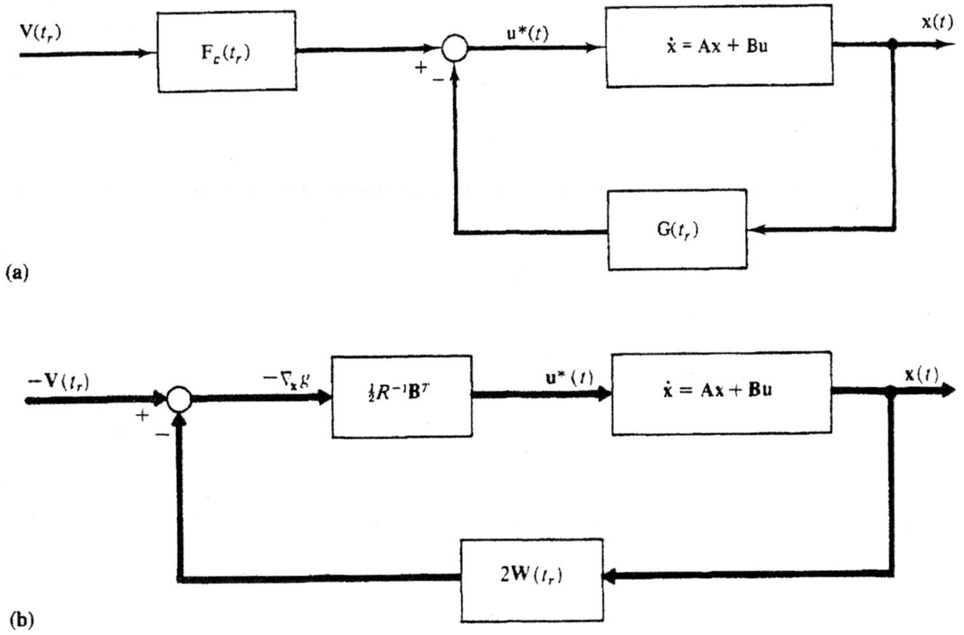

Figure 14.8

transformed into a pair of linear differential equations, in an analogous fashion to the treatment of the discrete-time Riccati equation in Sec. 14.3.3. Let $\mathbf{W} = \mathbf{E}\mathbf{F}^{-1}$. Since $\mathbf{F}\mathbf{F}^{-1} = \mathbf{I}$, $d\{\mathbf{F}\mathbf{F}^{-1}\}/dt = [\mathbf{0}]$, or

$$\dot{\mathbf{F}}\mathbf{F}^{-1} + \mathbf{F}d\{\mathbf{F}^{-1}\}/dt = [\mathbf{0}] \quad \text{or} \quad d\{\mathbf{F}^{-1}\}/dt = -\mathbf{F}^{-1}\dot{\mathbf{F}}\mathbf{F}^{-1}$$

From this, $\dot{\mathbf{W}} = \dot{\mathbf{E}}\mathbf{F}^{-1} - \mathbf{E}\mathbf{F}^{-1}\dot{\mathbf{F}}\mathbf{F}^{-1}$. Using this in Eq. (*14.40*) and then postmultiplying by $\mathbf{F}$ gives

$$-(\dot{\mathbf{E}} - \mathbf{E}\mathbf{F}^{-1}\dot{\mathbf{F}}) = \mathbf{E}\mathbf{F}^{-1}\mathbf{A}\mathbf{F} + \mathbf{A}^T\mathbf{E} - \mathbf{E}\mathbf{F}^{-1}\mathbf{B}\mathbf{R}^{-1}\mathbf{B}^T\mathbf{E} + \mathbf{Q}\mathbf{F}$$

If the terms linear in $\mathbf{E}$ and $\mathbf{F}$ are equated, that is,

$$-\dot{\mathbf{E}} = \mathbf{A}^T\mathbf{E} + \mathbf{Q}\mathbf{F} \tag{14.41}$$

then the remaining nonlinear terms give

$$\mathbf{E}\mathbf{F}^{-1}\dot{\mathbf{F}} = \mathbf{E}\mathbf{F}^{-1}\mathbf{A}\mathbf{F} - \mathbf{E}\mathbf{F}^{-1}\mathbf{B}\mathbf{R}^{-1}\mathbf{B}^T\mathbf{E}$$

When this is premultiplied by $[\mathbf{E}\mathbf{F}^{-1}]^{-1}$, the second linear equation is found to be

$$\dot{\mathbf{F}} = \mathbf{A}\mathbf{F} - \mathbf{B}\mathbf{R}^{-1}\mathbf{B}^T\mathbf{E} \tag{14.42}$$

Equations (*14.41*) and (*14.42*) can be stacked into one homogeneous linear system of coupled equations:

$$\begin{bmatrix}\dot{\mathbf{F}}\\ \dot{\mathbf{E}}\end{bmatrix} = \begin{bmatrix}\mathbf{A} & -\mathbf{B}\mathbf{R}^{-1}\mathbf{B}^T\\ -\mathbf{Q} & -\mathbf{A}^T\end{bmatrix}\begin{bmatrix}\mathbf{F}\\ \mathbf{E}\end{bmatrix} \tag{14.43}$$

Note that the superior dots indicate $d\{\ \}/dt$ rather than $d\{\ \}/dt_r$. The $2n \times 2n$ coefficient matrix on the right side of Eq. (*14.43*) is generally referred to as the Hamiltonian matrix **H**. Its form is different from—and should not be confused with—the discrete-time version of Sec. 14.3. However, similar to the discrete case, this **H** also has its eigenvalues occurring in stable-unstable pairs. That is, if λ is an eigenvalue of **H**, then so is $-\lambda$. Equation (*14.43*) has a solution in terms of the exponential matrix, that is,

$$\begin{bmatrix} \mathbf{F}(t) \\ \mathbf{E}(t) \end{bmatrix} = \exp\{(t - t_0)\mathbf{H}\} \begin{bmatrix} \mathbf{F}(t_0) \\ \mathbf{E}(t_0) \end{bmatrix} \tag{14.44}$$

for any t, t_0 pair. Since the "initial" conditions on **W**—and hence **E** and **F**—are given at time t_f and not t_0, Eq. (*14.44*) does not appear ready for use. Two possible modification methods exist, which show that it is possible to replace t_0 by t_f. The first method reverts to the time-remaining variable t_r and gives a sign change on the derivative terms. Since $t = t_f$ implies $t_r = 0$, this means that

$$\begin{bmatrix} \mathbf{F}(t_r) \\ \mathbf{E}(t_r) \end{bmatrix} = \exp\{-t_r\mathbf{H}\} \begin{bmatrix} \mathbf{F}(t_r = 0) \\ \mathbf{E}(t_r = 0) \end{bmatrix} \tag{14.45}$$

The other method evaluates Eq. (*14.44*) with the general time t replaced by the final time t_f and the initial time t_0 replaced by the current time t. This gives

$$\begin{bmatrix} \mathbf{F}(t_f) \\ \mathbf{E}(t_f) \end{bmatrix} = \exp\{(t_f - t)\mathbf{H}\} \begin{bmatrix} \mathbf{F}(t) \\ \mathbf{E}(t) \end{bmatrix} \tag{14.46}$$

Inverting the exponential matrix, which just changes the sign in its exponent, to solve for current values gives Eq. (*14.44*) with t_0 replaced by t_f. Equation (*14.45*) gives the same expression, provided it is recognized that the different methods of indexing time arguments both refer to the same physical time instant. Since $\mathbf{W}(t = t_f) = \mathbf{M}$, we set $\mathbf{E}(t = t_f) \equiv \mathbf{E}(t_r = 0) = \mathbf{M}$ and $\mathbf{F}(t_r = 0) = \mathbf{I}$. The similarity transformation relation between **H** and its Jordan form $\mathbf{J} = \text{Diag}[\mathbf{J}_s \quad \mathbf{J}_u]$ is $\mathbf{H} = \mathbf{TJT}^{-1}$, where, as before, $\mathbf{J}_s$ and $\mathbf{J}_u$ contain the stable and unstable blocks and where $\mathbf{T} = \begin{bmatrix} \mathbf{T}_{11} & \mathbf{T}_{12} \\ \mathbf{T}_{21} & \mathbf{T}_{22} \end{bmatrix}$ is the modal matrix of eigenvectors of **H**, with the first n columns being associated with the stable eigenvectors. Let $\mathbf{T}^{-1} = \begin{bmatrix} \mathbf{V}_{11} & \mathbf{V}_{12} \\ \mathbf{V}_{21} & \mathbf{V}_{22} \end{bmatrix}$. Since $\exp\{-t_r\mathbf{H}\} = \mathbf{T}\,\text{Diag}[\exp\{-t_r\mathbf{J}_s\}\ \exp\{-t_r\mathbf{J}_u\}]\mathbf{T}^{-1}$, the solution for **W** at a given time t (or the same corresponding time-remaining value t_r) is found using essentially the same steps which lead to Eq. (*14.29*)

$$\begin{aligned} \mathbf{W}(t_r) = {} & [\mathbf{T}_{21}\exp\{-t_r\mathbf{J}_s\}\mathbf{V}_{11} + \mathbf{T}_{22}\exp\{-t_r\mathbf{J}_u\}\mathbf{V}_{21} \\ & + (\mathbf{T}_{21}\exp\{-t_r\mathbf{J}_s\}\mathbf{V}_{12} + \mathbf{T}_{22}\exp\{-t_r\mathbf{J}_u\}\mathbf{V}_{22})\mathbf{M}] * \\ & [\mathbf{T}_{11}\exp\{-t_r\mathbf{J}_s\}\mathbf{V}_{11} + \mathbf{T}_{12}\exp\{-t_r\mathbf{J}_u\}\mathbf{V}_{21} \\ & + (\mathbf{T}_{11}\exp\{-t_r\mathbf{J}_s\}\mathbf{V}_{12} + \mathbf{T}_{12}\exp\{-t_r\mathbf{J}_u\}\mathbf{V}_{22})\mathbf{M}]^{-1} \end{aligned} \tag{14.47}$$

Once $\mathbf{W}(t_r)$ is determined, the control law is given by

$$\mathbf{u}^*(t) = -\mathbf{R}^{-1}\mathbf{B}^T\{\mathbf{W}(t_r)\mathbf{x}(t) + \tfrac{1}{2}\mathbf{V}(t_r)\} \tag{14.48}$$

The external input command $\mathbf{V}(t_r)$ is zero for the regulator problem. In the tracking problem it must be determined by solving Eq. (*14.38*) using the now-known matrix $\mathbf{W}(t_r)$ and the specified state trajectory $\boldsymbol{\eta}(t)$ which is to be tracked.

14.4.2 Infinite Time-to-Go Problem; The Algebraic Riccati Equation

A commonly used simplification of the previous LQ solution is to let $t_r \to \infty$. This is equivalent to letting the derivative $\dot{\mathbf{W}}$ go to zero, leaving the so-called algebraic Riccati equation (ARE)

$$\mathbf{A}^T\mathbf{W} + \mathbf{W}\mathbf{A} - \mathbf{W}\mathbf{B}\mathbf{R}^{-1}\mathbf{B}^T\mathbf{W} + \mathbf{Q} = \mathbf{0} \tag{14.49}$$

A number of ways of solving the ARE exist. For low-order problems, it may be feasible to write out the components explicitly. Using the known symmetry of $\mathbf{W}$ will yield $n(n+1)/2$ coupled quadratic equations. Since quadratics have multiple solutions, a question arises about which, if any, of these solutions is the correct one for the problem at hand. Under the assumptions that $\mathbf{R}$ is positive definite and that $\mathbf{Q}$ is at least positive semidefinite, it is known that one unique positive definite solution for $\mathbf{W}$ exists provided that the system $\{\mathbf{A}, \mathbf{B}\}$ is stabilizable and $\{\mathbf{A}, \mathbf{C}\}$ is detectible, where $\mathbf{C}^T\mathbf{C} = \mathbf{Q}$. A stronger and numerically safer set of requirements is sometimes given, namely: Either $\mathbf{A}$ is asymptotically stable or $\{\mathbf{A}, \mathbf{B}\}$ is controllable and $\{\mathbf{A}, \mathbf{C}\}$ is observable.

If a positive definite solution is known to exist because of satisfaction of these requirements, it could be found by numerical integration of Eq. (*14.39*), backward in time, until $\mathbf{W}$ approaches its constant final value. This approach is discussed in Problem 14.11, along with some suggestions for integration step-size selection. The general solution found in the last section can be used to develop another approach to finding the infinite-time-remaining solution $\mathbf{W}_\infty$. When $t_r \to \infty$, $\exp\{-t_r\mathbf{J}_s\} \to \infty$ and $\exp\{-t_r\mathbf{J}_u\} \to \mathbf{0}$. The general results then reduce to

$$\mathbf{W}_\infty = [\mathbf{T}_{21}\exp\{-t_r\mathbf{J}_s\}\mathbf{V}_{11} + \mathbf{T}_{21}\exp\{-t_r\mathbf{J}_s\}\mathbf{V}_{12}\mathbf{M}] *$$
$$[\mathbf{T}_{11}\exp\{-t_r\mathbf{J}_s\}\mathbf{V}_{11} + \mathbf{T}_{11}\exp\{-t_r\mathbf{J}_s\}\mathbf{V}_{12}\mathbf{M}]^{-1}$$

which reduces to

$$\mathbf{W}_\infty = \mathbf{T}_{21}\mathbf{T}_{11}^{-1} \tag{14.50}$$

It is of interest to note that if Eq. (*14.44*) is treated in a similar fashion but with $t \to \infty$, it is found that a "steady-state" solution for $\mathbf{W}$ is given by $\mathbf{W}_{ss} = \mathbf{T}_{22}\mathbf{T}_{12}^{-1}$. This is *not* the correct solution and explains why the commonly used terminology *steady-state solution* must be used with care.

In calculating Eq. (*14.50*) and/or Eq. (*14.30*), the eigenvalues and eigenvectors of $\mathbf{H}$ will often be complex. However, Problem 4.23 shows that only real arithmetic is needed to calculate $\mathbf{W}_\infty$. It is also pointed out that the n stable eigenvalues of $\mathbf{H}$ are exactly the closed-loop eigenvalues of the system when the constant feedback gain control law $\mathbf{u}(t) = -\mathbf{R}^{-1}\mathbf{B}^T\mathbf{W}_\infty\mathbf{x}(t) = -\mathbf{G}_\infty\mathbf{x}(t)$ is used. The following example uses simple second-order systems to illustrate the determination of the optimal feedback gains and several potential pitfalls.

EXAMPLE 14.10 Analyze the ARE for each of the following systems, with $\mathbf{Q} = \mathbf{I}$ and $\mathbf{R} = 1$.

(a) $\mathbf{A} = \begin{bmatrix} -3 & 1 \\ 0 & 2 \end{bmatrix}$, $\mathbf{B} = \begin{bmatrix} 1 \\ 0 \end{bmatrix}$. This system is in the Kalman canonical form, which shows it to be unstable, and the unstable mode is uncontrollable; hence it is not stabilizable. Ignoring this fact, it is found that the eigenvalues of $\mathbf{H}$ are ± 2 and ± 3.16227. The eigenvectors associated with the stable eigenvalues are found to be

$$\begin{bmatrix} \mathbf{T}_{11} \\ \hline \mathbf{T}_{21} \end{bmatrix} = \begin{bmatrix} 0 & -1 \\ 0 & 0 \\ \hline 0 & -0.16228 \\ -1 & -0.13962 \end{bmatrix}$$

Clearly $\mathbf{T}_{11}$ is singular, and the method of Eq. (*14.50*) fails, as well it should. If the numerical integration method is attempted, it also fails because the solution never settles to a constant matrix.

(b) $\mathbf{A} = \begin{bmatrix} -3 & 1 \\ 0 & -2 \end{bmatrix}$, $\mathbf{B} = \begin{bmatrix} 1 \\ 0 \end{bmatrix}$. The uncontrollable mode is now stable; hence this system is stabilizable. Using the eigenvalues-eigenvectors of $\mathbf{H}$, it is found that $\mathbf{W}_\infty = \begin{bmatrix} 0.16228 & 0.031435 \\ 0.031435 & 0.26547 \end{bmatrix}$, $\mathbf{G}_\infty = [0.16228 \quad 0.031435]$, and the closed loop eigenvalues are $\lambda = -2, -3.16228$. Note that the uncontrollable mode has its eigenvalue unchanged from the open-loop value of $\lambda = -2$. This solution is also verified by using fourth-order Runge-Kutta integration on Eq. (*14.39*).

(c) $\mathbf{A} = \begin{bmatrix} 2 & 1 \\ 0 & -3 \end{bmatrix}$, $\mathbf{B} = \begin{bmatrix} 1 \\ 0 \end{bmatrix}$. The same two eigenvalues occur here as in part (a), but now the unstable mode is controllable; hence the system is stabilizable. Routine application of both the eigenvector method and numerical integration give $\mathbf{W}_\infty = \begin{bmatrix} 4.23607 & 0.80902 \\ 0.80902 & 0.32725 \end{bmatrix}$, $\mathbf{G}_\infty = [4.2361 \quad 0.80902]$, and the closed-loop eigenvalues are at $\lambda = -3, -2.23607$. Again the uncontrollable mode has its eigenvalue unchanged.

(d) $\mathbf{A} = \begin{bmatrix} 0 & 1 \\ 0 & -3 \end{bmatrix}$, $\mathbf{B} = \begin{bmatrix} 1 \\ 0 \end{bmatrix}$. The Hamiltonian matrix is $\mathbf{H} = \begin{bmatrix} 0 & 1 & -1 & 0 \\ 0 & -3 & 0 & 0 \\ -1 & 0 & 0 & 0 \\ 0 & -1 & -1 & 3 \end{bmatrix}$. The eigenvalues are found to be at $\lambda = -1, 1, -3,$ and 3. The eigenvectors for -1 and -3 are $[-1 \quad 0 \quad -1 \quad -0.25]^T$ and $[0.375 \quad -1 \quad 0.125 \quad -0.145833]^T$, respectively. These could be used in Eq. (*14.50*) to find $\mathbf{W}_\infty$. Since this system is stabilizable, it must have a unique positive definite solution for $\mathbf{W}_\infty$. The direct solution of the ARE will be used here to find the result. $\mathbf{A}^T\mathbf{W} + \mathbf{W}\mathbf{A} - \mathbf{W}\mathbf{B}\mathbf{R}^{-1}\mathbf{B}^T\mathbf{W} + \mathbf{Q} = \mathbf{0}$ expands to

$$\begin{bmatrix} 0 & 0 \\ 1 & -3 \end{bmatrix}\begin{bmatrix} w_{11} & w_{12} \\ w_{12} & w_{22} \end{bmatrix} + \begin{bmatrix} w_{11} & w_{12} \\ w_{12} & w_{22} \end{bmatrix}\begin{bmatrix} 0 & 1 \\ 0 & -3 \end{bmatrix} + \begin{bmatrix} 1 & 0 \\ 0 & 1 \end{bmatrix}$$

$$-\begin{bmatrix} w_{11} & w_{12} \\ w_{12} & w_{22} \end{bmatrix}\begin{bmatrix} 1 & 0 \\ 0 & 0 \end{bmatrix}\begin{bmatrix} w_{11} & w_{12} \\ w_{12} & w_{22} \end{bmatrix} = \begin{bmatrix} 0 & 0 \\ 0 & 0 \end{bmatrix}$$

From the 1, 1 term, $w_{11}^2 = 1$. $\mathbf{W}$ must be positive definite, so $w_{11} = 1$. From the 1, 2 or 2, 1 terms, $w_{11} - 3w_{12} - w_{12}w_{11} = 0$, so $w_{12} = 0.25$. From the 2, 2 term, $2w_{12} - 6w_{22} - w_{12}^2 + 1 = 0$ gives $w_{22} = 0.23958$. From this, the constant gain matrix is $\mathbf{G}_\infty = [1 \quad 0.25]$, and the closed-loop eigenvalues are at $\lambda = -1$ and -3. Attempts to solve this problem with numerical integration gave mixed results. The correct answer was sometimes found and other times not, depending upon the integration step size and the stopping criteria.

(e) A similar example, with pure integrators in the open-loop system, has $\mathbf{A} = \begin{bmatrix} 0 & 1 \\ 0 & 0 \end{bmatrix}$, $\mathbf{B} = \begin{bmatrix} 0 \\ 1 \end{bmatrix}$. Note that $\mathbf{B}$ has been changed to make this system controllable. The Hamiltonian matrix is $\mathbf{H} = \begin{bmatrix} 0 & 1 & 0 & 0 \\ 0 & 0 & 0 & -1 \\ -1 & 0 & 0 & 0 \\ 0 & -1 & -1 & 0 \end{bmatrix}$. The eigenvalues are found to be at $\lambda = -0.86615 \pm 0.5j$ and $0.86615 \pm 0.5j$. Again, the clean separation into mirror-image stable and unstable eigenvalues is noted. It is easy to use the expanded components of the ARE, as in part (d), to find $\mathbf{W}_\infty = \begin{bmatrix} \sqrt{3} & 1 \\ 1 & \sqrt{3} \end{bmatrix}$, $\mathbf{G}_\infty = [1 \quad \sqrt{3}]$, and the closed loop eigenvalues are at $\lambda = -0.866 \pm 0.5j$. The same results are obtained by using the two stable eigenvectors $\xi_1 = [-0.5 - 0.866j \quad 0.866 + 0.5j \quad -j \quad 1]^T$ and its complex conjugate in Eq. (*14.50*). To avoid inverting a complex matrix, the result of Problem 4.23 can be used to write $\mathbf{W}_\infty = \begin{bmatrix} 0 & -1 \\ 1 & 0 \end{bmatrix} \begin{bmatrix} -0.5 & -0.866 \\ 0.866 & 0.5 \end{bmatrix}^{-1}$. This problem can also be solved using numerical integration, if appropriate step-size and stopping criterion are selected. The suggestions given in Problem 14.11 are not useful when all the open-loop poles are at the origin. ■

For simple low-order problems, a variety of solution methods can be used, which become impractical in realistic problems. The eigenvector method of Eq. (*14.50*) is capable of solving most ARE problems for which unique solutions are known to exist. There may be occasions where the $2n \times 2n$ Hamiltonian matrix has eigenvalues on the $j\omega$ axis. The clean separation into stable and unstable values thus breaks down. In these cases, a small change to $\mathbf{Q}$, $\mathbf{R}$, or even $\mathbf{A}$ might be used to allow the algorithm to proceed successfully. The system of Example 14.10(e), with $\mathbf{Q} = [\mathbf{0}]$, is one such case. Essentially the same answer would be obtained by changing to a very small nonzero matrix $\mathbf{Q}$ or by setting the diagonal terms of $\mathbf{A}$ to a small nonzero value ϵ. Other solution methods are also known [2].

Note that when the cost of control is very expensive—i.e., when $\mathbf{R} \to \infty$—the quadratic terms become vanishingly small and Eq. (*14.49*) reduces to a Lyapunov equation. Solutions of the Lyapunov equation were examined in detail in Sec. 6.10. This limiting result does not seem to be especially useful because it leads to a vanishingly small feedback gain $\mathbf{G}_\infty = \mathbf{R}^{-1}\mathbf{B}^T\mathbf{W}$ as well, so the system has no feedback control.

*.5 PONTRYAGIN'S MINIMUM PRINCIPLE

Optimal control problems can be analyzed from a number of alternative viewpoints. The continuous-time versions of Eqs. (*14.1*) and (*14.2*) are considered again. If the set of admissible controls is unrestricted, the calculus of variations [6] can be used to derive necessary conditions which characterize the optimal solution. When the admissible control set is bounded, unrestricted variations in $\mathbf{u}(t)$ are not allowed. This situation is analogous to the problem of finding the minimum of a function on a closed and bounded interval. If the minimum occurs at a boundary point, then it is not necessarily true that the first variation (analogous to the slope) vanishes at that minimal point. Pontryagin's minimum principle [5] is an extension of the methods of variational calculus to problems with bounded control and/or state variables. The simple version of the minimum principle presented next provides a set of necessary conditions for optimality (see Problems 14.12 and 14.13). This brief introduction to the theory is incomplete in that bounded states are not considered, sufficiency conditions are not treated, etc. A heuristic relation to dynamic programming results is given.

The pre-Hamiltonian is defined as the scalar function

$$\mathcal{H}(\mathbf{x}, \mathbf{u}, \mathbf{p}, t) \triangleq L(\mathbf{x}, \mathbf{u}, t) + \mathbf{p}^T(t)\mathbf{f}(\mathbf{x}, \mathbf{u}, t) \tag{14.51}$$

where $\mathbf{p}(t)$ is the $n \times 1$ *costate vector* and satisfies

$$\dot{\mathbf{p}} = -\left[\frac{\partial \mathbf{f}}{\partial \mathbf{x}}\right]^T \mathbf{p} - \nabla_{\mathbf{x}} L(\mathbf{x}, \mathbf{u}, t) \tag{14.52}$$

The minimum principle states that the optimal control $\mathbf{u}^*(t)$ is that member of the admissible control set U which minimizes $\mathcal{H}$ at every time. If $\mathbf{u}(t)$ has r components, then minimizing $\mathcal{H}$ gives r algebraic equations which allow the determination of $\mathbf{u}^*(t)$ in terms of the still unknown $\mathbf{p}(t)$ and $\mathbf{x}(t)$. Then $\mathbf{u}(t)$ can be eliminated from equations (*14.1*) and (*14.52*). These equations can also be written in canonical form as

$$\dot{\mathbf{x}} = \frac{\partial \mathcal{H}}{\partial \mathbf{p}}, \quad \dot{\mathbf{p}} = -\frac{\partial \mathcal{H}}{\partial \mathbf{x}} \tag{14.53}$$

Equation (*14.53*) consists of $2n$ first-order differential equations, so $2n$ boundary conditions are required for solution. The initial conditions $\mathbf{x}(t_0) = \mathbf{x}_0$ give n of them. The remaining n conditions will apply at the final time t_f. Their exact nature depends on the particular problem. $\mathbf{x}(t_f)$ will be specified directly in some cases. If $\mathbf{x}(t_f)$ is free, then $\mathbf{p}(t_f) = \nabla_{\mathbf{x}} S(\mathbf{x}(t), t)|t_f$. Combinations of these two kinds of conditions, as well as others, can arise. If the final time t_f is not specified, an additional equation is required to determine it (see Problem 14.13).

Equation (*14.53*), along with n boundary conditions at t_0 and n more at t_f, constitutes a two-point boundary value problem. Linear two-point boundary value problems are easily solved, at least in principle. Nonlinear two-point boundary value problems are generally difficult to solve, even numerically. Much of the effort in optimal control theory has been devoted to the development of efficient algorithms for computer solution of these problems (Problem 14.14).

Note that if $\mathbf{p}$ is defined as $\nabla_x g$, then Eq. (*14.35*) of dynamic programming also indicates that the pre-Hamiltonian must be minimized at each instant by proper choice of $\mathbf{u}(t)$. When the pre-Hamiltonian is evaluated with all its arguments optimally selected, it is called simply the Hamiltonian $\mathcal{H}^*$. (Often the * is omitted.) Equation (*14.35*) gives physical meaning to $\mathcal{H}^*$. It is the rate of change of the cost g as time-remaining changes, $\partial g/\partial t_r = \mathcal{H}^*$. If sufficient continuity properties are assumed for g, the gradient with respect to the state $\mathbf{x}$ gives

$$\nabla_x(\partial g/\partial t_r) = \partial\{\nabla_x g\}/\partial t_r = \nabla_x \mathcal{H}^*$$

By reverting to t instead of t_r, using the definition $\mathbf{p} = \nabla_x g$, and using the alternate notation $\partial\{\ \}/\partial\mathbf{x}$ for the gradient, the second of Eq. (*14.53*) is obtained. The first of this canonical pair is automatically satisfied because of the definition of $\mathcal{H}$. Thus the essential features of the minimum principle follow directly from the dynamic programming approach presented earlier. In some problems, especially those involving the exact satisfaction of terminal boundary conditions, the minimum principle formulation seems more convenient. This is explored in the examples and problems.

EXAMPLE 14.11 A simplified model of the linear motion of an automobile is $\dot{x} = u$, where $x(t)$ is the vehicle velocity and $u(t)$ is the acceleration or deceleration. The car is initially moving at x_0 ft/sec. Find the optimal $u(t)$ which brings the velocity $x(t_f)$ to zero in *minimum time* t_f. Assume that acceleration and braking limitations require $|u(t)| \le M$ for all t.

The minimum time performance criterion is $J = \int_0^{t_f} 1\ dt$, so the Hamiltonian is $\mathcal{H} = 1 + p(t)u(t)$. In order to minimize $\mathcal{H}$, it is obvious that $u(t) = -M$ if $p(t) > 0$ and $u(t) = M$ if $p(t) < 0$. That is, $u^*(t) = -M \text{ sign } (p(t))$. The optimal control has been found as a function of the unknown $p(t)$. The differential equation for p is $\dot{p} = -\partial\mathcal{H}/\partial x = 0$, so $p(t)$ is constant. The value of this constant must be determined from boundary conditions. (Actually only the sign of the constant is needed for this problem.) The available boundary conditions are $x(0) = x_0$, $x(t_f) = 0$. The form of the solution for $x(t)$ is $x(t) = x_0 + \int_0^t u(\tau)\,d\tau$, but $u(t)$ is a constant, either $+M$ or $-M$. Therefore, $x(t_f) = x_0 \pm Mt_f = 0$. Clearly, if $x_0 > 0$, then $u = -M$, maximum deceleration. If $x_0 < 0$, then $u = M$, maximum acceleration. That is, $u^*(t) = -M \text{ sign } (x_0)$. The minimum stopping time is $t_f = |x_0|/M$. Note that $u^*(t)$ is expressed in terms of $x(t_0)$, so this represents an open-loop control law. This is a typical result of using the minimum principle. ■

EXAMPLE 14.12 Consider the linear, constant system $\dot{\mathbf{x}} = \mathbf{A}\mathbf{x} + \mathbf{B}\mathbf{u}$ with $\mathbf{u}(t)$ unrestricted. Find $\mathbf{u}(t)$ which minimizes a trade-off between terminal error and control effort,

$$J = [\mathbf{x}(t_f) - \mathbf{x}_d]^T[\mathbf{x}(t_f) - \mathbf{x}_d] + \int_0^{t_f} \mathbf{u}^T(t)\mathbf{u}(t)\,dt$$

Time t_f is fixed.

The Hamiltonian is $\mathcal{H} = \mathbf{u}^T\mathbf{u} + \mathbf{p}^T[\mathbf{A}\mathbf{x} + \mathbf{B}\mathbf{u}]$. $\mathcal{H}$ is minimized by setting

$$\frac{\partial\mathcal{H}}{\partial\mathbf{u}} = \mathbf{0} = 2\mathbf{u} + \mathbf{B}^T\mathbf{p} \quad \text{or} \quad \mathbf{u}^*(t) = -\tfrac{1}{2}\mathbf{B}^T\mathbf{p}(t)$$

This is not yet a useful answer, since $\mathbf{p}(t)$ is unknown. However, $\dot{\mathbf{p}} = -\partial\mathcal{H}/\partial\mathbf{x} = -\mathbf{A}^T\mathbf{p}$. Combining this with the original system equation gives, after eliminating $\mathbf{u}(t)$,

$$\begin{bmatrix} \dot{\mathbf{x}} \\ \dot{\mathbf{p}} \end{bmatrix} = \left[\begin{array}{c|c} \mathbf{A} & -\tfrac{1}{2}\mathbf{B}\mathbf{B}^T \\ \hline \mathbf{0} & -\mathbf{A}^T \end{array}\right] \begin{bmatrix} \mathbf{x} \\ \mathbf{p} \end{bmatrix} \tag{14.54}$$

Note that the $2n \times 2n$ coefficient matrix is exactly the Hamiltonian matrix arising out of our solution to the Riccati equation of Sec. 14.4, in the special case of $\mathbf{R} = \mathbf{I}$ and $\mathbf{Q} = [\mathbf{0}]$. The $2n$ boundary conditions are $\mathbf{x}(0) = \mathbf{x}_0$ and $\mathbf{p}(t_f) = \nabla_\mathbf{x} S|_{t_f} = 2[\mathbf{x}(t_f) - \mathbf{x}_d]$. This linear two-point boundary value problem is treated in Problem 14.15. ■

6 THE SEPARATION THEOREM

In previous sections of this chapter it has been assumed that the entire state vector $\mathbf{x}$ can be measured and used in forming the feedback control signal. In Chapter 13 it was shown that if the system is *linear* and if the control law is *linear,* then a *linear* observer can be used to estimate **x** from the available outputs **y** without changing the closed-loop poles which the controller is designed to give.

Another version of the separation theorem is more commonly stated in conjunction with optimal control problems. It states that *if*

a. the system models are linear (both the dynamics of Eq. (*3.11*) or (*3.13*) and the output equation of Eq. (*3.12*) or (*3.14*)) and
b. all measurement errors and disturbances have Gaussian probability density functions, and
c. the cost function of Eq. (*14.2*) is quadratic,

then the expected value of the cost function J is minimized by

1. designing an optimal controller based on the assumption that all states are available,
2. designing an optimal estimator to provide an estimate $\hat{\mathbf{x}}$ of $\mathbf{x}$,
3. using $\hat{\mathbf{x}}$ in place of $\mathbf{x}$ in the control law of step 1.

These two versions of the separation principle are closely related because of the following facts: The optimal control law for a linear system with a quadratic cost function is a linear control law. The optimal estimator for a linear system subjected to Gaussian noise is a linear estimator of the same form as the linear observer of Chapter 13. While version one is true for *any* linear controller and linear estimator, regardless of how they were designed, and without any Gaussian assumptions, all that is guaranteed is that the desired closed-loop poles are still achieved. Version two guarantees the optimality of J in a stochastic sense, but demands somewhat stricter assumptions. The class of problems that meet these restrictions is referred to as *LQG* problems (linear, quadratic, Gaussian). Most successful applications of optimal control theory so far have been members of this class.

The implication of the separation theorem is that the assumption that the full state is available can be relaxed. The optimal controllers in this chapter can be cascaded with an observer (as in Chapter 13) or a Kalman filter (Problem 6.18) to obtain the overall design. This practice is generally not valid outside the *LQG* class, although it is still sometimes used as a rather effective suboptimal scheme.

Intuitively, one might expect that if control inputs could be tailored in some way, certain system modes would be more strongly excited, making their states easier to estimate. This idea is the basis for "probing"-type control signals. Likewise, it seems intuitive that if one or more states are only known to within some level of uncertainty, but the control system behaves as if it were exactly known, then overly bold or decisive control actions might incorrectly be taken at times. A more cautious controller which somehow takes into account its knowledge of the uncertainty in the state estimates might give better overall performance. The notions of probing and caution come into play in those cases where the separation theorem does not apply. This is still a subject of active research. To give any more than the above intuitive definitions would go beyond the scope of this book.

7 ROBUSTNESS ISSUES

Kalman [7] first showed that a single input control system designed with the *LQ* theory has a scalar return difference (first introduced in Sec. 2.4) which satisfies $|F_d(j\omega)| \geq 1$ for all ω. Using Nyquist frequency response methods, it can be shown [7] that this guarantees a phase margin $PM \geq 60°$ and an upward gain margin $GM = \infty$. The downward gain reduction margin is 0.5. This means that the *LQ* system will remain stable for any arbitrary gain increase and for all gains at least half the nominal design value.

These reassuring stability margins have also been shown to apply to the multiple-input case, at least for a diagonal **R** weighting matrix [8]. This was shown by using the return difference matrix $\mathbf{R}_d$ and by using its smallest singular value as the measure of how close it approaches singularity. The gain margin portion of this result is easily verified in the time-domain, state space domain by using Lyapunov stability theory of Sec. 10.6. Suppose the positive definite matrix $\mathbf{W}_\infty$ from the ARE is used to define a Lyapunov function

$$V = \mathbf{x}^T \mathbf{W}_\infty \mathbf{x}$$

for the closed-loop *LQ* regulator described by

$$\dot{\mathbf{x}} = (\mathbf{A} - \mathbf{B}\mathbf{G}_\infty)\mathbf{x}$$

(Since the presence of an external input for the *LQ* tracking problem does not alter the loop stability characteristics, only the regulator problem is considered for simplicity.) Then

$$\dot{V} = \dot{\mathbf{x}}^T \mathbf{W}_\infty \mathbf{x} + \mathbf{x}^T \mathbf{W}_\infty \dot{\mathbf{x}}$$

$$= \mathbf{x}^T\{\mathbf{A}^T \mathbf{W}_\infty + \mathbf{W}_\infty \mathbf{A} - \mathbf{G}_\infty^T \mathbf{B}^T \mathbf{W}_\infty - \mathbf{W}_\infty \mathbf{B}\mathbf{G}_\infty\}\mathbf{x}$$

Assume that the gain actually used in the feedback loop is $\mathbf{G}_\infty = \beta \mathbf{R}^{-1}\mathbf{B}^T\mathbf{W}_\infty$, that is, it equals the optimal *LQ* gain only when $\beta = 1$. Then

$$\dot{V} = \mathbf{x}^T\{\mathbf{A}^T \mathbf{W}_\infty + \mathbf{W}_\infty \mathbf{A} - 2\beta \mathbf{W}_\infty \mathbf{B}\mathbf{R}^{-1}\mathbf{B}^T\mathbf{W}_\infty\}\mathbf{x}$$

which can be regrouped into

$$\dot{V} = \mathbf{x}^T\{\mathbf{A}^T \mathbf{W}_\infty + \mathbf{W}_\infty \mathbf{A} - \mathbf{W}_\infty \mathbf{B}\mathbf{R}^{-1}\mathbf{B}^T\mathbf{W}_\infty + \mathbf{Q}\}\mathbf{x} - \mathbf{x}^T\{\mathbf{Q} + (2\beta - 1)\mathbf{W}_\infty \mathbf{B}\mathbf{R}^{-1}\mathbf{B}^T\mathbf{W}_\infty\}\mathbf{x}$$

Since $\mathbf{W}_\infty$ satisfies the ARE, the first quadratic form is zero. Since **Q** is at least positive semidefinite, the second quadratic form will be at least positive semidefinite, provided that $\beta \geq 0.5$. Lyapunov's theorem therefore guarantees that the gain-perturbed *LQ* system remains stable provided that $\frac{1}{2} \leq \beta < \infty$.

While this simple verification gives insight into the robustness issue, most thorough analyses of the subject are based on singular value analysis of the return difference matrix. It should be pointed out that singular value analysis can sometimes give overly conservative estimates of system robustness. The minimum singular value is a measure of how close the matrix is to the nearest singular matrix. Actual perturbations in a physical model may restrict the kinds of changes that can occur in a system matrix in such a way as to preclude the "nearest singular matrix" from ever occurring. For example, if a resistor value is the major unknown parameter, these model perturbations may cause only a single matrix element to vary. This points up the difference between unstructured model perturbations and structured perturbations [9, 10]. Obviously, an analysis that uses more information about the form which model perturbations might take is likely to give sharper answers.

A useful theorem [11, 12], which is valid for any arbitrary unstructured model perturbations, gives the following bounds on the stability margins. Let $\sigma(\omega)$ be the minimum singular value of a return difference matrix at any given frequency ω. If there exists a bound $\alpha \le 1$ such that $\sigma(\omega) \ge \alpha$ for all ω, then

$$PM \ge \pm 2 \sin^{-1}(\alpha/2) \quad \text{and} \quad GM = 1/(1 \pm \alpha)$$

These margins apply for gain or phase perturbations introduced in the control loop at the point for which the return difference matrix applies. Figure 14.9 shows the multivariable LQ regulator in block diagram form. The return difference of interest is computed at the input to the plant as $\mathbf{R}_d(s) = \mathbf{I} + \mathbf{G}_\infty(s\mathbf{I} - \mathbf{A})^{-1}\mathbf{B}$. The bound $\alpha = 1$ has been shown to apply to this return difference matrix. The preceding theorem then guarantees the previously stated stability margins for arbitrary gain-phase perturbations introduced at the plant input.

When perfect full state feedback is not available for use in the LQ control loop, the separation theorem allows estimates of the states to be used in forming the feedback control signals. Figure 14.10a through d shows block diagram representations of the controller-observer system. These are obtained, one from another, by standard block diagram manipulations. Figure 14.10 is the multivariable equivalent of Figure 13.15, except here the control gain $\mathbf{G}_\infty$ is explicitly identified.

If perfect models were available, the resulting system would perform as designed, except perhaps for a brief transient due to initial condition mismatches between the true and estimated states. In fact, the transfer functions from $\mathbf{v}$ to $\mathbf{y}$ are the same for Figures 14.9 and 14.10 for any arbitrary control gain $\mathbf{G}$. For Figure 14.9, $\mathbf{x}(s) = (s\mathbf{I} - \mathbf{A})^{-1}\mathbf{B}\mathbf{u}(s) = (s\mathbf{I} - \mathbf{A})^{-1}\{\mathbf{v} - \mathbf{G}\mathbf{x}\}$, so that $\mathbf{x} = [\mathbf{I} + (s\mathbf{I} - \mathbf{A})^{-1}\mathbf{B}\mathbf{G}]^{-1}(s\mathbf{I} - \mathbf{A})^{-1}\mathbf{B}\mathbf{v} = [s\mathbf{I} - \mathbf{A} + \mathbf{B}\mathbf{G}]^{-1}\mathbf{B}\mathbf{v}$. Since $\mathbf{y} = \mathbf{C}\mathbf{x}$, the full state transfer function is $\mathbf{H}_f = \mathbf{C}[s\mathbf{I} - \mathbf{A} + \mathbf{B}\mathbf{G}]^{-1}\mathbf{B}$. From Figure 14.10, $\mathbf{y} = \mathbf{C}(s\mathbf{I} - \mathbf{A})^{-1}\mathbf{B}\mathbf{u} = \mathbf{W}_1(s)\mathbf{u}$. But $\mathbf{u} = \mathbf{v} - \mathbf{G}(s\mathbf{I} - \mathbf{A}_c)^{-1}\mathbf{B}\mathbf{u} - \mathbf{G}(s\mathbf{I} - \mathbf{A}_c)^{-1}\mathbf{L}\mathbf{y} = \mathbf{v} - \mathbf{G}\mathbf{W}_3\mathbf{B}\mathbf{u} - \mathbf{G}\mathbf{W}_2\mathbf{y}$. Combining and solving for $\mathbf{y}$ leads to the input-output transfer function

$$\mathbf{H}_o = \{\mathbf{I} + \mathbf{W}_1[\mathbf{I} + \mathbf{G}\mathbf{W}_3]^{-1}\mathbf{G}\mathbf{W}_2\}^{-1}\mathbf{W}_1[\mathbf{I} + \mathbf{G}\mathbf{W}_3]^{-1}$$

The claim is that $\mathbf{H}_f = \mathbf{H}_o$ when exact models are used. The proof uses a result from Problem 4.4, which is restated here and is referred to as a gain rearrangement identity. This identity is true for any conformable matrices $\mathbf{F}$ and $\mathbf{G}$ for which the indicated inverses exist:

$$[\mathbf{I} + \mathbf{F}\mathbf{G}]^{-1}\mathbf{F} = \mathbf{F}[\mathbf{I} + \mathbf{G}\mathbf{F}]^{-1}$$

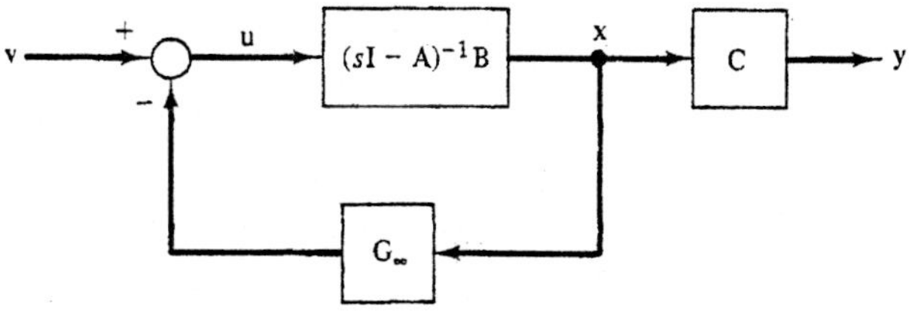

Figure 14.9

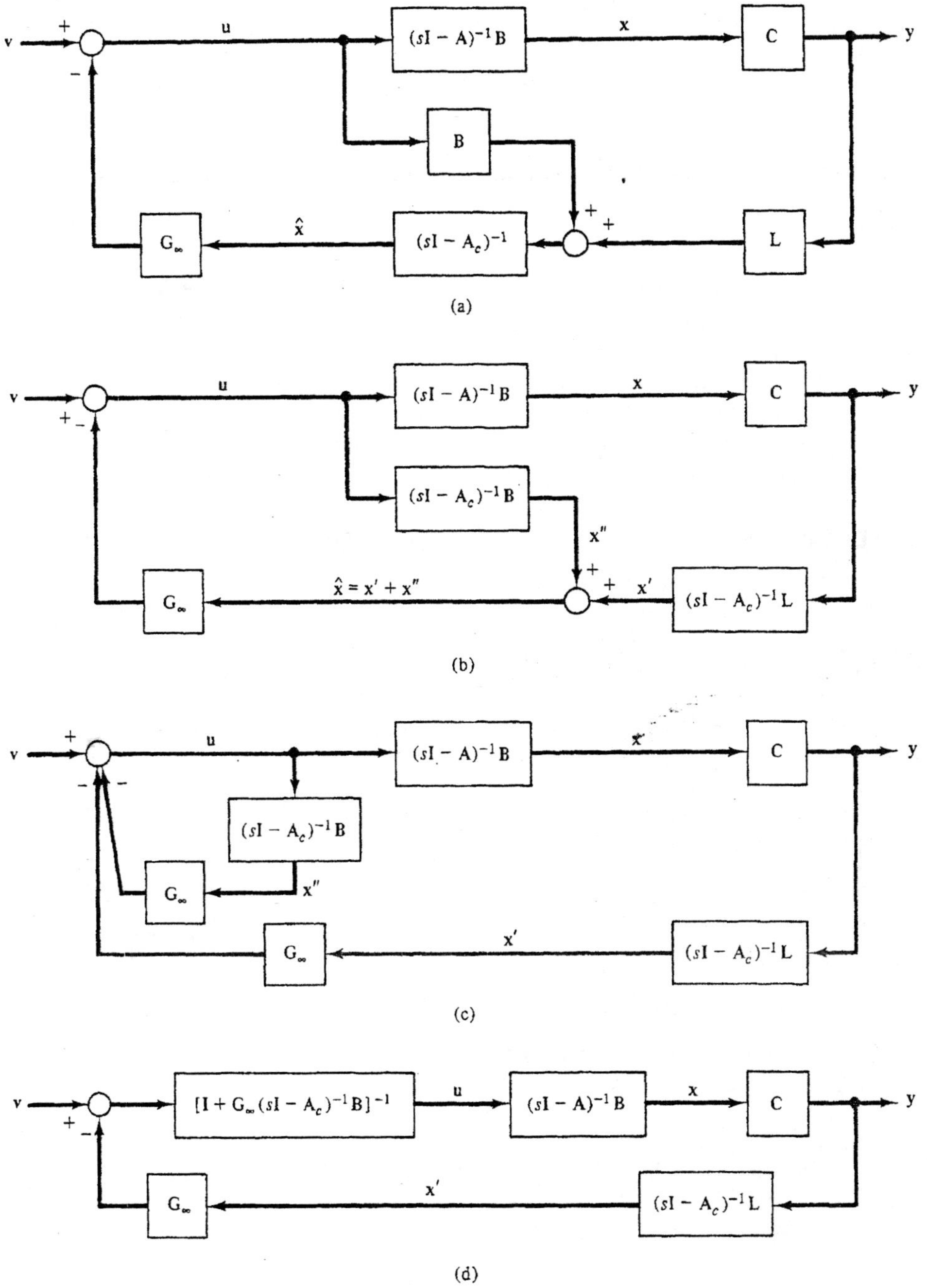
v
u
$(sI - A)^{-1}B$
x
C
y
B
L
$\hat{x}$
$(sI - A_c)^{-1}$
G_∞
(a)
$(sI - A_c)^{-1}B$
x″
$\hat{x} = x' + x''$
x′
$(sI - A_c)^{-1}L$
(b)
(c)
$[I + G_\infty(sI - A_c)^{-1}B]^{-1}$
(d)

Figure 14.10

Then a series of standard manipulations gives

$$\begin{aligned}\mathbf{H}_o &= \mathbf{W}_1\{\mathbf{I} + [\mathbf{I} + \mathbf{G}\mathbf{W}_3]^{-1}\mathbf{G}\mathbf{W}_2\mathbf{W}_1\}^{-1}[\mathbf{I} + \mathbf{G}\mathbf{W}_3]^{-1} \\ &= \mathbf{W}_1\{[\mathbf{I} + \mathbf{G}\mathbf{W}_3][\mathbf{I} + [\mathbf{I} + \mathbf{G}\mathbf{W}_3]^{-1}\mathbf{G}\mathbf{W}_2\mathbf{W}_1]\}^{-1} \\ &= \mathbf{W}_1\{\mathbf{I} + \mathbf{G}\mathbf{W}_3 + \mathbf{G}\mathbf{W}_2\mathbf{W}_1\}^{-1}.\end{aligned}$$

In terms of the original system matrices this becomes $\mathbf{H}_o = \mathbf{C}(s\mathbf{I} - \mathbf{A})^{-1}\mathbf{B}\{\mathbf{I} + \mathbf{G}(s\mathbf{I} - \mathbf{A}_c)^{-1}[\mathbf{I} + \mathbf{L}\mathbf{C}(s\mathbf{I} - \mathbf{A})^{-1}]\mathbf{B}\}^{-1}$. Replacing $\mathbf{I}$ by $(s\mathbf{I} - \mathbf{A})(s\mathbf{I} - \mathbf{A})^{-1}$ in the term in brackets allows an $(s\mathbf{I} - \mathbf{A})^{-1}$ to be factored out. Then using $\mathbf{A}_c = \mathbf{A} - \mathbf{L}\mathbf{C}$ gives $\mathbf{H}_o = \mathbf{C}(s\mathbf{I} - \mathbf{A})^{-1}\mathbf{B}\{\mathbf{I} + \mathbf{G}(s\mathbf{I} - \mathbf{A})^{-1}\mathbf{B}\}^{-1}$. Using the gain rearrangement identity to move $\mathbf{B}$ allows the two inverse terms to be combined, giving $\mathbf{H}_o = \mathbf{C}(s\mathbf{I} - \mathbf{A} + \mathbf{B}\mathbf{G})^{-1}\mathbf{B} \equiv \mathbf{H}_f$. It is clear from Figure 14.10 that the combined system is of order $2n$ (or higher with some rearrangements). Since the input-output transfer function equals that for an nth-order system, some modes must be either uncontrollable, unobservable, or both. In this case all the modes introduced by the observer are uncontrollable.

EXAMPLE 14.13 Consider the unstable system described by $\mathbf{A} = \begin{bmatrix} 0 & 1 \\ 2 & -3 \end{bmatrix}$, $\mathbf{B} = \mathbf{I}_2$, $\mathbf{C} = [1 \quad 0]$, and $\mathbf{D} = [0 \quad 0]$. Design a constant feedback LQ controller using $\mathbf{Q} = \mathbf{R} = \mathbf{I}_2$. Then design an observer by using the duality between controllers and observers. That is, replace $\mathbf{A}$ by $\mathbf{A}^T$ and $\mathbf{B}$ by $\mathbf{C}^T$, let $\mathbf{Q} = \mathbf{I}_2$, and $\mathbf{R} = r$, and then solve the ARE for $\mathbf{W}'_\infty$. The observer gain $\mathbf{L}$ is the transpose of the dual control gain matrix. That is, $\mathbf{L} = \mathbf{W}'_\infty\mathbf{C}^T\mathbf{R}^{-T}$. The separation theorem assures us that the final $2n$th-order system will have n poles determined by the $\mathbf{Q}$, $\mathbf{R}$ weights—i.e., the desired controller poles—and n poles determined by the observer. Select the scalar r so that the observer poles are faster—i.e., further to the left in the complex plane—than the closed-loop controller poles. Investigate the composite controller-observer system for controllability, observability, and eigenvalue locations. Then compare the step response of the full state feedback system with the controller-observer system.

Solving Eq. (*14.49*) gives $\mathbf{W}_\infty = \begin{bmatrix} 1.6012 & 0.4392 \\ 0.4392 & 0.26887 \end{bmatrix}$, and since $\mathbf{Q} = \mathbf{B} = \mathbf{I}$, $\mathbf{G}_\infty = \mathbf{W}_\infty$. Using this gain with full state feedback gives the closed-loop eigenvalues $\lambda = -1.1818$ and -3.6882. Solving the dual problem with three values of r gives:

$r = 10 \Rightarrow \mathbf{L} = [1.1919 \quad .66033]^T$ and $\lambda_0 = -0.62729, -3.5646$

$r = 1 \Rightarrow \mathbf{L} = [1.6350 \quad .8366]^T$ and $\lambda_0 = -1.0411, -3.5939$

$r = 0.1 \Rightarrow \mathbf{L} = [3.5876 \quad 1.4353]^T$ and $\lambda_0 = -2.487, -4.10$

Note that the observer gains get higher and its response gets faster as the value of r is decreased. The same effect occurs as $\mathbf{Q}$ is made larger, since only the ratio $\mathbf{Q}/r$ is significant in minimizing the quadratic performance function. The value $r = 0.1$ is accepted, since it gives observer eigenvalues somewhat faster than the controller eigenvalues. The composite system is described by

$$\begin{bmatrix} \dot{\mathbf{x}} \\ \dot{\hat{\mathbf{x}}} \end{bmatrix} = \begin{bmatrix} \mathbf{A} & -\mathbf{B}\mathbf{G}_\infty \\ \mathbf{L}\mathbf{C} & \mathbf{A} - \mathbf{B}\mathbf{G}_\infty - \mathbf{L}\mathbf{C} \end{bmatrix}\begin{bmatrix} \mathbf{x} \\ \hat{\mathbf{x}} \end{bmatrix} + \begin{bmatrix} \mathbf{B} \\ \mathbf{B} \end{bmatrix}\mathbf{v}$$

where $\mathbf{v}$ is an arbitrary external input and $\mathbf{y} = [1 \quad 0 \quad 0 \quad 0]\begin{bmatrix} \mathbf{x} \\ \hat{\mathbf{x}} \end{bmatrix}$. The 4×4 composite system matrix is

$$\begin{bmatrix} 0 & 1 & -1.6012 & -0.4392 \\ 2 & -3 & -0.4392 & -0.26887 \\ 3.5876 & 0 & -5.1888 & 0.5608 \\ 1.4353 & 0 & 0.1255 & -3.26887 \end{bmatrix}$$

Its eigenvalues are verified to consist of the desired controller and observer eigenvalues, $\lambda = -1.1818, -3.6882, -2.487$, and -4.100. The composite system is observable but not controllable. In fact, the Kalman controllable canonical form for this fourth-order system has

$$\mathbf{A}' = \left[\begin{array}{cc|cc} -1.6012 & 0.5608 & -0.557862 & 5.17739 \\ 1.5608 & -3.26887 & -0.311709 & 1.86785 \\ \hline 0 & 0 & -2.9645 & -0.550433 \\ 0 & 0 & -0.985733 & -3.62310 \end{array}\right],$$

$$\mathbf{B}' = \left[\begin{array}{cc} 1.41421 & 0 \\ 0 & 1.41421 \\ \hline 0 & 0 \\ 0 & 0 \end{array}\right]$$

and $\mathbf{C}' = [0.707107 \quad 0 \quad -0.016186 \quad 0.7070]$. The 2×2 lower right corner of $\mathbf{A}'$ describes the uncontrollable modes, and the eigenvalues of this partition are $\lambda = -2.4869$ and -4.100. These are the observer modes. To test the transient performance, it is necessary to specify the inputs $v_1(t)$ and $v_2(t)$ as well as the four initial conditions. Only one case is presented, with v_1 equal to a unit step function, $v_2 = 0$, and $x_1(0) = x_2(0) = \hat{x}_2(0) = 0, \hat{x}_1(0) = 1$. Figure 14.11*a* shows the response of x_1, $\hat{x}_1$, and x_1^*, where x_1^* is the response obtained with the same inputs and initial conditions $\mathbf{x}(0)$, but with full state feedback. Figure 14.11*b* gives the same comparisons for the second state variable and its estimate. The estimate, actual, and optimal curves all agree in the limit. There is a transient difference between $\mathbf{x}(t)$ and $\hat{\mathbf{x}}(t)$ due to initial condition mismatch. This

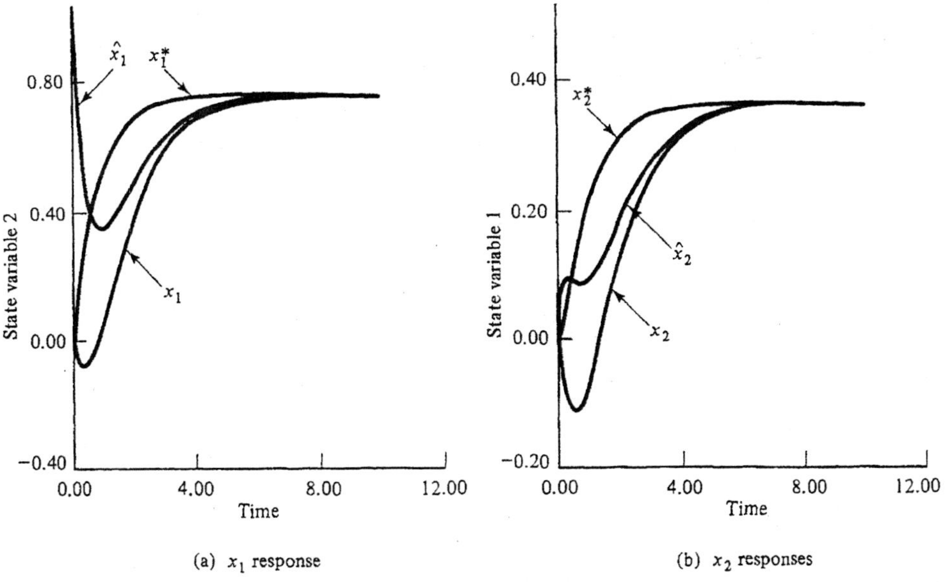

Figure 14.11 Comparison of full state step response with observer true and estimated.

difference decays according to the observer eigenvalue response. The composite system responses deviate from the optimal full state feedback case because of this transient. Even with zero input and all initial conditions zero except for $\hat{x}_1$, for example, the output y, which is just x_1, still shows a nonzero transient. This indicates that the observer modes' initial condition responses are observable in the output, but as stated earlier, they are not controllable. ■

Since all models have some degree of inaccuracy or oversimplification, the question of robustness needs to be considered here also. It has been shown [13, 14] that the introduction of a poorly designed observer can cause the comfortable stability margins of the *LQ* problem to be lost. The return difference matrix analysis is somewhat more complicated for the observer system shown in Figure 14.10. Consider a perturbation $\delta\mathbf{u}$ entering at the input to the plant. From Figure 14.10, it will affect $\mathbf{x}$, $\mathbf{x}'$, and $\hat{\mathbf{x}}$ but not $\mathbf{x}''$ if the loop is opened at $\mathbf{G}_\infty$, for example. A comparison of Figures 14.9 and 14.10 shows that any such $\delta\mathbf{u}$ will affect both $\mathbf{x}$ and $\mathbf{x}'$ (and hence $\hat{\mathbf{x}}$) in the same way, provided that

$$(s\mathbf{I}-\mathbf{A})^{-1}\mathbf{B} = (s\mathbf{I}-\mathbf{A}_c)^{-1}\mathbf{L}\mathbf{C}(s\mathbf{I}-\mathbf{A})^{-1}\mathbf{B} \qquad (14.55)$$

This condition, after modest algebraic manipulations, leads to Eq. (*14.56*), which is called the *Doyle-Stein robust observer condition.* It has been shown [13] that if the extra freedom which exists in specifying observer poles and finding the observer gain $\mathbf{L}$ is used to satisfy, or approach, the Doyle-Stein condition, then the stability robustness associated with the full state *LQ* system of Figure 14.9 can be asymptotically recovered in the *LQG* system of Figure 14.10. The original derivation of the Doyle-Stein condition assumed that there were r inputs and $m = r$ outputs, so that $\mathbf{C}(s\mathbf{I}-\mathbf{A})^{-1}\mathbf{B}$ is square. Then postmultiplying by the inverse of this matrix and premultiplying by $(s\mathbf{I}-\mathbf{A})$ alters Eq. (*14.55*) to

$$\mathbf{B}[\mathbf{C}(s\mathbf{I}-\mathbf{A})^{-1}\mathbf{B}]^{-1} = (s\mathbf{I}-\mathbf{A})(s\mathbf{I}-\mathbf{A}+\mathbf{L}\mathbf{C})^{-1}\mathbf{L}$$

where the definition $\mathbf{A}_c = \mathbf{A}-\mathbf{L}\mathbf{C}$ has been used. The left side is in the desired final form. The right side can be rearranged to give

$$(s\mathbf{I}-\mathbf{A})[\{\mathbf{I}+\mathbf{L}\mathbf{C}(s\mathbf{I}-\mathbf{A})^{-1}\}(s\mathbf{I}-\mathbf{A})]^{-1}\mathbf{L} = [(\mathbf{I}+\mathbf{L}\mathbf{C}(s\mathbf{I}-\mathbf{A})^{-1}]^{-1}\mathbf{L}$$

Again, using the gain rearrangement identity allows the two $\mathbf{L}$ terms to be rearranged, giving

$$\mathbf{B}[\mathbf{C}(s\mathbf{I}-\mathbf{A})^{-1}\mathbf{B}]^{-1} = \mathbf{L}[(\mathbf{I}+\mathbf{C}(s\mathbf{I}-\mathbf{A})^{-1}\mathbf{L}]^{-1} \qquad (14.56)$$

Other forms of Eq. (*14.55*) which do not require the assumption that $r = m$ can be written, such as

$$(s\mathbf{I}-\mathbf{A}_c)(s\mathbf{I}-\mathbf{A})^{-1}\mathbf{B} = \mathbf{L}\mathbf{C}(s\mathbf{I}-\mathbf{A})^{-1}\mathbf{B}$$

which leads to

$$[\mathbf{I}+\mathbf{L}\mathbf{C}(s\mathbf{I}-\mathbf{A})^{-1}]\mathbf{B} = \mathbf{L}\mathbf{C}(s\mathbf{I}-\mathbf{A})^{-1}\mathbf{B}$$

A full exploitation of the robust observer results is left to the references. One of the major results that comes from this analysis is that the placement of all observer poles far to the left in order to speed convergence of $\hat{\mathbf{x}}$ to $\mathbf{x}$ is not always the best strategy when the possibility of model errors exists. Rather, some of the observer poles should approach the stable plant zeros. The rest of the observer poles do migrate outward to the left along the classical Butterworth configuration.

EXTENSIONS

Lower-Order Controllers via Projective Controls [15, 16, 17]

From results of the past two chapters, it is known that the closed-loop eigenvalues of a completely controllable and observable system can be relocated as desired by using full state feedback. In most practical problems, all states are not available for use. The only signals that realistically can be assumed available for feedback are the outputs. Static output feedback can control the location only of a subset of m eigenvalues. The remaining $n - m$ eigenvalues may or may not take on acceptable values. An observer of order n, which is a dynamic feedback compensator, can be used with output feedback and total discretion about eigenvalue locations is again possible. A reduced order observer, which is a dynamic compensator of order $n - m$, where $\text{rank}(\mathbf{C}) = m$, can also be used. The order of these feedback compensators may be higher than desired in many cases. One extension to the LQ feedback theory provides for suboptimal lower-order controllers, which preserve some of the optimal eigenstructure associated with the full state LQ feedback solution. For simplicity the discussion is restricted to constant, strictly proper systems ($\mathbf{D} = [\mathbf{0}]$) which are both controllable and observable. Only the infinite control horizon is considered, so that the algebraic Riccati equation governs the optimal solution. The method of projective controls makes it possible to use output feedback through a dynamic feedback compensator of order p, with $0 \le p \le n - m$. The projective controls procedure allows for matching $p + m$ of the eigenvalues and eigenvectors of the suboptimal system with those of the optimal LQ system. The case of static output feedback is considered first. The optimal LQ closed-loop system dynamics, with optimal feedback gain $\mathbf{G}_0$, are governed by the matrix $\mathbf{A} - \mathbf{B}\mathbf{G}_0 = \mathbf{A} - \mathbf{B}\mathbf{R}^{-1}\mathbf{B}^T\mathbf{W}$. Let the eigenvalues of this matrix be $\{\lambda_i, i = 1, n\}$ and the corresponding eigenvectors be $\boldsymbol{\xi}_i$. Suppose that m of these eigenvalue-eigenvector pairs, perhaps the dominant modes, must be retained in the output feedback solution. Form the $n \times m$ matrix $\mathbf{X}_m$ with columns made up of the selected eigenvectors. The $n \times n$ projection matrix $\mathbf{P} = \mathbf{X}_m(\mathbf{C}\mathbf{X}_m)^{-1}\mathbf{C}$ (note $\mathbf{P}^2 = \mathbf{P}$; see Sec. 5.13) can be shown to project any arbitrary state $\mathbf{x}$ onto the subspace spanned by the m selected eigenvectors. If the feedback control law is

$$\mathbf{u} = -\mathbf{G}_0\mathbf{P}\mathbf{x} = -\mathbf{G}_0[\mathbf{X}_m(\mathbf{C}\mathbf{X}_m)^{-1}\mathbf{C}]\mathbf{x} = -\mathbf{G}_0[\mathbf{X}_m(\mathbf{C}\mathbf{X}_m)^{-1}]\mathbf{y}$$

then the suboptimal system will have the desired m eigenvalues and eigenvectors. The balance of the eigenvalues-eigenvectors is determined by the matrix $\mathbf{A}_{22} - \mathbf{N}\mathbf{A}_{12}$, where $\mathbf{A}_{22}$ is the lower right $(n - m) \times (n - m)$ partition of the original system matrix $\mathbf{A}$, $\mathbf{A}_{12}$ is the $m \times (n - m)$ upper right partition, and $\mathbf{N} = \mathbf{X}_b\mathbf{X}_a^{-1}$. The eigenvector matrix $\mathbf{X}_m$ is partitioned into the upper m rows, $\mathbf{X}_a$, and the lower $n - m$ rows, $\mathbf{X}_b$. The obvious restriction on choice of $\mathbf{X}_m$ applies. If $\boldsymbol{\xi}_i$ is complex and is selected for inclusion, then the conjugate column must also be selected. There are many instances where the static output controller will not be satisfactory. If only one output is available and both eigenvalues of an optimal second-order system are complex, the preceding procedure cannot be used. In other cases the closed loop system will be unstable or otherwise unsatisfactory due to the uncontrolled eigenvalues. In these cases a pth-order dynamic controller can be considered with

$$\dot{\mathbf{z}} = \mathbf{Hz} + \mathbf{Ly} \quad \text{and} \quad \mathbf{u} = -\{\mathbf{G}_z \mathbf{z} + \mathbf{G}_y \mathbf{y}\} \tag{14.57}$$

The controller parameters **H** and **L** as well as the two gain matrices must be determined. A composite $(n+p)$th-order system can be written as

$$\begin{bmatrix} \dot{\mathbf{z}} \\ \dot{\mathbf{x}} \end{bmatrix} = \begin{bmatrix} \mathbf{H} & \mathbf{LC} \\ \mathbf{0} & \mathbf{A} \end{bmatrix} \begin{bmatrix} \mathbf{z} \\ \mathbf{x} \end{bmatrix} + \begin{bmatrix} \mathbf{0} \\ \mathbf{B} \end{bmatrix} \mathbf{u}$$

shortened to $\dot{\mathbf{x}}_a = \mathbf{A}_a \mathbf{x}_a + \mathbf{B}_a \mathbf{u}$, and

$$\begin{bmatrix} \mathbf{z} \\ \mathbf{y} \end{bmatrix} = \begin{bmatrix} \mathbf{I} & \mathbf{0} \\ \mathbf{0} & \mathbf{C} \end{bmatrix} \begin{bmatrix} \mathbf{z} \\ \mathbf{x} \end{bmatrix}$$

notationally shortened to $\mathbf{y}_a = \mathbf{C}_a \mathbf{x}_a$. An augmented $\mathbf{Q}_a = \begin{bmatrix} \mathbf{0} & \mathbf{0} \\ \mathbf{0} & \mathbf{Q} \end{bmatrix}$ is also defined so that the augmented problem has the same cost function J as the original problem.

The standard ARE could be applied to the augmented problem if **H** and **L** were known. However, it can be shown that if **H** is asymptotically stable, the only possible solution to the ARE for the enlarged problem is $\mathbf{W}_a = \begin{bmatrix} \mathbf{0} & \mathbf{0} \\ \mathbf{0} & \mathbf{W} \end{bmatrix}$, where **W** is the solution to the parent ARE equation. This means that the determination of **H** and **L** can be postponed until later. The previous projection results for state output feedback can be formally applied to the enlarged problem by defining the matrix of $m+p$ desired eigenvectors,

$$\mathbf{X}_{m+p} = \begin{bmatrix} \mathbf{X}_{11} & \mathbf{X}_{12} \\ \mathbf{X}_{21} & \mathbf{X}_{22} \\ \mathbf{X}_{31} & \mathbf{X}_{32} \end{bmatrix}$$

The eigenvectors desired for the original nth-order system are contained in the partitions $\begin{bmatrix} \mathbf{X}_{21} & \mathbf{X}_{22} \\ \mathbf{X}_{31} & \mathbf{X}_{32} \end{bmatrix} = \{\xi_i,\ i = 1,\ m+p\}$ and are known. The pth-order controller requires augmentation with the still unknown blocks $\mathbf{X}_{11}$ $(p \times p)$ and $\mathbf{X}_{12}$ $(p \times m)$. It is convenient to assume that the states and measurements are arranged so that $\mathbf{C} = [\mathbf{I}_m \quad \mathbf{0}]$. Then the projection matrix for the augmented problem is

$$\mathbf{P}_a = \mathbf{X}_{m+p}[\mathbf{C}_a \mathbf{X}_{m+p}]^{-1} \mathbf{C}_a = \mathbf{X}_{m+p} \begin{bmatrix} \mathbf{X}_{11} & \mathbf{X}_{12} \\ \mathbf{X}_{21} & \mathbf{X}_{22} \end{bmatrix}^{-1} \begin{bmatrix} \mathbf{I}_p & \mathbf{0} & \mathbf{0} \\ \mathbf{0} & \mathbf{I}_m & \mathbf{0} \end{bmatrix}$$

Therefore, $\mathbf{P}_a \mathbf{x}_a = \begin{bmatrix} \mathbf{I}_{m+p} \\ \mathbf{N} \end{bmatrix} \begin{bmatrix} \mathbf{z} \\ \mathbf{x} \end{bmatrix}$, where

$$\mathbf{N} = [\mathbf{X}_{31} \quad \mathbf{X}_{32}] \begin{bmatrix} \mathbf{X}_{11} & \mathbf{X}_{12} \\ \mathbf{X}_{21} & \mathbf{X}_{22} \end{bmatrix}^{-1} = [\mathbf{N}_p \quad \mathbf{N}_m] \tag{14.58}$$

The projective feedback control is $\mathbf{G}_a \mathbf{x}_a = -\mathbf{R}^{-1} \mathbf{B}_a^T \mathbf{W}_a \mathbf{P}_a \mathbf{x}_a$, which reduces to $\mathbf{G}_a \mathbf{x}_a = -\{\mathbf{G}_z \mathbf{z} + \mathbf{G}_y \mathbf{y}\}$, where

$$\mathbf{G}_z = \mathbf{G}_0\begin{bmatrix}\mathbf{0}\\ \mathbf{N}_p\end{bmatrix} = \mathbf{G}_u\mathbf{N}_p \quad \text{and} \quad \mathbf{G}_y = \mathbf{G}_0\begin{bmatrix}\mathbf{I}_m\\ \mathbf{N}_m\end{bmatrix} = \mathbf{G}_m + \mathbf{G}_u\mathbf{N}_m$$

The original full state optimal gain $\mathbf{G}_0 = [\mathbf{G}_m \quad \mathbf{G}_u]$ has been partitioned into $\mathbf{G}_m$, which multiplies measured states, and $\mathbf{G}_u$, which multiplies unmeasured states. The projective control form is now known, but the matrix $\mathbf{N}$ of Eq. (*14.58*) used in forming the gains depends upon the still-unknown eigenvector blocks $\mathbf{X}_{11}$ and $\mathbf{X}_{12}$. These are directly related to the dynamics of the controller, $\mathbf{H}$ and $\mathbf{L}$ of Eq. (*14.57*), which have not yet been specified. As long as the controller matrix $\mathbf{H}$ is asymptotically stable, any $\mathbf{X}_{11}$ and $\mathbf{X}_{12}$ could be used, and the primary objective of retaining $m+p$ eigenvalues and eigenvectors of the optimal system will be met. The remaining $n-m$ eigenvalues, called the complementary, or residual, spectrum, are affected by the choice of the remaining unknowns, since these eigenvalues are determined by the matrix

$$\mathbf{A}_r = \mathbf{A}_{22} - \mathbf{N}\mathbf{A}_{12} = \mathbf{A}_{22} - [\mathbf{X}_{31} \quad \mathbf{X}_{32}]\begin{bmatrix}\mathbf{X}_{11} & \mathbf{X}_{12}\\ \mathbf{X}_{21} & \mathbf{X}_{22}\end{bmatrix}^{-1}\mathbf{A}_{12}$$

This is similar to the problem encountered in pole placement using output feedback in Sec. 13.5. Thus the selection of $\mathbf{X}_{11}$ and $\mathbf{X}_{22}$ can be used to give acceptable eigenvalues to $\mathbf{A}_r$. As a minimum, all the residual eigenvalues must be stable. With $\mathbf{X}_{11}$ and $\mathbf{X}_{12}$ determined, the gains $\mathbf{G}_z$ and $\mathbf{G}_y$ can be evaluated. The last task is to determine the control matrices $\mathbf{H}$ and $\mathbf{L}$. The augmented closed-loop eigenvector equation is $[\mathbf{A}_a - \mathbf{B}_a\mathbf{G}_a]\mathbf{X}_{m+p} = \mathbf{X}_{m+p}\mathbf{\Lambda}_{m+p}$. It has $m+p$ known, specified eigenvectors in $\mathbf{X}_{m+p}$ and eigenvalues in $\mathbf{\Lambda}_{m+p}$. By assumption we are dealing with the simple eigenvalues cases I or II_1 of Sec. 7.4, so that $\mathbf{\Lambda}_{m+p}$ will be diagonal. However, in the case of complex eigenvalues and eigenvectors, it may be more convenient to work with purely real arithmetic. Postmultiplication by a transformation $\mathbf{T}$ (see Problem 4.23) gives

$$[\mathbf{A}_a - \mathbf{B}_a\mathbf{G}_a]\mathbf{X}_{m+p}\mathbf{T} = \mathbf{X}_{m+p}\mathbf{T}\mathbf{T}^{-1}\mathbf{\Lambda}_{m+p}\mathbf{T}, \quad \text{or} \quad [\mathbf{A}_a - \mathbf{B}_a\mathbf{G}_a]\mathbf{X}'_{m+p} = \mathbf{X}'_{m+p}\mathbf{\Lambda}'_{m+p}$$

In the primed form only real numbers are required, but $\mathbf{\Lambda}_{m+p}$ will no longer be diagonal. The prime is dropped below, and the expanded partitioned form of these equations can be solved to give

$$[\mathbf{H} \quad \mathbf{L}] = [\mathbf{X}_{11} \quad \mathbf{X}_{12}]\mathbf{\Lambda}_{m+p}\begin{bmatrix}\mathbf{X}_{11} & \mathbf{X}_{12}\\ \mathbf{X}_{21} & \mathbf{X}_{22}\end{bmatrix}^{-1} \tag{14.59}$$

EXAMPLE 14.14 The third-order system with $\mathbf{A} = \begin{bmatrix}0 & 1 & 0\\ 0 & 0 & 1\\ -2 & -1 & -1\end{bmatrix}$, $\mathbf{B} = \begin{bmatrix}0\\0\\1\end{bmatrix}$, and $\mathbf{C}^T = \begin{bmatrix}1\\0\\0\end{bmatrix}$ and with $\mathbf{Q} = \mathbf{I}_3$, $R = 5$ has the optimal full state feedback gain $\mathbf{G}_0 = [0.04939 \quad 1.2591 \quad 0.92826]$ and the closed-loop eigenvalues are $\lambda_i = \{-1.37085, -0.2787 \pm 1.1905j\}$, and the corresponding eigenvectors are

$$\begin{bmatrix}-0.532106 & 0.599405 \pm 0.296929j\\ 0.729456 & 0.186430 \mp 0.79634j\\ -1 & -1\end{bmatrix}$$

If static output feedback is used, one eigenvalue-eigenvector from the optimal set can be retained, and it must be the real eigenvalue because complex pairs cannot be split. It is easy to calculate that $\mathbf{G}_y = \mathbf{G}_0\boldsymbol{\xi}_1[\mathbf{C}\boldsymbol{\xi}_1]^{-1} = 0.067813$ will in fact retain the first eigenvalue-eigenvector.

However, the other two eigenvalues will be in the right-half plane, so this solution is not useful. A dynamic controller of order $p = 1$ can be designed which will retain the complex conjugate eigenpair, as follows. The values $\mathbf{X}_{11} = -1$ and $\mathbf{X}_{12} = 1$ were found to give stable complementary eigenvalues. Using these, $\mathbf{N} = \begin{bmatrix} 1.76109 & 3.24908 \\ -0.981661 & -3.30605 \end{bmatrix}$, $\mathbf{G}_z = 1.30615$, $\mathbf{G}_y = 1.07144$, $\mathbf{H} = -3.806533$, and $\mathbf{L} = -7.871706$. It can be verified that these values do retain the complex eigenvalue-eigenvector pair in the suboptimal solution. The two remaining eigenvalues are at -0.23494 and -4.014. The first of these may be too close to the $j\omega$ axis. A systematic search for other values of $\mathbf{X}_{11}$ and $\mathbf{X}_{12}$ which give a more satisfactory residual spectrum can be carried out [16]. ■

Robustness Enhancement via Frequency-Weighted Cost Function [18, 19]

There are many situations where the system model used in control design is known to be inaccurate at high frequencies. The steady-state D.C. error may be particularly important in other cases. The suppression of certain vibration modes may be crucial in other circumstances. All these criteria suggest that errors at certain frequencies may be more costly than others. It is possible to extend the infinite time LQ optimization theory to frequency-sensitive cost functions by applying Parseval's theorem:

$$\int_0^\infty |f(t)|^2\,dt = \frac{1}{2\pi}\int_{-\infty}^{\infty} |F(j\omega)|^2\,d\omega \tag{14.60}$$

where $F(j\omega)$ is the Fourier transform of $f(t)$. The cost functional J can be written in a similar way by defining $\mathbf{f}^T(t) = [(\mathbf{Q}^{1/2}\mathbf{x})^T(\mathbf{R}^{1/2}\mathbf{u})^T]$, because then

$$\begin{aligned}\int_0^\infty |\mathbf{f}(t)|^2\,dt &= \int_0^\infty \{\mathbf{x}^T\mathbf{Q}\mathbf{x} + \mathbf{u}^T\mathbf{R}\mathbf{u}\}\,dt \\ &= \frac{1}{2\pi}\int_{-\infty}^{\infty} \{\mathbf{x}^*(j\omega)\mathbf{Q}(j\omega)\mathbf{x}(j\omega) + \mathbf{u}^*(j\omega)\mathbf{R}\mathbf{u}(j\omega)\}\,d\omega\end{aligned} \tag{14.61}$$

where $(\ )^*$ indicates the complex conjugate of the transpose of the quantity inside. Up until now the weighting matrices $\mathbf{Q}$ and $\mathbf{R}$ have been constants. In order to obtain frequency-dependent weighting while still being able to utilize the standard LQ results, we select $\mathbf{Q}(j\omega) = \mathbf{T}^*(j\omega)\mathbf{T}(j\omega)$ and $\mathbf{R}(j\omega) = \mathbf{U}^*(j\omega)\mathbf{U}(j\omega)$ and define two new vector variables whose transforms are related to $\mathbf{x}$ and $\mathbf{u}$ according to

$$\mathbf{z}(j\omega) = \mathbf{T}(j\omega)\mathbf{x}(j\omega) \quad \text{and} \quad \mathbf{w}(j\omega) = \mathbf{U}(j\omega)\mathbf{u}(j\omega) \tag{14.62}$$

The implication is that $\mathbf{z}$ is the output of a filter or dynamic system with transfer function $\mathbf{T}(j\omega)$ and input $\mathbf{x}(j\omega)$. If $\mathbf{T}$ is a proper transfer function, then there is a corresponding time domain relation involving new states $\mathbf{z}_s$,

$$\dot{\mathbf{z}}_s = \mathbf{F}_z\mathbf{z}_s + \mathbf{G}_z\mathbf{x} \quad \text{and} \quad \mathbf{z} = \mathbf{H}_z\mathbf{z}_s + \mathbf{D}_z\mathbf{x} \tag{14.63}$$

A similar relationship can be written between $\mathbf{u}$ and $\mathbf{w}$ by introducing additional states $\mathbf{w}_s$ for the transfer function $\mathbf{U}$ using the methods of Chapters 3 and 12:

$$\dot{\mathbf{w}}_s = \mathbf{F}_u\mathbf{w}_s + \mathbf{G}_u\mathbf{u}, \qquad \mathbf{w} = \mathbf{H}_u\mathbf{w}_s + \mathbf{D}_u\mathbf{u} \tag{14.64}$$

Equations (*14.63*) and (*14.64*) can be combined with the original system state equations to give

$$\begin{bmatrix} \dot{\mathbf{x}} \\ \dot{\mathbf{z}}_s \\ \dot{\mathbf{w}}_s \end{bmatrix} = \begin{bmatrix} \mathbf{A} & \mathbf{0} & \mathbf{0} \\ \mathbf{G}_z & \mathbf{F}_z & \mathbf{0} \\ \mathbf{0} & \mathbf{0} & \mathbf{F}_u \end{bmatrix} \begin{bmatrix} \mathbf{x} \\ \mathbf{z}_s \\ \mathbf{w}_s \end{bmatrix} + \begin{bmatrix} \mathbf{B} \\ \mathbf{0} \\ \mathbf{G}_u \end{bmatrix} \mathbf{u} \tag{14.65}$$

$$\begin{bmatrix} \mathbf{y} \\ \mathbf{z} \\ \mathbf{w} \end{bmatrix} = \begin{bmatrix} \mathbf{C} & \mathbf{0} & \mathbf{0} \\ \mathbf{D}_z & \mathbf{H}_z & \mathbf{0} \\ \mathbf{0} & \mathbf{0} & \mathbf{H}_u \end{bmatrix} \begin{bmatrix} \mathbf{x} \\ \mathbf{z}_s \\ \mathbf{w}_s \end{bmatrix} + \begin{bmatrix} \mathbf{D} \\ \mathbf{0} \\ \mathbf{D}_u \end{bmatrix} \mathbf{u} \tag{14.66}$$

The cost function is now $J = \int \{\mathbf{z}^T\mathbf{z} + \mathbf{w}^T\mathbf{w}\}\,dt$. The $\mathbf{z}^T\mathbf{z}$ term can be written as a quadratic in the new states $\mathbf{x}$ and $\mathbf{z}_s$ by using Eq. (*14.66*). If the same approach is used with the $\mathbf{w}^T\mathbf{w}$ term, cross products of state and control terms arise. It is not difficult to return to Sec. 14.4 and derive a modified Riccati equation to provide the solution in the presence of these cross terms. An alternate approach is first to solve for the control variable $\mathbf{w}$ that optimizes

$$J = \int \{\mathbf{x}_a^T \mathbf{Q}_a \mathbf{x}_a + \mathbf{w}^T \mathbf{R}_a \mathbf{w}\}\,dt \tag{14.67}$$

with the definitions $\mathbf{x}_a^T = [\mathbf{x}^T \quad \mathbf{z}_s^T \quad \mathbf{w}_s^T]$

$$\mathbf{Q}_a = \begin{bmatrix} \mathbf{D}_z^T\mathbf{D}_z & \mathbf{D}_z^T\mathbf{H}_z & \mathbf{0} \\ \mathbf{H}_z^T\mathbf{D}_z & \mathbf{H}_z^T\mathbf{H}_z & \mathbf{0} \\ \mathbf{0} & \mathbf{0} & \mathbf{0} \end{bmatrix}, \qquad \mathbf{R}_a = \mathbf{I} \tag{14.68}$$

Equation (*14.65*) must be rewritten in terms of $\mathbf{w}$ rather than $\mathbf{u}$. This is done by using the last partition of Eq. (*14.66*), assuming that $\mathbf{D}_u$ is invertible. This gives

$$\mathbf{u} = \mathbf{D}_u^{-1}(\mathbf{w} - \mathbf{H}_u\mathbf{w}_s) \tag{14.69}$$

$$\dot{\mathbf{x}}_a = \begin{bmatrix} \mathbf{A} & \mathbf{0} & -\mathbf{B}\mathbf{D}_u^{-1}\mathbf{H}_u \\ \mathbf{G}_z & \mathbf{F}_z & \mathbf{0} \\ \mathbf{0} & \mathbf{0} & \mathbf{F}_u - \mathbf{G}_u\mathbf{D}_u^{-1}\mathbf{H}_u \end{bmatrix} \mathbf{x}_a + \begin{bmatrix} \mathbf{B}\mathbf{D}_u^{-1} \\ \mathbf{0} \\ \mathbf{G}_u\mathbf{D}_u^{-1} \end{bmatrix} \mathbf{w} \tag{14.70}$$

Equations (*14.67*) and (*14.70*) are in the standard forms used in the earlier *LQ* problems, and the solution for the optimal $\mathbf{w}$ is obtained by solving an ARE for $\mathbf{W}_a$. Then $\mathbf{w} = -\mathbf{R}_a^{-1}\mathbf{B}_a^T\mathbf{W}_a\mathbf{x}_a$, and using Eq. (*14.69*) gives $\mathbf{u} = -\mathbf{K}_x\mathbf{x} - \mathbf{K}_z\mathbf{z}_s - \mathbf{K}_w\mathbf{w}_s$.

EXAMPLE 14.15 A fourth-order system with input-output transfer function $y(s)/u(s) = (s+5)/[s(s+1)(s^2/\omega_m^2 + 2\zeta/\omega_m s + 1)]$ is to be approximated by a second-order model $(s+5)/[s(s+1)]$, which is a good low-frequency approximation if $\omega_m >> 1$. Assume $\zeta = 0.5$ and $\omega_m = 10$. Then the state variable equations for the true system and the model have the following matrices.

$$\mathbf{A} = \begin{bmatrix} -11 & 1 & 0 & 0 \\ -110 & 0 & 1 & 0 \\ -100 & 0 & 0 & 1 \\ 0 & 0 & 0 & 0 \end{bmatrix} \quad \mathbf{B} = \begin{bmatrix} 0 \\ 0 \\ 100 \\ 500 \end{bmatrix} \quad \mathbf{A}_m = \begin{bmatrix} -1 & 1 \\ 0 & 0 \end{bmatrix} \quad \mathbf{B}_m = \begin{bmatrix} 1 \\ 5 \end{bmatrix}$$

The full-state feedback optimal *LQ* solution for the fourth-order system with $\mathbf{Q} = \text{Diag}[1\ 1\ 0\ 0]$ and $\mathbf{R} = 1$ has the feedback gains $\mathbf{G}_\infty = [-4.9368 \quad 0.4406 \quad 0.09376 \quad 0.00822]$, and this gives closed-loop $\lambda_i = \{-5.583 \pm 1.3338j; -6.659 \pm 11.102j\}$.

When the second-order approximate model is used with a cost function containing $\mathbf{Q} = \text{Diag}[1 \quad 1]$ and $\mathbf{R} = 1$, the resulting gains are $\mathbf{G}_{m\infty} = [0.41421 \quad 1]$, and the eigenvalues are $\lambda_i = \{-1.414, -5\}$. When these two approximate gains are used in the true fourth-order system, the step response of Figure 14.12 shows a very strong ringing at the frequency of the unmodeled modes. A frequency weighted cost function is selected in an attempt to suppress the oscillation. Various approaches could be studied, such as weighting both diagonals in **Q.** This would mean at least a second-order controller. Alternatively, a weighting on **R** could be introduced, and to keep the example simple $w(s) = [(10s + 1)/(s + 1)]u(s)$ is selected. As frequency (or s) gets large, this weights the fluctuations in u^2 100 times more than in the original $\mathbf{R} = 1$ case. The dynamics of this weighting filter are described by $\dot{w}_s = -w_s - 9u$ and $w = w_s + 10u$.

Augmenting the model as in Eqs. (*14.65*) through (*14.70*) gives the third-order design system (no extra $\mathbf{z}_s$ states are needed here):

$$\begin{bmatrix} \dot{x}_1 \\ \dot{x}_2 \\ \dot{w}_s \end{bmatrix} = \begin{bmatrix} -1 & 1 & 0 \\ 0 & 0 & -0.5 \\ 0 & 0 & -0.1 \end{bmatrix} \begin{bmatrix} x_1 \\ x_2 \\ x_3 \end{bmatrix} + \begin{bmatrix} 0.1 \\ 0.5 \\ -0.9 \end{bmatrix} w \qquad \mathbf{Q}_a = \text{Diag}[1 \quad 1 \quad 0], \qquad \mathbf{R}_a = 1$$

The ARE solution shows the optimal $w = -[0.19968 \quad 1.2145 \quad -0.63393]\mathbf{x}_a$. Converting back to the actual control gives $\mathbf{u} = -[0.019968 \quad 0.12145 \quad 0.036607][x_1 \quad x_2 \quad w_s]^T$. When this controller is used with the true fourth-order model, the high-frequency oscillation is replaced by a few cycles of a much lower frequency damped response. The step responses for (1) the true system with full optimal feedback, (2) the true system with static feedback of only the first two states (gains derived from the second-order model), and (3) the true system with the first-order dynamic controller just derived are compared in Figure 14.12. ■

The preceding example illustrates that the frequency weighting concept increases the order of the design model and the resulting controller, although not necessarily to

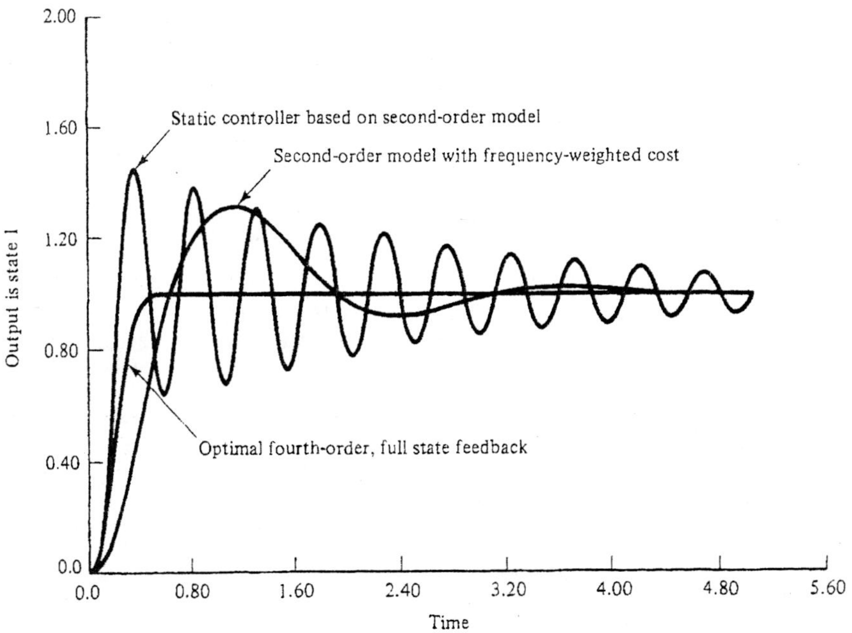

Figure 14.12

the full order of the true model, which is usually unknown. The projective controls approach reduces the order of the implemented controller. These two concepts were introduced for different fundamental reasons, one for controller order reduction and the other for enhancement of system robustness. Both concepts can be combined into a single design methodology [20].

9 CONCLUDING COMMENTS

Dynamic programming can be applied to a wide variety of optimization problems. An optimal assignment problem is described in Reference 21, and applications to control of systems described by partial differential equations are described in Reference 22. In principle, almost any optimal control problem can be solved numerically by using the dynamic programming approach. Let time be divided into a sufficiently small set of discrete points, and also quantize the allowable range of each state variable into a finite set of points. The set of allowable controls at each state is also quantized to a finite number of discrete possibilities. This means that the state-time space is then covered by a multidimensional mesh of node points similar to those in Figures 14.1 through 14.3 and in Problem 14.2. The solution process will normally require an interpolation step because an arbitrarily quantized set of controls at one time step will not necessarily give states at the next time step which fall on the discrete state values. Stated differently, the best control at a given state-time node will frequently fall between two of the discrete controls. Note that in this purely numerical approach, bounds on admissible controls and state variables are actually helpful, since they limit the range of values that need to be quantized and searched over. The biggest limitation to this method is what Bellman [1] called the "curse of dimensionality." If there are n state variables and if each is quantized into 100 points, there will be $(100)^n$ grid points for each time step. At each of these, the best control $\mathbf{u}[\mathbf{x}(k)]$ and the best cost $g[\mathbf{x}(k)]$ must be found (usually by direct enumeration) and stored. Computer storage and solution time become the limiting factors.

In terms of analytical solutions, dynamic programming yields the Hamilton-Jacobi-Bellman equation (*14.35*). This is a nonlinear partial differential equation, and general closed-form solutions are not known. When the set of admissible controls and states are not constrained, the minimization required in the HJB equation can be explicitly attempted. If the system is linear and if the cost is quadratic, this approach leads to the Riccati equation, which is still nonlinear but is an ordinary differential equation. The Riccati equation can be reduced to a coupled pair of linear differential equations, essentially equivalent to those obtained using the minimum principle. In the constant coefficient case, a closed-form solution can be written in terms of the exponential matrix. The infinite time-to-go solution can be found by solving the simpler algebraic Riccati equation, and the optimal control is expressed in terms of state feedback with constant gains.

Major emphasis here has been placed on the linear-quadratic problem and some interesting extensions to it. Notice that even here, computational aides are essential. Even a lowly second-order example becomes a fourth-order Hamiltonian system, which is not conducive to hand solution. In solving optimal control problems, it seems not to be a question of whether a computer will be used but how it will be used.

Tools for solving the Riccati equation are widely available. Solutions also can be obtained by using standard eigenvalue-eigenvector or numerical integration packages, which are even more widely available. Thus the mechanics of basic *LQ* theory can be applied by the controls practitioner in an easy fashion. Many extensions to the basic theory, such as those of Sec. 14.8, assume knowledge of and a facility with *LQ* theory as a starting point. Some adaptive and self-tuning control methods contain an inner loop consisting of an *LQ* optimal controller or a pole placement controller. The *LQ* theory is the most widely used aspect of optimal control theory, which is why it was stressed here.

There is a significant learning benefit derived from developing the software tools as needed rather than using canned packages. One reason for providing the many numerical examples here was that they allow the reader to calibrate his or her programs against known results.

REFERENCES

1. Bellman, R. and S. E. Dreyfus: *Applied Dynamic Programming,* Princeton University Press, Princeton, N.J., 1962.
2. Anderson, B. D. O. and J. B. Moore: *Linear Optimal Control,* Prentice Hall, Englewood Cliffs, N.J., 1971.
3. Bryson, A. E. and Y. C. Ho: *Applied Optimal Control,* Blaisdell (Xerox), Waltham, Mass., 1969.
4. Kirk, D. E.: *Optimal Control Theory,* Prentice Hall, Englewood Cliffs, N.J., 1970.
5. Pontryagin, L. S., V. G. Boltyanskii, R. V. Gamkrelidze, and E. F. Mishchenko: *The Mathematical Theory of Optimal Processes,* Interscience: John Wiley, New York, 1962 (translated from Russian edition).
6. Gelfand, I. M. and S. V. Fomin: *Calculus of Variations,* Prentice Hall, Englewood Cliffs, N.J., 1963 (translated from Russian edition).
7. Kalman, R. E.: "When Is a Linear Control System Optimal?" *Trans. of the ASME, Journ. of Basic Engineering,* Vol. 86D, March 1964, pp. 51–60.
8. Safonov, M. G. and M. Athans: "Gain and Phase Margins for Multiloop LQG Regulators," *IEEE Trans. Automatic Control,* Vol. AC-22, No. 2, Apr. 1977, pp. 173–179.
9. Doyle, J.: "Analysis of Feedback Systems with Structured Uncertainties," IEE Proceedings, Part D, Vol. 129, Nov. 1982, pp. 242–250.
10. Morari, M. and E. Zafiriou: *Robust Process Control,* Prentice Hall, Englewood Cliffs, N.J., 1989.
11. Lehtomaki, N. A., N. R. Sandell, Jr., and M. Athans: "Robustness Results on LQG Based Multivariable Control System Designs," *IEEE Trans. Automatic Control,* Vol. AC-26, No. 1, Feb. 1981, pp. 66–74.
12. Friedland, B.: *Control System Design,* McGraw-Hill, New York, 1986.
13. Doyle, J. C. and G. Stein: "Robustness with Observers," *IEEE Trans. Automatic Control,* Vol. AC-24, No. 4, Aug. 1979, pp. 607–611.
14. Doyle, J. C. and G. Stein: "Multivariable Feedback Design: Concepts for a Classical/Modern Synthesis," *IEEE Trans. Automatic Control,* Vol. AC-26, No. 1, Feb. 1981, pp. 4–16.
15. Hopkins, W. E. Jr., J. V. Medanic, and W. R. Perkins: "Output Feedback Pole Placement in the Design of Suboptimal Linear Quadratic Regulator," *Int. Journ. of Control,* Vol. 34, No. 3, 1981, pp. 593–612.

16. Medanic, J. V. and Z. Uskokovic: "The Design of Optimal Output Regulators for Linear Multivariable Systems with Constant Disturbances," *Int. Journ. of Control,* Vol. 37, No. 4, 1983, pp. 809–830.
17. Meo, J. A. C., J. V. Medanic, and W. R. Perkins: "Design of Digital PI+ Dynamic Controllers Using Projective Controls," *Int. Journ. of Control,* Vol. 43, No. 2, 1986, pp. 539–559.
18. Gupta, N. K.: "Frequency-Shaped Cost Functionals: Extension of Linear-Quadratic-Gaussian Design Methods," *AIAA J. Guidance and Control,* Vol. 3, No. 6, Nov.–Dec. 1980, pp. 529–535.
19. Anderson, B. D. O. and Mingori, D. L.: "Use of Frequency Dependence in Linear Quadratic Control Problems to Frequency-Shape Robustness," *AIAA J. Guidance,* Vol. 8, No. 3, May–June 1985, pp. 397–401.
20. Tharp, H. S., J. V. Medanic, and W. R. Perkins: "Parameterization of Frequency Weighting for a Two-Stage Linear Quadratic Regulator Based Design," *Automatic,* Vol. 24, No. 3, 1988, pp. 415–418.
21. Brogan, W. L.: "Algorithm for Ranked Assignments with Applications to Multiobject Tracking," *AIAA J. Guidance, Control and Dynamics,"* Vol. 12, No. 3, May–June 1989, pp. 357–364.
22. Brogan, W. L.: "Optimal Control Theory Applied to Systems Described by Partial Differential Equations," *Advances in Control Systems,* Vol. 6, C. T. Leondes, Ed., Academic Press, New York, 1968.
23. Perkins, W. R. and J. B. Cruz: "Feedback Properties of Linear Regulators," *IEEE Trans. Automatic Control,* Vol. AC-16, No. 6, Dec. 1971, pp. 659–664.
24. Cruz, J. B., J. S. Freudenberg, and D. P. Looze: "A Relationship Between Sensitivity and Stability of Multivariable Feedback Systems," *IEEE Trans. Automatic Control,* Vol. AC-26, No. 1, Feb. 1981, pp. 66–74.
25. Dorato, P.: "A Historical Review of Robust Control," *IEEE Control Systems Magazine,* Vol. 7, No. 2, Apr. 1987, pp. 44–47.
26. Brogan, W. L.: "Computer Control System Design for a Radar Camouflage Production Line," *Ninth Hawaii International Conference on System Sciences,* Jan. 1976.

ILLUSTRATIVE PROBLEMS

Linear, Time-Varying Minimum Norm Problem

1 Consider the general linear time-varying system $\dot{\mathbf{x}}(t) = \mathbf{A}(t)\mathbf{x}(t) + \mathbf{B}(t)\mathbf{u}(t)$. Find the control which transfers the state from $\mathbf{x}(t_0)$ to $\mathbf{x}_d$ at a fixed time t_f while minimizing the control effort, $J = \int_{t_0}^{t_f} \mathbf{u}^T(t)\mathbf{u}(t)\,dt$.

Since at the final time t_f, $\mathbf{x}(t_f)$ must equal $\mathbf{x}_d$, the solution to the state equation must satisfy

$$\mathbf{x}_d - \mathbf{\Phi}(t_f, t_0)\mathbf{x}(t_0) = \int_{t_0}^{t_f} \mathbf{\Phi}(t_f, \tau)\mathbf{B}(\tau)\mathbf{u}(\tau)\,d\tau \tag{1}$$

The right-hand side of equation *(1)* defines a linear operator $\mathcal{A}(\mathbf{u})$ such that $\mathcal{A}: \mathcal{U} \to \Sigma$. The left-hand side is a fixed vector in the state space Σ. The input function $\mathbf{u}_{[t_0, t_f]}$ is an element of the infinite dimensional input function space $\mathcal{U}$. The performance criterion is the norm squared of an element in $\mathcal{U}$. The problem is therefore one of finding the minimum norm solution to a linear operator equation.

Results of Problem 6.26b are directly applicable, $\mathbf{u}^*(t) = \mathcal{A}^*(\mathcal{A}\mathcal{A}^*)^{-1}[\mathbf{x}_d - \mathbf{\Phi}(t_f, t_0)\mathbf{x}(t_0)]$. The adjoint operator $\mathcal{A}^*$ is defined by $\langle \mathcal{A}(\mathbf{u}), \mathbf{v}\rangle_\Sigma = \langle \mathbf{u}, \mathcal{A}^*\mathbf{v}\rangle_{\mathcal{U}}$. Assuming $\mathbf{\Phi}$, $\mathbf{B}$, and $\mathbf{u}$ are real, the results of Sec. 11.6 give $\mathcal{A}^* = \mathbf{B}^T(t)\mathbf{\Phi}^T(t_f, t)$. The optimal control is

$$\mathbf{u}^*(t) = \mathbf{B}^T(t)\mathbf{\Phi}^T(t_f, t)\left[\int_{t_0}^{t_f} \mathbf{\Phi}(t_f, \tau)\mathbf{B}(\tau)\mathbf{B}^T(\tau)\mathbf{\Phi}^T(t_f, \tau)\, d\tau\right]^{-1} [\mathbf{x}_d - \mathbf{\Phi}(t_f, t_0)\mathbf{x}(t_0)]$$

Note that the indicated inverse exists if the system is controllable. By making use of $\mathbf{\Phi}(t_f, \tau) = \mathbf{\Phi}(t_f, t_0)\mathbf{\Phi}(t_0, \tau)$, the solution can be expressed in alternative but equivalent forms.

Other problems can be solved in the same way whenever the performance criterion J can be interpreted as some more general norm of a function $\mathbf{u}(t) \in \mathcal{U}$.

Discrete Dynamic Programming

14.2 Find the minimum cost path which starts at point a and ends at any one of the points h, j, k, l of Figure 14.13. Travel costs are shown beside each path segment and toll charges are shown by each node.

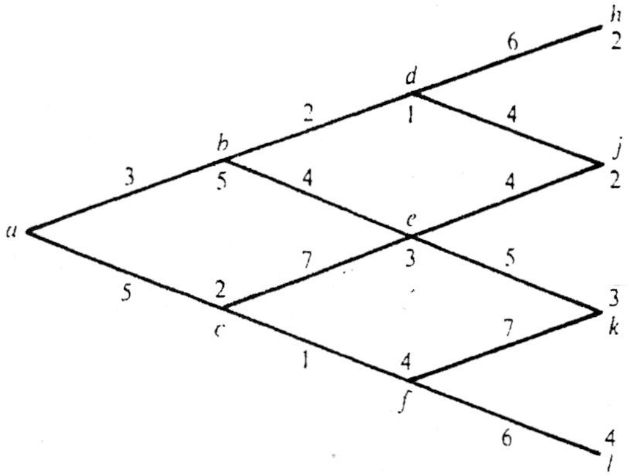

Figure 14.13

Let g_α be the minimum cost from node α to any one of the four possible termination points.

Last stage: $g_h = 2 \quad g_k = 3$
$g_j = 2 \quad g_l = 4$

Next stage: $g_d = 1 + \min[6 + g_h, 4 + g_j]$
$= 1 + \min[6 + 2, 4 + 2] = 7,$ from d to j
$g_e = 3 + \min[4 + g_j, 5 + g_k] = 9,$ from e to j
$g_f = 4 + \min[7 + g_k, 6 + g_l] = 14,$ from f to k or l

Next stage: $g_b = 5 + \min[2 + g_d, 4 + g_e] = 14,$ from b to d
$g_c = 2 + \min[7 + g_e, 1 + g_f] = 17,$ from c to f

Initial stage: $g_a = 0 + \min[3 + g_b, 5 + g_c] = 17,$ from a to b

The minimum total cost is $g_a = 17$ and is achieved on path a, b, d, j.

14.3 A two-stage discrete-time system is described by $x(k+1) = x(k) + u(k)$ with $x(0) = 10$. Use dynamic programming to find $u(0)$ and $u(1)$ which minimize $J = [x(2) - 20]^2 + \sum_{k=0}^{1} \{x^2(k) + u^2(k)\}$. There are no restrictions on $u(k)$.

$g[x(k)]$ is defined as the minimum cost from state $x(k)$ at time k to some final state $x(2)$. Then $g[x(2)] = [x(2) - 20]^2$ and

$$g[x(1)] = \min_{u(1)} \{x^2(1) + u(1)^2 + g[x(2)]\}$$
$$= \min_{u(1)} \{x^2(1) + u^2(1) + [x(1) + u(1) - 20]^2\}$$

Differentiating with respect to $u(1)$ and equating to zero gives $u(1) = [20 - x(1)]/2$, so that $g[x(1)] = x^2(1) + [20 - x(1)]^2/2$. Using this in $g[x(0)] = \min_{u(0)}\{x^2(0) + u^2(0) + g[x(1)]\}$ and setting $\partial\{\ \}/\partial u(0) = 0$ gives $u(0) = [20 - 3x(0)]/5$. This completes the backward pass. Since it is known that $x(0) = 10, u(0) = -2$. This, plus the difference equation, yields $x(1) = 8$. Using this in the expression found for $u(1)$ gives $u(1) = 6$. Finally, $x(2) = x(1) + u(1) = 14$. The minimum cost is $g[x(0)] = 240$.

Apply the general recursive algorithm of Sec. 14.3.3 to recalculate the solution of Problem 14.3.

For this case $\mathbf{A} = \mathbf{B} = \mathbf{M} = \mathbf{Q} = \mathbf{R} = 1$ and $\boldsymbol{\eta}(k) = 0$. The final time is $N = 2$. The initial conditions at $k = 2$ (or equivalently $k' = 0$) are $\mathbf{M}(0) = 1$ and $\mathbf{V}(0) = -20$. The calculations are tabulated from left to right in the order performed. The table also indicates the correct starting points and time sequencing relationships among the variables. The extra quantities $\mathbf{U}$ and $(\mathbf{I} - \mathbf{KB}^T)$ are given for convenience since they are both used more than once per line of table

k	$k' = N - k$	$\mathbf{u}(k) = -\mathbf{K}^T\mathbf{A}\mathbf{x}(k) - \mathbf{U}\mathbf{B}^T\mathbf{V}$	V	M	$\mathbf{U} = [\mathbf{B}^T\mathbf{M}\mathbf{B} + \mathbf{R}]^{-1}$	$\mathbf{K} = \mathbf{M}\mathbf{B}\mathbf{U}$	$(\mathbf{I} - \mathbf{K}\mathbf{B}^T)$	P
2	0	—	−20	1	1/2	1/2	1/2	1/2
1	1	$\mathbf{u}(1) = -1/2\mathbf{x}(1) + 10$	−10	3/2	2/5	3/5	2/5	3/5
0	2	$\mathbf{u}(0) = -3/5\mathbf{x}(0) + 4$	−4	8/5	—	—	—	—

entries in the optimal tracking problem. Once $k = 0$ is reached on the backward pass, the knowledge that $\mathbf{x}(0) = 10$ is used to give $\mathbf{u}(0) = -\frac{3}{5}(10) + 4 = -2$. This control gives $\mathbf{x}(1) = 8$ and then $\mathbf{u}(1) = -\frac{1}{2}(8) + 10 = 6$. The final state is $\mathbf{x}(2) = \mathbf{x}(1) + \mathbf{u}(1) = 14$. These all agree with Problem 14.3.

The general mechanization of the optimal controller for the optimal tracker is given in Figure 14.3.

5 Figure 14.14a gives a schematic for an industrial system [26] typical of many in the pulp and paper industries. A linearized model of the fluidic control aspects will be treated here. The three major system elements are (1) the variable speed pump, which controls total flow volume, (2) the dilution valve, which determines the fraction of dilution water to concentrate, and (3) the head box, which acts as a reservoir and applies the mixture to a moving substrate. Each of these can be approximated by a first-order time lag. This leads to the block diagram of Figure 14.14b. Use the values $K = 10$, $\tau_h = 2$, $\tau_p = 1$, and $\tau_v = 0.5$ and obtain a discrete-time state model for a sample time $T = 0.2$. Design a constant-gain LQ feedback controller which attempts to maintain the state near $\eta = [4 \quad 0.6 \quad 1]^T$. Maintaining the primary output x_1 near 4 is most important. It is also important to keep u_1 small to avoid valve saturation.

The continuous-time state model is

$$\dot{\mathbf{x}} = \begin{bmatrix} -1/\tau_h & -K/\tau_h & K/\tau_h \\ 0 & -1/\tau_v & 0 \\ 0 & 0 & -1/\tau_p \end{bmatrix}\mathbf{x} + \begin{bmatrix} 0 & 0 \\ 1/\tau_v & 0 \\ 0 & 1/\tau_p \end{bmatrix}\mathbf{u}$$

When numerical values are substituted, Eq. (*9.21*) and Problem 9.10 give the discrete model

$$\mathbf{x}(k+1) = \begin{bmatrix} 0.904837 & -0.781725 & 0.861067 \\ 0 & 0.67032 & 0 \\ 0 & 0 & 0.818731 \end{bmatrix}\mathbf{x}(k) + \begin{bmatrix} -0.16990 & 0.09056 \\ 0.32968 & 0 \\ 0 & 0.18127 \end{bmatrix}\mathbf{u}(k)$$

A complete parametric study of the effect of Q and R values is not possible here. Three sets are presented. The value of Q_{22} was successively increased in an attempt to force a positive value for x_2. The DARE was solved to find the matrix $\mathbf{W}_\infty$. Then the gains $\mathbf{G}_\infty$, and $\mathbf{F}_{c\infty}$, and the input $\mathbf{V}_\infty$ is calculated from Eq. (*14.31*). The total external input $\mathbf{v}_{ext} = \mathbf{F}_{c\infty}\mathbf{V}_\infty$ is tabulated along with $\mathbf{G}_\infty$.

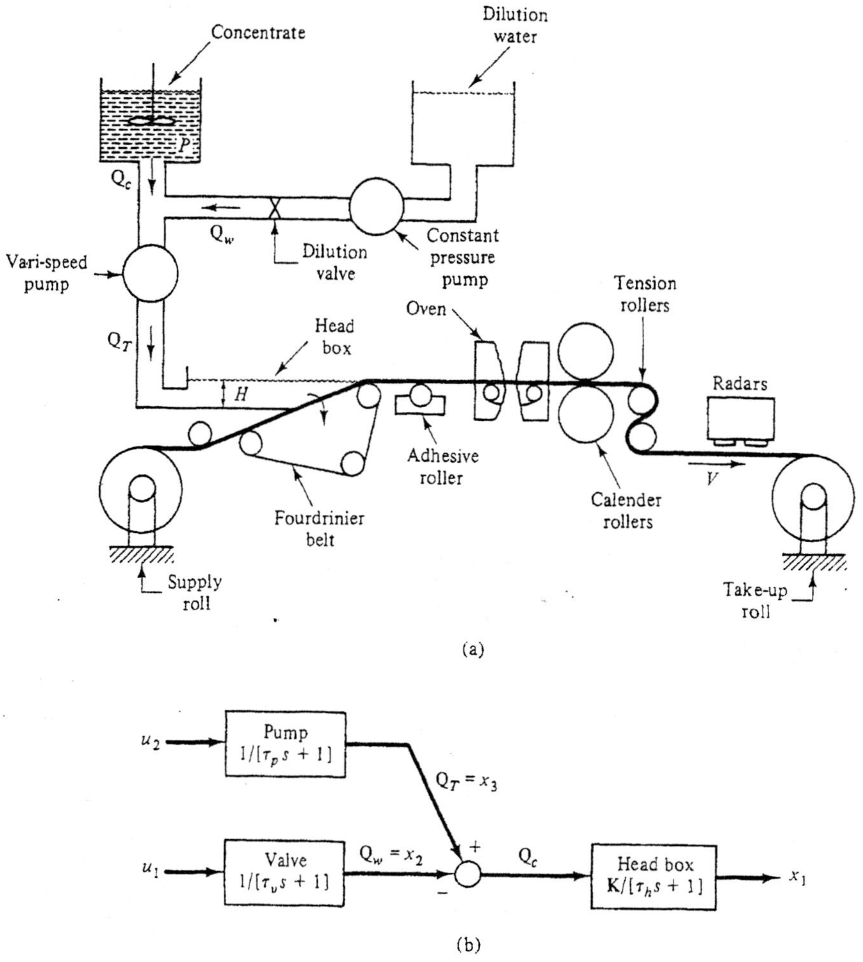

Figure 14.14

Case 1: $\mathbf{Q} = \text{Diag}[10, 1, 1]$, $\mathbf{R} = \text{Diag}[10, 4]$

$$\mathbf{v}_{\text{ext}} = \begin{bmatrix} -2.035 \\ 3.2415 \end{bmatrix} \qquad \mathbf{G}_\infty = \begin{bmatrix} -0.41905 & 0.71697 & -0.85416 \\ 0.57808 & -1.0044 & 1.4288 \end{bmatrix}$$

Case 2: $\mathbf{Q} = \text{Diag}[10, 4, 2]$, $\mathbf{R} = \text{Diag}[10, 4]$

$$\mathbf{v}_{\text{ext}} = \begin{bmatrix} -1.8476 \\ 3.5009 \end{bmatrix} \qquad \mathbf{G}_\infty = \begin{bmatrix} -0.40645 & 0.7620 & -0.80626 \\ 0.58728 & -0.96201 & 1.5118 \end{bmatrix}$$

Case 3: $\mathbf{Q} = \text{Diag}[10, 8, 2]$, $\mathbf{R} = \text{Diag}[10, 4]$

$$\mathbf{v}_{\text{ext}} = \begin{bmatrix} -1.6562 \\ 3.7437 \end{bmatrix} \qquad \mathbf{G}_\infty = \begin{bmatrix} -0.38764 & 0.80754 & -0.76125 \\ 0.60643 & -0.93136 & 1.5725 \end{bmatrix}$$

The transient responses for these cases are presented in Figure 14.15. The responses of x_1 are almost indistinguishable, but the effect of $\mathbf{Q}$ on x_2 and x_3 is clearly evident.

14.6 In order to compare quadratic optimal controllers with earlier methods, the dc motor system of Example 2.6 and Problems 2.22 through 2.24 is reconsidered. Using a sample time of $T = 1$ and

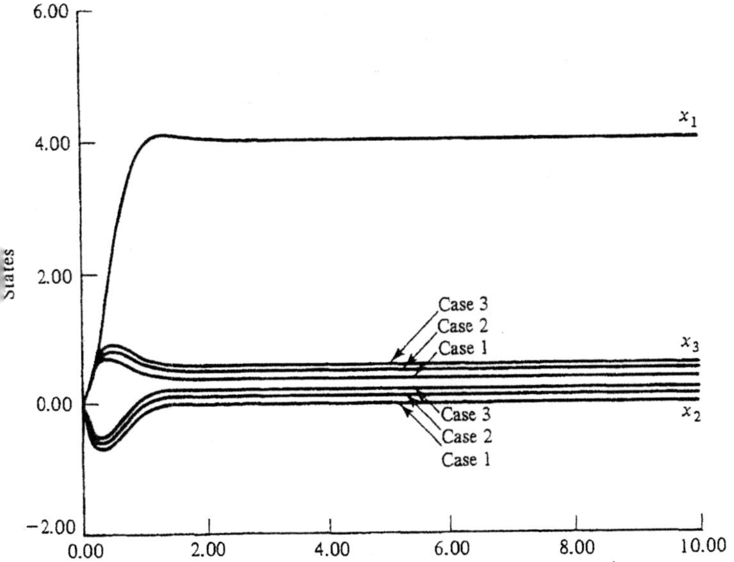

Figure 14.15

a motor gain of $K = 0.5$ allows the following observable canonical form state equations to be written for the open-loop system.

$$\mathbf{x}(k+1) = \begin{bmatrix} 1.6065 & 1 \\ -0.6065 & 0 \end{bmatrix} \mathbf{x}(k) + \begin{bmatrix} 0.1065 \\ 0.0902 \end{bmatrix} E(k)$$

The first component $\mathbf{x}_1(k)$ is the motor shaft position angle, and its measurement is assumed available for use. It can be seen that the second state can be determined as $\mathbf{x}_2(k) = -0.6065\mathbf{x}_1(k-1) + 0.0902E(k-1)$. Therefore, it is justifiable to assume that both states are available for use. Design a controller which minimizes the cost function J of Eq. (*14.18*).

Equations (*14.15*), (*14.16*), and (*14.17*), plus equations (*14.21*) and (*14.22*) were solved with $N = 15$ steps of $T = 1$ s each. The first group of cases studied all had $\boldsymbol{\eta}(k) = \mathbf{0}$ and $\mathbf{x}_d = \mathbf{0}$ (i.e., the regulator problem). For each case the steady-state feedback gain matrix is listed, along with the closed-loop eigenvalue locations. In each case steady state was reached in about seven steps, a little over three motor time constants.

Case	M	Q	R	$\mathbf{G}_{ss} = \mathbf{K}^T\mathbf{A}$	Closed-loop Eigenvalues
1	I	I	1	[2.106 1.9160]	$0.6057 \pm j0.1756$
2	0	I	1	[2.1060 1.9160]	$0.6057 \pm j0.1756$
3	20I	I	1	[2.1060 1.9160]	$0.6057 \pm j0.1756$
4	I	I	10	[0.8163 0.7834]	0.6305, 0.8184 (real)
5	I	I	0.1	[4.516 3.810]	$0.3909 \pm j0.2505$

From this it is seen that the final results are independent of the initial **M**. For relatively smaller **R** values, larger gains and a faster, more responsive system is obtained. Larger **R** values give slower, smoother response and smaller gains. Smaller motor inputs will be required. The function of the **R** weighting term is to prevent large control signals from being called for. The **Q** weighting term is intended to get the state near to $\boldsymbol{\eta}(k)$.

Another group of cases with $\boldsymbol{\eta}(k) = [1. \quad -0.6065]^T$ was run, and for simplicity **M** was set to zero and $\mathbf{R} = 1$ was held fixed. In addition to the steady-state feedback gains, the steady-state

(a) Move F

(b) Pick normalizing $\alpha = \dfrac{1}{v_{s.s.}}$

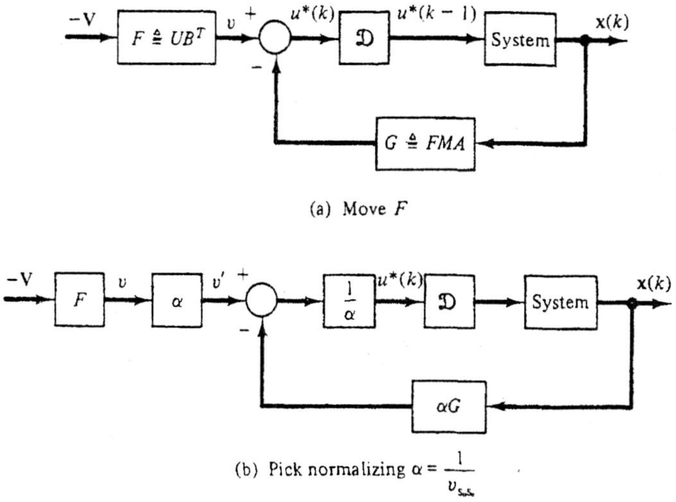

Figure 14.16

values of the external inputs $v = -UB' V$ are given, along with the scalar closed-loop transfer function $H(z)$ from $v(z)$ to $x_1(z)$ (normalized as described below). The eigenvalues are, of course, the denominator roots. The increase in Q is equivalent to a relative decrease in R. The previous observation remains true: as Q gets bigger relative to R the system response gets faster, but at the expense of larger control effort. The role of the external input needs to be clarified. This input V is a two-component vector. By shifting gains around, an equivalent normalized scalar input system can be obtained from Figure 14.4. The two steps used in shifting and normalizing are shown in Figure 14.16. The transfer functions given above are from v' to x. The final value theorem shows that each transfer function has a steady-state gain of one, meaning that there will be no steady state error in the position x_1 after a unit input at v'. This means that in steady state Gx must equal v.

Case	Q	Gain	v	Transfer Function
6	I	[same as case 1]	0.9439	$0.3971(z - 0.5324)/[(z - 0.6057 \pm j0.1756)]$
7	100I	[7.1525 5.6840]	3.7052	$1.2744(z - 0.4281)/[(z - 0.166 \pm j0.182)]$
8	10000I	[8.3299 6.4773]	4.4013	$1.4714(z - 0.4116)/[(z - 0.0068)(z - 0.1283)]$

The desired state $\eta(k)$ above happens to be a point of equilibrium where x_1 and x_2 are in balance when the input E is zero. Two more cases are solved where this is not true. x_1 is selected as unity and x_2 is chosen arbitrarily. In case 9 the errors in both states are equally weighted, and in case 10 the error in x_2 is ignored so η_2 could have been any value without changing the results. (This is not to say that the solution does not depend on Q_{22}, however.)

We see from the preceding results that the closed-loop system poles can be moved around by changing the relative importance of Q and R. If any one of the above sets of eigenvalues had been selected a priori, then the pole-placement methods of Chapter 13 would give the same feedback gain. Sometimes it may be more natural to select pole locations based on their relation

Case	Q	Gain	v	Transfer Function
9	100I	[7.1525 5.684]	1.06	$1.2766(z - 0.4260)/[z - 0.1650 \pm j0.1883]$
10	Diag[100, 0]	[7.7430 5.696]	4.288	$1.3384(z - 0.3698)/[z - 0.1340 \pm j0.3060]$

to a desired time response. Pole placement would easily give the gains. Sometimes it may seem more natural to express the design goals in terms of a cost function to be minimized. In either case, if extremes are requested, the magnitudes of the resultant control signals may become excessive or even ridiculous. It is not possible to make a turbogenerator set, which normally takes minutes to come up to rated speed, respond with a time constant in the microsecond range just because "the math in Chapter 13 says you can put the poles anywhere you want." Neither can state variable feedback transform a light pleasure aircraft into a high-performance military fighter. The fallacy is due to the validity of the linear models breaking down and due to state or control limits being exceeded. The linear models are intended for use over a limited range of values of the signals. Conductors tend to vaporize when their rated loads are drastically exceeded. Wings or control surfaces tend to get ripped off under excessive loads. Similar limitations exist in other situations.

The step responses for the designs of cases 6, 7, 8, and 10 are shown in Figure 14.17 for comparison with the Chapter 2 results. Perhaps case 7 gives the best response of any considered here or in Chapter 2. Further parametric studies may uncover a better response between case 6 and case 7.

Find the optimal feedback control law for the unstable scalar system $\dot{x} = x + u$ which minimizes $J = Mx(t_f)^2 + \int_0^{t_f} u(t)^2\,dt$. There are no restrictions on $u(t)$, and t_f is fixed.

Using the dynamic programming results of Sec. 14.4.1 with $A = 1$, $B = 1$, $x_d = 0$, $Q = 0$, $R = 1$, the optimal control is $u^*(t) = \frac{1}{2}\nabla_x g[x(t), t_r]$, where $\nabla_x g = 2W(t_r)x(t) + V(t_r)$.

The Riccati equation for $W(t_r)$ is $dW/dt_r = 2W - W^2$, with $W(t_r = 0) = M$.

The equation for $V(t_r)$ is $dV/dt_r = [1 - W(t_r)]V$, but since $V(t_r = 0) = 0$, $V(t_r) \equiv 0$ for all t_r.

Using $W(t_r) = E(t_r)/F(t_r)$ for this scalar case leads to $dE/dt_r = E$, $dF/dt_r = E - F$, with $E(t_r = 0) = M$, $F(t_r = 0) = 1$. Solving gives $E(t_r) = e^{t_r}M$ and $F(t_r) = e^{-t_r} + (e^{t_r} - e^{-t_r})M/2$.

Since $W(t_r) = E(t_r)/F(t_r)$, it is seen that $W \to 2$ for large t_f. This in turn gives $u^*(t) = -2x(t)$ and the stable closed-loop system $\dot{x} = -x(t)$.

The general feedback control system is given in Figure 14.18.

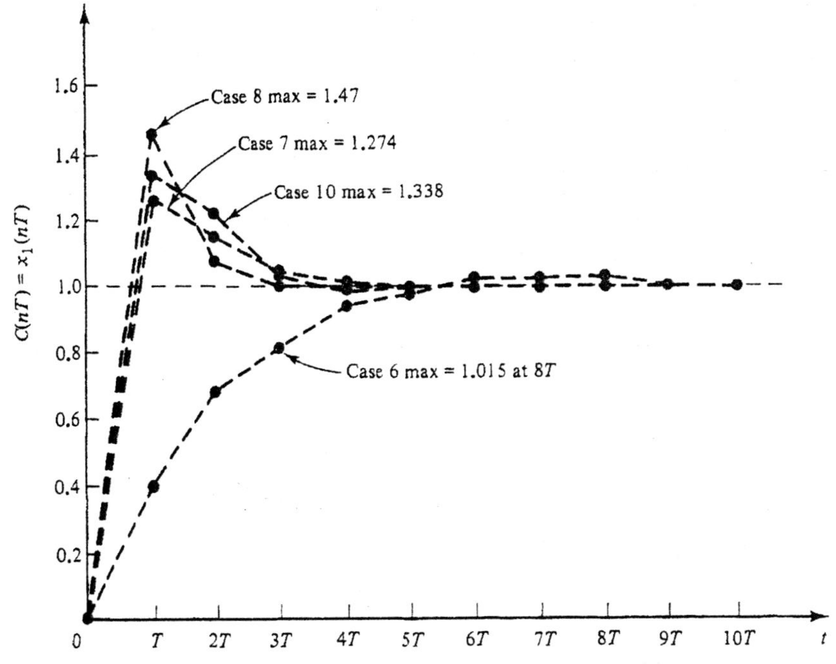

Figure 14.17 (Compare with Figures 2.29, 2.30, and 2.32.)

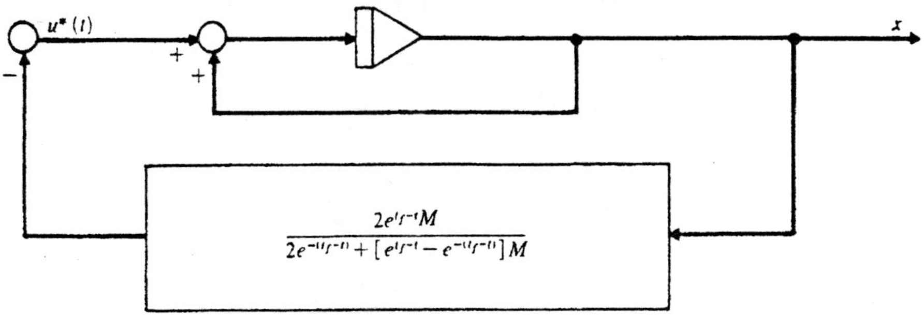

Figure 14.18

14.8 Consider the scalar control system $\dot{x} = Ax + u$. Investigate the constant state feedback control system which minimizes

$$J = x(t_f)^2 M + \int_0^{t_f} \{Qx^2 + u^2\}\, dt \qquad \text{as } t_f \to \infty$$

What is $u^*(t)$ if $Q = 0$? What if $Q = -A^2$?

The Riccati equation is $dW/dt_r = Q + 2WA - W^2$. In general, the steady-state solution can be found by numerically integrating until steady state is reached, or by setting the derivative equal to zero and solving a nonlinear algebraic equation for W. In this scalar case, setting $dW/dt_r = 0$ gives a simple quadratic equation for the steady-state values of W. Its solutions are $W = A \pm \sqrt{A^2 + Q}$.

Since $V(t_r) = 0$, $u^*(t) = -W(t_r)x(t)$, the steady-state feedback system satisfies $\dot{x} = Ax - [A \pm \sqrt{A^2 + Q}]x = \mp\sqrt{A^2 + Qx}$.

For a stable system, the plus sign in W must be selected, giving $\dot{x} = -\sqrt{A^2 + Qx}$. If $Q = 0$, $\dot{x} = -|A|\mathbf{x}$. If the original system was stable, $-|A| = A$, so this implies that $u^* = 0$. But $Q = 0$ means that it does not matter what $x(t)$ is, so the optimal control strategy is to do nothing. This gives a zero value to the integral part of J. Also, since $t_f \to \infty$, $x(t_f) \to 0$ for a stable system. If the original system is unstable, the feedback system is stabilized. If $Q = -A^2$, then $W = A$ in steady state, so $u^*(t) = -Ax$. This gives $\dot{x} = 0$, and the integral term in J is once again identically zero.

14.9 A building is divided into two zones, as shown in Figure 14.19a. Figure 14.19b shows the analogous electric network, using the through variable/across variable analogies of Sec. 1.3. A furnace with heat output q_f is located in zone 1. If T_0 is the ambient temperature and T_1 and T_2 are the zone temperatures, the heat flow equations can be written as

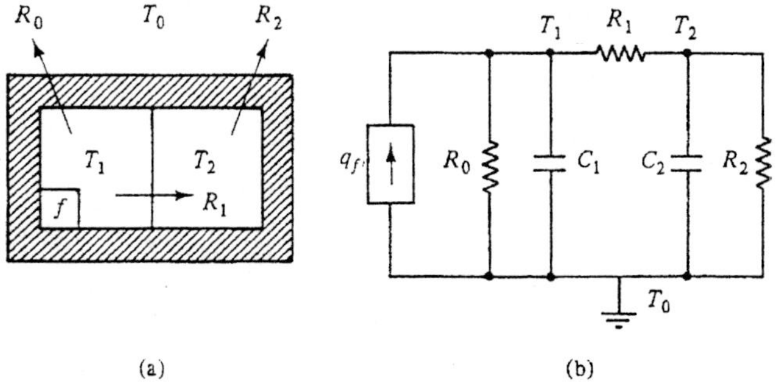

Figure 14.19

$$C_1\dot{T}_1 = q_f - (T_1 - T_0)/R_0 - (T_1 - T_2)/R_1$$

$$C_2\dot{T}_2 = (T_1 - T_2)/R_1 - (T_2 - T_0)/R_1$$

where C_i is the thermal capacitance of zone i. R_i are the lumped thermal resistances shown in Figure 14.19*a*. Let the control u be the furnace heat input q_f and let x_1 and x_2 be zone temperatures above ambient. The state equations become

$$\dot{\mathbf{x}} = \begin{bmatrix} -1/C_1[1/R_0 + 1/R_1] & 1/(C_1R_1) \\ 1/(C_2R_1) & -1/C_2[1/R_1 + 1/R_2] \end{bmatrix}\mathbf{x} + \begin{bmatrix} 1/C_1 \\ 0 \end{bmatrix}u$$

(a) Design a constant gain full state LQ controller which minimizes

$$J = \int_0^\infty \{(\mathbf{x} - \boldsymbol{\eta})^T\mathbf{Q}(\mathbf{x} - \boldsymbol{\eta}) + \mathbf{R}u^2\}\, dt.$$

where $\boldsymbol{\eta} = [1 \quad 1]^T$ is the vector of desired zone temperatures above ambient, i.e., the set points. For simplicity, let $C_1 = C_2 = 1$, $R_0 = 10$, $R_1 = .1$, and $R_2 = 2$. Let $\mathbf{R} = 1$ and consider $\mathbf{Q} = \alpha\mathbf{I}$ for four values of α: .1, 1, 10, and 100.

(b) Show that if only one state is a measured output, an equivalent feedback transfer function $\mathbf{H}_{eq}(s)$ can be found which gives the same feedback signal as does full state feedback through the gain matrix G_∞. In particular, if the thermostat measurement is in zone 2, H_{eq} is a PD-(proportional-derivative-) type lead compensator. If the thermostat is in zone 1, then H_{eq} is a lag compensator.

(a) For this tracking problem, an external input is required in addition to the feedback gains. Setting $\dot{V} = 0$ in Eq. (*14.38*) and solving for $\mathbf{V}$ gives

$$\mathbf{V} = 2[\mathbf{A}^T - \mathbf{W}\mathbf{B}\mathbf{R}^{-1}\mathbf{B}^T]^{-1}\mathbf{Q}\boldsymbol{\eta}$$

and since $\mathbf{F}_\infty = -(\frac{1}{2})\mathbf{R}^{-1}\mathbf{B}^T$, the composite term is

$$v_{ext} = \mathbf{F}_\infty\mathbf{V} = -\mathbf{G}_\infty[\mathbf{A}^T\mathbf{W} - \mathbf{W}\mathbf{B}\mathbf{R}^{-1}\mathbf{B}^T\mathbf{W}]^{-1}\mathbf{Q}\boldsymbol{\eta}$$

$$= \mathbf{G}_\infty[\mathbf{W}\mathbf{A} + \mathbf{Q}]^{-1}\mathbf{Q}\boldsymbol{\eta}$$

The last form used the ARE. It shows that if $\mathbf{Q}$ was large compared with $\mathbf{WA}$, then the control signal u depends on the total error $\mathbf{x} - \boldsymbol{\eta}$ as in many error-nulling servo systems; i.e., $u = \mathbf{G}_\infty(\boldsymbol{\eta} - \mathbf{x})$. This situation does not happen exactly because as $\mathbf{Q}$ increases, so does $\mathbf{W}$. The key results for the four values of $\mathbf{Q}$ are as follows:

α	$\mathbf{G}_\infty$	v_{ext}	λ_{cl}
0.1	[0.077102 0.07317]	0.27004	−0.374, −20.30
1.0	[0.48751 0.45427]	1.30478	−0.773, −20.30
10.0	[2.0883 1.8273]	4.43	−2.2645, −20.42
100.0	[7.7286 5.7925]	14.126	−6.714, −21.61

(b) Since all states may not be available as measured outputs, an observer may be needed to implement this controller. An alternative approximate method can often be used with single-input, single-output systems. The Laplace transform of the feedback signal under full state feedback is

$$F(s) = \mathbf{G}_\infty\mathbf{x}(s) = \mathbf{G}_\infty(s\mathbf{I} - \mathbf{A})^{-1}\mathbf{B}u(s)$$

If instead, $y(s)$ is fed back through a transfer function $H(s)$, the feedback signal is

$$F(s) = H(s)y(s) = Hs)\mathbf{C}\mathbf{x}(s) = H(s)\mathbf{C}(s\mathbf{I} - \mathbf{A})^{-1}\mathbf{B}u(s)$$

Equating gives $H(s) = \{\mathbf{G}_\infty(s\mathbf{I} - \mathbf{A})^{-1}\mathbf{B}\}/\{\mathbf{C}(s\mathbf{I} - \mathbf{A})^{-1}\mathbf{B}\}$. For this problem, if the thermostat is in zone 2, $\mathbf{C} = [0 \quad 1]$, and then $\mathbf{C}(s\mathbf{I} - \mathbf{A})^{-1}\mathbf{B} = 10/[(s + 0.298)(s + 20.302)]$; for the case $\alpha = 10$, $\mathbf{G}_\infty(s\mathbf{I} - \mathbf{A})^{-1}\mathbf{B} = 2.088(s + 19.25)/[(s + 0.298)(s + 20.302)]$, so that $H(s) = 0.2088(s + 19.25)$. Since this is not a proper transfer function—that is, the numerator degree is higher than the denominator—it is not physically realizable. It is clearly of the PD-type lead compensator. A small time constant τ, say $\tau = 0.01$, could be used to form an approximation which is physically realizable, $H(s) \approx 0.2088(s + 19.25)/(\tau s + 1)$. This can be shown to give a response due to the v_{ext} step input, which is very close to the full state feedback response. If the thermostat is located in zone 1, then $\mathbf{C} = [1 \quad 0]$ leads to the physically realizable transfer function $H(s) = 2.088(s + 19.25)/(s + 10)$. This is recognized as a lag compensator.

14.10 The longitudinal dynamics of an aircraft cruising at constant speed can be approximated by [3, pp. 171–172] (see Figure 14.20)

$$\dot{\theta} = q$$

$$\dot{q} = -\omega^2(\alpha - S\delta)$$

$$\dot{\alpha} = -\alpha/\tau + q$$

Assume the values $\tau = 0.25$, $\omega = 2.5$, and $S = 1.6$. Then

$$\dot{\mathbf{x}} = \begin{bmatrix} 0 & 1 & 0 \\ 0 & 0 & -6.25 \\ 0 & 1 & -4 \end{bmatrix}\mathbf{x} + \begin{bmatrix} 0 \\ 10 \\ 0 \end{bmatrix}u$$

Design a constant feedback LQ controller which attempts to maintain the velocity vector near horizontal while keeping control effort small. That is, it is desired to minimize

$$J = \int_0^\infty \{Q(\theta - \alpha)^2 + R\delta^2\}\,dt \qquad \textit{Note:} \text{ The matrix } \mathbf{Q} = Q\begin{bmatrix} 1 & 0 & -1 \\ 0 & 0 & 0 \\ -1 & 0 & 1 \end{bmatrix}$$

Select a Q/R ratio that gives fast response to a sudden 10° change in α due to a gust or wind shear but that requires a maximum $|\delta| \le 10°$.

In this case Q and R are scalars and only the ratio is important. By trial and error, it is found that a value $Q/R = 2.8$ is the maximum permissible value—i.e., the fastest response—for which δ stays within the specified limit for an initial perturbation $\mathbf{x}(0) = [10 \quad 0 \quad 10]^T$. The feedback gains for four trial cases are tabulated, and the transient responses for α, θ, and δ are shown in Figure 14.21*a*, *b*, and *c*.

Case	Q/R	$\mathbf{G}_\infty$	λ_{c1}
1	0.1	[0.31623 0.14943 −0.20457]	$-1.982, -1.756 \pm j1.816$
2	1.0	[1 0.33116 −0.45166]	$-3.429, -1.942 \pm j2.810$
3	2.8	[1.6733 0.44895 −0.66554]	$-4.1098, -2.1898 \pm j3.1389$
4	10.0	[3.1623 0.63169 −1.1671]	$-5.0872, -2.6148 \pm j4.2458$

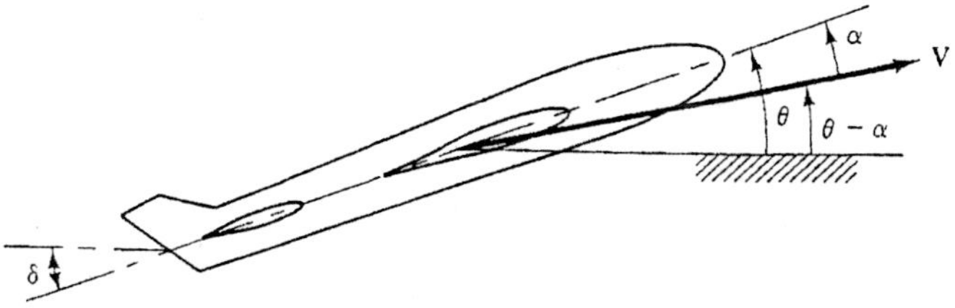

Figure 14.20

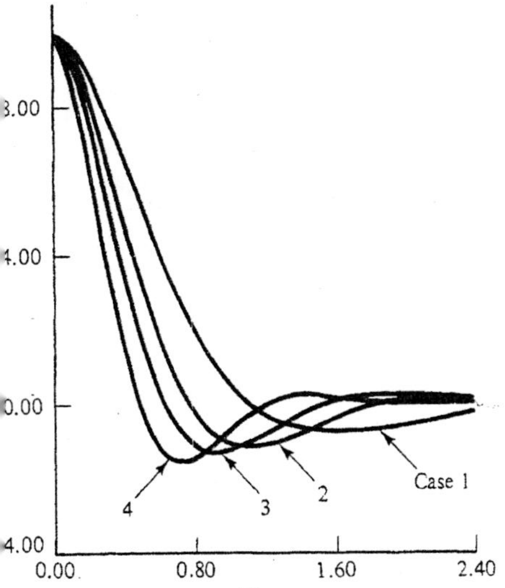

(a) Pitch attitude response for various Q/R ratios

Figure 14.21a

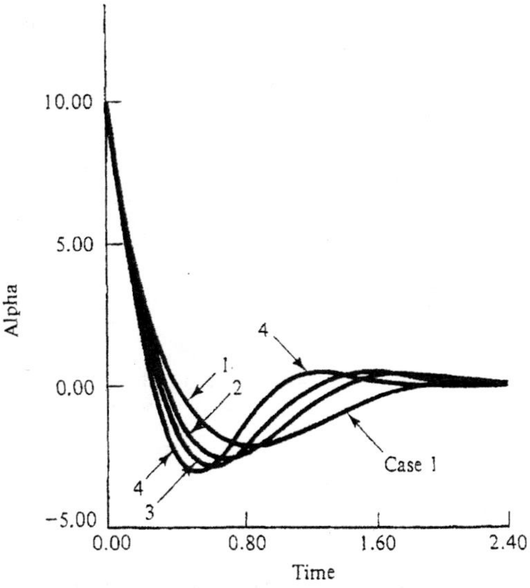

(b) Angle of attack for various Q/R ratios

Figure 14.21b

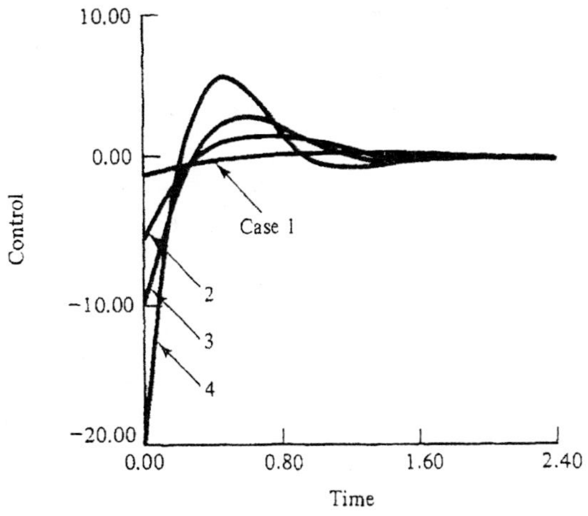

(c) Elevator deflection history for various Q/R ratios

Figure 14.21c

Applying the result of Problem 14.9 for an equivalent feedback transfer function to Case 3 and assuming that θ is the measured output gives

$$H(s) = \frac{4.4895(s + 3.122 + j2.2715)(s + 3.122 - j2.2715)}{s + 4}$$

This is not physically realizable (is not proper), since the numerator is of higher degree than the denominator. It may be acceptable to add a small time constant term $(\tau s + 1)$ in the denominator to give a realizable feedback transfer function. If τ is sufficiently small (i.e., the added pole is sufficiently far to the left), the contribution of the extra mode may be negligible.

14.11 A linearized aircraft model was introduced in Problem 11.9, where its controllability and observability properties were examined. Its transfer functions were obtained in Problem 12.5 and pole placement feedback controllers for it were found in Problem 13.35. Design a controller using the steady-state optimal regulator approach. For convenience, the system models are repeated here.

$$\mathbf{A} = \begin{bmatrix} -10 & 0 & -10 & 0 \\ 0 & -0.7 & 9 & 0 \\ 0 & -1 & -0.7 & 0 \\ 1 & 0 & 0 & 0 \end{bmatrix} \qquad \mathbf{B} = \begin{bmatrix} 20 & 2.8 \\ 0 & -3.13 \\ 0 & 0 \\ 0 & 0 \end{bmatrix}$$

The solution involves selecting the appropriate weighting matrices **M**, **Q**, and **R** and then solving the Riccati Equation 14.39 until steady state is reached. The controller is given by

$$\mathbf{u}^*(t) = -(\tfrac{1}{2})\mathbf{R}^{-1}\mathbf{B}^T[2\mathbf{W}\mathbf{x}(t) + \mathbf{V}]$$

as shown in Figure 14.8. Two issues remain. What method is to be used to solve the Riccati equation? Numerical integration of Eq. (*14.39*) is used until **W** effectively becomes constant. For an asymptotically stable constant coefficient system, a reasonable estimate is that this will occur within four time constants. The time constant is estimated as the reciprocal of the smallest nonzero real part of the eigenvalues of **A**. The eigenvalues of **A** are 0, $-0.7 \pm 3j$, and -10, so the time constant is estimated as $1/0.7 = 1.4$ sec. The integration stepsize ΔT must not be too large, or poor accuracy will result. If it is too small, excessive computer time is required. A method which seems satisfactory is to select ΔT as about $1/(20|\lambda_{max}|)$, which means that the number of integration steps might be on the order of $T/\Delta T = 80|\lambda_{max}|/|\lambda_{min}|$, which can easily exceed 1000. These are just estimates. The magnitudes of the changes in **W** must be monitored to determine when all components have essentially stopped changing.

The second issue is selection of the weighting matrices. For small problems with only a few parameters it may be feasible to parametrically examine the range of possibilities. For most problems a more focused approach is desirable. The expanded quadratic will contain terms of the form $x_i^2 Q_{ii} + u_i^2 R_{ii}$. Similar treatment of the final value terms can be done, but here $M = 0$ is selected because it will have no effect on steady-state answers. If x_i is a position variable with a magnitude of thousands of feet, and if u_i is an angle of say 0.01 radian, it is clear that u_i will have no effect on J unless $R_{ii} >> Q_{ii}$. The point is that scaling units and variable magnitudes are important, as well as the subjective choice of the importance of keeping u_i small compared to keeping x_i small. If all variables in the quadratic cost function are intended to be equally important, then one method is to estimate the maximum possible or allowable values of each variable and use these estimates to select the weights. For the airplane model, suppose that structural limits and prevention of pilot blackout require that the roll rate p be <300 deg/s and the yaw rate r be <18 deg/s. The maximum side-slip angle is estimated as $\beta < 15$ deg and the maximum roll angle as $\phi < 180$ deg. Likewise, there are maximums for the control surface deflection angles, and the hypothetical values assumed here are $\delta_a < 40$ deg and $\delta_r < 10$ deg. Then setting each term in J to unity when the variables are at their limits gives

$$\mathbf{Q} = \text{Diag}[0.0011, 0.308, 0.44, 0.003]$$

$$\mathbf{R} = \text{Diag}[0.0626, 1]$$

The relative magnitudes are all that matter. The above components have been scaled so the largest entry is unity. Considerable rounding off is probably justified because of the gross approximations involved in estimating the maximums. Using the above values for **Q** and **R**, the approximate steady-state feedback gain matrix and the closed-loop eigenvalues it yields are

$$\mathbf{G} = \begin{bmatrix} 0.0358 & 0.0197 & -0.0355 & 0.1942 \\ 0.0001 & -0.3072 & -0.1191 & -0.0001 \end{bmatrix}$$

$$\lambda_i = \{-0.376, -1.181 \pm 2.898j, -10.340\}$$

Cause		x_1	x_2	x_3	x_4
Commands	δ_a	10	0.3	0.5	35
	δ_r	0.03	5.5	1.7	0.01

The resultant system has moved the open-loop pole from the origin to -0.376, which means that the perturbations caused by any step disturbance will decay back to zero. The settling time and damping ratio of the dominant complex poles have been improved considerably and the remaining nondominant pole is not much changed.

If the maximum value of each state variable is multiplied by its two gain values, the resultant commands for δ_a and δ_r can be estimated. These values are valid if only one state variable is at its maximum and all others are zero. This shows that no maximum command is exceeded, unlike the results that can occur if arbitrarily selected **Q** or **R** values are used, or if extreme requests are made of a pole-placement design. In this particular example it is clear that δ_a is primarily controlled by the roll rate x_1 and roll angle x_4, and δ_r is primarily controlled by yaw rate x_2 and side-slip angle x_3 as expected from the physics of the problem.

A series of other sets of Q and R have been analyzed for comparison. With $\mathbf{Q}=\mathbf{I}$ and $\mathbf{R}=\mathbf{I}$,

$$\mathbf{G}=\begin{bmatrix} 0.6587 & 0.0768 & -0.2610 & 0.9909 \\ 0.0802 & -0.7184 & -0.2743 & 0.0729 \end{bmatrix}$$

$$\lambda_i=\{-0.889, -1.82\pm j2.6, -22.5\}$$

The dominant complex poles are about the same as the earlier case and the dominant real pole has an improved settling time, but the major difference is the shift in the nondominant pole, which will not be reflected much in system behavior. The required gains are much higher here and commanded deflection angles could easily exceed their limits.

With $\mathbf{Q}=10\mathbf{I}$ and $\mathbf{R}=0.1\mathbf{I}$ the resultant eigenvalues are all real $\{-0.995, -1.27, -30.7,$ and $-202.2\}$ and the gains are so high as to be ridiculous. Several are on the order of 9 or 10!

With $\mathbf{Q}=\text{Diag}[0.01, 0.1, 0.1, 0.01]$ and $\mathbf{R}=\text{Diag}[0.1, 1]$ the results are

$$\mathbf{G}=\begin{bmatrix} 0.1160 & 0.0275 & -0.0952 & 0.2956 \\ 0.0012 & -0.1148 & -0.0559 & 0.0004 \end{bmatrix}$$

$$\lambda_i=\{-0.5, -0.8798\pm 2.965j, -11.823\}$$

The dominant poles are less damped and have a poorer settling time than the first case in spite of the fact that gains are typically higher and command limits could be exceeded.

The Minimum Principle

2 A system is described by $\dot{\mathbf{x}}=\mathbf{f}(\mathbf{x},\mathbf{u},t)$, with $\mathbf{x}(t_0)$ given. Find the necessary conditions which $\mathbf{x}(t)$ and $\mathbf{u}(t)$ must satisfy if they are to minimize

$$J=S(\mathbf{x}(t_f),t_f)+\int_{t_0}^{t_f} L(\mathbf{x}(t),\mathbf{u}(t),t)\,dt$$

The admissible controls must satisfy $\mathbf{u}(t)\in U$. Assume t_f is fixed.

Let $\mathbf{x}^*(t), \mathbf{u}^*(t)$ be the optimal quantities and let $\mathbf{x}(t)$ be arbitrary and let $\mathbf{u}(t)$ be arbitrary but admissible. Let $\mathbf{p}(t)$ be an $n\times 1$ vector of Lagrange multipliers (also called the *costate* or *adjoint* variables). Adjoin the differential constraints to J and call the result J':

$$J'=S(\mathbf{x}(t_f),t_f)+\int_{t_0}^{t_f}\{L(\mathbf{x},\mathbf{u},t)+\mathbf{p}^T(t)[\mathbf{f}(\mathbf{x},\mathbf{u},t)-\dot{\mathbf{x}}]\}\,dt$$

Since $\mathbf{x}^*, \mathbf{u}^*$ are optimal, $\Delta J = J'(\mathbf{x},\mathbf{u}) - J'(\mathbf{x}^*,\mathbf{u}^*) \geq 0$ for $\mathbf{x}$ arbitrary and $\mathbf{u} \in U$.

Let $\mathbf{x} = \mathbf{x}^* + \delta\mathbf{x}$. Using Taylor series expansion gives

$$J'(\mathbf{x}^* + \delta\mathbf{x}, \mathbf{u}) = S(\mathbf{x}^*(t_f), t_f) + [\nabla_{\mathbf{x}} S]^T|_{t_f}\delta\mathbf{x}(t_f) + \int_{t_0}^{t_f}\Big\{L(\mathbf{x}^*,\mathbf{u},t) + [\nabla_{\mathbf{x}} L(\mathbf{x}^*,\mathbf{u},t)^T]\,\delta\mathbf{x} + \mathbf{p}^T\Big[\mathbf{f}(\mathbf{x}^*,\mathbf{u},t) + \frac{\partial\mathbf{f}}{\partial\mathbf{x}}\delta\mathbf{x} - \dot{\mathbf{x}}^* - \delta\dot{\mathbf{x}}\Big]\Big\}\,dt$$
$$+ \text{higher-order terms}$$

Therefore,

$$\Delta J = (\nabla_{\mathbf{x}} S)^T|_{t_f}\delta\mathbf{x}(t_f) + \int_{t_0}^{t_f}\{L(\mathbf{x}^*,\mathbf{u},t) + \mathbf{p}^T\mathbf{f}(\mathbf{x}^*,\mathbf{u},t) - L(\mathbf{x}^*,\mathbf{u}^*,t) - \mathbf{p}^T\mathbf{f}(\mathbf{x}^*,\mathbf{u}^*,t)\}\,dt + \int_{t_0}^{t_f}\Big\{[\nabla_{\mathbf{x}} L(\mathbf{x}^*,\mathbf{u},t)]^T\delta\mathbf{x} + \mathbf{p}^T\frac{\partial\mathbf{f}}{\partial\mathbf{x}}\delta\mathbf{x} - \mathbf{p}^T\delta\dot{\mathbf{x}}\Big\}\,dt$$
$$+ \text{higher-order terms.}$$

When the higher-order terms are dropped, the result is called δJ, the *first variation of J*. The condition for optimality is that $\Delta J \geq 0$ for arbitrary $\delta\mathbf{x}(t)$ and $\mathbf{u}(t) \in U$. If $\mathbf{x}, \mathbf{u}$ are sufficiently close to $\mathbf{x}^*, \mathbf{u}^*$, then the sign of ΔJ is the same as the sign of δJ. By defining the Hamiltonian as $\mathcal{H}(\mathbf{x}^*,\mathbf{u}^*,\mathbf{p},t) = L(\mathbf{x}^*,\mathbf{u}^*,t) + \mathbf{p}(t)^T\mathbf{f}(\mathbf{x}^*,\mathbf{u}^*,t)$ and using integration by parts, $\int_{t_0}^{t_f}\mathbf{p}^T\delta\dot{\mathbf{x}}\,dt = \mathbf{p}^T\delta\mathbf{x}\Big|_{t_0}^{t_f} - \int_{t_0}^{t_f}\dot{\mathbf{p}}^T\delta\mathbf{x}\,dt$, we obtain

$$\delta J = [\nabla_{\mathbf{x}} S - \mathbf{p}]_{t_f}^T\delta\mathbf{x}(t_f) + \int_{t_0}^{t_f}[\mathcal{H}(\mathbf{x}^*,\mathbf{u},\mathbf{p},t) - \mathcal{H}(\mathbf{x}^*,\mathbf{u}^*,\mathbf{p},t)]\,dt + \int_{t_0}^{t_f}\Big\{\Big[[\nabla_{\mathbf{x}} L(\mathbf{x}^*,\mathbf{u},t)]^T + \mathbf{p}^T\frac{\partial\mathbf{f}}{\partial\mathbf{x}} + \dot{\mathbf{p}}^T\Big]\delta\mathbf{x}\Big\}\,dt$$

In order for this to be nonnegative for arbitrary $\delta\mathbf{x}$, the coefficient of $\delta\mathbf{x}(t)$ inside the integral must vanish on the optimal trajectory.

$$\dot{\mathbf{p}} = -\left[\frac{\partial\mathbf{f}}{\partial\mathbf{x}}\right]^T\mathbf{p} - \nabla_{\mathbf{x}} L$$

The term involving $\delta\mathbf{x}(t_f)$ must also vanish. If $\mathbf{x}(t_f)$ is fixed, then $\delta\mathbf{x}(t_f) = \mathbf{0}$. If $\mathbf{x}(t_f)$ is not fixed, then $\delta\mathbf{x}(t_f) \neq \mathbf{0}$, so $\mathbf{p}(t_f) = \nabla_{\mathbf{x}} S|_{t_f}$. The remaining integral term must be nonnegative for any admissible $\mathbf{u}$, including a $\mathbf{u}$ which equals $\mathbf{u}^*$ for all except an infinitesimal time interval. Therefore, it is concluded that

$$\mathcal{H}(\mathbf{x}^*,\mathbf{u},\mathbf{p},t) \geq \mathcal{H}(\mathbf{x}^*,\mathbf{u}^*,\mathbf{p},t) \qquad \text{for all } t, \text{ all } \mathbf{u} \in U$$

This constitutes one version of the minimum principle, since $\mathbf{u}^*(t)$ is that $\mathbf{u}$ which minimizes $\mathcal{H}$. If there are no restrictions on $\mathbf{u}$, this reduces to $\partial\mathcal{H}/\partial\mathbf{u} = \partial L/\partial\mathbf{u} + \mathbf{p}^T(\partial\mathbf{f}/\partial\mathbf{u}) = \mathbf{0}$.

14.13 What modifications are necessary in the previous problem if t_f is not fixed in advance, but must be determined as part of the solution?

In forming ΔJ, variations in the final time must now be considered. Let t_f^* be the optimal stopping time and $t_f = t_f^* + \delta t_f$ is the perturbed time. Referring to Figure 14.22, it is apparent that two different variations in $\mathbf{x}$ at the final time must be considered. $\delta\mathbf{x}(t_f)$ is the variation used in Problem 14.12 and $\Delta\mathbf{x}(t_f)$ is the total variation.

$$\Delta J = J'(\mathbf{x}^* + \delta\mathbf{x}, \mathbf{u}, t_f^* + \delta t_f) - J'(\mathbf{x}^*,\mathbf{u}^*,t_f^*)$$
$$= S(\mathbf{x}^*(t_f^*) + \Delta\mathbf{x}_f, t_f^* + \delta t_f) - S(\mathbf{x}^*(t_f^*), t_f^*)$$

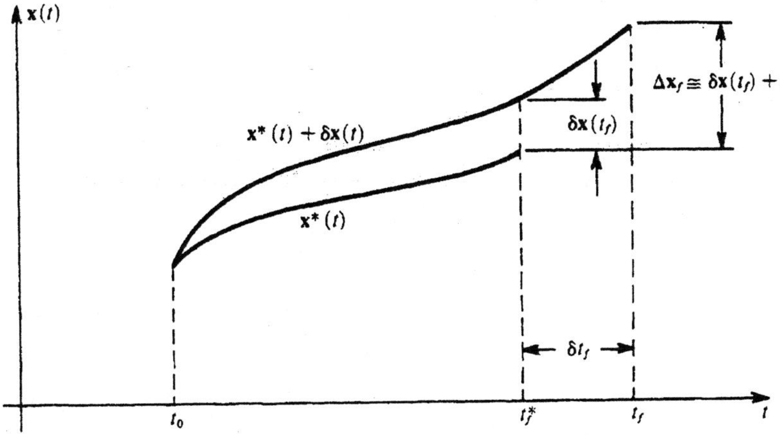

Figure 14.22

$$+\int_{t_0}^{t_f^*+\delta t_f} \{L(\mathbf{x}^* + \delta\mathbf{x}, \mathbf{u}, t) + \mathbf{p}^T[\mathbf{f}(\mathbf{x}^* + \delta\mathbf{x}, \mathbf{u}, t) - \dot{\mathbf{x}}^* - \delta\dot{\mathbf{x}}]\}\, dt$$

$$-\int_{t_0}^{t_f^*} \{L(\mathbf{x}^*, \mathbf{u}^*, t) + \mathbf{p}^T[\mathbf{f}(\mathbf{x}^*, \mathbf{u}^*, t) - \dot{\mathbf{x}}^*]\}\, dt$$

In computing δJ, the following relation is used:

$$\int_{t_0}^{t_f^*+\delta t_f} \{\ \}\, dt = \int_{t_0}^{t_f^*} \{\ \}\, dt + \int_{t_f^*}^{t_f^*+\delta t_f} \{\ \}\, dt \cong \int_{t_0}^{t_f^*} \{\ \}\, dt + \delta t_f L$$

Then

$$\delta J = (\nabla_{\mathbf{x}} S)^T|_{t_f} \Delta\mathbf{x}_f - \mathbf{p}^T(t_f)\delta\mathbf{x}(t_f) + \left[\frac{\partial S}{\partial t} + L(\mathbf{x}^*, \mathbf{u}, t)\right]\bigg|_{t_f^*} \delta t_f$$

$$+ \int_{t_0}^{t_f} [\mathcal{H}(\mathbf{x}^*, \mathbf{u}, t) - \mathcal{H}(\mathbf{x}^*, \mathbf{u}^*, t)]\, dt$$

$$+ \int_{t_0}^{t_f} \left\{\left[[\nabla_{\mathbf{x}} L(\mathbf{x}^*, \mathbf{u}, t)]^T + \mathbf{p}^T \frac{\partial \mathbf{f}}{\partial \mathbf{x}} + \dot{\mathbf{p}}^T\right] \delta\mathbf{x}\right\} dt$$

From this, conclusion regarding the two integral terms are the same as in Problem 14.12. The only changes are in the boundary terms. If $\mathbf{x}^*(t_f)$ is free, then using $\delta\mathbf{x}(t_f) = \Delta\mathbf{x}_f - \dot{\mathbf{x}}^*(t_f)\delta t_f$ leads to boundary terms $[(\nabla_{\mathbf{x}} S)^T|_{t_f^*} - \mathbf{p}^T(t_f)]\Delta\mathbf{x}_f + [\partial S/\partial t) + L(\mathbf{x}^*, \mathbf{u}, t) + \mathbf{p}^T\dot{\mathbf{x}}^*]|_{t_f} \delta t_f$. The conclusion is that $\mathbf{p}(t_f) = \nabla_{\mathbf{x}} S|_{t_f^*}$ as before, and the additional scalar equation required for determining t_f^* is $\partial S/\partial t + \mathcal{H}|_{t_f} = 0$.

If the final state had been restricted in some way, say to lie on a surface $\psi(\mathbf{x}(t_f), t_f)$, then $\Delta\mathbf{x}_f$ and δt_f are interrelated. Their coefficients cannot then be separately set equal to zero, but instead, additional restrictions must be imposed [3, 5].

14 Develop an iterative method of solving the two-point boundary value problem which results when the minimum principle is applied to:

$$\dot{\mathbf{x}} = \mathbf{f}(\mathbf{x}, \mathbf{u}), \qquad \mathbf{x}(t_0) \text{ known}, \qquad t_f \text{ fixed},$$

$$\text{minimize } J = \tfrac{1}{2}[\mathbf{x}(t_f) - \mathbf{x}_d]^T \mathbf{M}[\mathbf{x}(t_f) - \mathbf{x}_d] + \int_{t_0}^{t_f} L\, dt$$

Minimizing $\mathcal{H}$ allows $\mathbf{u}^*(t)$ to be found as a function of $\mathbf{p}(t)$. Then the two-point boundary value problem of Eq. (14.53) can be written as $\dot{\mathbf{x}} = \mathbf{f}(\mathbf{x}, \mathbf{p}), \dot{\mathbf{p}} = \mathbf{h}(\mathbf{x}, \mathbf{p}); \mathbf{x}(t_0) = \mathbf{x}_0, \mathbf{p}(t_f) =$

$\mathbf{M}[\mathbf{x}(t_f) - \mathbf{x}_d]$. If $\mathbf{p}(t_0)$ were known, the equations for $\mathbf{x}$ and $\mathbf{p}$ could be solved by numerical integration.

One way to proceed is to assume a $\mathbf{p}(t_0)^{(0)}$, numerically integrate to find $\mathbf{x}(t_f)^{(0)}$ and $\mathbf{p}(t_f)^{(0)}$. Then the terminal conditions can be checked, and the error can be used to estimate a new $\mathbf{p}(t_0)^{(1)}$. Figure 14.23 is used to illustrate the correction procedure [4]. Since $\mathbf{p}(t_f)^{(0)}$ and $\mathbf{x}(t_f)^{(0)}$ are determined by $\mathbf{p}(t_0)^{(0)}$, an unknown functional relation exists among these variables, as implied by the graph. The "slope" of the function multiplied by $\Delta\mathbf{p}(t_0)$ is set equal to the error in the terminal conditions. That is, a Newton-Raphson correction scheme is used.

The new estimate is

$$\mathbf{p}(t_0)^{(1)} = \mathbf{p}(t_0)^{(0)} - \Delta\mathbf{p}(t_0)$$

where

$$\left[\frac{\partial\mathbf{p}(t_f)}{\partial\mathbf{p}(t_0)} - \mathbf{M}\frac{\partial\mathbf{x}(t_f)}{\partial\mathbf{p}(t_0)}\right]^{(0)} \Delta\mathbf{p}(t_0) = \mathbf{p}(t_f)^{(0)} - \mathbf{M}[\mathbf{x}(t_f)^{(0)} - \mathbf{x}_d]$$

The two $n \times n$ sensitivity matrices $\mathbf{S}_\mathbf{p}(t_f) \triangleq \partial\mathbf{p}(t_f)/\partial\mathbf{p}(t_0)$ and $\mathbf{S}_\mathbf{x}(t_f) \triangleq \partial\mathbf{x}(t_f)/\partial\mathbf{p}(t_0)$ can be found by solving two sets of $n \times n$ matrix equations, obtained from the differential equations for $\mathbf{x}$ and $\mathbf{p}$ by interchanging $\partial/\partial\mathbf{p}(t_0)$ and d/dt. They are

$$\dot{\mathbf{S}}_\mathbf{x} = \frac{\partial\mathbf{f}}{\partial\mathbf{x}}\mathbf{S}_\mathbf{x} + \frac{\partial\mathbf{f}}{\partial\mathbf{p}}\mathbf{S}_\mathbf{p}, \qquad \mathbf{S}_\mathbf{x}(0) = [\mathbf{0}]$$

$$\dot{\mathbf{S}}_\mathbf{p} = \frac{\partial\mathbf{h}}{\partial\mathbf{x}}\mathbf{S}_\mathbf{x} + \frac{\partial\mathbf{h}}{\partial\mathbf{p}}\mathbf{S}_\mathbf{p}, \qquad \mathbf{S}_\mathbf{p}(0) = \mathbf{I}_n$$

These can be integrated along with the $\mathbf{x}$ and $\mathbf{p}$ equations. Only the terminal values are needed in the correction scheme, which generalizes to

$$\mathbf{p}(t_0)^{(k+1)} = \mathbf{p}(t_0)^{(k)} - [\mathbf{S}_\mathbf{p}(t_f)^{(k)} - \mathbf{M}\mathbf{S}_\mathbf{x}(t_f)^{(k)}]^{-1}\{\mathbf{p}(t_f)^{(k)} - \mathbf{M}[\mathbf{x}(t_f)^{(k)} - \mathbf{x}_d]\}$$

Success of the method depends on a good initial estimate for $\mathbf{p}(t_0)$ as well as the characteristics of the functions $\mathbf{f}$ and $\mathbf{h}$ and their derivatives.

14.15 Assuming that matrices $\mathbf{A}$ and $\mathbf{B}$ and vectors $\mathbf{x}(t_0) = \mathbf{x}_0$ and $\mathbf{x}_d$ are given, find the solution for the optimal control in Example 14.12.

When $\mathbf{A}$ and $\mathbf{B}$ are constant, the form of the solution for Eq. *(14.54)* is

$$\begin{bmatrix}\mathbf{x}(t)\\ \mathbf{p}(t)\end{bmatrix} = e^{\begin{bmatrix}\mathbf{A} & -1/2\mathbf{B}\mathbf{B}^T\\ \mathbf{0} & -\mathbf{A}^T\end{bmatrix}(t-t_0)}\begin{bmatrix}\mathbf{x}(t_0)\\ \mathbf{p}(t_0)\end{bmatrix}$$

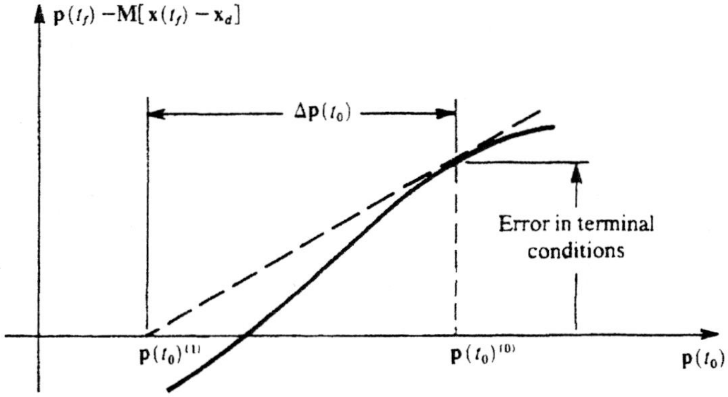

Figure 14.23

For convenience, the $2n \times 2n$ exponential matrix is written in partitioned form as

$$\begin{bmatrix} \boldsymbol{\phi}_{11}(t, t_0) & \boldsymbol{\phi}_{12}(t, t_0) \\ \boldsymbol{\phi}_{21}(t, t_0) & \boldsymbol{\phi}_{22}(t, t_0) \end{bmatrix}$$

This may be computed using the methods of Chapter 8. Actually $\boldsymbol{\phi}_{21}(t, t_0)$ will be the $n \times n$ null matrix for this problem, but the more general form is treated. At the final time,

$$\mathbf{x}(t_f) = \boldsymbol{\phi}_{11}(t_f, t_0)\mathbf{x}_0 + \boldsymbol{\phi}_{12}(t_f, t_0)\mathbf{p}(t_0)$$

$$\mathbf{p}(t_f) = \boldsymbol{\phi}_{21}(t_f, t_0)\mathbf{x}_0 + \boldsymbol{\phi}_{22}(t_f, t_0)\mathbf{p}(t_0)$$

Using the boundary condition $\mathbf{p}(t_f) = 2[\mathbf{x}(t_f) - \mathbf{x}_d]$ gives

$$2[\boldsymbol{\phi}_{11}(t_f, t_0)\mathbf{x}_0 + \boldsymbol{\phi}_{12}(t_f, t_0)\mathbf{p}(t_0) - \mathbf{x}_d] = \boldsymbol{\phi}_{21}(t_f, t_0)\mathbf{x}_0 + \boldsymbol{\phi}_{22}(t_f, t_0)\mathbf{p}(t_0)$$

The unknown $\mathbf{p}(t_0)$ can now be found:

$$\mathbf{p}(t_0) = [\boldsymbol{\phi}_{22}(t_f, t_0) - 2\boldsymbol{\phi}_{12}(t_f, t_0)]^{-1}\{[2\boldsymbol{\phi}_{11}(t_f, t_0) - \boldsymbol{\phi}_{21}(t_f, t_0)]\mathbf{x}_0 - 2\mathbf{x}_d\}$$

With $\mathbf{p}(t_0)$ known, $\mathbf{p}(t)$ and $\mathbf{u}^*(t)$ are given by

$$\mathbf{p}(t) = \boldsymbol{\phi}_{21}(t, t_0)\mathbf{x}_0 + \boldsymbol{\phi}_{22}(t, t_0)\mathbf{p}(t_0)$$

$$\mathbf{u}^*(t) = -\tfrac{1}{2}\mathbf{B}^T\mathbf{p}(t)$$

The optimal control law is open loop since $\mathbf{u}^*(t)$ is expressed as a function of $\mathbf{x}_0$ and $\mathbf{x}_d$.

6 Convert the control law of Problem 14.15 to a closed-loop control law.

Instead of writing $\begin{bmatrix} \mathbf{x}(t_f) \\ \mathbf{p}(t_f) \end{bmatrix} = \boldsymbol{\Phi}(t_f, t_0)\begin{bmatrix} \mathbf{x}_0 \\ \mathbf{p}(t_0) \end{bmatrix}$, use $\begin{bmatrix} \mathbf{x}(t_f) \\ \mathbf{p}(t_f) \end{bmatrix} = \boldsymbol{\Phi}(t_f, t)\begin{bmatrix} \mathbf{x}(t) \\ \mathbf{p}(t) \end{bmatrix}$, for a general time t.

Repeating much of Problem 14.15 but solving for $\mathbf{p}(t)$ instead of $\mathbf{p}(t_0)$ gives

$$\mathbf{p}(t) = [\boldsymbol{\phi}_{22}(t_f, t) - 2\boldsymbol{\phi}_{12}(t_f, t)]^{-1}\{[2\boldsymbol{\phi}_{11}(t_f, t) - \boldsymbol{\phi}_{21}(t_f, t)]\mathbf{x}(t) - 2\mathbf{x}_d\}$$

Once again $\mathbf{u}^*(t) = -\frac{1}{2}\mathbf{B}^T\mathbf{p}(t)$. The feedback control law is illustrated in Figure 14.24.

7 Use the minimum principle to find the input voltage $u(t)$ which charges the capacitor of Figure 14.25 from x_0 at $t = 0$ to x_d at a fixed t_f while minimizing the energy dissipated in R. There are no restrictions on $u(t)$.

The state equation is $\dot{x} = -x/RC + u/RC$. The energy dissipated is

$$J = \int_0^{t_f} i^2 R\, dt = \int_0^{t_f} \frac{[u(t) - x(t)]^2}{R}\, dt$$

The Hamiltonian is $\mathcal{H} = (u - x)^2/R + p(-x/RC + u/RC)$. Minimize $\mathcal{H}$ by setting $\partial\mathcal{H}/\partial u = 0$, or $2(u - x)/R + p/RC = 0$. Therefore, $u^*(t) = -p(t)/2C + x(t)$.

The costate equation is $\dot{p} = -\partial\mathcal{H}/\partial x = 2(u - x)/R + p/RC$.

Eliminating $u^*(t)$, and simplifying, the two-point boundary value problem is $\dot{x} =$

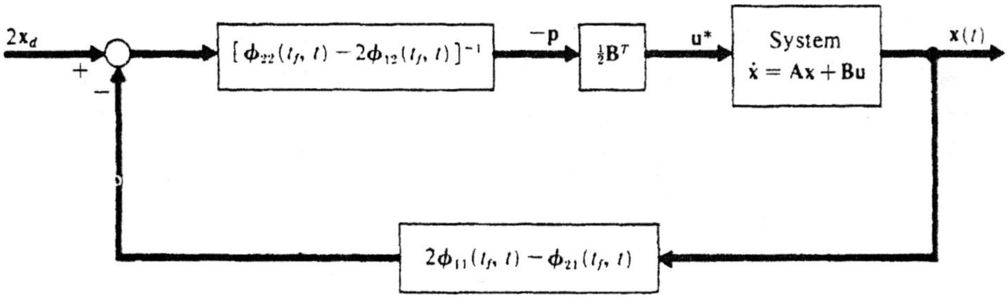

Figure 14.24

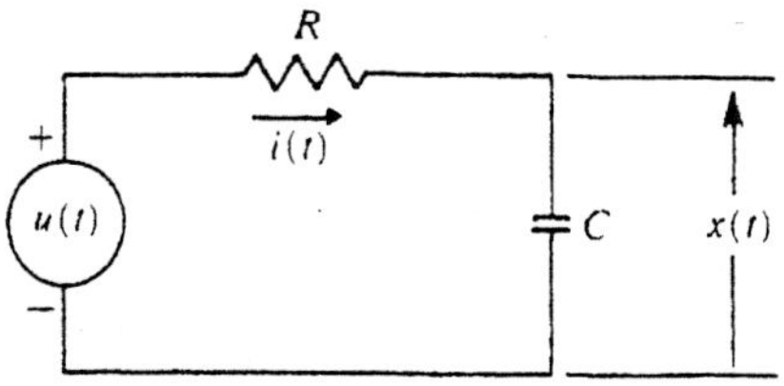

Figure 14.25

$-p/(2RC^2), \dot{p} = 0$, with $x(0) = x_0, x(t_f) = x_d$. This implies that $p(t) = $ constant, α. Therefore, $x(t) = x_0 - \alpha t/(2RC^2)$. From the terminal boundary condition, $\alpha = -2RC^2(x_d - x_0)/t_f$. Therefore, $u^*(t) = RC(x_d - x_0)/t_f + x(t)$. But $x(t) = x_0(1 - t/t_f) + x_d t/t_f$, so $u^*(t) = x_0 + (x_d - x_0)(RC + t)/t_f$.

14.18 A spin-stabilized satellite is wobbling slightly, with components of angular velocity due to the wobble being $x_1(t)$ and $x_2(t)$. The state equations are (see Problems 3.16, and 9.6).

$$\begin{bmatrix} \dot{x}_1 \\ \dot{x}_2 \end{bmatrix} = \begin{bmatrix} 0 & -\Omega \\ \Omega & 0 \end{bmatrix}\begin{bmatrix} x_1 \\ x_2 \end{bmatrix} + \begin{bmatrix} u_1(t) \\ u_2(t) \end{bmatrix}$$

Find the control $\mathbf{u}^*(t)$ which drives $\mathbf{x}(t_f)$ to $\mathbf{0}$ at a fixed t_f while minimizing the control energy $J = \int_0^{t_f} \mathbf{u}^T\mathbf{u}\,dt$. There are no restrictions on $\mathbf{u}(t)$.

The Hamiltonian is $\mathcal{H} = \mathbf{u}^T\mathbf{u} + \mathbf{p}^T[\mathbf{Ax} + \mathbf{u}]$. It is minimized by $\mathbf{u}^*(t) = -\frac{1}{2}\mathbf{p}(t)$, where the costate vector satisfies $\dot{\mathbf{p}} = -\partial\mathcal{H}/\partial\mathbf{x} = -\mathbf{A}^T\mathbf{p}$. Note that since $\mathbf{A} = \begin{bmatrix} 0 & -\Omega \\ \Omega & 0 \end{bmatrix}$, we have $-\mathbf{A}^T = \mathbf{A}$.

The two-point boundary value problem is

$$\begin{bmatrix} \dot{\mathbf{x}} \\ \dot{\mathbf{p}} \end{bmatrix} = \left[\begin{array}{c|c} \mathbf{A} & -\frac{1}{2}\mathbf{I}_2 \\ \hline \mathbf{0} & \mathbf{A} \end{array}\right]\begin{bmatrix} \mathbf{x} \\ \mathbf{p} \end{bmatrix} \qquad \text{with } \mathbf{x}(0) = \mathbf{x}_0, \mathbf{x}(t_f) = \mathbf{0}$$

From Problem 8.17, the 4 × 4 transition matrix for this system is

$$\mathbf{\Phi}(t, 0) = \left[\begin{array}{c|c} e^{\mathbf{A}t} & -\frac{t}{2}e^{\mathbf{A}t} \\ \hline \mathbf{0} & e^{\mathbf{A}t} \end{array}\right], \qquad \text{where } e^{\mathbf{A}t} = \begin{bmatrix} \cos\Omega t & -\sin\Omega t \\ \sin\Omega t & \cos\Omega t \end{bmatrix}$$

Using the terminal boundary condition gives $\mathbf{0} = e^{\mathbf{A}t_f}\mathbf{x}_0 - (t_f/2)e^{\mathbf{A}t_f}\mathbf{p}(0)$, from which $\mathbf{p}(0) = (2/t_f)\mathbf{x}_0$. Using $\mathbf{p}(0)$ gives $\mathbf{p}(t) = e^{\mathbf{A}t}\mathbf{p}(0)$, so $\mathbf{u}^*(t) = -(1/t_f)e^{\mathbf{A}t}\mathbf{x}_0$. When this control is used, the state satisfies $\mathbf{x}^*(t) = (1 - t/t_f)e^{\mathbf{A}t}\mathbf{x}_0$ and the minimum cost is

$$J = \frac{1}{t_f^2}\int_0^{t_f} \mathbf{x}_0^T[e^{\mathbf{A}t}]^T e^{\mathbf{A}t}\mathbf{x}_0\,dt = \mathbf{x}_0^T\mathbf{x}_0/t_f \qquad \text{since } [e^{\mathbf{A}t}]^T = [e^{\mathbf{A}t}]^{-1}$$

14.19 Consider a more general version of Example 14.11. Let the position of the automobile be $x_1(t)$ and the velocity be $x_2(t)$. The admissible controls must satisfy $|u(t)| \leq 1$ for all t. Find the $u^*(t)$ which drives the position and velocity to zero simultaneously, in mimimum time.

The system equations are $\dot{x}_1 = x_2, \dot{x}_2 = u$. The Hamiltonian is $\mathcal{H} = 1 + p_1\dot{x}_1 + p_2\dot{x}_2 = 1 + p_1x_2 + p_2u$. $\mathcal{H}$ is minimized by selecting $\mathbf{u}^*(t) = -\text{sign}[p_2(t)]$.

The equations for $\mathbf{p}(t)$ are $\dot{p}_1 = -\partial\mathcal{H}/\partial x_1 = 0$ and $\dot{p}_2 = -\partial\mathcal{H}/\partial x_2 = -p_1$. Therefore, $p_1(t) = $ constant, $p_1(0)$, and $p_2(t) = p_2(0) - p_1(0)t$. This indicates that $p_2(t)$ is a linear function of t and changes sign at most once (ignoring the exceptional case where $p_1(0) = p_2(0) = 0$). Therefore, $u^*(t)$ changes sign at most once.

Rather than attempting to determine $p_1(0)$ and $p_2(0)$, the behavior of the system is investigated for $u(t) = +1$ and $u(t) = -1$. With $u = +1$, $x_2(t) = x_2(0) + t$ and $x_1(t) = x_1(0) + x_2(0)t + t^2/2$, or $x_1(t) = \frac{1}{2}x_2^2(t) + x_1(0) - \frac{1}{2}x_2^2(0)$. In the x_1x_2 plane, this represents a family of parabolas open toward the positive x_1 axis. Similarly with $u = -1$, $x_1(t) = -\frac{1}{2}x_2^2(t) + x_1(0) + \frac{1}{2}x_2^2(0)$. This is a family of parabolas open toward the negative x_1 axis. Just one member of each family passes through the specified terminal point $x_1 = x_2 = 0$.

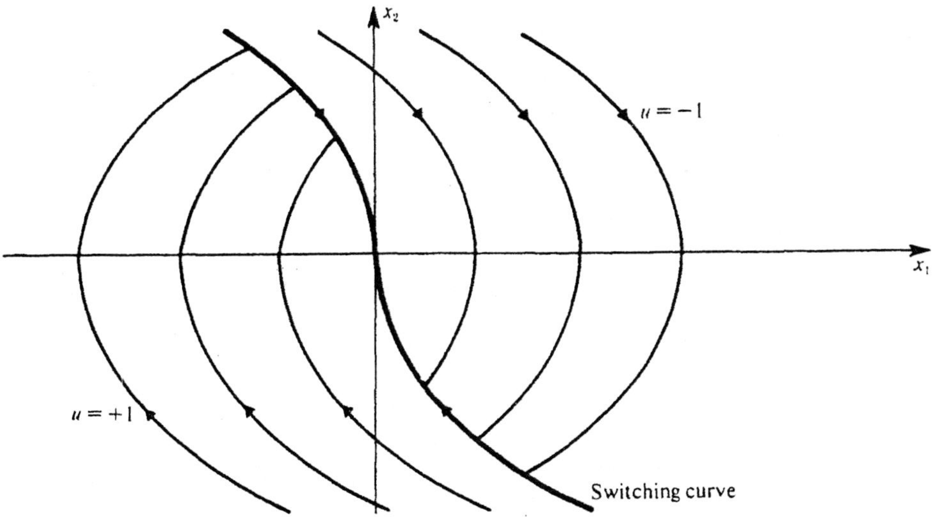

Figure 14.26

Figure 14.26 shows the portion of these two families that are of interest. Segments of the two parabolas through the origin form the switching curve. For any initial state above this curve, $u = -1$ is used until the state intersects the switching curve. Then $u = +1$ is used to reach the origin. For initial states below the switching curve, $u = +1$ is used first, then $u = -1$.

This problem is an example of bang-bang control. It illustrates that the fastest method of coming to a red light and stopping is to use maximum acceleration until the last possible moment, and then use full braking to stop (hopefully) at the intersection. This example also illustrates the remarks in Sec. 14.1. This control is certainly not optimal in terms of tire wear or the number of traffic tickets received.

PROBLEMS

Dynamic Programming

Use dynamic programming to find the path which moves left to right from point a to point z of Figure 14.27 while minimizing the sum of the costs on each path traveled.

A student has four hours available to study for four exams. He will earn the scores shown in Table 14.2 for various study times. Use dynamic programming to find the optimal allocation of time in order to maximize the sum of his four scores. Consider only integer numbers of hours.

2 Find the control sequence which minimizes $J = [x(2) + 2]^2 + \sum_{k=0}^{1} u(k)^2$ for the scalar system $x(k+1) = \frac{1}{2}x(k) + u(k), x(0) = 10$, no restrictions on $u(k)$. Also find the resulting sequence $x(k)$ and the minimum value of J.

3 The same system and initial conditions of Problem 14.22 are considered. Find the optimal control and state sequences which minimize $J = \sum_{k=0}^{4} u(k)^2$ and which give $x(5) = 0$.

4 A scalar system is described by $x(k+1) = \frac{1}{2}x(k) + 2u(k)$, with $x(0) = x_0$. Find $x^*(k)$ and $u^*(k)$ which minimize $J = \sum_{k=0}^{2} \{x^2(k) + u^2(k-1)\}$.

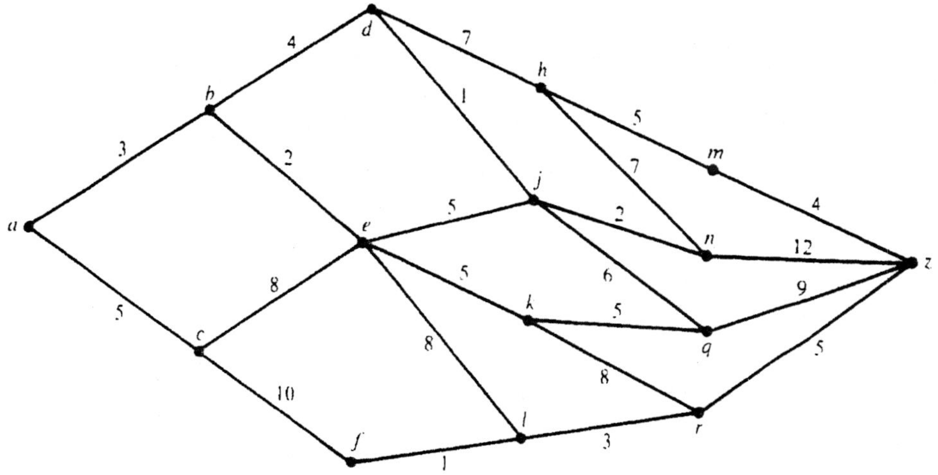

Figure 14.27

14.25 An unstable discrete-time system is described by

$$\mathbf{x}(k+1) = \begin{bmatrix} 1 & 2 \\ -2 & -3 \end{bmatrix}\mathbf{x}(k) + \begin{bmatrix} 1 \\ 0 \end{bmatrix} u(k) \qquad (1)$$

$$y(k) = [0 \quad 1]\mathbf{x}(k) \qquad (2)$$

Find the steady-state feedback gain matrix for the optimal regulator problem, assuming all states are available for use. Also determine the closed-loop eigenvalues that result.

Use the following weights

(a) $\mathbf{Q} = \mathbf{I}, \mathbf{R} = 1$

(b) $\mathbf{Q} = \mathbf{I}, \mathbf{R} = 10$

(c) $\mathbf{Q} = 10\mathbf{I}, \mathbf{R} = 1$

(d) $\mathbf{Q} = 1000\mathbf{I}, \mathbf{R} = 1$

14.26 The system of Problem 14.25 is now assumed to have additive white noises $\mathbf{w}(k)$ on equation (*1*) and $\mathbf{v}(k)$ on equation (2). The noise covariance matrices are

$$\mathbf{Q} = \begin{bmatrix} 9 & 0 \\ 0 & 4 \end{bmatrix} \quad \text{and} \quad \mathbf{R} = 16, \text{ respectively}$$

TABLE 14-2 TEST SCORES

Study Hours	Course No.			
	1	2	3	4
0	20	40	40	80
1	45	45	52	91
2	65	57	62	95
3	75	61	71	97
4	83	69	78	98

(a) A Kalman filter (recursive least-squares estimator) is to be used rather than a deterministic observer. Find the steady-state Kalman gain K and the pole locations of the filter's dynamics (i.e., equivalent to the observer poles).

(b) The above filter is used along with the controllers found in Problem 14.25 as indicated by the separation theorem. How do the presence of the filter and the use of estimated states rather than actual states affect the overall system performance in each case? Remember that the separation theorem guarantees that this approach is the best that can be done. But it does not guarantee that the results will be what the controller asked for or that the system will even work well.

Repeat Problem 14.26 if the measurement noise covariance is reduced to $\mathbf{R} = 0.1$.

Find the optimal control sequence and the resulting state sequence for the system

$$\mathbf{x}(k+1) = \begin{bmatrix} 1.0 & 0.0 \\ -0.5 & 0.5 \end{bmatrix} \mathbf{x}(k) + \begin{bmatrix} 1 \\ -1 \end{bmatrix} u(k)$$

with initial conditions $\mathbf{x}(0) = [4 \quad -2]^T$. The optimal control is to minimize the cost function J defined in Example 14.5, with final time $N = 5$ and

$$\mathbf{M} = \text{Diag}[1000, 1000], \qquad \mathbf{Q} = \text{Diag}[10, 10], \qquad \mathbf{R} = 1, \qquad \mathbf{x}_d = [8 \quad -7]^T,$$

$$\boldsymbol{\eta} = [-1 \quad 1]^T$$

Note the conflict being requested between $\mathbf{x}_d$ and $\boldsymbol{\eta}$. In general, fixed terminal conditions $\mathbf{x}(N) = \mathbf{x}_d$ can be closely approximated by using a sufficiently large $\mathbf{M}$ final weighting matrix.

Use the discrete-time approximate model of the airplane in Example 13.6 and design a steady-state optimal regulator with

$$\mathbf{Q} = \text{Diag}[0.1, 1.0, 1.0, 0.1] \quad \text{and} \quad \mathbf{R} = \mathbf{I}$$

A commonly used simplification of the previous system is that $\delta_a = 4\delta_r$. This allows the simpler single-input analysis to be carried out, using as the input matrix $\mathbf{B}$ four times column 1 plus column 2 of the original $\mathbf{B}$ matrix. Using this simplification, find the feedback gains for generating the equivalent control δ_r if

(a) $\mathbf{Q} = \mathbf{I}, \mathbf{R} = 1$

(b) $\mathbf{Q} = \mathbf{I}, \mathbf{R} = 0.1$

(c) $\mathbf{Q} = \mathbf{I}, \mathbf{R} = 25$

(d) $\mathbf{Q} = \text{Diag}[0.1, 1.0, 1.0, 0.1], \mathbf{R} = 1$

A linear, constant continuous-time system is described by $\mathbf{A} = \begin{bmatrix} 0 & 1 \\ -2 & -3 \end{bmatrix}, \mathbf{B} = \mathbf{I}$. Solve the algebraic Riccati equation to find $\mathbf{W}_\infty$ if $\mathbf{Q} = \mathbf{I}$ and $\mathbf{R} = r\mathbf{I}$, for five cases: $r = 50, 10, 4, 1,$ and 0.2. Also determine the constant full feedback gain matrices and the resulting closed-loop eigenvalues.

Solve the Lyapunov equation $\mathbf{A}^T\mathbf{W} + \mathbf{W}\mathbf{A} = -\mathbf{Q}$, using $\mathbf{Q}$ and $\mathbf{A}$ from the previous problem. Compare the result with the "large r" case $r = 50$.

A system with $\mathbf{A}$ of Problem 14.31 and $\mathbf{B} = [-1 \quad 1]^T$ is uncontrollable, but since the eigenvalues of $\mathbf{A}$ are $\lambda = -2, -1$, the system is stabilizable. Solve the ARE for $\mathbf{W}_\infty$ and then find $\mathbf{G}_\infty$ and the closed-loop eigenvalues. Use $\mathbf{Q} = \mathbf{I}$ and $\mathbf{R} = 1$.

An unstable but controllable continuous-time system has $\mathbf{A} = \begin{bmatrix} 0 & 1 \\ 2 & -3 \end{bmatrix}, \mathbf{B} = \mathbf{I}$. Let $\mathbf{Q} = \mathbf{I}$ and $\mathbf{R} = r\mathbf{I}$. Find the feedback gain matrix $\mathbf{G}_\infty$ for $r = 50$ and $r = 1000$. Also find the resulting closed-loop eigenvalues. What happens when the Lyapunov equation is used to approximate the ARE for large r in this unstable case?

For the scalar system $\dot{x} = x + u$, the ARE becomes a simple scalar quadratic. Let $Q = 1$ and

$R = \frac{1}{3}$, and find both roots of this quadratic. Then form the 2×2 Hamiltonian matrix $\mathbf{H}$ and use its eigenvectors to find both $\mathbf{W}_\infty$ and $\mathbf{W}_{ss}$ defined in Sec. 14.4.2. Compare the results.

14.36 Combine the pole placement controller of Example 13.1 and the observer of Example 13.10 into a composite fourth-order system. Find the Kalman observable canonical form to show that the observer modes are not controllable. Also verify that the composite system is observable.

The Minimum Principle

14.37 Use the minimum principle to find the optimal control $u^*(t)$ and the corresponding $x^*(t)$ for the system $\dot{x} = u(t)$ with $x(0) = 0, x(1) = 1$ and no restrictions on $u(t)$. The performance criterion is $J = \int_0^1 (x^2 + u^2)\, dt$.

14.38 Use the minimum principle to solve the optimization problem: $\dot{x}_1 = x_2$, $\dot{x}_2 = u$, $x_1(0) = 1$, $x_2(0) = 1$, $x_1(2) = 0$, $x_2(2) = 0$, minimize $J = \frac{1}{2}\int_0^2 u^2(t)\, dt$ with $u(t)$ unrestricted.

14.39 Solve the following two-point boundary value problem on the interval $0 \le t \le 1$: $\dot{x} = -2x - \frac{1}{2}p$, $\dot{p} = -2x + 2p$, $x(0) = 10$, $p(1) = 4x(1)$.

14.40 The equations of motion for a rocket flying in a vertical plane under the influence of constant gravity g and constant thrust T can be written as $\dot{x}_1 = x_3$, $\dot{x}_2 = x_4$, $\dot{x}_3 = (T/m)\cos u(t)$, $\dot{x}_4 = (T/m)\sin u(t) - g$, where x_1 and x_2 are horizontal and vertical position components, x_3 and x_4 are horizontal and vertical velocity components, and the control $u(t)$ is the thrust angle measured from the horizontal. The rocket is to be flown to a specified terminal altitude with zero vertical velocity at t_f and maximum horizontal velocity. Use the minimum principle to establish that the optimal thrust angle follows the linear-tangent steering law.

$$\tan u^*(t) = \alpha t - \beta \qquad \text{where } \alpha \text{ and } \beta \text{ are constants}$$

(*Hint:* Minimize $J = -x_3(t_f)$.)

14.41 Consider a system which is nonlinear with respect to $\mathbf{x}$, but linear with respect to $\mathbf{u}$, i.e., $\dot{\mathbf{x}} = \mathbf{f}(\mathbf{x}, t) + \mathbf{B}\mathbf{u}$. In each of the following cases, use the minimum principle to find the form of $\mathbf{u}^*(t)$ in terms of $\mathbf{p}(t)$.

(a) $J = \int_{t_0}^{t_f} \mathbf{u}^T\mathbf{u}\, dt$, $\mathbf{x}(t_f) = \mathbf{0}$, t_f fixed, $\mathbf{u}^T(t)\mathbf{u}(t) \le 1$ for all t.
(b) Drive $\mathbf{x}(t_f)$ to zero in minimum time, $\mathbf{u}^T(t)\mathbf{u}(t) \le 1$ for all t.
(c) Same as b except each component of $\mathbf{u}$ satisfies $|u_i(t)| \le 1$ for all t.
(d) $J = \int_{t_0}^{t_f} \Sigma|u_i(t)|\, dt$, with $|u_i(t)| \le 1$ for all t.

14.42 A scalar system is described by $\dot{x} = x + u$ with $x(0) = 2$. The admissible controls must satisfy $|u(t)| \le 1$. Use the minimum principle to find $u^*(t)$ which drives $x(t)$ to zero in minimum time.

14.43 Analyze an equivalent single-input model of the aircraft in Problem 14.11 by assuming that $\delta_a = 4\delta_r$. Use $\mathbf{Q} = \text{Diag}[0.1, 1.0, 1.0, 0.1]$ and $\mathbf{R} = 1$.

15

An Introduction to Nonlinear Control Systems

INTRODUCTION

Most of the text so far has dealt with linear systems. Yet when an engineer is faced with a "real problem," he or she invariably bumps into nonlinearities. Some typical examples are as follows:

1. In positioning a robotic device or in pointing a sensor at a target, the geometry of coordinate transformations comes into play. Sines and cosines are nonlinear functions of their arguments.
2. An actuating motor has inherent current—and hence torque—limitations. Saturation of the control commands is a common nonlinearity. Backlash, hysteresis, and dead zone are other commonly encountered nonlinearities.
3. Many important physical processes are described by nonlinear models. Drag on a moving vehicle is proportional to velocity squared, for example. The voltage-current characteristics of most electronic devices are nonlinear. Coulomb friction is of constant magnitude and always opposes motion, unlike the linear viscous friction model often assumed. Gravitational and electrostatic attraction are inversely proportional to distance squared.
4. Deliberate nonlinearities may be introduced by the control system. On-off relay controllers are common examples.

What are the implications of nonlinearities on the control system analysis and design techniques which have been presented thus far? Many aspects of former chapters are still applicable:

1. The physical modeling techniques using linear graphs apply to nonlinear systems. The individual elemental equations may be nonlinear, but the continuity and compatibility laws apply as before (Chapter 1).

2. The general form of the state equations

$$\dot{\mathbf{x}} = \mathbf{f}(\mathbf{x}, \mathbf{u}, t) \quad \text{and} \quad \mathbf{y} = \mathbf{h}(\mathbf{x}, \mathbf{u}, t)$$

   are still valid (Chapter 3).
3. Knowledge of matrix theory and linear algebra is still essential. A rigorous treatment of nonlinear systems would require much supplemental mathematics, however [1, 2].
4. The Lyapunov stability theory applies to nonlinear systems (Chapter 10).
5. The general formulation of the optimal control problems, using either dynamic programming or the minimum principle, is still valid (Chapter 14).

Many of the linear system results do not apply to nonlinear systems:

1. Superposition does not apply. Knowledge of the system response to initial conditions or to individual inputs does not allow prediction of the total response due to initial conditions plus several simultaneous inputs.
2. Homogeneity does not apply. The response to an input $\alpha\mathbf{u}(t)$ is not just α times the response to $\mathbf{u}(t)$. The response to $\beta\mathbf{x}(t_0)$ is not just β times the response to $\mathbf{x}(t_0)$. The whole concept of designing control systems based on typical test inputs (unit steps, sinusoids, and so on) and then predicting behavior to an actual input by scaling and superposition is generally invalid.
3. The nice correlation between transfer function pole and zero locations and time response behavior is generally invalid.
4. Stability of a system is no longer just a simple function of eigenvalue locations. In fact, it is not proper to speak about stability of a nonlinear system. Rather, the stability of equilibrium points must be investigated, and nonlinear systems may have multiple equilibrium points, some stable and others not.
5. An unforced nonlinear system can possess limit cycles and other behavior not predicted by linear theory.
6. A periodically excited nonlinear system is not restricted to yielding steady-state outputs of the same frequency as the input. Higher harmonics, subharmonics, and even continuous spectra (chaos) can occur in the output of a nonlinear system.

   Jump resonance, beat phenomenon, and other behaviors not predicted by linear theory can occur. Although these are of general interest in nonlinear system dynamics, the design of control systems is usually directed toward the avoidance of such behaviors.
7. Many linear system properties were derived based on full knowledge of the closed-form solution to the linear state equations. In the nonlinear case no known analytical solutions are available except in very rare exceptional cases. In fact the whole question of global existence and uniqueness of solutions to nonlinear differential equations cannot be taken for granted in general [3].
8. Properties such as controllability and observability can no longer be tested for on a global basis by using simple rank tests.

The goal of this chapter is to provide some insight into nonlinear system behavior and to give the engineer some useful tools for attacking certain classes of nonlinear problems. A complete treatment of nonlinear systems and control is not possible. Consideration is restricted to certain classes of nonlinear systems and certain kinds of problems:

1. Many systems with mild, sufficiently smooth nonlinearities can be treated by using a linear approximate model, obtained by linearizing about a known nominal solution or operating point. Most electronic circuit design is based on behavior in the vicinity of an "operating point." Many aircraft, rocket, and spacecraft control systems have been successfully designed using linear behavior in the neighborhood of a nominal trajectory.
2. Some nonlinear systems can be linearized by using a nonlinear state variable feedback controller. The principal advantage of this approach is that known techniques and results for linear systems analysis can then be applied.
3. An introduction to describing functions is presented. This constitutes a useful, albeit an approximate, approach for dealing with many of the unavoidable nonlinearities encountered in real system design.
4. Lyapunov stability theory can be used to analyze systems and even to design controllers of certain types. Application of these very general methods is often limited by the difficulties in finding suitable Lyapunov functions. By restricting attention to systems which are linear except for one nonlinear element, some easy-to-apply results such as the Popov criterion and the circle criterion can be derived. Stability is always a major concern in control system design. It is imperative that the effect of nonlinearities on system stability can be evaluated. The assumptions made in deriving these results are general enough to fit many engineering applications.

Phase-plane representations are used at various times because of the insight they provide. Although general phase-plane analysis techniques exist, they are fully effective only for second-order systems and are not pursued in detail here.

Because of the lack of analytical solutions for nonlinear system equations, simulation takes on much greater significance. Many numerical integration schemes will diverge if the integration step size is too large for the frequency of the signal being integrated. The major danger of applying fixed step-size integration schemes to nonlinear systems whose response frequencies may not be known in advance is that a numerical algorithm instability may be falsely interpreted as a system instability. One commonsense approach suggests that when a simulated response seems to be diverging, the simulation should be retried with a much smaller integration step size.

2 LINEARIZATION: ANALYSIS OF SMALL DEVIATIONS FROM NOMINAL

Consider the general nonlinear state variable model of Eqs. (*3.6*) and (*3.7*):

$$\begin{aligned}\dot{\mathbf{x}} &= \mathbf{f}(\mathbf{x}, \mathbf{u}, t)\\ \mathbf{y} &= \mathbf{h}(\mathbf{x}, \mathbf{u}, t)\end{aligned} \qquad (15.1)$$

Suppose a nominal solution $\mathbf{x}_n(t)$, $\mathbf{u}_n(t)$, and $\mathbf{y}_n(t)$ is known. The difference between these nominal vector functions and some slightly perturbed functions $\mathbf{x}(t)$, $\mathbf{u}(t)$, and $\mathbf{y}(t)$ can be defined by

$$\delta\mathbf{x} = \mathbf{x}(t) - \mathbf{x}_n(t)$$

$$\delta\mathbf{u} = \mathbf{u}(t) - \mathbf{u}_n(t)$$

$$\delta\mathbf{y} = \mathbf{y}(t) - \mathbf{y}_n(t)$$

Then Eq. (*15.1*) can be written as

$$\dot{\mathbf{x}}_n + \delta\dot{\mathbf{x}} = \mathbf{f}(\mathbf{x}_n + \delta\mathbf{x}, \mathbf{u}_n + \delta\mathbf{u}, t)$$

$$= \mathbf{f}(\mathbf{x}_n, \mathbf{u}_n, t) + \left[\frac{\partial \mathbf{f}}{\partial \mathbf{x}}\right]_n \delta\mathbf{x} + \left[\frac{\partial \mathbf{f}}{\partial \mathbf{u}}\right]_n \delta\mathbf{u} + \text{higher-order terms}$$

$$\mathbf{y}_n + \delta\mathbf{y} = \mathbf{h}(x_n + \delta\mathbf{x}, \mathbf{u}_n + \delta\mathbf{u}, t)$$

$$= \mathbf{h}(\mathbf{x}_n, \mathbf{u}_n, t) + \left[\frac{\partial \mathbf{h}}{\partial \mathbf{x}}\right]_n \delta\mathbf{x} + \left[\frac{\partial \mathbf{h}}{\partial \mathbf{u}}\right]_n \delta\mathbf{u} + \text{higher-order terms}$$

where $[\ ]_n$ means the derivatives are evaluated on the nominal solutions. Since the nominal solutions satisfy Eq. (*15.1*), the first terms in the preceding Taylor series expansions cancel. For sufficiently small $\delta\mathbf{x}$, $\delta\mathbf{u}$, and $\delta\mathbf{y}$ perturbations, the higher-order terms can be neglected, leaving the linear equations

$$\begin{aligned}\delta\dot{\mathbf{x}} &= \left[\frac{\partial \mathbf{f}}{\partial \mathbf{x}}\right]_n \delta\mathbf{x} + \left[\frac{\partial \mathbf{f}}{\partial \mathbf{u}}\right]_n \delta\mathbf{u} \\ \delta\mathbf{y} &= \left[\frac{\partial \mathbf{h}}{\partial \mathbf{x}}\right]_n \delta\mathbf{x} + \left[\frac{\partial \mathbf{h}}{\partial \mathbf{u}}\right]_n \delta\mathbf{u}\end{aligned} \tag{15.2}$$

If $\mathbf{x}_n(t) = \mathbf{x}_e = \text{constant}$ and if $\mathbf{u}_n(t) = 0 = \delta\mathbf{u}(t)$, then the stability of the equilibrium point $\mathbf{x}_e$ is governed by

$$\delta\dot{\mathbf{x}} = \left[\frac{\partial \mathbf{f}}{\partial \mathbf{x}}\right]_n \delta\mathbf{x} \tag{15.3}$$

For this case, the Jacobian matrix $[\partial\mathbf{f}/\partial\mathbf{x}]$ is constant and its eigenvalues determine system stability in the neighborhood of $\mathbf{x}_e$, in the following sense. If all λ_i have negative real parts, the equilibrium point is asymptotically stable for sufficiently small perturbations. If one or more eigenvalues have positive real parts, the equilibrium point is unstable. If one or more of the eigenvalues are on the $j\omega$ axis and all others are in the left-half plane, no conclusion about stability can be drawn from this linear model. Whether the actual behavior of the system is divergent or convergent will depend upon the neglected higher-order terms in the Taylor series expansion. Thus, except for the borderline $j\omega$ axis case, stability of the nonlinear Eq. (*15.1*) is the same as the linearized model Eq. (*15.2*), at least in a small neighborhood of the equilibrium point. Problem 15.1 proves these results and gives precise conditions under which they apply.

EXAMPLE 15.1 Find the equilibrium points for the system described by

$$\ddot{y} + (1 + y)\dot{y} - 2y + 0.5y^3 = 0$$

Then evaluate the linearized Jacobian matrix at each equilibrium point and determine the stability characteristics from the eigenvalues.

Letting $x_1 = y$ and $x_2 = \dot{y}$ gives the state variable model

$$\begin{bmatrix} \dot{x}_1 \\ \dot{x}_2 \end{bmatrix} = \begin{bmatrix} x_2 \\ 2x_1 - 0.5x_1^3 - (1 + x_1)x_2 \end{bmatrix} = \mathbf{f}(\mathbf{x})$$

Equilibrium points are solutions of $\mathbf{f}(\mathbf{x}) = \mathbf{0}$, so each must have $x_2 = 0$ and $2x_1 - 0.5x_1^3 = 0$. The three solutions are $x_{e1} = \begin{bmatrix} 0 \\ 0 \end{bmatrix}$, $x_{e2} = \begin{bmatrix} 2 \\ 0 \end{bmatrix}$ and $x_{e3} = \begin{bmatrix} -2 \\ 0 \end{bmatrix}$. The Jacobian matrix is

$$\frac{\partial \mathbf{f}}{\partial \mathbf{x}} = \begin{bmatrix} 0 & 1 \\ 2 - \dfrac{3x_1^2}{2} - x_2 & -(1 + x_1) \end{bmatrix}$$

so that

$$\left[\frac{\partial \mathbf{f}}{\partial \mathbf{x}}\right]_1 = \begin{bmatrix} 0 & 1 \\ 2 & -1 \end{bmatrix}$$

Its eigenvalues are at +1 and −2, so this is a saddle point (see Figure 10.3).

$$\left[\frac{\partial \mathbf{f}}{\partial \mathbf{x}}\right]_2 = \begin{bmatrix} 0 & 1 \\ -4 & -3 \end{bmatrix}$$

Its eigenvalues are at $-\frac{3}{2} \pm j\sqrt{7}/2$, so this point is a stable focus (see Figure 10.4).

$$\left[\frac{\partial \mathbf{f}}{\partial \mathbf{x}}\right]_3 = \begin{bmatrix} 0 & 1 \\ -4 & 1 \end{bmatrix}$$

Its eigenvalues are at $\frac{1}{2} \pm j\sqrt{15}/2$, so this point is an unstable focus (see Figure 10.4). If this system has initial conditions exactly at any one of the three equilibrium points, the state will remain there indefinitely in the absence of disturbances. For any other initial condition the state will eventually settle to $\mathbf{x} = [2 \quad 0]^T$. Figure 15.1 shows one representative phase plot which begins near the unstable focus, with $\mathbf{x}(0) = [-2, 0.1]^T$ and settles at the stable focus. ■

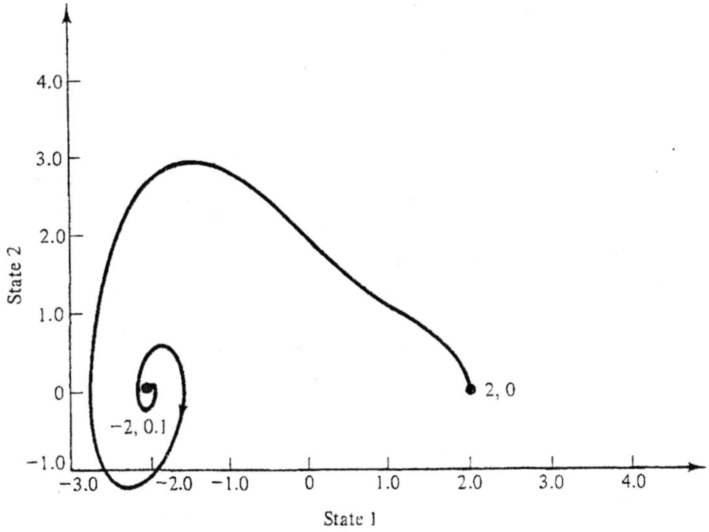

Figure 15.1

If a time-varying nominal solution $\{\mathbf{x}_n(t), \mathbf{u}_n(t), \mathbf{y}_n(t)\}$ is used (perhaps as obtained from numerical solution of Eq. (*15.1*)), then the Jacobian matrices of Eq. (*15.2*) will also be time-varying in general. As was pointed out in Sec. 10.7 the stability of linear time-varying systems is not as straightforward as the linear, constant case.

With $\delta u(t)$ restricted to zero, Eq. (*15.2*) can be used to investigate the passive behavior of perturbed trajectories. It is of interest to know whether a trajectory $\mathbf{x}(t)$ will passively return to $\mathbf{x}_n(t)$ (i.e., asymptotic stability) or will remain within some bounded neighborhood of it (i.e., stability i.s.L.) or will diverge from it (i.e., unstable). These types of analysis must always be used with caution because of the assumptions made regarding $\delta\mathbf{x}(t)$ remaining small.

The input perturbation $\delta\mathbf{u}(t)$ can be used to actively control the behavior of $\delta\mathbf{x}(t)$, thus forcing it to return to and remain at or near zero. Thus the linearizing assumption that $\delta\mathbf{x}(t)$ is small can be made somewhat self-fulfilling. A linear feedback control law, $\delta\mathbf{u}(t) = -\mathbf{K}\delta\mathbf{x}(t)$, could be used, and a typical implementation is shown in Figure 15.2. The overall goal is to maintain the trajectory near the known, precomputed nominal in spite of initial condition perturbations or input disturbances. The gain matrix $\mathbf{K}$ in the control law could be computed using pole placement techniques of Sec. 13.4. If the closed-loop poles are forced to be sufficiently stable, then $\delta\mathbf{x}(t)$ will rapidly return to $\mathbf{0}$ after any upset. Alternatively, the gain $\mathbf{K}$ could be found as the result of an optimal regulator design problem, as discussed in Sec. 14.4.

EXAMPLE 15.2 The equations for the orbit-plane motion of a satellite in orbit about a planet with an ideal inverse-square gravity field are

$$\ddot{r} - \dot{\theta}^2 r = -\frac{\mu}{r^2} + a_r$$

$$r\ddot{\theta} + 2\dot{r}\dot{\theta} = a_i$$

where a_r and a_i are the radial and in-track components of any acceleration terms due to thrust, drag, gravitational anomalies, and the like. These will be treated as components of the input vector $\mathbf{u}$. Determine the equations that describe small perturbations about a circular orbit of radius R.

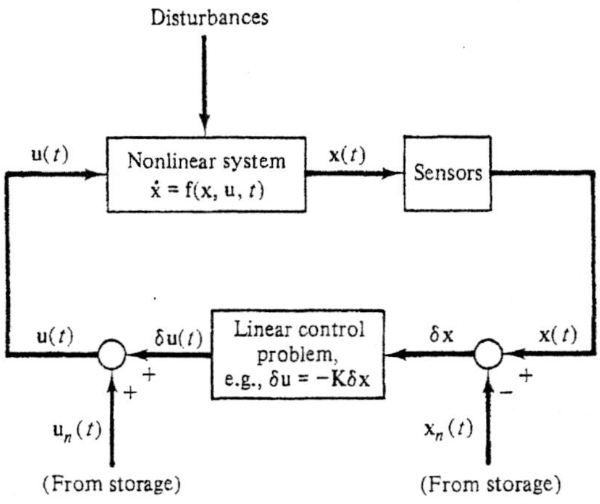

Figure 15.2

Let the state vector be $\mathbf{x} = [r \quad \theta \quad \dot{r} \quad \dot{\theta}]^T$. Then the nonlinear state equations are

$$\dot{\mathbf{x}} = \begin{bmatrix} x_3 \\ x_4 \\ x_4^2 x_1 - \mu/x_1^2 + u_1 \\ -\dfrac{2x_3 x_4}{x_1} + \dfrac{u_2}{x_1} \end{bmatrix} = \mathbf{f}(\mathbf{x}, \mathbf{u})$$

The Jacobian matrices are

$$\frac{\partial \mathbf{f}}{\partial \mathbf{x}} = \begin{bmatrix} 0 & 0 & 1 & 0 \\ 0 & 0 & 0 & 1 \\ x_4^2 + \dfrac{2\mu}{x_1^3} & 0 & 0 & 2x_4 x_1 \\ \dfrac{2x_3 x_4}{x_1^2} - \dfrac{u_2}{x_1^2} & 0 & \dfrac{-2x_4}{x_1} & \dfrac{-2x_3}{x_1} \end{bmatrix}$$

and

$$\frac{\partial \mathbf{f}}{\partial \mathbf{u}} = \begin{bmatrix} 0 & 0 \\ 0 & 0 \\ 1 & 0 \\ 0 & \dfrac{1}{x_1} \end{bmatrix}$$

For a circular nominal orbit, $x_{1n} = R$, $x_{3n} = \dot{R} = 0$, and $x_{4n} = \dot{\theta}_n = \omega$. To maintain R constant, $\dot{R}$ and $\ddot{R}$ must be zero, which leads to the relation $\omega = \mu/R^3$. x_{2n} does not appear explicitly in the linearized equations but clearly will increase linearly with time, since its derivative $x_{4n} = \omega$ is constant. The nominal values of both components of $\mathbf{u}$ are zero. Using these results gives the linear perturbation model

$$\delta\dot{\mathbf{x}} = \begin{bmatrix} 0 & 0 & 1 & 0 \\ 0 & 0 & 0 & 1 \\ 3\omega^2 & 0 & 0 & 2R\omega \\ 0 & 0 & \dfrac{-2\omega}{R} & 0 \end{bmatrix} \delta\mathbf{x} + \begin{bmatrix} 0 & 0 \\ 0 & 0 \\ 1 & 0 \\ 0 & \dfrac{1}{R} \end{bmatrix} \delta\mathbf{u}$$

The eigenvalues of $[\partial\mathbf{f}/\partial\mathbf{x}]_n$ can be found to be $\lambda_i = \{0, 0, j\omega, -j\omega\}$. Since they are all on the $j\omega$ axis, the linear model gives inconclusive results regarding stability of the nonlinear system. The linear model is very useful for studying the effect of perturbations away from the nominal circular orbit and remains accurate for sizable changes in altitude, as long as δr is small compared to the (large) nominal R value. Rocket thrusters can be used to actively drive observed perturbations back to zero. In order to predict the future effect of state perturbations, it is useful to know that the transition matrix is

$$\Phi(t, 0) = \begin{bmatrix} 4 - 3\cos\omega t & 0 & \dfrac{\sin\omega t}{\omega} & \dfrac{-2R(\cos\omega t - 1)}{\omega} \\ \dfrac{6(\sin\omega t - \omega t)}{R} & 1 & \dfrac{-2(\cos\omega t - 1)}{R\omega} & \dfrac{4\sin\omega t}{\omega} - 3t \\ 3\omega\sin\omega t & 0 & \cos\omega t & 2R\sin\omega t \\ \dfrac{6\omega(\cos\omega t - 1)}{R} & 0 & \dfrac{-2\sin\omega t}{R} & 4\cos\omega t - 3 \end{bmatrix}$$

The effect of the accelerations $\delta\mathbf{u}$ over a period T is given by $\int_0^T \Phi(T,\tau)[\partial\mathbf{f}/\partial\mathbf{u}]\delta\mathbf{u}(\tau)\,d\tau$, and for $\delta\mathbf{u}$ constant over the interval $[0, T]$, this can be integrated to obtain

$$\int_0^T \Phi(T,\tau)\left[\frac{\partial\mathbf{f}}{\partial\mathbf{u}}\right]\delta\mathbf{u}(\tau)\,d\tau = \begin{bmatrix} \dfrac{1-\cos\omega T}{\omega^2} & \dfrac{-2(\sin\omega T-\omega T)}{\omega^2} \\ \dfrac{-2(\sin\omega T-\omega T)}{\omega^2 R} & \dfrac{4(1-\cos\omega T)}{R\omega^2}-\dfrac{3T^2}{2R} \\ \dfrac{\sin\omega T}{\omega} & \dfrac{2(1-\cos\omega T)}{\omega} \\ \dfrac{-2(1-\cos\omega T)}{R\omega} & \dfrac{4\sin\omega T}{R\omega}-\dfrac{3T}{R} \end{bmatrix}\delta\mathbf{u}$$

■

15.3 DYNAMIC LINEARIZATION USING STATE FEEDBACK

In the previous section local linearization of a nonlinear system was investigated. We now consider the problem of synthesizing a control input $\mathbf{u}(t)$, which will cause the system

$$\dot{\mathbf{x}} = \mathbf{f}(\mathbf{x}, \mathbf{u}, t) \tag{15.4}$$

to have a response which matches some specified template system. That is, let $\mathbf{y}(t) = \mathbf{H}\mathbf{x}(t)$. It is desired that $\mathbf{y}(t)$ match as closely as possible the response of the specified template system

$$\dot{\mathbf{y}}_d = \mathbf{g}(\mathbf{y}_d, \mathbf{y}, t) \tag{15.5}$$

In a typical example, the $\mathbf{g}$ function might specify a linear system,

$$\dot{\mathbf{y}}_d = \mathbf{F}\mathbf{y}_d + \mathbf{G}\mathbf{v} \tag{15.6}$$

with $\mathbf{v}(t)$ being perhaps a step function input and with the response possessing certain desirable transient characteristics. If a control input $\mathbf{u}(t)$ can be found to achieve the goal $\dot{y} \equiv \dot{\mathbf{y}}_d$, then the original system will behave as a linear system. This is what we term *dynamic linearization.*

Define the error $\mathbf{e}(t) = \mathbf{H}\mathbf{x}(t) - \mathbf{y}_d(t)$. Then

$$\begin{aligned}\dot{\mathbf{e}}(t) &= \mathbf{H}\dot{\mathbf{x}}(t) - \dot{\mathbf{y}}_d(t) \\ &= \mathbf{H}\mathbf{f}(\mathbf{x}, \mathbf{u}, t) - \mathbf{g}(\mathbf{y}_d, \mathbf{v}, t)\end{aligned} \tag{15.7}$$

Suppose for the moment that $\mathbf{H} = \mathbf{I}$. When $\dot{\mathbf{e}}$ is set to zero, it may be possible to solve the resulting equation for the unknown input $\mathbf{u}$ in terms of known or measurable quantities $\mathbf{x}$, $\mathbf{y}_d$, and $\mathbf{v}$. If this is accomplished, the feedback-modified system will have the same derivative as the template system. If the template system is linear, then the original system will have been linearized. In essence the nonlinearities of the original system are canceled and replaced by the desired linear terms. This form of dynamic linearization has been known for many years [4, p. 560].

EXAMPLE 15.3 [5–7] It is desired that the first-order nonlinear system

$$\dot{x} = x + u + xu$$

behave like the linear system

$$\dot{y}_d = -\sigma y_d \qquad \text{with initial condition } y_d(0) = 10$$

For this scalar system let $y(t) = x(t)$. Setting $\dot{x} = \dot{y}_d$ leads to $u(t) = [-x(t) - \sigma y_d(t)]/[1 + x(t)]$, provided $x(t) \neq -1$. At least two potential problems exist with this scheme. (1) Even if the derivaties can be made to match exactly, the initial condition $x(0)$ may not match $y_d(0)$ for a variety of reasons. The exact initial conditions may not be known due to measurement error, or the desire may be to have the system respond like the template system regardless of its initial $x(0)$ value. Of course, matching derivatives does not mean matching response curves. This will be addressed in the sequel. (2) The resulting control law for $u(t)$ has a singularity at $x(t) = -1$. An infinite amount of control would be required at this point. Truxal [4] pointed out that forcing a nonlinear system to respond like a linear system generally means that components must be overdesigned to allow the avoidance of nonlinear behavior. The singularity of this example is an extreme case of this. ∎

The problem of initial condition mismatch, either deliberate or unintentional, can be addressed by adding a convergence factor matrix **S**, as follows. Instead of setting $\dot{\mathbf{e}} = \mathbf{0}$, we require that

$$\dot{\mathbf{e}} = \mathbf{S}\mathbf{e} \tag{15.8}$$

In a similar manner to the development of Chapter 13 for state variable observers, the matrix **S** is specified with asymptotically stable eigenvalues. Then $\mathbf{e}(t) \to \mathbf{0}$, and thus $\mathbf{Hx}(t) \to \mathbf{y}_d(t)$ at a rate controlled by choice of **S**. Note that the previous development is a special case with $\mathbf{S} \equiv [0]$. The equation for finding the control $\mathbf{u}(t)$ is thus

$$\mathbf{Hf}(\mathbf{x}, \mathbf{u}, t) = \mathbf{g}(\mathbf{y}_d, \mathbf{v}, t) + \mathbf{S}[\mathbf{Hx}(t) - \mathbf{y}_d(t)] \tag{15.9}$$

The existence of a solution of Eq. (*15.9*) for $\mathbf{u}(t)$ can be established in certain cases by using the implicit function theorem [8], which establishes sufficient conditions on the function **f**. The solvability is also influenced by the number of independent equations that need to be satisfied relative to the number of unknown control components. This explains the existence of the matrix **H**. It will not generally be possible to match all n components of **x** to an n-dimensional $\mathbf{y}_d$ vector when there are only $r < n$ components in the control vector **u**. General conditions for solvability are addressed in Problems 15.5, 15.6, and 15.7. For any specific problem, a direct attempt to solve for **u** will often be the most expedient method of determining whether or not such a solution can be found, and this is the approach presented in the example problems. Assuming the existence of a solution, the control law will be of the feedback form

$$\mathbf{u}(t) = \mathbf{u}(\mathbf{x}(t), \mathbf{y}_d(t), \mathbf{v}(t), \mathbf{H}, \mathbf{S}) \tag{15.10}$$

and the procedure is obviously a model-matching or model-tracking scheme, as shown in Figure 15.3.

Note that when a linear system is used as the template, then Eq. (*15.9*) becomes

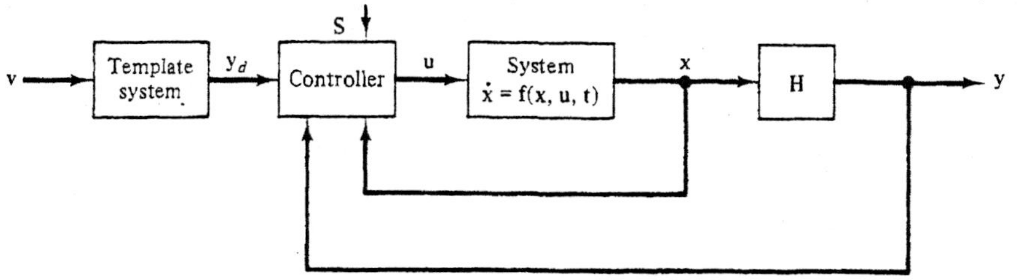

Figure 15.3

$$\begin{aligned}\mathbf{Hf}(\mathbf{x}, \mathbf{u}, t) &= \mathbf{F}\mathbf{y}_d + \mathbf{G}\mathbf{v} + \mathbf{S}[\mathbf{Hx} - \mathbf{y}_d] \\ &= [\mathbf{F} - \mathbf{S}]\mathbf{y}_d + \mathbf{SHx} + \mathbf{Gv}\end{aligned} \tag{15.11}$$

If the convergence matrix $\mathbf{S}$ is selected equal to $\mathbf{F}$, then $\mathbf{y}_d$ is not directly required, and the need to synthesize the template system is removed. If $\mathbf{S} = [\mathbf{0}]$ is selected, a major feedback path is eliminated. Figure 15.4a shows the general form, with $\mathbf{S}$ presumably being selected to be somewhat faster than $\mathbf{F}$—i.e., eigenvalues more negative. Figure 15.4b shows the result when $\mathbf{S} = \mathbf{F}$, and Figure 15.4c shows the configuration with $\mathbf{S} = [\mathbf{0}]$. In all cases the "solve" box refers to finding the input $\mathbf{u}$ which satisfies Eq. *(15.11)*.

EXAMPLE 15.4 The scalar system of Example 15.3 is reconsidered, but now the convergence factor S is included so that for all initial conditions, $x(0)$, $x(t)$ will ultimately approach the desired response $y_d(t)$. Set $\dot{e} = \dot{x} - \dot{y}_d = Se$, where S is a negative real number. Then

$$x + u + xu + \sigma y_d = S(x - y_d)$$

from which, if $x(t) \neq -1$,

$$u(t) = \frac{S[x(t) - y_d(t)] - x(t) - \sigma y_d}{1 + x(t)}$$

Substituting this back into the system equations gives the coupled pair

$$\begin{bmatrix} \dot{x} \\ \dot{y}_d \end{bmatrix} = \begin{bmatrix} S & -(\sigma + S) \\ 0 & -\sigma \end{bmatrix} \begin{bmatrix} x \\ y_d \end{bmatrix}$$

The 2×2 transition matrix is easily found using methods of Chapter 8:

$$\Phi(t, 0) = \begin{bmatrix} e^{St} & e^{-\sigma t} - e^{St} \\ 0 & e^{-\sigma t} \end{bmatrix}$$

Then the system response is

$$x(t) = e^{-\sigma t} y_d(0) + [x(0) - y_d(0)]e^{St}$$

This is the desired template response, $e^{-\sigma t} y_d(0)$, plus an initial condition mismatch term, which dies out at a rate determined by the convergence factor S. If $S << -\sigma$, this term will quickly die out, leaving only the desired response. ■

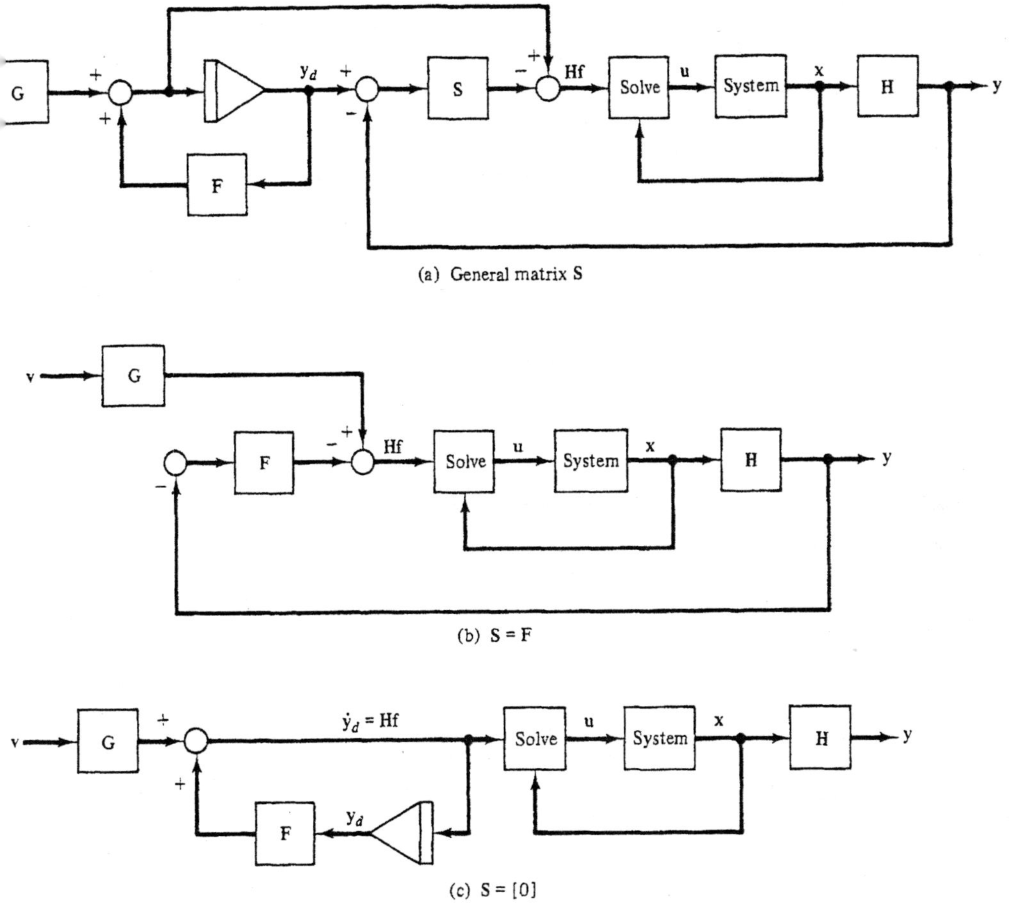

Figure 15.4

4 *HARMONIC LINEARIZATION: DESCRIBING FUNCTIONS* [4, 9–11]

Harmonic linearization is an approximate method of analyzing certain kinds of nonlinear systems using well-understood linear methods. It differs from the small perturbation linearization method of Sec. 15.2 in that signal amplitudes are not restricted to be small perturbations from known nominal values or equilibrium points. Rather, large amplitude signals can be treated, provided they are nearly sinusoidal and certain additional conditions are satisfied. Initially the systems under discussion will be limited to nth-order unforced linear systems with one scalar-valued nonlinearity, $n(x_i)$, which depends on a single state x_i as its input.

$$\dot{\mathbf{x}} = \mathbf{A}\mathbf{x} - \mathbf{B}n(x_i) \tag{15.12}$$

The minus sign is arbitrary, since it could have been absorbed into the definition of the input column matrix **B**. The chosen form allows representing Eq. (*15.12*) in the traditional single loop, negative-feedback block diagram form of Figure 15.5, with $G(s) = \mathbf{C}[s\mathbf{I} - \mathbf{A}]^{-1}\mathbf{B}$. $G(s)$ is the linear transfer function from the output u of the nonlinearity to x_i, which plays the role of the system output y, and thus $y = \mathbf{Cx}$ determines the appropriate matrix **C**. If the input e to the nonlinearity is *assumed* sinusoidal,

$$e(t) = E \sin \omega t$$

the output $u(t)$ will generally also be periodic but not just a pure sinusoid. The first few terms of its Fourier series expansion are given as

$$u(t) = b_0 + a_1 \sin \omega t + b_1 \cos \omega t + a_2 \sin 2\omega t + b_2 \cos 2\omega t + a_3 \sin 3\omega t + b_3 \cos 3\omega t + \cdots \tag{15.13}$$

The coefficients are given by

$$a_k = \frac{1}{T}\int_{-T}^{T} \sin(k\omega t) n(t)\, dt \qquad \text{where } T = \frac{\pi}{\omega}$$

$$b_k = \frac{1}{T}\int_{-T}^{T} \cos(k\omega t) n(t)\, dt \qquad \text{for } k > 0$$

$$b_0 = \frac{1}{2T}\int_{-T}^{T} n(t)\, dt$$

The output of the linear portion of the system—namely, $G(s)$—will likewise be the sum of terms, one from each component in the expansion for u. The *essential* assumption upon which the validity of harmonic linearization is based is that $G(s)$ is of a sufficiently low-pass nature, so that all harmonics are attenuated to a negligible level. Only the fundamental frequency terms survive the trip around the loop and have an effect on the input to the nonlinearity. Furthermore, many interesting nonlinearities have a zero average value, so that b_0 is zero. (This assumption can be removed.) Then the important part of the time-domain nonlinearity output caused by the input $E \sin \omega t$ is just $a_1 \sin \omega t + b_1 \cos \omega t$. The ratio of the Laplace transforms of the nonlinearity output and input is thus $N(s) = [a_1 \omega + b_1 s]/E\omega$. This ratio has no meaning except for sinusoidal signals, and when $s = j\omega$ it reduces to

$$N(j\omega) = \frac{a_1}{E} + \frac{b_1 j}{E} = \frac{\sqrt{a_1^2 + b_1^2}\, e^{j\varphi}}{E}$$

where

$$\varphi = \tan^{-1}\left(\frac{b_1}{a_1}\right) \tag{15.14}$$

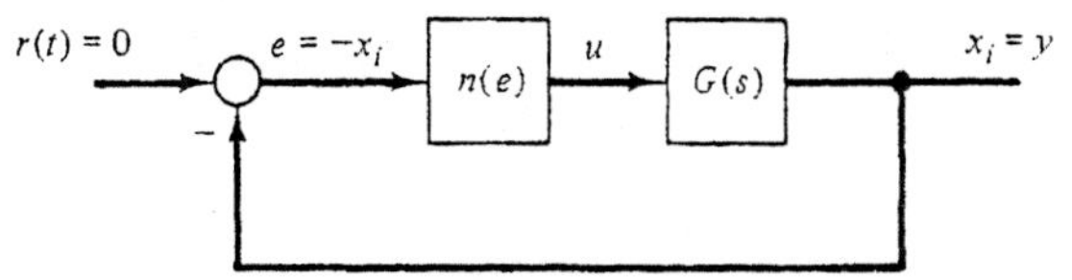

Figure 15.5

This is the *describing function* for the nonlinearity. It is the ratio of the magnitude of the fundamental output component to the mangitude of the input to the nonlinearity and as such plays the role of an equivalent gain. To emphasize that it depends on the input magnitude E as well as the frequency, we will write it as $N(E, \omega)$ from now on. Notice that the phase shift φ will be zero and the describing function N will be purely real for any single-valued odd symmetric nonlinearity, since then $b_k = 0$. In this case all even harmonic terms a_k are also zero, by symmetry. Thus the first neglected term is the a_3 term, which (1) is usually much smaller than a_1 to begin with and (2) is more strongly attenuated by $G(s)$ because of its higher frequency. Many odd symmetric nonlinearities are very accurately accounted for using the describing function approach.

For many nonlinearities, N is not really a function of frequency but only of input amplitude E. In those cases the nonlinearity is replaced by a real but signal-dependent gain, N. Viewing the describing function as an equivalent gain is most helpful in the applications of describing functions to stability analysis that follow.

EXAMPLE 15.5 Derivation of a Describing Function A large number of odd-symmetric nonlinearities of engineering significance can be represented by three linear segments, as shown in Figure 15.6*a* and as determined by three slopes K_1, K_2, and K_3 and two breakpoints α and β. For example, the dead zone nonlinearity in Figure 15.6*b* has $\alpha = 0$, $\beta = 1$, $K_2 = 0$, $K_3 = 1$, and K_1 arbitrary. The saturation characteristic of Figure 15.6*c* has $\beta = \infty$, $K_2 = 0$, and K_3 arbitrary. The relay of Figure 15.6*d* (which has the same shape as coulomb friction and preload) has $\alpha = 0$, $\beta = 0$ (or some small ϵ to prevent a discontinuity), K_1 arbitrary, $K_2 = 1/\epsilon$, and $K_3 = 0$. Many other combinations can be formed, such as a relay with dead zone, by proper choice of these five parameters. In order to derive the describing function for this nonlinearity, ten separate time increments must be considered for each full (normalized) cycle $t' = \omega t \in [0, 2\pi]$ in the most general case. (Sketch the input sine wave and identify when its magnitude falls in each segment of the nonlinearity.) In the following discussion, the prime is dropped from the normalized t' for convenience.

1. $0 < t < t_\alpha$, input sinusoid value $e(t) < \alpha$, output $n(e) = n_1(t) = K_1 E \sin t$
2. $t_\alpha < t < t_\beta, \alpha < e(t) < \beta, n(e) = n_2(t) = K_2 E[\sin t - \sin t_\alpha] + K_1 E \sin t_\alpha$
3. $t_\beta < t < t_{\beta'}, e(t) > \beta, n(e) = n_3(t) = K_3 E[\sin t - \sin t_\beta] + K_2 E[\sin t_\beta - \sin t_\alpha] + K_1 E \sin t_\alpha$
4. $t_{\beta'} < t < t_{\alpha'}, \alpha < e(t) < \beta$ (same as region 2)
5. $t_{\alpha'} < t < \pi, 0 < e(t) < \alpha$ (same as region 1)

The last five regions divide the negative half-cycle $[\pi, 2\pi]$ into the same regions as (1) through (5). The values for t_α and t_β are functions of the input amplitude E and are given by

$$t_\alpha = \sin^{-1}\left(\frac{\alpha}{E}\right) \text{ if } E > \alpha \quad \text{and} \quad t_\alpha = \frac{\pi}{2} \text{ otherwise}$$

$$t_\beta = \sin^{-1}\left(\frac{\beta}{E}\right) \text{ if } E > \beta \quad \text{and} \quad t_\beta = \frac{\pi}{2} \text{ otherwise}$$

Symmetry gives the other times, such as

$$t_{\beta'} = \pi - t_\beta \quad \text{and} \quad t_{\alpha'} = \pi - t_\alpha$$

The kth harmonic coefficient can be evaluated from

$$a_k = \left(\frac{4}{\pi}\right)\left\{\int_0^{t_\alpha} n_1(t)\sin(kt)\,dt + \int_{t_\alpha}^{t_\beta} n_2(t)\sin(kt)\,dt + \int_{t_\beta}^{\pi/2} n_3(t)\sin(kt)\,dt\right\}$$

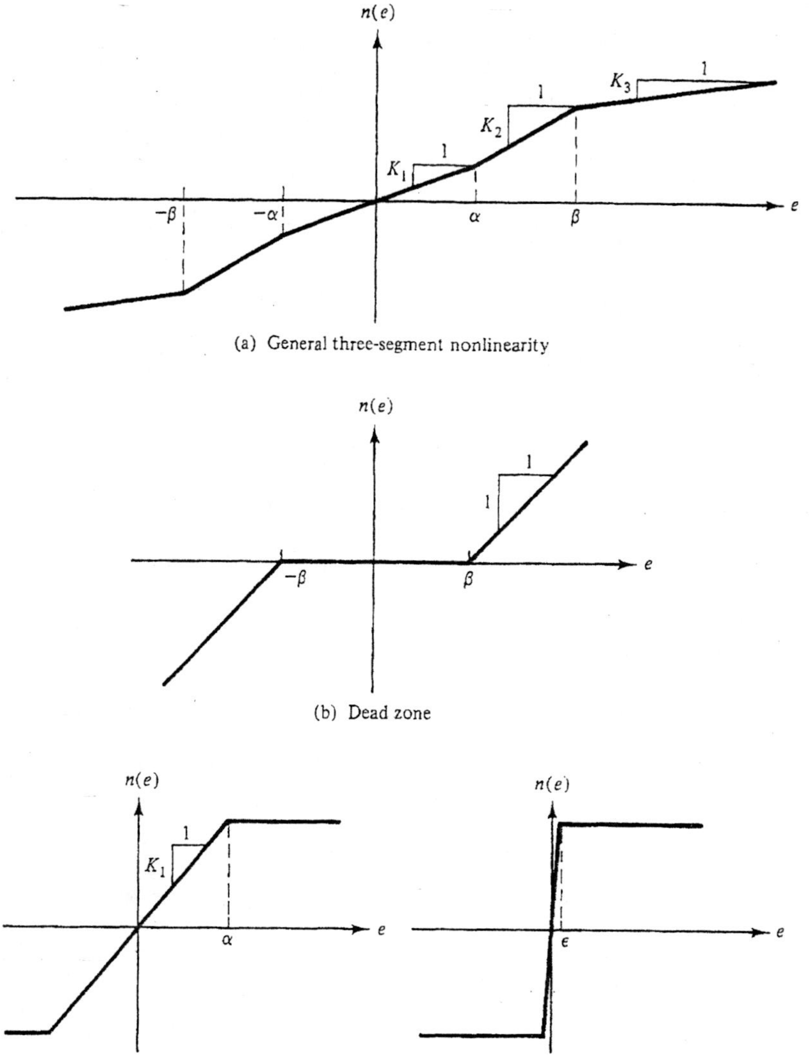

(a) General three-segment nonlinearity

(b) Dead zone

(c) Saturation characteristic

(d) Relay

Figure 15.6 Some common single-valued odd symmetric nonlinearities.

Evaluation of these integrals for $k = 1$ gives the describing function, which is independent of frequency ω:

$$N(E) = \frac{a_1}{E} = \frac{2(K_1 - K_2)[t_\alpha + \sin(2t_\alpha)/2]}{\pi}$$

$$+ \frac{2(K_2 - K_3)[t_\beta - \sin(2t_\beta)/2]}{\pi} + K_3 + \frac{2K_2 \sin(2t_\beta)}{\pi}$$

Note that if E is less than α, then both t_α and t_β are equal to $\pi/2$, and $N(E)$ reduces to just K_1 as it should. If $t_\alpha = 0$ and $t_\beta = \pi/2$, then $N(E)$ reduces to K_2. A computer is helpful in evaluating all except these special limiting cases. ∎

The describing functions for the preceding class of nonlinearities are real and independent of the excitation frequency. This is not always true, as demonstrated by the following multiple-valued nonlinearity and by the example in Problem 15.10.

EXAMPLE 15.6 Relay with Hysteresis Figure 15.7 shows a common model for a relay with hysteresis. Assuming $E > \alpha$, there are three critical time periods during a typical cycle of the input sine wave:

$0 < t < t_1$: $0 < e(t) < \alpha$, $e(t)$ rising to α, for which $N(e) = -M$

$t_1 < t < t_2$: $-\alpha < e(t) < \alpha$, $e(t)$ falling toward $-\alpha$, for which $N(e) = M$

$t_2 < t < 2\pi$: $e(t) < \alpha$, $e(t)$ again rising toward α, $N(e) = -M$

From Figure 15.7 it is seen that $t_1 = \sin^{-1}(\alpha/E)$ and $t_2 = \sin^{-1}(-\alpha/E) = \pi + t_1$. The two fundamental Fourier coefficients are thus

$$a_1 = \left(\frac{M}{\pi}\right)\left\{-\int_0^{t_1} \sin t\,dt + \int_{t_1}^{t_2} \sin t\,dt - \int_{t_2}^{2\pi} \sin t\,dt\right\} = \frac{4M\ \cos(t_1)}{\pi}$$

$$= \left(\frac{4M}{\pi}\right)\sqrt{1-\left(\frac{\alpha}{E}\right)^2}$$

$$b_1 = \left(\frac{M}{\pi}\right)\left\{-\int_0^{t_1} \cos t\,dt + \int_{t_1}^{t_2} \cos t\,dt - \int_{t_2}^{2\pi} \cos t\,dt\right\} = -\frac{4M\ \sin(t_1)}{\pi}$$

$$= -\frac{4M\alpha}{\pi E}$$

so that the describing function is

$$N(\omega, E) = \left(\frac{4M}{\pi E}\right)\left\{\sqrt{1-\left(\frac{\alpha}{E}\right)^2} - \frac{j\alpha}{E}\right\} \qquad \text{for } E > \alpha$$

When $E < \alpha$, a_1, b_1, and N are all zero. ■

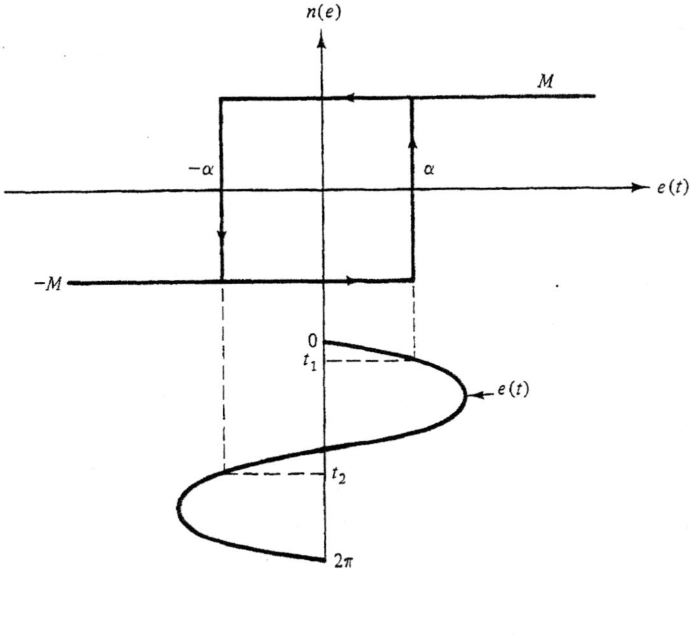

Figure 15.7

Other examples of describing functions are given in the problems. Extensive tabulations for every conceivable type of nonlinearity can be found in the references [9, 13]. For actual system components, describing functions can be determined experimentally.

15.5 APPLICATIONS OF DESCRIBING FUNCTIONS

A distinctive characteristic of nonlinear systems is the possibility of exhibiting limit cycles. A limit cycle is a self-excited, self-sustaining periodic oscillation. A stable limit cycle is one for which trajectories slightly perturbed from it are attracted back to it. An unstable limit cycle is one for which perturbed trajectories do not return to it. Rather, they approach other limit cycles, equilibrium points, grow without bound, or approach some other bounded but nonperiodic trajectories. The later case is now referred to as *chaos* [14–17]. The major uses of describing functions are (1) to investigate the stability of systems like the one in Figure 15.5, (2) to investigate the possible existence of limit cycles and to predict their amplitude and frequency when they exist, and (3) to modify or compensate the linear system to prevent the occurrence of undesirable limit cycles (or chaos?). A limit cycle is by definition periodic, although the waveform is not necessarily sinusoidal. If the limit cycle has a strong fundamental Fourier component, then describing functions can provide acceptably accurate results in their analysis. Consider the system described by Eq. (*15.12*), with the nonlinearity replaced by its describing function approximation, N. The unforced closed-loop system is described by

$$\dot{\mathbf{x}} = (\mathbf{A} - \mathbf{B}[\mathbf{N}]\mathbf{C})\mathbf{x} \tag{15.15}$$

The output matrix $\mathbf{C}$ is used to select the components of $\mathbf{x}$ which are input to the nonlinearities. At this point the factor $[\mathbf{N}]$ could be a matrix of describing functions representing multiple nonlinearities, $\mathbf{n}(\mathbf{x}) \approx [\mathbf{N}]\mathbf{C}\mathbf{x}$. The closed-loop eigenvalues satisfy

$$|\mathbf{I}\lambda - (\mathbf{A} - \mathbf{B}[\mathbf{N}]\mathbf{C})| = 0 \tag{15.16}$$

This can be rearranged as

$$\begin{aligned}|\mathbf{I}\lambda - \mathbf{A}|\cdot|\mathbf{I} + (\mathbf{I}\lambda - \mathbf{A})^{-1}\mathbf{B}\mathbf{C}[\mathbf{N}]| &= |\mathbf{I}\lambda - \mathbf{A}|\cdot|\mathbf{I} + [\mathbf{N}]\mathbf{C}(\mathbf{I}\lambda - \mathbf{A})^{-1}\mathbf{B}| \\ &= |\mathbf{I}\lambda - \mathbf{A}|\cdot|\mathbf{I} + [\mathbf{N}]G(\lambda)| = 0\end{aligned}$$

The whole concept of describing functions is based on periodic solutions, so this result has meaning only for complex conjugate sets of λ values. Thus the condition for existence of periodic solutions is that

$$|\mathbf{I} + [N]G(j\omega)| = 0 \tag{15.17}$$

In the simplest case of one nonlinearity, this determinant is just a scalar equation $1 + NG(j\omega) = 0$. Points at which $NG(j\omega) = -1$ identify potential limit cycles. These points can be determined either analytically or graphically (by using Nyquist, Bode, or log magnitude versus angle plots of $G(j\omega)$ and comparing them with the critical point $-1/N$ from the describing function, rather than the usual point -1). That is, $G(j\omega) = -1/N$ can be analyzed, or alternatively inverse Nyquist plots can be used with $N = -1/G(j\omega)$.

Real, Frequency-Independent Describing Functions

In cases such as Example 15.5, where N is purely real and independent of frequency, all the usual conclusions apply. In particular, the number of unstable closed-loop poles Z_r (called Z because they are zeros of the characteristic equation) can be determined from Nyquist's criterion as

$$Z_r = P_r - N_c \qquad (15.18)$$

where P_r is the number of right-half plane open-loop poles of G and N_c is the number of counterclockwise encirclement of the critical point $-1/N$ by the polar plot of $G(j\omega)$ as ω ranges over $-\infty$ to ∞. In the case of real, frequency-independent $N(E, \omega)$, root locus techniques can also be used (with caution) on $NG(s) = -1$. The describing function N plays the role of a varying gain. Since this is valid only for $s = j\omega$, the root locus results are helpful only in predicting behavior near the $j\omega$ axis crossover points. Attempts to correlate frequency response results with transient response behavior using known root locus methods are generally not very satisfactory.

The General Case

When $N(E, \omega)$ is complex and frequency-dependent, the Nyquist or Bode method can be applied without modification to find the limit cycle locations. Stability conclusions drawn from Eq. (*15.18*) can be wrong, however. The number of encirclements of $NG(j\omega)$ is no longer determined solely by the behavior of $G(j\omega)$. The safest approach is always to rely on the basic equation $NG(j\omega) = -1$, but this means the advantage of seeing separately the effects of the linear G and the nonlinearity is lost. In some cases it may be possible to separate $N(E, \omega)$ into a product $N(E, \omega) = N_1(E)N_2(\omega)$, with N_1 real. Then a modified $G'(j\omega) = G(j\omega)N_2(\omega)$ can be used in $G'(j\omega) = -1/N_1(E)$. Equation (*15.18*) can now be used again with $G'(j\omega)$, which will generally not have the same number of encirclements as $G(j\omega)$. See Problems 15.11 and 15.12 for an example. The root locus procedure must be similarly modified. The polar form of $N(E, j\omega) = Ke^{j\varphi}$ shows that if a root locus approach were to be attempted using $G(j\omega)$, the 180° locus would not suffice. It would need to be modified by the phase shift φ, which is generally frequency-dependent. This approach is usually not worth the required effort. In those cases where it is possible to form $G'(s)$, the root locus approach can be useful.

EXAMPLE 15.7 A single-loop system has $G(s) = 2/[s(s+p)]$ cascaded with the hysteresis-type relay of Example 15.6, with $\alpha = M = 1$. If this system can exhibit a limit cycle, find its approximate amplitude and frequency by using describing functions.

For this scalar case Eq. (*15.17*) gives the limit cycle requirement as $4/(\pi E)\{\sqrt{1-1/E^2} - j/E\} = -j\omega(j\omega + p)/2 = (\omega^2 - j\omega p)/2$. From the imaginary part we find that $\pi\omega p/8 = 1/E^2$. From the real part we find $\omega^2/2 = 4/(\pi E)\sqrt{1-1/E^2}$. Combining gives

$$\omega^3 + p^2\omega - \frac{8p}{\pi} = 0$$

Table 15.1 gives the results for three different pole locations p, as well as numerical results obtained by simulating the actual nonlinear system.

TABLE 15.1

p	4	1	0.5
Describing function predictions:			
ω (rad/s)	0.62161	1.1245	1.0071
Period $T = 2\pi/\omega$ (s)	10.1079	5.5875	6.239
E	1.012	1.5	2.2487
Simulation results:			
Period T	9.3	5.6	6.4
E	1.044	1.6	2.32
Percent error in:			
T	8.7	0.22	2.5
E	3.06	6.25	3.1

Figure 15.8 shows the time response of the input to the nonlinearity for the cases with $p = 4$ and $p = 0.5$. The later case is much more nearly sinusoidal, and this explains the higher accuracy of the describing function in predicting the period in this case. Figure 15.9 shows the phase-plane plot of the limit cycle behavior, with $p = 0.5$, for initial conditions inside and outside the limit cycle. Note that a simple root locus plot shows that this system is always asymptotically stable for any *real* gain N. A real equivalent gain cannot explain the sustained oscillations exhibited here. ■

Although some interesting and useful aspects of describing function analysis of nonlinear systems have been presented here and in the chapter-end problems, space

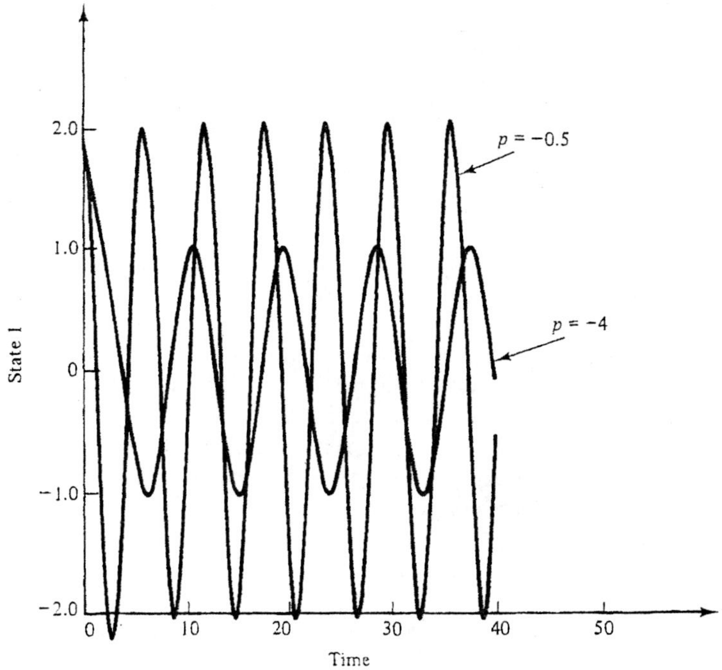

Figure 15.8

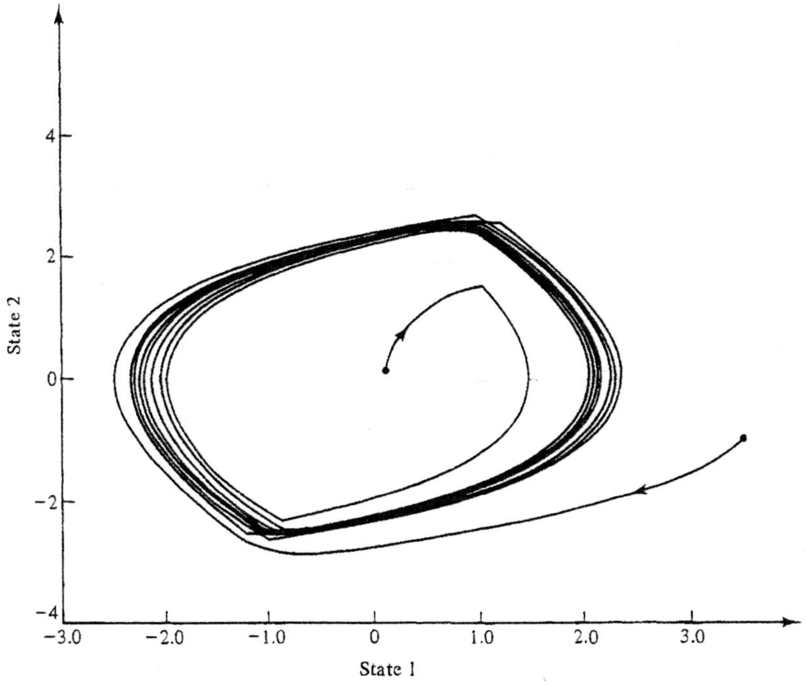

Figure 15.9

limitations have forced omission of several additional considerations, which can be found in references such as 9 and 11. Nonsymmetrical nonlinearities which yield a nonzero average value is one such topic. Another has to do with nonzero external inputs. Harmonic analysis techiques can be applied to such forced, closed-loop systems, and additional interesting phenomena such as jump resonances and subharmonic responses then arise. Significant space has been devoted in the problems to the application of describing functions to systems with (potentially) chaotic behavior. The field of chaos in dynamic systems is still evolving, and only a small segment of it has been discussed here. In control problems the usual goal would be to *avoid* such chaotic behavior (although it can inadvertently creep into various adaptive and self-learning control systems, which are invariably nonlinear [18, 19]). Describing functions are presented here as a tool for understanding and predicting the possibility of chaos in some nonlinear systems. Then, compensation or redesign techniques can be used to avoid such potentially troublesome behavior. Finally, the very useful topic of dual input describing functions has not been developed here. The concept involves the deliberate insertion of a second sinusoidal input, *dither,* to a nonlinearity. Dither is usually of a much higher frequency than the fundamental frequencies of interest. If the assumed low-pass nature of the linear system is present, the high-frequency dither component will have negligible direct effect on most of the system. It can have a pronounced effect on the behavior of the nonlinearity. For example a nonlinearity with deadband, as in Figure 15.6*b*, will have no output until the amplitude of the input $e(t)$ exceeds β. If a high-frequency dither signal $d(t)$ is superimposed onto $e(t)$, then some output will appear whenever $e(t) + d(t)$ exceeds β, so the effective gain of the nonlinearity has been changed. The low-pass filtered envelope of this $e(t) + d(t)$ output can provide positive benefits to system behavior [9, 11].

15.6 LYAPUNOV STABILITY THEORY AND RELATED FREQUENCY DOMAIN RESULTS

The direct method of Lyapunov [20–23] was presented in Sec. 10.6 with sufficient generality to allow application to nonlinear systems being considered here. Although all the stability and instability theorems apply without change, the determination of suitable Lyapunov functions is less straightforward. The solution of Lyapunov's equation, Eq. (*10.22*), to find quadratic Lyapunov functions is generally not adequate. A variety of methods for obtaining Lyapunov functions have been proposed [12, 24–26], and a few of these are illustrated in Problems 15.17 through 15.20. The variable gradient method [27, 28] introduced in Problem 10.14 and illustrated in Problem 10.15 is one fairly general approach. Problems 15.21 through 15.23 demonstrate the variable gradient method on several nonlinear systems.

Lyapunov analysis of nonlinear systems differs from linear systems in that linear system stability conclusions are global. In nonlinear problems, a given equilibrium point may have a finite or local domain of attraction. That is, initial states within that domain will converge to the equilibrium point. Initial points outside the domain of attraction will diverge from it, settle into some sort of limit cycle around it, or perhaps undergo some other more complex behavior (chaos). Lyapunov stability theorems can be used to estimate the extent of the domains of stable attraction. Lyapunov instability theorems can be used to estimate the extent of the unstable (repulsive) behavior around an equilibrium point. We now use the variable gradient method to illustrate this.

EXAMPLE 15.8 Consider the Van der Pol equation $\ddot{y} + \mu(y^2 - 1)\dot{y} + \beta y = 0$. In state variable form the system equations are

$$\dot{x}_1 = x_2 \quad \text{and} \quad \dot{x}_2 = -\mu(x_1^2 - 1)x_2 - \beta x_1$$

This system has a single equilibrium point at the origin. Assume that the gradient of the still unknown Lyapunov function V is

$$\nabla V = \begin{bmatrix} \alpha_{11}x_1 + \alpha_{12}x_2 \\ \alpha_{21}x_1 + \alpha_{22}x_2 \end{bmatrix}$$

The final coefficient α_{22} can be set to 2 without loss of generality, and this will be seen to ensure that V is quadratic in x_2. This will be done here. Assuming that V is not an explicit function of time, the time rate of change of V is

$$\dot{V} = [\nabla V]^T\dot{\mathbf{x}} = [\alpha_{11} - 2\beta - \mu\alpha_{21}(x_1^2 - 1)]x_1x_2 - \alpha_{21}\beta x_1^2$$
$$- [\alpha_{12} - 2\mu(x_1^2 - 1)]x_2^2$$

The troublesome cross terms can be eliminated by selecting

$$\alpha_{11} = 2\beta + \mu\alpha_{21}(x_1^2 - 1)$$

leaving

$$\dot{V} = -\alpha_{21}\beta x_1^2 + [\alpha_{12} - 2\mu(x_1^2 - 1)]x_2^2$$

One easy way to proceed would be to select $\alpha_{12} = \alpha_{21} = 0$. Then, if $\mu > 0$, there is a region defined by $x_1^2 < 1$ in which $\dot{V}$ will be positive. With these choices, $\nabla V^T = [2\beta x_1 \quad 2x_2]$. The so-called curl equations $\partial(\nabla V)_i/\partial x_j = \partial(\nabla V)_j/\partial x_i$ are automatically satisfied, and the line integral easily gives $V = \int (\nabla V)_1\, dx_1 + \int (\nabla V)_2\, dx_2 = \beta x_1^2 + x_2^2$. This is a positive definite function, and a given value of V defines an ellipse. The largest such ellipse which satisfies $x_1^2 < 1$ defines the region Ω, where V and $\dot{V}$ are both positive. The equation for the family of ellipses has no $x_1 x_2$ term so x_1 and x_2 are principal axes. Setting $x_2 = 0$ shows that the maximum value of V is β. Then setting $x_1 = 0$ shows that the maximum $x_2 = \sqrt{\beta}$. By the instability theorem, Theorem 10.6, the system is unstable and Ω is an estimate of the region of repulsion. If μ is negative, the origin is a point of stable equilibrium. The same region Ω would then be an estimate of the region of attraction, since at all points within it, we have $V > 0$ and $\dot{V} < 0$. ■

EXAMPLE 15.9 A different estimate of the regions of attraction or repulsion for the Van der Pol equation can be found by selecting different coefficients for the gradient of V. The same selection for α_{11} is again made to eliminate the cross terms. Now, a tentative choice for $\alpha_{12} = \alpha_{21} = -2\mu$ will give $\dot{V} = -2\mu[x_2 - \beta]x_1^2$. The selected gradient is

$$\nabla V = \begin{bmatrix} [-2\mu^2(x_1^2 - 1) + 2\beta]x_1 - 2\mu x_2 \\ -2\mu x_1 + 2x_2 \end{bmatrix}$$

The curl equations are again satisfied automatically, and a line integral gives

$$V = \left[\beta + \mu^2 - \frac{\mu x_1^2}{2}\right] x_1^2 - 2\mu x_1 x_2 + x_2^2$$

If we define $w = [\beta + \mu^2 - \mu x_1^2/2]$, then

$$V = \mathbf{x}^T \begin{bmatrix} w & -\mu \\ -\mu & 1 \end{bmatrix} \mathbf{x}$$

Using principle minors, this is seen to be positive definite provided $w > 0$ and $w - \mu^2 > 0$—that is, as long as $x_1^2 < 2\beta/\mu$. Likewise, (when $\mu > 0$) $\dot{V} \geq 0$ provided that $x_2^2 < \beta$. It is seen that the limit on x_1 is different here than in the previous example, and the limit on x_2 is the same. Here the figure determined by V = constant is no longer a simple ellipse. The estimate of Ω is given by the largest closed contour V = constant that satisfies the stated limits on x_1 and x_2. The actual boundary of Ω is the interior of a limit cycle. It is of irregular shape, somewhat similar to the limit cycle in Figure 15.9. See Reference 11 or 12 for the precise shape. ■

As demonstrated here and in the problems, finding a suitable Lyapunov function may require a bit of ingenuity. If attention is restricted to a certain subclass of nonlinear systems, then fairly general sufficient conditions for stability can be, and have been, derived by using Lyapunov methods. One such approach, valid for systems with a single nonlinearity, uses a Lyapunov function which is a quadratic (as in the linear system case) plus an integral of the nonlinearity. This approach is usually associated with the name Lur'e [20, 23, 28]. One simple demonstration is given.

Consider the system $\dot{\mathbf{x}} = \mathbf{A}\mathbf{x} - \mathbf{B}n(y)$, where $y = \mathbf{C}\mathbf{x}$ is a scalar input to the nonlinearity. Assume that $\mathbf{A}$ is asymptotically stable. It is also assumed that $n(0) = 0$, so that the origin is the only equilibrium point. Then by setting $\mathbf{Q}=\mathbf{I}$ and solving Lyapunov's equation (10.22), a positive definite matrix $\mathbf{P}$ can be found. In order to study the stability of the nonlinear system, a Lyapunov function is selected as $V=\mathbf{x}^T\mathbf{P}\mathbf{x}+\int_0^y n(\xi)\,d\xi$. Clearly, if the integral term is nonnegative for all $\mathbf{x}$, then V will be positive definite. The time-derivative is given by

$$\begin{aligned}\dot{V} &= \mathbf{x}^T[\mathbf{A}^T\mathbf{P}+\mathbf{P}\mathbf{A}]\mathbf{x}-2n(y)\mathbf{B}^T\mathbf{P}\mathbf{x}+n(y)\dot{y}\\ &= -\mathbf{x}^T\mathbf{x}-n(y)[2\mathbf{B}^T\mathbf{P}-\mathbf{C}\mathbf{A}]\mathbf{x}-\mathbf{C}\mathbf{B}n(y)^2\end{aligned}$$

The question of stability for this class of problems reduces to a determination of whether the nonlinear term $n(y)$ can cause $\dot{V}$ to lose the negative-definiteness which the quadratic term in $\mathbf{x}$ would give. Other strategies for selecting $\mathbf{P}$ may also be used.

EXAMPLE 15.10 Consider the system

$$\dot{\mathbf{x}}=\begin{bmatrix}0 & 1\\ -8 & -6\end{bmatrix}\mathbf{x}-\begin{bmatrix}0\\1\end{bmatrix}n(x_1)$$

The solution of $\mathbf{A}^T\mathbf{P}+\mathbf{P}\mathbf{A}=-\mathbf{I}$ gives

$$\mathbf{P}=\begin{bmatrix}\frac{9}{8} & \frac{1}{16}\\ \frac{1}{16} & \frac{3}{32}\end{bmatrix}$$

Since the input to $n(\ )$ is x_1, $\mathbf{C}=[1\quad 0]$, so $\mathbf{CB}=0$; then

$$\dot{V}=-x_1^2-x_2^2-\frac{n(x_1)x_1}{8}-x_2\left[x_2+\left(\frac{13}{16}\right)n(x_1)\right]$$

This can be rearranged into

$$\dot{V}=-\frac{n(x_1)x_1}{8}-\mathbf{x}^T\begin{bmatrix}1 & \dfrac{13n(x_1)}{32x_1}\\ \dfrac{13n(x_1)}{32x_1} & 1\end{bmatrix}\mathbf{x}$$

The first term is negative for all x_1 except 0 for any nonlinearity which lies in the first and third quadrant of $n(x_1)$ versus x_1 space and which satisfies $n(0)=0$. The quadratic-form term is negative definite, and hence the system is asymptotically stable, provided $n(x_1)^2<x_1^2(\frac{32}{13})^2$. ■

The preceding result is extremely conservative. Recall that Lyapunov theorems give sufficient conditions, not necessary conditions. Problem 15.20 uses the variable gradient method and includes this system as a special case, with $f(x_1)=6$ and $g(x_1)=8x_1+n(x_1)$. There it is shown that a sufficient condition for global asymptotic stability is that $n(x_1)x_1>0$—that is, any nonlinearity in the first and third quadrants. An even more general result is given in Reference 28, where the condition is that $\int n(\xi)\,d\xi>0$. This would include, for example, a nonlinearity such as $n(x_1)=e^{-x_1}\sin(x_1)$ which does not stay within the first and third quadrants. The results derived from the Lur'e approach of using a quadratic plus integral Lyapunov function obviously depend upon the particular quadratic selected. It has been shown that the strongest results that can be obtained by using a Lur'e-type Lyapunov function are those contained in the

Popov criterion [9, 12, 23]. The Popov criterion and the closely similar *circle criterion* are examples of frequency domain stability conditions. The interested reader should consult Reference 23 and the references therein for proofs and to learn about the difficult step of converting standard Lyapunov function results into frequency domain conditions for stability. There are several advantages of the frequency domain results to be given next: (1) They circumvent the difficult step of finding a Lyapunov function; (2) they apply without special regard to the order of the system; and (3) they rely on Nyquist-type stability analysis methods, which are well known from linear systems analysis. The principle disadvantage is that they do not always apply. These conditions are applicable to so-called sector nonlinearities—that is, linearities that are contained between two straight lines through the origin. Hysteresis-type nonlinearities and nonlinearities involving products of several state variables are notable examples that cannot be treated by these methods. To be specific, attention is restricted to systems like the one in Figure 15.5. The nonlinearity is assumed to be a single-valued, piecewise continuous function which satisfies $n(0) = 0$ and $K_l < n(e)/e < K_u$ for $e \neq 0$, where K_l and K_u are slopes of the straight lines which provide the lower and upper bounds of the nonlinearity. This is sometimes stated as $n(e)$ belongs to the sector $[K_l, K_u]$.

Popov's Stability Criterion

If the linear portion of the system $G(s)$ is a proper, asymptotically stable transfer function and if $K_l = 0$, $K_u < \infty$ and $n(e)$ is not an explicit function of time, then Popov's criterion can be applied. It ensures that the system is globally asymptotically stable if a real number q and an arbitrarily small positive δ can be found such that

$$\operatorname{Re}\{(1 + j\omega q)G(j\omega)\} + 1/K_u \geq \delta > 0 \tag{15.19}$$

If $G(s)$ has simple poles *on* the $j\omega$ axis, Popov's criterion remains valid provided the lower bounding slope K_l is larger than some arbitrarily small positive ϵ. Popov's criterion can be given a useful graphical interpretation. To do so we define a Popov locus for the system, which is a modification of the familiar Nyquist locus. The real part is unchanged, and the imaginary part is multiplied by ω. That is, define $G^*(j\omega) = \operatorname{Re}\{G(j\omega)\} + j\omega \operatorname{Im}\{G(j\omega)\}$. Then, Eq. (*15.19*) becomes $\operatorname{Re}\{G^*(j\omega)\} - q \operatorname{Im}\{G^*(j\omega)\} + 1/K_u \geq \delta > 0$. This indicates that a plot of the Popov locus must lie entirely to the right of a straight line through the point $-1/K_u$ and having a finite slope $1/q$.

EXAMPLE 15.11 Find the maximum upper sector bound K_u for the nonlinearity $n(e)$ of Figure 15.5 for which Popov's criterion can assure stability if $G(s) = [s^3 + 13s^2 + 72s + 160]/[s(s^4 + 6s^3 + 26s^2 + 56s + 80)]$.

A portion of the Popov locus is shown in Figure 15.10. The real-axis crossover occurs at $\omega = 2.866$, with $|G^*(j\omega)| = 0.965$. Thus any value of $-1/K_u$ more negative than -0.965 admits a finite positive sloping line that is totally to the left of the locus. That is, $K_u = 1/0.965 = 1.036$ is the maximum slope of the upper sector bound for which Popov can guarantee stability. Note that since $G(s)$ has one pole on the $j\omega$ axis, the lower sector slope must be larger than some positive ϵ. ■

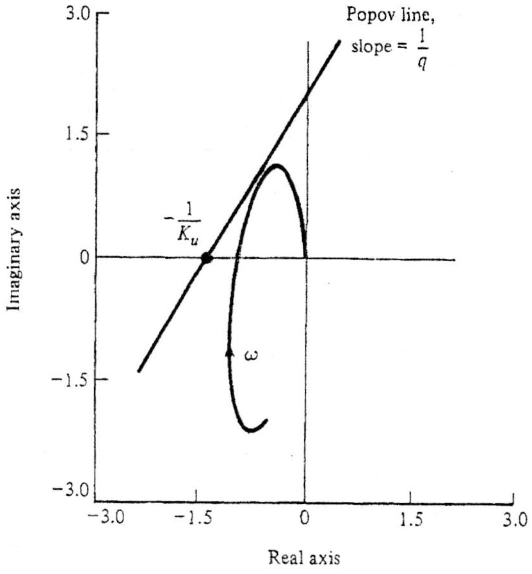

Figure 15.10 Popov locus.

It is pointed out [9] that the output of the system in Figure 15.5 will still asymptotically approach zero for nonzero reference inputs $r(t)$ into the system, provided that $r(t)$ is bounded, uniformly continuous, and square integrable. This rules out any type of constant or bias input component, but in such a case a zero asymptotic output would not be a reasonable expectation on purely intuitive grounds.

The Circle Criterion

The circle stability criterion applies to the same general type of system as discussed before, but with somewhat different specific conditions. The nonlinearity may now be a function of time, $n(e, t)$, but for all time it remains inside the bounding sector $[K_l, K_u]$. The linear system need not be stable, but it is assumed that $G(s)$ is strictly proper (i.e., the state model matrix $\mathbf{D}$ is zero) and there are no common numerator and denominator factors (i.e., the full-order state model is both controllable and observable). A critical circle or disk will be utilized in place of the critical point $-1 + j0$ of the Nyquist criterion for linear system stability. The critical circle cuts the real axis at points $-1/K_u$ and $-1/K_l$ and has a diameter given by $1/K_l - 1/K_u$. The usual Nyquist polar plot, and not the modified Popov locus, is used with the circle criterion. The circle criterion states that the polar plot of $G(j\omega)$ must not cut the critical circle, and the encirclements or lack of encirclements of the critical circle by the locus follow the usual rules associated with the point -1 for linear systems. Several subcases are possible:

(a) If both K_l and K_u are positive, the critical circle is entirely to the left of the origin.

(b) If both K_l and K_u are negative, the critical circle will be entirely to the right of the origin.

(c) When K_l and K_u have opposite signs, the critical circle will have the origin as an internal point.

If the open-loop system $G(s)$ has no right-half plane poles ($P_r = 0$ in Eq. (*15.18*)), then stability requires that the critical circle not be encircled. This means that in cases (a) and (b), the polar plot must stay outside the critical circle. In case (c) the polar plot must stay entirely within the critical circle. If $G(s)$ has unstable poles, then the polar plot still must not intersect the critical circle but must have the correct number of encirclements of it, $N_c \equiv P_r$ as given by Eq. (*15.18*), in order to give $Z_r = 0$ and thus assure asymptotic stability.

EXAMPLE 15.12 Examine the system of Example 15.11 using the circle criterion. The positive-frequency portion of the Nyquist polar plot is shown in Figure 15.11. Since the open-loop system has no unstable poles, the critical circle must not be encircled by the polar plot; i.e., $N_c = 0$ is required. The real-axis crossover is the same as in the previous example because $G^*(j\omega)$ and $G(j\omega)$ have equal real parts. Thus the limit on the upper-sector slope must be less than 1.036. The exact limit depends on the lower-sector bound K_l. If $K_l = 0$, then an infinite-radius circle (i.e., a vertical line) results. This vertical line would need to be positioned further to the left than -1.036 to avoid intersecting the locus. Also note that if both sector bounds have negative slopes, the locus will not intersect the critical circle but will encircle it once in the clockwise direction. The sufficient conditions for stability cannot be satisfied by this system with any negative slope. ■

Sector-type nonlinearities contain all linear gains between K_l and K_u as special cases. Stability for any nonlinearity in this class then necessarily requires that the system be stable for this range of linear gains. The Nyquist stability criterion must therefore hold for all critical points $[-1/K_l, -1/K_u]$. These are all the points along the real-axis diameter of the critical circle. *Aizerman's conjecture* [9, 12, 23] was that a

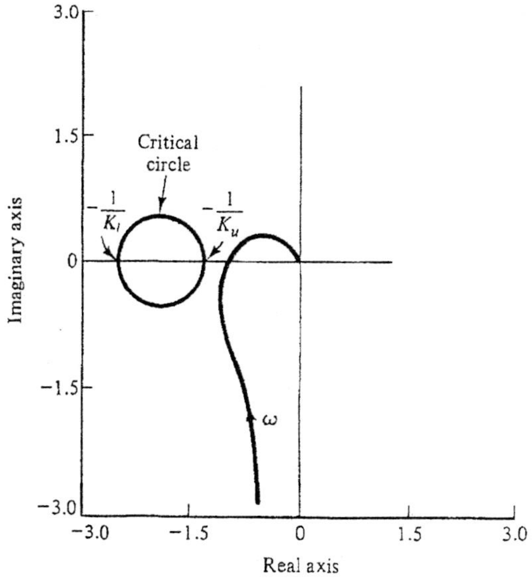

Figure 15.11 Nyquist plot.

nonlinear system with a sector nonlinearity would be stable if the linear system was stable for the entire range of gains in the sector. Aizerman's conjecture is generally false. The circle criterion indicates that not just the real-axis diameter but rather the entire critical circle must be checked for intersections with, or encirclements by, the polar plot.

Some extensions of these single nonlinearity results to systems with multiple nonlinearities are available. A brief survey may be found in Appendix C of Reference 9.

REFERENCES

1. Brockett, R. W.: "Nonlinear Systems and Differential Geometry," *Proceedings of the IEEE,* Vol. 64, No. 1, January 1976, pp. 61–72.
2. Isidori, A.: *Nonlinear Control Systems: An Introduction,* Springer-Verlag, Berlin, 1985.
3. Coddington, E. A. and N. Levinson: *Theory of Ordinary Differential Equations,* McGraw-Hill, New York, 1955.
4. Truxal, J. G.: *Automatic Control Systems Synthesis,* McGraw-Hill, New York, 1955, Chapters 10 and 11.
5. Boye, A. J. and W. L. Brogan: "A Method of Controlling Deterministic Nonlinear Systems," *Proc. American Control Conference,* Boston, Mass. June 1985.
6. Boye, A. J. and W. L. Brogan: "Application of a Nonlinear System Controller to Flight Control," *29th Midwestern Symposium on Circuits and Systems,* Lincoln, Nebraska, August 1986.
7. Boye, A. J. and W. L. Brogan: "A Nonlinear System Controller," *International Journal of Control,* Vol. 44, No. 5, December 1986, pp. 1209–1218.
8. Taylor, A. E. and R. W. Mann: *Advanced Calculus,* Xerox, Boston, Mass. 1972.
9. Atherton, D. P.: *Nonlinear Control Engineering,* Van Nostrand Reinhold Co., New York, 1975.
10. D'Azzo, J. J. and C. H. Houpis: *Feedback Control System Analysis and Synthesis,* 1st ed., McGraw-Hill, New York, 1960, Chapter 18.
11. Elgerd, O. I.: *Control Systems Theory,* McGraw-Hill, New York, 1967, Chapter 8.
12. Power, H. M. and R. J. Simpson: *Introduction to Dynamics and Control,* McGraw-Hill, New York, 1978, Chapter 9.
13. Gibson, J.: *Nonlinear Automatic Control,* McGraw-Hill, New York, 1963.
14. Guckenheimer, J. and P. Holmes: *Nonlinear Oscillations, Dynamical Systems, and Bifurcations of Vector Fields,* Springer-Verlag, New York, 1983.
15. Schuster, H. G.: *Deterministic Chaos,* Physik-Verlag, Weinheim, F. R. of Germany, 1984.
16. Chua, L. O. and R. N. Madan: "Sights and Sounds of Chaos," *IEEE Circuits and Devices Magazine,* January 1988, pp. 3–13.
17. Bartissol, P. and L. O. Chua: "The Double Hook," *IEEE Transactions on Circuits and Systems,* Vol. 35, No. 12, December 1988, pp. 1512–1522.
18. Mareels, I. M. Y. and R. R. Bitmead: "Bifurcation Effects in Robust Adaptive Control," *IEEE Transactions on Circuits and Systems,* Vol. 35, No. 7, July 1988, pp. 835–841.
19. Salam, F. M. A. and Shi Bai: "Complicated Dynamics of a Prototype Continuous-Time Adaptive Control System," *IEEE Transactions on Circuits and Systems,* Vol. 35, No. 7, July 1988, pp. 842–849.
20. LaSalle, J. and S. Lefschetz: *Stability by Liapunov's Direct Method,* Academic Press, New York, 1961.
21. Kalman, R. E. and J. E. Bertram: "Control System Analysis and Design Via the Second

Method of Lyapunov I. Continuous-Time Systems," *Trans. ASME Journal of Basic Engineering,* Vol. 82D, June 1960, pp. 371–393.

22. Kalman, R. E. and J. E. Bertram: "Control System Analysis and Design Via the Second Method of Lyapunov II. Discrete-Time Systems," *Trans. ASME Journal of Basic Engineering,* Vol. 82D, June 1960, pp. 394–400.
23. Willems, J. L.: *Stability Theory of Dynamical Systems,* John Wiley, New York, 1970.
24. Reiss, R. and G. Geiss: "The Construction of Liapunov Functions," *IEEE Transactions on Automatic Control,* Vol. AC-8, No. 4, October 1963, pp. 382–383.
25. Puri, N. N. and C. N. Weygandt: "Second Method of Liapunov and Routh's Canonical Form," *Journal of The Franklin Institute,* Vol. 276, No. 5, 1963, pp. 365–84.
26. Ku, Y. H. and N. N. Puri: "On Liapunov Functions of High Order Nonlinear Systems," *Journal of The Franklin Institute,* Vol. 276, No. 5, 1963, pp. 349–364.
27. Schultz, D. G. and J. E. Gibson: "The Variable Gradient Method for Generating Liapunov Functions," *AIEE Trans.,* Part II, Vol. 81, September 1962, pp. 203–210.
28. Ogata, K.: *State Space Analysis of Control Systems,* Prentice Hall, Englewood Cliffs, N.J., 1967, Chapter 8.
29. Baumann, W. T. and W. J. Rugh: "Feedback Control of Nonlinear Systems by Extended Linearization," *IEEE Transactions on Automatic Control,* Vol. AC-31, No. 1, January 1986, pp. 40–46.
30. Chang, L-C: *Probabilistic Error Model of Robot End-Effector's Dynamic State,* Unpublished Ph.D Dissertation, University of Nebraska, Lincoln, August 1987.
31. Matsumoto, T., L. O. Chua, and M. Komuro: "A Double Scroll," *IEEE Transactions on Circuits and Systems,* Vol. 32, No. 8, August 1985, pp. 797–818.

ILLUSTRATIVE PROBLEMS

Nonlinear System Stability

Let the origin be an equilibrium point of a slightly nonlinear system $\dot{\mathbf{x}} = \mathbf{f}(\mathbf{x})$. If this system is linearized about the origin, perhaps using Taylor's series, then $\dot{\mathbf{x}} = \mathbf{A}\mathbf{x} + \mathbf{h}(\mathbf{x})$, where $\mathbf{A}$ is the Jacobian matrix $[\partial\mathbf{f}/\partial\mathbf{x}]$ evaluated at $\mathbf{x} = 0$, and $\mathbf{h}(\mathbf{x})$ represents higher-order terms. Show that if $\|\mathbf{h}(\mathbf{x})\| \le \alpha\|\mathbf{x}\|$ for some positive constant α, then asymptotic stability of the linear equation $\dot{\mathbf{x}} = \mathbf{A}\mathbf{x}$ implies asymptotic stability of the nonlinear equation as well [21, 23].

Let $\mathbf{\Phi}(t, \tau) = e^{\mathbf{A}(t-\tau)}$ be the transition matrix of the linear equation. Then treating $\mathbf{h}(\mathbf{x})$ as a forcing term, the solution for the nonlinear system can be written as

$$\mathbf{x}(t) = \mathbf{\Phi}(t, t_0)\mathbf{x}(t_0) + \int_{t_0}^{t} \mathbf{\Phi}(t, \tau)\mathbf{h}(\mathbf{x}(\tau))\, d\tau$$

Therefore,

$$\|\mathbf{x}(t)\| \le \|\mathbf{\Phi}(t, t_0)\|\cdot\|\mathbf{x}(t_0)\| + \int_{t_0}^{t} \|\mathbf{\Phi}(t, \tau)\|\, \|\mathbf{h}(\mathbf{x}(\tau))\|\, d\tau$$

Asymptotic stability of the linear part ensures that $\|\mathbf{\Phi}(t, \tau)\|$ is bounded by a decaying exponential, $\|\mathbf{\Phi}(t, \tau)\| \le Me^{-k(t-\tau)}$ for all $t \ge \tau$, all $\tau \ge t_0$. Using this and the bound on $\|\mathbf{h}(\mathbf{x})\|$ leads to

$$\|\mathbf{x}(t)\| \le Me^{-k(t-t_0)}\|\mathbf{x}(t_0)\| + \int_{t_0}^{t} \alpha M e^{-k(t-\tau)}\|\mathbf{x}(\tau)\|\, d\tau$$

Multiplying by the positive function e^{kt} leaves the sense of the inequality unchanged, so

$$e^{kt}\|\mathbf{x}(t)\| \le Me^{kt_0}\|\mathbf{x}(t_0)\| + \int_{t_0}^{t} \alpha M e^{k\tau}\|\mathbf{x}(\tau)\|\, d\tau \qquad (1)$$

Call the right-hand side of equation (*1*) $U(t)$ for convenience. Note that $\dot{U}(t) = \alpha M e^{kt}\|\mathbf{x}(t)\| = \alpha M$ times left-hand side of equation (*1*). Hence

$$\dot{U}/(\alpha M) \le U \quad \text{or} \quad dU/U \le \alpha M\, dt$$

Integrating both sides gives $\ln(U(t)/C) \le \alpha M(t - t_0)$, where the integration constant C is the value of U at $t = t_0$, namely, $C = Me^{kt_0}\|\mathbf{x}(t_0)\|$. Then

$$U(t) \le Ce^{\alpha M(t - t_0)} \le Me^{kt_0}\|\mathbf{x}(t_0)\|e^{\alpha M(t - t_0)}$$

is an explicit upper bound for the right-hand side of equation (*1*). Therefore,

$$e^{kt}\|\mathbf{x}(t)\| \le Me^{kt_0}\|\mathbf{x}(t_0)\|e^{\alpha M(t - t_0)}$$

or

$$\|\mathbf{x}(t)\| \le Me^{-(k - \alpha M)(t - t_0)}\|\mathbf{x}(t_0)\|$$

Thus $\|\mathbf{x}(t)\| \to 0$ provided the bound on $\|\mathbf{h}(\mathbf{x})\|$ is sufficiently small, i.e., if $\alpha < k/M$. The constant α must satisfy $\alpha \ge \|\mathbf{h}(\mathbf{x})\|/\|\mathbf{x}\|$. Since $\mathbf{h}(\mathbf{x})$ is composed of second- or higher-order terms in components of $\mathbf{x}$, this restriction can be made as small as we please by restricting $\|\mathbf{x}\|$ to be sufficiently small.

Thus asymptotic stability of the linear part of the system implies asymptotic stability of the nonlinear system in a sufficiently small neighborhood of the origin. It can also be shown that if the Jacobian matrix has one or more right-half plane eigenvalues, then the equilibrium point is unstable. In summary, the stability of the nonlinear system in the neighborhood of an equilibrium point is the same as the linearized portion, with one exception. If one or more eigenvalues are on the imaginary axis and all others are in the left-half plane, the linearized system gives no definite information about stability. Second-order (and higher) terms need to be evaluated in this case.

15.2 Derive the nonlinear differential equation which relates the angle θ of the inverted pendulum [29] mounted on a cart to the input force f_x on the cart. See Figure 15.12*a* and the free-body diagrams of the two members in Figure 15.12*b* and *c*. The pendulum is of length L and mass m, and its moment of inertia about the center of gravity is J. The cart has mass M. The horizontal and vertical displacements of the center of gravity of the pendulum are $x = X + L/2 \sin\theta$ and $y = L/2 \cos\theta$. Letting the reaction forces at the pendulum support point be F_x and F_y, summing forces on the pendulum gives

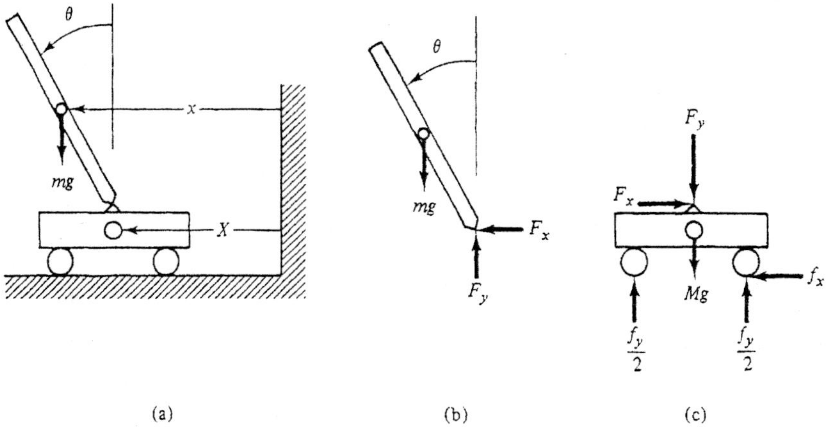

Figure 15.12

$$F_x = m\ddot{X} + \frac{mL}{2}\ddot{\theta}\cos\theta - \frac{mL}{2}\dot{\theta}^2\sin\theta \tag{1}$$

$$F_y - mg = -\frac{mL}{2}\ddot{\theta}\sin\theta - \frac{mL}{2}\dot{\theta}^2\cos\theta \tag{2}$$

$$\frac{F_y L}{2}\sin\theta - \frac{F_x L}{2}\cos\theta = J\ddot{\theta} \tag{3}$$

The external force imparted to the cart by its drive wheels is f_x. By summing horizontal forces on the cart, one obtains

$$\ddot{X} = \frac{f_x - F_x}{M} \tag{4}$$

Substituting Eq. (4) into (1) and then combining that result with Eq. (2) and (3) gives a nonlinear second-order differential equation relating the input force f_x to the pendulum angle θ,

$$\left\{J + \frac{mL^2}{4}\sin^2\theta + \frac{mML^2}{4(M+m)}\cos^2\theta\right\}\ddot{\theta} = \frac{mgL}{2}\sin\theta - \frac{m^2L^2}{2(m+M)}\dot{\theta}^2\sin(2\theta) - \frac{mL\cos\theta f_x}{2(M+m)}$$

If the pendulum has a uniform mass distribution, $J = mL^2/12$, and then Eq. (5) reduces to

$$\ddot{\theta} = \frac{\{2g\sin\theta - mL/[2(m+M)]\dot{\theta}^2\sin(2\theta) - 2\cos\theta f_x/(M+m)\}}{4L/3 - mL\cos^2\theta/(M+m)} \tag{6}$$

If the length L is written as twice the half-length, Eq. (6) becomes identical to the result in [29].

3 (a) Let $x_1 = \theta$, $x_2 = \dot{\theta}$, and $u = f_x$. Derive the state equations for linear perturbations of the pendulum in Problem 15.2. Use $\theta = \dot{\theta} = u = 0$ as the nominal status.

(b) Use the linearized constant-coefficient perturbations to design state feedback control gains that yield closed-loop poles at $\lambda = -4$ and -5. Use the parameter values [29] $m = 2$ kg, $M = 8$ kg, $L = 1$ m, and $g = 9.8$ m/s.

(c) Use the linear state feedback gains of part (b) with the actual nonlinear system given in part (a). Use computer simulation to investigate the transient response of θ, $\dot{\theta}$, and u for initial values of θ of 0.5, 1, 1.2, and 1.25 rad. Let $\dot{\theta}(0) = 0$.

(a) The nonlinear state equations are

$$\dot{\mathbf{x}} = \begin{bmatrix} x_2 \\ \dfrac{2g\sin x_1 - mL/[2(m+M)]x_2^2\sin(2x_1) - 2\cos x_1 f_x/(M+m)}{D} \end{bmatrix}$$

$$= \mathbf{f}(\mathbf{x}, u) \tag{1}$$

where the denominator is $D = 4L/3 - mL\cos^2\theta/(M+m)$. Then

$$\left[\frac{\partial \mathbf{f}}{\partial \mathbf{x}}\right]_n = \begin{bmatrix} 0 & 1 \\ \dfrac{2g}{4L/3 - mL/(M+m)} & 0 \end{bmatrix}, \qquad \left[\frac{\partial \mathbf{f}}{\partial u}\right]_n = \begin{bmatrix} 0 \\ \dfrac{-2}{4L(M+m)/3 - mL} \end{bmatrix}$$

(b) Using the pole placement algorithm of Sec. 13.4, the feedback gain matrix is found to be $\mathbf{K} = [-211.333 \quad -51]$. If, instead, an infinite horizon **LQ** optimum problem (Sec. 14.4.2) had been specified with weights $\mathbf{Q} = \mathbf{I}$ and $\mathbf{R} = 1$, the gain would then be $\mathbf{K} = [-196 \quad -47.142]$, and the resulting closed-loop poles of the linear system would be $\lambda = [-4.07$ and $-4.24]$.

(c) When the first set of gains is used to form $u(t) = -\mathbf{K}\mathbf{x}$ in Eq. (1) with values $x_1(0) = 0.5$, 1, and 1.2 and with $x_2(0) = 0$, the three response curves of Figure 15.13*a* are obtained for $x_1 = \theta(t)$. Figure 15.13*b* gives the corresponding $x_2(t) = \dot{\theta}(t)$. Figure 15.13*c* shows the input

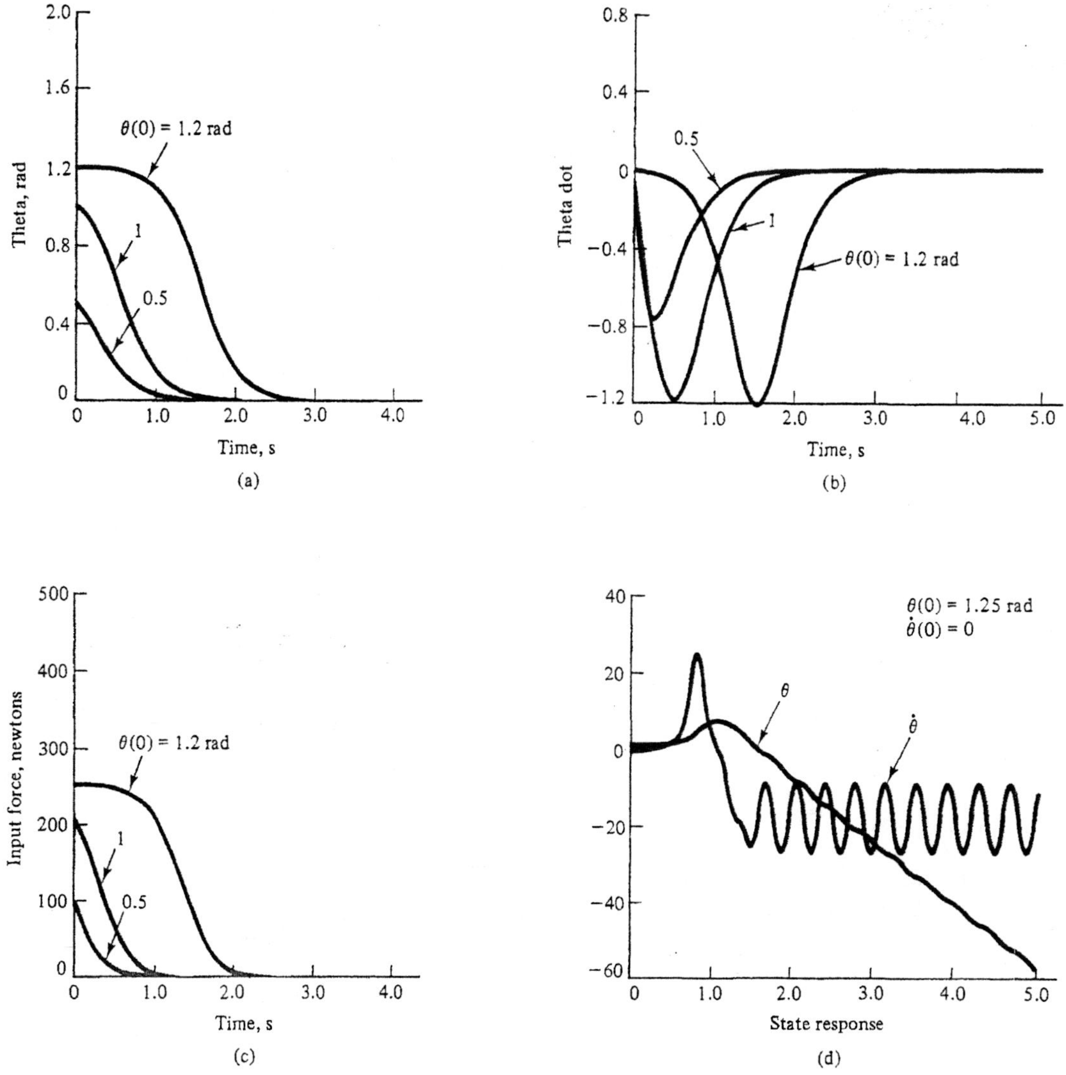

Figure 15.13

force $u(t) = f_x$ for these cases. When $\theta(0) = 1.25$ rad was attempted, the controller failed to balance the pendulum, as shown in Figure 15.13*d*. The maximum allowable control magnitude was set to 1000 N, and in the last case the commanded control saturated at this limit. A larger limit will allow a larger initial angle to be driven to zero successfully.

15.4 The two-link mechanism in Figure 15.14 demonstrates nonlinearities typical of many robot devices. The equations governing the motion as a function of the two input joint torques can be derived by any of several methods [30]. By summing forces and torques on each link it can be shown that

$$T_1 = \left[\frac{m_1}{2} + m_2\right] g L_1 \cos\theta + \frac{m_2 g L_2}{2} \cos\psi$$

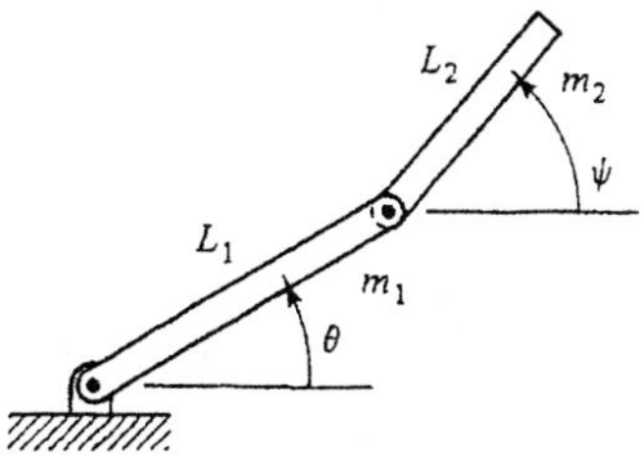

Figure 15.14

$$+\ddot{\theta}\left\{\frac{m_1 L_1^2}{3} + m_2\left[L_1^2 + \frac{L_1 L_2 \cos(\psi - \theta)}{2}\right]\right\} + \ddot{\psi}\left\{m_2\left[\frac{L_2^2}{3} + \frac{L_1 L_2 \cos(\psi - \theta)}{2}\right]\right\}$$

$$-\frac{\dot{\psi}^2 m_2 L_1 L_2 \sin(\psi - \theta)}{2} + \frac{\dot{\theta}^2 L_1 L_2 \sin(\psi - \theta)}{2} \qquad (1)$$

and

$$T_2 = \frac{m_2 g L_2 \cos(\psi)}{2} + \frac{\ddot{\theta} m_2 L_1 L_2 \cos(\psi - \theta)}{2} + \frac{m_2 L_2^2 \ddot{\psi}}{3} + \frac{m_2 L_1 L_2 \dot{\theta}^2 \sin(\psi - \theta)}{2} \qquad (2)$$

Let $x_1 = \theta$, $x_2 = \psi$, $x_3 = \dot{\theta}$, and $x_4 = \dot{\psi}$ and write equations (*1*) and (*2*) in state variable form.

Since $\dot{x}_3$ and $\dot{x}_4$ appear in both equations, we must first separate out these variables. Rewrite Eqs. (*1*) and (*2*) as

$$D_{11}(\mathbf{x})\ddot{\theta} + D_{12}(\mathbf{x})\ddot{\psi} = F_1(\mathbf{x}) + T_1$$

$$D_{21}(\mathbf{x})\ddot{\theta} + D_{22}(\mathbf{x})\ddot{\psi} = F_2(\mathbf{x}) + T_2$$

where the coefficients of the second derivatives have been redefined as D_{ij} and where all other terms except the torques have been combined into the definitions of the two F_i terms. The state equations can then be written as

$$\begin{bmatrix}\dot{x}_1\\ \dot{x}_2\\ \dot{x}_3\\ \dot{x}_4\end{bmatrix} = \begin{bmatrix} x_3 \\ x_4 \\ \begin{bmatrix} D_{11} & D_{12} \\ D_{21} & D_{22}\end{bmatrix}^{-1} \begin{bmatrix} F_1(\mathbf{x}) + T_1 \\ F_2(\mathbf{x}) + T_2\end{bmatrix}\end{bmatrix} = \mathbf{f}(\mathbf{x}, \mathbf{T})$$

These nonlinear equations can be linearized for analyzing small motions about some nominal conditions by using the methods of Sec. 15.2, leading to

$$\delta\dot{\mathbf{x}} = [\partial\mathbf{f}/\partial\mathbf{x}]_n\,\delta\mathbf{x} + [\partial\mathbf{f}/\partial\mathbf{T}]_n\,\delta\mathbf{T}$$

Details of evaluating the partial derivatives $\partial\mathbf{f}/\partial\mathbf{x}$ and $\partial\mathbf{f}/\partial\mathbf{T}$ are left as an exercise; they may also be found in [30].

.5 Consider an nth-order nonlinear system which is linear in the r control components, $\dot{\mathbf{x}} = \mathbf{f}(\mathbf{x}) + \mathbf{B}\mathbf{u}$. Investigate conditions under which it is possible to find a control function satisfying $\mathbf{f}(\mathbf{x}) + \mathbf{B}\mathbf{u} - \mathbf{F}\mathbf{y}_d - \mathbf{G}\mathbf{v} = \mathbf{S}(\mathbf{x} - \mathbf{y}_d)$ as required in dynamic linearization.

Rewriting this as $\mathbf{B}\mathbf{u} = \mathbf{w}$, where $\mathbf{w} = -\mathbf{f}(\mathbf{x}) + \mathbf{F}\mathbf{y}_d + \mathbf{G}\mathbf{v} + \mathbf{S}(\mathbf{x} - \mathbf{y}_d)$, the results of Chapter 6 indicate that a unique solution exists for any *arbitrary* $\mathbf{w}$ if and only if $\mathbf{B}$ is square and nonsingular. This is a very restrictive condition which is rarely met with control systems of order higher than one. Again from Chapter 6 results, for a *specific* $\mathbf{w}$, solutions will exist if rank$[\mathbf{B}]$ = rank$[\mathbf{B} \mid \mathbf{w}]$. A unique $\mathbf{u}$ will exist for $\mathbf{w}$ if, in addition, rank$[\mathbf{B}] = r$, the number of control components. Some influence can be exerted on $\mathbf{w}$ through choice of $\mathbf{F}$, $\mathbf{G}$, and $\mathbf{S}$. What is desired is an ability to find $\mathbf{u}$ for arbitrary $\mathbf{x}$ and $\mathbf{y}_d$, and this can be achieved under certain conditions if compatible choices are made for the forms of $\mathbf{F}$, $\mathbf{G}$, and $\mathbf{S}$. For example, consider an nth-order system in phase variable form, with one input, $\overset{(n)}{y} = f(y, \dot{y}, \ldots, \overset{(n-1)}{y}) + u$, which becomes in state variable form

$$\dot{\mathbf{x}} = \begin{bmatrix} x_2 \\ x_3 \\ x_4 \\ \vdots \\ x_n \\ f(\mathbf{x}) \end{bmatrix} + \begin{bmatrix} 0 \\ 0 \\ 0 \\ \vdots \\ 0 \\ 1 \end{bmatrix} u$$

Clearly rank$[\mathbf{B}] = 1$ and $[\mathbf{B} \mid (\mathbf{F} - \mathbf{S})\mathbf{y}_d + \mathbf{Sx} - \mathbf{f}(\mathbf{x}) + \mathbf{Gv}]$ will also have rank $= 1$ for all $\mathbf{x}$ and $\mathbf{y}_d$ if both $\mathbf{F}$ and $\mathbf{S}$ are in companion form and if the first $n-1$ rows of $\mathbf{Gv}$ are zero. This ensures that the first $n-1$ rows of $\mathbf{Bu} = \mathbf{w}$ are identically zero. The last row gives a scalar equation that can be solved for $u(t)$.

The preceding results easily generalize to any number r of coupled pth-order scalar differential equations. We illustrate with just two equations in terms of two input variables,

$$\dddot{y}_1 = f_1(\ddot{y}_1, \dot{y}_1, y_1, \dot{y}_2, y_2) + b_{11} u_1 + b_{12} u_2$$

$$\ddot{y}_2 = f_2(\ddot{y}_1, \dot{y}_1, y_1, \dot{y}_2, y_2) + b_{21} u_1 + b_{22} u_2$$

By picking state variables $x_1 = y_1$, $x_2 = \dot{y}_1$, $x_3 = \ddot{y}_1$, $x_4 = y_2$, and $x_5 = \dot{y}_2$, the state equations take the form

$$\dot{\mathbf{x}} = \begin{bmatrix} x_2 \\ x_3 \\ f_1(\mathbf{x}) + b_{11} u_1 + b_{12} u_2 \\ x_4 \\ f_2(\mathbf{x}) + b_{21} u_1 + b_{22} u_2 \end{bmatrix}$$

Since rows 1, 2, and 4 of the matrix $\mathbf{B}$ are zero, rank$[\mathbf{B}] \le 2$. Consideration of the solvability condition rank$[\mathbf{B}] =$ rank$[\mathbf{B} \mid \mathbf{w}]$ suggests a compatible choice for $\mathbf{F}$, $\mathbf{G}$, and $\mathbf{S}$ should make rows 1, 2, and 4 of $\mathbf{w}$ zero also. This can be accomplished by selecting

$$\mathbf{F} = \begin{bmatrix} 0 & 1 & 0 & 0 & 0 \\ 0 & 0 & 1 & 0 & 0 \\ F_{31} & F_{32} & F_{33} & F_{34} & F_{35} \\ 0 & 0 & 0 & 0 & 1 \\ F_{51} & F_{52} & F_{53} & F_{54} & F_{55} \end{bmatrix}, \quad \mathbf{S} = \begin{bmatrix} 0 & 1 & 0 & 0 & 0 \\ 0 & 0 & 1 & 0 & 0 \\ S_{31} & S_{32} & S_{33} & S_{34} & S_{35} \\ 0 & 0 & 0 & 0 & 1 \\ S_{51} & S_{52} & S_{53} & S_{54} & S_{55} \end{bmatrix}, \quad \mathbf{G} = \begin{bmatrix} 0 & 0 \\ 0 & 0 \\ G_{31} & G_{32} \\ 0 & 0 \\ G_{51} & G_{52} \end{bmatrix}$$

Thus there are only two nontrivial rows which contribute to the solution for $\mathbf{u}$, and a unique solution exists for all $\mathbf{x}$, $\mathbf{y}_d$, and $\mathbf{v}$ provided the 2×2 matrix $[b_{ij}]$ is nonsingular.

15.6 The system in Problem 15.5 is now generalized to allow it to be nonlinear in the control variables: $\dot{\mathbf{x}} = \mathbf{f}(\mathbf{x}, \mathbf{u})$. What can be said about the existence of a dynamic linearizing control law?

If the function $\mathbf{f}(\mathbf{x}, \mathbf{u})$ is continuous in some region $\mathcal{R}$ containing the point $\mathbf{x}_0$, $\mathbf{u}_0$, and if it is continuously differentiable in $\mathcal{R}$, the results of Problem 15.5 remain valid in the region $\mathcal{R}$, with $[\partial\mathbf{f}/\partial\mathbf{u}]_0$ replacing the previous matrix $\mathbf{B}$. This follows from the implicit function theorem [8]. If there are r input components and n states with $r < n$, then the $n \times r$ matrix $[\partial\mathbf{f}/\partial\mathbf{u}]_0$ must have full rank r. The augmented matrix $\mathbf{W}$ must also have rank r. This can sometimes be ensured by selecting $\mathbf{F}$, $\mathbf{G}$, and $\mathbf{S}$ in a form compatible with the given nonlinear state equations, just as in the case of linear control terms in Problem 15.5. This restricts the number and location of the nonlinear terms which can be dealt with in the state equations. Problem 15.7 shows a system where this condition cannot be met. When rank$[\mathbf{B}] \neq$ rank$[\mathbf{W}]$, the linearizing equations are inconsistent and no exact solution exists. Two choices may be considered in this situation. Certain equations can be ignored or combined so that a smaller set of consistent equations remains. This can be accomplished by premultiplying $\mathbf{f}(\mathbf{x}, \mathbf{u})$ by an $r \times n$ matrix $\mathbf{H}$. This means that only r states, or combinations of states, are being matched to an rth-order template system. Of course the resulting nth-order closed-loop system will generally not be linear in this case. The second possibility is to use a least-squares approximate solution to the full nth-order template-matching problem. This can be done provided $\mathbf{B}$ is of full rank, but the result will not be exactly linear, and its performance may not be satisfactory. Both possibilities are demonstrated in Problem 15.7.

.7 Consider the problem of applying dynamic linearization to the system

$$\dot{\mathbf{x}} = \begin{bmatrix} 0 & -1 & -1 \\ 1 & 0 & 0 \\ 0 & 0 & -4 \end{bmatrix}\mathbf{x} + \begin{bmatrix} 0 \\ ux_2 \\ 2 + x_1x_3 \end{bmatrix}$$

For this system, $\partial\mathbf{f}/\partial u = [0 \quad x_2 \quad 0]^T$ has rank 1, except at $x_2 = 0$. Clearly any choice for $u(t)$ will have no effect on the third component of $\dot{\mathbf{x}}$ in this case, so exact linearization by matching to a third-order linear template is not possible. By selecting

$$\mathbf{H} = \begin{bmatrix} 1 & 0 & 0 \\ 0 & 1 & 0 \end{bmatrix} \quad \text{or} \quad \mathbf{H} = \begin{bmatrix} 1 & 0 & 0 \\ 0 & 1 & 1 \end{bmatrix}$$

we can match a second-order template system to either the first two components of $\mathbf{x}$, or with the second $\mathbf{H}$, we would match x_1 to y_{d_1} and $x_2 + x_3$ to y_{d_2}. The second $\mathbf{H}$ is selected, along with a template system described by

$$\dot{\mathbf{y}}_d = \begin{bmatrix} 0 & -1 \\ 2 & -3 \end{bmatrix}\mathbf{y}_d + \begin{bmatrix} 0 \\ 1 \end{bmatrix}v$$

This is a stable system with eigenvalues at $\lambda = -1$ and -2. The steady-state solution for a step input of magnitude V is $\mathbf{y}_d = [-V/2 \quad 0]^T$. The convergence matrix is initially selected as

$$\mathbf{S} = \begin{bmatrix} 0 & -1 \\ 25 & -10 \end{bmatrix}$$

but $\mathbf{S} = \mathbf{F}$ will also be tested. Except when $x_2 = 0$, the control is given by

$$u(t) = \frac{2y_{d_1} - 3y_{d_2} + v + S_{21}(x_1 - y_{d_1}) + S_{22}(x_2 + x_3 - y_{d_2}) - x_1 + 4x_3 - 2 - x_1x_3}{x_2}$$

The nonlinear system equations were numerically integrated using fourth-order Runge-Kutta with a step size of 0.02. The magnitude of $u(t)$ was limited to 100. Figure 15.15a compares y_{d_1} and the state x_1 obtained with the two matrices $\mathbf{S}$ mentioned. Figure 15.15b shows y_{d_2} along with two sets of x_2, x_3 responses. The control command is shown in Figure 15.15c. It saturated immediately with the fast $\mathbf{S}$ and remained at either 100 or -100 for the first second and thereafter remained well within the bounds. When $\mathbf{S} = \mathbf{F}$ was selected, the system response was much slower and smoother. The commanded control signal gradually increased until saturation occurred at about 3 s.

Attempting to least-squares fit to a third-order template system in this problem is not productive, since $u = [\mathbf{B}\mathbf{B}^T]^{-1}\mathbf{B}^T\{\ldots\}$ amounts to ignoring the first and third components of the matching equations.

.8 Use the method of dynamic linearization to control the inverted pendulum of Problems 15.2 and 15.3. Compare the results with the linear perturbation controller of Problem 15.3.

The controller in Problem 15.3 was designed to yield closed-loop eigenvalues at $\lambda = -4$ and -5. For comparison purposes, a linear template system which has these same eigenvalues is specified, namely,

$$\dot{\mathbf{y}}_d = \begin{bmatrix} 0 & 1 \\ -20 & -9 \end{bmatrix}\mathbf{y}_d + \begin{bmatrix} 0 \\ 1 \end{bmatrix}v$$

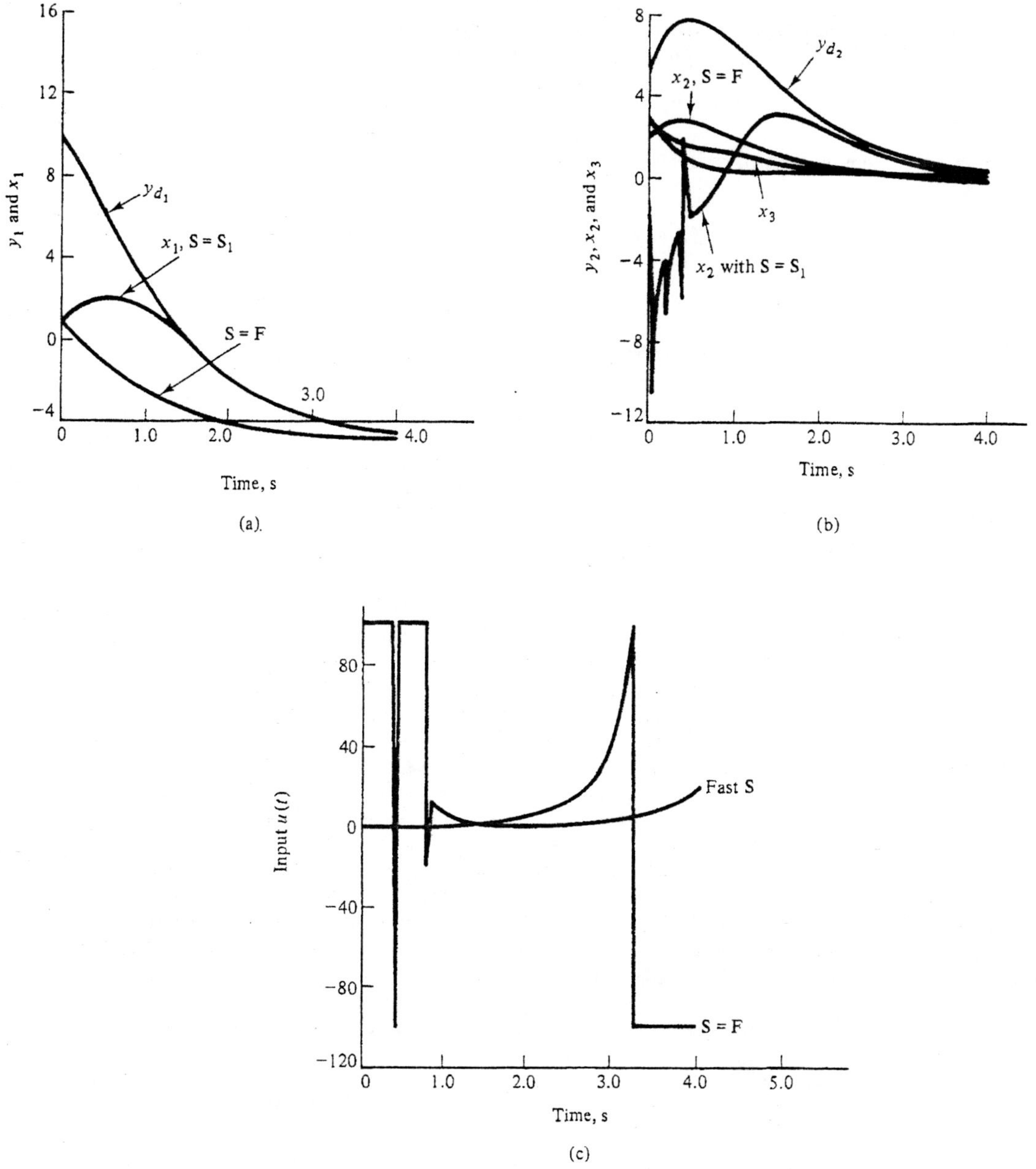

Figure 15.15

The nonzero convergence matrix **S** will also be specified to be in the companion form

$$\mathbf{S} = \begin{bmatrix} 0 & 1 \\ S_{21} & S_{22} \end{bmatrix}$$

This is not a totally arbitrary choice. The first component of Eq. (*15.11*) here requires that the first rows of **F** and **S** agree so that $x_2 = y_{d_2} + x_2 - y_{d_2}$. The second component of Eq. (*15.11*) is the only one containing the unknown control $u(t)$. If we define $w = (F_{21} - S_{21})y_{d_1} + (F_{22} - S_{22})y_{d_2} + S_{21}x_1 + S_{22}x_2 + v$ and the denominator D of Problem 15.3, this second component can be written as

$$2g \sin x_1 - \frac{mLx_2^2 \sin(2x_1)}{2(m+M)} - \frac{2 \cos x_1 u}{M+m} = Dw$$

from which, if $\cos x_1 \neq 0$, i.e., if $x_1 = \theta \neq \pi/2$, the control law is

$$u = \frac{(m+M)\{2g \sin x_1 - mL/[2(m+M)]x_2^2 \sin(2x_1) - Dw\}}{\cos x_1}$$

Unless stated otherwise, $v = 0$ in the following simulation tests of this control law. Three values of the matrix **S** are tested.

1. With $S_{21} = -100$, $S_{22} = -20$ ($\lambda = -10$ is a double root). The typical result of Figure 15.16*a* is obtained, using $\mathbf{x}(0) = [-0.5 \quad 0]^T$ and $\mathbf{y}_d(0) = [0.5 \quad 0]^T$. It is seen that $\mathbf{x}(t)$ and $\mathbf{y}_d(t)$ come together in a fraction of a second, as determined by **S**, and that they both settle to zero in less than 2 s, as determined by **F**.
2. With $\mathbf{S} \equiv \mathbf{F}$ but all other parameters as before, the response of Figure 15.16*b* is obtained. The settling times are noticeably longer.
3. With $\mathbf{S} = [0]$ and all other parameters as before, $x_2(t) \rightarrow y_{d_2}(t)$, but the first component of **x**—i.e., θ—does not go to zero, as shown in Figure 15.16*c*. The derivatives of **x** and $\mathbf{y}_d$ do match, and since the second components have equal initial conditions, these terms remain in perfect agreement. Having θ converge to -1 rad is not a satisfactory solution to the balancing problem. This points out the utility of the convergence matrix **S**.

The preceding three tests were repeated with the template input $v(t)$ being a periodic square wave. Results are shown in Figures 15.17*a, b, c*. Again, the first two cases successfully track the desired response, but (2) is considerably slower and allows a much bigger transient error to build up before settling. With $\mathbf{S} = [\mathbf{0}]$, the pendulum fails to achieve vertical balance. When the original angle error is allowed to increase to 1.48 rad, the controller is no longer able to balance the pendulum (the first **S** was used), but at $\theta(0) = 1.47$ rad, it performed correctly. Note that this is a much larger error that could be nulled by the linear controller of Problem 15.3. The actual response is shown in Figure 15.18, with θ oscillating around $\pi/2$; the input force history commanded by the controller is a square wave oscillating between the ± 1000-N limits.

Find the describing function for the general hysteresis-type nonlinearity in Figure 15.19*a*, which can be described by the five parameters α, β, K_1, K_2, and K_3. Note that the symmetry of the parallelogram gives the expression $\gamma = [K_1(\beta + \alpha) + K_2(\beta - \alpha)]/2$. Assume the amplitude E of the input sinusoid is larger than β.

There are seven time segments of interest in each cycle of the sine wave in Figure 15.19*b*.

1. $0 < t < t_1$; $0 < e(t) < \alpha$, $n(e) = -\gamma + K_1[\beta + e(t)]$
2. $t_1 < t < t_2$; $\alpha < e(t) < \beta$, $n(e) = -\gamma + K_1[\alpha + \beta] + K_2[e(t) - \alpha]$
3. $t_2 < t < t_3$; $\beta < e(t)$, $n(e) = \gamma + K_3[e(t) - \beta]$
4. $t_3 < t < t_4$; $-\alpha < e(t) < \beta$, $n(e) = \gamma + K_1[e(t) - \beta]$
5. $t_4 < t < t_5$; $-\beta < e(t) < -\alpha$, $n(e) = -\gamma + K_2[e(t) + \beta]$
6. $t_5 < t < t_6$; $e(t) < -\beta$, $n(e) = -\gamma + K_3[e(t) + \beta]$
7. $t_6 < t < 2\pi$; $-\beta < e(t) < 0$, $n(e) = -\gamma + K_1[e(t) + \beta]$

where $t_1 = \sin^{-1}(\alpha/E)$, $t_2 = \sin^{-1}(\beta/E)$, $t_3 = \pi - t_1$, $t_4 = \pi + t_1$, $t_5 = \pi + t_2$, $t_6 = 2\pi - t_2$. These times are all normalized times.

The actual times are obtained by dividing by the frequency ω. Using $e(t) = E \sin(t)$, the Fourier coefficient a_1 is composed of the sum of seven terms, each of the form

$$\left(\frac{1}{\pi}\right)\int \{[C_0 + C_1 \sin t] \sin t\}\, dt = \left(\frac{1}{\pi}\right)\left\{C_0(\cos t_a - \cos t_b) + C_1(t_b - t_a) + \frac{C_1}{4}[\sin(2t_a) - \sin(2t_b)]\right\}$$

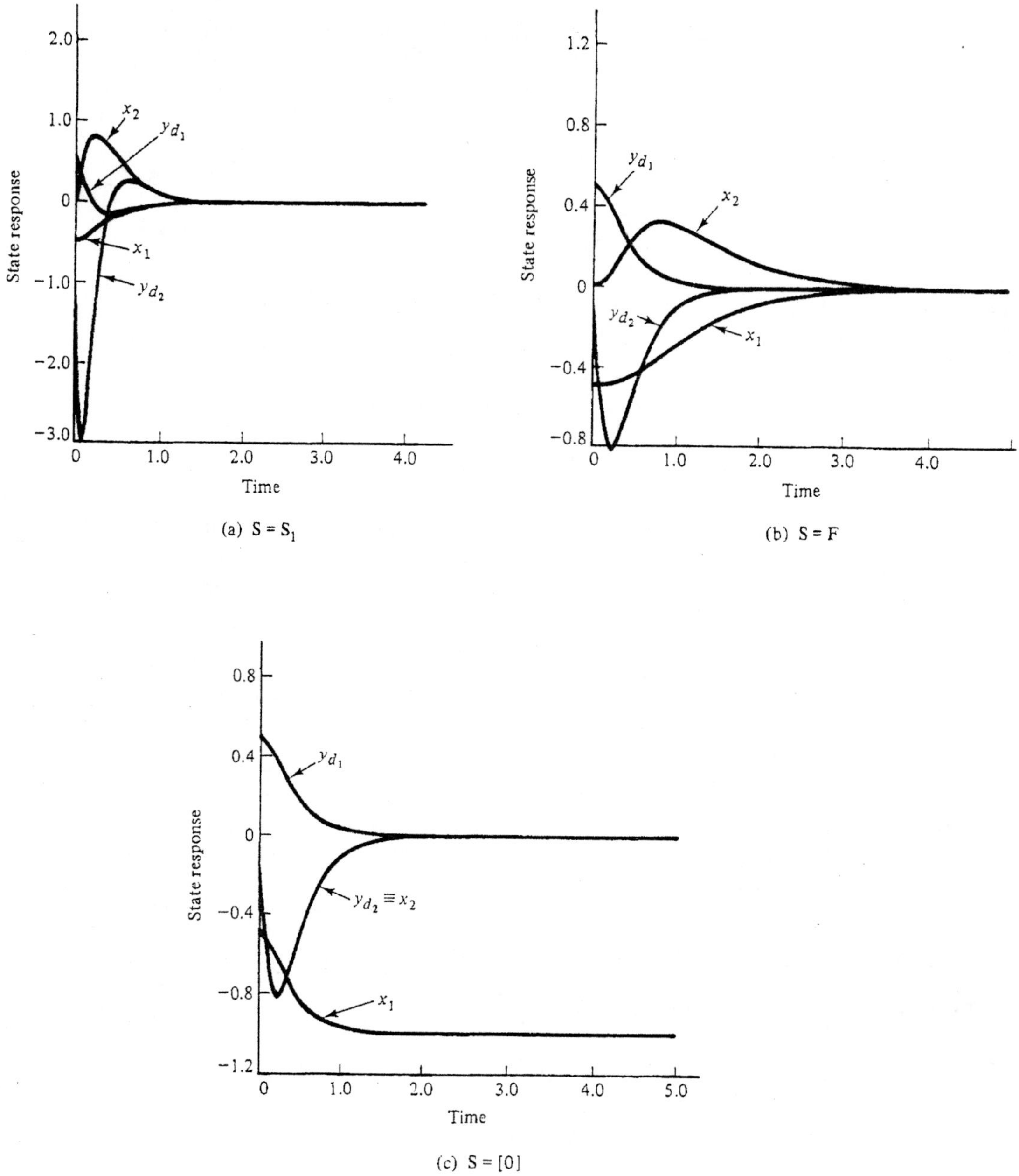

Figure 15.16

This will be defined as a function $S_f(t_a, t_b, C_0, C_1)$ to avoid repetition. Then a_1 is the sum of seven $S_f(\ )$ functions, each with appropriate arguments. For example, on the first interval, $t_a = t_1$, $t_b = t_2$, $C_0 = -\gamma + K_1\beta$, and $C_1 = K_1 E$. On interval 2, $t_a = t_2$, $t_b = t_3$, $C_0 = -\gamma + K_1(\alpha + \beta) - K_2\alpha$, and $C_1 = K_2 E$. The others are similar. The coefficient b_1 is given similarly by the sum of seven terms, each of the form

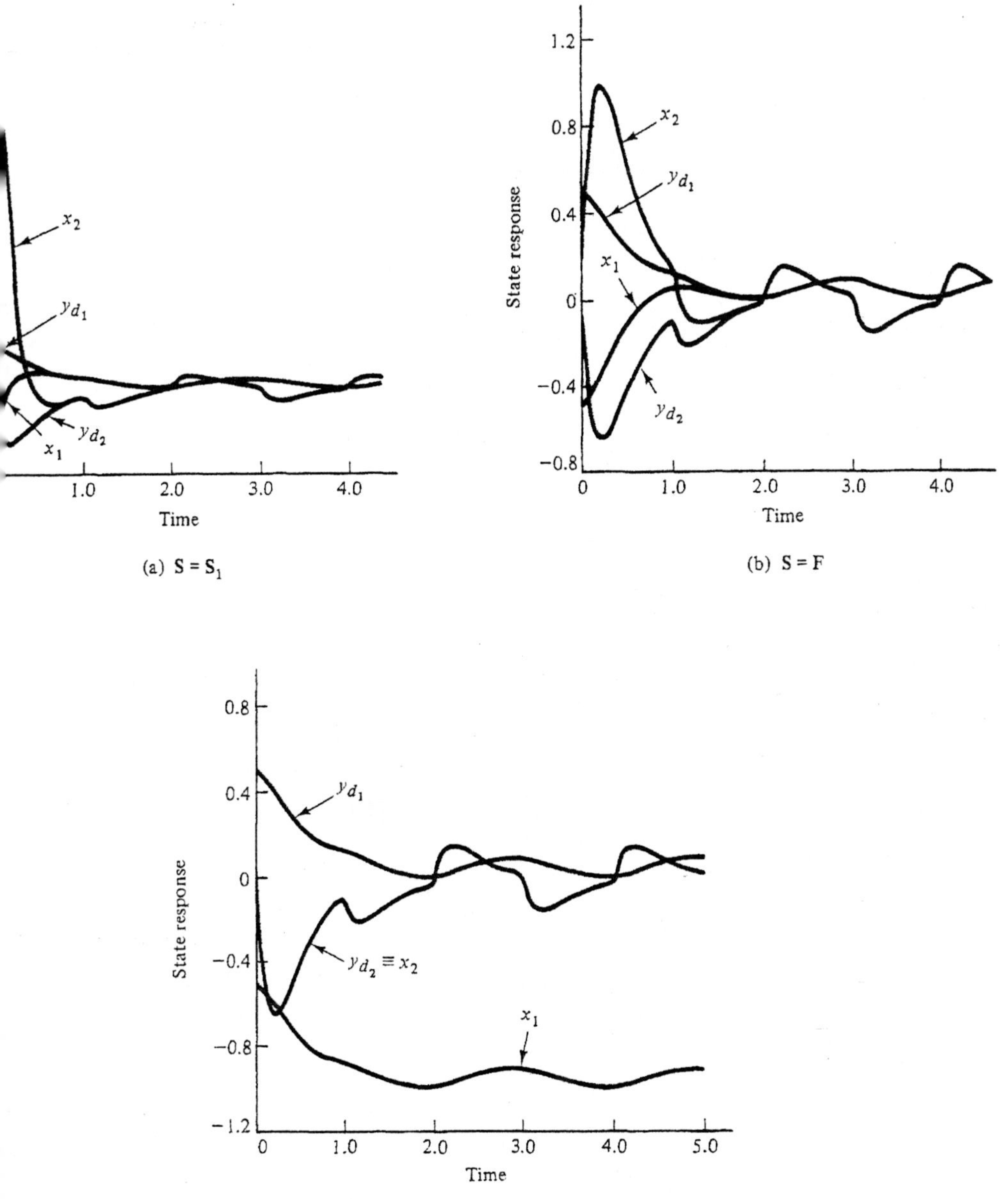

(a) $\mathbf{S} = \mathbf{S}_1$

(b) $\mathbf{S} = \mathbf{F}$

(c) $\mathbf{S} = [0]$

Figure 15.17

$$\left(\frac{1}{\pi}\right)\int\{[C_0 + C_1 \sin t] \cos t\}\, dt = \left(\frac{C_0}{\pi}\right)(\sin t_b - \sin t_a) + \frac{C_1}{(4\pi)}[\cos(2t_a) - \cos(2t_b)]$$

$$\equiv C_f(t_a, t_b, C_0, C_1)$$

The same arguments are used to evaluate the seven cosine coefficients $C_f(\ )$ and the sine coefficients $S_f(\ )$ on each interval.

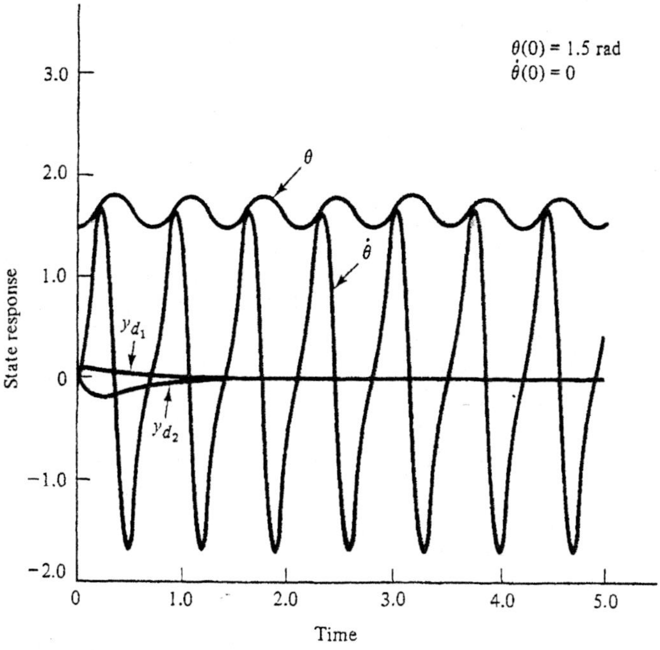

Figure 15.18

15.10 Certain nonlinearities with multiple but related inputs can be treated as if there were a single input, and the previous methods of finding describing functions can be applied. Demonstrate this by finding the describing function for $n(e, \dot{e}) = e^2 \dot{e}$.

If $e(t) = E \sin(\omega t)$, then $\dot{e}(t) = E\omega \cos(\omega t)$, so that $n(e, \dot{e})$ can be treated as having a single input $n(e) = E^3 \omega \sin^2(\omega t) \cos(\omega t)$. In this case no integration is required to find the describing function. Trigonometric identities give

$$n(e) = \left(\frac{E^3 \omega}{2}\right) \sin(\omega t) \sin(2\omega t) = \left(\frac{E^3 \omega}{4}\right)[\cos(\omega t) - \cos(3\omega t)]$$

The fundamental component is now obvious. It has a 90° phase shift relative to the input. Thus the describing function is

$$N(E, \omega) = \left(\frac{E^2 \omega}{4}\right) e^{j\pi/2} = \frac{E^2 \omega j}{4}$$

This example is frequency-dependent and purely imaginary.

Since the product of two sinusoids of frequencies ω_1 and ω_2 produces terms with frequencies $\omega_1 - \omega_2$ and $\omega_1 + \omega_2$, the foregoing procedure will fail to produce a valid describing function in many cases. For example $e(t)^2 = E^2 \sin^2(\omega t)$ gives a dc term and a double-frequency term but no fundamental component.

15.11 Use the describing function of Problem 15.10 to analyze the Van der Pol equation for possible limit cycles.

$\ddot{y} + \mu(y^2 - 1)\dot{y} + \beta y = 0$; both μ and β are positive.

Separating the linear and nonlinear parts gives $\ddot{y} - \mu\dot{y} + \beta y = -\mu y^2 \dot{y}$, which has the representation of Figure 15.5, with $n(e) = \mu y^2 \dot{y}$ and $G(s) = 1/[s^2 - \mu s + \beta]$. Thus the condition for existence of a limit cycle is that $-1/G(j\omega) = N(E, \omega)$. Equating imaginary parts gives $\mu\omega = \mu\omega E^2/4$, from which the approximate amplitude of the limit cycle is $E = 2$. By equating

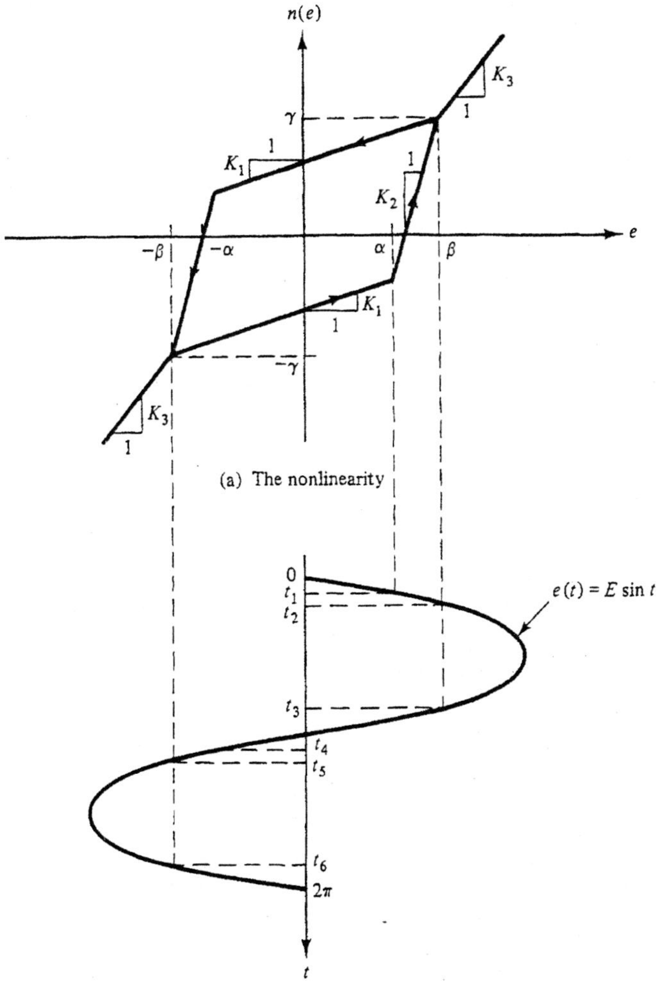

(a) The nonlinearity

(b) The input function

Figure 15.19

real parts, it is found that the limit cycle frequency (at least the fundamental) is $\omega = \sqrt{\beta}$. Sketches of the Nyquist plot and the $-1/N$ locus are given in Figure 15.20. The solution just found identifies the intersection of these two curves.

2 Is the limit cycle of Problem 15.11 stable or unstable?

Any critical point inside the closed contour formed by the plot of $G(j\omega)$ in Figure 15.20, for $-\infty < \omega < \infty$, is encircled once in the counterclockwise direction—i.e., $N_c = 1$. Any critical point outside the closed contour has $N_c = 0$. Since $G(s)$ has two open-loop right-half plane poles, $P_r = 2$. Thus application of Nyquist's criterion (Eq. (*15.18*)) to $G(j\omega)$, using $-1/N$ as the critical point, indicates the following:

1. The closed-loop system is unstable, with $Z_r = 2$ if $-1/N$ is outside the closed contour. This is *correct.*
2. The closed-loop system is unstable with $Z_r = 1$ if $-1/N$ is inside the closed contour. This is *incorrect.*

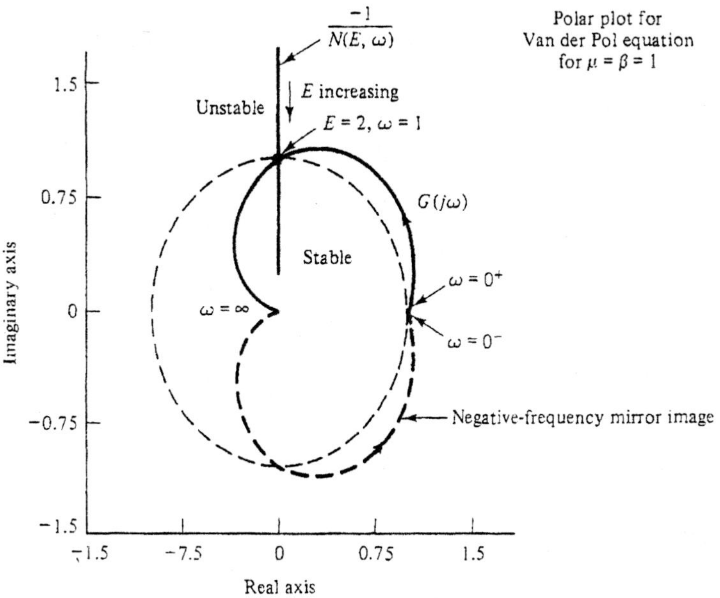

Figure 15.20

Note that $N(E, \omega)$ factors into $N_1(E)N_2(\omega)$. To analyze this system correctly, a modified

$$G'(j\omega) = \frac{j\omega}{-\omega^2 - \mu j\omega + \beta}$$

which incorporates the frequency-dependent part of $N(E, \omega)$, is defined. The remaining $N_1(E) = E^2/4$ will be used to define a real critical point $-1/N_1$. The polar plot of $G'(j\omega)$ is a double loop, one as ω varies from 0 to ∞ and the other, the mirror image for negative ω. Figure 15.21 shows the modified plot. Now critical points inside the contour have $N_c = 2$, and hence $Z_r = 0$, indicating stability. Critical points outside the contour are unstable. Thus stability of the equivalent gain closed-loop system requires that the critical point $-1/N$ must be *inside* the contour. (This is opposite the usual situation.) Now the stability of the limit cycle can be determined. Assume that E and ω are at the intersection point, i.e., we have a limit cycle. Then if the amplitude E increases slightly, $-1/N$ moves inside the contour, causing the system to become asymptotically stable and causing E to decrease back toward the limit cycle value. If E decreases a little, $1/N$ increases, and the critical point $-1/N$ moves outside the Nyquist contour. This gives an unstable system, causing E to grow back toward the intersection point. Thus the limit cycle is stable.

15.13 The circuit of Figure 15.22 has been widely used as an example of chaotic behavior [16, 17]. It is linear except for one nonlinear resistor. Its current will be modeled as a nonlinear function of the voltage across it, $i = n(v_{C1})$. The nonlinearity is piecewise linear, of the type treated in Example 15.5. Selecting states shown in the diagram, the state equations for this system are

$$\begin{bmatrix} \dot{x}_1 \\ \dot{x}_2 \\ \dot{x}_3 \end{bmatrix} = \begin{bmatrix} \left(\dfrac{1}{RC_1}\right)(x_2 - x_1) - \dfrac{n(x_1)}{C_1} \\ \left(\dfrac{1}{RC_2}\right)(x_1 - x_2) + \dfrac{x_3}{C_2} \\ -\left(\dfrac{1}{L}\right)x_2 \end{bmatrix}$$

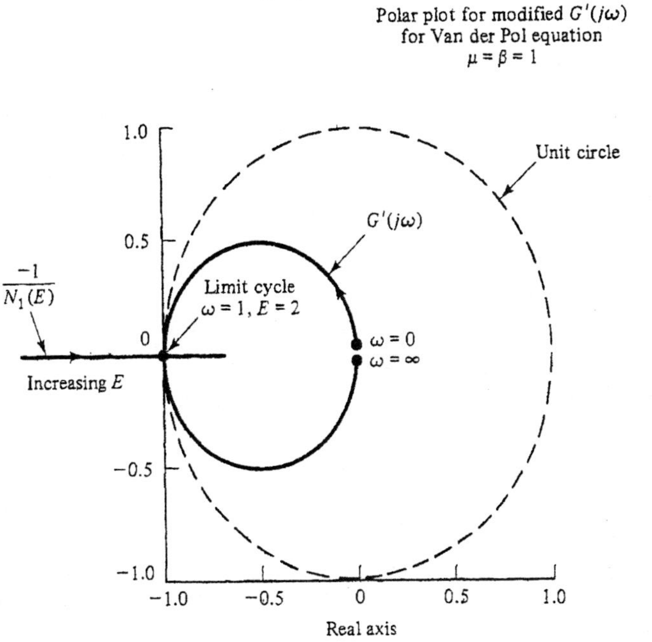

Figure 15.21

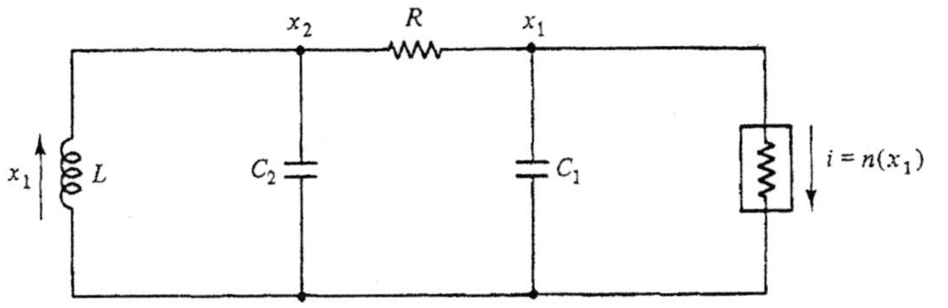

Figure 15.22 Chua's circuit example

Use Chua's values, $C_1 = \frac{1}{9}$, $C_2 = 1$, $L = \frac{1}{7}$, $\alpha = 1$, $K_1 = -0.8$, $R = 1/0.7$, $K_2 = -0.5$, $\beta = \infty$, and K_3 arbitrary. Find the equilibrium points of this nonlinear system and determine their stability (for small perturbations).

Using the given values, the state equations reduce to

$$\dot{\mathbf{x}} = \begin{bmatrix} -6.3 & 6.3 & 0 \\ 0.7 & -0.7 & 1 \\ 0 & -7 & 0 \end{bmatrix}\mathbf{x} - \begin{bmatrix} 9 \\ 0 \\ 0 \end{bmatrix} n(x_1) = \mathbf{A}\mathbf{x} - \mathbf{B}n(x_1)$$

$\dot{\mathbf{x}} = \mathbf{0}$ implies that $x_2 = 0$, $-6.3x_1 - 9n(x_1) = 0$ and that $0.7x_1 + x_3 = 0$. One equilibrium point is $\mathbf{x}_{e_1} = \mathbf{0}$. Other solutions are found from $x_1 = -9n(x_1)/6.3$. On the first linear segment, $n(x_1) = -0.8x_1$, and the only solution is $x_1 = 0$. On the second segment, $n(x_1) = -0.8 - 0.5(x_1 - 1)$ from which we find $x_1 = \pm 1.5$. This gives $\mathbf{x}_{e_2} = [1.5 \quad 0 \quad -1.05]^T$ and $\mathbf{x}_{e_3} = [-1.5 \quad 0 \quad 1.05]^T$. The Jacobian matrix is just $\mathbf{A}$ with A_{11} changed to $A_{11} - 9\partial n/\partial x_1$. Since $\partial n/\partial x_1$ is just K_1, the Jacobian at $\mathbf{x}_{e_1}$ is

$$[\partial \mathbf{f}/\partial \mathbf{x}]_1 = \begin{bmatrix} 0.9 & 6.3 & 0 \\ 0.7 & -0.7 & 1 \\ 0 & -7 & 0 \end{bmatrix}$$

and the eigenvalues are $\lambda = 1.552, -0.676 \pm j1.897$. At both $\mathbf{x}_{e_2}$ and $\mathbf{x}_{e_3}$, $\partial n(x_1)/\partial x_1 = K_2 = -0.5$, so the Jacobian at both these points is

$$\left[\frac{\partial \mathbf{f}}{\partial \mathbf{x}}\right]_2 = \begin{bmatrix} -1.8 & 6.3 & 0 \\ 0.7 & -0.7 & 1 \\ 0 & -7 & 0 \end{bmatrix},$$

and the eigenvalues are $\lambda = -2.759, 0.1297 \pm j2.1329$. All three equilibrium points are unstable, but their behavior would be expected to be different. Perturbations from the first would be expected to be dominated by a growing exponential $e^{1.552t}$, whereas the other two would be expected to exhibit growing oscillations of the form $e^{0.1297t}\sin(2.1329t)$.

15.14 Use the describing function approach to analyze the Chua circuit of Problem 15.13.

The results of Example 15.5 are used to calculate the describing function $N(E)$ for the nonlinearity $n(x_1)$. It is convenient here to examine $G(j\omega) = -1/N(E)$, and for that purpose $-1/N(E)$ is plotted in Figure 15.23. The transfer function from the output of the nonlinearity to x_1 is found by using $\mathbf{C} = [1 \quad 0 \quad 0]$, along with $\mathbf{A}$ and $\mathbf{B}$ of Problem 15.13. This gives

$$G(s) = \mathbf{C}[s\mathbf{I} - \mathbf{A}]^{-1}\mathbf{B} = \frac{9[s^2 + 0.7s + 7]}{[s^3 + 7s^2 + 7s + 44.1]}$$

The poles are at $s = -6.9105$ and $s = -0.0447 \pm j2.526$. The zeros are at $s = -0.35 \pm j2.6225$. Because of the lightly damped complex poles and zeros at frequencies rather close together, rapid changes in the magnitude and phase of the transfer function can be expected at nearby frequencies. The real-axis crossover points are crucial and are found analytically.

$$G(j\omega) = \frac{9\{(-\omega^2 + 7) + 0.7j\omega\}}{(44.1 - 7\omega^2) + j(7\omega - \omega^3)} \tag{1}$$

Multiplication by the complex conjugate of the denominator and setting the imaginary component to zero gives the equation for real axis crossover points, $0.7\omega(44.1 - 7\omega^2) - (-\omega^2 + 7)(7\omega - \omega^3) = 0$. One root of this equation is $\omega_1 = 0$. Factoring this term out and rearranging gives $\omega^4 - 9.1\omega^2 + 18.13 = 0$. This quadratic in ω^2 has two real roots, and their square roots give the positive frequencies at which $G(j\omega)$ crosses the real axis, $\omega_2 = 1.71642$ and $\omega_3 = 2.4807$. Using these in the real component of (*1*) gives the magnitudes at the crossover points as $G(0) = 1.42$, $G(j\omega_2) = 1.554$, and $G(j\omega_3) = 7.4455$. A sketch of the Nyquist polar plot

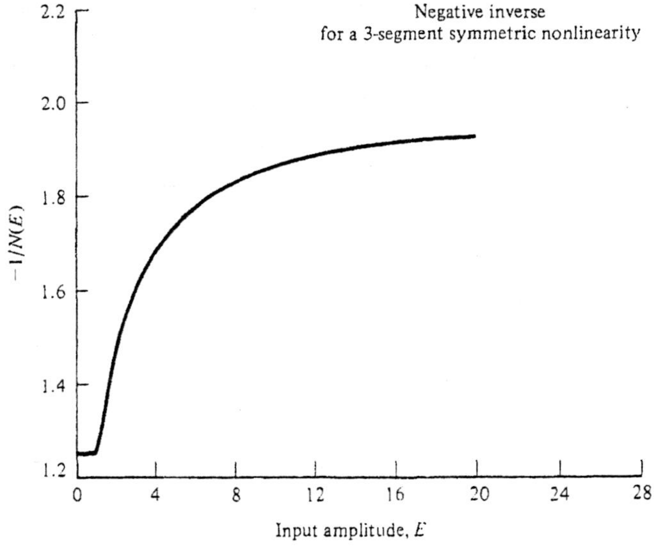

Figure 15.23

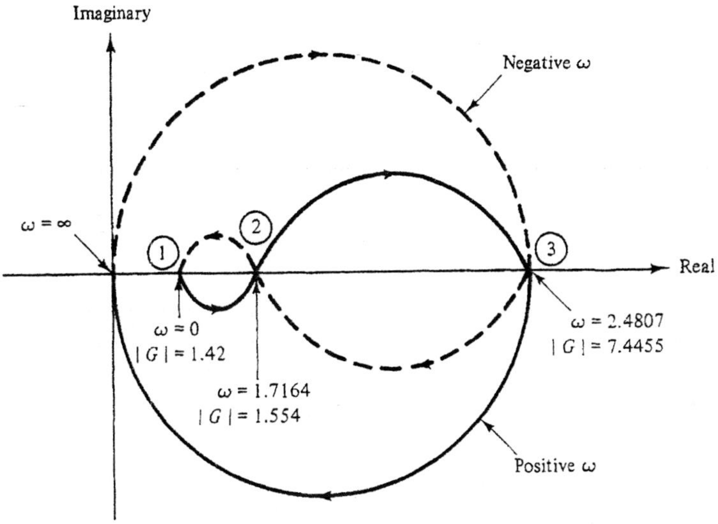

Figure 15.24 Polar plot of $G(j\omega)$

(not to scale) is given in Figure 15.24. The Nyquist criterion, Eq. (*15.18*), indicates that stability requires no encirclements of the critical point. For this system there are zero encirclements if the critical point is outside the entire plot or inside the small loop between crossover points 1 and 2. There are two clockwise encirclements ($N_c = -2$) for all critical points between crossover points 2 and 3, and $N_c = -1$ for all critical points between the origin and the first crossover point. For the Chua nonlinearity, the plot of $-1/N$ in Figure 15.23 shows a continuum of critical points between 1.25 and 2. In this range there are two intersections of $G(j\omega)$ and $-1/N$. The first one occurs for $\omega = 0$, and the second occurs for $\omega = 1.71642$. These intersections indicate potential limit cycles according to the describing function approximations. To check the stability of these two limit cycle points, assume we are at point 1 and a small increase occurs in the amplitude of x_1. This moves the critical point into a stable region, and the perturbed amplitude should decay back toward point 1. A small decrease in amplitude moves $-1/N$ into an unstable region, and the amplitude would be expected to increase back toward point 1. Thus point 1 appears to indicate a stable dc "limit cycle." Similar considerations at intersection point 2 show this to be an unstable limit cycle condition. In particular if the amplitude is perturbed away from point 2 to the right, it will not return to point 2, but neither can it proceed to another limit cycle, since point 3 is not an intersection of G and $-1/N$. Some more complicated behavior is indicated for a range of frequencies above $\omega = 1.7164$ but below $\omega = 2.4807$. The actual response to this system with the selected values is chaotic. The response contains a continuous spectrum of frequencies, not just discrete limit cycle frequencies. A small segment of the time response, starting with $\mathbf{x}(0) = [0.01 \quad 0 \quad 0]^T$, is shown in Figure 15.25. Although clearly not sinusoidal, counting peaks in various regions shows about 4 cycles in 10 s, or a period of 2.5 s. This crudely indicates some frequency content at about the frequency of crossover point 3. Superimposed are a range of lower modulating frequencies. Two-dimensional phase-plane plots of various pairs of states are given in Figures 15.26*a*, *b*, and *c*. Tendencies to oscillate about the equilibrium points $\mathbf{x}_{e_2}$ and $\mathbf{x}_{e_3}$ found in Problem 15.13 are clearly evident.

15 Add a modifying gain K to Chua's system of Problem 15.14 so that we now have $Kn(x_1)$ being fed into the same linear system. Use the describing function technique to predict behavior for various values of K.

The effect of K could be included in a modified $G(j\omega)$ plot or a modified $N' = KN$. We select the latter case for discussion. The real axis crossover frequencies are unchanged, and by selecting K we can cause $-1/N'$ to intersect with the crossover points in various ways:

1. If $K < 1.25/7.445 = 0.1679$, the entire interval $-1/N'$ is to the right of point 3, and stable behavior is predicted (no intersections, no limit cycles predicted).

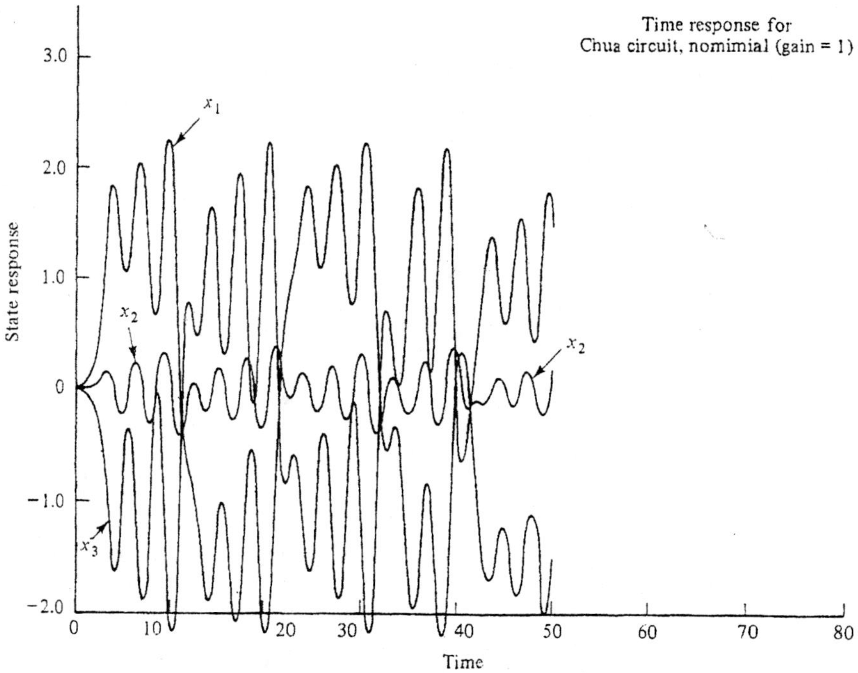

Figure 15.25

2. If $K > 2/1.42 = 1.408$, the entire interval $-1/N'$ falls between the origin and point 1, giving an unambiguous prediction of an unstable system (no intersections, no limit cycles predicted).
3. If $1.287 < K < 1.408$, the interval $-1/N'$ has just one intersection with $G(j\omega)$ at point 1, and some sort of stable dc ($\omega = 0$) behavior is predicted in steady state. The amplitude E required to make this happen can be predicted by using Figure 15.23 to find E for which $-1/N' = 1.42$. With $K = 1.3$, for example, this means $-1/N = 1.3(1.42) = 1.846$. This corresponds to $E \approx 8$ or 9, but it is not clear what this might mean for $\omega = 0$.
4. If $0.16789 < K < 0.2686$, $-1/N'$ has a single intersection with $G(j\omega)$ at point 3, and this predicts that a stable limit cycle with a frequency of $\omega = 2.4807$ rad/s. The amplitude E is also predicted by using Figure 15.23 to estimate the value of E at which $-1/N'(E)$ takes on the value 7.4455, that is, where $-1/N = 7.4455(K)$. With $K = 0.188$, the estimate is $E \approx 1.7$.

One simulation case in each of these categories is provided in Figures 15.27*a*, *b*, *c*, and *d*. In each case the general character of the describing function predictions are borne out reasonably well. The precision of the demarcation points is not clear, since a very slow growth or decay in amplitude takes a long time to notice. Also recall that the amplitude predictions are only for state x_1.

15.16 Use describing functions to investigate the possibility of limit cycles for the control loop of Figure 15.5 if $G(s) = (s + 20)/[s(s + 2)(s + 4)]$ and the nonlinearity is of the type given in Problem 15.9, with $\alpha = 0$, $\beta = 1$, $K_1 = 1$, $K_2 = 2$, and $K_3 = 1$.

The result in Problem 15.9 assumed that the amplitude E exceeded β. To analyze this system, the nonlinearity behavior must also be specified for amplitudes smaller than β. Here it is assumed that a set of smaller parallelograms nested inside the one shown in Figure 15.19*a* applies. The slope changes occur whenever $\dot{e}$ changes sign (at the maximum and minimum amplitudes) and whenever e changes sign. Under these assumptions the polar plot of $-1/N$ of

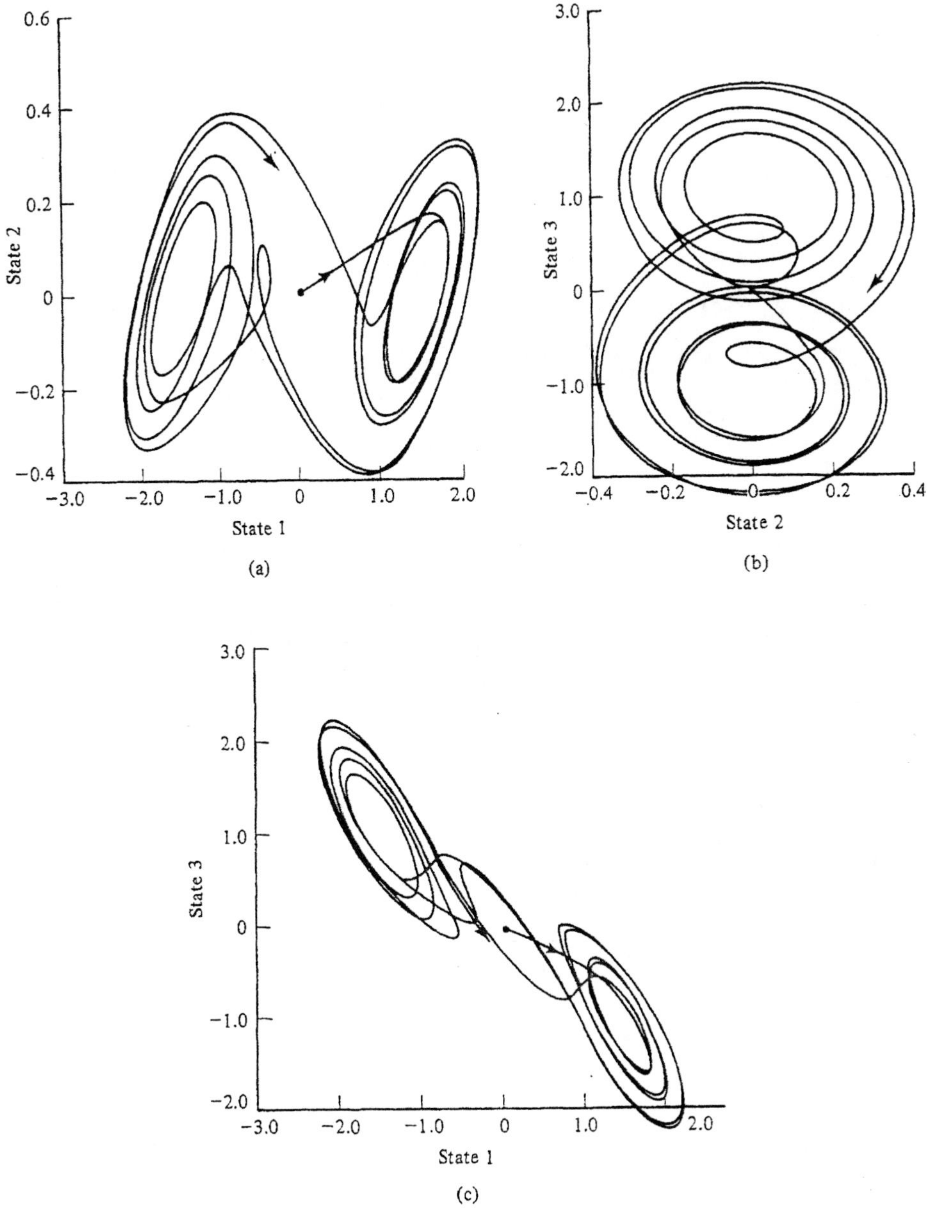

Figure 15.26

Figure 15.28 is obtained. Superimposed on it is a portion of the polar plot for $G(j\omega)$. There is one intersection, and it indicates existence of a stable limit cycle with an amplitude of $E \approx 0.8$ and a frequency of $\omega \approx 1.85$ rad/s. A simulation of this system generated the time response shown in Figure 15.29. This shows an x_1 amplitude of the expected magnitude, about 0.8, but the period of oscillation is 2.8 s, indicating $\omega \approx 2.24$, about 20% higher than predicted by the polar plot intersection. The probable reason is that the behavior of $n(e)$ is not really determined by e

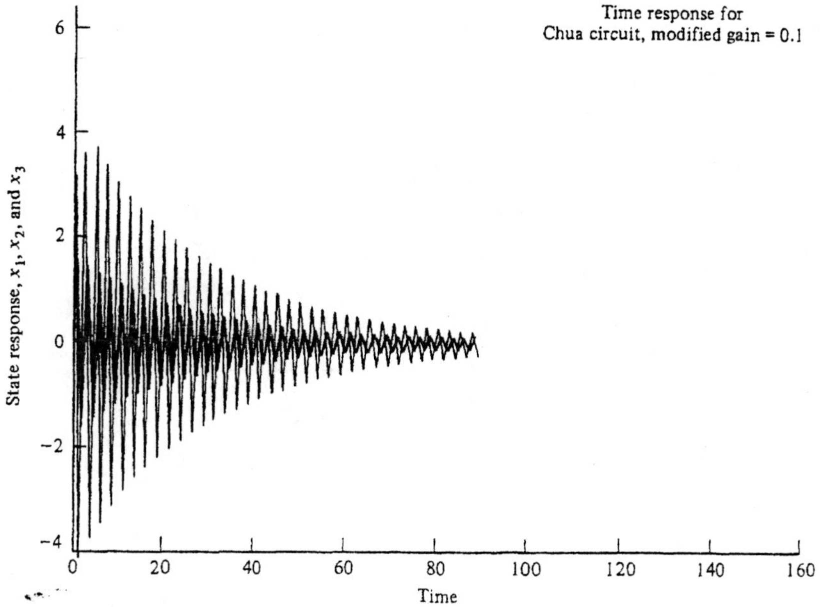

(a)

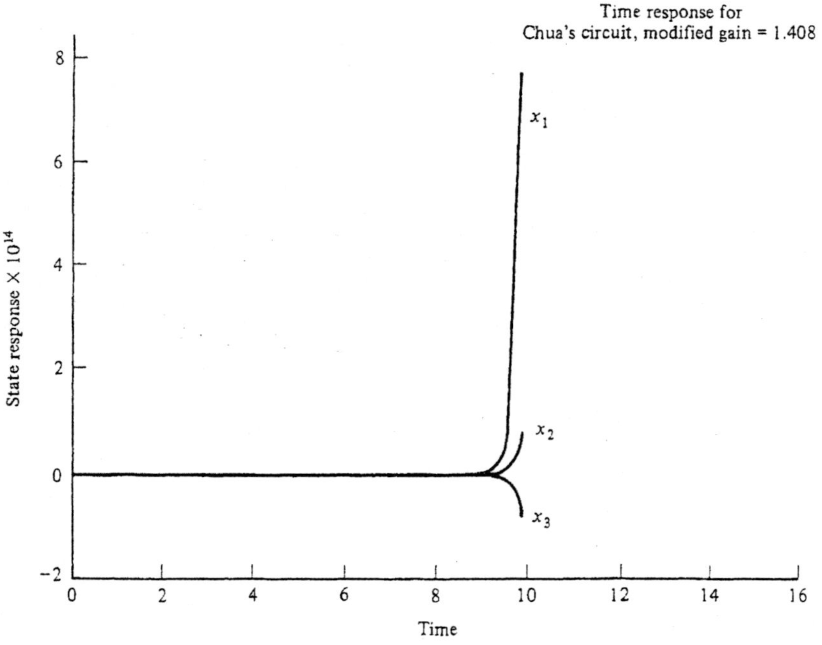

(b)

Figure 15.27

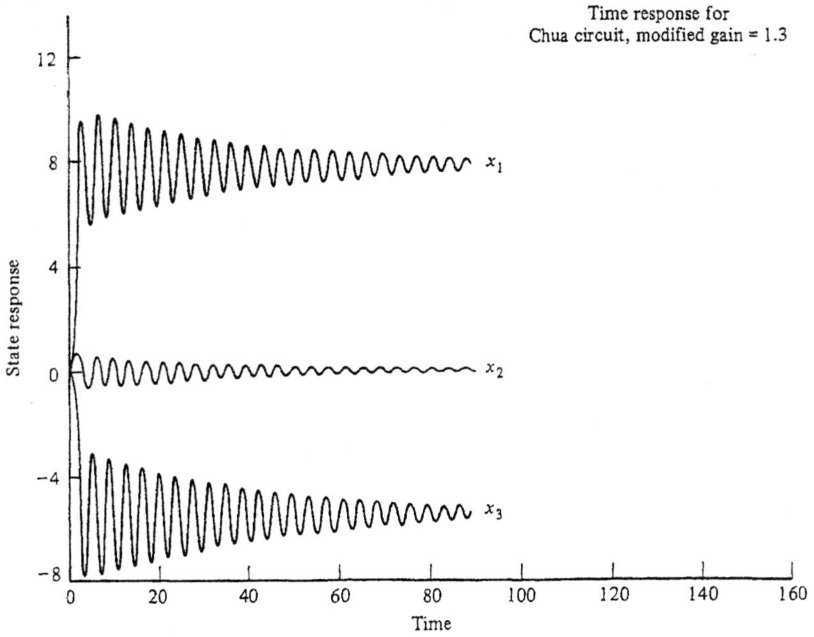

(c)

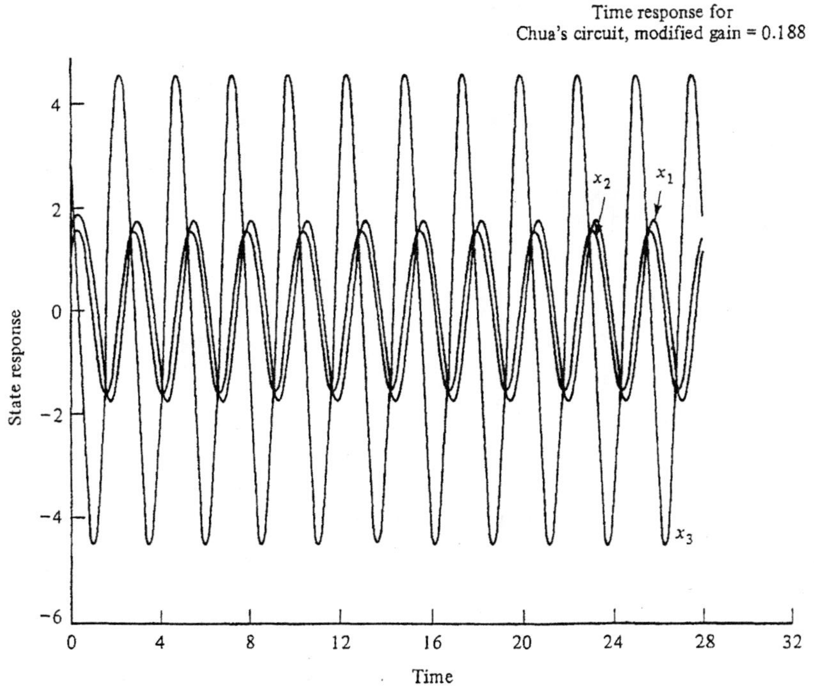

(d)

Figure 15.27 (Continued)

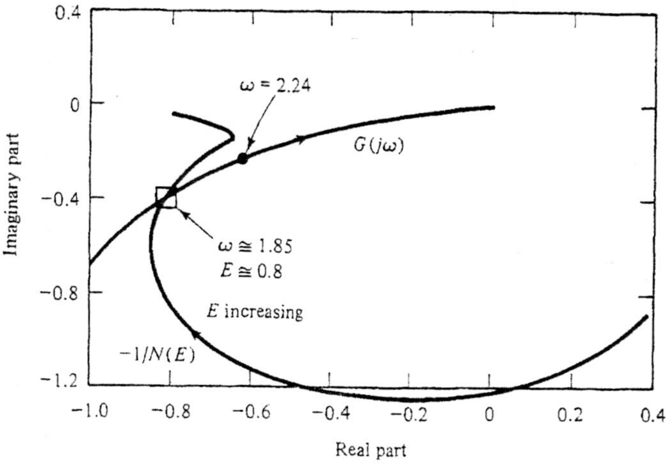

Figure 15.28

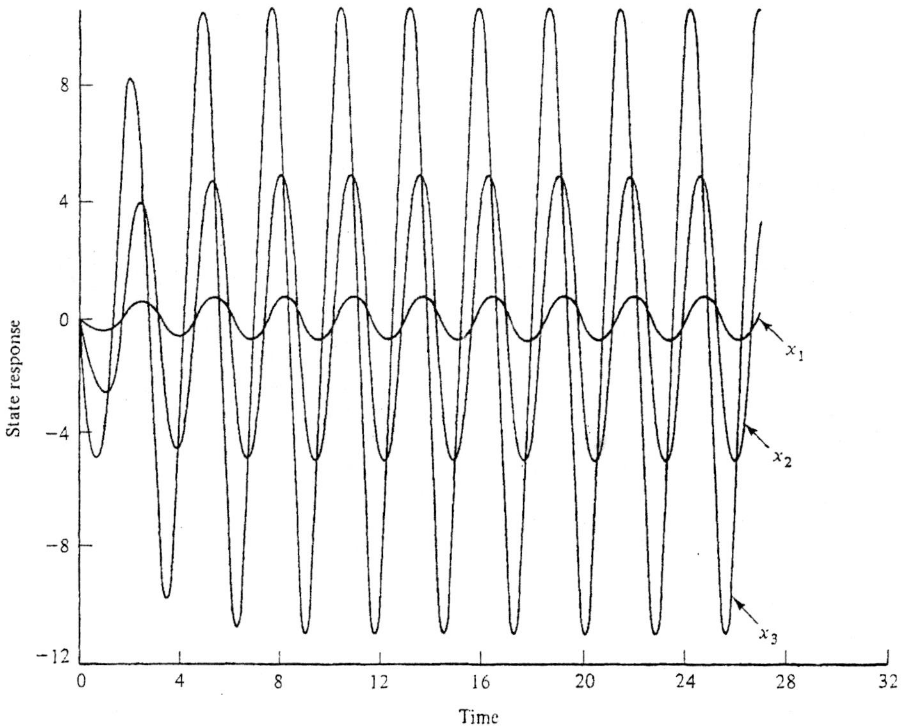

Figure 15.29

alone, nor even by both e and $\dot{e}$. Figure 15.30 shows four different signals, all having the same instantaneous value of $e(t)$. Two have positive $\dot{e}$ and two have negative $\dot{e}$. In order to determine correctly which parallelogram is being traversed, more history about $e(t)$ is required. In the simulation the most recent sign change in $\dot{e}(t)$ was used to determine the maximum amplitude of the current parallelogram being traversed. Other similar errors of approximation in hysteresis-type nonlinearities are discussed more fully in Reference 9.

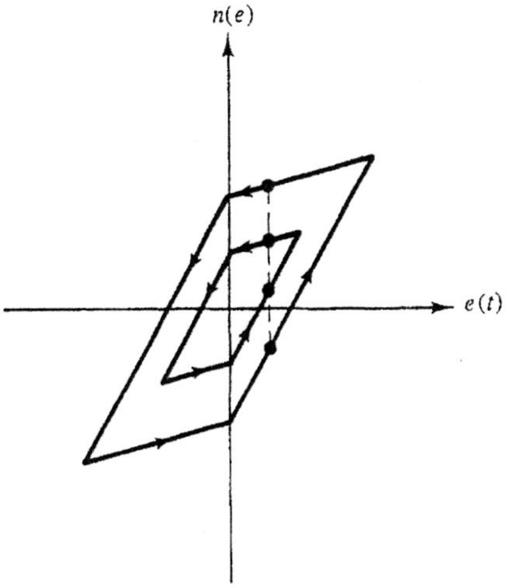

Figure 15.30

Investigate the following nonlinear system for stability:

$$\dot{x}_1 = x_2, \qquad \dot{x}_2 = -g(x_2) - f(x_1) \tag{1}$$

The stability of an *equilibrium point* must be investigated. Since there could be several, it is not really proper to speak of *system stability*. Equilibrium points satisfy $\dot{\mathbf{x}} = \mathbf{0}$, so $x_{2_e} = 0$ is required, and then $f(x_{1_e}) = -g(0)$. It is assumed that $g(0) = 0$ and that $f(x_1) = 0$ only at $x_1 = 0$. Thus the origin is the only equilibrium point, by assumption.

If x_1 is thought of as a position, then x_2 is a velocity, and the above equations might represent a unit mass connected to a nonlinear spring and damper. The spring force is $f(x_1)$ and the damper force is $g(x_2)$. This analogy suggests trying a Lyapunov function composed of a kinetic energy-like term with x_2^2 and a potential spring energy term (equal to work done by $f(x_1)$):

$$V(\mathbf{x}) = c_1 x_2^2 + c_2 \int_0^{x_1} f(\xi)\, d\xi$$

This term is positive definite if $c_1 > 0, c_2 > 0$ and if $f(x_1)$ always has the same sign as x_1, for example, any odd function of x_1. Then

$$\dot{V} = 2c_1 x_2 \dot{x}_2 + c_2 f(x_1)\dot{x}_1$$

Using equation (*1*) gives

$$\dot{V} = 2c_1 x_2[-g(x_2) - f(x_1)] + c_2 f(x_1) x_2$$

Selecting $c_2 = 2c_1$ gives $\dot{V}(\mathbf{x}) = -c_2 x_2 g(x_2)$. $\dot{V}$ is negative semidefinite if $g(x_2)$ always has the same sign as x_2. If this is true, stability i.s.L. is ensured by Theorem 10.1. Actually a slight generalization of Theorem 10.2 is possible. If, instead of $\dot{V}(\mathbf{x})$ being negative definite, $\dot{V}(\mathbf{x})$ can be shown to be always negative *along any trajectory of the system*, asymptotic stability can still be concluded [20].

In this problem, $\dot{V} = 0$ only if $x_2 = 0$, and is negative otherwise. But, if $x_2 \equiv 0$, then $\dot{x}_2 = 0$ also and this requires that $f(x_1) = 0$. By assumption, this means $x_1 = 0$, so $\dot{V} < 0$ for all possible $\mathbf{x}(t)$ trajectories, except at the equilibrium point $\mathbf{x} = \mathbf{0}$. It is concluded that this system is asymptotically stable if $f(x_1)x_1 > 0$ and $g(x_2)x_2 > 0$ for all $\mathbf{x} \neq \mathbf{0}$. Further, since $V(\mathbf{x}) \to \infty$ as $\|\mathbf{x}\| \to \infty$, the stability is global.

15.18 Use Lyapunov's direct method to study the stability of the origin $\mathbf{x} = \mathbf{0}$ for the system [21] described by

$$\dot{x}_1 = x_2 - ax_1(x_1^2 + x_2^2)$$
$$\dot{x}_2 = -x_1 - ax_2(x_1^2 + x_2^2) \tag{1}$$

A trial Lyapunov function is assumed as $V(\mathbf{x}) = c_1 x_1^2 + c_2 x_2^2$, with c_1 and c_2 unspecified but positive constants. Then $V(\mathbf{x})$ is positive definite and $V(\mathbf{x}) \to \infty$ as $\|\mathbf{x}\| \to \infty$. The time derivative is

$$\dot{V}(\mathbf{x}) = 2c_1 x_1 \dot{x}_1 + 2c_2 x_2 \dot{x}_2$$

Using Eq. (*1*), this becomes

$$\dot{V}(\mathbf{x}) = 2c_1 x_1[x_2 - ax_1(x_1^2 + x_2^2)] + 2c_2 x_2[-x_1 - ax_2(x_1^2 + x_2^2)]$$

If the selection $c_1 = c_2$ is made, then the troublesome $x_1 x_2$ product terms cancel, leaving $\dot{V}(\mathbf{x}) = -2ac_1(x_1^2 + x_2^2)^2$. If the constant a is positive, $\dot{V}(\mathbf{x})$ is negative definite and the origin is globally asymptotically stable by Theorem 10.3.

15.19 Derive conditions which ensure asymptotic stability of the origin for the system in Eq. (*1*) [24]. Use the integration technique of Problem 10.11 to determine a suitable Lyapunov function:

$$\dot{x}_1 = x_2, \qquad \dot{x}_2 = x_3, \qquad \dot{x}_3 = -(x_1 + cx_2)^n - bx_3 \tag{1}$$

Try $\dot{V}(\mathbf{x}) = -x_3^2$. Then $V(\mathbf{x}) = \int_{t_1}^{t} \dot{V}(\mathbf{x})\,dt = -\int_{t_1}^{t} x_3 \dot{x}_2\,dt$. Integration by parts gives

$$V(\mathbf{x}) = -x_3 x_2 + \int x_2 \dot{x}_3\,dt = -x_3 x_2 - \int x_2(x_1 + cx_2)^n\,dt - b\int x_2 x_3\,dt$$

But $x_3 = \dot{x}_2$, so $\int_{t_1}^{t} x_2 x_3\,dt = \int_0^{x_2} x_2\,dx_2 = x_2^2/2$. Adding and subtracting $\int cx_3(x_1 + cx_2)^n\,dt$ gives

$$V(\mathbf{x}) = -x_3 x_2 - \int (x_2 + cx_3)(x_1 + cx_2)^n\,dt + \int cx_3(x_1 + cx_2)^n\,dt - \frac{bx_2^2}{2}$$

From Eq. (*1*), $x_2 + cx_3 = \dot{x}_1 + c\dot{x}_2$ and $(x_1 + cx_2)^n = -\dot{x}_3 - bx_3$. Therefore,

$$V(\mathbf{x}) = -x_3 x_2 - \frac{(x_1 + cx_2)^{n+1}}{n+1} - \frac{bx_2^2}{2} - c\int x_3 \dot{x}_3\,dt - bc\int x_3^2\,dt$$
$$= -x_3 x_2 - \frac{(x_1 + cx_2)^{n+1}}{n+1} - \frac{bx_2^2}{2} - \frac{cx_3^2}{2} - bc\int x_3^2\,dt$$

This $V(\mathbf{x})$ is not positive definite; in fact, it can be made negative definite. Therefore, a modified function is selected as

$$V'(\mathbf{x}) = -V(\mathbf{x}) - bc\int_{t_1}^{t} x_3^2\,dt = \frac{(x_1 + cx_2)^{n+1}}{n+1} + \frac{bx_2^2}{2} + \frac{cx_3^2}{2} + x_2 x_3$$
$$= \frac{(x_1 + cx_2)^{n+1}}{n+1} + \frac{b}{2}\left(x_2 + \frac{x_3}{b}\right)^2 + \frac{(bc - 1)x_3^2}{2b}$$

This is positive definite if $b > 0$, $bc - 1 > 0$ and if $n + 1$ is any even positive integer. Also, $\dot{V}'(\mathbf{x}) = -\dot{V}(\mathbf{x}) - bcx_3^2 = -(bc - 1)x_3^2$. The same conditions ensure that $\dot{V}'(\mathbf{x}) \le 0$ for all $\mathbf{x}$. Since $x_3 \equiv 0$ requires $\dot{x}_3 = 0$ and $\dot{x}_2 = 0$, $x_1 =$ constant, it is seen that $x_3 = 0$ holds along a solution only at the point $\mathbf{x} = \mathbf{0}$. The above conditions thus ensure asymptotic stability.

For another direct-integration method of generating Lyapunov functions, see the discussion of Parks' method [12].

15.20 Consider the nonlinear system $\dot{\mathbf{x}} = \mathbf{f}(\mathbf{x})$, and assume that $\mathbf{f}(\mathbf{x})$ equals zero only at $\mathbf{x} = \mathbf{0}$. Then

$\mathbf{f}(\mathbf{x})^T\mathbf{f}(\mathbf{x})$ is a positive definite function of $\mathbf{x}$ and can serve as a potential Lyapunov function. Find sufficient conditions for asymptotic stability.

Since $V = \mathbf{f}(\mathbf{x})^T\mathbf{f}(\mathbf{x})$, $\dot{V} = \dot{\mathbf{f}}(\mathbf{x})^T\mathbf{f}(\mathbf{x}) + \mathbf{f}(\mathbf{x})^T\dot{\mathbf{f}}(\mathbf{x})$. But $\dot{\mathbf{f}}(\mathbf{x}) = [\partial\mathbf{f}/\partial\mathbf{x}]\dot{\mathbf{x}} = [\partial\mathbf{f}/\partial\mathbf{x}]\mathbf{f}$, so that

$$\dot{V} = \mathbf{f}^T\left\{\left[\frac{\partial\mathbf{f}}{\partial\mathbf{x}}\right]^T + \left[\frac{\partial\mathbf{f}}{\partial\mathbf{x}}\right]\right\}\mathbf{f}$$

Thus, by Theorem 10.3 the origin is asymptotically stable if $[\partial\mathbf{f}/\partial\mathbf{x}]^T + [\partial\mathbf{f}/\partial\mathbf{x}]$ is negative definite—i.e., has all its eigenvalues strictly in the left-half plane for all $\mathbf{x}$. Furthermore, if $\mathbf{f}(\mathbf{x})^T\mathbf{f}(\mathbf{x}) \to \infty$ as $\|\mathbf{x}\| \to \infty$, then the origin is *globally* asymptotically stable. This result is attributed to Krasovskii [23, 28].

Use the variable gradient technique to investigate the stability of the nonlinear system (see pages 59 and 67 of Reference 20) described by

$$\dot{x}_1 = x_2, \qquad \dot{x}_2 = -f(x_1)x_2 - g(x_1) \tag{1}$$

It is known that $g(0) = 0$ and that $\mathbf{x} = \mathbf{0}$ is the only equilibrium point. The gradient is assumed to be of the form

$$\nabla_\mathbf{x} V = \begin{bmatrix} \alpha_{11}x_1 + \alpha_{12}x_2 \\ \alpha_{21}x_1 + x_2 \end{bmatrix}$$

The curl equations require $\partial\nabla V_1/\partial x_2 = \partial\nabla V_2/\partial x_1$, or

$$x_1\frac{\partial\alpha_{11}}{\partial x_2} + \alpha_{12} + x_2\frac{\partial\alpha_{12}}{\partial x_2} = x_1\frac{\partial\alpha_{21}}{\partial x_1} + \alpha_{21}$$

Using the assumed gradient and Eq. (1) gives

$$\begin{aligned}\dot{V} &= (\nabla_\mathbf{x} V)^T\begin{bmatrix} x_2 \\ -f(x_1)x_2 - g(x_1)\end{bmatrix} \\ &= \alpha_{11}x_1x_2 + \alpha_{12}x_2^2 - \alpha_{21}f(x_1)x_1x_2 - f(x_1)x_2^2 - g(x_1)\alpha_{21}x_1 - g(x_1)x_2\end{aligned}$$

This expression should be made at least negative semidefinite.

One possible solution begins by setting $\alpha_{21} = 0$. Then the curl equations are satisfied if $\alpha_{12} = 0$ and if α_{11} is not a function of x_2. By setting $\alpha_{11} = g(x_1)/x_1$, we obtain $\dot{V} = -f(x_1)x_2^2$, which is negative semidefinite if $f(x_1) \geq 0$ for all x_1. Then $\nabla_\mathbf{x} V = [g(x_1) \quad x_2]^T$ and

$$V = \int_0^{x_1} g(\xi)\,d\xi + \int_0^{x_2} \xi\,d\xi = \int_0^{x_1} g(\xi)\,d\xi + \frac{x_2^2}{2}$$

Thus V is positive definite, and hence a Lyapunov function, if $g(x_1)x_1 > 0$ for all $x_1 \neq 0$.

The following conclusions regarding stability can be drawn:

1. Stable i.s.L. if $g(x_1)x_1 > 0, f(x_1) \geq 0$, for all $\mathbf{x} \neq \mathbf{0}$ by Theorem 10.1.
2. Asymptotically stable if, in addition, $f(x_1)$ and $g(x_1) = 0$ only at $x_1 = 0$. This ensures that $\dot{V} \neq 0$ on any solution of equation (1) except at $\mathbf{x} = \mathbf{0}$.
3. Globally asymptotically stable if, in addition, $\int_0^{x_1} g(\xi)\,d\xi \to \infty$ as $|x_1| \to \infty$.

Use Lyapunov's direct method to investigate the stability of the nonlinear time-varying system [27] $\ddot{x} + a\dot{x} + g(x, t)x = 0$.

The state equations are $\dot{x}_1 = x_2, \dot{x}_2 = -ax_2 - g(x_1, t)x_1$.

Assume that $\nabla_\mathbf{x} V = \begin{bmatrix} \alpha_{11}x_1 + \alpha_{12}x_2 \\ \alpha_{21}x_1 + x_2 \end{bmatrix}$. Then since $V(\mathbf{x})$ may be an explicit function of time,

$$\begin{aligned}\dot{V}(\mathbf{x}) &= (\nabla_\mathbf{x} V)^T\dot{\mathbf{x}} + \partial V/\partial t \\ &= \alpha_{11}x_1x_2 + \alpha_{12}x_2^2 - a\alpha_{21}x_1x_2 - g(x_1, t)x_1^2\alpha_{21} - ax_2^2 - g(x_1, t)x_1x_2 + \partial V/\partial t\end{aligned}$$

In order to remove the $x_1 x_2$ product terms, set $\alpha_{11} = a\alpha_{21} + g(x_1, t)$. The curl equations can then be satisfied if $\alpha_{21} = \alpha_{12} =$ constant. Then

$$V(\mathbf{x}) = \int_0^{x_1} [a\alpha_{21} + g(x_1, t)]x_1 dx_1 + \int_0^{x_2} [\alpha_{21} x_1 + x_2]\, dx_2 \Big|_{x_1 = \text{constant}}$$

$$= a\alpha_{21}\frac{x_1^2}{2} + \alpha_{21} x_1 x_2 + \frac{x_2^2}{2} + \int_0^{x_1} g(x_1, t)x_1\, dx_1$$

If $\int_0^{x_1} g(x_1, t)x_1\, dx_1 > 0$ for all x_1 and t, then $V(\mathbf{x}) > \frac{1}{2}(x_2 + \alpha_{21} x_1)^2 + \frac{1}{2}(a\alpha_{21} - \alpha_{21}^2)x_1^2$. Thus $V(\mathbf{x})$ is positive definite if $a > 0$ and $a\alpha_{21} > \alpha_{21}^2$. This is ensured by selecting $\alpha_{21} = a - \epsilon$, where ϵ is a small positive number. Checking the time derivative,

$$\frac{dV}{dt} = a\alpha_{21} x_1 \dot{x}_1 + \alpha_{21} \dot{x}_1 x_2 + \alpha_{21} x_1 \dot{x}_2 + x_2 \dot{x}_2 + g(x_1, t)x_1 \dot{x}_1 + \int_0^{x_1} \frac{\partial g(x_1, t)}{\partial t} x_1\, dx_1$$

Using $\alpha_{21} = a - \epsilon$ and the differential equations for $\dot{x}_1$ and $\dot{x}_2$ gives

$$\dot{V} = -\epsilon x_2^2 - (a - \epsilon)g(x_1, t)x_1^2 + \int_0^{x_1} \frac{\partial g(x_1, t)}{\partial t} x_1\, dx_1$$

This expression is negative definite if $g(x_1, t) > 0$ for all x_1, t and if the integral term is sufficiently small. That is, if

$$\int_0^{x_1} \frac{\partial g(x_1, t)}{\partial t} x_1\, dx_1 < a g(x_1, t)x_1^2$$

This will be true, for example, if $\max_{x_1, t} [\partial g(x_1, t)/\partial t] < 2ag(x_1, t)$ for all x_1 and t. Additionally, if $g(x_1, t)$ is bounded for all x_1 and t, then $V(\mathbf{x}, t)$ can be bounded as required by condition (4) of Theorem 10.5. If these conditions are satisfied, the system is uniformly globally asymptotically stable.

PROBLEMS

15.23 Use dynamic linearization to design a controller for the first order nonlinear system $\dot{x} = x^2 u$ so that the resulting response mimics the linear system $\dot{y}_d = -2y_d + v$, where v is a unit step function. Investigate various initial conditions for the actual system and for the template system. What impact do these have on potential singularities in the linearizing control law?

15.24 The desired response for the second-order nonlinear system described by

$$\dot{x}_1 = x_2^2 + u \quad \text{and} \quad \dot{x}_2 = x_2 u$$

is intended to mimic the uncoupled linear system

$$\dot{y}_{d_1} = -y_{d_1} + v, \qquad \dot{y}_{d_2} = -4y_{d_2} + v$$

Investigate the dynamic linearization procedure, noting that a single control variable is being asked to satisfy two conflicting equations. Use a least-squares approximation to find a control law, and test its response in a simulation. Let $v(t)$ be a unit step function and set $\mathbf{y}_d(0) = \mathbf{0}$.

15.25 Use dynamic linearization to design a controller for the nonlinear orbit equations of Example 15.2. Select a linear template system satisfying $\ddot{r} + \alpha\dot{r} + \beta r = \beta R$ and $\ddot{\theta} + \kappa\dot{\theta} + \sigma\theta = \sigma\mu t/R^3$. Select values for α, β, κ, and σ so that r and θ smoothly approach their steady-state values in about one-fourth of an orbit revolution.

15.26 Verify that the limit cycle found in Example 15.7 is stable.

15.27 Repeat the analysis of Problem 15.13 of the system equilibrium points, but reverse the order of the two slopes in the nonlinearity, so that $K_1 = -0.5$ and $K_2 = -0.8$.

Use describing functions to investigate the possibility of limit cycles (and chaotic behavior), as was done in Problem 15.14, but with the modified nonlinearity of Problem 15.27. Results derived in Example 15.5 yield the plot of $-1/N$ given in Figure 15.31.

Repeat the analysis of Problem 15.15 but with the nonlinearity of Problems 15.27 and 15.28.

Another version of Chua's chaotic circuit [31] has a three-segment nonlinearity of the type treated in Example 15.5. The breakpoints and slopes are now $\alpha = 1$, $\beta = 5$, $K_1 = -0.8$, $K_2 = -0.5$, and $K_3 = +2$. The linear part of the circuit is the same as in Problem 15.13, with all the same values. Find the equilibrium points and investigate their stability. Use describing functions to investigate the possibility of limit cycles or other strange oscillations. A plot of the describing function is given in Figure 15.32.

Repeat Problem 15.30, but with new parameters values $R = 0.71$, $L = 1$, $C_1 = C_2 = 0.1$, $\alpha = 1$, $\beta = 2$, $K_1 = -6$, $K_2 = -5$, and $K_3 = +10$. The describing function is plotted in Figure 15.33.

A system with the configuration of Figure 15.5 has a nonlinearity with dead zone and saturation, described by Figure 15.6*a* with $\alpha = 0.25$, $\beta = 2.5$, $K_1 = K_3 = 0$, and $K_2 = 1$. The linear portion is described by $G(s) = K(s+8)/[s(s+2)(s+4)(s+6)]$. Find the range of gain values K for which the linear system, without the nonlinearity, is unstable. Show that the presence of the nonlinearity can stabilize the system. Also find a value of K where limit cycles can exist. Determine the amplitude and frequency of the stable limit cycle.

Consider the nonlinearity of Problem 15.32 along with $G(s) = 10(s^2 + 4s + 68)/[s(s+2)(s+4)]$. Use root locus analysis to find the amplitude and frequency of a stable limit cycle by treating the describing function $N(E)$ as a part of the root locus gain $KN(E)$ and finding the $j\omega$ axis crossover points.

(a) Using precisely the same steps as in Problem 10.11c, find a Lyapunov function and its time derivative for the nonlinear system [20].

$$\dot{x}_1 = x_2, \qquad \dot{x}_2 = x_3, \qquad \dot{x}_3 = -F(x_2)x_3 - ax_2 - bx_1$$

(b) Determine sufficient conditions for asymptotic stability.

A special case of the Lorenz equation, widely used as an example of chaotic behavior, is given by [14, 15]

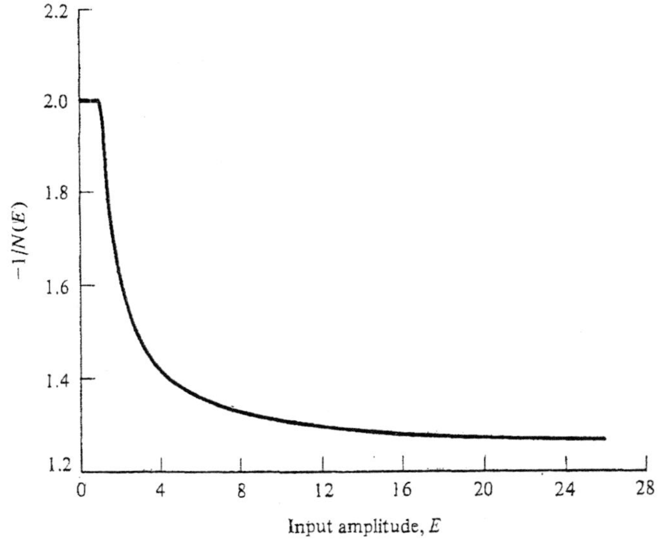

Figure 15.31

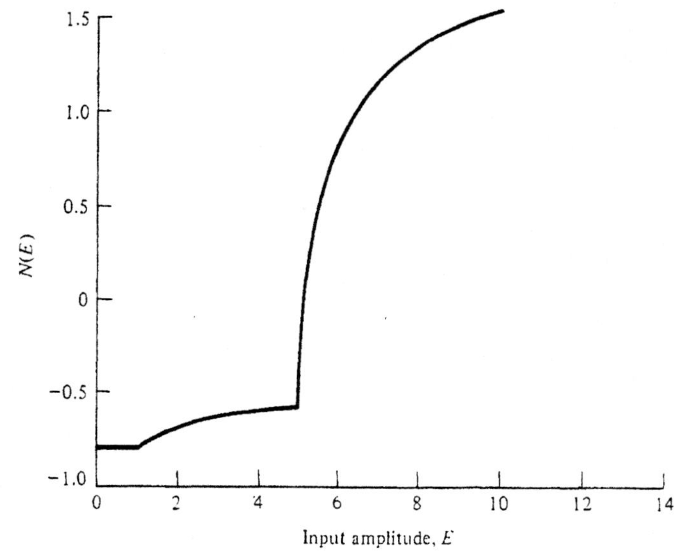

Figure 15.32

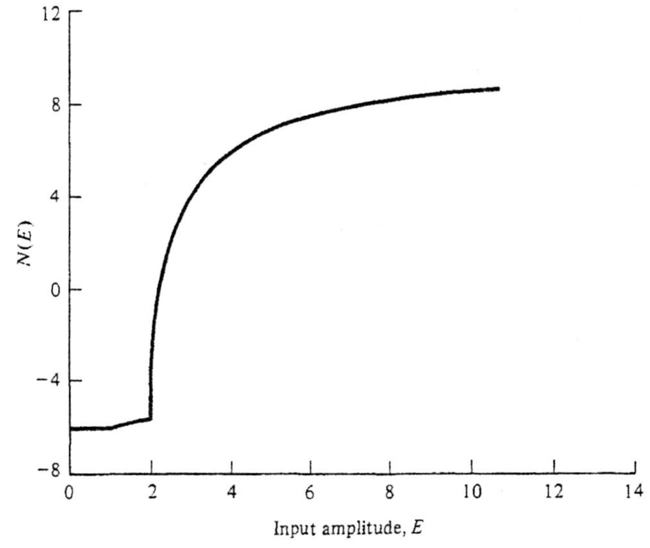

Figure 15.33

$$\begin{bmatrix} \dot{x}_1 \\ \dot{x}_2 \\ \dot{x}_3 \end{bmatrix} = \begin{bmatrix} -10 & 10 & 0 \\ r & -1 & 0 \\ 0 & 0 & -\frac{8}{3} \end{bmatrix} \begin{bmatrix} x_1 \\ x_2 \\ x_3 \end{bmatrix} + \begin{bmatrix} 0 \\ -x_1 x_3 \\ x_1 x_2 \end{bmatrix}$$

(a) Show that if $r < 1$, there is only one equilibrium point located at the origin, and it is locally stable.

(b) Use a suitable Lyapunov function to estimate the domain of stable attraction to this equilibrium point. Set $r = 0.5$ for simplicity.

(c) The more interesting, chaotic behavior occurs for values of r near 28. Show that there are then three equilibrium points, all of which are locally unstable.

(d) Estimate the domain of instability which surrounds the origin by using Lyapunov techniques.

The divergence theorem [8] states that for a sufficiently smooth vector field $\mathbf{f}(\mathbf{x})$ defined over a sufficiently smooth closed volume Y with surface area S composed of incremental area elements dA with outward pointing normal vectors $\mathbf{n}(\mathbf{x})$,

$$\iiint_{Y} \operatorname{div}(\mathbf{f})\, dV = \iint_{S} \mathbf{f}(\mathbf{x}) \cdot \mathbf{n}(\mathbf{x})\, dA$$

where $\operatorname{div}(\mathbf{f}) = \partial f_1/\partial x_1 + \partial f_2/\partial x_2 + \cdots + \partial f_n/\partial x_n$ is the divergence of $\mathbf{f}$. Let $\mathbf{f} = \dot{\mathbf{x}}$ and note that $\iint \mathbf{f}(\mathbf{x}) \cdot \mathbf{n}(\mathbf{x})\, dA = \iint \dot{\mathbf{x}}\Delta t \cdot \mathbf{n}(\mathbf{x})\, dA/\Delta t \to dY/dt$. Apply this to the vector $\mathbf{f}$ of the Lorenz equations and show that the volume Y containing an arbitrary set of initial states satisfies $dY/dt = -\frac{41}{3}Y$. This means that the volume that contains the solution trajectories through these initial states is always decreasing according to $Y(t) = Y(0)e^{-(41/3)t}$ [15].

Apply the divergence theorem of the previous problem to a linear system $\dot{\mathbf{x}} = \mathbf{A}\mathbf{x}$ and show that the trajectories which originally occupy a volume $Y(0)$ satisfy $Y(t) = Y(0)e^{\alpha t}$, where $\alpha = \operatorname{trace}\{\mathbf{A}\}$. From this show that "bounded volume" result for solution trajectories does not in any way imply bounded norms for $\mathbf{x}(t)$ and hence does not imply any kind of stability. *Hint:* Consider a specific unstable case such as $\mathbf{A} = \operatorname{diag}[2 + 2j,\ 2 - 2j,\ -10]$. Trajectories initially in a three-dimensional volume diverge to infinity, but in a planar subspace, with zero volume.

Answers to Problems

CHAPTER 1

1.21 $Q = \frac{A}{\rho g}\dot{P}$

1.22 $v(t) = v(t_0) + \int_{t_0}^{t} Q(\tau)\,d\tau$

1.23 $v_0 = L(df_3/dt) + Rf_3$

1.24 $y(s)/u(s) = [Cs + 1/R_1]/[Cs + (1/R_1 + 1/R_2)]$

1.26 $Y(z) = \sum_{n=0}^{\infty} e^{-0.1nT}e^{-nsT} = \frac{1}{1 - z^{-1}e^{-0.2}} = \frac{z}{z - e^{-0.2}}$

1.27 $H(z) = \frac{10z^2(z + 0.5)}{(z - 0.2)(z - 0.4)(z - 0.8)}$

Poles at $z = 0.2, 0.4, 0.8$
Zeros at 0., 0., −0.5
Stable

1.28 $y(nT) = 156.25 - 2.916(0.2)^n + 30(0.4)^n - 173.33(0.8)^n$

CHAPTER 2

2.25 $3 < K < 9$

2.26 $K = 20$, $\omega = 2$ rad/s

2.27 With $K = \sqrt{327{,}680} \cong 572$, the s row of Routh's array is zero. The auxiliary equation then is $44.2s^2 + 2288 = 0$, indicating the cross-over frequency is $\omega = 7.19$ rad/s.

2.28 (a) Type 0, $K_p = 20$ db $= 10$, $K_v = 0$, $K_a = 0$ (b) Type 1, $K_p = \infty$, $K_v = -20$ db $= 0.1$, $K_a = 0$ (c) Type 2, $K_p = \infty$, $K_v = \infty$, $K_a = -60$ db $= 0.001$

2.30 Stable for all K, but very low stability margins for small values of K.

2.31 For $K > 7.5$ there is one clockwise and one counterclockwise encirclement. Therefore, $N = 0$ and system is stable. There are two unstable roots for $0 < K < 7.5$.

2.32 Gain margin $\cong 1.76$, i.e., $K > 8800$ causes system instability. Phase margin $\cong 13°$. $\omega \cong 15$ rad/s for a $-180°$ phase angle.

2.33 $K = 9.65$, $\omega = 11.8$ rad/s (*Hint:* Dominant roots at $-\alpha \pm j\omega$; $e^{-\alpha} = 0.0015$.)

2.34 Nonminimum phase

2.35 Direct realization: $y(t_k) = -0.5y(t_{k-1}) + E(t_{k-1}) - 0.5E(t_{k-2})$

Parallel realization:
$$y_1(t_k) = E(t_{k-1})$$
$$y_2(t_k) = -0.5y_2(t_{k-1}) + E(t_{k-1})$$
$$y(t_k) = -y_1(t_k) + 2y_2(t_k)$$

Cascade realization:
$$E_1(t_k) = E(t_{k-1})$$
$$y(t_k) = E_1(t_k) - 0.5[E_1(t_{k-1}) + y(t_{k-1})]$$

2.36 $C(nT) \cong 13.3333 - 20(0.5)^n + 6.6667(-0.5)^n$

2.37 (a) $C(z) = 1 + (\alpha + \beta - a - b)z^{-1} + (ab + \alpha^2 + \alpha\beta + \beta^2 - a\alpha - b\alpha - a\beta - b\beta)z^{-2} + \cdots$. Therefore, $C(0) = 1$, $C(T) = (\alpha + \beta - a - b)$ and $C(2T) = ab + \alpha^2 + \alpha\beta + \beta^2 - a\alpha - b\alpha - a\beta - b\beta$.

(b) $$C(nT) = \frac{ab}{\alpha\beta}\delta_{0n} + \left[\frac{\alpha^2 - a\alpha - b\alpha + ab}{\alpha(\alpha - \beta)}\right]\alpha^n + \left[\frac{\beta^2 - a\beta - b\beta + ab}{\beta(\beta - \alpha)}\right]\beta^n$$

where $\delta_{0n} = \begin{cases} 0 & \text{if } n \neq 0 \\ 1 & \text{if } n = 0 \end{cases}$

(c) $C(nT) \to 0$ as $n \to \infty$

2.38 (a) $$\frac{C(z)}{R(z)} = \frac{0.004845K(z + 0.9672)}{z^2 - (1.9239 - 0.00484K)z + (0.90484 + 0.004686K)}$$

(b)

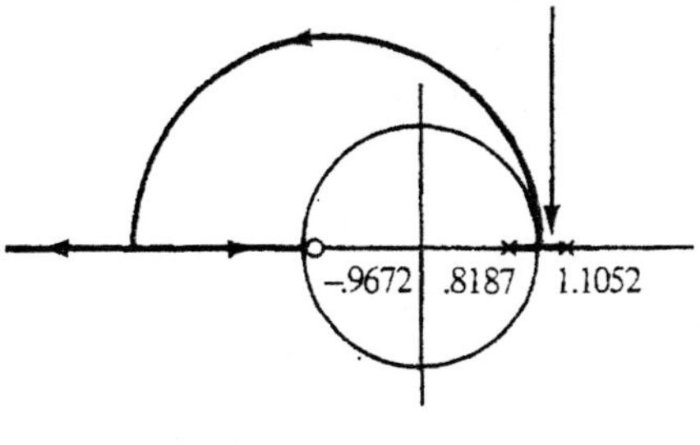

Figure A2.38

2.39 Let [ ]* indicate the Z-transform of whatever is inside [ ].

$$E(z) = \frac{[RG_1]^*}{1 + [G_1G_2H_1]^* + [G_1G_2G_3H_2]^*}$$

$$C(s) = \frac{[RG_1]^*G_2G_3G_5/[1 + G_5H_3]}{1 + [G_1G_2H_1]^* + [G_1G_2G_3H_2]^*} + \frac{RG_4G_5}{1 + G_5H_3}$$

$C(z)$ is the same except put an * on $[G_2G_3G_5/[1 + G_5H_3]]$ and $[RG_4G_5/[1 + G_5H_3]]$

2.40 $$G_c(z) = \frac{Az^2 + Bz + D}{(1 + \tau/T)z^2 - (2\tau/T + 1)z + \tau/T}$$

where
$$A = K_D/T + K_P(1 + \tau/T) + K_I(T + \tau)$$
$$B = -2(K_D/T + K_P\tau/T) - K_P - K_I\tau$$
$$D = K_D/T + K_P\tau/T$$

For simplicity, let $\tau \to 0$. Then one obtains an algorithm of the form

$$E_1(t_k) = E_1(t_{k-1}) + \alpha E(t_k) - \beta E(t_{k-1}) + \gamma E(t_{k-2})$$

CHAPTER 3

3.19 $\mathbf{A}_1 = \begin{bmatrix} 0 & 1 \\ -b & -a \end{bmatrix}$, $\mathbf{B}_1 = \begin{bmatrix} 0 \\ 1 \end{bmatrix}$, $\mathbf{C}_1 = [1 \quad 0]$, $D_1 = 0$

$\mathbf{A}_2 = \begin{bmatrix} -a & 1 \\ -b & 0 \end{bmatrix}$, $\mathbf{B}_2 = \mathbf{B}_1$, $\mathbf{C}_2 = \mathbf{C}_1$, $D_2 = D_1$

3.20 Let $x_1 = y_1$, $x_2 = y_2$, $x_3 = \dot{y}_2$, $\mathbf{x} = [x_1 \quad x_2 \quad x_3]^T$, $\mathbf{y} = [y_1 \quad y_2]^T$, $\mathbf{u} = [u_1 \quad u_2]^T$,

$$\mathbf{A} = \begin{bmatrix} -3 & -3 & 0 \\ 0 & 0 & 1 \\ 0 & -3 & -4 \end{bmatrix}, \mathbf{B} = \begin{bmatrix} 1 & 0 \\ 0 & 0 \\ 0 & 1 \end{bmatrix}, \mathbf{C} = \begin{bmatrix} 1 & 0 & 0 \\ 0 & 1 & 0 \end{bmatrix}, \mathbf{D} = \begin{bmatrix} 0 & 0 \\ 0 & 0 \end{bmatrix}$$

3.21 Of the many possible answers, one is obtained by setting $x_1 = y_1$, $x_2 = \dot{y}_1 - u_2$, $x_3 = y_2 - 2u_1$.

$$\text{Then } \mathbf{A} = \begin{bmatrix} 0 & 1 & 0 \\ -2 & -3 & 2 \\ 3 & 0 & -3 \end{bmatrix}, \mathbf{B} = \begin{bmatrix} 0 & 1 \\ 5 & -3 \\ -6 & 1 \end{bmatrix}, \mathbf{C} = \begin{bmatrix} 1 & 0 & 0 \\ 0 & 0 & 1 \end{bmatrix}, \mathbf{D} = \begin{bmatrix} 0 & 0 \\ 2 & 0 \end{bmatrix}$$

3.22
$$\begin{bmatrix} \dot{x}_1 \\ \dot{x}_2 \\ \dot{x}_3 \end{bmatrix} = \left[\begin{array}{cc|c} -1 & 1 & 0 \\ 0 & -1 & 0 \\ \hline 0 & 0 & -6 \end{array}\right] \begin{bmatrix} x_1 \\ x_2 \\ x_3 \end{bmatrix} + \begin{bmatrix} 0 \\ 1 \\ 1 \end{bmatrix} u \quad \text{and} \quad y = [\tfrac{1}{5} \quad -\tfrac{1}{25} \quad \tfrac{1}{25}]\mathbf{x}$$

3.23 With x_1 = voltage across C_1, x_2 = current through L_1, x_3 = current through L_2,

$$u = i_s, \quad \begin{bmatrix} \dot{x}_1 \\ \dot{x}_2 \\ \dot{x}_3 \end{bmatrix} = \begin{bmatrix} 0 & 1/(C_1 + C_2) & -1/(C_1 + C_2) \\ -1/L_1 & -R/L_1 & 0 \\ 1/L_2 & 0 & 0 \end{bmatrix} \begin{bmatrix} x_1 \\ x_2 \\ x_3 \end{bmatrix} + \begin{bmatrix} 0 \\ R/L_1 \\ 0 \end{bmatrix} u$$

3.24 x_1, x_2 = voltage across C_1, C_2. x_3 = current through L. Then

$$\begin{bmatrix} \dot{x}_1 \\ \dot{x}_2 \\ \dot{x}_3 \end{bmatrix} = \begin{bmatrix} \dfrac{-1}{C_1(R_1 + R_3)} & 0 & \dfrac{R_1}{C_1(R_1 + R_3)} \\ \dfrac{NK}{R_4 C_2} & -\dfrac{1}{R_4 C_2} & 0 \\ \dfrac{-R_1}{L(R_1 + R_3)} & 0 & \dfrac{-(R_1 R_2 + R_1 R_3 + R_2 R_3)}{L(R_1 + R_3)} \end{bmatrix} \begin{bmatrix} x_1 \\ x_2 \\ x_3 \end{bmatrix} + \begin{bmatrix} 0 \\ -\dfrac{NK}{R_4 C_2} \\ \dfrac{1}{L} \end{bmatrix} v_s$$

$$y = [0 \quad 1 \quad 0]\mathbf{x}$$

3.25
$$\dot{x}_1 = \frac{1}{C_1}\left[-\frac{x_1}{R_1 + R_2} - \frac{R_1}{R_1 + R_2} x_3 + \frac{u_1(t)}{R_1 + R_2}\right]$$

$$\dot{x}_2 = \frac{-1}{C_2} u_2(t)$$

$$\dot{x}_3 = \frac{1}{L}\left[\frac{R_1 x_1}{R_1 + R_2} - \frac{(R_1 R_2 + R_1 R_3 + R_2 R_3) x_3}{R_1 + R_2} + \frac{R_2}{R_1 + R_2} u_1(t) - R_3 u_2(t)\right]$$

3.26 $x_1(k) = y_1(k)$, $x_2(k) = y_1(k+1) + 10 y_1(k) - 2u_1(k) - y_2(k)$, $x_3(k) = y_2(k)$. Then

$$\mathbf{x}(k+1) = \begin{bmatrix} -10 & 1 & 1 \\ -3 & 0 & -2 \\ 4 & 0 & -4 \end{bmatrix} \mathbf{x}(k) + \begin{bmatrix} 2 & 0 \\ 1 & 0 \\ -1 & 2 \end{bmatrix} \mathbf{u}(k)$$

$$\mathbf{y}(k) = \begin{bmatrix} 1 & 0 & 0 \\ 0 & 0 & 1 \end{bmatrix} \mathbf{x}(k)$$

3.27 Let $x_1(k) = y(k), x_2(k) = y(k+1)$. Then

$$\mathbf{x}(k+1) = \begin{bmatrix} 0 & 1 \\ -2 & -3 \end{bmatrix}\mathbf{x}(k) + \begin{bmatrix} 0 \\ 1 \end{bmatrix}u(k), \quad y(k) = [1 \quad 0]\mathbf{x}(k)$$

Let $x_1(k) = y(k),\ x_2(k) = y(k+1) + 3y(k)$. Then

$$\mathbf{x}(k+1) = \begin{bmatrix} -3 & 1 \\ -2 & 0 \end{bmatrix}\mathbf{x}(k) + \begin{bmatrix} 0 \\ 1 \end{bmatrix}u(k), \quad y(k) = [1 \quad 0]\mathbf{x}(k)$$

PTER 4

4.35 $i_1 = \dfrac{(h_{22}R_L + 1)}{\Delta}v_1, i_2 = \dfrac{h_{21}}{\Delta}v_1, v_2 = \dfrac{-h_{21}R_L}{\Delta}v_1$, where $\Delta = (h_{11}h_{22} - h_{12}h_{21})R_L + h_{11}$

4.36 $v_1 = \dfrac{1}{\Delta}[h_{11}(1 + h_{22}R_L) - h_{12}h_{21}R_L]i_1, i_2 = \dfrac{h_{21}i_1}{\Delta}, v_2 = -\dfrac{1}{\Delta}h_{21}R_L i_1$, where $\Delta = 1 + h_{22}R_L$

4.37 $|\mathbf{A}| = -468$

4.38 $$\mathbf{A}^{-1} = \frac{1}{17}\begin{bmatrix} 3 & 4 & -3 \\ 7 & -19 & 10 \\ -2 & 3 & 2 \end{bmatrix}, \mathbf{B}^{-1} = \frac{1}{58}\begin{bmatrix} 59 & 2 & 3 & 4 & 5 \\ 1 & 60 & 3 & 4 & 5 \\ -4 & -8 & 46 & -16 & -20 \\ -2 & -4 & -6 & 50 & -10 \\ -8 & -16 & -24 & -32 & 18 \end{bmatrix}$$

$$\mathbf{C}^{-1} = \left[\begin{array}{cc|cc|cc} \frac{1}{5} & \frac{-2}{5} & 0 & 0 & 0 & 0 \\ \frac{2}{5} & \frac{1}{5} & 0 & 0 & 0 & 0 \\ \hline 0 & 0 & 0 & 1 & 0 & 0 \\ 0 & 0 & \frac{1}{7} & \frac{-5}{7} & 0 & 0 \\ \hline 0 & 0 & 0 & 0 & \frac{-3}{55} & \frac{8}{55} \\ 0 & 0 & 0 & 0 & \frac{8}{55} & \frac{-3}{55} \end{array}\right]$$

4.39 $$\mathbf{A}(s) = \begin{bmatrix} \frac{1}{s} & \frac{1}{s^2} \\ \frac{1}{s+a} & \frac{b\beta}{s^2+\beta^2} \\ \frac{2}{s^3} & \frac{s+1}{(s+1)^2+\beta^2} \end{bmatrix}, \quad \mathbf{B}(s) = \begin{bmatrix} \frac{s}{s^2-\beta^2} & \frac{\beta}{s^2-\beta^2} \\ \frac{1}{(s+a)^2} & \frac{s}{s^2+\beta^2} \end{bmatrix}$$

4.40 $\mathbf{A}(t) = [t^4 \quad t\cos\beta t], \mathbf{B}(t) = \begin{bmatrix} e^{-t} - e^{-2t} & -e^{-t} + 2e^{-2t} \\ 0 & \delta(t) \end{bmatrix}$

4.41 **(a)** $\mathbf{S} = \begin{bmatrix} 2 & -0.5 & 1 \\ 0 & 2.78388 & 1.61645 \\ 0 & 0 & 2.32101 \end{bmatrix}$ **(b)** Not possible, **A** not positive definite.

(c) $$\mathbf{S} = \begin{bmatrix} 4 & 1 & 0.25 & -0.25 & 0.75 \\ 0 & 3 & 1.25 & 0.75 & -0.91666 \\ 0 & 0 & 4.83477 & 0.64636 & 0.40505 \\ 0 & 0 & 0 & 3.15551 & 2.41267 \\ 0 & 0 & 0 & 0 & 3.10035 \end{bmatrix}$$

4.42 $\mathbf{H}(s) = \begin{bmatrix} 1 & 1 \\ 0 & 1 \end{bmatrix} \begin{bmatrix} (s+1)^2 & 0 \\ 0 & s+1 \end{bmatrix}^{-1} = \begin{bmatrix} (s+1)^2 & 0 \\ 0 & s+1 \end{bmatrix}^{-1} \begin{bmatrix} 1 & s+1 \\ 0 & 1 \end{bmatrix}$

4.43 $\mathbf{H}(s) = \begin{bmatrix} 1 & 1 \\ s+1 & -1 \end{bmatrix} \begin{bmatrix} (s+1)(s+2) & -2(s+3) \\ 0 & s+3 \end{bmatrix}^{-1}$. Other valid answers may be found.

4.44 Using the observable canonical form for each subsystem gives

$$\mathbf{A} = \begin{bmatrix} -3 & 1 & 0 & 0 \\ -6 & 0 & 0 & 0 \\ 1 & 0 & -4 & 1 \\ 2 & 0 & -3 & 0 \end{bmatrix}, \mathbf{B} = \begin{bmatrix} 1 \\ 5 \\ 0 \\ 0 \end{bmatrix}, \mathbf{C} = \begin{bmatrix} 1 & 0 & 0 & 0 \\ 0 & 0 & 1 & 0 \end{bmatrix}, \mathbf{D} = [0].$$

Using the controllable canonical form for each subsystem gives

$$\mathbf{A} = \begin{bmatrix} 0 & 1 & 0 & 0 \\ -6 & -3 & 0 & 0 \\ 0 & 0 & 0 & 1 \\ 5 & 1 & -3 & -4 \end{bmatrix}, \mathbf{B} = \begin{bmatrix} 0 \\ 1 \\ 0 \\ 0 \end{bmatrix}, \mathbf{C} = \begin{bmatrix} 5 & 1 & 0 & 0 \\ 0 & 0 & 2 & 1 \end{bmatrix}, \mathbf{D} = [0].$$

4.45

$$\mathbf{A} = \begin{bmatrix} 0 & 1 & 0 & 0 & 0 & 0 & 0 \\ -6 & -3 & 0 & 0 & 0 & 0 & 0 \\ 0 & 0 & 0 & 1 & 0 & 0 & 0 \\ 5 & 1 & -3 & -4 & 10 & 0 & 0 \\ 0 & 0 & 0 & 0 & -6 & 0 & 0 \\ 0 & 0 & 0 & 0 & 0 & 0 & 1 \\ 0 & 0 & -2 & -1 & 10 & -1 & -1 \end{bmatrix}, \mathbf{B} = \begin{bmatrix} 0 & 0 \\ 1 & 0 \\ 0 & 0 \\ 0 & 0 \\ 0 & 1 \\ 0 & 0 \\ 0 & 0 \end{bmatrix}, \mathbf{D} = [0], \text{ and}$$

$$\mathbf{C} = \begin{bmatrix} 5 & 1 & 0 & 0 & 0 & 0 & 0 \\ 0 & 0 & 2 & 1 & 0 & 0 & 0 \\ 0 & 0 & 0 & 0 & 10 & 0 & 0 \\ 0 & 0 & 0 & 0 & 0 & -2 & 1 \end{bmatrix}$$

4.46 $\mathbf{A} = \begin{bmatrix} 0 & 1 & 0 & 0 \\ -6 & -3 & -2 & -1 \\ 0 & 0 & 0 & 1 \\ 5 & 1 & -3 & -4 \end{bmatrix}, \mathbf{B} = \begin{bmatrix} 0 \\ 1 \\ 0 \\ 0 \end{bmatrix}, \mathbf{C} = \begin{bmatrix} 5 & 1 & 0 & 0 \\ 0 & 0 & 2 & 1 \end{bmatrix}$

4.47 $Q \approx Q_n + (c/2)\sqrt{\rho g/h_n}\,\delta h$, where $h = h_n + \delta h$ and $Q_n = c\sqrt{\rho g h_n}$

4.48 **(a)** $\theta = \cos^{-1}\left[\dfrac{\langle(\mathbf{r}_1 - \mathbf{x}), (\mathbf{r}_2 - \mathbf{x})\rangle}{\|\mathbf{r}_1 - \mathbf{x}\|\,\|\mathbf{r}_2 - \mathbf{x}\|}\right]$

(b) $\delta\theta = [\nabla_{\mathbf{x}}\theta]_n^T\,\delta\mathbf{x}$. Letting $\mathbf{r}_1 - \mathbf{x} = \mathbf{u}$ and $\mathbf{r}_2 - \mathbf{x} = \mathbf{v}$, the three components of $\nabla_{\mathbf{x}}\theta$ are given by

$$\frac{\partial\theta}{\partial x_i} = \frac{(u_i + v_i)\|\mathbf{u}\|\,\|\mathbf{v}\| - (v_i\|\mathbf{u}\|^2 + u_i\|\mathbf{v}\|^2)\cos\theta}{\|\mathbf{u}\|^2\|\mathbf{v}\|^2\sin\theta}$$

CHAPTER 5

5.39 **(a)** $|\mathbf{G}| = 16$

(b) $\hat{\mathbf{v}}_1 = \dfrac{1}{\sqrt{14}}[1 \;\; 2 \;\; 3]^T$, $\hat{\mathbf{v}}_2 = \dfrac{1}{\sqrt{35}}[1 \;\; -5 \;\; 3]^T$, $\hat{\mathbf{v}}_3 = \dfrac{1}{\sqrt{10}}[-3 \;\; 0 \;\; 1]^T$

5.40 $\mathbf{r}_1 = \begin{bmatrix} \frac{5}{4} & \frac{1}{4} & -\frac{1}{4} \end{bmatrix}^T$, $\mathbf{r}_2 = \begin{bmatrix} -\frac{1}{4} & -\frac{1}{4} & \frac{1}{4} \end{bmatrix}^T$, $\mathbf{r}_3 = [-3 \;\; 0 \;\; 1]^T$

5.41 $\mathbf{z} = \dfrac{5}{\sqrt{14}}\hat{\mathbf{v}}_1 - \dfrac{23}{\sqrt{35}}\hat{\mathbf{v}}_2 - \dfrac{21}{\sqrt{10}}\hat{\mathbf{v}}_3$

5.42 $\mathbf{z} = \frac{37}{4}\mathbf{x}_1 - \frac{13}{4}\mathbf{x}_2 - 21\mathbf{x}_3$

5.43 $\mathbf{r} = \frac{1}{10}[3 \quad -1 \quad 0]^T$, $\mathbf{r}_2 = \frac{1}{70}[-7 \quad 19 \quad -10]^T$, $\mathbf{r}_3 = \frac{1}{7}[0 \quad -1 \quad 2]^T$

5.44 $\mathbf{r}_1 = \begin{bmatrix} 0 & \frac{1}{2} & \frac{1}{2} \end{bmatrix}^T$, $\mathbf{r}_2 = \begin{bmatrix} 0 & -\frac{1}{2} & \frac{1}{2} \end{bmatrix}^T$, $\mathbf{r}_3 = [1 \quad 0 \quad -1]^T$, $\mathbf{z} = 2\mathbf{x}_1 - \mathbf{x}_2 + 5\mathbf{x}_3$

5.46

$$\mathbf{G} = \begin{bmatrix} 2.5784502E+03 & 6.3518805E+02 & -1.1518515E+02 & 3.3449918E+02 \\ 6.3518805E+02 & 6.2262292E+02 & 6.9457092E+01 & 8.6028557E+01 \\ -1.1518515E+02 & 6.9457092E+01 & 2.9569002E+01 & -1.9252985E+01 \\ 3.3449918E+02 & 8.6028557E+01 & -1.9252985E+01 & 1.1678545E+02 \end{bmatrix}$$

$|\mathbf{G}| = 3.12183 \times 10^8$

The orthonormal basis vectors $\mathbf{v}_i$ are

$$\begin{bmatrix} 8.7438685E-01 \\ 2.5207549E-01 \\ 2.9540095E-02 \\ -4.1356131E-01 \end{bmatrix}, \begin{bmatrix} -1.4671828E-01 \\ 8.4976387E-01 \\ 4.4605288E-01 \\ 2.3960790E-01 \end{bmatrix}, \begin{bmatrix} -3.4713072E-01 \\ 4.2903343E-01 \\ -6.5270102E-01 \\ -5.1904905E-01 \end{bmatrix}, \begin{bmatrix} 3.0564949E-01 \\ 1.7403936E-01 \\ -6.1167425E-01 \\ 7.0862055E-01 \end{bmatrix}$$

Then $\mathbf{x} = -0.6777\mathbf{v}_1 + 12.712\mathbf{v}_2 - 14.736\mathbf{v}_3 + 31.548\mathbf{v}_4$. Note that the sum of the squares of the components of $\mathbf{x}$ in the original and the new coordinates is 1374.5.

5.47 The Grammian is

$$\begin{bmatrix} 4.0000000E+00 & 4.0008001E+00 & -7.9900002E+00 \\ 4.0008001E+00 & 4.0016012E+00 & -7.9916024E+00 \\ -7.9900002E+00 & -7.9916024E+00 & 1.5960100E+01 \end{bmatrix}$$

Its determinant is not zero, but very small, because the three vectors are very nearly colinear. To proceed with these poorly conditioned vectors as a basis set would produce very unreliable results.

5.48 An orthonormal basis for the subspace is, from Gram-Schmidt,

$$\begin{bmatrix} 7.3127246E-01 \\ 3.6563623E-01 \\ -3.6563623E-01 \\ 7.3127240E-02 \\ 4.3876347E-01 \end{bmatrix}, \begin{bmatrix} -1.1037266E-01 \\ -2.8796402E-01 \\ -7.9833186E-01 \\ -4.9211112E-01 \\ -1.5933463E-01 \end{bmatrix}, \begin{bmatrix} 5.2741766E-01 \\ 1.2584069E-01 \\ 2.0743863E-01 \\ -2.8101894E-01 \\ -7.6419383E-01 \end{bmatrix}$$

The components of $\mathbf{x}_1$ and $\mathbf{x}_2$, expressed in these coordinates, are

$$\begin{bmatrix} 1.2431632E+00 \\ -1.8481143E+00 \\ -1.8451571E-01 \end{bmatrix}, \begin{bmatrix} 4.8995252E+00 \\ 3.3418725E+00 \\ -1.1238300E+01 \end{bmatrix}$$

5.49 Dimension is 2.

5.50 $\|\mathbf{x}_n\| = \frac{5}{\sqrt{14}}$, $x_1 = \frac{-5}{7}$, $x_2 = \frac{-15}{14}$, $x_3 = \frac{5}{14}$

5.51 $\mathbf{y}_p = \begin{bmatrix} \frac{1}{2} & 2 & \frac{9}{2} \end{bmatrix}^T$

5.52 The sine and/or cosine functions form an orthogonal basis set, $\{\mathbf{v}_i, i = 1, \infty\}$. The reciprocal basis $\{\mathbf{r}_i\}$ differs only by normalizing constants. The inner product is that defined in Problem 5.22 and the Fourier coefficients can be thought of as components of an infinite dimensional vector.

5.53 $\mathbf{A}$ must be an $n \times n$ real matrix which satisfies $\mathbf{x}^T\mathbf{A}\mathbf{x} > 0$ for all $\mathbf{x} \neq \mathbf{0}$. This last condition is the definition of a positive definite matrix.

5.55 $\langle \mathbf{f}, \mathbf{g} \rangle = \int_a^b \bar{\mathbf{f}}^T(\tau)\mathbf{g}(\tau)\,d\tau$, $\|\mathbf{f}\| = \langle \mathbf{f}, \mathbf{f} \rangle^{1/2}$

5.58 $\mathbf{A} = \begin{bmatrix} 0 & -1 & 2 \\ -3 & 2 & 2 \\ 0 & 2 & 3 \end{bmatrix}$

5.59 No, it depends upon the fact that $\mathbf{B}^{-1} = \mathbf{B}^T$.

5.60 $\mathbf{A}^{-1} = \dfrac{1}{\sin^2 \alpha} \begin{bmatrix} \hat{\mathbf{u}}^T - \hat{\mathbf{v}}^T \cos\alpha \\ \hline \hat{\mathbf{v}}^T - \hat{\mathbf{u}}^T \cos\alpha \\ \hline \mathbf{w}^T \end{bmatrix}$

5.61 $\dot{\mathbf{T}}_{BE} = \begin{bmatrix} 0 & \omega_z & -\omega_y \\ -\omega_z & 0 & \omega_x \\ \omega_y & -\omega_x & 0 \end{bmatrix} \mathbf{T}_{BE}$, where $\boldsymbol{\omega} = [\omega_x \quad \omega_y \quad \omega_z]^T$ is the angular velocity of the body coordinates with respect to the inertial coordinates, which could be measured with body-mounted rate gyros. (*Hint:* Let $\mathbf{x}_I = [1 \quad 0 \quad 0]^T$ be a fixed inertial vector. The same vector expressed in body coordinates is $\mathbf{x}_B$, the first column of $\mathbf{T}_{BE}$. Then use Euler's formula for differentiating a vector in rotating coordinates,

$$\left.\frac{d\mathbf{x}}{dt}\right|_I = \left.\frac{d\mathbf{x}}{dt}\right|_B + \boldsymbol{\omega} \times \mathbf{x}_B$$

Repeat for two other vectors $\mathbf{y}_I$ and $\mathbf{z}_I$ and use the result of Problem 5.6 for the cross product.)

5.62 $\mathbf{y}_r = \frac{1}{7}[6 \quad 19 \quad 32]^T$, $\mathbf{y}_p = \frac{1}{14}[34 \quad 5 \quad 53]^T$

5.65 **(a)** *Hint:* Result of Problem 7.37 will be helpful. **(b)** *Hint:* Results of Problem 7.26 will be helpful.

5.66 Ignoring the boundary terms, the formal adjoint equation is $\mathbf{y}(k-1) = \mathbf{A}_k^T\,\mathbf{y}(k)$. Note that the time index k on the adjoint equation runs backward.

CHAPTER 6

6.30 $\mathbf{x} = \frac{1}{27}[40 \quad -49 \quad 48]^T$

6.31 $\mathbf{x}_2(0) = \phi_{22}^{-1}(T)[\mathbf{x}_2(T) - \phi_{21}(T)\mathbf{x}_1(0)]$, where $\boldsymbol{\Phi}(T)$ is partitioned into $\left[\begin{array}{c|c} \phi_{11}(T) & \phi_{12}(T) \\ \hline \phi_{21}(T) & \phi_{22}(T) \end{array}\right]$.

6.33 $\mathbf{x} = [-4\alpha \quad 0 \quad \alpha]^T$, α an arbitrary scalar

6.34 $\mathbf{x}_1 = [-1 \quad 0 \quad 1]^T$, $\mathbf{x}_2 = [-1 \quad 1 \quad 0]^T$ and any linear combination of these.

6.35 Yes, since $r_A = 1$, $\mathbf{x} = [-2\alpha \quad \alpha]^T$, α arbitrary.

6.36 The degeneracy of $\mathbf{A}$ is 3.; the null space basis is

$$\begin{bmatrix} 0.0000000E+00 \\ -1.0000000E+00 \\ 1.1764708E-01 \\ 4.1176471E-01 \\ 0.0000000E+00 \end{bmatrix}, \begin{bmatrix} -1.0000000E+00 \\ 0.0000000E+00 \\ -7.6470608E-01 \\ 8.2352948E-01 \\ 0.0000000E+00 \end{bmatrix}, \begin{bmatrix} 0.0000000E+00 \\ 0.0000000E+00 \\ 6.4705873E-01 \\ 7.6470590E-01 \\ -1.0000000E+00 \end{bmatrix}$$

6.37 **(a)** $x_1 = \frac{13}{5}$, $x_2 = \frac{1}{5}$ **(b)** $x_1 = \frac{84}{35}$, $x_2 = \frac{14}{35}$

6.38 $x = 2$. Column space is one dimensional with basis $[2 \quad 1]^T$. $\mathbf{y} - \mathbf{A}x$ is perpendicular to this line, i.e., $\mathbf{A}x$ is the orthogonal projection of $\mathbf{y}$ onto this line.

6.40 $a = 1$, $b = 3$

6.41 $a = \frac{32}{37}$, $b = \frac{107}{37}$, $\|\mathbf{e}\|^2 = \frac{8}{37}$

6.42 $a = \frac{104}{109}$, $b = \frac{323}{109}$

6.43 $i(0) \cong 17.3$

6.44 $\mathbf{a} = [2.6286 \quad 0.082145 \quad -0.003572]^T$, $\|\mathbf{y}_e\| = 0.4326$, and the estimated final GPA is 3.057, which is probably more realistic even though the residual error is larger here.

6.45 $C = 5.3704$, $\alpha = 1.077$, estimated mile time is 5 min, 22 s

6.46 After processing first four equations $\mathbf{x}_4 = [-.566 \quad -.033 \quad .533]^T$, and after five $\mathbf{x}_5 = [-.472 \quad -.127 \quad .627]^T$. Roughly these same values were obtained using several initial estimates for $\mathbf{x}$, as long as the first $\mathbf{P}$ is very large. Disagreement with Example 6.7 after $k = 4$ would not be surprising since unreliable results can be given by the recursive equations whenever $\mathbf{A}$ is not full rank.

6.47 Introduction of a suitable weighting matrix can be handled by defining a weighted inner product, $\langle \mathbf{x}_1, \mathbf{x}_2 \rangle_Q = \mathbf{x}_1^T \mathbf{Q} \mathbf{x}_2$. This new inner product modifies the meaning of $\mathcal{A}^*$, since $\langle \mathbf{y}, \mathcal{A}(\mathbf{x}) \rangle_Q = \langle \mathcal{A}^*(\mathbf{y}), \mathbf{x} \rangle_Q$. With this definition for $\mathcal{A}^*$, the previous results still apply.

6.48 $\mathbf{X} = \begin{bmatrix} -0.5 & 0.25 \\ 0.25 & -0.25 \end{bmatrix}$.

6.49 No solution exists. $\mathbf{B}$ has $+1$ as an eigenvalue and $\mathbf{A}$ has -1 as an eigenvalue. The 4×4 $\mathbf{Q}$ matrix is singular, of rank 3 and the augmented $\mathbf{W}$ matrix has rank 4. The equations are inconsistent and no solution exists.

PTER 7

7.39 $\lambda_i = 1, -2, 3$. $\mathbf{x}_i = [-1 \quad 1 \quad 1]^T, [11 \quad 1 \quad -14]^T$, and $[1 \quad 1 \quad 1]^T$. $\mathbf{J} = \text{diag}[1, -2, 3]$

7.40 $\lambda_1 = \lambda_2 = 2$, $m = 2$, $q = 2$, $\mathbf{x}_1 = [1 \quad 0]^T$, $\mathbf{x}_2 = [0 \quad 1]^T$

7.41 $|\mathbf{A} - \mathbf{I}\lambda| = 0 \Rightarrow \lambda^3(2 - \lambda) = 0$. Therefore, $\lambda = 0$ is an eigenvalue with algebraic multiplicity 3 and index $k = 2$. $\text{Rank}(\mathbf{A} - \mathbf{I}\lambda)|_{\lambda=0} = 2$. Therefore, $q = 2$, so there are two eigenvectors $\mathbf{x}_1, \mathbf{x}_3$ and one generalized eigenvector associated with $\lambda = 0$, and they are solutions of $\mathbf{A}^2\mathbf{x} = \mathbf{0}$. $\lambda_4 = 2$ has the eigenvector $\mathbf{x}_4$. These are shown as columns of

$$\mathbf{M} = \begin{bmatrix} 0 & 0 & 1 & 1 \\ 1 & 0 & 0 & 0 \\ 0 & 0 & -2 & 0 \\ 0 & 1 & 0 & 0 \end{bmatrix}$$

7.42 $\lambda_1 = 3 + j$, $\lambda_2 = 3 - j$, $\lambda_3 = 6$, $\mathbf{x}_1 = [6 - 2j \quad 2 - 4j \quad 0]^T$, $\mathbf{x}_2 = \bar{\mathbf{x}}_1$, $\mathbf{x}_3 = [0 \quad 0 \quad 1]^T$, $\mathbf{J} = \text{diag}[3 + j,\ 3 - j,\ 6]$

7.43 No. Although both matrices have $\lambda_1 = \lambda_2 = 2$, the Jordan form for $\mathbf{A}$ is $\mathbf{J} = \begin{bmatrix} 2 & 1 \\ 0 & 2 \end{bmatrix}$. Since $\mathbf{B}$ is already in Jordan form and since $\mathbf{J} \neq \mathbf{B}$, they are not similar.

7.44 $\lambda_1 = 15$, $\mathbf{x}_1 = [1 \quad 1 \quad -1]^T$, $\lambda_2 = 9$, $\mathbf{x}_2 = [1 \quad -2 \quad -1]^T$, $\lambda_3 = 3$, $\mathbf{x}_3 = [1 \quad 0 \quad 1]^T$

7.45 $\lambda_1 = 5.049$, $\lambda_2 = 0.643$, $\lambda_3 = 0.308$, $\mathbf{x}_1 = [1 \quad 0.802 \quad 0.445]^T$, $\mathbf{x}_2 = [1 \quad -0.555 \quad -1.247]^T$, $\mathbf{x}_3 = [1 \quad -2.247 \quad 1.802]^T$

7.46 **(a)** *Hint:* Set $\lambda = 0$ in the definition of $\Delta(\lambda)$ and in $\Delta'(\lambda)$. **(b)** Use results of Problem 7.15 and note: (i) the only factor in $\Delta(\lambda)$ which involves all the diagonal elements a_{ii}, and which gives rise to all λ^{n-1} terms, is of the form $(a_{11} - \lambda)(a_{22} - \lambda) \cdots (a_{nn} - \lambda)$, and (ii) the form of the coefficient of λ^{n-1}, and (iii) $c_i' = (-1)^n c_i$.

7.47 $\lambda_i = \{-4.33945,\ 2.169727 \pm 2.4745j\}$, $\mathbf{x}_1 = [0.69505 \quad -0.355598 \quad -1]^T$, $\mathbf{x}_2 = [0.863702 - 0.181845j \quad -j \quad 0.1867736 - 0.116646j]^T$ and $\mathbf{x}_3 = \bar{\mathbf{x}}_2$. $\mathbf{J} = \text{Diag}[\lambda_1, \lambda_2, \lambda_3]$

7.48 $\lambda_i = \{1.48431,\ 6.257845 \pm 1.123484j\}$, $\mathbf{x}_1 = [0.97929 \quad -1 \quad -0.4636]^T$, $\mathbf{x}_2 = [-0.5107 - 0.41175j \quad -0.30786 - 0.17794 \quad -1]^T$, and $\mathbf{x}_3 = \bar{\mathbf{x}}_2$. $\mathbf{J} = \text{Diag}[\lambda_1, \lambda_2, \lambda_3]$

7.49 **(a)** Negative definite, **(b)** positive semidefinite, **(c)** negative semidefinite, **(d)** positive definite, **(e)** indefinite

7.50 $\mathbf{x} = (1/\sqrt{2})[-1 \quad 1]^T$ (an eigenvector), $\mathbf{Q}_{\max} = 4 = \lambda_{\max}$ = eigenvalue associated with $\mathbf{x}$.

7.51 From equation (7.5), $\mathbf{E}_i = \mathbf{x}_i\rangle\langle\mathbf{r}_i$. This is a projection since $\mathbf{E}_i\mathbf{E}_i = \mathbf{E}_i$ and $\mathbf{E}_i\mathbf{E}_j = 0$ with $i \neq j$.

7.52 Similar to the preceding problem except the eigenspaces are now mutually orthogonal.

CHAPTER 8

8.25 $\mathbf{A}^{-1} = \frac{1}{6}[-\mathbf{A}^2 + 2\mathbf{A} + 3\mathbf{I}] = \frac{1}{6}\begin{bmatrix} -2 & 4 & 0 \\ 2 & 2 & 0 \\ 3 & -3 & 3 \end{bmatrix}$

8.26 $\begin{bmatrix} e^{-t} & \frac{1}{4}(e^{t} - e^{-t}) \\ 0 & e^{t} \end{bmatrix}$

8.27 $\begin{bmatrix} e^{-3t} & 2te^{-3t} \\ 0 & e^{-3t} \end{bmatrix}$

8.29 $\begin{bmatrix} \frac{1}{2}(e^{7t} + e^{-t}) & \frac{5}{8}(e^{7t} - e^{-t}) \\ \frac{2}{5}(e^{7t} - e^{-t}) & \frac{1}{2}(e^{7t} + e^{-t}) \end{bmatrix}$

8.30 $\frac{1}{3}\begin{bmatrix} e^{-t} + 2e^{-4t} & e^{-t} - e^{-4t} \\ 2(e^{-t} - e^{-4t}) & 2e^{-t} + e^{-4t} \end{bmatrix}$

8.31 $\begin{bmatrix} (\frac{1}{2})^k & -k(\frac{1}{2})^k & k(2-k)(\frac{1}{2})^{k-1} \\ 0 & (\frac{1}{2})^k & k(\frac{1}{2})^{k-2} \\ 0 & 0 & (\frac{1}{2})^k \end{bmatrix}$

8.32 $\begin{bmatrix} -3e^{-t} + 4e^{-2t} & -6(e^{-t} - e^{-2t}) & 0 \\ 2(e^{-t} - e^{-2t}) & 4e^{-t} - 3e^{-2t} & 0 \\ 0 & 0 & e^{-3t} \end{bmatrix}$

8.33 $\begin{bmatrix} \alpha_0 & \alpha_1 & \alpha_2 \\ \alpha_2 & \alpha_0 - 3\alpha_2 & \alpha_1 + 3\alpha_2 \\ \alpha_1 + 3\alpha_2 & -3\alpha_1 - 8\alpha_2 & \alpha_0 + 3\alpha_1 + 6\alpha_2 \end{bmatrix}$ with $\begin{cases} \alpha_0 = e^{t} - te^{t} + \frac{1}{2}t^2 e^{t} \\ \alpha_1 = te^{t} - t^2 e^{t} \\ \alpha_2 = \frac{1}{2}t^2 e^{t} \end{cases}$

8.34

$$e^{\mathbf{A}t} = \begin{bmatrix} 0 & 0 & 0 & 0 \\ 0 & 0 & 0 & 0 \\ 0 & 0 & 0 & 0 \\ 0.1 & 0.1054 & -0.0738 & 0.1 \end{bmatrix} + e^{-t}\begin{bmatrix} 1 & 0.1047 & 0.9739 & 0 \\ 0 & 0 & 0 & 0 \\ 0 & 0 & 0 & 0 \\ -0.1 & -0.0105 & -0.0974 & 0 \end{bmatrix}$$

$$+ e^{-0.7t}\sin(3t)\begin{bmatrix} 0 & 0.3246 & -0.3142 & 0 \\ 0 & 0 & 3 & 0 \\ 0 & -0.333 & 0 & 0 \\ 0 & -0.0571 & -0.2847 & 0 \end{bmatrix}$$

$$+ e^{-0.7t}\cos(3t)\begin{bmatrix} 0 & -0.1047 & -0.9739 & 0 \\ 0 & 1 & 0 & 0 \\ 0 & 0 & 1 & 0 \\ 0 & -0.0949 & 0.1712 & 0 \end{bmatrix}$$

8.35 $\mathbf{A}^k = (0.5)^k\begin{bmatrix} 0 & 4 & -4 \\ 0 & 2 & -2 \\ 0 & 1 & -1 \end{bmatrix} + \begin{bmatrix} 0 & -1 & 2 \\ 0 & -1 & 2 \\ 0 & -1 & 2 \end{bmatrix}$

8.38

$$e^{\mathbf{A}t} = \begin{bmatrix} e^{3t}[\cos(t) + \sin(t)] & -2e^{3t}\sin(t) & 0 \\ e^{3t}\sin(t) & e^{3t}[\cos(t) - \sin(t)] & 0 \\ 0 & 0 & e^{6t} \end{bmatrix}$$

With $t = 0.2$ this gives $\mathbf{A}_1 \approx \begin{bmatrix} 2.147797 & -0.72399 & 0 \\ 0.361999 & 1.423799 & 0 \\ 0 & 0 & 3.320117 \end{bmatrix}$ with

$\lambda = \{3.320117, 1.785798 \pm 0.36199j\} = \{e^{1.2}, e^{0.6}[\cos(0.2) \pm \sin(0.2)j]\}$

8.41 An easy approximation is given by using the dominant eigenvalue, λ_d, that is, the eigenvalue with the smallest nonzero real part. This mode is called dominant since it will take longest to decay. Let the real part of λ be $-r$. Then, solving $e^{-rt} = 0.02$ gives $T_s \approx -\ln(0.02)/r \approx 4/r$. The term $1/r$ is an approximation of the dominant time constant τ_d. A 1% definition of settling is often used, and this changes the 4 to 4.6. This rough approximation may not be sufficiently accurate if the one mode is not truly dominant [4].

8.42 Nyquist's sampling theorem requires that $\omega_{max} T < \pi$, that is, there must be at least two samples per period of the highest frequency to avoid aliasing [4] and loss of information. This theoretical limit is almost never enough in controls problems. For adequate representation of states a rough rule of thumb is more like 6 to 10 samples per period, or $T < 2\pi/6\omega_{max}$. The same 6 to 10 rule is often applied to the number of samples per dominant time constant as well. The value used for ω_{max} is usually interpreted as the highest *significant* modal frequency to be retained in the model. The distinction is necessary because some physical problems involve many or even an infinite number of frequencies due to structural vibration or other causes [5]. Other modes may have such a large negative real part of λ that they rapidly decay to insignificance.

PTER 9

9.29 $y_1(t) = +\frac{9}{2}te^{-3t} + \frac{25}{4}e^{-3t} - \frac{21}{4}e^{-t}$; $y_2(t) = \frac{7}{2}e^{-t} - \frac{3}{2}e^{-3t}$

9.30 $y_1(t) = 11e^{-2t} - 3e^{-3t} - 7e^{-t}$; $y_2(t) = \frac{7}{2}e^{-t} - \frac{3}{2}e^{-3t}$

9.31 $\mathbf{x}(t) = \begin{bmatrix} (10 - 9t)e^{-t} \\ (1 - 9t)e^{-t} \end{bmatrix}$

9.32 $\mathbf{x}(t) = \begin{bmatrix} \frac{83}{9}e^{-t} - \frac{25}{3}te^{-t} + \frac{7}{9}e^{2t} \\ \frac{8}{9}e^{-t} - \frac{25}{3}te^{-t} + \frac{1}{9}e^{2t} \end{bmatrix}$

9.33 $\mathbf{x}(t_f) = \mathbf{0}$. Note that $\mathbf{u}(t) = -\frac{1}{t_f}\mathbf{\Phi}(t, 0)\mathbf{x}(0)$.

9.35 $\mathbf{\Phi}(t, 0) = e^{-t/2}\begin{bmatrix} \cos\frac{3}{2}t - \frac{1}{3}\sin\frac{3}{2}t & -\frac{5}{3}\sin\frac{3}{2}t \\ \frac{2}{3}\sin\frac{3}{2}t & \cos\frac{3}{2}t + \frac{1}{3}\sin\frac{3}{2}t \end{bmatrix}$

9.37 $x_1(k) = 2 - 5k + 5k(3 - k)$, $x_2(k) = 5 + 10k$, $x_3(k) = 10$

9.38 $\mathbf{x}(k) = \begin{bmatrix} 0 \\ 10 \end{bmatrix} + \sum_{j=1}^{k} (0.632)(0.368)^{k-j}\begin{bmatrix} 1 \\ -1 \end{bmatrix}$

9.39 $\mathbf{x}(k) = \begin{bmatrix} 1 & 0 \\ -1 & 1 \end{bmatrix}\mathbf{q}(k)$ gives $\mathbf{q}(k+1) = \begin{bmatrix} 1 & 0 \\ 0 & \frac{1}{2} \end{bmatrix}\mathbf{q}(k) + \begin{bmatrix} 1 \\ 0 \end{bmatrix}u(k)$

$\mathbf{y}(k) = [4 \quad 1]\mathbf{q}(k)$

9.40 No. The system matrix **A** is singular and consequently $\mathbf{\Phi}^{-1}$ does not exist.

9.41 $\dot{\mathbf{x}} = \begin{bmatrix} 0 & 1 \\ 0 & -1/\tau \end{bmatrix}\mathbf{x} + \begin{bmatrix} 0 \\ K/\tau \end{bmatrix}u(t)$

$$\mathbf{x}(k+1) = \begin{bmatrix} 1 & \tau(1 - e^{-\Delta t/\tau}) \\ 0 & e^{-\Delta t/\tau} \end{bmatrix}\mathbf{x}(k) + \begin{bmatrix} \Delta t - \tau(1 - e^{-\Delta t/\tau}) \\ 1 - e^{-\Delta t/\tau} \end{bmatrix}Ku(k)$$

9.42 $\mathbf{x}(k+1) = \begin{bmatrix} 1 - K\Delta t + K\tau(1-e^{-\Delta t/\tau}) & \tau(1-e^{-\Delta t/\tau}) \\ -K(1-e^{-\Delta t/\tau}) & e^{-\Delta t/\tau} \end{bmatrix} \mathbf{x}(k)$

$$+ \begin{bmatrix} \Delta t - \tau(1-e^{-\Delta t/\tau}) \\ 1 - e^{-\Delta t/\tau} \end{bmatrix} Kr(k)$$

$$\theta(k) = [1 \quad 0]\mathbf{x}(k)$$

9.43 $\mathbf{A} = \begin{bmatrix} 0.3666 & 0.1010 & 0.0110 \\ 0 & 0.6703 & 0.1341 \\ 0 & 0 & 0.6703 \end{bmatrix}, \mathbf{B} = \begin{bmatrix} 0.0136 \\ 0.1802 \\ 0.1648 \end{bmatrix}$

9.44 One of many possible answers, obtained by using a parallel realization, is

$$\mathbf{x}(k+1) = \begin{bmatrix} 0.04979 & 0 & 0 \\ 0 & 0.22313 & 0 \\ 0 & 0 & 0.60653 \end{bmatrix} \mathbf{x}(k) + \begin{bmatrix} 0.2121 \\ 0.1934 \\ 0.0649 \end{bmatrix} u(k)$$

$$y(k) = [0.04979 \quad -0.22313 \quad 0.60653]\mathbf{x}(k)$$

CHAPTER 10

10.16 **(a)** Eigenvalues are $\lambda_1 = -4, \lambda_2 = 2$. Every term in $\mathbf{\Phi} \to \infty$, as e^{2t} when $t \to \infty$. Therefore, $\|\mathbf{\Phi}(t, 0)\| \to \infty$ and the system is unstable. **(b)** Eigenvalues are $\lambda = -1, -2$, and -3. $\|\mathbf{\Phi}(t, 0)\| \to 0$ as $t \to \infty$ and the system is globally asymptotically stable. **(c)** Eigenvalues are $\lambda = -\frac{1}{4}$ and $-\frac{3}{4}$. The system is asymptotically stable.

10.17 **(a)** $\lambda_1 = 1, \lambda_2 = \frac{1}{2}$. Stable i.s.L. but not asymptotically stable; observable but not controllable. **(b)** $\lambda_1 = -1, \lambda_2 = -3, \lambda_3 = -5$. Asymptotically stable, uncontrollable and unobservable. **(c)** $\lambda = 1 \pm \sqrt{21}$. System is unstable, completely controllable and observable. These results illustrate that stability, controllability and observability are independent system properties. One property does not imply or require any of the others.

10.18 $u_1(t) = -M_1 \operatorname{sign}[7x_1(t) - x_2(t)]$, $u_2(t) = -M_2 \operatorname{sign}[-8x_1(t) + 19x_2(t)]$

CHAPTER 11

11.26 System is completely controllable and completely observable.

11.27 **(a)** Completely controllable, **(b)** completely controllable, **(c)** not completely controllable.

11.28 **(a)** Not controllable, not observable. This system will reappear in Example 12.2. **(b)** Not controllable; it is observable. This system is a special case of Problem 12.1b with $\alpha = 1$. **(c)** Both controllable and observable. See also Problems 12.8 and 12.9.

11.29 Yes. When put into state variable form, the system is completely controllable.

11.30 Completely controllable and completely observable for all finite values of k_1 and b_2 except when $k_1 b_2 = 1$.

11.31 Controllable, unless $R_3 C_2 = R_2 C_1$. Observable unless $R_1 = 0$. In these cases pole-zero cancellation occurs.

11.32 **(a)** Uncontrollable and unobservable. Two identical RLC systems in parallel. **(b)** Not completely controllable, due to pole-zero cancellation. Completely observable.

11.33 *Hint:* Use the semigroup property of transition matrices.

11.34 No. Rank of $\mathbf{P}$ is only 2.

11.35 The system *is* observable. The Kalman observable canonical form matrices are

$$\mathbf{A} = \begin{bmatrix} 0.5 & 0.866 & 2.4495 \\ -0.28867 & -0.5 & 2.8284 \\ -0.8165 & -1.4142 & 2.0 \end{bmatrix}, \mathbf{B} = \begin{bmatrix} 3.5355 & 2.1213 \\ 2.8577 & 1.2247 \\ 1.1547 & 0 \end{bmatrix}, \mathbf{C} = \begin{bmatrix} 1.4142 & 0 & 0 \\ 0.7071 & 1.2247 & 0 \end{bmatrix}$$

11.36 **(a)** Controllable. The Kalman controllable canonical matrices are

$$\mathbf{A} = \begin{bmatrix} -4. & -1.9466 & -1.4868 \\ -0.64888 & -6.1053 & -3.1355 \\ 0.84958 & 4.0657 & 1.1053 \end{bmatrix}, \mathbf{B} = \begin{bmatrix} 1.5275 & 0.43643 \\ 0 & 1.3452 \\ 0 & 0 \end{bmatrix},$$

$$\mathbf{C} = \begin{bmatrix} 0.6546 & 2.01778 & 3.0822 \\ 0.65465 & 4.99134 & 5.3533 \end{bmatrix}$$

(b) Not observable. Kalman observable canonical matrices are

$$\mathbf{A} = \begin{bmatrix} -1.0714 & 0.37115 & 0 \\ -4.925 & -3.9286 & 0 \\ -5.5549 & -1.06905 & -4 \end{bmatrix}, \mathbf{B} = \begin{bmatrix} 0.26726 & 0.80178 \\ -0.66865 & 0.87438 \\ 1.34715 & 0.7698 \end{bmatrix}, \mathbf{C} = \begin{bmatrix} 3.74165 & 0 & 0 \\ 7.216 & 1.3887 & 0 \end{bmatrix}$$

PTER 12

12.19 $\dot{\mathbf{x}} = \begin{bmatrix} -1/R_1C - 1/R_2C & -1/C \\ 1/L & 0 \end{bmatrix}\mathbf{x} + \begin{bmatrix} -1/R_2C \\ 1/L \end{bmatrix}u,\ y = [1 \quad 0]\mathbf{x} + u,$

$$H(s) = \frac{s(s + 1/R_1C)}{s^2 + (1/R_1C + 1/R_2C)s + 1/LC}$$

12.20 $\dot{\mathbf{x}} = \begin{bmatrix} -1/CR_1 & 0 & 1/C & 0 \\ 0 & -1/CR_1 & 0 & -1/C \\ -1/L & 0 & -R_2/L & -R_2/L \\ 0 & 1/L & -R_2/L & -R_2/L \end{bmatrix}\mathbf{x} + \begin{bmatrix} 0 & 0 \\ 0 & 0 \\ 1/L & 0 \\ 0 & 1/L \end{bmatrix}\begin{bmatrix} u_1 \\ u_2 \end{bmatrix},\ \mathbf{y} = \begin{bmatrix} 1 & 1 & 0 & 0 \\ 0 & 1 & 0 & 0 \end{bmatrix}\mathbf{x}$

12.21 $\mathbf{H}(s) = \begin{bmatrix} 0 & 0 \\ 1/(s+3) & 0 \end{bmatrix}$

12.22 First-order state equation $\dot{x}_1 = [-x_1/R_1 + u_1/R_1 + u_2]/C,\ y_1 = x_1,\ y_2 = x_1 + u_2R_2,$

$$\mathbf{H}(s) = \frac{\begin{bmatrix} 1/R_1 & 1 \\ 1/R_1 & 1 + R_2C(s + 1/R_1C) \end{bmatrix}}{(s + 1/R_1C)C}$$

12.23 $\mathbf{D} = \begin{bmatrix} 0 & 0 & 1 \\ 0 & 0 & 1 \end{bmatrix}$. One possible choice for $\{\mathbf{A}, \mathbf{B}, \mathbf{C}\}$ is

$$\mathbf{A} = \begin{bmatrix} -1 & 0 & 0 & 0 \\ 0 & -1 & 0 & 0 \\ 0 & 0 & -2 & 0 \\ 0 & 0 & 0 & -3 \end{bmatrix}, \mathbf{B} = \begin{bmatrix} 1 & 1 & -1 \\ 0 & 1 & 0 \\ 0 & 1 & 0 \\ 0 & 1 & 4 \end{bmatrix}, \mathbf{C} = \begin{bmatrix} 1 & 0 & -1 & 0 \\ 0 & \frac{1}{2} & 0 & -\frac{1}{2} \end{bmatrix}$$

12.24 **(a)** is reducible since it is completely observable but not controllable. **(b)** is reducible since it is completely controllable but not observable. **(c)** is irreducible. It is completely controllable and observable.

12.25 $\mathbf{D} = \begin{bmatrix} 1 & 0 & 0 \\ 0 & 0 & 0 \end{bmatrix}$. One possible solution for $\{\mathbf{A}, \mathbf{B}, \mathbf{C}\}$ is

$$\mathbf{A} = \begin{bmatrix} -1 & 0 & 0 & 0 \\ 0 & -2 & 0 & 0 \\ 0 & 0 & -3 & 0 \\ 0 & 0 & 0 & 0 \end{bmatrix}, \mathbf{B} = \begin{bmatrix} -1 & 1 & 0 \\ 0 & 1 & 0 \\ 0 & 0 & 1 \\ 0 & 0 & 1 \end{bmatrix}, \mathbf{C} = \begin{bmatrix} 1 & -1 & 1 & 0 \\ 1 & -1 & 0 & 1 \end{bmatrix}$$

12.26 $\mathbf{D} = \begin{bmatrix} 0 & 0 \\ 0 & 0 \end{bmatrix}$. One possible solution for $\{\mathbf{A}, \mathbf{B}, \mathbf{C}\}$ is

$$\mathbf{A} = \left[\begin{array}{cc|c|c} -1 & 1 & 0 & 0 \\ 0 & -1 & 0 & 0 \\ \hline 0 & 0 & -1 & 0 \\ \hline 0 & 0 & 0 & 0 \end{array}\right], \mathbf{B} = \left[\begin{array}{cc} 0 & 0 \\ 1 & 0 \\ \hline 1 & 1 \\ \hline 1 & 0 \end{array}\right], \mathbf{C} = \left[\begin{array}{cc|c|c} 1 & -2 & 1 & 0 \\ 0 & 0 & 1 & 1 \end{array}\right]$$

Another solution has the same $\mathbf{A}$, but $\mathbf{B} = \left[\begin{array}{cc} -1 & 1 \\ 1 & 0 \\ \hline 1 & 1 \\ \hline 1 & 0 \end{array}\right]$ and $\mathbf{C} = \left[\begin{array}{cc|c|c} 1 & 0 & 0 & 0 \\ 0 & 0 & 1 & 1 \end{array}\right]$. A third solution is given by Chen (see page 251 of Reference 4).

12.27 One eighth-order irreducible realization is

$$\dot{\mathbf{x}} = \left[\begin{array}{cccc|ccc|c} 0 & 1 & 0 & 0 & & & & \\ 0 & 0 & 1 & 0 & & \mathbf{0} & & \\ 0 & 0 & 0 & 1 & & & & \\ 0 & 0 & 0 & 0 & & & & \\ \hline & & & & 0 & 1 & 0 & \\ & \mathbf{0} & & & 0 & 0 & 1 & \\ & & & & 0 & 0 & 0 & \\ \hline & & & & & & & 0 \end{array}\right]\mathbf{x} + \left[\begin{array}{ccc} 0 & 0 & 0 \\ -1 & 0 & 1 \\ 0 & 0 & 0 \\ 1 & 1 & -2 \\ \hline 0 & 0 & 0 \\ 2 & 0 & -2 \\ 1 & 0 & 1 \\ \hline 1 & 0 & -1 \end{array}\right]\mathbf{u}$$

$$\mathbf{y} = \left[\begin{array}{cccc|ccc|c} 1 & 0 & 0 & 0 & 0 & 0 & 0 & 1 \\ 1 & 1 & 0 & 0 & \frac{1}{2} & 0 & 0 & 1 \\ 1 & 0 & -1 & 0 & -1 & -5 & 1 & 10 \end{array}\right]\mathbf{x}$$

See Reference 4 for a detailed solution.

12.28 $$\mathbf{T}(z) = \begin{bmatrix} \dfrac{0.5z^2 - 1.3z + 1.245}{(z-1)} & -(z+0.3) \\ \dfrac{1.5z^2 - 1.2z + 0.145}{(z-1)} & 2(z-0.5) \end{bmatrix} \frac{1}{(z^2 - z + 0.89)}$$

12.29 One possible answer is

$$\mathbf{A} = \begin{bmatrix} 1 & 0 & 0 & 0 \\ 0 & 0.3679 & 0 & 0 \\ 0 & 0 & 1 & 1 \\ 0 & 0 & -0.5 & 0 \end{bmatrix} \mathbf{B} = \begin{bmatrix} 1 \\ 1 \\ -0.9878 \\ -1.3761 \end{bmatrix} \mathbf{C} = \begin{bmatrix} 3.1641 & -2.1762 & 1 & 0 \\ 1.8984 & -0.8984 & 0 & 0 \end{bmatrix} \mathbf{D} = [0]$$

12.30 $$\mathbf{H}(z) = \begin{bmatrix} (z-1)(z-0.5) & 0 \\ 0 & (z-0.5)^2(z-0.7) \end{bmatrix}^{-1} \begin{bmatrix} 0.2z & 3z(z+0.3) \\ z+0.8 & z(z-0.5)(z-0.7) \end{bmatrix}$$

$$\mathbf{A} = \begin{bmatrix} 1.5 & 1 & 0 & 0 & 0 \\ -0.5 & 0 & 0 & 0 & 0 \\ 0 & 0 & 1.7 & 1 & 0 \\ 0 & 0 & -0.95 & 0 & 1 \\ 0 & 0 & 0.175 & 0 & 0 \end{bmatrix}, \mathbf{B} = \begin{bmatrix} 0.2 & 5.4 \\ 0 & -1.5 \\ 0 & 0.5 \\ 1 & -0.6 \\ 0.8 & 0.175 \end{bmatrix},$$

$$\mathbf{C} = \begin{bmatrix} 1 & 0 & 0 & 0 & 0 \\ 0 & 0 & 1 & 0 & 0 \end{bmatrix}, \mathbf{D} = \begin{bmatrix} 0 & 3 \\ 0 & 1 \end{bmatrix}$$

12.31 Original eighth-order system: Same as in Problem 12.10 except for the 2,2 block representation of h_{12}. Those blocks change to $\mathbf{A}_3 = \begin{bmatrix} 0 & 1 \\ -0.5 & 1.5 \end{bmatrix}$ and $\mathbf{C}_3 = \begin{bmatrix} -1.5 & 5.4 \\ 0 & 0 \end{bmatrix}$. **B** and **D** are unchanged. Original Rank(P) = 6, Rank(Q) = 5. Use **Q** to select fifth-order reduced system with

$$\mathbf{A} = \begin{bmatrix} 1.263911 & 0 & 0.150202 & 0 \\ 0 & 0.489418 & 0 & 8.13180 \\ -1.34220 & 0 & 0.236089 & 0 \\ 0 & -0.78092 & 0 & 1.220540 \\ 0 & 0.2418970 & 0 & 0.164580 \end{bmatrix},$$

$$\mathbf{B} = \begin{bmatrix} 0.035663 & 0.962905 \\ 0 & 0.363697 \\ 0.56055 & -0.267254 \\ 0.894503 & 0.004733 \\ 0.438318 & 0.174045 \end{bmatrix}, \mathbf{C}^T = \begin{bmatrix} 5.608 & 0 \\ 0 & 1.3748 \\ 0 & 0 \\ 0 & 0 \\ 0 & 0 \end{bmatrix},$$

and $\mathbf{D} = \begin{bmatrix} 0 & 3 \\ 0 & 1 \end{bmatrix}$. This realization is also controllable and hence irreducible.

12.32 $\mathbf{H}' = \begin{bmatrix} 3 & 2s(s^2+2s+10)(s+5) \\ s+1 & (s+1)(s^2+2s+10) \end{bmatrix} \begin{bmatrix} (s^2+2s+10)(s+5) & 0 \\ 0 & (s^2+2s+10)(s+5) \end{bmatrix}^{-1}$

can be reduced to $\mathbf{H}' = \begin{bmatrix} 3 & 2s \\ s+1 & s+1 \end{bmatrix} \begin{bmatrix} (s^2+2s+10)(s+5) & 0 \\ 0 & s+5 \end{bmatrix}^{-1}$ from which a minimal realization is

$$\mathbf{A} = \begin{bmatrix} 0 & 1 & 0 & 0 \\ 0 & 0 & 1 & 0 \\ -50 & -20 & -7 & 0 \\ 0 & 0 & 0 & -5 \end{bmatrix}, \mathbf{B} = \begin{bmatrix} 0 & 0 \\ 0 & 0 \\ 1 & 0 \\ 0 & 1 \end{bmatrix}, \mathbf{C}^T = \begin{bmatrix} 15 & 1 \\ 3 & 1 \\ 0 & 0 \\ -10 & -4 \end{bmatrix}, \mathbf{D} = \begin{bmatrix} 0 & 2 \\ 0 & 1 \end{bmatrix}$$

12.33 The original eighth-order realization is

$$\mathbf{A} = \begin{bmatrix} 0 & 1 & 0 & 0 & 0 & 0 & 0 & 0 \\ 0 & 0 & 1 & 0 & 0 & 0 & 0 & 0 \\ -50 & -20 & -7 & 0 & 0 & 0 & 0 & 0 \\ 0 & 0 & 0 & 0 & 1 & 0 & 0 & 0 \\ 0 & 0 & 0 & 0 & 0 & 1 & 0 & 0 \\ 0 & 0 & 0 & -50 & -20 & -7 & 0 & 0 \\ 0 & 0 & 0 & 0 & 0 & 0 & -5 & 0 \\ 0 & 0 & 0 & 0 & 0 & 0 & 0 & -5 \end{bmatrix}, \mathbf{B} = \begin{bmatrix} 0 & 0 \\ 0 & 0 \\ 1 & 0 \\ 0 & 0 \\ 0 & 0 \\ 1 & 0 \\ 0 & 1 \\ 0 & 1 \end{bmatrix}, \mathbf{C}^T = \begin{bmatrix} 3 & 0 \\ 0 & 0 \\ 0 & 0 \\ 0 & 1 \\ 0 & 1 \\ 0 & 0 \\ -10 & 0 \\ 0 & -4 \end{bmatrix}$$

and $\mathbf{D} = \begin{bmatrix} 0 & 2 \\ 0 & 1 \end{bmatrix}$. For this system, Rank(P) = 4 and Rank(Q) = 6. The fourth-order controllable realization is

$$\mathbf{A}=\begin{bmatrix}-7 & 0 & -2 & -5\\ 0 & -5 & 0 & 0\\ 1 & 0 & 0 & 0\\ 0 & 0 & 1 & 0\end{bmatrix},\ \mathbf{B}=\begin{bmatrix}1.4142 & 0\\ 0 & 1.4142\\ 0 & 0\\ 0 & 0\end{bmatrix},\ \mathbf{C}^T=\begin{bmatrix}0 & 0\\ -7.071 & -2.8284\\ 0 & 0.7071\\ 2.1213 & 0.7071\end{bmatrix}$$

and $\mathbf{D}=\begin{bmatrix}0 & 2\\ 0 & 1\end{bmatrix}$. This realization is observable and hence irreducible.

12.34 After reduction to the minimal determinant degree, one answer is

$$\mathbf{H}(s)=\begin{bmatrix}(s+1)(s+3) & -.25(s+2)(3s+5)\\ 0 & (s+1)(s+2)(s+5)\end{bmatrix}^{-1}\begin{bmatrix}-1.5(s+2) & .25(s-1)\\ 2(s+2)(s+5) & (s+1)^2\end{bmatrix}$$

From this, a fifth-order realization is

$$\mathbf{A}=\begin{bmatrix}-4 & 1 & -0.25 & 0 & 0\\ -3 & 0 & 0.25 & 0 & 0\\ 0 & 0 & -8 & 1 & 0\\ 0 & 0 & -17 & 0 & 1\\ 0 & 0 & -10 & 0 & 0\end{bmatrix},\ \mathbf{B}=\begin{bmatrix}-1.5 & 0.25\\ -3 & -0.25\\ 2 & 1\\ 14 & 2\\ 20 & 1\end{bmatrix},\ \mathbf{C}^T=\begin{bmatrix}1 & 0\\ 0 & 0\\ 0.75 & 1\\ 0 & 0\\ 0 & 0\end{bmatrix},\ \mathbf{D}=[0]$$

12.35 $\mathbf{H}(s)=\begin{bmatrix}(s+2)^2(s+5) & -1\\ 0 & (s+2)\end{bmatrix}^{-1}\begin{bmatrix}0 & (s+2)^2\\ 1 & 0\end{bmatrix}$

$$\mathbf{A}=\begin{bmatrix}-9 & 1 & 0 & 0\\ -24 & 0 & 1 & 0\\ -20 & 0 & 0 & 1\\ 0 & 0 & 0 & -2\end{bmatrix},\ \mathbf{B}=\begin{bmatrix}0 & 1\\ 0 & 4\\ 0 & 4\\ 1 & 0\end{bmatrix},\ \mathbf{C}=\begin{bmatrix}1 & 0 & 0 & 0\\ 0 & 0 & 0 & 1\end{bmatrix}$$

CHAPTER 13

13.22 $\mathbf{K}=\begin{bmatrix}1 & -5 & -1\\ 0 & 9 & 3\end{bmatrix}$. $\left(\textit{Hint:}\text{ Define } \alpha=\dfrac{1}{\lambda+3}\text{ and let } \lambda\to -3 \text{ after finding } \mathbf{G}^{-1}.\right)$

13.24 One solution is $\mathbf{K}'=[17\ \ 0]^T$, another is $\mathbf{K}'=[0\ \ 17]^T$.

13.25 One solution is $\mathbf{K}'=\begin{bmatrix}0 & 6\\ 9 & 0\end{bmatrix}$. For another solution using a slightly different method, see [6].

13.26 $\mathbf{K}'=[-1/\epsilon^2\ \ 1/\epsilon^2]$

13.27 No, because $\mathbf{N}=\begin{bmatrix}0 & 1\\ 0 & 1\end{bmatrix}$ is singular [11].

13.28 No, $\mathbf{N}=\begin{bmatrix}1 & 0\\ 1 & 0\end{bmatrix}$ is singular [11].

13.29 $\mathbf{F}_d=\begin{bmatrix}1 & -1\\ 0 & 1\end{bmatrix}$, $\mathbf{K}_d=\begin{bmatrix}-2 & -2 & 0\\ 0 & -3 & -4\end{bmatrix}$, $\mathbf{H}(s)=\begin{bmatrix}1/s & 0\\ 0 & 1/s^2\end{bmatrix}$

13.30 $\mathbf{K}=[1\ \ 0.5]$

13.31 $\mathbf{K}=[1.34\ \ 0.4]$

13.32 Two choices are

$$\mathbf{K}=\begin{bmatrix}0.4 & -0.3\\ 0.25 & 0.15\end{bmatrix}\text{ and }\begin{bmatrix}0.3 & -0.4\\ 0.15 & 0.25\end{bmatrix}$$

13.33 Using the nested integrator method (Example 3.6) gives

$$\mathbf{A} = \begin{bmatrix} -3 & 1 & 0 & 0 & 0 \\ -2 & -2 & 2 & 1 & 0 \\ 2 & 0 & -6 & 0 & 1 \\ 4 & -4 & 0 & 0 & 0 \\ 0 & 0 & -1 & 0 & 0 \end{bmatrix} \quad \mathbf{B} = \begin{bmatrix} 1 & 0 \\ 0 & 0 \\ 0 & 0 \\ 5 & 1 \\ 0 & 1 \end{bmatrix} \quad \mathbf{C} = \begin{bmatrix} 1 & 0 & 0 & 0 & 0 \\ 0 & 1 & 0 & 0 & 0 \\ 0 & 0 & 1 & 0 & 0 \end{bmatrix} \quad \mathbf{D} = [0]$$

One choice for **K** is

$$\begin{bmatrix} -2.1251 & -67.1349 & -26.9825 & -12.045 & -15.107 \\ 4.8752 & 171.368 & 70.0198 & 34.4558 & 40.8934 \end{bmatrix}$$

A choice for the full state observer is

$$\mathbf{L} = \begin{bmatrix} 6.3127 & 1.0443 & -0.0136 \\ -35.2083 & 13.0842 & 1.5313 \\ -43.6735 & -2.2322 & 9.6030 \\ -288.0065 & 49.2646 & -2.4314 \\ -367.7839 & -17.7946 & 59.789 \end{bmatrix}$$

13.36 $p(z) = z^2 + 0.3z - 0.009$; $q(z) = 0.3990z - 0.1515$; $e(z) = 0.3z - 0.001$
13.37 $p(z) = z + 0.09$; $q(z) = 0.31z - 0.035$; $e(z) = p(z) - d(z) = 0.01$

PTER 14

14.20 $g_a = 21$ is minimum cost. Optimal path is *a, b, e, l, r, z*.
14.21 2 hr on course 1, score = 65; 0 hr on course 2, score = 40; 1 hr on course 3, score = 52; 1 hr on course 4, score = 91. Total max. score = 248.
14.22 $u^*(0) = -1$, $u^*(1) = -2$, $x^*(0) = 10$, $x^*(1) = 4$, $x^*(2) = 0$, $J^* = 9$. For comparison, if no control is used, $u(k) = 0$, $x(0) = 10$, $x(1) = 5$, $x(2) = 2.5$, and $J = 20.25$.
14.23

$k =$	0	1	2	3	4	5
$u^*(k)$	−0.0147	−0.0293	−0.0587	−0.1173	−0.2346	—
$x^*(k)$	10.0	4.985	2.463	1.173	0.469	0.0

14.24 $u^*(0) = -21x_0/104$, $u^*(1) = -x_0/52$, $x^*(0) = x_0$, $x^*(1) = 5x_0/52$, $x^*(2) = x_0/104$
14.25 The gain matrices and eigenvalues are: **(a)** $[-1.620 \quad -1.959]$, $0.190 \pm j0.148$ **(b)** $[-1.199 \quad -1.413]$, $-0.4 \pm j0.26$ **(c)** $[-1.755 \quad -2.136]$, -0.033, -0.0212 (both real) **(d)** $[-1.775 \quad -2.162]$, 0.0044, -0.229 (both real)
14.26 **(a)** $\mathbf{K}_{ss}^T = [-0.5202 \quad 0.8588]$. Filter eigenvalues are $0.2319 \pm j0.2956$. **(b)** Based on the filter pole locations, the filter will be much slower than controllers c and d and a little slower than a. Therefore, the filter will have a major impact on system response, maybe even be the dominant effect. In case b the filter is somewhat faster than the controller, and the total system will act more nearly like the full state controller.
14.27 **(a)** $\mathbf{K}_{ss}^T = [-0.4561 \quad 0.9977]$. Eigenvalues are $0.0404 \pm j0.0257$. **(b)** Because of the lower noise level, the filter response is much faster. The effect on the control loop will be essentially negligible and the controller will work about as well as if it had the actual state to work with rather than estimate. Stated differently, with this low noise level the estimated states will agree very closely with the true values.

14.28 On backward pass find

k	k'	$\mathbf{G}^T$	$v = -\mathbf{FV}$
5	0	—	—
4	1	[0.74963 −0.24988]	7.4963
3	2	[0.72159 −0.23296]	−0.5684
2	3	[0.72140 −0.23305]	−0.9369
1	4	[0.72140 −0.23305]	−0.9537
0	5	[0.72140 −0.23305]	−0.9544

Then on the forward pass, starting with the given $\mathbf{x}(0)$, find

k	$\mathbf{x}(k)^T$	$u(k) = -\mathbf{G}(k)\mathbf{x}(k) + v(k)$
0	[4 −2]	−4.3061
1	[−0.3061 1.3061]	−0.4284
2	[−0.7345 1.2345]	−0.1192
3	[−0.8538 1.1038]	0.3048
4	[−0.5490 0.6740]	8.0762
5	[7.5273 −7.4648]	—

14.29 $\mathbf{G} = \begin{bmatrix} 0.0422 & 0.0484 & -0.1059 & 0.2630 \\ 0.0013 & -0.4759 & -0.7321 & -0.0005 \end{bmatrix}$

$\lambda_i = \{0.104, 0.898, 0.6068 \pm j0.3519\}$

The feedback matrix G and the resulting eigenvalues are:
(a) $[0.0271 \quad -0.0032 \quad -0.1099 \quad 0.1255]$, $\{0.0025, 0.8196, 0.721 \pm j0.479\}$
(b) $[0.0276 \quad -0.0039 \quad -0.1119 \quad 0.1292]$, $\{2.76 \times 10^{-4}, 0.818, 0.721 \pm j0.4788\}$
(c) $[0.0208 \quad 0.0004 \quad -0.0833 \quad 0.1052]$, $\{0.0447, 0.8437, 0.7197 \pm j0.4819\}$
(d) $[0.0227 \quad -0.1097 \quad -0.1507 \quad 0.1120]$, $\{0.0222, 0.8306, 0.699 \pm j0.455\}$. For reference the open-loop eigenvalues are $\{1.0, 0.135, 0.7175 \pm j0.4909]$

14.31

r	$\mathbf{W}_\infty$	$\mathbf{G}_\infty$	λ_{cl}
50	$\begin{bmatrix} 1.2276 & 0.2421 \\ 0.2421 & 0.24699 \end{bmatrix}$	$\mathbf{W}_\infty/50$	$-1.9896, -1.0398$ note: open-loop $\lambda = -2, -1$
10	$\begin{bmatrix} 1.1508 & 0.21573 \\ 0.21573 & 0.23686 \end{bmatrix}$	$\mathbf{W}_\infty/10$	$-1.9396, -1.1992$
4	$\begin{bmatrix} 1.04327 & 0.17995 \\ 0.17995 & 0.22322 \end{bmatrix}$	$\mathbf{W}_\infty/4$	$-1.6583, -1.6583$
0.2	$\begin{bmatrix} 0.42801 & 0.20482 \\ 0.20482 & 0.76754 \end{bmatrix}$	$5\mathbf{W}_\infty$	$-2.9535 \pm j1.10677$

14.32 $\mathbf{W} = \begin{bmatrix} 1.25 & 0.25 \\ 0.25 & 0.25 \end{bmatrix}$

14.33 $\mathbf{W}_\infty = \begin{bmatrix} 0.80127 & 0.14102 \\ 0.14102 & 0.212813 \end{bmatrix}$, $\mathbf{G}_\infty = [-0.66025 \quad 0.071797]$ and $\lambda = -2, -1.73205$.

14.34 With $r = 50$, $\mathbf{G}_\infty = \begin{bmatrix} 1.0582 & 0.29701 \\ 0.29701 & 0.08639 \end{bmatrix}$, $\lambda_{cl} = -0.5805, -3.564$. With $r = 1000$, $\mathbf{G}_\infty =$

$\begin{bmatrix} 1.0419 & 0.2925 \\ 0.2925 & 0.08229 \end{bmatrix}$, $\lambda_{cl} = -0.5625, -3.560$. Note: Open-loop eigenvalues are at $\lambda = +0.56155$, -3.5616

14.35 The roots of the quadratic are $\mathbf{W} = 1$ and $-\frac{1}{3}$. Only one solution is positive definite. The eigenvector method gives $\mathbf{W}_\infty = 1$ and $\mathbf{W}_{ss} = -\frac{1}{3}$.

14.36 The canonical **A** matrix is $\mathbf{A}' = \begin{bmatrix} \mathbf{A}_{11} & \mathbf{A}_{12} \\ 0 & \mathbf{A}_{22} \end{bmatrix}$. Also $\mathbf{B}' = \begin{bmatrix} \mathbf{B}_1 \\ 0 \end{bmatrix}$

and $\mathbf{C} = [0 \quad 0.7071 \quad -0.56904 \quad 0.41975]$

where $\mathbf{A}_{11} = \begin{bmatrix} -7 & -6 \\ 2 & 0 \end{bmatrix}$, $\mathbf{A}_{12} = \begin{bmatrix} -59.4519 & 31.42803 \\ -15.2902 & 11.2787 \end{bmatrix}$.

$\mathbf{A}_{22} = \begin{bmatrix} -39.19376 & 33.12394 \\ -29.37606 & 23.19376 \end{bmatrix}$ is uncontrollable and has $\lambda = -8, -8$. $\mathbf{A}_{11}$ has $\lambda = -3, -4$.

14.37 $u^*(t) = \dfrac{\cosh t}{\sinh(1)}$, $x^*(t) = \dfrac{\sinh t}{\sinh(1)}$

14.38 $u^*(t) = -\frac{7}{2} + 3t$, $x_1^*(t) = 1 + t - 7t^2/4 + t^3/2$, $x_2^*(t) = 1 - 7t/2 + 3t^2/2$

14.39 $x(t) \cong 10 \cosh\sqrt{5}t - 10.065 \sinh\sqrt{5}t$, $p(t) \cong 5.011 \cosh\sqrt{5}t - 4.462 \sinh\sqrt{5}t$

14.41 **(a)** $u^*(t) = -\frac{1}{2}\mathbf{B}^T\mathbf{p}(t)$ if $\|\frac{1}{2}\mathbf{B}^T\mathbf{p}(t)\| \le 1$; $\mathbf{u}^*(t) = -\mathbf{B}^T\mathbf{p}(t)/\|\mathbf{B}^T\mathbf{p}(t)\|$ otherwise. Linear control with saturation. **(b)** $u^*(t) = -\mathbf{B}^T\mathbf{p}(t)/\|\mathbf{B}^T\mathbf{p}(t)\|$ for all t unless $\mathbf{B}^T\mathbf{p}(t) \equiv \mathbf{0}$ on a finite time interval (singular control case). Optimal control is always on the boundary of U except in the case of singular control. **(c)** Each component satisfies $u_i^*(t) = -\text{sign}[\mathbf{B}^T\mathbf{p}(t)]_i$, where $[\mathbf{B}^T\mathbf{p}(t)]_i$ is the ith component of $\mathbf{B}^T\mathbf{p}(t)$. This is the so-called bang-bang control and is valid provided $[\mathbf{B}^T\mathbf{p}(t)]_i \ne 0$. If $[\mathbf{B}^T\mathbf{p}(t)]_i \equiv 0$ on a finite interval, we have singular control. **(d)** $u_i^*(t) = 0$ when $|[\mathbf{B}^T\mathbf{p}(t)]_i| \le 1$: $u_i^*(t) = -\text{sign}[\mathbf{B}^T\mathbf{p}(t)]_i$ otherwise.

14.42 No solution exists. It is not possible to reach the origin with this unstable system and the bounded admissible controls unless $|x(0)| < 1$. The minimum principle provides only necessary conditions. It does not guarantee existence of a solution.

14.43 $\mathbf{G} = [0.2215 \quad -0.3735 \quad 0.2227 \quad 0.3162]$; $\lambda_i = \{-0.926, -28.15, -0.9165 \pm j2.88\}$

PTER 15

15.23 With $e = x - y_d$, and using $\dot{e} = -\alpha e$, the control is $u(t) = \{-2y_d(t) + v(t) - \alpha[x(t) - y_d(t)]\}/x(t)^2$ provided $x \ne 0$. For $x(0) > 0$ the singularity at $x(t) = 0$ causes no problem in reaching the desired response.

15.24 The error equation $\dot{\mathbf{e}} = \mathbf{S}\mathbf{e}$, with diagonal S, gives $\begin{bmatrix} 1 \\ x_2 \end{bmatrix}\mathbf{u} = \begin{bmatrix} v - x_2^2 - y_{d_1} + S_{11}(x_1 - y_{d_1}) \\ v - 4y_{d_2} + S_{22}(x_2 - y_{d_2}) \end{bmatrix}$.

Least squares solution gives $u(t) = \{v - x_2^2 - y_{d_1} + S_{11}(x_1 - y_{d_1}) + x_2[v - 4y_{d_2} + S_{22}(x_2 - y_{d_2})]\}/(1 + x_2^2)$

15.25 Pick all $S_{ij} = 0$ except $S_{13} = S_{24} = 1$ and S_{31}, S_{33}, S_{42} and S_{44}. The first two error equations are automatically satisfied and the last two give

$$u_1(t) = S_{31}(x_1 - y_1) + S_{33}(x_3 - y_3) - x_4^2 x_1 + \mu/x_1^2 - \alpha y_3 - \beta y_1 + \beta R$$

$$u_2(t) = x_1\{S_{42}(x_2 - y_2) + S_{44}(x_4 - y_4) + 2x_3 x_4/x_1 - \kappa y_4 - \sigma y_2 - \sigma\mu t/R^3\}$$

Parameter values must be selected. The orbit period is $T = 2\pi R^3/\mu$. The time constants τ for the template system are selected to give $4\tau = T/4$. For simplicity set $\alpha = \kappa$ and $\beta = \sigma$. Damping of $\zeta = 0.707$ gives $\beta = \{8\sqrt{2}\mu/(\pi R^3)\}^2$ and $\alpha = 16\mu/(\pi R^3)$. The convergence factors are selected so that the error settles twice as fast as the template system. This gives $S_{31} = S_{42} = 4\beta$ and $S_{33} = S_{44} = 2\alpha$. There are many other acceptable answers.

15.27 $\mathbf{x}_{e1} = \mathbf{0}$, $\mathbf{x}_{e2}$ and $\mathbf{x}_{e3} = [\pm 3 \quad 0 \quad 2.1]^T$. The Jacobian matrix evaluated at $\mathbf{x}_{e1}$ here is the same as in Problem 15.13 evaluated at $\mathbf{x}_{e2}$, and likewise the Jacobian here for $\mathbf{x}_{e2}$ is the same as in Problem 15.13 at $\mathbf{x}_{e1}$.

15.28 The plot of $-1/N(E)$ still ranges from 1.25 to 2, but in the *opposite* direction as a function of E. Intersections with $G(j\omega)$ indicate a stable limit cycle at $\omega = 1.71642$ with $E \approx 2.7$ and an unstable limit cycle at $\omega = 0, E \approx 4$. Simulation indicates that for small initial conditions a response very similar to the predicted limit cycle (but with a small dc offset) develops. For $x_1(0)$ larger than about 1.7 an unstable growth ensues.

15.29 Refer to Figure 15.24.

Range of K	Intersections of $G(j\omega)$ and $-1/N$	Predicted Behavior
$K < 0.16788$	None. $-1/N$ to right of point 3	Stable for all I.C.
0.16788, 0.2686	Intersects point 3	Unstable limit cycle‡
0.2686, 0.804	None. $-1/N$ between pts. 2,3	Unstable for all I.C.
0.804, 0.880	Intersects point 2	Stable limit cycle $\omega = 1.7164$
0.880, 1.287	Intersects both 1 and 2	Chaotic?
1.287, 1.408	Intersects 1.	Unstable limit cycle‡
$K > 1.408$	None. $-1/N$ between 0 and 1	Unstable for all I.C.

‡ Whether actual behavior is stable or unstable is dependent upon initial conditions (I.C.)

15.30 The plot of $-1/N$ now continues along the positive real axis, approaching ∞ for an amplitude E somewhat larger than 5 (the point where the effective gain $N(E)$ crosses through 0). The $-1/N$ plot then resumes at $-\infty$ and proceeds to -0.5 as E continues to increase. One new intersection with $G(j\omega)$ at $\omega = 2.4807$ indicates a stable limit cycle. Simulation shows behavior is the same as that of Problem 15.14 for smaller initial conditions, because the second nonlinearity breakpoint occurs at a large amplitude. For initial conditions sufficiently large the predicted stable limit cycle is observed here, whereas the system of Problem 15.14 goes unstable.

15.31 The circuit still appears to be chaotic for some initial conditions.

15.32 Routh's criterion shows the system is unstable for $K > 30.69$. The auxiliary equation shows the frequency at which the $j\omega$ axis is crossed is $\omega = 2.56$. Example 15.5 results yield a describing function which increases from 0 at an amplitude $E = 0.25$ to a maximum of 0.873 at $E = 2.51$ and then decreases again. For $30.69 < K < 35.15$ the effective gain $N(E)K$ is less than 30.69 and the system is stabilized by the nonlinearity. For each $K > 35.15$ there are two amplitudes E which give the critical $N(E)K = 30.69$, and thus two potential limit cycles. The larger E, where $N(E)$ is decreasing, gives a stable limit cycle. One example has $K = 40$, $\omega = 2.56$ and $E = 3.3$ as a stable limit cycle, excited if $y(0) > 1.35$.

15.33 The root locus has two $j\omega$ axis crossings and at each there are two amplitudes E which give the critical gain values.

ω	$10N(E)$	$N(E)$	approximate E (see Example 15.5)
6.1127	7.372	0.7372	1.2 (stable) or 3.5 (unstable)
3.8096	1.628	0.1628	0.345 (unstable) or 17.0 (stable)

Two *stable* limit cycles are possible, as noted. Initial conditions will determine which one, if any, is excited.

15.34 (a) $V'(\mathbf{x}) = \frac{1}{2b}(ax_2 + bx_1)^2 + \frac{1}{2ab}(ax_3 + bx_2)^2 + \int_0^{x_2}\left[F(x_2) - \frac{b}{a}\right]x_2\,dx_2$

$$\dot{V}'(\mathbf{x}) = -\frac{a}{b}\left[F(x_2) - \frac{b}{a}\right]x_3^2$$

(b) Asymptotically stable if $a > 0$, $b > 0$, and $F(x_2) > b/a$ for all x_2.

15.35 (a) $\mathbf{f}(\mathbf{x}) = \mathbf{0} \Rightarrow (r - 1 - x_3)x_1 = 0$ and $8x_3/3 = x_1^2$. One solution is $\mathbf{x}_{e_1} = \mathbf{0}$. If $x_1 \neq 0$, then $x_3 = r - 1$ and $x_1 = \pm[8(r-1)/3]^{0.5}$. For $r < 1$ the only real solution is $x_{e_1} = \mathbf{0}$.

$$\partial\mathbf{f}/\partial\mathbf{x} = \begin{bmatrix} -10 & 10 & 0 \\ r - x_3 & -1 & -x_1 \\ x_2 & x_1 & -\frac{8}{3} \end{bmatrix}$$

At $\mathbf{x} = \mathbf{0}$, $|I\lambda - \partial\mathbf{f}/\partial\mathbf{x}| = (\lambda + \frac{8}{3})[\lambda^2 + 11\lambda + 10(1 - r)]$. All roots stable if $r < 1$.
(b) One answer, using variable gradient is $\nabla V = [x_1 + 0.1x_2 \quad 0.1x_1 + 2x_2 \quad 2x_3]^T$, $\dot{V}$ is negative definite if $x_1^2 < 20(\frac{16}{3})$. $V = x_1^2/2 + 0.1x_1x_2 + x_2^2 + x_3^2$ is positive definite for all $\mathbf{x}$.
(c) With $r = 28$, part (a) gives two more equilibrium points. $\mathbf{x}_e = [\pm 8.485 \quad \pm 8.485 \quad 27]^T$. Eigenvalues of $\partial\mathbf{f}/\partial\mathbf{x}$ are: at $\mathbf{x}_{e_1}$; $\lambda = -\frac{8}{3}, 11.828, -22.828$ (saddle point) at $\mathbf{x}_{e_2}$ and $\mathbf{x}_{e3}$; $\lambda = -13.845, 0.0892 \pm j10.165$
(unstable focus).

15.37 The trace of $\mathbf{A}$ can be negative while still having one or more unstable eigenvalues. The suggested example has $\mathrm{Tr}(\mathbf{A}) = -6$, yet the solution diverges as an unstable focus in the x_1, x_2 plane as e^{2t}. Even though $\|\mathbf{x}(t)\| \to \infty$, the infinite oscillatory growth is confined to a zero volume (plane).

Index

B

C

D

E

G

H

I

J

M

O

P

S